From Greenhouse to Icehouse

FROM GREENHOUSE TO ICEHOUSE

The Marine Eocene-Oligocene Transition

EDITED BY
DONALD R. PROTHERO, LINDA C. IVANY,
AND ELIZABETH A. NESBITT

COLUMBIA UNIVERSITY PRESS
NEW YORK

Columbia University Press
Publishers Since 1893
New York Chichester, West Sussex

Library of Congress Cataloging-in-Publication data
From greenhouse to icehouse : the marine Eocene-Oligocene transition / edited by Donald R. Prothero, Linda C. Ivany, and Elizabeth A. Nesbitt.
p. cm.
Includes bibliographic references and index.
ISBN 0-231-12716-2 (acid-free paper)
1. Eocene-Oligocene boundary. 2. Marine fauna—Gulf Coast Region (U.S.) 3. Marine fauna—Atlantic Coast Region. 4. Marine fauna—Pacific Coast Region. I. Prothero, Donald R. II. Ivany, Linda C. III. Nesbitt, Elizabeth R.
QE692.8 .F76 2003
560'.457—dc21 2002073645

∞

Columbia University Press books are printed on permanent and durable acid-free paper
Printed in the United States of America
c 10 9 8 7 6 5 4 3 2 1

This volume is dedicated to the memory of
Charles "Ray" Givens
(1938–1999)
for his many contributions to our understanding of
Eocene-Oligocene mollusks and stratigraphy

Contents

Participants in the Penrose Conference on the marine Eocene-Oligocene transition, held at Evergreen State College in Olympia, Washington, on August 17–22, 1999. *First row (left to right):* Clio Bitboul, Karina Hankins, Linda Donohoo, Ashley Streig, Elizabeth Sanger, Don Prothero, Louie Marincovich; *second row:* Ewan Fordyce, Rowan Lockwood, Manuel Iturralde, Karen Wetmore, James Whitehead, John Armentrout, Carol Tang, Anton Oleinik; *third row:* Paul Pearson, Paul Gammon, Ellen Thomas, Carole Hickman, Richard Squires, Franca Oboh-Ikuenobe, Nancy Buening, Noël Vandenberghe, Tom Yancey, Liz Nesbitt, Earl Manning, Linda Ivany; *fourth row:* John Hurley, Dave Scholl, Rick Fluegeman, Miklos Kazmer, Etienne Steurbaut, Yin Yanhong, Steve Pekar, Warren Allmon, Matt Campbell; *back row:* Wylie Poag, Burt Carter, David Campbell, Hiroshi Nishi, Steve Schellenberg. Photo by Anton Oleinik.

Preface

The Eocene-Oligocene transition was a critical turning point in Earth's climatic history, when the "greenhouse" world of the early Eocene changed into the "icehouse" conditions of the early Oligocene. In the past two decades, much research has focused on the details of the Eocene-Oligocene transition in both the terrestrial and marine records. In 1989, a Penrose Conference was held in Rapid City, South Dakota, which brought the terrestrial and marine records up-to-date and together for the first time in decades. That conference eventually led to three different books on the Eocene-Oligocene transition: *Eocene-Oligocene Climatic and Biotic Evolution*, edited by D. R. Prothero and W. A. Berggren (Princeton University Press, 1992); *The Eocene-Oligocene Transition: Paradise Lost*, by D. R. Prothero (Columbia University Press, 1994); and *The Terrestrial Eocene-Oligocene Transition in North America,* edited by D. R. Prothero and R. J. Emry (Cambridge University Press, 1996).

Since 1989, much additional new research has accumulated on the details of the marine Eocene-Oligocene transition, especially in the marine deposits of the Gulf, Atlantic, and Pacific Coasts of North America. These topics in particular were barely represented at the 1989 Penrose, because so few people were working on them at the time, and even fewer were available to participate. Since that time, magnetostratigraphic, biostratigraphic, sequence stratigraphic, and single-crystal $^{40}Ar/^{39}Ar$ dates have been completed in a number of sections, producing a high-resolution chronostratigraphy and precise correlations of these strata to each other and to the global time scale. Oxygen and carbon isotopes have been studied in numerous places, allowing a direct comparison with the pelagic isotopic records. Paleontologists have amplified the database of foraminifera, mollusks, and echinoids, and begun to examine patterns of evolutionary change in these faunas in response to global climatic changes. In particular, the details of the long deterioration from the warm, high-diversity late early Eocene and earliest middle Eocene (about 52–49 Ma) through the 12 million years of the middle Eocene (49–37 Ma) and the 3 million years of the late Eocene (37-34 Ma) are being studied with much higher resolution than ever before. This interval will be critical to testing climatic and evolutionary hypotheses about the Eocene deterioration, the so-called "doubthouse" conditions between the greenhouse and the icehouse.

Most of this new research was not yet published, and there is only limited communication among all the people who work on these fossils and deposits. Consquently, we organized another Penrose Conference on "The Marine Eocene-Oligocene Transition," which was held August 17–22, 1999, at Evergreen State College in Olympia, Washington. Forty-seven scientists from the United States, Canada, Cuba, China, Japan, New Zealand, Germany, Belgium, and Hungary participated; among them were 10 graduate students, and 4 undergraduate students. Together we hoped to broaden our understanding of the changes associated with the transition from Eocene warmth to Oligocene glaciers from oceanographic, climatic, and paleontological perspectives. We addressed such questions as: what is the nature of change during the transition? Was it synchronous or diachronous? Were there discrete events, or was the change protracted and cumulative? What was the nature of the extinctions at the end of the middle Eocene, and during the early Oligocene? Is there selectivity dependent upon latitude or ecology? How can we correlate and compare records?

By the end of the meeting, it was very clear that the interest level, excitement, and morale of the participants was very high, and we were all eager to learn more and make further connections. This momentum is now maintained by a Web site <www.washington.edu/burkemuseum/paleo.html, click on Geology, Invertebrate Paleo>. Shortly after the conference ended, we began making arrangements to publish the proceedings of the symposium, so that the rest of the geological community could benefit from these recent discoveries. Our efforts have resulted in the compilation of papers represented in this volume. Many of the questions listed above received at least partial answers, and we learned many unexpected things during the intense workshop atmosphere of a Penrose Conference. Of course, many questions still remain to be answered, but the papers contained in this volume represent an important starting point for all future research into this fascinating and critical interval of time.

Part I

THE PACIFIC RIM

Pacific Coast Eocene-Oligocene Marine Chronostratigraphy: A Review and an Update

Donald R. Prothero

ABSTRACT

Exposed in the Pacific Coast of Washington, Oregon, and California is a geographically extensive record of relatively thick sequences of Eocene-Oligocene marine sediments deposited across a broad bathymetric spectrum in forearc basins. These strata and their fossils have been studied for more than 140 years. They yield an important record not only of their mollusks and benthic foraminifera, but also geochemical proxies and other climatic indicators. Yet the dating and chronostratigraphic resolution of many of these rocks has been controversial and poorly understood, hampering the efforts to use these paleoclimatic indicators in our understanding of the marine Eocene-Oligocene transition. Recent magnetostratigraphic studies, calibrated with planktonic microfossils and $^{40}Ar/^{39}Ar$ dates, have greatly improved the dating of these beds, resolved the issue of their correlation with the global time scale and with each other, and allowed us to date many of them to the nearest 100,000 years. This allows precise calibration of the benthic foraminiferal stages of Mallory (1959) and Kleinpell (1938). The Ulatisian/Narizian boundary is diachronous from 46.5 to 47.5 Ma. The Narizian/Refugian boundary is dated at 36.5 Ma at its type section but may be slightly younger elsewhere. The Narizian spans most of the middle Eocene, and the Refugian most of the late Eocene. The Refugian/Zemorrian boundary occurs early in Chron C13r (32.8 Ma, earliest Oligocene), so the long-controversial Refugian Stage ranges from late Eocene through the early Oligocene (as originally suggested by Almgren et al., 1988). The Zemorrian spans almost the entire Oligocene and a portion of the earliest Miocene. Likewise, the Californian molluscan zonation of Clark and Vokes (1936) can be calibrated. The Domengine-Transition stage boundary occurs at 47.0 Ma, and the Transition-Tejon boundary at 45.3 Ma. The rest of the Eocene in California is occupied by the Tejon Stage, or informal unnamed stage of Addicott (1973). The molluscan stages of Armentrout (1975) allow slightly better biostratigraphic resolution of the upper Eocene and Oligocene in Washington.

INTRODUCTION

The first Eocene or Oligocene fossils found on the West Coast of the United States were discovered in 1862, when William More Gabb, paleontologist for the Geological Survey of California, described mollusks from a boulder found in "Cañada de las Uvas" (now "Grapevine Canyon" along Interstate 5, between the San Emigdio and Tehachapi Ranges of southern California) (Squires, 1999a). Gabb (1864) thought these fossils were Cretaceous in age, and in 1869 he assigned them to beds he called the "Tejon Group" (from Fort Tejon and the Tejon Pass, just south of the "Grapevine"). But in 1855, Timothy Conrad of the Academy of Natural Sciences of Philadelphia reassessed the "Tejon" fossils and assigned them to the Eocene. Eventually, the term "Tejon" became entrenched in the literature as a local term for Eocene in California.

In the 1870s and 1880s, Thomas Condon (then of the University of Oregon) discovered distinctive Eocene bivalves such as *Venericardia* (Orr and Orr, 1999). Based on similarities to *Venericardia* from the Paris Basin, he realized that some beds in Oregon (such as the Coaledo Formation) were correlative with Lyell's Eocene in Europe. The mollusks of the San

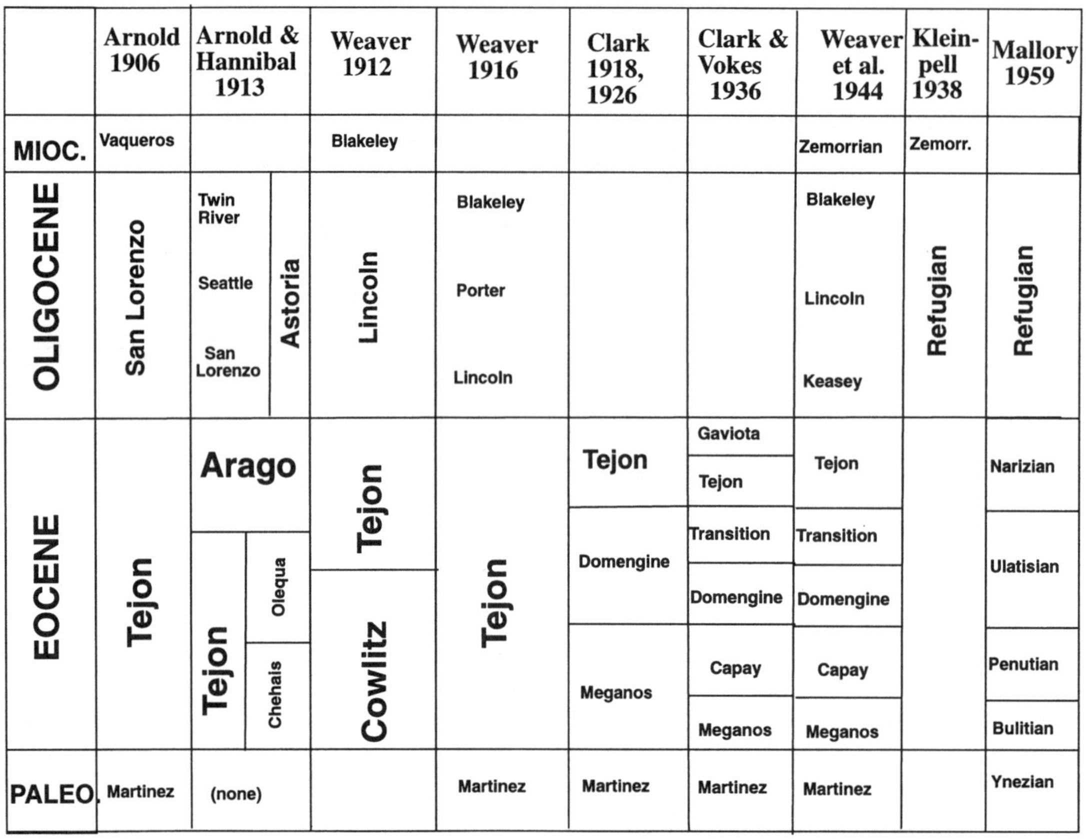

Figure 1.1. History of stratigraphic concepts for the Paleogene in the Pacific Coast. PALEO., Paleocene; MIOC., Miocene.

Lorenzo Formation in the Santa Cruz Mountains were also described, and their similarity to the Oligocene of Europe led early paleontologists to use the term "San Lorenzo" as equivalent of the California Oligocene. In 1875, the discovery of oil in California and the discovery of important coal deposits in Oregon and Washington further spurred biostratigraphic research on the Pacific Coast.

By the beginning of the twentieth century, many key stratigraphic units and their fossils had been named, described and mapped in all three states (Addicott, 1981). The first attempt to synthesize this understanding was the stratigraphic scheme of Ralph Arnold (1906), then at the U.S. Geological Survey (figure 1.1). He recognized very broad-scale stratigraphic divisions of the Cenozoic, with local names applied to each Lyellian epoch, so the entire Eocene was called "Tejon" and the entire Oligocene was the "San Lorenzo." This time scale was based largely on rock units and their pectinid bivalve fossils in California, although Arnold (1906:10) did indicate some possible correlatives in the Pacific Northwest. Like most scientists of his time, Arnold (1906) did not distinguish rock units and the time represented by their contained fossils, but treated rock units, time-rock units, and time units as interchangeable; the distinction among these concepts would not come for another 35 years. The names and concepts in Arnold's (1906) time scale have subsequently been refined and subdivided. Some parts have become obsolete, but the basic framework remains as the foundation for all later Cenozoic time scales on the Pacific Coast. Arnold (1906) also constructed the first paleogeographic maps for the Cenozoic, recognizing the difference between the Pacific Northwest and Californian molluscan provinces, and made the first paleoclimatic inferences based on mollusks.

A year after Arnold's pioneering effort was published, Charles Weaver finished his doctorate at the University of California, Berkeley, and took a position at the University of Washington. He then began a long-term field research program on the Cenozoic of the Pacific Northwest. Over the next few decades, he did extensive mapping and collecting, greatly enlarged the collections from such classic units as the Lincoln Creek, Cowlitz, and Coaledo Formations, improved the stratigraphic understanding of these units, and discovered and described many additional new fossils and formations. By 1912, Weaver had published his preliminary stratigraphic succession for the Pacific Coast, mixing Washington rock units (such as the Cowlitz Formation for middle Eocene) with California units (Tejon Formation as upper Eocene), and using the Washington Lincoln Creek Formation (then called the "Lincoln Formation," but later renamed "Lincoln Creek Formation" since "Lincoln Formation" was preempted) to represent the entire Oligocene (figure 1.1). Weaver later (1916a, 1916b, 1916c) modified his Eocene-Oligocene time scale, dropping the term "Cowlitz" and using "Tejon" as equivalent to the entire Eocene, but subdividing his "Lincoln" stage into successive Oligocene Lincoln, Porter, and Blakeley Stages.

While Weaver worked on Oregon and Washington strata, Arnold continued his research in California, collaborating with Harold Hannibal at Berkeley on a refined version of the Arnold (1906) time scale. Hannibal had first been trained at Stanford, but then worked with Weaver in Washington, so he was familiar with rocks and fossils from both regions. In 1913, Arnold and Hannibal published a refinement of the original Arnold (1906) stratigraphic column, mixing both Californian and Northwestern rocks and fossil terms (figure 1.1). They retained Arnold's "Tejon Stage" for the middle Eocene, but subdivided it based on Washington strata, and named the Chehalis and "Olequa" (= Cowlitz) substages for the middle Eocene. The "Arago" (= Coaledo) was considered to be of late Eocene age. The Oligocene still included the San Lorenzo, as well as Oregon-Washington units such as the Astoria, Seattle, and Twin River formations.

The next phase of research was initiated by Bruce Clark at Berkeley. He continued the study of the Eocene mollusks and formations in California, and by 1918 had subdivided the old "Tejon Stage" (spanning most of the Eocene) into the lower Eocene "Meganos Stage", middle Eocene "Domengine Stage", and upper Eocene "Tejon Stage" (figure 1.1). In 1936, Clark collaborated with his student Harold Vokes (who had just completed a dissertation on the Domengine Formation faunas), and elaborated on the Eocene molluscan stages of the Pacific Coast. Six successive Eocene stages were discussed: "Meganos," "Capay," "Domengine," "Transition," "Tejon," and "Gaviota." Although all of these named were based on California rocks and fossils, Clark and Vokes (1936) also indicated their ideas of correlative faunas in the Pacific Northwest.

Molluscan biostratigraphy had been the original basis for Cenozoic correlation and time scales in the Pacific Coast. By the 1920s the importance of microfossils was beginning to be appreciated. Some microfossils, especially benthic foraminifera, were abundant and distinctive in the marine rocks of this region, especially in the deeper-water shales that yielded few diagnostic mollusks. Their importance in correlation for the oil industry soon spurred intensive training and research by many micropaleontologists. The pioneering work by Joseph Cushman and others on the East Coast led to work on the West Coast. In 1924, Hanna and Hanna described the foraminifera of the Cowlitz Formation, and in 1928, Cushman and Hubert Schenck described two foraminiferal faunules from Oregon. Schenck (then at Stanford) was particularly influential in this regard, for he trained many of the micropaleontologists who carried on the next generation of foraminiferal research. He was also important for his influence in theoretical stratigraphy. Long bothered by the fact that geologists used rock and time terms interchangeably, he promoted the idea that rock units and time units were separate concepts, and required different nomenclature. In 1941, Schenck and Muller published their famous paper that established the separate hierarchies of time, time-rock, and rock units, and these concepts have been central to western stratigraphic thought ever since.

Schenck's most important student was Robert M. Kleinpell, who wrote a 1933 Stanford dissertation on the Miocene stratigraphy of California. Published as an AAPG monograph in 1938, this volume integrated his research on benthic foraminifera with his extensive knowledge of oil-company data from all over the state, and established the first benthic foraminiferal time scale for the California Miocene. Included in Kleinpell's 1938 time scale were units such as the Zemorrian (which he intially interpreted as early Miocene, but he later recognized as Oligocene), and the Refugian Stage, which he and Schenck had defined in 1936. In 1946, after serving in World War II and suffering internment in a Japanese prison camp, Kleinpell took a position at Berkeley, where he soon became the most influential micropaleontologist on the West Coast, and trained most of the next few generations of micropaleontologists for the oil industry and academia.

In the late 1930s, the Geological Society of America commissioned a number of committees of distinguished stratigraphers and paleontologists to formalize the stratigraphic nomenclature in North America. The task of heading up the Pacific Coast Cenozoic committee fell to Charles Weaver, but was divided up among nearly all the Cenozoic mollusk specialists (Clark, Durham, Grant, Putnam, Taliaferro, and others), as well as micropaleontologists (Kleinpell, Schenck, Bramlette, Kelley, Beck, Carlson, Forest, and Hill) who were active at the time. This report was eventually published in 1944 (Weaver et al., 1944) in the *Geological Society of America Bulletin.* The major part of the paper consisted of a large foldout correlation chart, divided by region, with each region being the responsibility of a different set of authors. In the text, Weaver et al. (1944) noted that a conflict had developed between the molluscan paleontologists and micropaleontologists over their concepts of correlation. The older, mollusk-based scheme tried to find common species with the Cenozoic mollusks of the type sections in Europe, so their local zonation could be compared to the global standard. The micropaleontologists were developing an independent scheme (with its own nomenclature) based on benthic foraminifera. As Weaver (1944:571) wrote, "The cleavage into two schools, due partly to the preference of some stratigraphers for one set of material and techniques and of others for a different set, springs largely from the inexorable fact that parts of the Cenozoic stratigraphic record in the West Coast province are rich in microfossils and almost lacking in mollusks or other large, shell-bearing groups, whereas other parts of the record are marked by abundant megafossils and few Foraminifera." The main conflict occurred when the two schools attempted to correlate their strata to the Lyellian epochs. Because both benthic foraminifera and benthic megainvertebrates are mostly endemic to the Pacific Coast, relatively few species could be confidently correlated to the type sections in Europe, so the position of the epoch boundaries on the chart was difficult to determine and highly controversial. On the right side of the chart (Weaver et al., 1944:figure 1), for example, are two different standard time scales (one based on mollusks, the other on foraminifera), which differ in many key points. For example, the molluscan Oligocene (based on the Lincoln and Blakeley Stages) was much shorter and more restricted than the foraminiferal Oligocene (more or less equivalent to the Refugian Stage). The foraminiferal time scale even had a transitional "Eo-Oligocene" for the early Refugian.

Although Pacific Coast molluscan biostratigraphic research almost ended with the Weaver Committee report, foraminiferal biostratigraphy continued to develop. In 1939, Boris Laiming published the first benthic foraminiferal zonation for the Eocene of the San Joaquin Basin of California. In 1959, Stan Mallory, one of Kleinpell's Berkeley students, published his dissertation research, which established a formal series of benthic foraminiferal stages for the Paleogene (figure 1.1). The Eocene was divided into four stages (Bulitian, Penutian, Ulatisian, Narizian), and the Refugian Stage was thought to span most of the Oligocene. Other authors placed the Refugian entirely within the Eocene, or thought it straddled the Eocene/ Oligocene boundary (see Prothero and Thompson, 2001, for a discussion). The problem of what was considered Oligocene in the West Coast was particularly acute. Some stratigraphers considered the Zemorrian Stage to span most or all of the Oligocene, while others correlated it only with the early Miocene. The problem was compounded by the fact that good continuous Oligocene or Zemorrian exposures are very rare in California and (except for the San Lorenzo Formation in central California) seldom yield mollusks. Kleinpell (1938) was forced to pick a very inadequate type section for the Zemorrian (see Prothero and Resseguie, 2001, for discussion), because better sections were known only from the subsurface. Tipton et al. (1973) tried to address the problem with a combination of surface and subsurface exposures, but they had difficulty deciding how much, if any, of these exposures were Oligocene in the Lyellian sense. Some authors even doubted whether there was any true Oligocene in the Pacific Coast at all. For example, Eames and others (1962) argued that "in the whole region we have considered (and even as far north as the state of Washington), there are no published records of stratigraphical successions of fossiliferous marine beds which can be dated as Oligocene." Lipps (1965) replied by showing that Oligocene planktonic foraminifera do occur in California, and since then the Oligocene age of many formations has been well documented.

As the stratigraphic controversies and confusion continued, the limits of zonations based on benthic foraminifera and mollusks became more and more apparent. The first problem was that benthic organisms tend to be highly provincial, so they seldom can be used for correlation to the stratotypes in Europe. Secondly, the stratigraphic distribution of benthic organisms is often affected by water depth. Consequently, their first and last occurrences may reflect changes in paleobathymetry rather than true first and last occurrences in time. Although some benthic foraminiferal workers (e.g., Kleinpell, 1938, 1972, 1980) knew of and debated the problem, its magnitude was not realized until a later generation of workers began to make detailed comparisons of the benthic foraminiferal correlations between California and Washington (e.g., McDougall, 1980). By the 1970s, years of work on the Deep Sea Drilling Project led to the realization that planktonic microfossils (especially foraminifera, coccoliths, radiolaria, and diatoms) were the best fossils for global correlation. Because microplankton live in the pelagic waters around the world and evolve rapidly, they are not controlled by differences in the paleobathymetry of the ocean floor beneath them, and can be found in marine sediments of almost any water depth in any region. In the 1960s, a number of studies utilized planktonic microfossils to resolve many of the long-standing debates. The most important of such studies was the solution of the controversies over the correlation of the European standards for the Lyellian epochs. Using planktonic microfossils from both deep-sea cores and the shallow marine stratotypes in Europe, Berggren and colleagues (Berggren, 1971; Hardenbol and Berggren, 1978; Berggren et al., 1978, 1985) ended the confusion of what was meant by Eocene and Oligocene. This in turn allowed the many concurrent studies on planktonic microfossils from Pacific Coast marine sections to determine the Lyellian epoch assignments of these rocks.

The planktonic microfossil research on the Pacific Coast Eocene-Oligocene took place largely in the 1960s and 1970s, and culminated in volumes edited by Armentrout (1981b) and Brabb (1983). Since that time, relatively few studies have been undertaken as the community of micropaleontologists largely moved on to other projects, or stopped working in this area due to changes in the oil industry and the U.S. Geological Survey (which once employed most of them). The most recent advance has been in the use of magnetic stratigraphy to correlate these previously studied sections to the time scale of Berggren et al. (1985, 1995). Although such studies have been published since the 1980s (e.g., Prothero and Armentrout, 1985; Flynn, 1986), the majority are published in a recent SEPM Pacific Section volume (Prothero, 2001a). Finally, the advent of $^{40}Ar/^{39}Ar$ geochronology has made high-precision dating of Pacific Coast marine Eocene-Oligocene rocks much more practical, and a number of recent dates have greatly improved the calibration and correlation of these sections.

CHRONOSTRATIGRAPHY OF THE BENTHIC FORAMINIFERAL STAGES

As discussed above, the designation of formal benthic foraminiferal stages for the Paleogene was a relatively late development. The Refugian was named in 1936 and the Zemorrian in 1938 (both largely due to Kleinpell's work). The first complete biostratigraphic zonation was that of Laiming (1939), but it wasn't until the publication of Mallory's (1959) zonation that formal biostratigraphic stages were named for the Paleocene and Eocene. As these zonation schemes became more widely adopted in the 1960s and 1970s, problems soon became apparent. McDougall (1980) compared the criteria for the Narizian, Refugian, and Zemorrian in Washington and California, and found that the zonations in one state were grossly time-transgressive when using the criteria of the other state. Poore (1980) pointed out some of the major difficulties in correlation to the Lyellian standard, based on his study of the planktonic foraminifera.

Almgren et al. (1988) reviewed the issues in detail, pointing out that many type sections were inadequate, or that the key indicator fossils did not occur as originally described. Using calcareous nannofossils, they recalibrated the zonation schemes of Laiming (1939) and Mallory (1959). The Bulitian Stage of Mallory (1959) is consistently earliest Eocene (55.0–53.5 Ma), but the successive Penutian Stage is grossly time-transgressive and facies-dependent. Although its beginning is probably synchronous within early Eocene nannofossil Zone CP9 (around 53.5 Ma), many of the Penutian indicator species are persistent deep-water taxa, so the Penutian ends as early as the middle of nannofossil Zone CP10 (52.0 Ma) in some places, but in others it overlaps the successive Ulatisian and even early Narizian Stages, ending as late as early Zone CP13 (47.5 Ma), or approximately 4.5 million years of diachrony. Almgren et al. (1988:figure 2) showed that Laiming's (1939) Zone D is interchangeable with the Bulitian, and Zone C with the Penutian, and that the latter zone is also plagued by a similar problem of deep-water index taxa that persist through the rest of the Eocene.

No paleomagnetic studies have been done on strata representing the Penutian or Bulitian Stages (largely because of this problem of time-transgression, and also due to poor exposures), but Ulatisian, Narizian, Refugian and Zemorrian strata have been sampled (figure 1.2). Prothero and Brabb (2001) analyzed the type section of the Ulatisian in the Vacaville Shale,

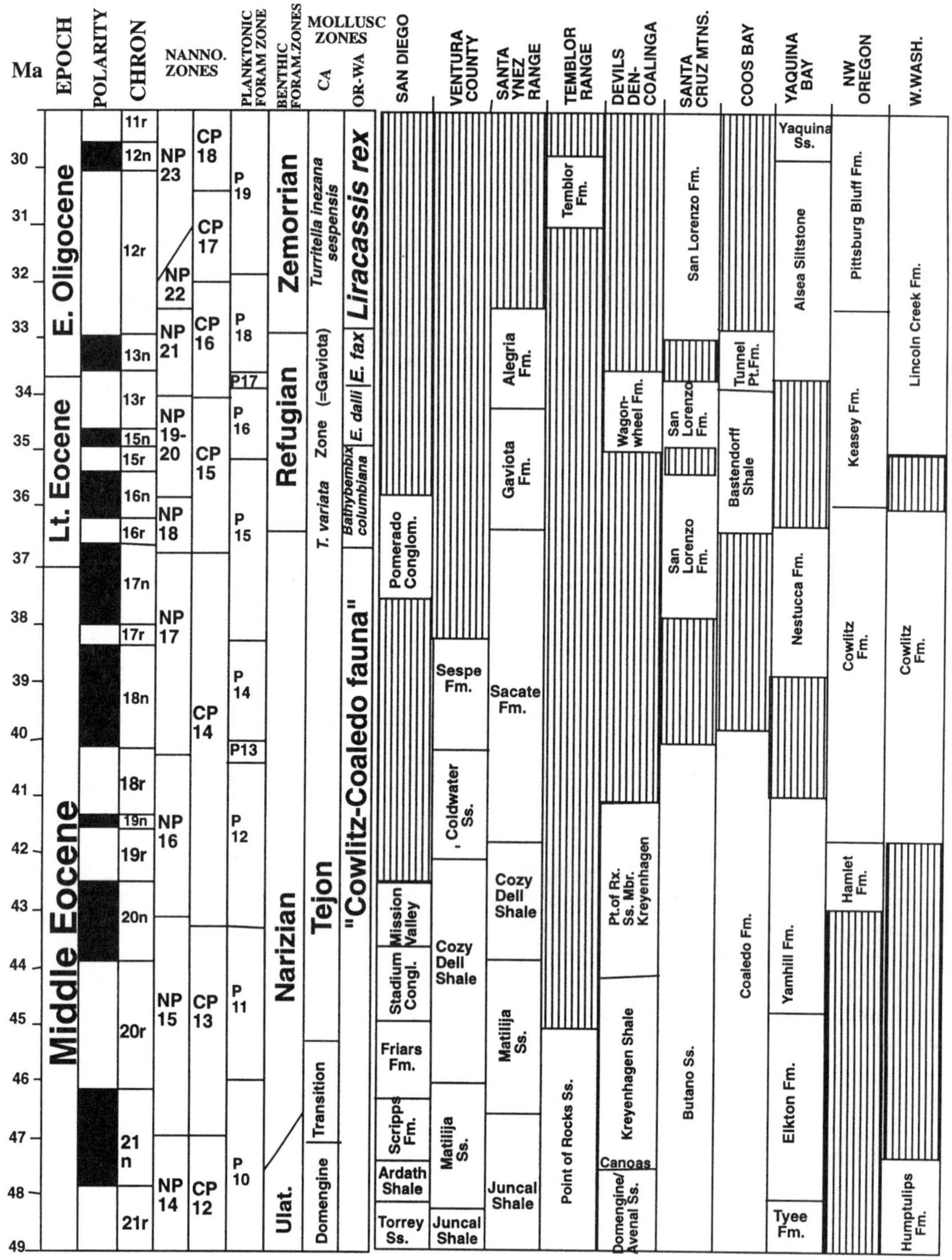

Figure 1.2. Correlation of key middle Eocene-early Oligocene formations in the Pacific Coast region, based on magnetostratigraphic studies cited in text, or in Prothero (2001a). Time scale after Berggren et al. (1995).

near Vacaville in Solano County, California. Based on the occurrence of CP11–CP13 nannofossils in the section, they correlated its magnetozones with Chrons C22r–C20r (45.0–50.0 Ma). This correlation places the Ulatisian/Narizian boundary within nannofossil Zone CP13 and Chron C21n (about 46.5 Ma). Elsewhere, Almgren et al. (1988:figure 3) showed that the Ulatisian/ Narizian boundary occurs at the Zone CP12/CP13 boundary in the Media Agua Creek section of the Temblor Range in the western San Joaquin Basin, and this is corroborated by its occurrence early in Chron C21n (about 47.5 Ma) in that section (Prothero, 2001b). Thus, the nannofossils and paleomagnetism show that this boundary is time-transgressive between the beginning and the middle of Zone CP13 and Chron C21n (46.5–47.5 Ma), or about a million years of demonstrable diachrony.

The rest of the 10–12 million years of the Narizian is less easy to calibrate. Prothero (2001b) analyzed Mallory's (1959) type section of the Narizian in the Point of Rocks Sandstone near Devil's Den, just north of the Temblor Range in the western San Joaquin Basin, but the exposures were too limited to produce much additional information. However, the magnetic stratigraphy of the available outcrops show that the type section does not span much of the long duration of the Narizian. In the San Lorenzo Formation of the Santa Cruz Mountains, California, Prothero, Sutton, and Brabb (2001) showed that the early/late Narizian boundary occurred within Chron C18r (about 41.0 Ma), as calibrated by Zone CP14a–14b nannofossils and Zone P14 planktonic foraminifera. Other Ulatisian-Narizian sections in California, such as the Matilija–Cozy Dell–Coldwater–Sespe sequence in the Transverse Ranges of Ventura County, California (Prothero and Vance, 1996; Prothero and Britt, 1998), or the San Diego middle Eocene sequence (Walsh et al., 1996), corroborate these interpretations.

Prothero and Thompson (2001) studied the paleomagnetism of the Narizian/Refugian boundary in the type section in the western Santa Ynez Range, Santa Barbara County, California. Using the nannofossil calibrations of Warren and Newell (1980) and Almgren and Filewicz (1984), they found that this boundary occurred in nannofossil Zone NP18, and in a reversed polarity interval correlative with Chron C16r (36.5 Ma). However, in the Lincoln Creek Formation in the southern Olympic Peninsula of Washington, Prothero and Armentrout (1985) found that the early Refugian is as old as nannofossil Zone NP19/20 and Chron C15r (35.3 Ma). Thus, the boundary is as old as 36.5 Ma in the type section, but could be younger in other regions. The early/late Refugian boundary falls with Chron C13r (about 34.0–34.2 Ma, slightly older than the Eocene/ Oligocene boundary) in both regions. With these correlations, the long debate (discussed above) over whether the Refugian Stage was late Eocene, early Oligocene, or both has been resolved (Prothero and Thompson, 2001). The early Refugian is late Eocene, but the latest Refugian is early Oligocene, so the stage crosses the epoch boundary, as suggested by Poore (1980) and Almgren and Filewicz (1984). This is corroborated by the magnetic stratigraphy of the Keasey Formation in northwest Oregon (Prothero and Hankins, 2000), where the early/late Refugian also straddles the Eocene/Oligocene boundary.

The Refugian/Zemorrian boundary has been calibrated in three places. In the Lincoln Creek Formation of Washington (Prothero and Armentrout, 1985), it is found in a continuous, conformable sequence and dates to early Chron C12r (32.8 Ma). In the Refugian type strata of the Santa Ynez Range (Prothero and Thompson, 2001), Zemorrian benthic foraminifera are lacking right above the latest Refugian fossils, but the latest Refugian foraminifera are associated with Zone CP16a nannofossils (Almgren and Filewicz, 1984) and found in reversed strata that also appear to correlate with early Chron C12r. In the San Lorenzo Formation (Prothero, Sutton, and Brabb, 2001), there is an unconformity at the Refugian/Zemorrian boundary, but early Zemorrian strata contain Zone CP18 nannofossils and Zone P19/20 planktonic foraminifera and are reversed in polarity, suggesting correlation with middle or late Chron C12r. The San Lorenzo Formation also calibrates the early/late Zemorrian as occurring late in Chron C12r (31.0 Ma).

The type section of the Zemorrian, however, turned out to be grossly inadequate for high-resolution biostratigraphy (figure 1.3). When Kleinpell (1938) designated as his type section the exposures of the lower Temblor Formation along Zemorra Creek in the Temblor Range, he picked what he thought were the best surface exposures representing the thick Zemorrian strata in the subsurface of the San Joaquin Basin. When these strata were paleomagnetically sampled and calibrated with calcareous nannofossils, however, they turned out to represent only a small portion of the Zemorrian (Prothero and Resseguie, 2001). The early Zemorrian Cymric Shale contains Zone CP16b–c nannofossils and is of reversed polarity, so it apparently spans early Chron C12r (32.5–33.0 Ma).

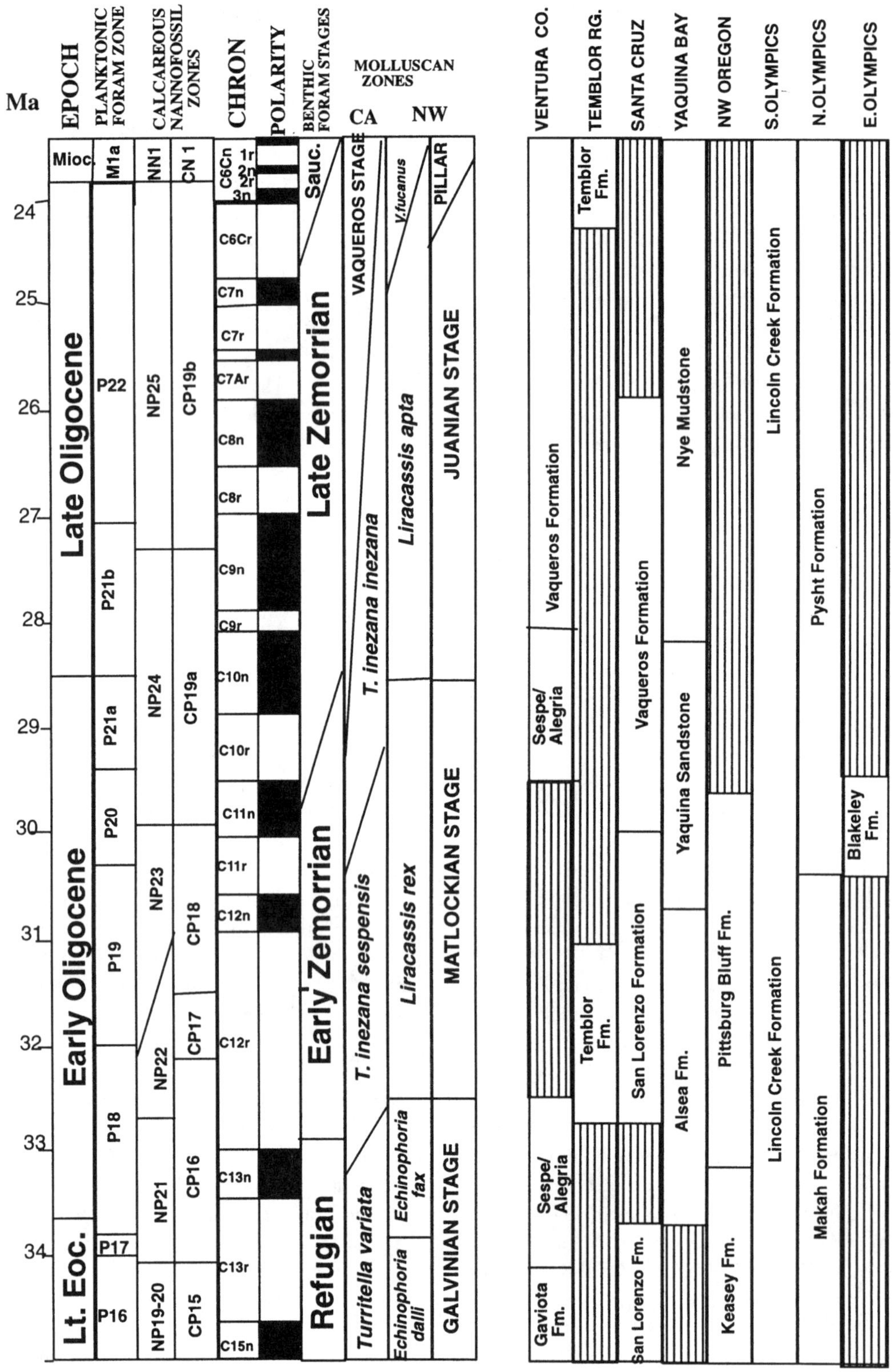

Figure 1.3. Correlation of key Oligocene formations in the Pacific Coast region, based on magnetostratigraphic studies cited in text, or in Prothero (2001a). Time scale after Berggren et al. (1995).

The overlying late Zemorrian Wygal Sandstone is also of reversed polarity and apparently correlates with some part of Chron C12r. The uppermost Zemorrian lower Santos Shale contains latest Oligocene Zone CP19b nannofossils, and passes conformably up into the Agua Sandstone, which contains early Miocene Zone CN1 nannofossils, so the rest of the type Zemorrian straddles the Oligocene/Miocene boundary (Chrons C6Cr–C6Cn, 23.8–24.4 Ma). Most of the Zemorrian, and most of the middle and late Oligocene (24.4–29.0 Ma) is unrepresented by strata in the type Zemorrian section (which has an unconformity spanning this interval). Much thicker, more continuous, and apparently conformable Zemorrian sections (figure 1.3) occur in the San Lorenzo Formation in California (Prothero, Sutton and Brabb, 2001), the Lincoln Creek Formation in Washington (Prothero and Armentrout, 1985), the Alsea-Yaquina-Nye Formations on Yaquina Bay in west-central Oregon (Prothero, Bitboul, Moore, and Niem, 2001), and in the Pittsburg Bluff Formation in northwestern Oregon (Prothero and Hankins, 2001).

CHRONOSTRATIGRAPHY OF THE MOLLUSCAN STAGES

The molluscan biostratigraphy was the original basis for most Pacific Coast marine Eocene-Oligocene time scales and correlation, so it is helpful to calibrate these zones as precisely as possible (Squires, 1988). As discussed above, five molluscan zones were well established in California for the Eocene by Clark and Vokes (1936). In California, a series of *Turritella* zones have been proposed (figures 1.2, 1.3).

By contrast, in the Pacific Northwest, only the long-ranging "Cowlitz–Coaledo fauna" represents the entire middle Eocene at this point, although there is the potential to subdivide the Cowlitz zonation further based on *Turritella* (E. Nesbitt, pers. commun.). But the late Eocene and Oligocene have been zoned by the use of the cassid gastropod lineage *Echinopho-ria-Liracassis* (Armentrout, 1975; Moore, 1984), so the Oligocene is better subdivided in the Pacific Northwest than it is in California (figures 1.2, 1.3).

The first two Eocene molluscan stages of Clark and Vokes (1936), the Meganos and Capay (figure 1.1), occur in very limited exposures in their type sections and there have been no planktonic microfossils recovered from these rocks (Squires, pers. commun.), so they are unsuitable for magnetic stratigraphy and cannot be dated any more precisely than early Eocene. Elsewhere, they have been more precisely dated using coccoliths in conjunction with mollusks (Squires, 1987; Squires and Demetrion, 1992; Squires, 1999b; Almgren et al., 1988). However, the next stage, the Domengine, has been sampled in several places. In its type section on Domengine Creek in Fresno County, California, Prothero (2001c) found that the classic Domengine faunas are associated with Zone CP12a nannofossils and reversed polarity, correlating these strata with early Chron C21r (48.5–49.0 Ma). The correlative Domengine Formation faunas in the Avenal Sandstone on Garza Creek on Reef Ridge in Kings County, California, also appear to correlate with Chron C21r. The Domegine/"Transition" molluscan zonal boundary occurs in Chron C21n (about 46.0–48.0 Ma) in both the Ventura County (Prothero and Britt, 1998) and San Diego County (Walsh et al., 1996) sections. The "Transition"/Tejon molluscan zonal boundary occurs in Chron C20r (about 45 Ma) in these same two regions as well.

Kleinpell and Weaver (1963) proposed an informal "*Turritella variata* Zone" for the type Refugian Gaviota Formation in the Santa Ynez Range, Santa Barbara County, California. This is apparently equivalent to the "Gaviota Stage" of Clark and Vokes (1936) that was later abandoned by Weaver et al. (1944). Prothero and Thompson (2001) showed that the Gaviota Formation spans Chrons C16r–C13r (34.5-36.5 Ma), so this zone is coextensive with the early Refugian. In fact, Schenck and Kleinpell's (1936) original definition of the Refugian included both benthic foraminifera and mollusks, but the later usage dropped the molluscan component, so today the Refugian is viewed strictly as a benthic foraminiferal stage. Addicott's (1973) early Oligocene "unnamed zone" (based on the *Phacoides* fauna of the Wygal Sandstone of the Temblor Formation, Temblor Range) correlates with early Oligocene Chron C12r (31–33 Ma) (Prothero and Resseguie, 2001).

In the Pacific Northwest, the molluscan zonation for the middle Eocene is much less well constrained than in California, but that for the late Eocene and Oligocene is much better known. The middle Eocene has been represented by the widespread Cowlitz–Coaledo fauna. However, paleomagnetic studies have shown that this fauna is extremely long-ranging. Prothero and Donohoo (2001) showed that the Coaledo Formation on Coos Bay in southwestern Oregon correlates with Chrons C18r–C20n (40.2–43.8 Ma), but the Cowlitz Formation correlates with Chron C17n-early Chron C19n (36.7–41.5 Ma), based on a $^{40}Ar/^{39}Ar$ date of 38.9 ± 0.01 Ma in the middle of the type section of the Cowlitz (Prothero, Sanger, Nesbitt, Niem, and Kleibacker, 2001). Thus, the two sections yielding the Cowlitz–Coaledo fauna only partly overlap in time.

Armentrout (1975) refined earlier work by Durham (1944) and proposed a series of molluscan stages and zones for the upper Eocene and Oligocene based on the Lincoln Creek Formation in southwestern Washington. These were calibrated paleomagnetically by Prothero and Armentrout (1985). The early Galvinian *Bathybembix columbiana* zone ends in Chron C15r, and may range as early as Chron C16r (35.3–36.5 Ma). The middle Galvinian *Echinophoria dalli* zone ranges from Chron C15r to mid C13r (34.2–35.3 Ma). The late Galvinian *Echinophoria fax* zone spans the interval from mid Chron C13r to early Chron C12r (32.0–34.2 Ma). The Matlockian *Liracassis rex* zone ranges from early Chron C12r to Chron C10n (32.0–28.5 Ma). The Juanian *Liracassis apta* zone spans the rest of the late Oligocene, from Chron C10n to the beginning of the Miocene (28.5-23.5 Ma).

DISCUSSION

The correlations of the major formations in this region are shown in figures 1.2 and 1.3. One of the most striking aspects of this chart can be seen by contrasting figures 1.2 and 1.3 with figure 1.1. In the early days of Pacific Coast Cenozoic marine stratigraphy, authors assumed that their stages were long-ranging and represented continuous intervals of time, with few gaps between them. Consequently, their typical representation on most time charts (such as those in Armentrout, 1981a, or the COSUNA charts) is as continuous, conformable boxes representing all of the time interval in question. However, as the actual stratigraphic basis for these stages has become more precisely dated, the overwhelming impression is one of discontinuity, and large stratigraphic gaps, hiatuses, and unconformities. Figures 1.2 and 1.3 show that relatively few regions have stratigraphic sequences that span more than a few million years continuously and conformably. Most of the sequences that were thought to be long-ranging (such as the type Zemorrian discussed above) turn out to represent relatively short intervals of time, with long hiatuses spanning most of the interval. This should come as no surprise to those who have followed contemporary stratigraphic thinking, such as Ager's (1963) assertion that the stratigraphic record is full of holes, and is "more gaps than record" (see Prothero, 1990). However, until the advent of precise (to the nearest 100,000 years) dating of these beds, it was not possible to appreciate just how incomplete the stratigraphic record of this region is. This is in spite of the fact that these are deposits of rapidly subsiding forearc basins, sometimes reaching more than 3000 m in thickness, and thus might be expected to be more complete than records in the slowly subsiding Gulf and Atlantic Coast passive margins, or on the craton. Yet even in this region of high sediment accumulation rates, the new chronostratigraphy indeed shows that there are "more gaps than record."

CONCLUSION

Magnetic stratigraphy, calibrated by planktonic microfossils and by occasional $^{40}Ar/^{39}Ar$ dates, has allowed us to date most of the important Eocene-Oligocene fossiliferous rocks in California, Oregon, and Washington much more precisely than ever before. Not only are their correlations to the global time scale and the Lyellian epochs much more secure, but with magnetic stratigraphy, it is possible to date individual events to the nearest 100,000 years. This new high-resolution chronostratigraphy has also pinpointed stratigraphic units that are facies-dependent and/or time transgressive.

With a correlation framework such as that shown in figures 1.2 and 1.3, it is now possible to precisely calibrate the regional paleoclimatic trends observed by Durham (1969) and Addicott (1970), and correlate them to global trends in climate. Such trends are discussed in other chapters in this volume.

ACKNOWLEDGMENTS

I thank A. Almgren, J. Armentrout, C. Hickman, E. Moore, E. Nesbitt, and R. Squires for helpful comments on this manuscript. This research was supported by National Science Foundation grants EAR 94-05942, 97-06046, 98-05071, and 00-00174 and by grants from the Donors of the Petroleum Research Fund, administered by the American Chemical Society.

REFERENCES

Addicott, W. O., 1970, Tertiary paleoclimatic trends in the San Joaquin Basin, California: U.S. Geological Survey Professional Paper, v. 644-D, p. D1-D19.

Addicott, W. O., 1973, Oligocene molluscan biostratigraphy and paleontology of the lower part of the type Temblor Formation, California: U.S. Geological Survey Professional Paper, v. 791.

Addicott, W. O., 1981, Brief history of Cenozoic marine biostratigraphy of the Pacific Northwest: Geological Society of America Special Paper, v. 184, p. 3-15.

Ager, D. V., 1963, The nature of the stratigraphical record: London, Macmillan.

Almgren, A. A., and Filewicz, M. V., 1984, Benthic foraminiferal and calcareous nannofossil biostratigraphy of the Markley Canyon fill, *in* Almgren, A. A., and Hacker, P. D., eds., Paleogene submarine canyons of the Sacramento Valley: Pacific Section AAPG Symposium Volume, p. 115-124.

Almgren, A. A., Filewicz, M. V., and Heitman, H. L., 1988, Lower Tertiary foraminiferal and calcareous nannofossil zonation of California: an overview: Pacific Section SEPM, v. 58, p. 83-105.

Armentrout, J. M., 1975, Molluscan biostratigraphy of the Lincoln Creek Formation, southwest Washington, *in* Weaver, D. W., et al., eds., Future Energy Horizons of the Pacific Coast—Paleogene symposium and technical papers: AAPG Pacific Section, SEPM, and SEG Annual Meeting, Long Beach, California, April 1975, pp. 14-48.

Armentrout, J. M., 1981a, Correlation and ages of Cenozoic chronostratigraphic units in Oregon and Washington: Geological Society of America Special Paper, v. 184, p. 137-148.

Armentrout, J. M., ed., 1981b, Pacific Northwest Cenozoic biostratigraphy: Geological Society of America Special Paper, v. 184.

Arnold, R., 1906, The Tertiary and Quaternary pectens of California: U.S. Geological Survey Professional Paper, v. 47, p. 1-264.

Arnold, R., and Hannibal, H., 1913, The marine Tertiary stratigraphy of the north Pacific Coast of North America: American Philosophical Proceedings, v. 52, p. 559-605.

Berggren, W. A., 1971, Tertiary boundaries, *in* Funnell, B. F., and Riedel, W. R., eds., The Micropaleontology of Oceans: Cambridge, Cambridge University Press, p. 693-808.

Berggren, W. A., Kent, D. V., and Flynn, J. J., 1985, Paleogene geochronology and chronostratigraphy: Geological Society of London Memoir, v. 10, p. 141-195.

Berggren, W. A., Kent, D. V., McKenna, M. C., Hardenbol, J., and Obradovich, J. D., 1978, Revised Paleogene polarity time scale: Journal of Geology, v. 86, p. 67-81.

Berggren, W. A., and Van Covering, J. A., 1978, Biochronology: American Association of Petroleum Geologists Memoir, v. 6, p. 39-55

Berggren, W. A., Kent, D. V., Swisher, C. C., III, and Aubry, M.-P., 1995, A revised Cenozoic geochronology and chronostratigraphy: SEPM Special Publication, v. 54, p. 129-212.

Brabb, E. E., ed., 1983, Studies in Tertiary Stratigraphy in the California Coast Ranges: U.S. Geological Survey Professional Paper 1213.

Clark, B. L., 1918, The San Lorenzo series of middle California: University of California Publications in Geological Sciences, v. 11, n. 2, p. 45-234.

Clark, B. L., and Vokes, H. E., 1936, Summary of the marine Eocene sequence of western North America: Geological Society of America Bulletin, v. 47, p. 851-878.

Conrad, T. A., 1855, Report on the fossil shells collected in California by W. P. Blake: U.S. Geological Survey Professional Paper, v. 59, p. 163-171.

Cushman, J. A., and Schenck, H. G., 1928, Two foraminiferal faunules from the Oregon Tertiary: University of California Publications in Geological Sciences, v. 27, p. 305-324.

Durham, J. W., 1944, Megafossil zones of the Oligocene of northwestern Washington: University of California Publications in Geological Sciences, v. 27, p. 84-104.

Durham, J. W., 1969, Tertiary climatic change in the marginal northeastern Pacific Ocean: Science, v. 165, p. 583-586.

Eames, F. E., Banner, F. T., Blow, W. H., and Clarke, W. J., 1962, Fundamentals of mid-Tertiary stratigraphic correlation: Cambridge, Cambridge University Press.

Flynn, J. J., 1986, Correlation and geochronology of middle Eocene strata from the western United States: Palaeogeography, Palaeoclimatology, Palaeoecology, v. 55, p. 335-406.

Gabb, W. M., 1864, Description of Cretaceous fossils; and index of localities of the Cretaceous fossils, *in* Gabb, W. M., Paleontology of California: Geological Survey of California, v. 1, sect. 4, p. 57-236.

Gabb, W. M., 1869, Description of new, and revision of previously described, Cretaceous fossils; and synopsis of all the described invertebrate Cretaceous fossils of California and the adjacent states, *in* Gabb, W. M., Paleontology of California: Geological Survey of California, v. 2, sect. 2, p. 127-254.

Hanna, G. D., and Hanna, M. A., 1924, Foraminifera from the Eocene of Cowlitz River, Lewis County, Washington: University of Washington Publications in Geology, v. 1, p. 57-62.

Hardenbol, J., and Berggren, W. A., 1978, A new Paleogene numerical time scale: American

Association of Petroleum Geologists Memoir, v. 6, p. 213-234.

Kleinpell, R. M., 1938, Miocene stratigraphy of California: Tulsa, Oklahoma, American Association of Petroleum Geologists.

Kleinpell, R. M., 1972, Some of the historical context in which a micropaleontological stage classification of the Pacific Coast middle Tertiary has developed, *in* Stinemeyer, E. H., and Church, C. C., eds., The Pacific Coast Miocene biostratigraphy symposium: Pacific Section SEPM, 47th Annual Meeting, p. 89-110.

Kleinpell, R. M., 1980, The Miocene stratigraphy of California revisited: American Association of Petroleum Geologists Studies in Geology, v. 11.

Kleinpell, R. M., and Weaver, D. W., 1963, Oligocene biostratigraphy of the Santa Barbara embayment: University of California Publications in Geological Sciences, v. 43, p. 1-77.

Laiming, B. 1939. Some foraminiferal correlations in the Eocene of the San Joaquin Valley, California: Sixth Pacific Science Conference Proceedings, v. 2, p. 535-638.

Lipps, J. H., 1965, Oligocene in California?: Nature, v. 228, p. 885-886.

Mallory, V. S., 1959, Lower Tertiary stratigraphy of the California Coast Ranges: Tulsa, Oklahoma, American Association of Petroleum Geologists.

McDougall, K., 1980, Paleoecological evaluation of late Eocene biostratigraphic zonations of the Pacific Coast of North America: Paleontological Society Monograph, v. 2.

Moore, E. J., 1984, Middle Tertiary molluscan zones of the Pacific Northwest: Journal of Paleontology, v. 58, p 718-737.

Orr, E. L., and Orr, W. N., 1999. Oregon Fossils: Dubuque, Iowa, Kendall/Hunt.

Poore, R. Z., 1980, Age and correlation of the California benthic foraminiferal stages: U.S. Geological Survey Professional Paper, v. 1162–C, p. C1–C18.

Prothero, D. R., 1990, Interpreting the Stratigraphic Record: New York, W. H. Freeman.

Prothero, D. R., 2001a, Magnetic Stratigraphy of the Pacific Coast Cenozoic: Pacific Section SEPM Special Publication, v. 91.

Prothero, D. R., 2001b, Magnetic stratigraphy of the middle Eocene (Ulatisian-type Narizian) Point of Rocks Sandstone, Kern County, California: Pacific Section SEPM Special Publication, v. 91, p. 65-74.

Prothero, D. R., 2001c, Magnetic stratigraphy of the lower-middle Eocene Domengine, Avenal, and Yokut Sandstones, western San Joaquin Basin, Fresno and Kings Counties, California: Pacific Section SEPM Special Publication, v. 91, p. 45-55.

Prothero, D. R., and Armentrout, J. M., 1985. Magnetostratigraphic correlation of the Lincoln Creek Formation, Washington: implications for the age of the Eocene/ Oligocene boundary: Geology, v. 13, p. 208-211.

Prothero, D. R., Bitboul, C. Z., Moore, G. W., and Niem, A., 2001, Magnetic stratigraphy and tectonic rotation of the Oligocene Alsea, Yaquina, and Nye formations, Lincoln County, Oregon: Pacific Section SEPM Special Publication, v. 91, p. 184-194.

Prothero, D. R., and Brabb, E. E., 2001, Magnetic stratigraphy of the lower middle Eocene (type Ulatisian) Vacaville Shale, Solano County, California: Pacific Section SEPM Special Publication, v. 91, p. 56-64.

Prothero, D. R., and Britt, J. R., 1998, Magnetic stratigraphy and tectonic rotation of the middle Eocene Matilija Sandstone and Cozy Dell Shale, Ventura County, California: implications for sequence stratigraphic correlations: Earth and Planetary Sciences Letters, v. 163, p. 261-273.

Prothero, D. R., and Donohoo, L. L., 2001, Magnetic stratigraphy and tectonic rotation of the middle Eocene Coaledo Formation, Coos County, Oregon: Geophysical Journal International, v. 145, p. 1-15.

Prothero, D. R., and Hankins, K. G., 2000, Magnetic stratigraphy and tectonic rotation of the upper Eocene-lower Oligocene Keasey Formation, Columbia County, Oregon: Journal of Geophysical Research, v. 105, n. B7, p. 16473-16480.

Prothero, D. R., and Hankins, K. G., 2001, Magnetic stratigraphy and tectonic rotation of the lower Oligocene Pittsburg Bluff Formation, Columbia County, Oregon: Earth and Planetary Sciences Letters, v. 91, p. 201-209.

Prothero, D. R., and Resseguie, J. L., 2001, Magnetic stratigraphy of the Oligocene type Zemorrian Stage, Temblor Formation, Kern County, California: Pacific Section SEPM Special Publication, v. 91, p. 210-223.

Prothero, D. R., and Thompson, M., 2001, Magnetic stratigraphy of the type Refugian Stage (late Eocene-early Oligocene), western Santa Ynez

Range, Santa Barbara County, California: Pacific Section SEPM Special Publication, v. 91, p. 119-134.

Prothero, D. R., and Vance, E. H., Jr., 1996, Magnetic stratigraphy of the upper middle Eocene Coldwater Sandstone, Ventura County, California, *in* Prothero, D. R., and Emry, R. J., eds., The Terrestrial Eocene-Oligocene Transition in North America: Cambridge, Cambridge University Press, p. 140-155.

Prothero, D. R., Sanger, E., Nesbitt, E., Niem, A. R., and Kleibacker, D., 2001. Magnetic stratigraphy and tectonic rotation of the upper middle Eocene Cowlitz and Hamlet Formations, western Oregon and Washington: Pacific Section SEPM Special Publication, v. 91, p. 75-95.

Prothero, D. R., Sutton, J. E., and Brabb, E. E., 2001, Magnetostratigraphy of the Eocene-Oligocene San Lorenzo and Vaqueros Formations, Santa Cruz Mountains, California: implications for California biostratigraphic zonations: Pacific Section SEPM Special Publication, v. 91, p. 154-168.

Schenck, H. G., and Kleinpell, R. M., 1936, Refugian Stage of the Pacific Coast Tertiary: American Association of Petroleum Geologists Bulletin, v. 20, p. 215-255.

Schenck, H. G., and Muller, S. W., 1941, Stratigraphic terminology: Geological Society of America Bulletin, v. 52, p. 1419-1426.

Squires, R. L., 1987, Eocene molluscan paleontology of the Whitaker Peak area, Los Angeles and Ventura Counties, California: Natural History Museum of Los Angeles County, Contributions in Science, v. 388, p. 1-93.

Squires, R. L., 1988, Geologic age refinements of West Coast Eocene marine mollusks: Pacific Section SEPM, v. 58, p. 107-112.

Squires, R.L., 1999a, William More Gabb, first paleontologist of the Geological Survey of California: California Geology, July/August 1999, p. 11-14.

Squires, R. L., 1999b, Upper Paleocene to lower Eocene ("Meganos Stage") marine megafossils in the uppermost Santa Susana Formation, Simi Valley, Southern California: Natural History Museum of Los Angeles County, Contributions in Science, v. 479, p. 1-38.

Squires, R. L., and Demetrion, R. A., 1992, Paleontology of the Eocene Batequa Formation, Baja California Sur, Mexico: Natural History Museum of Los Angeles County, Contributions in Science, v. 434, p. 1-55.

Tipton, A., Kleinpell, R. M., and Weaver, D. W., 1973, Oligocene biostratigraphy, San Joaquin Valley, California: University of California Publications in Geological Sciences, v. 105, p. 1-72.

Walsh, S. A., Prothero, D. R., and Lundquist, D. J., 1996, Stratigraphy and paleomagnetic correlation of middle Eocene Friars Formation and Poway Group in southwestern San Diego County, California, *in* Prothero, D.R., and Emry, R. J., eds., The Terrestrial Eocene-Oligocene Transition in North America: Cambridge, Cambridge University Press, p. 120-155.

Warren, A. D., and Newell, S. H., 1980, Plankton biostratigraphy of the Refugian and adjoining stages of the Pacific Coast Tertiary: Cushman Foundation for Foraminiferal Research Special Publication, v. 19, p. 233-251.

Weaver, C. E. 1912. A preliminary report on the Tertiary paleontology of western Washington: Bulletin of the Washington Geological Survey, v. 15, p. 1-18.

Weaver, C. E. 1916a. The post-Eocene formations of western Washington: Proceedings of the California Academy of Sciences, Series 4, v. 6, p. 28-41.

Weaver, C. E. 1916b. The Tertiary formations of western Washington: Washington State Geological Survey, v. 13, p. 319.

Weaver, C. E. 1916c. Tertiary faunal horizons of western Washington: University of Washington Publications in Geology, v. 1, p. 1-67.

Weaver, C. E., and 20 others, 1944, Correlation of the marine Cenozoic formations of western North America: Geological Society of America Bulletin, v. 55, p. 569-598.

CHAPTER 2

Turnovers in Marine Gastropod Faunas During the Eocene-Oligocene Transition, West Coast of the United States

Richard L. Squires

ABSTRACT

A detailed inventory of the genera and subgenera of marine gastropods known from Paleogene rocks in Washington, Oregon, and California, in combination with chronostratigraphic control provided by recent paleomagnetic studies of key formations, revealed a loss of many warm-water taxa and the addition of new cooler water taxa associated with global cooling during the Eocene-Oligocene transition (about 40–30 Ma). Only full-stage ranges are known for most of the taxa, thereby giving the appearance that this overall change in the fauna took place as three turnover events. More detailed biostratigraphies are needed to prove if the three turnovers (especially the last two) are truly coincident with stage boundaries, or if there was just one gradual turnover (about 12 million years-long).

From Paleocene through middle Eocene time, nearly all the marine gastropods on the West Coast of the United States and Baja California, Mexico, were shallow-water, warm-water forms. Faunal diversity reached its highest (143 genera/subgenera) for the West Coast Paleogene during the early Eocene. At the beginning of the late Eocene, about 36.5 Ma, the first turnover took place. Forty-eight percent of the previous taxa disappeared, but many warm-water taxa survived. There was also a major influx of cooler-water taxa, and it was the largest such influx for the region during the entire Paleogene. The first pteropod gastropods also appeared on the West Coast at this time, and bathyal, chemosynthetic cold-seep gastropods became more common, especially in Washington.

About 33 Ma and 28.5 Ma, second and third turnovers took place. Like the first one, warm-water-holdover taxa survived, and new arrivals of cooler-water taxa took place. Unlike the first turnover, however, faunal diversity was much lower, and the percentage of new arrivals was much lower. Following the third turnover, there was a record low number (40) of gastropod genera/subgenera for the West Coast Paleogene rock record.

Certain gastropods (e.g., *Calyptraea, Crepidula, Euspira, Glossaulax, Neverita, Sinum, Turritella*) survived all three turnovers and the arrival of cooler waters. Several of these were solely represented by long-lived species that persisted through most of the Paleogene.

INTRODUCTION

During the Eocene-Oligocene transition (about 40–30 Ma), a global lowering of atmospheric and oceanic temperatures initiated the oceanic psychrosphere and the Antarctic glaciation (Berggren et al., 1998). This interval was the most significant episode of climatic change and extinction since the end of the Cretaceous, and warm-adapted species were the primary victims (Berggren and Prothero, 1992). In-depth paleoclimatic studies have been made on Gulf Coast Eocene-Oligocene mollusks (Dockery, 1986; Hansen, 1987, 1992), but similar detailed studies on Pacific Coast Eocene-Oligocene mollusks have never been attempted. Late Paleogene marine gastropods from California, Oregon, and Washington have much potential for studying the changes in paleoclimate during the Eocene-Oligocene transition because the faunas are common to abundant, taxonomically diverse, span about 16° of latitudinal difference, and have been the subject of considerable research over the last 150 years.

Henceforth, the term "West Coast" refers to the coastal region from Seattle, Washington, to San Diego, California. If Baja California, Mexico, is to be included in any discussion, it will be mentioned by name. This present report is the first detailed study of the taxa-turnover events of marine gastropods that occurred on the West Coast during the Eocene-Oligocene transition. In order to more fully understand changes in the gastropod faunas during this transition, a background inventory was also made of all West Coast Paleogene marine-gastropod occurrences leading up to this transition. For this background inventory, Baja California, Mexico was also included. Unfortunately, the Eocene-Oligocene transition molluscan fauna of Baja California is very poorly known; thus, this geographic area could not be included in the analysis of that fauna.

PREVIOUS STUDIES OF EOCENE-OLIGOCENE TRANSITION MOLLUSKS

The tropical aspect of shallow-water, early Paleogene molluscan faunas of the West Coast has been discussed by several paleontologists (e.g., Arnold, 1909; Dickerson, 1917a; Smith, 1919; Clark and Vokes, 1936; Vokes, 1940; Durham, 1950; Addicott, 1969, 1970a, 1970b; Squires, 1984, 1987; Givens, 1989).

The comparatively cool aspect of Eocene-Oligocene transition faunas of the West Coast has been discussed by Dickerson (1917b), Durham (1950, 1959), Addicott (1969, 1970a, 1970b), Moore (1976), and Hickman (1969, 1972, 1974, 1976, 1980). Bathyal habitats for some of these gastropods were first reported by Watkins (1974) and Hickman (1974, 1976, 1980, 1984), and chemosynthetic cold-seep communities for some of the gastropods have been recognized by several paleontologists (e.g., Goedert and Squires, 1990: Goedert and Campbell, 1995; Goedert et al., 1995; Squires and Goedert, 1996; Goedert and Kaler, 1996; Goedert and Benham, 1999). Pteropods from West Coast Paleogene strata have been described by Squires (1989) and Squires et al. (1999). Brackish-marine gastropods from West Coast Paleogene strata have been described by Arnold (1909), Vokes (1939), Givens (1974), Givens and Kennedy (1976, 1979), and Squires (1991, 1998, 1999a).

Abrupt changes (taxon turnovers) between middle and late Eocene West Coast molluscan faunas were interpreted by Clark and Vokes (1936) and Hickman (1974, 1976, 1980).

WEST COAST OF NORTH AMERICA PALEOGENE MOLLUSCAN STAGES

Fossil gastropods are abundant and widely distributed in shallow-marine Paleogene strata on the West Coast of North America, so much so that they are the basis of the biostratigraphy of the provincial mega-invertebrate (molluscan) stages (figure 2.1) used in this area during the last 90 years. The Paleocene and most of the Eocene Pacific coast stages are informal ones and predominantly stem from work by Clark and Vokes (1936) and Weaver et al. (1944). As pointed out by Givens (1974), most of these stages and their boundaries are poorly defined, yet the recognized sequence of the traditional Paleocene and Eocene "stages" ("Martinez," "Meganos," "Capay," "Domengine," "Transition," and "Tejon") of the West Coast usually can be confidently recognized. Givens (1974) modified the use of the "Capay Stage," and it is in this modified sense that "Capay Stage" is used here. Saul (1983) and Squires (1988) correlated most of the Paleocene and Eocene stages to the standard planktonic foraminiferal and calcareous nannofossil biozones, but these correlations are now out of date because of better precision allowed by the integrated usage of current magnetic stratigraphy studies.

Early workers (e.g., Clark and Vokes, 1936; Weaver et al., 1944; Durham, 1950) used the stage names "Martinez," "Meganos," "Capay," "Domengine," "Transition," and "Tejon" for rocks distributed from southern California to the Pacific Northwest. Armentrout (1975) also believed that molluscan faunas of California and the Pacific Northwest constitute a single faunal province and that a single set of stages can be utilized throughout the region. Nevertheless, Armentrout (1981) did not use these stage names for the Pacific Northwest region. In this present report, all the previously recognized molluscan stages are used for the entire West Coast and Baja California, including a lower Paleocene "unnamed stage" recognized by Saul (1983).

The upper Eocene and Oligocene West Coast molluscan stages stem largely from work by Clark and Vokes (1936), Schenck and Kleinpell (1936), Durham (1944), Weaver et al. (1944), and Armentrout (1975). The latter worker proposed the Galvinian Stage (upper Eocene and lowermost Oligocene). This stage corresponds to all but the lowermost part of the Refugian Stage of Schenck and Kleinpell (1936), which was based on mollusks and benthic foraminifers. Over the years, however, the Refugian Stage "has taken on the character of a stage based exclusively on benthic foraminiferal zones"

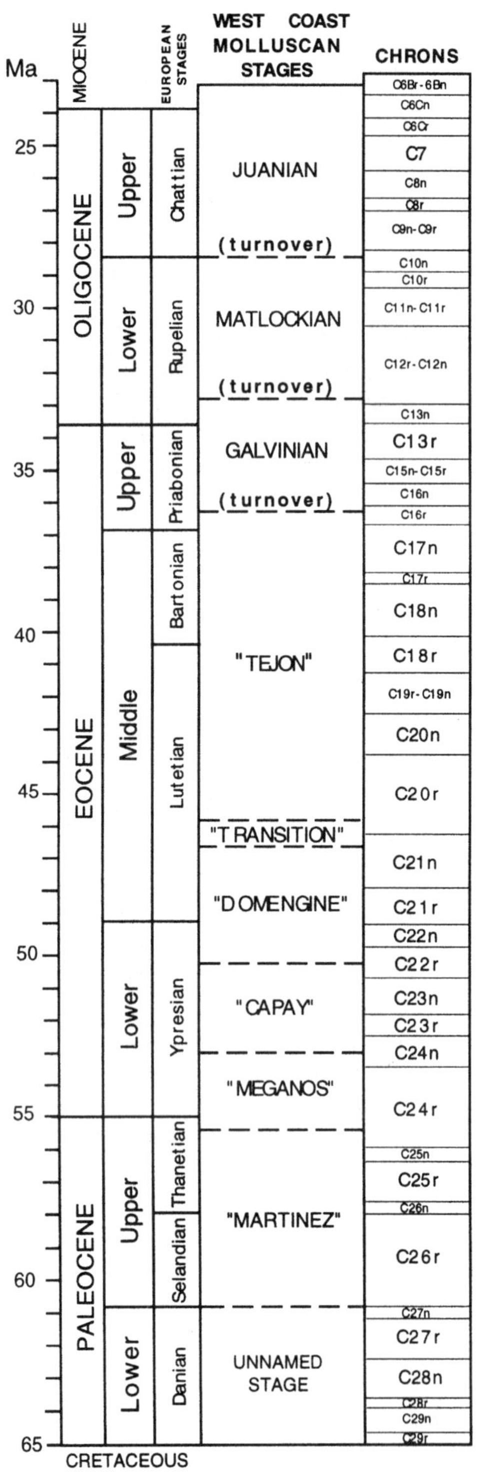

Figure 2.1. West Coast and Baja California, Mexico, Paleogene marine molluscan stages versus European stages and magnetic stratigraphy chrons (adapted from Berggren et al., 1995). Turnovers in marine gastropod genera/subgenera so noted.

(Moore, 1976:24). Armentrout (1975) also proposed the Matlockian Stage (upper Oligocene), but Armentrout (1981) revised this stage to include only the lower Oligocene. Addicott (1976a) proposed the Juanian Stage (upper Oligocene to lowermost Miocene).

PROCEDURE AND DEFINITION OF TERMS

The gastropod content of numerous marine Paleogene West Coast and Baja California, Mexico, formations and members (table 2.1) was used in this study. The list of formations in table 2.1 is not intended to be an exhaustive compilation of every mollusk-bearing formation in the study region. The listed formations are those containing abundant and stratigraphically representative molluscan faunas that are well documented in readily available publications. The ages of seventeen of these formations have been recently refined using magnetic stratigraphy studies (table 2.2), and this information was used to calibrate the ages of some of the other formations. In addition, the Simpson coefficient of faunal similarity was calculated for various formations. This coefficient is the number of taxa in common between two faunas, divided by the total number of taxa present in the smaller of the two faunas. (See Fallaw, 1979, for a discussion of this technique.) The most comprehensive and/or up-to-date references for the gastropod contents of the stages are as follow: "Martinez Stage"—Squires (1997); "Meganos Stage"—Squires (1999b); "Capay Stage"— Squires and Goedert (1994), Squires (2000); "Domengine Stage"—Squires (1984, 2000, 2001); "Transition Stage"—(Squires, 1999a); "Tejon Stage"—Clark and Anderson (1938), Givens and Kennedy (1979), Nesbitt (1995); Galvinian—Durham (1944), Weaver and Kleinpell (1963), Armentrout (1975), Hickman (1984); Matlockian—Hickman (1969), Moore (1976); Juanian—Addicott (1976a, 1976b), Moore (1984). Moore's (1984) recognition of molluscan zones of the cassid gastropod *Liracassis* in the Pacific Northwest has been very useful in distinguishing Matlockian versus Juanian Stage faunas. In that same paper, she also compiled the mollusks characteristic of these two stages and listed the formations that are correlative to each stage. Her compilation was incorporated into this present study (tables 2.4 and 2.5).

In the process of compiling published species lists of all the reported gastropods used in this present study, I employed updates in taxonomy whenever possible, and critically scrutinized each list for any obvious misidentifications (as to genus and/or sub-

Table 2.1. Formations Used in Obtaining the Stage Ranges of the Marine Gastropods

JUANIAN STAGE
Twin River Fm, WA; Lincoln Creek Fm (in part), WA

MATLOCKIAN STAGE
Lincoln Creek Fm (in part) WA; Blakeley Fm (in part) WA; Pysht Fm, WA, Makah Fm, WA; Pittsburg Bluff Fm, WA; Eugene Fm, OR; Alsea Fm (in part), OR; The Indians area, CA; Temblor Fm (Wygal Sandstone Mbr), CA

GALVINIAN STAGE
Lincoln Creek Fm, WA (in part); Gries Ranch Fm, WA; Quimper Fm, WA; Bear River deposit, WA; Keasey Fm, OR; Wheatland Fm, CA; Kreyenhagen Fm, CA; San Emigdio Fm, CA; Gaviota Fm, CA

"TEJON STAGE"
Cowlitz Fm, WA; Tukwila Fm, WA; transition zone between Crescent Fm & McIntosh Fm, Doty Hills, WA; Coaledo Fm, OR; Spencer Fm, OR; Sacate Fm/"Coldwater Sandstone", CA; Markley Fm, CA; Coldwater Sandstone, Sespe Creek, CA, Tejon Fm (Metralla Sandstone Mbr), CA; Tejon Fm (Liveoak Shale Mbr), CA; Tejon Fm at Edmonston Pumping Plant, CA (in part); Mission Valley Fm, CA; Stadium Conglomerate, CA; Bateque Fm, Baja California Sur, Mexico (in part)

"TRANSITION STAGE"
Friars Fm, CA; upper Scripps Fm, CA; Tejon Fm at Edmonston Pumping Plant, CA (in part)

"DOMENGINE STAGE" Crescent Fm at Pulali Point, WA; Humptulips Fm, WA; Tyee Fm, OR; Domengine Fm, CA; Juncal? Fm, Lockwood Valley, CA (in part); Matilija Sandstone, Matilija Hot Springs near Ojai, CA; Matilija Sandstone? & Juncal Fm (in part), Whitaker Peak, CA; Juncal Fm, Pine Mountain area, CA (in part); Llajas Fm, CA; Santiago Fm, Vista, CA; lower Scripps Fm, CA; Ardath Shale, CA; Delmar Fm, CA; Bateque Fm, Baja California Sur, Mexico (in part)

"CAPAY STAGE"
Crescent Fm, Little River, WA; White Tail Ridge fm., OR; Capay Fm, CA; Vacaville Shale, CA; lower and middle Juncal? Fm, Lockwood Valley, CA; Juncal Fm, Whitaker Peak, CA; Juncal Fm, Pine Mountain area, CA (in part); Maniobra Fm, CA; basal Llajas Formation, Ca; Bateque Fm, Baja California Sur, Mexico (in part)

"MEGANOS STAGE"
Meganos Fm, CA; Goler Fm, CA; upper 100 m of Santa Susana Fm, Simi Valley, CA

"MARTINEZ STAGE"
Martinez Fm, CA; Santa Susana Fm, Simi Hills, CA; Santa Susana Fm, Santa Monica Mtns, CA; Sepultura Fm, Baja California, Mexico

"UNNAMED STAGE"
Dip Creek "fm", CA (in part); Las Virgenes Sandstone, CA; San Francisquito Fm, Warm Springs Mtn, CA

genus). In compiling the geologic stage range of each genus/subgenus, I used an assumption of continuous occurrence (e.g., if a genus is recorded from the "Capay Stage" and the "Tejon Stage," it is also assumed to be present in the intervening stages, even though it has not been acutally reported from these intervening stages). In tables 2.4 and 2.5, the known occurrences versus the assumed occurrences are so indicated. This assumption of continuous presence had no impact on the number of genera/subgenera counted for the "Martinez" and Juanian Stages, negligible impact for the Galvinian and Matlockian Stages, minor impact for the "Capay" and "Tejon" Stages, moderate impact for the "Domengine Stage," and considerable to moderate impact for the "Meganos" and "Transition" Stages, respectively. These latter two stages are predictably the most affected because they are poorly represented in the rock record. The assumption of continuous occurrence for these two stages brings the number of genera/subgenera in line with what would be expected.

Totals of gastropod taxa per stage given in this

Table 2.2. Eocene-Oligocene Transition Formations Whose Magnetozones Are Known

Stage	Chrons	Formation	State	Reference
Matlockian	C12r-C11n	Pittsburg Bluff	WA	Prothero & Hankins, 2001
Matlockian	C12r	Eugene	OR	Prothero, pers. commun.
Galvinian	C13n-C12r	Alegria	So. CA	Prothero & Thompson, 2001
Galvinian	latest C16n-earliest C12r	Keasey	OR	Prothero & Hankins, 2000
Galvinian	C13R	Gries Ranch	WA	Prothero et al., 2001a
Galvinian	C16r-C13r	Gaviota	So. CA	Prothero & Thompson, 2001
"Tejon"	?C17n-C16r?	Sacate/"Coldwater" Sandstone	So. CA	Prothero & Thompson, 2001
"Tejon"	C17r-C17n	Cowlitz	WA	Prothero et al., 2001c
"Tejon"	C19r-C18r	Coldwater	So. CA	Prothero & Vance, 1996
"Tejon"	C20n-C19r	Mission Valley	So. CA	Walsh et al., 1996
"Transition"	C21n-C20r	Friars	So. CA	Walsh et al., 1996
"Transition"	C21n	Upper Scripps	So. CA	Walsh et al., 1996
"Domengine"	C21n	Ardath	So. CA	Walsh et al., 1996
"Domengine"	C21r	Humptulips	WA	Prothero et al., 2001b
"Domengine"	C21r	Domengine	So. CA	Prothero, 2001
"Domengine"	C22n-C20r	Matilija (Ojai area)	So. CA	Prothero & Britt, 1998
"Domengine"	C22n-C21n	Llajas (not basal)	So. CA	Bottjer et al., 1991

paper include taxa from all of the different habitats of the marine gastropods; namely, shallow-marine (less than 100 m paleodepth), brackish-marine, pelagic (i.e., pteropods), bathyal (greater than 100 m paleodepth), and bathyal-chemosynthetic communities. In order to avoid double or triple counting, differentiation as to habitat was not considered for the counting process, because some of the taxa lived in more than one habitat during any one stage. For the analysis of faunal changes at a stage boundary, the terms "first appearance datum" (FAD) and "last appearance datum" (LAD) are used. Here, these terms apply only to the Paleogene stages in the study region (table 2.3). The first appearance of a taxon could be related either to its being newly evolved or to its immigrating to the West Coast for the first time. The last appearance of a taxon could be related either to its final extinction or to its disappearance from the West Coast during the Paleogene, because the environment was no longer suitable. The FAD and LAD amounts are expressed as percent values of the total number of gastropod genera/subgenera present at a stage boundary. The term "turnover" refers to the number of first appearances versus the number of last appearances at a stage boundary. In making the counts of genera/subgenera present in each stage, I used full-stage ranges for the taxa because of the present state of our knowledge of the fossil record. Detailed biostratigraphies are not known for hardly any of the taxa; hence, only full-stage ranges are known for most the taxa, and any partial-stage ranges are lost in these data (see "Problems" section).

Table 2.3. Gastropod Turnover Data for West Coast Paleogene Stages

Stage	Number Genera/ Subgenera	LAD %	FAD %
JUANIAN	40	—	15
MATLOCKIAN	61	44	16
GALVINIAN	110	55	43
"TEJON"	122	48	22
"TRANSITION"	103	7	3
"DOMENGINE"	143	30	20
"CAPAY"	141	18	46
"MEGANOS"	78	4	14
"MARTINEZ"	85	22	—
UNNAMED	unknown	—	—

LAD = last appearance datum
FAD = last appearance datum

The terms "thermophilic," and "temperate" are used for paleoclimatic analysis. Here, thermophilic refers to tropical to subtropical temperatures, and temperate refers to intermediate temperatures between tropical-subtropical and cold.

Table 2.4. Paleogene Stage Ranges and Paleohabitats of Marine Gastropod Genera/Subgenera

GENUS (SUBGENUS)	Ma	Me	C	D	Tr	Te	G	Mt	J	GENUS (SUBGENUS)	Ma	Me	C	D	Tr	Te	G	Mt	J
Abderospira	X	x	x	x	x	X				Colus? (b)							X		
Abyssochrysos (ch)				X						Colwellia	X	X	X	X	X	X			
Acamptogenotia (s, b)						X	X			Comitas (Boreocomitas) (s, b)						X	X		
Acmaea (s, b)	X	x	X	X	x	x	X			Conomitra (s, b)	X	x	X	X	X	x	X		
Acteon = Kleinacteon	X	x	X	X	x	X	X	X	X	Conus (s, b)			X	X	X	X	X		
"Admete" (b, ch)							X	X		Corsania (Januncia)	X	X	X						
Aforia (b, ch)							X	X	X	Crepidula (s, bk)			X	X	X	X	X	X	X
Akera			X	X						Crommium (s, bk)	X	x	X	X					
Alvania			X	x	x	X	X			Cryptochorda	X	x	X	X					
Amaea (Scalina)=Acrilla	X	x	X	X	X	X	X	X	X	Cryptoconus		X	X	X					
Amaurellina			X	X	x	X				Cyclostremella (b)							X		
Amauropsis	X	x	X							Cylichnina	X	X	X	X	X	X	X	X	
Ammonicera			X							Cymatium (Septa)	X	x	x	X	x	X			
Amphissa									X	Cypraedia			X						
Ampullella			X	X						Cypraeogemmula			X	x	x	X	X		
Ancilla (Spirancilla) (s, b)	X	x	X	X	x	x	X			Depressigyra? (ch)				X					
Ancillarina?		X								Diodora	X	x	X	X	X	X			
Ancistrolepis (s, b)								X	X	Dirocerithium			X	X					
"Ancistrolepis?"				X						Discohelix	X	x	x	X					
Antiplanes (b)								X		Domenginella			X	X					
Apiotoma	X	x	X	X						Drepanochilus	X								
Araeodactylus?	X									Echinophoria (s, b)				X?	x	X	X		
Architectonica s.s.		X	X	X	x	X				Ectinochilus (Cowlitzia)			X	X	X	X			
Architectonica (Stellaxis)	X	X	X	X	x	X				Ectinochilus (Macilentos)			X	X	x	X			
"Architectonica?"								X		Elimia (bk)			X	x	x	X	X	X	
Arene			X	X						Emarginula			X	x	x	X			
Argobuccinum (b)							X	X	X	Eocernina	X	x	X	X	X	X			
Athleta			X							Eocithara	X	x	X	X					
Balcis (b)							X			Eocypraea	X	x	X	X					
Bathybembix (b)							X	X	X	Eopleurotoma			X	X					
Bathytoma				X						Eopleurotoma? (b)							X		
Benoistia			X	X	x	X				Eosiphonalia								X	
Bernaya s.s.				X						Eosurcula	X	x	X	X	X	X			
Bernaya (Protocypraea)			X							Epitonium (Asperiscala)									X
Bittium s.s.			X	X	X	X	X			Epitonium (Boreoscala) (s, b)						X	X	X	
Bonellitia s.s. (b)							X			Erato							X		
Bonellitia (Admetula) (s, b)	X	X	X	X	x	X	X			Eratotrivia			X						
Brachysphingus	X	X	X							Erginus (s, ch)			X	x	x	X	X	X	
Bruclarkia (s, b)							X	X	X	Euspira (s, b)	X	X	X	X	x	X	X	X	X
Bullia (Buccinanops)?							X			Exilia (s, b)	X	x	X	X	X	X	X	X	
Calicantharus								X		Falsifusus	X	x	X	X					
Calliostoma						X				Ficopsis		X	X	X	X	X			
Calliovarica			X							Ficus				X	x	X	X	X	X
Calorebama		X	X	X	X	X				Fulgurofusus (s, b)						X	X		
Calyptraea (s, b)	X	X	X	X	X	X	X	X	X	Fulgoraria	X								
Calyptraphorus	X									Fusinus	X	X	X	X	x	X	X		
Campanile	X	X	X							"Fusinus" (b)							X		
Cancellaria							X	x	X	"Fusitriton" (b)							X	x	X
"Cantharus"						X				Fusiturricula (Crenaturricula)			X	X					
Cantrainea = Nehalemia (s, b)						X				Galeodea (Caliagaleodea)			X	X					
Cellana							X			Galeodea (Mambrina)			X	X	X	X			
Certhiopsis			X	X	x	X	X			Galeodea s.l.		X	X	X	x	X			
Cerithium	X	x	X	X	X					Gemmula (s, b)	X	X	X	X	x	X	X	X	
Chedevillia			X							Gilbertina	X								
Cidarina						X				Gisortia (Megalocypraea)			X						
Cirsochilus							X			Glossaulax				X	x	X	X	X	X
Cirsotrema				X	x	x	x	x	X	Granula							X		
Clavilithes			X	X						Gyrodes s.s.	X								
Clavus						X				Haminoea							X		
Clio (p)									X	Haplocochlias			X	x	x	X			
Clivuloturris (b)							X			Hemitoma (Montfortia)			X	x	x	X			
Cochlespiropsis?				X						Hipponix	X	x	X	X	x	X	X		

(continued)

Table 2.4. (continued)

GENUS (SUBGENUS)	Ma	Me	C	D	Tr	Te	G	Mt	J
"Homalopoma"			X	X	x	X	X		
Homalopoma? (ch)				X					
Janiopsis			X	X					
Keilostoma				X					
Kelletia?								X	
Lacinia?			X	X					
Lacunaria	X								
Laevityphis				X	X	X			
Lapparia			X	X	X	X			
Lapsigyrus			X						
Latirus	X	x	X	X	x	X	X		
Limacina (p)							X	X	X
Liotia						X	X		
Liracassis							X	X	X
Littorina						X	X		
Loxotrema (bk)	X	x	X	X	X	X			
Lyria	X	x	X	X					
Lyrischapa			X	X					
"Mangelia?"						X			
Margarites s.s. (b)							X		
Margarites (Pupillaria?) (b, ch)							X	X	
Margarites s.l.							X		
"Margarites"?	X								
"Marginella"				X	x	X	X	X	X
Mazzalina		X							
Megasurcula? (b)									X
"Megathura?"						X			
Megistostoma		X	X	X	X				
"Melanatria" (bk)				X	x	X			
Melanella = Eulima			X	x	x	X	X		
Melanoides (bk)						X			
Mesalia s.s.	X								
Mesalia (Sigmesalia)	X								
Metacerithium	X								
Microglyphus (b)									X
Mitra	X	x	X	X	x	X			
"Mitramorpha"				X					
Mitrella			X						
Molopophorus							X	X	X
Monodonta (Incisilabium?)			X						
"Murex"	X	x	X	x	x	X			
Muricopsis				X					
Musashia s.l. (b, ch)							X	X	X
Natica s.s. (s, b)					X	X	X	X	X
Natica (Carinacca)				X	x	X			
Natica (Naticarius)			X	X	x	X	X	x	X
Nekewis	X	x	X	x	X	X	X	X	
Nerita (Bajanerita?)			X	x	x	x	X		
Nerita (Theliostyla) (s, bk)	X	x	X	X	X	X			
Neritina (bk?)						X			
Neverita s.s. (s, bk, b)	X	X	X	X	X	X	X	X	X
Niso	X	x	x	X	x	X			
Nucleolaria						X			
Odostomia		X	X	X	x	X	X	X	
Olequahia (s, b)	X	x	X	X	x	X	X	X	
Olivella s.l.	X	X	X	X	X	X			
Opalia s.l.						X	x	X	
Orbitestella			X						
Otostoma			X	X	X				
Ovatella			X						
Oxystele								X	
Pachycrommium		X	X	X	X	X	X		
Paraseraphs			X	X	X				
Parasyrinx s.l. (s, b)	X	x	x	x	x	x	X	X	
Parvisipho						X			
Pegocomptus	X								
Patelloida					X	X			
Perse				X	X	X	X	X	X
Phalium (Bezoardica)							X		
Phalium ("Semicassis")		X	X	X	X	X			
Phanerolepida (s, b)							X		
Phos						X			
Platyoptera			X	X					
Pleurofusia	X	x	X	X	X	X			
Pleuroliria (b)							X		
Pleurotomaria			X	x	x	X			
Polinices s.s. (s, b)	X	X	X	X	X	X			
Popenoeum	X								
Potamides (Potamidopsis) (bk)				X					
Potamides s.l. (bk)	X	x	X	X	X	X			
Praehyalocylis (p)							X	X	X
Priscoficus	X								
Priscofusus						X	X	X	X
Proadusta			X						
Procerapex (b)							X		
Propustularia	X								
Provanna (ch)							X	X	X
Proximitra?				X					
Pseudomelatoma (b)								X	
Pseudoperissolax	X	x	X	X	X	X			
Pseudovertagus (Pseudoaluco)			X						
Pterynotus s.s.				X	x	X			
Ptychosyrinx (b)							X	X	
Puncturella (Altrix)			X						
Puncturella s.l.							X		
Pyrgulifera (bk)				X					
Pyramidella			X	X	X	X			
Pyropsis = Heteroterma	X								
Ranella = Gyrineum	X	X	x	X	x	X	X	X	
Ranellina	X	x	x	X	X	X			
Ringicula	X	X	X						
Sassia	X	x	X	X	X				
Scaphander (Mirascapha)	X	X	X	X	X	X			
Scaphander s.l. (s, b, ch)						X	X	X	X
Searlesia							X		
Segmentina							X		
Seraphs?							X		
Serpulorbis?				X					
Serratocerithium	X								
Sinum (s, b)			X	X	X	X	X	X	X
Siphonalia	X	X	X	X	x	X	X		
Siphonaria							X		
Sissurella s.s.			X						
Solariella (Machaeroplax) (b)							X		
Solariella s.l.	X	X	X	X	X	X	x	x	X
Sphaerocypraea	X								
Spirotropis (b)							X	x	X
Sthenorytis (b)									X
Strepsidura			X	X	X	X			
Streptochetus			X						
Suavodrillia (b)							X	X	X
Sulcocypraea				X	x	X			
Surculites (s, b)	X	x	X	X	X	x	x	X	
Sveltella (b)							X		
Sycostoma	X	X							
?Symmetrocapulus					X				

GENUS (SUBGENUS)	Ma	Me	Ca	D	Tr	Te	G	Mt	J
Taranis								X	
Tegula								X	
Teinstoma	X	x	x	X	x	X			
Tejonia = ?Ampullospira			X	X	X	X			
Tenagodus	X	x	X						
Terebra			X	X	X	X			
Terebralia	X								
Terebrifusus				X	X				
Thalassonerita? (ch)				X					
Thiara (bk)							X		
Tibia (Eotibia)			X						
Tornatellaea			X	X					
"Trichotropis?"						X	X		
Trophon							X		
Trophonopsis (b)							X		
Typanotoma			X	X					
Turbella			X	x	x	x	X		

GENUS (SUBGENUS)	Ma	Me	Ca	D	Tr	Te	G	Mt	J
Turbonilla		X	X	X	X	X			
"Turricula" (s, b)	X	X	X	X	X	X	X	X	
Turrinosyrinx (b)							X	X	
"Turris"						X	X		
Turritella (s, b)	X	X	X	X	X	X	X	X	X
Turritellopsis?						X	X	x	X
Tympanotonos s.s.				X					
Umpquaia			X	X					
Urosalpinx						X	X		
Velates	X	X	X	X					
Vitrinella = ?Adeorbis			X	X	X	X			
Volutilithes	X	x	X	X					
Volutocorbis	X	x	X						
Volvulella						X	X		
Whitneyella	X	X	X	X	x	X	X		
Xenophora	X	x	X	X	X	X			

PROBLEMS

The lower Paloecene "unnamed stage" is represented by only a few formations, and the associated faunas remain largely undescribed. In most cases, these formations contain a conformable section across the Cretaceous-Tertiary boundary (Kirby and Saul, 1995) or contain strata deposited near this boundary (Squires and Saul, 1993). In the latter case, the exact age of the fossils is in question (Saul, 1986). This "unnamed stage" is not included in table 2.4 because this stage is not a well-established one, its fauna is poorly known, and it is unresolved whether some of the mollusks are latest Cretaceous or earliest Paleocene in age.

The "Meganos Stage" is represented by few formations, and its fauna is incompletely known. The "Transition Stage" might have lasted only a short time, and only a few formations can be confidently assigned to this stage. Consequently, the fauna of this stage is also poorly known, making this presumed time-rock interval very difficult to recognize.

The middle middle Eocene to late middle Eocene "Tejon Stage" has rather poor temporal resolution because the "Tejon Stage" spans 10 million years (figure 2.1). Moore (1992:figure 3) generally subdivided the "Tejon Stage" into a lower and upper part, as did Squires (1998). Although it would be immensely useful to do so, I did not subdivide the "Tejon Stage" in this paper, pending magnetostratigraphic studies of critically important mollusk-bearing stratigraphic units. Until more detailed biostratigraphic work is done on the "Tejon Stage" (see below), it would be premature to try to subdivide this stage in a rigorous fashion.

This study is hampered also by the following: scarcity of mollusk-bearing shallow-marine Oligocene rocks in California; need for more precise temporal correlation work on many of the upper Eocene and Oligocene West Coast formations; preservational and sampling biases against microgastropods, minute gastropods, and thin-shelled gastropods; need for detailed taxonomic investigations on nearly every genus and subgenus (e.g., *Colwellia, Conus,* "*Marginella*," *Molopophorus*, and "*Turricula*"); undetected misidentified gastropods and junior synonyms (literature before 1960 is especially prone to these kinds of problems); and need for magnetic stratigraphic and paleontologic studies of many of the faunas. Some of the most critical West Coast formations in need of magnetic stratigraphy studies, as well as in need of revised molluscan biostratigraphic studies, are the Liveoak Shale Member and Metralla Sandstone Member of the Tejon Formation in south-central California (see Nilsen, 1987), the "Matilija Sandstone" and "Coldwater Sandstone" in the Pine Mountain area in southern California (see Givens, 1974), and the Markley Formation in northern California. These studies will also greatly help in determining the much-needed details of the tentatively reported (Weaver et al., 1944; Givens, 1974; Givens and Kennedy, 1979) evolutionary lineage of *Turritella uvasana neopleura*, *T. uvasana* s.s., and *T. uvasana sargeanti* within the "Tejon Stage." The San Emigdio Formation in south-central California, the Wheatland Formation in northern California, and many of the Oligocene formations in Washington also need magnetostratigraphic studies. Some of the Matlockian Stage and Juanian Stage formations listed by Addicott (1976a) and Moore (1984) are currently being investigated by means of magnetic stratigraphic studies (Donald Prothero, pers. commun.). More precise depositional environments and taphonomic studies of many of the West Coast faunas, especially the Oligocene ones, are also badly needed. This kind of information is necessary in order to determine whether some of the gastropods might be part of mixed assemblages of shallow-marine taxa transported into bathyal depths. In addition, it is likely that some of the West Coast Paleogene genera and subgenera (and especially species) are oversplit, because too few specimens of them have been collected and only the "end points" of growth and ecologic series are ever seen. Furthermore, there are the problems of disjunct stratigraphic sections separated by unconformities, ground cover, and/or faults.

One of the most serious problems in this study is that full-stage ranges were used in determining FAD and LAD values. This was unavoidable, given the present state of biostratigraphic knowledge of Paleogene mollusks from the West Coast and Baja California, Mexico. With the ranges of taxa recorded as beginning or ending at the stage boundaries, any faunal changes are going to appear markedly stepped, whether they are or not. Only within-stage taxa ranges will be able to prove or disprove the apparent abruptness of each of the three turnovers, as well as to delineate exactly when the turnovers took place.

PALEOCENE THROUGH "TEJON STAGE" FAUNAS

For about 29 million years, from Paleocene to the end of the "Tejon Stage," there was a buildup of shallow-marine, thermophilic gastropod genera/subgenera on the West Coast and in Baja California, Mexico. Many were immigrants from the Old World Tethyan Sea region or from the Gulf Coast and the Caribbean, and they arrived via westward-flowing currents that passed through a low-latitude seaway (Squires, 1987).

The "Capay Stage" coincided with a flood of new mollusks and other marine organisms onto the West Coast and Baja California, and the number of gastropod genera/subgenera was 141. The gastropods had a FAD value of 46%, which was the highest for the entire Paleogene (table 2.3). About 30 genera are either Old World genera or show Tethyan affinities (Clark and Vokes, 1936; Vokes, 1939; Givens and Kennedy, 1976; Givens, 1978; Squires, 1984, 1987). Representative Tethyan genera are *Clavilithes*, *Conus*, *Echtinochilus*, *Galeodea*, and *Gisortia*. *Campanile*, *Lyria*, and *Velates*, all of which had arrived during the late Paleocene, continued to be important faunal components. The late Paleocene was the warmest time of the Cenozoic (Bains et al., 1999), and the influx of taxa was most likely linked to this warming event.

During the "Domengine Stage" through the "Tejon Stage," the number of genera/subgenera remained high, the disappearance of taxa (LAD) was low, and the influx of new taxa (FAD) was at a moderate level (table 2.3). Nearly all the gastropods were shallow, thermophilic gastropods. Representative taxa of the "Domengine" through "Tejon" Stages are *Architectonica*, *Eocernina*, *Eocypraea*, *Ficopsis*, *Lyria*, *Pseudoperissolax*, and *Xenophora*. All these were also present earlier on the West Coast.

A few of the newly appearing gastropod genera/subgenera on the West Coast during the "Domengine Stage" were members of bathyal chemosynthetic communities found in the Humptulips Formation in southwestern Washington (see Saul et al., 1996; Goedert and Kaler, 1996; Goedert and Benham, 1999). This formation has usually been assigned a middle to late Eocene age, but recent investigations on the magnetic stratigraphy of this formation (table 2.2) suggest correlation to the "Domengine Stage" (Prothero et al., 2001a, b).

Some of the newly appearing taxa in the Cowlitz Formation in the Pacific Northwest near the end of the "Tejon Stage" were ones that are more indicative of the cooler waters associated with the Galvinian Stage. Representatives include *Acamptogenotia*, *Cantrainea* [= *Phanerolepida*], *Comitas* (*Boreocomitas*), *Epitonium* (*Boreoscala*), *Opalia* (*Dentiscala*?), *Priscofusus*, and *Scaphander* s.l.

GALVINIAN STAGE FAUNA

The Galvinian Stage lasted about 4 million years and is represented by numerous formations (table 2.1), most of which are in the Pacific Northwest. The Quimper Sandstone in Washington, the San Emigdio Formation in south-central California, and the Wheatland Formation in northern California are three of these formations whose stage assignments have historically been inconsistent. During the course of this present investigation, all three are assigned to the Galvinian Stage based on their similar gastropod faunas. Central to this similarity study is the Gries Ranch Formation of Washington. This stratigraphic unit has a somewhat problematic nomenclatural history, which was reviewed by Weaver (1937), Effinger (1938), and Prothero et al. (2001a). Most workers have referred to the unit as the Gries Ranch beds, but Weaver (1937) suggested that it be termed the Gries Ranch Formation. Accordingly,

Moore (1976:figure 7) also used this name. Effinger (1938) studied the molluscan content of the Gries Ranch Formation, and Prothero et al. (2001a), using magnetic stratigraphy studies, assigned this formation to the Galvinian Stage.

Durham (1944) studied the molluscan fauna of the Quimper Sandstone. The Simpson coefficient of similarity is 62% between the gastropod genera/subgenera of the Gries Ranch Formation and that of the Quimper Sandstone. Both stratigraphic units also have the gastropods *Echinophoria dalli* and *Molopophorus stephensoni*, and these two species are characteristic of the Galvinian Stage (Armentrout, 1975). Wagner and Schilling (1923) studied the molluscan fauna of the San Emigdio Formation, and DeLise (1967) provided an emended megafaunal list. The Simpson coefficient of similarity coefficient is 68% between the gastropod genera/subgenera of the San Emigdio Formation and the Quimper Sandstone and is 53% between the San Emigdio Formation and the Gries Ranch Formation. Schenck and Kleinpell (1936) also correlated the San Emigdio Formation with the Gries Ranch Formation.

Clark and Anderson (1938) studied the molluscan content of the Wheatland Formation. The Simpson coefficient of similarity is 70% between the gastropod genera/subgenera of this formation and the San Emigdio Formation. Both formations also have *Bullia* (*Buccinanops*) *clarki* and *Turritella variata* s.l. Clark and Anderson (1938) correlated the Wheatland Formation to the Gries Ranch Formation, the Keasey Formation, and to the lower part of the Gaviota Formation. They also concluded that the Wheatland Formation is probably slightly older than the San Emigdio Formation.

There were 110 gastropod genera/subgenera on the West Coast during the Galvinian Stage. Sixty-four (58%) were holdovers from previous stages, and 49 (77%) of them wère thermophilic taxa that range far back into the Paleogene (table 2.5). Representatives are *Acteon*, *Calyptraea*, *Bonellitia* (*Admetula*), *Cerithiopsis*, *Conus*, *Cylichnina*, *Exilia*, *Ficus*, *Hipponix*, *Latirus*, *Olequahia*, *Pachycrommium*, *Ranella*, *Surculites*, *Turritella*, and *Whitneyella*.

Forty-three percent of the Galvinian Stage gastropod genera/subgenera were newly appearing taxa not previously found before on the West Coast (e.g., *Molopophorus*, *Ptychosyrinx*, *Turrinosyrinx*). Most of them occurred in the Keasey Formation of Oregon and Washington, and Hickman (1976, 1980, 1984) reported that they were cool-water-adapted bathyal gastropods. Two of these same taxa, *Bruclarkia columbiana* and *Scaphander* cf. *S. stewarti*, were reported by Watkins (1974) from the low-diversity and isolated *Delectopecten peckhami* community in the Kreyenhagen Formation [= Tumey Formation] of north-central California.

Pteropods appeared in West Coast Paleogene deposits for the first time during the Galvinian Stage. One of these, *Praehyalocylis cretacea*, was a cosmopolitan species (Janssen, 1989). It occurred both in Oregon (Keasey Formation) and in Washington (Quimper Sandstone) (Squires, 1989).

Bathyal cold-seep, chemosynthetic communities became more common on the West Coast, primarily in Washington, during the Galvinian Stage. Only a few gastropods are found in these communities, because in both the Pacific Northwest (Goedert and Squires, 1990; Campbell and Bottjer, 1993) and central California (Squires and Gring, 1996) the communities are bivalve dominated. *Provanna* is one of the chemosynthetic community gastropods, and it is found in Galvinian through Juanian Stage cold-seep limestones at various sites in southwestern Washington (Squires, 1995; Goedert and Campbell, 1995).

TURNOVER AT THE "TEJON STAGE"/ GALVINIAN STAGE BOUNDARY

Overall Pattern of Turnover

Sixty-four gastropod taxa co-occur in both the "Tejon" and Galvinian Stages, but at the end of the "Tejon Stage," many genera/subgenera of gastropods disappeared from the Paleogene West Coast (table 2.5). The LAD value is 48% (table 2.3). Nearly all of the casualities were thermophilic taxa (e.g., *Abderospira*, *Ectinochilus*, *Galeodea* s. l., *Nerita* [*Theliostyla*], *Strepsidura*). Of the many (58%) Galvinian Stage gastropods that were holdover taxa, 77% of them were thermophilic ones that range far back into previous Paleogene stages (e.g., *Acteon*, *Calyptraea*, *Conus*, *Cylichnina*, *Exilia*, *Ficus*, *Pachycrommium*, *Turritella*). Some of them were represented by long-lived species, such as *Calyptraea diegoana* and *Neverita* (*Neverita*) *globosa*. Hansen (1987, 1992) similarly reported that cool-water-tolerant genera such as *Calyptraea* and *Turritella*, as well as species of the family Naticidae, form a larger percentage of the fauna during low-temperature intervals in the rock record of the Gulf Coast of North America. Twenty-three percent of the Galvinian Stage holdover gastropods were warm-water forms that had arrived during the "Tejon Stage" (e.g., *Cidarina*, *Nucleolaria*).

The Galvinian Stage began with a flood of new

Table 2.5. Stage Ranges and Paleohabitats of Marine Gastropod Genera/Subgenera from the West Coast Eocene-Oligocene Transition

GENUS (SUBGENUS)	Ma	Me	C	D	Tr	Te	G	Mt	J
Abderospira	X	x	x	x	x	X			
Architectonica (Stellaxis)	X	X	X	X	x	X			
Colwellia	X	X	X	X	X	X			
Cymatium (Septa)	X	x	x	X	x	X			
Diodora	X	x	X	X	X	X			
Eocernina	X	x	X	X	X	X			
Eosurcula	X	x	X	X	X	X			
Loxotrema (bk)	X	x	X	X	X	X			
Mitra	X	x	X	X	x	X			
"Murex"	X	x	X	x	x	X			
Nerita (Theliostyla) (s, bk)	X	x	X	X	X	X			
Niso	X	x	x	X	x	X			
Olivella s.l.	X	X	X	X	X	X			
Pleurofusia	X	x	X	X	X	X			
Polinices s.s. (s, b)	X	X	X	X	X	X			
Potamides s.l. (bk)	X	x	X	X	X	X			
Pseudoperissolax	X	x	X	X	X	X			
Ranellina	X	x	x	X	X	X			
Scaphander (Mirascapha)	X	X	X	X	X	X			
Teinostoma	X	x	x	X	x	X			
Xenophora	X	x	X	X	X	X			
Architectonica s.s.		X	X	X	x	X			
Calorebama		X	X	X	X	X			
Ficopsis		X	X	X	X	X			
Galeodea s.l.		X	X	X	x	X			
Phalium ("Semicassis")		X	X	X	X	X			
Turbonilla		X	X	X	X	X			
Amaurellina			X	X	x	X			
Benoistia			X	X	x	X			
Ectinochilus (Cowlitzia)			X	X	X	X			
Ectinochilus (Macilentos)			X	X	x	X			
Emarginula			X	x	x	X			
Galeodea (Mambrina)			X	X	X	X			
Haplocochlias			X	x	x	X			
Hemitoma (Montfortia)			X	x	x	X			
Lapparia			X	X	X	X			
Pleurotomaria			X	x	x	X			
Pyramidella			X	X	X	X			
Strepsidura			X	X	X	X			
Tejonia = ?Ampullospira			X	X	X	X			
Terebra			X	X	X	X			
Vitrinella = ?Adeorbis			X	X	X	X			
Laevityphis				X	X	X			
Natica (Carinacca)				X	x	X			
Pterynotus s.s.				X	x	X			
"Melanatria" (bk)				X	x	X			
Patelloida					X	X			
Calliostoma						X			
Cantrainea = Nehalemia (s, b)						X			
"Cantharus"						X			
Cidarina						X			
Clavus						X			
"Mangelia?"						X			
"Megathura?"						X			
Melanoides (bk)						X			
Neritina (bk?)						X			
Nucleolaria						X			
Parvisipho						X			
Phos						X			
Acmaea (s, b)	X	x	X	X	x	x	X		
Ancilla (Spirancilla) (s, b)	X	x	X	X	x	x	X		

GENUS (SUBGENUS)	Ma	Me	C	D	Tr	Te	G	Mt	J
Bonellitia (Admetula) (s, b)	X	X	X	X	x	X	X		
Fusinus	X	X	X	X	x	X	X		
Hipponix	X	x	X	x	x	X	X		
Latirus	X	x	X	X	x	X	X		
Siphonalia	X	X	X	X	x	X	X		
Whitneyella	X	X	X	X	x	X	X		
Pachycrommium		X	X	X	X	X	X		
Alvania			X	x	x	X	X		
Bittium s.s.			X	X	X	X	X		
Certhiopsis			X	X	x	X	X		
Conomitra (s, b)	X	x	X	X	X	x	X		
Cypraeogemmula			X	x	x	X	X		
"Homalopoma"			X	X	x	X	X		
Melanella = Eulima			X	x	x	X	X		
Turbella			X	x	x	x	X		
Conus (s, b)			X	X	X	X	X		
Nerita (Bajanerita?)			X	x	x	x	X		
Echinophoria (s, b)				X?	x	X	X		
Sulcocypraea				X	x	X	X		
Acamptogenotia (s, b)						X	X		
Comitas (Boreocomitas) (s, b)						X	X		
Fulgurofusus (s, b)						X	X		
Liotia						X	X		
Littorina						X	X		
"Trichotropis?"						X	X		
"Turris"						X	X		
Turritellopsis?						X	X		
Urosalpinx						X	X		
Volvulella						X	X		
Balcis (b)							X		
Bonellitia s.s. (b)							X		
Bullia (Buccinanops)?							X		
Cellana							X		
Cirsochilus							X		
Clivuloturris (b)							X		
Colus? (b)							X		
Cyclostremella (b)							X		
Eopleurotoma? (b)							X		
Erato							X		
"Fusinus" (b)							X		
Granula							X		
Haminoea							X		
Margarites s.s. (b)							X		
Margarites s.l.							X		
Phalium (Bezoardica)							X		
Phanerolepida (s, b)							X		
Pleuroliria (b)							X		
Procerapex (b)							X		
Puncturella s.l.							X		
Searlesia							X		
Segmentina							X		
Seraphs?							X		
Siphonaria							X		
Solariella (Machaeroplax) (b)							X		
Sveltella (b)							X		
Thiara (bk)							X		
Trophon							X		
Trophonopsis (b)							X		
Cylichnina	X	X	X	X	X	X	X	X	
Gemmula (s, b)	X	X	X	X	x	X	X	X	
Nekewis	X	x	X	x	X	X	X	X	

GENUS (SUBGENUS)	Ma	Me	C	D	Tr	Te	G	Mt	J	GENUS (SUBGENUS)	Ma	Me	C	D	Tr	Te	G	Mt	J
Olequahia (s, b)	X	x	X	X	x	X	X	X		Natica (Naticarius)			X	X	x	X	X	x	X
Ranella = Gyrineum	X	X	x	X	x	X	X	X		Sinum (s, b)			X	X	X	X	X	X	X
"Turricula" (s, b)	X	X	X	X	X	X	X	X		Ficus				X	x	X	X	X	X
Exilia (s, b)	X	X	X	X	X	X	X	X		Glossaulax				X	x	X	X	X	X
Parasyrinx s.l. (s, b)	X	x	X	X	X	X	X	X		"Marginella"				X	x	X	X	X	X
Surculites (s, b)	X	x	x	x	x	x	X	X		Perse				X	X	X	X	X	X
Odostomia	X	x	X	X	X	x	x	X		Cirsotrema				X	x	x	x	x	X
Elimia (bk)			X	x	x	X	X	X		Natica s.s. (s, b)					X	X	X	X	X
Erginus (s, ch)			X	x	x	X	X	X		Priscofusus						X	X	X	X
Opalia s.l.						X	x	X		Turritellopsis?						X	X	x	X
Epitonium (Boreoscala) (s, b)						X	X	X		Scaphander s.l. (s, b, ch)						X	X	X	X
"Admete" (b, ch)							X	X		Aforia (b, ch)							X	X	X
Margarites (Pupillaria?) (b, ch)							X	X		Argobuccinum (b)							X	X	X
Ptychosyrinx (b)							X	X		Bathybembix (b)							X	X	X
Turrinosyrinx (b)							X	X		Bruclarkia (s, b)							X	X	X
Antiplanes (b)								X		Cancellaria							X	x	X
"Architectonica?"								X		"Fusitriton" (b)							X	x	X
Calicantharus								X		Limacina (p)							X	X	X
Eosiphonalia								X		Liracassis							X	x	X
Kelletia?								X		Molopophorus							X	X	X
Oxystele								X		Musashia s.l. (b, ch)							X	X	X
Pseudomelatoma (b)								X		Praehyalocylis (p)							X	X	X
Taranis								X		Provanna (ch)							X	X	X
Tegula								X		Spirotropis (b)							X	x	X
Acteon = Kleinacteon	X	x	X	X	x	X	X	X	X	Suavodrillia (b)							X	X	X
Amaea (Scalina)=Acrilla	X	x	X	X	x	X	X	X	X	Ancistrolepis (s, b)								X	X
Calyptraea (s, b)	X	x	X	X	X	X	X	X	X	Amphissa									X
Euspira (s, b)	X	X	X	X	X	X	X	X	X	Clio (p)									X
Neverita s.s. (s, bk, b)	X	X	X	X	x	X	X	X	X	Epitonium (Asperiscala)									X
Solariella s.l.	X	X	X	X	X	X	X	X	X	Megasurcula? (b)									X
Turritella (s, b)	X	X	X	X	X	X	X	X	X	Microglyphus (b)									X
Crepidula (s, bk)			X	X	X	X	X	X	X	Stenorytis (b)									X

Abbreviations: X=known occurrence; x=assumed occurrence; Ma="Martinez Stage;" Me="Meganos Stage;" C="Capay Stage;" D="Domengine Stage;" Tr="Transition Stage;" Te="Tejon Stage;" G=Galvinian Stage; Mt=Matlockian Stage; J=Juanian Stage; b=bathyal; bk=brackish; ch=chemosynthetic; p=pteropod; s=shallow marine (note: unless otherwise indicated, habitat is shallow marine); s.l.=sensu lato (in the broad sense); s.s.=sensu stricto (in the strict sense). Taxa listed so as to delineate turnovers.

taxa (FAD value = 43%) that rivaled the influx of new taxa during the "Capay Stage." There was a major difference between these two events. The "Capay Stage" influx involved only thermophilic taxa, whereas the beginning of the Galvinian Stage influx involved temperate taxa. The beginning of the Galvinian Stage also was coincident with the first West Coast Paleogene pteropod gastropods and the first West Coast Paleogene records of bathyal (nonchemosynthetic) mollucan communities. In addition, chemosynthetic cold-seep communities became more common on the West Coast. Representative newly appearing genera during the Galvinian were *Bruclarkia*, *Limacina* (pteropod), *Liracassis*, *Margarites* s.s., *Provanna* (chemosynthetic), *Ptychosyrinx* (bathyal), *Searlesia*, and *Trophon*.

Although a few brackish-marine gastropods were present during the Galvinian Stage, this habitat was greatly reduced relative to the "Tejon Stage." For example, *Potamides* s. l. ranges from Paleocene through the "Tejon Stage" and is geographically widespread, but it does not continue into the Galvinian Stage.

The turnover at the "Tejon Stage"/Galvinian Stage boundary is approximately coincident with the Bartonian/Priabonian boundary (figure 2.1). According to Hansen (1987, 1992), in the rock record of the Gulf Coast of North America, there was a severe extinction of warm-water mollusks at the Bartonian/Priabonian boundary. He correlated this extinction with a cooling event. Zinsmeister (1982) also recognized an abrupt decline of warm-water mollusks in New Zealand and Australia in the late Eocene, and he attributed this decline to the foundering of the South Tasman Rise in the Southern Ocean, the beginning of circum-Antarctic circulation, and the onset of glaciation.

Turnover Pattern Indicated by Local Stratigraphic Sections

Sacate Formation–"Coldwater" Sandstone–Gaviota Formation Section. Mollusk-bearing stratigraphic sections that are continuous or nearly continuous between the upper part of the "Tejon Stage" and the lower part of the Galvinian Stage are extremely rare on the West Coast. The only known fully continuous section is the Sacate Formation–"Coldwater" Sandstone and overlying Gaviota Formation in Santa Barbara County, California. These two formations are predominantly shallow-marine, lithologically gradational, and show a depositional continuum from the "Tejon Stage" to the Galvinian Stage (Weaver and Kleinpell, 1963; Prothero and Thompson, 2001). The gastropod diversity in each of these formations is anomalously low, especially considering the shallow-marine environment. The combined number of genera/subgenera of gastropods from these two formations is only about 25. Although this might be a function of unknown paleoenvironmental factors, it seems that a more detailed study of the molluscan fauna is warranted.

Eastward, the Sacate Formation interfingers with a stratigraphic unit called the "Coldwater" Sandstone by Weaver and Kleinpell (1963). Quote marks are used for this "Coldwater" Sandstone because it is not the actual Coldwater Sandstone, which crops out much farther east in Ventura County (in the Ojai area) and is considerably older (Prothero and Vance, 1996). Squires (1994) reported on the mollusks from the true Coldwater Sandstone in the Ojai area. Weaver and Kleinpell (1963) also used an "undifferentiated Sacate-Gaviota" Formation stratigraphic unit. Using their correlation charts or locality descriptions, however, most of the molluscan localities in this undifferentiated unit could be assigned to either the Sacate–"Coldwater" Sandstone or to the Gaviota Formation.

Clark and Vokes (1936) were the first to mention the faunal turnover between the lower Gaviota Formation and the underlying rocks, and my study corroborates this turnover. The Sacate Formation-"Coldwater" Sandstone fauna includes thermophilic forms, such as *Ficus*, *Ficopsis*, and large *Turritella*, all of which lived under nearly the same surface temperature as did the underlying "Tejon Stage" Eocene formations (i.e., Matilija Sandstone and Cozy Dell Shale). Of the 13 gastropod genera/subgenera reported by Weaver and Kleinpell (1963) as being present in the Sacate Formation–"Coldwater" Sandstone, 7 (54%) have their last (LAD) occurrence in the West Coast Paleogene at the end of the "Tejon Stage." These taxa are *Ectinochilus*, *Ficopsis*, *Galeodea*, *Nerita* (*Theliostyla*), *Ranellina*, *Strepsidura*, and *Tejonia*. Each ranges downward into the "Capay Stage" or older strata (table 2.4). All of them, as well as *Pachycrommium* and *Natica* (*Naticarius*), both of which occur only in the lowermost part of the Gaviota Formation, disappeared at or near the top of the "Tejon Stage" and were strictly warm-water forms represented by long-lived and widespread species. *Pachycrommium clarki* is very significant because it is very long-lived on the West Coast. Its geologic range is 18 million years, from the "Meganos Stage" to the lowermost part of the Galvinian Stage. Other than *Pachycrommium* and *Natica* (*Naticarius*), 11 other gastropod genera/subgenera were reported by Weaver and Kleinpell (1963) as being present in the Galvinian Stage Gaviota Formation, and most of these occur only in the middle part of the formation. All are present in the underlying "Tejon Stage" strata, but are present as different species, except for *Calyptraea diegoana*, *Glossaulax reclusiana*, and *Turritella variata* s.l.

Cowlitz Formation–Keasey Formation Section. The only other known mollusk-bearing stratigraphic section that represents the upper part of the "Tejon Stage" and most of the Galvinian Stage occurs in the southwestern Washington and northwestern Oregon area and involves the Cowlitz Formation and the conformably overlying Keasey Formation. The Cowlitz Formation was deposited at an estimated paleolatitude of 40°N to 45°N in warm tropical to subtropical, shallow-marine and marginal-marine environments associated with a prograding river-dominated delta. This formation has some of the best-preserved fossils found in the Paleogene rock record of the West Coast and is very diverse, with over 190 species of invertebrates (Nesbitt, 1995). The Cowlitz Formation, which is chronostratigraphically near the top of the "Tejon Stage," is correlative to Chrons C17r–C19n (Prothero et al.., 2001c), and a basalt flow in the lower part of the formation has been age dated at 38.9 Ma. The type section of the lower part of the Keasey Formation is near Vernonia, northwest Oregon (Armentrout, 1975). The benthic foraminifers in these rocks represent an outer shelf–upper slope (outer neritic–upper bathyal) environment, and the cold, deep-water species present were responding to cold water and not to depth (McDougall, 1975). Hickman (1980) reported that the formation was deposited at outer neritic to bathyal depths (200 m to 1000 m) in a deep basin. She also reported that global cooling coincided with the beginning of Keasey deposition and resulted in a "dramatic faunal change," reflected in the numerous generic- and species-level extinctions and appearances.

A comparison of the magnetostratigraphy of the Keasey Formation versus that of the Gaviota Formation (table 2.2) reveals that the lower part of the Galvinian Stage is probably slightly older than the lower Keasey Formation. By definition (Schenck and Kleinpell, 1936), the base of the Refugian Stage [= Galvinian Stage] corresponds to the base of the Gaviota Formation. Thus, the lower Keasey Formation, which was used by Armentrout (1975) to define the lowermost part of his Galvinian Stage is probably not complete. Approximately one-fourth, or 1 million years of the Galvinian Stage might not be represented by the Keasey Formation.

Nearly all the gastropod genera/subgenera in the Cowlitz Formation are thermophilic ones, and nearly all have been found elsewhere on the West Coast in older warm-water deposits. About 77% of these taxa range downward into the "Capay Stage"–"Domengine Stage" or older strata. Fifty-nine genera/subgenera of gastropods have been reported from the Cowlitz Formation by Weaver (1943), Hickman (1976), Marincovich (1977), and Nesbitt (1995). An additional 5 genera (*Alvania, Cylichnina, Teinostoma, Turbonilla,* and *Turritellopsis?*), all of which are minute-sized genera, were found by the author in collections of the Cowlitz fauna housed at the Natural History Museum of Los Angeles County Museum (Invertebrate Paleontology Section). Of these 64 taxa, the following have their last (LAD) West Coast and Baja California Paleogene occurrence in the Cowlitz Formation: *Calorebama, Cantharus, Clavus, Colwellia, Cymatium, Diodora, Ectinochilus* (*Cowlitzia*), *Ficopsis, Melanoides, Mitra,* "*Murex,*" *Nerita* (*Theliostyla*), *Neritina, Nucleolaria, Olivella, Parvisipho, Pleurofusia, Potamides, Pseudoperissolax, Pyramidella, Teinostoma,* and *Turbonilla.* Three of these, *Ectinochilus, Ficopsis,* and *Nerita* (*Theliostyla*), are the same ones that disappeared above the contact with the Sacate Formation/"Coldwater Sandstone" and the Gaviota Formation. The reason that many of the Cowlitz Formation gastropods do not continue into the Keasey Formation would seem to be facies controlled. The Cowlitz Formation was deposited in shallow-marine depths and the Keasey Formation was deposited in bathyal depths.

Nevertheless, 15 (23%) of the Cowlitz Formation genera/subgenera (e.g., *Conus, Conomitra, Echinophoria, Exilia, Olequahia, Turritella*) were warm-water-holdover taxa that continued into the Keasey Formation. These holdover taxa actually make up a significant portion of the gastropod faunas in shallow-marine formations of the Galvinian Stage on the West Coast (e.g., Gries Ranch Formation, Quimper Sandstone, San Emigdio Formation, Wheatland Formation). Several genera (e.g., *Euspira, Gemmula, Natica* s.s., *Turritella, Neverita* s.s., *Scaphander, Sinum*) were even more persistent holdover taxa and are present in the Matlockian Stage.

A few of the Cowlitz Formation gastropods, such as *Acamptogenotia, Comitas* (*Borecocomitas*), *Epitonium* (*Boreoscala*), and *Priscofusus,* appeared for the first time during deposition of this formation, and they became much more important later, during the temperate times associated with the Keasey and Pittsburg Bluff Formations.

Brackish-marine, warm-water gastropods like *Potamides* s.l. are locally important components of the Cowlitz Formation, but they do not occur in the Keasey Formation because its depositional setting did not involve deltaic environments and was associated with cooler waters.

The cool-water, outer neritic to bathyal gastropods in the Keasey Formation are very important because they coincided in time with the first appearance of these types of mollusks on the West Coast. Of the 48 gastropod genera/subgenera reported by Hickman (1976, 1980) for the Keasey Formation, 20 (42%) have their FAD in this formation: "*Admete,*" *Balcis, Bathybembix, Bonellitia* s.s., *Bruclarkia, Clivuloturris, Cyclostremella, Eopleruotoma?*, "*Fusinus,*" "*Fusitriton,*" *Granula, Margarites* s.s., *Margarites* (*Pupillaria*), *Pleuroliria, Procerapex, Ptychosyrinx, Solariella* (*Machaeroplax*), *Sveltella, Trophonopsis,* and *Turrinosyrinx.*

MATLOCKIAN STAGE

The Matlockian Stage lasted about 4.5 m.y. and is represented by several formations (table 2.1), most of which are in the Pacific Northwest. There were 61 gastropod genera/subgenera on the West Coast during this stage, and this number represents a significant decrease in diversity relative to all the other previous West Coast Paleogene stages. Fifty-one (84%) were holdovers from previous stages, and 28 (55%) of them were thermophilic taxa that range far back into the Paleogene (table 2.5). Representative thermophilic holdovers, some of which are the same as those of the Galvinian Stage, are *Acteon, Exilia, Ficus, Olequahia, Perse, Ranella, Surculites,* and *Turritella.*

Thirty-five percent of the holdovers were temperate forms that had first appeared in the Galvinian Stage (e.g., *Bathybembix*, *Limacina*, *Liracassis*, *Musashia*, *Provanna*, *Ptychosyrinx*).

Sixteen percent of the Matlockian Stage gastropod genera/subgenera were newly appearing temperate taxa not previously found before on the West Coast (e.g., *Antiplanes*, *Calicantharus*, *Oxystele*, *Taranis*). Cold-seep gastropods reached their highest diversity (7 genera) during the Matlockian Stage, and many of them are found in the Makah Formation, Olympic Peninsula, Washington. Hickman (1984) reported that the large volutid gastropod *Musashia* is a regular member of the *Thyasira-Lucinoma-Solemya* deep-water community in rocks that are temporally equivalent to the Matlockian Stage in Washington. Campbell and Bottjer (1993) recognized this community as being a cold-seep one. *Scaphander* has been reported (Goedert et al., 1995) as being a rarely occurring chemosynthetic community gastropod where it occurs with whale-fall remains in Matlockian Stage rocks in southwestern Washington.

Although having many fewer genera/subgenera of gastropods than the Galvinian Stage (61 versus 110), the Matlockian Stage is similar to the Galvinian Stage in having a high percentage of holdover taxa and many temperate taxa, as well as having pteropods, bathyal gastropods, and cold-seep gastropods. Representative genera of the Matlockian Stage are *Acteon*, *Amaea*, *Ancistrolepis* (bathyal), *Calyptraea*, *Epitonium* (*Boreoscala*), *Euspira*, *Ficus*, *Limacina* (pteropod), *Liracassis*, "*Marginella*," *Molopophorus*, *Musashia* (can be chemosynthetic), *Olequahia*, *Perse*, *Priscofusus*, *Ranella*, *Surculites*, and *Turritella*.

TURNOVER AT THE GALVINIAN STAGE/ MATLOCKIAN STAGE BOUNDARY

Overall Pattern of Turnover

Fifty-one gastropod taxa co-occur in both the Galvinian and Matlockian Stages, but at the end of the Galvinian Stage, a significant number of genera/subgenera of gastropods disappeared from the Paleogene West Coast (table 2.5). The LAD value is 55% (table 2.3) and is the highest LAD value for any of West Coast-Paleogene stage. Half of the casualities were thermophilic taxa (e.g., *Ancilla* [*Spirancilla*], *Conus*, *Cypraeogemmula*, *Latirus*, *Siphonalia*, *Whitneyella*), and the other half were temperate forms that had arrived on the West Coast during the Galvinian Stage (e.g., *Colus?*, *Procerapax*, *Trophonopsis*). Of the many (84%) Matlockian Stage gastropods that were surviving holdover taxa, the majority (55%) were thermophilic ones that range far back into previous Paleogene stages (e.g., *Acteon*, *Exilia*, *Ficus*, *Surculites*, *Turritella*), and some were cooler-water forms that had arrived during the Galvinian Stage (e.g., *Aforia*, *Ptychosyrinx*).

The Matlockian Stage had a moderate level (FAD value = 16%) of new taxa, which were temperate forms (e.g., *Ancistrolepis*, *Calicantharus*, *Oxystele*). The Matlockian Stage chemosynethic community gastropods are represented by only a few genera. one of these, *Provanna*, is also present in the Galvinian Stage.

The faunal turnover at the end of the Galvinian Stage was unlike the one at the end of the "Tejon Stage." The latter had a high LAD followed by a high FAD, but at the end of the Galvinian Stage there was a high LAD followed by a low FAD (table 2.3). The turnover at the Galvinian-Matlockian boundary is near the Eocene-Oligocene boundary. According to Hansen (1987, 1992), the rock record of the Gulf Coast of North America provides evidence of an extinction event at the end of the Eocene that he correlated to a cooling event.

Turnover Pattern Indicated by Local Stratigraphic Sections

Mollusk-bearing stratigraphic sections that are continuous from the Galvinian Stage into the overlying Matlockian Stage are extremely rare. The best-studied section is the Keasey Formation (discussed earlier) and the conformably overlying Pittsburg Bluff Formation in northwestern Oregon and southwestern Washington. Fossil mollusks in the Pittsburg Bluff Formation represent a mixed-depth assemblage consisting of taxa derived from offshore (about 100 m depth), shallow water (about 20 m depth), and nearshore to intertidal habitats (Moore, 1976).

The Keasey Formation has 48 gastropod genera/subgenera, and 40 (83%) are not in the Pittsburg Bluff fauna listed by Moore (1976:table 1). (See Hickman [1976:67] and Marincovich [1977:308] for some changes in taxonomy used by Moore). Only a few Pittsburg Bluff gastropod genera were new to the West Coast, and these are *Aforia*, "*Architectonica?*," *Eosiphonalia*, *Opalia*, and *Taranis*. The Pittsburg Bluff molluscan fauna has a co-occurrence of warm-water and cooler-water elements. The fauna is neither tropical nor boreal, rather it seems to indicate a warm-temperate sea (Moore, 1976). It is important to mention that Hickman (1969) similarly reported the occurrence of tropical and temperate molluscan genera in the Eugene Formation (Matlockian Stage) in Oregon. Furthermore, Hickman reported that the fauna in this shallow-marine formation lived in a cooling subtropical to warm-temperate sea.

Although the end of the Galvinian Stage is represented by the transition from the Gaviota Formation into the overlying Alegria Formation in Santa Barbara County, southern California (Prothero and Thompson, 2001), the Alegria is only sparsely fossiliferous, and, to the east, it grades into alluvial sediments.

JUANIAN STAGE

This stage lasted about 5 million years and is represented by several formations in Washington and Oregon (table 2.1). There were only 40 gastropod genera/subgenera on the West Coast during this stage, and this was the lowest diversity during the West Coast Paleogene. Thirty-four (85%) were holdovers from previous stages, and 16 (47%) of them were thermophilic taxa that range far back into the Paleogene. Representatives, all of which are the same as those in the Galvinian Stage, are *Acteon, Calyptraea, Ficus,* and *Turritella.* The other holdovers were those that had first appeared in either the temperate-water Galvinian or the Matlockian Stages (e.g., *Aforia, Bathybembix, Bruclarkia, Limacina, Liracassis, Molopophorus, Perse, Provanna, Suavodrilia*).

Fifteen percent of the fauna were newly appearing temperate forms not previously found before on the West Coast (e.g., *Amphissa, Clio* [pteropod], *Stenorytis*). Only a few cold-seep gastropods were present (e.g., *Provanna, Musashia*). Three pteropods, *Clio, Limacina,* and *Praehya-loclis cretacea,* representing the highest pteropod diversity for the Paleogene, have been found in Juanian Stage rocks.

Although much lower in number of genera/subgenera of gastropods than the Matlockian Stage (61 versus 40), the Juanian Stage is similar to the Matlockian Stage in having many holdover taxa from warm-water stages, as well as pteropods, bathyal gastropods, and cold-seep gastropods. The only brackish-marine gastropod was *Elimia.* Representative genera/subgenera of the Juanian Stage are *Acteon, Amaea, Ancistrolepis* (bathyal), *Calyptraea, Epitonium* (*Boreoscala*), *Euspira, Ficus, Limacina* (pteropod), *Liracassis, "Marginella," Molopophorus, Musashia* (can be chemosynthetic), *Perse,* and *Priscofusus.*

TURNOVER AT THE MATLOCKIAN STAGE/ JUANIAN STAGE BOUNDARY

Thirty-four gastropod taxa co-occur in both the Matlockian and the Juanian Stages, but at the end of the Matlockian Stage, a significant number of genera/subgenera of gastropods disappeared from the Paleogene West Coast (table 2.5). The LAD value is 44% (table 2.3). Nearly half of the casualities were thermophilic taxa that range far back into previous Paleogene stages (e.g., *Cylichnina, Exilia, Olequahia, Ranella, Surculites*), and the other half were those that had arrived during the Galvinian Stage (e.g, *Ptychosyrinx, Turrinosyrinx*) or during the Matlockian Stage (e.g., *Eosiphonalia, Pseudomelatoma*). Of the Juanian Stage surviving holdover taxa, 47% were thermophilic ones that range far back into previous Paleogene stages (e.g., *Acteon, Ficus, Turritella*). The Juanian Stage chemosynethic community gastropods are represented by only two genera; namely, *Provanna* and *Musashia.* Both are present in the Galvinian Stage.

At the beginning of the Juanian Stage, there was a relatively low number (FAD value = 15%) of new taxa, which were temperate forms (e.g., *Clio, Stenorytis*). This low number is similar to that of the Matlockian Stage. The faunal turnover at the end of the Matlockian was similar to the one at the end of the Galvinian Stage in that both had a high LAD value followed by a low FAD value.

CONCLUSIONS

For the 29 million years from the Paleocene through the "Tejon Stage," nearly all the marine gastropods on the West Coast of the United States and Baja California, Mexico, were shallow-marine, thermophilic forms, and many were immigrants from the Old World Tethys Sea. Their highest diversity occurred during the "Capay" and "Domengine" Stages and came just after the late Paleocene thermal maximum, which was the warmest time in the last 90 million years.

At about 36.5 Ma, near the boundary of the "Tejon Stage" and Galvinian Stage, a taxonomic turnover event began. Of the turnovers discussed in this paper, this first one has the best geologic age constraint. This turnover involved the disappearance of 48% of the gastropod genera/subgenera, nearly all of which were thermophilic, and most had been present on the West Coast since the late Paleocene or early Eocene. Forty-three percent of the gastropod genera/subgenera in the Galvinian Stage had their first appearance on the West Coast, and they were temperate forms. The Galvinian Stage corresponds to a time on the West Coast when the habitats of gastropods were greatly diversified. The first pteropods known in the Cenozoic rock record of the West Coast appeared in the Pacific Northwest during the Galvinian Stage, and this group continued throughout the Oligocene in this area. The first bathyal gastropods known in the Cenozoic rock record of the West Coast appeared in both the Keasey Formation in northwestern Oregon and the Kreyenhagen Formation [= Tumey Formation] in central California during the Galvinian Stage, and about one-third of the bathyal gastropods in the Keasey Formation are represented by genera never seen before on the West Coast. Chemosynthetic community gastropods also became more common on the West Coast during the Galvinian Stage, and these gastropods were associated with deep-sea, cold-seep limestones. This group of gastropods continued throughout the Oligocene, with some members becoming associated with whale-fall habitats.

The pattern of ongoing disappearance of thermophilic taxa and the appearance of new cooler-water taxa that began with the initiation of the Galvinian Stage continued into the Matlockian and Juanian Stages. Using the only available data, which consists of full-stage ranges of the taxa, predictably gives the appearance that this overall pattern of change was stepped. Resolution of whether it was truly stepped, or was a long gradual change, awaits better data.

A second turnover (apparent?) took place about 33 Ma at the Galvinian Stage/Matlockian Stage boundary. Fifty-five percent of the Galvinian Stage gastropod genera/subgenera disappeared. Again, the majority of these were holdover taxa that range into previous warm-water Paleogene stages, and the rest that disappeared were temperate-water taxa that had arrived on the West Coast during the Galvinian Stage. Sixteen percent of the gastropod genera/subgenera in the Matlockian Stage had their first appearance on the West Coast, and they were temperate forms.

A third turnover (apparent?) took place about 28.5 Ma, at the Matlockian Stage/Juanian Stage boundary. Forty-four percent of the Matlockian Stage gastropod genera/subgenera disappeared. Once again, the majority of these were holdover taxa that range into previous warm-water Paleogene stages, and the rest that disappeared were temperate forms that had appeared during either the Galvinian Stage or the Matlockian Stage. Fifteen percent of the taxa in the Juanian Stage had their first appearance on the West Coast, and they were also temperate forms. This third turnover, as well as the second one, differed from the first one in that there were fewer new appearances of taxa. The Juanian Stage had the lowest number (40) of marine gastropod genera/subgenera for the West Coast Paleogene.

The presence of some temperate-water taxa showing up within the "Tejon Stage" and Cowlitz Stage indicates that the initial arrival of cooler waters along the West Coast was not abrupt. Also, it is evident that the turnovers were not catastrophic, because after each turnover relatively high percentages (near or greater than 50%) of the remaining gastropod genera/subgenera were survivors from earlier warm-water Paleogene stages. Even 12 million years after the initial influx of cooler waters into the West Coast United States began and the diversity of marine gastropod genera/subgenera had declined by two thirds, half of the taxa present were warm-water ones that had been present earlier on the West Coast. A few of these very persistent holdover taxa were *Calyptraea*, *Crepidula*, *Turritella*, and the naticids *Euspira*, *Glossaulax*, *Neverita* s.s., and *Sinum*. Some of these genera persisted as single species for long intervals of time.

The co-occurrence of warm-water and cooler-water mollusks, and the relatively high percentages of warm-water holdover taxa present seem to indicate that the cooler waters that arrived during the Eocene-Oligocene transition were most likely associated with warm-temperate seas. This conclusion is in agreement with the work by Hickman (1969) and Moore (1976). Several other factors should be considered, as well. There might have been mixing of cooler and warmer faunal elements. For example, along the outer coast of modern Baja California, Mexico, warmer water species are found on the north side of points and cooler water species on the south side of points, with the number of warm-water species increasing to the south (Brusca, 1980). The warmer-water forms would not have to become cool-water adapted to show up among cooler taxa; they would just need to live in the right place to be mixed post-mortem.

The holdover taxa might be represented, at least in part, by extralimital species that can live beyond the normal habitat of the genus. For example, the modern gastropods *Turritella*, *Conus*, and *Cypraea* are normally warm-water gastropods (Keen, 1971), but at least one species of each of these genera can survive in warm-temperate waters along the West Coast of the United States (McLean, 1978). Periodic local and temporary current changes (e.g., El Niño–like currents) could have permitted the introduction of larvae of tropical mollusks into areas of cooler water temperatures. This type of scenario was recognized by

Zinsmeister (1974) as having affected molluscan assemblages of the California Pleistocene.

The holdover taxa might also represent taxa that were eurythermal and could live at a wide range of temperatures, but it is doubtful this explanation could be used to account for such a high number of holdover taxa. Of course, the presence of the holdover taxa might represent any combination of the above-mentioned possibilities. Certainly, more information concerning the paleoecology of Eocene-Oligocene-transition mollusks is needed.

ACKNOWLEDGMENTS

Donald Prothero kindly and generously shared information about his ongoing work on the magnetic stratigraphy of selected West Coast Paleogene formations. His help was invaluable in establishing the stage assignment of many of the key formations. Ellen J. Moore shared her considerable knowledge of West Coast Paleogene biostratigraphy and offered words of encouragement. Lindsey T. Groves, Natural History Museum of Los Angeles County (Invertebrate Paleontology Section) allowed access to the collections. Ellen J. Moore and LouElla R. Saul critically reviewed the manuscript and offered important insights that greatly improved it.

REFERENCES

Addicott, W. O., 1969, Tertiary climatic change in the marginal northeastern Pacific Ocean: Science, v. 165, p. 583-586.

Addicott, W. O., 1970a, Tertiary paleoclimatic trends in the San Joaquin basin, California: U. S. Geological Survey Professional Paper 644-D, 19 p.

Addicott, W. O., 1970b, Latitudinal gradients in Tertiary molluscan faunas of the Pacific coast: Palaeogeography, Paleoclimatology, Palaeoecology, v. 8, p. 287-312.

Addicott, W. O., 1976a, Neogene molluscan stages of Oregon and Washington, *in* Fritsche, A. E., Terbest, H., Jr., and Wornardt, W. W., eds., The Neogene symposium: SEPM, Pacific Section, Annual Meeting, p. 95-115.

Addicott, W. O., 1976b, New molluscan assemblages from the upper member of the Twin River Formation, western Washington: significance in Neogene chronostratigraphy: U. S. Geological Survey Journal of Research, v. 4, p. 437-447.

Armentrout, J. M., 1975, Molluscan biostratigraphy of the Lincoln Creek Formation, southwest Washington, *in* Weaver, D. W., Hornaday, G. R., and Tipton, A., eds., Paleogene symposium & selected technical papers, conference on future energy horizons of the Pacific Coast: Los Angeles, American Association of Petroleum Geologists, SEPM, and Society of Economic Geologists, Pacific Sections, Annual Meeting, p. 14-48.

Armentrout, J. M., 1981, Correlation and ages of Cenozoic chronostratigraphic units in Oregon and Washington, *in* Armentrout, J. M., ed., Pacific northwest Cenozoic biostratigraphy: Geological Society of America Special Paper 184, p. 137-148.

Arnold, R., 1909, Environment of the Tertiary faunas of the Pacific coast of the United States: Journal of Geology, v. 17, p. 509-533.

Bains, S., Corfield, R. M., and Norris, R. D., 1999, Mechanisms of climate warming at the end of the Paleocene: Science, v. 285, p. 724-727.

Berggren, W. A., and Prothero, D. R., 1992, Eocene-Oligocene climatic and biotic evolution: an overview, *in* Prothero, D. R., and Berggren, W. A., eds., Eocene-Oligocene climatic and biotic evolution: Princeton, New Jersey, Princeton University Press, p. 1-28.

Berggren, W. A., Kent, D. V., Swisher, C. C., and Aubry, M.-P., 1995, A revised Cenozoic geochronology and chronostratigraphy, *in* Berggren, W. A., Kent, D. V., Aubry, M. -P., and Hardenbol, J., eds., Geochronology, time scales, and global stratigraphic correlation: SEPM (Society for Sedimentary Geology) Special Publication 54.

Berggren, W. A., Lucas, S., and Aubry, M.-P., 1998, Late Paleocene-early Eocene climatic and biotic evolution: an overview, *in* Aubry, M. -P., Lucas, S., and Bergrren, W. A., eds., Late Paleocene-early Eocene climatic and biotic events in the marine and terrestrial record: New York, Columbia University Press, p. 1-17.

Bottjer, D. J., Lund, S. P., Powers, J. E., Steele, M. C., and Squires, R. L., 1991, Magnetostratigraphy of Paleogene strata in San Diego and the Simi Valley, southern California, *in* Abbott, P. L., and May, J. A., eds., Eocene geologic history San Diego region: Los Angeles, SEPM, Pacific Section, Book 68, p. 115-124.

Brusca, R. C., 1980, Common intertidal invertebrates of the Gulf of California (second edition): University of Arizona Press, Tucson, Arizona, 513 p.

Campbell, K. A., and Bottjer, D. J., 1993, Fossil cold

seeps: National Geographic Research & Exploration, v. 9, p. 326-343.

Clark, B. L., and Anderson, C. A., 1938, Wheatland Formation and its relation to early Tertiary andesites in the Sierra Nevada: Bulletin of the Geological Society of America, v. 49, p. 931-956.

Clark, B. L., and Vokes, H. E., 1936, Summary of marine Eocene sequence of western North America: Bulletin of the Geological Society of America, v. 47, p. 851-878.

DeLise, K. C., 1967, Biostratigraphy of the San Emigdio Formation, Kern County, California: University of California Publications in Geological Sciences, v. 68, p. 1-66.

Dickerson, R. E., 1917a, Climatic zones of Martinez Eocene time: Proceedings of the California Academy of Sciences, series 4, v. 7, p. 193-196.

Dickerson, R. E., 1917b, Climate and its influence upon the Oligocene faunas of the Pacific coast, with descriptions of some new species from the *Molopophorus lincolnensis* Zone: Proceedings of the California Academy of Sciences, series 4, v. 7, p. 157-192.

Dockery, D. T. III., 1986, Punctuated succession of Paleogene mollusks in the northern Gulf coastal plain: Palaios, v. 1, p. 582-589.

Durham, J. W., 1944, Megafaunal zones of the Oligocene of northwestern Washington: University of California Publications Bulletin of the Department of Geological Sciences, v. 27, p. 101-212.

Durham, J. W., 1950, Cenozoic marine climates of the Pacific coast: Bulletin of the Geological Society of America , v. 61, p. 1243-1264.

Durham, J. W., 1959, Paleoclimates, *in* Ahrens, L. H. and others, eds., Physics and Chemistry of the Earth, Vol. 3: London, Pergamon Press, p. 1-16.

Effinger, W. L., 1938, The Gries Ranch fauna (Oligocene) of western Washington: Journal of Paleontology, v. 12, p. 355-390.

Fallaw, W. C., 1979, A test of the Simpson Coefficient and other binary coefficients of faunal similarity: Journal of Paleontology, v. 53, p. 1029-1034.

Givens, C. R., 1974, Eocene molluscan biostratigaphy of the Pine Mountain area, Ventura County, California: University of California Publications in Geological Sciences, v. 109, p. 1-107.

Givens, C. R., 1978, An occurrence of the Tethyan genus *Volutilithes* (Gastropoda: Volutidae) in the Eocene of California: Journal of Paleontology, v. 52, p. 104-108.

Givens, C. R., 1989, First record of the Tethyan genus *Volutilithes* (Gastropoda: Volutidae) in the Paleogene of the Gulf coastal plain, with a discussion of Tethyan molluscan assemblages in the Gulf coastal plain and Florida: Journal of Paleontology, v. 63, p. 852-856.

Givens, C. R., and Kennedy, M. P., 1976, Middle Eocene mollusks form northern San Diego County, California: Journal of Paleontology, v. 50, p. 954-975.

Givens, C. R., and Kennedy, M. P., 1979, Eocene molluscan stages and their correlation, San Diego area, California, *in* Abbott, P. L., ed., Eocene depositional systems, San Diego: Los Angeles, SEPM, Pacific Section.

Goedert, J. L., and Benham, S. R., 1999, A new species of *Depressigyra*? (Gastropoda: Peltospiridae) from cold-seep carbonates in Eocene and Oligocene rocks of western Washington: The Veliger, v. 42, p. 112-116.

Goedert, J. L., and Campbell, K. A., 1995. An early Oligocene chemosynthetic community from the Makah Formation, northwestern Olympic Peninsula, Washington: The Veliger, v. 38, p. 22-29.

Goedert, J. L,, and Kaler, K., 1996, A new species of *Abyssochrysos* (Gastropoda: Loxonematoidea) from a middle Eocene cold-seep carbonate in the Humptulips Formation, western Washington: The Veliger, v. 39, p. 65-70.

Goedert, J. L., and Squires, R. L., 1990, Eocene deep-sea communities in localized limestones formed by subduction-related methane seeps, southwestern Washington: Geology, v. 18, p. 1182-1185.

Goedert, J. L., Squires, R. L., and Barnes, L. G., 1995, Paleoecology of whale-fall habitats from deep-water Oligocene rocks, Olympic Peninsula, Washington state: Palaeogeography, Palaeoclimatology, Palaeoecology, v. 118, p. 151-158.

Hansen, T. A., 1987, Extinction of late Eocene to Oligocene molluscs: relationship to shelf area, temperature changes, and impact events: Palaios, v. 2, p. 69-75.

Hansen, T. A., 1992, The patterns and causes of molluscan extinction across the Eocene/Oligocene boundary, *in* Prothero, D. R., and Berggren, W. A., eds., Eocene-Oligocene climatic and biotic evolution: Princeton, New Jersey, Princeton University Press, p. 341-348.

Hickman, C. S., 1969, The Oligocene marine molluscan fauna of the Eugene Formation in Oregon: University of Oregon Museum of Natural History Bulletin, v. 16, p. 1-112.

Hickman, C. S., 1972, Review of the bathyal gastropod genus *Phanerolepida* (Homalopomatinae) and description of a new species from the Oregon Oligocene: The Veliger, v. 15, p. 107-112.

Hickman, C. S., 1974, *Nehalemia hieroglyphica*, a new genus and species of archaeogastropod (Turbinidae: Homalopomatinae) from the Eocene of Oregon: The Veliger, v. 17, p. 89-91.

Hickman, C. S., 1976, Bathyal gastropods of the family Turridae in the early Oligocene Keasey Formation in Oregon, with a review of some deep-water genera in the Paleogene of the eastern Pacific: Bulletins of American Paleontology, v. 70, p. 1-119.

Hickman, C. S., 1980, Paleogene marine gastropods of the Keasey Formation in Oregon: Bulletins of American Paleontology, v. 78, p. 1-112.

Hickman, C. S., 1984, Composition, structure, ecology, and evolution of six Cenozoic deep-water mollusk communities: Journal of Paleontology, v. 58, p. 1215-1234.

Janssen, A., 1989, Pteropoda (Gastropoda, Euthecosomata) from the Australian Cenozoic: Scripta Geologica, v. 91, p. 1-76.

Keen, A. M., 1971, Sea Shells of Tropical West America, Marine Mollusks from Baja California to Peru (second edition): Stanford University Press, Stanford, California, 1064 p.

Kirby, M. S., and Saul, L. R., 1995, The Tethyan bivalve *Roudairia* from the Upper Cretaceous of California: Palaeontology, v. 38, p. 23-38.

McDougall, K., 1975, The microfauna of the type section of the Keasey Formation of northwestern Oregon, *in* Weaver, D. W., Hornaday, G. R., and Tipton, A., eds., Paleogene symposium & selected technical papers, conference on future energy horizons of the Pacific coast: Los Angeles, American Association of Petroleum Geologists, SEPM, and Society of Economic Geologists, Pacific Sections, Annual Meeting, p. 343-359.

McLean, J. L., 1978. Marine shells of southern California (revised edition): Natural History Museum of Los Angeles County, Science Series 24, 104 p.

Marincovich, L., Jr., 1977, Cenozoic Naticidae (Mollusca: Gastropoda) of the northeastern Pacific: Bulletins of American Paleontology, v. 70, p. 169-494.

Moore, E. J., 1976, Oligocene marine mollusks from the Pittsburg Bluff Formation in Oregon: U. S. Geological Survey Professional Paper 922, 66 p.

Moore, E. J., 1984, Middle Tertiary molluscan zones of the Pacific Northwest: Journal of Paleontology, v. 58, p. 718-737.

Moore, E. J., 1992, Tertiary marine pelecypods of California and Baja California: Erycinidae through Carditidae: U. S. Geological Survey Professional Paper 1228-E, 37 p.

Nesbitt, E. A., 1995, Paleoecological analysis of molluscan assemblages from the middle Eocene Cowlitz Formation, southwestern Washington: Journal of Paleontology, v. 69, p. 1060-1073.

Nilsen, T. H., 1987, Stratigraphy and sedimentology of the Eocene Tejon Formation, western Tehachapi and San Emigdio Mountains, California: U. S. Geological Survey Professional Paper 1268, 110 p.

Prothero, D. R., 2001, Magnetic stratigraphy of the lower-middle Eocene Domengine, Avenal, and Yokut Sandstones, western San Joaquin Basin, Fresno and King counties, California: Pacific Section SEPM Special Publication, v. 91, p.45-55.

Prothero, D. R., and Britt, J. R., 1998. Magnetic stratigraphy and tectonic rotation of the middle Eocene Matilija Sandstone and Cozy Dell Shale, Ventura County, California: implications for sequence stratigraphy correlations: Earth and Planetary Sciences, v. 163, p. 261-273.

Prothero, D. R., and Hankins, K. G., 2000, Magnetic stratigraphy and tectonic rotation of the upper Eocene-lower Oligocene Keasey Formation, northwest Oregon: Journal of Geophysical Research, v. 105 (B7), p. 16473-16480.

Prothero, D. R., and Hankins, K. G., 2001, Magnetic stratigraphy and tectonic rotation of the lower Oligocene Pittsburg Bluff Formation, Columbia County, Oregon: Pacific Section SEPM Special Publication, v. 91, p.201-209.

Prothero, D. R., and Thompson, M., 2001, Magnetic stratigraphy of the type Refugian Stage (late Eocene-early Oligocene), western Santa Ynez Range, Santa Barbara County, California: Pacific Section SEPM Special Publication, v. 91, p. 119-134.

Prothero, D. R., and Vance, E. H., Jr., 1996, Magnetostratigraphy of the upper midddle Eocene Coldwater Sandstone, central Ventura County, California, *in* Prothero, D. R., and Emry, R. J., eds., The terrestrial Eocene-Oligocene transition in North America: Cambridge, England, Cambridge University Press, p. 155-170.

Prothero, D. R., Burns, C., and Nesbitt, E. R., 2001a, Paleomagnetism of the upper Eocene Gries Ranch Formation, Lewis County, Washington: Pacific Section SEPM, v. 91, p. 179-183.

Prothero, D. R., Pearson, P., and Armentrout, J. M., 2001b, Magnetic stratigraphy of the upper middle Eocene (Ulatisian-Narizian) Humptulips Formation, southern Olympic Peninsula, Washington: Pacific Section SEPM, v. 91, p. 96-106.

Prothero, D. R., Sanger, E., Nesbitt, E. A., Niem, A., and Kleibacker, D., 2001c, Magnetic stratigraphy and tectonic rotation of the upper middle Eocene Cowlitz and Hamlet formations, western Oregon and Washington: Pacific Section SEPM Special Publication, v. 91, p. 75-95.

Saul, L. R., 1983, Notes on Paleogene turritellas, venericardias, and molluscan stages of the Simi Valley area, California, *in* Squires, R. L., and Filewicz, M. V., eds., Cenozoic geology of the Simi Valley area, southern California: Los Angeles, SEPM, Fall Field Trip Volume and Guidebook, p. 71-80.

Saul, L. R., 1986, Mollusks of latest Cretaceous and Paleocene age, Lake Nacimiento, California, *in* Grove, K., and Graham, S., eds., Geology of Upper Cretaceous and lower Tertiary rocks near Lake Nacimiento, California: Los Angeles, SEPM Book 49, p. 25-31.

Saul, L. R., Squires, R. L., and Goedert, J. L., 1996, A new genus of cryptic lucinid? bivalve from Eocene cold seeps and turbidite-influenced mudstone, western Washington: Journal of Paleontology, v. 70, p. 788-794.

Schenck, H. G., and Kleinpell, R. M., 1936, Refugian Stage of Pacific coast Tertiary: The Bulletin of the American Association of Petroleum Geologists, v. 20, p. 215-225.

Smith, J. P., 1919, Climatic relations of the Tertiary and Quaternary faunas of the California region: Proceedings of the California Academy of Sciences, series 4, v. 9, p. 123-173.

Squires, R. L., 1984, Megapaleontology of the Eocene Llajas Formation, Simi Valley, California: Natural History Museum of Los Angeles County Contributions in Science 350, 76 p.

Squires, R. L., 1987, Eocene molluscan paleontology of the Whitaker Peak area, Los Angeles and Ventura counties, California: Natural History Museum of Los Angeles County Contributions in Science 388, 93 p.

Squires, R. L., 1988, Geologic age refinements of West Coast Eocene marine mollusks, *in* Filewicz, M. V., and Squires, R. L., eds., Paleogene stratigraphy, West Coast of North America: Los Angeles, SEPM, Pacific Section, West Coast Paleogene Symposium, Book 58, p. 107-212.

Squires, R. L., 1989, Pteropods (Mollusca: Gastropoda) from Tertiary formations of Washington and Oregon: Journal of Paleontology, v. 63, p. 443-448.

Squires, R. L., 1991, A new middle Eocene potamidid gastropod from brackish-marine deposits, southern California: The Veliger, v. 34, p. 354-359.

Squires, R. L., 1994, Macropaleontology of Eocene marine rocks, upper Sespe Creek area, Ventura County, southern California, *in* Fritsche, A. E., ed., Sedimentology and paleontology of Eocene rocks in the Sespe Creek area, Ventura County, California: Los Angeles, SEPM (Society for Sedimentary Geology), Pacific Section, Book 74, p. 39-56.

Squires, R. L., 1995, First fossil species of the chemosynthetic-community gastropod *Provanna*: localized cold-seep limestones in upper Eocene and Oligocene rocks, Washington: The Veliger, v. 38, p. 30-36.

Squires, R. L., 1997, Taxonomy and distribution of the buccinid gastropod *Brachysphingus* from uppermost Cretaceous and lower Cenozoic marine strata of the Pacific slope of North America: Journal of Paleontology, v. 71, p. 847-861.

Squires, R. L., 1998, New information on morphology, stratigraphy, and paleoclimate implications of the Eocene brackish-marine gastropod *Loxotrema turritum* Gabb, 1868, from the West Coast of the United States: The Veliger, v. 41, p. 297-313.

Squires, R. L., 1999a, Middle Eocene brackish-marine mollusks from the Matilija Sandstone at Matilija Hot Springs, Ventura County, southern California: Natural History Museum of Los Angeles County Contributions in Science 459, 32 p.

Squires, R. L., 1999b, Upper Paleocene to lower Eocene ("Meganos Stage") marine megafossils in the uppermost Santa Susana Formation, Simi Valley, southern California: Natural History Museum of Los Angeles County Contributions in Science 479, 38 p.

Squires, R. L., 2000, Additions to the molluscan fauna of the Eocene Juncal? Formation,

Lockwood Valley, Ventura County, southern California [abs.]: American Association of Petroleum Geologists Bulletin, v. 84, p. 596.

Squires, R. L., 2001, Additions to the Eocene megafossil fauna of the Llajas Formation, Simi Valley, southern California: Natural History Museum of Los Angeles County Contributions in Science 489, 40 p.

Squires, R. L., and Goedert, J. L., 1994, Macropaleontology of the Eocene Crescent Formation in the Little River area, southern Olympic Peninsula, Washington: Natural History Museum of Los Angeles County Contributions in Science 444, 32 p.

Squires, R. L., and Goedert, J. L., 1996, A new species of *Thalassonerita*? (Gastropoda: Neritidae?) from a middle Eocene cold-seep carbonate in the Humptulips Formation, western Washington: The Veliger, v. 39, p. 270-272.

Squires, R. L., and Gring, M. P., 1996, Late Eocene chemosynthetic? bivalves from suspect cold seeps, Wagonwheel Mountain, central California: Journal of Paleontology, v. 70, p. 63-73.

Squires, R. L., and Saul, L. R., 1993, A new species of *Otostoma* (Gastropoda: Neritidae) from near the Cretaceous/Tertiary boundary at Dip Creek, Lake Nacimiento, California: The Veliger, v. 36, p. 259-264.

Squires, R. L., Goedert, J. L., and Benham, S. R., 1999, First fossil record of the pteropods *Limacina* from the Pacific coast of North America: Bulletin of the Southern California Academy of Sciences, v. 98, p. 80-89.

Vokes, H. E., 1939, Molluscan faunas of the Domengine and Arroyo Hondo Formations of the California Eocene: Annals of the New York Academy of Sciences, v. 38, p. 1-246.

Vokes, H. E., 1940, Paleoecology of the fauna of the Domengine Formation, middle Eocene, California: Proceedings of the Sixth Pacific Science Congress, v. 2, p. 597-605.

Wagner, C. M., and Schilling, K. H., 1923, The San Lorenzo Group of the San Emigdio region, California: University of California Publications, Bulletin of the Department of Geological Sciences, v. 14, p. 225-276.

Walsh, S. L., Prothero, D. R., and Lundquist, D. J., 1996, Stratigraphy and paleomagnetism of the middle Eocene Friars Formation and the Poway Group, southwestern San Diego County, California, *in* Prothero, D. R., and Emry, R. J., eds., The Terrestrial Eocene-Oligocene Transition in North America: Cambridge, England, Cambridge University Press, p.120-154.

Watkins, R., 1974, Palaeobiology of an offshore molluscan fauna from the California Oligocene: Palaeogeography, Palaeoclimatology, Palaeoecology, v. 15, p. 245-266.

Weaver, C. E., 1937, Tertiary stratigraphy of western Washington and northwestern Oregon: University of Washington Publications in Geology, v. 4, p. 1-266.

Weaver, C. E., 1943, Paleontology of the marine Tertiary formations of Oregon and Washington: University of Washington Publications in Geology, v. 5, p. 1-789.

Weaver, C. E., and others, 1944, Correlation of the marine Cenozoic formations of western North America: Bulletin of the Geological Society of America, v. 55, p. 569-598.

Weaver, D. W., and Kleinpell, R. M., 1963, Mollusca from the *Turritella variata* Zone and their chronologic and biogeographic significance. Part 2: Oligocene biostratigraphy of the Santa Barbara embayment, California: University of California Publications in Geological Sciences, v. 43, p. 81-161.

Zinsmeister, W. J., 1974, A new interpretation of thermally anomalous molluscan assemblages of the California Pleistocene: Journal of Paleontology, v. 48, p. 84-94.

Zinsmeister, W. J., 1982, Late Cretaceous-early Tertiary molluscan biogeography of southern circum-Pacific: Journal of Paleontology, v. 56, p. 84-102.

CHAPTER 3

Biotic Response to the Eocene-Oligocene Transition: Gastropod Assemblages in the High-Latitude North Pacific

Anton E. Oleinik and Louie Marincovich Jr.

ABSTRACT

Global changes in climate and ocean circulation from the middle Eocene through the early Oligocene caused a significant worldwide biotic turnover. A widely accepted scenario of climatic deterioration at the beginning of the Oligocene suggests that the most severe cooling and most significant biotic impact occurred in high-latitude regions. As a result of this turnover, high-latitude cool-water faunas, ancestral to the modern cold-water biotas in the North Pacific, replaced an Eocene marine cosmopolitan biota. Diversity patterns of middle Eocene through early Oligocene shallow-marine gastropod assemblages from Kamchatka and the Gulf of Alaska Tertiary Province provide insights into the biotic changes in high latitudes associated with climatic cooling at the end of the Paleogene.

Temporal changes in the diversity and the composition of shallow-marine gastropod assemblages indicate that faunal turnover in the northern high latitudes began in the middle Eocene. Shallow-marine Eocene through early Oligocene gastropod assemblages show a stepwise decline in cosmopolitan and warm-water taxa and an increase in northern Pacific temperate and cool-water elements. Temporal diversity patterns differ significantly between the northeastern Gulf of Alaska and western Kamchatka. Diversity changes in Paleogene faunas of the Gulf of Alaska Tertiary Province and their geologic occurrence clearly indicate that these faunas evolved at latitudes significantly lower than their present location and moved northward with the Yakutat tectonostratigraphic terrane. Patterns of diversification and extinction/survival rates during the Eocene to Oligocene transition in Kamchatka suggest that, although extinctions accelerated toward the Eocene/Oligocene boundary, they were accompanied by conspicuous radiations in a small number of families.

INTRODUCTION

The biotic response to Eocene-Oligocene climatic cooling in the North Pacific (figure 3.1) has traditionally been based on two major lines of evidence: changes in the taxonomic structure and isotopic composition of oceanic planktonic and benthic microfossils, and analyses of leaf floras and spores and pollen from terrestrial sections.

The Eocene/Oligocene boundary (figure 3.2) is marked by enrichment in $\delta^{18}O$ in the tests of benthic foraminifera. Planktonic foraminiferal $\delta^{18}O$ data show a latitudinal pattern in which the isotopic enrichment in high latitudes reflects greater surface-water cooling than in the low latitudes. Bottom-water temperatures in the equatorial Pacific dropped to about 7°C by the Eocene/Oligocene boundary, to 5°C in the late early Oligocene, and to a low of 4°C during the late Oligocene thermal minimum (Savin et al., 1975). Surface-water temperatures in the western Pacific (about 17°N) dropped from 17°C in the middle Eocene to 15°C during the late Eocene, and to 14°C in the early Oligocene (Monin and Shishkov, 1979). Enrichment of benthic foraminifers in $\delta^{18}O$ at the middle/late Eocene boundary reflects deep-water cooling of the same or larger magnitude (2–4°C) as at the Eocene/Oligocene boundary (Corliss and Keigwin, 1986).

Rapidly deteriorating climatic conditions led to well-documented extinction events, or to significant

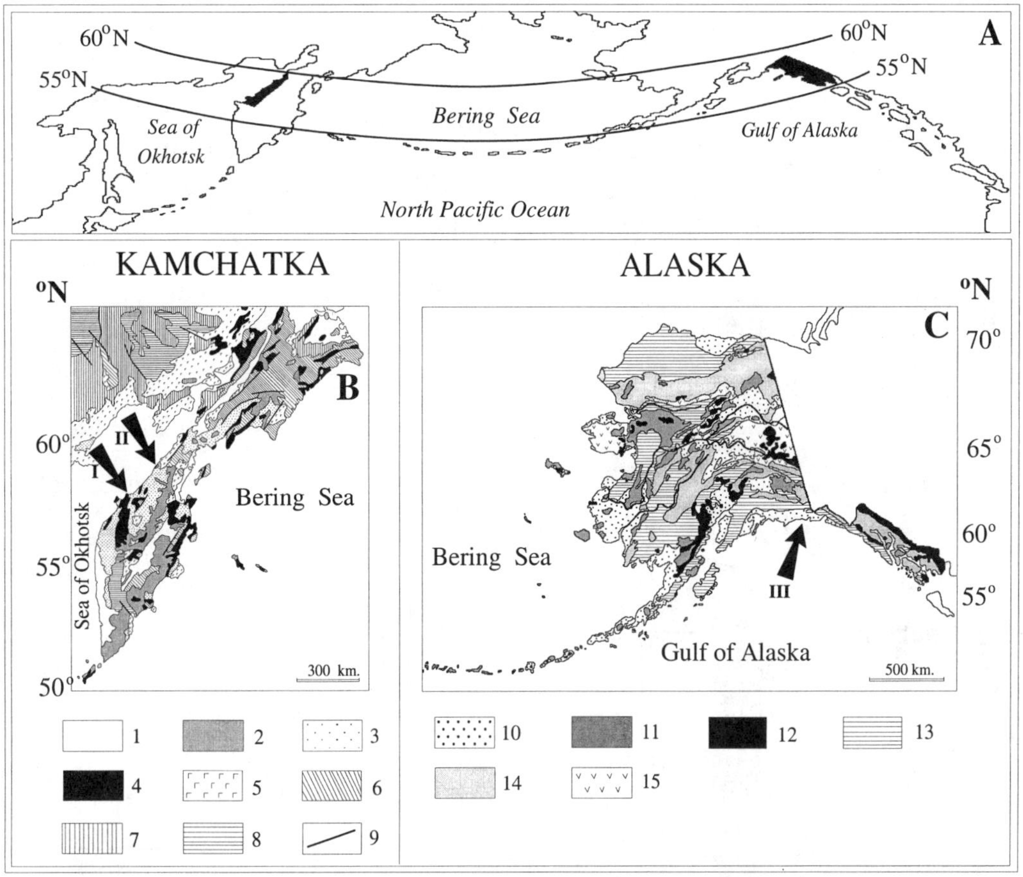

Figure 3.1. Location of study areas in the North Pacific (A), Kamchatka (B), and Alaska (C). Legend: I , Tigil District; II, Podkagernaya Bay, Palana District; III, Katalla district. Kamchatka: 1, Quaternary sediments; 2, Quaternary basalts; 3, Neogene rocks; 4, Paleogene rocks; 5, volcanic rocks of various ages; 6, Upper Cretaceous rocks; 7, Mesozoic (Triassic–Lower Cretaceous) rocks; 8, Paleozoic rocks; 9, major faults; Alaska: 10, Quaternary and Tertiary rocks; 11, Paleozoic to Quaternary volcanic rocks; 12, Paleozoic to Quaternary intrusive rocks; 13, Mesozoic rocks; 14, Paleozoic rocks; 15, Proterozoic through Mesozoic metamorphic rocks.

reductions in diversity in temperature-sensitive marine algae (Aubry, 1992), diatoms (Baldauf, 1992), dinoflagellates (Brinkhius, 1992), and foraminifera (Boersma et al., 1987; Thomas, 1992; Gaskell, 1991). Oligocene planktonic foraminiferal and nannofossil assemblages of the low latitudes were particularly impoverished in comparison with the highly diversified Eocene assemblages (Cavelier et al., 1981). Faunal turnover near the Eocene/Oligocene boundary in the oceans correlates with inferred bottom-water cooling, based on the enrichment in $\delta^{18}O$ observed in benthic foraminifera. It has been suggested, however, that cooling at the Eocene/Oligocene boundary did not have a major effect on the diversity of benthic foraminifera, because temperature-sensitive species had been eliminated during earlier cooling events, leaving species with wide environmental tolerances able to withstand a 2–3°C bottom-water temperature decrease (Corliss and Keigwin, 1986).

Profound climatic changes during the late Eocene and at the Eocene/Oligocene boundary in the northern Pacific has also been inferred from analysis of terrestrial floras, using both leaf physiognomy and palynology. Floristic analysis suggests that middle Eocene floras of southern Alaska were megathermal, living in a climate where the mean annual temperature exceeded 20°C, and was probably as high as 22°C (Wolfe, 1980). Climatic changes at the end of the Eocene and the beginning of the Oligocene in the middle and high latitudes resulted in a dramatic change in vegetation type within a geologically short time span. Vast areas of the Pacific Northwest and

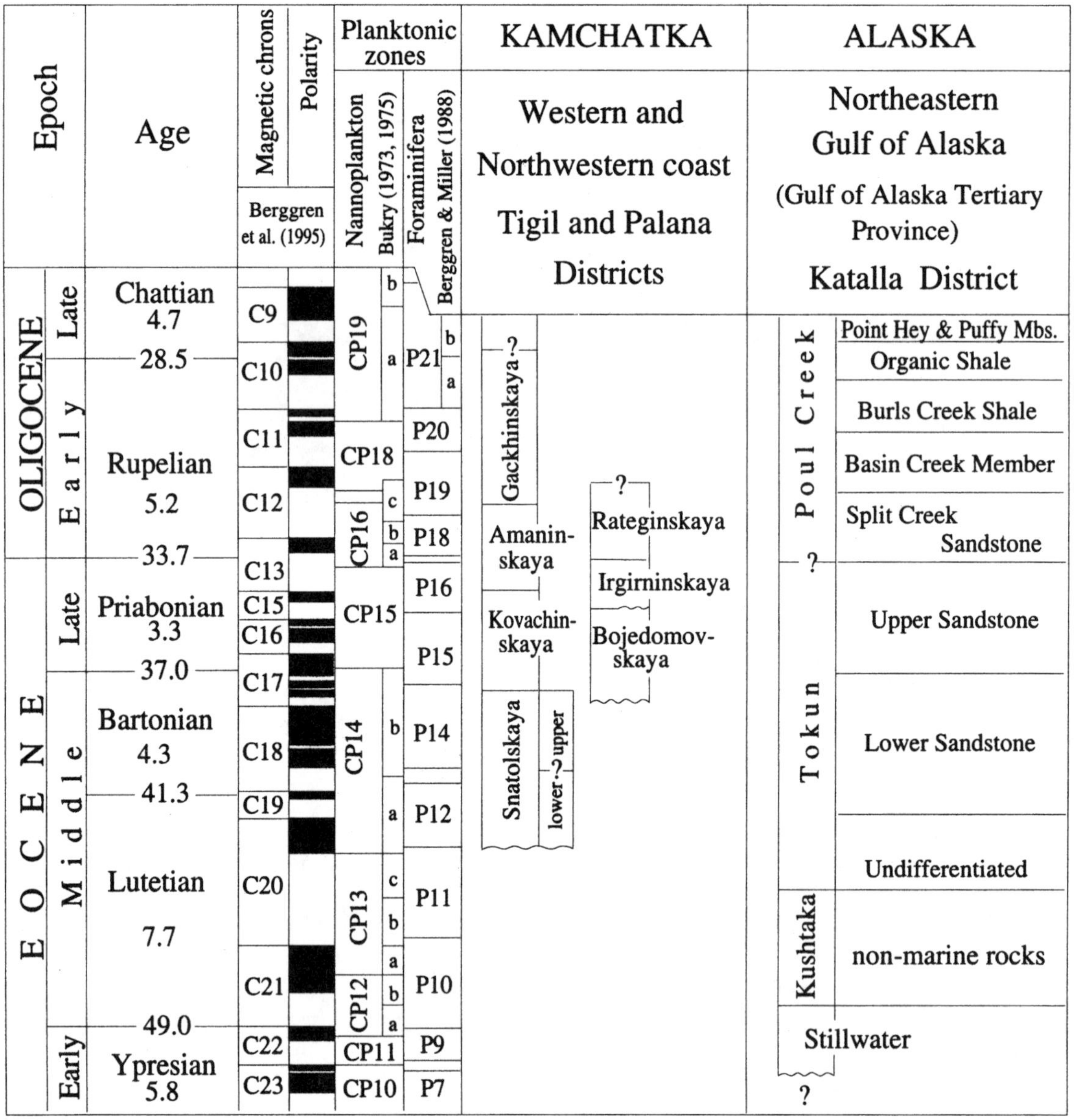

Figure 3.2. Preliminary correlation of Kamchatkan and Alaskan formations with the Paleogene time scale.

Alaska that had been a broad-leaved evergreen forest became a temperate broad-leaved deciduous forest. Based on this change in vegetation, the decline in mean annual temperature at the Eocene/Oligocene boundary was inferred to be 12–13°C at 60°N in Alaska, and about 10–11°C at 45°N in the Pacific Northwest. The shift in temperature equitability was as profound in the Pacific Northwest. The mean annual temperature range was as low as 3–5°C in the middle Eocene, and rose to at least 21°C, and probably to 25°C, in the Oligocene (Wolfe, 1971).

Middle Eocene (upper Lutetian–Bartonian) floral assemblages of western Kamchatka were mixed conifer–broadleaf deciduous forests that were dominated by angiosperms, and imply a warm-temperate climate. The late Eocene (Priabonian) floras of western Kamchatka contained even larger percentages of thermophylic taxa, including mesophylic entire-margined evergreen forms such as *Cinnamomum, Lindera, Pternandra,* and *Rhododendron.* These thermophylic forms made up over 55% of the floral assemblage, and the late Eocene climate of Kamchatka was certainly warmer than it was in the middle Eocene (Chelebaeva, 1991; Chelebaeva and Bratseva, 1985; Fotyanova, 1977).

Data from marine and terrestrial sections clearly

indicate that the deterioration of the warm, humid climate started sometime during the late Eocene, experienced a major drop at the Eocene/Oligocene boundary, and continued to decrease throughout the early Oligocene. This dramatic drop in temperature forced tropical and subtropical faunas to retreat southward and also stimulated the southward migration of temperate faunas. The rate and intensity of these changes appear to have been different in the oceanic and terrestrial realms in and around the northern Pacific Ocean. Analysis of the Eocene-Oligocene marginal-marine (shelf) faunas can, therefore, shed some additional light on the nature and intensity of biotic changes at the Eocene/ Oligocene boundary in the North Pacific.

The biotic responses of shallow-water (shelf), northern high-latitude benthic faunas to these events, however, have remained poorly known, largely owing to the fragmentary knowledge of these faunas. Geologic fieldwork in far-eastern Russia and Alaska has been impeded by geographic isolation, rugged topography, dense vegetation, and poor weather. Even today, the expense and distances involved are barriers to assembling comprehensive fossil collections from these regions. Many of our conclusions in this study are based on unpublished studies of mollusks from the former collections of the United States Geological Survey at Menlo Park, California, now housed in the Museum of Paleontology, University of California at Berkeley, and the ongoing research of the senior author on the Paleogene mollusks of Kamchatka.

The shallow-marine Paleogene molluscan record from the North Pacific rim, although incompletely collected, provides an excellent resource for examining changes in faunas across the Eocene/Oligocene boundary in high latitudes. Recent advances in the Paleogene biostratigraphy of Kamchatka allow more precise dating of the shallow-marine molluscan assemblages used in this study.

STUDY AREAS, MATERIALS, AND METHODS

Eocene and Oligocene gastropods used in this study (table 3.1) came from shallow-marine deposits along the western coast of Kamchatka (the Tigil and Palana districts) and the northeastern Gulf of Alaska, often referred to as the Gulf of Alaska Tertiary Province (Plafker, 1967). The Alaskan faunas in this study occur in the Katalla district in the western part of this province (figure 3.1). The choice of these two areas was not accidental. Both are located between 55 and 60°N (figure 3.1A) and have an extended history of geological studies due to intense petroleum exploration. Paleogene marine rocks in these areas are the most fossiliferous of any in Kamchatka or Alaska, and contain the northernmost record of the Eocene-Oligocene interval in the northern Pacific. Eocene and Oligocene sediments in both areas were deposited exclusively in shallow-shelf, normal-marine environments (Marincovich and McCoy, 1984; Shantser et al., 1991; Oleinik, 1999). These similarities make possible both the temporal and lateral comparisons of gastropod assemblages that lived in the same or similar habitats throughout the northern Pacific and across the Eocene/Oligocene boundary. Middle Eocene through early Oligocene gastropod faunas analyzed in this paper indicate a shallow-water, subtropical to warm-temperate environment. The studied genera (or their modern analogs) lived in the intertidal to lower subtidal zones. A water depth no greater than 50 m has been inferred herein for most assemblages, the only exceptions being some assemblages in the Kovachinskaya and Poul Creek formations, where the paleodepth may have ranged from 100 m to 200 m. Gastropod diversity patterns, and the number and percentage of extinct, carryover, and newly appearing species, as well as their biogeographic affinities within each stratigraphic unit, were analyzed to document changes in middle Eocene through Oligocene gastropod assemblages. Percentages of extinct and survivor taxa were calculated for assemblages below the boundary, and the percentage of newly appearing taxa for assemblages above the boundary. Each genus was assigned to a particular biogeographic category, based on its latitudinal distribution and on comparison with its worldwide occurrences during the Paleogene (Oleinik, in press). In all, taxa analyzed for this study include 101 species in 53 genera, belonging to 27 families of gastropods from Kamchatka, and 48 species in 26 genera, belonging to 18 families from Alaska.

STRATIGRAPHIC SETTING

Alaska

Fossiliferous Paleogene strata in the Katalla district of the Gulf of Alaska Tertiary Province (figure 3.1C) are at least 3000 m thick and consist of the Stillwater, Kushtaka, Tokun, and Poul Creek formations (figure 3.2). These strata and the fossils in them have never been well studied, so that their geologic, stratigraphic and faunal relationships are often uncertain. For example, it is worth noting that faunas once referred in the literature to the Katalla Formation have been reassigned to the Poul Creek Formation, and use of

Table 3.1. Distribution of Gastropods in the Eocene and Oligocene Formations of the Western Kamchatka and Northeastern Gulf of Alaska Regions

Region Formation	K Sn1	K Sn2	K Kv	K Rt	A St	A Tk1	A Tk2	A Tk3	A PCr1	A PCr2	A PCr3	A PCr4	A PCr5	A PCr6
Acmaea (*Acmaea*) new species	x			x										
Acrilla (*Ferminoscala*) cf. *dickersoni* Durham, 1937							x							
Acteon (*Acteon*) *microspiralis* Oleinik, 1991	x													
Acteon (*Acteon*) new species.	x													
Aforia new species				x										
Ancilla new species				x										
Ancilla kovatschensis Oleinik, 1991		x	x											
Ancilla meganosensis D. Palmer, 1923		x												
Ancilla sp.		x												
Ancistrolepis matchgarensis (Makiyama, 1934)				x										
Ancistrolepis rateginensis Titova, 1993				x										
Antimelatoma new species.			x											
Apiotoma new species.	x													
Arctomelon rateginense Oleinik, 1996				x										
Beringius (*Beringius*) *pressulus* Titova, 1986			x	x										
Bonellitia (*Admetula*) *paucivaricata* (Gabb, 1864)							x							
Bonellitia (*Bonellitia*) *microbonum* (Oleinik, 1991)			x											
Brucklarkia cf. *columbiana* (Anderson & Martin, 1914)									x					
Brucklarkia columbiana (Anderson & Martin, 1914)									x	x				
Calyptraea diegoana Conrad, 1855	x	x	x	x										
Cancellaria (*Bivetiella*) *snatolensis* (L. Krishtofovich, 1947)	x	x												
Cancellaria cf. *oregonensis* Dall, 1909							x							
Cancellaria sp.							x							
Clivoluturris spiraliformis (Oleinik, 1991)			x											
Colus (*Aulacofusus*) *asagaensis* Makiyama, 1934			x											
Comitas (*Boreocomitas*) (?) sp.			x											
Conominolia ovriensis Oleinik, 1991	x	x												
Conomitra new species				x										
Conus cf. *warreni* Hendon, 1938							x							
Conus sp.	x													
Creonella altus (Oleinik, 1991)	x													
Crepidula pileum (Gabb, 1869)	x	x	x	x										
Crepidula tigilensis (Oleinik, 1991)	x													
Crepidula ungana Dall, 1904				x										
Cylichnina sp.										x				
Diodora sp	x	x												
Echinophoria fax (Tegland, 1931)										x				
Echinophoria rex (Tegland, 1931)										x				
Eotrominina dispar (Takeda, 1953)			x	x										
Eotrominina new species 1				x										

the former formation name has been abandoned (Nelson et al., 1985). The Poul Creek Formation crops out most extensively in the Yakataga district, immediately to the east of the Katalla district, where its molluscan fauna has been partly described by Kanno (1971). The only non-marine Paleogene unit in the Katalla district is the Kushtaka Formation, which contains fossil plants that range in age from early to late Eocene (Wolfe, 1977). The Stillwater, Tokun, and Poul Creek formations are marine units that contain abundant, often well preserved mollusks. However, to emphasize how little these faunas have been studied, not a single species of mollusk from these formations has previously been figured (figure 3.3). The dating of these sequences is imprecise. Paleogene formations have been assessed on the basis of benthic foraminifers (Rau et al., 1983), but without notable success. In the absence of age control using planktonic microfossils or isotopes, the age assignments adopted in this paper are based almost entirely on mollusks.

Two approaches were used to assign ages to marine units within this Alaskan Paleogene sequence. The first approach is to use the succession of warmer-water versus cooler-water genera throughout the sequence and correlate the changes observed within this Alaskan sequence with presumably coeval changes in the better-dated Paleogene sequences of Kamchatka and the Pacific Northwest (figure 3.4). The second approach for age control is based on species-level correlations with better-dated Paleogene formations of western Washington and Oregon. Species shared with these regions are well represented in Gulf of Alaska faunas. Such age-diagnostic species as *Ficopsis cowlitzensis, Nekewis washingtoniana, Molopophorus californicus lonsdalei, Bonellitia*

Region Formation	K Sn1	K Sn2	K Kv	K Rt	A St	A Tk1	A Tk2	A Tk3	A PCr1	A PCr2	A PCr3	A PCr4	A PCr5	A PCr6
Eotrominina new species 2				x										
Epitonium (Boreoscala) cf. *condoni* Dall, 1909					x		x	x	x	x	x			
Epitonium (Boreoscala) keaseyense Durham, 1937				x	x				x					
Epitonium (Boreoscala) new species.	x	x	x											
Epitonium (Boreoscala) sp.					x									
Epitonuim (Boreoscala) cf. *condoni eugenense* Durham, 1937					x									
Exilia bentsonae Hickman, 1980	x	x												
Exilia new species.			x											
Exilia lincolnensis Weaver, 1916							x			x				
Ficopsis cowlitzensis (Weaver, 1912)							x							
Ficopsis meganosensis Clark and Woodford, 1927	x													
Ficopsis tuberculata Oleinik, 1991	x													
Ficus (Trophosycon) new species		x												
Fissurella sp.				x										
Fulgoraria (Kurodina) nagaoi (Shikama, 1967)				x										
Fulgoraria (Kurodina) new species				x										
Fulgoraria (Miopleiona) sp.													x	
Fulgoraria (Musashia) naibutiensis (L. Krishtofovich, 1954)				x										
Fulgoraria (Musashia) ochotensis Oleinik, 1991		x	x											
Fulgoraria (Musashia) new species 1				x										
Fulgoraria (Musashia) new species 2				x										
Fulgoraria (Musashia) new species 3			x	x										
Fulgoraria (Psephaea) new species 4				x										
Fulgoraria (Psephaea) new species 5				x										
Fulgoraria (Psephaea) new species 6	x			x										
Fusinus (Fusinus) new species 1			x	x										
Fusinus (Fusinus) new species 2	x													
Fusinus (Fusinus) willisi (Dickerson, 1915)	x													
Fusinus (Priscofusus) cf. *chechalisensis* (Weaver, 1912)									x	x				
Fusinus sp. 1														
Fusinus sp. 2							x							
Fusitriton sp.													x	
Gibbula (Gibbula) kahtanenesis Oleinik, 1991	x													
Hataiella (Kotakiella) hellenica (L. Krishtofovich, 1947)	x	x	x											
Hataiella (Kotakiella) poronaiensis (Takeda, 1953)			x	x										
Hataiella (Merriamella) kahtanenesis Titova, 1991	x	x												
Hataiella (Merriamella) rateginensis Titova, 1994				x										
Hataiella (Merriamella) snatolensis (L. Krishtofovich, 1947)	x													
Homalopoma (?) sp.			x	x										
Latirus cf. *roseburgensis* Hendon, 1938							x							
Lirosoma (?) new species.			x											
Lischkeia new species			x											
Makiyamaia decorata Oleinik, 1991	x													

(Admetula) paucivaricata, Siphonalia oregonensis, Turcicula columbiana, Echinophoria rex, Echinophoria fax, Trominina teglandae, Brucklarkia columbiana, and *Whitneyella washingtoniana* proved useful in this approach. The percentages of conspecific species within each Alaskan stratigraphic unit were used as a guide to determine its approximate age. Based on these data, the Stillwater Formation is herein assigned to the middle Eocene (most probably middle to upper Lutetian) (figure 3.2), the two lower members of the Tokun Formation are assigned to the middle and upper Bartonian, and the upper sandstone member of the Tokun Formation to the Priabonian (figure 3.2). Mollusks of the Poul Creek Formation occur stratigraphically throughout a considerable portion of the section. The uppermost members of the Poul Creek Formation contain cold-water North Pacific taxa, such as *Neptunea* and *Fusitriton* (figure 3.3). The molluscan fauna in the middle part of the formation is correlative with the Oligocene portion of the Poul Creek Formation in the Yakataga district. The latter formation is the best-known Paleogene marine unit in southern Alaska, with a mollusk fauna of over 60 species (Kanno, 1971; Addicott et al., 1978). The bulk of the Poul Creek Formation is most probably early Oligocene in age, although it is possible that the lowermost part of the formation is Eocene. In that case, the Eocene/Oligocene boundary would be within the lower members of the Poul Creek Formation. These age assignments should be treated as preliminary and are likely to change as our knowledge of stratigraphy and paleontology in southern Alaska advances.

Most mollusks in the marine part of the sequence suggest open-marine conditions at shelf depths, with *Corbicula* and lacustrine snails indicating

Table 3.1 (continued)

Region / Formation	K Sn1	K Sn2	K Kv	K Rt	A St	A Tk1	A Tk2	A Tk3	A PCr1	A PCr2	A PCr3	A PCr4	A PCr5	A PCr6
Margarites (*Margarites*) *uvutschensis* Oleinik, 1991	x	x	x											
Margarites (*Pupillaria*) *effingeri* (Durham, 1944)			x											
Margarites (*Pupillaria*) new species				x										
Marshallena tigilensis (Oleinik, 1991)			x											
Micropleurotoma new species.			x											
Mipus vinjukovi (L. Krishtofovich, 1954)			x											
Molopohorus aff. *tejonensis* Anderson & Hanna, 1925					x		x							
Molopohorus cf. *anglonana* (Anderson, 1905)							x		x					
Molopophorus californicus Clark & Woodford, 1927							x		x					
Molopophorus lincolnensis ovumiformis Oleinik, 1991	x	x												
Molopophorus longissimus Oleinik, 1991	x													
Molopophorus sp. 1			x				x		x	x				
Molopophorus sp. 2														
Nekewis costellatus Oleinik, 1991	x													
Nekewis washingtoniana (Weaver, 1912)					x		x							
Neptunea altispirata (Nagao, 1928)			x	x										
Neptunea cf. *lyrata* (Gmelin, 1791)														x
Neptunea sp.													x	
Nihonia gigas Oleinik, 1991			x											
Notoacmaea (*Conacmea*) *pseudoconica* Oleinik, 1991	x	x												
Notoacmaea (*Parvacmaea*) *snatolica* Oleinik, 1991	x	x												
Odostomia (*Doliella*) *hiltoni* (Van Winkle, 1918)	x	x	x		x						x			
Olivella matchewsonii umpquaensis Turner, 1938	x	x												
Optoturris new species				x										
Parasyrinx (*Lirasyrinx*) new species				x										
Plicifusus (*Retifusus*) new species				x										
Polinices (*E*.) cf. *nuciformis cowlitzensis* Dickerson, 1915							x		x					
Polinices (*Euspira*) *ashiyaensis* Nagao, 1928				x										
Polinices (*Euspira*) *lincolnensis* (Weaver, 1916)	x	x												
Polinices (*Euspira*) *nuciformis* (Gabb, 1864)	x													
Polinices (*Polinices*) *kamtschatica* L. Krishtofovich, 1947	x													
Polinices (*Polinices*) *snatolensis* L. Krishtofovich, 1947	x	x												
Polinices (*Polinices*) *tigilensis* L. Krishtofovich, 1947	x	x												
Problacmaea (?) sp. Oleinik, 1991		x												
Pseudoliomesus podkagernensis Titova, 1988	x	x		x										
Puncturella (?) sp.	x	x												
Scaphander (*Mirascapha*) *alaskensis* Clark, 1932	x	x	x											x
Scaphander (*Mirascapha*) *costatus* (Gabb, 1864)	x													
Siphonalia cf. *washingtoniensis* (Weaver, 1916)									x					
Siphonalia ex. gr. *sophenahensis* (Weaver, 1912)					x		x		x		x			
Siphonalia oregonensis (Dall, 1909)								x	x					
Siphonalia pseudofusoides Oleinik, 1988	x													
Siphonalia supragranulata Nagao, 1928				x										

a minor brackish- and freshwater component, which is not uncommon in nearshore-marine sections. An apparent cooling trend upsection correlates well with the cooling inferred for the western Gulf of Alaska based on Paleogene faunas from the Tolstoi and Stepovak Formations, and the so-called "Narrow Cape Formation" of Sitkinak Island (Marincovich and McCoy, 1984). The true Narrow Cape Formation crops out only on Kodiak Island and contains early middle Miocene marine mollusks (Marincovich and Moriya, 1992).

Paleogene strata in the Katalla district are complexly deformed and contain intraformational unconformities and hiatuses. The Stillwater Formation consists of sandstone and siltstone of unknown thickness, and mollusks have been collected from its stratotype and from equivalent unnamed rocks in the Katalla district (Miller, 1961, 1975). The Tokun Formation overlies the Stillwater and Kushtaka formations with local disconformity and consists of 330 to 1,050 m of interbedded sandstone and concretionary siltstone in its upper part, and mainly concretionary siltstone in its lower part. The Poul Creek Formation varies in lithology from calcareous siltstone with minor sandstone, glauconite, and pyroclastic rocks in its lower and middle parts to pebbly siltstone, pebbly conglomerate and sandstone in its upper part. Its thickness is estimated to be 960 to 1,830 m thick for its lower and middle parts, and at least 2,760 m for its upper part (Plafker, 1967)

Region Formation	K Sn1	K Sn2	K Kv	K Rt	A St	A Tk1	A Tk2	A Tk3	A PCr1	A PCr2	A PCr3	A PCr4	A PCr5	A PCr6
Siphonalia washingtonensis (Weaver, 1916)									x					
Snatolia elongata Oleinik, 1988	x													
Snatolia snatolensis (Ilyina, 1939)	x													
Solariella kamtschatica L. Krishtofovich, 1973			x											
Surculites (?) sp.					x									
Trigonostoma (*Trigonostoma*) new species				x										
Trominina new species.			x											
Trominina teglandae (Weaver, 1942)									x	x	x			
Turcicula cf. *arnoldi* (Weaver, 1912)									x					
Turcicula columbiana Dall, 1909								x		x				
Turcicula sp.														x
Turricula keaseyensis Hickman, 1976			x											
Turrinosyrinx new species.			x											
Turrinosyrinx snatolensis (L. Krishtofovich, 1947)	x													
Turrinosyrinx subelongata (Oleinik, 1991)	x	x	x											
Turriola tokunagai (Yokoyama, 1924)	x	x	x											
Turritella cf. *uvasana stewarti* Merriam, 1938							x		x					
Turritella cf. *uvasana washingtoniana* Weaver & Palmer, 1922							x							
Whitneyella cf. *lincolnensis* (Van Winkle, 1918)					x		x		x					
Whitneyella cf. *markleyensis* (Clark, 1938)					x	x	x							
Whitneyella cf. *oregonensis* Turner, 1938					x									
Whitneyella cf. *sinuata aragoensis* Turner, 1938					x	x	x		x				x	
Whitneyella cf. *washingtoniana* (Weaver, 1912)					x		x		x					

Abbreviations: K, Kamchatka; A, Alaska; Sn1, lower member of the Snatolskaya Formation; Sn2, upper member of the Snatolskaya Formation; Kv, Kovachinskaya Formation; Rt, Rateginskaya Formation; St, Stillwater Formation; Tk1, undifferentiated member of the Tokun Formation; Tk2, lower member of the Tokun Formation; Tk3, upper siltstone member of the Tokun Formation; PCr1, Split Creek Sandstone Member of the Poul Creek Formation; PCr2, Basin Creek Member of the Poul Creek Formation; PCr3, Burls Creek Member of the Poul Creek Formation; PCr4, organic shale member of the Poul Creek Formation; PCr5, Point Hey Member of the Poul Creek Formation; PCr6, Puffy Member of the Poul Creek Formation.

Kamchatka

Fossiliferous Paleogene shallow-marine rocks crop out in coastal bluffs and river valleys in the western coast of Kamchatka (figure 3.1B). The Eocene portion of this sequence has been subdivided into the Snatolskaya and Kovachinskaya Formations, which comprise the basal portion of a major transgressive cycle, and are separated from older Paleogene rocks by a regional unconformity (Gladenkov et al., 1991a). The dating of both formations is based on benthic and planktonic foraminifers, the radiometric ages of interbedded volcanic rocks and tuffs, leaf floras, and marine mollusks.

The Snatolskaya Formation consists of shallow-marine, medium- to fine-grained, laterally continuous sandstones, with abundant lagoonal and delta-front deposits, and contains a rich and diverse fauna of benthic mollusks. The thickness of the Snatolskaya Formation in the Tigil and Palana districts (a continuous and complete section with observable lower and upper contacts) varies from 10 to 295 m. The Snatolskaya Formation becomes thinner and coarser-grained, changing from fine-grained sandstone to medium- and coarse-grained sandstone and conglomerate toward exposures of underlying Cretaceous rocks.

Benthic foraminifers in the Snatolskaya Formation occur at two localities in the Tigil district. The lower and middle parts of the Snatolskaya Formation contain *Rhizammina indivisa, Bathysiphon nodosariaformis, Silicosigmolina* cf. *californica, Haplophragmoides glabratus, Asanospira* cf. *exavata, Asanospira* cf. *akkeshiensis, Glomospiracorona,* and *Recurvoides* sp. The upper part of the Snatolskaya Formation contains *Trochammina vitrea, Gyroidina kamtschatica, Elphidium californicum, Reophax difflugiformis, Haplopharagmoides snatolensis, Cyclammina ezoensis, Trochammina markini, Cyclammina tani, Dorothia amakusaensis, Textularia imariensis, Guttulina irregularis, Gyroidina kamtschatica, Elphidium asagiensis, Caucasina eocaenica kamtschatica, Globobulimina pacifica,* and *Bolivina jacksonensis.* These assemblages strongly imply a middle Eocene age for the Snatolskaya Formation (Gladenkov et al., 1991b). The planktonic foraminiferal assemblage is poorly preserved and can be generally assigned to the *Globigerina boweri* regional Zone (Serova, 1969; Krasheninnikov et al., 1988), which suggests a

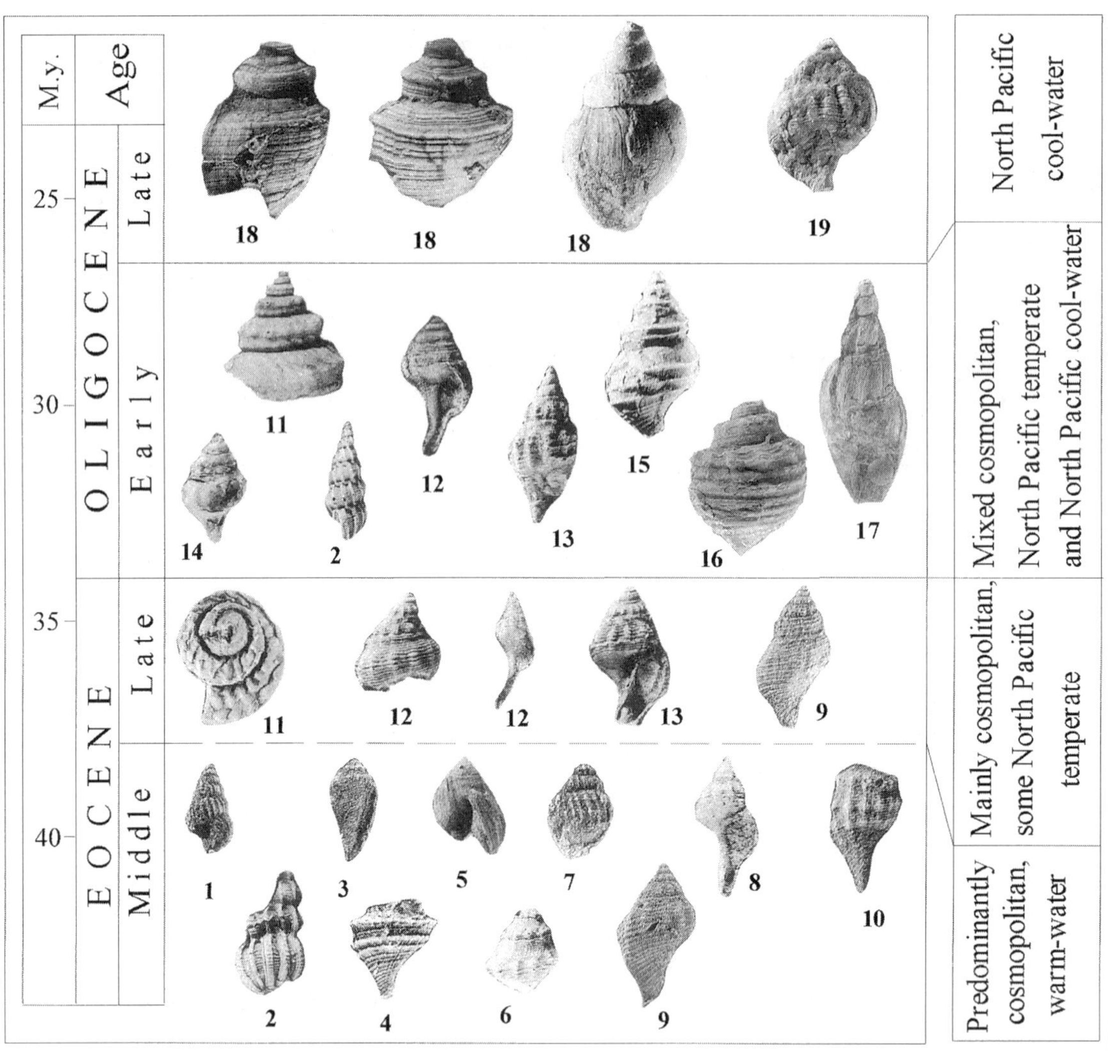

Figure 3.3. Representative middle Eocene through late Oligocene gastropod genera and subgenera from the Katalla District, northeastern Gulf of Alaska. Legend: 1, *Acrilla* (*Ferminoscala*); 2, *Epitonium* (*Boreoscala*); 3, *Conus*; 4, *Surculites*; 5, *Molopophorus*; 6, *Nekewis*; 7, *Cancellaria*; 8, *Latiru;*, 9, *Siphonalia*; 10, *Ficopsis*; 11, *Turcicula;* 12, *Whitneyella*; 13, *Fusinus*; 14, *Bruclarkia;* 15, *Trominina*; 16, *Echinophoria*; 17, *Fulgoraria* (*Musashia*); 18, *Neptunea*; 19, *Fusitriton.*

Lutetian to Bartonian age for the Snatolskaya Formation (figure 3.2). A number of conspecific benthic and planktonic foraminifers in the Snatolskaya Formation suggest correlation with the Narizian Stage of California (Mallory, 1959), the middle Eocene (Lutetian-Bartonian) Takisawa Formation (Honshu), and the lowermost part of the Poronai Formation (Upper Bartonian, Hokkaido) of Japan, which contain both benthic foraminiferal assemblages and nannoplankton (Saito et al., 1984).

The Kovachinskaya Formation conformably overlies the Snatolskaya Formation and consists of marine siltstone, claystone, and sandstone. It varies in thickness from 30 to 475 m and is conformably overlain by deep-water deposits of the Amaninskaya and Gakhinskaya formations of early Oligocene age. The Kovachinskaya Formation reaches its maximum thickness in areas most distant from outcrops of Cretaceous rocks. Sediments of this formation become progressively coarser-grained toward Cretaceous outcrops, where it reaches its minimal thickness and consists mostly of sandstone and conglomerate.

Because of its more open-marine depositional

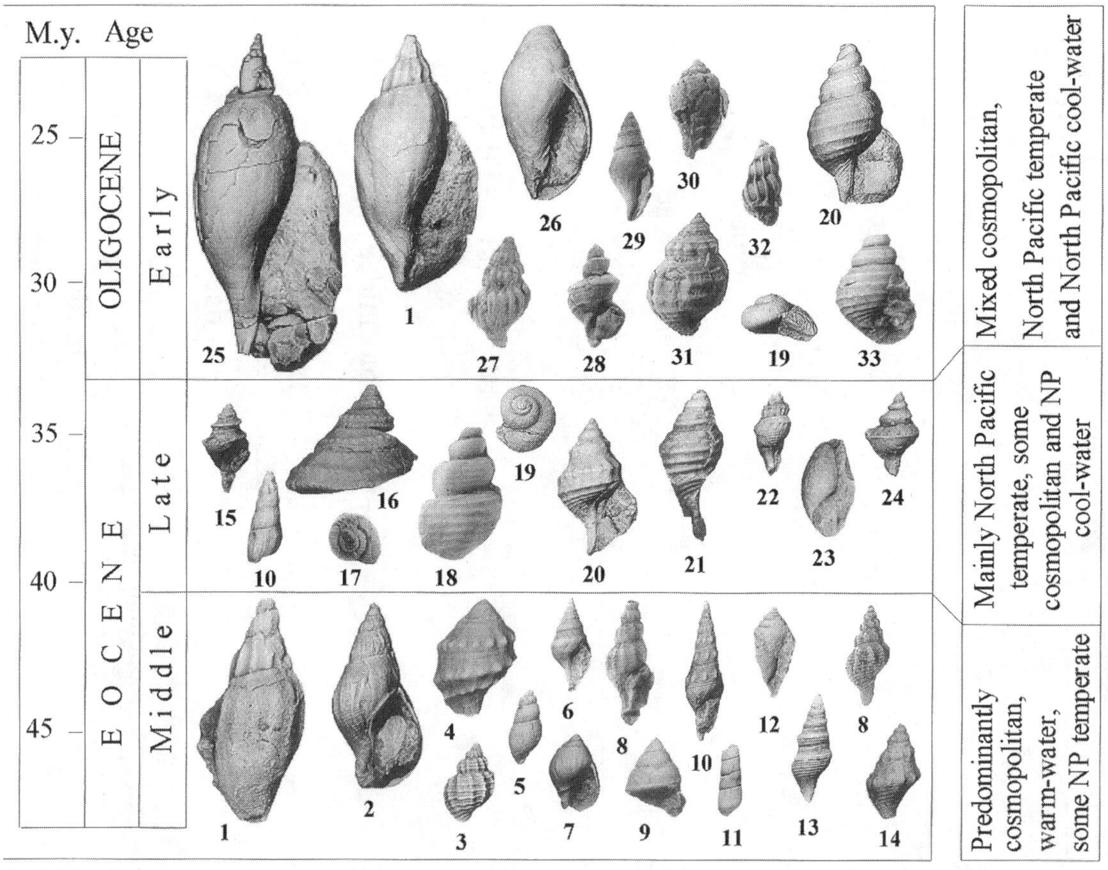

Figure 3.4. Representative middle Eocene through early Oligocene gastropod genera and subgenera from the Tigil and Palana districts of western Kamchatka. Legend: 1, *Fulgoraria* (*Musashia*); 2, *Snatolia*; 3, *Cancellaria*; 4, *Ficopsis*; 5, *Odostomia* (*Doliella*); 6, *Turrinosyrinx*; 7, *Molopophorus*; 8, *Fusinus*; 9, *Gibbula*; 10, *Exilia*; 11, *Creonella*; 12, *Conus*; 13, *Apiotoma*; 14, *Nekewis*; 15, *Clivoluturris*; 16, *Lishkeia*; 17, *Solariella*; 18, *Beringius*; 19, *Margarites*; 20, *Trominina*; 21, *Neptunea*; 22, *Marshallena*; 23, *Scaphander* (*Mirascapha*); 24, *Mipus*; 25, *Fulgoraria* (*Kurodina*); 26, *Arctomelon*; 27, *Plicifusus*; 28, *Trigonostoma*; 29, *Aforia*; 30, *Volutomitra*; 31, *Pseudoliomesus*; 32, *Epitonium* (*Boreoscala*); 33, *Ancistrolepis*.

setting, the Kovachinskaya Formation contains more abundant and diverse foraminiferal assemblages than does the Snatolskaya Formation. *Plectofrondicularia packardi packardi, Alabami-na californica, Gyrodina condoni, Bulimina debilis, Bathysiphon eocenicus, Ammodiscus pennyi, Haplophragmoides obliquicameratus, Budashevella deserta, Cyclammina cushmani, C. pacifica, Ammobaculites kamtschaticus, A. akabiroensis, Ammomarginulina stephensoni, Robulus inornatus, R. alatolimbatus, Dentalina plamerae, Alabamina californica, Cibicides pachecoensis, C. beckii, Nonion durchani, N. sorachiense, Melonis planatum, Bulimina debilis, Globobulimina pacifica,* and *Globocassidulina globosa* are among the most abundant benthic foraminifers in the Kovachinskaya Formation (Gladenkov et al., 1991a). According to Serova (1985), the foraminiferal assemblage of the Kovachinskaya Formation contains over 150 species, mostly benthic forms, whereas planktonic taxa are rare and represented solely by *Globigerina* (Serova, 1985). The foraminiferal assemblage characteristic of the Kovachinskaya Formation also occurs in eastern Kamchatka, Japan and western North America (Serova, 1985). Fotyanova and Serova (1976) referred the Kovachinskaya Formation to the *Bulimina corrugata* Zone. *Bulimina corrugata* co-occurs with *Globigerina frontosa* in the *Globigerina frontosa* Zone of the Ilpinskiy Peninsula, northeastern

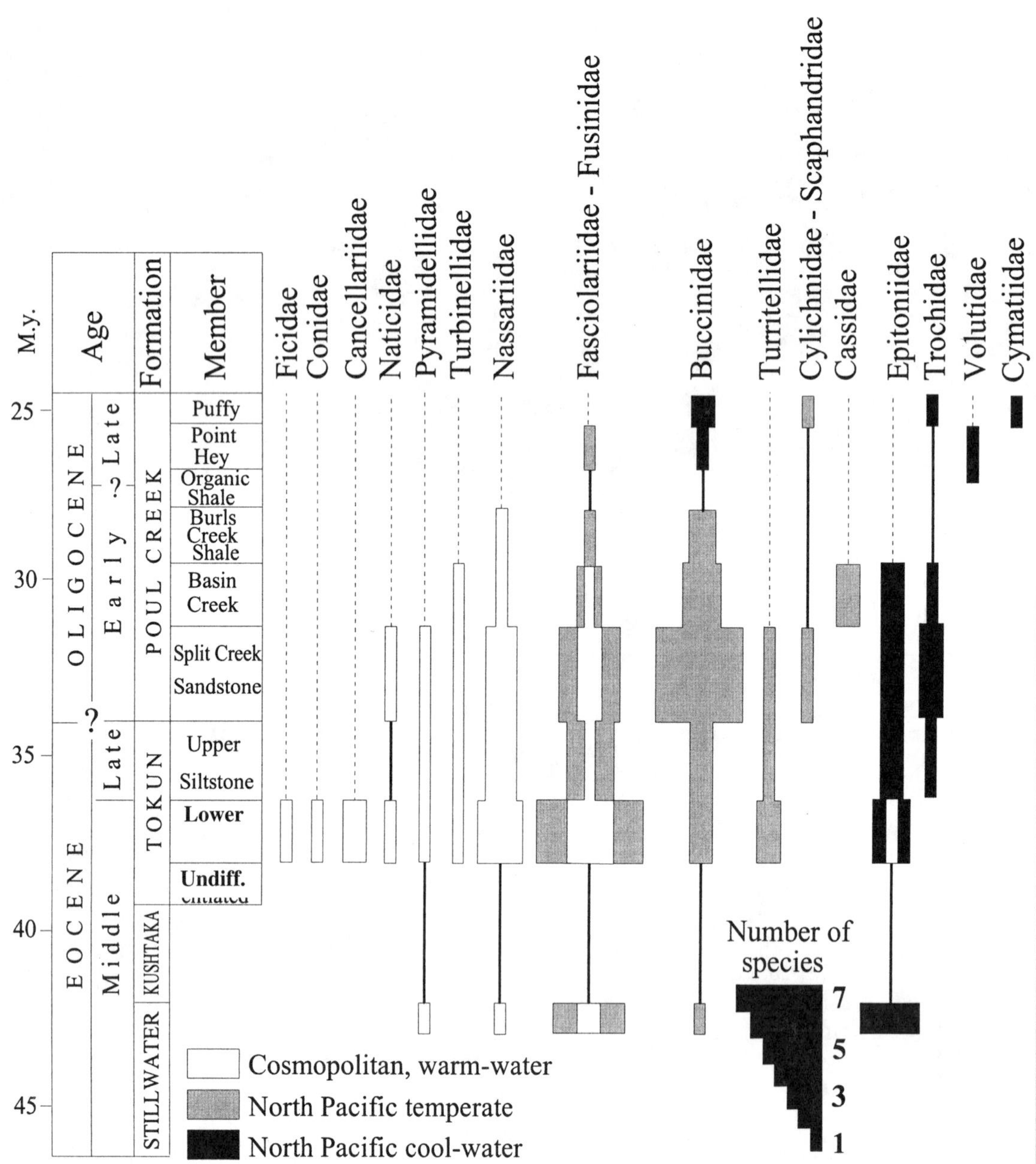

Figure 3.5. Diversity patterns of gastropod families from the Katalla District, northeastern Gulf of Alaska.

Kamchatka, as well as in the *Bulimina corrugata* Zone of California. Both the *Globigerina frontosa* and *Bulimina corrugata* zones are correlative with the P14 Zone of the standard planktonic foraminiferal sequence and indicate a Bartonian age (Krasheninnikov et al., 1988; Volobueva et al., 1994). This correlation is also supported by the co-occurrence of *Globigerina frontosa* with calcareous nannoplankton on the Ilpinskiy Peninsula, indicative of the NP17 Zone of Martini (1971) and the CP14 Zone of Bukry (1973, 1975), which suggests a late Bartonian age for at least the basal beds of the Kovachinskaya Formation (figure 3.2). Similar foraminiferal assemblages in western North America that contain *Buliminasculptilis, Plectofrondicularia packardi, Uvigerina atwili,* and *Cyclammina clarki* have been

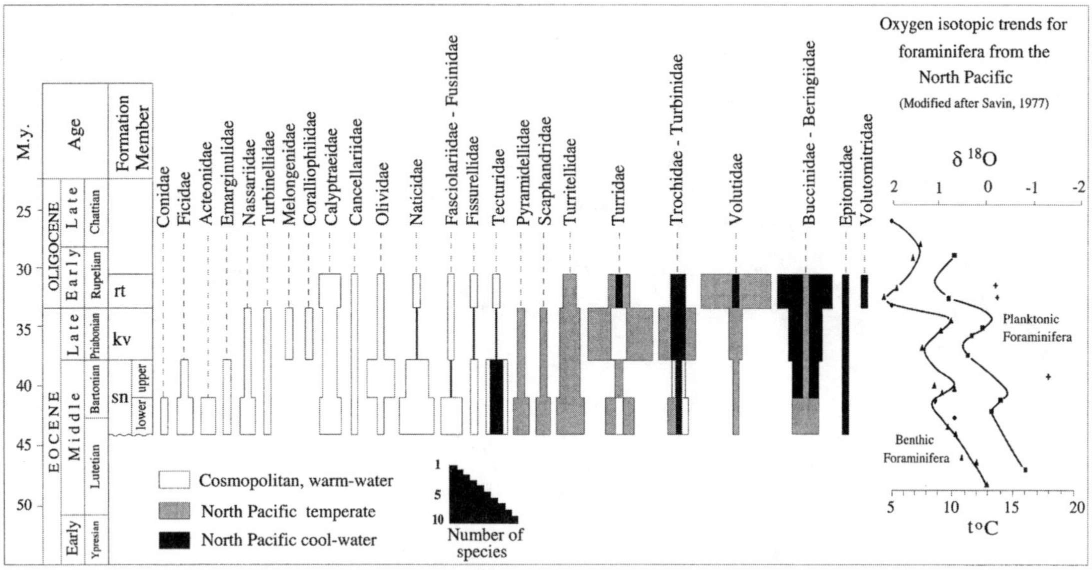

Figure 3.6. Diversity patterns of gastropod families from the Tigil and Palana districts of western Kamchatka, compared with the foraminiferal isotopic record for the northern Pacific.

assigned to the Refugian Stage (Serova, 1985; Krasheninnikov et al., 1988). Nannoplankton biochronozone NP20 within the Refugian Stage further suggests a Priabonian age for most of the Kovachinskaya Formation. Analogs of the Kovachinskaya Formation foraminiferal assemblage in Japan are found in the Okinoshiman Stage (Kyushu), which is dated as late middle to late Eocene based on planktonic foraminifers, and which contains a diverse benthic assemblage (50 species and subspecies) of the *Plectofrondicularia packardi–Bulimina ezoensis* Zone. This zone is also present in the upper part of the Poronai Formation of Hokkaido, where it is known as the *Plectofrondicularia packardi–Bulimina schwageri* or *Bulimina schwageri–Gyrodina yokoyamai* Zone (Gladenkov et al., 1991b) (figure 3.2).

Shallow-marine sandstone and siltstone of early Oligocene age crop out in the center of a broad syncline along the shores of Podkagernaya Bay in the northernmost part of the Palana district and are assigned to the Rateginskaya Formation. The thickness of the Rateginskaya Formation varies from 50 m in the northern part of the bay to 225 m in the southern part. Few mollusks are conspecific with those in the upper Eocene Kovachinskaya Formation of the Tigil district (Gladenkov et al., 1988; Sinelnikova et al., 1991). However, species-level correlations with northwestern Pacific (Sakhalin and Japan) faunas of benthic forminifers suggest an early Oligocene age for the Rateginskaya Formation. The benthic foraminiferal assemblage of the Rateginskaya Formation contains a number of age-diagnostic species: *Haplophragmoides laminatus, Pseudoglandulinaovata, Guttulina hantkeni, G. roemeri, Globulina minuta, G. gackhensis, Fissurina laevigata, Cribroelphidium sumitomoi, Cassidulina galvinensis,* and Asanospira carinata. These assemblages correlate well with the early Oligocene Aluginskaya Formation of the Ilpinskiy Peninsula, northeastern Kamchatka, which has been dated using planktonic foraminifera (Benyamovskiy, 1997a, 1997b). In addition, Fotyanova and Serova (1983) reported early Oligocene planktonic foraminifers in the uppermost part of the Rateginskaya Formation (figure 3.2).

PALEOGEOGRAPHIC SETTING

The shorelines, areal extent, and bathymetry of the northern Pacific Ocean in the Paleogene were quite different from those today. For example, Kamchatka during most of the Tertiary consisted of an archipelago along a volcanic arc. On an even larger scale, one of the major paleogeographic elements in the North Pacific was the Bering Land Bridge—the central part of the region known as Beringia—a land connection between Eurasia and North America that persisted from the middle Cretaceous to the late Tertiary

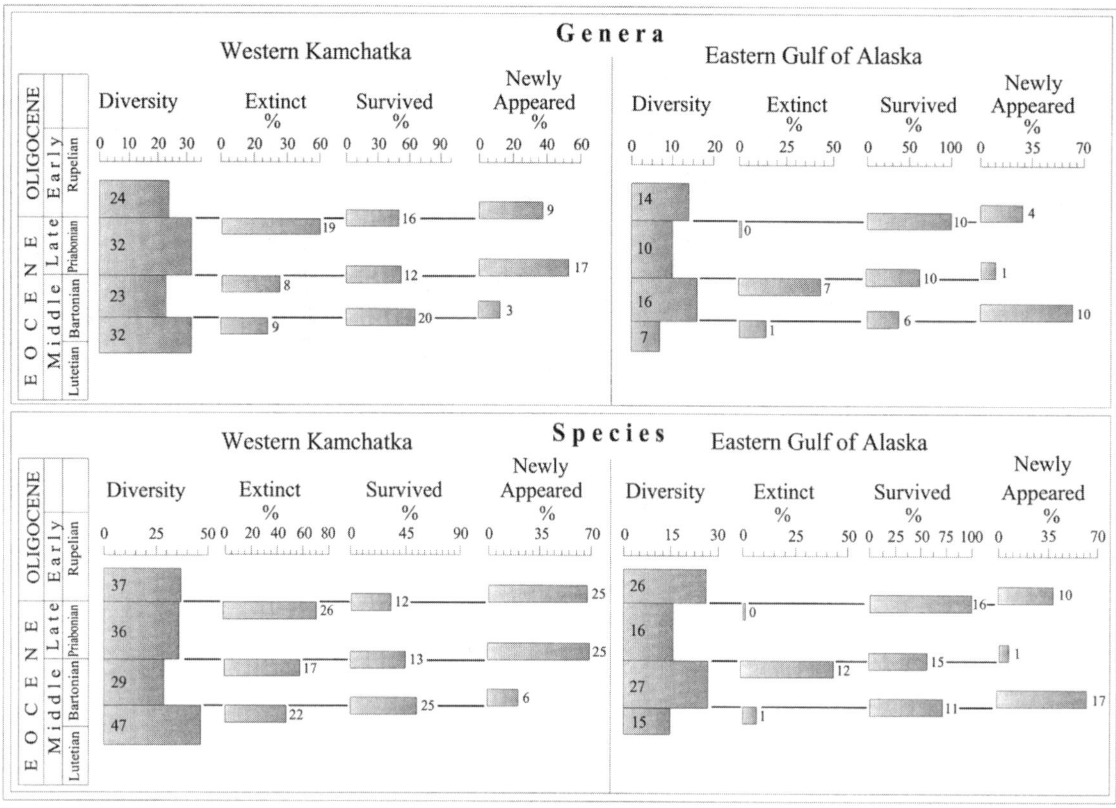

Figure 3.7. Gastropod diversity, extinction, survival, and origination patterns across the Eocene-Oligocene boundary in Alaska and western Kamchatka. The number of species is given next to the percentage bars.

(Durham and MacNeil, 1967; Hopkins, 1967; Marincovich and Gladenkov, 1999). The similarity of land mammals and plants suggests an unhindered terrestrial migration between the Old and New Worlds (Wolfe and Hopkins, 1967) during that 100-million-year length of time prior to the opening of the Bering Strait at 4.8-5.5 Ma (Marincovich and Gladenkov, 1999). The record of marine mollusks clearly indicates an absence of faunal interchange between the Pacific and Arctic basins during the Paleogene (Scarlato and Kafanov, 1976; Kafanov, 1979; Marincovich, 1985, 1993; Marincovich and Gladenkov, 1999). Paleogene unconformities recorded in seismic data at the western part of the Bering Sea further suggest the existence of large areas of dry land from the Late Cretaceous (Maastrichtian) through the middle Eocene (Goryachev, 1965).

MOLLUSCAN DIVERSITY AND EXTINCTION PATTERNS

Diversity changes in middle Eocene through early Oligocene gastropod assemblages of western Kamchatka and the northeastern Gulf of Alaska have been analyzed here at the family, genus, and species levels. Patterns of extinction, survival and origination were studied for three intervals: (1) within the middle Eocene, (2) across the middle Eocene to late Eocene boundary, and (3) across the late Eocene to early Oligocene boundary (figure 3.5).

Mollusks of the Stillwater Formation in the northeastern Gulf of Alaska are underrepresented in collections, and additional collecting is needed to obtain a more reliable understanding of diversity changes within this middle Eocene unit. Therefore, data from the Stillwater Formation are used only to show changes in general in the biogeographic composition of middle Eocene gastropod assemblages. All taxonomic categories clearly show a decrease in diversity from the middle Eocene to the early Oligocene. Individual patterns of decrease, however, differ between families and genera, as well as between Kamchatka and Alaska. The major decrease in diversity on all levels occurred at the Eocene/Oligocene boundary in Kamchatka (figures 3.6 and 3.7), where-

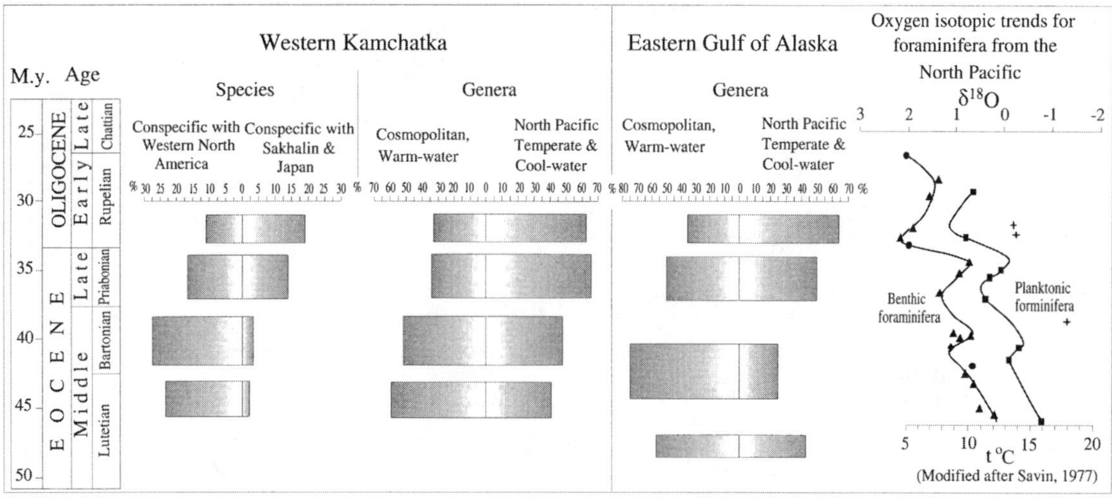

Figure 3.8. Changes in biogeographic composition of molluscan assemblages from the northeastern Gulf Alaska and western Kamchatka, compared with the foraminiferal isotopic record for the North Pacific.

as in Alaska the major decline, based on available data, occurred at the middle to late Eocene boundary. In fact, the Eocene/Oligocene boundary in Alaska does not seem to be marked by extinction at any level (figures 3.5 and 3.6). Instead, it shows diversification in Fasciolariidae and Buccinidae at the middle/late Eocene boundary (figure 3.6). The overall pattern for gastropods of the middle Eocene and across the Eocene/Oligocene boundary in Alaska, is one of high survival rates, with very minor extinction and little radiation. It is of particular interest that based on the currently available data, not a single species went extinct at the Eocene/Oligocene boundary. The number of taxa that survived across the boundary, as well as the number of newly appearing taxa, tend to increase upsection (figures 3.5 and 3.6). The scenario appears to be quite different in Kamchatka, where extinctions increased and survival decreased in magnitude upsection for both genera and species. Our data sets also clearly show that extinctions were accompanied by the appearance of new taxa in succeeding faunas. It is of particular interest that the number of newly appearing genera and species exceeded the number of taxa that became extinct at the middle to late Eocene boundary (figures 3.5, 3.6, and 3.7). It is evident from the data in figure 3.6 that the nature of these new appearances was different in the middle Eocene compared to the late Eocene to early Oligocene interval. The appearance of new genera and species within the middle Eocene is spread evenly among families, whereas late Eocene and early Oligocene radiations show an increase in species diversity that is confined almost exclusively to selected families. For the late Eocene, these families are Turridae, Trochidae, and Turbinidae, and for the early Oligocene are Buccinidae, Beringiidae, and Volutidae (figure 3.7). This pattern is, in part, correlative with the one observed in Alaskan faunas, where middle Eocene species diversity is greatest in the Fasciolariidae, and early Oligocene diversity is greatest in the Buccinidae (figure 3.6). The early Oligocene increase of diversity in the Buccinidae confirms the pattern that was observed previously for this family in the North Pacific (Titova, 1993, 1994). The overall pattern for Kamchatka can be described as stepwise extinctions with increasing magnitude, accompanied by conspicuous radiation in selected groups.

Changes in the biogeographic composition of gastropod assemblages in Alaska and Kamchatka are more similar than are diversity trends in these areas. Alaskan and Kamchatkan assemblages since middle Eocene have undergone a steady decline in cosmopolitan and warm-water genera, and an increase in the abundance of North Pacific temperate and cool-water genera (figures 3.5, 3.6, and 3.7). Many cosmopolitan and warm-water genera survived the Eocene-Oligocene transition, but the vast majority had a conspicuous decline in diversity. The radiation of cool-water genera of North Pacific affinities in the early Oligocene seems to be more conspicuous in Kamchatka than in Alaska, where most cool-water

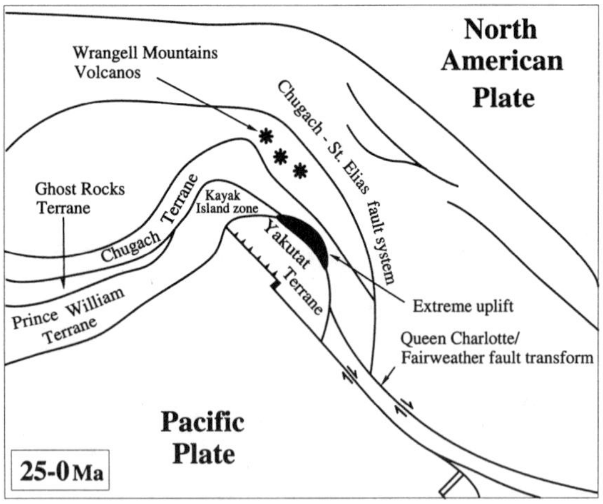

Figure 3.9 Inferred tectonic setting of the Yakutat Terrane (after Plafker, 1987).

genera appeared later in the Oligocene (figures 3.6 and 3.7). No species in common with Sakhalin or Japan were identified from Paleogene faunas in the northeastern Gulf of Alaska. All Alaskan mollusks identified to the species level are conspecific with Paleogene taxa of Oregon and Washington. Western Kamchatkan faunas experienced a decline in number of species in common with Washington and Oregon, and a corresponding increase in the number of species conspecific with Sakhalin and Japan (figure 3.8).

DISCUSSION

The pattern of an overall stepwise decline in diversity from the middle Eocene to the early Oligocene, and the gradual replacement of warm-water taxa with more temperate ones having more restricted latitudinal distributions, are consistent with changes seen in oceanic microfossils and terrestrial vegetation. Conspicuous differences in the temporal changes in diversity patterns of Kamchatka and the Gulf of Alaska Tertiary Province imply different causes for these patterns. The pattern observed in the Gulf of Alaska Tertiary Province, one of high survival rates, with minor extinction and little radiation, is more characteristic of temperate latitudes rather than high latitudes in the cooling world at the end of the Eocene and the beginning of the Oligocene. The logical reasoning behind this conclusion is that the cooling in a warm greenhouse world should have its most severe impact on the high-latitude faunas, whereas low-latitude regions will remain less- or non-affected. Available data on shallow-marine mollusks, even within the Pacific Northwest, seems to agree with that pattern. Gastropod assemblages of the Oligocene Eugene and Keasey formations in Oregon contain abundant cosmopolitan and warm-water genera (*Conus, Olequahia, Echinophoria, Ancilla, Bonellitia, Ficus, Gemmula, Molopophorus, Bruclarkia,* and *Exilia*) that survived from the late Eocene (Hickman, 1969; 1980). *Turritella, Ficus, Pseudoperissolax, Echinophoria, Bruclarkia, Fusinus, Exilia,* and *Hemipleurotoma* are also present in the early Oligocene Blakeley Formation in Washington (Weaver, 1942). Faunas located closer to the equatorial regions, like the Oligocene assemblages of the Gulf Coast, contain an even larger proportion of warm-water genera that survived the Eocene/Oligocene boundary (MacNeil and Dockery, 1984; Dockery, 1977, 1980; Hansen, 1992)

The presence of a "high survival" pattern at 60°N in Alaska can probably be attributed more to geologic and tectonic causes rather than have a purely biological or paleoceanographic explanation. The Katalla district of the Gulf of Alaska is part of the Yakutat tectonostratigraphic terrane that was transported to its present position during the late Cenozoic by dextral strike-slip movement along the Fairweather–Queen Charlotte transform fault system (figure 3.9) from a site along the continental margin at least 600 km to the southeast, and is the most recent terrane to dock with southern Alaska. This transform fault system was initiated during the late Oligocene, at about 25 Ma. After that, about 600 km of the Yakutat terrane were subducted northward beneath the Chugach and Prince William terranes along the Chugach–St. Elias fault system and the Kayak Island zone. Deep subduction of the Yakutat terrane produced arc volcanoes in the Wrangell Mountains, the oldest of which have been radiometrically dated at about 20 Ma (Jones et al., 1986; Stone and Wallace, 1987; Plafker, 1987). There are no reliable onshore paleomagnetic data that would reveal a more detailed displacement history of the Yakutat Terrane (Hillhouse and Grommé, 1984). Preliminary offshore data suggest 13° ± 3° post-Eocene latitudinal displacement, relative to cratonal North America, with no more than 20° rotation (Van Alstine et al., 1985). Existing paleontologic data point to a similar pattern of displacement of the terrane relative to the continental margin (Keller et al., 1984; Rau et al., 1983; Marincovich and McCoy, 1984; Wolfe and McCoy, 1984).

Therefore, the pattern of diversity change seen in Paleogene faunas of the Katalla district can be considered as yet more independent evidence for northward displacement of the Yakutat terrane. The original habitats of the Katalla district Eocene and early Oligocene faunas were located to the south of their present position, approximately at the latitude of present-day British Columbia, possibly northernmost Washington. Although the species-level composition of these Alaskan faunas remains poorly known, our preliminary studies uncovered no endemic species in these assemblages. The lack of endemic species suggests that these Alaskan Paleogene faunas were essentially identical with the Eocene and Oligocene faunas of Oregon and Washington. The currently unresolved ages of Paleogene rocks in the Gulf of Alaska region, however, precludes us from reaching thoroughly documented conclusions on the degree of similarity between Paleogene faunas of southern Alaska and the Pacific Northwest. Better dating of the Alaskan sections will allow us to compare the timing of faunal changes between Paleogene faunas of the Yakutat terrane and faunas of Oregon and Washington. Precise timing of biotic changes on the Yakutat terrane and the continental part of the Pacific Northwest would contribute significantly to our understanding of the initial latitudinal occurrence of the Gulf of Alaska Paleogene faunas.

Gastropod assemblages from western Kamchatka appear to be autochthonous and represent the remains of a true Paleogene high-latitude biota. The accretion of exotic blocks in Kamchatka took place predominantly from a southeastern direction and occurred along the front of the Kamchatka and/or "pre-Kamchatka" ancestral volcanic arc (Zonenshain et al., 1990). The accretion of exotic blocks on the eastern side of the arc evidently did not affect the western side of the arc (the western Kamchatka back arc basin). There is no geologic evidence of strike-slip motion in the western Kamchatka basin, which suggests that the accumulation of Paleogene sequences there took place at approximately the same latitude as today. This interpretation is confirmed by a high percentage of endemic genera and species of marine mollusks throughout the Paleogene of Kamchatka (Oleinik, 1987, 1994, in press).

Temporal changes in western Kamchatkan gastropod assemblages from the middle Eocene through early Oligocene show a pattern of stepwise extinctions of marine cosmopolitan taxa and taxa of warm-water affinities. These taxa were gradually replaced by more temperate and cool-water elements, often ancestral to cold-water taxa that inhabit shallow waters of the northern Pacific today (Oleinik, 1993, 1996). This pattern correlates well with the decline in global temperatures throughout the studied interval (figures 3.4, 3.7, and 3.8). A significant characteristic of Kamchatkan faunas is that it is often difficult to determine if extinctions and new occurrences were results of true extinction or of evolutionary radiation events, or merely reflected a continuous migration pattern. Poor preservation and incomplete knowledge of the Eocene faunas of Japan precludes us from delineating temporal changes in the latitudinal ranges of genera and species of shallow-marine gastropods along the coast of Asia. The continental margin of Eurasia, linked in the north with the margin of Beringia, made up a continuous shallow-marine habitat along the northwestern rim of the Pacific Ocean (Snavely and Wagner, 1963; Nilsen and McKee, 1978). The occurrence of Eocene species in common with western North America supports the existence of a counterclockwise anticyclonic gyre in the high-latitude North Pacific, similar to the one existing today. The predominant westward flow along the southern margin of Beringia provided an excellent migration route for molluscan faunas from western North America to Asia. Some clades, such as the subfamily Fulgorariinae (family Volutidae), and Buccinidae, suggest evolutionary radiation, which produced a number of closely related species. Others, such as Turridae, show a pattern of rapid shift at the genus level, which is probably more indicative of migration. The two most conspicuous patterns of the middle Eocene through early Oligocene transition in Kamchatka are the large number of newly appeared taxa restricted to only four families. Our study of diversity within these families shows that this "diversification" event is almost entirely explainable by the appearance of 9 species and 7 genera of turrids and 4 species and 2 genera of trochids (figure 3.6). This pattern can be (at least in part) explained by the fact that the upper Eocene Kovachinskaya Formation (figure 3.2) consists of facies ranging from shallow to deeper shelf, whereas the deeper facies are not known for the underlying Snatolskaya Formation. The short-term late Eocene warming that is expressed as slight changes in the composition of Kamchatkan terrestrial floras is less apparent (if at all) in shallow-marine gastropods. The appearance of a small number of warm-water groups (Coralliophylidae, Melongenidae) was accompanied by a much more impressive increase in intemperate and cool-water groups (figure 3.6). The second "explosive" radiation event, in the early Oligocene, largely occurred within the families Volutidae (subfamily Fulgorariinae) and Buccinidae, and did not exceed the number of extinct species at the Eocene/

Oligocene boundary. These families, which still inhabit the northern Pacific today, were evidently well pre-adapted to colder conditions. Detailed studies of Buccinidae by Gladenkov et al. (1988) and Titova (1993, 1994) and of Volutidae by Oleinik (1993, 1996, and work in progress) suggest that these groups underwent extensive radiation with the onset of cold conditions and probably migrated southward following continuing climatic cooling during the Neogene.

ACKNOWLEDGMENTS

We thank Carole Hickman and the staff of the Museum of Paleontology, University of California at Berkeley, for providing access to Alaskan Paleogene molluscan collections. Special thanks to Geerat Vermeij of the University of California in Davis, William Zinsmeister of the Purdue University, and Ellen Moore of Corvallis, Oregon for critical reviews and helpful comments that significantly improved the manuscript.

REFERENCES

Addicott, W. O., and Plafker, G., 1972, Paleocene mollusks from the Gulf of Alaska Tertiary province—a significant new occurrence on the North Pacific rim: U.S. Geological Survey Professional Paper, 750-B, p. 48-52.

Addicott,W. O., Winkler, G. R., and Plafker, G., 1978, Preliminary megafossil biostratigraphy and correlation of stratigraphic sections in the Gulf of Alaska Tertiary Province: U.S. Geological Survey Open-File Report, p. 78-491 (2 sheets).

Aubry, M.-P., 1992, Late Paleogene calcareous nannoplankton evolution: a tale of climate deterioration, *in* Prothero, D. R., and Berggren, W. A., eds., Eocene-Oligocene climatic and biotic evolution: Princeton, New Jersey, Princeton University Press, p. 245-271.

Baldauf, J. G., 1992, Middle Eocene through early Miocene diatom floral turnover, *in* Prothero, D. R., and Berggren, W. A., eds., Eocene-Oligocene climatic and biotic evolution: Princeton, New Jersey, Princeton University Press, p. 310-326.

Benyamovskiy, V. N., 1997a, Age and depositional environments of the Rateginskaya Formation of the western Kamchatka, based on foraminifera, Paper 1, Foraminiferal assemblages and zones: Transactions of Russian Academy of Sciences, no. 3, p. 37-42 [in Russian].

Benyamovskiy, V. N., 1997b, Age and depositional environments of the Rateginskaya Formation of the western Kamchatka, based on foraminifera, Paper 2, Paleogeography of Oligocene marginal basins of the Far Eastern region: Transactions of Russian Academy of Sciences, no. 4, p. 15–26 [in Russian].

Boersma, A., Premoli-Silva, I., and others, 1987, Atlantic Eocene planktonic foraminiferal paleohydrographic indicators and stable isotope paleoceanography: Paleoceanography, v. 2, p. 287-331.

Brinkhius, H., 1992, Late Paleogene dinoflagellate cysts with special reference to the Eocene/Oligocene boundary, *in* Prothero, D. R., and Berggren, W. A., eds., Eocene-Oligocene climatic and biotic evolution: Princeton, New Jersey, Princeton University Press, p. 327-340.

Bukry, D., 1973, Low-latitude coccolith biostratigraphic zonation: Initial reports of the Deep Sea Drilling Project, v. 15, p. 127-149.

Bukry, D., 1975, Coccolith and silicoflagellate stratigraphy, northwestern Pacific Ocean, Deep Sea Drilling Project Leg 32: Initial reports of the Deep Sea Drilling Project, v. 32, p. 677-701.

Cavelier, C., Chateaneuf, J. J., and others, 1981, The geological events at the Eocene/Oligocene boundary: Palaeogeography, Palaeoclimatology, Palaeoecology, 18, p. 223-248.

Chelebayeva, A. I., 1991, Floral characteristics of Snatolskiy and Kovachinskiy horizons, Snatolskaya Formation flora, Kovachinskaya Formation flora, phytohorizons, *in* Devyatkin, E. V., ed., Eocene of western Kamchatka: Transactionsof the U.S.S.R. Academy of Sciences Geological Institute, Moscow, Nauka Publishers, v. 467, p. 52-62. [in Russian, original title and contents translated].

Chelebayeva, A. I., and Bratseva, G. M., 1985, On the climatostratigraphy of the Paleogene based on paleofloras (North of the Pacific Region and adjacent regions), *in* Puscharovsky, Yu. M., and Gladenkov, Yu. B., eds., Correlation of the Cenozoic strata of the Far East, Part 2: U.S.S.R. Academy of Sciences Geological Institute, Moscow, p. 157-209. [in Russian].

Corliss, B. H., and Keigwin, L. D., 1986, Eocene-Oligocene paleoceanography, *in* Hsu, K. J., ed., Mesozoic and Cenozoic oceans: American Geophysical Union and Geological Society of America Geodynamics Series, vol. 15, Publication No. 0131 of the International Lithosphere Program, p.101-118.

Dockery, D. T., III, 1977, Mollusca of the Moodies

Branch Formation, Mississippi: Mississippi Geological Survey Bulletin 120, 212 p

Dockery, D. T., III, 1980, The invertebrate macropaleontology of the Clarke County, Mississippi, area: Mississippi Department of Natural Resources Bureau of Geology, Bulletin 122, 387 p.

Durham, J. W., and MacNeil, F. S., 1967, Cenozoic migrations of marine invertebrates through the Bering Strait region, in Hopkins, D. M., ed., The Bering Land Bridge: Stanford, California: Stanford University Press, p. 326-349.

Fotyanova, L. I., 1977, Cenozoic floras and climate of the North Pacific, in Fossil fauna and flora of the Far East and questions of the Phanerozoic stratigraphy: Vladivostok, p.66-85 [in Russian].

Fotyanova, L. I., and Serova, M. Ya., 1976, Position of Cenozoic floras in the zonal biostratigraphic scale of Japan and correlation with the floras of Sakhalin and Kamchatka: Papers on the geology and paleontology of the Far East, Vladivostok, U.S.S.R. Academy of Sciences Far Eastern Center, p. 104-107 [in Russian].

Fotyanova, L. I., and Serova, M. Ya., 1983, Biostratigraphy of the Paleogene of the northwestern Kamchatka (Podkagernaya Bay): Transactions of the U.S.S.R. Academy of Sciences, Geology Series, v. 10, p. 65-72 [in Russian].

Gaskell, B. A., 1991, Extinction patterns in Paleogene benthic foraminiferal faunas: relationship to climate and sea-level: Palaios 6, p.2-16.

Gladenkov, Yu. B., Sinelnikova, V. N., and Titova, L. V., 1988, Stages of formation of the fauna of Neogene shelf basins of Kamchatka (on the example of Buccinidae), in Timofeev, P. P., and Gladenkov, Yu. B., eds., Lithology and stratigraphy of the Mesozoic and Cenozoic of the eastern regions of the U.S.S.R.: Moscow, Nauka Publishers, p. 58-135 [in Russian].

Gladenkov, Yu. B., Shantser, A.E., and others, 1991a, Stratigraphy of the Eocene Snatolskaya and Kovachinskaya formations of the western Kamchatka, in Devyatkin, E. V., ed., Eocene of the western Kamchatka: Transactions of the U.S.S.R. Academy of Sciences Geological Institute, Moscow, Nauka Publishers, v. 467, p. 7- 32 [in Russian, original title translated].

Gladenkov, Yu. B., Sinelnikova, V. N., and others, 1991b, Marine paleocommunity analysis (mollusks and foraminifers), in Devyatkin, E. V., ed., Eocene of western Kamchatka: Nauka Publishers, Transactions of the U.S.S.R. Academy of Sciences Geological Institute, Moscow, v. 467, p. 41-48. [in Russian, original title and contents translated].

Goryachev, A. V., 1965, Alpine cycle of the tectonic development of the Kuril-Kamchatka folding zone, Paper 1, History of the tectonic movements: Proceedings of MOIP, Geology Series, v. 70, no. 1, p. 46-70 [in Russian].

Hansen, T. A., 1992, The patterns and causes of molluscan extinction across the Eocene/Oligocene boundary, in Prothero, D. R., and Berggren, W. A., eds., Eocene-Oligocene climatic and biotic evolution: Princeton, New Jersey: Princeton University Press, p. 341-348.

Hickman, C. S., 1969, The Oligocene marine molluscan fauna of the Eugene Formation in Oregon: Museum of Natural History, University of Oregon, Bulletin 16, 112 p.

Hickman, C. S., 1980, Paleogene marine gastropods of the Keasey Formation in Oregon: Bulletins of American Paleontology, 78(310), 112 p.

Hillhouse, J. W., and Gromme, C. S., 1984, Northward displacement and accretion of Wrangelia: new paleomagnetic evidence from Alaska: Journal of Geophysical Research, vol. 89, no. B14, p. 4461–4477.

Hopkins, D. M., 1967, The Cenozoic history of Beringia—a synthesis, in Hopkins, D. M., ed., The Bering Land Bridge. Stanford, California: Stanford University Press, p. 451-484.

Jones, D. L., Silberling, N. J., and others, 1986, Lithotectonic terrane map of Alaska (west of 41st meridian): U.S. Geological Survey Miscellaneous Field Studies Map, MF-1874, 1 sheet.

Kafanov, A. I., 1979, Cenozoic migrations of the shelf molluscan faunas in the northern Pacific, *in* Mollusks: Results and perspectives of molluscan research: Transactions of IVAll-Union Malacological Meeting. Nauka Publishers, Leningrad, p. 91-93 [in Russian].

Kanno, S., 1971, Tertiary molluscan fauna from the Yakataga district and adjacent areas of southern Alaska: Paleontological Society of Japan Special Papers, v. 16, p.1-154.

Keller, G., von Huene, R. E., Bruns, T. R., and McDougall, K., 1984, Paleoclimatic evidence for Cenozoic migration of Alaskan terranes: Tectonics, v. 3, p. 473–495.

Krasheninnikov, V. A., Serova, M. Ya, and Basov, I. A., 1988, Stratigraphy and planktonic foraminifera of the Paleogene of high-latitudes

Pacific Ocean: Transactions of the U.S.S.R. Academy of Sciences Geological Institute, v. 429, Nauka Publishers, Moscow, p. 1-119.

MacNeil, F. S. and Dockery, D. T., III, 1984, Lower Oligocene Gastropoda, Scaphopoda, and Cephalopoda of the Vicksburg Group in Mississippi: Mississippi Department of Natural Resources, Bureau of Geology, Bulletin 124, 415 p.

Mallory, V. S., 1959, Lower Tertiary biostratigraphy of the California Coast Ranges: American Association of Petroleum Geologists, Tulsa, Oklahoma, p. 1-416.

Marincovich, L., Jr., 1985. Early Tertiary marine fossils from northern Alaska: Implications for Arctic Ocean paleogeography and faunal evolution: Geology, v. 13, p. 770-773.

Marincovich, L., Jr., 1993, Danian mollusks from the Prince Creek Formation, northern Alaska, and implications for Arctic Ocean paleogeography: Journal of Paleontology, v. 67, no. 5, Paleontological Society Memoir 35, p. 1-35.

Marincovich, L., Jr., and Gladenkov, A. Yu, 1999, Evidence for early opening of the Bering Strait: Nature, v. 397, p. 149-151.

Marincovich, L., Jr., and McCoy, S., 1984, An overview of Paleogene molluscan biostratigraphy and paleoecology of the Gulf of Alaska region: Palaeogeography, Palaeoclimatology, Palaeoecology 47, p. 91-102.

Marincovich, L., Jr., and Moriya, S., 1992, Early middle Miocene mollusks and benthic foraminifers from Kodiak Island, Alaska: U.S. Geological Survey Bulletin 1991, p. 163-169.

Martini, E., 1971, Standard Tertiary and Quaternary calcareous nannoplankton zonation, *in* Farinacci, A., ed., Proceedings of the second planktonic conference, Roma, 1970, Technoscienzia, Roma, p. 739-785.

Miller, D. J., 1961, Geology of the Katalla district, Gulf of Alaska Tertiary Province, Alaska: U.S. Geological Survey open-file map, 2 sheets.

Miller, D. J., 1975, Geologic map and sections of the central part of the Katalla district, Alaska: U.S. Geological Survey Miscellaneous Field Studies Map, MF-772, 2 sheets.

Monin, A. S., and Shishkov, Yu. A., 1979. History of climate: Leningrad, Hydrometeoizdat, p. 1 -408. [in Russian].

Nelson, S. W., Dumoulin, J. A., and Miller, M. L., 1985, Geologic map of the Chugach National Forest, south-central Alaska: U.S. Geological Survey Miscellaneous Field Studies Map MF-1645-B, 1 sheet.

Nilsen, T. H., and McKee, E. H., 1978, Paleogene paleogeography of the western United States, *in* Armentrout, J. R., Colle, M. R., and Terberst, H., Jr., eds., Cenozoic paleogeography of the western United States. Pacific Section, SEPM, Pacific Coast Paleogeography Symposium, 3, p. 257-276.

Oleinik A. E., 1987, Formation of the western Kamchatka Eocene gastropod fauna, in Mollusks: Results and perspectives of molluscan research, Transactions of VIII All-Union Malacological Meeting. Nauka Publishers, Leningrad, p. 104-106 [In Russian, title translated].

Oleinik, A. E., 1993, The Genus Fulgoraria (Gastropoda: Volutidae) of the northeastern Kamchatka Peninsula and Sakhalin Island, with notes on the paleoecology and distribution of the subfamily Fulgorariinae in the Oligocene of the northern Pacific: The Nautilus, v. 106, p. 137-146.

Oleinik, A. E., 1994, Climatically induced diversifications and extinctions in the Paleogene gastropods from the western North Pacific: GSA Annual Meeting, Seattle, Washington, Abstracts with Programs, A-171.

Oleinik, A. E., 1996, Genus Arctomelon (Gastropoda, Volutidae) in the Tertiary of the northwestern Pacific: evolution and adaptations: Journal of Paleontology, v. 70, no. 2, p. 236-246.

Oleinik, A. E., 1999, A view from above: high-latitude North Pacific marine molluscan record of the Eocene-Oligocene transition: Geological Society of America Annual Meeting, Denver, Colorado, Abstracts with Programs, vol. 31, no. 7, p. A310.

Oleinik, A. E., in press, Eocene gastropods of western Kamchatka–implications for high-latitude North Pacific biostratigraphy and biogeography: Palaeogeography, Palaeoclimatology, Palaeoecology, Special Volume.

Plafker, G., 1967, Geologic map of the Gulf of Alaska Tertiary Province: U.S. Geological Survey Miscellaneous Geologic Investigations Map I-484, 1 sheet.

Plafker, G., 1987, Regional geology and petroleum potential of the northern Gulf of Alaska continental margin, *in* Scholl, D.W., Grantz, A., and Vedder, J. G., eds., Geology and resource potential of the continental margin of western North

America and adjacent ocean basins: Circum-Pacific council for energy and mineral resources, Houston, Texas, p. 229-268.

Rau, W. W., Plafker, G., and Winkler, G. R., 1983, Foraminiferal biostratigraphy and correlations in the Gulf of Alaska Tertiary Province: U.S. Geological Survey Oil and Gas Investigations Chart, OC-120, p. 1-11, 3 charts.

Saito, T., Okada, H., and Kaiho, K., eds., 1984, Biostratigraphy and international correlation of the Paleogene System in Japan: Faculty of Science, Yamagata University, Yamagata, Japan. p. 1-137 [in Japanese].

Savin, S. M., Douglas, R. G., and Stehli, F. G., 1975, Tertiary marine paleotemperatures: Geological Society of America Bulletin, v. 86, p. 1499-1510.

Scarlato, O. A., and Kafanov, A. I., 1976, Pre-Pliocene fauna of the Polar Basin and the question of primary autochthoneity of euarctic genera: Zoological Journal, v. 55, no. 12, p. 1765-1772 [in Russian].

Serova, M. Ya., 1969, Middle Eocene planktonic foraminifera of the western Kamchatka: Voprosy Mikropaleontologii, v. 12, p. 145-166. [in Russian].

Serova, M. Ya., 1983, Upper Paleogene of Sakhalin and Kamchatka (foraminifera, age, correlations): Proceedings of the U.S.S.R. Academy of Sciences, Geology Series, v. 4, p. 58-72. [in Russian].

Serova, M. Ya., 1985, Oligocene of the northwestern part of the Pacific province, *in* Pysharovsky, Yu. M. and Gladenkov, Yu. B., eds., Correlation of the Cenozoic strata of the Far East. Part II. U.S.S.R. Academy of Sciences, Geological Institute, Moscow, p. 210-243. [in Russian].

Shantser, A. E., Gladenkov, Yu. B., and Oleinik, A. E., 1991, Eocene paleogeography and geologic history of the western Kamchatka, *in* Devyatkin, E. V., ed., Eocene of the western Kamchatka: Transactions of the U.S.S.R. Academy of Sciences Geological Institute, Moscow, Nauka Publishers, v. 467, p. 69-75 [in Russian, original title translated].

Sinelnikova, V. N., Oleinik, A. E., and others, 1991, Systematic descriptions of mollusks, *in* Devyatkin, E. V., ed., Eocene of the western Kamchatka: Transactions of the U.S.S.R. Academy of Sciences Geological Institute, Moscow, Nauka Publishers, v. 467, p. 119-162 [in Russian, original title translated].

Snavely, P. D. Jr., and Wagner, H. C., 1963, Tertiary geologic history of western Oregon and Washington: Washington Division of Mines and Geology Report 22, p. 1-25.

Stone, D. B., and Wallace, W. K., 1987, A geological framework of Alaska: Episodes, v. 10, no. 4, p. 283-289.

Thomas, E., 1992, Middle Eocene–late Oligocene bathyal benthic foraminifera (Weddell Sea): Faunal changes and implications for oceanic circulation, *in* Prothero, D. R., and Berggren, W. A., eds., Eocene-Oligocene climatic and biotic evolution: Princeton, New Jersey, Princeton University Press, p. 245-271.

Titova, L. V., 1993, The early history of the North Pacific Ancistrolepidinae (Gastropoda: Buccinidae): Ruthenica, The Russian Malacological Journal, v. 3, no. 1, p. 1-16.

Titova, L. V., 1994, Cenozoic history of Turritelloidea and Buccinoidea (Mollusca: Gastropoda) in the North Pacific: Palaeogeography, Palaeoclimatology, Palaeoecology 108, p. 319-334.

Van Alstine, D. R., Bazard, D. R., and Witney, J. B., 1985, Paleomagnetism of cores from the Yakutat well: American Geophysical Union Transactions, v. 66, no. 46, p. 865.

Volobueva, V. I., Gladenkov, Yu. B., and others, 1994, Key section of marine Paleogene of the northern Far East (Ilpinskiy Peninsula), Parts 1 and 2: Northeastern Research Center of the Russian Academy of Sciences, p. 1-120. [in Russian].

Weaver, C. E., 1942, Paleontology of the marine Tertiary formations of Oregon and Washington: University of Washington Publications in Geology, v. 5, p.1-789.

Wolfe, J. A., 1971, Tertiary climatic fluctuations and methods of analysis of Tertiary floras: Paleogeography, Paleoclimatology, Paleoecology 9, p. 27-57.

Wolfe, J.A., 1977, Paleogene floras from the Gulf of Alaska. U.S. Geological Survey Professional Paper 997, p. 1-108.

Wolfe, J. A., 1980, Tertiary climates and floristic relationships at high latitudes in the Northern Hemisphere: Palaeogeography, Palaeoclimatology, Palaeoecology 30, p. 313-323.

Wolfe, J. A., and Hopkins, D. M., 1967, Climatic changes recorded by Tertiary land floras in northwestern North America, *in* Hatai, K., ed., Tertiary correlations and climatic changes in the Pacific: Sasaki Publishers, Sendai, Japan, p. 67-76.

Wolfe, J. A., and McCoy, S., Jr., 1984, Comment on "Model for the origin of the Yakutat block, an accreting terrane in the northern Gulf of Alaska": Geology, v. 12, no. 9, p. 564-565.

Zonenshain L. P., Kuzmin, M. I., and Natapov, L. M., 1990, Geology of the U.S.S.R.: a plate-tectonic synthesis: American Geophysical Union, Geodynamics Series, v. 21, Washington D.C., p. 149-167.

CHAPTER 4

Changes in Shallow-Marine Faunas from the Northeastern Pacific Margin Across the Eocene/Oligocene Boundary

Elizabeth A. Nesbitt

ABSTRACT

Shallow-marine molluscan faunas of the northeastern Pacific margin change from middle Eocene through the early Oligocene as the global climatic conditions changed from greenhouse to global icehouse. Faunal changes are described in terms of paleocommunity reorganization over four biozones during this critical 10-million-year period. Although there is rapid species turnover at each of the four biozone boundaries under investigation, the molluscan community structure changed at a much slower pace, taking 5 to 6 million years. The middle Eocene/late Eocene boundary is notable for the immediate loss of tropical taxa and drop in species richness. Cool-water gastropod taxa, endemic to the northern Pacific, slowly replace the locally extinct genera. This response to the rapid cooling events, previously characterized in deep-sea studies, is concomitant with a rapid change in tectonic setting and subsequent depositional environment. In contrast to the gastropods, changes in bivalve taxa reflect the replacement of craton-derived feldspathic and arkosic sediments in the shallow-shelf substrate, to magmatic arc-derived tuffaceous sediments. The molluscan paleocommunity analysis reveals two different environmental signatures.

INTRODUCTION

The Terminal Eocene Event was described by Wolfe in 1978, and the intervening twenty years have revealed that the climatic shift from global greenhouse to icehouse is more complex and severe, and extends from late middle Eocene through early Oligocene time (e.g. Prothero and Berggren, 1992; Prothero, this volume). Four global events mark progressively deteriorating climates over this transition, and the biozonal boundaries based on different taxa illustrate the protracted temperature shifts. The global climate changed from very wide tropical and subtropical belts that prevailed through the Cretaceous and early Tertiary to the narrowed tropics, wide temperate and boreal zones that characterized the Neogene. This change coincides with major marine extinctions depicted in studies of open-ocean, deep-sea organisms (e.g. Thomas, 1989; Schellenberg, 1999), but very few studies have documented these events in relation to nearshore organisms. Diverse and abundant molluscan faunas from Paleogene outcrops along the margin of the northeastern Pacific Ocean provide a different and independent database for studying the effects of this major global climate change. This study presents a preliminary investigation of the changes in paleocommunity structure from 40 million to 30 million years ago.

The Eocene–Oligocene boundary is placed at the last occurrence of the planktonic Foraminifera genus *Hantkenina* in the marine section at Massignano, northern Italy, that is dated at 33.7 Ma (Berggren et al., 1995). The epoch boundary is 0.1 million years younger than the worldwide planktonic foraminiferal Zone boundary P17/18 based on the LAD of the *Turborotalia cerroazulensis/cunialensis* group, and neither boundary coincides with nannoplanktonic zonal boundaries (Berggren and Miller, 1988; Berggren et al., 1995) (figure 4.1).After the early Eocene temperature maxima, global cooling is first recorded at the end of the middle Eocene (37 Ma), when the deep-sea, benthic foraminiferal $\delta^{18}O$ isotope record shows a 1‰ drop, corresponding to a 4°C temperature

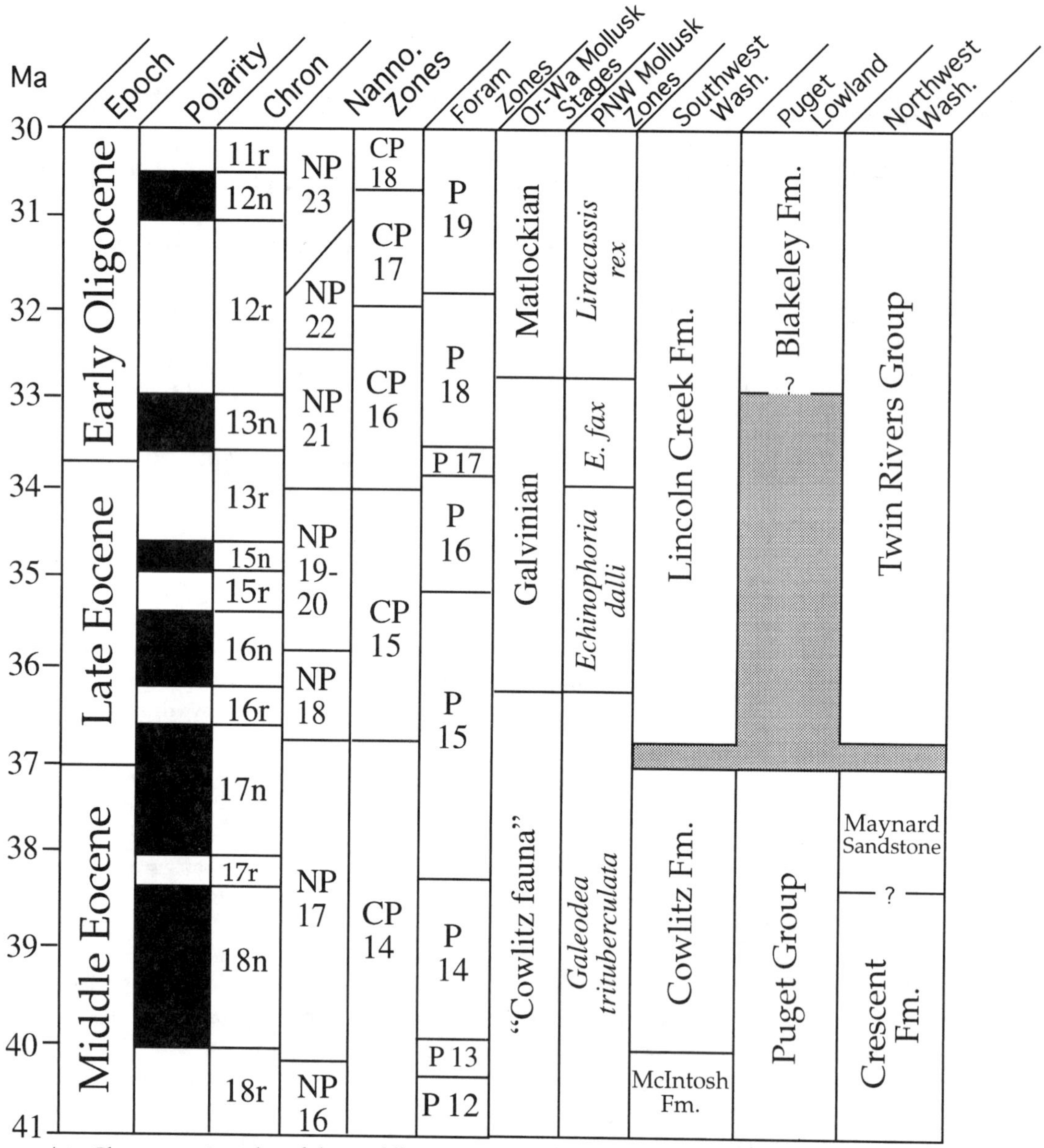

Figure 4.1. Chronostratigraphy of the middle Eocene to early Oligocene biostratigraphic units and formations of western Washington, modified from Prothero and Armentrout (1985) and Prothero (2001).

decrease in the open ocean (Miller et al., 1987; Miller, 1992; Zachos et al., 1999). A second precipitous temperature drop of around 5°C in the ocean at 33.2 Ma (Oi1, the early Oligocene event) is marked by another 1‰ benthic and planktic foraminiferal $\delta^{18}O$ isotope shift (Miller, 1992). This is also concurrent with changes in global tectonic regimes: the closure of the Tethys and opening of the circum-Antarctic seaway. Nonsynchronous extinctions of different marine groups occurred over this time period (Prothero and Berggren, 1992). Extinction patterns of mollusks across the Eocene and Oligocene have been documented only for the U.S. Gulf Coast (Haasl and Hansen, 1996; Dockery, 1998) and they appear to be contradictory. Interpretations of changes in diversity were attributed to change in climate or to change in shelf environments.

High-resolution chronostratigraphy of middle Eocene through Oligocene marine sedimentary rocks of western Washington and Oregon (Prothero, 2001,

Table 4.1. Formation Names, Stratigraphic Thickness, Sediment Types, and Depositional Environments for the Upper Middle Eocene Through Lower Oligocene Fossiliferous Units in Western Washington

Era	Biozone	Stratigraphic units and thickness	Sediments and depositional environments of fossiliferous units
Early Oligocene	*Liracassis rex*	Lincoln Creek Fm. – Blakeley Fm. – 1,800m Upper Pysht Fm. (Twin Rivers Group) – 400m	Massive, tuffaceous siltstones, with alternating thin feldspathic and quartz sandstones, some turbidites, numerous concretionary layers. Shallow to deep shelf, and marine embayments.
Eocene - Oligocene boundary	*Echinophoria fax*	Lincoln Creek Fm. – Upper Pysht Fm. (Twin Rivers Group) – 500m	Thick sequences of tuffaceous siltstones and fine-grained sandstone, with strata containing abundant fossiliferous concretions. Shallow to deep shelf
Late Eocene	*Echinophoria dalli*	Lincoln Creek Fm. – Gries Ranch beds of Toutle Fm. – 20 m Jensen Creek Mb. of Makah Fm. – 200m	Coarser-grained, basaltic sandstones, glauconitic sandstones and sandy siltstones deposited on basaltic basement on shallow shelf
Late middle Eocene	*Galeodea trituberculata*	Cowlitz Fm. – 2,000m Marine beds of the Tukwila Fm. (Puget Group) – 200m	Feldspathic sandstones and siltstones, basaltic sandstones, and inter-bedded carbonaceous strata, deposited in a variety of deltaic depositional environments

Sources: Compiled from Weaver (1937), Durham (1944), Henriksen (1956), Roberts (1958), Armentrout (1973, 1975), Nesbitt (1995, 1998), and unpublished UWBM data.

and this volume) now allows detailed studies of temporal changes in faunal composition in shelf sequences. The purpose of this study is twofold: (1) to determine if shelf faunas reflect the global climatic shift over the Eocene-Oligocene transition, as documented in the deep-sea record, and (2) to ascertain the cause of that reorganization in the composition and character of shallow-shelf faunas on the northeastern Pacific margin. A preliminary investigation these was undertaken using new collections to supplement existing museum holdings of Eocene and Oligocene mollusks from western Washington that are housed in the Burke Museum, University of Washington.

On the active plate margin of the northeastern Pacific, eustatic sea-level curves of Haq et al. (1988) appear to be overridden by the local tectonic effects of subducting plate angle and geographic position of the active volcanic arc (Armentrout, 1987; Cheney, 1994). Hence, stratigraphic hierarchical models for paleoecological reconstructions, such as those of Rollins et al. (1990), are not yet applicable. However, this first-order investigation of the temporal paleocommunity dynamics provides a measure of the nearshore marine biotic response to the global cooling event across the Eocene-Oligocene transition.

REGIONAL SETTING

Stratigraphy

The marine Cenozoic strata of western Washington State (table 4.1) are richly fossiliferous and provide ideal data to test faunal responses to global cooling events. Invertebrate fossils have been collected systematically since Charles E. Weaver became the first paleontologist in the state in 1909. Weaver (1937, 1943, and references therein) described the formations and assigned them to time periods, described the invertebrate faunas, published paleogeographic maps, and his extensive collections established the Paleontology Division of the Burke Museum. Weaver's work in the Pacific Northwest culminated in his chairing the Geological Society of America stratigraphic commission that established a correlation chart of all the Cenozoic marine sedimentary formations in California, Oregon, Washington, and southern Vancouver Island, British Columbia (Weaver et al., 1944). This correlation chart was entirely regional in scope and was not dependent on any correlations with European biostratigraphy. Two problems were noted by the committee—that of faunal provinciality, and that the very thick stratigraphic sections "show marked changes in lithology within relatively short horizontal distances" (Weaver et al., 1944: 571).

Armentrout (1975, 1981) refined the stratigraphic units in western Washington, based on restricted and overlapping ranges of numerous molluscan fossils from the Lincoln Creek Formation (late Eocene to early Miocene) in southwestern Washington. This scheme has provided the stratigraphic framework for paleontological investigations. High-resolution correlations of Paleogene fossiliferous sections along the northeastern Pacific margin have been difficult because of paucity in planktonic microfossils, lack of continuous sections, and increasing biogeographic provinciality along the West Coast of North America through the Cenozoic. Over the last few years much more precise chronostratigraphic correlations have been possible with magnetostratigraphic studies on key sections (Prothero, 2001). Paleomagnetic signatures of type sections for both molluscan and benthic foraminiferal stages and zones in western Washington are tied to the rare planktonic foraminifera and diatom index fossils, and to radiometric dates (Irving et al., 1996; Prothero, 2001; Prothero et al., 2001; Prothero, this volume). For the first time these sections can be correlated with the worldwide timescale of Berggren et al. (1995). The resulting chronology for Washington is shown in figure 4.1.

Data for this study spans four biozones (figure 4.1) that range from late middle Eocene through early Oligocene, a period of 10 million years. Table 4.1 gives a summary of depositional environments and source rocks of the fossils. Two amendments to the published biostratigraphic chart are included here. First, Armentrout (1981) used species of large and distinctive cymatiid gastropods (*Echinophoria* and *Liracassis*) as zonal indicators for the late Eocene through Oligocene of the Pacific Northwest. The cymatiid *Galeodea trituberculata* is restricted to the Cowlitz and Tukwila strata, and in keeping with the zonal nomenclature of Armentrout (1981) this has been used here as an informally designated zone name for the late middle Eocene (figure 4.1). The late Eocene–earliest Oligocene Galvinian Stage (Armentrout, 1973, 1975) was divided into three biozones; however, they are characterized by faunas from different depositional environments. The oldest, the *Bathybembix columbiana* zone, was defined by cool-water or deep-water mollusks and microfossils that inhabited finer-grained sediment. This biozone is based on outcrops of the Keasey Formation, Oregon, two stream-cut sections of the lower Lincoln Creek Formation, Washington, and it is equivalent to the "Keasey Stage" of Weaver et al. (1944). *Bathybembix columbiana* fossils are not included herein, and this author suggests that the zone is the coeval, deep-water facies of the shallow-shelf *Echinophoria dalli* zone.

Tectonic Framework

Temporal patterns of depositional environments and fossil preservation along the Pacific Northwest coast of North America are influenced by being on an active tectonic margin. This has been a zone of active subduction from the Jurassic, and the Cenozoic tectonic history of the continental margin has been characterized by oblique convergence (Miller et al., 1992; Snavely and Wells, 1996). Plate reconfiguration of the oceanic Kula and Farallon plates during the late Eocene resulted in decreased convergence rates and inferred steepening of the subducting plate (Miller et al., 1992). Changes in sedimentary basin morphology and in sediment sources resulted in a variety of depositional regimes along the forearc through the critical Eocene-Oligocene time period. In the Pacific Northwest, the Paleocene and early Eocene volcanic arc (the Challis arc) was situated across northeastern Washington, Idaho, and western Montana (Armentrout, 1987; Wells and Heller, 1988). Along the coastal margins of Washington and Oregon, large volumes of feldspathic sediments, derived from the continental sources, were deposited during the early and middle Eocene onto a basement of Eocene submarine and subaerial basalts (Henriksen, 1956; Heller and Ryberg, 1983; Heller et al., 1985). Extensive middle Eocene delta systems prograded across the basaltic basement, and these formations now contain minor gas fields and economically viable coal deposits (Snavely and Wells, 1996, and references therein). Continued eruptions of small basalt flows and breccias—for example, the Grays River and the Goble volcanic units—locally interfingered with these marine sediments (Armentrout, 1973; Wells et al., 1983; Snavely and Wells, 1996) and provided source rocks for basaltic sandstone.

During the early and middle Eocene the active volcanic arc migrated progressively westward, concurrent with crustal extension of eastern Washington and Oregon (Wells and Heller, 1988). The oldest-known volcanic tuffs of the Western Cascades yield $^{40}Ar/^{39}Ar$ (single-crystal laser fusion) dates of 38.9 ± 0.1 Ma in southwestern Washington (Irving et al., 1996), and 38.9 ± 0.03 Ma in north-central Oregon (Retallack et al., 1996). Andesitic lava flow complexes of the Western Cascades first began accumulating around 37 Ma, and thick tuff bed sequences are present from approximately 35 Ma (Hammond, 1998). The

Table 4.2. Relative Numbers, as a Percentage, of Each of the Assigned Higher-order Taxa, (A) Bivalvia and (B) Gastropoda, for the Four Biozones that Span the Late Middle Eocene Through Early Oligocene of the Northeastern Pacific Margin

	Late middle Eocene	**Late Eocene**	**Eocene-Oligocene**	**Early Oligocene**
	Galeodea trituberculata Zone	*Echinophoria dalli* Zone	*Echinophori a fax* Zone	*Liracassis rex* Zone
A) BIVALVIA				
Cryptodonta	15%	10%	44%	70%
Pteriomophia	15%	7%	8%	5%
Veneroida	73%	83%	48%	25%
B) GASTROPODA				
Trochidae	–	–	–	12%
Turritellidae	42%	–	16%	12%
Naticidae	21%	25%	26%	24%
Cymatiacea + Tonnacea	5%	–	4%	38%
Buccinacea	28%	75%	49%	12%
Conacea	4%	–	5%	2%

change in tectonic regime of the northeastern Pacific margin towards the end of the Eocene is evident in the depositional environments. The change in tectonic regime of the northeastern Pacific margin toward the end of the Eocene is also evident in the depositional environments. Fossiliferous units change from thick continent-derived, marginal-marine, and marine-deltaic units and fluvial deposits, to tuffaceous marine sequences derived from the magmatic arc (Weaver, 1937; Armentrout, 1973, 1975; Hickman 1984; Niem et al., 1992). Petrofacies of late Eocene and Oligocene coastal marine deposits are characterized by unaltered volcanic lithic fragments that increased volumetrically as the Western Cascades developed (Heller and Ryberg, 1983). Rapid subsidence of the forearc basin during the Oligocene resulted in very thick units of tuffaceous siltstone that reached bathyal depths. These packages of Eocene and Oligocene marine sediments became part of the accretionary wedge, and are thus structurally disconnected and are overlain with younger sedimentary and volcanic units.

METHODS

Half a century ago, Durham (1950) used the molluscan components of Cenozoic marine faunas to reconstruct paleoclimates along the West Coast of North America. By comparing the fossil genera with analogous living taxa, he estimated a change in the positions of the 19°C isotherm in February (midwinter) and recognized that the tropical belts narrowed gradually over the Paleogene and Eocene. This molluscan data showed that temperate climates prevail in the Pacific Northwest only later in the Oligocene (Durham, 1950). The study presented here is based on the same faunas that Durham used, but with a substantially augmented database and well-constrained stratigraphical units. Collections of fossil mollusks from the four biostratigraphic units that span the Eocene–Oligocene transition in Washington are housed in the Burke Museum, University of Washington (UWBM). To ensure legitimate faunal comparisons, this data included only those assemblages characteristic of the inner neritic (to a water depth of approximately 100 m).

Table 4.3. Higher-order Taxa, Their Life Habits and Trophic Mode of the Numerically Abundant Bivalves and Gastropods in the Eocene-Oligocene Marine Faunas of the Northeastern Pacific Margin

Higher order taxa	Families represented in Eocene Oligocene on eastern Pacific margin	Life habit and trophic mode of living analogues
A) BIVALVIA		
Cryptodonta	Order Nuculoida Order Solemyacea	• primitive, infaunal, sessile or mobile detritus feeders, generally with thin-shells. • infaunal detritus feeders or with endosymbiotic chemosynthetic bacteria
Pteriomorphia	Orders Arcoida, Pteroida, and Mytiloida	• nestling, epibyssate, or partially endobyssate, filter feeders.
Heterodonta	Order Veneroida including subfamilies Crassatellacea, Carditacea, Lucinacea, Veneracea, Mactracea, Tellinacea, and Solenacea	• largely filter-feeding, shallow and deeply infaunal bivalves • some Tellinacea are deposti feeders in fine-grained substrates • *Thyasira* and *Lucinoma* (Luciniacae) with endosymbiotic chemosynthetic bacteria
B) GASTROPODA		
Trochidae	Genus *Bathybembix*,	• mobile epifaunal deposit feeder that today live at bathyal depths of the northern Pacific
Turritellidae	Genus *Turritella*	• gregarious semi-infaunal, sedentary habit in soft sediment, and they are ciliary suspension feeders
Naticidae	Family Nacticidae (moon snails)	moon snails are active predators • *Neverita, Natica (Natica)* and *Polinices (Polinices)* generally world-wide tropical distributed, with a few temperate seas • *(Cryptonatica)* and *P. (Euspira)* are common in temperate to Arctic shallow, subtidal regions.
Cymatiacea	Family Cymatiacea,	• Cymatiacea feed on polychaetes and bivalves; nektonic larvae results in widely spread distributions.
Tonnacea	Families Ficidae (fig shells), Cassidae (helmit shellls), and Tonnidae(tun-shells)	• active predators feeding on on echinoids, asteroids, and holothurians
Buccinidae	Family Neptunidae, Nassariidae, Fasciolaridae and Buccinidae.	• mostly active predatory gastropods, some scavingers, • distributed worldwide, and the most abundant large gastropods in the northern Pacific and circum-boreal regions • many of them inhabit intertidal to shallow subtidal depths, on coarse-grained sediments.
Conacea	Conidae (cone shells), turrids.	• living members possess harpoon -like teeth and buccal poison glands, and are species-specific in their food preferences • in the Eocene-Oligocene sediments, a diversity of turrids been used to indicate cool, deep-water depositional environments.

Sources: Compiled from Cox et al. (1969), Smith (1970), Keen (1971), McPhereson (1971), Stanley (1970, 1972), Marincovich (1977), Hickman (1976, 1980, 1984), Taylor et al. (1980), Nesbitt (1982, 1995, 1998), and Allmon (1988, 1990).

Benthic molluscan assemblages, typical of Cretaceous and Tertiary coastal marine settings, are substrate dependent, and biotic interactions can be described in terms of their trophic structure in combination with sedimentological features (e.g. see discussions in Scott, 1976; Stanton and Dodd, 1976; Aberhan, 1994,) which reflect the prevailing environmental conditions, such as energy level, relative temperature, and position on the shelf/slope. Defining paleocommunities by a combination of biological and nonbiological characteristics broadens the scope and allows for higher resolution in paleoecological reconstructions. Although benthic paleocommunities preserved in soft sediment are now known to be time-averaged assemblages, the composition does reflect the persistence of environmental constraints over geologic time periods. In fact, the persistence of paleocommunities is encoded into each biostratigraphic unit: recognition of strata by an enduring co-occurrence of species characterizes a single biozone, followed by reorganization of species abundance distributions and assemblage characteristics at biozonal boundaries. Reorganization of paleocommunities indicates the shifting environmental parameters, and is an alternate perspective to species turnover.

The first evidence of rapid faunal changes over the middle Eocene through Oligocene on the northeastern Pacific margin is the timespan of each biozones. There is a temporal restriction of biozones around the boundary, preceded and followed by wider zonal periods (figure 4.1). The pre-Cowlitz fauna spans almost 4 million years and the late Oligocene *Liracassis apta* biozone spans 6 million years. The slower pace of taxonomic turnover in the later Oligocene indicates stabilization of the environment after the Eocene-Oligocene perturbations.

To determine community structure within each temporal unit in this study, individual taxa were counted from bulk samples that were collected from multiple localities. Cenozoic strata in the Pacific Northwest are relatively thick (table 4.1) but are generally exposed only in road cuts and creek beds, and frequently stratigraphic sections are interrupted by covered areas at a scale of >1 kilometer. However, the recurrence of assemblage composition in species individual ranking is regarded as the same paleocommunity repeated in time and space. The fact that these assemblages are repetitious infers community stasis, but that hypothesis is still to be tested. Individual specimens of gastropods and bivalves were counted (550–1,200 individuals per fauna) and assigned to higher-order taxonomic groups (table 4.2). Taphonomic features favor whole shells, excellent preservation of both calcitic and aragonitic shell structure, and range from large adults to small and juvenile shells. In many localities, exposure results in fracturing of thin shells, but careful collecting can preserve these, or they can be counted on the outcrop. At this time, statistical testing has not been applied to the data. Paleoecological analysis of the middle Eocene Cowlitz Formation molluscan paleocommunities shows that a few taxa are numerically dominant, and that most of the taxa are rare (Nesbitt, 1982, 1995). Rare species, each accounting for less than 2% of the fauna, were not included.

After a review of paleocommunities, community stasis, and ecological locking, Stanton and Dodd characterized a paleocommunity as "essentially a group of organisms that occupied a specific habitat at a specific time and persisted through a particular time span" (1997:240). In addition, use of the word "community" infers that the species composition and numerical abundance of each species remains approximately.

RESULTS

For ready comparison, the data is presented here in higher-order taxonomic categories grouped by trophic guild: infaunal and epifaunal suspension feeders, infaunal and epifaunal deposit feeders, and predatory/scavenging carnivores (tables 4.2 and 4.3). This is feasible because mollusks are classified by characteristics associated with feeding modes, which in turn dictates their habitat. Thus family- and order-level designations embody attributable autecological data for each animal (e.g. see Hickman, 1984; Allmon, 1988; Nesbitt, 1982, 1995). The trophic structure of a paleocommunity is characterized by the sum of feeding habits of each individual; therefore, these data are plotted as number of individuals counted, and not species presence or absence. Not included in this study are the almost monospecific beds of shallow-subtidal to intertidal bivalves, usually mytilids or ostraeids that occur in all formations (e.g. Nesbitt, 1995).

The three most abundant trophic groups are the carnivorous gastropods, the infaunal filter-feeding bivalves (Veneroida), and a group consisting of deposit-feeding bivalves and those that harbor endosymbiotic chemosynthetic bacteria (tables 4.2 and 4.3). The latter group includes the genera *Solemya* (Cryptodonta), *Lucinoma* (Veneroida), and *Thyasira* (Veneroida), which are associated with methane seeps and high-sulfur muds in the North Sea and the Mediterranean (Dando et al., 1986; Dando and Spiro 1993; Dando et al., 1994). Other trophic groups that are minor components of these Eocene-Oligocene communities are deposit-feeding gastropods, semi-infaunal filter-feeding gastropods, and epifaunal filter-feeding bivalves.

Late Middle Eocene Fauna

Late middle Eocene molluscan faunas, from the *Galeodea trituberculata* zone, are characterized by a high diversity (>150 molluscan species) of warm-water taxa that includes the gastropod families Strombidae, Cypraeidae, Muricidae, and Conidae, numerically abundant Turritella, as well as the very thick-shelled bivalves *Venericardia* and *Acanthocardia* (Nesbitt, 1995, 1998). Genera, and a few species, are in common with late middle Eocene faunas of California, Oregon, and southern Alaska. Thus, at this time, biogeographic provinces covered a very wide latitudinal span along the eastern Pacific margin.

The shallow-subtidal assemblages of the *Galeodea trituberculata* zone are preserved in coarse- to fine-grained sandstone and are numerically dominated by infaunal suspension-feeding individuals. Bivalves in this trophic group account for 65% of the total fauna, and the semi-infaunal, filter-feeding gastropod *Turritella uvasana* comprises up to 40% of the number of individuals per counted sample (Nesbitt, 1995). The most common bivalves are the veneroid genera *Venericardia, Acanthocardia, Pitar, Tivelina,*

Spisula, and *Callista* (table 4.2). Shell morphology of the common bivalves indicates two types: the robust, rounded, highly ornamented shells that would have been shallow-burrowers; and the slim, minimally ornamented, and thin-shelled genera that indicate a deep-burrowing habit. Epifaunal filter-feeders are <15% of the total fauna.

Numerically important gastropods include a high species diversity of buccinid genera, and those naticid genera and subgenera that are typical of tropical seas: *Neverita, Polinices* (*Polinices*), and *Natica* (*Natica*). Distribution of gastropod taxa across the deltaic front shows that buccinids and naticids dominate on the shallow shelf, and were gradually replaced on the outer shelf and upper bathyal slope by three species of *Conus*, a diversity of turrids, and shelled opisthobranchs (Nesbitt, 1995).

Late Eocene Fauna

The Lincoln Creek Formation unconformably overlies the type Cowlitz Formation in southwestern Washington (Armentrout, 1987; Payne and Niem, 1996) and has faunas ranging from late Eocene to early Miocene in age (Durham, 1944; Armentrout, 1973, 1975; Prothero and Armentrout, 1985). The late Eocene *Echinophoria dalli* zone (figure 4.1) is characterized by a shallow-subtidal to inner-shelf molluscan fauna (Armentrout, 1975; unpublished UWBM data). The base of the Lincoln Creek Formation is characterized by basaltic sand, derived locally from the exposed basement rocks. In the middle of this biozone, tuffaceous sediments begin to accumulate, and rapidly overwhelm those derived from local basaltic sources. Included in this study are two other shallow-subtidal assemblages, the Gries Ranch and the Jensen Creek faunas. The Gries Ranch Beds, are exposed in one short section along the Cowlitz River and are mapped as Toutle Formation (Roberts, 1958). Mollusks from this highly fossiliferous locality, described by Van Winkle (1918) and Armentrout (1973), occur immediately below a thick layer of densely packed oysters (not included in the study) and are preserved in a fine-grained sandstone with pebble-size pumice clasts. A remarkably similar fauna is found in the Jensen Creek Member of the Makah Formation, lowermost Twin Rivers Group, preserved within a glauconitic basaltic sandstone (J. Goedert, pers. commun., and unpublished UWBM data).

The shallow-shelf molluscan fauna of the E*chinophoria dalli* zone is reduced in diversity (95 species) compared with the older *Galeodea tritaberculata* zone, especially in gastropod taxa. The common Eocene genera *Venericardia* and *Turritella* are present, but are relatively thin-shelled forms that are rare in the late Eocene. Regional extinctions of the tropical index taxa, at the late middle/late Eocene boundary, resulted in reduced diversity.

Typically, Eocene gastropods that become extinct at the boundary or within the late Eocene are muricids, strombids, and a diversity of buccinid and cymatiid genera. Naticidae and Buccinidae continue to be the most abundant and diverse gastropod groups in these paleocommunities (table 4.2), but most of the common genera are new to this zone—for example, *Bruclarkia, Priscofusus, and "Molopophorus"* (six distinct species that probably belong in three nassariid genera, see Allmon, 1990). The zonal index genus *Echinophoria* (Tonnacea) and the cymatiid genera are generally large in body size but are uncommon. Cones and the opisthobranchs *Scaphander* and *Cylichnella* persist through this period but are progressively more rare.

These late Eocene paleocommunities are numerically dominated by veneroid bivalves similar in taxonomic composition as in the previous zone: *Pitar, Spisula, Tellina, Callista*, and *Solen*. There is, however, an almost complete turnover in species. As in the *Galeodea trituberculata* fauna, these bivalves have shell forms that indicate substrate depth tiering. Although the *Echinophoria dalli* biozone has been characterized by a variety of pteriomorph genera—for example *Lima, Barbatia, Arca*, and *Glycymeris* (Van Winkle, 1918; Effinger, 1938; Armentrout, 1973) —they occur in low numbers in each counted assemblage. In addition, the abundant Oligocene veneroid genus *Lucinoma* and the large-shelled trochid gastropod *Bathybembix* make their first appearances here, but are not common in this biozone.

Overall, the change in fauna between the late middle Eocene and late Eocene is a loss of the high diversity of warm-water gastropods, but the persistence of the numerically dominant infaunal bivalves.

Eocene/Oligocene Boundary Fauna

The *Echinophoria fax* zone, upper Galvinian Stage, in western Washington crosses the Eocene/Oligocene boundary (Prothero et al., 2001) (figure 4.1). Fossiliferous sediments consist of thick sequences of highly bioturbated, tuffaceous siltstones and fine-grained sandstones. The *E. fax* paleocommunities consist of almost entirely new species. Gastropod taxa reflect cooler water temperatures, including the appearance of a variety of Epitoniidae, Marginellidae, and Turridae, and return of the genus *Turritella.* More gastropod taxa in this zone originated in the North Pacific, and fewer taxa ranged into California (Hickman, this volume; Oleinik and Marincovich, this volume; Squires, this volume). The stratigraphic index species, such as *Echinophoria, Musashia,* and *Pseudoperissolax* are rare.

There was another rapid turnover in bivalve species at the *Echinophoria dalli-E. fax* boundary, and a decline in dominance and diversity of the infaunal filter-feeding families Veneracea, Tellinacea, and Mactridea to 48%. Increasing numbers of nuculanids, and the genera *Solemya, Thyasira,* and *Lucinoma* give these assemblages a novel structure (table 4.2). These bivalves are characteristic of cold, methane-seep ecology and have been recorded in such environments in the Pacific Northwest (Campbell, 1992; Goedert and Squires, 1990; Nesbitt et al., 1994; Goedert and Campbell, 1995). All three genera living today rely on the nutrition supplied by endosymbiotic chemosynthetic bacteria; however, where they are found in the thick stratigraphic sequences of this biozone, there are very few obvious seep-carbonate sediments. Living *Lucinoma* are also associated with seagrass and substrates with high organic content. In the early Oligocene, marine sediments of Washington that contain numerous *Lucinoma, Thyasira,* and *Solemya* are frequently found in substrates that include abundant organic matter as well as larger pieces of wood.

Early Oligocene Fauna

The *Liracassis rex* zone is found in the upper Lincoln Creek, Blakeley, and lower Pysht formations (table 4.1). Sediments of these fossiliferous formations consist of fine-grained, tuffaceous mudstone that in some places is part of turbidite sequences, and in others is interbedded with occasional laminae of clean quartz sand. Calcite cement concretions are common and frequently formed around fossil bone and crustacean remains. Diversity increased in these molluscan assemblages (120 species) compared with diversity on the previous two zones. Dominant groups of gastropods are markedly different from previous faunas. *Liracassis rex* and the three species of the large, thin-shelled trochid, *Bathybembix,* are common (table 4.2). Naticids in these paleocommunities belong in typically cold-water subgenera *Natica* (*Cryptonatica*) and *Polinices* (*Euspira*) (Marincovich, 1977). And the large, circum-boreal carnivorous gastropods such as the cymatiids Argobuccinum and Fusitriton, as well as neptunids, give the fauna a northern temperate to boreal aspect. The numerically dominant bivalves (70%) are the chemosynthetic trophic grouping, *Solemya, Thyasira,* and *Lucinoma.* Filter-feeding veneroids, including the genera *Macoma, Solen,* and *Pitar* comprise only 25%; and epifaunal filter-feeders are 5% of these bivalve faunas.

CONCLUSIONS

Results of this preliminary study show significant changes in the higher-level taxonomic composition of paleocommunities over the Eocene–Oligocene transition. Shallow-shelf molluscan paleocommunities along the Washington coast experienced a complete rearrangement of numerically dominant taxa across the Eocene/Oligocene boundary. The climate shift, previously documented in the deep-sea record, is clearly evident in the shallow-shelf molluscan paleocommunities of the northeastern Pacific margin, and the living communities of the northeastern Pacific have their roots in the changes over the Eocene/Oligocene boundary.

However, the most notable result of this study is that the molluscan components reflect changes in two different physical features of their environment. Gastropod taxa responded to a drop in water temperatures, and the bivalves (largely infaunal) responded to a change in substrate composition (figure 4.2).

On the northeastern Pacific margin a major climate shift from tropical temperatures in the middle Eocene to cool water in the Oligocene is evident in the almost 100% extinction of molluscan species at each zonal boundary (Hickman, this volume; UWBM unpublished data). The loss of Eocene index species such as *Venericardia* and cypraeid, strombid, muricid, and cone gastropods occurred primarily at the middle Eocene/late Eocene boundary (37 Ma), and continued through the late Eocene period. These extinctions resulted in a loss of species diversity in the late Eocene, and diversity did not increase for another 3–4 million years, until the *Liracassis rex* zone (30–32 Ma). Over the 10-million-year interval studied here, gastropod replacement within paleocommunities was overwhelmingly temperate in composition, by genera that were endemic northern Pacific. These new taxa include the large-shelled *Bathybembix,* as well as Cassidae, Cymatiidae, Fasciolariidae, and Neptunidae genera, Naticidae subgenera, and a radiation of small-shelled turrids.

However, the change in bivalve components of the paleocommunities signifies a completely different environmental response. They reflect a change in substrate, rather than a temperature drop (figure 4.2). As

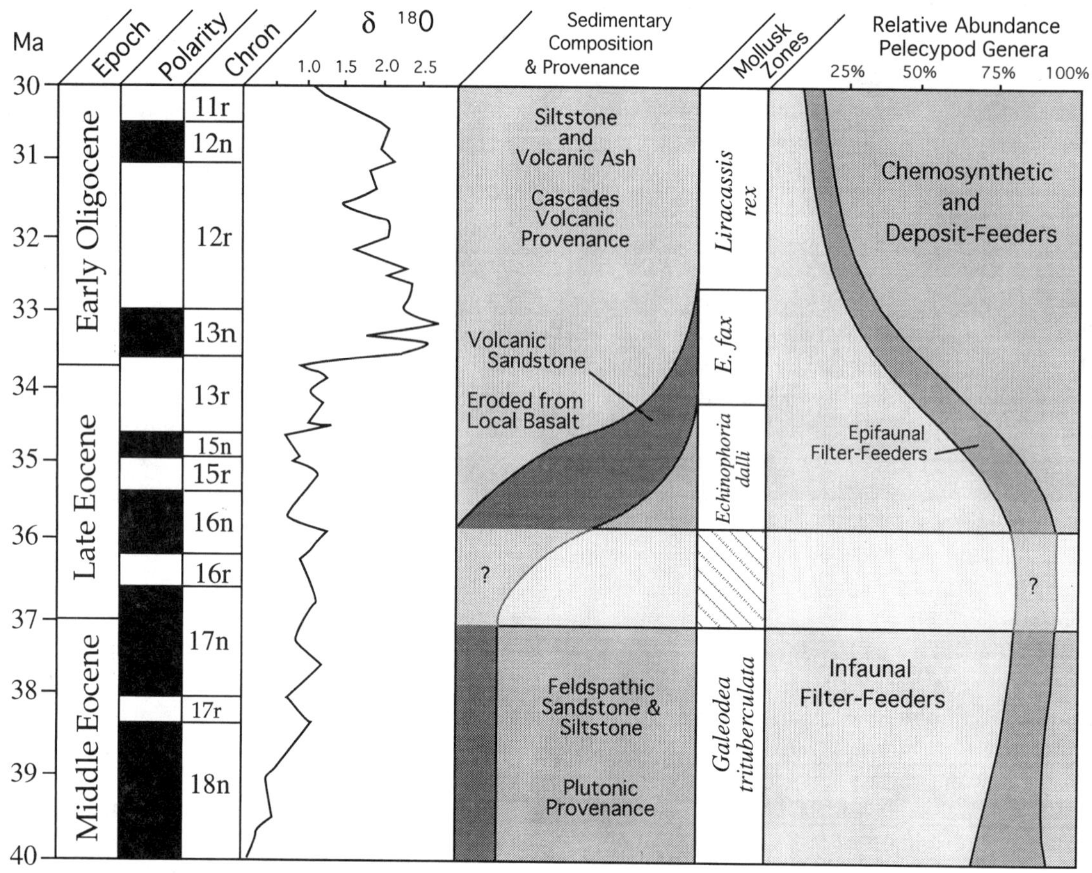

Figure 4.2. Relative abundances of bivalve trophic groups from the four Eocene-Oligocene biozones of the northeastern Pacific margin, plotted, benthic foraminiferal $\delta^{18}O$ isotope record, and the changing sediment on the shallow shelf. Deep-sea, benthic foraminiferal $\delta^{18}O$ isotope curves from the southern oceans modified from Miller et al. (1987) and Zachos et al. (1999).

the Western Cascades magmatic arc developed very close to the northeastern Pacific margin in the late Eocene, basaltic sands (locally derived), feldspathic sand, and silt substrates (continental bloc provenance) of the nearshore shelf were overlain by voluminous, organic-rich, tuffaceous mud substrates. The shallow-infaunal, robust, veneroid bivalve genera *Pitar* and *Spisula*, with the deeper burrowing, thin-shelled *Tivelina, Tellina*, and *Solen*, are the most common components of the late middle Eocene sediments and are present but in diminishing abundance through the late Eocene and early Oligocene. Rapid species changes of those genera that fall within the infaunal filter-feeding trophic group occurred at each biozonal boundary, but the same genera remain significant paleocommunity components through the entire 10 million years. As the shallow-shelf substrate changed, the dominance of filte- feeders is replaced by abundant bivalves that harbored endosymbiotic chemosynthetic bacteria. For this study, these are put into the same grouping as the nuculanid deposit-feeders, which similarly become more numerous in the fine-grained, organic-rich sediment. From this result, it is difficult to discern a response to cool water by the bivalve community, although this may have been an additional factor. The next step in this study is to investigate the $\delta^{18}O$ and $\delta^{13}C$ isotope signatures of these faunas.

Bambach and Bennington (1996) questioned the thesis that communities evolve, and suggested that the patterns documented in the literature still require explanation. Documenting changes in inner-neritic, soft-sediment, benthic paleocommunities through the 10 million years of the last major global greenhouse to icehouse event provides some of the answers. Bivalves and gastropods have different responses. Although most of the abundant bivalve genera persist through this time period, their species evolve rapidly. Composition of the replacement fauna shows that substrate is as important a physical stimulus to change as is water temperature. Gastropods, being mobile, are less controlled by substrate types and much better indicators of water temperatures.

ACKNOWLEDGMENTS

John Armentrout's unpublished dissertation and associated collection of the Lincoln Creek fauna formed the foundation of this study, and I appreciate this valuable resource. Years of successive class collections and associated field notes accumulated and curated in the Burke Museum, under the direction of V. S. Mallory from 1957 to 1985, provided additional data. I thank Don Prothero for all the work he has accomplished, and continues to pursue, on West Coast Cenozoic chronostratigraphy, which makes this study possible. I greatly appreciate the stimulating discussions generated at the Penrose Conference, especially among the molluscan paleontologists. I thank Kathy Campbell, Carol Tang, and Ellen Moore for their reviews, and Catherine Townsend for discussions and graphic design.

REFERENCES

Aberhan, M., 1994, Guild-structure and evolution of Mesozoic benthic shelf communities: Palaios, v. 9, p. 516–545.

Allmon, W. D., 1988, Ecology of recent turritelline gastropods (Prosobranchia, Turritellidae) current knowledge and paleontological implications: Palaios, v. 3, p. 259–248.

Allmon, W. D., 1990, Review of the *Bullia* Group (Gastropoda: Nassariidae) with comments on its evolution, biogeography and phylogeny: Bulletins of American Paleontology, v. 99, 171 p.

Armentrout, J. M., 1973, Molluscan biostratigraphy and paleontology of the Lincoln Creek Formation (late Eocene-Oligocene), southwestern Washington: [Unpubl. Ph.D. diss.]: Seattle, University of Washington, p. 749.

Armentrout, J. M., 1975, Molluscan biostratigraphy of the Lincoln Creek Formation southwestern Washington, *in* Weaver, D. W., Hornaday, G. R., and Tipton, A., eds., Paleogene Symposium and Selected Technical Papers: American Association of Petroleum Geologists, Pacific Section, p. 14–48.

Armentrout, J. M., 1981, Correlation and ages of Cenozoic chronostratigraphic units in Oregon and Washington, *in* Armentrout, J. M., ed., Pacific Northwest Cenozoic Biostratigraphy: Geological Society of America, Special Paper, v. 184, p. 137–148.

Bambach, R. K., and Bennington, J. B., 1996, Do communities evolve? A major question in evolutionary paleoecology, *in* Jablonski, D., Erwin, D. H., and Lipps, J. H., eds. Evolutionary paleobiology: Chicago, University of Chicago Press, p. 123–160.

Berggren, W. A., and Miller, K. G., 1988, Paleogene tropical planktonic foraminiferal biostratigraphy and magnetochronology: Micropaleontology, v. 34, p. 362–380.

Berggren, W. A., Kent, D. V., Swisher, C. C. III, and Aubry, M. P., 1995, Revised Cenozoic geochronology and chronostratigraphy: SEPM Special Publication, v. 54, p. 129–212.

Campbell, K. A., 1992, Recognition of a Mio-Pliocene cold seep setting from the Northwest Pacific convergent margin, Washington, U.S.A.: Palaios, v. 7, p. 422–433.

Cheney, E. S., 1994, Cenozoic unconformity-bounded sequences in central and eastern Washington: Washington Division of Geology and Earth Resources Bulletin, v. 89, p. 115–139.

Cox, L. R., and 24 others, 1969, Bivalvia, *in* Moore, R. C., ed., Treatise on paleontology, Part N, 952 p.

Dando, P. R., and Spiro, B., 1993, Varying nutritional dependence of thyasirid bivalves Thyasira sarsi and T. equalis on chemoautotrophic symbiotic bacteria, demonstrated by large isotope rations of tissue carbon and shell carbonate: Marine Ecology Progress Series, v. 92, p. 151–158.

Dando, P. R., Southward, A. J., and Southward, E. C., 1986, Chemoautotrophic symbionts in the gills of the bivalve mollusc Lucinoma borealis and the sediment chemistry of its habitat: Proceedings of the Royal Society of London, Series B Biological Sciences, v. 227, p. 227–247.

Dando, P. R., Ridgeway, S. A., and Spiro, B. 1994, Sulphide "mining" by lucinid bivalves molluscs, demonstrated by sulphur stable isotope measurements and experimental models: Marine Ecology Progress Series, v. 1107, p. 169-175.

Dockery, D. T., 1998, Molluscan faunas across the Paleocene Eocene boundary in the North American Gulf coast plain, *in* Aubry, M. P., Lucas, S, and Berggren, W. A., eds., Late

Paleocene–early Eocene climatic and biotic events in the marine and terrestrial record: Columbia University Press, New York, p. 269–322.

Durham, J. W., 1944, Megafaunal zones of the Oligocene of northwestern Washington. University of California Publications of the Department of Geological Sciences, v. 27, no. 5, p. 101–212.

Effinger, W. L., 1938, The Gries Ranch fauna (Oligocene) of western Washington: Journal of Paleontology, v. 12, p. 355–390.

Goedert, J. L., and Campbell, K. A., 1995, An early Oligocene chemosynthetic community from the Makah Formation, northwestern Olympic Peninsula, Washington: The Veliger, v. 38, p. 22–29.

Goedert, J. L., and Squires, R. L., 1990, Eocene deep-sea communities in localized limestones formed by subduction-related methane seeps, southwestern Washington: Geology, v. 18, p. 1182–1185.

Haasl, D. M., and Hansen, T. A., 1996, Timing of latest Eocene molluscan extinction patterns in Mississippi: Palaios, v. 11, p. 487–494.

Haq, B. U., Hardenbol, J., and Vail, P. R., 1988. Mesozoic and Cenozoic chronostratigraphy and cycles of sea-level changes, *in* Wilgus, C. K., et al., eds., Sea-level changes: an integrated approach: SEPM Special Paper 42, p. 71–180.

Heller, P. L., and Ryberg, P. T., 198,. Sedimentary record of subduction to forearc transition in the rotated Eocene basin of western Oregon: Geology, v. 11, p. 380–383.

Heller, P. L., Peterman, Z. E., O'Niel, J. R., and Shafiqllah, M., 1985, Isotopic provenance of sandstones from the Eocene Tyee Formation, Oregon Coast Range: Geological Society of America Bulletin, v. 96, p. 770–780.

Henriksen, D. A., 1956, Eocene stratigraphy of the lower Cowlitz valley-eastern Willapa Hills area, southwestern Washington: Washington Division of Mines and Geology Bulletin, 43, 122 p.

Hickman, C. S., 1976, Bathyal gastropods of the Family Turridae in the early Oligocene Keasey Formation in Oregon, with a review of some deep water genera in the Paleocene of the eastern Pacific: Bulletins of American Paleontology, v. 70, no. 292, 106 pp.

Hickman, C. S., 1980, Paleogene marine gastropods of the Keasey Formation in Oregon: Bulletins of American Paleontology, v. 78, no. 310, 112 p.

Hickman, C. S., 1984, Composition, structure, ecology and evolution of six Cenozoic deep water mollusk communities: Journal of Paleontology, v. 58, p. 1215–1234.

Irving, A. J., Nesbitt, E. A., and Renne, P. R., 1996, Age constraints on earliest Cascade arc volcanism and Eocene marine biozones from a feldspar rich tuff in the Cowlitz Formation, southwestern Washington: EOS, v. 77, p. F814.

Keen, A. M., 1971, Sea shells of tropical west America: Stanford, California, Stanford University of Press, 1064 p.

MacPherson, E., 1971. The marine molluscs of Arctic Canada: National Museum of Canada, National Museum of Natural Sciences Publications in Biological Oceanography, no. 3, 149 p.

Miller, D. M., Nilsen, T. H., and Bilodeau, W. L., 1992, Late Cretaceous to early Eocene geological evolution of the U.S. Cordillera, *in* Burchfiel, B. C., Lipman, P. W. and Zoback, M. L., eds., The Cordilleran orogen: conterminous U.S.: Geological Society of America, The Geology of North America, v. G-3, p. 205–222.

Miller, K. G., 1992, Middle Eocene to Oligocene stable isotopes, climates and deep-water history: the Terminal Eocene event, *in* Prothero D. R., and Berggren, W. A., eds., Eocene-Oligocene climatic and biotic evolution: New York, Princeton University Press, p. 160–77.

Miller K. G., Fairbanks R. G., and Mountain, S. G., 1987, Tertiary oxygen isotope synthesis, sea level history and continental margin erosion: Paleoceanography, v. 2, p. 1–9.

Nesbitt, E. A., 1982, Paleoecology and biogeography of Eocene marine assemblages from western North America [Unpubl. Ph.D. diss.]: Berkeley, University of California, 215 p.

Nesbitt, E. A., 1995, Paleoecological analysis of molluscan assemblages from the middle Eocene Cowlitz Formation, southwestern Washington: Journal of Paleontology, v. 69, p. 1060–1073.

Nesbitt, E. A., 1998, Marine fauna of the middle Eocene Tukwila Formation, King County: Washington Geology, v. 25, p. 13–19.

Nesbitt, E. A., Campbell, K. A., and Goedert, J. L., 1994, Paleogene cold-seeps and macroinvertebrate faunas in a forearc sequence of Oregon and Washington, in Swanson, D. A., and Haugerud, R. A., eds., Geologic field trips of the Pacific Northwest. Geological Society of America, pp. D1–11.

Payne, C. W., and Niem, A. R., 1996, Stratigraphy, lithofacies and bounding formations of the Eocene Cowlitz Formation, southwestern

Washington. Cordilleran Section Geological Society of America, Abstracts with Programs, v. 28, no. 3, p. 99.

Prothero, D. R., 2001, Chronostratigraphic calibration of the Pacific Coast Cenozoic, *in* Prothero, D. R., ed., Magnetic stratigraphy of the Pacific Coast Cenozoic: Pacific Section SEPM, no 91, p. 377–394.

Prothero, D. R., and Armentrout, J. M., 1985, Magnetostratigraphic correlation of the Lincoln Creek Formation, Washington: implications for the age of the Eocene/Oligocene boundary: Geology, v. 13, p. 208–211.

Prothero, D. R., and Berggren, W. A., eds., 1992, Eocene-Oligocene climatic and biotic evolution. Princeton New Jersey, Princeton University Press, 568 p.

Prothero, D. R., Jaquette, C. D., and Armentrout, J. M., 2001, Magnetic stratigraphy of the Eocene–Oligocene Lincoln Creek Formation, Porter Bluffs, Washington, *in* Prothero, D. R., ed., Magnetic stratigraphy of the Pacific Coast Cenozoic: Pacific Section SEPM, no 91, p. 169–178.

Roberts, A. E., 1958, Geology and Coal Resources of the Toledo–Castle Rock District Cowlitz and Lewis Counties, Washington: U.S. Geological Survey Bulletin, v. 1062, 71 p.

Rollins, H. B., West, R. R., and Busch, R. M., 1990, Hierarchical genetic stratigraphy and marine paleoecology, *in* Miller, W. III, ed., Paleocommunity temporal dynamics: the long-term development of multispecies assemblages: Paleontological Society Special Publication, n. 5, p. 273-308.

Schellenberg, S. A., 1999, Deep-ocean ostracod response to the Eocene-Oligocene transition: high-resolution records from the southern ocean ODP sites 744 and 689: Geological Society of America Abstracts with Programs, v. 31, p. 310.

Scott, R. W., 1976, Trophic classification of benthic communities, *in* Scott, R. W., and West, R. R., eds., Structure and classification of paleocommunities: Stroudsburg Pennsylvania, Dowden Hutchinson and Ross Inc., p. 29–66.

Smith, J. T., 1970. Taxonomic distribution and phylogeny of the cymatiid gastropods *Argobuccinum, Fusitriton, Mediargo* and *Priene*: Bulletins of American Paleontology, v. 56, no. 254, p. 287–447.

Snavely, P. D., and Wells, R. E., 1996, Cenozoic evolution of the continental margin of Oregon and Washington: U.S. Geological Survey Professional Paper, v. 1560, p.161–182.

Stanley, S. M., 1970, Relation of shell form to life habits of the Bivalvia (Mollusca): Geological Society of America Memoir, v. 125, 296 p.

Stanley, S. M., 1972, Functional morphology and evolution of byssaly attached bivalve mollusks. Journal of Paleontology, v. 46, p. 165–212.

Stanton, R. J. Jr., and Dodd, J. R., 1976, The application of trophic structure of fossil communities in paleoenvironmental reconstruction: Lethaia, v. 9, p. 327–342.

Stanton, R. J. Jr., and Dodd, J. R., 1997, Lack of stasis in late Cenozoic marine faunas and communities, central California: Lethaia, v. 30, p. 239–256.

Taylor, J. D., Morris, N. J., and Taylor, C. N., 1980, Food specialization and the evolution of predatory gastropods: Palaeontology, v.23, p. 357–409.

Thomas, E. 1989. Late Eocene-Oligocene faunal changes in deep-sea benthic Foraminifera, Maud Rise (Antarctica): Geological Society of America Abstracts with Programs, v. 21, no. 6, p. A88.

Van Winkle, K. E, 1918, Paleontology of the Oligocene of the Chehalis Valley, Washington: University of Washington Publication in Geology, v. 1, p. 69–97.

Weaver, C. E., 1937, Tertiary stratigraphy of western Washington and northwestern Oregon: University of Washington Publication in Geology, v. 4, 266 p.

Weaver, C. E., 1943, Paleontology of marine formations of Oregon and Washington: University of Washington Publications in Geology v. 5, 789 p.

Weaver, C. E., and 20 others, 1944, Correlation of the marine Cenozoic formation of western North America: Geological Society of America Bulletin, v. 55, p. 569–598.

Wells, R. E., and Heller, P .L., 1988, The relative contribution of accretion, shear and extension to Cenozoic tectonic rotation in Pacific Northwest: Geological Society of America Bulletin, v. 100, p. 325–338.

Wells, R. E., Niem, A. R., MacLeod, N. S., Snavely, P. D., and Niem, W. A, 1983, Preliminary geologic map of the west half of the Vancouver (Washington-Oregon) 1° by 2° quadrangle. U.S. Geological Survey Open-File Report 83–591, scale 1:250,000.

Wolfe, J. A., 1978, A paleobotanical interpretation of Tertiary climates in the northern hemisphere: American Scientist, v. 66, p. 694–703.

Zachos, J. C., Opdyke, B. N., Quinn, T. M., Jones, C. E., and Halliday, A. N., 1999, Early Cenozoic glaciation, Antarctic weathering, and seawater $^{87/86}Sr$; is there a link?: Chemical Geology, v. 161, p. 165–180.

CHAPTER 5

Evidence for Abrupt Eocene-Oligocene Molluscan Faunal Change in the Pacific Northwest

Carole S. Hickman

ABSTRACT

Abrupt changes occurred in marine molluscan faunas of the Pacific Northwest during the latest Eocene and earliest Oligocene. New data show that turnover is the important measure of faunal change rather than species diversity. Superimposed on a pattern of gradual decline in diversity over the 10-million-year transition are two precipitous regional extinctions followed immediately by rapid bursts of origination and/or first appearances of taxa. The two episodes of regional extinction separate three distinctive molluscan faunas: (1) a tropical Eocene fauna, (2) a turnover fauna, and (3) a recovery fauna. The first episode is at 36.5 Ma between the tropical Eocene fauna of the Cowlitz Formation and the turnover fauna of the Keasey Formation. The second is at approximately 32.5 Ma, between the turnover fauna of the Keasey Formation and the recovery faunas of the Pittsburg Bluff, Eugene, and Scappoose Formations. The most dramatic event is at 36.5 Ma between the Cowlitz (145 species) and Keasey (109 species) faunas, where there is 100% regional extinction and 100% regional origination, respectively. Other features of the overall pattern of rapid faunal turnover at 36.5 Ma include (1) loss of thermophilic taxa; (2) appearance of cryophilic precursors of the modern temperate fauna; (3) appearances and disappearances of taxa characteristic of specific active tectonic settings; (4) origins of four chemosymbiotic mollusk communities associated with ancient methane seeps; (5) heightened faunal exchange with Asia; (6) appearance of regional endemic genera; (7) appearance of a distinctive bathyal facies fauna; and (8) changes in faunal abundance and structure.

Although global climate deterioration may play an important role in the faunal transition in the Pacific Northwest, the main turnover event precedes the negative oxygen and positive carbon isotopic excursions at 33.5 Ma that are inferred to mark abrupt global cooling. Regional tectonic change is more strongly and directly reflected in the structural and sedimentary features of strata across the transition interval, and more closely correlated with the observed changes in fauna.

INTRODUCTION

A molluscan faunal transition from warm-water Eocene faunas to cold-water Oligocene faunas has been recognized on the West Coast of North America for nearly a century. Ideas about the tempo and mode of transition have changed with the rise of modern macroevolutionary theory, and paleoecological interpretations have changed with new systematic knowledge of the molluscan faunas. For paleontologists, the most serious impediment to understanding the transition has been the lack of a framework in which biostratigraphic and lithostratigraphic units have an numerical chronostratigraphic (radiometric) basis. In the Pacific Northwest, progress in dating and correlation has been impeded further by poor exposures of thick marine sequences, deposited in localized basins that promoted the evolution of endemic faunal elements.

Although correlation of Oligocene stratigraphic units on the Pacific Coast has always been problematic in contrast to the Eocene and Miocene (Moore, 1976), the greatest deterrent to documenting the Eocene-Oligocene transition has been dissension over the position of the boundary, and, indeed, dissension as to whether there are any rocks at all of Oligocene age on the Pacific Coast. William Healey Dall (*in* Diller, 1896) was the first to suggest that there were faunas of Oligocene age, and Arnold (1906), Clark (1918), and others subsequently assigned the unique faunas between the "Tejon" (upper Eocene) and Monterey (Miocene) to the Oligocene. However, the matter was not then settled, and Eames et al. (1962:49), in a major work on worldwide Paleogene correlation, stated that there is no Oligocene on the West Coast of North America. Subsequent biostratigraphic data generated by monographs of "Oligocene" faunas and taxa from the Keasey (Hickman, 1976a), Eugene (Hickman, 1969) and Pittsburg Bluff (Moore, 1976) Formations, and biostratigraphic revision of the Lincoln Creek Formation (Armentrout, 1973, 1975), could not be used to explore patterns of faunal change in a global context until calibrated by independent stratigraphies. The evidence for faunal change was good, but the events were not fixed in time.

The redefined Cenozoic biostratigraphic framework of Armentrout (1981) was a first step in refined calibration. It was followed by the Berggren et al. (1985, 1992, 1995) time scales, based on magnetostratigraphic calibrations that further increased the precision of correlation of provincial biochronologies to the global time scale. Higher resolution was required for rocks of the Eocene-Oligocene transition, and that has been provided by Donald Prothero and his students, using magnetostratigraphy to calibrate the key sections containing the fossil record of the transition in Washington and Oregon (Prothero and Armentrout, 1985; Prothero and Donohoo, 2001; Prothero and Hankins, 1999, 2001; Prothero et al., 2001a, 2001b, 2001c; and Prothero, 2001).

The primary objective of this paper is to characterize the Eocene-Oligocene molluscan faunal transition in the Pacific Northwest in a global temporal context using the magnetostratigraphic framework. The patterns of faunal diversity and turnover are based on data from the Cowlitz, Keasey, Eugene, Pittsburg Bluff, and Scappoose Formations, spanning an interval of approximately 10 million years. Data are compiled here for the first time for entire faunas in order to capture the broad outlines of change. Nesbitt (1999 and this volume) provides a contrasting perspective for a shorter time interval that is restricted to shallow-water faunas in the Pacific Northwest.

The secondary objective is to integrate the new perspective on faunal change (1) with changing perspectives on climatic deterioration in the northeastern Pacific during the Eocene-Oligocene transition and (2) with changing perspectives on the tectonic setting in which the evidence for faunal change is preserved. The Eocene-Oligocene molluscan fossil record is best preserved and most intensively studied in passive margin sequences. The active margin of the Pacific Northwest provides an excellent opportunity to examine faunal change in the context of major regional tectonic events that are strongly and directly reflected in the petrofacies and sedimentary features of the rocks that contain the molluscan faunas.

The paper begins with a brief review of changing perspectives on the nature of climatic cooling and its faunal consequences in the northeastern Pacific, followed by a brief review of changing perspectives on the tectonic setting of the Eocene-Oligocene transition in the Pacific Northwest. The faunal results are presented both as quantitative data on changes in diversity and rates of extinction, origination, and turnover, and as qualitative patterns. Some of the qualitative patterns implicate a strong role for climatic deterioration, while others implicate a predominance of tectonic control.

CHANGING PERSPECTIVES ON CLIMATIC AND FAUNAL TRANSITION IN THE NORTHEASTERN PACIFIC

Ideas of Progressive Cooling and Gradual Loss of Faunal Diversity

Throughout the world, molluscan faunas have been noted for their high diversity and tropical appearance during the Eocene, followed by a drop in diversity in the Oligocene, with concomitant loss of tropical faunas at higher latitudes. On the Pacific Coast of North America, Arnold (1909), Smith (1910, 1919), and Dickerson (1917) were among the first to infer an onset of climatic cooling at the close of the Eocene, reflected in the loss of what they considered to be tropical faunal elements, decrease in faunal diversity, loss of cosmopolitan Eocene taxa, and the disappearance of geographically widespread faunal associations. Durham (1950, 1954) further expanded the inference of post-Eocene cooling. These earlier interpretations viewed the cooling as gradual and progressive, beginning at high latitudes and increasingly restricting thermophilic taxa to low latitudes. Progressive latitudinal differentiation of faunas has been expressed as an increase in the number of faunal provinces (Hall, 1964). The single Pacific Coast Eocene molluscan fauna recognized by Clark and Vokes (1936) presents a striking contrast to as many as eight modern molluscan provinces and subprovinces between 20° and 60°N (Hall, 1964; Valentine, 1966).

Addicott (1968, 1970a, b) altered the model of continuous, progressive intensification of the latitudinal thermal gradient by noting a prominent reversal of the Oligocene cooling trend and a return in the Miocene to marine climates substantially warmer than at present. Although Addicott's interpretation of climatic change was gradualistic, he clearly identified the Oligocene as an anomalous interval of low temperatures in shallow water. He was especially perplexed by the fact that the anomaly was more pronounced in central California (34–37°N), where generic diversity drops below that of Oregon and Washington (Addicott, 1970a). To explain the latitudinal component of the anomaly, he invoked a local oceanographic and paleogeographic hypothesis of upwelling along a straight segment of coast (Addicott, 1970a).

The Shift to Ecological Thinking and a Punctuational View of Global Cooling

The notion of an abrupt and potentially global cooling event emerged in the 1970s, when a new effort was launched (1) to document key undescribed Pacific Coast mollusk faunas, (2) to monograph key molluscan evolutionary lineages, and (3) to designate molluscan communities and characterize ecological relationships. Hickman (1969) had noted the abrupt origin of cold-water faunal lineages, the invasion of Asian cold-water lineages, and the origin and diversification of several endemic genera during what was identified as an interval of climatic instability and transition. This late Eocene to early Oligocene interval was represented in the Pacific Northwest in both the Eugene Formation (Hickman, 1969) and the Keasey Formation (Hickman, 1976a, 1980). Abrupt disappearances and appearances of generic-level taxa emerged in other monographic studies. For example, Hickman's (1976a) review of Paleogene gastropods of the family Turridae documented the sudden appearance of 10 genus-level taxa, 4 of them new. In his treatment of Cenozoic gastropods of the family Naticidae, Marincovich (1977) noted the greatest extinction of taxa at the end of the Eocene, when the number of species dropped from 13 to 5. Smith (1970) documented abrupt appearances of modern cryophilic genera of cymatiid gastropods at the beginning of the Oligocene.

Paleoecological studies further supported the hypothesis of a more abrupt transition. Changes in community structure and architecture during critical intervals in life history are often more revealing than are more traditional tabulations of taxa and their stratigraphic ranges. The late Eocene communities designated by Nesbitt (1995) differ dramatically in physiognomy, trophic structure, and inferred interactions from the superposed late Eocene to early Oligocene communities defined by Hickman (1984). Abrupt change at the community level is so profound that it can be recognized intuitively in the field without detailed knowledge of the taxa.

CHANGING PERSPECTIVES ON THE TECTONIC SETTING OF THE EOCENE-OLIGOCENE TRANSITION

The Cenozoic tectonic history of the active continental margin of the Pacific Northwest has been a subject of intensive study, debate, and periodic review (e.g., Dickinson, 1976; Ewing, 1980; Heller et al., 1987; Duncan and Kulm, 1989; Yeats et al., 1996; and Snavely and Wells, 1996). The overall tectonic trend that affects the Eocene-Oligocene transition is Paleogene convergence and oblique subduction of the oceanic Kula and Farallon plates beneath the North American plate (figure 5.1A,B). The modern Coast Range Block (figure 5.1C) contains the transition faunas and sediments that accumulated in a forearc basin between the ancestral Cascade volcanic arc and the subducting Farallon and Kula plates. The sediments were deposited on an accreted basaltic basement terrane that formed offshore during the Paleocene and early Eocene as an archipelago of oceanic islands and seamounts capping oceanic basalts (62–56 Ma). Although several models have been proposed for formation of the Coast Range basement basalts (e.g,. Duncan, 1982; Wells et al., 1984), most invoke a combination of rifting along the Kula-Farallon plate boundary and the presence of a hotspot close to the margin of the North American plate (Duncan and Kulm, 1989). The island chain is inferred to have accreted by 50 Ma (Heller and Ryberg, 1983; Snavely and Wells, 1996) (figure 5.1A).

Following accretion of the basaltic basement, a subsiding forearc basin developed to the west of the pre-Cascade Challis volcanic arc (figure 5.1B). Massive deltaic sequences accumulated in this basin from source areas in the newly uplifted Klamath Mountains in the southwest and the Idaho Batholith to the northeast. Petrologic features of early and middle Eocene sediments record a dramatic change at 50 Ma from subduction zone lithic facies to forearc-basin arkosic facies (Heller and Ryberg, 1983; Heller

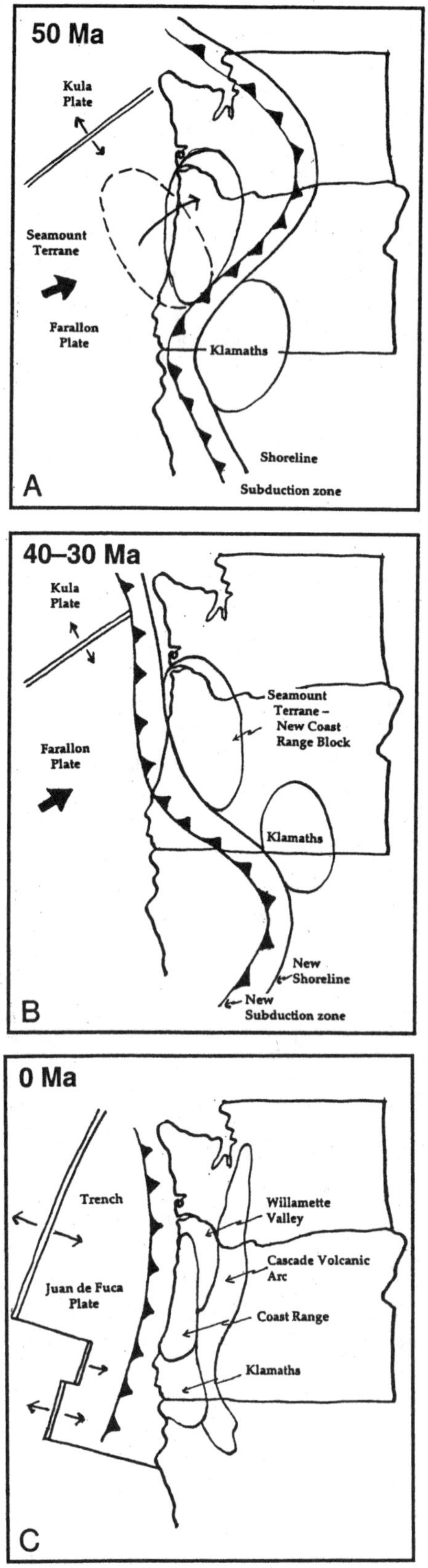

et al., 1987). The late Eocene tropical fauna of the Pacific Northwest is best preserved in the Cowlitz Formation (Nesbitt, 1995), a thick deltaic sequence that formed as the delta prograded into the forearc basin (Niem et al., 1992). The estimated thickness of the sequence is between 1,300 and 1,500 m (Henriksen, 1956).

The early Eocene trench and subduction zone shut down at approximately 48 Ma, after accretion of the oceanic seamount terrane (Heller et al., 1987). Two subsequent tectonic events affected the sedimentary habitats of the transition faunas: activation of a new middle to late Eocene spreading center and trench farther offshore (figure 5.1B) and an associated westward shift in the position of the volcanic arc (Heller et al., 1987). The shift coincides with the change from arkosic to volcaniclastic deposition (Heller and Ryberg, 1983) and the onset of Cascade arc volcanism (Snavely and Wells, 1996). The earliest well-dated evidence of Cascade arc volcanism is a discrete ash bed in the upper Eocene Cowlitz Formation in Washington at 38.9 Ma (Irving et al., 1996) in the deltaic complex inhabited by the tropical fauna.

The volume of ash was not great at the outset, and the most significant change in lithology occurs with the beginning of voluminous Cascade volcanism between 36 and 37 Ma. The deposition of massive tuffaceous siltstones coincides with the disappearance of the tropical fauna and appearance of the turnover fauna. The massive tuffaceous units containing the turnover fauna are the Eugene (Hickman, 1969) and Keasey Formations (Hickman, 1976a) in Oregon and the Lincoln Creek Formation (Armentrout, 1975) in Washington. Rapid subsidence and continued accu-

Figure 5.1. Generalized tectonic history of the Pacific Northwest, showing major features at 50 Ma, between 30 and 40 Ma, and at present. (A) Collision and subsequent rotation of the early Eocene seamount terrane to form the ancestral Coast Range Block. (B) Establishment of the new trench to the west of the ancestral Coast Range Block. (C) Modern setting with generalized positions of the Cascade arc, Willamette Valley, Coast Range Block, Klamath terrane, and subducting Juan de Fuca Plate. Modified from Dickinson, 1976; Heller and Ryberg, 1983; Wells and Coe, 1985; Heller et al., 1987; Wells and Heller, 1988; Duncan and Kulm, 1989; and Wells, 1990.

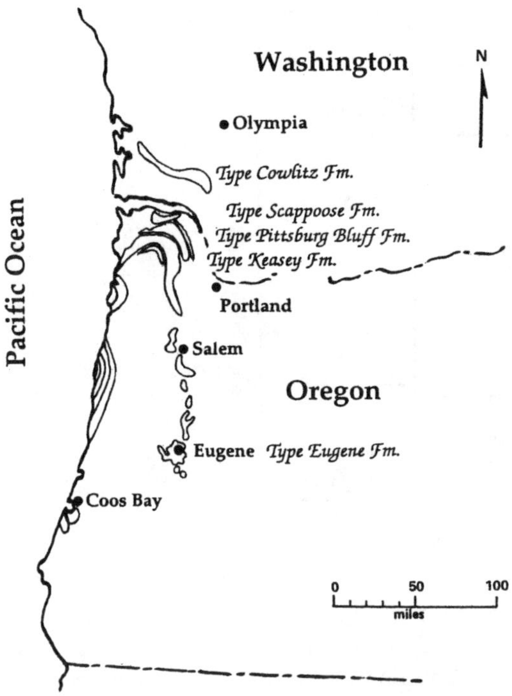

Figure 5.2. Geographic setting of the Eocene-Oligocene formations in Oregon and Washington containing the tropical, turnover, and recovery faunas. The areas containing the type sections of the Cowlitz, Keasey, Eugene, Pittsburg Bluff, and Scappoose Formations are labeled. Unlabeled coastal outcrops include the Nye, Yaquina, Alsea, Nestucca, Tunnel Point, and Bastendorf Formations as well as undifferentiated Eocene and Oligocene rocks. Modified from Hickman, 1969.

mulation of thick tuffaceous siltstone and pumiceous sandstone continued into the Oligocene and is characteristic of sedimentary settings in which the temperate recovery faunas are preserved, including the Pittsburg Bluff and Scappoose Formations in Oregon.

This brief summary of the tectonic evolution of the Pacific Northwest emphasizes events with consequences for basin development: changes in sedimentary settings, rates, and depositional facies that are correlated directly with faunal turnover. There are other features of the complex accretionary history of the Pacific Northwest that may eventually be linked with faunal change. It is therefore appropriate to note that unresolved aspects of tectonic history have led some authors to identify the entire Cascade arc subduction complex as "an extreme end member in the spectrum of convergent plate margins" (Duncan and Kulm, 1989). Of particular interest is the evidence of widespread and substantial (15–80°) clockwise tectonic rotation during the Cenozoic history of the arc. At least three models have been proposed for the tempo and mode of rotation (Simpson and Cox, 1977; Magill et al., 1981). Many more models have been proposed to explain the underlying mechanisms (see Wells and Coe, 1985; Wells and Heller, 1988; Wells, 1990), although oblique subduction appears to be the untimate driving force (Wells and Heller, 1988). There is debate about the number of crustal blocks that were involved as well as the extent to which they may have been translated, as well as rotated (Bates et al., 1981). Although tectonic rotation would not have been significant on an ecological time scale or had a direct ecological effect, too little is known about the processes involved to discount a role in faunal turnover through indirect effects.

A NEW CONCEPTUAL FRAMEWORK FOR ANALYSIS OF FAUNAL TRANSITION

Patterns of faunal change during critical intervals in Earth history are complex and varied. A new conceptual framework is proposed for analysis of Eocene-Oligocene molluscan faunal change in the Pacific Northwest. The framework contains five elements. The first is a set of five fossiliferous formations that are relatively well exposed in northwestern Oregon and southwestern Washington and which span the 10-million-year interval of the Eocene-Oligocene transition. The geographic distribution of the five formations is illustrated in figure 5.2. The second is recognition of three distinctive faunas, separated by two faunal turnover events. The third is correlation of the formations containing the three faunas with one another and to the global time scale using magnetostratigraphic data. The fourth is the recognition and chronostratigraphic fixation of regional tectonic events that are recorded in these formations. The fifth is correlation of the global isotopic excursions that have been invoked as climatic events to which faunas have responded in other parts of the world. The five formations, three faunas, magnetic polarity and absolute time scales, tectonic events, and isotopic excursions are illustrated in figure 5.3.

The Three Faunas and Their Occurrence

Three distinctive faunas are recognized here for the first time: an Eocene tropical fauna, a turnover fauna, and an Oligocene recovery fauna. They are defined

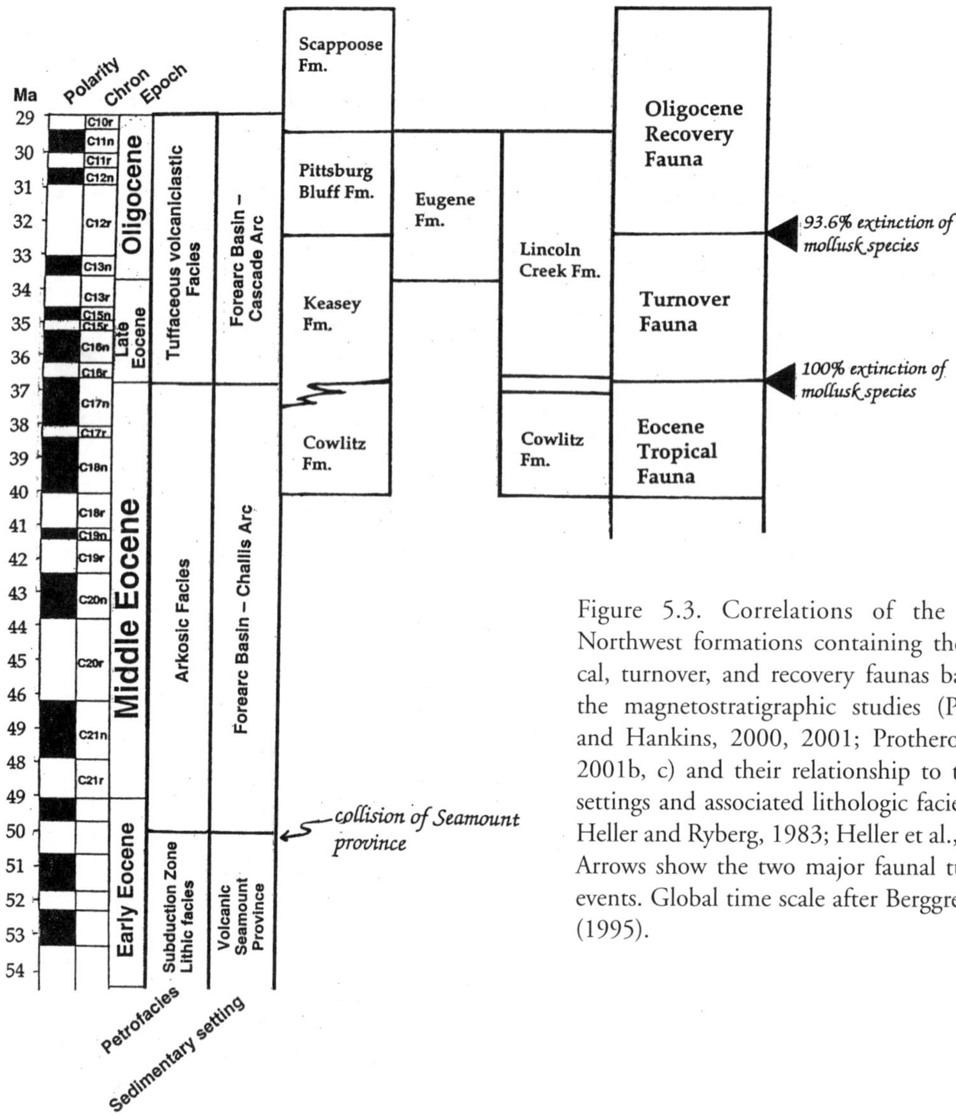

Figure 5.3. Correlations of the Pacific Northwest formations containing the tropical, turnover, and recovery faunas based on the magnetostratigraphic studies (Prothero and Hankins, 2000, 2001; Prothero et al., 2001b, c) and their relationship to tectonic settings and associated lithologic facies (after Heller and Ryberg, 1983; Heller et al., 1987). Arrows show the two major faunal turnover events. Global time scale after Berggren et al. (1995).

on the basis of large turnover of species (>>75%) documented below. The three faunas and their relationship to the formations in which they occur is illustrated in figure 5.3. The Cowlitz Formation in southwestern Washington and northwestern Oregon contains the Eocene tropical fauna. The overlying Keasey Formation in Oregon contains the turnover fauna and above the Keasey, the Pittsburg Bluff, Eugene, and Scappoose Formations in Oregon contain the Oligocene recovery fauna.

In the Pacific Northwest, the Eocene tropical fauna is not well represented outside of the Cowlitz Formation. It occurs in the Tejon Formation in California, and other formations including a portion of the Kreyenhagen, the Markley, and the Coldwater Sandstone. The turnover fauna is also present in the basal beds of the Eugene Formation, the lower portion of the Lincoln Creek Formation, and the Gries Ranch Formation of southwestern Washington and northwestern Oregon. In California the turnover fauna occurs in the Rices Mudstone Member of the San Lorenzo Formation. Elements of the turnover fauna also occur in the Nestucca Formation on the Oregon coast. The recovery fauna is clearly represented in the Sooke Formation on Vancouver Island, the Blakeley Formation in Washington, and the Tunnel Point Sandstone and the Yaquina and Alsea Formations in Oregon.

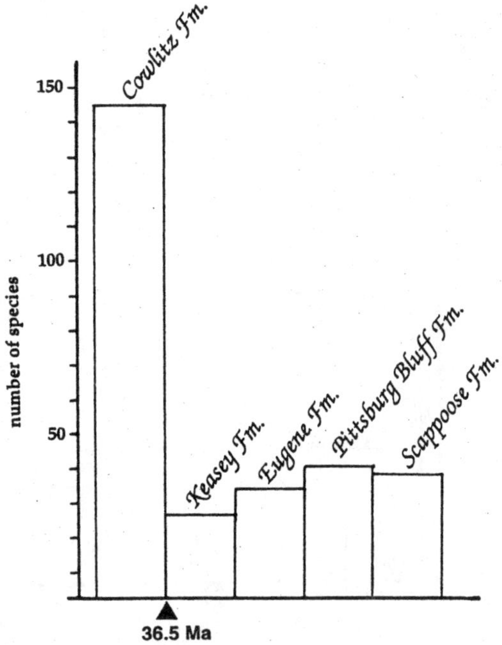

Figure 5.4. Molluscan species diversity of the Cowlitz, Keasey, Pittsburg Bluff, and Scappoose Formations compiled from sources published prior to 1950, showing a crash in diversity at 36.5 Ma, followed by gradual recovery to a much lower level.

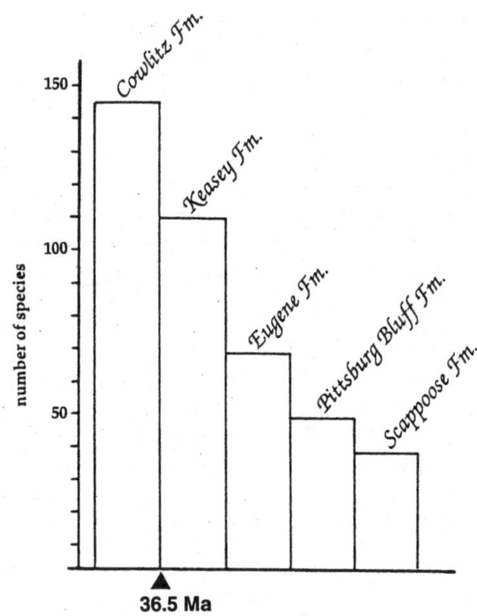

Figure 5.5. Molluscan species diversity of the Cowlitz, Keasey, Pittsburg Bluff, and Scappoose Formations compiled to include post-1950 data from monographs and unpublished sources, showing a progressive stepwise decline in diversity through he Eocene-Oligocene transition.

Magnetostratigraphy and Chronostratigraphy

The new chronostratigraphic element of the conceptual framework for assessment of faunal change is based on magnetostratigraphic data from the five key formations in the Pacific Northwest that contain the three distinctive faunas. Magnetostratigraphy of the Cowlitz Formation is based on Prothero et al. (2001c). Magnetostratigraphy of the Keasey Formation is based on Hankins and Prothero (1999), and Prothero and Hankins (2000). Magnetic stratigraphy of the Pittsburg Bluff Formation is based on Prothero and Hankins (2001). Additional magnetostratigraphic data for the Lincoln Creek, Tunnel Point, Alsea, Yaquina, and Nye Formations also are from the work of Prothero and his colleagues (Prothero and Armentrout, 1985; Prothero et al, 2001a, 2001b; Prothero and Donohoo, 2001). Correlation of the normal and reversed polarity intervals recorded in these formations with the global time scale is shown in figure 5.3.

Correlation of Faunal and Tectonic Events with Isotopic Excursions

Three distinctive tectonic regimes and three distinctive lithofacies are closely correlated with the tropical, turnover, and recovery faunas (figure 5.3). The major abrupt global climate shift at 33.5 Ma, inferred from the high-resolution foraminiferal oxygen isotope curves of Zachos et al. (1996), does not coincide with the main tectonic shift and main faunal turnover event in the Pacific Northwest, both of which occurred at 36.5 Ma, approximately 3 million years earlier (figure 5.3). Because the $\delta^{18}O$ and corresponding $\delta^{13}C$ excursions occur after the major episode of regional extinction and origination, they can be eliminated from theories of causation.

ANALYSIS OF PATTERNS OF FAUNAL CHANGE IN THE PACIFIC NORTHWEST

Quantitative Data

Species diversity. Data compiled for this paper from diverse sources published prior to 1950 show a precipitous drop in species diversity between the middle Eocene fauna of the Cowlitz Formation and the upper Eocene to lower Oligocene faunas of the Keasey, Eugene, and Pittsburg Bluff Formations (figure 5.4). As of 1950, 145 mollusk species had been described or reported from the Cowlitz Formation, and it was the only fauna that had been studied intensively. Only 26 species had been reported from the overlying Keasey Formation. Literature records for the Eugene (33 species) and Pittsburg Bluff (30 species) Formations suggested a slight rise in diversity following the initial drop.

How well has this pattern held up with subsequent research? No new species have been described from the Cowlitz Formation, but monographic studies beginning in the 1960s altered the initial impression of low diversity in the Eugene, Keasey, and Pittsburg Bluff Formations. Hickman's (1969) monograph of the Eugene Formation more than doubled the number of species, bringing the total to 67. Even more surprising is the effect of intensive study of the sparsely fossiliferous underlying Keasey Formation. Hickman (1976a) described 12 new species of turrid gastropods from the Keasey Formation, and subsequently (Hickman, 1980) named an additional 23 new species of gastropods, bringing the total number of Keasey gastropods to 68. A monographic study of the Keasey bivalves is now complete (Hickman, unpublished), and there are 40 bivalve species, 22 of them new. This increases the known diversity of the Keasey molluscan fauna to 109 species. The pattern of diversity now shows a gradual decline from the tropical Eocene fauna to the Oligocene, rather than a precipitous drop during the late Eocene transition period (compare figures 5.4 and 5.5).

Although intensive collecting and monographic treatment of the Keasey and Eugene faunas resulted in marked increases in species diversity, an equally intensive monographic effort on the fauna of the Pittsburg Bluff Formation (Moore, 1976) resulted in a more modest increase from 39 to 48 mollusk species. The fauna of the Scappoose Formation, which is currently known to contain 37 mollusk species, has not been monographed. However, large bulk samples of the fauna from new localities (Hickman, unpublished) suggest that the old number is not likely to change considerably. The tropical, turnover, and recovery faunas show a gradual decline in species diversity.

Generic diversity. Another surprising result from new research on the molluscan faunas of the Eocene-Oligocene transition is that the tropical Cowlitz

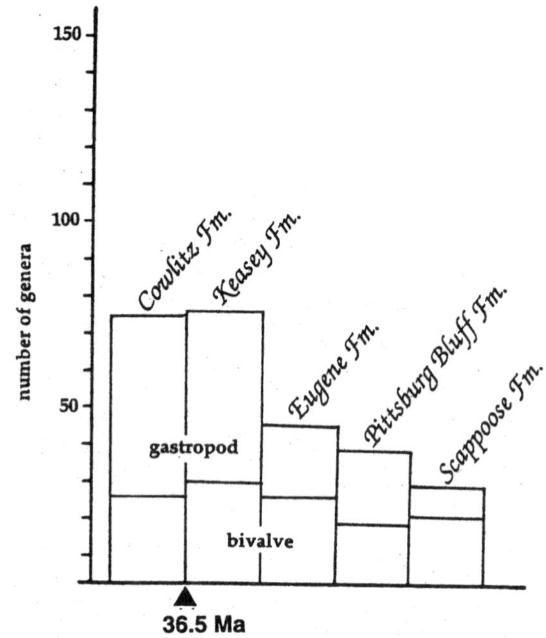

Figure 5.6. Molluscan generic diversity as it is currently known, illustrating a pattern that differs from that of species diversity. There is no decline in generic diversity at 36.5 Ma between the tropical Eocene fauna and the turnover fauna, but there is a subsequent decline in generic diversity in the recovery faunas.

fauna and late Eocene–early Oligocene Keasey fauna are almost equivalent in generic diversity (figure 5.6). The Keasey fauna has the highest bivalve diversity of any of the faunas during the transition. The drop in generic diversity is between the Keasey fauna and the faunas of the Eugene, Pittsburg Bluff, and Scappoose Formations. Most of the decline in generic diversity is in the gastropod component.

Extinctions, originations, and faunal turnover. Species diversity is not providing the critical information about faunal change; instead, the dramatic event between the Cowlitz and Keasey is one of turnover. At the species level there is 100% turnover. Even at the generic level there is a massive turnover. Only 4 genera of bivalves and 15 genera of gastropods persist from the Cowlitz Formation into the Keasey Formation. This results in 74% generic extinction for mollusk genera in the Cowlitz Formation (85% for the bivalves and 62% for the gastropods). In the Keasey Formation 87% of the bivalve genera and

67% of the gastropods are new, and the total for mollusks is 75% new genera.

Turnover is also striking between the Keasey and the Pittsburg Bluff Formations. Only seven mollusk species persist from the Keasey into the Pittsburg Bluff (94% extinction). Thirteen species persist into the Eugene Formation (88% extinction), and the only reason that persistence is this high is that the basal beds of the Eugene Formation are coeval with the upper portion of the Keasey Formation. Several species that are present in the basal beds of the Eugene Formation do not persist into the Oligocene portion of the formation where the typical recovery fauna occurs.

The turnover between the Pittsburg Bluff and overlying Scappoose Formations is substantially less. Fourteen of the 18 bivalve genera in the Pittsburg Bluff persist into the Scappoose (22% extinction), although only 4 of the 20 gastropod genera survived (80% extinction). The pattern of turnover in the gastropods is particularly interesting because it is accompanied by a shift in the Scappoose fauna to low diversity and high abundance of nonherbivorous caenogastropods (discussed further below).

Qualitative Data

Demise of the tropical Eocene fauna. The disappearance of tropical, thermophilic taxa at the end of the Eocene was first noted by Smith (1919), and has been reiterated by many subsequent authors. Vokes (1935) and Clark and Vokes (1936) saw connections between the tropical Eocene faunas of California and those of the Paris Basin in genera such as *Harpa*, *Voluta*, *Nerita*, and *Pseudoliva*, in the presence of the neritid species *Velates perversus* in both faunas, and the occurrence species of "species pairs," congeners that were so similar in their morphology as to indicate a close phylogenetic relationship. Species pairs between the Paris Basin and western North America were recognized in the following genera: *Potamides*, *Turritella*, *Typhus*, *Cuneocorbula*, *Cryptochordia*, *Cryptoconus*, *Eocypraea*, *Keilostoma*, and *Lyria*.

Other authors associated the demise of the tropical fauna primarily with a decline in taxonomic diversity (Keen and Bentson, 1944; Addicott, 1970b). Watkins (1974) recognized a separate "Northern Province" in the Pacific Northwest by the beginning of the Oligocene, noting that the fauna of the Pittsburg Bluff Formation contained fewer than 17% thermophilic genera.

The demise of the tropical Eocene fauna in the Pacific Northwest is marked by a change in patterns of calcification. As noted by Vermeij (1978) tropical taxa often have "large, thick, heavily sculptured shells." Large thick-shelled *Venericardia* disappear, along with spinose and delicately ornamented gastropods, including the frilly muricids, the Eocene genus *Ficopsis*, and three endemic subgenera of the stromboidean genus *Ectinochilus*. *Ficopsis* and *Ectinochilus* are two of the most distinctive elements of the Cowlitz fauna. The late Eocene tropical fauna also contained a diverse assemblage of cerithoidean species that disappeared with the demise of the tropical fauna. Extinction of large-shelled taxa included the genus *Pachychromium*, which Marincovich (1977) considered one of the most distinctive Cenozoic naticids of western North America.

Origins of cryophilic taxa. The Keasey Formation marks the origin of the turnover fauna in the Pacific Northwest, with the appearance of genera that typically occur in cool-temperate habitats today, and are therefore considered indicative of cooling. Cryophilic bivalve originations include *Truncacila*, *Panope*, *Thracia*, *Crenella*, *Cardita*, and *Moerella*. Cryophilic gastropod originations include the cymatiid genera *Fusitriton* and *Argobuccinum*, the trochid genus *Margarites*, and the boreal muricid genus *Trophonopsis*.

Additional cryophilic genera make their first appearance in the Oligocene recovery fauna, and several of the cool-water taxa that originated in the turnover fauna diversify and become abundant in the recovery fauna. These include *Cryptonatica* and striking, large-shelled turrids of the genus *Aforia*. Other genera that make their first appearance in the turnover fauna, such as *Bathybembix*, *Phanerolepida*, and *Acesta*, are often considered more indicative of deep (bathyal) water rather than cold water (Hickman, 1972, 1984). A method of distinguishing deep-water communities from communities living in cold shallow water is outlined by Hickman (1984).

Associations with tectonic settings. During the Eocene-Oligocene transition, there are two peculiar molluscan faunal elements that are associated with changing tectonic settings in the Pacific Northwest. The first is the association of large pleurotomariid gastropods with the early to middle Eocene oceanic basaltic terrane that was accreted onto the North American continent at 50 Ma. The occurrence of pleurotomariids in limestone interbeds in oceanic basalt of the Coast Range Block is reviewed by Hickman (1976b), who interpreted the limestones as sedimentary cappings on seamounts and submarine banks and noted the similarity of occurrence to the settings in which pleurotomariids occur off eastern Japan today (Hickman, 1976b; Hickman and Lindberg, 1984).

A second molluscan peculiarity correlated with tectonic change is the brief appearance in the turnover fauna of an entire guild of limpets that no longer occurs on the northeastern Pacific margin. Acmaeid limpets comprise the Neogene and Holocene guilds on rocky shores, but the Keasey Formation and coeval Gries Ranch Formation contain all three elements of the modern Australian limpet guild: *Patelloida* (a patelloidid), *Cellana* (a nacellid), and *Siphonaria*. (Hickman and Lindberg, 1984; Lindberg and Hickman, 1986). *Siphonaria*, an air-breathing pulmonate limpet, occurs in the Panamic Province today, but species of *Cellana* are almost exclusively associated with basalt substrates, and the nearest living fauna in which nacellids and siphonariids co-occur is in the Hawaiian chain (Lindberg and Hickman, 1986).

Origins of chemosymbiotic mollusk communities. Cold-seep carbonates are a prominent feature of many forearc settings in western North America, and there is a prominent clustering of seep occurrences in the Pacific Northwest during the Eocene-Oligocene transition (Goedert and Squires, 1990; Squires and Goedert, 1991; Campbell, 1992; Campbell and Bottjer, 1993; Nesbitt et al. 1994; Goedert and Campbell, 1995; Squires, 1995). The formation of carbonate mounds, chimneys, and small nodular carbonate masses is correlated with the onset of Cascade arc volcanism (Nesbitt et al., 1994; Niem et al., 1994), with inferred redirection and intensification of plate convergence, and with strong pulses of regional fluid venting through forearc accretionary prisms (Campbell and Bottjer, 1993, 1995; Nesbitt et al., 1994). Similar deposits are forming today off the Oregon coast, where methane-rich fluid venting is concentrated at compression-related faults at approximately 2,000 m water depth over the subduction zone complex (Kulm et al., 1986; Kulm and Suess, 1990). Associated with these authigenic carbonates are assemblages of relatively abundant mollusks in lineages whose modern representatives all house chemosynthetic bacteria within their gills. Four distinctive chemosymbiotic community types or associations make their first appearance in the Pacific Northwest as part of the turnover fauna.

Thyasira-Lucinoma-Solemya communities (Hickman, 1984) are characterized by low diversity and abundance of species of three bivalve genera in which nutrition is partially or exclusively provided by endosymbiotic bacteria in the gills. Modern communities dominated by these bivalves are typically chemoautotrophic and dependent upon a source of reduced sulfur. The communities that appear in the Eocene-Oligocene transition fauna occur at inferred methane seeps (Campbell and Bottjer, 1993; Nesbitt et al., 1994; Niem et al., 1994), suggesting that they housed methanotrophic bacteria (i.e., heterotrophic nutrition based on an organic compound, methane) rather than thiotrophic bacteria (i.e., autotrophic nutrition based on reduced sulfur, an inorganic compound) (see Hickman, 1994, for a discussion).

Thyasira-Lucinoma-Solemya communities seldom contain all three genera and tend to be dominated by species of either *Lucinoma* or *Thyasira*. In the turnover fauna, these communities are particularly well developed in the Keasey and Lincoln Creek Formations, where the carbonate mounds, lenses, nodules, and associated fauna occur in tuffaceous siltstones reflecting nearby volcanic activity. One of the most remarkable features of these communities is the large size of both the thyasirids and the solemyids. Thyasirid bivalves elsewhere are typically small-shelled (<10 mm), but the species that characterize the Eocene-Oligocene transition are giant thyasirids of the genus *Conchocele*, attaining lengths up to 130 mm. The solemyids in the turnover fauna also attain lengths up to 130 mm and are classified in the subgenus *Archarax*, in contrast to typical *Solemya*, which are generally less than 40 mm in length.

A second type of chemosymbiotic community that appears as part of the late Eocene turnover fauna was documented by Goedert and Squires (1990) from authigenic subduction-related carbonates, formed at inferred methane seeps, at three sites in western Washington. It is designated here as the *Modiolus* community, based on the observation that "modiolid bivalves are always present with either vesicomyid bivalves or thyasirid bivalves" (Goedert and Squires, 1990). The *Modiolus* community occurs in autochthonous limestone bodies, where the modiolids may form dense concentrations. The *Modiolus* community lacks a living counterpart in the modern subduction zone complex off Oregon, but modiolids are the dominant organisms at many hydrothermal vents and sulfide seeps elsewhere (Turner and Lutz, 1984).

The third type of chemosymbiotic community contains vesicomyid bivalves of the genus *Calyptogena*. *Calyptogena* communities occur in turbidity flows within the subduction complex as well as in subduction-related limestone bodies (Goedert and

Squires, 1990; Squires and Goedert, 1991), and these communities persist into the Oligocene recovery interval (Goedert and Squires, 1993). *Calyptogena* is also present at methane-venting sites in the modern Oregon subduction zone complex (Suess et al., 1985; Kulm et al., 1986).

The fourth type of chemosymbiotic community is characterized by large numbers of provannid gastropods. The first recognition of the genus *Provanna* in the fossil record was in the Eocene-Oligocene turnover and recovery faunas of the Pacific Northwest (Goedert and Campbell, 1995; Squires, 1995). At three localities more than 50 specimens were extracted from seep limestones. Although none of the living species of *Provanna* are chemosymbiotic, they are all are restricted to chemosynthetically nourished communities (Warén and Bouchet, 1993). The family Provannidae (Warén and Ponder, 1991) is known from both hydrothermal vents and cold seeps, and the only gastropods known to have chemosynthetic bacteria are the provannid genera *Alviniconcha* and *Ilfremeria*, which occur only at hot vent sites (Stein et al., 1988; Warén and Bouchet, 1993).

Heightened faunal exchange with Asia. Strong connections between the Pacific Northwest and Asia first become apparent in the turnover fauna, a connection which is seen earlier in Alaska (Marincovich, 1988). Whereas the tropical Eocene Cowlitz fauna was most comparable with the Eocene fauna of the Paris Basin (Clark and Vokes, 1936), the Keasey and Lower Lincoln Creek faunas have strong affinities with coeval and slightly older faunas in Alaska and Asia. The influx of Asian taxa, coinciding with inferred onset of cooling, was first emphasized by MacNeil et al. (1961) and MacNeil (1965). MacNeil provided groundbreaking studies of the Cenozoic faunas of Alaska and observed that several Asian taxa, notably the *Acila shumardi* lineage, appear first in the Alaskan Paleogene, and subsequently in the Pacific Northwest (MacNeil, 1971). Kanno (1960) reported connections at the species level, observing that two species typical of the Oligocene recovery fauna in the Pacific Northwest, *Cyclocardia hannibali* and *Lucinoma columbiana*, are present in the Oligocene Ushikubitoga Formation in Japan. Both species also occur in the Oligocene portion of the Poul Creek Formation in Alaska, a faunal similarity first noted by Clark (1932).

The direction of faunal exchange in the North Pacific during the transition was not always from Asia to North America via Alaska. Hickman (1972, 1980) noted that the distinctive genus *Phanerolepida* appears suddenly in the Keasey Formation and at the base of the Lincoln Creek Formation, with a precursor at the top of the Cowlitz Formation in Oregon (Hickman, 1974). The sole living species of *Phanerolepida* is restricted to the Japanese bathyal fauna, and the fossil record of the genus in the western Pacific is restricted to Neogene rocks in Japan.

Alaskan Paleogene faunas are poorly studied and will undoubtedly provide increasingly important evidence of the pattern of Eocene-Oligocene faunal response to global cooling and the relative roles of climate change and regional tectonic change in the Pacific Northwest faunal turnover event.

Origination of regional endemic taxa. The turnover and recovery faunas of the Pacific Northwest are most easily recognized by the high proportion of now-extinct endemic Pacific Coast genera. Once they appeared, the regional endemics not only tended to speciate rapidly, but also became some of the most frequent and abundant elements of the turnover and recovery faunas.

The best examples are among the caenogastropods and include *Eosiphonalia* and *Bruclarkia* (both neptuneids), *Molopophorus* (a buccinid), *Priscofusus* (a fusinid), and *Perse* (a fasciolariid). *Priscofusus* and *Perse* originated in the tropical fauna but did not become prominent faunal elements until the turnover, and *Molopophorus*, *Eosiphonalia*, and *Bruclarkia* are hallmark genera of the turnover and recovery faunas. In the Pittsburg Bluff Formation *Perse*, *Bruclarkia*, and *Molopophorus* not only are very abundant, but also are remarkable for the broad range of intraspecific variation in sculpture (Moore, 1976).

One of the most dramatic faunal changes at the Eocene-Oligocene transition is an abrupt turnover of the turrid gastropod fauna (Hickman, 1976a). Four endemic genera were proposed by Hickman (1976a) for species that characterize the turnover and recovery faunas: *Turrinosyrinx*, *Clivuloturris*, *Procerapex*, and *Boreocomitas*. A single species of *Boreocomitas* appears in the Cowlitz fauna, but the other three genera are restricted to latest Eocene and Oligocene faunas of the Pacific Northwest. Other turrid genera that have longer stratigraphic ranges and broader geographic distributions make their first appearances on the Pacific Coast in turnover faunas. They include *Turricula*, *Parasyrinx*, *Ptychosyrinx*, and *Acamptogenotia*. Five distinct stratigrapic intervals in rocks representing the turnover and recovery faunas contain five distinctive species of *Ptychosyrinx*.

Trochid gastropods, like turrids, show rapid turnover at the species level during the transition. The large-shelled trochid *Bathybembix* s.l. makes its first appearance at the base of the Keasey Formation, and the genus shows a high rate of species replacement in the turnover and recovery faunas, where is it represented by at least seven unnamed species (Hickman, unpublished).

Appearance of the modern temperate bathyal fauna. The major elements of the modern bathyal (>200 m) fauna appear abruptly in the Pacific Northwest as part of the turnover fauna, accompanying the shift from arkosic sediments of the Cowlitz Formation to volcaniclastic sediments of the Keasey Formation. The six bathyal communities that originate in the Keasey Formation all have close counterparts (shared genera) in the modern fauna. Each community tends to be dominated by one or a few species, and none is represented by abundant fossil material. In contrast, the Cowlitz bathyal facies contains a more diverse fauna that Nesbitt (1995) designated as the *Nuculana* assemblage, based on the most abundant bivalve.

Although the genera are different, and faunal diversity and abundances are different, the Keasey and Cowlitz offshore faunas share a common taxonomic and trophic structure that identifies them as bathyal (Hickman, 1984). The key structural similarities are the predominance of deposit-feeding protobranch bivalves and carnivorous caenogastropods over suspension-feeding bivalves and herbivorous caenogastropods.

Prominent elements of the modern bathyal fauna that make their first northeastern Pacific appearance in the turnover fauna include assemblages dominated by deposit-feeding archaeogastropods, assemblages dominated by mud pectens (Pectinidae and Propeamussiidae), and assemblages dominated by giant limids (*Acesta*) and siliceous sponges (Hickman, 1984).

Changes in faunal abundance and structure. Accompanying the progressive decline in species diversity from the tropical Eocene to the Oligocene recovery fauna, there is a shift from species-rich assemblages to species-dominant assemblages. The Cowlitz fauna contains eight distinctive assemblages (Nesbitt, 1995). Four of these assemblages are fresh to brackish water, and four are fully marine. In a landmark quantitative study of the structure and paleoecology of the Cowlitz mollusk fauna, Nesbitt (1995) provided the first data on the relative abundances of tropical marine Eocene species from shallow to middle bathyal settings in the eastern Pacific. Her four marine assemblages, from shallow to deep, are a *Turritella-Solena* assemblage (34 species), a *Turritella-Tivilena* assemblage (72 species), a *Pitar* assemblage (41 species), and a *Nuculana* assemblage (50 species). In contrast, the most diverse assemblages in the Keasey Formation contain a maximum of 19 species (Hickman, 1984; and unpublished data). In the Cowlitz Formation, 17 gastropod species are represented by ≥10 specimens in a single bulk sample, whereas in the Keasey Formation only 7 gastropod species are represented by ≥10 specimens from a single, intensively collected locality.

Nesbitt's quantitative data were obtained by laborious disaggregation of bulk samples by hand in the field in order to count specimens that disintegrated even with the most careful handling. Hickman's data for the Keasey Formation are based on collecting every fossil found at a given locality. Although the data acquisition methods are necessarily different, there are two robust qualitative conclusions. The first is that fossils are much more abundant in the Cowlitz Formation, and the second is that abundance is more evenly distributed among species in Cowlitz assemblages. For example, in the *Turritella-Tivilena* assemblage, there were 12 mollusk species represented by >20 specimens and 22 mollusk species represented by >10 specimens.

The recovery fauna of the Pittsburg Bluff, Eugene, and Scappoose Formations resemble those of the turnover fauna in the prevalence of species-dominant assemblages. However, the recovery faunas contain a much greater abundance of fossil material. Bulk samples from the Scappoose Formation (Hickman, unprocessed) should provide data on relative abundances comparable to data from the Cowlitz Formation.

One of the peculiarities of the fauna of the Pittsburg Bluff Formation is the great abundance of specimens of 6 species of carnivorous (or scavenging) gastropods. Moore (1976) estimated frequency and relative abundance of species observed in the field and in collections prepared from slabs of rock. She ranked 6 gastropod species as abundant (>20 specimens). Three are naticids, and 3 are species of the endemic turnover and recovery genera *Bruclarkia*, *Molopophorus*, and *Perse*.

A final example of the change in dominance structure and abundance across the three faunas is the abrupt decline in *Turritella*-dominated assemblages. Two of Nesbitt's (1995) marine assemblages are dominated by *Turritella* specimens. *Turritella* is a suspension-feeding gastropod that typically occurs in dense aggregations (Hickman, 1985). In her quantitative sample of the *Turritella-Tivilena* assemblage, Nesbitt counted 297 specimens of *Turritella uvasana olequahensis.* Hickman (1980) described *T. keaseyensis* as a new species from the turnover fauna of the Keasey Formation from a total of 11 specimens from four localities. When Moore (1976:14) described *T. pittsburgensis* from the recovery fauna of the Pittsburg Bluff Formation, she characterized it as scarce, "found at several localities, usually singly or only a few together." Although *Turritella* is present in all three faunas, it is a relatively insignificant element after the demise of the tropical fauna.

CONCLUSION

With a new conceptual framework for assessing marine molluscan faunal change at the Eocene-Oligocene transition in the Pacific Northwest, particularly an ability to date the fossiliferous sequences using a calibrated magnetic stratigraphy, it is clear that the most significant episode of regional extinction and origination occurs in the late Eocene, prior to the abrupt global cooling event inferred from isotopic proxies for temperature. The turnover event is more closely correlated with regional tectonics and a prominent coincidental change in sedimentary setting and lithofacies accompanying the onset of voluminous Cascade arc volcanism. A second, but less profound, episode of regional extinction and origination occurs early in the Oligocene, but postdating the abrupt cooling event by several million years. The two episodes of regional extinction in the Pacific Northwest serve to define a late Eocene tropical fauna, a turnover fauna, and an Oligocene recovery fauna. Climatic deterioration and intensification of the latitudinal thermal gradient occurred across the entire 10-million-year interval under investigation. The effects of global cooling are reflected in some of the patterns reported here and are part of a latitudinal trend in the timing of climate-related faunal change that emerged during the conference generating this volume (Carter et al., this volume). Teasing apart the relative effects of climate change and regional tectonic events may be difficult, but the new framework illustrates how increased temporal resolution of faunal data, proxies for climate change, and increasingly detailed reconstructions of the complex tectonic history of the northeastern Pacific margin have already helped constrain interpretations of the Eocene-Oligocene transition.

ACKNOWLEDGMENTS

I thank Don Prothero for the opportunity to participate in the Penrose Conference leading to this volume as well as for the inducement that his research has provided to return to study of molluscan faunas of the Pacific Northwest. I am grateful to Liz Nesbitt and Carol Tang for discussions of this work, to all of the paleontologists and geologists who have stimulated my interest in and enthusiasm for the geology and faunas of the Pacific Northwest, and to Ellen Moore, Kathy Campbell, and Liz Nesbitt for helpful reviews of the manuscript. This is contribution no. 1786 of the University of California Museum of Paleontology.

REFERENCES

Addicott, W. O., 1968, Tertiary climatic change in the marginal northeastern Pacific Ocean: Science, v. 165, p. 583-586.

Addicott, W. O., 1970a, Latitudinal gradients in Tertiary molluscan faunas of the Pacific Coast: Palaeogeography, Palaeoclimatology, Palaeoecology, v 8, p. 187-312.

Addicott, W. O., 1970b, Tertiary paleoclimatic trends in the San Joaquin Basin, California: U.S. Geological Survey Professional Paper 644-D, 19 p.

Armentrout, J. M., 1973, Molluscan paleontology and biostratigraphy of the Lincoln Creek Formation, late Eocene-Oligocene, southwestern Washington [Unpubl. Ph.D. dissert.]: University of Washington, Seattle, 478 p.

Armentrout, J. M., 1975, Molluscan biostratigraphy of the Lincoln Creek Formation, southwest Washington, *in* Weaver, D. E., ed., Future energy horizons of the Pacific Coast: American Association of Petroleum Geologists, Long Beach, California, p. 14–28.

Armentrout, J. M., 1981, Correlation and ages of Cenozoic chronostratigraphic units in Oregon and Washington: Geological Society of America Special Paper, no. 184, p. 137–148.

Arnold, R., 1906, The Tertiary and Quaternary pectens of California: U.S. Geological Survey Professional Paper 47, 264 p.

Arnold, R., 1909, Paleontology of the Coalinga District, Fresno and Kings Counties, California: U.S. Geological Survey Bulletin 396, 173 p.

Bates, R. G., Beck, M. E., and Burmeister, R. F., 1981, Tectonic rotations in the Cascade Rnage of southern Washington: Geology, v. 9, p. 184–189.

Berggren, W. A., Kent, D. V., and Flynn, J. J., 1985, Paleogene geochronology and chronostratigraphy, *in* Snelling, N. J., ed., The chronology of the geological Record: Geological Society of America Memoir 10, p. 141–195.

Berggren, W. A., Kent, D. V., Obradovich, J. D., and Swisher, C. C., III, 1992, Towards a revised Paleogene geochronology, *in* Prothero, D. R., and Berggren, W. A., eds., Eocene-Oligocene climatic and biotic evolution: Princeton, New Jersey, Princeton University Press, p. 29-45.

Berggren, W. A., Kent, D. V., Swisher, C. C., III, Aubry, M.-P., and Hardenbol, J., 1995, A revised Cenozoic geochronology and chronostratigraphy: SEPM Special Publication 54, p. 129–212.

Campbell, K. A., 1992, Recognition of a Mio-Pliocene cold seep setting from the northeast Pacific convergent margin, Washington, U.S.A.: Palaios, v. 7, p. 422–433.

Campbell, K. A., and Bottjer, D. J., 1993, Fossil cold seeps (Jurassic–Pliocene) along the convergent margin of western North America: National Geographic Research & Exploration, v. 9, p. 326–343.

Campbell, K. A., and Bottjer, D. J., 1995, Brachiopods and chemosymbiotic bivalves in Phanerozoic hydrothermal vent and cold seep environments: Geology, v. 23, p. 321–324.

Clark, B. L., 1918, The San Lorenzo Series of middle California: University of California Publications, Bulletin of the Department of Geology, v. 11, p. 45-235.

Clark, B. L., 1932, Fauna of the Poul and Yakataga Formations (upper Oligocene) of southern Alaska: Geological Society of America Bulletin, v. 43, p. 797–846.

Clark, B. L., and Vokes, H. E., 1936, Summary of marine Eocene sequence of western North America: Geological Society of America Bulletin, v. 17, p. 63-142.

Dickerson, R. E., 1917, Climate and its influence upon the Oligocene faunas of the Pacific Coast, with descriptions of some new species from the *Molopophorus lincolnensis* Zone: California Academy of Sciences, Proceedings, ser. 4, v. 7, no. 6, p. 157–192.

Dickinson, W. R., 1976, Sedimentary basins developed during evolution of Mesozoic–Cenozoic arc-trench system in western North America: Canadian Journal of Earth Sciences, v. 13, p. 1268–1287.

Diller, J. S., 1896, A geological reconnaissance of northwestern Oregon, U.S. Geological Survey 17th Annual Report, p. 441-520.

Duncan, R. A., 1982, A captured island chain in the Coast Range of Oregon and Washington: Journal of Geophysical Research, v. 87, no. B13, p. 10827–10837.

Duncan, R. A., and Kulm, L. D., 1989, Plate tectonic evolution of the Cascades arc-subduction complex, *in* Winterer, E. L., Hussong, D. M., and Decker, R. W., eds., The Eastern Pacific Ocean and Hawaii: Boulder, Colorado, Geological Society of America, The Geology of North America, V. N, p. 413–438.

Durham, J. W., 1950, Cenozoic marine climates of the Pacific Coast: Geological Society of America Bulletin, v. 61, p. 1243-1264.

Durham, J. W., 1954, The marine Cenozoic of southern California: California Division of Mines Bulletin, v. 170, p. 23-31.

Eames, F. R., Banner, F. T., Blow, W. H., Clarke, W., and Cox, L. R., 1962, Fundamentals of mid-Tertiary stratigraphical correlation: Cambridge, England, Cambridge University Press, 163 p.

Ewing, T. E., 1980, Paleogene tectonic evolution of the Pacific Northwest: Journal of Geology, v. 88, no. 6, p. 619–638.

Goedert, J. L., and Campbell, K. A., 1995, An early Oligocene chemosynthetic community from the Makah Formation, northwestern Olympic Peninsula, Washington: Veliger, v. 38, p. 22–29.

Goedert, J. L., and Squires, R. L., 1990, Eocene deep-sea communities in localized limestones formed by subduction-related methane seeps, southwestern Washington: Geology, v. 18, p. 1182–1185.

Goedert, J. L., and Squires, R. L., 1993, First Oligocene records of *Calyptogena* (Bivalvia: Vesicomyidae): The Veliger, v. 36, p. 72–77.

Hall, C. A., Jr., 1964, Shallow-water marine climates and molluscan provinces: Ecology, v. 45, p. 226-234.

Hankins, K. G., and Prothero, D. R., 1999, Magnetic stratigraphy and tectonic rotation of the Oligocene Pittsburg Bluff Formation, northwest Oregon: Geological Society of America, Abstracts with Programs, v. 31, no. 7, p. A289.

Heller, P. L., and Ryberg, P. T., 1983, Sedimentary record of subduction to forearc transition in the rotated Eocene basin of western Oregon: Geology, v. 11, p. 380–383.

Heller, P. L., Tabor, R. W., and Suczek, C. R., 1987, Paleogeographic evolution of the United States Pacific Northwest during Paleogene time: Canadian Journal of Earth Sciences, v. 24, p. 1652–1667.

Henriksen, D. A., 1956, Eocene stratigraphy of the

lower Cowlitz River–eastern Willapa Hills area, southwestern Washington: Washington Division of Mines and Geology Bulletin, v. 43, p. 1–122.

Hickman, C. S., 1969, Oligocene marine molluscan fauna of the Eugene Formation in Oregon: Bulletin 16 of the Museum of Natural History, University of Oregon. 112 p.

Hickman, C. S., 1972, Review of the bathyal gastropod genus *Phanerolepida* (Homalopomatinae) and a description of a new species from the Oregon Oligocene: The veliger, v. 15, p. 107–112.

Hickman, C. S., 1974, *Nehalemia hieroglyphica*, a new genus and species of archaeogastropod (Turbinidae: Homalopomatinae) from the Eocene of Oregon: The veliger, v. 17, p. 89–91.

Hickman, C. S., 1976a, Bathyal gastropods of the family Turridae in the early Oligocene Keasey Formation in Oregon, with a review of some deep-water genera in the Paleogene of the eastern Pacific: Bulletins of American Paleontology, v. 70, no. 292, 119 p.

Hickman, C. S., 1976b, *Pleurotomaria* (Archaeogastropoda) in the Eocene of the northeastern Pacific: A review of Cenozoic biogeography and ecology of the genus: Journal of Paleontology, v. 50, p. 1090–1102.

Hickman, C. S., 1980, Paleogene marine gastropods of the Keasey Formation in Oregon: Bulletins of American Paleontology, v. 78, no. 310, 112 p.

Hickman, C. S., 1984, Composition, structure, ecology, and evolution of six Cenozoic deep-water mollusk communities: Journal of Paleontology, v. 5, p. 1215–1234.

Hickman, C. S., 1985, Comparative morphology and ecology of free-living suspension-feeding gastropods from Hong Kong: *in* Morton, B., and Dudgeon, D., eds., The Malacofauna of Hong Kong and of Southern China: Hong Kong University Press, p. 217-234.

Hickman, C. S., 1994, The genus *Parvilucina* in the Eastern Pacific: Making evolutionary sense of a chemosymbiotic species complex: The Veliger, v. 37, p. 43–61.

Hickman, C. S., and Lindberg, D. R., 1984, Relationship of molluscan biogeographic anomalies to changing settings on an active continental margin: Geological Society of America Abstracts with Programs, v. 16, p. 289.

Irving, A. J., Nesbitt, E. A., and Renne, P. R., 1996, Age constraints on earliest Cascade arc volcanism and Eocene marine biozones from a feldspar-rich tuff in the Cowlitz Formation, southwestern Washington: EOS, v. 77, p. F814.

Kanno, S., 1960, Palaeontology, Part II, *in* Arai, J., and Kanno, S., The Tertiary system of the Chichibu Basin, Saitama Prefecture, central Japan: Tokyo, Ueno, Japan Society for the Promotion of Science, p. 123–396.

Keen, A. M., and Bentson, H., 1944, Checklist of California Tertiary marine Mollusca: Geological Society of America Special Paper 56, 280 p.

Kulm, L. D., and Suess, E., 1990, Relationship between cabonate deposits and fluid venting: Oregon accretionary prism: Journal of Geophysical Research, v. 95, p. 8899–8915.

Kulm, L. D., and 13 others, 1986, Oregon subduction zone: venting, fauna, and carbonates: Science, v. 231, p. 561-566.

Lindberg, D. R., and Hickman, C. S., 1986, A new anomalous giant limpet from the Oregon Eocene (Mollusca: Patellida): Journal of Paleontology, v. 60, p. 661–668.

MacNeil, F. S., 1965, Appendix D, *in* Burk, C. A., Geology of the Alaska Peninsula—Island arc and continental margin: Geological Society of America Memoir 99, p. 1, p. 221–228.

MacNeil, F. S., 1971, Selected mollusks from the Poul Creek and Yakataga Formations, Yakataga and Malaspina districts, showing tentative identification and stratigraphic range, *in* Miller, D. J., Geologic map of Yakataga District, Gulf of Alaska: U.S. Geological Survey Misc. Geol. Inv. Map I-610, scale l:125,000, p. 3-5.

MacNeil, F. S., Wolfe, J. A., Miller, D. J., and Hopkins, D. M., 1961, Correlation of Tertiary formations of Alaska: American Association of Petroleum Geologists Bulletin, v. 45, p. 1801–1809.

Magill, J. R., Cox, A. V., and Duncan, R. A., 1981, Tillamook volcanic series: Further evidence for tectonic rotation of the Oregon Coast Range: Journal of Geophysical Research, v. 86, p. 2953–2970.

Marincovich, L., Jr., 1977, Cenozoic Naticidae (Mollusca: Gastropoda) of the northeastern Pacific: Bulletins of American Paleontology, v. 70, no. 294, 494 p.

Marincovich, L., Jr., 1988, Late middle Eocene mollusks of the Tolstoi Formation, Alaska Peninsula, Alaska, and correlations with faunas from California to the far-eastern U.S.S.R., *in* Filewicz, M. V., and Squires, R .L., eds., Paleogene stratigraphy, West Coast of North America: Pacific Section, SEPM, West Coast Paleogene Symposium, 58, p. 265–281.

Moore, E. J., 1976, Oligocene marine mollusks from the Pittsburg Bluff Formation in Oregon: U.S. Geological Survey Professional Paper 922, 66 p.

Nesbitt, E. A., 1995, Paleoecological analysis of molluscan assemblages from the middle Eocene Cowlitz Formation, southwestern Washington: Journal of Paleontology, v. 69, p. 1060–1073.

Nesbitt, E. A., 1999. Changes in marine neritic paleoassemblages of the northeastern Pacific margin in response to the Eocene-Oligocene global climate shift: Geological Society of America Abstracts with Programs, v. 31, no. 7, p. A310.

Nesbitt, E. A., Campbell, K. A., and Goedert, J. L., 1994, Paleogene cold seeps and macroinvertebrate faunas in a forearc sequence of Oregon and Washington, *in* Swanson, D.A., and Haugerud, R.A., eds., Geologic field trips in the Pacific Northwest: Geological Society of America Annual Meeting, v. 1, p. 1D-1 –1D-11.

Niem, A. R., MacLeod, N. S., Snavely, P. D., Jr., Huggins, D., Fortier, J. D., Meyer, H. J., and Seeling, A., 1992, Onshore-offshore geologic cross-sections, northern Oregon Coast Range to continental slope: Oregon Department of Geology and Mineral Industries Special Publication 26, 38 p.

Niem, A. R., McKnight, B. K., Meyer, H. J., and Campbell, K. A., 1994, Sedimentary, volcanic, and tectonic framework of forearc basins and the Mist Gas Field, northwest Oregon, *in* Swanson, D. A., and Haugerud, R. A., eds., Geologic field trips in the Pacific Northwest: Geological Society of America Annual Meeting, v. 1, p. 1F-1–1F-42.

Prothero, D. R., 2001, Magnetic Stratigraphy of the Pacific Coast Cenozoic: Pacific Section SEPM Special Publication, v. 91.

Prothero, D. R., and Armentrout, J. M., 1985, Magnetostratigraphic correlation of the Lincoln Creek Formation, Washington: Implication for the age of the Eocene/Oligocene boundary: Geology, v. 13, p. 208–211.

Prothero, D. R., and Donohoo, L. L., 2001, Magnetic stratigraphy and tectonic rotation of the lower Oligocene Tunnel Point Formation, Coos County, Oregon: Pacific Section SEPM Special Publication, v. 91, p. 195-200.

Prothero, D. R., and Hankins, K. G., 2000, Magnetic stratigraphy and tectonic rotation of the upper Eocene-lower Oligocene Keasey Formation, Columbia County, Oregon: Journal of Geophysical Research, v. 105, n. B7, p. 16473-16480.

Prothero, D. R., and Hankins, K. G., 2001, Magnetic stratigraphy and tectonic rotation of the lower Oligocene Pittsburg Bluff Formation, Columbia County, Oregon: Earth and Planetary Sciences Letters, v. 91, p. 201-209.

Prothero, D. R., Bitboul, C. Z., Moore, G. W., and Niem, A., 2001, Magnetic stratigraphy and tectonic rotation of the Oligocene Alsea, Yaquina, and Nye formations, Lincoln County, Oregon: Pacific Section SEPM Special Publication, v. 91, p. 184-194.

Prothero, D. R., Jaquette, C., and Armentrout, J. M., 2001, Magnetic stratigraphy of the upper Eocene-upper Oligocene Lincoln Creek Formation, Porter Bluffs, Washington: Pacific Section SEPM Special Publication, v. 91, p. 169-178.

Prothero, D. R., Sanger, E., Nesbitt, E., Niem, A. R., and Kleibacker, D., 2001. Magnetic stratigraphy and tectonic rotation of the upper middle Eocene Cowlitz and Hamlet Formations, western Oregon and Washington: Pacific Section SEPM Special Publication, v. 91, p. 75-95.

Simpson, R. W., and Cox, A. V., 1977, Paleomagnetic evidence for tectonic rotation of the Oregon Coast Range: Geology, v. 5, p. 485–589.

Smith, J. P., 1910, The geologic record of California: Journal of Geology, v. 18, p. 216-227.

Smith, J. P., 1919, Climatic relations of the Tertiary and Quaternary faunas of the California region: California Academy of Sciences Proceedings, ser. 4, v. 9, p. 123-173.

Smith, J. T., 1970, Taxonomy, distribution, and phylogeny of the cymatiid gastropods *Argobuccinum, Fusitriton, Mediargo,* and *Priene*: Bulletins of American Paleontology, v. 56, no. 254, p. 445–573.

Snavely, P. D., Jr., and Wells, R. E., 1996, Cenozoic evolution of the continental margin of Oregon and Washington: U.S. Geoogical Survey Professional Paper 1560, p. 161–182.

Squires, R. L., 1995, First fossil species of the chemosynthetic-community gastropod Provanna: Localized cold-seep limestones in upper Eocene and Oligocene rocks, Washington: Veliger, v. 38, p. 30–36.

Squires, R. L., and Goedert, J. L., 1991, New late Eocene mollusks from localized limestone deposits formed by subduction-related methane seeps, southwestern Washington: Journal of Paleontology, v. 63, p. 412-416.

Stein, J. L., Cary, S. C., Hessler, R. R., Ohta, S., Vetter, R. D., Childress, J. J., and Felbeck, H., 1988, Chemoautotrophic symbiosis in a hydrothermal vent gastropod: Biological Bulletin, v. 174, p. 373–378.

Suess, E., Carson, B., Ritger, S. D., Moore, J. C., Jones, M. L., Kulm, L. D., and Cochrane, G. R., 1985, Biological communities at vent sites along the subduction zone off Oregon, *in* Jones, M.S., ed., The hydrothermal vents of the eastern Pacific: An overview: Biological Society of Washington Bulletin 6, p. 475–484.

Turner, R. D., and Lutz, R. A., 1984, Growth and distribution of mollusks at deep-sea vents and seeps: Oceanus, v. 27, p. 55–62.

Valentine, J. W., 1966, Numerical analysis of marine molluscan ranges on the extratropical northeastern Pacific shelf: Limnology and Oceanography, v. 11, p. 198-211.

Vermeij, G. J., 1978, Biogeography and adaptation: Cambridge, Massachusetts, Harvard University Press.

Vokes, H. E., 1935, The genus *Velates* in the Eocene of California, University of California Publications, Bulletin of the Department of Geological Sciences, v. 23, p. 381–390.

Warén, A., and Bouchet, P., 1993, New records, species, genera, and a new family of gastropods from hydrothermal vents and hydrocarbon seeps: Zoologica Scripta, v. 22, p. 1–90.

Warén, A., and Ponder, W. F., 1991, New species, anatomy, and systematic position of the hydrothermal vent and hydrocarbon seep gastropod family Provannidae fam. n. (Caenogastropoda): Zoologica Scripta, v. 20, p. 27–56.

Watkins, R., 1974, Paleobiology of an offshore molluscan fauna from the California Oligocene: Palaeogeography, Palaeoclimatology, Palaeoecology, v. 15, p. 245–266.

Wells, R. E., 1990, Paleomagnetic rotations and the Cenozoic tectonics of the Cascade Arc, Washington, Oregon, and California: Journal of Geophysical Research, v. 95, no. B12, p. 19409–10417.

Wells, R. E., and Coe, R. S., 1985, Paleomagnetism and geology of Eocene volcanic rocks of southwest Washington, implications for mechanisms of tectonic rotation: Journal of Geophysical Resarch, v. 90, no. B2, p. 1925–1947.

Wells, R. E., and Heller, P. L., 1988, The relative contribution of accretion, shear, and extension to Cenozoic tectonic rotation the Pacific Northwest: Geological Society of America Bulletin, v. 100, p. 325-338.

Wells, R. E., Engebretson, C. C., Snavely, P. D., Jr., and Coe, R. S., 1984, Cenozoic plate motions and the volcano-tectonic evolution of western Oregon and Washington: Tectonics, v. 3, p. 275–294.

Yeats, R. S., Graven, E. P., Werner, K. S., Goldfinger, C., and Popowski, T. A., 1996, Tectonics of the Willamette Valley, Oregon: U.S. Geological Survey Professional Paper 1560, p. 183–198.

Zachos, J. D., Quinn, T. M., and Salamy, K. A., 1996, High-resolution (10^4 years) deep-sea foraminiferal stable isotope records of the Eocene-Oligocene climate transition: Paleoceanography, v. 11, p. 251–266.

CHAPTER 6

An Overview of Eocene-Oligocene Echinoderm Faunas of the Pacific Northwest

Casey Burns and Rich Mooi

ABSTRACT

We summarize current knowledge of Eocene and Oligocene echinoderm faunas of the Pacific Northwest within the context of the Eocene-Oligocene transition. Echinoderms, notably irregular echinoids, present a diversity of preservational scenarios and unresolved systematic issues. Some spatangoid echinoids (such as schizasterids) range widely both before and after the transition, but are taxonomically oversplit and as yet unhelpful in assessing major patterns of faunal change. Although clypeasteroid echinoids occur at fewer localities, better estimates of their biodiversity make them more reliable in such analyses. Crinoids, ophiuroids, and asteroids are rare, but new search paradigms relying on present knowledge should minimize some of the gaps in the record. We also present new data on occurrence and morphology of previously known, stratigraphically important taxa (such as *Schizaster, Brisaster, Kewia, Eoscutella,* and *Isocrinus*) and on as yet unnamed but significant new finds (including the first western North American protoscutellid sand dollar and uniquely preserved asteroids). With ongoing reappraisal of outdated systematics, it will be feasible to develop a scenario, based on the echinoderm record, for the evolution of Eocene-Oligocene marine environments of the region.

INTRODUCTION

Echinoderms are mostly benthic, deuterostome animals with distinctive pentameral symmetry. They are entirely marine, and can be found in varying abundance in virtually every type of environment in today's oceans. With few exceptions, echinoderms possess within the body wall an endoskeleton constructed of individual, commonly articulated elements. These elements, which can be connected to each other in loose meshwork or in a rigid body wall skeleton known as a test, are made of stereom—a modified, lightweight but very durable type of calcite. In the fossil record, echinoderms such as echinoids (sea urchins), are relatively easily preserved as intact body fossils because the test is made of partially interlocking plates. However, even among sea urchins, differences in preservation potential can skew perceptions of their abundance in a given paleoenvironment. As shown by Kier (1977), the so-called "regular echinoids," long-spined sea urchins with a globose test, have a poor fossil record compared to that of the so-called "irregularia," which includes the burrowing, bilaterally symmetric heart urchins (spatangoids) and greatly flattened sand dollars (clypeasteroids). Spatangoids, clypeasteroids, and even some regular echinoids can survive intact in a higher energy environment of deposition than echinoderms with only soft tissue connections between plates, such as crinoids (sea lilies and feather stars), asteroids (starfish), and ophiuroids (brittle stars).

With respect to the Oligocene-Eocene transition in the Pacific Northwest, variation in taphonomic factors has not been well studied in echinoderms of this region, and virtually nothing has been published. The enduring nature of the echinoid test has led to taphonomic bias, and taxonomic studies have resulted in oversplitting within some echinoid clades. This can create difficulties in developing an accurate picture of faunal compositions by overestimating echinoid diversity. In contrast, our knowledge of the more readily disarticulated asteroids, ophiuroids, and crinoids, tends to lag behind that of the echinoids. An even larger gap is evident in our understanding of the importance of holothuroids (sea cucumbers), echinoderms that have soft body walls that typically contain only minute ossicles. In spite of their abundance in many habitats today, only extraordinary Lagerstätten or careful examination of microfossil assemblages will provide evidence of holothuroids in paleoenvironments.

The occurrence of intact individuals of the more easily disarticulated echinoderm groups indicates extraordinary site-specific preservational conditions such as quiet and/or deeper water, absence of scavengers or burrowers, and repeated sediment flows. Due at least in part to their rarity, fewer non-echinoid taxa have been described, and their taxonomy is considerably less controversial. Paradoxically, many non-echinoid species remain unnamed, which is in distinct contrast to the overabundance of names for echinoids, a significant proportion of which will probably turn out to be invalid as fuller taxonomic revisions are developed.

Major taxonomic monographs of the northwestern Pacific fossil echinoderms include Kew (1920) and Grant and Hertlein (1938). Important contributions from Weaver (1942) and Moore and Vokes (1953) added considerably to our knowledge of both the faunas and the specific strata in which they were found. However, most of these important works, such as Zullo et al. (1964) and Linder et al. (1983), have a relatively narrow taxonomic focus, and no comprehensive overview for echinoderm occurrences in the Pacific Northwest currently exists.

Taxonomic oversplitting of more commonly preserved taxa further complicates studies of past diversity. Regrettably, there has been a tendency to name species on the basis of what are now recognized as insufficient morphological grounds, especially when modern analogues are considered. Naming new species, or synonymizing with an established species, becomes an overwhelming task that often requires complete revision of a family. Unfortunately, and all too often, this exhaustive and exhausting revisionary work is left incomplete in favor of the expediency and recognition that comes with naming new taxa. This is understandable, given that such scholarship is impractical for every major taxon, especially where potentially synonymous type material yields few relevant diagnostic features. Nevertheless, the addition of new names without due consideration of prior work and type material continues to plague assessments of past biodiversity. This problem is, of course, not restricted to the realm of echinoderm systematics.

Rather than concern ourselves with naming species, we can propose to deal with difficult taxa from a given region by morphotyping them as useful indicators of local diversity and populations. In this way, obfuscatory taxonomic proclivities can be disregarded. In the case of the schizasterids, for example, morphotyping may give us a clearer picture of true local diversity and population without forcing a complete revision of the family.

Here, we review the various occurrences of echinoderms in the Eocene and Oligocene of the Pacific Northwest by formation, and we discuss the depositional environment of each unit. We then discuss what, if any, trends we can discern from components of this Pacific Northwest fauna. With additional data points, we may eventually be able to uncover differential trends in the shallower-water and deeper-water echinoderm faunas across the Eocene/Oligocene boundary. A side benefit of this attempt will be our ability to point out problematic taxa, as characterized above.

Except where attributed to a specifically referenced source, all reported echinoderm occurrences are based on specimens collected by the authors, or examined from repositories or private collections. We examined material and field data from the following institutions: Burke Museum of Natural History and Culture, Seattle, Washington (BM); geological collections of the California Academy of Sciences, San Francisco, California (CASG); Museum of Comparative Zoology, Harvard University, Cambridge, Massachusetts (MCZ); National Museum of Natural History, Washington, D.C. (NMNH); Natural History Museum of Los Angeles County, Los Angeles, California (LACM); California State University, Northridge, California (CSUN); University of California Museum of Paleontology, Berkeley, California (UCMP).

EOCENE DEPOSITS AND ECHINODERM FAUNAS

Crescent Formation

Middle Eocene shallow-water sedimentary interbeds composed of dark green basaltic sandstone between tholeiitic basalt flows in the Little River and Satsop River areas of Grays Harbor and Mason Counties in Washington have yielded several species of echinoids. The first author has also collected one sample of a

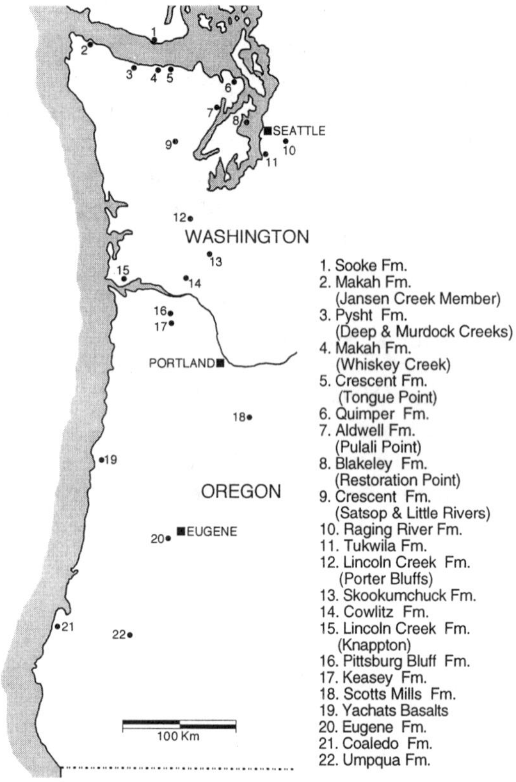

Figure 6.1. Map showing important localities discussed in paper.

poorly preserved, but articulated series of crinoid columnals. According to Squires and Goedert (1994), these deposits were produced by high-energy erosion of pillow basalts that were extruded into shallow water, thereby providing both hard substrates (on the basalts themselves) and soft substrates (the erosion of the basalts producing local sands and muds). The condition of the preserved invertebrate fauna indicates minor postmortem transport and rapid preservation.

An undescribed species of sand dollar, currently under study by the second author, has been collected from one of these sedimentary interbeds exposed on the middle fork of the Satsop River (see figure 6.1 for general locations). These echinoids are most commonly preserved as solitary individuals, although groups of as many as 20 individuals have been found in a few instances. The species is notable not only because it represents a new Eocene clypeasteroid, but also because it appears to belong to the Protoscutellidae, a group considered to be among the most basal members of the scutelline clade (Durham, 1955; Mooi, 1987; Mooi et al., 2000). Because this family is otherwise known only from Eocene deposits on the east coast of North America, phylogenetic analyses including the Satsop River species could considerably revise current views of the early evolution of scutelline sand dollars.

Echinoids tentatively ascribed to the spatangoid genus *Schizaster* have also been found at a number of places in the Crescent Formation (Squires and Goedert, 1994), most commonly at BM locality B666. No articulated spines were found with the echinoids, indicating possible postmortem transport. Larger spatangoid echinoids and several burrow traces have also been found in the Crescent Formation type area at Tongue Point, in Clallam County, Washington (J. Goedert, UWBM, pers. commun., 1999).

At Pulali Point in Jefferson County, Washington, unidentified spatangoids and cidaroid spines have also been reported in similar subtidal deposits of the Crescent Formation (Squires et al., 1992). A spatangoid tentatively identified as *Schizaster diabloensis* (Kew, 1920) and figured by Squires et al. (1992) was collected from deposits in the overlying Aldwell Formation from locality CSUN 1518. The Pulali Point *S. diabloensis* resembles *Schizaster* from BM locality B666, especially in terms of petaloid length ratios and in the slight curvature of the anterior petaloids. Further study will determine if these are indeed synonymous with *S. diabloensis.* In spite of the uncertainty in the systematics of this family (see below), *S. diabloensis* was considered by Squires et al. (1992) to be stratigraphically important in correlating this occurrence to the Californian Eocene.

Examination of the schizasterids of the Pacific Coast has indicated much taxonomic instability and almost certain oversplitting. Most Californian Eocene and Oligocene species are regrettably based on poorly preserved individual holotypes, of which at least one has been further damaged in recent transport (L. Groves, LACM, pers. commun., 1999). These problems, along with recent success in using a morphometric approach to differentiate other schizasterid species (Hood and Mooi, 1998), compelled us to use petaloid dimensions in an attempt to uncover some order among the nominal *Schizaster* taxa in the Northwest (table 6.1). Ontogeny, phenotypic variation, and taphonomic distortion undoubtedly lie at the heart of the taxonomic uncertainty, and could undermine some of the results given here. However, these comparisons remain important in efforts to sort out all of the North American West Coast *Schizaster*

Table 6.1. Ratio of Anterior Petaloid Length to Posterior Petaloid Length of Pacific Coast *Schizaster*

Specimen	Locality	Formation and age	Ratio
Schizaster cordifomis Kew, 1920, holotype	Mount Diablo, CA	Martinez Fm., Eocene	1.45
Schizaster cordiformis Kew, 1920, holotype	Mount Diablo, CA	Martinez Fm., Eocene	1.45
Schizaster lecontei Merriam, 1899, holotype	Contra Costa County, CA	Martinez Fm., Eocene	1.48
Schizaster martinezensis Kew, 1920, holotype	Contra Costa County, CA	Martinez Fm., Paleocene	2.6
Schizaster diabloensis Kew, 1920, holotype	Mount Diablo, CA	Martinez Fm., Eocene	2.5
Schizaster cf. *S. diabloensis*	Jefferson County, WA	Aldwell Fm., Eocene	2.5
Schizaster stalderi Weaver, 1908, holotype	Humboldt County, CA	Wild Cat Fm., Pliocene	1.9
"*Brisaster*" *maximus* H. L. Clark, 1937, holotype	Washington County, OR	Pittsburg Bluff Fm., Oligocene	1.5
Schizaster sp.	BM locality B666, WA	Crescent Fm. upper Eocene	2.75
Schizaster sp.	King County, WA	Tukwila and Raging River Fms., upper Eocene	2.0
Schizaster sp.	Coal Creek locality, WA	Cowlitz Fm., upper Eocene	1.6

species. Additional *Schizaster* specimens from other shallow water and deeper water Eocene and Oligocene formations in Oregon and Washington have been collected or reported, but they have not yet been studied to determine their affinities.

Tentatively identified regular echinoid spines have also been documented from these rocks (Squires et al., 1992). Additional collection should yield information concerning the number of species represented by these spines, even if test material remains unavailable. Float containing a well cemented coquina of mostly molluscan origin yielded a poorly preserved yet articulated crinoid columnal of uncertain affinity.

Age assignments of the Crescent Formation in the Little River Area point toward a middle early Eocene age, on the basis of co-occurring macrofossils and benthic foraminifers. These are correlated with the provincial Californian molluscan "Capay Stage" as used by Weaver et al. (1942). The macrofauna also argues for warm-water tropical conditions (Squires and Goedert, 1994). The age assignments in the Aldwell Formation overlying the Crescent Formation at Pulali Point imply a middle Eocene age on the basis of benthic foraminifers and macrofossils, correlative to the Narizian Stage (as used by Mallory, 1959) (Squires et al., 1992).

Cowlitz and Tukwila Formations

The upper Eocene Cowlitz Formation in southwest Washington has produced rare echinoids at two localities only, which is disappointing considering the spectacular preservation of the diverse molluscan fauna of that formation (Nesbitt, 1995). It can be argued that the presence of diverse mollusks, especially of benthic forms, essentially prohibited the whole body preservation of echinoderms because of bioturbation. Given the warm-water molluscan diversity this fauna represents, a starfish fauna may yet appear as disarticulated elements.

A poorly preserved, single, large spatangoid of uncertain affinities was found at the junction of Olequa and Stillwater Creeks. Echinoids of the genus *Schizaster* have been found at one locality in the Coal Creek section, associated with the large mollusks *Venericardia* and *Turritella.* Given that the strata represent nearshore deposition, and given the diversity and body size of this mostly molluscan fauna, it seems serendipitous that these echinoids were preserved as well as they are. In general, these are relatively undistorted but lack spines, suggesting post-mortem transport.

A single specimen of spatangoid echinoid from the upper Eocene Tukwila Formation (age equivalent to the Cowlitz Formation) from King County, Washington was examined and identified as "*Brisaster*" *maximus* Clark, 1937, by the first author and reported as such by Nesbitt (1998). We use the quotation marks for the genus name because neither Clark's (1937) description nor our own examination of the type material allows unambiguous attribution to the genus *Brisaster* as diagnosed by Mortensen (1951) and Hood and Mooi (1998). Recent work by the first author suggests that the Tukwila Formation

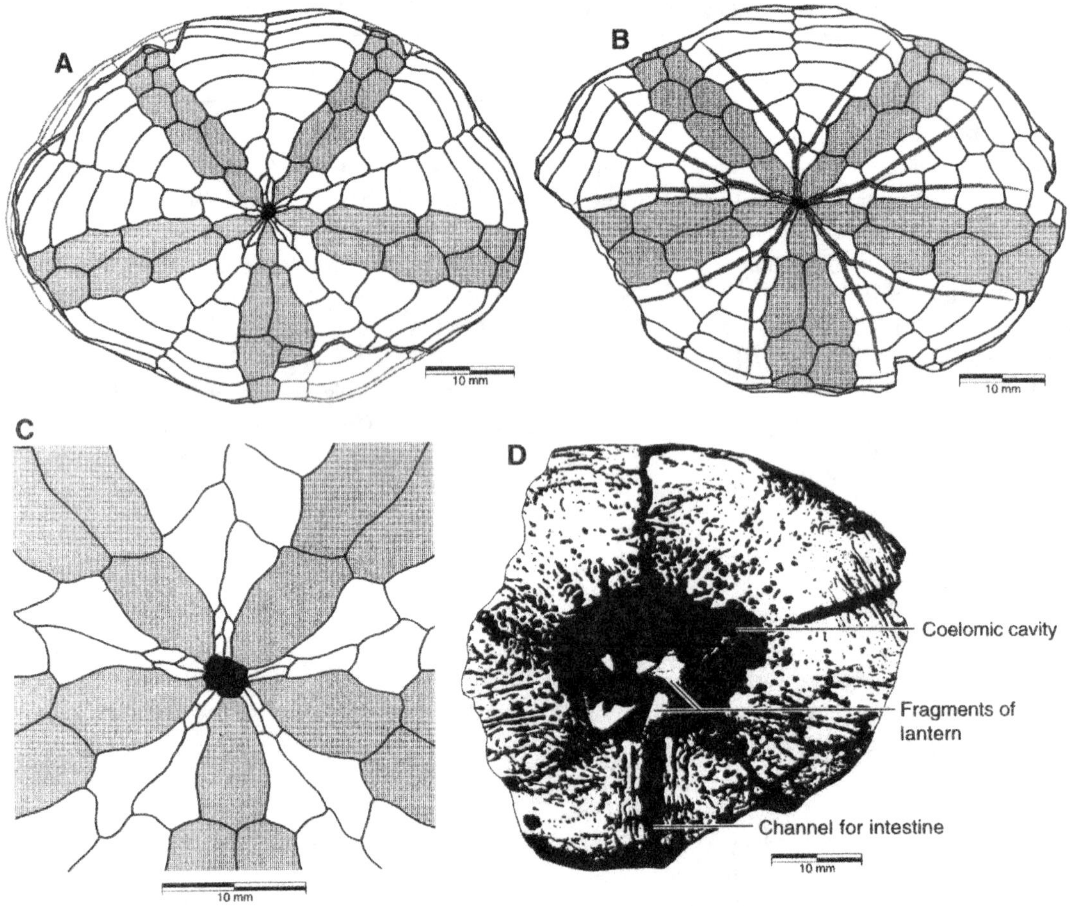

Figure 6.2. *Eoscutella coosensis* Kew, 1920: (A) oral surface plate architecture of CASG 69495, shaded lines indicate reconstructed areas; (B) oral surface plate architecture of CASG 68494, main branches of food grooves indicated by shaded lines; (C) plate architecture of adoral region of CASG 68494; (D) aboral view of UCMP A-860, aboral surface polished off to show internal structure (white) and coelomic cavities (black). For 1-3, mouth is in solid black, interambulacral plates are shaded.

Schizaster specimen is more like the Crescent Formation form and that it is not attributable to "*Brisaster*" *maximus* or the species represented by the Coal Creek *Schizaster* (which might, in fact, also be "*Brisaster*" *maximus*) (table 6.1). It is also unlikely that other Eocene records of "*B.*" *maximus* such as those cited in Linder (1988) are correct. We deal more extensively with "*B.*" *maximus* in the sections on Keasey Formation faunas below.

Sparsely fossiliferous shallow water gritty sandstones of the Cowlitz-age-equivalent upper Eocene Raging River Formation in the vicinity of Tiger Mountain, King County, Washington, produce rare molds of a small *Schizaster*-like echinoid. Vine (1969:15-16) reported these as *Eupatagus* n. sp. aff. *E. carolinensis* Clark and *E. ocalensis* Clark, but the poor preservation and overall appearance of these fossil spatangoids cast serious doubt that such precise determinations are possible. The Raging River specimens superficially resemble the Crescent Formation *Schizaster*. However, the ratio of anterior to posterior petaloid length of the Raging River Formation *Schizaster* is considerably lower than that of the Crescent Formation echinoids, but higher than in the Cowlitz Formation echinoids (table 6.1).

Coaledo Formation

Shallow-water upper middle Eocene deposits of the Coaledo Formation near Coos Bay, Oregon, contain abundant remains of the clypeasteroid *Eoscutella coosensis* (Kew, 1920). These deposits were derived from a deltaic system similar to that present in southwestern Washington during Eocene time. *Eoscutella* (Grant and Hertlein, 193), is very difficult to relate to other scutelline sand dollar genera and has therefore been placed in its own family, Eoscutellidae (Durham, 1955). Part of the problem in determining affinities of this unusual clypeasteroid is that Durham's (1955) figures of the plate architecture are incorrect. As shown, Durham's (1955:96, 98) specimens (along with several other species depicted in his monograph) appear to violate the ambulacral basicoronal plate pattern called Lovén's Rule (see explanation in David et al., 1995), a pattern found in all other echinoids. We were able to prepare several well-preserved CASG specimens of *E. coosensis*, compare them to the holotype, CASG 9, and reveal that the type species of the genus does not violate Lovén's Rule (figure 6.2). The second author is currently researching the phenomenon of Lovén's Rule in the clypeasteroids and has already discovered unusual plate patterns similar to that of *Eoscutella* in other fossil forms. The significance of oral surface plate architecture in the phylogenetic systematics of the northeastern Pacific sand dollars has still to be worked out in detail, but these patterns are already suggesting that taxa such as the genus *Kewia* (Nisiyama) listed in Linder et al. (1983) and Hickman (1969) will need to be reassessed (see below).

As Durham (1955) noted, *Eoscutella* is among the earliest recorded scutellines and it is "primitive" in having continuous interambulacra. However, *Eoscutella* also has advanced features shared with much more crownward members of the scutelline clade, including an extremely flattened test and a highly developed peripheral buttress system (figure 6.2). *Eoscutella. coosensis* is unique in that the test is much wider than long (figure 6.2), a feature shared with Argentine Eocene species placed in the same genus (Parma, 1985). Nothing is known of the plate pattern in the Argentine *Eoscutella*. Because the pattern is so distinctive in *E. coosensis*, it is imperative that the plate architecture of the Argentine specimens be determined for comparison, especially in light of the enormous range implied by the present generic assignment of the South American *Eoscutella*. Our analysis reveals that *Eoscutella* has features found in other early scutellines such as protoscutellids, including food grooves that bifurcate distally to the peristome (figure 6.2), not close to the peristome as suggested by Durham (1955). The combination of plesiomorphic and apomorphic characteristics in *Eoscutella*, as well as its early appearance in the fossil record, suggest that in an evolutionary context, it might be among the most important Pacific Northwest sand dollars.

Other significant taxa recorded from the Coaledo include a *Schizaster* sp. reported by Turner (1938). Current information does not permit elaboration on this record. In addition to these echinoids, Blake and Allison (1970) recorded dense accumulations of the ophiuroid, *Ophiocrossota baconi* (Blake and Allison, 1970), from deposits at Shore Acres State Park. These represent the first eastern Pacific Ocean record of this modern Australian ophiuroid genus. Another species, *Ophiocrossota oweni* Blake, 1975, has been reported from the Miocene of central California's coast ranges (Blake, 1975).

Eocene-Oligocene *Hypalocrinus* in Washington

A new species of isocrinid tentatively identified as a member of the extant genus *Hypalocrinus* (T. Oji, University of Tokyo, pers. commun., 1999) is occasionally found in upper Eocene and Oligocene deposits in western Washington. A complete crown was collected by J. Goedert from the deep-water upper Eocene Makah Formation in Clallam County, Washington, and is currently under study by C. Franzen-Bengston of the Swedish Natural History Museum. An additional crown from deep-water deposits of the Lincoln Creek Formation at Porter Bluffs, Washington, has been located at the Burke Museum. Although the exact stratigraphic position of this specimen cannot be ascertained, additional collection of *Hypalocrinus* columnals at Porter Bluffs points toward a latest Eocene age. *Hypalocrinus* columnals have also been found in the upper Eocene sections of the Lincoln Creek Formation along the Satsop River. A substantial set of pluricolumnals has also been collected from the Eocene McIntosh Formation ("Crescent-McIntosh transitional zone") in southwest Washington.

Miscellaneous Eocene Occurrences

Additional echinoderms reported or collected in Eocene rocks of Washington and Oregon are mentioned here for the completeness of this record and to suggest future avenues of research. Some of these fossils require additional documentation because there are serious questions concerning their provenance and systematic assignment.

Comatulid centrodorsals and arm parts have been collected from the upper Eocene Skookumchuck Formation in southwestern Washington. A possible centrodorsal has also been found in a slab of mostly disarticulated spatangoids from the Eocene Umpqua Formation in southwestern Oregon. These two finds may represent the oldest fossil comatulids on the Pacific Coast.

Holothuroid ossicles have been reported from the Roseburg, Lookingglass, and Flournoy Formations of Oregon (Orr and Miles, 1978). Linder (1988) listed *Ophiocrossota* from the Siletz River Volcanics. These volcanics are part of the Siletz terrane, which includes the Crescent Formation, and are older than the Coaledo or Cowlitz Formations. The specimens are probably not conspecific with named members of this genus from the younger Coaledo Formation (Blake and Allison, 1970). Linder (1988) listed undetermined clypeasteroids from the Tyee Formation, but these need further study. Linder (1988) also reported *Eoscutella* from the Umpqua Formation. Thoms (1965) recorded poorly preserved regular echinoid spines from the Elkton Formation, which appears to be an age equivalent of the Coaledo but laid down in deeper water. Spatangoids have also been reported from the Yamhill Formation (Linder, 1988). Some localities of the Yamhill Formation, which is probably equivalent or older than the Cowlitz Formation, yield extremely well preserved selachian fossils (Welton, 1979), suggesting that these localities be reviewed for possible echinoderm faunas, notably asteroids and holothuroids. Baldwin and Roberts (1952) indicated the presence of spine and test fragments in the silty sandstones of the upper Eocene Burpee Formation west of McMinnville. The Spencer Formation, possibly a deeper-water deposit equivalent in age to the Cowlitz Formation near Eugene, contains poorly preserved test and spine fragments, as well as ophiuroids listed as *Ophiocrossota* by Linder (1988). A spatangoid identified as *Spatangus* sp. has been reported from tuffaceous sandstones near Dallas, Polk County (Stokesbary, 1933).

OLIGOCENE DEPOSITS AND ECHINODERM FAUNAS

Eugene Formation

In her review of the fauna of the lower Oligocene Eugene Formation, Hickman (1969) noted that deposition likely occurred in relatively shallow seas, from intertidal to about 200 meters deep. These waters appeared to be subtropical or temperate and indicative of a gradually cooling environment in which faunas intermediate between the Keasey and Pittsburg Bluff Formations can be found. Hickman (1969) figured and briefly described some fairly well preserved but unidentified ophiuroids, as well as some echinoid fragments that she attributed to the genus *Salenia.* Several specimens of a small undetermined species of *Kewia* have also been collected in the Eugene Formation. Hickman (1969) noted that her material was probably conspecific with the material described and figured in Durham (1957) as *Kewia* sp. indet. and similar to the Oligocene *Kewia newcombei* (Kew, 1920) from the Sooke Formation (see below). Features suggesting to Durham (1957) that this was a species of *Kewia* include the aboral periproct and the simple internal buttresses on the interior of the test. The latter feature is easily seen in Hickman's (1969) specimens as well. However, several other clypeasteroid taxa exhibit these features, and it will be necessary to examine additional material to determine diagnostic features (as discussed below) of the oral surface plate pattern to make possible a full identification and description of these specimens.

Scotts Mills Formation

Linder et al. (1983) described a diverse echinoid fauna from shallow-water deposits of the upper Oligocene upper Marquam unit of the Scotts Mills Formation near Mollala, Oregon. This fauna includes representatives of the regular echinoids *Salenia, Glyptocidaris, Arbacia, Lytechinus,* and *Gagaria,* and a minute clypeasteroid assigned to the genus *Kewia.* According to Linder et al. (1983) these occur in storm and swell lag deposits among sea stacks of older basalts that formed a natural strand, providing optimal conditions for preservation. A late Oligocene *Kewia, K. marquamensis,* was described by Linder et al. (1983), who also analyzed the occurrence of what appears to be a magnetite weight belt in many specimens (see review in Mooi and Chen, 1995). The deposits of the Scotts Mills Formation may represent a slightly shallower gradient, quieter-water upper Oligocene analogue to the Eocene Crescent formation discussed above. These upper Oligocene deposits cannot shed much light on what happened to specific echinoderm taxa as they went from the "greenhouse" to the "icehouse" except to extend our knowledge of cladogenesis in echinoid lineages in the period after the Eocene-Oligocene transition.

Keasey Formation at Mist, Oregon

The upper middle Keasey Formation at Mist, Oregon, which measures approximately 250 meters in length and 50 meters in height,exposed in a large bluff along the Nehalem River, contains one of the most spectacular Tertiary echinoderm faunas of the Pacific Coast. It is famous primarily for its superbly preserved articulated stalked crinoids of the genus *Isocrinus.* Other echinoderms found at this site include several starfish, an ophiuroid, and at least three echinoids. A number of these echinoderms remain in open nomenclature pending future investigations. Several beautifully preserved representatives of other phyla are also found at Mist, along with a large molluscan fauna and large flora derived from nearby shorelines. Recent magnetostratigraphic work by Prothero and Hankins (2000) suggests an early Oligocene age for this locality. Its position close to the Eocene/Oligocene boundary warrants its detailed treatment in the context of this volume.

Early attention was given to the paleoecology of this site owing to its novelty among other Pacific Coast Tertiary faunas. In their paper describing the Keasey crinoids, Moore and Vokes (1953) recognized this novelty and compared the Mist fauna with other Tertiary crinoid occurrences. They noted that except for two or three species that were represented by cups, all described forms of Tertiary crinoids in the world are identified on the basis of fragments of the stems, the arms or both. On the basis of physical features of the tuffaceous matrix presumably derived from nearby andesitic volcanic sources, as well as bathymetric correlation of the diverse molluscan fauna and the presence of a well-preserved terrestrial flora, they suggested a depth ranging from 183 m to greater than 914 m, at a site only a few tens of kilometers from the shoreline.

Zullo et al. (1964) further reviewed bathymetric data and concluded that the Keasey Formation was deposited at a depth of 365 meters, but supported Moore and Vokes' (1953) conclusion that the shoreline was not far away. These data were heavily influenced by the abundance of the coral genus *Flabellum,* whose modern species have a reported optimal depth of 324 to 401 meters, and the echinoid *Salenia,* which has been reported from 91 to 456 meters in modern oceans. Additional depth interpretations utilizing bathypelagic fishes (David, 1956), mollusks (Hickman, 1976, 1980), selachians (Welton, 1979), and microfossils (McDougall, 1980) suupport this depth interpretation.

The crinoids, the abundant *Salenia,* and the corals occur at different stratigraphic levels representing three temporally isolated depositional and preservational scenarios. Work in progress by K. Campbell (University of Auckland) and the first author postulates a potential cold seep derivation for the *Isocrinus* Lagerstätte at the base of the outcrop, a second potential cold seep locality a few meters upsection dominated primarily by *Flabellum* and mollusks, surmounted several meters upsection by a time averagedtime-averaged assemblage dominated by *Salenia.* At this near-vertical outcrop, rocks from the *Salenia* layer have only been found as float commonly mixed with float derived from active quarrying of the crinoid Lagerstätte. No successful attempts have been made to pinpoint the source of this material, due to the hazardous conditions at this site. In the densest section of the Lagerstätte, the number of crinoids approaches 75 to 100 individuals per square meter for a thickness of approximately 1.5 meters and an approximate exposed area of 3-4 square meters. This density quickly tapers off to one or two individuals per square meter lateral to the area of highest concentration. In places, layers of apparently bioturbated crinoid coquina are present.

Even among weathered material, the preservation of the crinoids is excellent, although specimens showing postmortem disarticulation are not uncommon. Many crinoid specimens are preserved in near-living positions with filtration fans outstretched. It seems likely that these were buried in situ, presenting a unique opportunity to compare this Oligocene crinoid population with modern ones. Evidence of arm regeneration is plentiful and isolated; autotomized body parts abound. A number of bathypelagic fishes (David, 1956) and selachians (Welton, 1979) were described in association with this fauna. However, predation rates for and potential predators upon the Keasey Formation crinoids remain to be studied.

The depth, distance from shore, and faunal composition of the Mist locality are strongly reminiscent of sites off Grand Bahama visited by the second author in the Johnson Sea Link submersible. This similarity may carry with it implications that are only now being explored. Spines of a deep-water cidaroid resembling those of an extant *Calocidaris* species occurring among isocrinids at the Grand Bahama locality are also found associated with the Keasey Formation crinoids, although not commonly. *Calocidaris* has recently been observed to consume columnals at the distal ends of extant stalked crinoids (Baumiller et al., 1999). A search is under way at the

Mist locality for more complete specimens of these spines in order to confirm the generic assignment, as well as for evidence of predation. If this material is in fact *Calocidaris*, it will represent the first Pacific occurrence of this otherwise Atlantic genus. Direct evidence of cidaroid predation in the fossil record, if found, will likewise have significant paleobiological implications.

Moore and Vokes (1953) named two species of crinoids, *Isocrinus oregonensis*, and *I. nehalemensis* found together at Mist. Assignment to the genus *Isocrinus* was based on the nature of the stellate columnals and the structure of the rays, including the isotomous mode of branching, union of two primibrachs of each ray by synarthry, and the nature of pinnulation. Sieverts-Doreck (1971) later reassigned these species to *Chladocrinus*, which is considered by some to be a subgenus of *Isocrinus*. In a review of mostly Asian occurrences of the Isocrinidae, Klikushin (1982) named a new genus, *Raymondicrinus*, to hold the two Pacific Northwest species. Klikushin based his new genus on a muscular articulation between IIBr1 and IIBr2, which is different from all other *Isocrinus*, but is found in the genera *Metacrinus* and *Saracrinus*. In the latter, the articulation between IBr1 and IBr2 is syzygial, but it is muscular in the Oregon specimens. Hess et al. (1999:233-235) figured examples of these articulations and agreed with Klikushin's (1982) assignment, further speculating that the smaller *I. nehalemensis* was a juvenile form of *I. oregonensis*. After consulting with active crinoid taxonomists (T. Oji, University of Tokyo and C. Messing, University of Florida, pers. commun., 1999), we remain dubious of the validity of *Raymondicrinus*, and we prefer to use the more universally recognized assignment to *Isocrinus* of Moore and Vokes (1953) pending fuller revision of the family.

Zullo et al. (1964) described the echinoid *Salenia schencki*, noting excellent preservation including attached primary and secondary spines and completely articulated lanterns, indicating minimal, if any, transport. The first author has observed that these occur on the top and/or bottom faces of a 15-cm layer that is mostly composed of current-aligned spines as the primary matrix component, indicating the possibility that some of these may have been transported a short distance, with survivors or immediate recruitment of young *Salenia*. From the Oligocene Lincoln Creek Formation exposed on the north side of Bear River Ridge in southwestern Washington, *Salenia schencki* has also been collected from a fine grained slaty siltstone that was deposited in deep water.

As reported by Zullo et al. (1964), excellently preserved asteroids and ophiuroids are associated with the crinoids (as opposed to the *Salenia* layers). These asteroids were described by Blake (1973), but two, a species of *Astropecten* and a species of goniasterid, were left unnamed. Two new monotypic genera were described based on the new species *Mistia spinosa* and *Nehalemia delicata*. *Mistia spinosa*, a benthopectinid, was previously misidentified by Zullo et al. (1964) and cited again as such by Linder (1988) as a possible brisingid (D. Blake, University of Illinois, pers. commun., 2000). The ophiuroids are currently under study by Gordon Hendler (LACM).

At the same stratigraphic level, but approximately 10 to 20 m south of the Lagerstätten at Mist, specimens of *Flabellum* are frequently found along with diverse mollusks and other faunal elements. Partially disarticulated spatangoids identified as "*Brisaster*" *maximus* are not uncommon and were noted by Zullo et al. (1964). Few, if any, crinoid remains are found at this location.

Elsewhere at the Mist locality, the matrix is remarkably devoid of macrofossils in comparison to the concentration of fossils in the Lagerstätten. Observations that modern stalked crinoids prefer slightly elevated hardgrounds (Meyer, 1997) seem to be echoed in the conditions found here. K. Campbell and the first author are investigating the hypothesis that a cold seep may have created a slightly elevated hardground that provided ideal habitat for the crinoids. Cold seep activity may have also inhibited benthic organisms during and after deposition, as well as cemented the matrix, thereby enhancing the preservation. The presence of the bivalve *Solemya*, often associated with cold seeps, and suspected pogonophoran worm tubes (Adegoke, 1967) supports this possibility. Cold seeps are known elsewhere from the Keasey Formation, including a well-preserved cold-seep locality that includes a large chimney structure exposed in cross section along the Timber-Vernonia road (Nesbitt, et al. 1994). Crinoids have not been found at this locality, however. In extant cold seeps, echinoderms are rarely encountered, presumably because isosmotic body fluids make waters near cold seeps toxic to these organisms. Therefore, the presence of any echinoderm at a cold seep is unusual.

Unfortunately, most of the crinoid Lagerstätte at Mist was removed by illegal commercial quarrying in the 1980s, making future study difficult. However, large amounts of material were collected from the Lagerstätte by the first author in 1970-1974, and may prove adequate for ongoing analyses. Fortunately, at the deposit characterized by the coral *Flabellum*, another presumed cold seep site remains intact and uncollected. This site might provide a valuable close analogue to the cold seep associated with the crinoid Lagerstätte if data from the latter are found to be lacking.

Fragmentary *Isocrinus* remains and occasional crowns are found elsewhere in the Keasey Formation, including in strata of late Eocene age. Similar remains have also been found in other formations, including in a basaltic sandstone at Yachats, Oregon, and from the upper Oligocene Lincoln Creek Formation at Knappton, Washington. Shallow-water occurrences include the upper Eocene basal Lincoln Creek Formation at the Oakville Quarry in Oakville, Washington, and suspected shallow-water deposits of the upper Eocene Tukwila Formation (Cowlitz Formation equivalent) at Dumas Bay, Washington. The Tukwila Formation specimens are found in a gritty sandstone float sourced nearby. These rocks have produced disarticulated *Isocrinus* stems and cidaroid spines, along with shallow-water bivalves such as *Mytilus*. The crinoid fragments might represent the only shallow water occurrence of Tertiary stalked crinoids from the eastern Pacific Ocean. These particular examples are the property of a private collector and were examined by the first author in 1995. They are no longer available for study, leaving us to hope that additional sampling will yield more material. According to Nesbitt (1995) these deposits represent a northern analog to the Cowlitz Formation of southwestern Washington and are derived from a large deltaic system that dominated most of western Washington in the late Eocene.

Linder and Orr (1984) also reported on other excellently preserved material in the Yachats basalts, including test and spine fragments of regular echinoids, two species of undetermined cidaroids, and specimens identified as *Heterocentrotus, Lytechinus*, and "*Brisaster" maximus* (see below). We have as yet been unable to examine this material, but we expect it to yield faunas comparable to the Keasey Formation, given that the Yachats unit seems to be an approximate age equivalent.

Other Keasey Formation Echinoderm Faunas

Another early Oligocene locality yielding significant echinoderm specimens is located 1.3 km south of Vernonia, Oregon, along Oregon Highway 47. This locality was pinpointed on a map of crinoid localities in Moore and Vokes (1953:114), but not mentioned in the text. Collections from recent quarrying have resulted in a diverse invertebrate fauna that rivals the Mist locality and includes hexactinellid sponges in excellent preservation. Disarticulated entire *Isocrinus* individuals as well as stems are locally abundant, and a few crinoid coquina layers have been noted. A poorly preserved asteroid resembling *Nehalemia delicata* Blake, an ophiuroid, a possible comatulid and scattered spines and test fragments tentatively identified as *Stylocidaris* have also been collected. Spines of the latter have also been collected elsewhere in the Keasey Formation. The most frequently collected echinoderm is a schizasterid spatangoid discussed below. Several crushed but complete tests and molds of tests, commonly with external features such as spines still articulated have been prepared for analysis by the first author.

"*Brisaster" maximus* is an enigmatic spatangoid whose name has been used in the literature on more than one occasion as a "garbage can category" in a similar manner to that for some species of *Schizaster*. The type series of "*Brisaster" maximus*, somewhat inadequately described and figured by Clark (1937), consists of a mold of the aboral surfaces of one entire and of five partial specimens, collected on one slab from the younger Oligocene Pittsburg Bluff Formation of northwestern Oregon. The apical system, petaloids, fascioles, and tuberculation suggest that the very large "*B.*" *maximus* of the Pittsburg Bluff Formation and the Keasey Formation specimens are conspecific, permitting us to make additional observations of the species represented by this material. Possession of three gonopores is integral to the diagnosis of *Brisaster* (Hood and Mooi, 1998). Mortensen (1951) stated that the number of gonopores in "*B.*" *maximus* was unknown. We have been able to determine that several of the new Keasey Formation specimens have four gonopores, and our re-examination reveals that the holotype of "*Brisaster*" *maximus* does as well (figure 6.3).

So, what is "*Brisaster*" *maximus*? The holotype has well-developed peripetalous and lateral fascioles typical of schizasterids, as well as a single regular series of pores in the anterior, aboral ambulacrum (figure 6.3) noted in some keys to the schizasterid genera (e.g., Mortensen, 1951). Ratios of the length of anterior to length of posterior petaloids are within the range of *Schizaster cordiformis* and *S. lecontei* (table 6.1), but the poor preservation of the holotypes of these species prevents further comparison. However, we can say that the petaloids in relatively uncrushed specimens of "*B.*" *maximus* from the Keasey do not

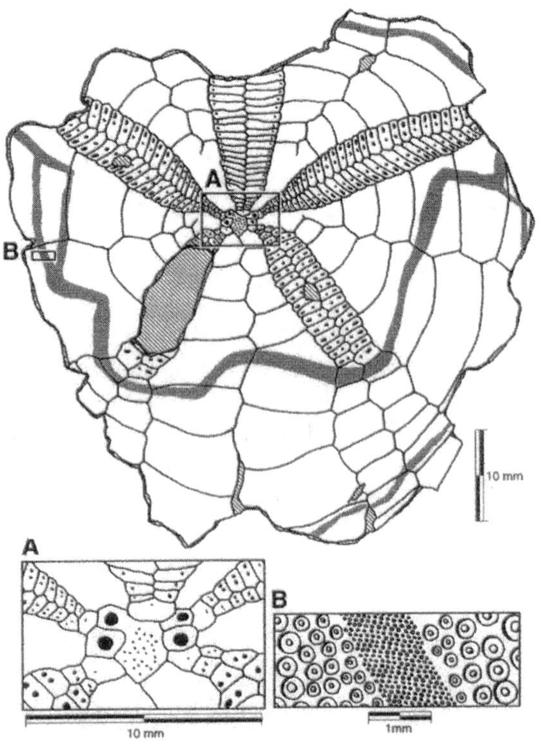

Figure 6.3. Holotype of "*Brisaster*" *maximus* H. L. Clark, 1937, MCZ 102088. Because specimen is a mold of outer surface, symmetry of drawing is reverse of natural orientation. Fascioles are shaded, damaged areas are hatched. Inset A shows shows detail of apical system. Inset B shows detail of tuberculation in peripetalous fasciole.

exhibit the curvature typical of the Eocene *Schizaster*, a genus that is also supposed to have only two gonopores. As suggested above, "*Brisaster*" *maximus* cannot be a *Brisaster* because of its straight petaloids and number of gonopores. The incomplete holotype of "*Brisaster*" *maximus* does not show the condition of the anterior notch or the periproct. The additional specimens from the Keasey Formation show that the anterior notch is well developed, and that the periproct seems to be elongate rather than transversely oval. Therefore, according to Mortensen's (1951) keys, "*Brisaster*" *maximus* is most like *Paraster* Pomel or *Linthia* Desor. Although the latter genus is the most likely candidate among the known genera with fossil representatives, a full revision of northeastern Pacific schizasterid spatangoids will be necessary before we can unequivocally refer the holotype and new material of "*Brisaster*" *maximus* to *Linthia.*

We had hoped to collect additional specimens from the type locality, described by Clark (1937:368-369) as "Washington County, Oregon, from the center of the south line of Sec. 12, T. 3 N. R. 4 W.; in a dug well along road at C. H. Bonham Farm; 1000 feet south of the coal seam exposure; Pittsburg Bluff Formation (J. T. Holman, coll., Stanford University) (Schenck)." We were unable to pinpoint this collecting site because of an unusual situation in which it is the type locality, not the type specimen that is lost. The authors speculated that the holotype was possibly collected from an isolated outcrop of the Keasey Formation, given the abundance of "*Brisaster*" *maximus* in that formation. However, the matrix of the holotype is of a different lithology from most of the Keasey Formation in containing more fine carbonaceous material, in having a gritty texture as opposed to the tuffaceous nature of the middle and upper members of the Keasey Formation, and in completely lacking foraminifera (D. Peterson, CAS, pers. commun., 1999).

Searching through historical maps and records at the Oregon Historical Society in Portland, Oregon, we failed to locate any "Bonham Farm" in Washington or Columbia Counties, including at the topographic coordinates. This locality is not listed in Schenck's otherwise complete listing of Holman's localities (Schenck, 1936), nor is it mentioned in Schenck's field notes so far available to us. However, while examining historic plat maps of the area, we did locate one potential candidate, a farm with name "C. H. Brown" in section 34, T5N, R4W in Columbia County along Oregon Highway 47 a few kilometers north of Vernonia, and a few kilometers south of the Pittsburg Bluff Formation type section. Nearby are fresh exposures of the basal Pittsburg Bluff Formation with a lithology that matches the holotype slab (A. Niem, Oregon State University, pers. commun., 1999), and that have also yielded one poorly preserved echinoid fragment. We suggest that this location is the type locality and that "*Brisaster*" *maximus* was collected from basal Pittsburg Bluff Formation rocks that are stratigraphically adjacent and above the Keasey Formation occurrences at Mist and just south of Vernonia.

"*Brisaster*" *maximus* has also been recorded from the lower Oligocene strata of the Lincoln Creek Formation in southwest Washington (Armentrout, 1973). Examination of Armentrout's material tentatively confirms this identification. As noted above, the first author's identification of a single specimen from the Tukwila Formation and reported by Nesbitt (1998) was later determined to be similar to the Crescent Formation *Schizaster,* rather than "*Brisaster*" *maximus.*

Another unstudied collection of echinoderms was assembled by R. Van Atta from Robinson Creek, Oregon, near Oregon Highway 26, during fieldwork associated with his doctoral thesis (Van Atta, 1971). This fauna was extracted from a gritty, limonitic, loosely consolidated sand and contains crinoid stems, echinoid spines, and several beautifully preserved asteroids. The authors have yet to examine this material. However, recent field work in this area (A. Niem, Oregon State University, pers. commun., 2000) indicates this matrix may be basal Keasey Formation or Clark and Wilson Sandstone of late Eocene age. Comparison of these late Eocene asteroids with the nearby early Oligocene asteroids from Mist will undoubtedly yield insights into the evolution of this important but understudied part of the echinoderm fauna.

Pysht Formation

Deeper-water concretionary sandstones, siltstones, and mudstones of the Oligocene Pysht Formation in Clallam County have produced a small but important echinoderm fauna. These have been the focus of recent collection efforts by the first author. Durham (1944) indicated the presence of an echinoid fauna at Durham's UCMP locality A-3687 in the lower part of the Pysht section. Echinoids of the brissid genus *Brissopsis* are not uncommon at this locality and they are preserved in abundance in layers at UCMP locality A-3689. These echinoids have been collected elsewhere in the Pysht Formation. Spines and test fragments commonly occur in samples washed for foraminiferal studies (J. Grieff, UWBM, pers. commun., 1999). *Brissopsis* has also been found in a deep-water facies of the Oligocene Blakeley Formation at Restoration Point in Kitsap County, Washington, and reported by Tegland (1933) and Grant and Hertlein (1938) as *Schizaster* sp. This echinoid has been also found at the Lincoln Creek Formation at Porter Bluffs, Washington. Throughout the Pysht Formation, abundant ichnofossils suggesting burrowing activity, possibly by *Brissopsis,* can be found.

Preservation of these echinoids ranges from good to excellent, especially in the Blakeley Formation. A few specimens have attached spines. Apical system details are similarly excellent, even in fragmentary specimens. In spite of the excellent preservation, specific identification of this echinoid will rely upon as yet incomplete comparison with other modern and fossil species of *Brissopsis.* The poor quality of type material of a similar Japanese Miocene species described by Morishita (1957) will undoubtedly complicate these comparisons. Modern *Brissopsis* were revised by Chesher (1968) on the basis of 41 morphometric characters. We have not yet attempted to place the Pysht Formation *Brissopsis* within this morphometric framework, and we remain skeptical that enough characteristics are preserved to do so.

Occurring with the *Brissopsis* at UCMP localities A-3687 and A-3689 are rare goniasterid asteroids and infrequent ophiuroids. These are preserved in a single layer containing crushed echinoid tests and frequent, well-preserved leaves from a nearby onshore source. The matrix contains several layers of upward-fining, fine-grained, rhythmic turbidite. Where bioturbated by the mud shrimp, *Callianassa,* or by *Brissopsis,* neither echinoids nor the asteroids can be found.

Two new species of asteroids and dense accumulations of ophiuroids occur near Deep Creek, Washington, in the upper Oligocene western part of the Pysht section near UCMP locality A-3673. At the same locality, layers containing many mytilid bivalves and flattened logs supporting abundant barnacles suggest shallow water. Loading structures, minor scouring and crossbedding are also suggestive of shallow-water deposition. The echinoderms occur at the top of a debris flow layer dominated by the large bivalve *Panopea,* aligned shells of the scaphopod *Dentalium* and other mollusks, sediment clasts, and wood fragments. Above this coarse layer the sediment fines upward into a very fine-grained matrix containing abundant well-preserved leaves and organic matter.

Two preservational scenarios seem to apply at this locality. In one layer, abundant *Pisaster*-like forcipulate asteroids with occasional ophiuroids are preserved in a very hard, resistant, and highly indurated layer. In a second layer, an astropectinid similar to *Dipsacaster* is frequently encountered with very abundant ophiuroids of an unidentified species, and rare individuals of another, larger species tentatively identified as a species of *Ophiocrossota* by the first author. Assignment of the unidentified ophiuroid will depend on discovery of additional specimens. This ophiuroid is occasionally found elsewhere in the upper Pysht Formation. Similar ophiuroids have also been collected from the deep-water Jansen Creek Member of the Makah Formation. All the Pysht

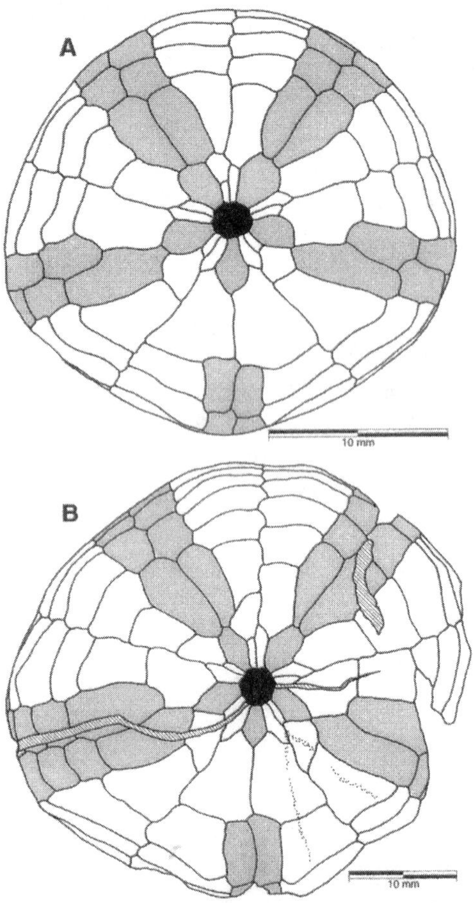

Figure 6.4. Comparison of two species of *Kewia*: (A) oral surface plate architecture of *K. blancoensis* (Kew, 1920), CASG 67237.01; (B) oral surface plate architecture of *K. newcombei* (Kew, 1920), holotype, UCMP 5164. Mouth in solid black, interambulacral plates shaded, damaged areas hatched. In B, some portions of food grooves are indicated by stippling in left posterior ambulacrum.

Formation ophiuroids are under study by Gordon Hendler (LACM).]The matrix of the second layer remains relatively unaltered and soft, sometimes complicating collection efforts. Post-depositional diagenesis may account for the differing lithologies of these two layers. However, there seems to be some correlation with species, possibly as a result of their substrate preferences.

In both layers, preservation of these asteroids and ophiuroids seems remarkable in light of the generally bioturbated nature of the Pysht Formation. However, vertical bioturbation is lacking in these beds, which together with the generally high organic content of the matrix and the presence of well preserved leaves, suggest that substrate conditions were anoxic and that these echinoderms lived at the sediment-water interface. Here, they were subjected to smothering by debris flows of anoxic nature.

Preservation of the asteroids suggests rapid burial and/or quiet water conditions either in a protected setting or below the wave base. Microscopic observations reveal that some of the astropectinids are preserved in the initial stages of disarticulation. Loose elements occur immediately adjacent to the specimens, further supporting the suggestion of quiet water during preservation as well as the passage of some time between mortality and burial. An influx of low salinity or fresh water from a nearby riverine source, followed by deposition of anoxic sediments with high organic content, could account for this type of preservation. This would also explain the abundance of leaves and other plant material. Similar taphonomic scenarios involving modern asteroids at the mouth of San Francisco Bay during flood stages of the Sacramento River have been observed by the second author.

Many of the arms of the astropectinids appear broken and thus suggest transport. However, the presence of loose elements adjacent to the specimens suggests that transport was minor and that this might represent postmortem trauma. In contrast, the abundant ophiuroids are generally preserved entire. In the bed containing the astropectinids and numerous ophiuroids, it can be seen that where the astropectinids are abundant, the ophiuroids are rare, and vice versa. Further work is necessary to determine if this is a preserved predation signature in which the astropectinids preyed upon the ophiuroids.

Another dense accumulation in a limestone matrix composed predominantly of *Ophiocrossota* occurs as float from a source above the beds containing the asteroids near Deep Creek. These dense accumulations resemble ophiuroid coquinas reported from the California Miocene (Blake, 1975). Aronson and Blake (1997) discussed the paleobiological significance of such dense accumulations of ophiuroids with respect to predational scenarios, especially in shallow water.

Sooke Formation

Kewia newcombei (Kew, 1920) is occasionally collected from the upper Oligocene shallower water deposits of the Sooke Formation on Vancouver Island. Although this formation is temporally distant from the Eocene/Oligocene boundary, *Kewia* is of great interest in analysis of echinoid evolution in the northeast Pacific because it is among the few clypeasteroid genera with purported Oligocene representatives. An understanding of its phylogenetic position is critical for analysis of the diversification of sand dollars after the Eocene.

Re-examination of type material at UCMP and CASG permits us to add some data about *K. newcombei* that confirms its placement in *Kewia.* As Durham (1955) noted, and contrary to Kew's (1920) original observations, *K. newcombei* clearly has a supramarginal periproct, a diagnostic feature of the genus. The plate pattern on the oral surface of the type species of the genus, *K. blancoensis* (Kew, 1920), is peculiar in that the first post-basicoronal plates in each interambulacrum are greatly elongated, but in the posterior paired interambulacra, only the plate in the posteriormost half-interambulacrum is continuous with the basicoronal. This unusual pattern is also found in *K. newcombei* (figure 6.4). Many *Kewia,* including the type species of the genus, are described as having "simple" food grooves that do not branch. However, the Sooke Formation material is similar to at least one of the species placed in *Kewia* by Wagner (1974) in having bifurcated food grooves (figure 6.4). A complete revision of the genus will be necessary in order to determine the ubiquity of this pattern and whether the oral surface plate architecture can be used to diagnose the genus.

Work by the second author is beginning to show that a significant number of the species listed as *Kewia* by Linder et al. (1988) do not have the salient features of either *Kewia blancoensis* or *K. newcombei.* For example, *K. fairbanksi* (Arnold, 1907) has trifurcated food grooves and an oral surface plate pattern typical of the echinarachniids, which also have trifurcated food grooves.

Miscellaneous Oligocene Occurrences

Echinoids are present at several Oligocene localities in the Pacific Northwest. Regular echinoids preserved as crush tests and resembling *Strongylocentrotus* have been collected from deep-water deposits of the lowest Oligocene Jansen Creek Member of the Makah Formation. Ophiuroids of uncertain affinities have also been collected from the Jansen Creek Member, the lower Oligocene Eugene Formation (Hickman, 1969); the lower Oligocene Quimper Formation on Discovery Bay, Jefferson County, Washington; from the Oligocene Pittsburg Fluff Formation type area near Vernonia, Oregon; from shallow-water upper Oligocene rocks near Seal Rock, Oregon; as well as in the Lincoln Creek Formation at Porter and Knappton, Washington (figure 6.1). Echinoid test fragments and spines have been collected from shallow-water deposits of the early Oligocene Gries Ranch fauna of southwestern Washington. A complete regular echinoid test was collected from shallow water deposits of the lower Oligocene Quimper Formation at Discovery Bay, Washington. In the lower Oligocene Bastendorff Formation, indeterminate test fragments and spines of echinoids, along with an unidentified clypeasteroid, have been reported by Linder (1988).

As yet unstudied spatangoids are known from the Oswald West mudstones (A. Niem, Oregon State University, pers. commun., 1999), a deep-water equivalent to the Keasey Formation. These mudstones contain an exceptionally diverse and well preserved ichnofauna and need to be fully studied in order to determine the presence of "*Brisaster*" *maximus,* isocrinids, or other members of the fauna already known to be characteristic of the Keasey Formation.

An Oligocene occurrence of crinoid stems was reported by Durham (1944) in the Blakeley Formation at Beans Point, Kitsap County, Washington. These specimens have not been located, but are presumed to be *Hypalocrinus* or *Isocrinus.* Disarticulated individuals and stems of *Hypalocrinus* are also occasionally found in upper Oligocene beds of the Lincoln Creek Formation at Knappton, Washington, and occur at some localities with *Isocrinus* columnals.

Deep-water Miocene echinoderm faunas in Oregon and Washington are notably rare, when compared with Eocene and Oligocene faunas. This is due mostly to the absence of preserved deeper-water deposits of this age in this region. A few asteroid fragments, ophiuroids, and echinoids have been collected from shallower-water deposits of the Astoria Formation of coastal Oregon (Moore, 1963). One superbly preserved astropectinid has been collected from the Astoria Formation at Beverly Beach, Oregon. Dense accumulations of ophiuroids have been collected from the Astoria Formation near Artic, Washington, and the Nye Formation near Newport, Oregon. Similarly, a few species of clypeasteroids have been found in the shallow-water upper Miocene to lower Pliocene Montesano, Empire and Astoria Formations.

Finally, large echinoids tentatively identified as the extant species *Brisaster latifrons* (Agassiz, 1898) have been found in shallow-water bioturbated deposits of the Pliocene Quinault Formation.

Additional specimens might confirm this identification and will potentially offer valuable evolutionary insights for this modern species. Should "*Brisaster*" *maximus* be reassigned to another genus, then these specimens will represent the first true fossil representatives of the genus *Brisaster*.

DISCUSSION

In the Eocene and Oligocene of the Pacific Northwest, clypeasteroids, spatangoids with robust tests, and fragmentary remains of cidaroids are the only echinoderms found in deposits from nearshore, higher-energy environments. More delicate echinoids are found in strata deposited in quieter-water environments. Crinoids are mostly associated with deeper water environments. However, except for one extraordinary occurrence, crinoids are commonly found disarticulated or, more commonly, as columnals only. A few shallow-water occurrences of stalked crinoids are noted. Asteroids, if found at all, are found in deeper-water deposits or between debris flows containing a high organic content, indicating possible anoxic conditions.

Of all the stalked crinoid occurrences in the Eocene and Oligocene of the Pacific Northwest, only two possible upper Eocene shallow-water occurrences have been found. No other occurrences are known from northeastern Pacific Ocean shallow water deposits from the Oligocene and later. Therefore, it could be suggested that stalked crinoids disappeared from shallow water in the Pacific Northwest at the end of the Eocene. Conversely, fossil analogues of the modern comatulid *Florometra seratissima*, which ranges from shallow water in Canada and Alaska to deeper water in Oregon and Washington, should be in the record, unless *Florometra* has only recently colonized this area. However, identification of the few comatulid centrodorsals found so far remains dubious. Absence of recorded comatulid remains in the shallow-water record suggests that the strata may simply not have preserved these taxa. Applying this admittedly weak stratigraphic control to the stalked crinoids will tend to undermine the idea of their Eocene disappearance from shallow water. A search paradigm for crinoidal elements, especially in microfaunal studies of Oligocene to recent sediments has the potential to yield information about the appearance of *Florometra*, and to shed new light on the departure of shallow-water stalked crinoids.

The disappearance of isocrinid genera from the deeper waters of Oregon and Washington can be timed with even less precision, since there are fewer deeper-water formations after the late Oligocene in the Pacific Northwest. To date there are no published occurrences of isocrinid stems occurring in deeper-water Miocene formations in California. In contrast, *Isocrinus* has been collected in the Miocene of Japan (Oji, 1990), and *Hypalocrinus* still inhabits the western Pacific.

The robust test architecture of the echinoids has enhanced their preservation potential with respect to other echinoderm classes. The available data for spatangoids suggest that *Schizaster* disappeared from the Pacific Northwest at the end of the Eocene. Without implying a causal connection, it would appear that taxa such as "*Brisaster*" *maximus* and *Brissopsis* replaced *Schizaster* in the Oligocene. There is a remarkable conservatism in morphotypes in these spatangoids, and even the Eocene schizasterids are virtually indistinguishable from modern forms.

Protoscutellid and eoscutellid sand dollars are known only from the Eocene, both in the Pacific Northwest and elsewhere. The undescribed protoscutellid from the Crescent Formation has no obvious close relatives in the Eastern Pacific, but preliminary evidence suggests that it is related to Eocene protoscutellids in the southeastern United States. *Eoscutella* appears to have representatives in eastern South America. The implications are that both the Protoscutellidae and Eoscutellidae had much wider distributions than formerly realized, and that these sand dollars, with their high preservation potential, have much to tell us about paleobiogeography before and after the Eocene-Oligocene transition.

The Oligocene and later sand dollars are morphologically and conspicuously distinct from the Eocene forms, especially in plate pattern, location of the periproct, and food groove branching pattern. Without a complete phylogenetic analysis of the scutelline sand dollars, it is difficult to say exactly to which clades the Oligocene taxa are most closely related, but it would seem that neither protoscutellids nor eoscutellids are the closest relatives of forms such as *Kewia*. This implies that the Oligocene taxa are derived from sand dollars that originated elsewhere. Indeed, several described species of *Kewia* from Japan and Alaska suggest a more northern and western origin of the *Kewia* and echinarachniid clades. Presumably these taxa were well situated to move southward during the post-Eocene cooling trend.

Studies of echinoderm taphonomy (e.g., Nebelsick, 1995a, 1995b) illustrate that taphonomy must be taken into account when studying overall diversity trends. Asteroids illustrate this very well because the northeast Pacific Ocean currently hosts a large diversity of sea star taxa. However, since the time during which the Pysht Formation was deposited (late Oligocene), the fossil record of sea stars is meager. Except for material from the Keasey and Pysht Formations, one specimen from the Lincoln Creek Formation, one from the Miocene Astoria Formation, and scattered California Tertiary occurrences, sea stars are generally absent from the record. Taphonomic factors besides depth must also bear on the frequency with which whole sea stars are preserved, because asteroids are absent from several deeper-water deposits in Oregon and Washington such as the Eocene Humptulips Formation of the southwestern Olympic Peninsula.

The rarity of these forms means that less can be said about trends of asteroid diversity in the Eocene and Oligocene of the Pacific Northwest compared to the present. To aid in characterizing starfish diversity in the Pacific Northwest Tertiary, Blake (1973) proposed a search for sea star skeletal elements. To date there have been no systematic searches for these elements in the microfaunal record of the Pacific Northwest. However, such studies are planned by the first author. Echinoderm elements as small and delicate as holothurian ossicles have been reported (Orr and Miles, 1978), suggesting that both these and starfish ossicles have been overlooked but will be found with future concerted effort.

It seems clear that once outstanding taxonomic issues are addressed, echinoids represent our best hope for using echinoderms to analyze Eocene-Oligocene faunal change. However, ongoing work with echinoderms less amenable to whole body preservation adds valuable data points that we hope will contribute to a more global scenario. Search protocols that can find individual elements are not now widely used, but these should yield additional data. Similarly, although the low preserved diversity of ophiuroids yields few data points of regional significance, dense accumulations of ophiuroids will become more important. This will become especially true as these accumulations are correlated to depth of deposition and stratigraphic occurrence, yielding information on predation pressures and movement of dense ophiuroid communities into deeper waters.

ACKNOWLEDGMENTS

J. Wood, C. Gar, K. Smejkal, and S. Meyer provided access to important Keasey localities. J. Goedert, J. Grieff, P. Grieff, J. Cornish, the Healy family, K. Reed, J. Thorsen, L. Nesbitt, R. Berglund, D. Palmiter, B. Smith, C. Hedeen, R. Gaspers, S. Burns, M. Kernan, N. Ball, and L. Burns assisted in the field, and collected and donated many of the specimens used in this study. D. Blake, C. Mah, K. Campbell, C. Hickman, E. Moore, D. Squires, T. Oji, E. Nesbitt, A. Niem, J. Goedert, J. Cornish, B. Orr, and L. Orr participated in many useful conversations. E. Nesbitt (BM), J. DeMouthe (CASG), D. Squires (CSUN), L. Groves (LACM), F. Collier and F. d'Escrivan (MCZ), K. Wetmore-Grycewicz (UCMP), and T. Rice of the Sea and Shore Museum in Port Gamble, Washington, kindly provided access to collections for study and loaned important comparative or type material. D. Blake and C. Messing reviewed the manuscript. G. Hendler, J. Goedert and K. Reed provided additional useful suggestions and corrections. Finally, we extend thanks to D. Prothero for inviting us to participate in this volume.

REFERENCES

Adegoke, O. S., 1967, A probable pogonophoran from the early Oligocene of Oregon: Journal of Paleontology, v. 41, p. 1090-1094.

Agassiz, A., 1898, Reports on the dredging operations off the west coast of Central America to the Galápagos, to the west coast of Mexico, and in the Gulf of California, in charge of Alexander Agassiz, carried on by the U.S. Fish Commission Steamer "Albatross," during 1891, Lieut. Comm. Z.L. Tanner, U.S.N., Comm., 23, Preliminary report on the Echini: Bulletin of the Museum of Comparative Zoology, v. 32, p. 71-86.

Armentrout, J. M., 1973, Molluscan paleontology and biostratigraphy of the Lincoln Creek Formation, late Eocene–Oligocene, southwestern Washington [Unpubl. Ph.D. dissert.]: Seattle, Washington, University of Washington, 479 p.

Arnold, R., 1907, New and characteristic species of fossil mollusks from the oil-bearing Tertiary formations of southern California: Proceedings of the U.S. National Museum, v. 32, p. 525-546.

Aronson, R. B., and Blake, D. B., 1997, Evolutionary paleoecology of dense ophiuroid populations, *in* Waters, J. A. and Maples, C. G., eds., Geobiology of echinoderms, Paleontological Society Papers, v. 3, p. 107-119.

Baldwin, E. M., and Roberts, A. E., 1952, Geology of Spirit Mountain Quadrangle, Oregon: U.S. Geological Survey Oil and Gas Investigations Map OM 129, scale 1:500,000, 1 sheet.

Baumiller, T. K., Mooi, R., and Messing, C., 1999,

Predator-prey interactions between cidaroids and crinoids and their ecological and evolutionary implications: Geological Society of America Abstracts with Programs, v. 31, no. 7, p. A-172.

Blake, D. B., 1973, Ossicle morphology of some Recent asteroids and description of some west American fossil asteroids: University of California Publications in Geological Sciences, v. 104, 59 p.

Blake, D. B., 1975, A new west American Miocene species of the recent Australian ophiuroid *Ophiocrossota*: Journal of Paleontology, v. 49, p. 501-507.

Blake, D. B. and Allison, R. C., 1970, A new west American Eocene species of the recent Australian ophiuroid *Ophiocrossota*: Journal of Paleontology, v. 44, p. 925-927.

Chesher, R. H., 1968, The systematics of sympatric species in West Indian spatangoids: A revision of the genera *Brissopsis, Plethotaenia, Paleopneustes,* and *Saviniaster*: Studies in Tropical Oceanography, 168 p.

Clark, H. L., 1937, New sea urchin from the "Oligocene" of Oregon: Transactions of the San Diego Society of Natural History, v. 8, p. 367-374.

David, B., Mooi, R., and Telford, M., 1995, The ontogenetic basis of Lovén's Rule clarifies homologies of the echinoid peristome, *in* Emson, R. H., Smith, A. B., and Campbell, A. C., eds., Echinoderm Research 1995: Rotterdam, Netherlands, A. A. Balkema, p. 155-164.

David, L. R., 1956, Tertiary anacanthin fishes from California and the Pacific Northwest; their paleoecological significance: Journal of Paleontology, v. 30, p. 568-607.

Durham, J. W., 1944, Megafaunal zones of the Oligocene of northwestern Washington: University of California Publications, Department of Geological Sciences Bulletin, v. 27, p. 101-211.

Durham, J. W., 1955, Classification of clypeasteroid echinoids: University of California Publications in Geological Sciences, v. 31, p. 73-198.

Durham, J. W., 1957, Notes on echinoids: Journal of Paleontology, v. 31, p. 625-631.

Fisher, W. K., 1911, Asteroidea of the North Pacific and adjacent waters: U.S. National Museum Bulletin, v. 76, 419 p.

Grant, U. S., and Hertlein, L. G., 1938, The west American Cenozoic Echinoidea: Publications of the University of California at Los Angeles in Mathematical and Physical Sciences, v. 2, 225 p.

Hess, H., Ausich, W. I., Brett, C. E., and Simms, M. J., 1999, Fossil crinoids: Cambridge, Cambridge University Press, 275 p.

Hickman, C. S., 1969, The Oligocene marine molluscan fauna of the Eugene Formation in Oregon: University of Oregon, Museum of Natural History Bulletin, v. 16, 112 p.

Hickman, C. S., 1976, Paleogene marine gastropods of the Keasey Formation in Oregon: Bulletins of American Paleontology, v. 78, p. 1-112.

Hickman, C. S., 1980, Bathyal gastropods of the family Turridae in the early Oligocene of the Keasey Formation in Oregon, with a review of some deep-water genera in the Paleogene of the eastern Pacific: Bulletins of American Paleontology, v. 70, p. 1-119.

Hood, S., and Mooi, R., 1998, Taxonomy and phylogenetics of extant *Brisaster* (Echinoidea: Spatangoida), *in* Mooi, R. and Telford, M., eds., Echinoderms: San Francisco: Rotterdam, Netherlands, A.A. Balkema, p. 681-686.

Kew, W. S., 1920, Cretaceous and Cenozoic Echinoidea of the Pacific Coast of North America: University of California Publications in Geological Sciences, v. 12, p. 23-236.

Kier, P. M., 1977, The poor fossil record of the regular echinoid: Paleobiology, v. 3, p. 168-174.

Klikushin, V. G., 1982, Taxonomic survey of fossil isocrinids with a list of the species found in the USSR: Geobios, v. 15, p. 299-325.

Linder, R. A., 1988, Mid-Tertiary echinoids and Oligocene shallow marine environments in the Oregon central western Cascades [M.Sc. dissert.]: Eugene, Oregon, University of Oregon, 196 p.

Linder, R., and Orr, W. N., 1984, Oregon Eocene echinoids: Oregon Academy of Science, 41st Annual Conference, Abstracts, p. 19.

Linder, R., Durham, J. W., and Orr, W. N., 1983, New Late Oligocene echinoids from the central Western Cascades of Oregon: Journal of Paleontology, v. 62, p. 945-958.

Mallory, V. S., 1959, Lower Tertiary biostratigraphy of the California Coast Ranges: Tulsa, Oklahoma, American Association of Petroleum Geologists, 416 p.

McDougall, K., 1980, Paleoecological evaluation of late Eocene biostratigraphic zonations of the Pacific Coast of North America: SEPM, Paleontological Monograph 2, 46 p.

Merriam, J. C., 1899, The Tertiary sea-urchins of

middle California: Proceedings of the California Academy of Sciences, series 3, v. 1, p.161-170.

Meyer, D. L., 1997, Implications of research on living stalked crinoids for paleobiology, *in* Waters, J. A. and Maples, C. G., eds., Geobiology of Echinoderms, Paleontological Society Papers, v. 3, p. 31-43.

Mooi, R., 1987, A cladistic analysis of the sand dollars (Clypeasteroida: Scutellina) and the interpretation of heterochronic phenomena [Ph. D. dissert.]: Toronto, Ontario, Canada, University of Toronto, 208 p.

Mooi, R., and Chen, C. P., 1995, Weight belts, diverticula, and the phylogeny of the sand dollars (Clypeasteroida: Echinoidea): Bulletin of Marine Science, v. 57, p. 186-195.

Mooi, R., Martínez, S., and Parma, S. G., 2000, Phylogenetic systematics of Tertiary monophorasterid sand dollars (Clypeasteroida: Echinoidea) from South America: Journal of Paleontology, v. 74, p. 263-281.

Moore, E. J., 1963, Miocene marine mollusks from the Astoria Formation in Oregon: U.S. Geological Survey Professional Paper 419, 109 p.

Moore, E. J., 1976, Oligocene marine mollusks from the Pittsburg Bluff Formation in Oregon: U.S. Geological Survey Professional Paper 922, 66 p.

Moore, R. C., and Vokes, H. E., 1953, Lower Tertiary crinoids from northwestern Oregon: U.S. Geological Survey Professional Paper 233-E, p. 113-148.

Morishita, A., 1957, On some fossil echinoids of Japan, *Brissopsis* and *Eupatagus*: Memoirs of the College of Science, University of Kyoto, Series B, v. 24, p. 161-164.

Mortensen, T., 1951, A monograph of the Echinoidea, 5.2, Spatangoida II: Copenhagen, C. A. Reitzel, 593 p.

Nebelsick, J. H., 1995a, Uses and limitations of actuopalaeontological investigations of echinoids: Geobios Memoire Speciale 18, p. 329-336.

Nebelsick, J. H., 1995b, Comparative taphonomy of clypeasteroids: Eclogae Geologicae Helvetiae, v. 88, p. 685-693.

Nesbitt, E. A., 1995, Paleoecological analysis of molluscan assemblages from the middle Eocene Cowlitz Formation, Southwestern Washington: Journal of Paleontology, v. 69, p. 1060-1073

Nesbitt, E. A., 1998, Marine fauna of the middle Eocene Tukwila Formation, King County: Washington Geology, v. 26, p. 13-19.

Nesbitt, E. A., Campbell, K. A., and Goedert, J. L., 1994, Paleogene cold seeps and macroinvertebrate faunas in a forearc sequence of Oregon and Washington, *in* Swanson, D. A., and Haugerud, R. A., eds., Geologic field trips in the Pacific Northwest: Geological Society of America Annual Meeting, p. 1D-11.

Oji, T., 1990, Miocene Isocrinidae (stalked crinoids) from Japan and their biogeographic implication: Transactions and Proceedings of the Paleontological Society of Japan, New Series, v. 157, p. 412-429.

Orr, W. N., and Miles, G. A., 1978, Holothurian microfossils and Eocene stratigraphy of southwest Oregon: American Association of Petroleum Geologists Bulletin, v. 62, p. 2358.

Parma, S. G., 1985, *Eoscutella* Grant and Hertlein (Echinodermata: Clypeasteroida) in the Patagoniano (Early Tertiary) of the Province of Santa Cruz, República Argentina: Ameghiniana, v. 22, p. 35-41. [in Spanish]

Prothero, D. R., and Hankins, K. G., 2000, Magnetic stratigraphy and tectonic rotation of the upper Eocene-lower Oligocene Keasey Formation, Columbia County, Oregon: Journal of Geophysical Research, v. 105, n. B7, p. 16473-16480.

Schenck, H. G., 1936, Nuculid bivalves of the genus *Acila*: Geological Society of America Special Paper 4, 149 p.

Sieverts-Doreck, H., 1971, Über *Chladocrinus* Agassiz (Isocrinidae) und die nomenklatorische Verankerung dieser Gattung: Neues Jahrbuch für Geologie und Palaontologie, Monatshifte, p. 314-320.

Squires, R. L., and Goedert, J. L., 1994, Macropaleontology of the Eocene Crescent Formation in the Little River Area, southern Olympic Peninsula, Washington: Los Angeles County Museum Contributions in Science 444, 32 p.

Squires, R. L., Goedert, J. L., and Kaler, K. L., 1992, Paleontology and stratigraphy of Eocene rocks at Pulali Point, Jefferson County, eastern Olympic Peninsula, Washington: Washington Division of Geology and Earth Resources, Report of Investigations 31, 27 p.

Stokesbary, W. A., 1933, A faunal study of the Dallas area in an attempt to determine the faunal horizon or horizons present and to correlate with other Oregon and California horizons [Unpubl. M.Sc. dissert.]: Corvallis, Oregon, Oregon State University, 73 p.

Tegland, N. M., 1933, The fauna of the type Blakeley,

upper Oligocene of Washington: University of California Publications in Geological Sciences, v. 23, p. 81-174.

Thoms, R. E., 1965, Biostratigraphy of the Umpqua Formation, southwest Oregon [Unpubl. Ph.D. dissert.]: Berkeley, California, University of California, 215 p.

Turner, F. E., 1938, Stratigraphy and mollusca of the Eocene of western Oregon: Geological Society of America Special Paper 10, 130 p.

Van Atta, R. O., 1971, Sedimentary petrology of some Tertiary Formations, upper Nehalem River Basin, Oregon [Unpubl. Ph.D. dissert.]: Eugene, Oregon, Oregon State University, 245 p.

Vine, J. D., 1969, Geology and coal resources of the Cumberland, Hobart and Maple Valley Quadrangles, King County, Washington: U.S. Geological Survey Professional Paper 624, 67 p.

Wagner, C. D., 1974, Fossil and recent sand dollar echinoids of Alaska: Journal of Paleontology, v. 48, p. 105-123.

Weaver, C. E., 1908, New echinoids from the Tertiary of California: University of California Publications, Department of Geological Sciences Bulletin, v. 5, p. 273-283.

Weaver, C. E., 1942, Paleontology of the marine Tertiary formations of Oregon and Washington: University of Washington Publications in Geology, v. 5, 790 p.

Welton, B. J., 1979, Late Cretaceous and Cenozoic Squalimorphii of the northeast Pacific Ocean [Ph.D. dissert.]: Berkeley, California, University of California, 553 p.

Zullo, V. A., Kaar, R. F., Durham, J. W., and Allison, E. C., 1964, The echinoid genus *Salenia* in the eastern Pacific: Palaeontology, v. 7, p. 331-349.

CHAPTER 7

Upper Eocene Greenhouse Deposits in Southern Australia: Paleoclimatology and Paleoceanography

Paul R. Gammon, Yvonne Bone, Jonathan D. A. Clarke, and Noel P. James

ABSTRACT

The earliest Oligocene opening between the South Tasman Rise and Antarctica was a primary factor in the global deterioration to Cenozoic icehouse climates. The geographic position of southern Australian late Eocene shelf successions was ideal to record the paleoclimatology and paleoceanography of this area immediately prior to this important event. The late Eocene Bremer, Eucla and St. Vincent basins record biosiliceous sponge-rich sedimentation as a fringe along 2000 km of the late Eocene southern Australian coastline. These deposits indicate that neritic paleoenvironments received high nutrient fluxes, especially silica, and their proximity to terrestrial environments implies a causal linkage to the Australian landmass. A humid late Eocene climate and flat hinterland peneplain probably preferentially partitioned precipitation into groundwater flow and concomitant deep chemical weathering. Silica-rich groundwater discharge into neritic marine environments provided optimal conditions for prolific biosiliceous sponge growth. Biosiliceous deposits grade offshore to extensive deep shelf chalk and marl indicating nutrient-rich, warm-temperate to possibly subtropical waters. Pollen and General Climate Model data indicate temperate late Eocene paleoclimates, with low seasonality and strong offshore winds. The hypothesized wind direction and Coriolis surface-forcing probably promoted upwelling along the entire southern Australian late Eocene shelf margin, especially in the Eucla Basin. The bottom-water source for upwelling currents was the Indian Ocean, which in the late Eocene was relatively warm (10°C) and nutrient-rich. This ensured year-round warm shelf temperatures and low seasonality, and the nutrients as indicated by the shelf successions. The data suggest that oceanic rather than atmospheric heat transport was probably more significant for the late Eocene Australian sector. Nutrient-rich upwelling and groundwaters produced a eutrophic and equable warm-temperate to subtropical late Eocene southern Australian marginal marine environment.

INTRODUCTION

The late Eocene was the last phase of a world that had enjoyed predominantly greenhouse climates since at least Jurassic times. An early Eocene Tertiary climatic optimum was followed by a cooling of average global climates by both steady decline and steps, jumps, and the odd reversal until the earliest Oligocene climatic crash introduced Oligocene "doubthouse." Following a Miocene climatic optimum the world entered the Miocene-Recent "icehouse" climates (Shackleton and Kennett, 1975; Prothero, 1994; Zachos et al., 1994). Eocene greenhouse climate is interpreted to have operated differently from the present icehouse climate (Barron, 1987), but no consensus has emerged as to the differing roles for oceans and atmosphere in such a world (e.g. Lyle, 1997, plus many others).

The earliest Oligocene climatic deterioration is hypothesized to be due to the opening of the Tasmanian gateway between the South Tasman Rise and Antarctica. This tectonic event precipitated a major change in oceanic circulation around Antarctica, which severely decreased heat transport and led to the growth of ice sheets on that continent (i.e., developed icehouse conditions; Shackleton and Kennett, 1975; Kennett, 1977). During this time the east-west oriented southern Australian shelf was situated in the gulf between Australia and Antarctica, a prime geographic location to record these changes in Southern Ocean surface currents (figures 7.1 and 7.2). The aim of this paper is to synthesise and extend the current paleoclimatic, sedimentologic, and paleoceanographic interpretations for the late Eocene southern Australian shelf successions.

GEOLOGICAL SETTING

The southern Australian margin has been in a drift phase since the Late Cretaceous separation between Antarctica and Australia (although minor extensional block faulting was possibly still occurring in St. Vincent Basin; Lowry, 1970; Veevers, 1984; Hocking, 1990a, b; Royer and Rollet, 1997; James and Bone, 2000). Cretaceous-Paleogene spreading was slow (~5

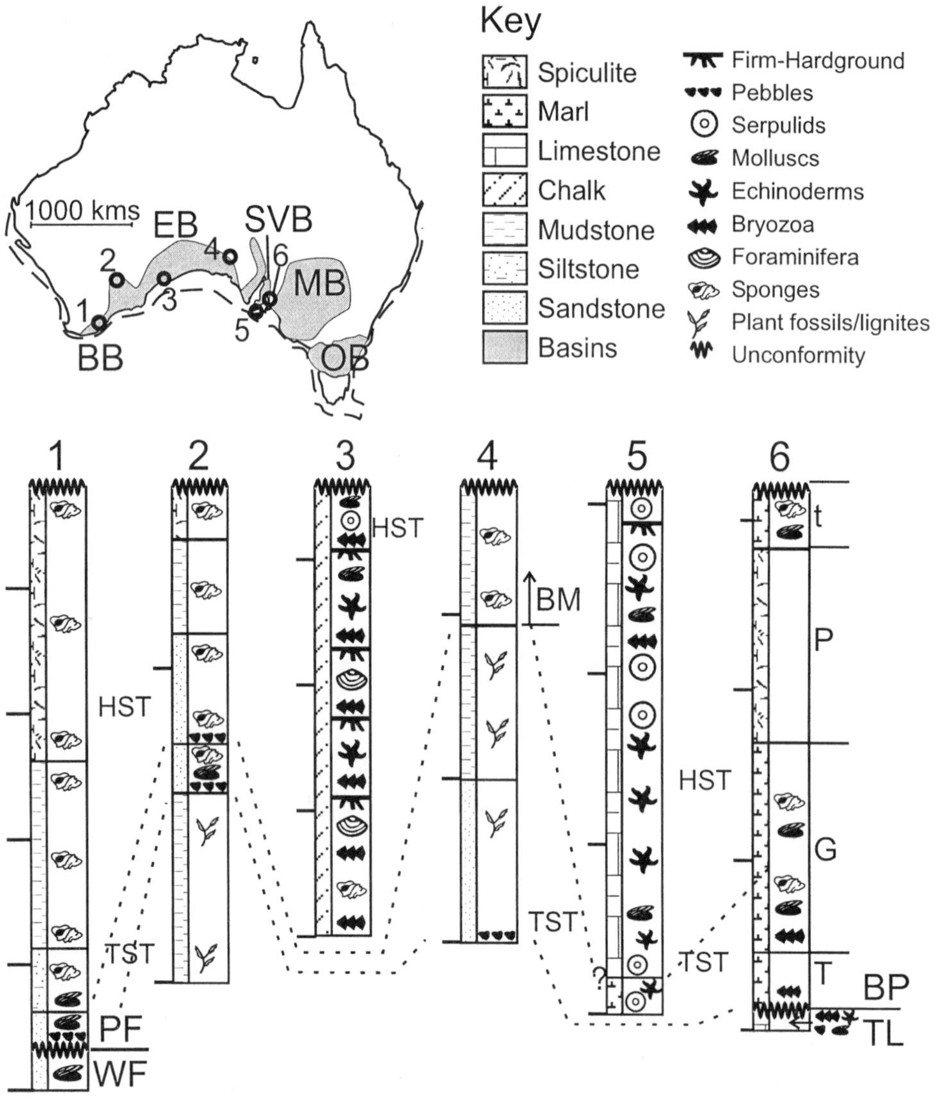

Figure 7.1. Location of southern Australian late Eocene basins (*shaded*) and representative stratigraphic successions for basins with biogenic sediments. Dashed line is the shelf edge, solid line is the land edge. Symbols: BB, Bremer Basin; EB, Eucla Basin; SVB, St. Vincent Basin; MB, Murray Basin; OB, Otway-Gippsland Basin. Numbered sections: 1, Pallinup Formation (WF, Werrilup Formation; PF, Pallinup Formation); 2, Princess Royal Spongolite; 3, Wilson Bluff Limestone; 4, Pidinga Formation (BM, Bring Member); 5, Kingscote Limestone; 6, Blanche Point Formation (TL, Tortachilla Limestone; BP, Blanche Point Formation; T, Tjuketja Member; G, Gull Rock Member; P, Perkana Member; t, Tuit Member). Sections 1, 2, 3 redrawn from Gammon et al. (2000a); 4 redrawn from Baohong et al. (2000); 5, 6 redrawn from James and Bone (2000). Tick marks to the side of each stratigraphic column indicate 10 m.

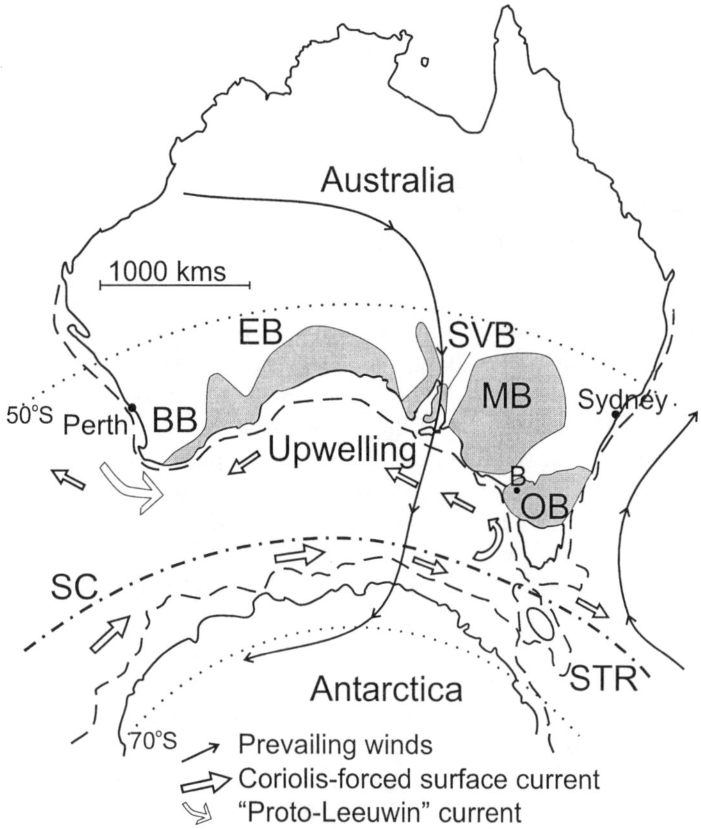

Figure 7.2. Paleogeographic setting of the southern Australian late Eocene margin, with Australian basins (from Veevers, 1984; Lawver et al., 1992). Coriolis circulation from Buening et al. (1998), and "proto-Leeuwin" current from McGowran et al. (1993). Prevailing wind (drawn along the main axis of modeled wind strength) from Barron and Peterson (1991). Other symbols and abbreviations as for figure 1; plus, SC, proto Sub-Antarctic convergence; STR, South Tasman Rise; B, Browns Creek Clay locality for the isotopic data of Kamp et al. (1990). The SC and water flow between Tasmania and Antarctica probably did not develop until the Oligocene (Murphy and Kennett, 1986).

mm/yr) until approximately 42 Ma when rearrangement of spreading ridges led to a rapid increase in spreading rate (33 mm/yr; Lawver et al., 1992; McGowran et al., 1997b; Royer and Rollet, 1997), which renewed subsidence in the marginal southern Australian basins facing the embryonic Southern Ocean (figures 7.1 and 7.2). Several sequences were deposited during this phase of subsidence, which have been termed the Wilson Bluff (Bartonian), Tortachilla (Bartonian-Priabonian), Tjuketja (Priabonian) and Tuit (Priabonian) transgressions (P12-P15; McGowran et al., 1997b). On sedimentological grounds James and Bone (2000) dispute the evidence for the Tuit transgression in the St. Vincent Basin, its type locality. Studies of other southern Australian basins have yet to report correlative sequences to the Tuit transgression (Gallagher and Holdgate, 2000; Gammon et al., 2000a). Therefore, the Tuit member is not regarded here as a transgressive-regressive cycle, but rather part of the Tjuketja transgressive-regressive cycle. This paper focuses on this particular late Eocene transgressive-regressive cycle within the biogenic successions of the Bremer, Eucla and St. Vincent basins (figure 7.1).

PALEOGEOGRAPHY AND PALEOHYDROLOGY

Southern Australia is an old landscape, and by late Eocene time peneplanation, and associated deep weathering, was well advanced (e.g., Ollier et al.,

1988; Clarke, 1994b; Twidale, 1994; Sircombe and Freeman, 1999, and many references therein). Rivers that were possibly incised as early as the Jurassic, were of low gradient (Clarke, 1994a). This hinterland paleogeography probably resulted in low runoff velocity that would (1) only be capable of delivering a fine-grained suspended sediment load (mostly clay from the deep weathering); and (2) favor chemical over physical weathering.

The late Eocene transgressive-regressive cycle produced the highest relative sea level in the Paleogene for these three basins (figure 7.2; Bunting et al., 1973; Clarke, 1993; McGowran et al., 1997b; Gammon et al., 2000a; James and Bone, 2000). In the Bremer and Eucla basins this transgression flooded paleovalleys and inner shelf embayments protected but not separated from the open ocean by basement island archipelagos (Gammon and James, 2001). The Kangaroo Island–Investigator Strait basement topographic high produced a similar neritic embayment paleogeography for the St. Vincent Basin (McGowran and Beecroft, 1986a; James and Bone, 2000). These gave way southwards to an open shelf facing the large oceanic gulf between Australia and Antarctica (figure 7.2; Lawver et al., 1992).

STRATIGRAPHY

In the Bremer Basin, prolific biosiliceous sponge facies dominate late Eocene depositional environments of the paleo-coastal Pallinup Formation, and paleovalley deposits of the Princess Royal Spongolite (figure 7.2; Cockbain, 1968; Clarke, 1993). Open marine shelf deposits are not represented in onshore Bremer Basin exposures. Eucla Basin paleovalley sediments are contained within the Princess Royal Spongolite in Western Australia, and within the middle-late Eocene Pidinga Formation in South Australia (Clarke, 1993; Benbow et al., 1995). The Pidinga Formation contains paleovalley and littoral terrigenous sandstone-siltstone and lignite, with the siliceous sponge spicule-rich Bring Member at the top (figure 7.1). Southward, late Eocene Eucla Basin open-marine sediments are contained in the aerially extensive outer-shelf bryozoan chalks of the Wilson Bluff Limestone (Lowry, 1970; Benbow et al., 1995).

The upper Eocene succession of the St. Vincent Basin contains the sponge-dominated Blanche Point Formation in the northeast, the terrigenous-biosiliceous Rogue Formation in the west, and the calcareous Kingscote Limestone to the south (James and Bone, 2000). The contact relationships between these three St. Vincent Basin formations are not exposed, but drilling data suggests that they are gradational (James and Bone, 2000). Terrigenous clastic sediments dominate late Eocene basins farther east (Shafik, 1996; McGowran et al., 1997b; Gallagher and Holdgate, 2000).

In littoral areas the late Eocene transgressive-regressive cycle is generally a simple threefold succession (figure 7.1; Cockbain, 1968; Lowry, 1970; Clarke, 1993; Benbow et al., 1995; McGowran and Li, 1997; Gammon et al., 2000a; James and Bone, 2000). Transgressive deposits are reworked terrigenous clastic sediments that fine-upward to fine-grained biota-dominated highstand deposits. Regressive deposits generally become less siliciclastic and more purely biogenic upward, with grain size, and faunal taphonomy and assemblage indicating shallowing-upward conditions. In the outer-shelf Wilson Bluff Limestone, biogenic limestones were deposited continuously throughout the transgressive-regressive cycle.

FAUNAL ASSEMBLAGE

Siliceous sponges within the Pallinup Formation comprise the most diverse assemblage yet reported from the Tertiary, and are closely related to Tethyan European taxa (Gammon et al., 2000b). Bryozoa, echinoderms, serpulids and mollusks dominate Wilson Bluff Limestone macroscopic biotic assemblages, with 10% siliceous sponges in some localities (figures 7.1 and 7.2; Lowry, 1970; Gammon et al., 2000b). The bryozoa and mollusk macrofauna are also biogeographically related to Mesozoic Tethyan ancestors (McGowran et al., 1997b). Marl-chalk lithologies contain abundant coccoliths and common benthic foraminifera (McGowran and Beecroft, 1986a; James and Bone, 2000). The faunal assemblage within biosiliceous facies has been interpreted to represent eutrophic conditions (Gammon et al., 2000b; James and Bone, 2000).

FACIES DISTRIBUTION

In all basins there is a north-south lateral gradation from biosiliceous paleovalley and protected embayment facies to open marine bryozoan limestone and chalk (Gammon et al., 2000b; James and Bone, 2000; Gammon and James, 2001). This facies distribution–paleogeographic relationship persisted for ~2,000 km along the late Eocene southern Australian margin. The lateral gradation from protected inner-shelf biosiliceous to open-shelf calcareous assemblages is centered on outer archipelago areas of the protected embayments (Gammon et al., 2000b).

The outer-shelf chalk coarsens upward to massive mollusk-serpulid-bryozoan rudstone that probably represents open-ocean shoreface conditions (Lowry, 1970; Gammon and James, 2001). Rudstone components are only weakly broken and abraded, with poor sorting and no sedimentary structures (probably because of bioturbation). This indicates that shoreface hydrodynamic conditions were probably relatively calm in comparison to the high-energy open-ocean swells and very well sorted sandy beaches of the modern Southern Ocean (figure 7.2).

PALEOCLIMATOLOGY

Current palynological knowledge indicates that late Eocene southern Australia terrestrial environments contained temperate rain forest floral assemblages with minor sclerophyllous elements (Carpenter and Pole, 1995; Alley et al., 1996; Alley, 1998). Modern relatives to these floras occur in cool tropical highlands, which suggests cooler than tropical temperatures with low seasonality. This is also suggested by the presence of *Nypa* palms, whose modern relatives inhabit tropical areas where the cold-month mean temperature is above 5°C (Greenwood and Wing, 1995).

The most comprehensive general circulation model (GCM) for the late Eocene suggests that the southern Australian margin was influenced by moderate offshore winds (~2 dynes/cm^2), had mean annual atmospheric temperatures of 10-15°C (i.e., a temperate climate regime), and was relatively humid (precipitation minus evaporation ~ 0.05 cm/day; Barron and Peterson, 1991).

PALEOCEANOGRAPHY

The South Tasman Rise cleared Antarctica in the earliest Oligocene, initiating deep-water flow through the Tasmanian gateway and the onset of an extensive early Oligocene Antarctic glaciation (figure 7.2; Kennett, 1977). Prior to this, late Eocene basins around the South Tasman Rise (STR) accumulated siliciclastic shelf sediments, with some organic-rich, dysaerobic benthic environments induced by high productivity due to upwelling (Hampton, 1975). The South Tasman Rise only subsided to its present deep-water position after it had cleared Antarctica. Kamp et al. (1990) suggest, on the basis of shallow-marine foraminiferal carbon and oxygen isotopes from the Otway Basin (figure 7.2), that late Eocene coastal waters were around 15-20°C, and that a proto-Sub-Antarctic Convergence only developed after the earliest Oligocene initiation of deep-water flow through the Tasmanian gateway (figure 7.2). In an earlier paper, Murphy and Kennett (1986) suggest that late Eocene deep and shallow waters entered the gulf at the southwestern corner and circulated anticlockwise (in keeping with Coriolis forcing) around the gulf, exiting northward at the northwestern corner (figure 7.2; see also Kennett, 1977; Kamp et al., 1990; Lazarus and Caulet, 1993).

Neritic southern Australian sponges, both late Eocene and modern, are taxa of Mesozoic Tethyan derivation, implying a tropical to subtropical heritage (Wiedenmayer, 1989, 1994; Pisera, 1999; Gammon et al., 2000b). The modern relatives of southern Australian late Eocene sponges live in 10-20°C, 100-800 m deep fore-reef waters off tropical island coral reefs (e.g., New Caledonia and West Indies; Lévi and Lévi, 1983, 1988; van Soest and Stentoft, 1988; Lévi, 1991; Wiedenmayer, 1994; Gammon et al., 2000b). Late Eocene neritic shallow-marine benthic diatoms in the Bremer Basin are also dominantly tropical forms (Gammon and James, 1998). "Extratropical" large foraminifera have also been interpreted as warm-water indicators in the St. Vincent Basin (McGowran and Beecroft, 1986b). Together with the isotopic data, the faunal data indicates that late Eocene shallow neritic waters were warm-temperate to subtropical. The cool-water carbonate bryozoan fauna of the deeper shelf Wilson Bluff Limestone perhaps suggests cooler waters in open-marine conditions, possibly due to deeper-water thermocline effects, although the isotopic 15-20°C bottom-water temperature range is also suitable for this faunal assemblage (James, 1997).

DISCUSSION

Paleoclimatology

Despite late Eocene rain forest flora indicating humid conditions, there are a number of sedimentologic and geomorphic indicators of low runoff: (1) Paleovalleys up to 200 km straight-line distance from the paleocoast have a faunal assemblage indicating fully marine salinities (Clarke et al., 1996), suggesting little dilution by freshwater runoff. Trough cross-bed direction indicates that paleovalleys were flood-tide-dominated estuaries, but the small size (30 cm wavelength by 15 cm amplitude) of the cross-beds indicates relatively low strength tidal currents, although still enough to maintain stable marine salinities (Gammon and James, 2001). (2) The change from spicular marine to fluvial sedimentation occurs over approximately 1 kilometer in the upper reaches of paleovalleys, indicating minimal fluvial influence beyond the riverhead and therefore low runoff volumes (Clarke, 1993;

Egoe, 1998). (3) Falling base level should rejuvenate rivers, resulting in increased fluvial erosion and valley incision (Posamentier and Vail, 1988). The earliest Oligocene climatic deterioration induced a large eustatic sea-level drop (Haq et al., 1988; Abreu and Anderson, 1998), but there is no evidence for equivalent hinterland drainage incision (the lack of incision and implication of low-velocity runoff or aridity continues to this day; Twidale, 1994). This suggests that runoff volumes were low by at least the later stages of deposition. (4) The upward decreasing terrigenous percentages in late Eocene neritic successions indicates that rivers had low fluvial competence throughout at least the later times of these successions.

Together, the geomorphic and sedimentologic factors indicate low volumes of freshwater runoff. The runoff that did occur was from a flat topography, and consequently was of low velocity, as also indicated by the suspended-load, mud-dominated siliciclastic sediment within neritic successions. It is possible to reconcile the humid palynology with the arid sedimentology by assuming that the terrestrial rain forest vegetation was concentrated around rivers and is overrepresented in late Eocene sampling. This would imply that sclerophyllous pollen are substantially underrepresented in all samples. However, rain forest floras have been reported from across late Eocene southern Australia (Carpenter and Pole, 1995), and it seems unlikely that all samples would be so heavily biased. Other alternatives are possible: (1) Evapotranspiration may have played a part, although this cannot yet be reconstructed and substantial changes in plant evapotranspiration are unlikely; and/or (2) a high percentage of precipitation was debouched to the oceans via subsurface groundwater flow rather than runoff (Gammon et al., 2000a). High groundwater flow would be aided by (a) sluggish, low-volume rivers due to low gradients; (b) increased soil moisture residence times due to low gradients (such high soil moisture may also have helped maintain a rain forest flora); and (c) the late Eocene being the highest relative sea-level for the Paleogene would have significantly decreased fluvial competence, and raised the water table. Both low gradients and high relative sea level were present in the late Eocene southern Australian landscape (van de Graaff et al., 1977; Ollier et al., 1988; McGowran et al., 1997b; Sircombe and Freeman, 1999).

Groundwater flows along permeability gradients, with the greatest permeability being river valleys and the coastline (loose sediment and free air versus regolith and rock; i.e. the same reason wetlands are usually found in valleys and coastal plains). Groundwaters would therefore have exited via paleovalleys and the coastline during the late Eocene (paleovalley systems in south-central and western Australia are the same as the valley system of the Eocene; van de Graaff et al., 1977). Groundwater flow results in subsurface chemical leaching (e.g., Twidale, 1982), which leads to nutrient rich waters that are especially high in dissolved silica (such weathering is often termed desilicification; Bland and Rolls, 1998). Siliceous sponges can only utilise dissolved silica for spicule formation. The rim of paleovalley and protected embayment sponges surrounding late Eocene southern Australia is here interpreted to indicate a causal connection between this faunal assemblage and dissolved-silica-rich groundwaters entering from the terrestrial realm.

Such a link would resolve the apparent contradiction of humid paleoclimatic and low runoff rate, with much of the indicated precipitation discharging via continental groundwater outflow rather than physical runoff. The volume of run-off water plus groundwater entering the paleovalleys and nearshore areas is still the same, but through leaching groundwater salinity helps maintain marine salinities within the long estuaries. The high relative late Eocene sea level would also promote marine saltwater penetration into groundwaters (which in turn would enhance the silica solubility of the groundwaters). In summary, the rim of biosiliceous sedimentation within late Eocene paleovalley estuaries and nearshore protected embayments along 2,000 km of the late Eocene southern Australian margin was probably maintained by two principle features: (1) increased nutrient fluxes due to chemical rather than physical weathering, producing waters sufficiently rich in nutrients, especially silica, to enable sponges to outcompete calcareous biota in neritic environments; and (2) lowered turbidity and normal marine salinity due to sluggish, low-volume rivers.

Paleoceanography

To maintain warm waters at high southern Eocene paleolatitudes (~55°S for the southern Australian margin) requires a transfer of tropical heat (Shackleton and Kennett, 1975; Barron, 1987; Greenwood and Wing, 1995; Sloan et al., 1995; Lyle, 1997; Sloan and Morrill, 1998). The mechanism of heat transfer (oceans versus atmosphere) remains controversial (Barron, 1987 outlined the problem). Pollen and GCM data indicate temperate paleoclimates with low seasonality, while paleoceanographic data indicate that coastal southern Australian waters were warm temperate to subtropical (Carpenter and Pole, 1995; Sloan et al., 1995). This suggests that the southward transfer of heat was oceanic rather than atmospheric. The circulation envisaged for the late Eocene Australia-Antarctica gulf has southern Indian Ocean-derived warm oceanic currents passing northern Antarctica before turning north and then west, bathing the southern Australian margin (figure 7.2). The Antarctic paleoflora during the late Eocene was an impoverished cool temperate flora (Hill and Scriven, 1995), suggesting that surface waters of the region did not lead to substantial warming, despite GCM data indicating onshore directed winds (although more data are required since other controls were possibly important for this flora, e.g., light constraints at ~70°S, wind velocities; figure 7.2; Barron and Peterson, 1991; Sloan et al., 1995). The present data suggests that surface Coriolis circulation was incapable of producing warm paleoclimatic conditions across Antarctica, and therefore could only have played a minor role in warming southern Australia.

The biotic assemblage, biotic ancestry and marl-chalk couplets of the Wilson Bluff Limestone are very similar to coccolith-rich European Mesozoic chalks, which also has an associated "Sponge Megafacies" (Pisera, 1991, 1997; Leinfelder et al., 1994; Wiedenmayer, 1994; Mitchell and Carr, 1998). The Tethyan Cretaceous chalks represent upwelling-influenced eutrophic, deep tropical-water shelf environments, with marl-chalk couplets related to changes in productivity modulated by paleoceanographic changes with Milankovitch periodicities (Mitchell and Carr, 1998 and references therein; Grafe, 1999). Sites of periodic upwelling along the modern southern Australian shelf edge also support rich bryozoan meadows (James et al., 1992, 1994; Boreen et al., 1993). The extensive Wilson Bluff Limestone bryozoan chalk represents late Eocene depositional environments similar to both Tethyan chalk and modern outer shelf–upper slope bryozoan meadows.

GCM data (Barron and Peterson, 1991) suggest that Australia had strong offshore winds during the late Eocene, which would reinforce offshore movement of east to west shelf surface water caused by the Coriolis effect. Combined with a blocked eastern end to the Australia-Antarctica gulf, these are ideal oceanographic conditions for the upwelling of bottom waters (Parrish, 1995), as also reported from some late Eocene South Tasman Rise basins (Hampton, 1975). The extensive Wilson Bluff Limestone chalk is interpreted to be the result of late Eocene upwelling along the southern Australian margin, supplemented by nutrient-rich continental groundwaters.

Isotope data for late Eocene Indian Ocean bottom waters, the probable source of Australian-Antarctic gulf deep waters, suggest that bottom-water temperatures were around 10°C (Barrera and Huber, 1993), with high productivity (and therefore high nutrients; Wright and Miller, 1993; Diester-Haass, 1996). Mixing of upwelled waters with sun-warmed shallow waters would result in shelf temperatures significantly warmer than 10°C, consistent with faunal and isotopic data for the shelf. An upwelling system is more likely to buffer the winter minimum temperature than a surface circulation, which is probably the real control on extratropical foraminifera and reduced seasonality. It would also explain the higher marine temperature indicators in comparison to terrestrial records. The Wilson Bluff Limestone, GCM, and isotope data is consistent with the upwelling of warm, nutrient-rich bottom waters. The late Eocene temperate, low-seasonality terrestrial floras are consistent with a warm temperate to possibly subtropical shelf water temperature that is combined with offshore winds, producing a loose coupling between the shelf and terrestrial paleoclimates.

McGowran et al. (1993) identify "extratropical" foraminifera within the St. Vincent Basin and equate this to the influence of a late Eocene "proto-Leeuwin Current" entering the Australian-Antarctic gulf from the northwestern corner (analogous to the modern Leeuwin Current; figure 7.2). The modern Leeuwin Current is a product of the Indonesian flow-through of the modern thermohaline circulation (McGowran et al., 1993). Late Eocene reconstructions do not indicate an equivalent late Eocene current constriction to that of the modern Indonesian Archipelago (e.g., Barron and Peterson, 1991; Lawver et al., 1992). The Eocene greenhouse thermohaline circulation was probably considerably different to the modern icehouse system (Barron, 1987). The three known localities of late Eocene extratropical foraminifera are spread over a distance of ~8,000 km of Australian coastline (McGowran et al., 1997a). Their presence indicates waters were continuously warm, but the present rock record is insufficient to determine whether or not a warm proto-Leeuwin Current entered the late Eocene Australia-Antarctic gulf from the northwest corner and circulated clockwise. If such a current did exist, its driving mechanism was probably substantially different to that of the modern Leeuwin Current. In the absence of more comprehensive data, the counterclockwise late Eocene surface circulation of Murphy and Kennett (1986:figure 2) is preferred. However, neither of these two possible surface currents are interpreted to be the main paleoclimatic temperature control for the late Eocene southern Australian margin, rather an upwelling current of Indian Ocean bottom waters reconciles the apparent data conflicts.

SUMMARY AND CONCLUSIONS

The earliest Oligocene opening of the Tasmanian gateway between the South Tasman Rise and Antarctica led to the onset of Cenozoic icehouse climates. Southern Australian basins therefore occupied an important paleogeographic position to record the changes leading up to and during this time. The southern Australian late Eocene Bremer, Eucla, and St. Vincent basin shelf successions record biosiliceous sponge-rich sedimentation within protected neritic environments, and calcareous limestone, chalk and marl sedimentation in open-shelf depositional environments. Terrigenous shelf sediments accumulated in basins farther east and southeast, including basins surrounding the South Tasman Rise.

Terrestrial rain forest pollen indicate high humidity, but sedimentological data indicate low runoff. Due to a flat peneplained paleogeography, precipitation infiltrated into continental groundwater rather than discharged into rivers. This induced deep chemical regolith weathering, and the discharge of these nutrient-rich groundwaters promoted biosiliceous sponge growth along 2,000 km of the late Eocene southern Australian margin.

Terrestrial pollen data and a climate model indicate temperate late Eocene paleoclimates, with low seasonality and strong offshore winds. However, shelf faunal and isotopic data indicate shelf waters were at least warm-temperate and probably subtropical. Warm waters most likely did not originate from Coriolis-forced surface waters circulating anticlockwise around the late Eocene Australian-Antarctic gulf. Offshore winds and Coriolis surface-forcing probably combined to promote upwelling of warm, nutrient-rich bottom waters along the entire southern Australian late Eocene shelf margin. The Eucla Basin was the focus of upwelling, with the resultant lithology being the extensive Wilson Bluff Limestone, a unit directly comparable to the high-nutrient, deep tropical water chalks of Mesozoic Europe. In combination with solar-warmed surface waters, the upwelled waters ensured warm shelf temperatures and low seasonality. The southern Australian late Eocene shelf was probably eutrophic due to nutrient-rich upwelling waters and groundwater outflow. The warm upwelling waters maintained year-round temperate, low-seasonality climates across the adjacent hinterland.

With respect to the ongoing debate about ocean versus atmosphere greenhouse poleward heat transport (Barron, 1987), the terrestrial and marine data indicate that for the Australian southern margin of the late Eocene planet, oceanic heat transport was probably the more significant. The upwelling of warm, nutrient-rich bottom waters resolves some of the outstanding paleoceanographic and paleoclimatic problems for the late Eocene southern Australian margin.

ACKNOWLEDGMENTS

We wish to thank Don Prothero for putting the Penrose Conference together, and for the huge task of editing this conference volume. We thank many people for the debates over the Australian late Eocene; they are all grist for the mill. We especially thank Drs. Larry Frakes and Roland Twidale. Thanks to Drs. N. Exon, P. Quilty and B. McGowran for their peer reviews; all their suggestions were welcomed. We thank the Canadian National Science and Engineering Research Council for research support (Grant No. OGP0002028).

REFERENCES

Abreu, V. S., and Anderson, J. B., 1998, Glacial eustasy during the Cenozoic: sequence stratigraphic implications: American Association of Petroleum Geologists Bulletin, v. 82, p. 1385-1400.

Alley, N. F., 1998, Cainozoic stratigraphy, palaeoenvironments and geological evolution of the Lake Eyre Basin: Palaeogeography, Palaeoclimatology, Palaeoecology, 144, p. 239-263.

Alley, N. F., Krieg, G. W., and Callen, R. A., 1996, Early Tertiary Eyre Formation, lower Nelly Creek, southern Lake Eyre Basin, Australia: palynological dating of macrofloras and silcrete, and paleoclimatic interpretations: Australian Journal of Earth Sciences, v. 43, p. 71-84.

Baohong, H., Frakes, L., Alley, N., Stamoulis, V., and Rowett, A., 2000, The application of geological and geophysical technology to the study of Tertiary palaeochannels draining the Gawler Craton, S. A.: Primary Industries and Resources South Australia (PIRSA), Adelaide, South Australia.

Barrera, E., and Huber, B. T., 1993, Eocene to

Oligocene oceanography and temperatures in the Antarctic Indian Ocean, *in* Kennett, J. P., and Warnke, D. A., eds., The Antarctic paleoenvironment: a perspective on global change. Part 2. Antarctic Research Series: Washington, D.C., American Geophysical Union, pp. 49-65.

Barron, E. J., 1987, Eocene equator-to-pole surface ocean temperatures: a significant climate problem? Paleoceanography, v. 2(6), p. 729-739.

Barron, E. J., and Peterson, W. H., 1991, The Cenozoic ocean circulation based on ocean General Circulation Model results: Palaeogeography, Palaeoclimatology, Palaeoecology, v. 83, p. 1-28.

Benbow, M. C., Alley, N. F., Callen, R. A., and Greenwood, D. R., 1995, Geology and Paleoclimate, *in* Drexel, J. F., and Preiss, W. V., eds., The Geology of South Australia, v. 2 The Phanerozoic: Geological Survey of South Australia, Adelaide, pp. 208-217.

Bland, W., and Rolls, D., 1998, Weathering: an introduction to the scientific principles: Arnold, London, 271 pp.

Boreen, T., James, N., Wilson, C., and Heggie, D., 1993, Surficial cool-water sediments on the Otway continental margin, southeastern Australia: Marine Geology, v. 112, p. 35-56.

Buening, N., Carlson, S. J., Spero, H. J., and Lee, D. E., 1998, Evidence for the early Oligocene formation of a proto-subtropical convergence from oxygen isotope records of New Zealand Paleogene brachiopods: Palaeogeography, Palaeoclimatology, Palaeoecology, v. 138(1-4), p. 43-68.

Bunting, J. A., van de Graaff, W. J. E., and Jackson, M. J., 1973, Palaeodrainages and Cainozoic palaeogeography of the Eastern Goldfields, Gibson Desert and Great Victoria Desert: Bureau of Mineral Resources Annual Report, p. 45-50.

Carpenter, R. J., and Pole, M., 1995, Eocene plant fossils from the Lefroy and Cowan paleodrainages, Western Australia: Australian Systematic Botany, v. 8, p. 1107-1154.

Clarke, J. D. A., 1993, Stratigraphy of the Lefroy and Cowan palaeodrainages, Western Australia: Journal of the Royal Society of Western Australia, v/ 76, p. 13-23.

Clarke, J. D. A., 1994a, Evolution of the Lefroy and Cowan palaeodrainages, Western Australia: Australian Journal of Earth Sciences, v. 41, p. 55-68.

Clarke, J. D. A., 1994b, Geomorphology of the Kambalda region, Western Australia: Australian Journal of Earth Sciences, v. 41, p. 229-239.

Clarke, J. D. A., Bone, Y., and James, N. P., 1996, Cool-water carbonates in an Eocene palaeoestuary, Norseman Formation, Western Australia: Sedimentary Geology, v. 101, p. 213-226.

Cockbain, A. E., 1968, The stratigraphy of the Plantagenet Group, Western Australia: Geological Survey of Western Australia Annual Report, v. 1967, p. 99-101.

Diester-Haass, L., 1996, Late Eocene-Oligocene paleoceanography in the southern Indian Ocean (ODP Site 744): Marine Geology, v. 130(1-2), p. 99-119.

Egoe, D. P., 1998, The origin and variability of gold in the Africa palaeochannel, Kambalda, Western Australia. [Unpublished B.Sc.(hons) Thesis], Melbourne , Latrobe University.

Gallagher, S. J., and Holdgate, G., 2000, The palaeogeographic and palaeoenvironmental evolution of a Palaeogene mixed carbonate-siliciclastic cool-water succession in the Otway Basin, southeast Australia: Palaeogeography, Palaeoclimatology, Palaeoecology, v. 156, p. 19-50.

Gammon, P. R., and James, N. P., 1998, Shallow, warm-water spiculites? Geological Society of America Abstracts with Programs, 30(7), p. 315.

Gammon, P. R., and James, N. P., 2001, Palaeogeographical influence on late Eocene biosiliceous sponge-rich sedimentation, southern Western Australia: Sedimentology, v. 48, p. 559-584.

Gammon, P. R., James, N. P., Bone, Y., and Clarke, J. D. A., 2000a, Sedimentology and lithostratigraphy of a late Eocene sponge-dominated sequence, southern Western Australia: Australian Journal of Earth Sciences, v. 47(6), p. 1087-1103.

Gammon, P. R., James, N. P., and Pisera, A., 2000b, Eocene spiculites and spongolites in southwestern Australia: not deep, not polar, but shallow and warm: Geology, v. 28(9), p. 855-858.

Grafe, K. U., 1999, Foraminiferal evidence for Cenomanian sequence stratigraphy and palaeoceanography of the Boulonnais (Paris Basin, northern France): Palaeogeography, Palaeoclimatology, Palaeoecology, v. 153(1-4), p. 41-70.

Greenwood, D. R., and Wing, S. L., 1995. Eocene continental climates and latitudinal temperature gradients: Geology, 23(11), p. 1044-1048.

Hampton, M. A., 1975, Detrital and biogenic sediment trends at DSDP sites 280, and 281, and evolution of middle Cenozoic currents: Initial Reports of the Deep Sea Drilling Program, pp. 1071-1076.

Haq, B. U., Hardenbol, J., and Vail, P. R., 1988, Mesozoic and Cenozoic chronostratigraphy and cycles of sea-level change, *in* Wilgus, C.K., et al., eds., Sea-level changes: an integrated approach: SEPM Special Publication 42, pp. 71-108.

Hill, R. S., and Scriven, L. J., 1995, The angiosperm-dominated woody vegetation of Antarctica: A review: Review of Palaeobotany and Palynology, v. 86, p. 175-198.

Hocking, R. M., 1990a, Bremer Basin: Geology and mineral resources of Western Australia: Geological Survey of Western Australia Memoir 3, pp. 561-563.

Hocking, R. M., 1990b, Bremer Basin: Geology and mineral resources of Western Australia: Geological Survey of Western Australia Memoir 3, pp. 548-560.

James, N. P., 1997, The cool-water carbonate depositional realm, *in* James, N. P., and Clarke, J. D. A. eds., Cool-water carbonates: SEPM Special Volume 56, pp. 1-20.

James, N. P., and Bone, Y., 2000, Eocene cool-water carbonate and biosiliceous sedimentation dynamics, St. Vincent Basin, South Australia: Sedimentology, v. 47(4), p. 761-786.

James, N. P., Bone, Y., Von Der Borch, C. C., and Gostin, V. A., 1992, Modern carbonate and terrigenous clastic sediments on a cool-water, high-energy, mid-latitude shelf: Lacepede, southern Australia: Sedimentology, v. 39, p. 877-903.

James, N. P., Boreen, T. D., Bone, Y., and Feary, D. A., 1994, Holocene carbonate sedimentation on the west Eucla Shelf, Great Australian Bight: a shaved shelf: Sedimentary Geology, v. 90, p. 161-177.

Kamp, P. J. J., Waghorn, D. B., and Nelson, C. S., 1990, Late Eocene–early Oligocene integrated isotope stratigraphy and biostratigraphy for paleoshelf sequences in southern Australia: paleoceanographic implications: Palaeogeography, Palaeoclimatology, Palaeoecology, v. 80, p. 311-323.

Kennett, J. P., 1977, Cenozoic evolution of Antarctic glaciation, the Circum-Antarctic Ocean, and their impact on global paleoceanography: Journal of Geophysical Research, v. 82(27), p. 3843-3860.

Lawver, L. A., Gahagan, L. M., and Coffin, M. F., 1992, The development of paleoseaways around Antarctica, *in* Kennett, J. P., and Warnke, D. A., eds., The Antarctic Paleoenvironment: A Perspective on Global Change: Antarctic Research Series, v. 56, pp. 7-30.

Lazarus, D., and Caulet, J. P., 1993, Cenozoic Southern Ocean reconstructions from sedimentologic, radiolarian, and other microfossil data, *in* Kennett, J. P., and Warnke, D. A., eds., The Antarctic paleoenvironment: a perspective on global change. Part 2. Antarctic Research Series: Washington, D.C., American Geophysical Union, pp. 145-174.

Leinfelder, R. R., et al., 1994, The origin of Jurassic reefs: current research developments and results: Facies, v. 31, p. 1-56.

Lévi, C., 1991, Lithistid sponges from the Norfolk Rise: Recent and Mesozoic genera, *in* Reitner, J., and Keupp, H., eds., Fossil and recent sponges: Berlin, Heidelberg, Springer-Verlag, pp. 72-82.

Lévi, C., and Lévi, P., 1983, Éponges Tétractinellides et Lithistides bathyales de Nouvelle-Calédonie: Bulletin du muséum national d'histoire naturelle., (4th series, section A), v. 5(1), p. 101-168.

Lévi, C., and Lévi, P., 1988, Nouveaux spongiaires Lithistides bathyaux à affinités crétacées de la Nouvelle-Calédonie.: Bulletin du muséum national d'histoire naturelle., (4, A), v. 10(2), p. 241-263.

Lowry, D. C., 1970, Geology of the Western Australian part of the Eucla Basin: Geological Survey of Western Australia Bulletin, v. 122, 201 pp.

Lyle, M., 1997, Could early Cenozoic thermohaline circulation have warmed the poles? Paleoceanography, v. 12(2), p. 161-167.

McGowran, B., and Beecroft, A., 1986a, Foraminiferal biofacies in a silica-rich neritic sediment, Late Eocene, South Australia: Palaeogeography, Palaeoclimatology, Palaeoecology, v. 52, p. 321-345.

McGowran, B., and Beecroft, A., 1986b, Neritic, southern extratropical foraminifera and the terminal Eocene event: Palaeogeography, Palaeoclimatology, Palaeoecology, v. 55, p. 23-34.

McGowran, B., and Li, Q., 1997, Stratigraphic excursion to Maslin and Aldinga Bays. Marine Geology 1997 Field Guide: Department of Geology and Geophysics, The University of Adelaide, Adelaide, South Australia, 41 pp.

McGowran, B., Qianyu, L., Almond, D., and Cann,

J., 1993, The Leeuwin Current at Cainozoic time scales: biogeographic impact on southern Australia since the later Eocene: Abstracts of the Second Australian Marine Geoscience Workshop, p. 45-46.

McGowran, B., et al., 1997a, Biogeographic impact of the Leeuwin Current in southern Australia since the late Eocene: Palaeogeography, Palaeoclimatology, Palaeoecology, v. 136, p. 19-40.

McGowran, B., Li, Q., and Moss, G., eds., 1997b, The Cenozoic neritic record in southern Australia: the biogeohistorical framework: SEPM Special Publication. v. 56, 185-204 pp.

Mitchell, S.F., and Carr, I.T., 1998, Foraminiferal response to mid-Cenomanian (Upper Cretaceous) palaeoceanographic events in the Anglo-Paris Basin (Northwest Europe): Palaeogeography, Palaeoclimatology, Palaeoecol-ogy, v. 137, p. 103-125.

Murphy, M. G., and Kennett, J. P., 1986, Development of latitudinal thermal gradients during the Oligocene: oxygen isotope evidence from the southwest Pacific, *in* Roberts, D.G., Schnitker, D., et al., eds., Initial Reports of the Deep Sea Drilling Program, pp. 1347-1360.

Ollier, C. D., Chan, R. A., Craig, M. A., and Gibson, D. L., 1988, Aspects of landscape history and regolith in the Kalgoorlie region, Western Australia. BMR Journal of Geology and Geophysics, 10, p. 309-321.

Parrish, J. T., 1995, Paleogeography of C_{org}-rich rocks and the preservation versus production controversy, *in* Auc, A.-Y., ed., Paleogeography, paleoclimatology, and source rocks: AAPG Studies in Geology, v. 40, pp. 1-20.

Pisera, A., 1991, Upper Jurassic sponge megafacies in Spain: preliminary report, *in* Reitner, J., and Keupp, H., eds., Fossil and Recent sponges: Berlin, Heidelberg, Springer-Verlag, pp. 486-497.

Pisera, A., 1997, Upper Jurassic siliceous sponges from the Swabian Alb: taxonomy and paleoecology: Palaeontologia Polonica, Warszawa , Publishing Department of the Institute of Paleobiology, v. 57, p. 1-216.

Pisera, A., 1999, Postpaleozoic history of the siliceous sponges with rigid skeleton, *in* Hooper, J. N. A., ed., Memoirs of the Queensland Museum, Brisbane , pp. 463-472.

Posamentier, H.W., and Vail, P.R., 1988, Eustatic controls on clastic deposition 2-Sequence and systems tract models, *in* Wilgus, C.K.. et al., eds., Sea-level changes: an integrated approach: SEPM Special Publication 42, pp. 125-154.

Prothero, D. R., 1994, The Eocene-Oligocene transition: paradise lost: New York, Columbia University Press, 291 pp.

Royer, J.-Y., and Rollet, N., 1997, Plate-tectonic setting of the Tasmanian region: Australian Journal of Earth Sciences, v. 44, p. 543-560.

Shackleton, N., and Kennett, J. P., 1975, Paleotemperature history of the Cenozoic and the initiation of Antarctic glaciation: Oxygen and carbon isotope analyses in DSDP sites 277, 279, 281: Initial Reports of the Deep Sea Drilling Project, v. 29, p. 801-807.

Shafik, S., 1996, Calcareous microplankton biostratigraphy of the Eocene Browns Creek Clay in the Aire District, Otway Basin of southeastern Australia: an update: AGSO Journal of Australian Geology and Geophysics, v. 16(3), p. 333-344.

Sircombe, K. N., and Freeman, M. J., 1999, Provenance of detrital zircons on the Western Australia coastline—Implications for the geologic history of the Perth basin and denudation of the Yilgarn craton: Geology, v. 27(10), p. 827-882.

Sloan, L. C. and Morrill, C., 1998, Orbital forcing and Eocene continental temperatures: Palaeogeography, Palaeoclimatology, Palaeoecol-ogy, v. 144(1-2), p. 21-35.

Sloan, L. C., Walker, J. C. G., and Moore, T. C., Jr., 1995, Possible role of oceanic heat transport in early Eocene climate: Paleoceanography, v. 10(2), p. 347-356.

Twidale, C. R., 1982, Role of subterranean water in landform development in tropical and subtropical regions, *in* LaFleur, R.G., ed., Groundwater as a geomorphic agent: Winchester, Allen and Unwin, pp. 91-134.

Twidale, C. R., 1994, Gondwanan (Late Jurassic and Cretaceous) palaeosurfaces of the Australian craton: Palaeogeography, Palaeoclimatology, Palaeoecology, v. 112, p. 157-186.

van de Graaff, W. J. E., Crowe, R. W. A., Bunting, J. A., and Jackson, M. J., 1977, Relict Early Cainozoic drainages in arid Western Australia: Zeitschrift fur Geomorphologie, v. 21, p. 379-400.

van Soest, R. W. M., and Stentoft, N., 1988, Barbados deep-water sponges: Studies on the fauna of Curacao and other Carribean Islands:

Amsterdam, Foundation for Scientific Research in Surinam and the Netherlands Antilles, v. 70, 175 pp.

Veevers, J. J., 1984, Phanerozoic earth history of Australia: Oxford geological sciences series No. 2. Oxford, Clarendon Press, 418 pp.

Wiedenmayer, F., 1989, Demospongiae (Porifera) from northern Bass Strait, southern Australia: Memoirs of the Museum of Victoria, v. 50, 242 pp.

Wiedenmayer, F., 1994, Contributions to the knowledge of post-Palaeozoic neritic and archibenthal sponges (Porifera): Schweizerische Paläontologische Abhandlungen, v. 116. Basel, Kommission der Schweizerischen Paläontologischen Abhandlungen.

Wright, J. D., and Miller, K. G., 1993, Southern Ocean influences on late Eocene to Miocene deepwater circulation, *in* J.P. Kennett and D.A. Warnke, eds., The Antarctic paleoenvironment: a perspective on global change. Part 2. Antarctic Research Series: Washington, D.C., American Geophysical Union, pp. 1-25.

Zachos, J. C., Stott, L. D., and Lohmann, K. C., 1994, Evolution of early Cenozoic marine temperatures: Paleoceanography, v. 9(2), p. 353-387.

CHAPTER 8

The Meiji Drift Body and Late Paleogene-Neogene Paleoceanography of the North Pacific-Bering Sea Region

David W. Scholl, Andrew J. Stevenson, Marlene A. Noble, and David K. Rea

ABSTRACT

Bottom water connected to a Bering Sea circulation pattern has deposited a lengthy (>1,500 km), wide (as much as 300 km), and southeastward thinning sediment body—the Meiji drift—in the northwestern corner of the Pacific Basin. Deposition took place along the north-facing flank of the Emperor Seamount Ridge. Drift deposition began in the early Oligocene and continues today. The drift body is thickest, ~1,800 m, just south of the 125-km-wide, ~4200-m-deep Kamchatka Strait that connects the Bering Sea and North Pacific Basins around the western end of the Aleutian Ridge.

Two current-driving forces, one operating external and the other local or internal to the North Pacific–Aleutian–Bering Sea (PAC-A-BERS) region (45–60°N) can account for the Meiji drift current. Both involve cells of thermohaline circulation or convection (THC). The external mechanism is the import to the North Pacific of THC-generated North Atlantic and Antarctic water that together form Pacific Bottom Water (PBW) and its return flow along the western side of the Kamchatka Trench below the Kamchatka Current. This western-boundary current exits the Bering Sea at Kamchatka Strait. It can be inferred that the deeper part of the outflow, which entrains Bering debris, impinges on the northeastern side of the Emperor Ridge and turns southeastward to flow along the Meiji drift body.

A local or Bering Sea cell of THC could also produce the Meiji bottom current by the southward flushing of densified bottom water around the western end of the Aleutian Ridge. Because the salinity of modern surface water is too low (32–33 ppt, or parts per thousand), no THC of significance presently takes place in the Bering Sea, a condition that has probably existed since at least the middle Pliocene. However, during the past ~50–55 Ma, left-oblique, plate boundary tectonism has steadily lengthened the ridge westward by transporting blocks of arc massif toward Kamchatka. Restoring the displaced blocks to the east opens a ~1000-km-wide, deep-water gateway along the southwestern side of Bering Sea Basin. A Bering Sea Basin open to the southwest could have allowed salty subtropical water to enter the basin from this direction and, if chilled, initiate THC and the flushing of abyssal waters over the Meiji drift.

Trace element and stable isotope data are unavailable to determine if the abyssal current system responsible for the deposition of the Meiji drift was generated by THC external to the North Pacific, locally within the Bering Sea Basin, or, at times, forced from both areas. Notwithstanding the oceanographic importance of addressing this wonderment, it is supposed that the formation of the drift body in the northwestern Pacific with its Bering Sea connection signals the movement through the PAC-A-BERS region of cold, thermohaline bottom water generated somewhere. It is thus conjectured that formation of the Meiji drift is a North Pacific manifestation of the post-Eocene icehouse world.

INTRODUCTION

The Pacific-Aleutian-Bering Sea (PAC-A-BERS) region occupies an anomalously low latitudinal position, 45-60°N, for the subboreal climate that it hosts, one that is characterized by cold, intense winter storms. The raw climes of the PAC-A-BERS region

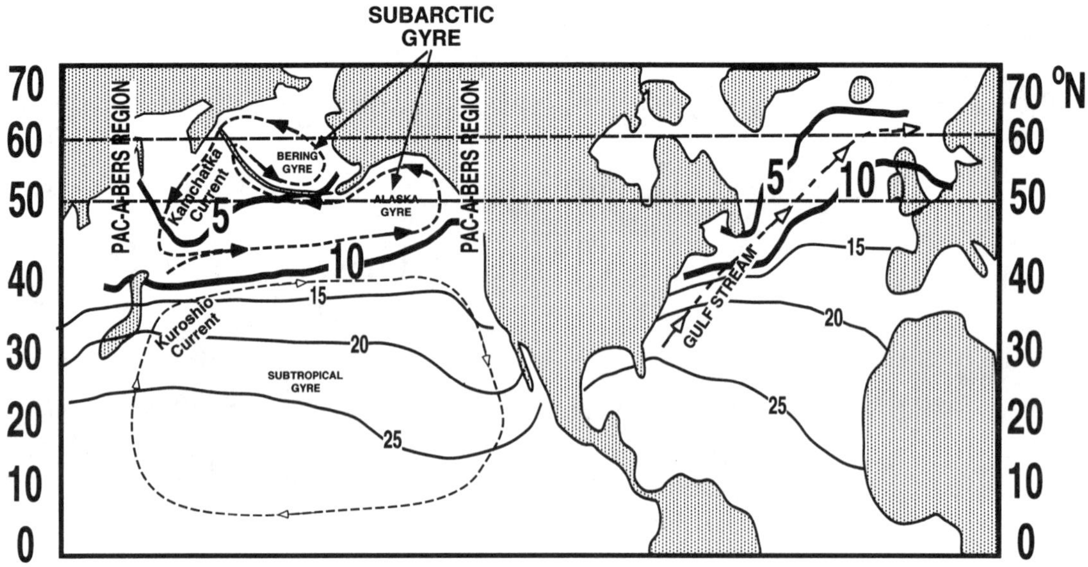

Figure 8.1. Mean sea-surface temperature, in °C, for the time period 1946–1993 (after Harrison and Larkin, 1998). Contours tracking the 5° and 10° isotherms are bold. Dashed horizontal lines identify the 50° and 60°N parallels. The subtropical waters of the latitude-crossing Gulf Stream keeps the high North Atlantic relatively warm. In comparison, the Pacific's Kuroshio western boundary current does reach the North Pacific-Aleutian-Bering Sea region (PAC-A-BERS), which is kept frigid by the virtually closed CCW circulating loop of the Subarctic gyre and its poor mixing with the Subtropical gyre south of about 40°N.

stand in contrast to the warmer, relatively salutary environment of the North Atlantic, despite its higher latitude (50-70°N; figure 8.1). This incongruity reflects the presence in the high North Atlantic of relatively warm and salty surface waters delivered there by the north-flowing Gulf Stream. No similar counterpart stream reaches the PAC-A-BERS realm, which is bathed by the cold, counterclockwise rotating, and relatively isolated Subarctic gyre (figure 8.2; Dodimead et al., 1963; Warren, 1983). This virtually closed circulation loop, which courses through the Bering Sea Basin, is driven by the powerful atmospheric depression of the Aleutian Low (figure 8.3).

An important factor contributing to the relative warmth of the high North Atlantic is winter cooling of salty surface waters and their consequent sinking to abyssal depths. This vertical cell of thermohaline circulation (THC) forms North Atlantic Deep Water (NADW), which, at depth, flows southward away from the North Atlantic. THC in the North Atlantic thus generates a positive feedback mechanism that draws warm and salty Gulf Stream waters to the North Atlantic to replace waters that chilled and exported southward as NADW.

Figure 8.2. (*facing page*) (A) Surface and deep-water circulation patterns for the North Pacific–Aleutian–Bering Sea Region (PAC-A-BERS). The Alaska and Bering Sea gyres together constitute the Subarctic gyre. Surface currents are from Dodimead et al. (1963), Stabeno et al. (1999), and Coachman et al. (1999). (B) Deep-water circulation patterns for the North Pacific–Aleutian–Bering Sea Region (PAC-A-BERS). Currents are based on information from Owens and Warren (2001) for the North Pacific and by Stabeno et al. (1999) and Coachman et al. (1999) for the Bering Sea Basin. The insert diagram (after Gordon, 1991) shows the import at abyssal depths of Antarctic Bottom Water (AABW) and North Atlantic Deep Water (NADW), which together form Pacific Bottom Water (PBW), to the North Pacific Basin. Upwelling and vertical mixing of PBW with low salinity upper waters produces North Pacific Deep Water (NPDW) that is exported southward from both the Pacific and Bering Sea Basins.

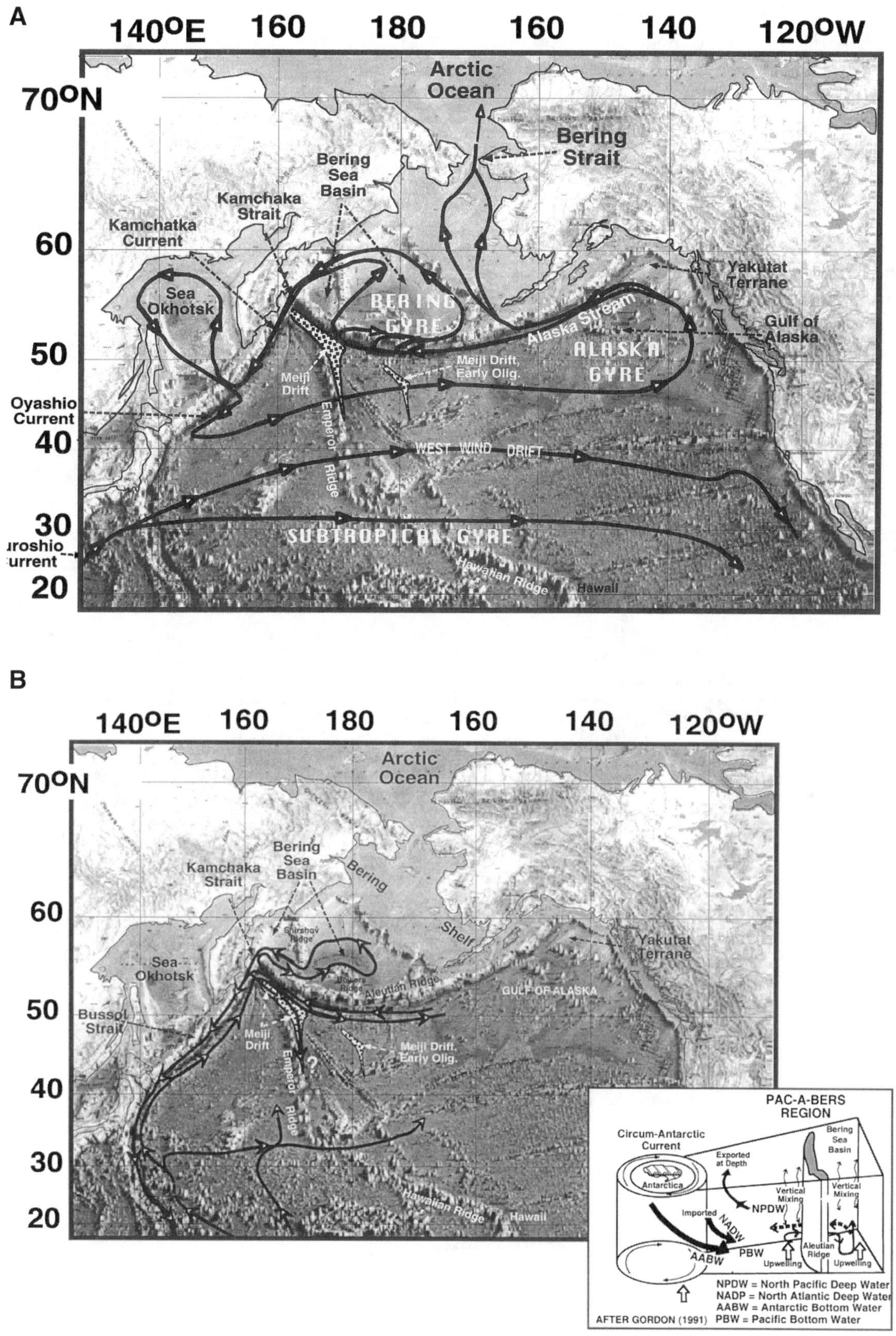
A
140°E
160
180
160
140
120°W
70°N
60
50
40
30
20
Arctic Ocean
Bering Strait
Bering Sea Basin
Kamchaka Strait
Kamchatka Current
Sea Okhotsk
BERING GYRE
Alaska Stream
ALASKA GYRE
Yakutat Terrane
Gulf of Alaska
Meiji Drift
Meiji Drift, Early Olig.
Emperor Ridge
Oyashio Current
WEST WIND DRIFT
SUBTROPICAL GYRE
Hawaiian Ridge
Hawaii
B
140°E
160
180
160
140
120°W
70°N
60
50
40
30
20
Arctic Ocean
Bering Sea Basin
Kamchaka Strait
Bering
Shelf
Shirshov Ridge
Aleutian Ridge
Yakutat Terrane
GULF OF ALASKA
Sea Okhotsk
Bussol Strait
Meiji Drift
Meiji Drift, Early Olig.
Emperor Ridge
Hawaiian Ridge
Hawaii
PAC-A-BERS REGION
Circum-Antarctic Current
Antarctica
Exported at Depth
Vertical Mixing
Vertical Mixing
Bering Sea Basin
Imported
NPDW
NADW
PBW
AABW
Upwelling
Aleutian Ridge
Upwelling
NPDW = North Pacific Deep Water
NADP = North Atlantic Deep Water
AABW = Antarctic Bottom Water
PBW = Pacific Bottom Water
AFTER GORDON (1991)

Conversely, on a regional scale, THC does not take place in the PAC-A-BERS region. The reason for this circumstance is that the salinity of surface water circulating through the Subarctic gyre is too low to support thermohaline sinking (Warren, 1983). As a consequence, winter-cooled surface water is not flushed southward as North Pacific deep water nor is warmish and salty replacement water imported from the western, subtropical Pacific. In fact, it is somewhat remarkable to recognize that abyssal waters in the North Pacific are imported from the Southern Hemisphere as a blend of NADW and Antarctic Bottom Water (AABW), commonly referred to as Pacific Bottom Water (PBW) (figure 8.2B). The slow mixing ascent of this exogenous water mass contributes to maintaining the coldness and low salinity of the surface water circulating in the Subarctic gyre (Warren, 1983; Broecker, 1987, 1991, 1997; Boyle, 1990; Gordon, 1991; Schmitz, 1995).

The occurrence in the northwestern Pacific of a lengthy (>1,500 km) and bathymetrically confined sediment body, the Meiji sediment tongue (Scholl et al., 1977), can be viewed as evidence that a vigorous, THC-generated flow of abyssal water formerly took place between the Bering Sea and North Pacific Basins (Mammerickx, 1985; Rea et al., 1995a,b; figure 8.2). The outflowing bottom waters apparently jetted around or through a deep-water pass in the Aleutian Ridge, which has separated the two ocean basins since at least 55 Ma.

Although the origin of the Meiji drift can be attributed to a local or Bering Sea cell of THC, the initiation and sustained formation of the Meiji body can as well be ascribed to the bottom-sweeping actions of an abyssal spur of the south-flowing, western-boundary current, the Kamchatka Current (Scholl et al., 1977). The upper ~4200 m of this strong water stream exits the Bering Sea at Kamchatka Strait, which separates the western end of the Aleutian Ridge and Kamchatka (figures 8.2 and 8.4). South of the strait, the basal flow of this vigorous western-boundary current system is driven by the

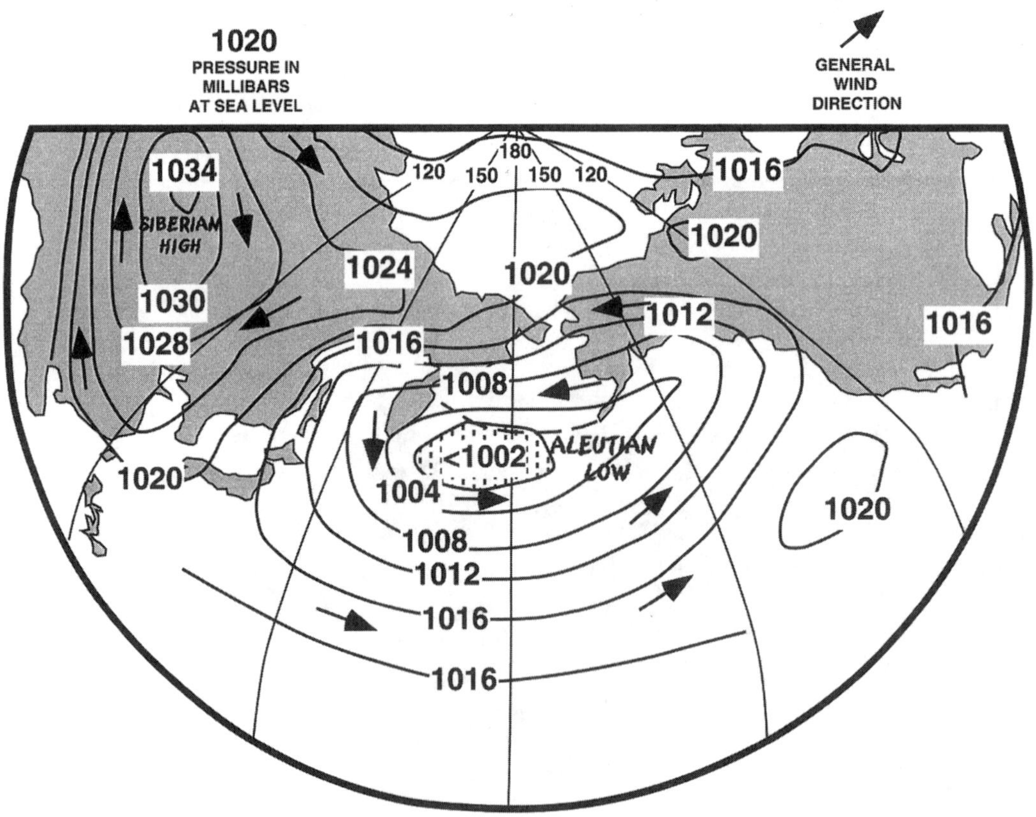

Figure 8.3. Sea level pressure (mb) for winter months (December–March), 1947–1994. Shaded area outlines center of the statistical atmospheric depression known as the Aleutian Low. After Niebauer et al. (1999).

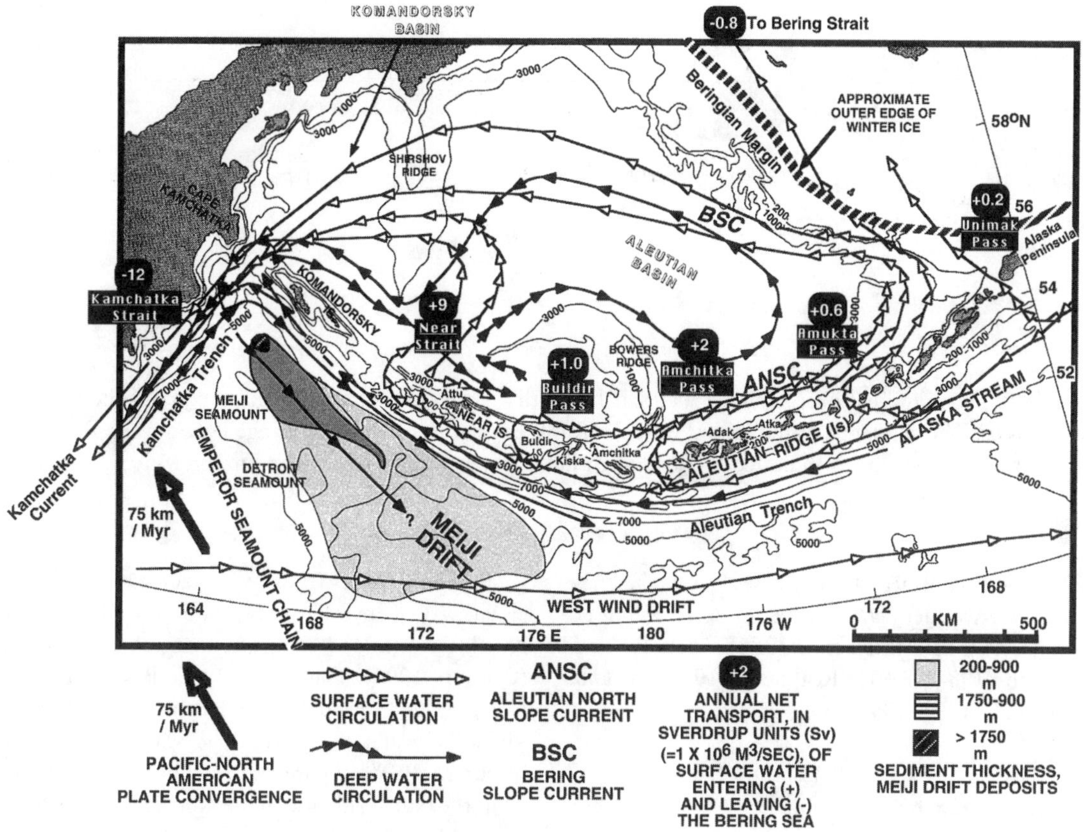

Figure 8.4. Deep and surface water circulation paths between the Bering Sea and the North Pacific Basins. The bottom flow over Meiji drift is conjectured. Flow Paths are based on the compilations of Warren and Owens (1988), Stabeno et al. (1999), Coachman (1999), and Owens and Warren (2001).

northward import of the externally- and THC-generated PBW (Owens and Warren, 2001).

DSDP and ODP cores establish that the Meiji drift body began to accumulate in the early Oligocene (Creager et al., 1973; Scholl and Creager, 1973; Rea et al.,1995a, 1995b; Tarduno et al., 2002). The initial formation of the Meiji drift can thus be seen as a PAC-A-BERS manifestation of the production of thermohaline bottom water in response to the Earth's transition from the greenhouse to the icehouse world (Miller et al., 1987; Miller et al., 1991; Miller, 1992). Beyond the global-scale climatic impact of gateway tectonism in the Southern Hemisphere and equatorial region (Kennett, 1985), local or northern Pacific-rim tectonic factors have also contributed to the oceanographic regime of the PAC-A-BERS region. Regional tectonism linked to Pacific–North American plate-boundary processes has affected drainage runoff, the course and debouchment of subarctic rivers, and determined the location and transport capacity of passes through which surface and bottom water exchange takes place between the Bering and Pacific Basins.

Evidence is presently unavailable to determine if a Bering Sea cell of THC, an externally driven, western-boundary current system, or an in-tandem combination of these forcing mechanisms account for the origin of the Meiji drift. The purpose of this paper is to assemble sedimentological, stratigraphic, and tectonic information about the Meiji drift body and to explore their oceanographic and tectonic ramifications. Because of the oceanographic implications involved, we probe the Bering THC scenario most thoroughly. To better view the past, we also explore how continuing tectonism of the Pacific's northern rim should, within the coming 1-2 million years, lead to a significant if not dramatic modification of the oceanographic regime of the PAC-A-BERS region

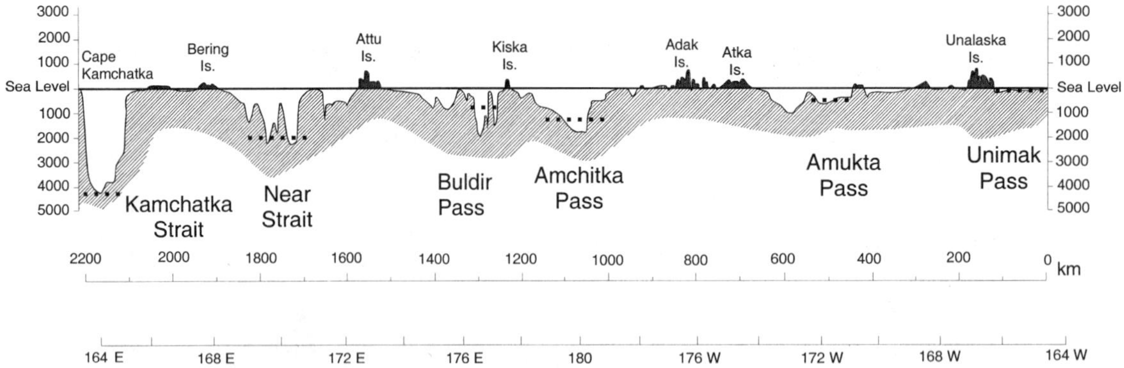

Figure 8.5. Longitudinal profile of the Aleutian Ridge that passes through the deepest parts of the between-island passes. The sill depth of the passes, which controls the depth of water exchange between the Bering Sea and North Pacific Basins, is indicated by the dashed lines. The horizontal scale is in km.

and adjacent continental landmasses. Early sketches about these wonderments can be found in Scholl et al. (1977), Mammerickx (1985), Scholl and Stevenson (1987, 1991), Rea et al. (1995a,b), and Scholl et al. (1998, 1999).

BACKGROUND INFORMATION

Modern Oceanographic Setting

General situation. Speaking generally, water moves through the PAC-A-BERS region along two principal circulation systems, a circuit of rapidly moving surface and upper waters (figure 8.2B), and, at depths exceeding about 2,500 m, the northward spread, lateral dispersal, and vertical ascent of abyssal waters imported from the Antarctic circumpolar region (figure 8.2B). The surface circulation system is vigorous, generally deep (1,500–2,500 m), gyral, and counterclockwise or cyclonic in nature. The gyre is powered by the winds associated with the barometric depression of the Aleutian Low (figure 8.3). Circulating surface waters of the major cell, the Subarctic gyre, are cool and relatively fresh (~4.0°C, and ~32-33 ppt, respectively). This gyre lies well north of, and mixes poorly with, the northern arm of the clockwise rotating Subtropical gyre to the south of the Subarctic front (~ 40°N; figure 8.2A; Warren, 1983, Sancetta and Silvestri, 1986). As noted, because of the persistence of relatively fresh surface waters in the Subarctic gyre, THC at any regionally important scale does not occur in the PAC-A-BERS region (Warren, 1983; Roden, 1995; Stabeno et al., 1999; Coachman et al., 1999).

In contrast to the isolated nature of circulating surface water, the underlying water mass of North Pacific Deep Water (NPDW) is connected to a global pathway of abyssal waters produced by THC in the North Atlantic and along the margins of Antarctica (Gordon, 1991). The North Pacific Basin is thus the global destination for a large mass of salty, cold, thermohaline water that makes its way northward as imported PBW (figure 8.2B). The transport time from the North Atlantic to the North Pacific is roughly 1,000 years (Roemmich and McCallister, 1989; Keigwin et al., 1992; Gordon, 1991; Broecker, 1987, 1991, 1997). The continued arrival and upwelling ascent of these abyssal waters, in both the North Pacific and Bering Sea Basins, contributes, in particular at depth, to the vigorous boundary-current circulation along the northern and western edges of the PAC-A-BERS region (Owens and Warren, 2001).

At least since the early Oligocene, when THC probably began adjacent to Antarctica (Carter and McCave, 1994; Carter et al., 1999), the oceanographic effects of the inflow of far-traveled abyssal water has been felt in the PAC-A-BERS region. As argued below, between the Pacific and Bering Sea Basins, patterns of both bottom- and surface-water circulation may have been much different in the early and middle Cenozoic than now.

Surface circulation. Following the descriptions of Dodimead et al. (1963), and as shown on figures 8.2A and 8.4, north of about 45°N, the Subarctic gyre is the main circulation system in the PAC-A-BERS region. The Subarctic gyre is constructed of two stable and interconnected subgyres, the Alaska gyre in the northeastern, or Gulf of Alaska, corner of the Pacific Basin, and, north of the Aleutian Ridge, the Bering gyre churning counterclockwise over the

Table 8.1. Transport of Water Through the Bering Sea

	Sill depth (m)	Area (km^2)	Flow volume (Sverdrup units, Sv, = 1 x 10^6 m^3/sec)
FLOW INTO BERING SEA BASIN AT:			
1) Unimak Pass (166ºW)	80	2	0.2 Sv North
2) Amukta Pass (172ºW)	430	19	0.6 Sv North
3) Amchitka Pass (180ºW)	1200	46	2.0 Sv North
4) Buldir Pass (175ºE)	650	28	1.0 Sv North
5) Near Strait (170ºE)	2000	240	9.0 Sv North
6) Kamchatka Strait (165ºE)	4200*	?	<0.1 Sv North
			Total in ~12.8 Sv
FLOW OUT OF BERING SEA BASIN AT:			
1) Bering Strait (170ºW)	45	3	0.8 Sv North
2) Kamchatka Strait (165ºE)	4200	335	12.0 Sv South
			Total out ~12.8 Sv

* Inflow is at depths >3500 m along east side of strait

deep-water Bering Sea Basin. The Alaska and Bering gyres are thus the principal circuits involved in the movement of the upper-water column (~1,500–2,500 m). The Bering Sea gyre can be viewed as a western and northern continuation of the North Pacific's cyclonic circulation pattern (Stabeno et al., 1999; Coachman et al.,1999; Owens and Warren, 2001)

As noted, the driving force behind the Subarctic gyre are the cyclonic winds of the atmospheric Aleutian Low (figure 8.3; Hughes et al., 1974; Bond et al. 1994; Stabeno et al., 1999; Niebauer et al., 1999). During winter months, storms, which can reach hurricane strength (Businger, 1991), average 3-5 per month, and their centers move eastward along a track just south of the Aleutian chain. Variations in transport of surface water through the Bering gyre, including the intensity of vertical mixing, are tied to variations in the position and intensity of the Aleutian Low (Bond et al., 1994, Stabeno et al., 1999). In combination with the more stable, clockwise-circulating Siberian High centered over northeastern Asia and the Sea of Okhotsk (figure 8.3), the Aleutian Low draws strong, frigid, northeast Arctic winds across the Bering Sea (Sancetta et al., 1985; Sancetta and Silvestri, 1986, Stabeno et al., 1999; Niebauer et al., 1999).

The northern arm of the Alaska gyre is the Alaska Stream, a swift, well-defined, offshore coastal current that flows westward from the Gulf of Alaska along the southern flank of the Aleutian Ridge. To nourish the Bering gyre, surface water from the Alaska Stream pours northward through bathymetric gaps in the Aleutian Ridge (figures 8.2A and 8.4). Alaska Stream water entering the Bering Sea is relatively warm and fresh (~4–6°C, 32–33 ppt). Although 14 water-exchange passes cross the ridge, significant water transport into the Bering from the Alaska Stream occurs at only 5 of them, and the flow generally increases westward along the ridge as does the cross-section area and sill depth of these passes (figure 8.5; Scholl et al., 1987). Drawing on the data assembled by Stabeno et al. (1999), Reed and Stabeno (1999), and Coachman et al. (1999), the mean transport and general dimension of the principal exchange passes are summarized in table 8.1.

From east to west, the significant gateways are Unimak (166°W), Amukta (172°W), Amchitka (180°W), and Buldir (175°E) Passes, and Near

(170°E) Strait (figures 8.4 and 8.5; table 8.1; Favorite, 1974; Stabeno et al. 1999; Reed and Stabeno, 1999). At the far western end of the ridge, along the eastern side of the Kamchatka Strait (165°E), a small volume of deep North Pacific water (PBW?) also enters the Bering Sea at depths between about 3,500 and 4,200 m (figure 8.2B).

Surface, mid-depth, and deep waters leave the Bering Sea at Kamchatka Strait—the sixth major exchange pass—as the south-bound Kamchatka Current (figure 8.2). This deep (~4,200 m) and wide (~125 km at 200 m isobath) waterway (figure 8.5) separates the Aleutian Ridge and Kamchatka through which ~95% of the Alaska Stream water entering the Bering Sea to the east is returned to the Pacific Basin. Surface water also leaves the Bering Sea northward at its Bering Strait passage to the Arctic Ocean. Only a relatively small volume (~5%) of all exported water flows northward through this arctic gateway (figures 8.2A and 8.4).

Circulation inside the Bering Sea begins at the eastern (Alaska) end of the Aleutian Ridge where a relatively small water transport (0.2 Sv, = sverdrups) occurs across the shallow (~100 m) Unimak Pass. This inflow tends to move northward across the Bering shelf to leave the Bering Sea at Bering Strait. Farther westward along the ridge, larger volumes (~0.6 and 2.0 Sv, respectively) of Alaska Stream water enter the Bering Sea via the deeper Amukta (~430 m) and Amchitka (~1,200 m) Passes. Water moving through these openings turns right to nourish the east-flowing Aleutian North Slope Current (ANSC), which forms the southern arm of the Bering gyre (figure 8.4; Reed and Stabeno, 1994, 1999). Still farther westward, water entering the Bering Sea via the fairly shallow (~650 m) but wide (~200 km) Buldir Pass (1.0 Sv) bolsters the east-flowing ANSC. This southern-boundary current directly feeds the Bering Slope Current, the westward-flowing, northern arm of the Bering gyre (figure 8.4).

Near the western end of the ridge, the largest transport of Alaska Stream water (9 Sv) enters the Bering Sea through the deep (~2,000 m) and wide (~450 km) Near Strait (figure 8.5). The bulk of this flow moves northward across the western Bering Sea Basin to join the Bering Slope Current and circle back counterclockwise as the western arm of the Bering gyre, which, as the south-flowing Kamchatka Current, exits the Bering through the Kamchatka Strait (figures 8.2A and 8.4; Stabeno et al., 1999; Reed and Stabeno, 1999, Coachman et al., 1999). The outflowing volume of the Kamchatka Current, a vigorous western-boundary current (Owens and Warren, 2001), is substantial, ranging from 12–14 or more Sv, and extends from the surface to at least a depth of 2500–3000 m (Verkhunov and Tkachenko, 1992; Stabeno et al., 1999), and most likely to the strait's sill depth at ~4,200 m (Owens and Warren, 2001). Annually, outflowing surface water is about 1° C cooler (4°C vs. >5°C) and 0.5 ppt less salty (32.5 vs. 33.1 ppt) than the Alaska Stream entering the Bering Sea (Luchin et al., 1999).

In summary, surface water provided by the west-flowing Alaska Stream enters the Bering Sea through 5 major between-island passes (figure 8.4). The volume of entering water increases westward along the ridge, with nearly 75% (9.0 Sv) flowing through the westernmost pass of Near Strait (table 8.1). Inside the Bering Sea, surface water moves counterclockwise over the Bering Sea Basin as the Bering gyre, and exits the basin southward through the deep and wide Kamchatka Strait as the Kamchatka Current. The flow volume of the Bering gyre is 12–13 Sv; only a small volume (0.8 Sv) of which is discharged northward to the Arctic ocean through the Bering Strait. The driving energy for the Bering gyre is provided by the intense barometric depression of the Aleutian Low (figure 8.3), which also powers the cyclonic movement of the larger Subarctic gyre (figure 8.2A).

Deep circulation. It was noted that abyssal water enters the North Pacific from the south mainly as a fusion of NADW and AABW that together form Pacific Bottom Water (PBW) (figure 8.2B). The bulk of PBW evidently moves first into the northwestern Pacific, then spreads eastward through three deep-water gateways into the northeastern Pacific Basin and the Gulf of Alaska region (Roemmich and McCallister, 1989, Corliss et al.,1986; Owens and Warren, 2001). Based on the bed-form features reported by Damuth et al. (1983), it seems likely that these abyssal currents are vigorous. Northward spreading is attended by lateral advective mixing, diffusion, and vertical blending with upper-layer water, an in-situ modification process that forms North Pacific Deep Water (NPDW) at a volume rate of about 10 Sv (Gordon, 1991; Schmitz, 1995; figure 8.2B). It is important to recognize that NPDW is created by very different processes than that involved in forming NADW. Thermohaline convection forms NADW, but NPDW is produced by a vertical mixing and diffusion of upper-surface water with an imported and upwelling mass of PBW. The upper part of NPDW advects vertically, reaching and chilling surface water circulation through the Alaska gyre of the

Subarctic gyre system. In a similar fashion, NPDW or PBW entering the Bering Sea at depth along the eastern side of Kamchatka Strait advectively mixes with upper water of the overlying Bering gyre (Coachman et al., 1999). Large quantities of NPDW leave the North Pacific by spreading southward at midocean depths to flow into both the Atlantic and the Indian oceans (figure 8.2B; Gordon, 1991; Schmitz, 1995; Corliss et al., 1986; Kwiek and Ravelo, 1999).

Part of the mixing that forms NPDW involves a counterclockwise flow of abyssal water. This circulation, forced by the interaction of the northward flooding PBW with the bathymetric relief of the Aleutian and Kamchatka-Kuril Arc-Trench systems, is confined to the northern end of the Pacific Basin (figure 8.2B; Warren and Owens, 1988; Keigwin et al., 1992; Owens and Warren, 2001). In the northwest corner of the Pacific, along the axis and seaward slope of the Kuril-Kamchatka Trench, the deep path flows swiftly (8 cm/s) northward toward Kamchatka Strait. Just south of the strait, the flow turns eastward along the southern, or seaward, side of the Aleutian Trench (Owens and Warren, 2001). On the landward side of the Kuril-Kamchatka Trench, a return abyssal current transports water parallel to but beneath the south-flowing, western-boundary Kamchatka Current. In a parallel way, a north-boundary abyssal current moves westward toward the Kamchatka Strait along the Aleutian Trench and the south flank of the Aleutian Ridge. At Kamchatka Strait, below about 3500 m, some of this deep, west-running current of NPDW or PBW turns northward to enter the Bering Sea Basin (figure 8.2B; Stabeno et al., 1999; Coachman et al., 1999; Owens and Warren, 2001).

Inside the Bering Sea, at depths of 3,500–4,200 m, NPDW or PBW flows eastward then counterclockwise around the Bering Sea Basin. Along this path, invading abyssal water is modified by vertical mixing and diffusion with overlying warmer and more dilute surface waters of the Bering gyre and the rain of organic debris from surface water productivity (Roden, 1995; Coachman et al., 1999). These mixing processes account for the observation that Bering Sea deep water is warmer, less salty, less dense, less oxygenated, but richer in silica then NPDW south of the Aleutian Ridge. These characteristics demonstrate long residence time (~350–500 years) of abyssal and deep water in the Bering Sea Basin and also that THC does not significantly ventilate the basin (Roden, 1995; Coachman et al., 1999). However, the detection of chlorofluorocarbons in bottom water documents that presently, and periodically, small volumes of shelf water densified by winter icing is flushed to abyssal depths (Warner and Roden, 1995).

Deep Bering Sea water also exits southward through Kamchatka Strait along its western side (Stabeno et al., 1999) where it joins a larger volume of PBW flowing southward along the landward slope of the Kuril-Kamchatka Trench. It is inferred by Scholl et al. (1977) and Owens and Warren (2001) that this deep part of the Kamchatka Current may be deflected eastward by the western end of the Meiji-Detroit sector of the Emperor Ridge to follow, as a western-boundary current jet, the ridge's northeast-facing flank toward the deeper north-central Pacific (Owens and Warren, 2001). The Meiji drift lies along this conjectured path.

Absence of thermohaline circulation. It has been emphasized that thermohaline sinking of surface water to abyssal depths does not take place in the Bering Sea Basin or anywhere in the PAC-A-BERS region. As explained by Warren (1983), the reason for this circumstance is the low salinity of surface water circulating within the Subarctic gyre (see also Gordon, 1983, 1991). The average salinity of surface water is ~32.7 ppt, but in order to sink to abyssal depths and displace cold, salty water residing there, winter-chilled surface water must start out with a salinity of 34–35 ppt (Warren, 1983). These salinities are never observed over the Bering Sea Basin, and only occur locally and occasionally near polynyas of ice formation on the shallow northern shelf (Coachman et al., 1999).

Warren (1983) cites a number of factors that combine to maintain low surface water salinity in the PAC-A-BERS region. But the key determinant is cold surface water temperature, which at 6–7°C, strongly suppresses evaporative losses. Even a small net increase in evaporation or precipitation has a strong effect on surface salinity (Roemmich and McCallister, 1989). As a consequence, over the PAC-A-BERS region, dilution of surface water salinity by precipitation and runoff greatly exceeds salinity gains caused by evaporation and advective import of saltier, deeper waters. For the PAC-A-BERS oceanographic realm, the annual rainfall (~90 cm/yr) exceeds evaporation (~55 cm) by about 35 cm. It is important to emphasize that surface waters remain cold in the North Pacific owing to both the upwelling of abyssal water and the poor horizontal mixing of the Subarctic gyre with the warmer, salty waters of the Subtropical gyre south of about 40°N (figure 8.1). The fact that THC does not take place further thwarts the import of subtropical water into the PAC-A-BERS region.

Aleutian Ridge, Tectonic Settings, and Fragmentation

General situation. As originally conceived in his prescient and paradigm-establishing paper, Coats (1962) recognized that supra-subduction-zone magmatism is the process that built the ~2500-km-long, ~150-km-wide, and 25-40 km-thick crustal mass of the Aleutian Ridge. Initial arc magmatism occurred, problematically, in the early Eocene, that is, 50-55 Ma. Grow (1973), Scholl et al. (1975, 1986, 1987), Vallier et al. (1994), and Fliedner and Klemperer (1999) describe the ridge's growth and tectonic evolution since this time.

Construction of the Aleutian Ridge physiographically and oceanographically cordoned off the northernmost sector of the early Tertiary Pacific Basin, forming the deep-water basin of the Bering Sea (figures 8.2A and 8.3). Prior to this event, circulating surface and deeper waters in the North Pacific had direct access to the area of the Bering Sea Basin, which, when corrected for the present thickness and weight of its sediment-column (3–10 km), was then approximately 5,000 m deep. It is presumed that as construction of the intraoceanic Aleutian "dam" progressed, reaching sea level by the early middle Eocene (Scholl et al., 1987), circulation between the Pacific and Bering Sea Basins was increasingly encumbered, especially in the Aleutian Basin area (figure 8.4). As documented on table 8.1, only 6 waterways—five inlets and one outlet—now conduct the major flow of water between the two basins (figure 8.4).

Each of these passes is a structural depression caused by fault-related shearing and disruption of the ridge's along-arc continuity. They are transitory products of the ridge's tectonic position along the Aleutian sector of the Pacific–North American plate boundary (figure 8.6). General information about the principal plate-boundary processes involved in building the Aleutian Ridge and its water-exchange passes can be found in Scholl et al. (1975, 1986, 1987), Cooper et al. (1977, 1987a, 1987b, 1992), Geist et al. (1988), Ryan and Scholl (1989, 1993), Geist and Scholl

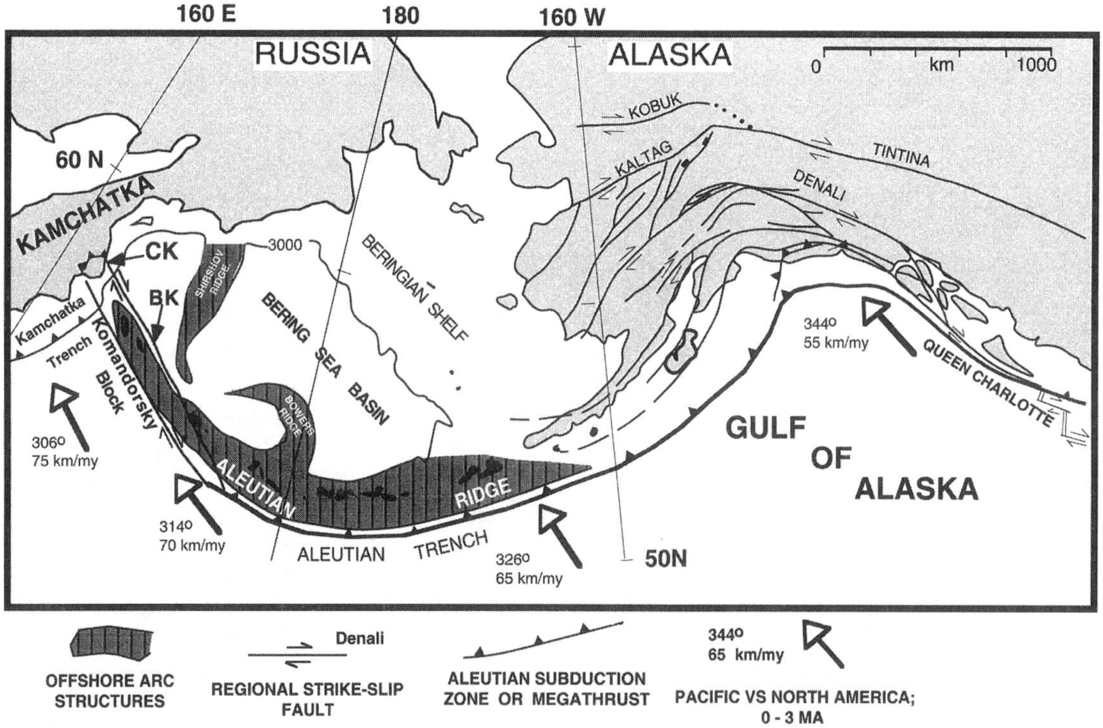

Figure 8.6. Regional tectonic setting map of the north rim of the Pacific Basin. Westward increasing obliquity of convergence transports blocks of the Aleutian Ridge westward. The Bering-Kresta (BK) shear zone guides the collision of the far western of Komandorsky block toward collision with the Kamchatka subduction zone (Trench) at Cape Kamchatka (CK).

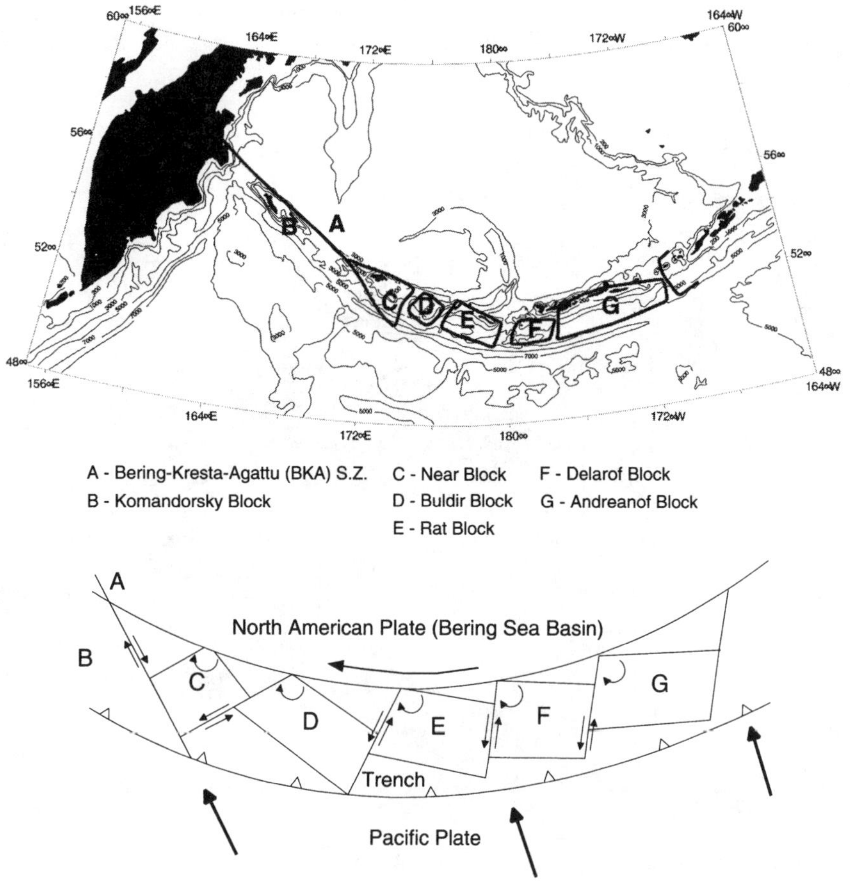

Figure 8.7. Upper graphic is a bathymetric base of the Aleutian Ridge with schematic outline of major clockwise rotating and westward translating crustal blocks. The far western or Komandorsky block (B) lies south of the Bering-Kresta-Agattu shear zone (A) and appears to be largely attached to the Pacific plate. Lower diagram idealizes the mechanics involved in the plate-boundary process of fragmenting the Aleutian Ridge and dispersing major crustal blocks westward toward the Kamchatka subduction zone (Geist et al., 1987, 1988; Geist and Scholl, 1992; Ryan and Scholl, 1989; Lallemant and Oldow, 2000).

(1992, 1994), Vallier et al. (1994), Ave Lallemant (1996), and Ave Lallemant and Oldow (2000).

Creation and evolution of between-island passes. Figure 8.7 displays the bathymetric configuration of the Aleutian Ridge and the relation of first-order physiographic elements to known or inferred faults and fault zones. This figure also graphically links these observations to a model of ridge shearing and disruption. This model ascribes the observed style of ridge shearing and fragmentation to the consequences of right-lateral, oblique plate-boundary coupling or shearing, which increases in magnitude westward (Geist et al., 1988; Geist and Scholl, 1992). Along the ridge's far western or Komandorsky sector the relative direction of plate contact is virtually parallel to the ridge's trend (figures 8.6 and 8.7).

The overall style of ridge disruption is thus one of extension or elongation parallel to its length and increasing in magnitude to the west (Geist et al., 1988; Ave Lallemant and Oldow, 2000). Extension, a manifestation of subduction-zone-displacement partitioning, is effected by westward movement of clockwise-rotating blocks, and in the Komandorsky sector, by westward shuttling of elongate splinters of arc crust (figure 8.7). Crustal fragments move in this direction along and between strike-slip faults that trend parallel or subparallel to the ridge's axial strike or elongation. Water-exchange passes form between

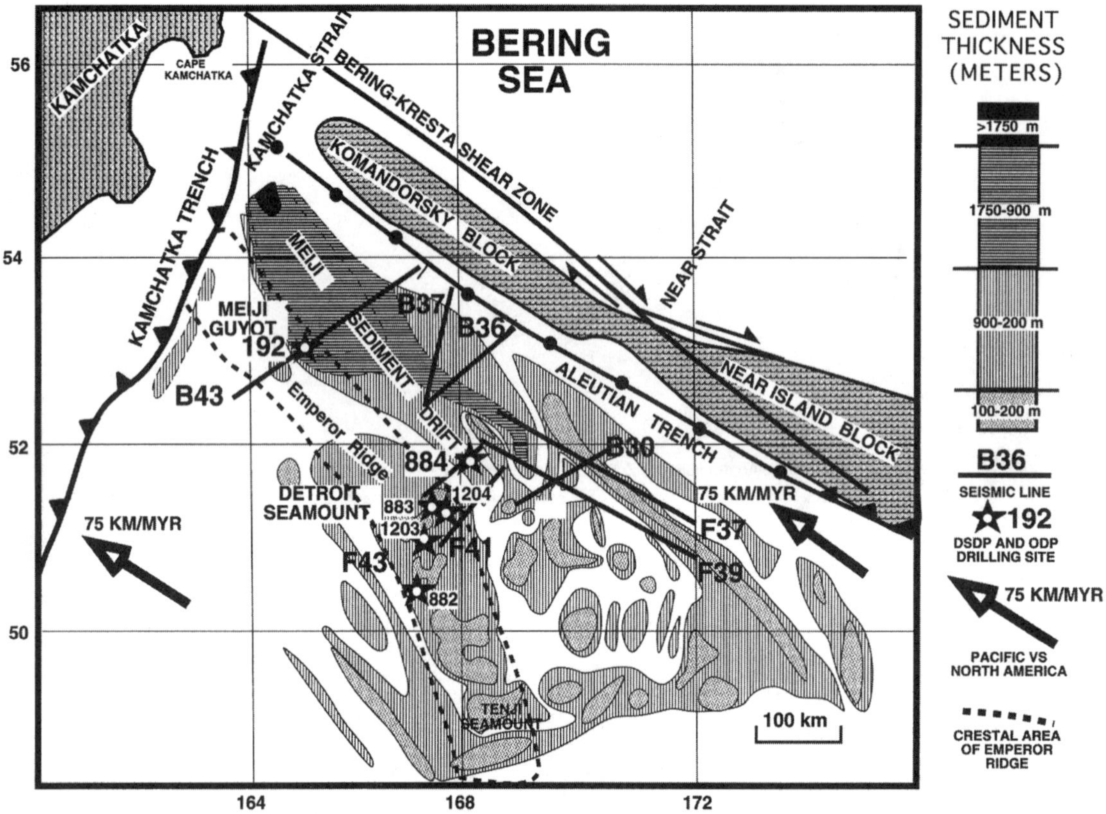

Figure 8.8. Thickness isopachs, based on seismic reflection and drilling data, of pelagic and hemipelagic beds in the northwestern corner of the Pacific Basin (Dadisman and Scholl, 1995). Areas without pattern either lack sufficient data or where the total sediment thickness is less than about 150 m. More than 90 percent of the sediment thickness is contributed by Meiji drift deposits of terrigenous silt and clay and diatomaceous debris or Oligocene and younger age.

the edges of differentially moving crustal fragments, which are tens to hundreds of km in length, 50–100 km in width, and travel at speeds of tens of km/million years (Ave Lallemant and Oldow, 2000). Their movement westward is directed at the north-south-trending Kamchatka subduction zone, toward which the blocks are drawn by the subducting Pacific plate (figures 8.6 and 8.7).

The general westward increasing speed of translation probably accounts for the observation that the depth and width of all the major transport passes (e.g., Amukta, Amchitka, Buldir, Near, and Kamchatka) increase in this direction, and contributes to the fact that the Near and Kamchatka Straits, the widest and deepest passes, handle, respectively, ~75% of all surface waters entering and ~95% of all surface waters exiting the Bering Sea Basin (figures 8.4 and 8.5). Newly acquired field and GPS data document strain partitioning and the westward increasing deformation and speed of block translation toward the Kamchatka subduction zones (Ave Lallemant, 1996; Ave Lallemant and Oldow, 2000).

Earthquake seismicity along the right-lateral Bering-Kresta strike-slip shear zone, which runs along the base of the northern side of the ridge's far western, or Komandorsky, sector (figures 8.6 and 8.7) implies that the plate boundary here lies dominantly on the Bering Sea side of the Aleutian Ridge, which means that effectively the ~800-km long Komandorsky sector is a part of the Pacific plate moving toward the Kamchatka subduction zone. In support of this interpretation, Geist and Scholl (1994) present geologic evidence that at least one former and more western "Komandorsky" island has collided with Kamchatka's landward trench slope to form Cape Kamchatka (figures 8.2, 8.6, and 8.7). This hypothesized collision potentially closed a former more western "Kamchatka Strait" of Tertiary age through which exiting deep water could have moved across the Meiji drift (Dadisman and Scholl, 1995).

In conjunction with the magmatic growth of the ridge, its concurrent tectonic dismemberment has constantly altered its effectiveness as a barrier to the exchange of water between the Pacific and Bering Sea Basins. Linked to the present style of whole-ridge fragmentation and block movement is the likelihood that, with time, the ridge has lengthened westward while central and eastern waterways widened and deepened. These shifting and changing geometries have altered, and will continue to alter, the locations of waterways and the cross-sectional area and depth of where water can enter and exit the Bering Sea. This situation is also true for the structurally controlled Bering Strait, which opened roughly 5 Ma (Marincovich and Gladenkov, 1999; Takahashi, 1999, Marincovich, 2000). As explored elsewhere below, past (and future) changes in the location and exchange capacity of the Aleutian passes have (and will) imposed commensurate changes on the physical and thermal properties of water circulating through, and created within, the Bering Sea.

The Meiji Sediment Body

In the far northwestern corner of the Pacific Basin, an elongate body of hemipelagic and pelagic sediment—the Meiji sediment tongue—overlies the northeast-facing flank of the seamount-crested Emperor Ridge (figure 8.2; Ewing et al., 1968; Scholl et al., 1977; Ludwig and Houtz, 1977; Dadisman and Scholl, 1995). Cores recovered at DSDP Site 192 (Leg 19) and ODP Site 884 (Leg 145, figure 8.8) determined that the Meiji body is constructed of a sequence of claystone and diatom-rich clay and clay-rich diatomaceous beds of Oligocene to at least latest Cenozoic age (Creager et al., 1973; Rea et al., 1995a, b). On the summit platform of Detroit Seamount (figure 8.8), drilling at ODP Sites 883 and 882 (Leg 145, Rea et al., 1995 a, b) and Site 1023 (Leg 197, Tarduno et al., 2002) further describe the physical characteristics of the Meiji sediment body. At these sites, the lower part of the diatom-rich Meiji section is more calcareous than noted for Site 884 at the base of the eastern side of the seamount.

Bathymetrically, the major (northern) mass of the Meiji body lies along the southern side of the wide (~250 km) geomorphic trough lying between the parallel ridges of the WNW trending Meiji-Detroit sector of the Emperor Ridge to the south and the far western, or Komandorsky, sector of the Aleutian Ridge to the north (figure 8.2A). Along the Meiji-Detroit sector, the Meiji drift body extends from the Aleutian Trench southward, upslope to the summit of the Emperor Ridge and perhaps along the southern slope of the ridge as well (figures 8.8 and 8.9A). The distance from the trench to the ridge crest is approximately 250 km, and the vertical relief over which the Meiji body is draped is roughly 2,000 m. At Detroit Seamount the Meiji body turns southeastward and follows the eastern flank of the main Emperor Ridge as well as along its summit. Mammerickx (1985) traced the Meiji body at least an additional 1,000-1,500 km. The seismic sections of Lonsdale et al. (1993) document that the body is well formed in the vicinity of Tenji Seamount about 250 km southwest of Detroit.

The tongue is thickest, approximately 1800 m, at the northwestern end just south of the outflow side of Kamchatka Strait (figure 8.8). At Site 192 (53.0°N), on the summit of Meiji Seamount, the Meiji sequence is 912 m thick. Along the axis of the sediment tongue, thinning is southeastward toward Detroit Seamount, along the eastern flank of which the drift section at ODP Site 884 (51.5°N) is about 700 meters thick (Rea et al.,1995a, b). The Meiji section is at least this thick at the summit of Detroit Seamount (Rea et al., 1993; Tarduno et al., 2002). At DSDP Site 192 and ODP Site 884 the Meiji sediment body overlies approximately 100 meters of older oceanic pelagic and seamount-derived debris. Based on preliminary studies of the sedimentary sections cored at Sites 883 (Rea et al., 1993), 1203 and 1204 (Tarduno et al., 2002), pre-drift deposits are also about 100 m thick on the summit of Detroit Seamount. Approximately 100 km south of Detroit, a 500-700 thick sequence of Meiji sediment overlies the eastern flank of Wayne Seamount (50.6°N; Lonsdale et al., 1993). At Tenji, Seamount (48.5°N), ~250 km southeast of Detroit, the body of sediment thins to about 300 m (Lonsdale et al., 1993). It presumably progressively thins southeastward to its distal end explored by Mammerickx (1985).

INTERPRETATIONS

Meiji Sediment Body as a Drift Deposit Involving the Export of Bering Sea Water

Based on depositional, nondepositional, and erosional bed forms, Mammerickx (1985) identified the Meiji sediment tongue as a sediment drift, a seaward-thinning body of contourite beds created by the outflow of bottom waters from the Bering Sea Basin. Her

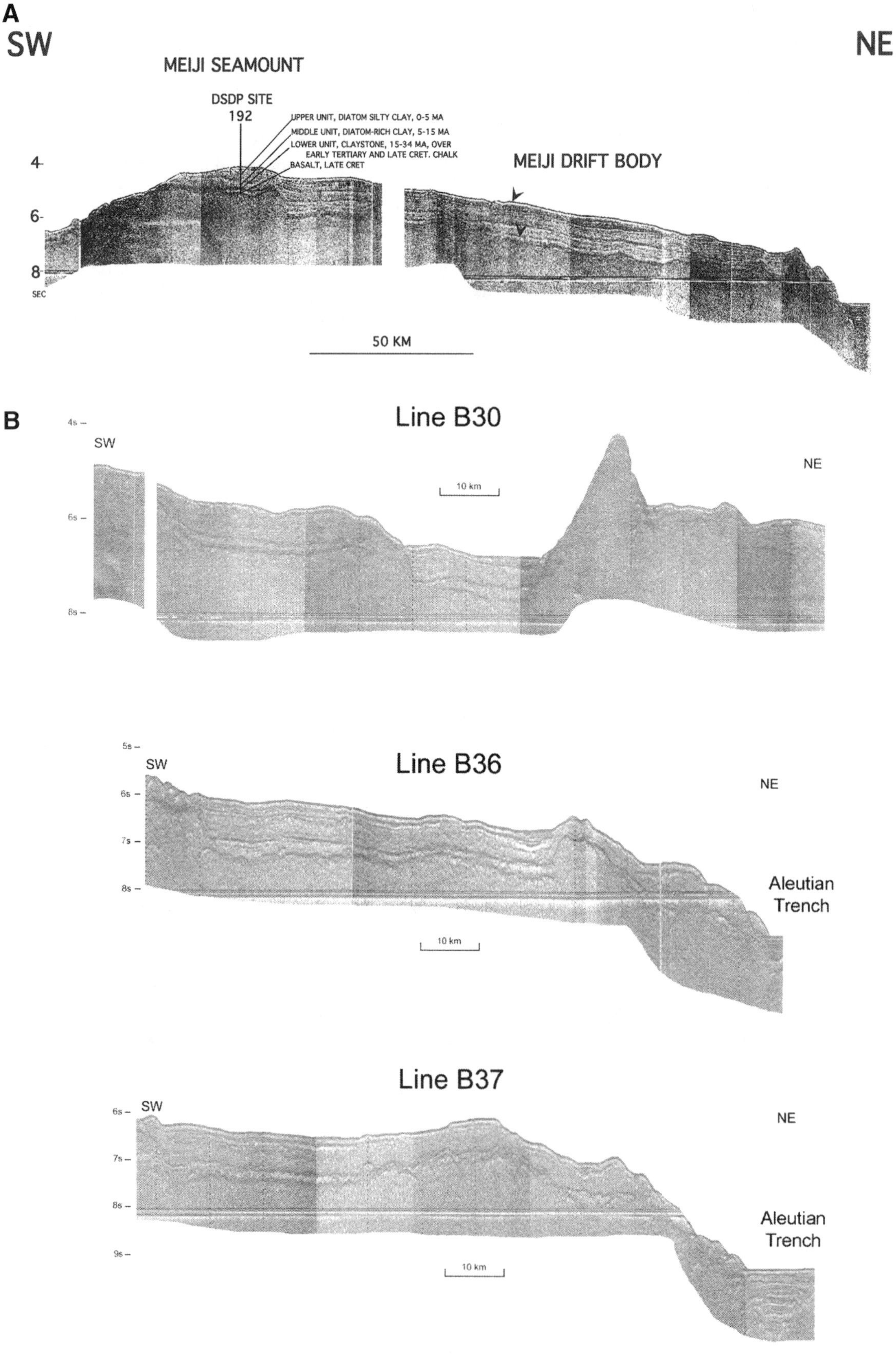
A
SW
NE
MEIJI SEAMOUNT
DSDP SITE
192
UPPER UNIT, DIATOM SILTY CLAY, 0-5 MA
MIDDLE UNIT, DIATOM-RICH CLAY, 5-15 MA
LOWER UNIT, CLAYSTONE, 15-34 MA, OVER
EARLY TERTIARY AND LATE CRET. CHALK
BASALT, LATE CRET
MEIJI DRIFT BODY
4
6
8
SEC
50 KM
B
Line B30
4s
6s
8s
SW
NE
10 km
Line B36
5s
6s
7s
8s
SW
NE
Aleutian
Trench
10 km
Line B37
6s
7s
8s
9s
SW
NE
Aleutian
Trench
10 km

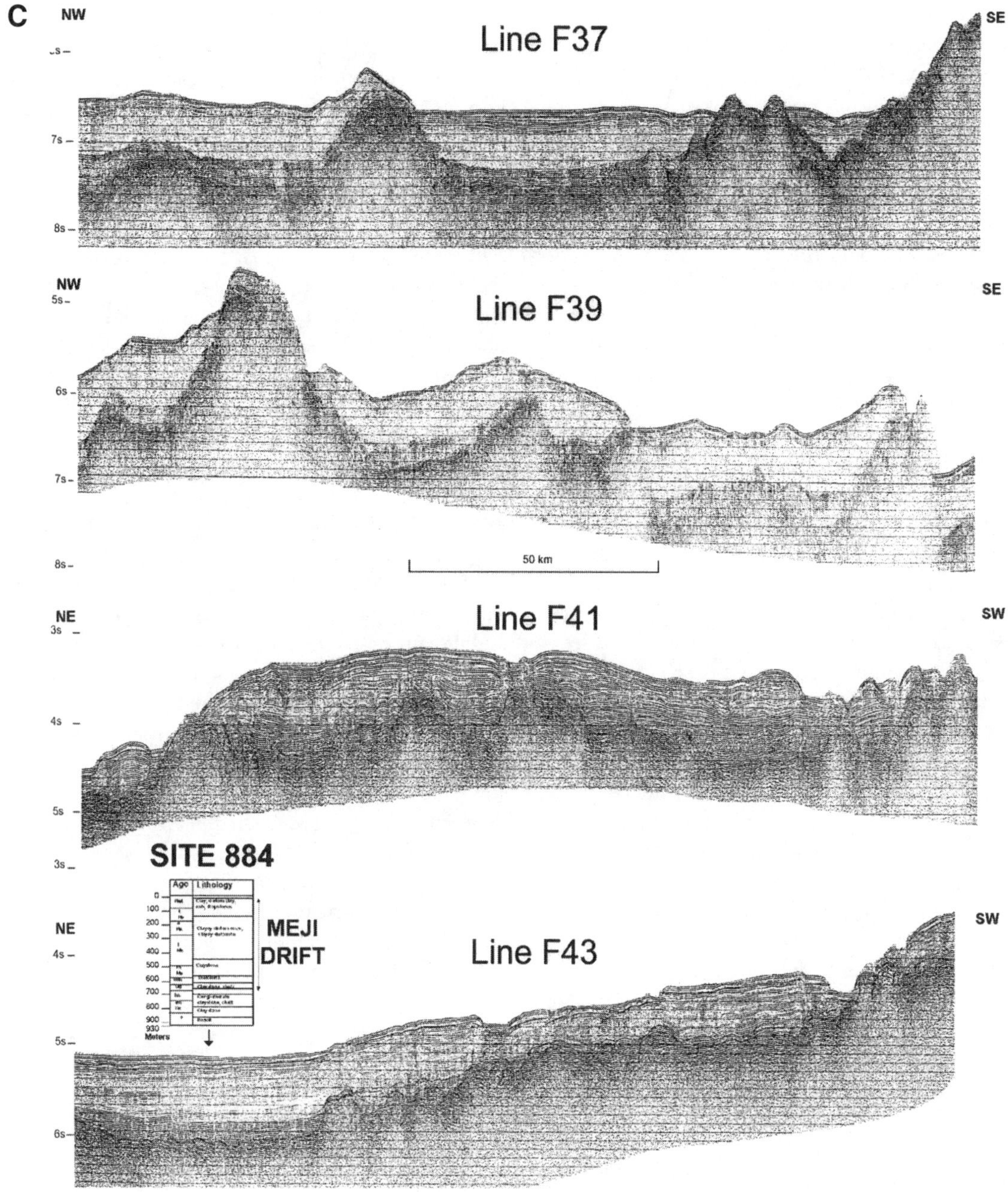

Figure 8.9. (A) Seismic line B43, which passes through DSDP Site 192, crosses the axis of the thickest part of the Meiji drift body overlying the northeastern flank of the Emperor Ridge. The body rapidly thins to the south across the crest of Meiji Seamount (guyot), the most northerly of the Emperor Seamounts. See figure 8.8 for location. (B) Seismic lines B30, B36, and B37 cross the axis the Meiji sediment body between Meiji and Detroit Seamounts. Line B36 and B37 show that the body thickens in basement depressions and thins across basement highs. See figure 8.8 for location. (C) Seismic lines F37 and F 39 are parallel to the NW-SE axis of the Meiji sediment body. Thickening to the left of basement highs, and thinning over them to the right, document a bottom current flowing from left to right, or toward the southeast and the deeper north central Pacific Basin. Profile F37 exhibits large wave forms. Seismic sections F41 and F43 display evidence of strong bottom currents interacting with basement relief. See figure 8.8 for location.

inference that the Meiji sediment body is a drift deposit was confirmed by subsurface samples recovered at ODP Site 884 (figure 8.8; Rea et al. 1993, 1995a, b). The principal observations that suggest the Meiji sediment body was formed and sculptured by abyssal waters exiting the Bering Sea can be grouped into two categories: (1) physical location and depositional geometries, and (2) sedimentological components and textural properties.

Location and geometries. The physical location and the southeastward thinning geometry of the Meiji body along the northeast-facing flank of the Emperor Ridge are consistent with the notion that bottom water connected to the Bering Sea Basin and flowing seaward along this flank laid out the Meiji sediment mass as a drift deposit. The seismic reflection records presented on figures 8.9A, 8.9B, and 8.9C, and by Lonsdale et al. (1993:figures 3 and 6) display depositional and erosional patterns of thickening, thinning, scouring, and moating expected of a drift deposit. On the flanks of the ridge, a widespread depositional pattern is that Meiji sediment fills bathymetric lows, where the section is thickest, but thins across highs or rapidly toward them. The upper two seismic sections of figure 8.9C, which were recorded along the southeast axis of the drift body, record sedimentary and erosional forms expected to be produced by a bottom current with a southeast component of flow. The southeastern end of record F37 reveals large sediment waves (~10 km crest to crest and ~100 m in height), documenting that at least locally the flow velocity has been strong (see also Mammericks, 1985 and Lonsdale et al., 1993). At the crest of Detroit Seamount, seismic records collected by ODP Leg 197 (Tarduno et al., 2002) image scouring and erosion scars cut into the Meiji body and also subsurface sediment waves that appear to increase in relief with depth. It is apparent that the Meiji current not only coursed over the northern and northeastern flanks of the Emperor Ridge but also along its summit.

Sedimentological and textural properties. The stratigraphy of the Meiji body has been explored at DSDP Site 192, located at depth of ~3,000 m atop the summit of Meiji Seamount, and at ODP Site 884, positioned at ~3,800 m at the eastern foot of Detroit Seamount (figure 8.8). Site 192 is closer to where the drift is thickest, but the drill site is positioned where the Meiji body thins rapidly across the top of Meiji Seamount (figure 8.9). The more distally located Site 884 is in a broad basement swale (figure 8.10).

At DSDP Site 192, the dominant component of the Meiji drift is terrigenous clay and fine silt. A major secondary constituent is diatomaceous debris and near the top of the section, volcanic ash is an important fraction (Creager et al., 1973; Scholl et al., 1977). The drift's initial deposits of lower Miocene and probably Oligocene age are firm mudstone or siltstone beds. The stratigraphy documented at Site 192 is also observed at Site 884, except that the initial beds here are more diatomaceous and calcareous, and basal units incorporate mass wasting debris tumbled from Detroit Seamount (Rea et al., 1995a, b).

At Site 192, the clay mineral component of the Meiji section is dominantly (75%) a mixture of illite, chlorite, and kaolinite. The mineralogy observed at Site 884 is uniform throughout the Meiji section. Quartz and plagioclase dominate the 2- to 20-micron-size fraction, with subordinate illite and chlorite and minor smectite (Arnold, 1995; Rea et al., 1995a). The Meiji suite of clay minerals signals derivation from a terrestrial source terrane. In contrast, at Site 192, the underlying beds of early Tertiary and Cretaceous deposits are a pelagic sequence rich (85%) in smectite derived from altered volcanic debris. Similar stratigraphic relations were noted at ODP Sites 883, 1203 and 1204 (Rea et al., 1995a; Tarduno et al., 2002), but the sedimentary sections cored at these sites are much less studied that those of Site 192 and 884 (figure 8.8).

Meiji's terrigenous clay mineral assemblage is virtually identical to that recovered from the surface sediment of the Komandorsky Basin of the western Bering Sea Basin (figure 8.4; Hein et al., 1976; Scholl et al., 1977). The Kamchatka Current forms over this basin before jetting through Kamchatka Strait into the northwestern Pacific. Eastern Siberian drainages, and subordinately Alaska drainages, provide the fine-grained component of the terrigenous beds that accumulate in the Komandorsky Basin (Creager et al., 1973; Stewart, 1977).

The fine-grained terrigenous component of the Meiji sequence deposited along the Meiji-Detroit sector of the Emperor Ridge is thus consistent with that provided by Bering Sea drainages and transported to the northwestern Pacific through Kamchatka Strait. This interpretation is supported by the occurrence at the more distal Site 884 of an Arctic-boreal taxa of Bering Sea diatoms, including shallow-water forms, entrained throughout the Meiji sedimentary section (Barron and Gladenkov, 1995). It is likely, as noted by Mammerickx (1985), that farther to the southeast the much thinner Meiji drift body is composed largely of bottom-current-distributed pelagic material and debris shed from the Emperor Ridge than from the Bering Sea Basin.

High rates of sediment accumulation are also characteristic of Meiji sediment. At Site 192, accumulation rates (corrected for compaction) of clay and fine-silt size material alone (diatoms excluded) are 30–45 m/million years. At Site 884, total sediment accumulation rates (uncorrected for compaction) are at least 50 m/million years. For hemipelagic deposits in oceanic settings, accumulation rates of these magnitudes are typical of drift deposits (Kidd and Hill, 1987).

Since the drillings at Site 884 (Leg 145), a technique that combines the results of grain size distribution and magnetic fabric has been developed to differentiate ice-rafted, hemipelagic, eolian, distal turbidite and drift deposits (Rea and Hovan, 1995; Joseph et al., 1998). The technique utilizes the anisotropy of magnetic susceptibility, or magnetic fabric, to determine whether sediment grains are randomly oriented, the result of hemipelagic or eolian processes, or have been aligned by a dominant flow direction along the bottom. These techniques were applied to a set of Site 884 samples from the top 270 meters (0–4.8 Ma) of the Meiji drift deposit. The results, shown on figure 8.10 demonstrate that seven of the eight samples are fine-grained, well-sorted, silty clay that exhibit a modest magnetic fabric. A similar texture and fabric characterizes drift deposits from the North Atlantic (Joseph et al., 1998).

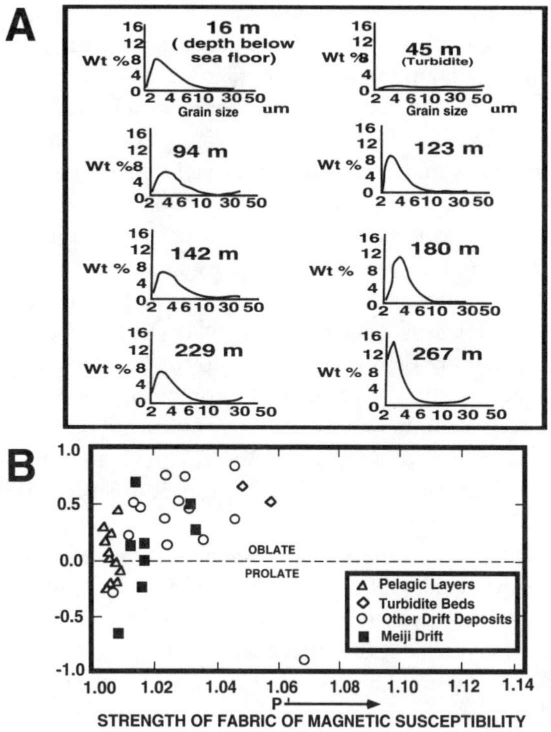

Figure 8.10. (A) Grain-size distribution of terrigenous beds from the upper 267 m of the Meiji sediment body. Skewness demonstrates winnowing expected of drift deposits. Top right plot is for a known turbidite. (B) Magnetic susceptibility fabric of beds formed by specific depositional processes. Meiji beds demonstrate a fabric typical of drift deposits shaped by contour-following currents (Joseph et al., 1998).

Oceanographic Implications of Meiji Drift

Rea et al. (1995a) summarize the principal information concerning the age and origin of the Meiji drift and what its oceanographic implications might be. Importantly, they recognize evidence that the Meiji drift formed more or less continuously during late Cenozoic time (figure 8.10). Scholl et al. (1977) speculate that the Meiji drift might be tied to an eastward splitting of the western-boundary current, the Kamchatka Current, south of its passage through Kamchatka Strait. Owens and Warren (2001), drawing on modern oceanographic observations, also hypothesize that a deep part of the boundary current might well be deflected along the northeastern side of the Emperor Ridge to wash along and shape the drift's continuing formation. Mammerickx (1985) emphasizes an alternative mechanism, that of drift formation caused by thermohaline circulation in the Bering Sea Basin and the export of abyssal water to the North Pacific Basins.

However, presently, THC of any volumetric significance does not take place in the Bering Sea, or anywhere in the PAC-A-BERS regions (Warren, 1983; Gordon, 1986, 1991; Roden, 1995; Schmitz, 1995: Stabeno et al., 1999; Coachman et al., 1999). Despite this reality, the detection of chlorofluorocarbons in Bering bottom water documents that, periodically, shelf-ice production creates surfaces waters of sufficient density to descend to the basin floor (Warner and Roden, 1995). Gardner et al. (1982) cite evidence of oxidized core sediment that during glacial episodes ventilation of the abyssal basin floor occurred periodically, presumably during times of surface-ice make up. Diatom taxa also support the likelihood that much of the surface water of the Bering Sea Basin became frozen during glacial episodes (Sancetta, 1979, 1983; Sancetta and Silvestri, 1986).

Based on regional information and modeling studies, Corliss et al. (1986) propose that significant volumes of thermohaline bottom water were produced in the northwestern Pacific during glacial episodes. Cadmium nutrient-minimum evidence also implies North Pacific formation of intermediate or deeper water in the late Quaternary (Ohkuochi et al., 1994). The record of Siberian coastal glaciation is consistent with warmish, evaporative summer months and thus more saline surface water in the Bering Sea (Gualtieri et al., 2000). Gorbarenko (1996) argues that carbon isotopic data from Detroit Seamount document the glacial-age production of intermediate water in the Bering Sea Basin. This water pushed southward to a depth of about 2,600 m over the Meiji-Detroit sector of the Emperor Ridge (figures 8.2A and 8.4). This conclusion is not supported by observations reported by Keigwin (1987) and Keigwin et al. (1992) from nearby Meiji Seamount.

Although evidence is inconclusive that significant thermohaline circulation occurred in the Bering Sea during the Quaternary, it is known that winter icing presently forms North Pacific Intermediate Water (NPIW) in the neighboring Sea of Okhotsk. Across the Kuril Ridge, NPIW escapes into the Pacific via the 2,300–m deep Bussol Strait (figure 8.2; Talley, 1991, 1993). Off southern California, the oceanographic impact of Okhotsk-made intermediate water has been identified within the late Quaternary sedimentary deposits of the Santa Barbara Basin (Behl and Kennett, 1996; Kennett et al., 2000). Thus, potentially, during the warmth of the early Pliocene (Barron, 1992a, 1992b, 1995), evaporation-supported production of intermediate water in the Bering Sea Basin may have contributed to higher d^{13} C values reported by Kwiek and Ravelo (1999) for the northeastern Pacific. Possibly, production of intermediate water in the Bering Sea Basin continued to the end of the Pliocene as well as during glacial episodes.

Despite the confused and vastly incomplete trace element and stable isotope record that might document THC in the PAC-A-BERS region, the Tertiary existence and continued Quaternary formation of the Meiji drift implies a strong bottom flow of water that has a Bering Sea connection. Presently, a small volume of deep water exiting the Bering Sea at Kamchatka Strait is joined by a larger current flowing westward along the northern slope of the Aleutian Trench toward the Pacific side of the strait. The combined flow, presumably enriched with the overlying fallout of Bering Sea terrigenous and biogenic detritus, turns southward and travels as the basal flow of the Kamchatka Current along the landward (eastern) slope of the Kuril-Kamchatka Trench. The southern reach of the deep current has been detected off northern Japan (Owens and Warren, 2001). However, as conjectured by Owens and Warren (2001), if the deep south-flowing water encounters the western end of the Emperor Ridge, then flow to the left along its northeastern side toward the deeper central Pacific Basin (>5,000–5,500 m) might be expected. This circumstance would be predicted if the abyssal Meiji-forming current was driven in the past by a nearby Bering Sea center of thermohaline circulation (THC) or in response to deep boundary-current flow forced by the arrival of imported THC-generated PBW

DISCUSSION

The message delivered by the depositional location, geometry, and composition of the main body of the Meiji drift strongly implicates the Bering Sea as a major center of THC during most of Oligocene, Miocene, and Pliocene time. It is also recognized that drift deposits first began to form in the North Atlantic in the early Oligocene (Kidd and Hill, 1987; Davies et al., 2001). The same is also true along the base of the Pacific margin of New Zealand (Carter and McCave, 1994; Carter et al., 1996, 1999). Both North Atlantic and New Zealand drifts are linked to bottom currents generated by nearby centers of THC. In the North Atlantic, the generation area for NADW is the Norwegian and Labrador Seas. In the Southern Hemisphere, the Weddell-Adelie coastal sector of Antarctica parents the THC-driven Deep Western Boundary Current (DWBC) (Carter et al., 1996). Initiation of drift deposition in both hemispheres is linked to the consequences of the global "icehouse" cooling of the early Oligocene and the consequent production of large volumes of thermohaline bottom water.

In consideration of the global implications of North Atlantic and New Zealand drift deposits, and the seeming strength of the Bering Sea signal recorded in the Meiji drift, we explore most fully below the oceanographic ramifications of a proximal or Atlantic-type scenario for a THC-generated, Bering Sea origin of the Meiji-depositing current. This hypothesis was first voiced by Mammerickx (1985).

More succinctly, the implications of the alternative or distal THC model is also probed. It is particularly important to look into the viability of this model because in the early and middle Tertiary the southwestern side of the Bering Sea was probably not blockaded by the Aleutian Ridge as it is now but was open to paths of deep-water circulation then traveling through the North Pacific Basin.

Exploring the Bering Sea Scenario

The coupled oceanographic-atmospheric regime of the modern Bering Sea is not sustained or altered by near-field factors alone. External controls that force climatic changes on a global or Pan-Pacific scale impacted the PAC-A-BERS region. These influences alter the temperature, salinity, and circulation paths of water flowing toward and through the North Pacific Basin, both at the surface and at depth. Internal or local factors specific to the PAC-A-BERS region have equally affected the physical properties and transport paths of surface and deeper waters. We comment below on information about these external and internal factors, how they may have interacted to influence the oceanic regime of the Bering Sea and the climatology of the North Pacific more generally, and how they could have lead to the start-up and later shutdown of the Bering Sea as an important center of thermohaline circulation.

Far-field Controlling Factors

Surface waters. The Bering Sea Basin, as a physiographic feature, is probably no older than the earliest Eocene, i.e., about 50-55 Ma (Scholl et al., 1986, 1987). Since this time oceanographic conditions in the Bering Sea are presumed to have shifted in sympathy with external factors that caused first-order changes in global climate. In particular, prior to the early Oligocene (~34 Ma), the transformation from the warmth of the so-called Late Cretaceous-early Tertiary greenhouse world to the oscillating but progressive cooling of the icehouse world that culminated in episodic continental glaciation during the past ~2.6-2.7 Ma (Barron and Baldauf, 1990; Miller et al., 1991; Prueher and Rea, 1998; Driscoll and Haug, 1998; Haug and Tiedemann, 1998). The post-Eocene beginning of global cooling is generally understood to be a consequence of the tectonically induced thermal isolation of Antarctica and the progressive reduction of atmospheric CO_2 in response to enhanced continental erosion and sea floor sedimentation (see, for example, Kennett, 1985; Miller, 1992; Miller et al. 1991, 1987; Flower and Kennett, 1994; France-Lanord and Derry, 1997). In the PAC-A-BERS region, evidence of a warm Paleocene and Eocene clime is documented by Burk (1965), Marincovich (1988), Beniamovskii and Gladenkov (1996) and Gladenkov (2000).

In the North Pacific, a progressive global cooldown would be expected to strengthen the gyral circulation pattern, in particular the Subarctic gyre and the atmospheric depression of the Aleutian Low that drives it (Kennett, 1985; figure 8.3). In the Bering Sea, the increasing chill, interrupted by important episodes of warming, is recorded in both onshore and offshore stratigraphic sections (Creager et. al., 1973; Gladenkov, 1990; Gladenkov and Shantser ,1990; Lagoe et al., 1993; Leg 145 Scientific Party, 1993; Rea et al., 1995a, 1995b; Barron, 1992a, 1992b; 1998; Haug and Tiedemann, 1998; Gladenkov, 2000). It can be supposed that the early Oligocene formation of the Meiji drift is core evidence of the arrival of global icehouse conditions in the North Pacific.

As global cooling progressed, and with specific respect to the paleoceanographic regime of the PAC-A-BERS region, two remote mid- to late Tertiary tectonic events in the Pacific Basin would be expected to have modified the thermal and salinity structures of the surface water entering the Bering Sea Basin. The first was the effective closing of the Indonesian seaway along the western edge of the equatorial Pacific, and, later, along the eastern equatorial sector, the gradual sealing of the Panama seaway.

The effect of restricting the Indonesian seaway was to deflect warm, salty, equatorial waters northward along the western margin of the Pacific toward high Pacific latitudes (Kennett, 1985; Kennett et al., 1985). Closing began about 16 Ma, in the early middle Miocene, and was largely complete by 10 Ma. The northward delivery of subtropical water toward the northwestern PAC-A-BERS region was presumably effected by the mid Tertiary Kuroshio Current, the western Pacific's counterpart to the North Atlantic's Gulf Stream, and, toward the northeastern Pacific, by the paleo-West Wind Drift (figure 8.2A). The flow of subtropical waters into the North Pacific, and potentially directly into the Bering Sea Basin (see figure 8.11) would have had a warming effects on the PAC-A-BERS region. This event seems to be recorded along the Pacific's northern rim (see, for example, Gladenkov, 1990, 2000; Gladenkov and Shantser, 1990; White et al., 1997; Lagoe et al., 1993).

The Central American or Panama seaway, which had long permitted the exchange of surface and deeper waters between the Pacific and Caribbean Basins, began to close in the early middle Miocene (~12 Ma), significantly restricting deep-water exchange by the early Pliocene (5.0–4.0 Ma), and fully ending interocean surface-water exchange in the middle to late Pliocene (~3.5–5 Ma) (Keigwin, 1982; Romine,

1985; Romine and Lombari, 1985; Duque-Caro, 1990; Driscoll and Haug, 1998; Collins et al., 1996; Haug et al., 1998; Haug and Tiedemann, 1998; Frank et al., 1999). In the Atlantic, closing deflected warm, salty, Caribbean equatorial waters that had formerly flowed into the Pacific to the North Atlantic, warming the Northern Hemisphere generally and enhancing the production of NADW and its export to the North Pacific (Haug and Tiedemann, 1998; Driscoll and Haug, 1998).

Sealing of the Panama gateway might be expected to have cooled the PAC-A-BERS region by the reduction of both the subtropical surface water available for transport to the North Pacific and the outflow of Pacific bottom water to the Caribbean Basin (Murdock et al., 1997). A larger chilling effect might be anticipated from the increased production and export of NADW to the high North Pacific and the upwelling and vertical mixing of PBW there (figure 8.2B; see Haug and Tiedemann, 1998; Haug et al., 1998; Billups et al., 1999). But, the isotopic data of Kwiek and Ravelo (1999) and Frank et al. (1999) testify that closing the Panama gateway had a complicated oceanographic impact on the North Pacific. One unexpected consequence appears to have been the reversal of water flow through the Bering Strait, from southward into the Bering Sea prior to closing to a closing-initiated northward flow into the Arctic Ocean (Marincovich, 2000; figure 8.2A). Although the linkage is convoluted and intricate, the progressing closure is matched in the PAC-A-BERS region by a Pliocene climatic deterioration punctuated by significant warming events prior to the onset of Northern Hemisphere glaciation at ~2.7 Ma was (Barron and Baldauf, 1990; Barron, 1992a, 1992b, 1995, 1998; Dowsett et al., 1995). As emphasized by Prueher and Rea (1998), it is likely that volcanic events in the PAC-A-BERS region contributed to triggering, or triggered, hemisphere icing that, arguably, had been conditioned by the sealing of the Panama gateway.

Thus episodically, but gradually, from the middle Miocene to the middle Pliocene (~15–3 Ma), external tectonic and climatic factors increasingly chilled and isolated surface waters circulating through the PAC-A-BERS region from interchange with the Subtropical gyre to the south and the tropical Atlantic and Caribbean to the east. Contrastingly, via the so-called global conveyor belt, these circumstances increasingly connected the abyssal waters of the North Pacific to the import of the cold discharges of the Atlantic and Southern Oceans and their eventual upwelling to chill surface waters.

Most probably, the present cold, fresh, and isolated Subarctic gyre that swings through the Bering Sea was fully formed by about ~2–3 Ma (Sanchetta and Silvestri, 1986). The late Neogene formation of the modern, and isolated circulation of PAC-A-BERS surface waters thus gradually allowed near-field tectonic factors (see below) to have a stronger say in what the final nature of the oceanographic regime of the Bering Sea would be. These factors include the rapid, late Tertiary construction of coastal mountain systems along the northern rim of the Gulf of Alaska, and, farther west, an accelerated change in location, depth, and width of major water-exchange passages along the Aleutian Ridge.

Deep water. External factors linked to the source regions and circulation of deep waters (i.e., >2,500–3,000 m) were involved in shaping the oceanographic environment of the PAC-A-BERS region. In the late early Miocene and middle Miocene (i.e., ~17 Ma to about 14 Ma), two external events, both in response to tectonic processes, directed the flow of abyssal waters to the North Pacific. One, in the North Atlantic, involved the subsidence of the Greenland-Scotland Ridge (see Davies et al., 2001), which allowed the southward flushing of North Atlantic Deep Water (NADW) to the Southern Ocean and from there into the North Pacific Basin. This event was signaled by the "silica shift" described by Keller and Barron (1983) and Barron (1986, 1992a,b), when surface water productive of opaline plankton, (i.e., mostly diatoms and radiolarians) switched from the low-latitude Atlantic to the Pacific Basin. The corresponding arrival in the North Pacific of subsurface water rich in silica, and its upwelling to the surface, is under way today (Stabeno et al., 1999; Coachman et al., 1999) and is documented in the geologic past by scientific drilling on DSDP Legs 18 (Kulm et al., 1973) and 19 (Creager et al., 1973), ODP Leg 145 (Rea et al., 1995a, b), and the widespread exposure of siliceous deposits along the rim of the Pacific Basin (Barron, 1986; Barron and Baldauf, 1990).

A second contributing circumstance to the flushing of silica-rich bottom waters to the North Pacific was Miocene icing and chilling of Antarctica and the presumed strengthening of the formation of Antarctic Bottom Waters (AABW), which, mingled with NADW, increasingly entered the North Pacific Basin as PBW (Gordon, 1991). As noted, the vertical mixing and upwelling of these deep, silica-rich waters served not only to fertilize but to significantly chill surface water of the Subarctic gyre (figure 8.1; Warren 1983; Barron and Baldauf, 1990).

The production and northward transport of cold, silica-charged abyssal water to the PAC-A-BERS region was evidently enhanced in the late Miocene and Pliocene in response to the progressing closure of the Panama seaway and increasing production of NADW (Haug et al., 1995, 1998; Haug and Tiedemann, 1998; Barron, 1998; Frank et al., 1999; Maier-Reimer et al., 1990). A general consequence was the eventual cooling and freshening of surface water circulating through the PAC-A-BERS region. These conditions are unfavorable to a Bering Sea center of deep thermohaline convection. But, perhaps during much of the Pliocene, conditions may have been favorable to the production of intermediate waters and their movement into the North Pacific as water enriched in δ^{13} C (Kwiek and Ravelo, 1999).

Near-field Controlling Factors

Surface waters. As emphasized earlier, tectonically driven changes in the location, depth, and width of water exchange passes that cross or go around the Aleutian Ridge can influence the source and nature of surface water circulating in the Bering gyre. Surface-water characteristics can also be transformed by changes in continental drainage systems, in particular by increased runoff to the Gulf of Alaska and the Alaska Stream that flows into the Bering Sea (figures 8.2A and 8.4). Moisture trapping along the northern rim of the Gulf of Alaska area was significantly increased beginning about 6 Ma as mountain building and the growth of alpine and piedmont glaciers got under way (Plafker et al., 1992; Fitzgerald et al., 1993, 1995; White et al., 1997). Mountain building was localized here by the progressing collision of a continental fragment, the Yakutat terrane or block, with southern Alaska (Bruns and Calrson, 1987; Plafker, 1987; Lagoe et al., 1993; Plafker and Berg, 1994; Plafker et al., 1994; figures 8.2 and 8.6).

Currently, via the Alaska Stream, surface water enters the Bering Sea only from the east (figures 8.2A and 8.4). To reach the Bering Sea, western Pacific water must now cross the width of the North Pacific Basin before turning back on the north side of the Alaska gyre to pour through one of five major Aleutian passes (figure 8.4). This long, trans-Pacific transport assures surface-water chilling and dilution through the combined impact of the cold surface winds of the Aleutian Low, the influx of slowly upwelling deep waters, the suppression of surface-water evaporation, and the dilution of Alaska Stream water by high precipitation over, and runoff from, the mountainous and glaciated drainages of the northern Gulf of Alaska. These circumstances sustain the low surface-water salinity of the Subarctic gyre and doom any chance of regional-scale thermohaline circulation taking place here (Warren, 1983; Roemmich and McCallister, 1989; Roden, 1995; Stabeno et al., 1999; Coachman et al., 1999).

Speculatively, if a northern continuation of the Kuroshio Current or a northern branch of the West Wind Drift could reach the Bering Sea from the southwest, then sufficiently salty and warm surface water might enter the basin and allow a deep-water cell of THC to operate there. However, the far western Aleutian exchange pass, the Kamchatka Strait, is presently the dominant gateway for the southward outflow of a deep-running western-boundary current, the Kamchatka Current (figures 8.2 and 8.4). This current guarantees that no significant volume of subtropical surface water can enter the Bering from the southwest. A second barrier to the direct entrance of western Pacific water is the low-salinity water of the Alaska Stream, the bulk of which enters the Bering Sea through the Near Strait immediately east of the Kamchatka Strait (table 8.1, figure 8.5).

Deep waters. The Meiji sediment body has been cited by Mammerickx (1985) as evidence for the former outflow of thermohaline water from the Bering Sea to the North Pacific (figure 8.2A). The central floor of the Bering Sea Basin is presently between 3,500 and 4,200 m. It is likely that prior to about 3 Ma, when rapid deposition of a glacial age terrigenous turbidite section began across the basin floor (Ewing , 1968; Cooper et al., 1987a), the central area was deeper (~300 m) than now. Thus for thermohaline circulation to export bottom water to the Pacific, one or more deep-water gaps (i.e., >3,000–4,000 m) that cross over or go around the Aleutian Ridge are needed.

The only contemporary deep-water pass is that of Kamchatka Strait (sill at ~4,200 m), just south of which lies the beginning and thickest part of the Meiji drift (figures 8.5 and 8.8). It seems likely that older "Kamchatka Straits," through which abyssal waters could have reached Meiji drift, have been tectonically eliminated by collision of arc fragments, and the passes between them, with western Kamchatka (Geist and Scholl, 1994). It is also likely that a lengthy (hundreds of km) sector of the Meiji drift has been subducted beneath or accreted to eastern Kamchatka (Dadisman and Scholl, 1995; Rea et al., 1995a). In a similar fashion the Hikurangi drift body,

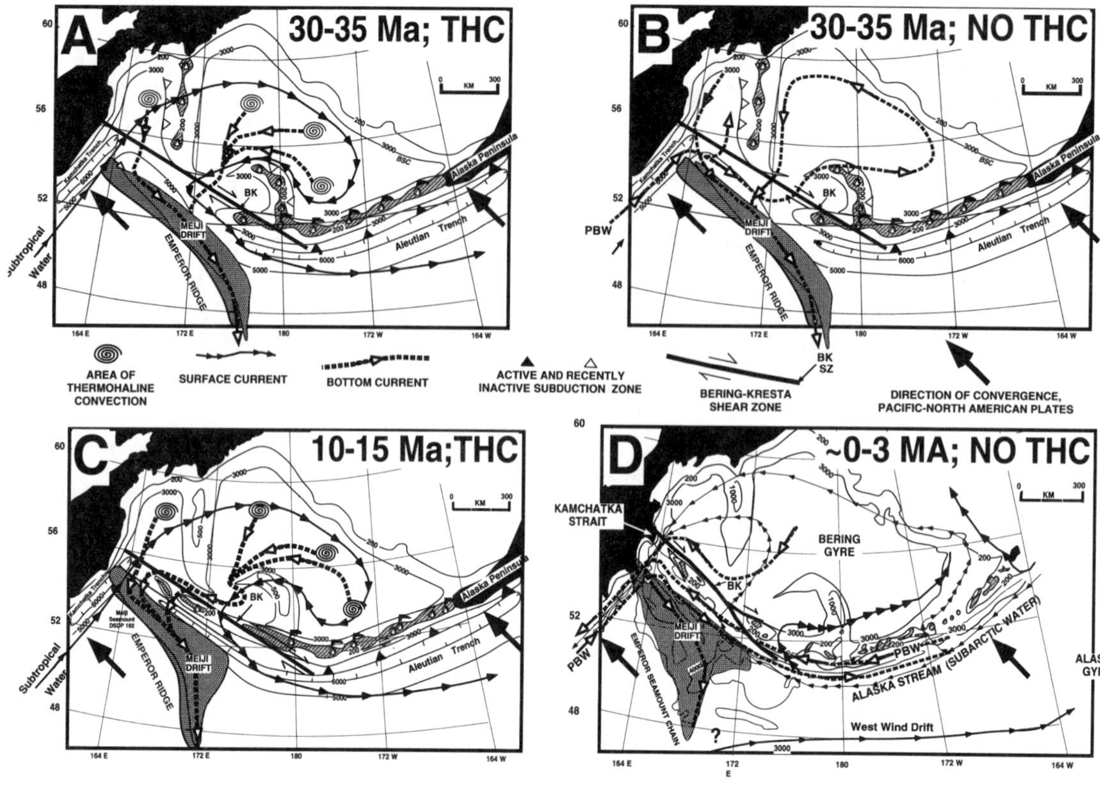

Figure 8.11. Diagram shows that, since the Eocene, the westward transport of crustal blocks of the Aleutian Ridge has progressively closed-off the southwestern side of the Bering Sea Basin. Along the far western or Komandorsky sector of the ridge, the main transform boundary has been the Bering-Kresta-Agattu shear zone (BKA).

Panel A is an early Oligocene reconstruction of the Aleutian Ridge showing a wide, ocean-depth gateway between the ridge's western end and Kamchatka. The panel infers that if salty, subtropical surface water was able to enter the Bering Sea from the western Pacific, that thermohaline circulation (THC) would take place during the winter months. Southward export path of this bottom water deposits Meiji drift along the northeastern flank of the Emperor Ridge.

Panel B is an early Oligocene reconstruction showing THC-generated Pacific Bottom Water (PBW), supplied from the North Atlantic and Antarctica, enters the Bering Sea Basin via the western gateway. The southward return, which courses along the Meiji drift, is at a shallower depth but also through the Aleutian-Kamchatka gateway. No THC in the Bering Sea is required of this model. Presumably the scenarios of Panels A and B could have worked in parallel to produce a strong Meiji current.

Panel C, an early Neogene reconstruction of the Aleutian Ridge showing a partly closed western gap but the continuation of the inflow of subtropical water from the western Pacific, THC, and the sustained growth of the Meiji drift.

Panel D displays present physiographic relations and a surface water circulation pattern that only allows subarctic surface water to enter the Bering Sea Basin from the east and mainly exit to the west via Kamchatka Strait. No THC of significance takes place in the basin.

Residual bottom flow over Meiji drift is maintained by an east-deflected basal jet of PBW underrunning the Kamchatka Current. Presumably, Bering-Sea sourced terrigenous and biogenic debris descending from the body of the flow is incorporated in bottom water passing over the Meiji drift body. This inferred circulation pattern has probably prevailed over the past 2-3 Ma.

formed by the northward-flowing Deep Western Boundary Current off eastern New Zealand, is now being subducted westward beneath North Island (Carter and McCave, 1994; Carter et al., 1996; Carter et al., 1999).

As outlined below, when the Meiji drift body was rapidly accumulating in the Oligocene and Miocene, a Bering Sea Basin substantially open to the southwest is likely to have existed. Discharging THC bottom waters from the Bering Sea could have found an easy way out by flowing around the Aleutian Ridge's western end, then located at least 1000-1200 km ESE of its present location at Kamchatka Pass. The Meiji body overlies crust of the Pacific plate, which when back-tracked to its early Oligocene position places the basal Meiji beds recovered at DSDP Site 192 and ODP Site 884 at least 1,200–1,500 km to the ESE of their present locations (figures 8.2A and 8.11). As recognized by Rea et al. (1995a), the implication of this reconstruction is that the north-northeast-trending Meiji-Detroit sector of the Emperor Ridge, which is supposed to have deflected outflowing Bering bottom water to the southeast, could well have formerly extended across much of the width of the northwestern Pacific (figure 8.11).

Tectonic and Oceanographic Implications of the Bering Sea THC Model

General situation and the open model. Given adequate winter chilling, the Bering Sea THC scenario for the formation of the Meiji drift only requires that surface water about 1.5–2.0 ppt saltier than now circulated over the Bering Sea Basin. In effect, the Bering THC scenario calls for Paleogene oceanographic conditions similar to those now existing in the North Atlantic (figure 8.1). As noted by Warren (1983), these two areas are presently very different because the pattern of wind stress in the North Pacific separates the Subtropical and Subarctic gyres. Hence the Kuroshio Current, the Pacific's counterpart to the Atlantic's Gulf Stream, does not reach the Aleutian–Bering Sea region. But even if salty Kuroshio water could reach the northwestern corner of the Pacific Basin, the lateral continuity of the Aleutian Ridge would keep subtropical water out of the Bering Sea Basin.

Since the early Eocene, the style of ridge tectonism has been that of left-oblique convergence, a setting that promotes the westward transport of blocks of arc crust toward Kamchatka (Geist et al., 1987, 1988; Lonsdale, 1988; Geist and Scholl, 1992, 1994). This block shuffling process, which has been observationally confirmed by the GPS studies of Ave Lallemant and Oldow (2000) would, over time, tend to close off the western side of the Bering Sea Basin and open access to it along the ridge's central and eastern sectors (figures 8.6 and 8.7; Geist et al., 1988; Ryan and Scholl, 1989).

Based on the plate-boundary configurations of Lonsdale (1988), and, in response to a counterclockwise rotation of the underthrusting Kula plate, ridge-parallel transportation of arc fragments most likely got under way in the early middle Eocene. Eocene and younger arc fragmentation is supported by the occurrence along the length of the ridge of uplifted sequences of basinal sedimentary successions of Paleogene and Neogene age (Scholl et al., 1987). It can be inferred from the late Cenozoic formation of most of the ridge's modern summit basins and water-exchange passes that the rate of arc fragmentation increased markedly during the past 5–7 Ma. Prior to this rapid breakup the ridge was higher and more continuous than now, an observation based on the existence of a guyot-like, wave-base-cut platform carved across the submerged summit of the ridge (Scholl et al., 1987; Geist et al., 1988; Ryan and Scholl, 1989).

Two operations are involved in reconstructing the length of the Paleogene Aleutian Ridge:

1. Internally closing up and laterally repositioning the ridge's Komandorsky sector ~700 km eastward to the southern, or forearc, side of the Near Islands sector. This restoration assumes that during much of the past 40-50 Ma, along the far western Aleutian Ridge, the Bering-Kresta shear zone has not only been the principal plate-boundary fault separating the Pacific and North America plates, but that east of Near Pass slip along this transform was transferred across the ridge to run along the Aleutian forearc south of Near Islands (Vallier et al., 1994) (figures 8.7, 8.8, and 8.11). Evidence for this possibly is provided by Rostovtseva and Shapiro (1998) who identify a northern—Near Island—provenance for a ~3,000-m -thick sequence of Eocene fan deposits exposed on the Komandorsky Islands. If correctly repositioned, then in the early Tertiary the Komandorsky sector formed the forearc of the Near Island sector. It is recognized that the kinematic pattern of fragmentation and movement of blocks of the arc massif, in particular how motion has been distributed along and across the width of the ridge during much of the Cenozoic, is very complicated.

2. Within and east of the Near Island sector, closing up the ~300–400 km of tear canyons associated

with the clockwise rotation and westward translation of the ridge's major tectonic blocks (figure 8.7; Geist et al., 1988, Geist and Scholl, 1992, 1994).

The combined repositioning and closings shorten the length of the Paleogene ridge by about 1,000–1,200 km, placing its western end ~50 Ma near 176°E, in the vicinity of Kiska Island (figures 8.4 and 8.11).

Initiation of THC in the Bering Sea Basin. With an open Bering Sea Basin, as depicted on figure 8.11, warm, salty, subtropical water spun off northward from the paleo-Kuroshio Current or West Wind Drift (figure 8.2A) could have directly entered the Bering Sea by flowing into its open, western side. Miocene closing of the Indonesian seaway would have enhanced the delivery of salty subtropical waters to the North Pacific. This hypothetical path of western entrance is very different than that of today when subarctic surface water can only enter the Bering Sea from the east and downstream of a full west to east crossing of the North Pacific and an east to west counterclockwise backswing through the Gulf of Alaska (figures 8.2A, 8.4, and 8.11).

The conjecture is thus made that thermohaline circulation was initiated ~34 Ma when, in response to global cooling, salty western Pacific water entering the Bering Sea from the southwest was sufficiently chilled to sink to abyssal depths. Southward spreading of the THC bottom water, which would have been capable of conveying an abundance of Beringian terrigenous and biogenic debris to the Pacific Basin, occurred west of the Aleutian Ridge or via one or more far western deep-water passes, one of which is the present Kamchatka Strait (figure 8.11). As thermohaline circulation got under way, in positive feedback fashion, additional subtropical water flowing northward along the western rim of the Pacific would have been drawn into the Bering Sea to replace that lost by southward export at depth. THC in the Bering Sea thus implies warmer, more evaporative conditions than now prevailing. The wind swept, fog-shrouded Aleutian Islands are presently covered with tundra, but unequivocal evidence for Neogene warmer climes is provided by the fossilized remnants of a crowning coniferous forest that disappeared in the Pliocene (see Scholl et al., 1987:137). In effect, during much of the Tertiary, mapping in the Aleutian Islands could have engendered a sunburn rather than the Nordic pallor gained by modern mappers and sailors.

Shutting down THC in the Bering Sea Basin. The open model states that since the late Eocene, the slow but gradual westward translation of blocks of the Aleutian Ridge progressively impacted the inflow of subtropical water into the Bering Sea Basin. Potentially, the long tectonic process of obstructing the western entrance to the basin may have contributed to, or dampened, some of the cooling and warming events documented within and along the rims of the North Pacific Basin (Beniamovskii and Gladenkov, 1966; Barron, 1992a, 1992b, 1986; Barron and Baldauf, 1990; Gladenkov, 1990; Gladenkov and Shantser, 1990; White et al., 1997; Kwiek and Ravelo, 1999).

In keeping with the open scenario, it is posited that beginning in the latest Miocene and early Pliocene (8–4 Ma), three roughly synchronous tectonic events combined to intensify the Subarctic gyre and the Aleutian Low that spins it, lower the gyre's surface-water salinity, and isolate the gyre from mixing with the warm, saltier waters of the Subtropic gyre to the south (figure 8.2). These events, which led to thermal isolation of the Subarctic gyre and the wind stress pattern that keeps it this way, include:

1. the late Miocene-early Pliocene closing of the Panama seaway and the eventual chilling of surface waters in the PAC-A-BERS region;
2. the late Miocene to present elevation of mountain belts along the northern rim of the Gulf of Alaska and the resulting increased precipitation over, and runoff to, the Alaska Stream; and
3. the late Cenozoic increased rate of closing-off of the western Bering Sea Basin and opening the eastern Aleutian passes.

The grand effect of these distal and local factors would have been to dilute surface water circulating through the Subarctic gyre and thus either shutting down or greatly diminishing THC in the Bering Sea. It is supposed that the onset of Northern Hemisphere glaciation at about 2.7 Ma would have finally ended any significant scale of deep THC in the Bering Sea Basin, including, possibly, that needed to contribute intermediate water to the North Pacific (Kwiek and Ravelo, 1999). It is important to re-emphasize that since this time the Meiji drift has continued to be swept by abyssal currents (figure 8.10; Rea et al., 1995a, b). Because Bering Sea THC cannot be called upon to explain this circumstance, the modern engine driving the Meiji bottom flow is best linked to the abyssal flooding of the North Pacific with imported PBW and the southward boundary current return of some of this water—with added debris from the Bering Sea Basin—across the Meiji drift body (figures 8.2A, 8.4, and 8.11). If re-circulating PBW is involved, then the path of the deflected deep western boundary flow is connected to the northern boundary current moving westward (toward Kamchatka Strait) along the base of the landward slope of the Aleutian Trench. The studies of Underwood and Hathon (1989) and Hathon and Underwood (1991) imply that this northern-boundary current will include fine-grained mineral matter shed from both Alaska and the Aleutian Ridge. Presumably, the water presently flowing over the Meiji drift body will also include Aleutian and Alaskan material.

Exploring the External THC Scenario

If, in the early Tertiary, the Bering Sea Basin was open to the southwest, then this circumstance could have favored the existence of an externally driven deep-current system inside the Bering Sea Basin. At that time the basin was probably at least 4,500–5,000 m deep, including the sill depth of the seaway connecting it with the Pacific (figure 8.11; Scholl, 1975, 1986). A deep, western entrance to the Bering Sea Basin could have allowed THC water generated from the North Atlantic and Antarctic margins to enter the Bering Sea Basin along the Pacific side of the early Oligocene Kamchatka Trench. This, in a much reduced way, happens today (figure 8.2B). Presumably, the imported bottom water would force northern and western boundary currents to flow in a CCW pattern around the abyssal edges of the basin, entraining sediment along its path. The southward return to the Pacific at a depth shallower than the imported abyssal water would have been guided by the eastern side of Shirshov Ridge or the basin's Kamchatka margin (figure 8.11). The return flow to the Pacific, if guided by the Emperor Ridge, could have nourished formation of the Meiji drift along it exiting path. In a way, this model for the initiation of drift deposition in the northwestern Pacific Basin is similar to that described by Carter,and McCave (1994) and Carter et al. (1996) for the drift-forming consequence of the Deep Western Boundary Current presently flowing northward east of New Zealand.

The likelihood that imported PBW reached the high North Pacific and generated abyssal boundary currents there does not mean that cells of THC were not also operating in the Bering Sea Basin. Conceivably, both distal and proximal THC combined to force deep boundary flow along the northwestern margins of the PAC-A-BERS region. A system of THC-forced currents can be imagined that allowed the return flow of imported PBW to mingle with exported Bering Sea Bottom water and the combined stream to course along the northeastern flank of the Emperor Ridge. This basal current would have been capable of transporting large quantities of Bering Sea material into the Pacific Basin.

SUMMARY WITH COMMENTS

In the high North Pacific, the 2,200-km-long Aleutian Ridge separates the Bering Sea Basin from that of the Pacific. Seismic reflection records and DSDP (Leg 19, Site 192) and ODP (Leg 145, Site 884) cores document that bottom water connected to the Bering Sea Basin deposited, in the North Pacific, a lengthy (>1500 km), wide (as much as 300 km), and seaward or southeast thinning body of sediment—the Meiji drift—along the northeastern flank of the Emperor Ridge. The drift body is best formed along the northwestern most or Detroit-Meiji sector of the Emperor Ridge. Drift deposits are thickest, ~1,800 m, just south of the 125-km wide and ~4,200 m deep Kamchatka Strait, which separates the western end of the Aleutian Ridge and Kamchatka (figures 8.2 and 8.4). The clay mineralogy and diatom taxa of these deposits fingerprint a Bering Sea source region. Drift deposition began in the early Oligocene, roughly 34 Ma, and continues today.

Two current-forcing mechanisms, one acting external to the North Pacific and the other internal or local to the Pacific–Aleutian–Bering Sea (PAC-A-BERS) region, can account for the abyssal current that shaped the Meiji drift body. Both mechanisms involve thermohaline circulation (THC) and the export at depth of densified surface water. The externally driven flow is in response to the influx to the North Pacific of the blended mass of North Atlantic Deep Water (NADW) and Antarctic waters that forms Pacific Bottom Water (PBW). The import of PBW forces strong, abyssal boundary-current flow along the northern and western edges of the modern PAC-A-BERS region. The Kamchatka Current, a largely wind-driven western boundary current, flows southward out of the Bering Sea via Kamchatka Strait. South of the strait, the deeper part of this outflow (>~3,500 m) is underrun by a southbound current of PBW that probably entrains some deep Bering water escaping through the strait (Stabeno et al., 1999; Owens and Warren, 2001). The combined flow, bearing fallout debris from the main body of the overlying Kamchatka Current, is inferred to impinge on the northeastern side of the Meiji-Detroit sector of the Emperor Ridge and, moving southeastward along this flank toward the central Pacific, distribute sediment to the Meiji drift (Scholl et al., 1977; Owens and Warren, 2001).

Local generation of the Meiji bottom current can be caused by THC in the Bering Sea Basin and the consequent venting of densified bottom water southward around the western end of the Aleutian Ridge through the deep-water pass (~4,200 m) of Kamchatka Strait (Mammerickx, 1985). Owing to low surface water salinity, no THC of any significance presently takes place in the Bering Sea Basin (Warren, 1983). Because sedimentological evidence documents that the Meiji drift continues to form (Rea et al., 1995 a, b; figure 8.10), only the inferred southeastward deflection of the keel of the Kamchatka Current can be called upon to effect late Cenozoic growth (Scholl et al., 1977; Owens and Warren, 2001).

However, it is imaginable that the deep-circulation system of the PAC-A-BERS region was very different in the early and middle Tertiary than now, in particular if saltier surface water entered the Bering Sea and nourished a regional-scale (~10^6 km^2) cell of thermohaline sinking. Water of appropriate salinity (34–35 ppt) to initiate THC can now only be found at the northern end of the Kuroshio Current off northern Japan and within the subtropical gyre south of about 40°N (figure 8.2A; Dodimead et al., 1963). These subtropical waters are presently constrained by global wind patterns to mix poorly with the Subarctic gyre and thus to stay well south of the Bering Sea (figure 8.2A). An additional contributor to isolating the Bering gyre from access to subtropical water is the lateral continuity of the modern Aleutian Ridge. The greater part of the water exchanged (~12 Sv) between the Bering Sea and the North Pacific is effected only across and around the far western sector of the Aleutian Ridge (table 8.1, figure 8.4). But good reasons exist to suppose that in the Oligocene and Miocene the Bering Sea Basin was open to the southwest.

This inference is based on the oblique to transform plate-boundary setting of the Aleutian Ridge, which causes blocks of the arc massif to rotate clockwise and move westward toward the Kamchatka subduction zone (Geist et al., 1988; Geist and Scholl, 1992; Ave Lallemant and Oldow, 2000). This process has been under way for at least the past 50-55 Ma, but probably accelerated during the past 5–7 Ma (Scholl et al., 1987). The consequence of block rotation and westward shuttling has been to open waterways connecting the Bering and North Pacific across the more continuous central and eastern sectors of the Aleutian Ridge, but progressively close off between-basin circulation around the western end of the ridge, i.e., through the deep-water Kamchatka Strait (figures 8.4 and 8.5). Restoring the westward displaced blocks to their early Oligocene positions contracts the length of the Aleutian Ridge and opens up the southwestern side of the Bering Sea Basin to more direct water exchange with the Pacific. Reconstructions imply that both surface- and deep-water circulation could have been conducted across a ~1,000-km-wide, 4,500–5,000-m deep seaway separating the Aleutian Ridge and Kamchatka (present distance is ~150 km; figure 8.11).

A Bering Sea Basin open to the southwest could have allowed relatively salty subtropical water from the western Pacific to enter the basin from this direction (figure 8.11). Winter cooling would then be expected to initiate THC and the flushing of cold, salty, sediment-laden water southward out of the basin, either around the western end of the Aleutian Ridge or through one or more deep-water gaps located along its distal end. In a self-sustaining or feedback fashion, THC would have drawn replacement western Pacific surface water into the Bering Sea Basin.

Because the Meiji drift resides on the west-northwest-moving Pacific plate, the location of the depositional site of the oldest drift beds lying along the Meiji-Detroit sector of the Emperor Ridge was originally 1200-1500 km to the east-southeast. This backtracked location places the forming Meiji drift immediately south of the restored western end of the early Oligocene ridge (figures 8.2A and 11). If correctly restored, then Pacific plate motion would have carried both the drift and the western end of the lengthening Aleutian Ridge toward the Kamchatka subduction zone (Fig. 11). This likelihood can explain why the Meiji drift is thickest immediately south of Kamchatka Strait, and why this relation is not evidence, as supposed by Scholl et al.(1977), for limited subduction of the Pacific plate during the past 30-40 Ma.

Looking at the Past from a Future Scenario

A perspective can be gained about the paleoceanographic effects of tectonism along the Pacific's northern rim by exploring what the likely impact might be of the continuation of these same tectonic processes on the future oceanographic setting of the PAC-A-BERS region. For example, sustained coastal mountain building adjacent to the Gulf of Alaska seems assured because the colliding Yakutat block elevating them will continue to underthrust southeastern Alaska (Bruns and Carlson, 1987; Plafker, 1987; Plafker et al., 1994; Plafker and Berg, 1994; Fuis, 1998). These high ranges will thus continue to trap atmospheric moisture streaming across the northern Pacific toward the North American interior, diverting it back to help maintain diluted surface-water conditions in the Bering Sea.

The main boundary between the Pacific plate and the far western Aleutian Ridge appears not to have been the Bering-Kresta shear zone, which in a complicated way splays northwestward from the Aleutian forearc to run along the Bering Sea base of the ridge's Komandorsky sector (figures 8.6, 8.7, and 8.8). It can thus be inferred that continuing WNW motion of the Pacific plate will transport the length of the Komandorsky block toward and into the Kamchatka subduction zone. At the current rate of relative plate motion, the 125-km-wide strait (between the facing 200-m isobaths) will close within 1-2 million years. The base (~4,000 m) of the Komandorsky block will impact Kamchatka's landward trench slope in about half this time. Movement of the Komandorsky block toward Kamchatka at near Pacific plate speed may not greatly widen the size of the Near Pass because, as documented by Ave Lallemant (1996) and Ave Lallemant and Oldow (2000), the Near Island block is moving westward at a speed close to half that of the Pacific plate.

A likely consequence of sealing Kamchatka Pass will be forcing the Kamchatka Current to exit 400 km to the east through the western side of Near Strait (figures 8.4 and 8.5). Presumably, if Near Strait is to handle the greater volume of both inflowing and outflowing surface waters, the cross-ridge exchange between the Pacific and the Bering Basins would be impeded, especially if the average depth and cross-sectional area of the future Near Strait is not substantially increased during the coming 1–2 million years (table 8.1).

The coming closure of the Kamchatka Strait has the potential of significantly altering the surface-water properties and general oceanographic setting of the Bering Sea Basin. It can be expected that sealing of the strait will restrict the inflow of warmish Pacific water across the Aleutian Ridge, in particular at Near Strait, a consequence of which could be the winter freeze-over of the Bering Sea Basin. Winter ice formation in the Bering Sea is presently limited to the area of the Beringian shelf, or about 37% of the total area of the Bering Sea (figure 8.4; LaBelle et al., 1983; Niebauer et al., 1999). Unless the sill depth of the future Near Strait is substantially deepened (it is now ~2000 m), closing of the ~4,200-m-deep Kamchatka Strait will end the direct exchange of deep water between the Bering and Pacific Basins and, presumably, lessen the delivery of Bering Sea debris to a remaining deep current running across the Meiji drift.

It is likely that full icing would foster a limited volume of thermohaline circulation in the Bering Sea Basin (Gardner et al., 1982; Takahashi, 1999). However, similar to what now happens in the Sea of Okhotsk (Talley, 1991), winter freeze up of the Bering Sea Basin can be expected to generate sufficiently dense water to sink to intermediate depths. Even with a sealed Kamchatka Strait, hypothetical Bering Intermediate Water would be capable of spreading into the North Pacific Basin via the western side of the future Near Strait. The presence of Bering Intermediate Water in the North Pacific could play an important role in ventilating the water columns bathing the margins of the North Pacific Basin (Gorbarenko, 1996; Behl and Kennett, 1996; Talley, 1991, 1993; Kennett et al., 2000).

It can be anticipated that full icing of the Bering Sea Basin will prominently affect ocean-atmospheric coupling and energy exchange in the PAC-A-BERS region. A full ice cover would reduce winter-months wind stress and thus the wind-driven transport of surface water into, across, and out of the Bering Sea. The center of the Aleutian Low, the north-central quadrant of which presently hovers over the ice-free Bering Sea Basin (figure 8.3), would move elsewhere for open-water nourishment. Movement of the center southeastward toward the Gulf of Alaska or southwestward toward the Kuril Islands is most likely, and presently happens in response to outside forcing by equatorial El Nino and La Nina oscillations (Niebauer et al., 1999). Presumably, an ice and snow covered Bering Sea Basin would allow the intensely cold Siberian High to expand southeastward and embrace the entire area of the Bering Sea (figure 8.3). This likelihood would tend to refrigerate the continental and watery domains of Beringia (Hopkins, 1967; Gualtieri et al., 2000).

It is intriguing to wonder what the more outreaching consequences of closing the Kamchatka Strait, and the commensurate redirecting and restricting of the western boundary current system, will have on the oceanographic realm of the North Pacific and the climatic regimes of its bordering continental areas.

Addressing the Remaining Questions

Returning to the past, if the greater bulk of the Meiji drift was formed by a Bering Sea-muddied, externally driven deep-water jet of the Kamchatka Current, then the water involved would have been old and long traveled since last at the sea surface, as would be true today. However, if the principal engine, or a contributing one, was THC in the Bering Sea, then newly formed bottom water would have been involved in fashioning the drift. At the present, trace element, nutrient, and stable isotope evidence from the Meiji body or from the Bering Sea Basin are unavailable to determine if distally or locally formed bottom waters, or both, were involved in starting the drift deposit and nourishing its growth through much of its history. It is unlikely that significant production of thermohaline bottom water has occurred in the Bering Sea Basin since the middle Pliocene. But properly selected drilling sites, core samples, and bottom samples should allow this question, and the larger one of the relative age and origin(s) of the bottom water that formed the greater volume of the Meiji drift body, to be resolved (see Keigwin, 1987; Keigwin et al., 1992; Ohkuochi et al., 1994; Gorbarenko, 1996; Kwiek and Ravelo, 1999). Field measurements can also verify if an inferred spur of basal western boundary flow presently passes over Meiji drift.

With respect to Pacific Basin paleoceanography, the issue of whether a local or distal source of THC, or a fusion of these source regions, created the bulk of the Meiji drift is an important question to tackle. But from the perspective of global climate, answering this question is not critical to the implications of the conjecture explored in this paper; i.e., that the formation of the Meiji drift is a high North Pacific manifestation of the onset of the post-Eocene icehouse world. It is the movement through the PAC-A-BERS region of cold, thermohaline bottom water generated somewhere that is hypothesized to have initiated and sustained the growth of the Meiji drift during the past ~34 Ma.

ACKNOWLEDGMENTS

The project to extract a paleoceanographic message from a history of North Pacific rim tectonism was originally suggested to us by John Barron, U.S. Geological Survey. John read early fragments and completed drafts of the manuscript and, drawing on his paleoceanographic insights and knowledge, provided the counsel need to best shape and direct its message. Jim Ingle, Stanford University, gave us special encouragement and guidance about how to wring likely oceanographic consequences out of a history of North Pacific tectonic processes. Jim also read the first complete draft of the manuscript and significantly contributed to improving its information content and focus. The authors are exceptionally grateful for the many messages of information, guidance, and counsel provided by Bruce Warren, Woods Hole Oceanographic Institution, concerning paths and driving forces of deep-water circulation in the high North Pacific. His curiosity and interest in the oceanographic implications of the Meiji drift body identified a viable mechanism for the drift's continuing history of growth.

REFERENCES

Arnold, E., 1995, Data report: Mineralogy of Site 884, *in* Rea, D. K., Basov, I. A, Scholl, D. W., Allan, J., et al., Proceedings of the Ocean Drilling Program, Scientific Results, v. 145, p. 657-660.

Ave Lallemant, H. G., 1996, Displacement partitioning and arc-parallel extension in the Aleutian volcanic island arc: Tectonophysics, v. 256, p. 279-293

Ave Lallemant, H. G., and Oldow, J. S., 2000, Active displacement partitioning and arc-parallel extension of the Aleutian volcanic arc based on GPS geodesy and kinematic analysis: Geology, v. 28, p. 739-742

Barron, J. A., 1986, Paleoceanographic and tectonic controls on deposition of the Monterey Formation and related siliceous rocks in California: Palaeogeography, Palaeoclimatology, Palaeoecology, v. 53, p. 27-45.

Barron, J. A, 1992a, Paleoceanography and tectonic controls on the Pliocene diatom record of the northeastern Pacific, *in* Tsuchi, R.. and Ingle, J. C., Jr., eds., Pacific Neogene Environment, Evolution and Events: Tokyo, Tokyo University Press, p. 25-41.

Barron, J. A., 1992b, Pliocene paleoclimatic interpretations of DSDP Site 580 (NW Pacific) using diatoms: Marine Micropaleontology, v. 20, p. 23-44.

Barron, J. A., 1995, High-resolution diatom paleoclimatology of the middle part of the Pliocene in the northwestern Pacific, *in* Rea, D. K., Basov, I. A., Scholl, D.W., Allan, J., et al., Proceedings of the Ocean Drilling Program, Scientific Results, v. 145, p. 43-53

Barron, J. A, 1998, Late Neogene changes in diatom sedimentation in the North Pacific: Journal of Asian Earth Sciences, v. 16, p. 85-95

Barron, J. A., and Baldauf, J. G., 1990, Development of biosiliceous sedimentation in the North Pacific during the Miocene and early Pliocene, *in* Tsuchi, R., ed., Pacific Neogene Events—their timing nature and interrelationship: Tokyo, University of Tokyo Press, p. 43-63.

Barron, J. A., and.Gladenkov, A. Y., 1995, Early Miocene to Pleistocene diatom stratigraphy of Leg 145, *in* Rea, D. K., Basov, I. A, Scholl, D. W., Allan, J., et al., Proceedings of the Ocean Drilling Program, Scientific Results, v. 145, p. 3-19.

Behl, R. J., and Kennett, J. P., 1996, Brief interstadial events in the Santa Barbara Basin, northeast Pacific during the past 60 Kyr: Nature, v. 379, p. 243-256.

Beniamovskii, V. N. and Gladenkov, Y. B., 1996, Paleogene climatic fluctuations and biotic migrations in the North Pacific, *in* Stratigraphy and Geological Correlations: Geological Institute of Russian Academy of Sciences, Moscow. v. 4, p. 480-495.

Billups, K., Ravelo, A. C., Zachos, J. C., and Norris, R. D., 1999, Link between oceanic heat transport, thermohaline circulation, and intertropical convergence zone in the early Pliocene Atlantic: Geology, v.27, p. 319-322.

Bond, N. A., Overland, J. E., and Turet, P., 1994, Spatial and temporal characteristics of the wind forcing of the Bering Sea: Journal of Climate, v. 7, p. 1119-1130.

Boyle, E., 1990, Quaternary deep water paleoceanography: Science, v. 249, p. 863-869.

Broecker, W. S., 1987, Unpleasant surprises in the greenhouse? Nature v. 328, p 123-126.

Broecker, W. S., 1991, The great ocean conveyor: Oceanography, v. 4, p. 79-89.

Broecker, W. S., 1997, Thermohaline circulation, the Achilles heel of our climate system: will man-made CO_2 upset the current balance? Science, v. 278, p. 1582-1588.

Bruns, T. R., and Carlson, P. R., 1987, Geology and petroleum potential of the southeast Alaska continental margin, *in* Scholl, D. W., Grantz, A., and Vedder, J. G., eds., Geology and resource potential of the continental margin of western North America and adjacent ocean basins—Beaufort Sea to Baja California: Houston, Texas, Circum-Pacific Council for Energy and Mineral Resources, Earth Science Series, v. 6, p. 269-282.

Burk, C. A., 1965, Geology of the Alaska Peninsula—island arc and continental margin: Geological Society of America Memoir 99, p. 1-280.

Businger, S., 1991, Arctic Hurricanes; American Scientist, v. 79, p. 18-33

Carter, L., and McCave, I. N., 1994, Development of sediment drifts approaching an active plate margin under the southwest Pacific deep western boundary current: Paleoceanography, v. 9., p. 1061-1085

Carter, L., Carter, R. M., McCave, I.N., and Gamble, J., 1996, Regional sediment recycling in the abyssal southwest Pacific Ocean: Geology, v. 24, p. 735-738.

Carter R. M., Carter, L., McCave, I. N., et al., 1999, The DWBC sediment drift record from Leg 181: drilling in the Pacific gateway for the global thermohaline circulation: JOIDES Journal, v. 25, p. 8-13.

Coachman, L. K., Whitledge, T. E., and Goering, J. J., 1999, Silica in Bering Sea deep and bottom water, *in* Loughlin, T. R., and Ohtani, K., eds., Dynamics of the Bering Sea: A summary of Physical, Chemical, and Biological Characteristics, a Synopsis of Research on the Bering Sea: North Pacific Marine Science Organization (PICES), University of Alaska Sea Grant, Fairbanks, p. 285-309.

Coats, R. R., 1962, Magma type and crustal structure in the Aleutian Arc: American Geophysical Union Monograph, v. 6, p. 92-10.

Collins, L. S., Coats, A. G., Berggen, W. A., Aubry, M.-P., and Zhang, J., 1996, The late Miocene Panama Isthmus Strait: Geology, v. 24, p. 687-690.

Cooper, A. K., Marlow, M. S., and Scholl, D. W., 1977, The Bering Sea—a multifarious marginal basin: American Geophysical. Union., Maurice Ewing Series, v. 1, p. 437-450.

Cooper, A. K., Scholl, D. W., and Marlow, M. S., 1987a, Structural framework, sedimentary sequences, and hydrocarbon potential of the Aleutian and Bowers Basins, Bering Sea, *in* Scholl, D. W., Grantz, A., and Vedder, J. G., eds., Geology and resource potential of the continental margin of western North America and adjacent ocean basins—Beaufort Sea to Baja California: Houston, Texas, Circum-Pacific Council for Energy and Mineral Resources, Earth Science Series, v. 6, p. 476-502.

Cooper, A. K., Marlow, M. S., and Scholl, D. W., 1987b, Geologic framework of the Bering Sea crust, *in* Scholl, D. W., Grantz, A., and Vedder, J. G., eds., Geology and resource potential of the continental margin of western North America and adjacent ocean basins—Beaufort Sea to Baja

California: Houston, Texas, Circum-Pacific Council for Energy and Mineral Resources, Earth Science Series, v. 6, p. 73-102.

Cooper, A. K., Marlow, M. S., Scholl, D. W., and Stevenson, A. J., 1992, Evidence for Cenozoic crustal extension in the Bering Sea: Tectonics, v. 11, p. 719-731

Corliss, B. H., Martinson, D. G., and Keffer, T., 1986, Late Quaternary deep-ocean circulation: Geological Society of America Bulletin, v. 97, p. 1106-1121

Creager, J. S., Scholl, D. W., et al., 1973, Initial Report of the Deep Sea Drilling Project: Washington, D.C., U.S. Government Print. Office, v. 19, 913 p.

Dadisman, S. V., and Scholl, D. W., 1995, Subduction of the Meiji sediment drift, cause of isotopic enrichment and excessive arc magmatism at Kamchatka's Klyuchevskoy volcano: EOS, v. 76, p. F538.

Damuth, J. E., Jacobi, R. D., and Hayes, D. E., 1983, Sedimentation processes in the northwestern Pacific Basin revealed by echo-character mapping studies: Geological Society of America Bulletin, v. 94, p. 381-395.

Davies, R., Cartwright, J., Pike, J., and Line, C., 2001, Early Oligocene initiation of North Atlantic Deep Water formation: Nature, v. 410, p. 917-920.

Dodimead. A. J., Favorite, F., and Hirano, T., 1963, Review of Oceanography of the subarctic Pacific Region; Salmon of the North Pacific Ocean, Part II: Vancouver, B.C., Canada, International North Pacific Fisheries Commission, Bulletin 13, 190 pp

Dowsett, H., Barron, J. A., and Poore, R. Z., 1995, Middle Pliocene sea surface temperatures: a global reconstruction: Marine Micropaleontology, v. 27, p. 13-25

Driscoll, N. W., and Haug, G. H., 1998, A short circuit in thermohaline circulation: A cause for Northern Hemisphere glaciation: Science, v. 282, p. 436-438.

Duque-Caro, H., 1990, Neogene stratigraphy, paleoceanography and paleobiology in northwest South America and the evolution of the Panama Seaway: Palaeoceanography, Palaeoclimatology, Palaeoecology, v. 77, p. 203-234.

Ewing, J., Ewing, M., Aitken, T., and Ludwig, W. J., 1968, North Pacific sediment layers measured by seismic profiling, *in* Knopoff. L., Drake, C. L., and Hart, P. J., eds, The Crust and Upper Mantle of the Pacific Area: American Geophysical Union Geophysical Monograph, v. 12, p. 147-173.

Favorite, F., 1974, Flow into the Bering Sea through Aleutian Island passes, *in* Hood, D. W., and Kelley, E. J., eds., Oceanography of the Bering Sea: Occasional Publication of the Institute of Marine Science, University of Alaska, Fairbanks, No. 2, p. 3-37

Fitzgerald, P. G., Stump, E., and Redfield, T. F., 1993, Late Cenozoic uplift of Denali and its relation to relative plate motion and fault morphology: Science, v. 259, p.497-499.

Fitzgerald, P. G., Sorkhabi, R. B., Redfield, T. F., and Stump, E., 1995, Uplift and denudation of the central Alaska Range: A case study in the use of apatite fission track thermoochronology to determine absolute uplift parameters: Jounal of Geophysical Research, v. 100, p. 20,175-20,191.

Fliedner, M. M., and Klemperer, S. L., 1999, Structure of an island-arc: wide-angle seismic studies in the eastern Aleutian Islands, Alaska: Journal of Geophysical Research, v. 104-p. 10,667-10,694

Flower, B. P., and Kennett, J. P., 1994, The middle Miocene climatic transition: East Antarctic sheet development, deep ocean circulation and global carbon cycling: Palaeogeography, Palaeoclimatology, Palaeoecology, v. 108, p. 537-555.

France-Lanord, C., and Derry, L. A., 1997, Organic carbon burial forcing of the carbon cycle from Himalayan erosion: Nature, v. 390, p. 65-67.

Frank, M., Reynolds, B. C., and O'Nions, R. K, 1999, Nd and Pb isotopes in Atlantic and Pacific water masses before and after closure of the Panama gateway: Geology, v. 27, p. 1147-1150.

Fuis, G. S., 1998, West margin of North America—a synthesis of recent seismic transects: Tectonophysics, v. 288, p. 265-292

Gardner, J. V., Dean, W. E., Klise, D. H., and Baldauf, J. G., 1982, A climate-related oxidizing event in deep-sea sediment from the Bering Sea: Quaternary Research, v. 18, p. 91-107

Geist, E. L., and Scholl, D. W., 1992, Application of continuum models to deformation of the Aleutian Island Arc: Journal of Geophysical Research, v. 97, p. 4953-4967.

Geist, E. L., and Scholl, D. W., 1994, Large-scale deformation related to the collision of the Aleutian Arc with Kamchatka: Tectonics, v. 13, p. 538-560

Geist, E. L., Childs, J. R., and Scholl, D. W., 1987, Evolution and petroleum geology of Amlia and

Amukta intra-arc summit basins, Aleutian Ridge: Marine and Petroleum Geology, v. 4. p. 334-352.

Geist, E. L., Childs, J. R., and Scholl, D. W., 1988, The origin of summit basins of the Aleutian Ridge: implications for block rotation of an arc massif: Tectonics, v. 7, p. 327-341.

Gladenkov, Y. B., 1990, Ecostratigraphy and the North Pacific Neogene holarctic; Palaeogeography, Palaeoclimatology, Palaeoecology, v. 77, p. 195-197.

Gladenkov, Yu. B., and Shantser, A. Y., 1990, Neogene of Kamchatka: stratigraphy and correlation of geological event, *in* Tsuchi, R., ed. Pacific Neogene Events: Tokyo, University of Tokyo Press, p. 173-181.

Gorbarenko, S., 1996, Stable isotope and lithologic evidence of late glacial and Holocene oceanography of the northwestern Pacific and marginal seas: Quaternary Research, v. 46, p. 230-250.

Gordon, A. L., 1983, A clever answer to a simple question: Nature, v. 305, p. 385-386.

Gordon, A. L., 1986, Interocean exchange of thermocline water: Journal of Geophysical Research, v. 91, p. 5037-5046

Gordon, A. L., 1991, The role of thermohaline circulation in global climate change: Lamont-Doherty Geological Observatory 1990 and 1991 Report, Lamont-Doherty Geological Observatory of Columbia University, p. 44-51

Grow, J. A., 1973, Crustal and upper mantle structure of the central Aleutian Arc: Geological Society of American Bulletin, v. 84, p. 2169-2192

Gualtieri, L., Glushkova, O., and Bringham-Grette, J., 2000, Evidence for restricted ice extend during the last glacial maximum in the Koryak Mountains of Chukotka, far eastern Russia: Geological Society of America Bulletin, v. 112, p. 1106-1118.

Hathon, E. G., and Underwood, M. B., 1991, Clay mineralogy and chemistry as indicators of hemipelagic sediment dispersal south of the Aleutian Arc: Marine Geology, v. 97, p. 145-166.

Haug, G. H., and Tiedemann, R., 1998, Effect of the formation of the Isthmus of Panama on Atlantic Ocean thermohaline circulation: Nature v. 393, p. 673-676.

Haug, G. L., Maslin, M. A., Sarnthein, M., Stax, R., and Tiedemann, R., 1995, Evolution of northwest Pacific sedimentation patterns since 6 Ma: Site 882: Proceedings of the Ocean Drilling Program, Scientific Results, v. 145, p 293-314.

Haug, G. L., Sigman, D. M., Tiedemann, R., and Zahn, R., 1998, The Panamanian Isthmus formation, Atlantic Ocean thermohaline circulation, and subarctic Pacific productivity: Potential effects on Plio-Pleistocene climate: EOS, v. 70, p. F25.

Hein, J. R., Scholl, D. W., and Gutmacher, C. E., 1976, Neogene clay minerals of the far northwest Pacific and southern Bering Sea: sedimentation and diagenesis, *in* Bailey, S.W., ed., Proceedings of the International Clay Conference, 1975: Applied Publishing Ltd., p. 71-80.

Hopkins, D. W., 1967, The Bering Land Bridge: Stanford, California, Stanford University Press, 495 p.

Hughes, F. W., Coachman, L. K., and Aaraard, K., 1974, Circulation, transport, and water exchange in the western Bering Sea, *in* Hood, D. W., and Kelley, E. J., eds., Oceanography of the Bering Sea: Occasional Publication of the Institute of Marine Science, University of Alaska, Fairbanks, No. 2, p. 59-98.

Johnson, R. G., 1997a, Ice age initiation by an ocean-atmospheric circulation change in the Labrador Sea: Earth. Planetary Sciences Letters, v. 148, p. 367

Joseph, L. H., Rea, D .K., and van der Pluijm, B. A, 1998, Use of grain size and magnetic fabric analysis to distinguish among depositional environments: Paleoceanography, v. 13, p. 491-501.

Keigwin, L .S., 1982, Isotopic paleoceanography of the Caribbean and east Pacific: role of Panama uplift in late Neogene time: Science, v. 217, p. 350-352.

Keigwin, L. S., 1987, North Pacific deep water formation during the latest glaciation: Nature, v. 330, p. 362-364

Keigwin, L. D., Jones, G. A., and Froelich, P. N., 1992, A 15,000 year paleoenvironmental record from Meiji Seamount far northwestern Pacific: Earth and Planetary Science Letters, v. 111, p. 425-440.

Keller, G., and Barron, J. A., 1983, Paleoceanographic implications of Miocene deep-sea hiatuses: Geological Society of America Bulletin, v. 94, p. 590-613.

Kennett, J. P., ed., 1985, The Miocene Ocean: Paleoceanography and Biogeography: Geological Society of America Memoir 163, 337 p.

Kennett, J. P., Keller, G., and Srinivasan, M. S., 1985, Miocene planktonic foraminifera biogeography and paleoceanographic development of the Indo-

Pacific region, *in* Kennett, J. P., ed., The Miocene Ocean: Paleoceanography and Biogeography: Geological Society of America Memoir 163, p. 197-236

Kennett, J. P., Cannariato, K. G., Hendy, I. L., and Behl, R. J., 2000, Carbon isotopic evidence for methane hydrate instability during Quaternary interstadials: Science, v.288, p. 128-132.

Kwiek, P .B., and Ravelo, A. C., 1999, Pacific Ocean intermediate and deep water circulation during the Pliocene: Palaeogeography, Palaeoclimatology, Palaeoecology, v. 154, p. 191-217.

Kidd, R. B., and Hill, P. R., 1987. Sedimentation on Feni and Gardar sediment drifts, *in* Ruddiman, W. F, Kidd, R. B., et al., Initial Reports of the Deep Sea Drilling Project, v. 94(2), p.1217-1244.

Kulm, L. D., von Huene, R., et al., 1973, Initial Reports of the Deep Sea Drilling Project, v. 18, 1077 p.

LaBelle, J. C., Wise, J. L., Voelker, R. P., Schulze, R. H., and Wohl, G .M., 1983, Alaska Marine Ice Atlas: Arctic Environmental Information and Data Center, University of Alaska, Fairbanks, 302 p.

Lagoe, M. B., Eyles, C. H., Eyles, N., and Hale, C., 1993, Timing of late Cenozoic tidewater glaciation in the far North Pacific: Geological Society of America Bulletin, v. 105, p. 1542-1560.

Leg 145 Scientific Party, 1993, Paleoceanographic record of North Pacific quantified: EOS, v. 74, p. 406-411.

Luchin, V. A., Menovshchikov, V. A., Lavrentiev, V. M., and Reed, R. K., 1999, Thermohaline structure and water masses in the Bering Sea, *in* Loughlin, T. R., and Ohtani, K., eds., Dynamics of the Bering Sea: A aummary of physical, chemical, and biological characteristics, and a synopsis of research on the Bering Sea: North Pacific Marine Science Organization (PICES), University of Alaska Sea Grant, Fairbanks, p. 61–91.

Ludwig, W .J., and Houtz, R. E., 1977, Isopach map of sediments in the Pacific Ocean Basin and marginal sea Basins: Tulsa, Oklahoma, American Association of Petroleum Geologists.

Lonsdale, P., 1988, Paleogene history of the Kula plate: Offshore evidence and onshore implications: Geological Society of America Bulletin, v. 100, p. 733-754

Lonsdale, P., Dieu, J., and Natland, J., 1993, Posterosional volcanism in the Cretaceous part of the Hawaiian hotspot trail; Journal of Geophysical Research, v. 98, p. 4081-4098.

Mammerickx, J., 1985, A deep-sea thermohaline flow path in the northwestern Pacific: Marine Geology, v. 65, p. 1-19.

Maier-Reimer, E., Mikolajewicz, U., and Crowley, T., 1990, Ocean general circulation model sensitivity experiment with an open Central American isthmus: Paleoceanography, v. 5., p. 349-366.

Marincovich, L., Jr., 1988, Late middle Eocene mollusks of the Tolstoi formation, Alaska Peninsula, Alaska, and correlations with faunas from California to the far-eastern USSR, *in* Filewizc, M. V., and Squires, R. L., eds, Paleogene Stratigraphy, West Coast of North America: Pacific Section SEPM., West Coast Paleogene Symposium, v. 58, p. 265-281.

Marincovich, L., Jr., 2000, Central American paleogeography controlled Pliocene Arctic Ocean molluscan migrations: Geology, v. 28, p. 551-554

Marincovich, L., Jr., and Gladenkov, A. Y., 1999, Evidence for an early opening of the Bering Strait: Nature, v. 397, p. 149-151.

Miller, K. G., 1992, Middle Eocene to Oligocene stable isotopes, climate, and deep-water history: The terminal Eocene Event, *in* Prothero, D. R., and Berggren, W. A., eds., Eocene-Pliocene climatic and biotic evolution: Princeton, New Jersey, Princeton University Press, p. 160-177.

Miller, K. G., Fairbanks, R. G., and Mountain, G. S., 1987, Tertiary oxygen isotope synthesis, sea level history, and continental margin erosion: Paleoceanography. v. 2, p. 1-19.

Miller, K. G., Wright, J. D., and Fairbanks, R. G., 1991, Unlocking the icehouse: Oligocene-Miocene oxygen isotopes, eustacy, and margin erosion: Journal of Geophysical Research, v. 96, p. 6829-6848.

Murdock, T. Q., Weaver, A. J., and Fanning, A. G., 1997, Paleoclimatic response of the closing of the Isthmus of Panama in a coupled ocean-atmosphere model: Geophysical Research Letters, v. 24, p. 253-256.

Niebauer, H. J., Bond, N. A., Yakunin L. P., and Plotnikov, V. V., 1999, An update on the climatology and sea ice of the Bering Sea, *in* Loughlin, T. R., and Ohtani, K., eds, Dynamics of the Bering Sea: A summary of physical, chemical, and biological characteristics, a synopsis of research on the Bering Sea: North Pacific Marine Science Organization (PICES), University of Alaska Sea Grant, Fairbanks, p. 29-59

Owens, W. B., and Warren, B. A., 2001, Deep circulation in the northwest corner of the Pacific Ocean: Deep Sea Research, Part 1, v. 38, p. 959-993.

Ohkuochi, N., Kawahata, H., Okada, M., Murayama, M., Nakamura, T., and Taira, A., 1994, Was deep water formed in the North Pacific during the late Quaternary?: Cadmium evidence from the northwest Pacific: Earth and Planetary Sciences Letters, v. 124, p. 185-194.

Plafker, G., 1987, Regional geology and petroleum potential of the northern Gulf of Alaska continental margin, *in* Scholl, D. W., Grantz, A., and Vedder, J. G., eds., Geology and resource potential of the continental margin of western North America and adjacent ocean basins—Beaufort Sea to Baja California: Houston, Texas, Circum-Pacific Council for Energy and Mineral Resources, Earth Science Series, v. 6, p. 229-268.

Plafker, G., and Berg, H. C., 1994, Overview of the geology and tectonic evolution of Alaska, *in* Plafker, G., and Berg, H. C., eds., The Geology of North America v. G-1, The Geology of Alaska: Boulder, Colorado, Geological Society of America, p. 989-1021.

Plafker, G., Moore, J. C., and Winkler, G. R., 1994, Geology of the southern Alaska margin, *in* Plafker, G., and Berg, H. C., eds., The Geology of North America v. G-1, The Geology of Alaska: Boulder, Colorado, Geological Society of America, p. 389-449.

Plafker, G., Naeser, C. W., Zimmermann, R. A., Luss, J. S., and Hudson, T., 1992, Cenozoic uplift history of the Mount McKinley area in the central Alaska Range basin on fission-track dating: U.S. Geological Survey Bulletin 2041, p. 202-212.

Prueher, L. M., and Rea, D. K., 1998, Rapid onset of glacial conditions in the subarctic North Pacific region at 2.67 Ma: clues to causality: Geology, v. 26, p. 1027-1030.

Raymo, M. E., and Ruddiman, W. F., 1992, Tectonic forcing of late Cenozoic climate: Nature, v. 359, p. 117-122.

Rea, D. K., and Hovan, S. A., 1995, Grain size distribution and depositional processes of the mineral component of abyssal sediments: Lessons from the North Pacific: Paleoceanography, v. 10, p. 251-158.

Rea, D. K., Basov, I. A., Janecek, T. R., Palmer-Julsun, A., et al., 1993. Proceedings of the Ocean Drilling Program: Initial Reports, v. 145, 1040 p.

Rea, D. K., Basov, I. A., Krissek, L. A., and the Leg 145 Scientific Party, 1995a, Scientific results of drilling the North Pacific transect: Proceedings of the Ocean Drilling Program, Scientific Results, v. 145, p. 577-596.

Rea, D. K, Basov, I. A., Scholl, D. W., and Allan, J. F., 1995b, Proceedings of the Ocean Drilling Program, Scientific Results, v. 145: 711 p.

Reed R. K., and Stabeno, P .J., 1994, Flow along and across the Aleutian Ridge: Journal of Marine Research, v. 52, p. 639-648.

Reed, R. K., and Stabeno, P. J., 1999, The Aleutian North Slope Current, *in* Loughlin, T. R., and Ohtani, K., eds, Dynamics of the Bering Sea: A summary of physical, chemical, and biological characteristics, a synopsis of research on the Bering Sea: North Pacific Marine Science Organization (PICES), University of Alaska Sea Grant, Fairbanks, p. 177-192.

Roden, C. L., 1995, Aleutian Basin of the Bering Sea: Thermohaline, oxygen, nutrient, and current structure in July 1993: Journal of Geophysical Research, v. 100, p. 13,539-13,554.

Roemmich, D., and McCallister, T., 1989, Large scale circulation of the North Pacific Ocean: Progress in Oceanography, v. 22, p. 171-204.

Romine, K, 1985, Radiolarian biogeography and paleoceanography of the North Pacific at 8 Ma, *in* Kennett, J. P., ed., The Miocene Ocean: paleoceanography and biogeography: Geological Society of America Memoir 163, p. 237-272

Romine, K, and Lombari, G., 1985, Evolution of Pacific circulation in the Miocene: Radiolarian evidence from DSDP Site 289, *in* Kennett, J. P., ed., The Miocene Ocean: paleoceanography and biogeography: Geological Society of America Memoir 163, p.273-290.

Rostovtseva, Yu. V, and Shapiro, M. N., 1998, Provenance of the Palaeocene-Eocene rocks of the Komandorsky Islands: Sedimentology, v. 45, p. 201-216

Ryan, H. F., and Scholl, D. W., 1989, The evolution of forearc structures along an oblique convergent margin, central Aleutian Arc: Tectonics, v. 8, p. 497-516.

Ryan, H. F., and Scholl, D. W., 1993, Geologic implications of great interplate earthquakes along the Aleutian Arc: Journal of Geophysical Research, v. 98, p. 22,135-22,146

Sancetta, C., 1979, Oceanography of the North Pacific during the past 18,000 years: evidence from fossil diatoms: Marine Micropaleontology, v. 4, p. 103-123.

Sancetta, C., 1983, Effect of Pleistocene glaciation upon oceanographic characteristics of the North Pacific Ocean and Bering Sea: Deep Sea Research, v. 30, p. 851-869.

Sancetta, C., and Silvestri, S., 1986, Pliocene-Pleistocene evolution of the North Pacific Ocean-atmosphere system, interpreted from fossil diatoms: Paleoceanography, v. 1, p. 163-180.

Sancetta, C., Heusser, L., Labreyrie, L., Naidu, A. S., and Robinson, S. W., 1985 Wisconsin-Holocene paleoenvironment of the Bering Sea: evidence from diatoms, pollen, oxygen isotopes and clay minerals: Marine Geology, v. 62, p. 55-68.

Schmitz, W. J., Jr., 1995, On the interbasin-scale thermohaline circulation: Reviews of Geophysics, v. 33, p. 151-173.

Scholl, D. W., and Creager, J. S., 1973, Geologic synthesis of Leg 19 (DSDP) results; far North Pacific and Aleutian Ridge and Bering Sea, *in* Creager, J. S., Scholl, D. W., et al.: Initial Reports of the Deep Sea Drilling Project, v. 19, p. 897-913.

Scholl, D. W., and Stevenson, A. J., 1987, Ocean margin tectonics and North Pacific paleoceanography—wonderments about the effects of the Pacific-Bering Sea connection: Proceedings of the Fourth International Congress of Pacific Neogene Stratigraphy, Berkeley, California, July 29-31, 1987, p 95-97.

Scholl, D. W., Buffington, E. C., and Marlow, M. S., 1975, Plate tectonics and the structural evolution of the Aleutian–Bering Sea region, *in* Forbes, R. R., ed., Contributions to the geology of the Bering Sea basin and adjacent regions: Geological Society of America Special Paper 151, p. 1-31.

Scholl, D. W., Hein, J. R., Marlow, M. S., and Buffington, E. C., 1977, Meiji sediment tongue: North Pacific evidence for limited movement between the Pacific and North American plates: Geological Society of America Bulletin, v. 88, p. 1567-1576.

Scholl, D. W., Vallier, T. L., and Stevenson, A. J., 1986, Terrane accretion, production, and continental growth—a perspective based on the origin and tectonic fate of the Aleutian-Bering Sea region: Geology, v. 14, p. 43-47.

Scholl, D. W., Vallier, T. L., and Stevenson, A. J., 1987, Geologic evolution and petroleum geology of the Aleutian Ridge, *in* Scholl, D. W., Grantz, A., and Vedder, J. G., eds., Geology and resource potential of the continental margin of western North America and adjacent ocean basins—Beaufort Sea to Baja California: Houston, Texas, Circum-Pacific Council for Energy and Mineral Resources, Earth Science Series, v. 6, p. 124-155.

Scholl, D. W., and Stevenson, A. J., 1991, Transpressive tectonism along the Aleutian margin of the Pacific Basin—a major factor affecting the paleoceanographic history of the far northern Pacific, *in* Anderson, D. M., and Eaton, G. P., eds., Geological Perspectives on Global Change: Proceedings of the Geodynamics Research Institute Symposium, Texas A & M University, College Station, Texas, 3 p.

Scholl, D. W., Stevenson, A. J., Noble, M. A., and Rea, D. K., 1998, Turn-on and turn-off oceanographic, climatic, and tectonic conditions for the Tertiary formation and southward discharge, via an Aleutian gateway, of thermohaline water from the Bering Sea to the North Pacific Basin: EOS, v. 79, p. 32.

Scholl, D. W., Stevenson, A. J., Noble, M. A., and Rea, D. K., 1999, The Meiji sediment body—a drift implying Oligocene initiation of thermohaline-driven circulation in the North Pacific-Bering Sea region: Geological Society of America Abstracts with Program, v.31, p. A311.

Stabeno, P. J., Schumacher, J. D., and Ohtani, K., 1999, Physical Oceanography of the Bering Sea, *in* Loughlin, T. R., and Ohtani, K., eds., Dynamics of the Bering Sea: A summary of Physical, Chemical, and Biological Characteristics, a Synopsis of Research on the Bering Sea: North Pacific Marine Science Organization (PICES), University of Alaska Sea Grant, Fairbanks, p. 1-28.

Stewart, R. J., 1977, Neogene turbidite sedimentation in Komandorsky Basin, western Bering Sea: American Association of Petroleum Geologists Bulletin, v. 61, p.192-206

Takahashi, K., 1999, Paleoceanography changes and the present environment of the Bering Sea, *in* Loughlin, T. R., and Ohtani, K., eds., Dynamics of the Bering Sea: A summary of Physical, Chemical, and Biological Characteristics, a Synopsis of Research on the Bering Sea: North Pacific Marine Science Organization (PICES), University of Alaska Sea Grant, Fairbanks, p. 365-383.

Talley, L. D., 1991, An Okhotsk Sea water anomaly: implications for ventilation in the North Pacific: Deep Sea Research, v. 38. Supplementary Issue 1A, p. S171-S190

Talley, L. D., 1993, Distribution and formation of North Pacific Intermediate Water: Journal Physical Oceanography, v. 23, p. 517-537.

Tarduno, J. A., Duncan, R. A., Scholl, D. W., et al., 2002, Proceedings Ocean Drilling Program, Initial Reports, v. 197 [CD-ROM]. Available from Ocean Drilling Program, Texas A&M University, College Station, Texas 77845.

Underwood, M. B., and Hathon, E. G., 1989, Provenance and dispersal of muds within the Aleutian forearc, North Pacific Ocean: Geo-marine Letters, v. 9, p. 76-75.

Vallier, T. L., Scholl, D. W., Fisher, M. A., Bruns, T. R., Wilson, F. H., von Huene, R., and Stevenson, A. J., 1994, Geologic framework of the Aleutian arc, Alaska, *in* Plafker, G., and Berg, H. C., eds., The Geology of North America, Vol. G-1, The Geology of Alaska: Geological Society of America, p. 367-388.

Verkhunov, A. V., and Tkachenko, Y. Y., 1992, Recent observations on variability in the western Bering Sea current system: Journal of Geophysical Research, v. 97, p. 14,369-14,376

Warren, B. A., 1983, Why is no deep water formed in the North Pacific; Journal of Marine Research, v. 41, p. 327-347.

Warren, B. A., and Owens, W. B., 1988, Deep currents in the Central Subarctic Pacific Basin: Journal of Physical Oceanography, v. 18, p. 529-551.

Warner, M. J., and Roden, G. I., 1995, Chlorofluorocarbon evidence for the recent ventilation of the deep Bering Sea: Nature, v. 372, p. 409-412.

White, J. M., Ager, T. A., Adam, D. P., Leopold, E. P., Liu, G., Jette, H., and Schweger, C. E., 1997, An 18-million-year record of vegetation and climate change in northwestern Canada and Alaska: tectonic and global climatic correlates: Palaegeography, Palaeoclimatology, Palaeoecology, v. 130, p. 293-306.

CHAPTER 9

Cetacean Evolution and Eocene-Oligocene Oceans Revisited

R. Ewan Fordyce

ABSTRACT

The structure, behavior and global distribution of many living cetaceans (whales, dolphins) is linked strongly to food and in turn water masses and global climate patterns. Similar relationships may be inferred for fossil Cetacea, based on trends in diversity and disparity, and evolution of major functional complexes. There were two major radiations of Cetacea in the Eocene-Oligocene. Firstly, stem Cetacea (Archaeoceti) originated late in the early Eocene (Ypresian) and diversified mainly in the Tethys in the middle Eocene, with diversity peaks about early Lutetian and early Bartonian. In the late Eocene (later Bartonian-Priabonian), diversity and disparity decreased, concurrent with expansion of basilosaurids into temperate latitudes. These changes might reflect later Eocene cooling; they seem unrelated to impact events. Secondly, crown-Cetacea (Neoceti—whales, dolphins) arose from amongst archaeocetes by the latest Eocene, with the evolution of filter-feeding (mysticetes, e.g. late Priabonian ≥34 Ma *Llanocetus*) and echolocation-assisted predation (odontocetes). The origin and Oligocene radiation of Neoceti links closely with the development of Antarctic ice and the inferred restructuring of global oceans. The early Oligocene (Rupelian) record is still poor. Late Oligocene (Chattian) Cetacea, including many undescribed species, are dramatically more diverse and disparate than earlier. Chattian Cetacea provide important early records for living groups, including sperm whales, true (delphinoid) dolphins, platanistoid ("river") dolphins, and rorqual-like baleen whales (stem-balaenopterids; "cetotheres" of past use). However, extinct clades, such as the Aetiocetidae (mysticetes) and *Xenorophus*-group (odontocetes) were also significant. Taxonomic, functional and geographic patterns suggest that cetacean ecology, while not truly "modern," was significantly more complex than for the Eocene.

INTRODUCTION

In the last 25 years, many articles have explored the links between the evolution of the Cetacea—whales and dolphins—and the history of the oceans. In general, these articles have focused on the Oligocene and younger radiations of cetaceans; contributions include those of Orr and Faulhaber (1975), Lipps and Mitchell (1976), Barnes (1977), Fordyce (e.g., 1980, 1992), Gaskin (1982), and Rothausen (1985, 1994). The history of the "modern" groups Odontoceti (toothed whales and dolphins) and Mysticeti (baleen whales and relatives) was probably influenced greatly by events of the late Eocene and Oligocene (Fordyce, 1992): restructuring of ocean circulation and food chains associated with the breakup of Gondwanaland, opening of the Southern Ocean, development of the Antarctic ice cap, and consequent global climate change. Articles on the Eocene history of early cetaceans—archaeocetes—have stressed the role of productive shallow epicontinental seas, particularly the Tethys between India and Eurasia, in the initial radiation of the group (Gingerich et al., 1983, Gingerich, 1998).

Advances in three areas allow some of the topics addressed previously (e.g. Fordyce, 1992) to be updated:

1. Many new finds of Archaeoceti—basal or stem-cetaceans that include the putative ancestors of

"modern" forms—have advanced understanding of taxonomy, anatomy and functional morphology, stratigraphy and geographic distribution. Important contributions were made by authors in Thewissen (1998), and in more-recent articles cited below. Most of the recognized Eocene cetacean species have been described, but a few remain to be named formally (e.g., Williams, 1998:figure 1).

2. New Oligocene cetaceans greatly expand knowledge of Odontoceti and Mysticeti. In contrast to archaeocetes, most of the new taxa are still unpublished or at best mentioned in abstract (e.g. for the annual meeting of the Society of Vertebrate Paleontology; see supplement to the *Journal of Vertebrate Paleontology*), with a few exceptions mainly in symposium volumes (Berta and Deméré, 1994, Barnes et al., 1995a).

3. Refinements in stratigraphy (especially bio- and sequence stratigraphy; e.g., Graciansky et al., 1998), and in understanding of oxygen isotopic records, origin of the Antarctic ice cap, oceanic circulation and productivity (e.g., Dingle and Lavelle, 1998b; Salamy and Zachos, 1999; Zachos et al., 1999; Pearson and Palmer, 2000), all help put cetacean evolution into broader perspective.

TAXONOMIC AND STRATIGRAPHIC FRAMEWORK

Cetaceans are widely recognized as an order within the Mammalia. In the last decade, molecular studies have placed the Cetacea close to living hippopotamus or to other artiodactyls (e.g. Gatesy, 1998, Nikaido et al., 1999). Limb structure in protocetid archaeocetes (Gingerich et al., 2001) further supports relationships with artiodactyls rather than, as previously supposed, with the extinct archaic ungulate group Mesonychia. Conventionally, three suborders are used within the Cetacea (e.g. Kellogg, 1928; Barnes and Mitchell, 1978; Fordyce and Barnes, 1994; Rice, 1999):

1. a paraphyletic Archaeoceti (Eocene only);
2. a monophyletic Odontoceti: toothed whales and dolphins (possibly Eocene/Oligocene boundary to Recent);
3. a monophyletic Mysticeti: baleen whales and relatives (probably latest Eocene to Recent).

Modern odontocetes and, by inference, fossil species, are unified by the ability to produce high-frequency sounds, which are used in echolocation to navigate and detect prey. Modern mysticetes, conversely, are not known to echolocate, and all filter-feed, a behavior inferred for most or all fossil mysticete). Together, Odontoceti + Mysticeti form the crown-group Cetacea (also termed Neoceti or Autoceta; see McKenna and Bell, 1998; Fordyce, 2002), while Archaeoceti forms the stem-group Cetacea. Archaeocetes lack features that clearly indicate the ability to echolocate or filter-feed.

Figure 9.1 shows the currently recognized families of Cetacea—both clades and grades—known from Eocene and Oligocene strata. For Archaeoceti, the definition, content and ranges of families were reviewed by Williams (1998), Uhen (1998a) and other contributors in Thewissen (1998), supplemented by contributions from Uhen (1998b, 1999), Luo and Gingerich (1999) and Gingerich et al. (2001); see also Kellogg (1936). Families of Odontoceti and Mysticeti were summarized by Fordyce and Barnes (1994), Fordyce and Muizon (2001), and references therein.

For archaeocetes, stratigraphic ranges are documented in some detail for Tethyan and North Atlantic species (Williams, 1998; Uhen, 1998a, 1998b, 1999; Gingerich et al. 2001). The sequence-stratigraphic framework used by Gingerich (1992) and Gingerich et al. (1997) is particularly useful in placing species by relative age. Nevertheless, cited numerical ages (in Ma) for a particular archaeocete species may vary widely (compare, for example, Williams, 1998:figure 1 with Gingerich et al., 1997:figure 14), so that caution is needed when interpreting ranges shown here in figures 9.1 and 9.3. For the Oligocene Cetacea, there have been fewer useful stratigraphic advances, and few species have been placed in sequence-stratigraphic framework. Currently, for crown-group Cetacea, the oldest mysticete is the toothed *Llanocetus denticrenatus* from about the Eocene/Oligocene boundary of Seymour Island, Antarctica (Mitchell, 1989; Fordyce, 1989a), while the oldest reported odontocetes are undescribed and unnamed species from about the Eocene/Oligocene boundary of Washington State, U.S.A. (Goedert et al., 1995; Goedert and Barnes, 1996).

Cladistic analyses are available for various levels within the Cetacea, although the Archaeoceti and many of its constituent taxa archaeocetes are still acknowledged grades. Figure 9.2 shows recently proposed cladistic relationships for some archaeocetes, slightly simplified from Uhen (1999) and Uhen and Gingerich (2001); some families are clearly grade taxa. For the Odontoceti, family-level cladistic relationships proposed by Muizon (1987, 1988, 1991) are supported by recent discoveries of early odontocetes. The higher-level branching sequence of

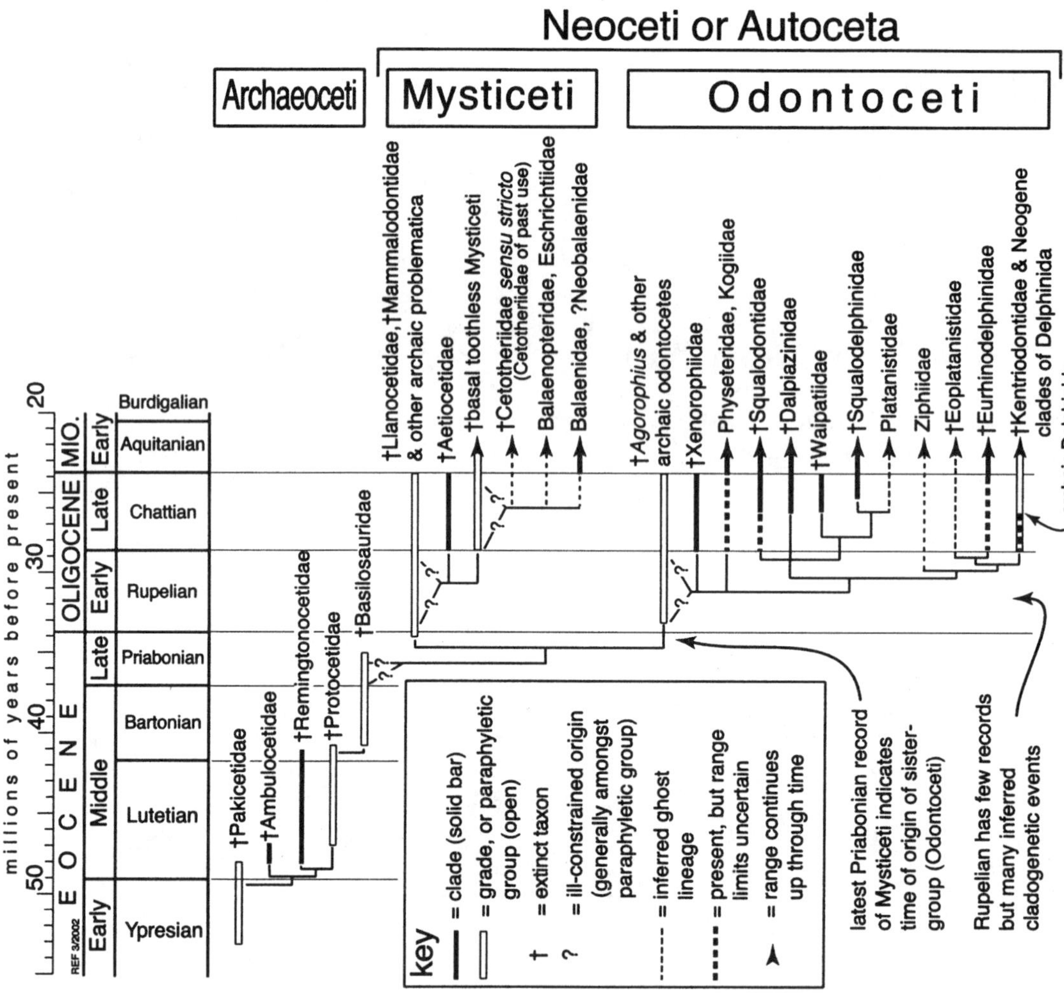

Figure 9.1. Currently used families of Cetacea from the Eocene and Oligocene. Based on taxa, inferred cladistic relationships, concepts of grades, and stratigraphic ranges documented by Williams (1998), Uhen (1998a, 1998b), Uhen and Gingerich (2001), Fordyce and Muizon (2001), and Nikaido et al. (2001).

Physeteridae + Kogiidae-Platanistoidea-Ziphiidae-Delphinida is based on the molecular results of Nikaido et al. (2001). A supposed link between sperm whales, Physeteridae, and rorquals, Balaenop-teridae, indicated by molecules (Milinkovitch, 1995) has been thoroughly dismissed (e.g. Heyning, 1997; Messenger and McGuire, 1998). The few cladograms produced for the mysticetes consider a limited number of taxa and elements (e.g. Geisler and Luo, 1996) and/or have yet to be tested by computer algorithms (e.g. Barnes and McLeod, 1984; McLeod et al., 1993; Barnes et al., 1995b). Overall, the cladistic framework for Neoceti is still poor at superfamilial to subordinal level, that is, toward the base of the Neoceti. Within the Neoceti, family-level relationships and stratigraphic ranges are understood well enough to identify some Neogene groups as ghost lineages with origins in the Oligocene; examples are the Ziphiidae—beaked whales (middle Miocene to Recent); and various mysticetes including Balaenidae—right whales, and Cetotheriidae *sensu stricto.*

There are important provisos to consider for the stratigraphic record, especially if species-level patterns are considered (figure 9.3). Cetaceans are large and conspicuous in outcrop but generally rather rare as fossils, so one cannot be sure how well the observed record approximates the real historical pattern. For plots of species diversity against time, biases are read-

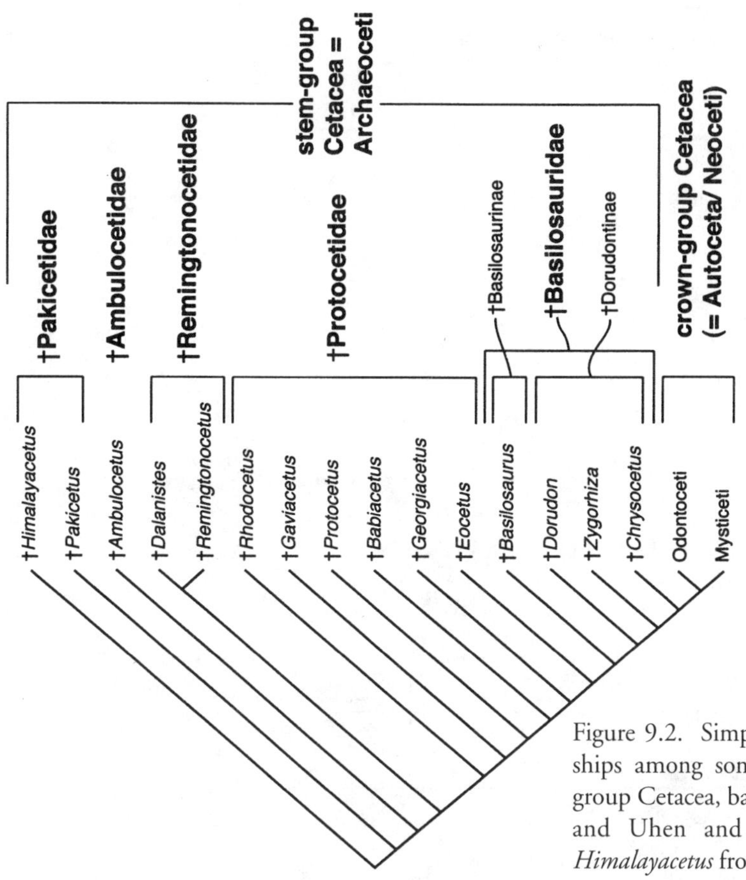

Figure 9.2. Simplified concept of cladistic relationships among some archaeocete genera and crown-group Cetacea, based mainly on Uhen (1998a, 1999) and Uhen and Gingerich (2001); position of *Himalayacetus* from Bajpai and Gingerich (1998).

ily identified. They arise from such factors as variation in volume of preserved rocks and area of exposed fossiliferous rocks (for the cetacean record, poor for Rupelian, better for other stages); prospecting and collecting effort (probably biased toward the Eocene, well-exposed in northern localities accessed by large numbers of paleontologists); different lengths of stages; reliability of stratigraphy (probably better for the Eocene than Oligocene); and approaches to taxonomy (probably more consistent for Eocene archaeocetes than for Oligocene Neoceti). For the Eocene, most individual species' occurrences can be ranked within stages to a resolution of about 1 million years (e.g. Gingerich et al., 1997, 2001) (figure 9.3). However, numerical ages cited in the last 20 years have varied because of changing local correlations and because of changing concepts of the international time scale (correlations between foraminiferal zones, inferred sea level changes, European stage sections, absolute dating, and paleomagnetic time scale). This is particularly the case with fossils from the eastern Tethys; compare, for example the widely cited Ypresian age of pakicetids (e.g. Williams, 1998:figure 1) with the recent assessment of early Lutetian (Gingerich et al., 2001). Most Oligocene cetaceans are from thin condensed sequences with poor internal resolution, or from shelf sequences which lack the oceanic (micro-) fossils needed for reliable biostratigraphy. At present, there is a great inbalance between the Eocene, with almost all known cetacean species actually described and named, and the Oligocene, with many species known but unnamed.

TAXONOMIC PATTERNS OVER TIME

A plot of families against time (figure 9.1) indicates the rapid appearance of successive grades (pakicetids and protocetids) and clades (Ambulocetidae and Remingtonocetidae) in the late Ypresian-Lutetian. These higher groups signal a rapid increase in disparity (variation in basic design), in turn an index of ecological partitioning. This pattern is consistent with rapid radiation by the early Lutetian, also revealed by species-level plots (figure 9.3). Disparity falls in the Bartonian, with the disappearance of non-basilosaurid archaeocetes (M. Uhen, pers. commun.).

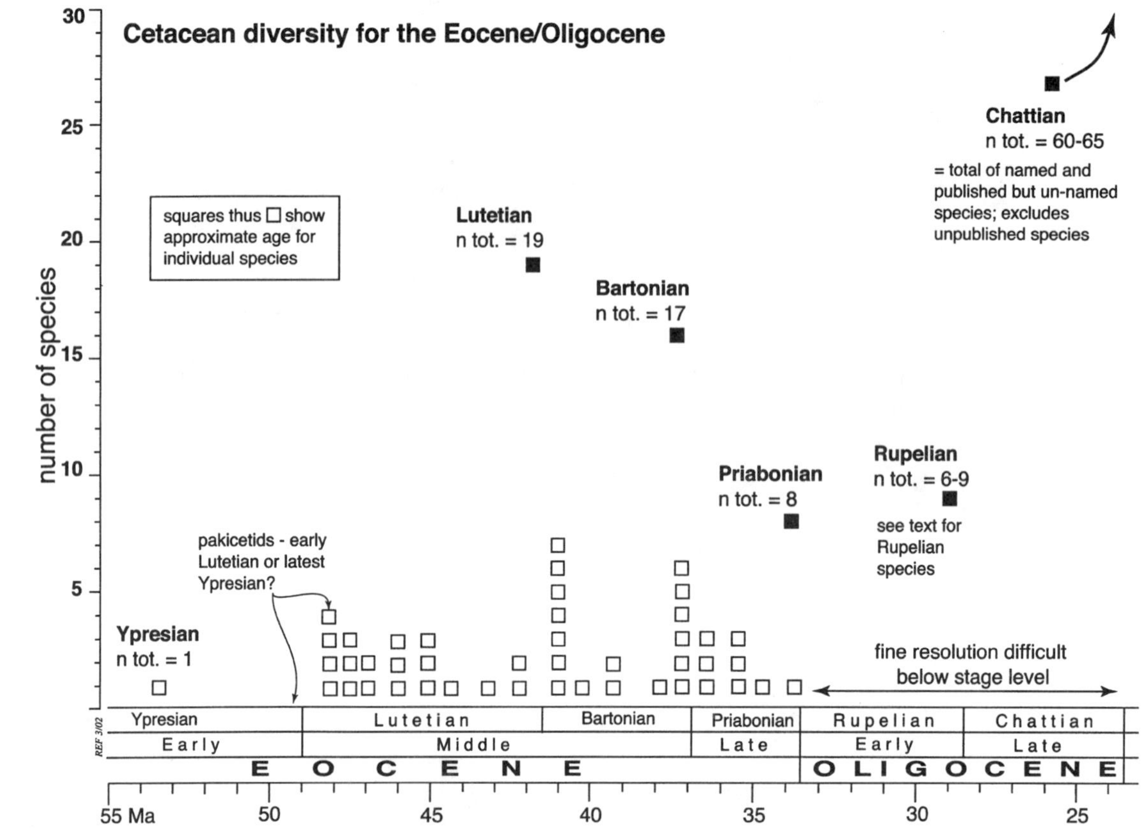

Figure 9.3. Cetacean species diversity during the Eocene and Oligocene. Numbers of species and stratigraphic ranges for archaeocetes are based on Williams (1998), Uhen (1998a, 1998b), Uhen and Gingerich (2001), and Fordyce and Muizon (2001). Oligocene species are based on references reviewed by Fordyce (1992; Fordyce and Barnes, 1994, and Fordyce and Muizon, 2001), on the present text, and on a few unpublished observations.

The second major radiation for the Cetacea (figure 9.1) started at about the Eocene/Oligocene boundary, 33.7 Ma. *Llanocetus denticrenatus* is the earliest reported Mysticeti, and the suborder Odontoceti (sister-taxon to Mysticeti) is presumed to have arisen by the same time (see comments under monophyly of Basilosauridae, below). Paradoxically, the record is poor for the ~5 million years of the Rupelian (early Oligocene). Eustatic sea-level fall, marked by a widespread unconformity at 29–30 Ma, probably caused widespread erosion and reworking of shallow-water bone-rich Rupelian strata. The Chattian record, above the unconformity, shows high diversity for species, genera and families, supporting the notion of an earlier dramatic radiation.

The third major radiation for Cetacea occurred later in the Neogene, and is not shown on figure 9.1. Then, the dolphins and relatives (Delphinida, including Delphinoidea of traditional use) diversified, eventually to become the most speciose group in modern seas (Fordyce and Muizon, 2001). This radiation is signalled by 11-12 Ma records of true dolphins and porpoises (Delphinidae, Phocoenidae) and relatives (e.g. Barnes, 1977; Fordyce and Barnes, 1994), possibly related to ocean restructuring associated with middle Miocene cooling (e.g. reviewed by Pearson and Palmer, 2000).

For species-level trends, known occurrences of cetaceans are plotted against Eocene and Oligocene stages to show changing species diversity over time (figure 9.3). The raw cumulative totals per stage provide a crude index of longer-term (smoothed) diversity changes. The Eocene shows a dramatic increase in species diversity from Ypresian to Lutetian, but this spans a gap of perhaps 4 million years and could be interpreted as showing either stepwise or gradual change. Diversity is high in the Lutetian, drops in the Bartonian and especially the Priabonian, rises slightly in the Rupelian, and rises dramatically in the Chattian. For the Oligocene, species-level and family-level diversity trends parallel one other (figures 9.1 and 9.3). Ghost lineages at family level (figure 9.1), and biases inferred from erosion of the Rupelian, suggest that Rupelian diversity was originally much higher.

To consider numbers, standing crops of 4-7 species occur around 48, 41 and 37 Ma. Judging from New Zealand Oligocene assemblages, Chattian standing crops were probably significantly higher, with 10 to 15+ species per fauna, but problems of time-averaging hamper more reliable totals. What do the figures mean? Globally, there are about 84 living species today (Rice, 1998); Barnes (1977) listed about 48 species for modern tropics, about 55 species for temperate regions, and about 28 species for cool-polar regions. Regional diversity patterns include about 30 species for the eastern North Pacific (Barnes, 1977), 29 species for tropical Sri Lanka (Leatherwood and Reeves, 1989), and 12 regularly south of the Antarctic convergence (Laws, 1977; Nishiwaki, 1977). In summary, Eocene standing crops were significantly less than today, while Chattian figures start to approach modern levels.

Related to issues of cooling and late Eocene bolide impacts (e.g. as reviewed by Vonhof et al,. 2000), species diversity in the Priabonian is lower than the Bartonian, dropping from 6 species at ~37 Ma (or 17 total for Bartonian) to 1-3 species at 35-36 Ma (and 8 total for Priabonian). Only basilosaurid archaeocetes are involved, and one species, *Dorudon atrox*, ranges from Bartonian into early Priabonian (see Uhen, 1998a). Diversity change around 37 Ma need not relate to known late Eocene impact craters, but might reflect later Eocene cooling. The geographic distribution of archaeocetes (below) is consistent with the notion that they were warm-water stenotherms, but to what extent might cooling be invoked as influencing diversity via direct physiological affects? It seems unlikely that cooling would cause extinction through loss of habitat, for warm stenothermal cetaceans might be expected to track water masses that contracted toward the tropics. It seems more likely that cooling and ocean restructuring influenced archaeocete ecology, and indirectly through food resources, diversity.

GEOGRAPHIC RANGES OVER TIME

Major occurrences of cetaceans for each stage are shown on generalized Eocene and Oligocene paleogeographic maps. For the Eocene (figure 9.4), northern localities in the former Tethys and central to North Atlantic predominate (localities reviewed by Williams, 1998; Uhen, 1998a; and others in Thewissen, 1998; see also Uhen, 1998b, 1999, 2000; Uhen and Gingerich, 2001; Thewissen et al., 2001; and Gingerich et al., 2001). There are notable occurrences from the central and South Pacific–marginal Southern Ocean (Fordyce, 1989a; Köhler and Fordyce, 1997; Fordyce and Muizon, 2001). From the North Pacific, one reportedly Eocene record of an archaeocete (a single vertebra from Nootka Island, British Columbia; Kellogg, 1936) is probably late Oligocene (Barnes et al., 1995b: table 9.1—Vancouver Island), and is indeterminate. More recently, unspecified Eocene cetaceans have been reported from Oregon (Heeden, 1997).

Oligocene records (figure 9.5) have expanded considerably in the 1990s. Notable new records are mostly from New Zealand, Baja California Sur, Oregon, Washington, and Japan (e.g. Fordyce, 1994; Fordyce and Watson, 1998; Goedert et al., 1995; Okazaki, 1995; Barnes et al., 1995b; Cruz-Marin and Barnes, 1996; Sanders, 1996; Barnes, 1998). Many potentially important Oligocene fossils are mentioned only in abstract. Early Oligocene species are considered further below.

The known geographic ranges of taxa, mostly families (both grades and clades), appear on generalized Eocene and Oligocene paleogeographic maps. For the Eocene (figure 9.6), the basal archaeocete families Pakicetidae, Ambulocetidae and Remingtonocetidae are recognized only in the eastern Tethys (Pakistan, India). Species of Protocetidae occur more widely, in mid-Tethyan regions (Egypt, Nigeria, Togo) and in the western Atlantic (Atlantic Coastal Plain/Gulf Coast) (reviewed by Williams, 1998; Uhen, 1998a, 1998b, 1999, 2000; Uhen and Gingerich, 2001). These distributions indicate increasingly pelagic habits, an inference supported also by changes in major functional complexes involving teeth and jaws, position of blowhole, development of sinuses in the skull base, the hearing apparatus, and reduction in pelvis and associated hind limbs (Thewissen and Fish, 1997; Gingerich, 1998; Uhen, 1998a). Oxygen isotopes from teeth have also been used to identify the rise of marine habits (Thewissen

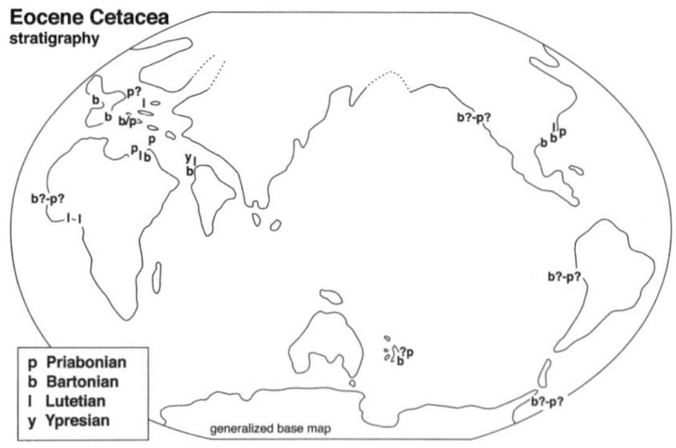

Figure 9.4. Key localities for Eocene Cetacea. Localities and ages are based on Williams (1998), Uhen (1998a, 1998b, 1999), Uhen and Tichy (2000), Gingerich et al. (2001), Pilleri (1989), Pilleri and Cigala-Fulgosi (1989), and Fordyce and Muizon (2001 and references therein).

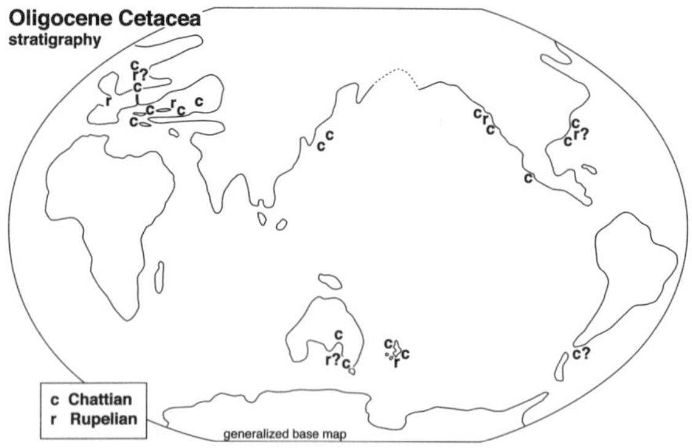

Figure 9.5. Key localities for Oligocene Cetacea. Localities and ages are based on summaries of Whitmore and Sanders (1977), Fordyce and Barnes (1994), Barnes et al. (1995), and Fordyce and Muizon (2001 and references therein).

et al., 1996; Roe et al., 1998). Basilosaurids are more widespread. Records around the Pacific and proto-Southern Ocean imply that basilosaurids were able to spread beyond the Tethys across open ocean and that they might have had oceanic lifestyles.

Oligocene Cetacea show more complex patterns that indicate radiation into most of the world's oceans (figure 9.7). Some families are reported from single localities or even single occurrences (e.g. *Kekenodon, Xenorophus*; Whitmore and Sanders, 1977; Sanders, 1996; Fordyce and Muizon, 2001) which, given the wide distribution patterns of most living Cetacea, reveal little about original ranges. Taxa so far reported only from the North Pacific include the toothed small mysticetes of the family Aetiocetidae (see Barnes et al., 1995b), and archaic *Agorophius*-like odontocetes (Whitmore and Sanders, 1977:figure 2; Kimura, 1989:figures 7 and 8). Many other important North Pacific species are unnamed, and not indicated on figure 9.7. In the south, the late Oligocene-early Miocene *Prosqualodon* (?Squalodontidae, or Prosqualodontidae; Cozzuol, 1996) appears to have a circum-Southern Ocean distribution, and one fragmentary *Llanocetus*-like archaic mysticete is known from the New Zealand basal Oligocene as well as from the latest Eocene of Seymour Island. The family Squalodontidae *sensu stricto* clearly has northern and southern temperate distributions similar to the "antitropical" distributions of many living species (Davies, 1963; see also Cipriano, 1997). The few records for the Waipatiidae (New Zealand *Waipatia*, and *Sulakocetus* from Caucasus; also possibly *Microcetus* from northern Germany; Fordyce, 1994), and provisionally the Eurhinodelphinidae (*Iniopsis* from the Caucasus, unnamed freshwater species from Central Australia) also indicate northern and southern temperate distributions. Other taxa straddling the Oligocene/Miocene boundary further expand distribution patterns: *Kentriodon*-like dolphins (archaic Delphinoi-dea) have been reported from the Caucasus (*Oligodelphis*), New Zealand (?*Kentriodon*, Kentriodontidae), and the basal Miocene or possible Oligocene of Patagonia (Ichishima et al., 1995). Dolphins in the platanistoid family Squalodelphinidae include important early Miocene records in Italy (*Squalodelphis*) and Patagonia (*Notocetus*), and latest Oligocene (to Oligocene/Miocene boundary) records from New Zealand (e.g. *Notocetus marplesi, sensu* Fordyce, 1994). The taxonomy of archaic baleen whales, Cetotheriidae *sensu lato* of many authors, is too confused to identify real patterns.

CLOSING THE GAP BETWEEN ARCHAEOCETES AND CROWN-GROUP CETACEA

New interpretations of long-known specimens, refinements in stratigraphy, and new discoveries of fossils help to close the phylogenetic and/or stratigraphic gaps between archaeocetes, mysticetes, and odontocetes. Key recent advances are summarized below.

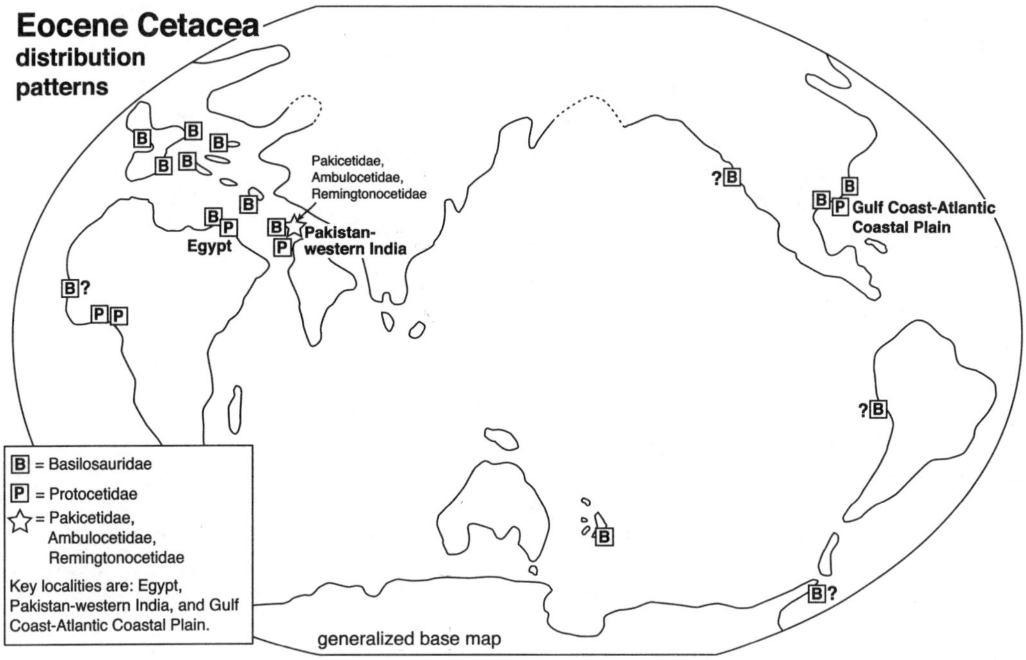

Figure 9.6. Geographic distribution of families of Eocene Cetacea. See figure 9.4 caption for references.

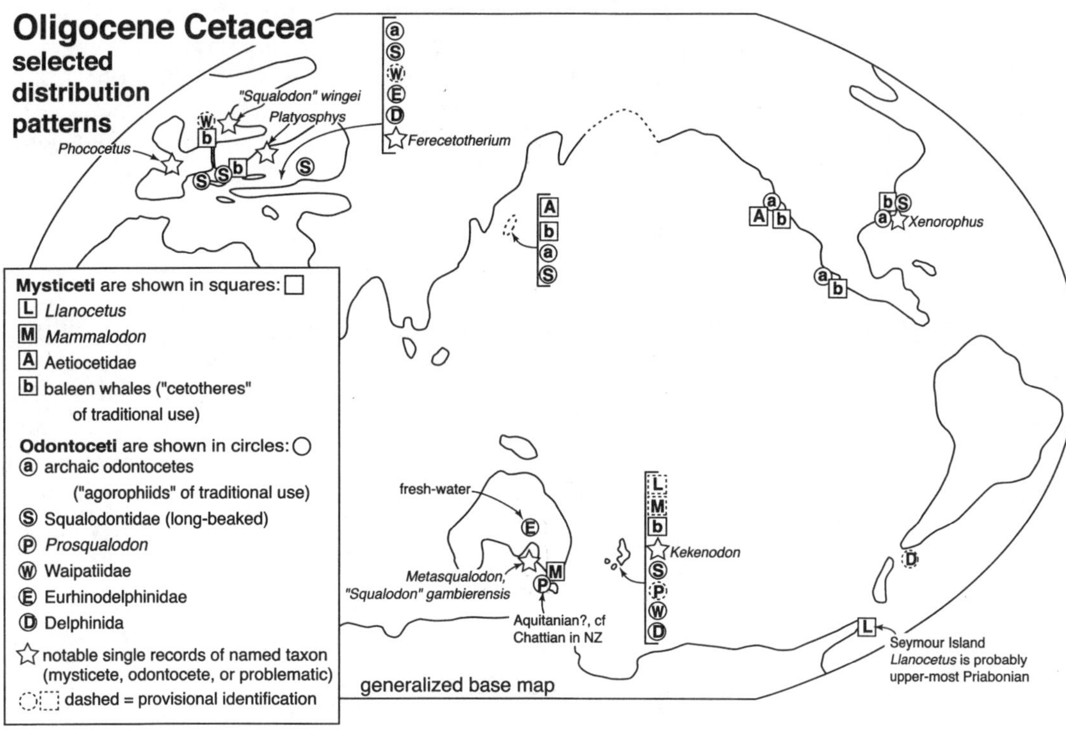

Figure 9.7. Simplified geographic distribution of families of Oligocene Cetacea. See figure 9.5 caption for references. Not all known occurrences are shown; see text for details.

The Last Archaeocetes

Uhen and Gingerich (2001) identified the dorudontine archaeocete *Chrysocetus healeyorum* (age: 36-38.5 Ma) as the sister taxon to Odontoceti + Mysticeti, because of the inferred style of tooth succession in *C. healeyorum*. Previously, geologically younger species of *Zygorhiza* and *Saghacetus* seemed likely candidates as sister taxa (Uhen, 1998a:figure 1; 1999:figure 7; see also Luo and Gingerich, 1999:figure 26). The monotypic *Zygorhiza kochii* is from the Priabonian of the Atlantic Coastal Plain (Uhen, 1998a), while the smaller monotypic *Saghacetus osiris* (= *Dorudon osiris* of older literature; Gingerich, 1992) is from the upper, but not top, Priabonian of Fayûm, Egypt. Judging from the time scale of Graciansky et al. (1998:chart 1), the LAD [last appearance datum, or extinction level] for *S. osiris* is >34-35 Ma. Despite named Oligocene "archaeocetes" based on teeth or postcranial remains, no skull has yet been described from the Oligocene. For example, the putative archaeocete *Platyosphys paulsonii*, which Kellogg (1936) identified as a possible basilosaurid, is known only from vertebrae that do not reliably indicate suborder. The species is of possible but not certain early Oligocene age. Another supposed archaeocete of possible early Oligocene age, *Phococetus vasconum*, is known only from an isolated tooth (Kellogg, 1936). *Platyosphys* and *Phococetus* are best placed in the Cetacea *incertae sedis* until more-diagnostic remains are available.

Monophyly of Basilosauridae and Crown-Group Cetacea

Synapomorphies involving the teeth, tympanoperiotics (ear bones), and pterygoid sinus fossae in the skull base indicate that the Basilosauridae, Mysticeti, and Odontoceti form a clade (Barnes and Mitchell, 1978; Luo and Gingerich, 1999; Fordyce and Muizon, 2001; Fordyce, 2002). There is no need to postulate that the crown-group cetaceans arose among more basal archaeocetes such as the protocetids. But, are the Odontoceti and Mysticeti sister-taxa, sharing an immediate common ancestor amongst basilosaurid archaeocetes? Or, did each arise from a different basilosaurid archaeocete? A few synapomorphies seen in both fossil and living species unite the Odontoceti and Mysticeti (Fordyce, 2002). For example, the premaxillae are partly or wholly separated by an open mesorostral groove dominated by the vomer, which, primitively, is exposed ventrally as well as dorsally. Posteriorly, a concave facial fossa on the maxilla has multiple dorsal infraorbital foramina (for infraorbital nerves and vessels), associated with a groove or notch for the facial nerve. The fossa is developed on the base of the rostrum in mysticetes and the most archaic odontocetes, but is supraorbital in more crownward odontocetes. More posteriorly, the jugal is delicate, and the squamosal and exoccipital exclude the mastoid process of the periotic from the skull wall (that is, the skull is amastoid). The other possible synapomorphies of note are tooth polydonty and monophyodonty, but these features are problematic (see discussion by Barnes et al., 1995b; Uhen 2000; Uhen and Gingerich, 2001).

Early and Basal Mysticeti

The oldest described putative mysticete is *Llanocetus denticrenatus*, named and described by Mitchell (1989) from a partial mandible and natural endocranial cast from the La Meseta Formation of Seymour Island, Antarctica. The large skull recovered from Seymour Island by Fordyce and others in 1987 (Fordyce, 1992:373), and prepared over a decade to the late 1990s, definitely represents the rest of the holotype. The specimen is from the top of unit Telm 7 of the upper La Meseta Formation (see Sadler, 1988) on the southeast side of Seymour Island west of the Argentine base of Marambio. According to Dingle and Lavelle (1998a), $^{87}Sr/^{86}Sr$ dates from mollusks from the uppermost La Meseta Formation are 34.2 ± 0.87 Ma, equivalent to uppermost Priabonian, just pre-dating the first (Oi1) Antarctic ice advance of ~33.6 Ma. In contrast to previous suggestions, none of the La Meseta Formation is demonstrably Oligocene.

New Zealand fossils include some presumed archaic mysticetes from the lower Whaingaroan local stage (*Subbotina angiporoides* Zone), spanning top Priabonian to middle or upper Rupelian (upper-most Eocene to lower Oligocene) (Morgans et al., 1996). Biostratigraphically, the lower Whaingaroan postdates the LAD of *Globigerinatheka index*, while its top is marked by the local LAD of *Subbotina angiporoides*. At outcrop level, the lower Whaingaroan underlies the conspicuous "mid-Oligocene" unconformity that presumably marks the major sea level fall of ~28.5 Ma, of international sequence stratigraphy (e.g. Graciansky et al., 1998:table 1). Four cetaceans from the lower Whaingaroan indicate a possible higher diversity in the Rupelian. Of these, Keyes (1973) described two teeth from the lower Ototara Limestone of North Otago, identifying them as from an unnamed "protosqualodont." Other elements were recognized during later preparation, and the specimen now comprises skull fragments including dorsal parts of the supraoccipital, parietals, left supraorbital process of frontal, a premaxilla, posterior parts of both maxillae, and 4 teeth. One upper posterior premolar (Keyes, 1973:figures 3-6) is phenetically similar to a loose presumed upper posterior premolar from the slightly older *Llanocetus denticrenatus* of Antarctica, differing in smaller absolute size and in having lower and less palmate denticles. The Ototara Limestone specimen is thus identified provisionally as *Llanocetus* species (Mysticeti).

Second, Fordyce (1989b) described a partial mandible of a toothed supposed Mysticeti from the Amuri Limestone of North Canterbury. Fragmentary teeth are more delicate, with smaller denticles, than in dorudontine archaeocetes, and the mandible is gracile and oval in section. Tooth proportions and ornament rule out close relationship with *Llanocetus*. The specimen lacks diagnostic features that would allow it to be named.

Third, Fordyce and Watson (1998) described a pathological cetacean, possibly a mysticete, from the basal Wharekuri Greensand of North Otago. The animal comprises large vertebrae and ribs of a ~6-m-long cetacean that show two conspicuous pathologies: congenital abnormality of incomplete vertebral arch ("spina bifida"), and bubbly proliferative bone probably caused by chronic periostitis. There is marked pathology-related bilateral asymmetry. The specimen lacks diagnostic features that would allow it to be named. Another specimen (OU 22371), newly recovered from the same locality, is a juvenile or neonate presumed archaic mysticete. Fragments of the disarticulated skull include apparent posterior of a maxilla, indicating a broad and possibly toothed rostrum, and an inflated tympanic bulla (~68 mm long), which, judging from other early mysticetes, indicates a cranial length around 300 mm (condyloid notch to base of rostrum). The bulla is superficially like that of dorudontine archaeocetes, and lacks the more rounded profile of mysticete bullae from higher in the New Zealand Oligocene. Not enough is preserved to determine the relationship to *Llanocetus* sp. (Ototara Limestone), above, and the young ontogenetic age would make comparisons difficult even if the specimen were complete. This is one of few juvenile or neonate fossil Cetacea reported.

The ecology and structure of living mysticetes is linked strongly with hydrographic conditions and the distribution of food. Such links suggest that patterns of mysticete evolution in the past, particularly the initial Oligocene radiation, may be a proxy for changing ocean structure (Fordyce, 1992). In summary, modern mysticetes are obligate microphagous filter-feeders that eat mainly plankton (Pivorunas, 1979). Mysticetes show distinctive structural adaptations for filter-feeding; the upper jaws are broad and flat-based (e.g., Kellogg, 1928), forming an origin for the hair-fringed baleen plates that form the filter, while the toothless lower jaws are bowed outward, increasing mouth volume, with a flexible joint rather than fixed bony contact (symphysis) anteriorly. Comparable jaw structures occur in Oligocene mysticetes, even in small toothed mysticetes (e.g. Aetiocetidae: Barnes et al., 1995b) which probably lacked baleen but fed using the teeth as a mechanical sieve. Different species of living mysticetes differ particularly in structure of the feeding apparatus and style of food capture (Nemoto, 1959) and distribution (Laws, 1977), presumably limiting niche overlap and competition (see Clapham and Brownell, 1996, for discussion). The distribution of modern mysticetes is governed by summer concentrations of food, mostly at mixing zones (e.g., convergences) and upwellings, and generally in cool-temperate to polar latitudes; mysticete distribution is widely assumed to reflect hydrographic controls on the diversity and abundance of living plankton (Nemoto, 1959; Gaskin, 1982; Kenney and Wishner, 1995; Tynan, 1998). Mysticete structure thus is linked strongly in terms of feeding ecology to plankton and hydrographic conditions. This relationship is assumed to apply to even the earliest mysticetes.

Early and Basal Odontoceti

In contrast to mysticetes, the candidate for the oldest odontocete is less clear. In terms of systematic position, the most basal named odontocete, athough not demonstrably the oldest, is *Archaeodelphis patrius* Allen, 1921. This species is known only from an incomplete skull from an uncertain horizon in the Atlantic Coastal Plain (Whitmore and Sanders, 1977), possibly from South Carolina. *Archaeodelphis patrius* could be Oligocene or latest Eocene in age. Beyond *A. patrius,* the oldest well-dated and named odontocetes known from skulls are from the Chattian (for reviews, see, e.g., Whitmore and Sanders, 1977; Fordyce, 1992). Many specimens of definite late Oligocene age are now known, but for most, the exact age within the Chattian is less certain, as the source horizon is rarely specified in terms of lithostratigraphy (e.g., in meters above the "mid-Oligocene" unconformity) or internationally recognized biostratigraphy (P, N or NP Zones). In the shelf deposits that produce most of the Chattian cetaceans, planktic foraminifera give rather poor biostratigraphic resolution. For example, in New Zealand, this interval includes the important cetacean horizon of the Kokoamu Greensand, which postdates the LAD of *Subbotina angiporoides*, predates the FAD of *Globoquadrina dehiscens*, and has no reliable planktic foraminiferal zonation (Hornibrook et al., 1989).

Among other possibly older records of Odontoceti, some supposedly Rupelian fossils from Japan (Fordyce, 1992:373) are now thought to be from the Chattian (Oishi and Hasegawa, 1995). The presumed odontocete "*Squalodon* (*Microzeuglodon*)" *wingei* Ravn, 1926, is based on fragments including teeth from the Branden Formation of Denmark. This unit is known to span Zones NP24a and NP24b, or later Rupelian to earlier Chattian (Michelsen et al., 1998:113). Pledge (1994) described an incomplete and indeterminate cheek-tooth, of supposed early Oligocene age, from near Wellington in South Australia. The suborder was not identifiable.

Goedert et al. (1995) reported lower Oligocene cetaceans from the upper Eocene to upper Oligocene Makah and Pysht Formations of Washington State. The mode of preservation indicates that specimens were buried in lower to mid bathyal depths, in basin plain to outer fan-fringe settings. Later, Goedert and Barnes (1996) gave a brief report of an odontocete cetacean from the lower part of the Lincoln Creek Formation, reportedly close to the Eocene/Oligocene boundary (planktic foraminiferal zone not specified), of Washington State. Important synapomorphies of the skull were mentioned, and the odontocete was stated to be of agorophiid affinities, implying relationships with *Agorophius pygmaeus* (Agorophiidae; Chattian?, South Carolina). The Lincoln Formation specimen appears to be the oldest reported odontocete.

Unlike the situation for mysticetes, the ecology and structure of living odontocetes is more difficult to relate to evolutionary patterns and oceanic change. Nevertheless, odontocetes radiated rapidly in the early Oligocene, presumably in response to changing ocean structure (Fordyce, 1992). As with mysticetes, odontocete evolution can be elucidated by considering behavior and functional morphology in living species. A few living odontocete species definitely use high frequency sounds to echolocate and many other living odontocetes are known to produce such sounds although they have not been shown experimentally to echolocate (Wartzok and Ketten, 1999); the consensus is that all odontocetes are echolocators, while mysticetes are not. Echolocation is used in navigation and hunting, presumably to great advantage below the photic zone and in turbid waters within the photic zone. Echolocation sounds seem to be produced within the soft tissues of the face, in diverticula of the nasal passages anterodorsal to the facial region of the skull. The nasal tissues are manipulated by enlarged nasofacial muscles (Mead, 1975; Heyning, 1989; Cranford et al., 1996), which have distinct origins from the enlarged and usually dished supraorbital process of the maxilla. Most of the nasal diverticula lie within soft tissue, but one bilateral diverticulum, the premaxillary sac, has a distinct fossa and associated foramina on the premaxilla. Other bony concomitants of the enlarged facial muscles include multiple dorsal infraorbital foramina and a prominent antorbital notch for the facial nerve. The enlarged supraorbital process and premaxillary sac fossa are present in all fossil members of the crown-group Odontoceti, implying that fossil species also could echolocate, but comparable structures are lacking in archaeocetes and mysticetes. These patterns of structure and inferred behavior across the Cetacea led Fordyce (e.g., 1980) to postulate that the evolution of echolocation was a defining character of the odontocetes which, like the unique feeding method of mysticetes, was tightly linked to the evolution of new oceanic structure and food resources in the Oligocene. Judging from diverse patterns of skull bones and soft tissues, living odontocetes probably employ a wide range of taxonomically specific styles of echolocation, but there is little understanding of how particular structures relate to particular sound systems, to ecological partitioning between species, and to oceanic structures such as convergences and upwellings.

DIRECTIONS FOR THE FUTURE

Taxonomy

It is axiomatic that if cetacean history is to be understood better, and related to the evolution of oceans, higher-level taxonomy must be refined through detailed description and interpretation of well-preserved specimens. Advances in archaeocete systematics in the last decade could provide a model for much-needed work on crown-group Cetacea (Neoceti). However, archaeocetes seem easier to consider than Neoceti in terms of taxonomy (diversity) and structure (disparity); there has been general acceptance of formal taxa based on fragmentary remains, and more routine use of acknowledged grades in family-level taxonomy. The strong evidence for rapid—indeed, explosive—radiation of Neoceti in the Oligocene should alert workers to a likely complex taxonomic pattern, for which simple interpretations are unlikely to endure. The pattern of cetacean history will become clearer as traditional grade taxa are redefined as clades, with incomplete or enigmatic taxa being placed elsewhere. This has been done for some taxa, for example, the odontocete groups Agorophiidae and Squalodontidae (see Fordyce and Muizon, 2001, for summary). Mysticetes still need attention.

Grades and Clades Among Basal Toothless Mysticetes

The wide use of mysticete grade taxa poses major problems in understanding possible links between the Oligocene evolution of mysticetes and oceans. Traditionally, the family Cetotheriidae has been used as a grade for toothless fossil mysticetes that do not belong with gray whales (Eschrichtiidae), right whales (Balaenidae), or rorquals and relatives (Balaenopteridae). Many archaic Oligocene mysticetes have, therefore, been identified as cetotheres—obscuring details of the early radiation of mysticetes. The family Cetotheriidae is based on the genus *Cetotherium*, with the type species being the late Miocene *Cetotherium rathkii* Brandt, 1843. The latter is known from a unique partial skull (figured by Van Beneden and Gervais 1868-80; Kellogg, 1928; Pilleri, 1986) in which the upper jaw or rostrum thrusts back into bones of the braincase with a sharp narrow triangular apex, and the posteromedial parts of the maxillae narrowly constrict the premaxillae and nasals. It is proposed here that the family Cetotheriidae be restricted formally as a clade for *C. rathkii*; it may also include the closely similar Pliocene *Herpetocetus* from the Yushima Formation of Japan (Oishi and Hasegawa, 1995). Basal toothless mysticetes reported from the Oligocene (e.g., species of *Mauicetus* and *Cetotheriopsis*) are not demonstrably in the same family-level clade as *C. rathkii*, and should be removed from the Cetotheriidae and placed in the Mysticeti incertae sedis. Here, such action is formally proposed for all named species of *Mauicetus*. The moves toward use of clades will elucidate mysticete history in the longer term, and will allow systematists to consider whether some of the archaic baleen-bearing mysticetes might be stem Balaenopteridae or stem Balaenidae.

Isotopes

Isotopes from Oligocene (and other) Cetacea offer great promise in helping elucidate the evolution of global ice and thus paleoclimates. Patterns of oxygen isotopes from bone hydroxyapatite in modern Cetacea (Yoshida and Miyazaki, 1991; Roe et al., 1998) can be related to different habitats (marine or fresh-water) used by different species. Cetacean bone, as is the norm for homoithermic mammals, is deposited at relatively warm constant body temperatures, and oxygen isotopes from core bone apparently lack any fluctuations attributable to ambient environmental temperature as seen in the carbonate skeletons of invertebrates. Thus, bone isotopes may provide a good proxy for isotopic composition of environmental water—as already noted in the literature. For the Eocene, oxygen isotopes have revealed habitat use in archaeocetes, indicating a change from brackish to marine settings (Roe et al., 1998). Surprisingly, the oldest reported cetacean, *Himalayacetus subathuensis*, produces an isotopic signal suggestive of both marine and freshwater habits (Bajpai and Gingerich, 1998). For the Miocene, Cetacea have provided independent evidence for ocean water oxygen isotopic composition (and, thus, global ice volumes), allowing adjustment of isotopic paleotemperatures from other, non-homoithermic sources (Barrick et al., 1992; see also LeCuyer et al., 1996). A caveat for early Miocene samples is that signal size may be close to analytical error.

Isotopes from Oligocene Cetacea are largely unstudied. These animals lived at a time of dramatic change in global climates, sea levels, ocean circulation, and oceanic ecosystems related to the opening of the Southern Ocean and growth of the Antarctic ice cap (e.g. Diester-Haass, 1996; Diester-Haass et al., 1996; Dingle and Lavelle, 1998b; Salamy and Zachos, 1999; Pearson and Palmer, 2000, and references therein). These changes in ocean structure may have been a major driving force in cetacean evolution. A key factor in Eocene-Oligocene climate change is Antarctic ice volume (e.g., Zachos et al., 1999:figure 3), and here isotopes from Oligocene cetacean bone phosphate offer a proxy for the oxygen isotopic composition of ancient seawater and ice volume estimates. In contrast to the early Miocene, ice volume signals are likely to be markedly larger than analytical limits. This indicates a significant new direction for research.

ACKNOWLEDGMENTS

Research was supported by the National Geographic Society (most recently grant 5381-94) and the Otago Research Committee. I thank A. Grebneff for his help in the field and lab program, R. D. Frew, A. Grebneff, H. Ichishima, and D. E. Lee for comments on an early version of text, M. D. Uhen and an anonymous referee for helpful comments, and participants in the

1999 Penrose Conference for their stimulating discussion.

REFERENCES

Allen, G. M., 1921, A new fossil cetacean (*Archaeodelphis patrius* gen. et sp. nov.): Bulletin of the Museum of Comparative Zoology, Harvard, v. 65, p. 1-14.

Bajpai, S., and Gingerich, P. D., 1998, A new Eocene archaeocete (Mammalia, Cetacea) from India and the time of origin of whales: Proceedings of the National Academy of Sciences of the United States of America, v. 95, p. 15464-15468.

Barnes, L. G., 1977, Outline of eastern North Pacific fossil cetacean assemblages: Systematic Zoology, v. 25, p. 321-343.

Barnes, L. G., 1998, The sequence of fossil marine mamal assemblages in México: Universidad Autonoma Del Estado De Hidalgo, Publicación Especial, v. 1, p. 26-79.

Barnes, L. G., and McLeod, S. A., 1984, The fossil record and phyletic relationships of gray whales, *in* Jones, M. L., Swartz, S. L., and Leatherwood, S., eds., The gray whale: *Eschrichtius robustus*: Orlando, Academic Press, p. 3-32.

Barnes, L. G., and Mitchell, E. D., 1978, Cetacea, *in* Maglio, V. J., and Cooke, H. S. B., eds., Evolution of African mammals: Cambridge, Mass., Harvard University Press, p. 582-602.

Barnes, L. G., Inuzuka, N., and Hasegawa, Y., eds., 1995a, Evolution and biogeography of fossil marine vertebrates in the Pacific realm: The Island Arc, v. 3, p. 243-537.

Barnes, L. G., Kimura, M., Furusawa, H., and Sawamura, H., 1995b, Classification and distribution of Oligocene Aetiocetidae (Cetacea; Mysticeti) from western North America and Japan: The Island Arc, v. 3, p. 392-431.

Barrick, R. E., Fischer, A. G., Kolodny, Y., Luz, B., and Bohaska, D., 1992, Cetacean bone oxygen isotopes as proxies for Miocene ocean composition and glaciation: Palaios, v. 7, p. 521-531.

Berta, A. and Deméré, T., eds., 1994, Contributions in marine mammal paleontology honoring Frank C. Whitmore, Jr: Proceedings of the San Diego Museum of Natural History, v. 29, 1-268.

Brandt, J. F., 1843, De Cetotherio, novo Balaenarum familae genre in Rossia meridionali ante aliquot annos effosso: Bulletin Scientifique de L'Académie (Impériale) des Sciences de Saint-Pétersbourg, v. series 2, 1, p. 145-148.

Cipriano, F., 1997, Antitropical distributions and speciation in dolphins of the genus *Lagenorhynchus*: a preliminary analysis: The Society for Marine Mammalogy Special Publication, v. 3, p. 305-316.

Clapham, P. J., and Brownell, R. L., 1996, Potential for interspecific competition in baleen whales: Reports of the International Whaling Commission, v. 46, p. 361-367.

Cozzuol, M. A., 1996, The record of the aquatic mammals in southern South America: Münchner Geowissenschaftliche Abhandlungen A, Geologie Und Paläontologie, v. 30, p. 321-342.

Cranford, T. W., Amundin, M., and Norris, K. S., 1996, Functional morphology and homology in the odontocete nasal complex: implications for sound generation: Journal of Morphology, v. 228, p. 223-285.

Cruz-Marin, A., and Barnes, L. G., 1996, A new agorophiid from San Juan de la Costa, Baja California Sur: the oldest odontocete from Mexico: 6th North American Paleontological Congress, Abstracts, Paleontological Society Special Publication, v. 8, p. 91.

Davies, J. L., 1963, The antitropical factor in cetacean speciation: Evolution, v. 17, p. 107-116.

Diester-Haass, L., 1996, Late Eocene-Oligocene paleoceanography in the southern Indian Ocean (ODP Site 744): Marine Geology, v. 130, p. 99-119.

Diester-Haass, L., Robert, C., and Chamley, H., 1996, The Eocene-Oligocene preglacial-glacial transition in the Atlantic sector of the Southern Ocean (ODP Site 690): Marine Geology, v. 131, p. 123-149.

Dingle, R. V., and Lavelle, M., 1998a, Antarctic Peninsular cryosphere: Early Oligocene (c. 30 Ma) initiation and a revised glacial chronology: Journal of the Geological Society, v. 155, p. 433-437.

Dingle, R. V., and Lavelle, M., 1998b, Late Cretaceous–Cenozoic climatic variations of the northern Antarctic Peninsula: new geochemical evidence and review: Palaeogeography, Palaeoclimatology, Palaeoecology, v. 141, p. 215-232.

Fordyce, R. E., 1980, Whale evolution and Oligocene Southern Ocean environments: Palaeogeography, Palaeoclimatology, Palaeoecology, v. 31, p. 319-336.

Fordyce, R. E., 1989a, Origins and evolution of Antarctic marine mammals: Special Publication of the Geological Society of London, v. 47, p. 269-281.

Fordyce, R. E., 1989b, Problematic Early Oligocene toothed whale (Cetacea, ?Mysticeti) from Waikari, North Canterbury, New Zealand: New Zealand Journal of Geology and Geophysics, v. 32, p. 385-390.

Fordyce, R. E., 1992, Cetacean evolution and Eocene/Oligocene environments, *in* Prothero, D., and Berggren, W. A., eds., Eocene-Oligocene climatic and biotic evolution: Princeton, New Jersey, Princeton Univ. Press, p. 368-381.

Fordyce, R. E., 1994, *Waipatia maerewhenua*, new genus and new species (Waipatiidae, new family), an archaic late Oligocene dolphin (Cetacea: Odontoceti: Platanistoidea) from New Zealand, *in* A. Berta and T. Deméré, eds., Contributions in marine mammal paleontology honoring Frank C. Whitmore, Jr. Proceedings of the San Diego Museum of Natural History, v. 29, p. 147-176.

Fordyce, R. E., 2002, Neoceti, *in* Perrin, W. F., Würsig, B., and Thewissen, J. G., eds., Encyclopedia of marine mammals: San Diego, Academic Press, p. 787-791.

Fordyce, R. E., and Barnes, L. G., 1994, The evolutionary history of whales and dolphins: Annual Review of Earth and Planetary Science, v. 22, p. 419-455.

Fordyce, R. E., and Muizon, C. de, 2001, Evolutionary history of cetaceans: a review, *in* Buffrenil, V. de, and Mazin, J., eds., Secondary adaptation to life in water: Munich, Pfeil Verlag, p. 169-234.

Fordyce, R. E., and Watson, A. G., 1998, Vertebral pathology in an Early Oligocene whale (Cetacea: ?Mysticeti) from Wharekuri, North Otago, New Zealand, *in* K.I. Grimm, M.C. Grimm, and M. Morlo, eds., Festschift zum 70 Geburtstag von Karlheinz Rothausen: Mainz, Naturhistoriches Museum Mainz, p. 161-176.

Gaskin, D. E., 1982, The ecology of whales and dolphins: London, Heinemann, 459 p.

Gatesy, J., 1998, Molecular evidence for the phylogenetic affinities of Cetacea, *in* Thewissen, J. G. M., ed., The emergence of whales. Evolutionary patterns in the origin of Cetacea: New York, Plenum, p. 63-111.

Geisler, J. H., and Luo, Z., 1996, The petrosal and inner ear of *Herpetocetus* sp. (Mammalia: Cetacea) and their implications for the phylogeny and hearing of archaic mysticetes: Journal of Paleontology, v. 70, 1045-1066.

Gingerich, P. D., 1992, Marine mammals (Cetacea and Sirenia) from the Eocene of Gebel Mokattam and Fayum, Egypt: stratigraphy, age and paleoenvironments: University of Michigan Papers on Paleontology, v. 30, p. 1-84.

Gingerich, P. D., 1998, Paleobiological perspectives on Mesonychia, Archaeoceti, and the origin of whales, *in* Thewissen, J. G. M., ed., The emergence of whales. Evolutionary patterns in the origin of Cetacea: New York, Plenum, p. 423-449.

Gingerich, P. D., Arif, M., Bhatti, M. A., Anwar, M., and Sanders, W. J., 1997, *Basilosaurus drazindai* and *Basiloterus hussaini*, new Archaeoceti (Mammalia, Cetacea) from the middle Eocene Drazinda Formation, with a revised interpretation of ages of whale-bearing strata in the Kirthar Group of the Sulaiman Range, Punjab (Pakistan): Contributions From the Museum of Paleontology, the University of Michigan, v. 30, p. 55-81.

Gingerich, P. D., Ul Haq, M., Zalmout, I. S., Khan, I. H., and Malkani, M. S., 2001, Origin of whales from early artiodactyls: hands and feet of Eocene Protocetidae from Pakistan: Science, v. 293, p. 2239-2242.

Gingerich, P. D., Wells, N. A., Russell, D. E., and Shah, S. M., 1983, Origin of whales in epicontinental remnant seas: new evidence from the early Eocene of Pakistan: Science, v. 220, p. 403-406.

Goedert, J. L., and Barnes, L. G., 1996, The earliest known odontocete: a cetacean with agorophiid affinities from latest Eocene to earliest Oligocene rocks in Washington State: 6th North American Paleontological Congress, Abstracts, Paleontological Society Special Publication, v. 8, p. 148.

Goedert, J. L., Squires, R. L., and Barnes, L. G., 1995, Paleoecology of whale-fall habitats from deep-water Oligocene rocks, Olympic Peninsula, Washington State: Palaeogeography, Palaeoclimatology, Palaeoecology, v. 118, p. 151-158.

Graciansky, P.-C., Hardenbol, J., Jacquin, T., and Vail, P. R., 1998, Mesozoic and Cenozoic sequence stratigraphy of European basins: SEPM (Society for Sedimentary Geology) Special Publication, v. 60, pp. 1-786.

Hedeen, C. D., 1997, Keasey Formation; a window into the bathyal late Eocene of Oregon: Geological Society of America Abstracts with Programs, v. 29 (6), p. 100-101.

Heyning, J. E., 1989, Comparative facial anatomy of beaked whales (Ziphiidae) and a systematic revision among the families of extant Odontoceti: Contributions in Science, Natural History Museum of Los Angeles County, v. 405, p. 1-64.

Heyning, J. E., 1997, Sperm whale phylogeny revisited: analysis of the morphological evidence: Marine Mammal Science, v. 13, p. 596-613.

Hornibrook, N. de B., Brazier, R. C., and Strong, C. P., 1989, Manual of New Zealand Permian to Pleistocene foraminiferal biostratigraphy: New Zealand Geological Survey Paleontological Bulletin, v. 56, p. 1-175.

Ichishima, H., Barnes, L. G., Fordyce, R. E., Kimura, M., and Bohaska, D. J., 1995, A review of kentriodontine dolphins (Cetacea: Delphinoidea: Kentriodontidae): systematics and biogeography: The Island Arc, v. 3, p. 486-492.

Kellogg, A. R., 1928, History of whales—their adaptation to life in the water: Quarterly Review of Biology, v. 3, p. 29-76 and p. 174-208.

Kellogg, A. R., 1936, A review of the Archaeoceti: Carnegie Institute of Washington Publication, v. 482, p. 1-366.

Kenney, R. D. and Wishner, K.F., eds., 1995, The South Channel ocean productivity experiment: SCOPEX: Continental shelf research, v. 15, p. 373-611.

Keyes, I. W., 1973, Early Oligocene squalodont cetacean from Oamaru, New Zealand: New Zealand Journal of Marine and Freshwater Research, v. 7, p. 381-390.

Kimura, M., 1989, Research on whales from Ashoro, *in* Anon., A record of studies on the Ashoro fossil fauna: Ashoro, Hokkaido, Board of Education, Ashoro, p. 77-117.

Köhler, R., and Fordyce, R. E., 1997, An archaeocete whale (Cetacea: Archaeoceti) from the Eocene Waihao Greensand, New Zealand: Journal of Vertebrate Paleontology, v. 17, p. 574-583.

Laws, R. M., 1977, The significance of vertebrates in the Antarctic marine ecosystem, *in* Llano, G., ed., Adaptations within Antarctic ecosystems: Washington D.C., Smithsonian Institution, p. 411-438.

Leatherwood, S., and Reeves, R. R., 1989, Marine mammal research and conservation in Sri Lanka, 1985-1986: Nairobi, Kenya, United Nations Environment Programme, Marine Mammal technical report number 1, 138 p.

LeCuyer, C., Grandjean, P., Paris, F., Robardet, M., and Robineau, D., 1996, Deciphering "temperature" and "salinity" from biogenic phosphates: the $\delta^{18}O$ of coexisting fishes and mammals of the middle Miocene sea of western France: Palaeogeography, Palaeoclimatology, Palaeoecology, v. 126, p. 61-74.

Lipps, J. H., and Mitchell, E. D., 1976, Trophic model for the adaptive radiations and extinctions of pelagic marine mammals: Paleobiology, v. 2, p. 147-155.

Luo, Z., and Gingerich, P. D., 1999, Terrestrial Mesonychia to aquatic Cetacea: transformation of the basicranium and evolution of hearing in whales: University of Michigan Papers on Paleontology, v. 31, 1-98.

McKenna, M. C. and Bell, S. K., 1998, Classification of mammals: above the species level: New York, Columbia University Press, 631 p.

McLeod, S. A., Whitmore, F. C., and Barnes, L. G., 1993, Evolutionary relationships and classification, *in* Burns, J. J., Montague, J. J., and Cowles, C. J., eds., The bowhead whale: Lawrence, Kansas, Society for Marine Mammalogy, p. 45-70.

Mead, J. G., 1975, Anatomy of the external nasal passages and facial complex in the Delphinidae (Mammalia: Cetacea): Smithsonian Contributions to Zoology, v. 207, p. 1-72.

Messenger, S. L., and McGuire, J. A., 1998, Morphology, molecules, and the phylogenetics of cetaceans: Systematic Biology, v. 47, p. 90-124.

Michelsen, O., Thomsen, O., Heilmann-Clausen, C., Jordt, H., and Laursen, G. V., 1998, Cenozoic sequence stratigraphy in the eastern North Sea: SEPM Special Publication, v. 60, p. 91-118.

Milinkovitch, M. C., 1995, Molecular phylogeny of cetaceans prompts revision of morphological transformations: Trends in Ecology and Evolution, v. 10, p. 328-334.

Mitchell, E. D., 1989, A new cetacean from the late Eocene La Meseta Formation, Seymour Island, Antarctic Peninsula: Canadian Journal of Fisheries and Aquatic Science, v. 46, p. 2219-2235.

Morgans, H. E. G., Scott, G. H., Beu, A. G., Graham, I. J., Mumme, T. C., St. George, W., and Strong, C. P., 1996, New Zealand Cenozoic time scale: Institute of Geological and Nuclear Sciences Science Report, v. 96/38, p. 1-7.

Muizon, C. de, 1987, The affinities of *Notocetus vanbenedeni*, an Early Miocene platanistoid (Cetacea, Mammalia) from Patagonia, southern Argentina: American Museum Novitates, v. 2904, p. 1-27.

Muizon, C. de, 1988, Les relations phylogénétiques des Delphinida (Cetacea, Mammalia): Annales de Paléontologie, v. 74, p. 159-227.

Muizon, C. de, 1991, A new Ziphiidae (Cetacea) from the early Miocene of Washington State

(USA) and phylogenetic analysis of the major groups of odontocetes: Bulletin du Muséum Nationale d'Histoire Naturelle, Paris, 4 Série, v. 12 C (3-4), p. 279-326.

Nemoto, T., 1959, Food of baleen whales with reference to whale movements: Scientific Reports of the Whales Research Institute, v. 14, p. 149-290.

Nikaido, M., Rooney, A. P., and Okada, N., 1999, Phylogenetic relationships among cetartiodactyls based on insertions of short and long interpersed elements: hippopotamuses are the closest extant relatives of whales: Proceedings of the National Academy of Sciences of the United States of America, v. 96, p. 10261-10266.

Nikaido, M., Matsuno, F., Hamilton, H., Brownell, R. L., Cao, Y., Ding, W., Zuoyan, Z., Shedlock, A. M., Fordyce, R. E., Hasegawa, M., and Okada, N., 2001, Retroposon analysis of major cetacean lineages: The monophyly of toothed whales and the paraphyly of river dolphins: Proceedings of the National Academy of Sciences of the United States of America, v. 98, p. 7384-7389.

Nishiwaki, M., 1977, Distribution of toothed whales in the Antarctic Ocean, *in* Llano, G., ed., Adaptations within Antarctic ecosystems: Washington D.C., Smithsonian Institution, p. 783-791.

Oishi, M., and Hasegawa, Y., 1995, A list of fossil cetaceans in Japan: The Island Arc, v. 3, p. 493-505.

Okazaki, Y., 1995, A new type of primitive baleen whale (Cetacea: Mysticeti) from Kyushu, Japan: The Island Arc, v. 3, p. 432-435.

Orr, W. N., and Faulhaber, J., 1975, A middle Tertiary cetacean from Oregon: Northwest Science, v. 49, p. 174-181.

Pearson, P. N., and Palmer, M. R., 2000, Atmospheric carbon dioxide concentrations over the past 60 million years: Nature, v. 406, p. 695-699.

Pilleri, G., 1986, Beobachtungen an den fossilen Cetaceen des Kaukasus: Ostermundigen, Switzerland, Hirnanatomisches Institut, 40 p.

Pilleri, G., 1989, An archaeocete vertebra from the Upper Eocene of Taradell, Catalonia, Spain, *in* Pilleri, G., ed., Contributions to the paleontology of some Tethyan Cetacea and Sirenia (Mammalia). II: Berne, University of Berne, Hirnanatomisches Institut, p. 7-12.

Pilleri, G., and Cigala-Fulgosi, F., 1989, First archaeocete record from the Eocene of Italy (Varano, Northern Appenines), *in* Pilleri, G., ed., Contributions to the paleontology of some Tethyan Cetacea and Sirenia (Mammalia). II: Berne, University of Berne, Hirnanatomisches Institut, p. 87-101.

Pivorunas, A., 1979, The feeding mechanisms of baleen whales: American Scientist, v. 67, p. 432-440.

Pledge, N. S., 1994, Cetacean fossils from the lower Oligocene of South Australia: Records of the South Australian Museum, v. 27, p. 117-123.

Ravn, J. P. J., 1926, On a cetacean, *Squalodon* (*Microzeuglodon*?) *wingei* nov. sp., from the Oligocene of Jutland: Meddelelser Fra Dansk Geologisk Forening, v. 7, p. 45-54.

Rice, D. W., 1998, Marine mammals of the world. Systematics and distribution: Lawrence, Kansas, Society for Marine Mammalogy, 231 p.

Roe, L. L., Thewissen, J. G. M., Quade, J., O'Neil, J. R., Bajpai, S., Sahni, A., and Hussain, S. T., 1998, Isotopic approaches to understanding the terrestrial-to-marine transition of the earliest cetaceans, *in* Thewissen, J. G. M., ed., The emergence of whales. Evolutionary patterns in the origin of Cetacea: New York, Plenum, p. 399-422.

Rothausen, K., 1985, The early evolution of Cetacea: Fortschritte der Zoologie, v. 30, p. 143-147.

Rothausen, K., 1994, Die Schritte der Tetrapoden in die Meere des frühen Känozoikums: Berliner Geowissenschaften Abhandlungen, v. E13, p. 99-112.

Sadler, P. M., 1988, Geography and stratification of uppermost Cretaceous and Paleogene units on Seymour Island, northern Antarctic Peninsula: Geological Society of America Memoir, v. 169, p. 303-320.

Salamy, K. A., and Zachos, J. C., 1999, Latest Eocene–early Oligocene climate change and Southern Ocean fertility: inferences from sediment accumulation and stable isotope data: Palaeogeography, Palaeoclimatology, Palaeoecology, v. 145, p. 61-77.

Sanders, A. E., 1996, The systematic position of the primitive odontocete *Xenorophus sloanii* (Mammalia: Cetacea) and two new taxa from the late Oligocene of South Carolina, USA: 6th North American Paleontological Congress, Abstracts. Paleontological Society Special Publication, v. 8, p. 338.

Thewissen, J. G. M., ed., 1998, The emergence of whales: evolutionary patterns in the origin of Cetacea: New York, Plenum, 477 p.

Thewissen, J. G. M., and Fish, F. E., 1997, Locomotor evolution in the earliest cetaceans: functional model, modern analogues, and paleontological evidence: Paleobiology, v. 23, p. 482-490.

Thewissen, J. G. M., Roe, L. J., O'Neil, J. R., Hussain, S. T., Sahni, A., and Bajpai, S., 1996, Evolution of cetacean osmoregulation: Nature, v. 381, p. 379-380.

Tynan, C. T., 1998, Ecological importance of the southern boundary of the Antarctic Circumpolar Current: Nature, v. 392, p. 708-710.

Uhen, M. D., 1998a, Middle to late Eocene basilosaurines and dorudontines, *in* Thewissen, J. G. M., ed., The emergence of whales. Evolutionary patterns in the origin of Cetacea: New York, Plenum, p. 29-61.

Uhen, M. D., 1998b, New protocetid (Mammalia, Cetacea) from the late middle Eocene Cook Mountain Formation of Louisiana: Journal of Vertebrate Paleontology, v. 18, p. 664-668.

Uhen, M. D., 1999, New species of protocetid archaeocete whale, *Eocetus wardii* (Mammalia: Cetacea) from the middle Eocene of North Carolina: Journal of Paleontology, v. 73, p. 512-528.

Uhen, M. D., 2000, Replacement of deciduous first premolars and dental eruption in archaeocete whales: Journal of Mammalogy, v. 81, p. 123-133.

Uhen, M. D. and Gingerich, P. D., 2001, New genus of dorudontine archaeocete (Cetacea) from the middle to late Eocene of South Carolina: Marine Mammal Science, v. 17, p. 1-34.

Uhen, M. D., and Tichy, G., 2000, A new basilosaurid archaeocete from Austria: Journal of Vertebrate Paleontology, Supplement, v. 20 (3), p. 74-75.

Van Beneden, P. J. and Gervais, P., 1868-1880, Ostéographie des Cétacés Vivants et Fossiles, Comprenant la Description et l'Iconographie du Squelette et du Système Dentaire de ces Animaux; ainsi que des documents relatifs à leur histoire naturelle: Paris, Arthus Bertrand, 2 volumes, Text, 634 pp., and Atlas, 64 pls.

Vonhof, H. B., Smit, J., Brinkhuis, H., Montanari, A., and Nederbragt, A. J., 2000, Global cooling accelerated by early late Eocene impacts?: Geology, v. 28, p. 687-690.

Wartzok, D., and Ketten, D. R., 1999, Marine mammal sensory systems, *in* Reynolds, J. E., and Rommel, S. A., eds., Biology of marine mammals: Washington DC, Smithsonian Institution, p. 117-175.

Whitmore, F. C., and Sanders, A. E., 1977, Review of the Oligocene Cetacea: Systematic Zoology, v. 25, p. 304-320.

Williams, E. M., 1998, Synopsis of the earliest cetaceans: Pakicetidae, Ambulocetidae, Remingtonocetidae, and Protocetidae, *in* Thewissen, J. G. M., ed., The emergence of whales. Evolutionary patterns in the origin of Cetacea: New York, Plenum, p. 1-28.

Yoshida, N., and Miyazaki, N., 1991, Oxygen isotope correlation of cetacean bone phosphate with environmental water: Journal of Geophysical Research, v. 96 (C1), p. 815-820.

Zachos, J. C., Opdyke, B. N., Quinn, T. M., Jones, C. E., and Halliday, A. N., 1999, Early Cenozoic glaciation, Antarctic weathering, and seawater $^{87}Sr/^{86}Sr$: is there a link?: Chemical Geology, v. 161, p. 165-180.

CHAPTER 10

Terrestrial Eocene-Oligocene Vegetation and Climate in the Pacific Northwest

Jeffrey A. Myers

ABSTRACT

Stratigraphically dense and well-dated paleofloras from sequences throughout the U.S. West Coast permit recalibration of the timing and magnitude of vegetational and paleoclimate events during the Eocene-Oligocene transition in western North America. Paleoclimate data from these assemblages indicate that a major cooling step of ~3–4°C in mean annual temperature (MAT) took place between 39 and 38.5 Ma. The vegetational response to cooling was the geologically rapid replacement of evergreen broad-leaved forest by deciduous broad-leaved forest in many areas of the interior Pacific Northwest, and the entry of diverse temperate lineages into near-sea level vegetation. A latest Eocene warming event produced a ~3–4°C increase in mean annual temperature (MAT) and returned much of the Pacific Northwest to warm temperate conditions by ~37 Ma, permitting the reentry of thermophylic Goshen-type indicator species to low and moderate elevations. A second major pulse of climatic cooling took place between < 34.0 and > 33.6 Ma, and produced no more than a ~3–6°C decline in MAT, a moderate increase in seasonality, and a pronounced decrease in dry month precipitation. This "boundary event" resulted in the widespread replacement of subtropical evergreen broad-leaved vegetation by deciduous broad-leaved vegetation dominated by diverse cool temperate lineages, although many thermophylic lineages persisted well into the Oligocene near sea level.

Revised estimates of the pattern, magnitude, and timing of late middle Eocene to early Oligocene climate change agree more closely with evidence from marine isotopes and with extinction and radiation events of mammals and shallow marine invertebrates than previous estimates. It is now clear that cooling near the Eocene/Oligocene boundary was only the last of a series of cooling and warming steps that impacted continental western North America, and that middle-late Eocene climate and vegetational change was at least as severe those of the more famous boundary event. However, the boundary cooling event crossed the minimum temperature tolerance of most subtropical lineages and resulted in the most catastrophic and permanent continental biotic crisis of the Cenozoic.

INTRODUCTION

Paleoclimate estimates from continental paleofloras indicate that continental climate cooled in a series of steps from the late Paleocene thermal maximum through the early Oligocene. The record is punctuated by shorter term warming and cooling steps (Wolfe, 1971, 1992, 1994), by far the most frequently cited and most catastrophic of which was interpreted by Wolfe (1992, 1994) to have taken place over a span of ~1 million years near 33 Ma, just after the Eocene/Oligocene boundary. The continental event is mirrored by an approximately 1‰ positive marine oxygen isotopic shift centered between 33.6 and 33.5 Ma (Zachos et al., 1996; Alroy et al., 2000). This event, termed the Terminal Eocene Event (TEE) by Wolfe (1978), had profound effects on vegetation, particularly at high and mid latitudes, resulting in the replacement of warm subtropical, predominantly evergreen, vegetation with predominantly deciduous temperate vegetation. In many areas of the interior Pacific Northwest and Alaska, cold-month mean temperature declined to a level between 1°C and –2°C, causing a severe regional dieoff of woody evergreen dicots.

Surprisingly few Eocene-Oligocene transitional paleofloras have been directly dated using modern single-crystal $^{40}Ar/^{39}Ar$ methods, and this paucity of well-dated assemblages has, until recently, prohibited the resolution of climate and vegetational trends spanning less than a few million years. Additionally, older K-Ar dates with dubious provenance and large errors have become entrenched in the literature, and have only recently begun to be replaced by more reliable $^{40}Ar/^{39}Ar$ ages. $^{40}Ar/^{39}Ar$ chronology, accompanied by precision stratigraphic analysis using magnetostratigraphic and/or physical correlation has in many cases allowed correlation of Eocene-Oligocene paleofloras with the standard marine and continental chronology in critical sections including the John Day Basin (Manchester, 1990, 1994; Bestland et al., 1994 a, b), the San Diego area, California (Walsh et al., 1996), western Oregon (Kester, 2001; Prothero and Retallack, 2001; Myers et al., 2002), and the Washington and Oregon coast (Prothero and Hankins, 2001). The purpose of this paper is to examine vegetational and climatic information from well-dated paleofloras of middle Eocene through early Oligocene age on the West Coast in order to bring the paleobotanical climate and vegetational record into line with the marine isotopic/invertebrate record of these events. The focus is on successions of paleofloras within more or less continuous and well-dated geological sequences spanning critical intervals of the record.

APPROACH

This paper will emphasize fossil floras directly dated using $^{40}Ar/^{39}Ar$ methods or physically correlated with dated horizons. Corrected (Dalrymple, 1979) older K-Ar ages are also cited but identified. Climate information from fossil leaf assemblages was obtained using Wolfe's (1993) CLAMP (Climate Leaf Analysis Multivariate "Program") database. CLAMP uses the statistical correlation between specific features of leaf physiognomy (morphology/anatomy) and specific climate variables in modern vegetation as a proxy for estimating paleoclimate from fossil samples of greater than 20 species of woody dicot leaves (Wolfe, 1993). This report considers the climate variables of mean annual temperature (MAT), cold-month mean temperature (CMMT), mean annual precipitation (MAP), and dry-season precipitation (DSP). Some assemblages cited from the literature were examined using only traditional univariate leaf-margin analysis of MAT (Wolfe, 1979), and are so noted.

Both site variability of standing vegetation and taphonomic sorting processes (decay, water and wind transport) directly or indirectly influence morphological features such as leaf size, leaf shape, and leaf-margin type of leaves in a fossil assemblage. Because these variables are predictors for interpreting paleoclimate, taphonomic sorting and site variability can significantly impact climate estimates from leaf assemblages (Greenwood, 1992; Burnham et al., 1994, 2001; Myers, 1998). These biases are particularly significant in assemblages derived from mixed evergreen and deciduous broad-leaved forests like those of the middle Tertiary West Coast. Where possible in this paper, estimates of climate change are made from assemblages preserved in the same depositional setting (table 10.1). In cases where this was not possible, a "corrected" MAT value of +3°C for lacustrine assemblages has been assigned (Myers, 1998), although corrections for other climate variables have not been quantified.

An additional problem when examining assemblages inland from a paleocoastline is that many interior assemblages grew at moderate elevation and cannot be directly compared with near-sea-level assemblages without a MAT correction (figure 10.1). Previous studies (Meyer and Manchester, 1997; Wolfe et al., 1998) indicate that central Oregon lay at ~1000 m during the middle Tertiary, while northeastern California may have lain about 500 m or more higher. The approximate paleoelevation of each flora examined is noted in table 10.1, and where it as been necessary to directly compare MAT information from sea level and upland assemblages, an altitude correction of +3.3°C/km has been added to the higher-elevation assemblage (Wolfe et al., 1998).

THE EOCENE-OLIGOCENE PALEOBOTANICAL RECORD OF THE U.S. WEST COAST

Middle Eocene Paleofloras

Middle Eocene paleofloras are known from four primary regions of the West Coast (figure 10.1). Sea level fluviodeltaic assemblages of the Torrey and Delmar Formations, San Diego (Myers, 1990, 1991; Frederiksen, 1991), co-occur with Bridgerian mammals, and are approximately 45.5–46 Ma in age (Prothero and Emry, 1996; T. A. Deméré, pers. commun., 1999). Slightly younger middle Eocene floras from the San Diego region occur in the Mission Valley and Friars Formation, stratigraphically above a horizon $^{40}Ar/^{39}Ar$ dated at 42.6 Ma, associated with Uintan vertebrates (Walsh et al., 1996). The floras have not been described, although they have been scored for climate analysis (Myers, 1999). The lowermost "Franklinian" and "Fultonian" assemblages of the Puget Group, Washington (Wolfe, 1968), also grew near sea level, share key floristic elements with the San Diego flora, and have been K-Ar dated at ~45 Ma (Turner et al., 1983; Brandon and Vance, 1992). Slightly younger floras from the interior West Coast include the ~43.5–44.5 Clarno Flora of central Oregon (Manchester, 1990, 1994) and the ~45 Ma Chalk Bluffs Flora of the Sierra Nevada (MacGinitie, 1941), both of which grew on the seaward-facing slope of the Paleogene Cordillera at elevations of ~1,000 m.

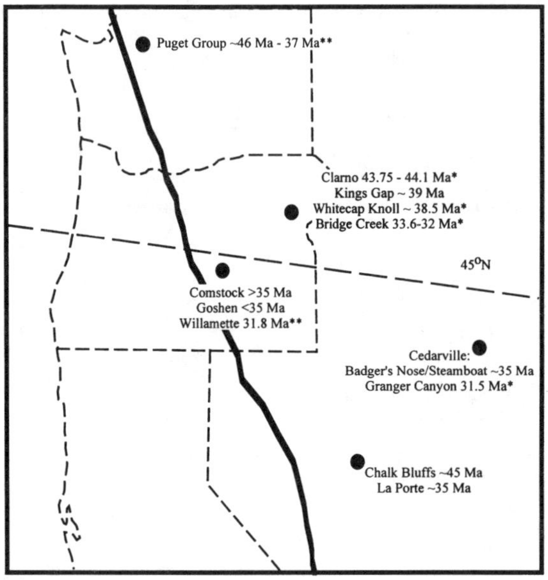

Figure 10.1. (*above*) Map of western North America indicating the location and age of floras discussed in this study, at the time of deposition with respect to the estimated Eocene-early Oligocene paleocoastline and paleolatitude. * indicates floras directly dated by $^{40}Ar/^{39}Ar$. ** indicates floras directly dated using K-Ar methods. (*below*) Chronology of West Coast floras discussed in the text. Chronology is from Berggren et al. (1995). Paleobotanical stages are modified from Wolfe (1968, 1981).

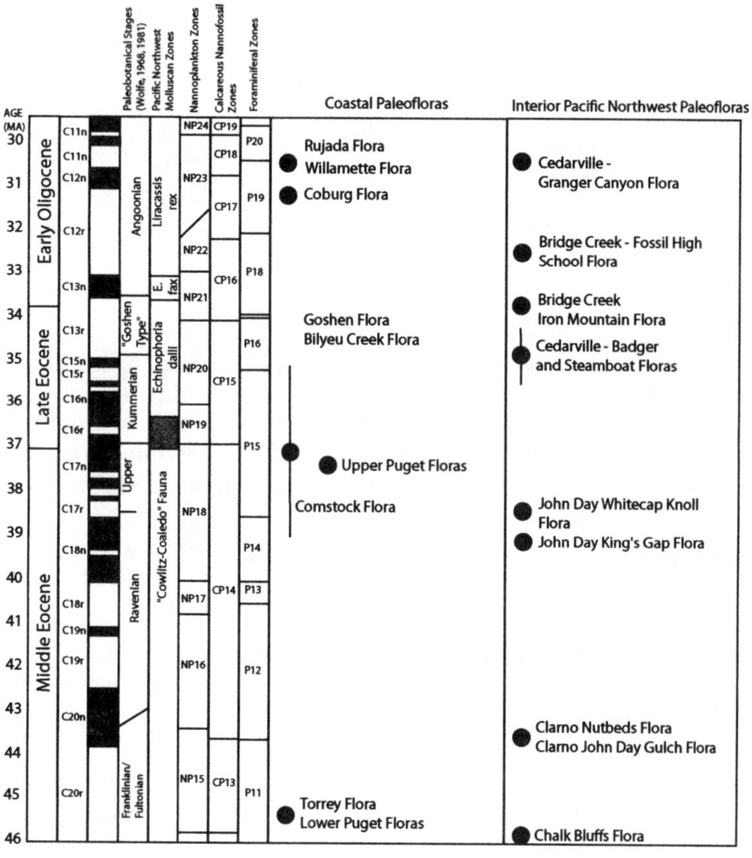

Table 10.1. Physiognomic Climate Estimates from Eocene and Oligocene West Coast Paleofloras

Flora	Age (Ma)	#spp.	Facies	Paleo Elev. (m)	MAT °C	CMMT °C	WMMT °C	MAP mm	3DMP mm	Reference
Standard deviation					2.0	3.6	2.9	58	9	Wing/Greenwood, 1993
Willamette*	30.6	N/A	Lacustrine Near shore	0	13.0	1.8	24.2			Kester, 2001 Myers et al., 2002
Cedarville Granger Canyon	31.5	17	Lacustrine Near shore	>1000	9.6	-0.4	21.1	2588	145	Myers 1998
Bridge Creek Fossil High School	32.6	44	Mid-Lacustrine	<1000	12.1	3.7	19.0	1675	230	Meyer and Manchester, 1997
Bridge Creek Dugout Gulch	33.6	33	Mid-Lacustrine	<1000	10.3	0.9	17.7	1800	162	Meyer and Manchester, 1997
Cedarville Badger's Nose	34 -35	32	Mid-Lacustrine	>1000	12.5	2.6	21.1	1780	246	Myers, 1998
Cedarville Steamboat	34-35	21	Flood plain Leaf litter	>1000	17.1	5.6	24.3	2340	314	Myers, 1998
Bilyeu Creek**	34-35	36	Flood plain Rapid Burial	0	20.7	6.6	24.8	4140	493	Klucking, 1964
Goshen**	34-34.8	46	Flood plain Leaf Litter	0	19.7	6.8	25.1	3790	469	Chaney and Sanborn, 1933; Kester, 2001
Comstock**	<39 >35	23	Flood plain Leaf Litter	0	22.4	7.0	26.5	3370	481	Sanborn, 1935
Whitecap Knoll*	~38.5	19	Lacustrine	unknown	9.0					Manchester, 2000
Kings Gap*	~39	21	Flood plain	unknown	15.5					Smith et al. 1998
Clarno WestBr. Ck.	43.5-44.5	24	Mid-Lacustrine	~1000	14.5	2.4	22.1	1740	286	Myers, 1998
Clarno - John Day Gulch	43.5-44.5	41	Flood plain Rapid Burial	~1000	19.0	6.3	25.1	3020	360	Myers, 1998

The number assigned to each assemblage in column 1 identifies data in Figures 2, 3, and 4. * Indicates climate results cited from indicated paper. ** Indicates floras that were scored from the literature. Standard errors do not take into account sample size or habitat/taphonomic biases.

The Eocene-Oligocene Transitional Sequence near Eugene, Oregon

Macrofloras from the vicinity of Eugene, Oregon, provide the most complete currently available picture of near-sea level floristic and vegetational change across the Eocene/Oligocene boundary (figure 10.1). At least four assemblages are known from the Eugene area, preserved in fluvial and lacustrine mudstone and sandy mudstone: (1) the undated Comstock Flora (Sanborn, 1937) recognized by Hergert (1954) to occur near the base of the Fisher Formation; (2) the Goshen Flora (Chaney and Sanborn, 1933); (3) a flora informally identified as the "Coburg" Flora high in the Fisher Formation; and (4) the Willamette Flora (Brown, 1959) within the basal deposits of the Little Butte Volcanics.

K-Ar and $^{40}Ar/^{39}Ar$ dates from rocks below the Comstock Flora indicate that the flora is younger than ~39 Ma in age. Rocks of the Eugene Formation interfinger with the upper Fisher Formation (Vokes and Snavely, 1948). Molluscan faunas from near the base of the Eugene Formation are tied to the Keasey West Coast local biozone. Recent magnetostratigraphic work by Prothero and Hankins (2001) places the Keasey Formation within Chrons C15n to C12r, approximately 35 million to 32.8 Ma, or latest Eocene/earliest Oligocene in age. This constrains the age of the upper Fisher Formation to <35 Ma, an interpretation that is confirmed by $^{40}Ar/^{39}Ar$ dates ranging from 34.0 Ma ± 0.7 Ma to 34.85 Ma ± 0.2 Ma from the Bond Creek Tuff within the upper Fisher Formation (Kester, 2001). The tuff stratigraphically underlies the Goshen Flora horizon (G.J. Retallack, pers. commun., 2000), but overlies the Comstock Flora. Hence, the Comstock Flora is most likely late middle Eocene in age, younger than ~39 Ma, but older than latest Eocene. The Goshen Flora is no older than latest Eocene (34.85 Ma), and no younger than early Oligocene (32.8 Ma). The Willamette Flora is Oligocene in age (>30.6 Ma), based on a new $^{40}Ar/^{39}Ar$ date from reworked tuff stratigraphically above the Willamette Flora sequence. Ongoing work in the Eugene region may further resolve stratigraphic and chronological relationships between the units and the enclosed paleofloras (Kester, 2001; Myers et al., 2002).

Additional sea level floras of latest Eocene-early Oligocene age include the youngest Kummerian floras of the Puget sequence, which were initially thought to be as young as the Eocene/Oligocene boundary, but are now known to be no younger than 37 Ma (Nesbitt, this volume). Other undated or imprecisely-dated near sea level floras of latest Eocene to early Oligocene age include the late Eocene LaPorte (Potbury, 1935), Sweet Home (Brown, 1950), Scio (Sanborn, 1947), and Bilyeu Creek (Klucking, 1964) floras. Oligocene floras approximately coeval with the Willamette Flora include the Gumboot Mountain, Washington, (Meyer and Manchester, 1997) and Rujada, Oregon, (Lakhanpal, 1958) floras.

The Eocene-Oligocene Transitional Sequence in the Interior Pacific Northwest

Late Eocene and earliest Oligocene paleofloras are also known from two regions of the interior Pacific Northwest: the Clarno and John Day Formations of central Oregon and the Lower Cedarville Formation of northeast California (figure 10.1). At least four small paleofloras occur near the base of the John Day Formation, bracketed by the basal John Day Ignimbrite A with an $^{40}Ar/^{39}Ar$ age of 39.17 Ma (Smith et al., 1998) and the Whitecap Knoll Tuff $^{40}Ar/^{39}Ar$ dated at ~38.5 Ma (Bestland et al., 1994a, b). These are the ~39 Ma Kings Gap Flora preserved in floodplain siltstone (Smith et al., 1998), the 38.5 Ma Whitecap Knoll Flora (Manchester, 2000), and the ~38 Ma Sumner Spring Flora (Smith et al., 1998), both from lacustrine shale. The lowermost radiometrically dated Oligocene assemblage in the John Day sequence is the Dugout Gulch assemblage of the Bridge Creek Flora, $^{40}Ar/^{39}Ar$ dated at 33.6 Ma (Meyer and Manchester, 1997). The Dugout Gulch and numerous other Bridge Creek assemblages treated by Meyer and Manchester (1997) are preserved in lacustrine shale unconformably overlying rocks containing the Whitecap Knoll Flora, and range in age from 33.62 to ~32 Ma.

The only latest Eocene assemblages from the interior Pacific Northwest (figure 10.1) are the Badger's Nose and Steamboat Canyon assemblages of the Cedarville Flora of northeastern California (Myers and Greenwood, 1997; Myers, 1998), which are bracketed by a 40.0 ± 3 Ma K-Ar age (Axelrod, 1966) from rocks unconformably underlying the assemblage, and by a 33.9 Ma K-Ar age well up section (Duffield and McKee, 1986). Although it should be emphasized that these K-Ar dates have large error bars, and the sequence requires additional radiometric age control, the Cedarville assemblages are of critical importance because the assemblages are regionally correlated, and are preserved in a range of depositional facies (Myers and Fisher, 1994; Myers, 1998), permitting the examination of regional-scale vegetational relationships. The Steamboat Canyon assemblage represents broad-leaved evergreen vegetation preserved in slowly aggraded floodplain and channel facies, while the Badger's Nose assemblage represents *Metasequoia*-dominated lake margin vegetation. Additionally, the Granger Canyon assemblage of the Cedarville Flora, $^{40}Ar/^{39}Ar$ dated at 31.5 Ma, unconformably overlies latest Eocene assemblages, and provides a basis for evaluating the Boundary Event in the Pacific Northwest interior (Myers, 1998).

VEGETATIONAL TRENDS AND BIOSTRATIGRAPHY OF LATE EOCENE FLORAS

Wolfe (1968, 1977, 1981) erected a traditional biostratigraphic zonation of Cenozoic paleobotanical stages in the Puget Group sequence of western Washington, which at the time was thought to represent a continuous section spanning the latest early Eocene through Oligocene. For the most part, Wolfe's paleobotanical biozones are not formally defined by the lowermost and uppermost occurrence of individual species, but are in the strict sense assemblage zones. Modern plant communities as well as the ranges of specific plant species are highly facies specific and limited by altitude, local ecology, local climate, migrational pathways, and other variables. As a result, any biostratigraphic zonation using plant index taxa or plant assemblages is likely to be diachronous and local. Many fossil plant taxa erected on the basis of foliage are form taxa that have long geological records, e.g., *Plafkeria, Fagus oregona*, and *"Cinnamomum" dilleri*, and in many instances do not have wide enough occurrences to be biostratigraphically useful. For example *Macclintockia pugetensis*, which marks the upper Kummerian stage, is thought to be a beach strand plant, and hence is extremely limited in occurrence. Four vegetational and floristic trends outlined below can be directly calibrated to Wolfe's paleobotanical stages, and clearly document climatically driven biotic trends during the Eocene-Oligocene transition.

The Uppermost Occurrence of *Macginitiea*

Macginitiea (Platanaceae) is ubiquitous among Eocene floras, often as a dominant element. The uppermost known occurrence of *Macginitiea* near sea level may be from the LaPorte Flora (B. H. Tiffney, pers. commun., 2000), although the latest Eocene-early Oligocene age assigned to this flora may be incorrect, and the *Macginitiea* fossils may have originated from middle Eocene strata underlying the LaPorte tuff. Elsewhere, *Macginitiea* occurs in abundance in the Comstock Flora of the far southern Willamette Valley, Oregon (Sanborn, 1937). K-Ar and $^{40}Ar/^{39}Ar$ dates from rocks both above and below the flora suggest the flora is younger than 38 to 40 Ma in age. *Macginitiea* is known from floras in central Oregon as young as the ~38 Ma Sumner Spring Flora (Smith et al., 1998), but the taxon has not been recovered from younger assemblages in the interior. Based on current evidence, the uppermost occurrence of *Macginitiea* is approximately synchronous throughout the Pacific Northwest near 38-39 Ma, and marks the boundary between Wolfe's (1968) middle and upper Ravenian paleobotanical zones (figure 10.2).

The Uppermost Occurrence of Diverse Subtropical Broad-leaved Evergreen Lineages

The late Eocene marks the uppermost occurrence of the majority of thermophylic evergreen broad-leaved dicot lineages that form Wolfe's Boreotropical Paleogene vegetation type. Although some characteristic subtropical forms extend into the early Oligocene, including the common tropical vine *Paleophytocrene*, the presence of *diverse* members of the families Magnoliaceae, Annonaceae, Lauraceae, Menispermaceae, and Euphorbiaceae is limited to the Eocene. Coastal latest Eocene Goshen-type assemblages, including the LaPorte, Bilyeu Creek, and Goshen floras, all include a rich representation of thermophylic evergreen forms. In the interior Pacific Northwest, most thermophylic forms appear to have suffered regional extinction due to changing climate during the late middle Eocene prior to 38.5 Ma. The majority of subtropical evergreen dicots represented in the Clarno Nut Beds and equivalent assemblages have not been recovered from younger assemblages of the Clarno and John Day Formations; and although *Magnolia, Kadsura*, Theaceae, and other evergreen broad-leaved forms have been recovered in low abundance from the ~35 Ma Goshen-type Cedarville Flora, diverse evergreen broad-leaved lineages with tropical affinities have not. The uppermost occurrence of diverse Boreotropical vegetation in the interior thus predates the uppermost occurrence of *Macginitiea*, while diverse Boreotropical vegetation may have persisted to the Eocene/Oligocene boundary in sea-level regions.

A number of distinctive evergreen broad-leaved taxa are limited to uppermost Eocene floras recognized by Wolfe (1981) as defining a separate biostratigraphic interval informally identified as the Goshen-type stage (figure 10.1). Taxa that characterize the Goshen-type paleobotanical stage, including *Allophylus wilsoni, Meliosma aesculifolia*, and *Staphylea kennetti*, occur in age-equivalent floras from different vegetation types in the interior and coastal Pacific Northwest at ~35 Ma. In the Cedarville Flora, Goshen-type indicators co-occur with the diagnostic Bridge Creek taxa *Parrotia brevipetiolata, Plafkeria obliquifolia*, and *Amalanchier covea*, along with a few species that belong to subtropical lineages (e.g. *Kadsura, Magnolia dayana*, Theaceae). The near-sea-level Bilyeu Creek–Thomas Creek Flora includes Goshen-type indicators along with other diverse subtropical families. Current evidence indicates that Goshen-type assemblages span a narrow geological range between ~35 Ma and the Eocene/Oligocene boundary at 33.7 Ma, and that Goshen-type indicators do not significantly overlap in range with Kummerian and Angoonian indicators. Because the assemblages are relatively easy to recognize, the Goshen-type stage is among the most useful continental biostratigraphic marker for a narrow but critical interval of the latest Eocene.

The Lowermost Occurrence of *Metasequoia*-dominated Vegetation

Metasequoia is present in low abundance in early Paleogene assemblages, but the tree is a dominant element of nearly all early Oligocene lacustrine assemblages of the interior Pacific Northwest. The taxon appears to have dominated well-drained lake-margin settings, but was not common in poorly drained floodplain or swamp facies (e.g., the Cedarville Steamboat Canyon assemblage). *Metasequoia* is absent in Eocene central Oregon assemblages, including lacustrine floras of the Clarno and lower John Day Formations; and large-scale migration of *Metasequoia* to central Oregon appears to have postdated the latest middle Eocene Whitecap Knoll Flora (38.5 Ma). *Metasequoia* was a dominant in the lacustrine Badger's Nose assemblage of the Cedarville Formation, where it co-occurs with Goshen-type indicators. These relationships suggest that large-scale migration of *Metasequoia* into central Oregon took place between ~38.5 and 35 Ma. With the exception of a few assemblages, *Metasequoia* is an uncommon element of low elevation coastal assemblages prior to the middle Oligocene. However, *Metasequoia* is a dominant element of the Goshen-type Bilyeu Creek Flora and its lateral equivalent the Scio Flora in coastal Oregon. Both these floras are preserved in coarse immature volcaniclastics where *Metaseqoia* may have been an early successional colonizer.

The Lowermost Occurrence of Diverse Arctotertiary Lineages

Neogene and modern temperate humid and mesic forests are characterized by a diverse species within the "Arctotertiary" families Pinaceae, Aceraceae, Betulaceae, Fagaceae, Juglandaceae, Rhamnaceae, Rosaceae, Tiliaceae, Ulmaceae, and others. The initial diversification of some Arctotertiary lineages took place in the higher-elevation regions of the Cordillera. Middle Eocene assemblages such as the Republic Flora (Wolfe and Wehr, 1987) and Thunder Mountain Flora (Axelrod, 1998) include diverse Pinaceae and Rosaceae, while the 34 Ma Florissant Flora, Colorado (MacGinitie, 1953), includes species of Pinaceae, Fagaceae, Betulaceae, Ulmaceae, Rosaceae, Tiliaceae, and Aceraceae that are closely related to forms found in latest Eocene and younger assemblages from the Pacific Northwest (Wolfe and Wehr, 1987; Meyer and Manchester, 1997).

Most Paleogene floras from the Pacific Northwest include one or a few species belonging to lineages that are taxonomic dominants in modern temperate vegetation. Cool temperate families such as Aceraceae, Betulaceae, Rosaceae, Juglandaceae, Pinaceae, and others are found in assemblages of the middle Eocene Clarno Flora (e.g. John Day Gulch, West Branch Creek), although they are not diverse at the genus and species level (Manchester, 1990, 1994). Myers and Fisher (1994) observed that these forms were in many cases opportunists ("woody weeds") that occupied episodically disturbed or less favorable habitats within predominantly evergreen broad-leaved subtropical forest.

A few Arctotertiary lineages had begun to build diversity at middle to low elevations of the Pacific Northwest by late Eocene time. For example, the Cedarville Badger's Nose assemblage includes five species in four genera of Rosaceae (*Rosa, ?Peraphyllum, Amalanchier, Rubus*). Both *?Peraphyllum* and *Amalanchier* appear to be conspecific with or similar to Florissant forms. However other cool-temperate families have not been recovered (e.g. Aceraceae) or are represented by a single species (e.g. Betulaceae, Ulmaceae, and Tiliaceae) in the Cedarville Flora. At Cedarville, cool-temperate groups were colonizers of marginal habitats, and did not yet intergrade with the regional evergreen oak-laurel forest (Myers, 1998). Assemblages of the younger Bridge Creek Flora include diverse members of all Arctotertiary families as numerical and taxonomic dominants. Thus it appears that important Bridge Creek Arctotertiary genera such as *Fagus pacifica, Tilia, Ostrya,* and *Paracarpinus* along with diverse Pinaceae arrived in the interior Pacific Northwest between ~35 and 33.62 Ma.

The arrival of diagnostic Arctotertiary lineages marks the base of Wolfe's (1981) Angoonian stage in lowland paleofloras. However, lowland Angoonian floras are fundamentally different from early Oligocene floras of the interior Pacific Northwest in that, aside from Pinaceae and Taxodiaceae, they do not include the rich species-level diversity of Arctotertiary families found the Bridge Creek Flora, and retain a larger proportion of evergreen broad-leaved plants than have been recovered from coeval floras of the interior. For example, the 31.6 Ma (Kester, 2001) Willamette Flora and approximately coeval Rujada Flora (Lakhanpal, 1958) include the Bridge Creek forms *Platanus condonii, Platanus exaspera, Crataegus merriamii, Plafkeria obliquifolia,* and others, in addition to *Paleophytocrene,* Lauraceae,

Annonaceae?, and other evergreen broad-leaved forms. It appears that lowland Oligocene vegetation retained a few late Eocene subtropical taxa, but did not accrue diverse cool temperate lineages until the early Miocene.

EOCENE-OLIGOCENE TRANSITIONAL CLIMATE AND VEGETATION

Middle Eocene Climate and Vegetation

The middle Eocene West Coast was clothed in uniform evergreen broad-leaved paratropical rain-forest at low elevation (figure 10.2). Although Bridgerian floras from San Diego are not diverse, they yield a MAT estimate of 21°C and ~2400 mm of growing season rainfall (Myers, 1990, and unpublished), with little temperature or precipitation seasonality. Approximately coeval Franklinian– Fultonian assemblages of the Puget Flora were equally humid, and yield MAT estimates between 18°C and 21°C (Wolfe, 1994). Vegetation at higher elevations of the interior Pacific Northwest included a significant component of deciduous temperate lineages. In many cases deciduous broad-leaved groups were opportunists colonizing volcanically disturbed habitats (Myers and Fisher, 1994). Both the Clarno and Chalk Bluffs floras yield MAT estimates of between 13°C and 19°C (table 10.1; figure 10.2), although the range of estimates is primarily due to preservational facies differences. Precipitation was in the range of 2500–2750 mm/yr, and dry season precipitation was substantial. Evidence of seasonality, in response to either rainfall or temperature, is evident from growth rings in wood from the Clarno Nut Beds Flora (Wheeler et al., 1996).

A dramatic decline in pollen diversity and an increase in species turnover between >43 Ma and 42 Ma floras of the Gulf Coast and San Diego sequence has been referred to by Frederiksen (1991) as the

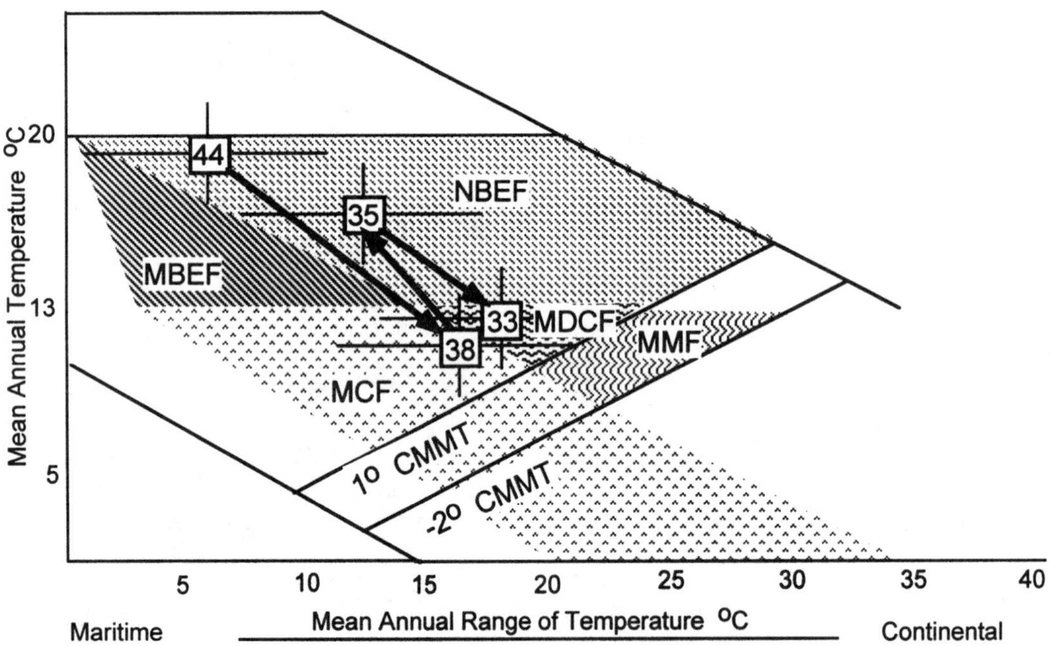

Figure 10.2. MAT/MART nomograph showing climate/vegetation trend of middle Eocene–earliest Oligocene paleofloras of the interior West Coast, with respect to major vegetational thresholds. MAT estimates from lacustrine assemblages include a +3°C facies correction. Floras are identified by number from table 10.1. Nomograph showing the mean annual temperature and mean annual range of temperature of major Southeast Asian forest communities is adapted from Wolfe (1979). Vegetational formations are identified as follows (from Wolfe, 1979): NBEF, Notophyllous Broadleaved Evergreen Forest; MBEF, Microphyllous Broadleaved Evergreen Forest; MDCF, Mixed Deciduous and Coniferous Forest; MCF, Mixed Coniferous Forest; MMF, Mixed Mesophytic Forest. The Palm Line is from Wing and Greenwood (1993). Floras are identified with numbers from table 10.1. Climate estimates include error bars from Wing and Greenwood (1993).

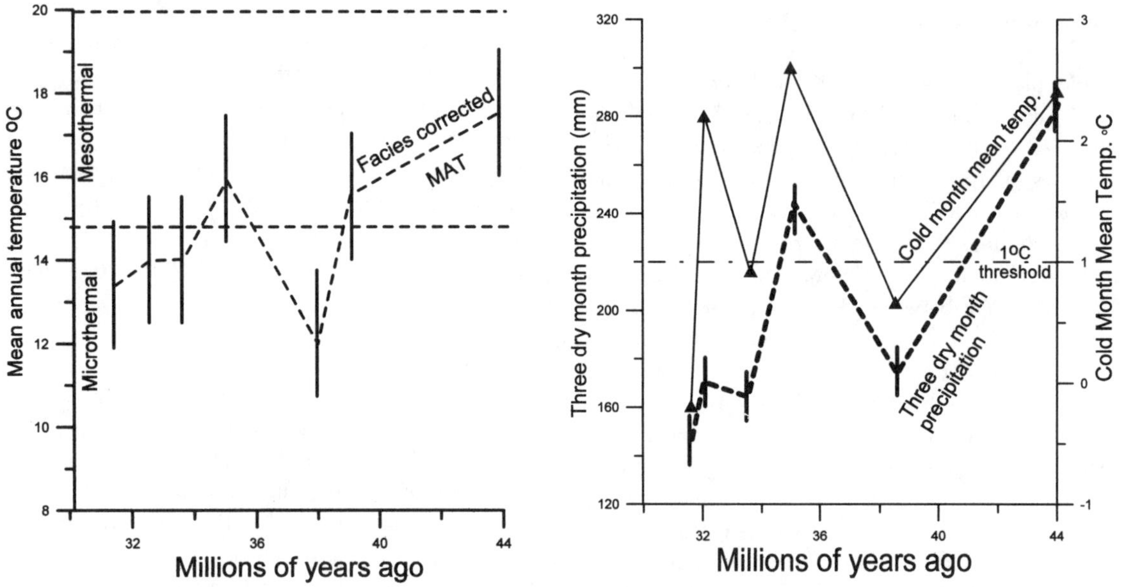

Figure 10.3. (*left*) MAT estimates from floras of the interior West Coast identified in table 10.1. The lower curve shows minimum MAT estimates from lacustrine assemblages. The "Facies Corrected" curve is an approximation of facies corrected MAT (Myers, 1998). 13°C identifies the boundary between mesothermal and microthermal climate (Wolfe, 1979). Error bars are from Wing and Greenwood (1993).

(*right*) Cold month mean temperature and three driest month precipitation estimates for interior Pacific Northwest lacustrine assemblages from table 10.1. Error bars are not indicated for CMMT estimates because they exceed the scale of the plot. CMMT and TDM estimates were not calculated by Smith et al. (1998) for the Kings Gap Flora. The 1°C CMMT threshold is the point below which most evergreen dicots are unable to grow.

"Middle Eocene Diversity Decline". The MEDD event marked a significant increase in seasonal drought, and a transition from humid paratropical forest to seasonally dry savanna vegetation (Frederiksen, 1991). Low-diversity ~42 Ma assemblages from the Scripps and Mission Valley formations, San Diego (Myers, 1999), have not been examined using CLAMP; however, leaf-margin analysis indicates a MAT of ~20°C, which is not significantly different from the Bridgerian Torrey assemblage. Although lineages typical of the dry tropics (Burseraceae, Sapindaceae, Anacardiaceae, etc.) are present in the assemblages, the composition and physiognomy of the leaf floras does not indicate pronounced seasonal drought. Approximately coeval late middle Eocene assemblages from the Puget Group indicate humid conditions with no significant dry season. In the interior, a marked paleosol transition in the Clarno Formation bracketed by $^{40}Ar/^{39}Ar$ ages of 43.75 Ma (Clarno Nut Beds) and 39.17 Ma (basal John Day Fm. Ignimbrite A) may mark a significant rainfall decline in the interior Pacific Northwest (Retallack et al., 2000).

The Middle-Late Eocene Event

The earliest of two major late Eocene MAT declines documented the Pacific Northwest took place near 38.5 Ma, based on comparison of the floodplain ~39 Ma Kings Gap Flora with the lacustrine ~38.5 Ma Whitecap Knoll, both from the lowermost John Day Formation in central Oregon (Smith et al., 1998; Manchester, 2000). A leaf-margin analysis MAT estimate of 15.5°C from the Kings Gap Flora is close to estimates from the Clarno Nut Beds and equivalent assemblages. Estimates for the ~38.5 Ma Whitecap Knoll Flora are ~9°C, indicating a decline in MAT of ~6.5°C over less than 1 million years (figure 10.3, left). Seasonality and precipitation data have not been calculated. Habitat and taphonomic biases cause physiognomic estimates from lacustrine assemblages to read "cool" by at least 3°C (Myers, 1998), and the "facies corrected" MAT for the Whitecap Knoll Flora would be closer to ~12°C (figure 10.3, left). This would place the Whitecap Knoll Flora within the MAT range of modern Mixed Mesophytic Forest (figure 10.2), which closely agrees with vegetational affinities interpreted from species found in the flora (Manchester, 2000). Comparison of MAT estimated from the Kings Gap Flora with corrected estimates from the Whitecap Knoll Flora indicates a decline in MAT of ~3.5°C between ~39 Ma and ~38.5 Ma. The magnitude of the climatic shift is comparable to estimates from coeval sea level Upper Ravenian floras of the Puget Lowlands (Wolfe, 1972, 1994).

The marine oxygen isotopic record indicates a major cooling step at ~37 Ma reflecting an approximately 4°C decline in global MAT (Miller, 1992), which may correspond to the 38.5 Ma paleobotanical event in the Pacific Northwest. Major extinction events near 38 Ma are also recorded by the shallow marine molluscan fauna (Nesbitt, this volume), and by continental vertebrate faunas of upper Duchesnean age (Janis, 1993; Prothero, 1994). Wolfe (1972, 1992) reported a severe decline in MAT during the Upper Ravenian paleobotanical stage from floras in Washington State and Alaska, and suggested that lowland assemblages may have experienced as much as a ~10°C decline in MAT between ~42 Ma and 40 Ma. Given the limited geochronological control of the assemblages studied, it is likely that the event cited by Wolfe is the same cooling event recognized in the interior Pacific Northwest.

In the interior of the Pacific Northwest, the ~38.5 Ma cooling event pushed cold-month temperatures below the threshold of tolerance of almost the majority of thermophylic evergreen taxa (figure 10.3, right), and may have produced significant seasonal drought. The few known late Eocene assemblages from the Pacific Northwest interior lack the diverse subtropical taxa represented in the Clarno Flora and other middle Eocene assemblages, but invariably include a major (although not diverse) component of cool temperate lineages, which were either already present in the regional vegetation of the middle Eocene or were new entries from higher elevations or northern sites. Hence, regional extinctions in response to climatic cooling produced a species-depauperate flora comprised of relict thermophylic plants, along with a limited diversity of temperate deciduous plants. At sea level, Upper Ravenian assemblages from the Puget Group retain most subtropical families, but also include temperate lineages not found in older coastal assemblages (Wolfe, 1971).

Because few known West Coast paleofloras are known from this period, little emphasis has been placed on the ~38.5 Ma cooling event. Preliminary information from floras near the base of the John Day Formation indicate that ~38.5 Ma cooling was of a magnitude approaching that of the Boundary Event, and that climatic cooling produced major vegetational and floristic turnover in the interior Pacific Northwest.

The Latest Eocene Warm Period

Wolfe (1972, 1977, 1994, 1995) has repeatedly emphasized that Kummerian and Goshen-type floras in the Pacific Northwest record a return to near tropical conditions during latest Eocene time; however, the rate and magnitude of late Eocene continental climate change on the West Coast between ~38.5 Ma and 35 Ma is difficult to evaluate because few fossil floras of that age are known. Kummerian paleofloras of the uppermost Puget Group are no younger than ~37 Ma, based on estimates of the maximum age of the stratigraphically higher Lincoln Creek Formation (Nesbitt, this volume). Hence, the ~38.5 Ma cool interval appears to have had a duration of less than 1 million years. Kummerian assemblages from the Puget Group produce MAT estimates of ~21°C, indicating an approximately 6°C increase in MAT relative to slightly older Upper Ravenian assemblages within the same sequence (Wolfe, 1995). At low elevation, warming resulted in a return to warm subtropical conditions near sea level, which permitted the reestablishment of diverse subtropical evergreen vegetation seen in the latest Eocene Goshen, and coeval assemblages.

Floras of ~38.5–35 Ma in age are unknown from central Oregon; however, the Badger's Nose and Steamboat Canyon assemblages of the Cedarville Flora (northeastern California) are Goshen-type assemblages, approximately 35 Ma in age. The lacustrine Badger's Nose assemblage yields a facies-corrected MAT estimate of ~15.5°C which is approximately comparable to an estimate ~17.1°C for the coeval Steamboat Canyon floodplain assemblage (table 10.1). Subtropical elements such as Magnoliaceae, Annonaceae, Lauraceae, and others are present in the flora, but they are of much lower diversity than in the middle Eocene Clarno assemblages, and may represent thermophylic lineages that migrated back into the interior following the late Eocene cool interval. After latitude correction, comparison of Cedarville assemblages with the ~38.5 Ma Whitecap Knoll Flora (~12°C corrected MAT) indicates an approximately 3.5°C MAT increase between ~38.5 Ma and ~35 Ma (figure 10.3, left), although possible elevation differences between the two localities have not been considered here.

An important aspect of late Eocene Kummerian and Goshen-type floras is the presence of thermophylic groups with humid warm temperate modern distributions, including *Meliosma* and *Staphylea*. Physiognomic analysis of Goshen-type assemblages yields mean annual precipitation estimates in excess of ~2,300 mm/year for the Cedarville Badger's Nose assemblage, and over 4,000 mm/yr near the coast, with little rainfall seasonality (table 10.1). Deeply weathered paleosols in the lower John Day Formation in the Painted Hills area (Retallack et al., 2000) also record the return of humid conditions to central Oregon. Hence, both coastal and interior Goshen-type assemblages indicate abundant rainfall and the absence of significant seasonal drought during latest Eocene time (figure 10.3, right). It is likely that abundant rainfall as much as increased MAT is responsible for the reexpansion of subtropical vegetation during the latest Eocene Goshen interval.

Vegetational Thresholds and the Boundary Event

The Boundary Event in the continental Pacific Northwest is bracketed by < 34.8 Ma pre-cooling Goshen Flora and the 33.6 Ma earliest Oligocene Dugout Gulch assemblage of the Bridge Creek Flora. The 34.0 Ma (Evanoff et al., 2001) Florissant Flora, Colorado is interpreted to be a pre-cooling assemblage, and would further restrict the timing of the event to a period between 34 Ma and 33.6 Ma in the western U.S. This places the cooling event close to the Eocene/Oligocene boundary at 33.7 Ma, and at least 0.6 million years earlier than estimates by Wolfe (1992). High-resolution marine oxygen isotopic chronology indicates that the marine isotopic record of the Boundary Event spans an interval from ~33.6 Ma to ~33.5 Ma, which is the initiation of the Oi-1a event of Zachos et al.(1996) and Salamy and Zachos (1999). The Oi-1 shift itself spans 400,000 years, although the rate of isotopic change marking the first stage of the event appears to have dramatically increased between 33.55 and 33.5 Ma (Salamy and Zachos, 1999). At least within the limits of systematic error, the continental and marine events appear to be isochronous.

Oligocene assemblages older than ~31.2 Ma have not been recovered from western Oregon, and the immediate short-term impacts of the Boundary Event cannot be determined from a paleofloras within a single stratigraphic sequence. However, later Oligocene near-sea level floras retain a low diversity of subtropical lineages (e.g., Magnoliaceae, Lauraceae, Annonaceae, Icacinaceae, Sabiaceae and the Goshen-type indicator *Fagus oregona*) which formed the most numerically abundant members of the local vegetation. Cool temperate lineages are not diverse in early Oligocene coastal assemblages, either, although some Bridge Creek dominants—including *Platanus condonii*—are found (Meyer and Manchester, 1997). Comparison of the Goshen and Willamette Floras indicates no more than a 6°C decline in MAT, accompanied by moderately depressed cold-month temperature and a significant decline in rainfall (table 10.1; figure 10.3, left). The absence of cold season frost probably accounts for the presence of thermophylic taxa in the post-cooling lowland assemblages, while relatively cool summer temperature and the absence of a pronounced dry season permitted the growth of diverse conifers.

Unusual low diversity Oligocene paleofloras of the Willamette-type reflect two consequences of the Boundary Event: (1) Cold-month temperature during the Oi-1 event were probably cooler than the 1.8°C CMMT estimate for the post- Oi-1 Willamette Flora. Cold-season temperatures below ~1°C would have extirpated the vast majority of thermophylic taxa in the vegetation. (2) The composition of Oligocene vegetation reflects variable responses of different dicot and conifer lineages to cold-month temperature threshold near 1°C, below which frost-intolerant plants cannot survive (figure 10.3). As cold-month temperature approached ~1°C during the Boundary Event, species diversity declined sharply in the flora, and many thermophylic lineages found in older Goshen-type floras became regionally extinct, while surviving forms became numerical dominants in the regional vegetation. As a result, vegetational change in response to the Boundary Event appears dramatic in the fossil record, although actual climatic cooling in the Pacific Northwest between the latest Eocene and early Oligocene appears to have been less severe than has traditionally been thought.

The Boundary Event in the interior Pacific Northwest appears to have produced a decline in MAT significantly smaller than that of coastal regions. Comparison of the latest Eocene Badgers Nose and early Oligocene Granger Canyon assemblages of the Cedarville Flora indicates no more than a ~3°C decline in MAT (figures 10.2 and 10.3, right).

However, because cold-month temperatures declined well below 1°C, vegetational and floristic turnover across the boundary is dramatic, and nearly all thermophylic taxa became regionally extinct. Post-cooling assemblages are dominated by *Acer, Ulmus, Platanus,* and *Pinus,* and lack evergreen broad-leaved plants found in late Eocene Cedarville assemblages.

Although deciduous angiosperms with cool temperate affinities dominated early Oligocene vegetation in central Oregon, the Bridge Creek Flora retain a few thermophylic taxa, including the conifer *Keteleeria,* and evergreen broad-leaved angiosperms *Terminalia,* and *Paleophytocrene* (Meyer and Manchester, 1997). Cold-month estimates for the Bridge Creek Flora are near or above the 1°C threshold, and some thermophylic plants may have survived the Boundary Event or migrated back into the regional vegetation following the event (figure 10.3, right). Comparison of the 33.6 Ma Dugout Gulch assemblage of the Bridge Creek Flora with the younger 32.6 Ma Fossil High School assemblage suggests a post-Boundary Event rebound of ~1.8°C in MAT, with a concomitant increase in annual and dry season rainfall. Marine oxygen isotopic data clearly document that after temperature rebounded significantly near 33 Ma (Salamy and Zachos et al., 1999).

SUMMARY

Previous assessments of Eocene and Oligocene vegetational and climate change relied on imprecisely dated paleofloras from a variety of different preservational facies. Recent treatment of new and previously known latest middle Eocene to early Oligocene paleofloras from the coastal and interior Pacific Northwest, combined with improved $^{40}Ar/^{39}Ar$ geochronology and magnetostratigraphic correlation produces a more precise record of climate and vegetation during the Eocene-Oligocene transition. The leaf physiognomic analysis of well-dated paleofloras from comparable preservational facies within continuous stratigraphic sequences indicate that previous interpretations of Eocene-Oligocene transitional climate and vegetation (Wolfe, 1971, 1978, 1992, 1994, 1995) are correct, although new evidence indicates that:

1. The 40 Ma "Upper Ravenian" cooling event described by Wolfe (1971, 1972, 1994) actually took place near ~38.5 Ma and correlates with major extinction events among land mammals and shallow marine invertebrates. The event produced as much as a ~3.5°C decline in MAT, and resulted in the regional replacement of evergreen broad-leaved subtropical vegetation in the interior Pacific Northwest by predominantly deciduous forest.

2. An ~3.5°C increase in MAT took place before ~37 Ma and resulted in the return of mesothermal climate to the interior Pacific Northwest. The resulting vegetation, which is reflected the Cedarville Flora of the interior Pacific Northwest, and by the coastal Goshen Flora, defines Wolfe's (1968) Goshen-type paleobotanical stage. The stratigraphically narrow occurrence of Goshen-type indicator taxa is a useful latest Eocene regional datum.

3. Comparison of floras from similar preservational facies indicates that, at least in the Pacific Northwest, the Boundary Event produced a far lower magnitude of cooling than suggested by previous studies (e.g. Wolfe, 1992). In the interior Pacific Northwest, the Boundary Event may have produced no more than a ~2-3°C decline in MAT concomitant with dramatically increased rainfall seasonality and the reduction in cold-month mean temperature below the critical 1°C threshold throughout much of the interior Pacific Northwest. The result was the coastward and elevationally downward migration of diverse cool temperate lineages comprising Bridge Creek vegetation (Wolfe's Angoonian stage). Coastal vegetation remained mesothermal, and was comprised of a mix of relict subtropical thermophylic taxa and Bridge Creek species, along with a major coniferous component that suggests mesic and equable coastal climatic conditions.

ACKNOWLEDGMENTS

This manuscript benefited from thoughtful reviews by Steven R. Manchester and Gregory J. Retallack and by the editors of this volume. This research includes results from NSF Graduate Fellowship DEB 9520826 to Richard V. Fisher and Jeffrey A. Myers. Additional support from the Geological Society of America, Sigma Xi, and the Department of Geological Sciences, UC Santa Barbara, is gratefully acknowledged.

REFERENCES

Alroy, J., Koch, P. L., and Zachos, J. C., Global climate change and North American mammalian evolution: Paleobiology, v. 26, p. 259-288.

Axelrod, D. I., 1966, The Eocene Copper Basin Flora of northeastern Nevada: University of California Publication in Geological Sciences, v. 59 p. 1-83.

Axelrod, D. I., 1998, The Eocene Thunder Mountain Flora of central Idaho: University of California

Publication in Geological Sciences, v. 142, p. 1-61.

Berggren, W. A., Kent, D. V, Swisher, C. C., and Aubry, M.-P., 1995, A revised Cenozoic geochronology and chronostratigraphy,. *in* Berggren, W.A. et al. eds., Geochronology, time scales, and global stratigraphic correlation: SEPM, Special Publication, v. 54, p. 129-212.

Bestland, E. A., Retallack, G. J., and Fremd, T., 1994a, Geology and Paleoenvironments of the late Eocene Clarno Unit, John Day Fossil Beds National Monument, central Oregon: Unpubl. Report, John Day Fossil Beds National Monument, p. 1-160.

Bestland, E. A., Retallack, G. J., and Fremd, T. J., 1994b, Geology and Paleoenvironments of the Painted Hills Unit John Day Fossil Beds National Monument, central Oregon: Unpubl. Report, John Day Fossil Beds National Monument, p. 1-211.

Brandon, M. T., and Vance, J. A., 1992, Tectonic evolution of the Cenozoic Olympic subduction complex, Washington state, as deduced from fission track ages for detrital zircons: American Journal of Science, v. 292, p. 565-636.

Brown, R. W., 1950, An Oligocene evergreen cherry from Oregon: Journal of the Washington Academy of Science, v. 40, p. 321-324.

Brown, R. W., 1959, A bat and some plants from the upper Oligocene of Oregon: Journal of Paleontology, v. 33, p. 125-129.

Burnham, R. J., 1994, Paleoecological and floristic heterogeneity in the plant-fossil record—an analysis based on the Eocene of Washington: U.S. Geological Survey Bulletin, v. 2085B, p. 1-36.

Burnham, R. J. Pitman, N. C. A., Johnson, K. R., and Wilf, P., 2001, Habitat-related error in estimating temperatures from leaf margins in a humid tropical forest: American Journal of Botany, v. 88, p. 1096-1102.

Chaney, R. W. and Sanborn, E. I., 1933, The Goshen Flora of central Oregon: Carnegie Institute of Washington Publications, v. 439 p. 1-237.

Dalrymple, G. B., 1979, Critical tables for conversion of K-Ar ages from old to new constants: Geology v. 7, p. 558-560.

Duffield, W. A., and McKee, E. H., 1986, Geochronology, structure, and basin-range tectonism of the Warner Range, northeastern California: Geological Society of America Bulletin, v. 97, p. 213-221.

Evanoff, E., McIntosh, W. C., and Murphey, P. C., 2001. Stratigraphic summary and $^{40}Ar/^{39}Ar$ geochronology of the Florissant Formation, Colorado, *in* Evanoff, E., Gregory-Wodzicki, K. M., and Johnson K. R., eds.. Fossil flora and stratigraphy of the Florissant Formation, Colorado: Proceedings of the Denver Museum of Nature and Science, series 4, no. 1, p. 1-16.

Frederiksen, N. O., 1991, Pulses of middle Eocene to earliest Oligocene climatic deterioration in southern California and the Gulf Coast: Palaios, v. 6, p. 564-571.

Greenwood, D. R., 1992, Taphonomic constraints on foliar physiognomic interpretations of Late Cretaceous and Tertiary paleoclimates: Reviews of Palaeobotany and Palynology, v. 71, p. 149-190.

Hergert, H. L., 1954, Tertiary floras of Western Oregon and their geological significance: [Unpubl. Ph.D. dissert.]: Corvallis, Oregon State University, p. 1-87.

Hickman, C. J., 1969, The Oligocene marine molluscan fauna of the Eugene Formation in Oregon: Bulletin of the Oregon Museum of Natural History, v. 16, p. 1-112.

Janis, C. M., 1993, Tertiary mammal evolution in the context of changing climates, vegetation, and tectonic events, Annual Review of Ecology and Systematics, v. 24, p. 467-500.

Kester, P. R., 2001, Paleoclimatic interpretation from the early Oligocene Willamette flora, Eugene, Oregon [unpubl. M.S. Thesis]: Seattle, University of Washington, 62 p.

Klucking, E., 1964, An Oligocene flora from the Western Cascades, [unpubl. Ph.D dissert.]: Berkeley, University of California, p. 1-241 .

Lakhanpal, R. N., 1958, The Rujada Flora of west-central Oregon: University of California Publication in Geological Sciences, v. 35, p. 1-66.

MacGinitie, H. D., 1941, A middle Eocene flora from the central Sierra Nevada: Carnegie Institute of Washington Publication, v. 534, p. 1-178 .

MacGinitie, H. D., 1953, Fossil plants of the Florissant beds, Colorado: Carnegie Institute of Washington Publication, v. 599, p. 1-198.

Manchester, S. R., 1990, Eocene to Oligocene floristic changes recorder in the Clarno and John Day formations, Oregon, U.S.A., *in* Knobloch, E., and Kvacek, Z., eds., Symposium proceedings, paleofloristic and paleoclimatic changes in the

Cretaceous and Tertiary: Prague, Geological Survey Press, p. 183-187.

Manchester, S. R., 1994, Fruits and seeds of the middle Eocene Nut Beds Flora, Clarno Formation, Oregon: Palaeontographica Americana, v. 58, p. 1-205.

Manchester, S. R., 2000, Late Eocene fossil plants of the John Day Formation, Wheeler County, Oregon: Oregon Geology, v. 62, p. 51-63.

Meyer, H. W., 1973, The Oligocene Lyons Flora of northwestern Oregon: Ore Bin, v. 35, p. 37-51.

Meyer, H. W., and Manchester, S. R., 1997, Revision of the Oligocene Bridge Creek floras of Oregon: University of California Publication in Geological Sciences, v. 141, p. 1-195.

Miller, K. G., 1992, Middle Eocene to Oligocene stable isotopes, climate, and deep-water history: The terminal Eocene event? *in* Prothero, D. R., and Berggren, W. A., eds., Eocene-Oligocene climatic and biotic evolution: Princeton, New Jersey, Princeton University Press, p. 160-177.

Miller, K. G., Fairbanks, R.G., and Mountain, G.S., 1987, Tertiary oxygen isotope synthesis, sea level history, and continental margin erosion: Paleooceanography, v. 2, p. 1-19.

Myers, J. A., 1990, A Bridgerian age flora from Del Mar, California [Unpubl. M.S. thesis]: San Diego State University, p. 1-164.

Myers, J. A., 1991, The early middle Eocene Torrey Flora, Del Mar, *in* Abbott, P. L., and May, J. A., eds., Eocene Geological History, San Diego Region: Pacific Section SEPM, Guidebook, v. 68, p. 201-216.

Myers, J. A., 1998, Paleovegetational heterogeneity and the record of Eocene-Oligocene climate change in the interior Pacific Northwest [Unpubl. Ph.D. dissert.]: Santa Barbara, University of California, p. 1-502.

Myers, J. A., 1999, Middle Eocene paleofloras, San Diego: A West Coast sea level datum: Geological Society of America Abstracts with Program, v. 31, p. A124.

Myers, J. A., and Fisher, R. V., 1994, Explosive volcanism as a regulator of middle Tertiary floral and vegetational change: Western North America: Geological Society of America Abstracts with Program, v. 26, p. A521.

Myers, J. A., and Greenwood, D. R., 1997, Landscape-scale vegetational heterogeneity within a late Eocene intra-arc basin: Cedarville Flora NE California, U.S.A: Geological Society of America Abstracts with Program, v. 29, p. 431.

Myers, J. A., Kester, P. R., and Retallack, G. J. 2002, Paleobotanical record of Eocene-Oligocene climate and vegetational change near Eugene, Oregon: Oregon Geology, in press.

Potbury, S. S., 1935, The La Porte Flora of Plumas County, California: Carnegie Institute of Washington Publication, v. 465, p. 29-81.

Prothero, D. R., 1994, The late Eocene-Oligocene extinctions: Annual Reviews of Earth and Planetary Sciences, v. 22, p. 145-165.

Prothero, D. R., and Emry, R. J., 1996,. Summary, *in* Prothero, D.R., and Emry, R.J. eds., The Terrestrial Eocene-Oligocene Transition in North America: Cambridge, Cambridge University Press, p. 664-683.

Prothero, D. R., and Hankins, K. G. 2001, Magnetic stratigraphy and tectonic rotation of the Eocene-Oligocene Keasey Formation, northwest Oregon: Journal of Geophysical Research, B, v. 105, no. 7, p. 16,473-16,480.

Prothero, D. R., and Retallack, G. J., 2001, Magnetic stratigraphy of the Eocene-Oligocene Eugene/Fisher formations, western Oregon: Implications for the "Terminal Eocene Event": North American Paleontological Congress, Berkeley.

Retallack, G. J., Bestland, E. A., and Fremd, T. J., 2000, Eocene and Oligocene paleosols of central Oregon: Geological Society of America Special Paper, v. 344, p. 1-192.

Salamy, K. A., and Zachos, J. C., 1999, Latest Eocene-early Oligocene climatic change and Southern Ocean fertility: inferences from sediment accumulation and stable isotope data: Palaeogeography, Palaeoclimatology, Palaeoecology, v. 145, p. 61-77.

Sanborn, E. I., 1937, The Comstock Flora of west central Oregon,. Carnegie Institute of Washington Publication, v. 465, p. 1-28.

Sanborn E. I., 1947, The Scio Flora of western Oregon: Oregon State College (Corvallis) Studies in Geology, v. 4, p. 1-29.

Smith G. A., Manchester, S. R., Ashwill, M., McIntosh, W. C., and Conrey, R. M., 1998, Late Eocene-early Oligocene tectonism, volcanism, and floristic change near Gray Butte, central Oregon: Geological Society of America Bulletin, v. 110, p. 759-778.

Turner, D. L., Frizell, V. A., Triplehorn, D. M., and Naeser, C. W., 1983,. Radiometric dating of ash partings in coal of the Eocene Puget Group, Washington: Implications for paleobotanical stages: Geology, v. 11, p. 527-531.

Vokes, H. E., and Snavely, P. D., 1948, The age relationships of the Eugene and Fisher Formations, Geological Society of the Oregon Country Newsletter, v. 14, p. 38-41.

Walsh, S. L., Prothero, D. R., and Lundquist, D. J., 1996, Stratigraphy and paleomagnetism of the middle Eocene Friars Formation and Poway Group, southwestern San Diego County, California, *in* Prothero, D. R., and Emry, R. J. eds., The Terrestrial Eocene-Oligocene Transition in North America: Cambridge, Cambridge University Press, p. 120-154

Wheeler, E. A., Wiemann, M., and Manchester, S. R., 1996, Eocene dicotyledonous woods of western North America: International Organization of Paleobotany, 5th Quadrennial Conference, Santa Barbara, California, Abstracts, p. 111.

Wing S. L., and Greenwood, D. R., 1993, Fossils and fossil climate: The case for equable continental interiors in the Eocene: Philosophical Transactions of the Royal Society of London, Series B, v. 341, p. 243-252.

Wolfe, J. A., 1968, Paleogene biostratigraphy of nonmarine rocks in King County, Washington: U.S. Geological Survey Professional Paper, v. 571, p. 1-33.

Wolfe, J. A., 1971, Tertiary climatic fluctuations and methods of analysis of Tertiary floras: Palaeogeography, Palaeoclimatology, Palaeoecology, v. 9, p. 27-57.

Wolfe, J. A., 1972, An interpretation of Alaskan Tertiary floras, *in* Graham, A., ed., Floristics and paleofloristics of Asia and eastern North America: Amsterdam, Elsevier, p. 201-231.

Wolfe, J. A., 1977, Paleogene floras from the Gulf of Alaska region: U.S. Geological Survey Professional Paper, v. 997, p. 1-108 .

Wolfe, J. A., 1978, A paleobotanical interpretation of Tertiary climates in the Northern Hemisphere: American Scientist, v. 66, p. 694-703.

Wolfe, J. A., 1979, Temperature parameters of humid to mesic forests of eastern Asia and relation to forests of other regions of the Northern Hemisphere and Australasia: U.S. Geological Survey Professional Paper, v. 1106, p. 1-37.

Wolfe, J. A., 1981,. A chronologic framework for Cenozoic megafossil floras of northwestern North America and its relation to marine geochronology: Geological Society of America Special Paper, v. 184, p. 39-47.

Wolfe, J. A., 1992, Climatic, floristic, and vegetational changes near the Eocene/Oligocene boundary in North America, *in* Prothero, D. R., and Berggren, W. A., eds., Eocene-Oligocene Climatic and Biotic Evolution: Princeton, New Jersey, Prince-ton Univ. Press, p. 421-436.

Wolfe, J. A., 1993, A method of obtaining climatic parameters from leaf assemblages: U.S. Geological Survey Bulletin, v. 2040, p. 1-71.

Wolfe, J. A., 1994, Tertiary climatic changes at middle latitudes of western North America: Palaeogeography, Palaeoclimatology, Palaeoecology, v. 108, p. 195-205.

Wolfe, J. A. 1995. Paleoclimatic estimates from Tertiary leaf assemblages: Annual Reviews of Earth and Planetary Sciences, v. 23, p. 119-142.

Wolfe, J. A., and Wehr, W., 1987, Middle Eocene dicotyledonous plants from Republic, northeastern Washington: U.S. Geological Survey Bulletin, v. 1597, p. 1-25 .

Wolfe, J. A., Forest, C. E., and Molnar, P., 1998, Paleobotanical evidence of Eocene and Oligocene paleoaltitudes in midlatitude western North America: Geological Society of America Bulletin, v. 110, p. 664-678.

Zachos, J. C., Quinn, T. M., and Salamy, K. A., 1996, High-resolution (10^4 years) deep-sea foraminiferal stable isotope records of the Eocene-Oligocene climate transition: Paleocean-ography, v. 11, p. 251-266.

Part II

THE ATLANTIC, GULF, AND CARIBBEAN

CHAPTER 11

Sequence Stratigraphy of the Eocene/Oligocene Boundary Interval: Southeastern Mississippi

Ronald J. Echols, John M. Armentrout, Stephan A. Root, Larry B. Fearn, James C. Cooke, B. Kirby Rodgers, and Peter R. Thompson

ABSTRACT

Important sections for study of the Eocene/Oligocene boundary in North America are located in the eastern Gulf Coastal Plain within the states of Mississippi and Alabama. Pronounced east-to-west facies change from relatively thin, carbonate-dominated sediment to thick, overwhelmingly siliciclastic sediment complicates stratigraphic analysis of these sections. This multidisciplinary study of three cores and two nearby outcrops integrates data to address the facies complexity. The study (1) identifies the local Eocene/Oligocene boundary; (2) clarifies the origin of specific depositional facies; and (3) provides an internally consistent local sequence stratigraphic framework for latest Oligocene to earliest Eocene strata. Data sets include foraminifers, calcareous nannofossils, marine and terrestrial palynomorphs, palynofacies, paleomagnetostratigraphy, strontium isotopes, and kerogen analysis.

The Eocene/Oligocene boundary, recognized by the last occurrence of in situ *Hantkenina alabamensis*, occurs within the time transgressive lithofacies of the uppermost Shubuta Clay and Red Bluff Clay. This interpretation is supported by both paleomagnetostratigraphic and published dinocyst graphic correlation data.

Two regionally significant condensed section lithofacies occur within the studied interval. A regional maximum flooding surface datum is interpreted within a condensed section located within the middle part of the Shubuta Clay. This condensed section is marked by distinct high gamma ray readings and is accompanied by an interval of fossil abundance, maximum water depth, and dominance of marine kerogen. Within the overlying Red Bluff Clay, an interval of burrowed horizons, shell hash, authigenic minerals and elevated foraminiferal abundance is interpreted as a composite-set of sediment-starved intervals associated with a distal delta foreset setting characterized by sediment bypass.

The youngest Eocene sequence boundary in the eastern Gulf Coastal Plain is located at the base of the Cocoa Sand member of the Yazoo Clay, within nannofossil Zone NP19/20 according to our data. The oldest Oligocene sequence boundary is located within nannofossil Zone NP21 at the base of the Mint Spring Marl Member of the Yazoo Clay. The Eocene/Oligocene boundary is located within the highstand systems tract between these two sequence boundaries.

INTRODUCTION

Some of the most important sections for study of the Eocene/Oligocene boundary in North America are located in the eastern Gulf Coastal Plain within the states of Mississippi and Alabama (MacNeil and Dockery, 1984; Toulmin, 1977). Although these sections have been studied intensively, there has been no consensus about the exact stratigraphic location of the Eocene/Oligocene (E/O) boundary within them or of their sequence stratigraphic interpretation. Marked differences in interpretation are possible because pronounced facies changes complicate the stratigraphy

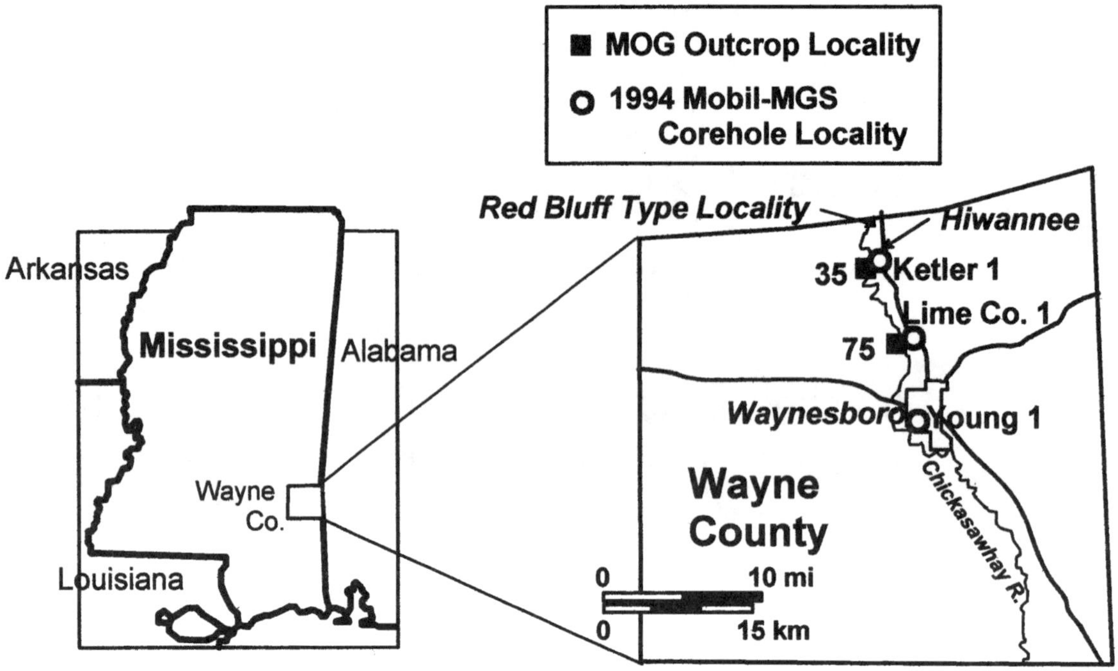

Figure 11.1. Area location map. MOG, Mississippi Office of Geology.

around the E/O boundary along the outcrop belt from southwestern Alabama to western Mississippi (Dockery, 1982, 1990; Mancini, 2000; Tew and Mancini, 1995). From east to west, upper Eocene and lowest Oligocene sediments increase dramatically in thickness and change from carbonate-rich in Alabama to overwhelmingly siliciclastic in central and western Mississippi (Murray, 1961). Stratigraphic sections within the Chickasawhay River Valley in Wayne County, southeastern Mississippi, are located in a pivotal position between these two very different facies successions of southwestern Alabama and western Mississippi (figure 11.1). Understanding these intermediate sections is important to reconciling differences in interpretation of the E/O boundary interval in the eastern Gulf Coastal Plain. Here, we discuss aspects of the chronostratigraphic and sequence stratigraphic setting of the upper Eocene and basal Oligocene based on our study of three core holes and two outcrop sections from southeastern Mississippi (figure 11.1). Our multidisciplinary study gathered data on foraminifers, calcareous nannofossils, marine and terrestrial palynomorphs, palynofacies, paleomagnetostratigraphy, strontium isotopes, and basic organic geochemistry (Total Organic Carbon and Rock-Eval Pyrolysis). This report complements the palynostratigraphic study on the same cores by Jaramillo and Oboh-Ikuenobe (1999).

Stratigraphy

The stratigraphic units considered in this study comprise the upper part of the Jackson Group and the lower part of the Vicksburg Group (Murray, 1961) (figure 11.2). From west to east the Yazoo Clay of the Jackson Group thins markedly from about 181 m (595 feet) thick in west-central Mississippi to 53 m (175 feet) in southeastern Mississippi to only 22 m (72 feet) in southwestern Alabama (Dockery, 1982; Dockery and Siesser, 1984; Mancini et al., 1987). The thin Yazoo Clay of southwestern Alabama is richer in carbonate, glauconite, and phosphate than sections of this formation farther west (Mancini et al., 1987).

In southeastern Mississippi and southwestern Alabama, the Yazoo Clay is divided into four members. In ascending order, these are the North Twistwood Creek Clay, the Cocoa Sand, the Pachuta Marl, and the Shubuta Clay (Murray, 1961) (figure 11.2). The distribution of the Cocoa Sand is very limited. From a thickness 10 to 19 m (33 to 62 feet) near the

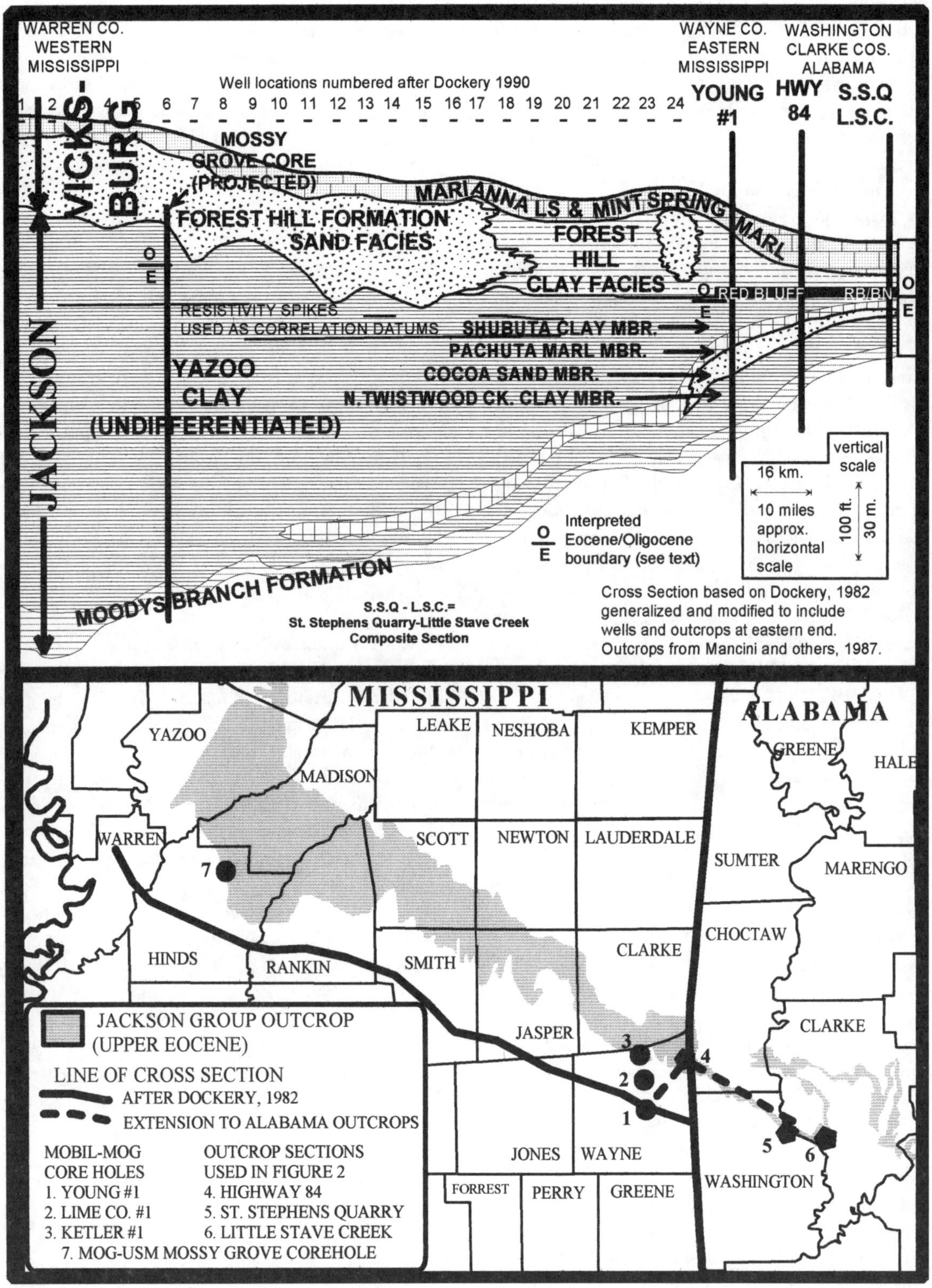

Figure 11.2. Generalized strike-oriented cross section of Jackson and Lower Vicksburg Groups (upper Eocene to basal Oligocene) in the eastern Gulf Coastal Plain.

Mississippi-Alabama state line (May, 1974), it thins and undergoes facies change toward both the northwest and east. In central Mississippi it is replaced by its facies equivalent, the lower part of the Pachuta Marl. In west-central Mississippi none of the members are developed, and there the Yazoo Clay is undifferentiated.

The lower part of the Vicksburg Group includes the Red Bluff Clay, the Bumpnose Limestone, the Forest Hill Formation, the Mint Spring Marl, and the Marianna Limestone (figure 11.2). The Red Bluff Clay is best developed in southeastern Mississippi (Wayne County) where it is 3.5 to 10 m (11 to 32 feet) thick (Dockery, 1982; May, 1974). It consists of ferruginous, glauconitic clay and marl, with abundant fossils. Northwest of Wayne County, it thins and disappears, possibly through facies change to the lower Forest Hill Formation (MacNeil, 1944). Toward the east in southwestern Alabama, the Red Bluff Clay is intertongued with the Bumpnose Limestone, the facies equivalent of the Red Bluff Clay and lower Forest Hill Formation in south-central Alabama and northwest Florida (Huddlestun and Toulmin, 1965; Jaramillo and Oboh-Ikuenobe, 1999). The stratigraphic interval of intertongued clay and limestone is referred to as Red Bluff/Bumpnose (RB/BN) (Mancini et al., 1987).

In central and west-central Mississippi, the Forest Hill Formation overlies the Yazoo Clay, and in southeastern Mississippi and southwestern Alabama, it overlies the Red Bluff Clay and Red Bluff/Bumpnose (figure 11.2). The Forest Hill Formation varies from about 22 to 44 m (72 to 145 feet) from west-central to southeastern Mississippi (Dockery, 1982; May, 1974). It thins abruptly near the border with Alabama and pinches out in southern Clarke County, Alabama (Mancini et al., 1987). The Forest Hill changes facies from silt and sand, with beds of lignite, in west-central and central Mississippi (sand facies) to predominantly clay, with variable amounts of dispersed lignite, in southeastern Mississippi and southwestern Alabama (clay facies) (Dockery, 1982; Dockery et al., 1991; May, 1974). MacNeil (1944), Dockery (1982, 1990), and Tew and Mancini (1995) interpreted the facies changes within the Forest Hill Formation to indicate delta progradation from west-central and central Mississippi toward the south and southeast. Our stratigraphic treatment of the Forest Hill Formation and Red Bluff Clay (or formation) generally follows practices of the Mississippi Office of Geology (Dockery, 1982; Dockery, et al., 1994). However, other workers (e.g., MacNeil, 1944; Pasley and Hazel, 1995) restrict Forest Hill Sand to the silt and sand-rich, largely non-marine facies that predominates in central and western Mississippi and expand Red Bluff Clay to include the clay-rich, marine beds, with dispersed lignite debris, that are here treated as the clay facies of the Forest Hill Formation.

The Mint Spring Marl consists of sandy, clayey, glauconitic fossiliferous marl and some fossiliferous, cross-bedded sand (Dockery, 1982; MacNeil, 1944; May, 1974; Tew and Mancini, 1995). It disconformably overlies the Forest Hill Formation. Thickness is variable, but in general, it thins from less than 7.5 m (25 feet) in west-central Mississippi to 0.6 to 0.9 m (2 to 3 feet) in southeastern Mississippi to 0.5 m (1.5 feet) at St. Stephens Quarry in southwestern Mississippi (MacNeil, 1944). The Mint Spring Marl is considered a formation in Mississippi and the basal member of the Marianna Limestone in Alabama. The Marianna Limestone varies from 12 to 18 m (40 to 60 feet) thick in Alabama and northwest Florida, but thins westward above the Mint Spring Marl to only 2.1 m (7 feet) at Vicksburg in west-central Mississippi (MacNeil, 1944).

Previous Sequence Stratigraphic Studies

In Alabama and Mississippi the Eocene/Oligocene boundary is located within a third order stratigraphic sequence designated TE 3.3 (Baum and Vail, 1987, 1988; Baum et al., 1982, 1994; Mancini and Tew, 1991; Mancini et al., 1987) and, more recently, TAGC 4.3 (Tew and Mancini, 1995). We will use locally derived designation TE 3.3 rather than TA 4.3, the globally derived designation used by Loutit et al. (1988). This will help to avoid prejudging the local sequence interpretation based on the ages assigned to the TA 4.4 sequence boundary in Europe (Haq et al., 1988; Michelsen et al., 1998; Vakarcs et al., 1998) or to sequence boundaries inferred from DSDP oxygen isotope studies (Miller et al., 1993; Baum et al., 1994; Abreu and Haddad, 1998). Jiang (1997) differs from these authors in that he correlates local sequence TE 3.3 and 3.2, and part of 3.1, to sequence TA4.4 of Haq et al. (1988).

The stratigraphers that published the now classic studies on the sequence stratigraphy of TE 3.3 in the 1980s generally agreed in their stratigraphic interpretations (Baum et al., 1982; Baum and Vail, 1987, 1988; Mancini et al., 1987; Loutit et al., 1988). This early consensus is summarized as follows (figure 11.3A): The sequence boundary is located at the base of the Cocoa Sand Member of the Yazoo Clay, a shelf-margin systems tract (Tew and Mancini, 1995). The

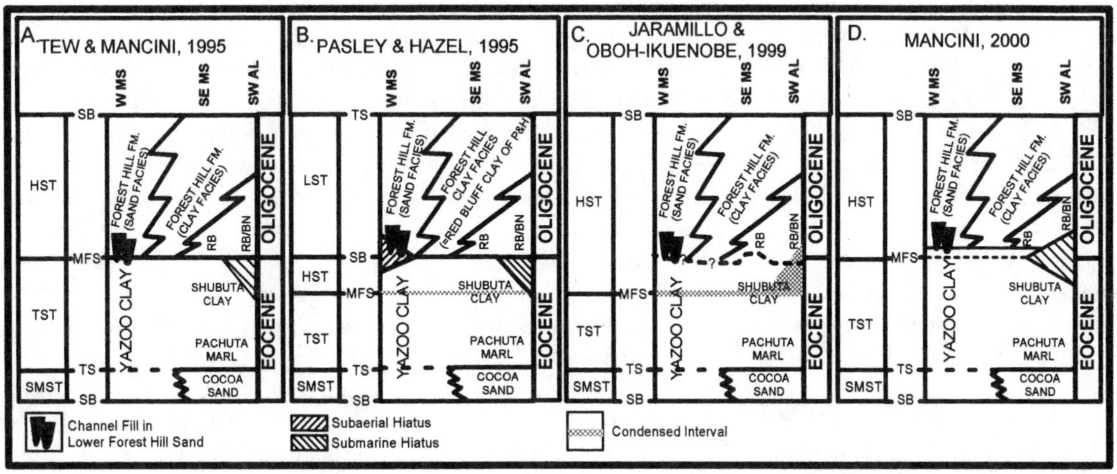

Figure 11.3. Three models proposed to reconcile observations appearing to indicate that the Eocene/Oligocene boundary is (or approximates) a maximum flooding surface in southeastern Alabama (SE AL), but a sequence boundary in western Mississippi (W MS). The southeastern Mississippi (SE MS) coreholes and outcrops provide an intermediate link between these contrasting stratigraphic successions. RB/BN, intertonguing of Red Bluff Clay and Bumpnose Limestone in southwestern Alabama; SMST, Shelf Margin Systems Tract; TST, Transgressive Systems Tract; HST, Highstand Systems Tract; SB, Sequence Boundary; TS, Transgressive Surface; MFS, Maximum Flooding Surface.

transgressive systems tract is made up of the Pachuta Marl and Shubuta Clay Members of the Yazoo Clay. The highstand systems tract consists of the Forest Hill delta deposits, and shelf sediments that correlate to the lower Forest Hill, the Red Bluff Clay and Bumpnose Limestone. In their seminal paper on condensed sections associated with maximum flooding surfaces, Loutit et al. (1988) selected the TE 3.3 condensed section at the St. Stephens Quarry section as one of their type examples. They interpreted the condensed section there to include the upper part of the Shubuta Clay and the lower part of the Red Bluff/Bumpnose.

Two reports challenged the prevailing consensus that the contact between the Shubuta Clay and Red Bluff Clay or Red Bluff/Bumpnose is a maximum flooding surface. Dockery (1990) correlated the base of the Red Bluff in southeastern Mississippi to the base of the sand facies of the Forest Hill Formation in western and central Mississippi using well log correlation and molluscan biostratigraphy (MacNeil, 1944; Dockery, 1982; Dockery and Siesser, 1984). He showed that lignite and sand of the basal Forest Hill in western and central Mississippi fill channels eroded into the Yazoo Clay (figure 11.2). He interpreted a sequence boundary at this eroded surface in western Mississippi and he pointed out features of the Red Bluff Clay in southeastern Mississippi that suggested a down shift in facies indicative of a sequence boundary there also. Miller et al. (1993) concluded that the contact between the Shubuta Clay Member and the Bumpnose Limestone should be a sequence boundary merged with a transgressive surface, because an important increase in global ice volume, indicated by the oxygen isotope record, caused a fall in sea level at the time represented by this contact.

Subsequent to the challenges by Dockery (1990) and Miller et al. (1993), four publications have proposed models to reconcile the interpreted maximum flooding surface within the Shubuta to Red Bluff/Bumpnose transition to the interpreted sequence boundary at the base of the western sand facies of the Forest Hill Formation (figure 11.3A-D).

As a part of thorough regional analysis of the sequence stratigraphy of TE 3.3 (their TAGC 4.3), Tew and Mancini (1995) reconfirmed the now "traditional" interpretation that the Red Bluff/Bumpnose contact with the Shubuta is a maximum flooding surface (figure 11.3A). They discounted the significance of the unconformable surface at the base of the Forest Hill Formation, explaining it as distributary channels eroded below sea level into underlying marine deposits during highstand.

Pasley and Hazel (1995) interpreted the surface at the base of the Forest Hill, Red Bluff, and Bumpnose as a sequence boundary, and they interpreted these lithostratigraphic units as facies of a lowstand systems tract of the oldest Oligocene sequence (figure 11.3B).

They interpreted the maximum flooding surface to occur within the Shubuta Clay at one of the several correlated resistivity peaks on wire-line logs (Dockery, 1982) (also see figure 11.2). In their interpretation, the upper part of the Shubuta Clay is the complete highstand of the youngest Eocene sequence. However, this upper Shubuta highstand is missing at St. Stephens Quarry because of a submarine hiatus, according to their graphic correlation. Therefore, the condensed section at the maximum flooding surface within the Shubuta is juxtaposed with the sequence boundary at the base of the Red Bluff/Bumpnose (figure 11.3B).

Like Pasley and Hazel (1995), Jaramillo and Oboh-Ikuenobe (1999) interpret the maximum flooding surface within the Shubuta Clay (figure 11.3C). However, unlike Pasley and Hazel (1995), Dockery (1990), and Miller et al. (1993), they find no evidence for a sequence boundary at the base of the Red Bluff Clay and Bumpnose Limestone. They quote the views of Tew and Mancini (1995) in dismissing evidence for a sequence boundary at the base of the sand facies of the Forest Hill Formation in western Mississippi. Their graphic correlation of dinocyst range data indicates to them that the base of the Red Bluff is diachronous. They believe it is an intra-highstand event of little significance for sequence analysis (figure 11.3C). As in the "traditional" model (figure 11.3A), they interpret the TE 3.3 highstand to consist of the Forest Hill, Red Bluff, and Bumpnose formations, and to these they add the upper Shubuta Clay. They interpret the top of the Shubuta Clay at St. Stephens Quarry as early highstand, highly condensed, rather than submarine hiatus (Pasley and Hazel, 1995). As they pointed out, their sequence analysis of the St. Stephens Quarry section is very much like the traditional interpretation summarized above. However, their analysis of the southeastern Mississippi sections is quite different (figure 11.3C).

The most recent synthesis (Mancini, 2000) proposes that the TE 3.3 maximum flooding surface occurs at approximately the same level as the E/O boundary (figure 11.3D). It integrates recently published information that the E/O boundary occurs within the uppermost Yazoo Clay, 16.5 m (54 feet) below the base of the Forest Hill Formation, in the Mossy Grove core, west-central Mississippi (Fluegeman, 1996) (see figure 11.2). In this model the stratigraphic horizon that is both E/O boundary and maximum flooding surface is located within the upper Yazoo in west-central Mississippi (Fluegeman, 1996), but it converges with the Yazoo-Red Bluff boundary to the southeast because highstand progradation (Forest Hill delta and uppermost Yazoo) occurred in that direction. Southeastward progradation caused the period of maximum sediment starvation associated with the maximum flooding surface to end progressively later to the southeast. This synthesis locates the base Forest Hill erosion surface in west-central and central Mississippi stratigraphically above earliest Oligocene marine highstand sediments that exhibit evidence of shallowing and coarsening upward in the Mossy Grove core (Fluegeman, 1996).

As these four interpretations emphasize, resolution of the sequence stratigraphy of the Eocene/Oligocene boundary interval in the eastern Gulf Coastal Plain involves a separate, although related, set of problems for western and central Mississippi, for southeastern Mississippi, and for southwestern Alabama. This paper aims to contribute to a resolution of issues for southeastern Mississippi. It has three principal objectives. The first is documentation of the local chronostratigraphy of the Eocene/Oligocene boundary interval. The second is bringing a diverse set of data and criteria to bear on the question of the stratigraphic location of the maximum flooding surface and the distinction of transgressive from highstand systems tracts in the southeastern Mississippi coreholes and sections. The third is proposition of a sedimentological model for the Red Bluff Clay taking into account the features that have led some workers to interpret the base of it as a downlap surface (Baum and Vail, 1988), others to interpret the base of it as a sequence boundary (Dockery, 1990), and Jaramillo and Oboh-Ikuenobe (1999) to dismiss it as a minor feature within the highstand.

Methods

Coring operations and lithostratigraphy of the three coreholes are described in Dockery et al. (1994). Each was logged for gamma ray (GR), spontaneous potential (SP), and resistivity (16 in. and 64 in.). The most downdip of the stratigraphic sections, the Young #1 corehole, has a complete TE 3.3 section and the Lime Company #1 and Ketler #1 coreholes composite to a complete section of TE 3.3 (figure 11.4). The stratigraphic overlap of the Lime Company #1 and Ketler #1 cores includes the Shubuta–Red Bluff Clay contact, identified as key horizon during the design of this study. Outcrop sections, previously measured and described in part in Dockery (1980), were photographed, described, and sampled by Mobil person-

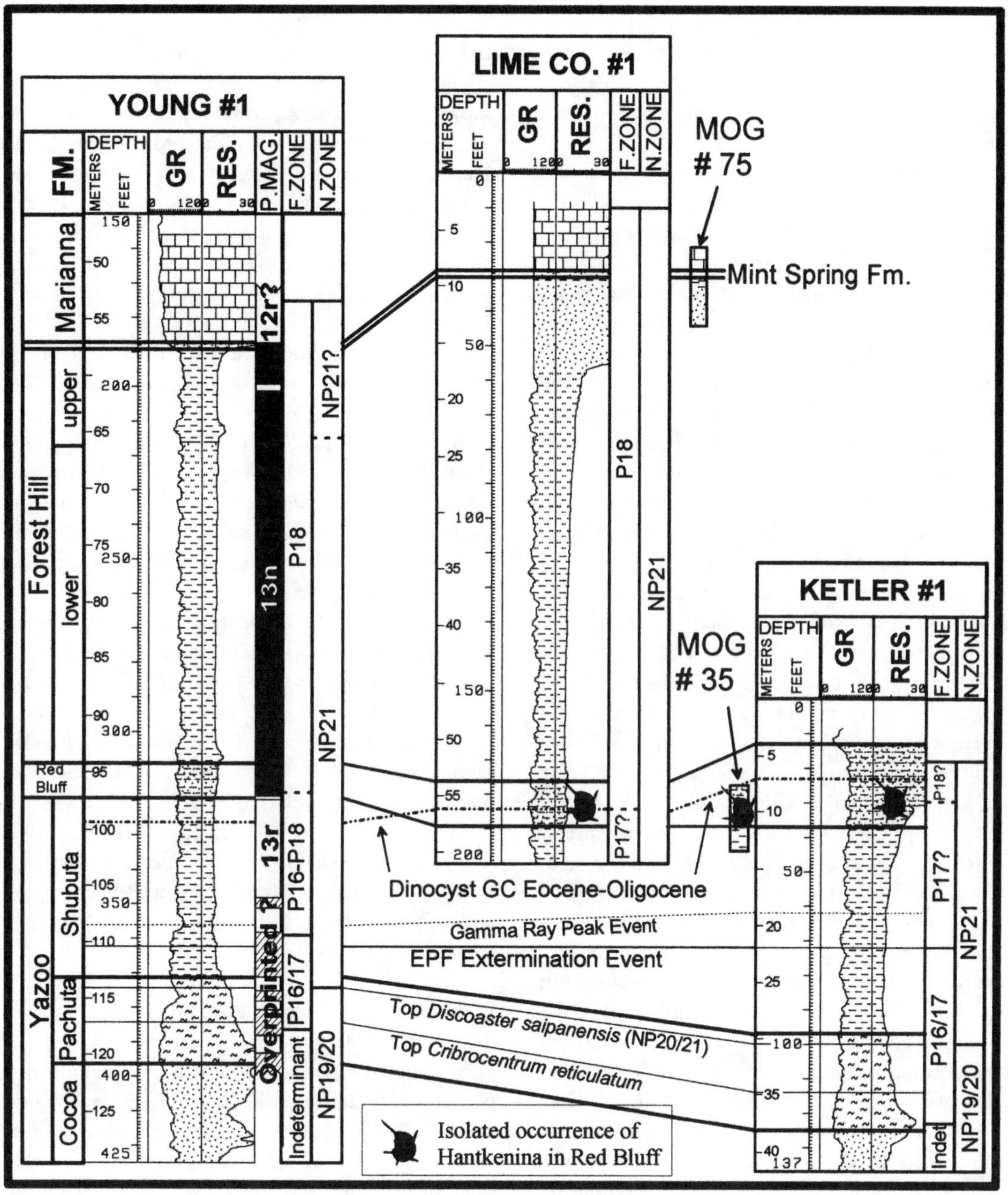

Figure 11.4. Correlation, biostratigraphic zonation, and magnetostratigraphy of three coreholes and two outcrops. Eocene planktonic foraminifers disappear up section at the Eocene Planktonic Foraminifer (EPF) Decimation Horizon except for isolated occurrences of *Hantkenina alabamensis* in the Red Bluff Clay. The Red Bluff *Hantkenina* occurrences in Lime Company #1 and Ketler #1 are stratigraphically below the dinocyst graphic correlation (GC) of the Eocene/Oligocene boundary by Jaramillo and Oboh-Ikuenobe (1999). Outcrop Locality MOG #35 was not included in the dinocyst graphic correlation study.

nel in 1993 (S. A. Root, J. M. Armentrout, D. Guido). Outcrops were logged with a hand-held gamma ray scintillometer (see Jordan et al., 1990). Section MOG #35 also includes the contact between Shubuta and Red Bluff Clays (MacNeil and Dockery, 1984).

Samples for biostratigraphic analyses were taken a minimum of 1.52 m (5 feet) apart in the core holes and 0.3 m (1 foot) apart in the measured sections. Each biostratigraphic sample was analyzed for foraminifers, calcareous nannofossils, and palynomorphs. Foraminiferal samples of estimated equal volume, about 35–40 cm^3, were disaggregated and washed through a sieve with 74 μm openings (200 mesh) to remove silt and clay. The processed samples were separated into size fractions that were then spread onto a tray marked with a surface grid for counting of foraminiferal taxa and estimating abundance of glauconite, macrofossil shell debris, and lignite clasts. Foraminiferal population statistics are based on sums of total or proportional partial counts of taxa on each sieve of 74 μm mesh size and larger. When necessary, a microsplitter was used to subdivide very large samples prior to foraminiferal analysis. However, the entire sample was always examined for rare, chronostratigraphically significant planktonic foraminifers. Nannofossil slides were prepared using a standard pipette strew slide technique (Bown and Young, 1998). The counts are based on one traverse of a 4-cm cover-slip. The total number of abundant species is calculated with a multiplication factor based on the number of fields of view. The entire nannofossil slide was examined for marker species. Standard palynological processing procedures on samples of approximately equal volume (10 cm^3) were employed, and both stained and unstained, mildly oxidized palynological residues (15–200 mm fraction) were examined. A fixed count method (200 total palynomorphs) was used and represents a compromise between generating statistically meaningful results and expediency (Traverse, 1988). Palynofacies analyses were performed on the unoxidized, <200 mm fraction kerogen residue. The biostratigraphic analyses were made by Edelman, Percival, and Associates (palynomorphs: D. W. Edelman and R. Stapleton; foraminifers: R.T. Guerra), by F. Oboh-Ikuenobe (palynofacies), by M .M. Jiang (calcareous nannofossils), and by Mobil (calcareous nannofossils of Young #1: S. A. Root). Core Lab, Inc., Plano, Texas, made the Rock-Eval pyrolysis and total organic carbon analyses.

BIOSTRATIGRAPHY AND CHRONOSTRATIGRAPHY

Calcareous Nannofossil Zonation

Highest occurrences of *Discoaster saipanensis, D. barbadiensis,* and *Cribrocentrum reticulatum,* associated with *Isthmolithus recurvus,* are recorded in the Pachuta Marl member of the Yazoo Clay in Young #1 and Ketler #1 (figure 11.5; table 11.1), thereby identifying the upper part of upper Eocene calcareous nannofossil Zone NP19/20 (figure 11.4). Rare, sporadic occurrences of *I. recurvus* in the underlying North Twistwood Creek Clay in Young #1 and Ketler #1 indicate that the oldest sediment considered in this report, the Cocoa Sand, is also within NP19/20 as previously reported by Jiang (1997). However, other workers believe the North Twistwood Creek Clay is older than NP19/20 (Siesser, 1983; Mancini and Tew, 1991; also see Fluegeman, 1996), and these differences are currently unresolved. Uppermost Eocene to lowest Oligocene calcareous nannofossil Zone NP21 extends upsection from the last occurrence of *Discoaster saipanensis* in the upper part of the Pachuta Marl Member to the highest stratigraphic occurrence of *Ericsonia formosa* in the basal part of the Marianna Formation of Lime Company #1. The last observed occurrence of *E. formosa* is near the stratigraphically highest sample available; therefore, it may be below the last occurrence datum of the species. Bybell's (1982) local zonation of neritic nannofossils is useful for correlation within NP21, particularly the nearly synchronous highest occurrences of *Pemma papillon* and *Chiasmolithus titus* within the Shubuta Clay (table 11.1).

Planktonic Foraminiferal Zonation

The Pachuta Marl and basal Shubuta Clay members yield abundant and relatively diverse assemblages of planktonic foraminifers including *Globigerina ampliapertura, Cribrohantkenina inflata, Hantkenina alabamensis,* and *Globorotalia (T.) cerroazulensis.* This assemblage is diagnostic of upper Eocene planktonic foraminiferal Zone P16 (figure 11.4). P17 could also be present above the highest observed *Cribrohantkenina inflata,* but this species occurs too sporadically in the studied sections for its absence to be considered meaningful. *Hantkenina alabamensis* and *Globorotalia (T.) cerroazulensis* are common in upper Pachuta Marl and lower Shubuta Clay samples, but both species are absent in our samples from the upper 13.7 m (45 feet) of the Shubuta in Young #1 and the upper 11.3 m (37 feet) of the Shubuta in Ketler #1. Isolated sin-

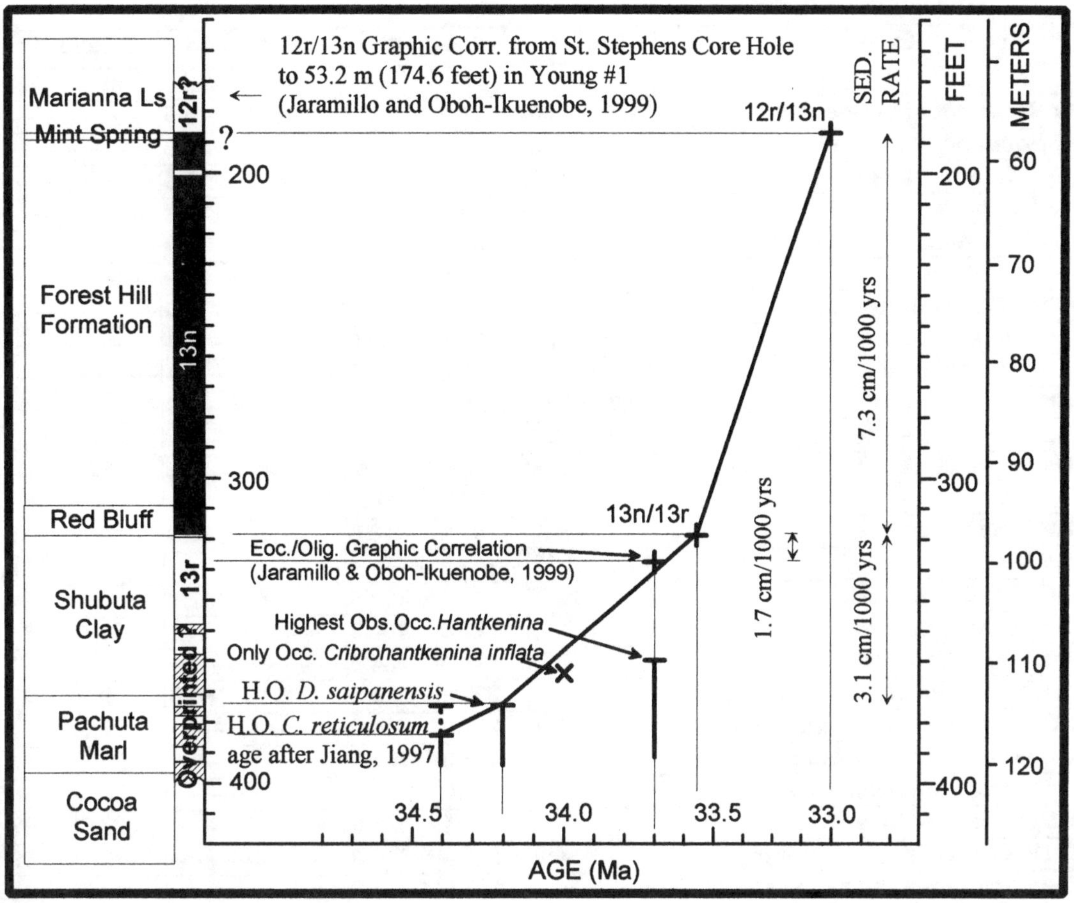

Figure 11.5. Magnetochronology and biochronostratigraphy of part of the Young #1 core-hole and rates of deposition (SED. RATE) of selected intervals.

gle occurrences of *Hantkenina alabamensis* in the Red Bluff of Lime Company #1, Ketler #1, and outcrop locality MOG #35 are difficult to evaluate because they may be reworked (see Bybell and Poore, 1983). Reflecting the uncertain status of occurrences of these species in the Red Bluff Clay, we assign the stratigraphic interval between the highest common and consistent occurrences of the Eocene index foraminifers in the lower Shubuta Clay and these isolated occurrences of *Hantkenina* sp. in the Red Bluff Clay to P17(?). Our results are similar to those of others who have reported sporadic occurrences of late Eocene index planktonic foraminifers above their highest common and consistent occurrences in Mississippi and Alabama sections (Deboo, 1965; Frederick, 1994; Keller, 1985; Miller et al., 1993; Waters and Mancini, 1982). The specialists who studied these occurrences cannot agree on whether they should be interpreted as reworked, and consequently no consensus exists on the stratigraphic location of the Eocene/Oligocene boundary here. Miller et al. (1993) attempted to avoid the problem of possible reworking of Eocene species into the Oligocene by using the first occurrence of *Cassigerinella chipolensis* to locate the base of the Oligocene. They found small and quite rare specimens of this species down to the base of the Shubuta Clay in the St. Stephens Quarry corehole. Pasley and Hazel (1995) challenged this correlation on the basis that the first occurrence datum of *C. chipolensis* is not well established. We did not find *C. chipolensis* in our Shubuta samples, although we did find *C. winnianna*, a possible precursor to *C. chipolensis,* in the Pachuta Marl and North Twistwood Creek Clay Member of the Yazoo Clay.

Table 11.1. Chronostratigraphy of the Mobil-MOG Coreholes

Event	Ma[1]	Young #1		Lime Co.#1		Ketler #1	
		m	feet	m	feet	m	feet
LO *Pseudohastigerina micra*[3] (f)	**32.00**	**54.9**	**180.0**	**4.7**	**15.5**	---------	---------
LO *Ericsonia formosa* (n)	**32.800**	---------	**low**	**3.0**	**10.0**	---------	---------
^{87}Sr/^{86}Sr isotope age	**32.89±0.23**	---------	---------	**4.1**	**13.5**	---------	---------
^{87}Sr/^{86}Sr isotope age	**33.16±0.23**	---------	---------	**5.9**	**19.2**	---------	---------
BASE: Marianna Limestone		**57.0**	**187.0**	**8.5**	**28.0**	---------	---------
Chron 12r/13n?	**33.048**	**57.0**	**187.0**	---------	---------	---------	---------
BASE: Mint Springs Marl		**57.7**	**189.4**	**9.1**	**30.0**	---------	---------
^{87}Sr/^{86}Sr isotope age	**33.54±0.23**	**71.6**	**235.0**	---------	---------	---------	---------
^{87}Sr/^{86}Sr isotope age	**33.59±0.23**	**77.4**	**254.0**	---------	---------	---------	---------
^{87}Sr/^{86}Sr isotope age	**33.76±0.35**	**88.2**	**289.5**	---------	---------	---------	---------
^{87}Sr/^{86}Sr isotope age	**33.83±0.20**	**96.8**	**317.5**	---------	---------	---------	---------
Chron 13n/13r	**33.545**	**96.8-97.4**	**317.5-319.5**	---------	---------	---------	---------
BASE: Red Bluff Clay		**97.2**	**319.0**	**57.6**	**189.0**	**11.2**	**36.7**
^{87}Sr/^{86}Sr isotope age	**33.21±0.32**	---------	---------	---------	---------	**21.2**	**69.5**
EOCENE-OLIGOCENE BNDY.[4]	**33.7**	**99.5**	**326.5**	**55.7**	**182.8**	**7.5**	**24.6**
LO *Pemma papillatum*[2] (n)	**(33.9)**	**111.0**	**364.0**	---------	---------	**22.6**	**74.0**
LO *Chiasmolithus titus*[2] (n)	**(33.9)**	**111.0**	**364.0**	---------	---------	**22.6**	**74.0**
LO *Hantkenina alablamensis* (f)		**109.6**	**359.5**	**56.1**	**184.0**	**9.0**	**29.5**
LCO *Hantkenina alabamensis* (f)	**33.7**	**111.0**	**364.0**	---------	---------	**22.6**	**74.0**
LO *Turborotalia cocoaensis* (f)	**33.8**	**111.0**	**364.0**	---------	---------	**22.6**	**74.0**
LO *Cribrohantkenina inflata* (f)	**34.0**	**111.0**	**364.0**	---------	---------	**30.3**	**(99.5)**
BASE: Shubuta Clay Member		**113.1**	**371.0**	---------	---------	**29.6**	**97.0**
LO *Discoaster saipanensis* (n)	**34.2**	**114.0**	**374.0**	---------	---------	**30.3**	**99.5**
LO *Reticulofenestra reticulata* (n)	**34.4**	**117.0**	**384.0**	---------	---------	**34.7**	**114.0**
BASE: Pachuta Marl Member		**120.8**	**396.2**	---------	---------	**38.1**	**125.0**
BASE: Cocoa Sand Member		**129.8**	**425.7**	---------	---------	**41.8**	**137.0**
BASE: N. Twistwood Creek Mbr.		**147.5**	**483.9**	---------	---------	**61.9**	**203.0**
BASE: Moodys Branch Fm.		**153.6**	**504.0**	---------	---------	**65.7**	**215.7**
FO *Isthmolithus recurvus* (n)	**36.0**	**167.2**	**548.5**	---------	---------	**65.7**	**215.5**

LO, Last Occurrence; LCO , Last Common Occurrence; FO, First Occurrence; (n), Calcareous Nannofossil; (f), Planktonic foraminifer. [1]Time scale of Berggren and others (1995); [2]local ages estimated by Jiang (1997); [3]last occurrence datum much higher than highest available sample; [4] after Jaramillo and Oboh-Ikuenobe (1999)

The Eocene Planktonic Foraminifer (EPF) Local Decimation Datum

The uppermost Eocene index foraminifers, *Hantkenina alabamensis* and *Globorotalia (T.) cerroazulensis*, and the local calcareous nannofossil markers *Pemma papillatum* and *Chiasmolithus titus* disappear up section virtually at the same stratigraphic horizon in Young #1 and Ketler #1 (table 11.1). All four species are common and they occur consistently in the upper part of the Pachuta Marl and lower Shubuta Clay. *Globorotalia (T.) cerroazulensis* is absent, and the other three have only very rare, isolated, occurrences stratigraphically higher in the upper Shubuta Clay or Red Bluff Clay. Regardless of whether any of these species became extinct at this stratigraphic level, an important change in the calcareous plankton occurred here. We will refer to it as the Eocene Planktonic Foraminifer (EPF) Local Decimation Datum (figures 11.4 and 11.5). We view it as a locally useful intra-Shubuta biostratigraphic correlation datum in southeastern Mississippi. It may correlate to the horizon below the type section of the Red Bluff Clay (figure 11.1) that Keller (1985) interpreted as the Eocene/Oligocene boundary.

Magnetostratigraphy and Strontium Isotope Dates
Magnetostratigraphic and strontium isotope analyses were performed at Lamont Geological Observatory and Rutgers University, under the direction of D. V. Kent and K .J. Miller (table 11.1). All three coreholes were analyzed for magnetostratigraphy. The magnetostratigraphic data appears easily interpreted for the Forest Hill to upper Shubuta Clay interval of the Young #1 core-hole (figures 11.4 and 11.5). However, the magnetostratigraphic record of the lower part of this core-hole, and all of the Lime Company #1 and Ketler #1 coreholes, is difficult to interpret and appears to have been overprinted (Miller, written communication, 1994). In Young # 1, the interpretable data identifies the Chron C13n/C13r boundary within a 0.6-meter (2-foot) sample gap that includes the basal 0.46 m (1.5 feet) of the Red Bluff Clay and the uppermost 0.15 m (0.5 feet) of the Shubuta Clay Member. These results are similar to those from the St. Stephens Quarry corehole in Alabama where the Chron C13n/C12r boundary is located 0.46 m (1.5 feet) above the Shubuta Clay member in the Red Bluff/Bumpnose (Miller et al., 1993). Strontium isotope dates for Young #1 are slightly older than dates for the same stratigraphic level based on magnetostratigraphy and biostratigraphy evidence (table 11.1). Similar discrepancies were noted for strontium isotope dates from the St. Stephens Quarry corehole and attributed to diagenesis (Miller et al., 1993). Suspecting diagenesis here also, we have not used these dates in our analysis.

The Eocene/Oligocene Boundary
Jaramillo and Oboh-Ikuenobe (1999) used dinocysts and the graphic correlation technique to correlate the E/O boundary at St. Stephens Quarry, as interpreted by Pasley and Hazel (1990), to specific stratigraphic levels in the coreholes of this study (figures 11.4 and 11.5). They correlated this interpreted E/O boundary to a location 2.3 m (7.5 feet) below the Shubuta–Red Bluff contact in Young #1, but 2.1 m (7.2 feet) above the contact in Lime Company #1 and 3.7 m (12.1 feet) above it in Ketler #1 (table 11.1). They concluded, therefore, that the Shubuta–Red Bluff contact is diachronous. The data reported here supports this conclusion.

The E/O boundary can be determined graphically in Young #1 assuming relatively constant sedimentation rate for Shubuta Clay, and little or no break in sedimentation at the Shubuta–Red Bluff contact in southeastern Mississippi (see Jaramillo and Oboh-Ikuenobe, 1999; Pasley and Hazel, 1990) (figure 11.5). Given these assumptions, the E/O boundary in Young #1 occurs within the upper Shubuta Clay in good agreement with the position of the E/O boundary determined from dinocyst graphic correlation (Jaramillo and Oboh-Ikuenobe, 1999).

Although the Eocene/Oligocene Global Section Stratotype and Point (GSSP) is located at the level of extinction of Eocene Hantkeninidae in the Massignano near Ancono, Italy (Premoli Silva and Jenkins, 1993), correlation of the E/O boundary in the eastern Gulf Coastal Plain by this criterion is difficult. *Hantkenina* and other late Eocene planktonic foraminiferal indices occur rarely and sporadically above the EPF Decimation Datum. However, rare occurrences of *Hantkenina* are not at all unusual in the Red Bluff and Red Bluff/Bumpnose (Bandy, 1949; Bybell and Poore, 1983; Cushman and Todd, 1948; Deboo, 1965; Eames et al., 1962; Mancini and Waters, 1986; Mornhinveg, 1941; Pasley and Hazel, 1995). Although several workers interpret these as reworked from underlying beds (Bybell and Poore, 1983; Deboo, 1965; Eames et al., 1962; Mancini and Waters, 1986), Pasley and Hazel (1995) disagree. They believe the highest in place occurrence of *Hantkenina alabamensis* is within the Red Bluff. Although our Red Bluff *Hantkenina* may indeed be reworked, such an interpretation is unnecessary if Jaramillo and Oboh-Ikuenobe (1999) correctly correlated the E/O boundary. In our data, the *Hantkenina* in the Red Bluff Clay occur below the dinocyst graphic correlation E/O boundary in Lime Company #1 and Ketler #1, but they do not occur in the Red Bluff of Young #1 that is Oligocene in age both by graphic correlation and magnetostratigraphic evidence (figure 11.4). Thus, our data are consistent with dinocyst graphic correlation evidence that the E/O boundary occurs within the Red Bluff Clay in the Ketler #1 and Lime Company #1 coreholes, but within the upper Shubuta Clay in Young #1.

PALEOBATHYMETRY
Interpretation of paleobathymetry is a key element in biostratigraphic sequence analysis (figures 11.6–11.8). The bathymetric units are standard biozones with approximate depth boundaries as follows: inner neritic (0 to 20–50 m), middle neritic (20–50 to 80–100 m), outer neritic (80–100 to 200 ± m), and upper bathyal (200± to 500± m) (see Tipsword et al., 1966). Other than in the most general way, no particular continental margin morphology is implied by these biozones. For example, the shelf-slope bound-

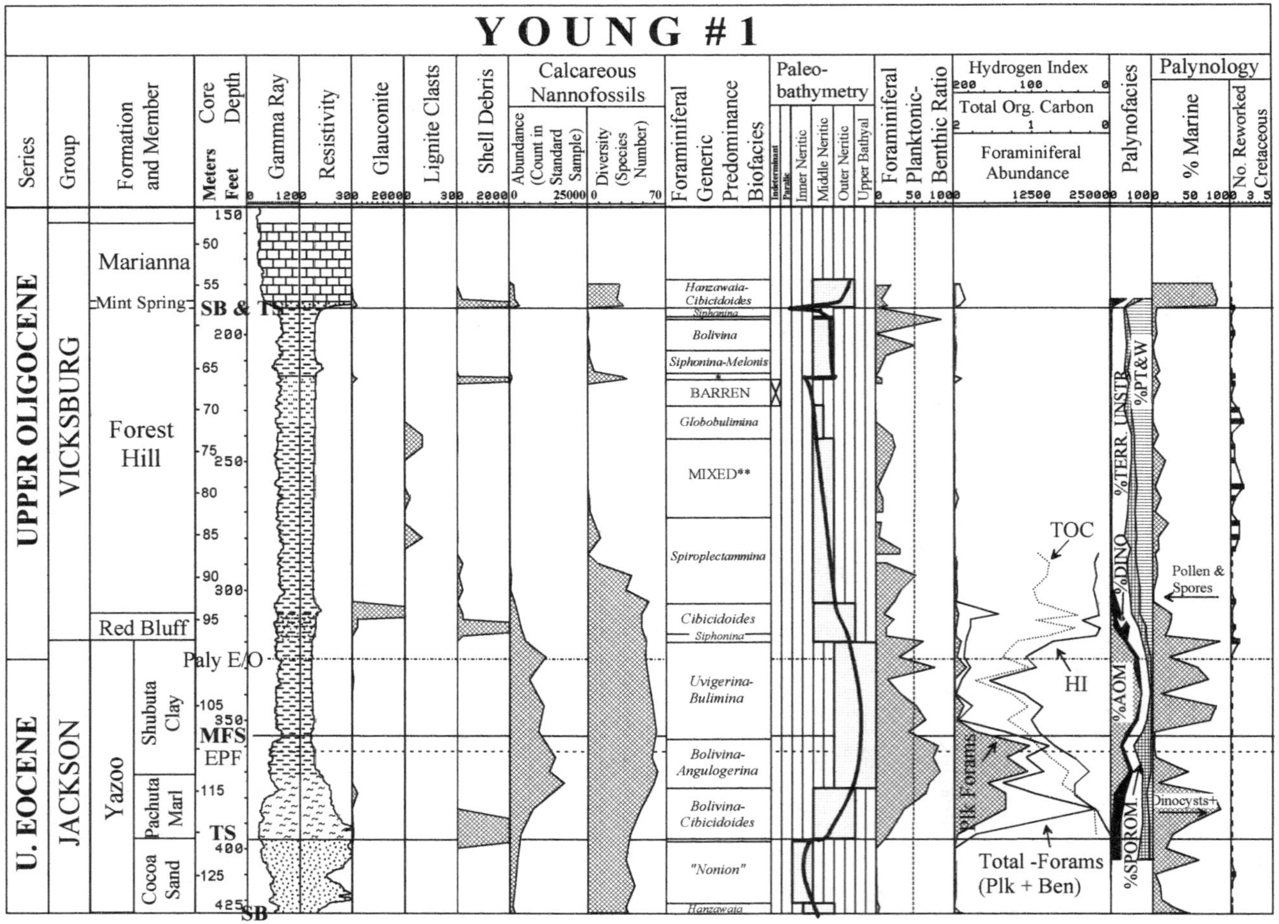

Figure 11.6. Sequence stratigraphy of the Eocene/Oligocene boundary interval in the Young #1 core-hole together with selected lithostratigraphic and biostratigraphic curves used to interpret sequence stratigraphy. GR, gamma ray curve; RES, resistivity curve; Paly E/O, dinocyst graphic correlation of Eocene/ Oligocene boundary (Jaramillo and Oboh-Ikuenobe, 1999); EPF, Eocene planktonic foraminifer decimation horizon; TOC, total organic carbon; HI, hydrogen index; Plk, Planktonic; Ben, Benthonic; for sequence stratigraphy abbreviations see figure 11.3. Palynofacies abbreviations are as follows: %AOM, Percent Aquatic Amorphous Organic Matter; %DINO, aquatic structured organic matter (dinocysts and foraminiferal test linings; %TERR. UNSTR., terrigenous unstructured; %SPOROM., terrigenous structured: sporomorphs; %PT&W, terrigenous structured: primarily plant tissue and woody debris. Sequence strati-graphy abbviations as in figure 11.3.

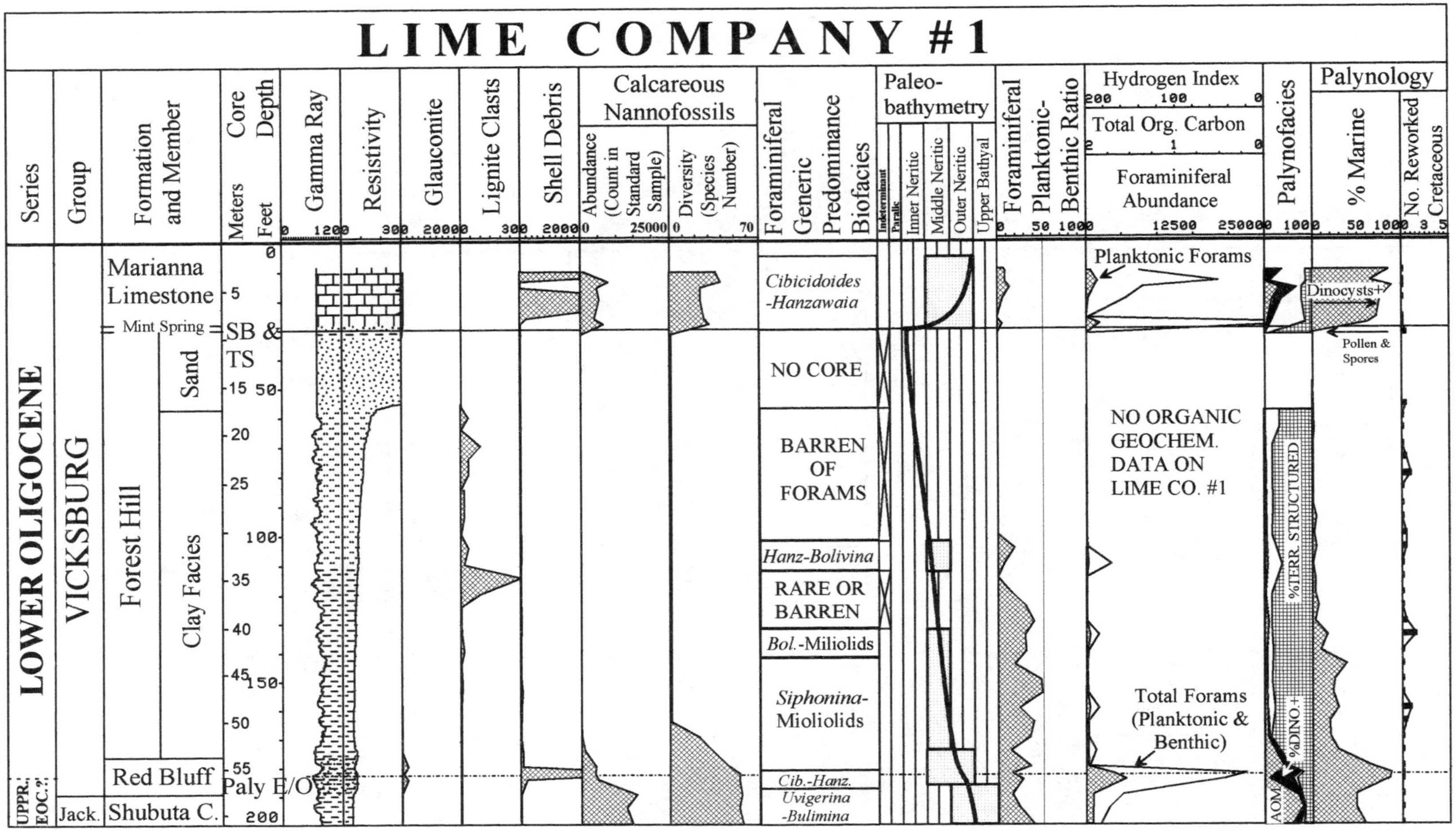

Figure 11.7. Sequence stratigraphy of the Eocene/Oligocene boundary interval in the Lime Company #1 core-hole together with selected lithostratigraphic and biostratigraphic data curves used to interpret sequence stratigraphy. For abbreviations, see figure 11.6.

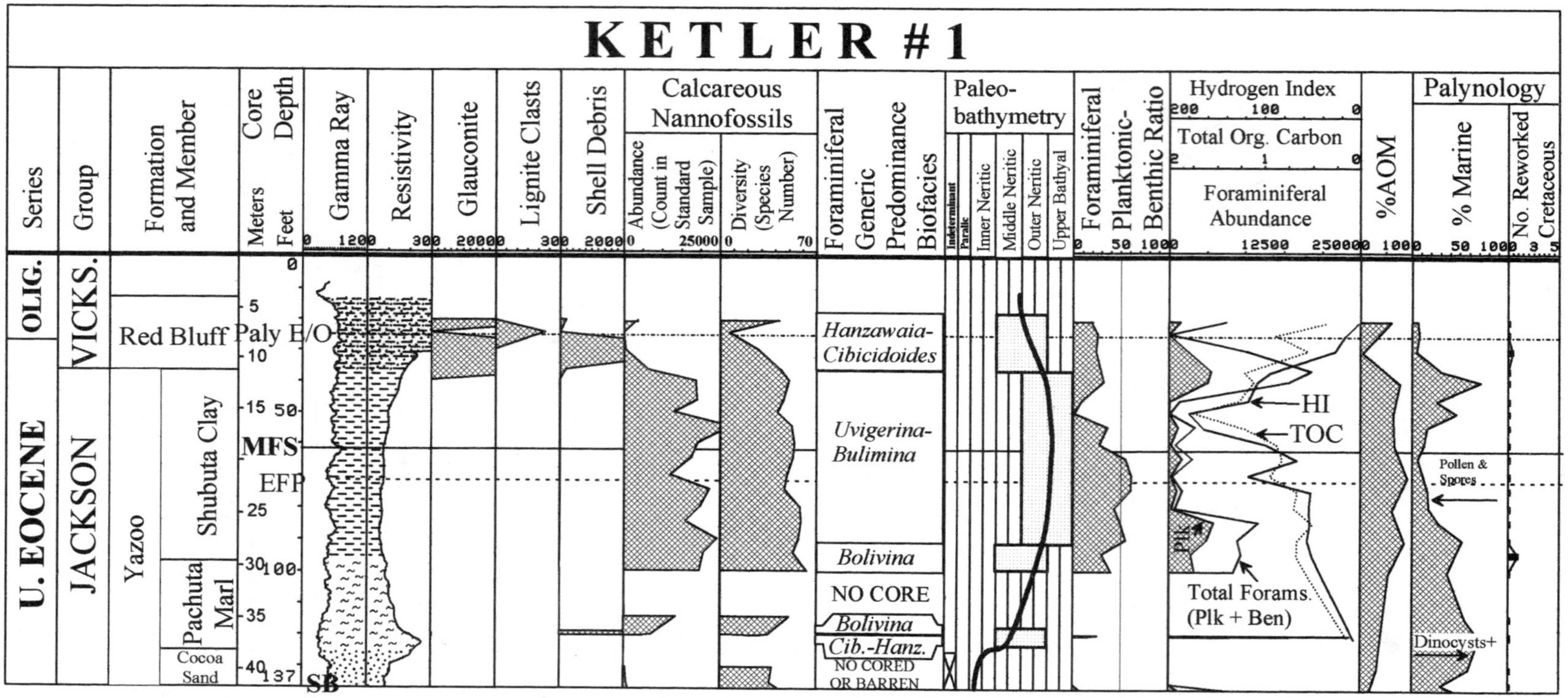

Figure 11.8. Sequence stratigraphy of the Eocene/Oligocene boundary interval in the Ketler #1 core-hole together with selected lithostratigraphic and biostratigraphic data curves used to interpret sequence stratigraphy. For abbreviations, see figure 11.6.

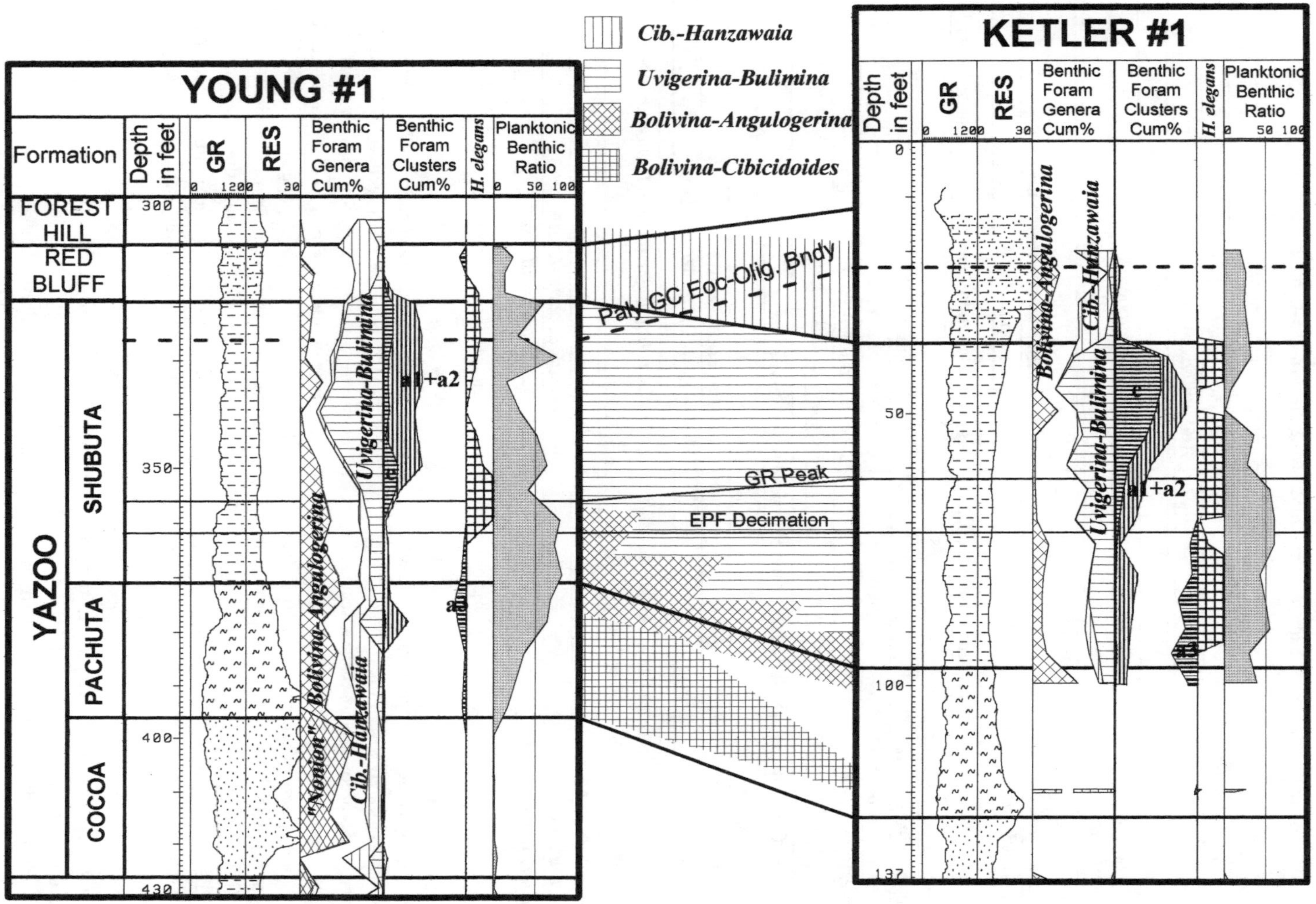

Figure 11.9. Quantitative distribution of foraminiferal biofacies in the lower part of Sequence TE 3.3. Species comprising benthonic foraminiferal cluster groups are documented in table 11.2.

Table 11.2. Benthic Foraminiferal Species Groups Identified by Cluster Analysis that Have Distributions Concentrated Within the Indicated Stratigraphic Intervals

Red Bluff Formation, Forest Hill Formation, and Marianna Limestone - Primary generic predominance facies associations are *Cibicidoides, Cibicidoides-Hanzawaia, Hanzawaia-Cibicidoides*, but present in others.		
	Custer Group **d**	
	Articulina terquemi	*Hanzawia susanaensis*
	Gyroidina octocamerata	*Anomalina umbonata*
	Cluster Group **j1**	
	Hanzawaia lobatulus	*Cibicidoides pippeni*
	Melonis planatus	*Gyroidina danvillensis*
	Cluster Group **j2**	
	Cibicides tallahattensis	*Cibicidoides pseudoungerianus*
	Cibicidoides vicksburgensis	*Cibicidoides mimulus*
Upper Pachuta Marl, Shubuta Clay, and basal Red Bluff. *Uvigerina-Bulimina* generic predominance facies		
	Cluster Groups **a1** and **a2**	
	Hoeglundina elegans	*Pullenia quinqueloba*
	Uvigerina cocoaensis	*Uvigerina glabrans*
	Uvigerina gardnerae	*Nodosaria latejugata*
	Bulimina jacksonensis	*Gaudryina jacksonensis*
Upper Shubuta Clay. *Uvigerina-Bulimina* generic predominance facies		
	Cluster Group **e**	
	Uvigerina vicksburgensis	*Discorbis cocoaensis*
	Bolivina dalli	*Planulina sp. cf. P. venezuelana*
Pachuta Marl and lower Shubuta Clay. *Uvigerina-Bulimina, Bolivina-Angulogerina*, and *Cibicidoides-Hanzawaia* generic predominance facies		
	Cluster Group **a3**	
	Cibicides cocoaensis	*Anomalina danvillensis*
	Cibicidoides cooki	*Hanzawaia danvillensis*

Note: The primary associations of species cluster groups with generic predominance facies are also indicated.

ary is within the outer neritic of the modern Gulf of Mexico, but it was located within the middle to inner neritic at some times of low glacio-eustatic sea level in the Pleistocene (Armentrout, 1991).

We interpret paleobathymetry primarily from benthonic foraminiferal biofacies and the planktonic-benthonic (PB) foraminiferal ratio (figures 11.6-11.9). In the paleobathymetry columns of figures 11.6-11.8, paleobathymetry interpreted from foraminiferal biofacies is shown as a bar graph. The superimposed paleobathymetry curve is constrained by the bar graph, but is further interpreted from sediment type, dinocyst paleoecology and palynofacies. Benthonic foraminiferal biofacies were differentiated in two separate analyses, one at the taxonomic level of genus and the other at the species level. Interpretation of paleobathymetry from foraminiferal biofacies must be done primarily at the generic level because most of the Paleogene species are extinct. Our approach is to compare generic predominance facies in the coreholes and outcrops to their closest counterparts in the modern Gulf of Mexico (see Poag, 1981; Walton, 1964). Most species associations are not interpretable by comparison to modern biofacies as are generic associations, but they may more precisely define the limits of paleoenvironments. This is so because environmentally sensitive species have more restricted distributions than most genera represented by two or more species. Species associations were recognized using the unweighted pair group method of cluster analysis (table 11.2). The cluster analysis used the entire data set including stratigraphically lower parts of the Young #1 and Ketler #1 coreholes that are not addressed in this report (see Dockery et al., 1994). Primarily because of the practical difficulties of counting some species of which many of the individuals are small (down to 74 μm), some taxa were lumped into groups identified to genus only. These genus-taxa were not used in the cluster analysis. Two separate cluster analyses were run using closely related coefficients, the Ochai and Dice (Romesburg, 1990). The final cluster groups consist of only species that joined the groups in analyses run with both coefficients. Although the cluster analysis used presence-absence coefficients, distributions of the cluster groups in the coreholes was investigated using quantitative data expressed as a percentage of the total number of benthonic foraminiferal tests in each sample (figures 11.9 and 11.10).

The planktonic-benthonic (PB) ratio [(P/P+B) * 100] is an effective empirical tool for interpreting *minimum* paleobathymetry. Gibson's (1989) extensive documentation showed that PB ratios of 50% or more indicate outer neritic to bathyal depths because ratios this high are very unusual at depths of less than 80–100 m in the modern ocean. However, the PB ratio curve is not a proxy for a paleobathymetry curve because low PB ratios are common over a very wide range of water depths.

Using foraminiferal biofacies and PB ratios we interpret that water depths increased from inner neritic to middle to outer neritic in the lower half of the sequence, peaked at outer neritic to upper bathyal in the Shubuta Clay, then decreased in the upper half of the sequence (figures 11.6-11.8, 11.10). Our paleobathymetric interpretations are generally similar to those of others for the same stratigraphic units in outcrop using evidence from foraminifers (Mancini et al., 1987), ostracodes (Hazel et al., 1980; Huff, 1970), and dinocyst and palynofacies data (Jaramillo and Oboh-Ikuenobe, 1999).

The "*Nonion*" predominance facies in the Cocoa Sand indicates an inner neritic environment. The "*Nonion*" predominance facies appears to be an upper Eocene equivalent of the modern inner neritic *Elphidium* predominance facies. Indeed, Bandy (1949) illustrated *Elphidium*-like cribrate apertures, but no sutural pores, for the upper Eocene "*Nonion*" species that occur in our samples (e.g.,, *N. advena, N. inexcavatus,* and *N. rolshauseni*). We have left the generic assignment open because apertures of the "*Nonion*" species are so obscured by a granular secondary layer that they are difficult to observe.

From inner-neritic depths that persisted through-

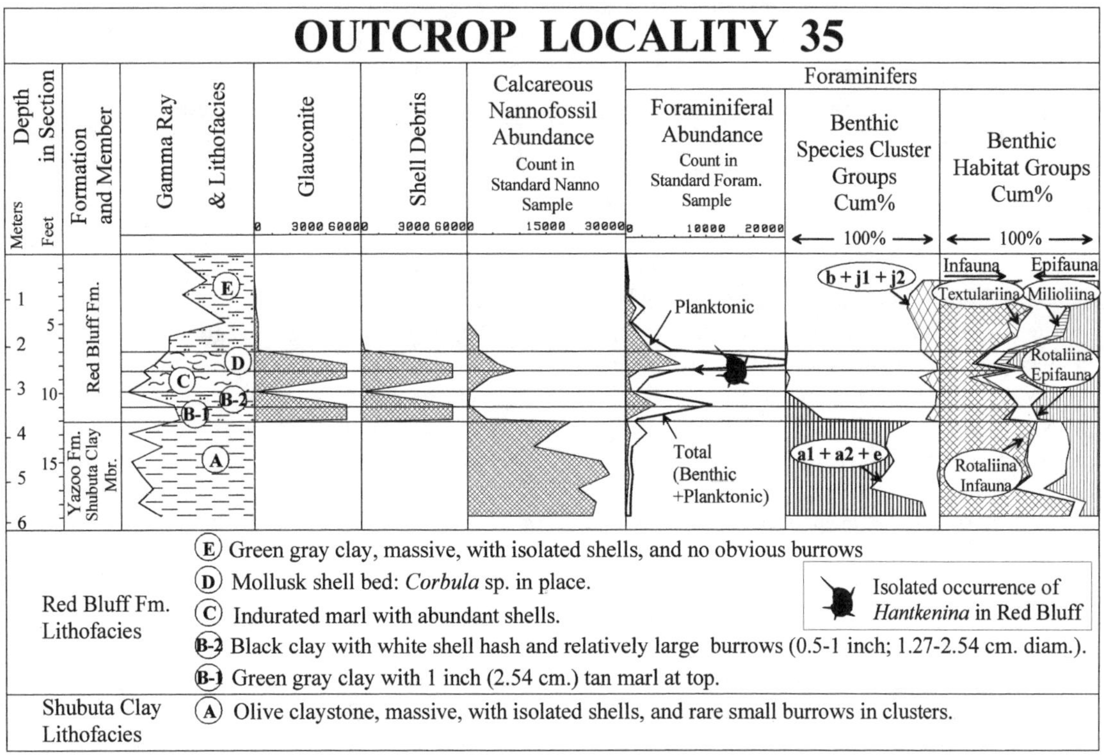

Figure 11.10. Biofacies and lithofacies changes across the boundary between the Shubuta Clay and Red Bluff Clay at outcrop locatity MOG #35.

out deposition of the Cocoa Sand, water depths increased to outer neritic during deposition of the Pachuta Marl, reached a maximum that may have been as deep as upper bathyal for Sequence TE 3.3 during deposition of the lower Shubuta Clay, and remained deep throughout deposition of the upper Shubuta Clay. The evidence for possible upper-bathyal water depths comes from comparison of the *Uvigerina-Bulimina* predominance facies to the modern *Uvigerina* predominance facies of the northwestern Gulf of Mexico (see Fluegeman, 1996, for a similar interpretation). The *Bolivina-Angulogerina* predominance facies in Young #1 occurred basinward of the *Uvigerina-Bulimina* facies in Ketler #1 during deposition of the lower Shubuta Clay, therefore we interpret possible upper bathyal for this facies also (figure 11.9). The important species of the *Uvigerina-Bulimina* biofacies are the eight of cluster groups a1 and a2, which range through the lower and upper Shubuta, and cluster group e, which is virtually confined to the upper Shubuta (table 11.2). One of the species of cluster group a1 is particularly noteworthy for calibrating paleobathymetry. The extant species *Hoeglundina elegans* occurs consistently throughout most of the Shubuta Clay in Young #1 and Ketler #1, but it disappears up section abruptly within the lowermost part of the Red Bluff Clay (figure 11.9). Although this species occasionally occurs in the middle neritic of the modern Gulf of Mexico, it is typical of outer-neritic and bathyal zones (Denne and Sen Gupta, 1993; Grünig and Herb, 1980; Pflum and Frerichs, 1976; Phleger, 1960). The PB ratio confirms that the lower Shubuta was deposited at depths at least as great as outer neritic. This ratio increases regularly up section in the Pachuta Marl and reaches maximum values of 70% to 80% in the lower Shubuta Clay in Young #1, the most basinward of our coreholes (figures 11.6 and 11.9). PB ratios are somewhat lower in Ketler #1, but still greater than 50% in the lower Shubuta (figures 11.8 and 11.9). The PB ratio is lower in the upper Shubuta although intermittently greater than 50% in Young #1.

The lithostratigraphic boundary between the Shubuta Clay and Red Bluff Formation coincides with change in benthonic foraminiferal generic predominance from *Uvigerina-Bulimina* to *Cibicidoides* or *Cibicidoides-Hanzawaia*. Comparing this change to the change in modern biofacies at the shelf break in the northwestern Gulf of Mexico, but allowing for a greater range of uncertainty based on evolutionary changes in faunas and differences in physiographic setting, we interpret the *Uvigerina-Bulimina* facies as outer neritic to upper bathyal and the *Cibicidoides* facies as middle to outer neritic.

Diverse middle-neritic biofacies are present in the Forest Hill Formation. Some of these, such as the *Siphonina*-Miliolid biofacies in Lime Company #1, include species that may be displaced downslope from inner-neritic environments (figure 11.7). Other species in the same biofacies (*Globobulimina ovata, Bifarina vicksburgensis* and *Cibicidoides pseudoungerianus*, for example) indicate middle neritic or greater water depths (Tipsword et al., 1966). Those parts of the upper Forest Hill Formation that are barren of foraminifers, or within the interval of the Forest Hill sand facies in Lime Company #1 that were not successfully cored, were probably deposited at inner neritic to paralic depths (figures 11.6 and 11.8). The evidence for this interpretation is discussed later in this report.

SEQUENCE STRATIGRAPHY

Biostratigraphic, wire-line log, and sedimentologic criteria for sequence analysis are described in a number of sources (Armentrout, 1991; Armentrout et al., 1990; Coleman and Roberts, 1988; Loutit et al., 1988; Mancini et al., 1987; Mitchum et al., 1991; Pasley et al., 1991; Schaffer, 1990; Vail and Wornardt, 1990). We apply these criteria using an assortment of data curves for the Young #1 (figure 11.6), Ketler #1 (figure 11.7), Lime Company #1 (figure 11.8) and outcrop locality MOG #35 (figure 11.10).

Identification of sequence boundaries is an important initial step in sequence analysis. Previous sequence stratigraphic analyses of the E/O boundary interval of the eastern Gulf Coastal Plain identified candidates for sequence boundaries at the base of the Cocoa Sand, the base of the Red Bluff Clay, and the base of the Mint Spring Marl (figure 11.3A–D). In the text that follows, we support arguments that the base of the Cocoa Sand is the TE 3.3 sequence boundary and the base of the Mint Spring Marl is the sequence boundary of overlying sequence TO 1.1. We do not interpret the base of the Red Bluff Clay as a sequence boundary although we recognize that the juxtaposition of relatively shallower biofacies and lithofacies of the Red Bluff Clay over deeper facies of the uppermost Shubuta Clay, with very little transition between them, could suggest falling sea level (Dockery, 1990; Pasley and Hazel, 1995). We reject interpretation of the Shubuta–Red Bluff contact as a sequence boundary primarily because of evidence reported by Jaramillo and Oboh-Ikuenobe (1999), and supported by this study, that the Shubuta–Red

Bluff contact is measurably younger in the Young #1 core-hole than in the other two coreholes. If falling sea level were the cause of missing transitional deposits between Shubuta and Red Bluff facies we would expect this contact to be virtually synchronous throughout the study area. Therefore, instead of invoking falling sea level, we will call upon progradation and normal filling of accommodation space by sediment to explain decreasing water depth. We will propose that different facies are sharply juxtaposed at the Shubuta–Red Bluff contact because incremental decreases in water depth caused crossing of a threshold between spatially adjacent environments that supported dissimilar biofacies.

TE 3.3 Sequence Boundary and Shelf-margin Systems Tract

We interpret the TE 3.3 sequence boundary at the base of the Cocoa Sand because its contact with the underlying North Twistwood Creek Clay is known to be a subaerial unconformity in southwestern Alabama (Tew and Mancini, 1995). To our knowledge, this unconformity has not been reported from southeastern Mississippi (see, e.g., Hendy, 1948; May, 1974). Furthermore, Dockery et al. (1994) interpreted the Cocoa Sand to grade both downward and laterally into the North Twistwood Creek Clay in Young #1 and Ketler #1. If they are correct, the base of the Cocoa Sand in this study does not precisely correlate between these two coreholes, leaving exact correlation to the sequence boundary in Alabama in question as well. Perhaps these localities lay basinward of the subaerial unconformity in an area of conformable deposition, such as could occur with a shelf-margin systems tract. If so, the sequence boundary in the southeastern Mississippi coreholes could vary above or below the North Twistwood Clay–Cocoa Sand contact at some undetected stratigraphic level rather than at the base of the Cocoa Sand. However, this is not an issue that can be solved with the data of this study. Therefore, we have adopted the prevailing view that all of the Cocoa Sand is a shelf-margin systems tract (Baum and Vail, 1988; Tew and Mancini, 1995). As defined by Van Wagoner et al. (1988), a shelf-margin systems tract is a set of weakly prograding to aggrading beds bound below by a "Type II" sequence boundary, and above by a transgressive surface.

In Young #1, the Cocoa Sand shelf-margin systems tract is comprised of inner-neritic, open-marine, clayey sand, 8.9 m (29.2 feet) thick, with little or no glauconite or macrofossil shell material (figure 11.6). In Ketler #1, it is only 3.7 m (12 feet) thick and its properties are less well known because of poor core recovery. It has an inner-neritic foraminiferal biofacies. Calcareous nannofossils are present in low abundance, but moderately high diversity, consistent with the open-marine, shallow-water environment. The Cocoa Sand palynoflora is overwhelmingly terrigenous. The impoverished dinocyst assemblage is dominated by *Spiniferites*, a cosmopolitan genus.

Transgressive Surface and Transgressive Systems Tract

The transgressive surface is interpreted at the Cocoa Sand–Pachuta Marl contact where the inner neritic "*Nonion*" predominance facies is abruptly replaced up section by the middle- to outer-neritic benthonic foraminiferal *Bolivina-Cibicidoides* predominance facies. The transgression that begins here continues up section through the Pachuta Marl and peaks in the lower Shubuta Clay where we locate our candidate for maximum flooding surface. The stratigraphic interval from the transgressive surface to the maximum flooding surface is the transgressive systems tract.

The TE 3.3 transgression is manifested in an orderly series of changes in sediment properties and fossil biota. The lower part of the Pachuta Marl contains abundant siliciclastic sand and macrofossil shell debris. These decrease then disappear up section concurrent with increasing abundance of benthonic foraminifers, planktonic foraminifers, and calcareous nannofossils (figure 11.6). The PB ratio increases regularly upsection along with increases in the percentage of marine palynomorphs. Increasing water depth during the transgression is interpreted from foraminiferal biofacies. From inner-neritic depths that persisted throughout deposition of the Cocoa Sand, water depths increased to middle to outer neritic during deposition of the Pachuta Marl, reaching a maximum of outer neritic to upper bathyal during deposition of the lower Shubuta Clay.

Maximum Flooding Surface Candidates

In eastern Mississippi, sequence TE 3.3 passes from transgressive retrogradation to highstand progradation through a series of gradual and complex changes. Reflecting that complexity, two candidates for maximum flooding surface (MFS) have emerged from previous local interpretations of this sequence. The stratigraphically lower of the two candidates was recognized recently on biostratigraphic evidence (Jaramillo and Oboh-Ikuenobe, 1999) and is the preferred candidate of this study. It caps the clearly transgressive deposits of the Pachuta Marl and lower

Shubuta Clay, described above. The other candidate is the contact between the Shubuta Clay and Red Bluff Clay (Baum and Vail, 1988; Mancini et al., 1987; Tew and Mancini, 1995). At this surface the local sedimentological regime changed from predominantly hemipelagic to delta-influenced. The stratigraphic interval from this surface to the overlying sequence boundary is interpreted as a highstand systems tract in this and most previous interpretations (Baum and Vail, 1988; Jaramillo and Oboh-Ikuenobe, 1999; Mancini, 2000; Mancini et al., 1987; Tew and Mancini, 1995). Our interpretation supports the status of the Shubuta Clay-Red Bluff Clay contact as an intra-highstand lithofacies event (Jaramillo and Oboh-Ikuenobe, 1999), which is time-transgressive in an offshore direction (figure 11.4). We interpret the upper Shubuta Clay above the intra-Shubuta MFS and below the Shubuta–Red Bluff lithofacies boundary as early highstand. However, we also recognize that some of its characteristics—its contained organic matter, in particular—are similar to those of some transgressive systems tracts. These characteristics suggest that the early highstand systems tract was deposited in a relatively distal setting.

In Young #1 and Ketler #1, the intra-Shubuta Clay candidate for maximum flooding surface is supported by three lines of evidence: dinocyst paleoecology, the foraminiferal PB ratio, and a gamma ray peak event. Our unpublished dinocyst data confirm observations by Jaramillo and Oboh-Ikuenobe (1999) that the Pachuta Marl and lower Shubuta Clay have dinocyst assemblages with a more offshore aspect than does the upper Shubuta. Jaramillo and Oboh-Ikuenobe (1999) quantified this trend with their Dinocyst Paleobathymetry Index (DPI) and used it to interpret an MFS within the lower to middle parts of the Shubuta Clay. Their MFS is 1 m or less above the gamma ray peak event and 2 m or less above the beginning of a trend toward lower PB ratios in these two coreholes. This pattern suggests a thin condensed section at the gamma ray peak followed upsection by a trend from more offshore to more nearshore dinocyst assemblages and PB values. The outer-neritic to upper-bathyal foraminiferal assemblages that persist to the top of the Shubuta Clay show that no great decrease in water depth is implied by the change in DPI and PB. Although DPI values in the upper Shubuta Clay may be in a range typical of inner-neritic environments, Jaramillo and Oboh-Ikuenobe (1999) recognize from their integrated analysis that the Shubuta Clay was deposited in water depths that were never less than deep middle neritic. Given that PB and DPI indicate that the most offshore aspect of sequence TE 3.3 was developed during deposition of the lower Shubuta Clay, the large terrigenous element in the palynomorph and palynofacies data seems a paradox (figure 11.6). However, the terrestrial component is accounted for by sporomorphs, a group that is readily transported to offshore environments by wind and water currents (see Jaramillo and Oboh-Ikuenobe, 1999). Sporomorphs, therefore, may not necessarily be out of place in late transgressive deposits in a deep-shelf setting seaward of a major fluvial system (see Tew and Mancini, 1995).

Early Highstand Systems Tract

We interpret the TE 3.3 highstand in southeastern Mississippi to consist of the upper part of the Shubuta Clay, the Red Bluff Clay, and the clay facies of the Forest Hill Formation. The depositional environments of these formations are very different. The early highstand part of the Shubuta Clay is similar in most respects to the late transgressive part below the MFS. It is massive hemipelagic clay with rare, widely dispersed, chalky, bivalve shells (Dockery et al., 1994). Glauconite is very rare in our Shubuta samples. The *Uvigerina-Bulimina* generic predominance facies occurs to the very top of the Shubuta Clay, underscoring the offshore, deep-water aspect of this lithostratigraphic unit (figures 11.6–11.9). Throughout, calcareous nannofossils are abundant and diverse, contributing to its hemipelagic aspect (figures 11.6–11.8). In Young #1, high percentage of aquatic amorphous organic matter (%AOM), and correspondingly low percentages of terrestrial organic matter, give the palynofacies of the early highstand an aspect that is even more offshore than that of the late transgressive (figure 11.6). The low concentrations of foraminifers in the upper Shubuta, relative particularly to the late transgressive Shubuta of Young #1, is noteworthy but of uncertain significance. Dilution by clastic clay sediment is probably not the principal cause because calcareous nannofossils should have been diluted to a similar degree, and they were not according to our data. Dissolution is another possible cause for the low concentrations of foraminifers in the highstand Shubuta because foraminifers are more solution susceptible than calcareous nannofossil, at least under some conditions (Honjo and Erez, 1978). Although dissolution may have reduced foraminiferal numbers locally, the good state of preservation of most upper Shubuta foraminifers indicates that dissolution is not the sole cause of their low numbers.

Early highstand deposits of the upper Shubuta

Clay are distinguished by maximum values of TOC and HI in Young #1 and Ketler #1, and maximum percent aquatic amorphous organic matter (%AOM) and minimum percent of structured terrestrial organic matter (%TOM) in Young #1 (figures 11.6, 11.8). The weight percent of the sediment that is organic carbon (%TOC) and the hydrocarbon content of the organic matter (HI) covary through the Pachuta Marl and Shubuta Clay members of Young #1 and Ketler #1 (figures 11.6, 11.8). Both increase through the late transgressive systems tract and peak within the early highstand Shubuta Clay, and then decrease toward the top of this unit. Although this early highstand Shubuta sediment has much more organic carbon than the late transgressive Shubuta, and the overlying Red Bluff, it is still not very organic-rich. The highest TOC values are similar to maximum values for the modern upper slope in the Gulf of Mexico (Jones, 1983; McKee et al., 1978; Phlum and Frerichs, 1976; Richards and Redfield, 1953). After correction for clay adsorption of pyrolyzable hydrocarbons, according to the graphical method of Langford and Blanc-Valleron (1990), the upper Shubuta kerogen assemblage is classified within the lower range of Type II, consistent with characterization of its palynofacies as oil-prone (Jaramillo and Oboh-Ikuenobe, 1999). However, because of the small amount of organic matter present, the oil source rock potential is only poor to fair (see Peters and Cassa, 1994). Published observations on the sequence stratigraphic significance of organic matter have tended to emphasize potential oil source rocks having very high TOC and HI and high %AOM and very low %TOM. These potential source rocks have been linked to transgressive flooding by oxygen-depleted water (Creaney and Passey, 1994; Pasley et al., 1991). Locally, at St. Stephens Quarry in southwestern Alabama, maximum %AOM and minimum %TOM characterize the (non-source rock) upper transgressive systems tract (Shubuta Clay) and maximum flooding surface (basal Red Bluff/Bumpnose) (Pasley and Hazel, 1990). Thus, the sedimentary organic matter of the upper Shubuta Clay has a more offshore aspect than the lower Shubuta Clay, contrasting with the DPI and PB ratios, which have a less offshore aspect in the upper Shubuta Clay than in the upper Pachuta Marl and lower Shubuta Clay. However, these modest quantities of relatively hydrogen-rich organic matter, linked to high frequency of aquatic amorphous organic matter, may not be inconsistent with interpretation of early highstand deposits. In the modern early highstand of the northern Gulf of Mexico, oxidation of the high organic production around the Mississippi Delta depresses the dissolved oxygen content to depths of 400 m, and the sediments that underlie this water have a higher content of organic carbon than sediments farther from the delta (Denne and Sen Gupta, 1991, 1993).

The Red Bluff Clay is a condensed shelf deposit related to the prodelta clay facies of the Forest Hill Formation by the nature of its contacts and its contained organic matter. In our coreholes and outcrop section MOG #35, its lower contact is sharp and burrowed, clearly delineating it from the underlying hemipelagic Shubuta Clay (Dockery et al., 1994). Its upper contact is transitional linking it with the clay facies of the Forest Hill Formation (see especially, Gilliland, 1980). With regard to properties of the organic matter, woody debris and plant tissue begin to increase upsection within the Red Bluff, suggesting that the first influence of the Forest Hill delta upon sedimentation in the area of southeastern Mississippi is to be found there (figure 11.6). Also, the lowest stratigraphic occurrence of lignite clasts in a microfossil sample is within Forest Hill clay facies interbedded with typical upper Red Bluff Clay in the Ketler #1 core (figure 11.8). Further distinguishing the Red Bluff Clay from the hemipelagic Shubuta Clay, the hydrocarbon content (HI) of the organic matter decreases precipitously up section in the lower Red Bluff Clay concurrent with (and presumably linked to) a decrease in %AOM in the palynofacies data. Although %AOM decreases here, the marine character of the palynofacies is retained through the Red Bluff of Young #1 by a high percentage of dinocysts and other structured marine palynodebris (Jaramillo and Oboh-Ikuenobe, 1999) (figure 11.6).

In other respects as well, the Red Bluff Clay marks an important change in the local sedimentological regime from offshore hemipelagic deposition of the Shubuta Clay to prodelta deposition of the overlying Forest Hill Formation. These changes are documented in all three coreholes (Figures 6-8) and especially in outcrop locality MOG #35 (figure 11.10). At the outcrop locality, 30 cm (1 foot) sample spacing, decreased to only 6 cm (0.2 foot) spacing at the Shubuta–Red Bluff contact, provides better data resolution than does the more widely spaced samples in the cores. The lower contact of the Red Bluff is marked by a sharp, although subtle, color change (Dockery et al., 1994) and by abrupt appearance of glauconite or macrofossil shell debris, or both, in abundance at the base of the Red Bluff (figures 11.6–11.8, 11.10). Some burrows that originate in the Red Bluff and penetrate the underlying Shubuta Clay are reportedly filled with Red Bluff sediment and fossils (MacNeil, 1944). Relatively large burrows, about 1.3 to 2.5 cm (1/2 to 1 inch) in diameter, are common in parts of the Red Bluff at outcrop locality MOG #35, contrasting with rare burrows that are small and pencil-thin in the underlying Shubuta Clay (figure 11.10). Calcareous nannofossils are abundant in the Shubuta Clay, decrease sharply in abundance at the base of the Red Bluff, and virtually disappear in the Forest Hill (figures 11.6–11.8, 11.10). Foraminiferal biofacies also change dramatically in the lower Red Bluff. The *Uvigerina-Bulimina* generic predominance facies of the Shubuta Clay is abruptly replaced by the *Cibicidoides-Hanzawaia* facies in the Red Bluff. The abruptness of the foraminiferal biofacies change is most evident at the species level (see benthonic species cluster groups in figure 11.10). From these foraminiferal biofacies changes, we interpret decreasing water depth stratigraphically upward from the Shubuta–Red Bluff contact. Supporting this interpretation, the outer neritic to bathyal species *Hoeglundina elegans* occurs consistently in the Shubuta Clay, but very rarely in the Red Bluff Clay. Also, the replacement of the *Uvigerina-Bulimina* facies by the *Cibicidoides-Hanzawaia* facies across the Shubuta–Red Bluff contact is similar to change in generic predominance facies that occurs from the upper slope to shelf-edge in the modern northwestern Gulf of Mexico (figures 11.6–11.10). In our comparison of generic predominance biofacies to modern ones, we imply that the environment to which the fossil assemblage was adapted and the environment to which the recent assemblage is adapted are similar. We do not imply a shelf-edge. We believe that the entire study area was located on the physiographic shelf evidenced, for example, by the locations of both middle Eocene and Oligocene lowstand wedge deposits far south of our study area (Coleman, 1990). Multiple factors, some not directly related to water depth, are likely to have been involved in causing the replacement of one foraminiferal biofacies by another. Further insight into this biofacies replacement is provided by Frederick's (1994) observation that the Shubuta–Red Bluff boundary is marked by a change in predominance from infaunal benthonic foraminifers in the Shubuta Clay to epifaunal predominance in the Red Bluff. We (and Frederick) interpret habitats of foraminiferal genera on the basis of their test shape, coiling patterns and pore distribution (Corliss, 1985, 1991; Corliss and Chen, 1988; Corliss and Fois, 1990). The abrupt benthonic foraminiferal biofacies change at the Shubuta–Red Bluff boundary is probably related to some degree to the different feeding strategies and substrate requirements of these two habitat groups (Kitazato, 1994) and not entirely to changes in vertically stratified water masses (see Denne and Sen Gupta, 1991, 1993). Internally, the Red Bluff Clay is very heterogeneous with different layers having highly variable content of glauconite, shell debris, in situ mollusk shell beds, foraminifers, and calcareous nannofossils, all in a clay and silt matrix (figure 11.10). Dinocyst assemblages underscore the heterogeneous nature of the Red Bluff Clay. The Red Bluff dinocysts in Young #1 are dominated by *Deflandrea* followed by *Spiniferites,* whereas those of the Lime Company #1 core and MOG #35 outcrop are dominated by *Cordosphaeridium* followed by *Operculodenium.* Ketler #1 samples are overwhelmingly terrigenous, to a degree that precludes characterization of the dinocyst assemblage.

The association of authigenic mineralization, concentration of shell hash and foraminifers, and common burrows are basis for interpreting the Red Bluff as a condensed section. For example, condensed sections associated with late Pleistocene maximum flooding surfaces of the Louisiana continental shelf share this association, albeit with different expression of authigenic mineralization (pyrite and carbonate overgrowths on shells rather than glauconite) (Coleman and Roberts, 1988).

Late Highstand Systems Tract: The Forest Hill Delta

Transition from Red Bluff Clay condensed deposits to prodelta clay deposits of the Forest Hill Formation is marked by change from marine-dominated to terrestrial-dominated palynomorph assemblages and palynofacies, and by absence in the clay facies of the Forest Hill of significant occurrences of shell hash, glauconite, burrows, or calcareous plankton. Beds with characteristics of the Red Bluff Clay may occur interbedded with the clay facies of the Forest Hill as high as 9.8 m (32 feet) above the Shubuta–Red Bluff contact in Wayne County, Mississippi (May 1974). In both Young #1 and Lime Company #1, the Forest Hill Formation can be divided into an upper and a lower part (figures 11.6 and 11.7). Most of the thickness of the Forest Hill is contained within the lower part, characterized by changes up section that signal progradation from a distal to more proximal prodelta environment (figure 11.11). Progradation is indicated by changes in sediment texture and foraminiferal

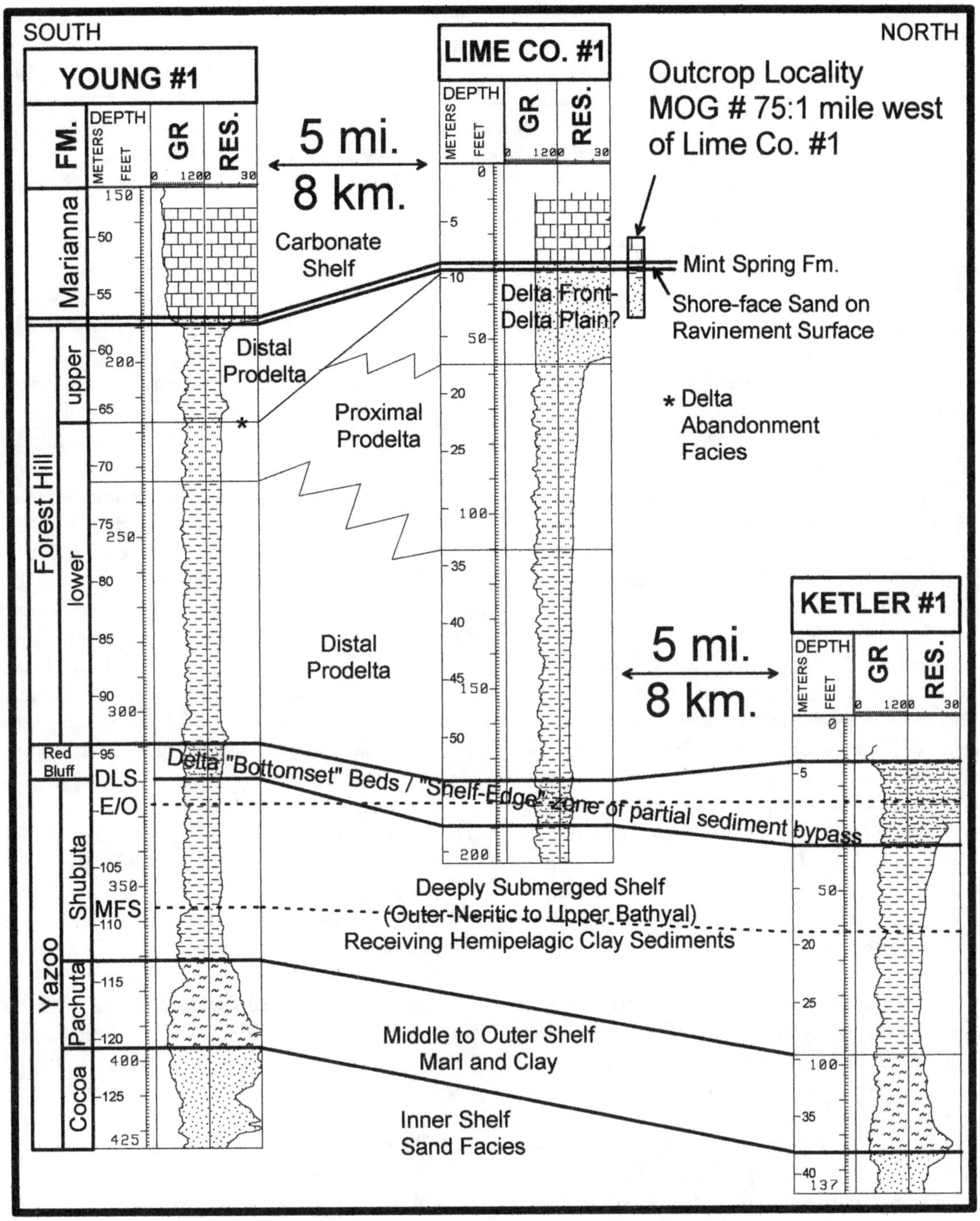

Figure 11.11. Depositional environments of the Eocene/Oligocene boundary interval (Sequence TE 3.3) in the three Mobil-MOG coreholes from southeastern Mississippi. E/O, Eocene/Oligocene boundary as determined by dinocyst graphic correlation (Jaramillo and Oboh-Ikuenobe (1999); MFS, Maximum Flooding Surface; DLS, Downlap Surface.

abundance and biofacies. The lower Forest Hill is olive gray clay, nearly homogeneous in its lower part that is here interpreted as distal prodelta, but with frequent irregular light gray silt stringers in the upper part that is interpreted as proximal prodelta. The silt stringers suggest temporal variation in the suspended load that is more likely in a proximal than a distal prodelta environment. The calcareous biogenic content is low. Macrofossil shell material and calcareous nannofossils are very rare. Although much less abundant than in the Pachuta Marl, Shubuta Clay, and Red Bluff Clay, foraminifers are common in the distal prodelta environment, but rare to absent in the proximal prodelta environment. Although Young #1 is located only 5 miles farther seaward of the paleoshoreline than Lime Company #1 (see Tew and Mancini, 1995), the distal prodelta environment is dominant in Young #1 and the proximal prodelta environment is dominant in Lime Company #1 (figure 11.11). For example, the uppermost part of the lower Forest Hill that is barren of foraminifers in our samples is only 3 m (10 feet) thick in Young #1, but 13.7 m (45 feet) thick in Lime Company #1 (figures 11.6 and 11.7). As observed in core photographs, silt stringers are frequent in the upper 5.2 m (17 feet) of the lower Forest Hill of Young #1 and intermittently frequent over the upper 15-18 m (50–60 feet) of the lower Forest Hill of Lime Company #1.

The base of the upper part of the Forest Hill Formation in Young #1 and Lime Company #1 is marked by an influx of siliciclastic sand (figures 11.6 and 11.7). Above that influx of sand, however, resemblance ceases between the upper parts of the Forest Hill in these two coreholes. The sand in Young #1 is only 0.2 m (0.6 feet) thick, dominantly siliciclastic, but with common glauconite grains, and rich in macrofossil debris. A small nannofossil peak (figure 11.6) also characterizes it. This thin sand is overlain by 0.3 m (0.9feet) of highly indurated fossiliferous sandy limestone or limey sandstone. The latter is overlain in turn by 7.8 m (25.6 feet) of clay (Dockery et al., 1994). This upper Forest Hill clay facies in Young #1 represents a return to a depositional environment similar to the lower distal deltaic part of the lower Forest Hill. It is olive gray in color and lacks macrofossil debris. Light gray silt stringers are rare through most of its thickness but become common in the upper 0.8 m (2.6 feet). Foraminifers, absent in samples from the upper 3.0 m (10 feet) of the lower Forest Hill, reappear in the sand and continue upsection to the top of the Forest Hill. Foraminiferal species that occur in the Red Bluff, but disappear upsection above the lower 6.1 m (20 feet) of Forest Hill, reappear in the thin sand at the base of the upper Forest Hill deposit. These include *Cibicidoides pippeni* and *Liebuella byramensis.* In Wayne County, Mississippi, a thin, fossiliferous sand layer subdividing the clay facies of the Forest Hill Formation, apparently similar to the one in Young #1, has been reported in three other cores (May, 1974) and one outcrop MOG #88 (MacNeil and Dockery, 1984). This upper Forest Hill unit in Young #1 may be a product of autocyclic delta lobe switching, but allocyclic causes, such as a "fourth-order" eustatic sea-level cycle, cannot be ruled out with the evidence at hand.

The sand that appears at the base of the upper Forest Hill Formation in Lime Company #1, is 7.6 m (25 feet) thick and the overlying clay only 0.6 m (2 feet) thick (figure 11.7). Cores of the upper Forest Hill sand unit were not recovered, but the SP and resistivity curves indicate that the sand is sharp-based, suggesting a channel fill deposit (G. R. Baum, pers. commun.). Upper Forest Hill sand sediment can be inspected in outcrop a mile west of Lime Company #1, along depositional strike, at outcrop localities MOG #74 and MOG #75 (Dockery, 1982; MacNeil and Dockery, 1984). At locality MOG #75, the upper 4.3 m (14 feet) of the Forest Hill was exposed above river level when the section was measured, described, and collected by Mobil personnel in October 1993. The lower 3.4 m (11 feet) of this exposure is unfossiliferous sandstone. Individual sand beds are 0.3 to 0.6 m (1 to 2 feet) thick and separated by clay drapes. Some of the beds are trough cross-bedded. Internal structure in other sand beds varies from parallel laminated to climbing ripples. The heterolithic character of the upper Forest Hill sand at MOG #75 indicates a tidally influenced environment, perhaps a tidal delta (R. Moiola, pers. commun.). Unfortunately, the lower boundary of the sand cannot be observed at this locality. Evidence of the "normal" upward facies succession from prodelta clay through interbedded sand and clay of the delta front facies is rare in Wayne and Clarke counties, Mississippi. Although most cores and wells of the Forest Hill Formation from these counties are exceedingly clay-rich (Gilliland, 1980; May, 1974), there are exceptions. MacNeil (1944) described a Forest Hill section in a well at Hiwannee (figure 11.1) that has a gradual upward progression from prodelta clay to "typical, laminated Forest Hill sand and clay" at the top. The Sohio #109-A and #142 wells from western Wayne County may have sand-rich delta front to delta plain facies as well (see Dockery, 1982).

The upper surface of the sand at MOG #75 is mottled and burrowed and it is overlain by almost 1 m (3 feet) of sandy clay with mollusk shells. Only one of the samples from the overlying sand yielded a foraminiferal fauna. It is dominated by *Hanzawaia* sp. and includes a few planktonic foraminifers indicating an open-marine, shallow-water environment. Dockery (1982) has also noted marine fossils in the silt and clay deposits that overlie the Forest Hill delta. He suggests these may have been washed into an estuarine environment by storms. However, we suggest an alternative hypothesis, namely that these fossils represent the updip expression of the flooding event in the upper Forest Hill Formation of Young #1, reflecting either an allocyclic fourth-order transgression or an autocyclic switching of deltaic distributary systems (figure 11.11).

TO 1.1 Sequence Boundary

In agreement with most previous studies, we place the TO 1.1 sequence boundary coincident with the base of the Mint Spring Marl, a prominent ravinement surface well described in the literature (Dockery, 1982; MacNeil, 1944; MacNeil and Dockery, 1984; Tew and Mancini, 1995). At this surface at outcrop locality #75, and in the Lime Company #1 core, the underlying Forest Hill Formation is gently scoured and penetrated by burrows filled with Mint Spring sediment. The transgressive Mint Spring Marl at these localities is shelly sand containing rounded clay clasts. In the Young #1 core, our most basinward locality, the Mint Spring Marl is glauconitic, sandy limestone with a 12-cm basal layer of glauconitic sand and clay (Dockery et al., 1994). The basal part of the Mint Spring Marl is time transgressive along its outcrop belt, falling within NP21 in southwestern Alabama and southeastern Mississippi and within NP22 in central and western Mississippi (Bybell, 1982; Pettway and Dunn, 1990; Siesser, 1983). Locally, we interpret no lowstand deposits, or incised valley fill, although we recognize that they may indeed be present. For example, Baum (pers. commun.) has suggested that the upper Forest Hill sand unit in Lime Company #1 may be incised valley fill citing the blocky profile of the log motif at its base as evidence. Although we acknowledge that this sand may be channel fill, we believe that it is more likely part of the normal delta front facies because the underlying sediments are proximal prodelta facies (figure 11.11).

Highstand Depositional Model

Figure 11.12 is a two-dimensional conceptual model for the transgressive and highstand systems tracts of sequence TE 3.3 in southeastern Mississippi. It proposes a sedimentological mechanism to explain the following observations: 1) the uppermost Shubuta Clay at Young #1 is a facies of the Red Bluff Clay at Lime Company #1 and Ketler #1; 2) the Red Bluff Clay is a condensed section, although not necessarily associated with the maximum flooding surface; and 3) at times during Red Bluff sedimentation the sea floor experienced fluid energy capable of maintaining fine clastic sediment in suspension. The model addresses differences in the patterns of concentration of biogenic and authigenic sediment between the Shubuta Clay, Red Bluff Clay, and clay facies of the Forest Hill Formation.

This report supports the dinocyst graphic correlation of uppermost Shubuta Clay in Young #1 to Red Bluff Clay in Lime Company #1 and Ketler #1 (Jaramillo and Oboh-Ikuenobe, 1999) because the results of that correlation appear to provide the most plausible explanation for the magnetostratigraphic data and planktonic foraminiferal occurrences in the same coreholes (figure 11.4). In each of the southeastern Mississippi coreholes, the Shubuta–Red Bluff boundary signals the end of hemipelagic sedimentation and the beginning of progradation of the highstand delta. Internal facies relationships within the Forest Hill Formation indicate that the direction of progradation included a strong north-to-south component from Lime Company #1 toward Young #1, although the true direction of progradation may have been more nearly southeast (Tew and Mancini, 1995). The Young #1 core-hole shows that the delta overrode hemipelagic Shubuta Clay deposited contemporaneous with the earliest phases of the delta in the other two coreholes (figure 11.12). More than 2.5 m of uppermost Shubuta Clay (from the Eocene/Oligocene boundary to the Chron C13n/C13r boundary) were deposited at Young #1 contemporaneous with sedimentation of the Red Bluff Clay at Lime Company #1. The thickness of contemporary Shubuta Clay was more than this because part of the Red Bluff of Lime Company #1 and Ketler #1 predates the Eocene/Oligocene boundary. In order to reach the location of Young #1, this clay sediment traveled over or around the Red Bluff Clay that was accumulating as a condensed section 8 km (5 miles) and 16 km (10 miles) away at Lime Company #1 and Ketler #1, respectively.

The Red Bluff Clay is a condensed shelf deposit influenced by proximity to the delta. True prodelta sedimentation, represented by the clay facies of the Forest Hill Formation, was still farther to the north and west at this time. Beds with very high concentrations of glauconite, and other characteristics of condensed sections previously enumerated, indicate

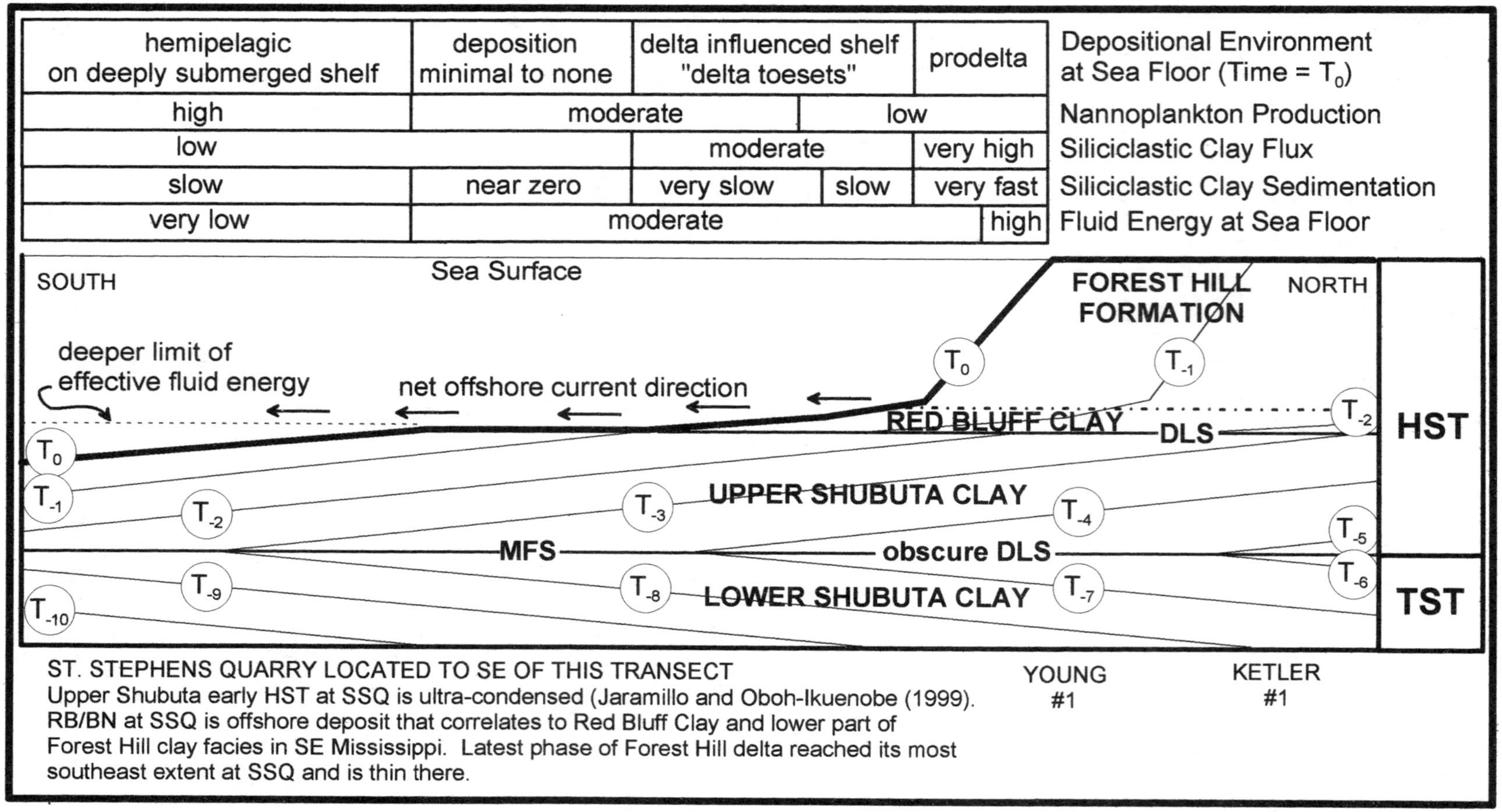

Figure 11.12. Conceptual model relating interpreted stratal geometry and depositional processes of transgressive systems tract (TST) and highstand systems tract (HST) of TE 3.3 in southeastern Mississippi.

that at least some of the Red Bluff Clay was deposited very slowly, perhaps more slowly than most of the Shubuta Clay in the coreholes where glauconite is insignificant. Although lacking glauconite or biogenic attributes of a maximum condensed section, the rate of deposition of the uppermost Shubuta Clay in Young #1 was slow in absolute terms, only 1.7 cm/1000 years (figure 11.6). This is within the range that has been postulated for condensed sections (<1-2 cm/1000 years, see Baum and Vail, 1988). It is slightly more than one-half of the average sedimentation rate calculated for all of the Shubuta Clay, 3.1 cm/1000 years (figure 11.6). Even the average rate for the Shubuta Clay differs little from the average sedimentation rate for condensed sections associated with late Pleistocene maximum flooding on the Louisiana shelf. This was calculated at 2.7 cm/1000 years after removal of outliers (Coleman and Roberts, 1988).

Glauconite and shell debris in the Red Bluff are concentrated in lenses that appear to rest on diastems (MacNeil and Dockery, 1984). Dockery (1990) inferred fluid energy at the seafloor from such observations, suggesting that storm waves may have been the responsible agent. Tew and Mancini (1995) interpret these features of the Red Bluff as shoaling, progradational parasequences, capped by scoured marine flooding surfaces. Their interpretation of scoured surfaces seems to acknowledge the significance of fluid energy during some phases of Red Bluff deposition. The diastems or scoured surfaces indicate intermittent resuspension of fine-grained sediment. Clay sediment resuspended or maintained in suspension by this fluid energy would have been available for offshore transport and final deposition as Shubuta Clay at greater water depths than that reached by the agent supplying the fluid energy.

Gravity acting on the prodelta slope and the transport mechanism described above, along with nearshore to offshore and shallow to deep environmental gradients, shaped the concentrations and relative proportions of authigenic and biogenic sediment that accumulated in each depositional setting. Shells of microfauna (including foraminifers) and macrofauna moved down the prodelta slope, accounting for the shallow-water biofacies found in some distal prodelta sediment (figure 11.7). Some of this downslope transport reached the apron of Red Bluff shelf sediment that accumulated next to the delta, accounting for the freshwater gastropod found in the Red Bluff Clay at Hiwannee (MacNeil, 1944). Unlike the assemblages of transported shallow water mollusks and foraminifers that occur in sand interbedded with uppermost Yazoo Clay at Cynthia Quarry in west-central Mississippi (Smith and Zumwalt, 1987), little siliciclastic sand is associated with transported shells in the clay facies of the Forest Hill. Probably, little siliciclastic sand was available for transport because the entire delta is so clay-rich in southeastern Mississippi. Glauconite, shell debris, and foraminifers are largely autochtonous in the Red Bluff. Although occasionally moved around, they are too large for significant transport by the deep-shelf currents. They were concentrated in glauconite and shell beds because much of the clay- and silt-sized sediment that could have diluted them was transported through the area to deeper environments. The properties of the clay- and silt-sized siliciclastic and biogenic material that accumulated in the Red Bluff are partly inherited from the delta and partly modified by nearshore and mid-shelf processes. The inherited properties include terrestrial palynodebris and organic carbon that has very little hydrocarbon content. Because calcareous nannofossils are in the size range of coarse clay to very fine silt, they are readily transported with the siliciclastic clay through Red Bluff depositional sites. The extreme scarcity of them in Forest Hill clay, and the low to moderate concentrations of them in the Red Bluff Clay can be attributed to a high flux of siliciclastic clay relative to local production of calcareous nannoplankton. Extremely low concentrations of calcareous nannofossils are typical of prodelta sediments (Everett, 1982). Their production is undoubtedly lower in surface waters of variable salinity and turbidity around the delta than on parts of the shelf distant from the delta. However, dinocysts seem to flourish on the mid-shelf in the proximity of deltas (Jaramillo and Oboh-Ikuenobe, 1999), if their relative proportions in palynodebris and palynomorph assemblages from the Red Bluff are any indication of their productivity.

Only the smallest of biogenic particles, and those most resistant to degradation, are transported offshore. Therefore, the Shubuta biogenic fraction is dominated by calcareous nannofossils and algal-sourced kerogen (AOM) produced in offshore surface waters, but stimulated by nutrient-rich water in the vicinity of the delta (see Denne and Sen Gupta, 1991, 1993).

The lithofacies and biofacies of the Red Bluff Clay can be understood as products of processes similar to those affected by modern shelf and shelf-edge dynamics in the northern Gulf of Mexico. The following quote from McGrail and Carnes (1983:263)

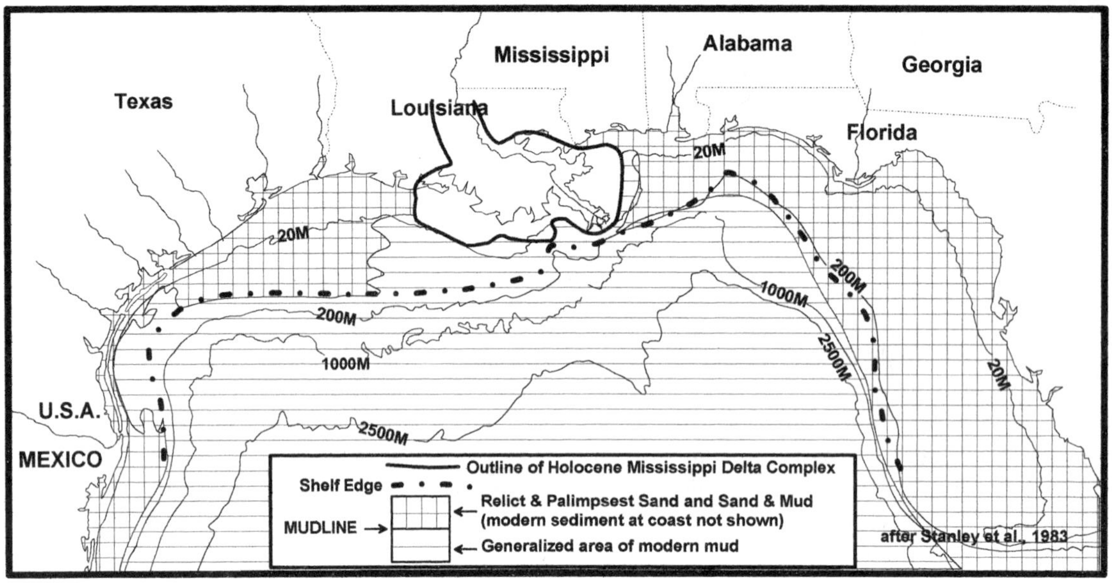

Figure 11.13. Northern Gulf of Mexico showing areas of mud depostion on shelf, slope, and basin and areas of relict and palimpsest sediment on shelf. Offshore oriented bottom currents sweep the shelf and prevent mud deposition except in areas of high sediment flux.

summarizes these dynamics: "The waters of the shelf-break are subject to motions with many time and length scales. These are superimposed on one another, producing energetic currents, which vary in time and space. Many of these motions produce oscillations in the flow that are directed across the trend of the shelf isobaths and sweep fine sediment off the shelf and into deep water. The flow is frequently of such magnitude that it could move sand if it were available at the shelfedge. These processes provide a very efficient mechanism for supplying quantities of sediment to the slope, rise, and deep sea." The currents are of several types including internal waves and inertial currents. Surface waves, including those forced by winter storms and hurricanes and by tidal currents are not important on the outer shelf of the Gulf of Mexico (McGrail and Carnes, 1983). At least in part, these processes apply to the middle and outer shelf as well as the shelf-edge. The efficiency of these processes is such that little or no net accumulation of silt and clay sediment occurs on much of the modern shelf of the Gulf of Mexico (figure 11.13). The fine grain sediment that passes over the outer shelf and shelf-edge is deposited on the slope below the mudline. As documented by Stanley et al. (1983), the mudline is the depth of substantially increased silt and clay content and the level below which deposition prevails on continental margins. It often occurs near, but is only rarely coincident with, the shelf-to-slope transition. Three of the most important factors controlling the position of the mudline are shelf width, sediment supply, and magnitude of fluid energy on the shelf, at the shelf-break, and on the upper slope.

The interpretation presented here visualizes that the Shubuta Clay was deposited on a deeply submerged shelf below the depth of effective currents and below the contemporary mudline (figure 11.12). When water depth decreased sufficiently to bring the seafloor to within the depth range of effective currents, the initial effect may have been complete sediment bypass and cessation of sediment accumulation for a while, conditions then being comparable to those on the modern Gulf of Mexico shelf and shelf-edge above the mudline. Such a break in sedimentation, too brief to be detected at the resolution of the graphic correlation studies, might help to explain the sharp, burrowed contact between the Shubuta Clay and Red Bluff Clay. Progradation of the Forest Hill delta brought with it an increase in flux of clay and silt, eventually overcoming the capacity of currents to transport all of this fine-grained sediment through the local area.

CONCLUSIONS

The data of this study do not uniquely determine the Eocene/Oligocene boundary because they cannot resolve whether sedimentation in the boundary interval is continuous or interrupted and whether isolated, rare, occurrences of marker species are in place or reworked. However, this data become significant when integrated with the dinocyst graphic correlation of the E/O boundary by Jaramillo and Oboh-Ikuenobe (1999). For example, their graphic correlation successfully predicts the stratigraphic level of the Chron 13n/13r boundary in Young #1. It also successfully predicts the highest stratigraphic occurrences of *Hantkenina* spp. in the Lime Company #1 and Ketler #1 coreholes suggesting that these very rare and stratigraphically isolated occurrences are in place rather than reworked. Therefore, we have accepted the dinocyst graphic correlation within the stratigraphic interval that includes the E/O boundary and have used it for sequence stratigraphic analysis.

In southeastern Mississippi, acmes of the latest Eocene index foraminifers *Hantkenina alabamensis* and *Globorotalia (T.) cerroazulensis* terminate abruptly and concurrently well below the E/O boundary. This very severe, local Eocene planktonic foraminiferal (EPF) decimation event occurred about 350,000 years before global extinction of these two species (figure 11.5).

The E/O boundary in the Mobil-MOG coreholes occurs in stratigraphic proximity to the lithostratigraphic boundary between the Shubuta Clay member of the Yazoo Clay and the overlying Red Bluff Clay confirming most previous studies. However, it does not coincide with this boundary (Jaramillo and Oboh-Ikuenobe, 1999). Rather, the Shubuta–Red Bluff lithostratigraphic boundary climbs stratigraphically in the offshore direction from uppermost Eocene in the Ketler #1 and Lime Company #1 coreholes to a location low within the Oligocene in Young #1. This conclusion has important implications for sequence stratigraphy because the lithofacies and biofacies changes at the clay-on-clay Shubuta–Red Bluff contact can be reasonably interpreted as evidence for an abrupt decrease in water depth, although we do not support this interpretation. This apparent downshift in facies at the Shubuta–Red Bluff contact has been used to support interpretation of a sequence boundary at the E/O boundary. However, if we are correct that the Shubuta–Red Bluff contact climbs section stratigraphically, it cannot be a sequence boundary because it does not separate older from younger beds everywhere. Thus, the sequence stratigraphic chronology of the E/O boundary interval in eastern Gulf Coastal Plain that we support here is out of phase with widely used global and supraregional sequence chronostratigraphy charts (Haq et al., 1987, 1988; Vakarcs et al., 1998) that correlate the E/O boundary to a sequence boundary.

Similar to the earliest interpretations of the E/O boundary sequence of the southeastern Gulf Coastal Plain, we interpret shelf margin, transgressive, and highstand systems tracts. However, in our interpretation, two distinct condensed sections complicate the transition from transgressive to highstand system tracts. The stratigraphically lower of the two interpreted condensed sections is located in the lower Shubuta Clay Member of the Yazoo Formation where it caps the strongly transgressive Pachuta Marl member. Peak foraminiferal concentrations, maximum foraminiferal planktonic-benthonic ratios, maximum water depth, the most offshore dinocyst facies, and a gamma ray peak are associated with this lower condensed section. The maximum flooding surface of the sequence may occur here, perhaps at the gamma ray peak. The upper part of the Shubuta Clay member between the two interpreted condensed sections is a deep-water, hemipelagic deposit characterized by maximum total organic carbon, kerogen with maximum hydrogen index, and maximum percent aquatic amorphous organic matter. A more shelfal aspect to the dinocyst populations and lower planktonic-benthonic foraminiferal ratios are the basis for including it within the earliest part of the highstand. Its depositional rate was scarcely more than that of some condensed sections. The upper condensed section is the Red Bluff Clay. It overlies upper Shubuta hemipelagic deposits and is transitional with lower Forest Hill prodelta clay. The Red Bluff Clay is an interval of burrowed horizons, shell hash, authigenic minerals, elevated foraminiferal abundance, and reduced calcareous nannofossil abundance. We interpret it as a composite set of sediment-starved intervals associated with a distal delta foreset setting characterized by sediment bypass.

The entire early highstand, that is, the upper Shubuta Clay and Red Bluff Clay, is highly condensed at St. Stephens Quarry, according to the dinocyst graphic correlation of Jaramillo and Oboh-Ikuenobe (1999).

ACKNOWLEDGMENTS

We thank ExxonMobil for permission to publish. We would especially like to thank David Dockery III and the Mississippi Office of Geology for guidance in setting up this project and for carrying out drilling of the coreholes and collection of the cores. During the course of this research, the first author consulted with David Dockery III, Gerald Baum and Richard Fluegeman on a number of technical topics. He wishes to thank them for freely sharing their expertise. We gratefully acknowledge the very helpful suggestions made by Nick Tew and Ernest Mancini in their reviews of an earlier draft of this manuscript.

REFERENCES

Abreu, V. S., and Haddad, G. A., 1998, Glacioeustatic fluctuations; the mechanism linking stable isotope events and sequence stratigraphy from the early Oligocene to middle Miocene, *in* de Graciansky, P.-C., Hardenbol, J., Jacquin, T., and Vail, P. R., eds., Mesozoic and Cenozoic sequence stratigraphy of European basins: SEPM Special Publication, v. 60, p. 245-259.

Armentrout, J. M., 1991, Paleontologic constraints on depositional modeling: examples of integration of biostratigraphy and seismic stratigraphy, *in* Weimer, P., and Link, M. H., eds., Seismic facies and sedimentary proceses of submarine fans and turbidite systems: Frontiers in sedimentary geology: New York, Springer Verlag, p. 137-170.

Armentrout, J. M., Echols, R. J., and Lee, T. D., 1990, Patterns of foraminiferal abundance and diversity: implications for sequence stratigraphic analysis, *in* Perkins, B. F., ed., Sequence stratigraphy as an exploration tool, concepts and practices in the Gulf Coast: Eleventh Annual Research Conference, Gulf Coast Section, SEPM Foundation, p. 15-23.

Bandy, O. L., 1949, Eocene and Oligocene Foraminifera from Little Stave Creek, Clarke County, Alabama: Bulletins of American Paleontology, v. 32, no. 131, p. 210.

Baum, G. R., and Vail, P. R., 1987, Sequence stratigraphy, allostratigraphy, isotope stratigraphy and biostratigraphy; putting it all together in the Atlantic and Gulf Paleogene, *in* McNulty, C. L., and Waterman, A. S., eds., Innovative biostratigraphic approaches to sequence analysis; new exploration opportunities: Eighth annual research conference, Gulf Coast Section, SEPM Foundation: Houston, Texas, p. 15-23.

Baum, G. R., and Vail, P. R., 1988, Sequence stratigraphic concepts applied to Paleogene outcrops, Gulf and Atlantic basins, *in* Wilgus, C. K., Hastings, B. S., Ross, C. A., Posamentier, H., Van Wagoner, J., and Kendall, C. G. S. C., eds., Sea-level changes; an integrated approach: SEPM Special Publication, v. 42, p. 309-327.

Baum, G. R., Vail, P. R., and Hardenbol, J., 1982, Unconformities and depositional sequences in relationship to eustatic sea level change, Gulf and Atlantic Coastal Plains, *in* International Geological Correlation Program Project 714, International Field Conference, Louisiana State University, Baton Rouge, p. 1-3.

Baum, J. S., Baum, G. R., Thompson, P. R., and Humphrey, J. D., 1994, Stable isotopic evidence for relative and eustatic sea-level changes in Eocene to Oligocene carbonates, Baldwin County, Alabama: Geological Society of America Bulletin, v. 106, p. 824-839.

Berggren, W. A., Kent, D. V., Swisher, C. C., III, and Aubry, M.-P., 1995, A revised Cenozoic geochronology and chronostratigraphy, *in* Berggren, W. A., Kent, D. V., Aubry, M.-P., and Hardenbol, J., eds., Geochronology, time scales, and global dtratigraphic vorrelation: SEPM Special Publication, v. 54, p. 129-212.

Bown, P. R., and Young, J. R., 1998, Techniques, *in* Bown, P. R., ed., Calcareous nannofossil biostratigraphy: British Micropalaeontological Society Publication Series, Chapman & Hall, p. 314.

Bybell, L. M., 1982, Late Eocene to early Oligocene calcareous nannofossils in Alabama and Mississippi: Gulf Coast Association of Geological Societies Transactions, v. 32, p. 295-302.

Bybell, L. M., and Poore, R. Z., 1983, Reworked *Hantkenina* specimens at Little Stave Creek, Alabama: Gulf Coast Association of Geological Societies Transactions, v. 33, p. 253-256.

Coleman, J. L., Jr., 1990, Sequence stratigraphy of the Tertiary of the North Central Gulf Coastal Plain: what goes on between the seismic reflectors, *in* Sequence Stratigraphy as an exploration tool: Eleventh Annual Research Conference, Gulf Coast Section, SEPM Foundation, Houston, p. 87-98.

Coleman, J. M., and Roberts, H. H., 1988, Sedimentary development of the Louisiana continental shelf related to sea level cycles: Part I—Sedimentary sequences: Geo-Marine Letters, v. 8, no. 2, p. 63-108.

Corliss, B. H., 1985, Microhabitats of benthic foraminifera within deep-sea environments: Nature, v. 314, p. 435-436.

Corliss, B. H., 1991, Morphology and microhabitat

preferences of benthic foraminifera from the Northwest Atlantic Ocean: Marine Micropaleontology, v. 17, no. 3-4, p. 195-236.

Corliss, B. H., and Chen, C., 1988, Morphotype patterns of Norwegian Sea deep-sea benthic foraminifera and ecological implications: Geology, v. 16, no. 8, p. 716-719.

Corliss, B. H., and Fois, E., 1990, Morphotype analysis of deep-sea benthic foraminifera from the Northwest Gulf of Mexico: Palaios, v. 5, no. 6, p. 589-605.

Creaney, S., and Passey, Q. R., 1994, Recurring patterns of total organic carbon and source rock quality within a sequence stratigraphic framework: American Association of Petroleum Geologists Bulletin, v. 77, p. 386-401.

Cushman, J. A., and Todd, R., 1948, Foraminifera from the Red Bluff-Yazoo section at Red Bluff, Mississippi: Contributions from the Cushman Laboratory for Foraminiferal Research, v. 24, p. 1-12.

Deboo, P. B., 1965, Biostratigraphic correlation of the type Shubuta Member of the Yazoo Clay and Red Bluff Clay with their equivalents in southwestern Alabama: Geological Survey of Alabama Bulletin 80, 84 p.

Denne, R. A., and Sen Gupta, B. K., 1991, Association of bathyal foraminifera with water masses in the northwestern Gulf of Mexico: Marine Micropaleontology, v. 17, no. 3-4, p. 173-193.

Denne, R. A., and Sen Gupta, B. K., 1993, Matching of benthic foraminiferal depth limits and watermass boundaries in the northwestern Gulf of Mexico: an investigation of species occurrences: Journal of Foraminiferal Research, v. 23, no. 2, p. 108-117.

Dockery, D. T., III, 1980, The invertebrate macropaleontology of the Clarke County, Mississippi, area: Mississippi Department Natural Resources, Bureau of Geology, Bulletin No. 122, 387 p.

Dockery, D. T., III, 1982, Lower Oligocene Bivalvia of the Vicksburg Group in Mississippi: Mississippi Department Natural Resources, Bureau of Geology, Bulletin 123, 261 p.

Dockery, D. T., III, 1990, The Eocene/Oligocene boundary in the northern Gulf—A sequence boundary, *in* Perkins, B. F., ed., Sequence stratigraphy as an exploration tool, concepts and practices in the Gulf Coast: Eleventh Annual Research Conference, Gulf Coast Section, SEPM Foundation, p. 141-150.

Dockery, D. T., III, and Siesser, W. G., 1984, Age of the upper Yazoo Formation in central Mississippi: Mississippi Geology, v. 5, no. 1, p. 1-10.

Dockery, D. T., III, Stover, C. W., Weathersby, P., Stover, C. W., Jr., and Ingram, S. L., 1991, A continuous core through the undifferentiated Yazoo Clay (late Eocene, Jackson Group) of central Mississippi: Mississippi Geology, v. 12, no. 3,4, p. 21-27.

Dockery, D. T., III, Thompson, D. E., and Ingram, S. L., 1994, The Mobil–Mississippi Office of Geology core-hole project: Mississippi Geology, v. 15, no. 1, p. 8-15.

Eames, F. E., Blow, W. H., Clarke, W. J., and Cox, L. R., 1962, Fundamentals of mid-Tertiary stratigraphical correlation: Cambridge, England, Cambridge University Press, 163 p.

Everett, R., 1982, Using nannofossil counts in the interpretation of subsurface deltas: Gulf Coast Association of Geological Societies Transactions, v. 32, p. 579-591.

Fluegeman, R. H., 1996, Preliminary paleontological report on the foraminifera of the Mossy Grove Core, Hinds County, Mississippi: Mississippi Geology, v. 17, no. 1, p. 9-15.

Frederick, D. L., 1994, Eocene and Oligocene benthic foraminifera in the eastern Gulf Coast [Unpubl. Ph.D. dissert.]: Knoxville, The University of Tennessee, 340 p.

Gibson, T. G., 1989, Planktonic, benthonic foraminiferal ratios: modern patterns and Tertiary applicability: Marine Micropaleontology, v. 15, p. 29-52.

Gilliland, W. A., 1980, Clarke County Geology, Clarke County Geology and Mineral Resources: Mississippi Bureau of Geology Bulletin 121, p. 11-147.

Grünig, A., and Herb, R., 1980, Paleoecology of late Eocene benthonic foraminifera from Possagno (Treviso—northern Italy), *in* Sliter, W. V., ed., Studies in marine micropaleontology and paleoecology: a memorial volume to Orville L. Bandy: Cushman Foundation for Foraminiferal Research Special Publication No. 19, p. 68-85.

Haq, B. U., Hardenbol, J., and Vail, P. R., 1987, Chronology of fluctuating sea levels since the Triassic: Science, v. 235, no. 4793, p. 1156-1167.

Haq, B. U., Hardenbol, J., and Vail, P. R., 1988, Mesozoic and Cenozoic chronostratigraphy and eustatic cycles, *in* Wilgus, C. K., Hastings, B. S., Posamentier, H., Van Wagoner, J., Ross, C. A., and Kendall, C. G. S. C., eds., Sea-level changes: an integrated approach: SEPM Special Publication, v. 42, p. 71-109.

Hazel, J. E., Mumma, M. D., and Huff, W. J., 1980, Ostracode biostratigraphy of the lower Oligocene (Vicksburgian) of Mississippi and Alabama: Gulf Coast Association of Geological Societies Transactions, v. 30, p. 361-401.

Hendy, W. J., 1948, Notes on the stratigraphy of northeastern Wayne County, Mississippi, Guide Book, Sixth Field Trip, June 18, 19, 20, Upper Eocene, Oligocene, and lower Miocene of Central Mississippi: Mississippi Geological Survey, p. 25-36.

Honjo, S., and Erez, J., 1978, Dissolution rates of calcium carbonate in the deep ocean; an in-situ experiment in the North Atlantic Ocean: Earth and Planetary Science Letters, v. 40, p. 287-300.

Huddlestun, P. F., and Toulmin, L. D., 1965, Upper Eocene–lower Oligocene stratigraphy and paleontology in Alabama: Gulf Coast Association of Geological Societies Transactions, v. 15, p. 155-159.

Huff, W. J., 1970, The Jackson Eocene Ostracoda of Mississippi: Mississippi Geological Survey Bulletin 114, 389 p.

Jaramillo, C., and Oboh-Ikuenobe, F. E., 1999, Sequence stratigraphic interpretations from palynofacies, dinocyst and lithological data of upper Eocene–lower Oligocene strata in southern Mississippi and Alabama, U.S. Gulf Coast: Palaeogeography, Palaeoclimatology, Palaeoecology, v. 145, p. 259-302.

Jiang, M. M., 1997, Middle Eocene through basal Miocene sequence bio-stratigraphy of the western Gulf Coast region: Marine and Petroleum Geology, v. 14, no. 7/8, p. 855-866.

Jones, R. W., 1983, Organic matter characteristics near the shelf-slope boundary, *in* Stanley, D. J., and Moore, G. T., eds., The shelfbreak: critical interface on continental margins. SEPM Special Publication, v. 33 p. 391-405.

Jordan, D. W., Slatt, R. M., Gillespie, R. H., D'Agostino, A. E., and Scheihing, M. H., 1990, Applications of outcrop gamma-ray logging to field development and exploration, The Integration of Geology, Geophysics, Petrophysics and Petroleum Engineering in Reservoir Delineation, Description and Management: Proceedings of the First Archie Conference, October 22–25, 1990, Houston, Texas, U.S.A.: Tulsa, Oklahoma, American Association of Petroleum Geologists, p. 104-122.

Keller, G., 1985, Eocene and Oligocene stratigraphy and erosional unconformities in the Gulf of Mexico and Gulf Coast: Journal of Paleontology, v. 59, no. 4, p. 882-903.

Kitazato, H., 1994, Foraminiferal microhabitats in four marine environments around Japan: Marine Micropaleontology, v. 24, no. 1, p. 29-41.

Langford, F. F., and Blanc-Valleron, M.-M., 1990, Interpreting rock-eval pyrolysis data using graphs of pyrolizable hydrocarbons vs. total organic carbon: American Association of Petroleum Geologists Bulletin, v. 74, no. 6, p. 799-804.

Loutit, T. S., Hardenbol, J., Vail, P. R., and Baum, G. R., 1986, Condensed sections: the key to age dating of continental margin sequences: Geological Society of America Abstracts with Programs, v. 18, no. 6, p. 676.

Loutit, T. S., Hardenbol, J., Vail, P. R., and Baum, G. R., 1988, Condensed sections: the key to age determination and correlation of continental margin sequences, *in* Wilgus, C. K., Hastings, B. S., Ross, C. A., Posamentier, H., Van Wagoner, J., and Kendall, C. G. S. C., eds., Sea-level changes; an integrated approach: SEPM Special Publication No. 42, p. 183-213.

MacNeil, F. S., 1944, Oligocene stratigraphy of southeastern United States: American Association of Petroleum Geologists Bulletin, v. 28, no. 9, p. 1313-1354.

MacNeil, F. S., and Dockery, D. T., III, 1984, Lower Oligocene Gastropoda, Scaphopoda, and Cephalopoda of the Vicksburg Group in Mississippi: Mississippi Department of Natural Resources, Bureau of Geology, Bulletin 124, 415 p.

Mancini, E. A., 2000, Sequence stratigraphy and chronostratigraphy of upper Eocene and lower Oligocene strata, eastern Gulf Coastal Plain: Gulf Coast Association of Geological Societies Transactions, v. 50, p. 379-388.

Mancini, E. A., and Waters, L. A., 1986, Planktonic foraminiferal biostratigraphy of upper Eocene and lower Oligocene strata in southern Mississippi and southwestern and south-central Alabama: Journal of Foraminiferal Research, v. 16, no. 1, p. 24-33.

Mancini, E. A., and Tew, B. H., 1991, Relationships of Paleogene stage and planktonic foraminiferal zone boundaries to lithostratigraphic and allostratigraphic contacts in the eastern Gulf Coastal Plain: Journal of Foraminiferal Research, v. 21, no. 1, p. 48-66.

Mancini, E. A., Tew, B. H., and Waters, L. A., 1987, Eocene/Oligocene boundary in southeastern Mississipi and southwestern Alabama: A stratigraphically condensed section of a Type 2 depositional sequence, *in* Ross, C. A., ed., Timing and depositional history of Eustatic sequences: constraints on seismic stratigraphy: Cushman Foundation Special Publication No. 24, p. 41-50.

May, J. H., 1974, Wayne County Geology, Wayne County Geology and Mineral Resources: Mississippi Geological Survey Bulletin, v. 117, p. 13-194.

McGrail, D. W., and Carnes, M., 1983, Shelfedge dynamics and the nepheloid layer, *in* Stanley, D. J., and Moore, G. T., eds., The shelfbreak: critical interface on continental margins. SEPM Special Publication 33, p. 251-264.

McKee, T. R., Jeffrey, L. M., Presley, B. J., and Whitehouse, U. G., II 1978, Holocene sediment geochemistry of continental slope and intraslope basin areas, northwest Gulf of Mexico, *in* Bouma, A. H., Moore, G. T., and Coleman, J. M., eds., Framework, facies, and oil-trapping characteristics of the upper continental margin: American Association of Petroleum Geologists, Studies in Geology 7, p. 313-326.

Michelsen, O., Thomsen, E., Danielsen, M., Heilmann-Clausen, C., Jordt, H., and Laursen, G. V., 1998, Cenozoic sequence stratigraphy in the eastern North Sea, *in* de Graciansky, P.-C., Hardenbol, J., Jacquin, T., and Vail, P. R., eds., Mesozoic and Cenozoic sequence stratigraphy of European basins: SEPM Special Publication, v. 60, p. 91-118.

Miller, K. G., Thompson, P. R., and Kent, D. V., 1993, Integrated late Eocene/Oligocene stratigraphy of the Alabama coastal plain: correlation of hiatuses and stratal surfaces to glacioeustatic lowerings: Paleoceanography, v. 8, no. 2, p. 313-331.

Mitchum, R. M., Jr., 1977, Seismic stratigraphy and global changes of sea level, Part I: Glossary of terms used in seismic stratigraphy, *in* Payton, C. E., ed., Seismic stratigraphy-applications to hydrocarbon exploration: American Association of Petroleum Geologists Memoir 26, p. 205-212.

Mitchum, R. M., Jr., Sangree, J. B., Vail, P. R., and Wornardt, W. W., 1991, Integrated well-log, seismic, and biostratigraphic approach to sequence stratigraphy in late Cenozoic expanded sections, Gulf of Mexico (abstract): American Association of Petroleum Geologists Bulletin, v. 75, no. 3, p. 637.

Mornhinveg, A. R., 1941, The foraminifera of Red Bluff: Journal of Paleontology, v. 15, p. 431-435.

Murray, G. E., 1961, Geology of the Gulf and Atlantic Coastal Province of North America: New York, Harper and Brothers, 692 p.

Pasley, M. A., and Hazel, J. E., 1990, Use of organic petrology and graphic correlation of biostratigraphic data in sequence stratigraphic interpretations: example from the Eocene/Oligocene boundary section, St. Stephens Quarry, Alabama: Gulf Coast Association of Geological Societies Transactions, v. 40, p. 661-683.

Pasley, M. A., and Hazel, J. E., 1995, Revised sequence stratigraphic interpretation of the Eocene/Oligocene boundary interval, Mississippi and Alabama, Gulf Coast Basin, U.S.A: Journal of Sedimentary Research, Section B: Stratigraphy and Global Studies, v. 65, no. 1, p. 160-169.

Pasley, M. A., Gregory, W. A., and Hart, G. F., 1991, Organic matter variations in transgressive and regressive shales: Organic Geochemistry, v. 17, no. 4, p. 483-509.

Peters, K. E., and Cassa, M. R., 1994, Applied source rock geochemistry, *in* The petroleum system—from source to trap: American Association of Petroleum Geologists Memoir 60: p. 93-120.

Pettway, W. C., and Dunn, D. A., 1990, Paleoenvironmental analysis of the lower Oligocene Mint Spring and Marianna Formations across Mississippi and southwestern Alabama: Gulf Coast Association of Geological Societies Transactions, v. 40, p. 701-709.

Pflum, C. E., and Frerichs, W. E., 1976, Gulf of Mexico deep-water foraminifers: Cushman Foundation for Foraminiferal Research Special Publication No. 14, 108 p.

Phleger, F. B., 1960, Ecology and distribution of recent foraminifera: Baltimore, John Hopkins Press, 297 p.

Poag, C. W., 1981, Ecologic atlas of benthic foraminifera of the Gulf of Mexico: Stroudsburg, Penn., Hutchinson Ross Publishing Co, 175 p.

Premoli Silva, I., and Jenkins, D. G., 1993, Decision on the Eocene/Oligocene boundary stratotype: Episodes, v. 16, no. 3, p. 379-382.

Richards, F. A., and Redfield, A. C., 1953, A correlation between the oxygen content of water and the organic content of marine sediments: Deep-Sea Research, v. 1, p. 279-281.

Romesburg, H. C., 1990, Cluster analysis for researchers: Malabar, Florida, Robert E. Krieger Publishing Co., 334 p.

Schaffer, B. L., 1990, The nature and significance of condensed sections in Gulf Coast late Neogene sequence stratigraphy: Gulf Coast Association of Geological Societies Transactions, v. 40, p. 186-195.

Siesser, W. G., 1983, Paleogene calcareous nannoplankton biostratigraphy: Mississippi, Alabama, and Tennessee: Mississippi Department of Natural Resources, Bureau of Geology, Bulletin No.125, 61 p.

Smith, S. M., and Zumwalt, G. S., 1987, Gravity flow introduction of shallow-water microfauna into deep water depositional environments: Mississippi Geology, v. 8, no. 2, p. 1-8.

Stanley, D. J., Addy, S. K., and Behrens, E. W., 1983, The mudline: variability of its position relative to shelfbreak, *in* Stanley, D. J., and Moore, G. T., eds., The shelfbreak: critical interface on continental margins: SEPM Special Publication, v. 33, p. 279-298.

Tew, B. H., and Mancini, E. A., 1995, An integrated stratigraphic method for paleogeographic reconstruction: examples from the Jackson and Vicksburg Groups of the eastern Gulf Coastal Plain: Palaios, v. 10, no. 2, p. 133-153.

Tipsword, H. L., Setzer, F. M., and Smith, F. L., Jr., 1966, Interpretation of depositional environment in Gulf Coast petroleum exploration from paleoecology and related stratigraphy: Gulf Coast Association of Geological Societies Transactions, v. 16, p. 119-130.

Toulmin, L. D., 1977, Stratigraphic distribution of Paleocene and Eocene fossils in the eastern Gulf Coast region: Geological Survey of Alabama Monograph 13, 602 p.

Traverse, A., 1988, Paleopalynology: Boston, Unwin Hyman Ltd., 600 p.

Vail, P. R., and Wornardt, W. W., 1990, Well log–seismic sequence stratigraphy: an integrated tool for the 90's, *in* Perkins, B. F., ed., Sequence stratigraphy as an exploration tool, concepts and practices in the Gulf Coast: Eleventh Annual Research Conference, Gulf Coast Section, SEPM Foundation, p. 379-388.

Vakarcs, G., Hardenbol, J., Abreu, V. S., Vail, P. R., Varnai, P., and Tari, G., 1998, Oligocene-middle Miocene depositional sequences of the central Paratethys and their correlation with regional stages, *in* de Graciansky, P.-C., Hardenbol, J., Jacquin, T., and Vail, P. R., eds., Mesozoic and Cenozoic sequence stratigraphy of European basins: SEPM Special Publication, v. 60, p. 209-231.

Van Wagoner, J. C., Posamentier, H. W., Mitchum, R. M. Jr., Vail, P. R., Sarg, J. F., Loutit, T. S., and Hardenbol, J., 1988, An overview of the fundamentals of sequence stratigraphy and key definitions, *in* Wilgus, C. K., Hastings, B. S., Ross, C. A., Posamentier, H. W., Van Wagoner, J., and Kendall, C. G. S. C., eds., Sea-level changes; an integrated approach: SEPM Special Publication, v. 42, p. 39-45.

Walton, W. R., 1964, Recent foraminiferal ecology and paleoecology, *in* Imbrie, J., and Newell, N., eds., Approaches to paleoecology: New York, John Wiley and Sons, Inc., p. 151-237.

Waters, L. A., and Mancini, E. A., 1982, Lithostratigraphy and biostratigraphy of upper Eocene and lower Oligocene strata in southwest Alabama and southeast Mississippi: Gulf Coast Association of Geological Societies Transactions, v. 32, p. 303-307.

CHAPTER 12

Late Middle Eocene Glacioeustasy: Stable Isotopes and Foraminifera from the Gulf Coastal Plain

John V. Hurley and Richard H. Fluegeman

ABSTRACT

A glacioeustatic control of the oxygen isotope record, as obtained from the late middle Eocene (Bartonian), is evident from the comparison of such a record with an independent proxy for paleobathymetry. Both $\delta^{18}O$ and planktonic: benthic foraminiferal ratios, as well as $\delta^{13}C$, were determined from samples taken every two feet from the 31.70-m section of the Moody's Branch Formation and lower Yazoo Clay as they occur in a section restrained to the *Truncorotaloides rohri* Biozone (P14) of the Mossy Grove Core, Hinds County, Mississippi. Isotope analyses were completed on tests of the benthic genus *Cibicidoides* while planktonic: benthic ratios were measured from processed and picked sample populations of 300 foraminifera tests.

Results of both the oxygen isotope analyses and the planktonic: benthic ratio determinations reveal a relationship of inverse correlation, whereby maximum planktonic: benthic values coincide with minimum $\delta^{18}O$ values and vice versa. Were late middle Eocene $\delta^{18}O$ records indicative only of temperature, as has been suggested, then the relationship determined here between the oxygen isotopes and the planktonic: benthic ratios would not be expected, as the latter is indicative of relative sea-level fluctuations, whereby higher values correspond to deeper water.

Seventeen coincident cycles, interpreted here as glacioeustatic, are apparent in the correlation of the $\delta^{18}O$ and planktonic: benthic curves during the 1.7-million-year P14 planktonic foraminiferan Zone. Sea-level changes are inferred from oxygen isotope fluctuations, during periods of known significant global ice volume. The findings reported here suggest that glacioeustacy may have been a significant controlling factor of sea-level fluctuations, by the Bartonian.

INTRODUCTION

During periods of widespread glacial advance and retreat, the oxygen isotopic composition of the world's oceans ($\delta^{18}O_{sw}$) fluctuate, causing changes in sea level and the oxygen isotopic compositions of benthic foraminifera ($\delta^{18}O_{bf}$) (Matthews and Poore, 1980). The history of sea level may be inferred from records of $\delta^{18}O_{bf}$, and our understanding of Quaternary climate and stratigraphy owes much to our knowledge of this relationship (Shackleton and Opdyke, 1973). However, during periods of no significant global ice volume, such as the Cretaceous, oxygen isotopic compositions of the oceans foraminifera are primarily controlled by temperature, and thus implications about sea-level changes may not be drawn from such data. Consequently, uncertainty about the timing of the onset of Cenozoic "ice house" Earth conditions has led to confusion about the history of the oxygen isotope record. Early interpretations of Miocene $\delta^{18}O$ records (Shackleton and Kennett, 1975), under the assumption of an ice-free Earth, have given way to the recognition of lower Oligocene ice-rafted debris (IRD) (Wise et al., 1991) and the realization that this implies an ice volume/sea level control of the $\delta^{18}O$ record (Keigwin and Corliss, 1986; Miller et al., 1993; Pekar and Miller, 1996). More recently, correlation of Atlantic Coastal Plain hiatuses to late middle Eocene $\delta^{18}O$ records have been interpreted as being indicative of glacioeustacy (Browning et al., 1996). However, a lack of signifi-

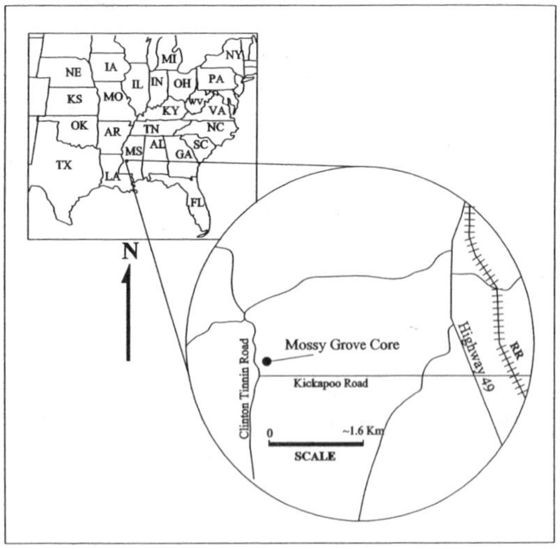

Figure 12.1. Location map of the Mossy Grove Core, in northwestern Hinds County, western Mississippi. Smaller scale figure modified from Dockery et al. (1991).

cant ice rafted debris in marine sediment, older than late Eocene (Zachos et al., 1992), around Antarctica warrants caution when interpreting earlier Paleogene oxygen isotope records as sea-level indicators.

An assertion that late middle Eocene oxygen isotope fluctuations, as recorded in benthic foraminifera, imply sea-level/ice-volume changes, rather than bottom water temperature changes, requires an independent line of evidence, suggestive of paleobathymetric or relative sea-level conditions. Such a line of evidence are the ratios of planktonic to benthic foraminifera, a ratio directly proportional to water depth (Van der Zwaan et al., 1990).

Purpose

Browning et al. (1996) document evidence for late middle Eocene icehouse conditions from the New Jersey Atlantic Coastal Plain by linking positive $\delta^{18}O$ excursions to stratigraphic hiatuses. The study presented here is an attempt to determine whether or not there is evidence from the Gulf Coastal Plain to corroborate the interpretation of benthic $\delta^{18}O$ values, near the Eocene/Oligocene boundary, as ice-volume/sea-level proxies. The upper middle Eocene sediments of the Mossy Grove Core, Hinds County, Mississippi (figure 12.1), compared to those from that period in the open ocean, were relatively rapidly deposited. The moderate sampling frequency of these Mossy Grove sediments and their respective benthic $\delta^{18}O$ values and planktonic :benthic foraminiferal ratios used here provide an opportunity to study short-term fluctuations. Assuming continuous and constant-rate sedimentation throughout the studied 1.7 million years, sampling frequency on average is approximately 30,000 years.

Zachos et al. (1996) recognized 41,000-year Milankovitch periodicity, ascribed to changes of the Earth's axial tilt (Imbrie and Imbrie, 1980), in a high resolution Oligocene oxygen isotope study of the Deep Sea Drilling Project's (DSDP) Site 522. Norris and Rohl (1999) recognized a signature of the 21,000-year Milankovitch precession cycles in high resolution Paleocene carbon isotope records from the Ocean Drilling Program's Site 1051. Global composite records of oxygen isotopes from the deep-sea were used by Browing et al. (1996) to identify evidence for

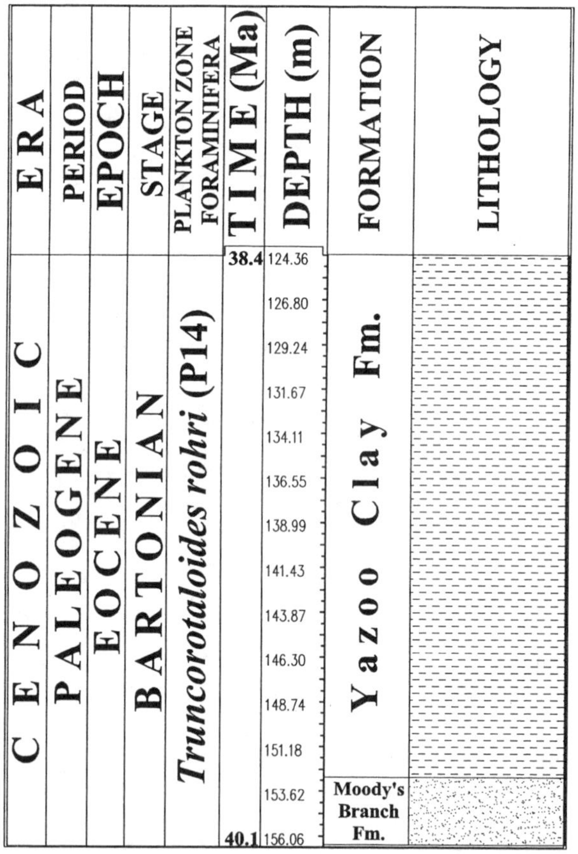

Figure 12.2. Stratigraphic column of the Planktonic Foraminiferan Zone P14 from the Mossy Grove Core. Age constraints are Fluegeman's (1996) biostratigraphy referenced to the Berggren et al. (1995) Cenozoic time scale. Lithologic symbols indicate sand for the Moody's Branch and clay for the Yazoo Clay.

late middle Eocene glacioeustacy. High-resolution stable isotope records from the Paleogene provide understanding of short-term climatic and oceanographic fluctuations. This work presented here offers another isotope record, from shallower waters than those from which deep-sea records are typically obtained, and seeks to provide interpretations of this record, so as to better understand climatic and oceanographic conditions of the Eocene-Oligocene transition.

The Mossy Grove Core

The Mossy Grove Core retrieved Eocene-Oligocene sediment of the Creola Member of the Cockfield Formation, Moody's Branch Formation, Yazoo Clay, and the Forest Hill Formation (Dockery et al., 1991). The sediment of the cored interval, 161.54 m deep and sampled every 0.61 m, represent planktonic foraminiferal Zones P14 through P18, which span the time interval from the late middle Eocene to lower Oligocene.

Upper middle Eocene (Bartonian) sediments occur in the core from a depth of 156.06 m up to 124.36 m, a section constrained to Zone P14 (*Truncorotaloides rohri*) and represented lithologically by the Moody's Branch Formation and the overlying lower section of the Yazoo Clay (figure 12.2). The former unit, a greenish, glauconitic, and foraminifera-bearing sand (Dockery et al., 1991), was deposited in outer neritic waters and the latter unit, a 'greyish-green montmorillonitic clay' (Dockery et al., 1991), was deposited in outermost neritic to upper bathyal depths. These paleodepths are suggested by the respective benthic foram assemblages of *Discorbis-Siphonina* and *Uvigerina-Bulimina* (Fluegeman, 1996).

SAMPLES AND METHODOLOGY

Mossy Grove Core samples were dried at 100°C, wash-sieved to remove the <60 μm size fraction, dried again, then dry-sieved to retain the >250 μm, the >150 μm, and the <150 μm size fractions. Following dry-sieving, samples were processed through microsplitters and sample populations of 300 foraminiferan tests were picked (Fluegeman, 1996). The processed sample populations of 300 tests from the >150 μm size fraction were used in this study for the determination of planktonic: benthic ratios. Additional *Cibicidoides* were picked from the remainder of the processed >150 μm size fraction and used for the stable isotope analyses. Tests of *Cibicidoides*, the foraminiferan genus used for the stable isotope analyses, were inspected for diagenetic crystal overgrowths, which may obscure a stable isotopic signature (Baum et al., 1994; Allan and Matthews, 1977) with a scanning electron microscope (SEM). Upon inspection of the foram tests with the SEM, no overgrowths were observed.

Planktonic to Benthic Ratios

The number of both planktonic and benthic foraminifera represented in each sample population of approximately 300 was determined and expressed as a ratio. This ratio is representative of paleobathymetric conditions, whereby a higher ratio or a relatively larger percentage of planktonics, would be implicit of deeper water, and vice versa (Loutit et al., 1988; Gibson, 1989).

Stable Isotope Analyses

Stable isotope ratio measurements were completed on tests of the benthic genus *Cibicidoides* at the Stable Isotope Laboratory of the University of South Carolina Department of Geological Science with a Micromass Optima Stable Isotope Ratio Mass Spectrometer. Ratios were calibrated to the PeeDee Belemnite (PDB) standard and results are referenced in the standard δ notation as ‰ (per mil, parts per thousand). $\delta^{18}O$ results were increased by 0.64‰, so as to correct for the vital effect of *Cibicidoides* (Woodruff et al., 1980; Miller et al., 1988).

RESULTS

Oscillations are characteristic of all three data sets: $\delta^{18}O$, $\delta^{13}C$, and planktonic: benthic foraminiferal ratios (figure 12.3). The latter two of these data sets, $\delta^{18}O$ and planktonic: benthic ratios, correlate to one another by a relationship of inverse proportionality. $\delta^{13}C$ does not correlate to either one of the other data sets, and unlike $\delta^{18}O$, $\delta^{13}C$ does display a trend with sample depth. Carbon isotope ratios generally decrease throughout the sampled interval. The data is tabulated in table 12.1.

Essentially 19 cycles occur in the obtained oxygen isotope curve (figure 12.3). Values, upon correction for the vital effect of *Cibicidoides*, range from -2.66‰ to -0.10‰ PDB, across a mean of -1.17‰ PDB and the standard deviation of the data set is 0.710 PDB. The standard deviation, as reflected by cycle amplitudes, of the upper ~60% of the curve is markedly higher than that of the lower portion of the curve. These $\delta^{18}O$ results are approximately 1–2‰ lower than typical deep-sea benthic $\delta^{18}O$ values from the middle Eocene and some cycle amplitudes from

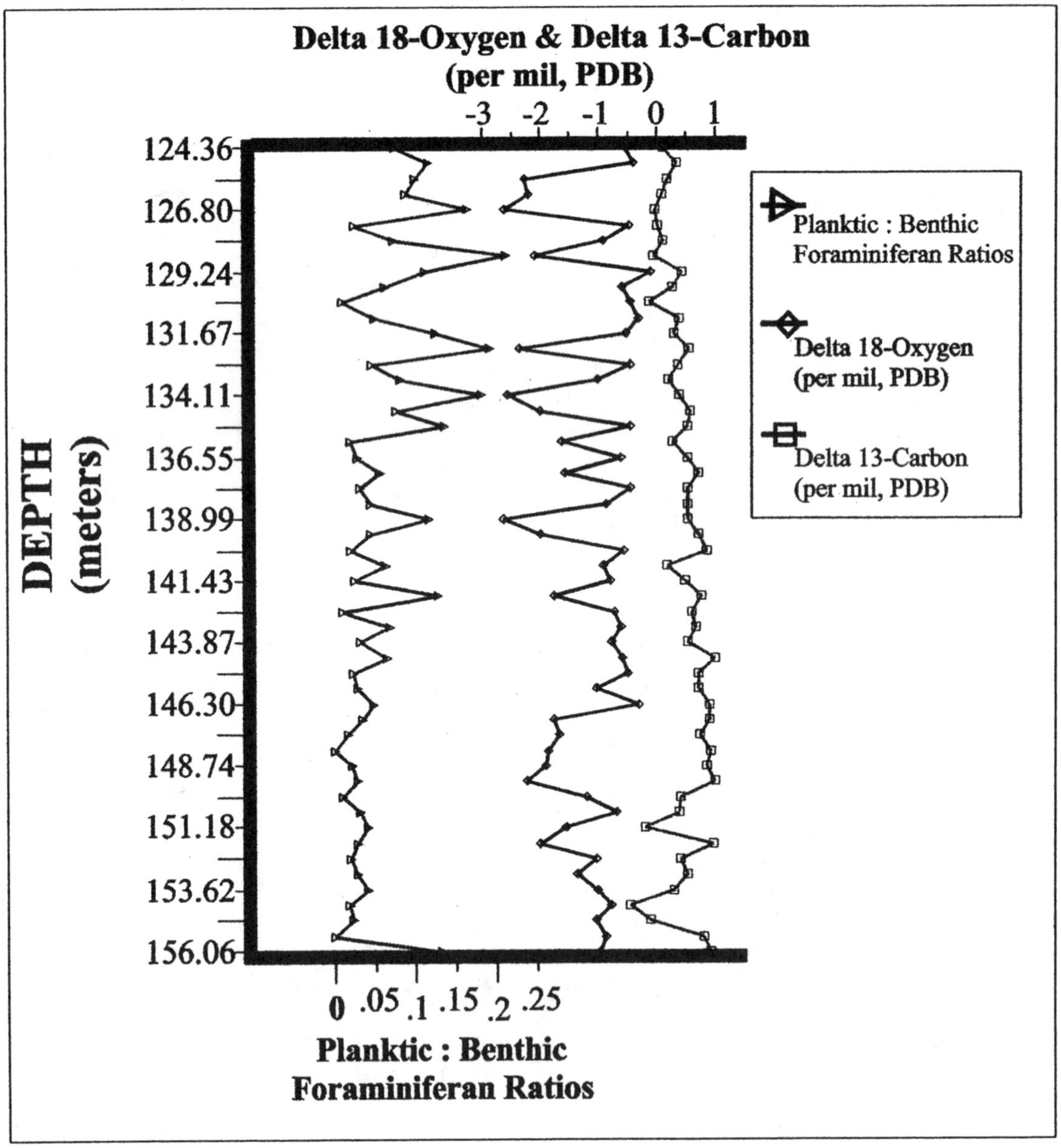

Figure 12.3. Stable isotope, $\delta^{18}O$ and $\delta^{13}C$, results and planktonic:benthic foraminiferan ratios from the P14 Zone of the Mossy Grove Core.

this data set are 1‰ to 2‰ higher than those from oxygen isotope curves generated from deep-sea Eocene records. Common deep-sea benthic $\delta^{18}O$ values, from the middle Eocene, range from 1‰ to 2‰ and typical cycle amplitudes are approximately 0.2–0.5‰ (Bralower et al., 1995; Zachos et al., 1993). In fact, variability of benthic $\delta^{18}O$ from the beginning to the end of the middle Eocene is approximately 1‰ (Zachos et al., 1993).

Carbon isotope values range from 0.990 to –0.46‰ PDB, with a mean of 0.43‰ PDB (figure 12.3). These $\delta^{13}C$ results are comparable in magnitude to benthic $\delta^{13}C$ values typically obtained from deep-sea drill sites. Typical deep-sea benthic $\delta^{13}C$ values range from approximately –0.5‰ to 1‰ and so are in good agreement with the carbon isotope values obtained from the Mossy Grove Core.

Eighteen cycles occur throughout the plankton-

Table 12.1. Stable Isotope and Planktonic to Benthic Ratios, Mossy Grove Core

Depth meters	Oxygen isotope ratios* (per mil, PDB)	Carbon isotope ratios (per mil, PDB)	Planktic : benthic ratios
124.36	-0.55	0.11	0.07
124.97	-0.40	0.34	0.11
125.58	-2.27	0.18	0.10
126.19	-2.18	0.09	0.09
126.80	-2.61	-0.04	0.16
127.41	-0.46	0.02	0.02
128.02	-0.91	0.11	0.07
128.63	-2.08	-0.05	0.21
129.24	-0.10	0.44	0.11
129.84	-0.58	0.26	0.06
130.45	-0.45	-0.13	0.01
131.06	-0.32	0.38	0.05
131.67	-0.51	0.29	0.12
132.28	-2.36	0.54	0.19
132.89	-0.45	0.37	0.05
133.50	-1.00	0.20	0.08
134.11	-2.57	0.38	0.18
134.72	-1.98	0.57	0.08
135.33	-0.45	0.54	0.14
135.94	-1.62	0.26	0.02
136.55	-0.61	0.52	0.03
137.16	-1.57	0.72	0.06
137.77	-0.44	0.54	0.03
138.38	-0.87	0.52	0.04
138.99	-2.63	0.53	0.12
139.60	-1.99	0.72	0.04
140.21	-0.58	0.86	0.02
140.82	-0.91	0.19	0.06
141.43	-0.79	0.47	0.02
142.04	-1.74	0.77	0.13
142.65	-0.72	0.60	0.01
143.26	-0.60	0.66	0.07
143.87	-0.77	0.54	0.03
144.48	-0.59	0.98	0.06
145.08	-0.49	0.71	0.02
145.69	-1.03	0.72	0.03
146.30	-0.32	0.90	0.05
146.91	-1.74	0.89	0.03
147.52	-1.65	0.74	0.02
148.13	-1.85	0.93	0.00
148.74	-1.89	0.86	0.02
149.35	-2.22	0.99	0.03
149.96	-1.20	0.40	0.01
150.57	-0.69	0.39	0.03
151.18	-1.54	-0.20	0.04
151.79	-1.99	0.98	0.03
152.40	-1.02	0.40	0.02
153.01	-1.37	0.52	0.03
153.62	-1.00	0.28	0.04
154.23	-0.79	-0.46	0.02
154.84	-1.02	-0.11	0.02
155.45	-0.86	0.80	0.00
156.06	-0.97	0.93	0.13

*Oxygen isotope values, shown here, were increased by 0.64 per mil to account for the vital effect of Cibicidoides.

ic: benthic foraminiferan ratio curve (figure 12.3). Values range from zero to 0.21, across a mean of 0.06, and the standard deviation of this data set is 0.05. As with the $\delta^{18}O$ data set, yet to a larger degree, cycle amplitudes of the planktonic: benthic ratio curve are larger in the latter ~60% of the section. These values, measured from a section deposited in neritic to upper bathyal conditions (Fluegeman, 1997), are lower than average, modern, planktonic:benthic ratios from comparable water depths in the Gulf of Mexico (Gibson, 1989; van der Zwaan et al., 1990).

The behavior of both the oxygen isotope and planktonic: benthic ratio curves is similar in that both exhibit higher amplitude cycles in the latter two-thirds of their sequences, as exemplified by the increased standard deviations of the parameters within the upper 19 m of the core section. Likewise, the number of complete cycles is nearly identical with the oxygen isotope curve independently containing 19 and the planktonic: benthic curve independently containing 18.

Another aspect of these two curves, with respect to one another is that boundaries of the complete cycles per curve, which are defined here as the peaks of the curves, coincide in sample depth with peaks on the other curve, the relationship is one of inverse proportionality. This is apparent from the curves that exhibit near mirror images one another. Twelve low peaks of the oxygen curve coincide, in sample depth, with high peaks of the planktonic: benthic curve. Likewise, ten high peaks of the oxygen curve coincide, in sample depth, with low peaks of the planktonic: benthic curve. There were coeval occurrences of high $\delta^{18}O$ values and low ratios of planktonic to benthic foraminifera in the water column. Likewise, there were coeval occurrences of low $\delta^{18}O$ values and high ratios of planktonic to benthic foraminifera. This aspect of the results has allowed the inverse correlation of these two curves and the recognition of 17 cycles in the late middle Eocene.

DISCUSSION AND SIGNIFICANCE

The curve of planktonic: benthic foraminiferan ratios, a paleobathymetric proxy, is inversely correlated to a $\delta^{18}O$ record obtained from the late middle Eocene. Seventeen cycles are interpreted from the inverse correlation of these curves. These cycles are restricted stratigraphically to the P14 plankton foraminiferal Zone, a 1.7-million-year zone (Berggren et al., 1995). A curious fact of this inverse correlation is that mathematically the cycles have an average period of 100,000 years, which is a period associated with Milankovitch's eccentricity of the earth's orbit cycles (Imbrie and Imbrie, 1980). This aspect of the data, the 100,0000-year average, may be an artifact of the sampling frequency, since many of the identified cycles contain no more than three points. 100,0000-year cyclicity, to the knowledge of the authors, has only been recognized from Pleistocene records. Milankovitch's ~20,000-year cycles have been recognized from the Paleocene (Norris and Rohl, 1999) and the 41,000-year cycles were recognized from Eocene-Oligocene records (Zachos et al., 1996), but the 100,0000-year eccentricity phenomenon has not been recognized in the Paleogene, let alone outside of the Quaternary.

The good correlation between the oxygen isotope curve and the planktonic:benthic curve does not completely explain the data as both sets are in contrast with expected values. $\delta^{18}O$ values are significantly lighter and more variable than middle Eocene deep-sea values (Bralower et al., 1995; Zachos et al., 1993). Planktonic:benthic ratios are low for the water depths, neritic to upper bathyal, implied by benthic foraminiferan assemblages of the studied section (Fluegeman, 1997). Modern day planktonic:benthic ratios, as measured in the Gulf Coast (Gibson, 1989; van der Zwaan et al., 1990) at water depths comparable to neritic to upper bathyal are, on average, greater than the maximum value obtained here of 0.21. The benthic assemblages, implicit of paleobathymetry, do not incidentally, alternate at frequencies comparable to the stable isotope and planktonic:benthic data sets. A larger order of sea-level change is inferred by the benthics which, at the Moody's Branch and Yazoo Clay contact, go from neritic to outer neritic/upper bathyal and remain as such through and above the studied interval.

$\delta^{18}O$ varies by up to 2‰. These variations, assuming the Pleistocene sea-level / $\delta^{18}O$ relationship of about 10m / 0.1‰, correspond to sea-level changes of up to 200 m. Sea level fluctuations this frequent have not been recorded by the benthic foraminiferan assemblages. The invocation of ice volume fluctuations that would accommodate 200 m of sea-level change would necessitate a very large late middle Eocene ice sheet on Antarctica. However, current estimates of Eocene-Oligocene transition ice do not exceed 40% of the current volume and what ice did exist at that time is thought to have been restricted to the southern hemisphere. Two-hundred-meter ice volume-induced sea-level changes in the late middle Eocene could very well call for an ice sheet on Antarctica larger than that of today's, as comparable

Pleistocene sea-level changes did occur, but did so during periods with both northern and southern hemisphere ice caps.

Aside from the implications of the data provided here, there is no current geologic evidence for an enormous Antarctic ice sheet in the late middle Eocene. Consideration should be given here to the possible notion that not only is there a glacioeustatic component to the $\delta^{18}O$ record, but that there is also a temperature signature on the data. In their studies on the Atlantic Coastal Plain, Browning et al. (1996) ascribed at most only 0.3‰ of the late middle Eocene $\delta^{18}O$ record to glacioeustacy. Paleotemperatures were calculated with the Erez and Luz (1983) equation, using a $\delta^{18}O_{seawater(sw)}$ of –0.8‰ PDB. This value for $\delta^{18}O_{sw}$ was chosen as a compromise between –0.6‰, PDB and -0.98‰, PDB. The former value, referenced here in the PDB scale, was determined, in the standard mean ocean water (SMOW) scale for early Oligocene (Zachos et al., 1996) ice house conditions. The latter value of -0.98‰ is for an ice-free earth and has been used to calculate paleotemperatures of the late Paleocene through middle Eocene (Bralower et al., 1995). Temperatures were determined to range from 13.9°C to 25.7°C, with an average of 18.8°C. If there was a very large, late middle Eocene, Antarctic ice sheet then our estimate of $\delta^{18}O_{SW}$ is low and these estimates of paleotemperatures are accordingly incorrect.

There have been arguments that $\delta^{18}O$ older than late Eocene should not be interpreted as a glacioeustatic proxy, but rather as temperature (Zachos et al., 1993, 1994) indices. However, Browning et al. (1996), demonstrated that late middle Eocene $\delta^{18}O$ data could be interpreted as indicative of ice volume/sea level. The inverse correlation made here, of the paleobathymetric proxy to the $\delta^{18}O$ record, shows that there is evidence from the Gulf Coastal Plain that glacioeustacy, or ice-volume-controlled sea-level change, was occurring prior to the Eocene-Oligocene transition.

The pattern of the $\delta^{13}C$ curve does not correlate to either the planktonic: benthic curve or the $\delta^{18}O$ curve, thereby supporting the defense against a nutrient- or terrestrial-influx-oriented explanation for the correlation between the planktonic: benthic curve and the $\delta^{18}O$ curve. If the oxygen isotope or planktonic: benthic data, or their inversely comparable relationship, could be explained by fluctuations of a nutrient system, resulting from either terrestrial water influx or temperature variations, then one would expect a correlation between the carbon isotope results and those of both or either of the other two data sets. There is no correlation obvious from comparison of the carbon data to oxygen or foraminiferal data. It should be noted, however, that isotopically light meteoric waters could have affected the oxygen isotope record without having done so to the carbon isotope record. Had this occurred, one might not expect such good inverse correlation between the oxygen and planktonic: benthic ratios.

CONCLUSIONS AND SUMMARY

The behaviors of the oxygen isotope curve and planktonic: benthic curve are inversely similar. The oxygen isotope curve contains 19 cycles and the planktonic: benthic curve contains 18 cycles. Coeval peaks of the curves correlate inversely, whereby high $\delta^{18}O$ peaks correspond to low planktonic: benthic peaks and vice versa. Both curves also display a marked increase in cycle amplitudes in the upper two-thirds of the section. The two curves are inversely correlated and 17 cycles are identified from this correlation.

Seventeen glacioeustatic cycles are discerned from the 1.7-million-year planktonic foraminiferal Zone P14 sampled at a frequency of approximately 31,000 years. The average temporal order of magnitude of these cycles is comparable to the durations of Quaternary glacioeustatic cycles.

Moderately high-resolution late middle Eocene oxygen isotope data is here interpreted as a proxy for glacioeustacy. The occurrence of glacioeustatic events in the late middle Eocene suggests significant ice buildup prior to the Eocene-Oligocene transition.

ACKNOWLEDGMENTS

This manuscript summarizes findings of the senior author's Master's Thesis research while at Ball State University. Stable isotopic analyses were completed at the the University of South Carolina Department of Geological Sciences. I thank the Ball State University Geology Department, the the University of South Carolina Department of Geological Sciences Stable Isotope Laboratory, and the Geological Society of America. I greatly appreciate reviews of earlier editions of this paper made by Stephan Pekar and Ellen Thomas. I thank the co-editors of this volume and the co-conveners of the Penrose Conference Linda Ivany, Liz Nesbitt, and Don Prothero. I thank my family for their support. This research was funded by a Geological Society of America Student Research Grant and Ball State University's Geology Department.

REFERENCES

Allan, J. R., and Matthews, R. K., 1977, Carbon and oxygen isotopes as diagenetic and stratigraphic tools: Surface and subsurface data, Barbados, West Indies: Geology, v. 5, p. 16-20.

Baum, J. S., Baum, G. R., Thompson, P. R., and Humphrey, J. D., 1994, Stable isotope evidence for relative and eustatic sea-level changes in Eocene to Oligocene carbonates, Baldwin County, Alabama: Geological Society of America Bulletin, v. 106, p. 824-839.

Berggren, W. A., Kent, D. V., Swisher, C. C., III, and Aubry, M.-P., 1995, A revised Cenozoic geochronology and chronostratigraphy: SEPM Special Publication, v. 54, p. 129-212.

Bralower, T. J., Zachos, J. C., Thomas, E., Parrow, M., Paull, P. K., Kelly, D. C., Premoli Silva, I., Sliter, W. V., and Lohmann, K. C., 1995, Late Peleocene to Eocene paleoceanography of the equatorial Pacific Ocean: Stable isotopes recorded at Ocean Drilling Program Site 865, Allison Guyot: Paleoceanography, v. 10, p. 841-865.

Browning, J. V., Miller, K. G., and Pak, D. K., 1996, Global implications of lower to middle Eocene sequence boundaries on the New Jersey coastal plain: the icehouse cometh: Geology, v. 24, p. 639-642.

Dockery, D. T,. III, Stover, D. T., Weathersby, P., Stover, C. W. Jr., and Ingram, S. L., 1991, A continuous core through the undifferentiated Yazoo Clay (Late Eocene, Jackson Group) of central Mississippi: Mississippi Geology, v. 12, p. 21-27.

Erez, J., and Luz, B., 1983, Experimental peleotemperature equation for planktonic foraminifera: Geochimica et Cosmochemica Acta, v. 47, p. 1025-1031.

Fluegeman, R. H., 1996, Preliminary paleontological report on the foraminifera of the Mossy Grove Core, Hinds County, Mississippi: Mississippi Geology, v. 17, p. 9-15.

Gibson, T. G., 1989, Planktonic benthonic foraminiferal ratios: Modern patterns and Tertiary applicablility: Marine Micropaleontology, v. 15, p. 29-52.

Imbrie, J., and Imbrie, J. Z., 1980, Modeling the climatic response to orbital variations: Science, v. 207, p. 943-953.

Keigwin, L. D., and Corliss, B. H., 1986, Stable isotopes in late middle Eocene to Oligocene foraminifera: Geological Society of America Bulletin, v. 97, p. 335-345.

Kennett, J. P., 1977, Cenozoic evolution of Antarctic glaciation, the circum-Antarctic Ocean, and their impact on global paleoceanography: Journal of Geophysical Research, v. 82, n. 27, p. 3843-3860.

Kennett, J. P., and Shackleton, N. J., 1976, Oxygen isotopic evidence for the development of the psychrosphere 38 Myr ago: Nature, v. 26, p. 513-515.

Kennett, J. P., and Stott, L. D., 1991, Abrupt deep-sea warming, paleoceanographic changes and benthic extinctions at the end of the Paleocene: Nature, v. 353, p. 225-229.

Loutit, T. S., Hardenbol, J., Vail, P .R., and Baum, G. R., 1988, Condensed sections: the key to age determination and correlation of continental margin sequences: SEPM Special Publication, v.42, p. 183-213.

Matthews, R. K., and Poore, R. Z., 1980, Tertiary $\delta^{18}O$ record and glacioeustatic sea-level fluctuations: Geology, v. 8, p. 501-504.

Miller, K. G., Fairbanks, R. G., and Mountain, G. S., 1987, Tertiary oxygen isotope synthesis, sea level history, and continental margin erosion: Paleoceanography, v. 2, p. 1-19.

Miller, K. G., Feigenson, M. D., Kent, D. V., and Olsson, R. K., 1988, Upper Eocene to Oligocene isotope ($^{87}Sr/^{86}Sr$, $\delta^{18}O$, $\delta^{13}C$) standard section, Deep-Sea Drilling Project site 522: Paleoceanography, v. 3, p. 223-233.

Miller, K. G., Kent, D. V., Brower, A. N., Bybell, L. M., Feigenson, M. D., Olsson, R .K., and Poore, R. Z., 1990, Eocene-Oligocene sea-level changes on the New Jersey coastal plain linked to the deep-sea record: Geological Society of America Bulletin, v. 102, p. 331-339.

Miller, K. G., Mountain, G. S., Browning, J. V., Kominz, M., Sugarman, P. J., Christie-Blick, N., Katz, M. E., and Wright, J. D., 1998, Cenozoic global sea level, sequences, and the New Jersey Transect: results from coastal plain and continental slope drilling: Reviews of Geophysics, v. 36, p. 569-601.

Miller, K. G., Thompson, P. R., and Kent, D. V., 1993, Integrated late Eocene-Oligocene stratigraphy of the Alabama coastal plain: correlation of hiatuses and stratal surfaces to glacioeustatic lowerings: Paleoceanography, v. 8, p.313-331.

Mix, A. C., 1987, The oxygen-isotope record of glaciation, *in* Ruddiman, W. F., and Wright, H. E., Jr., eds., North America and adjacent oceans during the last deglaciation: Boulder, Colorado,

Geological Society of America, The Geology of North America, v. K-3, p. 111-135.

Norris, R. D., and Rohl, U., 1999, Carbon cycling and chronology of climate warming during the Palaeocene/Eocene transition: Nature, v. 401, p. 775-778.

Oberhansli, H., McKenzie, J., Toumarkine, M., and Weissert, H., 1984, A paleoclimatic and paleoceanographic record of the Paleogene in the central South Atlantic (Leg 73, sites 522, 523, and 524): Initial Reports of the Deep Sea Drilling Project, v. 73, p. 737-747.

Pekar, S., and Miller, K. G., 1996, New Jersey Oligocene "icehouse" sequences (ODP Leg 150X) correlated with global $\delta^{18}O$ and Exxon eustatic records: Geology, v. 24, p. 567-570.

Prentice, M. L., and Matthews, R. K., 1988, Cenozoic ice volume history: development of a composite oxygen isotope record: Geology, v. 16, p. 963-966.

Prothero, D. R., 1994, The Eocene-Oligocene transition: paradise lost: New York, Columbia University Press, 291 p.

Savin, S. M., Douglas, R. G., and Stehli. F., 1975, Tertiary marine paleotemperatures: Geological Society of America Bulletin, v. 86, p. 1499-1510.

Shackleton, N. J., 1967, Oxygen isotope analyses and Pleistocene temperatures re-assessed: Nature, v. 215, p. 15-17.

Shackleton, N. J., 1987, Oxygen isotopes, ice volume and sea level: Quaternary Science Reviews, v. 6, p. 183-190.

Shackleton, N. J., and Boersma, A., 1981, The climate of the Eocene ocean: Journal of the Geological Society of London, v. 138, p. 153-157.

Shackleton, N. J., and Kennett, J. P., 1975, Paleotemperature history of the Cenozoic and the initiation of Antarctic glaciation: oxygen and carbon isotope analyses in DSDP Sites 277, 279, and 281: Initial Reports of the Deep Sea Drilling Project, v. 29, p. 743-756.

Shackleton, N. J., and Opdyke, N. D., 1973, Oxygen isotope and palaeomagnetic stratigraphy of equatorial pacific core V28-238: oxygen isotope temperatures and ice volumes on a 10^5 year and 10^6 year scale: Quaternary Research, v. 3, p. 39-55.

Shackleton, N. J., Hall, M. A., and Boersma, A., 1984, Oxygen and carbon isotope data from Leg 74 foraminifers: Initial reports of the Deep Sea Drilling Project, v. 74, p. 599-612.

Van der Zwaan, G. J., Jorissen, F. J., and de Stigter, H. C., 1990, The depth dependency of planktonic/benthic foraminiferal ratios: Constraints and applications: Marine Geology, v. 95, p. 1-16.

Wei, W., Pospichal, J. J., and Wise, S. W., 1991, Cenozoic calcareous nannofossil magnetobiochronology of the southern oceans: a summary: Antarctic Journal of the United States, v. 26, p. 134-138.

Wise, S. W., Jr., Breza, J. R., Harwood, D. M., and Wei, W., 1991, Paleogene glacial history of Antarctica, *in* McKenzie, J. A., Muller, O. W., and Weissert, H., eds., Controversies in modern geology: New York, Academic Press, p. 133-171.

Woodruff, F., Savin, S.M., and Douglas, R. G., 1980, Biological fractionation of oxygen and carbon isotopes by recent benthic foraminifera: Marine Micropaleontology, v. 5, p. 3-11.

Zachos, J. C., Breza, J. R., and Wise, S. W., 1992, Early Oligocene ice-sheet expansion on Antarctica: stable isotope and sedimentological evidence from Kerguelen Plateau, southern Indian Ocean: Geology, v. 20, p. 569-573.

Zachos, J. C., Lohmann, K. C, Walker, J. C. G., and Wise, S. W., 1993, Abrupt climate change and transient climates during the paleogene: a marine perspective: Journal of Geology, v. 101, p. 191-213.

Zachos, J. C., Stott, L. D., and Lohmann, K. C, 1994, Evolution of early Cenozoic marine temperatures: Paleoceanography, v. 9, p. 353-387.

Zachos, J. C., Quinn, T. M., and Salamy, K. A., 1996, High-resolution (10^4 years) deep-sea foraminiferal stable isotope records of the Eocene-Oligocene climate transition: Pale-oceanography, v. 11, p. 251-266.

CHAPTER 13

Paleogene Temperature History of the U.S. Gulf Coastal Plain Inferred from $\delta^{18}O$ of Fossil Otoliths

Linda C. Ivany, Kyger C Lohmann, and William P. Patterson

ABSTRACT

The selective loss of warm-water molluscan taxa from the U.S. Gulf Coastal Plain during the middle Eocene through Oligocene transition suggests that cooling temperatures played a role in these extinctions. To test this hypothesis, and to shed light more broadly on environmental changes that may have occurred on the shelf during this interval of time, we sectioned and incrementally microsampled fish otoliths ("ear stones") collected throughout the Paleogene in the Gulf Coast. Stable oxygen isotopic compositions of aragonite subsamples provide a proxy record for temperature during the lifetime of each fish (Iacumin et al., 1992; Kalish, 1991a, 1991b; Patterson et al., 1993). Moreover, high-resolution analyses of accretionary otolith aragonite enable the approximation not only of mean annual temperature but also of the seasonal range of temperature variation.

We find that mean early Eocene temperatures were, on average, about 5°C warmer than later in the Eocene or Oligocene; Paleocene temperatures may have been even warmer, but data is limited. Cooling occurred sometime before the middle of the middle Eocene, and mean temperatures remained roughly constant through the remainder of the middle Eocene and late Eocene. Mean temperature change across the middle/late Eocene boundary and the Eocene/Oligocene boundary was minimal. Seasonal range of temperature variation in the Paleocene and early Eocene was comparable to that later in the Eocene. There is a suggestion of reduced seasonality in the middle Eocene that has yet to be confirmed with further analyses. Late Eocene seasonality was roughly 5°C, with values fluctuating between 15°C and 20°C. Across the Eocene/Oligocene boundary, seasonality increased in association with a 4°C drop in minimum winter temperatures; summer temperatures were not affected. This increase in seasonality correlates with the molluscan extinction at the boundary, hence we suggest a causal link between temperature variability and faunal turnover.

INTRODUCTION

The transition from the warm equable climates of the early Eocene to the cooler conditions of the Oligocene is one of the fundamental turning points in Cenozoic earth history. The beginning of the Oligocene records the first major expansion of glacial ice on Antarctica (Zachos et al., 1992) and a steepening of latitudinal temperature gradients (Zachos et al., 1994), both of which led to a strengthening of thermohaline circulation (Kennett and Stott, 1990; Miller, 1992) and increases in upwelling and productivity in the oceans (Diester-Haass, 1995; Salamy and Zachos, 1999; Thomas and Gooday, 1996). Broadly coincident with this transition are the following: a significant turnover in the shallow marine and terrestrial biota, including regional extinctions in mollusks (Dockery, 1986; Haasl and Hansen, 1996), benthic foraminiferans (McGowran, 1987; Gaskell, 1991; McGowran et al., 1992), and echinoids (McKinney et al., 1992); the transition from archaic whales to modern odontocete and mysticete whales (Fordyce, 1992; Fordyce and Barnes, 1994); significant reshuffling of the terrestrial mammalian fauna (Legendre and Hartenberger, 1992; Meng and McKenna, 1998); and turnover in terrestrial plants (Wolfe, 1992).

The U.S. Gulf Coastal Plain preserves one of the

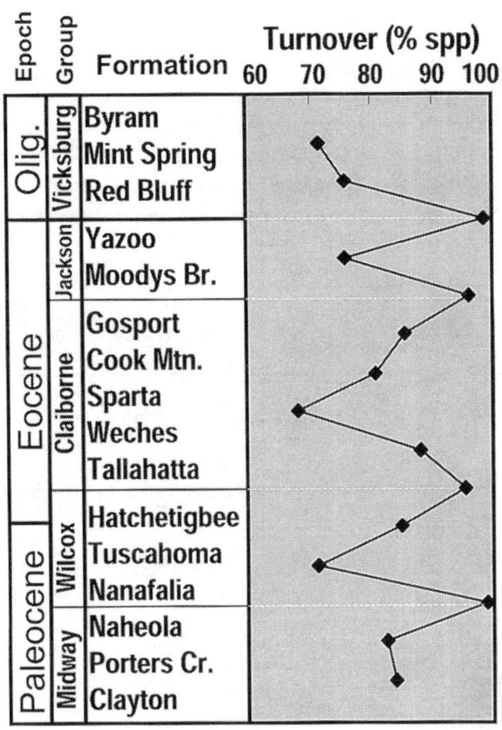

Figure 13.1. Percent turnover of Gulf Coast mollusk species across formation boundaries (from Dockery, 1986). Turnover represents proportion of the fauna involved in either origination or extinction between formations. Stratigraphic resolution is at formation level.

best-studied Paleogene marine shelf sections in the world. Preservation of original aragonite is commonplace, and the generally unconsolidated nature of the sediments makes collection of very large numbers of specimens practicable. The molluscan fossil record is especially well constrained due to well over a century of intensive collecting and the compilation of a standardized taxonomy (Dockery, 1982; Harris, 1895; Harris, 1919; Harris and Palmer, 1946-47; MacNeil and Dockery, 1984; Palmer, 1937; Palmer and Brann, 1965, 1966; Toulmin, 1977). The record is characterized by a series of four large turnovers that define the Gulf Coast group boundaries and are marked by high percentages of both first and last appearances (figure 13.1). Turnover between these episodes is relatively lower, and Dockery (1986) noted a pattern of decreasing first appearances and increasing last appearances through time within each group. Turnover at the Eocene/Oligocene boundary in particular is very high, with fewer than 10% of late Eocene Yazoo Clay species surviving into the Oligocene Red Bluff Formation (Dockery, 1986; Haasl and Hansen, 1996). This event is not unique to the Gulf Coast, nor to mollusks (e.g., papers in this volume and Prothero and Berggren, 1992), and is counted as one of the more significant global mass extinctions in the last 200 million years, and the largest in the Cenozoic (Raup and Sepkoski, 1984, 1986).

Hansen (1987) investigated the relationship between Gulf Coast diversity and various environmental parameters. He concluded that changes in shelf area, favored by Dockery (1986, 1987), did not correlate as well to turnovers as did changes in temperature, at least for middle Eocene to Oligocene faunas. He maintained that the general correlation between temperature and diversity suggests that temperature played a defining role in the diversity history of the Gulf (Hansen, 1992, 1987), as it seems to have in other times and settings (Stanley, 1984, 1986). Hansen supported this contention with the observation that warm-water taxa suffer disproportionately in middle Eocene through Oligocene extinctions. However, the temperature data available at the time and upon which Hansen drew for his analysis were based on equatorial and Southern Hemisphere (mainly Indo-Pacific) foraminiferan assemblage composition (Keller, 1983) and Northern Hemisphere plant morphology (Wolfe, 1978), neither of which are specific to the Gulf Coast. The few proxies available that are based on Gulf Coast material seem not to support the idea of significant temperature change, at least at the Eocene/Oligocene boundary (see below). While the selective extinction of thermophilic taxa clearly suggests some role for temperature in the extinctions in the Gulf, there must be some independent demonstration of correlated temperature change in this region if cooling is to remain a viable causal mechanism.

Here, we take steps toward evaluating the hypothesis that episodes of temperature change led to some of the observed faunal turnovers in the Gulf Coast section, in particular the Eocene/Oligocene boundary extinction. We construct a temperature record for the Gulf based on the oxygen isotopic composition of biogenic (otolith) aragonite through the section that can then be compared to the record of faunal change. It is hoped that this data set will make assessing the relationship between temperature and diversity a more tractable endeavor.

Table 13.1. Qualitative and Quantitative Paleoclimatic Interpretations for the Paleogene Derived from Gulf Coastal Plain Sediments and Low-latitude Deep-sea Cores

TIME INTERVAL	NORTH ATLANTIC, GULF, AND CARIBBEAN	GULF COASTAL PLAIN					
		Nanno-plankton[3]	$\delta^{18}O$ of gastropod shells[4]	Foliar Phys.[5]	Palyno-floras[6]	Palyno-floras[7]	Sporo-morphs[8]
Early Oligocene	18-22[1]	temperate late			Warm temperate	Cool temperate?	
Late Eocene	19-22[1]	Sub-tropical		20-28	Warm temperate	Cool – warm temperate	Cooler
Middle Eocene	15.5-23[2]	Sub-tropical	19-28 seasonal range	24-30		Warm temperate to subtropical	Cool early, warm late
Early Eocene	26[1]	Sub-tropical		27-30		Cool – warm temperate	Very warm
Late Paleocene	26[1]	temperate		19-20			

[1] (Zachos et al., 1994)
[2] (Wade et al., in press)
[3] (Seisser, 1984)
[4] (Andreasson and Schmitz, 2000)
[5] (Wolfe and Poore, 1982)
[6] (Oboh et al., 1996)
[7] (Elsik, 1974)
[8] (Frederiksen, 1988)

Temperature ranges are in °C. North Atlantic, Gulf, and Caribbean data are from low latitude (≤30°) deep sea sites and derived from oxygen isotopic measurements of planktonic foraminiferan test calcite.

EXISTING CLIMATE RECORDS FROM THE GULF COAST, AND AN ALTERNATIVE APPROACH

Existing climate data specific to the Gulf Coastal Plain (table 13.1) are limited and generally qualitative. Most are based on floral or faunal diversity to begin with, so circularity is a significant concern: testing the effects of temperature change on a fauna when temperature itself is estimated by changes in faunal composition poses a serious problem! Wolfe and Poore (1982) used physiognomic characteristics of leaf assemblages (Wolfe, 1993) to produce semi-quantitative temperature estimates for the Paleogene. However, no information is given about the data set or methods, so their numerical values are difficult to evaluate. More recently, Andreasson and Schmitz (2000) looked at the $\delta^{18}O$ of gastropod shells from two middle Eocene units (the Stone City Beds, Texas, and the Doby's Bluff Tongue, Mississippi) and inferred seasonal ranges of temperature variation from them, but there was no temporal component to their study. Planktonic foraminiferans collected from deep-sea cores in the Caribbean and low-latitude North Atlantic provide some insight into climate change through the Paleogene in the region. $\delta^{18}O$ values of test carbonate suggest warm surface waters in the Paleocene and early Eocene, and somewhat cooler waters in the late Eocene and Oligocene (Zachos et al., 1994). Late middle Eocene data from Wade et al. (in press) suggest values more akin to those in the late Eocene. In general, the data sets seem to agree that the early and perhaps middle Eocene were the warmest times in the Gulf, but records are inconsistent about the timing of cooling trends in the late Eocene and Oligocene.

Climatic events surrounding the Eocene/Oligocene boundary in the Gulf Coast are of particular interest because of the associated severe molluscan extinction (Haasl and Hansen, 1996). Despite suspicions that temperature played a role in the turnover (Hansen, 1992, 1987), these records do not support the occurrence of any significant climate change at the boundary (table 13.1). Temperatures derived from the $\delta^{18}O$ of planktonic foraminiferans from the Gulf of Mexico, Caribbean, and low-latitude North Atlantic show no change in temperature across the boundary (Zachos et al., 1994). Palynomorph assemblages from Gulf Coast sections suggest a progressive cooling and drying on land beginning in the late Eocene and continuing into the early Oligocene (Frederiksen, 1988; Oboh et al., 1996; Yancey and Elsik, 1999), but detailed paleofloristic work could resolve no significant climate change at the boundary (Oboh et al., 1996). Calcareous nannoplankton assemblages suggest that cooling began only later in the early Oligocene, after the boundary extinctions (Seisser, 1984). The suggestion that temperature did not change significantly across the boundary is also more broadly supported by recent Mg/Ca data from benthic foraminifera (Lear et al., 2000) and benthic ostracodes (Schellenberg, 1999) sampled from deep-sea cores. From these data alone, it is difficult to maintain that temperature was a factor in the Eocene-Oligocene extinction.

Because of the commonplace preservation of primary carbonate in biogenic materials from the Gulf Coast, the potential exists to retrieve quantitative estimates of temperature throughout the Paleogene. The well-known relationship between the stable isotopic composition of oxygen in carbonate minerals and the temperature conditions under which the carbonate is precipitated (Epstein et al., 1953; Friedman and O'Neil, 1977; Grossman and Ku, 1986) has permitted successful reconstruction of temperature from the $\delta^{18}O$ of carbonate from many groups, particularly foraminifera. Temperature inference based on stable oxygen isotopic analysis of bivalve shell carbonate is common practice now and generally yields robust and reliable results (Dettman and Lohmann, 1993; Jones and Allmon, 1995; Jones and Quitmyer, 1996; Klein et al., 1996; Purton and Brasier, 1999). Inferences from gastropods have been fewer but work on Eocene taxa shows promise (Andreasson and Schmitz, 1996, 2000; Purton and Brasier, 1997), and Eocene-Oligocene brachiopod shell carbonate has been used to infer temperature change associated with changing circulation patterns in New Zealand (Buening et al., 1998).

Most paleotemperatures reported in the literature are derived from whole-organism $\delta^{18}O$ values, be they single foraminifera or homogeneous bulk shell values. The advantage of using accretionary biogenic carbonate for temperature inference, however, is that one can not only estimate mean annual temperature, but also derive the seasonal range of temperature variation. Seasonality is a particularly important variable in controlling modern biogeographic distributions (Valentine, 1973; Vermeij, 1978), and this was likely true in the past, as well. Organisms that produce accretionary carbonate and live for more than a single year therefore provide an advantage over foraminifera for temperature reconstructions because they can be microsampled across growth bands, allowing resolution at subannual scales and thus retrieval of this important climate variable. Only one study (Andreasson and Schmitz, 2000) at this scale of resolution has been conducted in the Coastal Plain Paleogene, and this involved only 5 shells from one stratigraphic horizon.

Mollusks provide an obvious choice for analysis in the Gulf Coast because they are so plentiful. Another possibility, however, are fish otoliths, or "ear stones." Like most mollusks, they are aragonitic, accretionary, and relatively abundant in the section. However, otoliths tend to be preserved more consistently than mollusks in some parts of the section. Localities where mollusks are chalky and fragile still produce robust, intact otoliths. This may be a function of differences in the overall ratio of organic matrix to carbonate or in the way the aragonite lathes are packed. We believe that $\delta^{18}O$ of otolith aragonite therefore provides an excellent proxy for temperature change within this section.

OTOLITH BIOLOGY AND CHEMISTRY

Otoliths are small accretionary aragonite structures produced in the inner ear of fishes. All vertebrates produce micron-scale calcareous crystals in their inner ears, but only in the bony fishes do these crystals form a solid mass (Nolf, 1985), the "ear stones." There are generally three paired structures, only one of which (usually the saccular otolith, or sagitta) is large enough to be practically used in taxonomy and recoverable as fossils from the sedimentary record (figure 13.2). Otoliths are formed in each of the three membranous chambers of the labyrinth (inner ear), a structure originally derived from the lateral line system and housed within the auditory capsule near the base of the skull. They appear to be secreted by the walls of the labyrinth and rest against ciliated cells that line the chambers (Norman and Greenwood, 1975). Not surprisingly, otoliths function in a sensory capacity, and are primarily associated with orientation and sound reception. Morris and Kittleman (1967) also suggest a role in depth perception based on their piezoelectric properties.

Growth of otoliths occurs through accretion of layers of aragonite crystals laid down over a proteinaceous organic template. If growth is rapid, the crystals are large and the ratio of carbonate to protein is high, producing a light-colored growth band in reflected light (translucent in transmitted light).

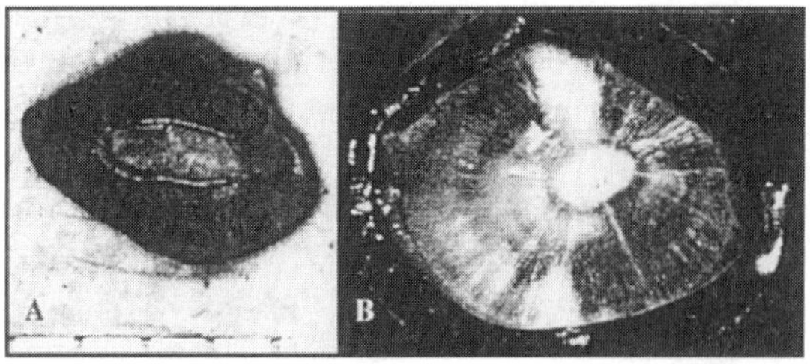

Figure 13.2. (A) Left sagittal otolith of *Paraconger* from the upper Eocene Yazoo Clay. Sculptured groove is the sulcus, upon which sensory nerves rest. Scale bar in mm. (B) Polished and mounted otolith of *Paraconger* showing growth bands; otolith is about 6.5 mm across.

When growth slows, the proportionately higher amount of organic matrix in the carbonate results in a darker growth band (opaque in transmitted light) (Morales-Nin, 1987; Nolf, 1985; Beckman and Wilson, 1995) (figure 13.2B). Growth banding usually results from some physiological stress on the fish, and may result from seasonal temperature drops, spawning, or some other environmental disturbance (Beckman and Wilson, 1995). Counting otolith growth bands in order to estimate the age of fish is a technique that goes back at least a century, and finer-scale banding has more recently been linked to the production of daily growth increments (Pannella, 1971; Stevenson and Campana, 1992).

Because otoliths are calcareous, they preserve well in sediments and have a good fossil record at least back into the Jurassic, with isolated occurrences as far back as the Devonian (Nolf, 1985, 1995). Their morphologies are distinct and taxon-specific, and so they can be readily distinguished and many of them allied to extant families and genera. Their aragonite mineralogy, good preservation potential, and distinct morphologies make them excellent candidates for paleoclimate studies. The oxygen isotopic composition of otolith aragonite can be related to the temperature of the water in which the carbonate was precipitated. And in addition, because they are accretionary, a record of temperature variation experienced throughout the life of the fish can be extracted by subsampling individual growth bands. One can therefore produce a record not only of mean temperature, but also of seasonal range of temperature variation, from subsampling and isotopic analysis of otolith carbonate.

Determining Paleoclimate from Otolith Carbonate

Key in assessing the usefulness of otolith isotope data in paleoclimate work is determining whether otolith aragonite is produced in isotopic equilibrium with the surrounding seawater, and what the nature of the relationship is with temperature. Because otoliths are precipitated inside the heads of fish, there is no a priori reason to expect equilibrium. Devereux (1967) was the first to suggest that otoliths are formed in oxygen isotopic equilibrium with seawater and that calculated temperatures, using the equation of Epstein et al. (1953) for biogenic calcite, fall within the expected range. This was later confirmed by Kalish (1991a,b) for a range of marine taxa using the equation of Grossman and Ku (1986) for biogenic aragonite. Iacumin et al. (1992) obtained similar results, but cautioned that it is unclear why the endolymph fluid from which otoliths are precipitated is in apparent oxygen isotopic equilibrium with seawater when other body fluids are clearly not. Despite this enigma, Weidman and Millner (2000) have recently produced high-resolution subsampled isotopic records from North Atlantic cod that closely approximate measured seasonal variation in water temperature in the regions in which the fish were captured, again suggesting equilibrium and emphasizing that accurate seasonal reconstructions are possible using otoliths.

Controlled experimental work to verify the relationship between the isotopic composition of the water, the otolith, and the temperature of the water are few at present. Patterson et al. (1993) analyzed the otoliths of freshwater fishes as well as the composition and temperature of the water in which the fish were growing. Fish included free-living, hatchery, and laboratory-reared specimens to yield a wide range of temperature conditions. Fractionation factors were calculated for each otolith using equation 1, and plotted against temperature. A regression line through the data gives the paleotemperature equation, equation 2 (figure 13.3). All isotope ratios are reported relative to Vienna standard mean ocean water (VSMOW).

$$\alpha = \delta CaCO_3 + 1000/\delta H_20 + 1000 \quad (1)$$

$$10^3 \ln\alpha = 18.56\ (\pm\ 0.319) \times (10^3)\ T^{-1}{}^\circ K - 33.49\ (\pm\ 0.307) \quad (2)$$

They found that otoliths are precipitated in oxygen isotopic equilibrium with ambient water, and that there appears to be no species-specific vital effect.

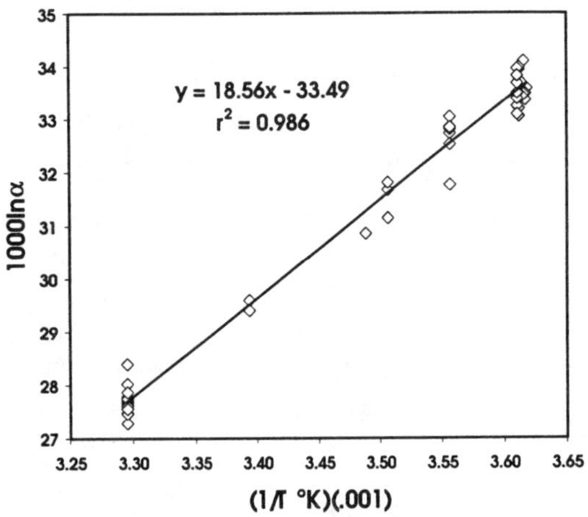

Figure 13.3. Data for temperature vs. fractionation factor constraining the otolith paleotemperature equation of Patterson et al. (1993). The equation of the regression line is the paleotemperature equation.

Moreover, earlier published data from marine fishes are described equally well by the equation, suggesting that it is robust to differences in taxonomic affiliation and ecology (Patterson et al., 1993). The slope of the equation is statistically indistinguishable from that of Grossman and Ku (1986), where $1000\ \ln\alpha = 18.07 \times 10^3T^{-1} - 31.08$, but the intercept is offset by approximately +2.5. Because the Grossman and Ku equation was derived from bivalve aragonite, the offset between the two may be due to vital effects associated with bivalve shell precipitation. Alternatively, it may be due to the vagaries of estimating environmental conditions (temperature and water composition) associated with collecting specimens in the wild. The studies of Grossman and Ku (1986) and to a lesser extent Patterson et al. (1993) were both susceptible to this potential source of error. However, because the Patterson et al. (1993) equation is very close to thermodynamic equilibrium (Tarutani et al., 1969), it is likely that this is the more robust equation for otoliths. This is the equation that we use in the remainder of this paper to constrain paleotemperatures.

METHODS

Otoliths were collected from lower Eocene to Oligocene strata exposed in western Alabama, Mississippi, Louisiana, and east Texas (table 13.2). Sediments here are generally fossiliferous, glauconitic, marly, fine-grained sands deposited on an open marine shelf. The section becomes thicker and muddier to the west due to the proximity of sediment sources. Otoliths were collected from horizons that also produce mollusks, and in many cases were drawn from bulk samples collected originally for their molluscan component. In total, representatives of 8 different families and 14 genera were collected. Individual assemblages generally contained 6 to 8 different species. Additional samples were graciously provided by D. Nolf at the Institute Royal des Sciences Naturelles de Belgique in Brussels, from his extensive collection of fossil otoliths. We sampled 37 otoliths for the present study.

Choice of taxa for use in temperature reconstruction is an important consideration. It is necessary to select fish with ecologies appropriate to the questions being asked. In this case, we are interested in reconstructing temperature change experienced by the associated benthic molluscan fauna, and hence we have chosen to use otoliths from fish that live on the bottom—the families Congridae (conger eels), subfamily Bathymyrinae (Smith, 1989a), and Ophidiidae (cusk eels), subfamily Ophidiinae (Nolf, 1980) (figure 13.4). Modern adults in these groups live on the bottom in burrows they have excavated using their distally stiffened tails to wisk sediment away. They are fully marine in relatively deep water, and pan-tropical in their distributions today (McEachran and Fech-helm, 1998; Smith, 1989a; Whitehead et al., 1985). Adults are nonmigratory, neither geographically nor vertically in the water column. All are desirable traits for recording temperatures experienced by the benthic shelf biota from which most of the fossil record comes. Species belonging to extant genera were chosen in order to minimize concerns about assumptions of ecological consistency through time.

Otoliths were polished flat on one side, mounted on glass slides, and polished again on the other to produce a thin layer through the center. The outer edge, center, and any obvious growth bands were digitized, and interpolated growth-line-parallel sampling paths calculated so that sample masses exceed roughly 20 μg (figure 13.5A). Micromilling was done on a computer-controlled, three-dimensional positioning stage set under a fixed high-precision dental drill (Wurster et al., 1999). Stable oxygen and carbon isotopic compositions of resulting powders were determined using the Finnegan MAT 252 mass spectrometer at Syracuse University or the Finnegan MAT 251 at the University of Michigan, both coupled with automated carbonate preparation systems. Values are reported relative to the Vienna Pee Dee belemnite

Table 13.2. Taxonomic, Stratigraphic, and Locality Information for Paleogene Otoliths Used in this Study

Otolith	Genus	Family	Samples	Age	Age* (Ma)	Member / Formation	Location
LCI 113	*Ariosoma*	Congridae	20	Chattian	27.75	Chickasawhay Lmst.	Waynesboro, MS
LCI 158	*Paraconger*	Congridae	43	Rupelian	29.75	Byram Fm.	Sylvarena, MS
LCI 110	*Ariosoma*	Congridae	53	Rupelian	30.3	Glendon Limestone	Warren Co., MS
LCI 104	*Ariosoma*	Congridae	64	Rupelian	31.1	Rosefield Clay	Rosefield, LA
LCI 100	*Ariosoma*	Congridae	45	Rupelian	32.9	Mint Spring Fm.	Vicksburg, MS
LCI 99	*Ariosoma*	Congridae	37	Rupelian	32.9	Mint Spring Fm.	Vicksburg, MS
LCI 87	*Paraconger*	Congridae	85	Priabonian	34.95	Yazoo Clay	Copenhagen, LA
LCI 88	*Paraconger*	Congridae	26	Priabonian	34.95	Yazoo Clay	Copenhagen, LA
LCI 92	*Ariosoma*	Congridae	25	Priabonian	34.95	Yazoo Clay	Copenhagen, LA
LCI 93	*Ariosoma*	Congridae	25	Priabonian	34.95	Yazoo Clay	Copenhagen, LA
LCI 94	*Ariosoma*	Congridae	20	Priabonian	34.95	Yazoo Clay	Copenhagen, LA
LCI 83	*Ariosoma*	Congridae	33	Priabonian	36.1	Moodys Branch Fm.	Jackson, MS
LCI 150	*Paraconger*	Congridae	90	Priabonian	36.1	Moodys Branch Fm.	Jackson, MS
LCI 79	*Paraconger*	Congridae	37	Bartonian	37.25	Gosport Sand	Jackson, AL
LCI 139	*Paraconger*	Congridae	55	Bartonian	37.25	Gosport Sand	St. Stephens, AL
LCI 141	*Paraconger*	Congridae	83	Bartonian	37.25	Gosport Sand	St. Stephens, AL
LCI 23	*"Lepophidiinarum"*	Ophidiidae	7	Bartonian	38	upper Lisbon Fm.	Jackson, AL
LCI 24	*"Lepophidiinarum"*	Ophidiidae	8	Bartonian	38	upper Lisbon Fm.	Jackson, AL
LCI 29	Myripristidae	Myripristidae	21	Bartonian	38	upper Lisbon Fm.	Jackson, AL
LCI 19	*"Lepophidiinarum"*	Ophidiidae	18	Bartonian	39.15	upper Lisbon Fm.	Coffeeville, AL
LCI 128	*Paraconger*	Congridae	17	Bartonian	39.15	upper Lisbon Fm.	Coffeeville, AL
LCI 129	*Ariosoma?*	Ophidiidae	4	Bartonian	39.15	upper Lisbon Fm.	Coffeeville, AL
LCI 14	*"Lepophidiinarum"*	Ophidiidae	18	Bartonian	40.2	Archusa Marl	Newton, MS
LCI 21	*"Lepophidiinarum"*	Ophidiidae	7	Bartonian	40.35	Wheelock Mbr.	Stone City, TX
LCI 10	*"Lepophidiinarum"*	Ophidiidae	10	Bartonian	41.15	upper Lisbon Fm.	Coffeeville, AL
LCI 11	*"Lepophidiinarum"*	Ophidiidae	37	Bartonian	41.15	Archusa Marl	Hickory, MS
LCI 125	*Paraconger*	Congridae	27	Bartonian	41.15	upper Lisbon Fm.	Coffeeville, AL
LCI 9	*"Lepophidiinarum"*	Ophidiidae	9	Lutetian	41.4	Doby's Bluff Tongue	Quitman, MS
LCI 7	*"Lepophidiinarum"*	Ophidiidae	6	Lutetian	44.8	Weches Fm.	Cladwell, TX
LCI 6	*"Lepophidiinarum"*	Ophidiidae	8	Lutetian	45.35	Weches Fm.	Burleson Bluff, TX
LCI 70	*Gnathophis*	Congridae	45	Ypresian	51.55	Hatchetigbee Fm.	Washington Co., AL
LCI 30	*Gnathophis*	Congridae	14	Ypresian	54.05	Bashi Marl	Elba Dam, AL
LCI 31	Congridae	Congridae	20	Ypresian	54.05	Bashi Marl	Elba Dam, AL
LCI 152	*Gnathophis*	Congridae	49	Ypresian	54.05	Bashi Marl	Campbell, AL
LCI 75	*Gnathophis*	Congridae	16	Ypresian	54.05	Bashi Marl	Meridian, MS
LCI 62	Congridae	Congridae	35	Selandian	59.8	Coal Bluff Marl	Wilcox Co., AL
LCI 64	*Rhynchoconger*	Congridae	22	Selandian	61	Matthew's Landing Marl	Wilcox Co., AL

Number of samples is the number of successful isotopic analyses per otolith.

*Age is estimated from timescale in Berggren et al. (1995) based on extrapolations from planktonic foraminiferan zones given in Baum et al. (1994) and Yancey and Davidoff (1994). Where stratigraphic information is unresolved, the age represents the average between maximum and minimum estimates. Absolute values may change, but relative stratigraphic ordering of samples will not.

Specific locality information is available from the authors.

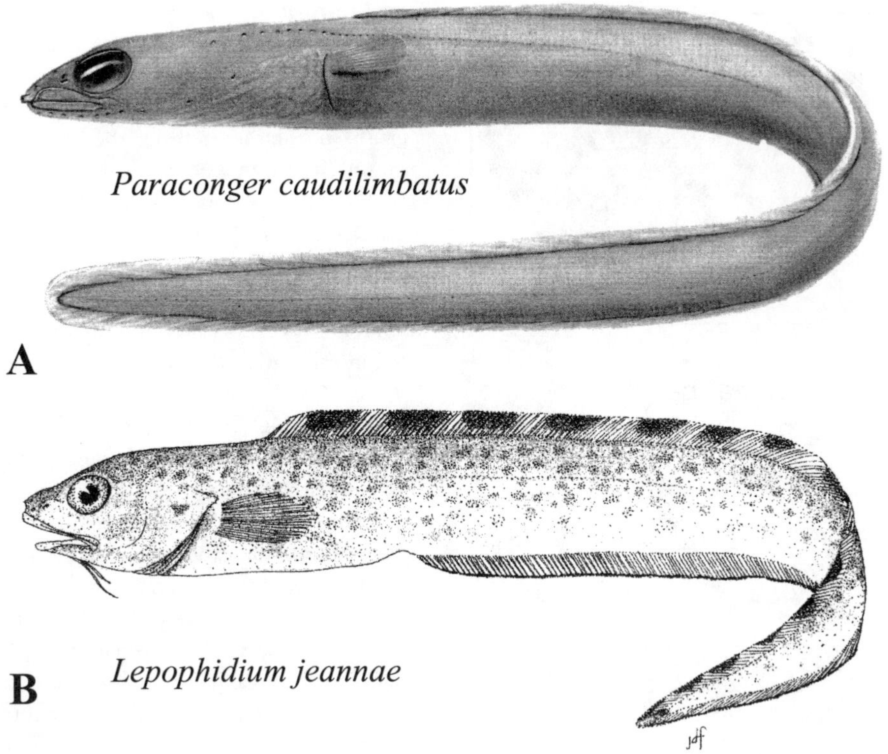

Figure 13.4. Modern representatives of the two families from which otoliths were used in this study. (A) *Paraconger caudilimbatus,* a congrid, 159 mm in length (from Smith, 1989a); (B) *Lepophidium jeannae,* an ophidiid, roughly the same length (from McEachran and Fechhelm, 1998).

(VPDB) standard. This sampling methodology provides a high-resolution profile of isotopic variation recorded by the otolith throughout the lifetime of the animal, with individual microsamples representing as little as 1.5–2 weeks of growth (figure 13.5B).

Paleotemperatures are calculated from the stable oxygen isotopic compositions of otolith aragonite using the otolith paleotemperature equation of Patterson et al. (1993) given above. Otolith isotopic compositions relative to VPDB are converted to VSMOW using the equation VPDB x 1.03091 + 30.91 = VSMOW. Calculated paleotemperatures are dependent upon the $\delta^{18}O$ value assumed for seawater ($\delta^{18}O_{sw}$). Zachos et al. (1994) provide estimates of seawater $\delta^{18}O$ for the Paleogene based on benthic foraminiferan $\delta^{18}O$ and independent proxies for the volume of glacial ice on Antarctica (Zachos et al., 1993) as follows:

early Eocene–early middle Eocene (Lutetian)	-1.0‰ (ice free world)
late middle Eocene (Bartonian)	-0.9‰
late Eocene	-0.8‰
Oligocene	-0.5‰

With revisions of these assumptions, absolute values for temperature will change, but the ranges of seasonal variation will not be affected. Lear et al. (2000) also provide estimates of Paleogene seawater $\delta^{18}O$ based on insights from the Mg/Ca ratio of benthic foraminiferans through time, and we consider the effects of using these alternate values later in the text.

The mean value of all analyses from all otoliths of a given age provides an approximation of mean annual temperature for that time. Calculated mean temperatures may be biased toward temperatures in the season of fastest growth because growth bands are thicker and hence more microsamples are possible, but this offset should be consistent through time. Mean $\delta^{18}O$ values are based on both ophidiid and congrid otoliths, and are well constrained for the early Eocene and middle middle Eocene through Oligocene. Data are currently unavailable for the early middle Eocene due to the poor preservation of aragonite during this interval (Tallahatta Formation and its equivalents).

Paleoseasonality can be estimated through a comparison of mean minimum and maximum tem-

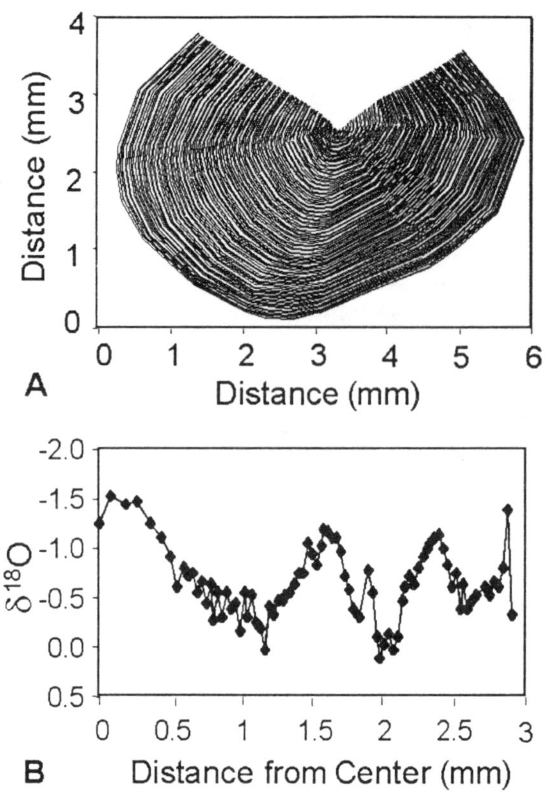

Figure 13.5. (A) Computer-generated sampling paths for LCI 87, a late Eocene congrid otolith from the Yazoo Clay, also shown in figure 13.2A. 113 intermediate paths were calculated, which yielded 92 discrete samples of carbonate after sampling paths were combined near the otolith center; 85 isotopic analyses were successful and are plotted in B. (B) Variation in oxygen isotopic composition of microsamples taken across the otolith (LCI 87), showing clear seasonal signals. Note that the scale is reversed. Isotopically light values correspond to warm temperatures.

perature values for each time interval, and by an examination of the ranges of temperature manifest within individual otoliths. Seasonality can be assessed with confidence for early Eocene, late Eocene and Oligocene congrid otoliths. Middle Eocene samples are dominated by ophidiids, and while co-occurring congrids and ophidiids show equivalent mean $\delta^{18}O$ values, lower otolith precipitation rate by the ophidiids may increase time-averaging and lead to an underestimate of seasonal range. Estimates of middle Eocene seasonal range based on congrids are pending at this time.

Error bars of one standard deviation around the mean composition and temperature are placed on the secular plots. The width of these bars does not represent the error associated with estimating the mean per se, but rather reflects the range of variation around the mean due to seasonal and interannual variability. These could be thought of as crude proxies for seasonality, however, this should be cautioned against due to the potential differences in time-averaging between ophidiids and congrids mentioned above, and because differing numbers of otoliths are used in calculating the mean at each time horizon.

RESULTS AND DISCUSSION

Secular Trends in Mean $\delta^{18}O$ and $\delta^{13}C$

The secular trend in mean $\delta^{18}O$ values of congrid and ophidiid otoliths is presented in figure 13.6. Lowest mean values are seen in the Paleocene and early Eocene, with a general increase into the middle and late Eocene to an early Oligocene maximum, and a slight return toward somewhat lower values later in the Oligocene. Values are relatively constant through the middle Eocene, increase slightly into the upper Eocene, and increase by ~0.6‰ across the Eocene/Oligocene boundary. Mean otolith $\delta^{13}C$ values average about -4‰ through the Eocene and decrease into the early Oligocene (figure 13.7).

Two apparent outliers exist in the data, in the earliest and latest Bartonian, and are indicated by dashed lines on figures 13.6 and 13.7. The early Bartonian point represents the mean of analyses from three different otoliths, one from the base of the Archusa Marl in Hickory, Mississippi, and the other two from the base of the upper Lisbon Formation at Coffeeville, Alabama. One of the Coffeeville samples falls in the same range as other middle Eocene samples. The Archusa sample and the second sample from Coffeeville, however, are significantly more negative than other otoliths in $\delta^{18}O$, reaching values of –2.1‰ and –3‰, respectively. These two samples are likewise quite negative in their $\delta^{13}C$ values, yet the range of seasonal variation in $\delta^{18}O$ is similar to other middle Eocene otoliths. We suspect that the compositions of these otoliths may have been influenced by waters of reduced salinity. The latest Bartonian mean value represents data from 3 otoliths from the Gosport Sand. In each case, the range of seasonal variation recorded in the otoliths is greater than those from samples above or below. The unusual nature of the Gosport Sand (medium-coarse grained, very fossiliferous, occasionally cross-bedded, and very restricted geographically) suggests that this

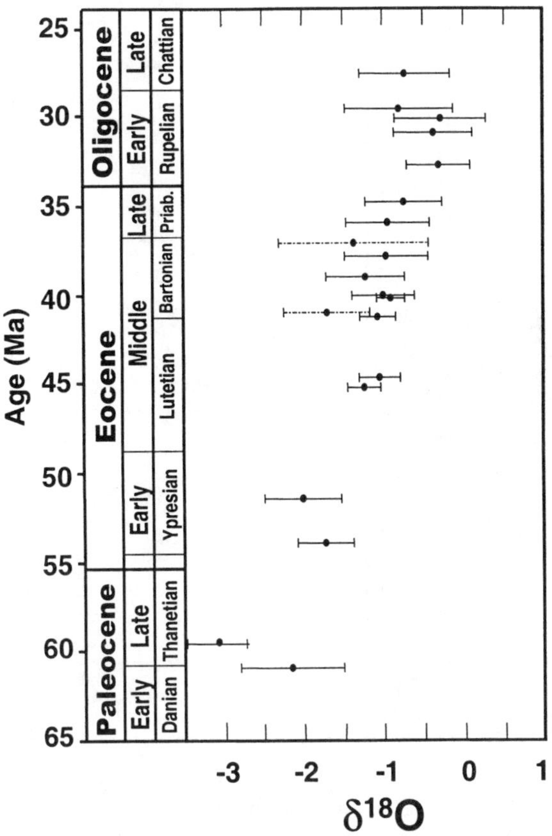

Figure 13.6. Secular trend in $\delta^{18}O$ based on means values of otolith carbonate. Means are the arithmetic mean of all microsamples for a given age. Error bars are one standard deviation around the mean, and represent a crude proxy for seasonal range (but see text for a cautionary note). Dashed samples are those that may have been affected by freshwater influx (see text). Time scale is based on Berggren et al. (1995), and sample ages are from table 13.2

unit was deposited closer to shore than the other units sampled in this study. If this were true, the inferred paleodepth might be substantially less than for other samples and hence the range of seasonal variation could be expected to be greater. In addition, proximity to shore might also make samples more susceptible to fluctuating salinity.

The Paleocene $\delta^{18}O$ values are also conspicuously negative. The means reflect samples taken from only two otoliths from the Matthews Landing Marl, and the other from the Coal Bluff Marl.

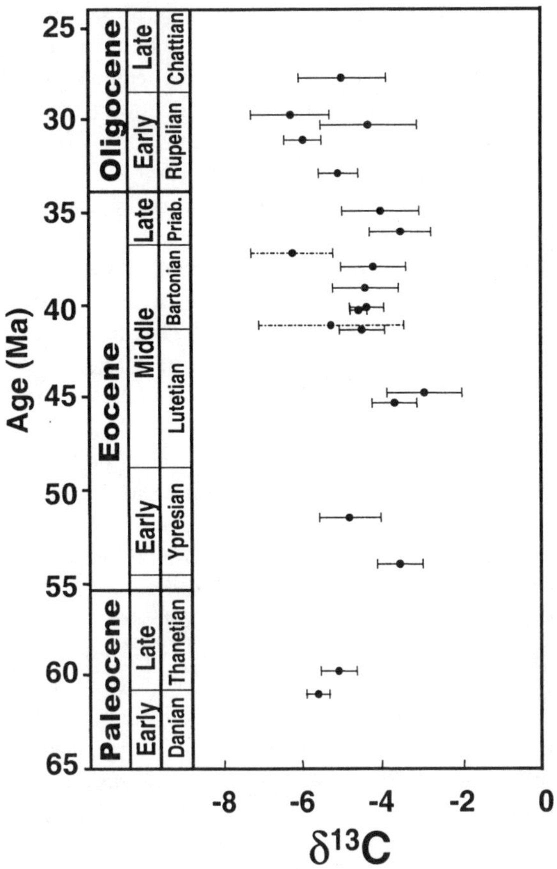

Figure 13.7. Secular trend in the $\delta^{13}C$ record based on mean values of otolith carbonate. Means are the arithmetic means of all microsamples of a given age.

Salinity may be a factor here also, although the carbon values do not show substantial depletion. Oxygen values and resulting paleotemperatures should be interpreted with caution due to the limited sampling from this interval of time, yet the possibility that the middle Paleocene was even warmer than the early Eocene in the Gulf Coast should be investigated further.

Secular Trend in Paleotemperature

Observed increases in $\delta^{18}O$ of otolith aragonite through the Eocene and into the Oligocene record some combination of cooling temperatures and increasing ice volume. Teasing apart these two variables is less than straightforward, but estimates of $\delta^{18}O_{SW}$ (and hence ice volume) are necessary before paleotemperatures can be calculated from $\delta^{18}O_{CaCO_3}$. Zachos et al. (1993) estimated Antarctic ice volume by assuming minimum temperature thresholds for ocean bottom water of 4°C and 1°C (conservative and extreme estimates). Shifts in $\delta^{18}O$ of benthic foraminifers that required bottom water temperatures to fall below these minima must be due to increases in ice volume and $\delta^{18}O_{SW}$. Their estimates of $\delta^{18}O_{SW}$ and ice volume based on $\delta^{18}O$ of benthic foraminifers are minimum estimates, but agree well with the distribution of glaciomarine sediments in the late Eocene and Oligocene (Zachos et al., 1993, 1994). Times not represented by ice rafted debris in marine sediments were assumed to be essentially ice-free ($\delta^{18}O_{SW}$ = -1‰); limited alpine glaciers are assumed to be present by the late middle Eocene (Bartonian).

When mean paleotemperatures are calculated using the Zachos et al. (1994) seawater values (figure 13.8), the data indicate warm temperatures in the Paleocene and early Eocene, with cooling of about 4°C occurring sometime after the early Eocene and before the middle of the middle Eocene. From the mid-Lutetian on, the data reveal little change all the way up through the Oligocene. A t-test shows that the difference between the pooled middle Eocene and late Eocene samples is statistically insignificant (0.3°C, p = 0.613 in 20 otoliths). The difference between pooled late Eocene and early Oligocene samples is somewhat larger, but again statistically insignificant (1.1°C , p = 0.186, in 12 otoliths). The maximal change across the Eocene/Oligocene boundary, as indicated by a comparison of latest Eocene and earliest Oligocene samples, is 1.8°C, but again is statistically insignificant (p = 0.202 in 7 otoliths) (Ivany et al., 2000).

More recently, Lear et al. (2000) arrived at estimates of $\delta^{18}O_{SW}$ through time by independently assessing temperature change in the deep ocean using Mg/Ca of benthic foraminiferans. They were then able to remove the temperature component from the $\delta^{18}O$ $CaCO_3$ record of the deep ocean. Their resulting curve of $\delta^{18}O_{SW}$ shows more variation than the Zachos estimates, with a much more substantial change at the Eocene/Oligocene boundary. Recalculation of mean otolith paleotemperatures using the Lear estimates results in a similar overall pattern to that in figure 13.8, but mean temperature across the Eocene/Oligocene boundary increases.

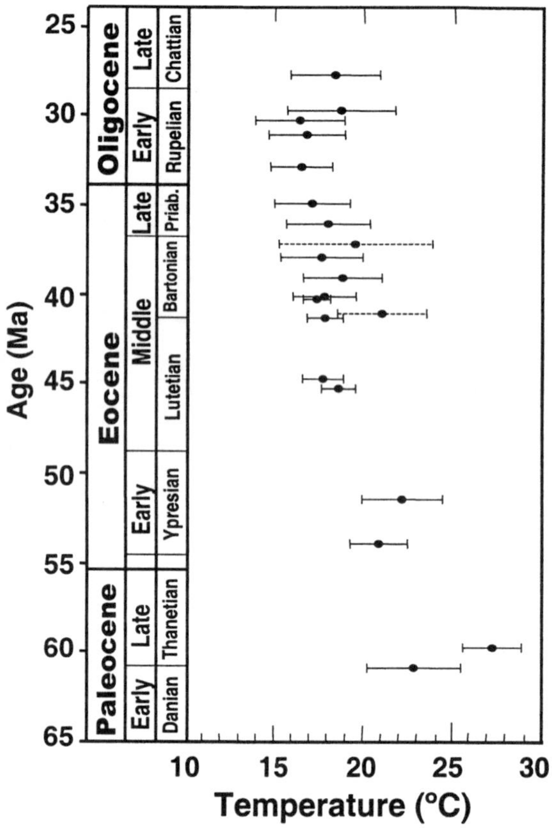

Figure 13.8. Secular trend in temperatures based on mean values of otolith carbonate. Means are based on the arithmetic mean of all microsamples from otoliths of a given age. Error bars are one standard deviation around the man, and represent a crude proxy for seasonality (but see text for a cautionary notes). Seawater $\delta^{18}O$ values are based on Zachos et al. (1994).

Comparisons with Other Records

The pattern of variation in mean $\delta^{18}O$ is broadly similar to that reported in Zachos et al. (1993, 1994; hereafter referred to as "the Zachos et al. data"). Several important differences exist between the two records, however, owing to differences in source data and what they record. In the Gulf Coast data set, the overall magnitude of the shift from early Eocene minimum (about –2.0‰; or even less if the Paleocene values are considered) to early Oligocene maximum (about 0‰) is smaller and the values shifted toward more negative numbers than in the Zachos et al. data. The latter are derived from measurements of open-ocean benthic foraminifera, which record the temperature and composition of bottom waters. Bottom-water temperature reflects the temperature at the surface where it formed, and hence the Zachos et al. data provide a signal of high-latitude conditions through time (providing bottom water had its origin throughout the Paleogene primarily at high latitudes due to the sinking of cold, dense water). Both data sets record some combination of the effects of fluctuations in temperature and ice volume, but since high latitudes were more strongly affected by Paleogene cooling than low (Barron, 1987; Zachos et al., 1994), the Zachos et al. data are dominated by the cooling signal and hence show a larger magnitude shift in isotopic ratios. In addition, higher latitudes and bottom waters are generally cooler overall than low-latitude shelf sites, and hence $\delta^{18}O$ of benthic foraminiferal carbonate is shifted toward more positive values.

In addition, the Zachos et al. data set shows more or less continuous change throughout the Eocene, while the Gulf Coast data show near stability throughout most of the middle and late Eocene. This again is most likely a function of the low-latitude position of the Gulf Coast, and reflects the relative stability of low-latitude climates during the cooling of the poles (Zachos et al., 1994; Schrag, 1999). The Gulf does show a $\delta^{18}O$ minimum in the Paleocene and early Eocene, but a significant positive shift of about 1‰ (~4°C of cooling) occurs sometime before the middle of the middle Eocene and little change occurs subsequently (figure 13.6). This is especially evident once $\delta^{18}O$ values are converted to paleotemperature (figure 13.8). Dockery's (1986) molluscan data indicate a significant faunal turnover near the early-middle Eocene boundary (Wilcox–Claiborne Group transition, figure 13.1) suggesting another possible link between turnover and cooling. Unfortunately, however, otolith paleotemperature data from the early middle Eocene needed to more precisely establish the timing of cooling during this transition are unavailable due to poor preservation of aragonite in this interval.

An increase in $\delta^{18}O$ at the Eocene/Oligocene boundary similar to that reported here (although of a somewhat higher magnitude) has been observed in numerous records of benthic and planktonic foraminiferan $\delta^{18}O$ in the deep ocean (Miller, 1992; Zachos et al., 1992, 1993, 1994). That this shift is manifest, if to a lesser degree, in the lower-latitude shelf waters of the Gulf Coast suggests that increasing ice volume and concomitant change in the $\delta^{18}O$ value of the global ocean is more likely to account for it than temperature change alone. By the middle to late Oligocene, isotopic compositions in the Gulf once again approximated those of the late Eocene, suggesting either that the ice sheet was not long-lived and/or that temperatures warmed into the late Oligocene.

Minimal change in mean annual temperature across the Eocene/Oligocene boundary inferred here is consistent with earlier work in the Gulf Coast and with Mg/Ca ratios of deep-sea benthic foraminiferans (Lear et al., 2000). This information on mean annual temperature alone does not particularly support the causal link between cooling and extinction suggested earlier.

The trend in $\delta^{13}C$ from the Gulf Coast otoliths also broadly mirrors that of Zachos et al. (1993), though it is more variable and absolute values are several per mil lower on average than those from the deep-sea sites. The short-lived positive shift in $\delta^{13}C$ documented by Salamy and Zachos (1999) from benthic foraminifera is not recognized in these data, yet the long-term trend toward more negative values in the Oligocene is also present in the benthic data from Zachos et al. (1993). This shift likely reflects the exhumation and oxidation of buried organic matter associated with sea-level fall near the boundary.

Trends in Seasonality

Beyond estimates of mean annual temperature, $\delta^{18}O$ values of individual, serially microsampled otoliths also record the seasonal temperature variation experienced throughout lifetimes of the fish. Figure 13.9 shows a selection of temperature profiles from early and late Eocene and early Oligocene congrid otoliths calculated using the $\delta^{18}O_{SW}$ values of Zachos et al. (1994).

Early Eocene temperatures are, on average, about 4-5°C warmer than those in the late Eocene, reflecting peak warm conditions on the planet at that time. Limited data from Paleocene otoliths suggest even warmer conditions in the Gulf. In both cases, however, the range of seasonal temperature variation is comparable or even greater than that seen later on (5-9*C), a result perhaps unexpected given our limited but growing understanding of warm equable climates.

Only limited data exist at this time for the middle Eocene based on congrid otoliths. Samples come only from the Bartonian and include those discussed earlier with reference to unusually negative isotopic values and warm paleotemperatures. The middle Eocene is better characterized by ophidiid otoliths. Lutetian temperature minima based on ophidiids are roughly 16°C, while maxima fall about 19°C. Bartonian values are generally comparable. The range of temperature variation within individual ophidiid otoliths in the middle Eocene is markedly smaller, however, as mentioned earlier in the text, and it is unclear whether this is a real phenomenon or an artifact of sampling (see discussion below). For this reason, the suggestion of reduced middle Eocene season-

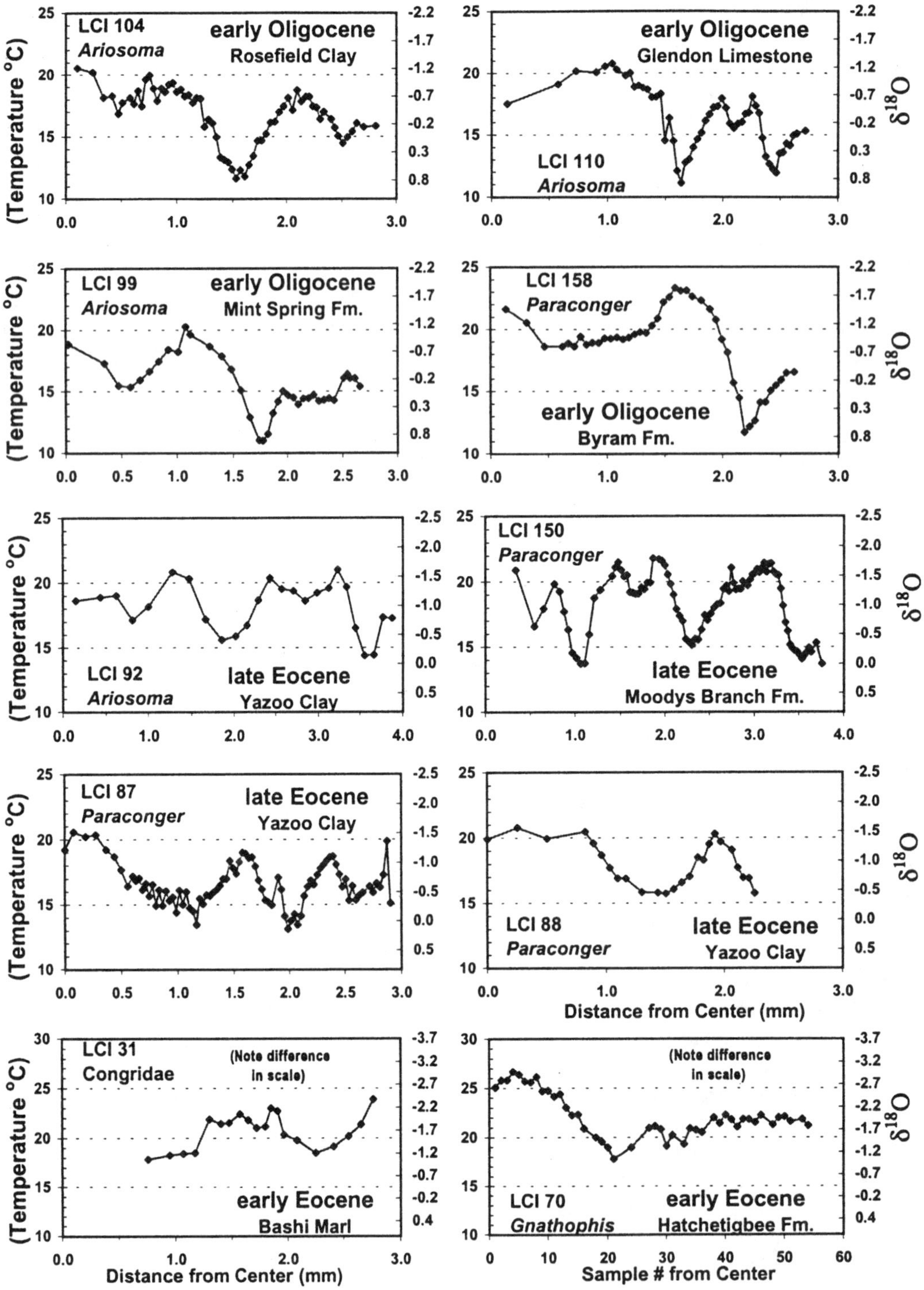

Figure 13.9. Individual temperature records from microsampled congrid otoliths from the early and late Eocene and early Oligocene. Sample number, genus, formation, and age are indicated on each plot. Note that (1) the early Eocene is on average 5°C warmer than later in the Eocene and Oligocene, but that seasonal range is comparabl, and (2) early Oligocene winters are cooler than late Eocene winters, yet summer temperatures do not differ appreciably.

ality must be regarded as tentative at this time. Andreasson and Schmitz (2000) report $\delta^{18}O$ from Gulf Coast gastropods at the top of the Lutetian that maximally range from about –2.0‰ to 0.0‰. This corresponds to a temperature range of roughly 13°C to 22°C when corrected for $\delta^{18}O_{SW}$ of –1.0‰.* Four of their five shells, however, recorded somewhat lower amplitudes and fell on the warm side of that range.

The range of temperatures experienced in the late Eocene generally falls between 15° and 20°C, with winter minima not extending below about 13.5°C. The Oligocene congrids record similar warm month temperatures of around 20°C, but winter temperatures consistently fall below those of the Eocene samples, with values as low as 11.0°C (Ivany et al., 2000). The scalloped nature of the Oligocene curves also suggests that otolith growth may have slowed significantly or stopped during the coldest month(s) (Dettman and Lohmann, 1993), indicating that winter temperatures may have been even colder than those recorded in the otolith carbonate. The trend in minimum and maximum temperatures from the late Eocene into the Oligocene is depicted in figure 13.10. When all local maximum and minimum values of temperature are compiled, a t-test demonstrates no significant difference in summer temperatures (p = .800, n = 21 summers in 12 otoliths), while Oligocene winters were significantly colder than late Eocene winters (p = .025, n = 21 winters in 12 otoliths) by as much as 4°C (Ivany et al., 2000). Such a decrease in winter minimum temperature on the shelf would have a significant effect on the biogeographic ranges of taxa, and could be responsible for the loss of stenothermal tropical taxa from the Gulf Coast.

Using Lear estimates for $\delta^{18}O_{SW}$, the pattern across the Eocene/Oligocene boundary changes. In this case, summer temperatures show warming, while winters remain the same. This is inconsistent with faunal data that suggest preferential extinction of warm-water taxa (Hansen, 1987, 1992), hence we prefer the Zachos et al. (1994) estimates of $\delta^{18}O_{SW}$.

Re-evaluation of ages of otolith-bearing strata may have a more significant impact on patterns using the Lear et al. estimates because of the greater variability in $\delta^{18}O_{SW}$ through time. It is important to emphasize, however, that patterns in the seasonal range of variability will not be affected by variation in $\delta^{18}O_{SW}$ through time. The pattern of increasing seasonality across the Eocene/Oligocene boundary is robust, regardless of whether summers get warmer, winters get colder, or some combination of both.

If our interpretations of $\delta^{18}O_{SW}$ across the boundary are correct, then the pattern of cooling winters with no change in summer temperatures might provide some explanation for the lack of temperature change demonstrated by isotopic analyses of planktonic foraminiferans (Zachos et al., 1994). There are indications that planktonic foraminifera are more abundant and larger in the summer months during times of higher phytoplankton productivity (Hemleben et al., 1989). If this were true in the Paleogene as well, and because $\delta^{18}O$ values of biogenic carbonate are biased toward temperatures during the season of maximum growth, we would expect to see little if any change in inferred temperature across the Eocene/Oligocene boundary based on analyses of foraminifera. Microsampling of biogenic carbonate from organisms that grow more or less continuously year-round can therefore provide not only a more complete picture of seasonal temperature variation, but also more robust estimates of mean annual temperature.

Sampling and Resolution

Several caveats about these data and the conclusions drawn from them should be mentioned. First and perhaps most serious is the question of how representative of an interval of time encompassed by an individual sampling horizon (on the order of 10^3–10^4 years) is the record of seasonal variation preserved in one or several otoliths, from fish that lived for 2–3 years? While this is a potential limitation, samples from otoliths of a similar age taken from different formations at different localities show a generally consistent pattern in the temperatures they generate; hence, the trends appear to be robust.

Other potential limitations are associated with interpreting seasonal ranges of temperature from dis-

*Andreasson and Schmitz (2000) used values of $\delta^{18}O_{SW}$ of only -0.4‰, and hence their calculated temperatures were much warmer (table 13.1). We believe their estimate of $\delta^{18}O_{SW}$ to be too heavy, and prefer -1.0‰ for the Lutetian in the Gulf. They justify their use of heavier values by suggesting that evaporation would have a significant impact on $\delta^{18}O_{SW}$. However, surface-water salinity of the Gulf today is roughly comparable to the global mean (NOAA/PMEL World Ocean Atlas, 1994), suggesting that this is not a concerning factor. The Gulf is more restricted today than it was in the Eocene, and therefore more likely to show effects of evaporation now if they were important.

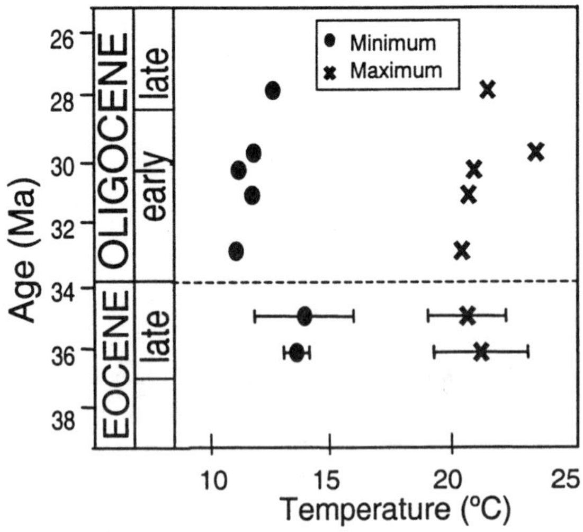

Figure 13.10. Change in seasonality across the Eocene/Oligocene boundary as recorded by the minimum and maximum values from congrid otolithsfrom the given horizon (from Ivany et al., 2000). the latest Eocene values represent means from 5 otoliths. Temperatures were calculated using $\delta^{18}O$ of seawater from Zachos et al. (1994).

cretely sampled carbonate. By necessity, the degree of annual temperature variation will correlate with sampling resolution. The more finely resolved the subsampling, the greater the range in inferred temperature; conversely, the coarser the sampling, the greater the amount of time averaged within each sample, and the more that seasonal amplitudes will be reduced. This relationship must be considered in future studies of paleoseasonality. As mentioned above, this is a real concern in comparing patterns based on ophidiid otoliths with those from congrids. Until the suggested pattern of reduced seasonality in the middle Eocene can be confirmed with congrid otoliths sampled comparably to later Eocene otoliths, this pattern must remain unsubstantiated. We do not believe sampling resolution to be a problem with our inferences concerning seasonality across the Eocene/Oligocene boundary. Resolution does vary to some degree among the late Eocene and Oligocene congrid otoliths presented here, yet sampling intensity is always high enough to record a significant number of points from each seasonal cycle. As a result, it is unlikely that inferred seasonal ranges of variation are compromised.

Paleodepth

Our data reflect temperature variations experienced at the bottom on a marine shelf. Paleodepth of these samples is poorly constrained; however, seasonal range of temperature variation in the northern Gulf at 50 m today is about 5°C (NOAA/PMEL World Ocean Atlas), roughly that observed in our data. Weidman et al. (1994) have shown that temperatures recorded on the bottom of the Atlantic shelf in 50 m accurately reflect surface temperatures from November to May, when the water column is isothermic. Stratification during the summer months prevents maximum bottom-water temperatures from reaching those at the surface, but turnover in the fall mixes warm surface water to the bottom. This suggests that while our warmest bottom temperatures may underestimate those reached at the surface, cold-month temperatures accurately record surface temperatures as long as Gulf waters experienced turnover.

Modern congrids have a planktonic leptocephalus larval stage (Smith, 1989b), suggesting the potential for resolving differences between surface- and deep-water temperatures using fossil otoliths. This phenomenon may be reflected in the profiles from several otoliths where the first summer is slightly warmer than those subsequent, but signs of distinctly different surface-water conditions could not be detected in the data. This suggests little if any complicating effect on our interpretation of shelf temperatures through time.

CONCLUSIONS

With the ability to reconstruct seasonal maximum and minimum temperatures through a substantial interval of earth history, we have added another variable to the list of climate parameters accessible from the rock record. Seasonality is an important variable to constrain in order to better understand the patterns and mechanisms of climate change and to interpret turnovers in the record of faunal composition.

Insight from this approach provides a potential resolution to the conundrum of selective loss of warm-water taxa without demonstrable evidence for change in mean annual temperature, and lends support to Hansen's (1987, 1992) suspicions that temperature likely played an important role in determining the diversity history of the Gulf Coastal Plain. At least for the Eocene/Oligocene boundary, and potentially for the middle/late Eocene boundary, we see evidence for an increase in seasonal range of temperature variation despite little change in mean annual temperature. We find that across the Eocene/

Oligocene boundary, warm-month temperatures do not significantly change, but early Oligocene winters are markedly cooler than those in the late Eocene. The increase in seasonality associated with a drop in cold-month mean temperature likely had a significant effect on the distributions of marine organisms in the Gulf, and may be the primary cause for the large faunal turnover associated with that boundary.

The correlation between increasing seasonality and faunal turnover emphasizes the link between biology and climate. When climate changes, taxa not capable of long-distance seasonal migration are left with three options: adapt, move, or go extinct. The Eocene-Oligocene tunover in the Gulf likely reflects a combination of all these factors. Climate change associated with the Eocene/Oligocene boundary was a transition that may have occurred over several hundred thousand years (Zachos et al., 1992). Faunal turnover was therefore probably not effected by actual death of individuals, as one might expect for an abrupt catastrophic event, but rather by a decrease in their ability to sustain viable population sizes. Physiological stress associated with cooling winter temperatures more likely resulted in a "failure to thrive" in the Gulf, which in turn led to regional and in some cases, global extinction. Whether the increase in seasonality recorded in the Gulf Coast is a more general phenomenon and relates to Eocene-Oligocene faunal turnover on a global basis remains to be seen. At high latitudes, turnover is more likely to be associated with a decrease in mean annual temperature at the boundary, but at lower latitudes this climate transition may be manifest only as an increase in seasonality. Microsampling of accretionary biogenic carbonate from taxa that grow for several years may be the only way to quantitatively resolve climate change of this nature.

ACKNOWLEDGMENTS

This paper has benefited greatly from discussions with participants of the Geological Society of America Penrose Conference on the Eocene–Oligocene transition held in August 1999, and with Jim Zachos, Roger Buick, Dan Schrag, Gerald Smith, and Bruce Wilkinson. Doug Jones, Judy Parrish, Liz Nesbitt, and Richard Ivany read the manuscript and offered many helpful suggestions. We thank Chris Wurster, Lora Wingate, and Geraldean Hourigan for running samples and assisting with microsampling, and Mark Suchter, Lisa Vosburgh, and Beth Reimer for assistance in microsampling and data entry. Dirk Nolf is gratefully acknowledged for providing additional Paleogene otoliths from the Gulf Coast to augment our samples. Ivany was supported by the Michigan Society of Fellows and National Science Foundation Postdoctoral Grant EAR-9704806.

REFERENCES

Andreasson, F. P., and Schmitz, B., 1996, Winter and summer temperatures of the early middle Eocene of France from *Turritella* $\delta^{18}O$ profiles: Geology, v. 24, p. 1067-1070.

Andreasson, F. P., and Schmitz, B., 2000, Temperature seasonality in the early middle Eocene North Atlantic region: evidence from stable isotope profiles of marine gastropod shells: Geological Society of America Bulletin, v. 112, p. 628-640.

Barron, E. J., 1987, Eocene equator-to-pole surface ocean temperatures: A significant climate problem?: Paleoceanography, v. 2, p. 729-739.

Baum, J. S., Baum, G. R., Thompson, P. R., and Humphrey, J. D., 1994, Stable isotopic evidence for relative and eustatic sea-level changes in Eocene to Oligocene carbonates, Baldwin County, Alabama: Geological Society of America Bulletin, v. 106, p. 824-839.

Beckman, D. W., and Wilson, C. A., 1995, Seasonal timing of opaque zone formation in fish otoliths, *in* Secor, D. H., Dean, J. M., and Campana, S. E., eds., Recent developments in fish otolith research, v. 27-43: Columbia, S.C., University of South Carolina Press.

Berggren, W. A., Kent, D. V., Swisher, C. C., and Aubry, M.-P., 1995, A revised Cenozoic geochronology and chronostratigraphy: SEPM Special Publication 54, p. 129-212.

Buening, N., Carlson, S. J., Spero, H .J., and Lee, D. E., 1998, Evidence for the Early Oligocene formation of a proto-subtropical convergence from oxygen isotope records of New Zealand Paleogene brachiopods: Palaeogeography, Palaeoclimatology, Palaeoecology, v. 138, p. 43-68.

Dettman, D. L., and Lohmann, K. C., 1993, Seasonal change in Paleogene surface water $\delta^{18}O$: Fresh-water bivalves of western North America, *in* Swart, P .K., Lohmann, K. C., McKenzie, J., and Savin, S., eds., Climate Change in Continental Isotopic Records, Geophysical Monograph 78, American Geophysical Union, p. 153-163.

Devereux, I., 1967, Temperature measurements from oxygen isotopic ratios of fish otoliths: Science, v. 155, p. 1684-1685.

Diester-Haass, L., 1995, Middle Eocene to early Oligocene paleoceanography of the Antarctic Ocean (Maud Rise, ODP Leg 113, Site 689): change from a low to a high productivity ocean: Palaeogeography, Palaeoclimatology, Palaeoecology, v. 113, p. 311-334.

Dockery, D. T., III, 1982, Lower Oligocene Bivalvia of the Vicksburg Group: Mississippi Department of Natural Resources Bureau of Geology, v. 123, p. 1-261.

Dockery, D. T., III, 1986, Punctuated succession of Paleogene mollusks in the Northern Gulf Coastal Plain: Palaios, v. 1, p. 582-589.

Dockery, D. T., III, 1987, Eocene-Oligocene molluscan extinctions: Comment to Hansen (1987): Palaios, v. 2, p. 620-621.

Elsik, W. C., 1974, Characteristic Eocene palynomorphs in the Gulf Coast, U.S.A.: Palaeontographica, Abteilung B, v. 149, p. 90-111.

Epstein, S., Buchsbaum, H. S., Lowenstam, H. A., and Urey, H. C., 1953, Revised carbonate-water isotopic temperature scale: Geological Society of America Bulletin, v. 64, p. 1315-1326.

Fordyce, R. E., 1992, Cetacean evolution and Eocene/Oligocene environments, *in* Prothero, D. R., and Berggren, W. A., eds., Eocene-Oligocene climatic and biotic evolution: Princeton, New Jersey, Princeton University Press, p. 368-381.

Fordyce, R. E., and Barnes, L. G., 1994, The evolutionary history of whales and dolphins: Annual Review of Earth and Planetary Sciences, v. 22, p. 419-455.

Frederiksen, N. O., 1988, Sporomorph biostratigraphy, floral changes,and paleoclimatology, Eocene and Earliest Oligocene of the Eastern Gulf Coast: U.S. Geological Survey Professsional Paper, v. 1448, p. 1-68, 16 pl.

Friedman, I., and O'Neil, J. R., 1977, Compilation of stable isotope fractionation factors of geochemical intereset., *in* Fleisher, M., ed., Data of Geochemistry, Sixth Edition, p. Chapter 440-KK, 12 pages, 49 figures.

Gaskell, B. A., 1991, Extinction patterns in Paleogene benthic foraminiferal faunas: Relationship to climate and sea level: Palaios, v. 6, p. 2-16.

Grossman, E. L., and Ku, T.-L., 1986, Oxygen and carbon isotope fractionation in biogenic aragonite: temperature effects: Chemical Geology, v. 59, p. 59-74.

Haasl, D. M., and Hansen, T. A., 1996, Timing of latest Eocene molluscan extinction patterns in Mississippi: Palaios, v. 11, p. 487-494.

Hansen, T., 1992, The patterns and causes of molluscan extinction across the Eocene/Oligocene boundary, *in* Prothero, D. R., and Berggren, W. A., eds., Eocene-Oligocene climatic and biotic evolution: Princeton, New Jersey, Princeton University Press, p. 341-348.

Hansen, T. A., 1987, Extinction of Late Eocene to Oligocene molluscs: relationship to shelf area, temperature changes, and impact events: Palaios, v. 2, p. 69-75.

Harris, G. D., 1895, Claiborne fossils: Bulletins of American Paleontology, v. 1, p. 1-52.

Harris, G. D., 1919, Pelecypoda of the St. Maurice and Claiborne Stages: Bulletins of American Paleontology, v. 6, p. 1-268.

Harris, G. D., and Palmer, K. V., 1946-47, The Mollusca of the Jackson Eocene of the Mississippi Embayment (Sabine River to the Alabama River), Parts 1 and 2, 563 p.

Hemleben, C., Spindler, M., and Anderson, O. R., 1989, Modern planktonic foraminifera: New York, Springer-Verlag, 363 p.

Iacumin, P., Bianucci, G., and Longinelli, A., 1992, Oxygen and carbon isotopic composition of fish otoliths: Marine Biology, v. 113, p. 537-542.

Ivany, L. C., Patterson, W. P., and Lohmann, K.C, 2000, Increase in seasonality across the Eocene-Oligocene boundary inferred from fossil otoliths: Nature, v 407, p. 887-890.

Jones, D. S., and Allmon, W. D., 1995, Records of upwelling, seasonality and growth in stable-isotope profiles of Pliocene mollusk shells from Florida: Lethaia, v. 28, p. 61-74.

Jones, D. S., and Quitmyer, I. R., 1996, Marking time with bivalve shells: oxygen isotopes and season of annual increment formation: Palaios, v. 11, p. 340-346.

Kalish, J. M., 1991a, ^{13}C and ^{18}O isotopic disequilibria in fish otoliths: metabolic and kinetic effects: Marine Ecology Progress Series, v. 75, p. 191-203.

Kalish, J. M., 1991b, Oxygen and carbon stable isotopes in the otoliths of wild and laboratory-reared Australian salmon (*Arripis trutta*): Marine Biology, v. 110, p. 37-47.

Keller, G., 1983, Biochronology and paleoclimatic implications of middle Eocene to Oligocene planktic foraminiferal faunas: Marine Micropaleontology, v. 7, p. 463-486.

Kennett, J. P., and Stott, L. D., 1990, Proteus and Proto-Oceanus: ancestral Paleogene oceans as revealed from Antarctic stable isotopic results, ODP Leg 113, *in* Barker, P. F., and Kennett, J. P. et al., eds., Proceedings of the Ocean Drilling Program, Scientific Results, v. 113, p. 865-880.

Klein, R. T., Lohmann, K. C., and Thayer, C. W., 1996, Bivalve skeletons record sea-surface temperature and $\delta^{18}O$ via Mg/Ca and $^{18}O/^{16}O$ ratios: Geology, v. 24, p. 415-418.

Lear, C. H., Elderfield, H., and Wilson, P. A., 2000, Cenozoic deep-sea temperatures and global ice volumes from Mg/Ca in benthic foraminiferal calcite: Science, v. 287, p. 269-272.

Legendre, S., and Hartenberger, J.-L., 1992, Evolution of mammalian faunas in Europe during the Eocene and Oligocene, *in* Prothero, D. R., and Berggren, W. A., eds., Eocene-Oligocene climatic and biotic evolution: Princeton, New Jersey, Princeton University Press, p. 516-528.

MacNeil, F. S., and Dockery, D. T., III, 1984, Lower Oligocene Gastropoda, Scaphopoda, and Cephalopoda of the Vicksburg Group in Mississippi: Mississippi Department of Natural Resources Bureau of Geology, v. 124, p. 1-415.

McEachran, J. D., and Fechhelm, J. D., 1998, Fishes of the Gulf of Mexico: Austin, University of Texas Press, 1112 p.

McGowran, B., 1987, Late Eocene perturbations: foraminiferal biofacies and evolutionary overturn, Southern Australia: Paleoceanography, v. 2, p. 715-727.

McGowran, B., Moss, G., and Beecroft, A., 1992, Late Eocene and early Oligocene in Southern Australia: Local neritic signals of global oceanic changes, *in* Prothero, D. R., and Berggren, W. A., eds., Eocene-Oligocene climatic and biotic evolution: Princeton, New Jersey, Princeton University Press, p. 178-201.

McKinney, M. L., McNamara, K. J., Cater, B. D., and Donovan, S. K., 1992, Evolution of Paleogene echinoids: a global and regional view, *in* Prothero, D. R., and Berggren, W. A., eds., Eocene-Oligocene climatic and biotic evolution: Princeton, New Jersey, Princeton University Press, p. 349-367.

Meng, J., and McKenna, M. C., 1998, Faunal turnovers of Paleogene mammals from the Mongolian Plateau: Nature, v. 394, p. 364-367.

Miller, K. G., 1992, Middle Eocene to Oligocene stable isotopes, climate, and deep-water history: the Terminal Eocene Event?, *in* Prothero, D. R., and Berggren, W. A., eds., Eocene-Oligocene climatic and biotic evolution: Princeton, New Jersey, Princeton University Press, p. 160-177.

Morales-Nin, B., 1987, Ultrastructure of the organic and inorganic constituents of the otoliths of sea bass, *in* Robert, C.S., and Hall, G.E., eds., The age and growth of fish: Ames, The Iowa State University Press, p. 331-343.

Morris, R. W., and Kittleman, L. R., 1967, Piezoelectric property of otoliths: Science, v. 158, p. 368-370.

NOAA/PMEL, World Ocean Atlas 1994, http://ferret.wrc.noaa.gov/fbin/climate_server, accessed November 8, 1999.

Nolf, D., 1980, Etude monographique des otolithes des ophidiiformes actuels et revision des especes fossiles (Pisces, Teleostei): Meded. Werkgr. Kwart. Geol., v. 17, p. 71-195.

Nolf, D., 1985, Handbook of Paleoichthyology: Otolithi Piscium: Stuttgart, Gustav Fischer Verlag, 145 p.

Nolf, D., 1995, Studies on fossil otoliths—the state of the art, *in* Secor, D. H., Dean, J. M., and Campana, S. E., eds., Recent developments in fish otolith Research, Columbia, University of South Carolina Press, p. 513-544.

Norman, J. R., and Greenwood, P. H., 1975, A history of fishes: London, Ernst Benn Ltd., 467 p.

Oboh, F. E., Jaramillo, C. A., and Reeves-Morris, L. M., 1996, Late Eocene–early Oligocene paleofloristic patterns in southern Mississippi and Alabama, U.S. Gulf Coast: Review of Palaeobotany and Palynology, v. 91, p. 23-34.

Palmer, K. V. W., 1937, The Claibornian Scaphopoda, Gastropoda and Dibranchiate Cephalopoda of the Southern United States: Bulletins of American Paleontology, v. 7, p. 1-731.

Palmer, K. V. W., and Brann, D. C., 1965, Catalogue of the Paleocene and Eocene Mollusca of the Southern and Eastern United States: Part I. Pelecypoda, Amphineura, Pteropoda, Scaphopoda, and Cepholopoda: Bulletins of American Paleontology, v. 48, no. 218, p. 1–443.

Palmer, K. V. W., and Brann, D. C., 1966, Catalogue of the Paleocene and Eocene Mollusca of the Southern and Eastern United States: Part II: Bulletins of American Paleontology, v. 48, no. 218, p. 471-1057.

Pannella, G., 1971, Fish otoliths: daily growth layers and periodical patterns: Science, v. 173, p. 1124-1127.

Patterson, W. P., Smith, G. R., and Lohmann, K. C, 1993, Continental paleothermometry and seasonality using the isotopic composition of aragonitic otoliths of freshwater fishes, *in* Swart, P., Lohmann, K. C, McKenzie, J., and Savin, S., eds., Continental Climate Change from Isotopic Indicators—American Geophysical Union Monographs, Volume 78, p. 191-202.

Prothero, D. R., and Berggren, W. A., eds., 1992, Eocene-Oligocene climatic and biotic evolution: Princeton , New Jersey, Princeton University Press, p. 568.

Purton, L. M. A., and Brasier, M. D., 1997, Gastropod carbonate $\delta^{18}O$ and $\delta^{13}C$ values record strong seasonal productivity and stratification shifts during the late Eocene in England: Geology, v. 25, p. 871-874.

Purton, L. M. A., and Brasier, M. D., 1999, Giant protist *Nummulites* and its Eocene environment: life span and habitat insights from $\delta^{18}O$ and $\delta^{13}C$ data from *Nummulites* and *Venericardia*, Hampshire basin, UK: Geology, v. 27, p. 711-714.

Raup, D. M., and Sepkoski, J. J., 1984, Periodicity of extinctions in the geologic past: Proceedings of the National Academy of Sciences USA, v. 81, p. 801-805.

Raup, D. M., and Sepkoski, J. J., 1986, Periodic extinction of families and genera: Science, v. 231, p. 833-836.

Salamy, K. A., and Zachos, J. C., 1999, Latest Eocene–early Oligocene climate change and Southern Ocean fertility: inferences from sediment accumulation and stable isotope data: Palaeogeography, Palaeoclimatology, Palaeoecology, v. 145, p. 61-77.

Schellenberg, S. A., 1999, Deep ocean ostracode response to the Eocene-Oligocene transition: high-resolution records from the Southern Ocean (ODP Sites 744 and 6): Geological Society of America Abstracts with Programs, 1999 Annual Meeting, v. 31, p. A310.

Schrag, D. P., 1999, Effects of diagenesis on the isotopic record of late Paleogene tropical sea surface temperatures: Chemical Geology, v. 161, p. 215-224.

Seisser, W. G., 1984, Paleogene sea levels and climates: U.S.A. Eastern Gulf Coastal Plain: Palaeogeography, Palaeoclimatology, Palaeoecology, v. 47, p. 261-275.

Smith, D. G., 1989a, Family Congridae: Conger Eels, *in* Böhlke, E. B., ed., Fishes of the western North Atlantic, Part 9, Volume 1: New Haven, Connecticut, Yale University, p. 460-567.

Smith, D. G., 1989b, Family Congridae: Leptocephali, *in* Böhlke, E. B., ed., Fishes of the western North Atlantic, Part 9, Volume 2: New Haven, Connecticut, Yale University, p. 723-763.

Stanley, S. M., 1984, Temperature and biotic crises in the marine realm: Geology, v. 12, p. 205-208.

Stanley, S. M., 1986, Anatomy of a regional mass extinction: Plio-Pleistocene decimation of the Western Atalantic bivalve fauna: Palaios, v. 1, p. 17-36.

Stevenson, D. K., and Campana, S. E., 1992, Otolith microstructure examination and analysis: Canadian Special Publication of Fisheries and Aquatic Sciences, v. 117: Ottawa, Department of Fisheries and Oceans, p. 126.

Tarutani, T., Clayton, R. N., and Mayeda, T. K., 1969, The effect of polymorphism and magnesium substitution on oxygen isotope fractionation between calcium carbonate and water: Geochimica et Cosmochimica Acta, v. 33, p. 987-996.

Thomas, E., and Gooday, A. J., 1996, Cenozoic deep-sea benthic foraminifers: Tracers for changes in oceanic productivity?: Geology, v. 24, p. 355-358.

Toulmin, L. D., 1977, Stratigraphic distribution of Paleocene and Eocene fossils in the eastern Gulf Coast region, Volumes 1 and 2: Geological Survey of Alabama, Monograph 13, 602 p.

Valentine, J. W., 1973, Evolutionary paleoecology of the marine biosphere: Englewood Cliffs, New Jersey, Prentice-Hall, Inc., 511 p.

Vermeij, G. J., 1978, Biogeography and adaptation: Cambridge, Massachusetts, Harvard University Press, 332 p.

Wade, B. S., Kroon, D., and Norris, R. D., in press, Orbitally forced climate change in the late middle Eocene at Blake Nose (Leg 171B): evidence from stable isotopes in foraminifera, Western North Atlantic Palaeogene and Cretaceous Palaeoceanography: Geological Society of London Special Publications.

Weidman, C. R., Jones, G. A., and Lohmann, K. C, 1994, The long-lived mollusc *Arctica islandica*: A new paleoceanographic tool for the reconstruction of bottom temperatures for the continental shelves of the Northern Atlantic Ocean: Journal of Geophysical Research, v. 99, p. 18,305-18,314.

Weidman, C. R., and Millner, R., 2000, High-resolution stable isotope records from North Atlantic cod: Fisheries Research, v. 46, p. 327-342.

Whitehead, P. J. P., Bauchot, M.-L., Hureau, J.-C., Nielsen, J., and Tortonese, E., 1985, Fishes of the North-eastern Atlantic and the Mediterranean, Unesco.

Wolfe, J. A., 1978, A paleobotanical interpretation of Tertiary climates in the northern hemisphere: American Scientist, v. 66, p. 694-703.

Wolfe, J. A., 1992, Climatic, floristic, and vegetational changes near the Eocene/Oligocene boundary in North America, *in* Prothero, D. R., and Berggren, W. A., eds., Eocene-Oligocene climatic and biotic evolution: Princeton, New Jersey, Princeton University Press, p. 421-436.

Wolfe, J. A., 1993, A method of obtaining climatic parameters from leaf assemblages: U.S. Geological Survey Bulletin, v. 2040, p. 1-71.

Wolfe, J. A., and Poore, R. Z., 1982, Tertiary marine and nonmarine climatic trends: Climate and earth history: Studies in Geophysics: Washington, D.C., National Academy Press, p. 154-158.

Wurster, C. M., Patterson, W. P., and Cheatham, M. M., 1999, Advances in computer-based microsampling of biogenic carbonates: Computers and Geosciences, v. 25, p. 1155-1162.

Yancey, T. E., and Davidoff, A. J., 1994, Paleogene sequence stratigraphy of the Brazos River section, Texas, Gulf Coast Association of Geological Societies, 104 p.

Yancey, T. E., and Elsik, W. C., 1999, The palynological record of late Eocene climate change and terrestrial palynostratigraphy, northwest Gulf of Mexico: Abstracts with Programs, Geological Society of America Annual Meeting, v. 31, p. A309.

Zachos, J. C., Breza, J. R., and Wise, S. W., 1992, Early Oligocene ice-sheet expansion on Antarctica: Stable isotope and sedimentological evidence from Kerguelen Plateau, southern Indian Ocean: Geology, v. 20, p. 569-573.

Zachos, J. C., Lohmann, K. C, Walker, J. C. G., and Wise, S. W., 1993, Abrupt climate change and transient climates during the Paleogene: A marine perspective: Journal of Geology, v. 101, p. 191-213.

Zachos, J. C., Stott, L. D., and Lohmann, K. C , 1994, Evolution of early Cenozoic marine temperatures: Paleoceanography, v. 9, p. 353-387.

CHAPTER 14

The Palynological Record of Late Eocene Climate Change, Northwest Gulf of Mexico

Thomas E. Yancey, William C. Elsik, and Recep H. Sancay

ABSTRACT

Palynofloras of east Texas show major changes in regional lowland plant assemblages through the middle Eocene–early Oligocene interval. In marine deposits containing a mixture of pollen assemblages, most of the Bartonian contains a *Cupuliferoipollenites*-dominated regional assemblage, the Priabonian generally contains a *Momipites*-dominated regional assemblage, and the early Oligocene contains a *Carya* and *Quercus* regional assemblage. This succession indicates longterm progressive change to cooler, drier land climates, tracking the global late Eocene cooling trend. The succession is well developed in the Texas segment of the northwestern Gulf region, where it occurs earlier than in the Mississippi-Alabama segment of the Gulf region. The diachronous succession of regional pollen assemblages eastward along the northern Gulf indicates the presence of a late Eocene moisture gradient across southern North America that influenced the timing of turnover during temperature-driven floral succession. Palynomorph changes in east Texas show continual, progressive late Eocene climate change, without the major climatic fluctuations suggested by leaf-margin analysis and without long intervals of uniformity.

Palynomorph ranges are enumerated against a framework of volcanic ash $^{40}Ar/^{39}Ar$ radiometric dates. Several palynomorph biostratigraphic assemblage zones based upon a cooling temperature-driven succession of floral elements are recognized within the section. Marine dinoflagellate and diatom microfloras are also present in the east Texas section, showing that most of the strata were deposited in coastal marine environments, which contain regional assemblages composed of a mixture of land-derived pollen and spores. Regional palynomorph assemblages are distinct from environmentally controlled palynomorph assemblages and do not carry information about depositional environment or plant associations; they contain a general sample of palynomorphs in a region and are sensitive to climate change. Examples of environmentally controlled assemblages occur in lignites and in nonmarine deposits of the late late Eocene in east Texas.

INTRODUCTION

The record of late Eocene climate change shows a rapid decline in global temperatures (Berggren and Prothero, 1992), culminating in an earliest Oligocene cold event accompanying the beginning of continental glaciation in Antarctica (Robert and Kennett, 1997). The most detailed records of late Eocene climate change are shown by shifts in oxygen isotope ratios of marine waters recorded in deep-sea sediments (Miller et al., 1987; Miller, 1992; Zachos et al., 1994), while continental records of climate change come from records of leaf floras (Upchurch and Wolfe, 1987; Wolfe, 1992; Greenwood and Wing, 1995) and palynomorphs (Leopold and MacGinitie, 1972) in nonmarine sediments preserved in continental interior basins. Together, these settings provide indications of oceanic and continental climate change and global cooling following the early Eocene high-temperature interval (Greenwood and Wing, 1995; Andreasson and Schmitz, 2000). Associated with the global cooling climate changes are major biotic turnovers at the middle/late Eocene boundary and the Eocene/Oligocene boundary (the Terminal Eocene

Event of Wolfe, 1978), corresponding with land mammal and land plant extinction events (Prothero and Swisher, 1992).

The record of Eocene climate change is examined mostly in the context of changing temperature, which is the most important control in ocean waters and readily determined from the isotopic composition of calcareous shell. Continental climate is also examined to determine trends in changing temperature, although continental climate reflects strong control by precipitation as well as temperature and calcareous shell is much less available or reliable for paleotemperature determination. Plant fossils provide the most useful record of continental climate, particularly paleotemperature indices based on leaf-margin analysis of forest plants (Wolfe, 1978, 1992; Greenwood and Wing, 1995), and provide a means of calculating paleotemperatures that are directly comparable with marine paleotemperature records. Because plants are sensitive indicators of precipitation, they also can provide records of precipitation change, either on temporal or geographic gradients, once the factor of paleotemperature condition is determined. The dual goals of interpreting paleotemperature change and identifying precipitation gradient are undertaken in this study, utilizing new data from a well-documented stratigraphic section in east Texas.

Late Eocene climate change is well recorded in the east Texas stratigraphic section, centered on the Brazos Valley, where palynology provides a detailed record of temporal change in shorezone depositional settings. This considerably extends the documentation of late Eocene climate change by adding a record from the land-sea interface (the shore zone), in settings with high rates of sediment accumulation and expanded section, at a site located geographically between the central-eastern Gulf Coast and east Pacific portions of North America. The initial work in this study builds a detailed record of climate change based on palynomorphs, which provide a means for determining climate change in marginal marine/coastal zone settings with siliciclastic sediments. In siliciclastics, carbonate has frequently been removed by dissolution, leaving organic-walled microfossils as the most available fossils. The land-sea margin is also important in providing a record of climate change from an interface where lowland continental climates are moderated by marine influences, useful in tying together oceanic and continental records of climate change. The culminating work of this study is a cross-longitude comparison of climate change across the southern portions of North America, where a gradient in precipitation exerts an important control on climate regime during the late Eocene (Frederiksen, 1991). This is done by comparison of the palynological record in east Texas (northwestern part of the Gulf of Mexico) with palynological records of the north-central and northeastern portions of the Gulf of Mexico (primarily Frederiksen, 1980a, 1988) and the palynological record of the Pacific margin in southern California (Frederiksen, 1991). The identification of co-occurring climate gradients (temporal and geographic) is based on data sets obtained from similar depositional environments, so local environmental control on plant assemblages is effectively removed as a factor in making comparisons of the data.

The Brazos Valley section of east Texas contains thick intervals of shallow, coastal marine deposits, interspersed with thinner intervals of open-shelf marine deposits and nonmarine deposits. Although the presence of lignite beds and lack of carbonate in the post-Crockett parts of the section (Yegua Formation and Jackson Group) have prompted many workers to assume that this section is dominantly nonmarine or upper deltaic in origin, strata in most intervals contain a variety of marine fossils indicative of open-ocean environments. The presence of marine dinoflagellates in strata once thought to be nonmarine—for example, the uppermost Yegua Formation at Koppes Bridge section (Atlee et al., 1967)—is proving to be the rule rather than the exception. Dinoflagellates are present in many palynology samples (if not screened out during sample preparation), test linings of marine foraminifera are present, marine trace fossil assemblages are common, and even marine diatoms are present in some samples (Yancey, 1997). Also, strata in most parts of the section contain common tidalite depositional structures.

PALYNOSTRATIGRAPHY

This study of climate change in east Texas is based on palynomorph occurrences compiled from data obtained in the study of 209 samples in 15 stratigraphic intervals (outcrop sections and drilled cores) centered around the Brazos River Valley (figure 14.1; appendix). Strata in this section are cyclic in character, deposited in parasequences that contain sediments of shallow marine and coastal plains environments. In the composite section, there are many range terminations and appearances of palynomorphs (figure 14.2). Several palynomorph assemblage zones are present in the section that are important for regional

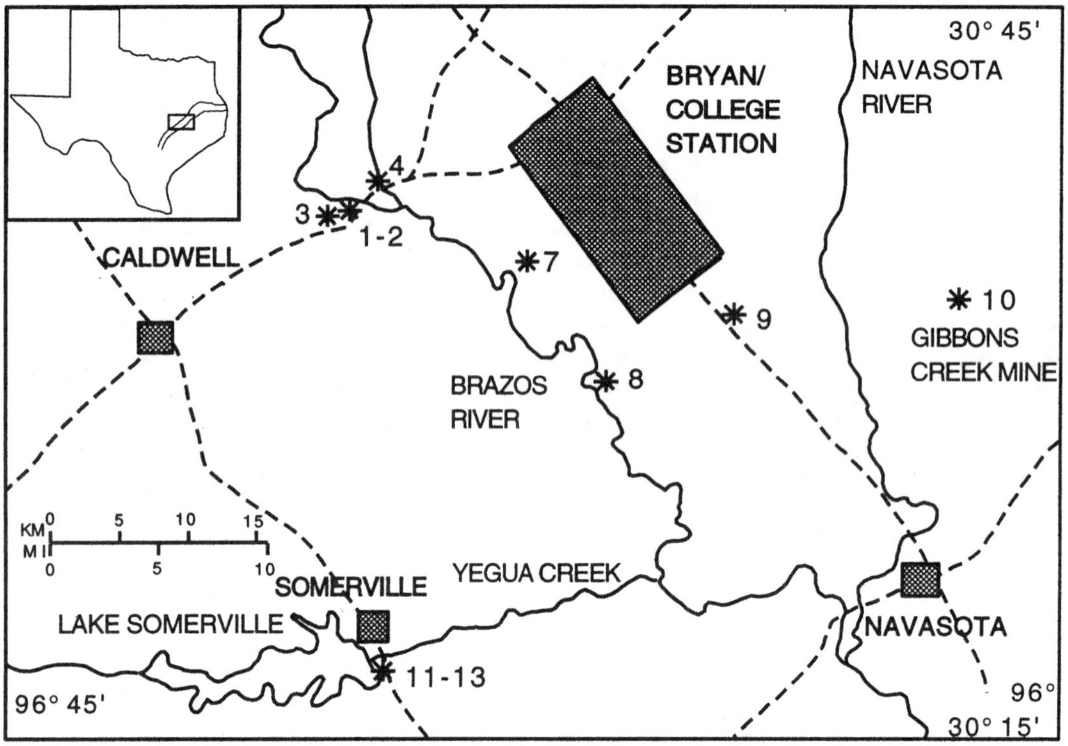

Figure 14.1. Location map of sampled stratigraphic sections in the Brazos River Valley, east-central Texas. Late Eocene outcrop belt of east Texas is shown on inset map. Sections: 1-2, Stone City Bluff; 3, Rocky Branch; 4, Little Brazos; 7, Turkey Creek; 8, Koppes Bridge; 9, Greens Prairie core; 10, Gibbons Creek mine; 11-13, Lake Somerville. Sections at Alabama Ferry, Hurricane Bayou, and Riverside quarries are located east of the Bryan/College Station area.

correlation (Elsik and Yancey, 2000). These assemblage zones add greatly to the Gulf Coast palynostratigraphy developed by Tschudy (1973), Elsik (1974) and Frederiksen (1980a). Age control for the palynomorph assemblage zones in this section is provided by radiometric $^{40}Ar/^{39}Ar$ determinations of several volcanic ash beds present in the section (Guillemette and Yancey, 1996; Yancey and Guillemette, 1998).

Although strata throughout the section yield well-preserved and diverse palynomorph populations, the microfloras are relatively unknown, as indicated by the number of grains assigned to the "angiosperm pollen undifferentiated" category in many samples. These rich residues contain many undescribed taxa of dinoflagellates, fungi, pollen, and spores that will yield additional useful index taxa. Several taxa that range throughout the section show cyclic variations in abundance. These include *Cicatricosisporites* spp., which become very abundant in the upper Manning Formation and are cyclic in abundance through the Oligocene of the Gulf Coast, and both *Momipites* spp. and *Cupuliferoipollenites* spp., which vary in abundance throughout the section (figure 14.3).

BARTONIAN PALEOCLIMATE: STARTING CONDITION

The east Texas climate record presented in this study begins in the Bartonian, at the end of the early middle Eocene high-temperature interval. This was a time of major marine highstand in the Gulf Coast region, providing a rich fossil record for paleotemperature determination to serve as a baseline for late Eocene climate comparison. Three studies have calculated shelf ocean water temperatures from oxygen isotope ratios in fossils of the Crockett Formation (Stone City and Wheelock members) in the Brazos River Valley.

Using marine gastropod shells, Andreasson and Schmitz (2000) report seasonal paleotemperature variation ranging from 19°C to 28°C (with mean annual temperature about 24°C) for the Wheelock Member in Burleson County. Their Texas data points

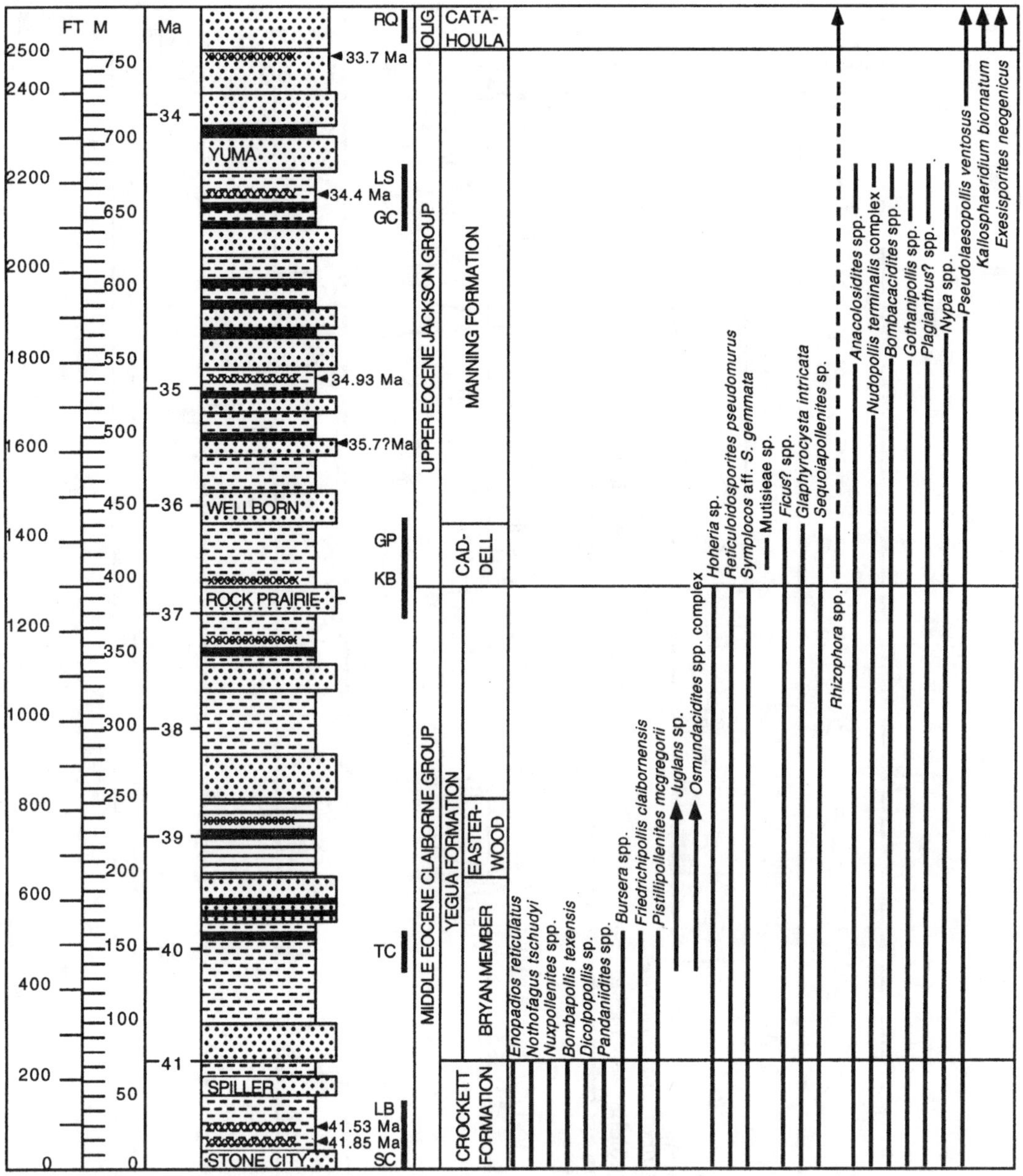

Figure 14.2. Composite section, late middle Eocene to early Oligocene, with ranges of stratigraphically important palynomorphs and radiometric dates providing age control. Bars indicate the range of taxon in section and arrows on end of bars indicate range continuing into younger intervals above top of section; dashes indicate interval in which taxon has not been recovered but probably present in the east Texas section. Lithologic section compiled from strata in Brazos County, Texas. Approximate position of sections and cores shown by heavy bars: SC, Stone City Bluff core and outcrop section, 47 samples; LB, Little Brazos River and equivalent sections, 14 samples; TC, Turkey Creek section, 10 samples; KB, Koppes Bridge section, 17 samples; GP, Greens Prairie core, 20 samples; LS, Lake Somerville dam and spillway outcrop sections and core, 40 samples; RQ, Riverside quarries, Huntsville, 3 samples. Radiometric dates are from Guillemette and Yancey (1996) and Yancey and Guillemette (1998), except the 35.7 Ma level, which indicates the probable source level of bediasite tektites derived from the late Eocene Chesapeake impact event. Age marks (Ma) to left of the lithologic column are not on a linear time scale, but are inferred from radiometric dates and tied to section thickness, showing a high rate of sediment accumulation in the upper part of Jackson Group.

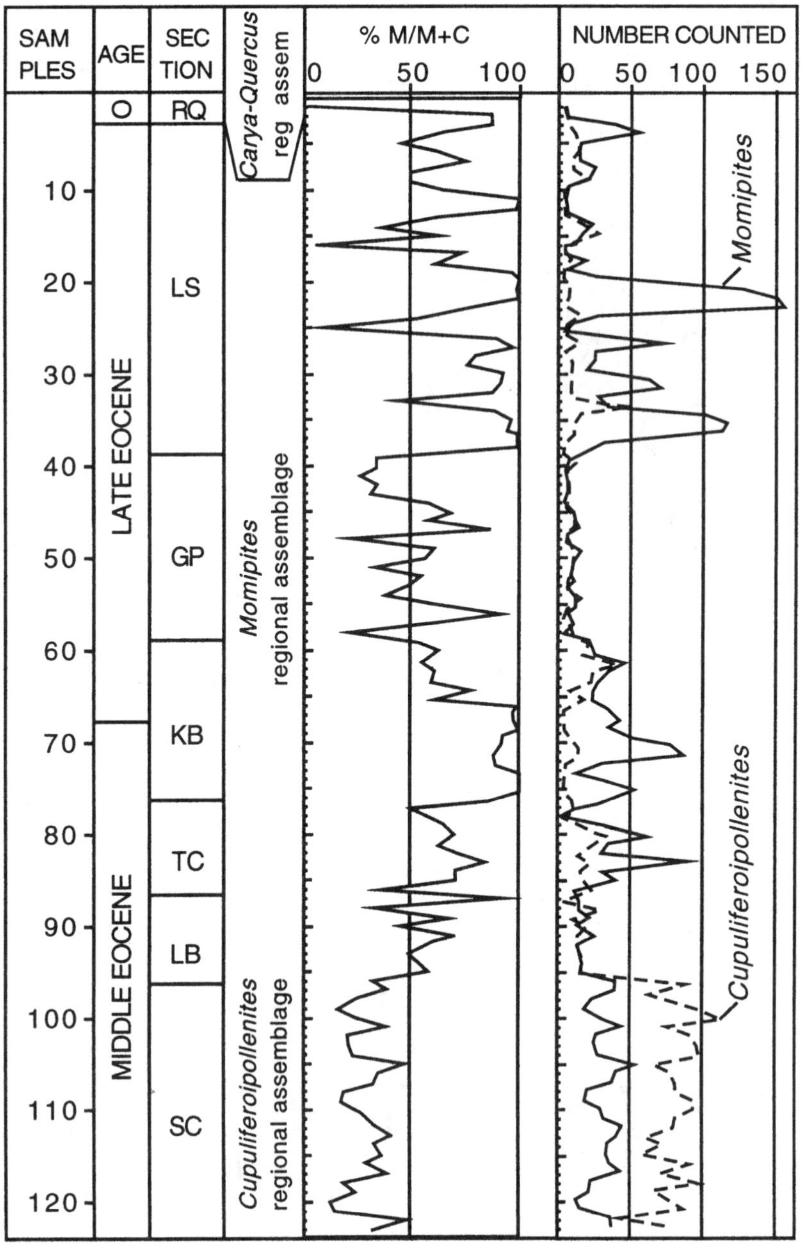

Figure 14.3. Stratigraphic positions of regional pollen assemblages in late middle Eocene to early Oligocene, Brazos Valley section, east Texas. The ratio M/M+C shows proportion of *Momipites* to combined total of *Momipites* plus *Cupuliferoipollenites* through the Eocene and illustrates the change to dominance of *Momipites* over *Cupuliferoipollenites.* Plots on the right side of diagram show grain counts of the two taxa. Tick marks in the % M/M+C and Number Counted columns indicate individual samples. Note that the low percentage peaks of *Momipites* in late Eocene portion of the % M/M+C plot correspond with samples having very low numbers of both *Momipites* and *Cupuliferoipollenites* pollen grains, indicating that these low percentages are artifices of the plotting method and not reliable indicators of dominance. Data compiled from 28 samples in Stone City Bluff core (SC), 11 samples in Little Brazos and equivalent sections (LB), 10 samples in Turkey Creek section (TC), 17 samples in Koppes Bridge section (KB), 20 samples in Greens Prairie core (GP), 35 sample in Lake Somerville core (LS), and 2 samples in Riverside quarries (RQ), a total of 123 samples.

come from shells collected in the Rocky Branch section (our section #3), which is Bartonian in age instead of Lutetian as reported in their paper (calcareous nannoplankton in the underlying Stone City Bluff section were assigned to zone NP16 by Mark Jiang in Yancey and Davidoff, 1994). Nearly identical paleotemperatures are indicated for marine gastropods from the Wheelock Member at the Highway 21 bridge on the Little Brazos River, 3 km distant from the Rocky Branch site, by Takuro Kobashi (pers. commun., 1999) of Texas A&M University: a mean annual temperature of 24°C and seasonal paleotemperature variation of 20°–28°C. This is representative of paleotemperatures determined from marine gastropods for other Bartonian sites in Mississippi and Alabama (Kobashi et al., 2001). Using fish otoliths, Ivany et al. (2000) report a mean annual temperature of about 18°C for the slightly older Stone City Member at Stone City Bluff in Burleson County, Texas, but the temperature calculations were done using a different estimate of seawater oxygen isotopic composition than the Andreasson and Schmitz (2000) and Kobashi et al. (2000) results. When recalculated in the same manner, the Stone City Bluff paleotemperature is 21°C (T. Kobashi, pers. commun.). A seasonal variation in temperature of 20°–28°C is a little warmer than modern ocean temperatures for coastal Texas (Rezak et al., 1985), which have winter water temperatures several degrees cooler than 20°C.

These temperatures place the Bartonian of east Texas within the paratropical climate zone, corresponding to an interval having mean annual temperatures between 20° and 25°C (Wolfe and Poore, 1982; Morley, 2000). Paleobotanical determinations of middle Eocene paleoclimate (Gray, 1960; Dilcher, 1973; Wolfe, 1978; Upchurch and Wolfe, 1987; Greenwood and Wing, 1995) and of late Eocene paleoclimate (Frederiksen, 1980b; Wolfe, 1992) for the northern Gulf Coast all infer winter dry-tropical to humid paratropical climate on lowlands and most infer no freezing. These results are similar to gastropod paleotemperature determinations, although results of leaf-margin analysis suggest two times of 28°C mean annual temperatures in the middle and late Eocene (Wolfe, 1978). Frederiksen (1980a,b) suggested a regime of winter-dry tropical climate on lowlands adjacent to the coastline and a marginal humid paratropical climate on the upper coastal plain, in agreement with results of leaf foliar physiognomy analysis (Wolfe, 1978) that indicate a drying trend through the Eocene. Petrified wood from the Yegua Formation (late Bartonian) of the Brazos River Valley is dominantly angiosperm hardwood, with minor amounts of palm wood. This wood contains an association of trees found today in Central America, along with wood of *Taxodium* (cypress). Most petrified wood in the Brazos Valley section is found in shallow coastal marine deposits (personal observation) and probably was transported to the ocean by river floods.

THE TEMPORAL RECORD OF CLIMATE CHANGE IN EAST TEXAS

Palynofloras of east Texas show major changes in land plant taxa and assemblages through the middle Eocene–early Oligocene interval, both in the appearance or disappearance of individual taxa and in the change in dominance of regional assemblages (figures 14.2, 14.3). Many warm-climate elements disappear or decline in abundance through time, concurrent with the appearance of taxa tolerant of cooler, drier climates. Tropical/paratropical indicators *Alangiopollis* spp., *Amanoa?* spp., *Bursera* spp., *Calamuspollenites* spp., *Cycadopites?* spp., *Ficus?* spp., *Friedrichipollis claibornensis*, *Maceopolipollenites quietus* group, *Nypa* spp., *Symplocos* aff S. *gemmata*, and *Triatriopollenites aroboratus* decline or disappear, as grassland indicators (Gramineae and Mutisieae) and grasslands dung fungi (*Delitschia* spp. and *Sporormiella* spp.), along with comminuted fungal hyphae from arthropod coprolites, appear and increase, indicating a change to drier land climates (table 14.1). There is a decline in abundance of *Tricolpo(ro)pollenites microhenricii* that parallels the decline of tropical/paratropical elements, although *T. microhenricii* is not a typical tropical/paratropical indicator. The warm-climate taxa listed here are usually present in the subtropical as well as the tropical/paratropical climate zone, but are rare or absent in cooler or drier climates of the subtropical zone. (Note: The climatological definition of subtropics is a temperature range in which occasional freezing events occur [Trewartha, 1968; Morley, 2000].) Warm-climate taxa were common in east Texas during the Bartonian, in climates with mean annual temperatures of 24°C (Andreasson and Schmitz, 2000).

The appearance of abundant grains of *Quercus* (mainly *Quercoidites*) and *Carya* spp. (especially the larger, modern types of grains) and the occurrence of abundance peaks of *Pinus* suggests change to cooler/drier climate conditions across the Eocene/Oligocene boundary. *Quercus* (mainly *Quercoidites*) is generally rare throughout the middle and late

Table 14.1. Stratigraphic Abundances and Ranges of Palynomorph Climatic Indicator Taxa in Sections 1-15, Late Middle Eocene to Early Oligocene, East Texas

SECTION	1. Stone City Bluff	2. Stone City core	3. Rocky Branch	4. Little Brazos	5. Alabama Ferry	6. Hurricane Bayou	7. Turkey Creek	8. Koppes Bridge	9. Greens Prairie core	10. Gibbons Creek mine	11. Lake Somerville dam	12. Lake Somerville core	13. L. Somerville spillway	14. Riverside new quarry	15. Riverside quarry	
TROPICAL/PARATROPICAL INDICATORS																
Alangiopollis spp.	C	X	S		S	X	S	S	S	X		R		S		Tropical: Alangiaceae
Amanoa? spp.	C	X	S		R	X	S	S	S	S				C		Tropical/paratropical mangrove
Anacolosidites spp.	X	X								X						Tropical
Annona? spp.	S		R			X	R	S	S		S	R	X	R		Tropical/paratropical
Arecipites spp.	S	X	S				S	S	S	X	S	S	S	S		Palm/palmetto
Bombacacidites spp.		S					S	R	S	X		S	S			Tropical/paratropical
Bombapollis spp.	S		R	R		X								cf		Tropical
Bursera spp.	X					X	S									Tropical/paratropical
Calamuspollenites spp.	C		S	S	R	X	C	C	S	X	S	S	S			Palm/palmetto
Cicatricosisporites spp.	X	S	R		C	X	S	A	S	S	A	VA	S	VA	R	Tropical/paratropical fern
Cycadopites? spp.	S		R	R					S							Palm/palmetto
Ficus? spp.	S		R				S		?S							Tropical/paratropical fig
Gleicheniidites spp.	R			R			R				S					Tropical fern
Maceopolipollenites /*Momipites* spp.	S		R	R	C	X	R	S			A					Tropical/paratropical
Momipites spp.	VA	S	VA	A	C	X	A	VA	C	S	A	VA	S	C	S	Tropical/paratropical
Nypa spp.		X				X	R	R		X		S	X			Tropical
Pandaniidites spp.					R	X										Tropical/paratropical
Pelliceria spp. aff.									S							Tropical/paratropical mangrove
Platycarya spp.	S		R		S		?S	S	?R		C		X	VA	C	Tropical/paratropical
Polypodiaceoisporites (*Pteris*) spp.							R	R	S							Pantropical fern
Rhizophora spp.									S					S		Tropical/paratropical mangrove
Symplocos aff. *S. gemmata*	R		R			X	R	S								Tropical/paratropical
Tetracolporopollenites spp.	VA	S	VA	C	C	X	C	S	C	S	S		S	C	S	Tropical/paratropical
Triatriopollenites aroboratus	R	?	R	R		X										Trop/paratropical by association
GRASSLAND INDICATORS																
Aglaoreidia spp.										X			X			Marsh herb
arid fungal epiphytes														R		Also dry grassland
Carex spp.	S		R			X	S	S		X				S		Sedge
Comminuted fungal hyphae					S			S	VA		S			VA	R	
Curvularia spp.									R							
Delitschia spp.									S					?C		Grassland/woodland dung fungus
Graminidites sp. (Gramineae)								S		X		S	X			Grass
Malvacipollis sp.								S		X						Malvaceae?
Milfordia spp.	X							R		S		R	X			Aquatic herb
Mutisieae								R	S							early Asteraceae
Sporormiella spp.									S							Grassland animal dung fungus
TEMPERATE - COOL INDICATORS																
Abies spp.	X	S			R				S					S		Fir
Alnus spp.	S	S	R	R	R	X	R	S	S	X	R	R				Mainly cool wet
Cedripites spp.					?				S					S		Cedar
Picea spp.	S	X	R		S	X	S	S	R		R	S		S		Spruce
Pinus spp.	C		S	R	S	X	S	S	R			A	X	A		Pine
Tsuga spp.			R						R							Hemlock

Note: Climatic trends are seen in the decline of tropical/paratropical taxa and increase in grasslands-savanna taxa as the paleoclimate shifted from warmer-wetter to cooler-drier. Cool indicator taxa are long distance transport components of assemblages and are rare, but their abundance variations correspond with minor climatic variation in this section. Abundance code: VA, very abundant; A, abundant; C, common; S, sparse; R, rare; X, present. Dashed lines delimit groups of sections from same formation.

Eocene, becoming common in the early Oligocene, and abundant large pollen of *Carya* spp. appear in the early Oligocene. The abundant pecan/hickory (*Carya*) content of early Oligocene deposits suggests a cooler paleoclimate than the tropical/paratropical climate of the Eocene. Pine (*Pinus*) is common in one sample in the late Eocene Lake Somerville core and in samples from the early Oligocene Riverside quarry.

The tropical creeping palm *Nypa*, whose growth habitat is in marginal marine, brackish water environments, is of special interest because it is an indicator of ocean water temperatures. It does not extend above the Eocene in the Brazos Valley section or other areas of the Gulf Coast (Frederiksen, 1980b) and is most common in the Yegua and Manning formations, which contain large amounts of shallow- and marginal-marine deposits. Similarly, the occurrence of several mangrove taxa (*Rhizophora*, *Pseudolaesopollis ventosus*, *Avicennia*, *Pelliceria*, and *Amanoa*?) in the late Eocene indicates that mean annual air temperatures remained within the tropical/paratropical climate range and that there were few frost days during winter months, based on comparisons to modern climatic tolerances of these genera (Chapman, 1977). Although some mangrove pollen are present in the early Oligocene, their diversity and abundance are low, suggesting marine temperatures were close to the limits for mangrove tolerance.

Long-term change in lowland climate is also shown by changes in dominant terrestrial pollen types preserved in marine deposits (figure 14.3). Marine deposits contain a composite of the land plant assemblages that are present in the regional flora, combining local environmental assemblages into a single regional assemblage that is a good indicator of average coastal zone climate. During the Bartonian there is a *Cupuliferoipollenites*-dominated regional assemblage present, which is replaced by a *Momipites*-dominated regional assemblage during the Priabonian, which in turn is replaced by an early Oligocene *Carya* and *Quercus*–dominated regional assemblage (figure 14.3). The plants producing this pollen remain in the regional flora throughout the interval, so changes in abundance are related primarily to climate change. Elsik (1969, 1974) first noted the change from a *Momipites*-dominated regional assemblage to *Carya* and *Quercus*–dominated regional assemblages at the Eocene/Oligocene boundary, a change that has been recognized across the Gulf Coast. Although *Cupuliferoipollenites* pollen is associated with trees similar to *Castanea* (chestnut), presently paratropical to temperate in distribution, this is the dominant pollen taxon in marine shelf sediments during the late middle Eocene—a time of global high temperatures. As suggested by Gray (1960), some Eocene species may not have had the same physiological adaptations to climate as related modern species. The change to the *Momipites*-dominated regional assemblage occurs during times of declining global temperatures and parallel development of drier climates.

The terminations of palynomorph ranges shown in figure 14.3 for the late middle Eocene to early Oligocene of east Texas indicate continual change in lowland climate through the late Eocene, supporting the conclusions of the cooling/drying trend presented by Elsik (1974) and providing the basis for recognition of a sequence of palynomorph biostratigraphic assemblage zones (Elsik and Yancey, 2000). This record of east Texas climate change is generally progressive and shows the same trend as records of late Eocene climate change determined from planktic foraminifera (Zachos et al., 1994). It does not show the major climatic fluctuations of the late Eocene suggested by leaf-margin analysis (Wolfe, 1992), nor does it show a long interval of late Eocene climate stability suggested for the eastern Gulf Coast (Frederiksen, 1980a, 1991). The amount of climate change shown in the east Texas section shows that major late Eocene climate change can be recognized in more southern latitudes than previously documented and in the northern margin of the tropical/paratropical climate zone.

REGIONAL PALYNOMORPH ASSEMBLAGES

Palynological changes, especially the change in dominant pollen type, present in samples of late Eocene marine strata along the northern Gulf Coast show a correspondence of floral change with declining paleotemperatures determined from marine mollusk shells. This is best seen in the regional palynomorph assemblages deposited in marine strata, which contain an averaged palynomorph assemblage of both wind-transported and water-transported grains derived from co-occurring plant populations of a large geographic area, not a specific local environment. The depositional site may be many kilometers distant from land, and the sediments are usually bioturbated, mixing the seasonal variations into a single composite sample representative of annual to decadal climate. The regional assemblage is therefore not determined by specific environments or floral communities, but is representative of averaged lowland climatic conditions and is a good indicator of average coastal zone climate.

Regional assemblages are different from environmentally controlled assemblages, such as those recovered from lignite deposits. Lignites tend to include in situ assemblages indicative of local environment, which can change during accumulation of the peat deposit (Frederiksen, 1981; Raymond et al., 1997). A stratigraphic interval containing local environmentally controlled assemblages from specific nonmarine environments is known from one part of the Brazos Valley section: the Lake Somerville section of the late Eocene. In that section, dense sampling in known depositional environments reveals the presence of three environmentally controlled palynomorph assemblages (Raymond et al., 1997; Sancay, herein). A *Momipites*-dominated environmental assemblage was deposited in peat swamp lignites and is similar to the transported regional assemblage of marine deposits. A *Cupuliferoipollenites*-dominated environmental assemblage occurs in the upper margins of lignite beds overlain by marine sediments, but is not present in the upper margin of lignites overlain by nonmarine strata. A fern-dominated environmental assemblage (*Cicatricosisporites, Laevigatosporites, Verrucatosporites*, etc.), probably produced by ground cover vegetation, occurs in nonmarine stream, lake, and alluvial deposits. The fern-dominated environmental assemblage even occurs in strata with dense rooting by large trees and in standing-water deposits containing common angiosperm leaf fossils, suggesting that the ground cover fern vegetation overwhelms the pollen contribution from forest trees and shrubs. The Lake Somerville section contains many leaf fossils and has petrified hardwood and palmwood, so large vegetation was common in the area and was dominantly angiosperm in character. The abundant fern spores indicate that ferns were a common ground cover, even in forested areas, but relatively few fern spores were transported to marine depositional environments.

A regional palynomorph assemblage is a composited assemblage and is produced in a different manner than an assemblage present in a local environment. It is possible that a regional palynomorph assemblage may be compositionally similar to one of the local environmental assemblages, but the similarity may be the result of a single environment on adjacent land being so widespread that it dominates the palynomorphs deposited in marine waters, or it may be the result of mixing of palynomorphs from multiple local environments in which the resemblance is entirely coincidence. The two situations cannot be distinguished in a marine sample, so the regional palynomorph assemblage is not a reliable indicator of local environmental assemblages present in the area. Similarly, the determination of statistical clusters in marine or regional assemblages does not necessarily indicate the presence of a real plant association. The recognition of local environmental assemblages and plant associations is dependent on sampling in sedimentologically-determined known depositional environments. The regional palynomorph assemblage is a good indicator of the general character of flora present, not an indicator of plant associations or communities.

CLIMATE CYCLES

Cyclic variations in climate can be recognized in parts of the section. Temperature variation is suggested by the periodic appearance of short-duration peaks of *Picea, Alnus, Ulmus*, and *Pinus* pollen (figure 14.4). *Picea* (spruce) is a good indicator, because it is a cool-climate upland plant that was probably not a part of the lowland coastal plains flora. Its appearance in any samples of an Eocene assemblage dominated by warm-climate plants is a useful indicator of times of cooling, and abundance peaks of this taxon provide evidence of brief episodes of regional cooling, when temperatures drop on adjacent highland areas. Dilcher (1973) suggested that some cool-adapted taxa could be from highlands of the Ozarks and Appalachian Mountains or more distant regions. The more abundant *Pinus* (pine) pollen, a plant that is presently common on warm midlatitude coastal plains, tracks the occurrence of *Picea* in this Eocene-Oligocene section and provides a useful indicator of temperature fluctuations. In this instance, abundance peaks of *Pinus* pollen indicate cooler-temperature events. *Picea* and *Pinus*–defined cool-climate peaks occur in the latest Bartonian (palynomorph zone; ~37.0 Ma) and late Priabonian (palynomorph zone; ~34.4 Ma) in the east Texas section, about 2.5 million years apart. A *Pinus* abundance peak is also present in the basal Oligocene of the Little Stave Creek, Alabama, section (figure 14.5), at a level just below the appearance of abundant *Quercus*. This occurrence is about 1 million years later than the Lake Somerville peak. Although the record is incomplete, the intermittent occurrence of these taxa suggests an intermediate-scale climatic cyclicity.

Short-duration climatic cyclicity is apparent in the offshore marine deposits of the Caddell Formation, where closely spaced samples of the Koppes Bridge section and in the Greens Prairie core show systematic cyclic variation (figure 14.6) in the

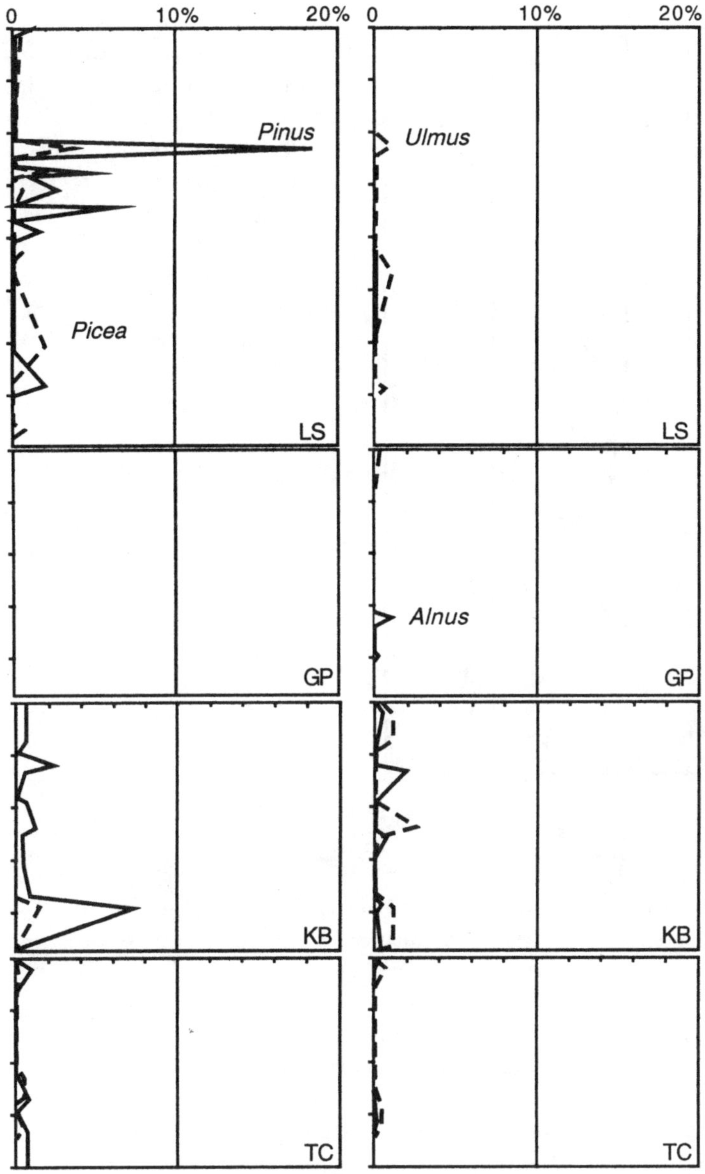

Figure 14.4. Relative abundance frequencies of *Pinus* and *Picea* (*left column*) and *Alnus* and *Ulmus* (*right column*) in late middle Eocene to late Eocene strata at Turkey Creek section (TC), Koppes Bridge section (KB), Greens Prairie core (GP), and Lake Somerville core (LS). Note that *Pinus* abundance is similar to the abundance of *Picea*. These cool temperature indicator taxa are not part of a local flora, but are added to the pollen assemblage by long distance transport; abundance peaks indicate climate fluctuations of distant highland habitats. Vertical scale corresponds to number of samples.

ratio of *Momipites* and *Cupuliferoipollenites* abundances, taxa common in the local flora. These cyclic variations are present in stratigraphic intervals where neither taxon is highly dominant in pollen assemblages. *Momipites* pollen was produced by trees typical of a more tropical, humid environment than present conditions on the Texas coastal plain, while *Cupuliferoipollenites* pollen was probably produced by a tree similar to *Castanea* (chestnut), which presently has a paratropical to temperate distribution. Comparisons to modern relatives would suggest that *Cupuliferoipollenites* lived in drier habitats, but the general trend of increasing *Momipites* dominance during the late Eocene suggests the possibility of different climatic tolerances for these taxa during the late Eocene.

The *Momipites-Cupuliferoipollenites* abundance cycles occur as variations in the regional palynology assemblages, deposited in lithologically similar marine sediments, so they are not related to cyclic changes in depositional environment. The Greens Prairie core samples are from a single parasequence containing one transgressive-regressive cycle of deposition and the maximum flooding horizon of the

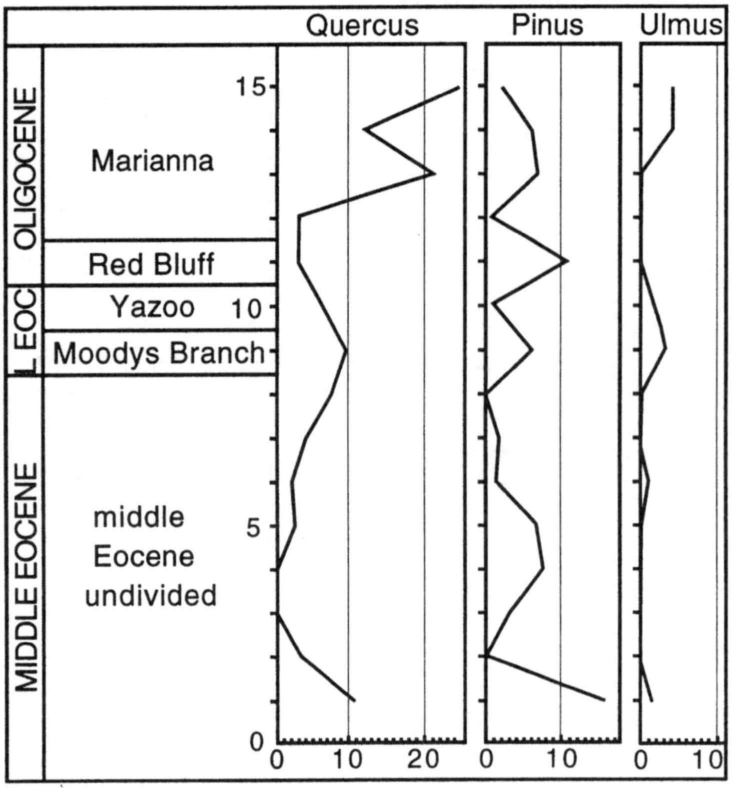

Figure 14.5. Relative frequency of *Quercus*, *Pinus*, and *Ulmus* against the sum of all nonmarine palynomorphs (except fungi) at Little Stave Creek, Alabama (Elsik, study in progress). Cooling event signified by *Pinus* spike just above base of the Oligocene, followed by dominance of *Quercus* pollen. Vertical scale is number of samples; horizontal scale is percentage.

parasequence lies within the lower climatic cycle. The boundary between climatic cycles occurs within a fairly uniform mudstone interval. The asymmetric character of the cycles, with rapid increase in *Momipites* followed by slow decline in abundance, is reminiscent of Pleistocene climate cycles and probably occurs on a similar time scale: durations of tens or hundreds of thousands of years; within the range of Milankovich cyclicity. The Caddell Formation was deposited over an interval of a half-million years or more, indicating the climatic cycles probably have a duration of 50,000 to 100,000 years each.

EAST-WEST (LONGITUDINAL) CLIMATIC GRADIENT

The changes from *Cupuliferoipollenites*-dominated regional assemblage to *Momipites*–dominated regional assemblage and from *Momipites*–dominated regional assemblage to *Carya* and *Quercus*–dominated regional assemblage present in the east Texas section also occur in other sections across the northern reaches of the Gulf of Mexico (Yancey and Elsik, 1999) and in southern California (Frederiksen, 1991, as shown by his pollen event 3—the time of appearance of abundant Quercoideae pollen). It is present in Mississippi and Alabama and can be seen in data from two independent studies of the Little Stave Creek section in Alabama (figure 14.7). Therefore, this is a typical succession of regional pollen assemblages during the interval of late Eocene cooling and earliest Oligocene thermal minimum. However, the timing of changes is different in each section, with succession of regional pollen assemblages occurring earlier to the west than to the east (figure 14.8). The *Cupuliferoipollenites* to *Momipites* succession change is diachronous, ranging from late Bartonian to late Priabonian across the northern Gulf Coast lowlands. The *Momipites* to *Carya* and *Quercus* succession occurs during the Priabonian in southern California, but occurs near the Eocene/Oligocene boundary across the entire Gulf Coast. Frederiksen (1991) concludes that this change occurred 2.5 million years earlier in California than in the Gulf Coast. Eocene lowland climates in southern California were summer-wet (Frederiksen, 1991, and references therein), like Eocene lowland climates in the Gulf Coast. Although the pollen succession occurs at different times in the sections studied, this does not preclude the use of other pollen events for chronostratigraphic correlation, as done by Frederiksen (1991) for the late Eocene.

The diachronous changes show that an east-west (longitudinal) climate gradient was already present in the southern part of North America during the late Eocene. The occurrence of caliche horizons and dry-climate vertebrate faunas in Lutetian strata of southern California (Frederiksen, 1991, and references therein) indicates an early development of dry to semi-arid climate in the region. This means that a rainfall gradient similar to the modern moisture gradient, ranging from western dry to eastern wet, was present during the late Eocene. The western continental region would be dry because of the presence of stable air masses and cold upwelling ocean currents, rain shadow effects and continental interior conditions, while moisture coming from the Gulf of Mexico produced increasing amounts of rainfall to the east along the coastal plain of the northern Gulf of Mexico. Another aspect would be greater climate extremes to the west and less variable (more equable) conditions to the east.

The east to west climate gradient is more directly related to moisture control than to temperature. Mean annual temperatures should be generally similar along the coastal plain of the northern Gulf of Mexico, lying at the same latitude and bordering the same oceanic body. The diachronous succession of late Eocene regional pollen assemblages from eastern Texas to western Alabama is therefore indicative of influence by the moisture gradient, the other great factor in terrestrial climate. During the late Priabonian, lowland floras producing the *Cupuliferoipollenites*-dominated regional assemblage, the *Momipites*-dominated regional assemblage and the *Quercus* and *Carya*–dominated regional assemblage lived at the same time along the moisture gradient.

ACKNOWLEDGMENTS

Judith A. Gennett shared palynology data from many late Eocene samples and has been very helpful at various stages of this study. Richard P. Zingula supplied samples from the Riverside quarries and other localities used in the ongoing study of the Brazos Valley section. Anne Raymond and graduate students at Texas A&M University have contributed thoughts and helped in acquiring core at the Lake Somerville spillway section. Most samples were processed at the Palynology Lab at Texas A&M University, under the direction of Vaughn Bryant. Lloyd Morris and David White of the Geology and Geophysics Department,

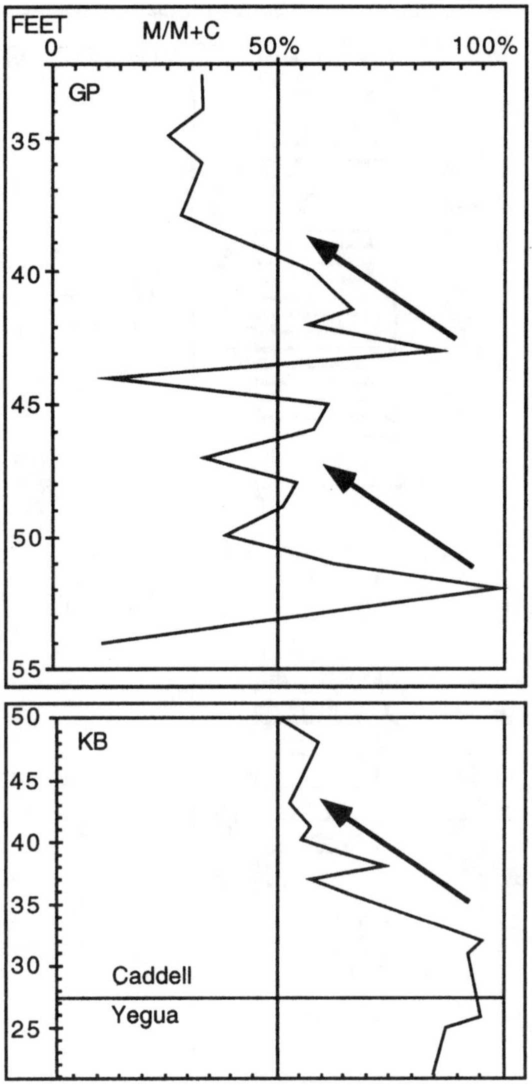

Figure 14.6. Short-term climate cyclicity in the late Eocene Caddell Formation, Koppes Bridge section (KP; 12 samples) and Greens Prairie core (GP; 20 samples). Cooling cycles shown by variation in ratio of *Momipites* to *Momipites* + *Cupuliferoipollenites*. Arrows indicate trend towards cooler and drier paleoclimate, based on comparison to habitat preference of modern *Engelhardtia (Momipites)*, listed by Graham and Jarzen (1969) as part of the upland warm temperate to subtropical community. Although both *Momipites* and *Cupuliferoipollenites* lived in lowland forests, *Momipites* appears to have a broader climatic tolerance than *Cupuliferoipollenites*. Cycles are present in offshore marine mudstones and the two cycles present in the Greens Prairie core occur in the cycle-center portion of a single parasequence.

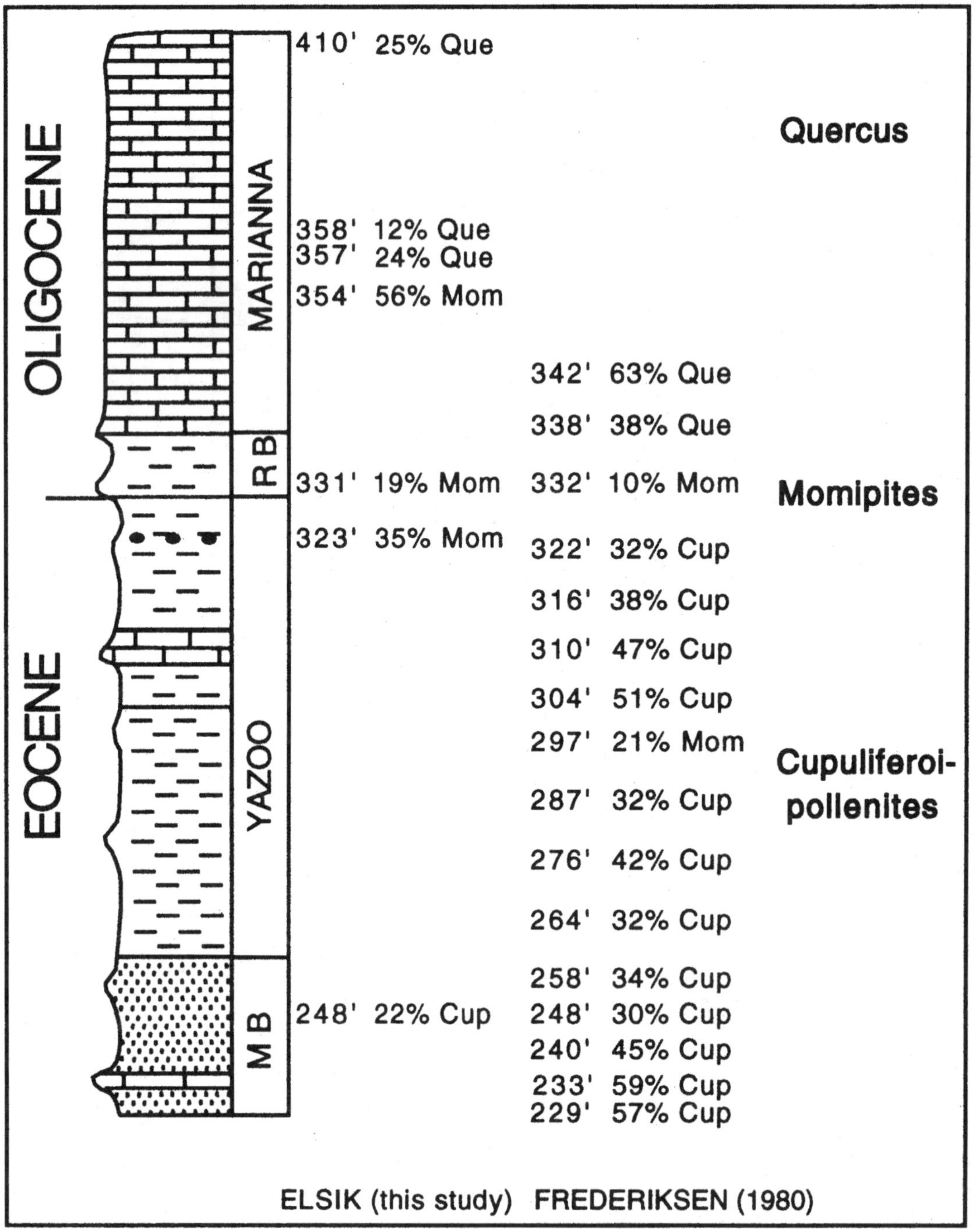

Figure 14.7. Stratigraphic succession of regional palynomorph assemblages in the Little Stave Creek section, Alabama, as determined from data set of Elsik (unpublished) and frequency diagrams of Frederiksen (1980a). Samples collected at different times and identified by different workers, but both data sets clearly show the succession of dominant regional pollen taxa. Stratigraphic levels standardized to measured section of Bandy (1949), which has a datum level of 331 feet at base of the Red Bluff. Formation code: MB, Moodys Branch; RB, Red Bluff.

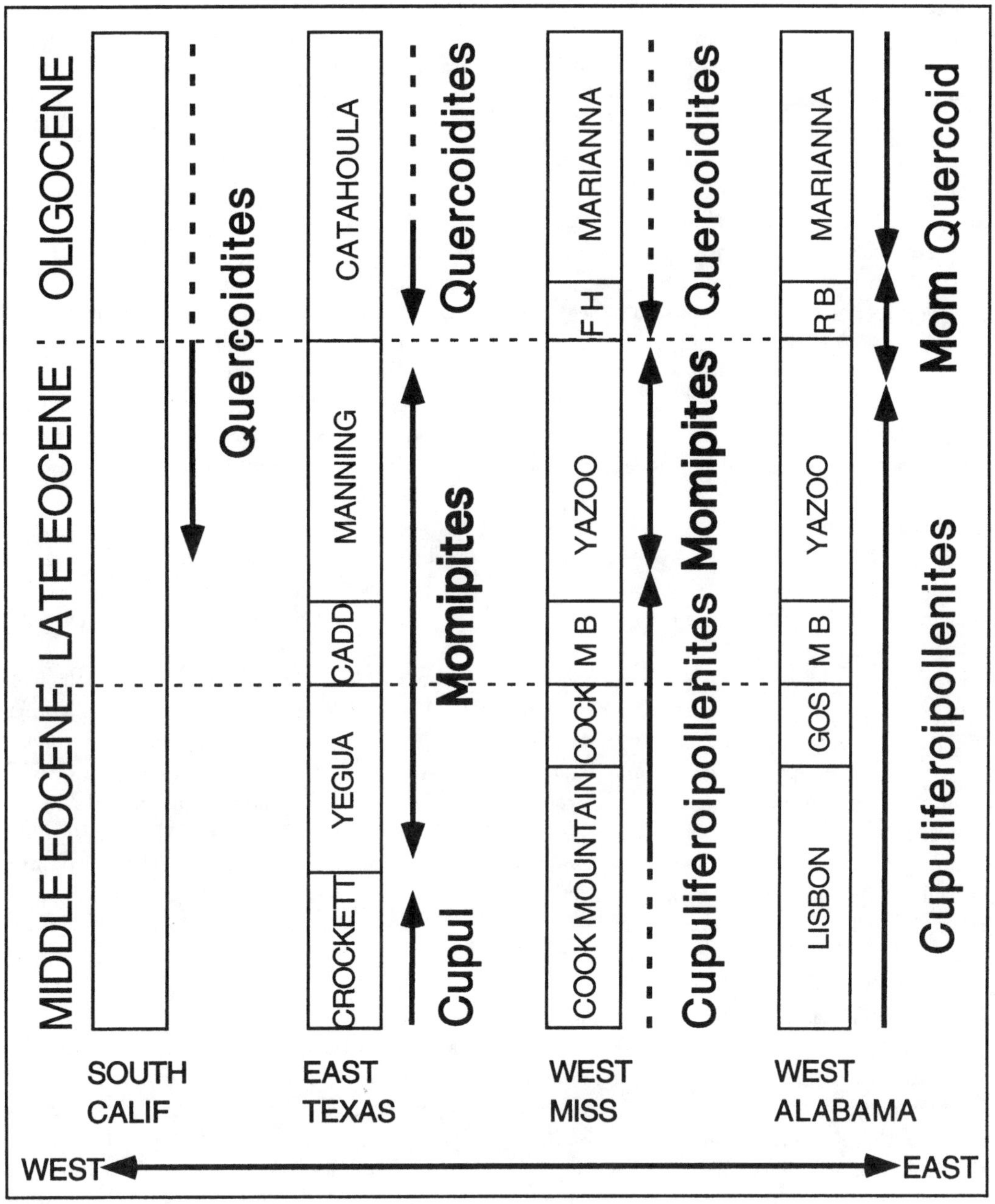

Figure 14.8. Diachronous stratigraphic succession of *Cupuliferoipollenites*-dominated regional assemblage to *Momipites*-dominated regional assemblage to *Carya* and *Quercus*-dominated regional assemblage across longitude in the late Eocene of southern North America. The name *Quercoidites* is equivalent to *Quercus* and indicates the *Carya* and *Quercus*-dominated regional assemblage of this report. Pollen data sources: southern California, Frederiksen (1991); east Texas, this report; Mississippi, Frederiksen (1980a); Alabama, this report and Frederiksen (1980a). Frederiksen (1991) does not provide percentage data or counts, but clearly identifies timing of abrupt rise in relative frequency of Quercoideae pollen in samples. Correlation of sections follows Yancey and Davidoff (1994), Ewing (1997) and Mancini (2000). Formation code: CADD, Caddell; COCK, Cockfield; FH, Forest Hill; GOS, Gosport; MB, Moodys Branch; RB, Red Bluff. Assemblage abbreviations: Cupul, *Cupuliferoipollenites*; Mom, *Momipites*; Quercoid, *Quercoidites*.

Texas A&M University, were drillers for the Lake Somerville spillway core and Greens Prairie core, respectively. This is a contribution of the Paleoecology Research Program at Texas A&M University.

REFERENCES

Andreasson, F. P., and Schmitz, B., 2000, Temperature seasonality in the early middle Eocene North Atlantic region: Evidence from stable isotope profiles of marine gastropod shells: Geological Society of America Bulletin, v. 112, p. 628-640.

Atlee, W. A., Loep, K. J., Elsik, W. C., Zingula, R. P., Pointer, G. N., and Ogden, J. C., 1967, Selected Cretaceous and Tertiary depositional environments: Gulf Coast Section SEPM, Field Trip Guidebook, 50 p.

Bandy, O. L., 1949, Eocene and Oligocene foraminifera from Little Stave Creek, Clarke County, Alabama: Bulletins of American Paleon-tology, v. 32, no. 131, 210 p.

Berggren, W. A., and Prothero, D. R., 1992, Eocene-Oligocene climatic and biotic evolution: an overview, *in* Prothero, D. R., and Berggren, W. A., eds., Eocene-Oligocene climatic and biotic evolution: Princeton, New Jersey, Princeton University Press, p. 1-28.

Chapman, V. J., 1977, Ecosystems of the world: wet coastal ecosystems: Amsterdam, The Netherlands, Elsevier Publishing Company, p. 1-27.

Dilcher, D. L., 1973, A paleoclimatic interpretation of the Eocene floras of southeastern North America, *in* Graham, A., ed., Vegetation and vegetational history of northern Latin America: Amsterdam, The Netherlands, Elsevier Publishing Company, p. 39-59.

Elsik, W. C., 1969, An evaluation of *Engelhardtia* in the Gulf Coast Tertiary: Geological Society of America, Abstracts with Program, v. 1, p. 10-11.

Elsik, W. C., 1974, Characteristic Eocene palynomorphs in the Gulf Coast, U.S.A.: Palaeontographica, Abteilung B, v. 149, p. 90-111.

Elsik, W. C., and Yancey, T. E., 2000, Palynomorph biozones in the context of changing paleoclimate, Middle Eocene to lower Oligocene of the northwest Gulf of Mexico: Palynology, v. 24, p. 177-186.

Ewing, T. E., 1997, Correlation of Yegua/Cockfield genetic cycles, Texas and Louisiana: Gulf Coast Association of Geological Societies Transactions, v. 47, p. 631-633.

Frederiksen, N. O., 1980a, Sporomorphs from the Jackson Group (upper Eocene) and adjacent strata of Mississippi and western Alabama: U.S. Geological Survey Professional Paper 1084, 75 p.

Frederiksen, N. O., 1980b, Mid-Tertiary climate of southeastern United States: the sporomorph evidence: Journal of Paleontology, v. 54, p. 728-739.

Frederiksen, N. O., 1981, Middle Eocene to early Oligocene plant communities of the Gulf Coast, *in* Gray, J. et al., eds., Communities of the past: Stroudsburg, Pennsylvania, Hutchinson Ross Publishing Company, p. 493-549.

Frederiksen, N. O., 1988, Sporomorph biostratigraphy, floral changes, and paleoclimatology, Eocene and earliest Oligocene of the eastern Gulf Coast: U.S. Geological Survey Professional Paper 1448, 68 p., 16 pl.

Frederiksen, N. O., 1991, Pulses of middle Eocene to earliest Oligocene climatic deterioration in southern California and the Gulf Coast: Palaios, v. 6, p. 564-571.

Gennett, J. A., 1996, Comparative palynology of clastics and lignites from the Manning Formation, Jackson Group, Upper Eocene, Grimes Co., Texas: Gulf Coast Association of Geological Societies Transactions, v. 46, p. 149-158.

Graham, A., and Jarzen, D. M., 1969, Studies in neotropical paleobotany. 1. The Oligocene communities of Puerto Rico: Annals of the Missouri Botanical Gardens, v. 56, p. 308-357.

Gray, J., 1960, Temperate pollen genera in the Eocene (Claiborne) flora, Alabama: Science, v. 132, p. 808-810.

Greenwood, D. R., and Wing, S. L., 1995, Eocene continental climates and latitudinal temperature gradients: Geology, v. 23, p. 1044-1048.

Guillemette, R., and Yancey, T. E., 1996, Composition and provenance of volcanic glass in late Eocene Manning Formation, east-central Texas: Gulf Coast Association of Geological Societies Transactions, v. 46, p. 159-166.

Ivany, L. C., Patterson, W. P., and Lohmann, K. C., 2000, Cooler winters as a possible cause of mass extinctions at the Eocene/Oligocene boundary: Nature, v. 407, p. 887-890.

Jones, J. G., and Gennett, J. A., 1991, Pollen and spores from the type section of the middle Eocene Stone City Formation, Burleson Co., Texas: Gulf Coast Association of Geological Societies Transactions, v. 41, p. 348-352.

Kobashi, T., Grossman, E. L., Yancey, T. E., and

Dockery, D.T., III, 2001, Reevaluation of conflicting tropical temperature estimates: molluscan oxygen isotope evidence for warm low latitudes: Geology, v. 29, p. 983-986.

Leopold, E. B., and MacGinitie, H. D., 1972, Development and affinities of Tertiary floras in the Rocky Mountains, *in* Graham, A., ed., Floristics and paleofloristics of Asia and eastern North America: Amsterdam, The Netherlands, Elsevier Publishing Company, p. 147-200.

Mancini, E. A., 2000, Sequence stratigraphy and chronostratigraphy of upper Eocene and lower Oligocene strata, eastern Gulf Coastal Plain: Gulf Coast Association of Geological Societies Transactions, v. 50, p. 379-388.

McMahon, J. M., 1997, Palynology and paleoecology of the Stone Creek Member, Crockett Formation, middle Eocene, Burleson County, Texas [unpubl. M.S. thesis]: College Station, Texas A& M Univerristy, 245 p.

Miller, K. G., 1992, Middle Eocene to Oligocene stable isotopes, climate, and deep-water history: the terminal Eocene event?, *in* Prothero, D. R., and Berggren, W. A., eds., Eocene-Oligocene climatic and biotic evolution: Princeton, New Jersey, Princeton University Press, p. 160-177.

Miller, K. G., Fairbanks, R. G., and Mountain, G. S., 1987, Tertiary oxygen isotope synthesis, sea level history, and continental margin erosion: Paleoceanography, v. 2, p. 1-19.

Morley, R. J., 2000, Origin and evolution of tropical rain forests: Chicester, England, John Wiley and Sons, 362 p.

Prothero, D. R., and Swisher, C. C., III, 1992, Magnetostratigraphy and geochronology of the terrestrial Eocene-Oligocene transition in North America, *in* Prothero, D. R., and Berggren, W. A., eds., Eocene-Oligocene climatic and biotic evolution: Princeton, New Jersey, Princeton University Press, p. 46-73.

Raymond, A. R., Phillips, M. K., Gennett, J. A., and Comet, P. A., 1997, Palynology and paleoecology of lignites from the Manning Formation (Jackson Group) outcrop in the Lake Somerville spillway of east-central Texas: Coal Geology, v. 34, p. 195-223.

Rezak, R., Bright, T. J., and McGrail, D. W., 1985, Reefs and banks of the northwestern Gulf of Mexico: New York, John Wiley and Sons, 259 p.

Robert, C., and Kennett, J. P., 1997, Antarctic continental weathering changes during Eocene-Oligocene cryosphere expansion: clay mineral and oxygen isotope evidence: Geology, v. 25, p. 587-590.

Trewartha, G. T., 1968, An introduction to climate: New York, McGraw-Hill Co., 408 p.

Tschudy, R. H., 1973, Stratigraphic distribution of significant Eocene palynomorphs of the Mississippi Embayment: U.S. Geological Survey Professional Paper 743B, 24 p.

Upchurch, G. R., and Wolfe, J. A., 1987, Mid-Cretaceous to early Tertiary vegetation and climate: evidence from fossil leaves and woods, *in* Friis, E. M., Chaloner, W. G., and Crane, P. H., eds., The origins of angiosperms and their biological consequences: Cambridge, England, Cambridge University Press, p. 75-105.

Wolfe, J. A., 1978, A paleobotanical interpretation of Tertiary climates in the northern hemisphere: American Scientist, v. 66, p. 694-703.

Wolfe, J. A., 1992, Climatic, floristic, and vegetational changes near the Eocene/Oligocene boundary in North America, *in* Prothero, D. R., and Berggren, W. A., eds., Eocene-Oligocene climatic and biotic evolution: Princeton, New Jersey, Princeton University Press, p. 421-436.

Wolfe, J. A., and Poore, R. Z., 1982, Tertiary marine and nonmarine climatic trends, *in* Climate in earth history: Washington, D.C., National Academy Press, p. 154-158.

Yancey, T. E., 1997, Depositional environments of late Eocene lignite-bearing strata, east-central Texas: Coal Geology, v. 34, p. 261-275.

Yancey, T. E., and Davidoff, A. J., 1994, Paleogene sequence stratigraphy of the Brazos River section, Texas: Gulf Coast Association of Geological Societies, Fieldtrip Guidebook, 104 p.

Yancey, T. E., and Elsik, W. C., 1994, Paleoclimatological analysis of upper Eocene core, Manning Formation, Brazos County, Texas: Gulf Coast Association of Geological Societies Transactions, v. 44, p. 776.

Yancey, T. E., and Elsik, W. C., 1999, The palynological record of late Eocene climate change and terrestrial palynostratigraphy, northwest Gulf of Mexico: Geological Society of America, Abstracts with Program, v. 31, no. 7, p. 309-310.

Yancey, T. E., and Guillemette, R., 1998, Major volcanic ash units in the late Eocene of east Texas: Gulf Coast Association of Geological Societies Transactions, v. 48, p. 511-516.

Zachos, J. C., Stott, L. D., and Lohmann, K. C., 1994, Evolution of early Cenozoic marine temperatures: Paleoceanography, v. 9, p. 353-387.

APPENDIX: SAMPLED SECTIONS

The east Texas data sets used in this study are centered on documentation of the Brazos River Valley section, compiled from several previously unpublished studies of Elsik and supplemented by published and unpublished thesis work done at Texas A&M University. Following is a list of intervals for which palynological data is available.

1. Stone City Bluff; Crockett Formation, Stone City Member. A data set of 19 samples from exposures along the southwest bank of the Brazos River in Burleson County, in the type section of the Stone City Member. Population counts by Elsik. Jones and Gennett (1991) published a similar study of this section

2. Stone City core; Crockett Formation, Wheelock Member and Stone City Member. A data set of 28 samples from a 26 m core drilled close to the riverbank exposures of the type section of the Stone City Member. The cored interval bottoms at about the same horizon as the base of the riverbank exposures. Population counts by McMahon (1997 MS thesis, Geology Dept., Texas A&M University).

3. Rocky Branch; Crockett Formation, Wheelock Member. A data set of 7 samples from exposures along Rocky Branch, located near Stone City Bluff in Burleson County. Population counts by Elsik.

4. Little Brazos River; Crockett Formation, Wheelock Member. A data set of 2 samples from the east bank of the Little Brazos River, upstream from the Highway 21 bridge crossing, Brazos County. The samples are underlain by a bentonite bed dated at 41.85 Ma. Population counts by Elsik.

5. Alabama Ferry landing; Crockett Formation, Wheelock Member. A data set of 2 samples from exposures along the banks of the Trinity River, downstream from the Alabama Ferry landing site, Houston County. The samples are bracketed by bentonite beds dated at 41.66 Ma and 41.53 Ma. Population counts by Elsik.

6. Hurricane Bayou; Crockett Formation, Hurricane Lentil. A data set of 3 samples from Hurricane Bayou, in the type section of the Hurricane Lentil, near the west edge of the city of Crockett, Houston County. Population counts by Elsik.

7. Turkey Creek; Yegua Formation, Bryan Member. A data set of 10 samples from the west bank of Turkey Creek, downstream from Highway 46 crossing, near the southwest edge of the city of Bryan, Brazos County. Population counts by Elsik.

8. Koppes Bridge; top of Yegua Formation (includes the Rock Prairie Member) and base of Caddell Formation. A data set of 1 sample from the bottom of Hope Creek and 16 samples from exposures along the east bank of the Brazos River, south of College Station, Brazos County. Population counts by Elsik.

9. Greens Prairie core; upper part of Caddell Formation. A data set of 20 samples from core drilled beside the Greens Prairie Road interchange on Highway 6, the south edge of the city of College Station, Brazos County. Population counts by Elsik; presented in a preliminary report by Yancey and Elsik (1994).

10. Gibbons Creek mine; upper part of Manning Formation. A data set of 56 samples from exposures and core in Gibbons Creek lignite mine, east of the Navasota River and south of Carlos, Grimes County. Sample interval includes the 3,500-, 5,500-, and 5,600-level lignite seams and a clastic cycle in Piedmont interval above the 5,600 seam. The lowest samples are underlain by a bentonite bed dated at 34.53 Ma. Population counts by Gennett (1996).

11. Lake Somerville dam; upper part of Manning Formation. A data set of 4 samples from exposures on the south bank of the Lake Somerville dam outlet channel, Yegua Creek, Washington County. Population counts by Elsik.

12. Lake Somerville core; upper part of Manning Formation. A data set of 12 outcrop samples and 35 samples from core drilled on the south slope of the Lake Somerville spillway channel, Washington County. Population counts by Sancay.

13. Lake Somerville spillway; Manning Formation, upper part. A data set of 36 samples from exposures on the south slope of the Lake Somerville spillway channel, Washington County. Population counts by Gennett and Phillips in Raymond et al. (1997).

14. Riverside new quarry; lower part of Catahoula Formation. A data set of 1 sample from exposures in quarry beside Route 274, north of Huntsville, Walker County. Population count by Elsik.

15. Riverside quarry; lower part of Catahoula Formation. A data set of 2 samples from quarry exposures in the old Riverside quarry (now Blue Lagoon diving park), north side of Huntsville, Walker County. Population counts by Elsik.

CHAPTER 15

Palynological Patterns in the Uppermost Eocene to Lower Oligocene Sedimentary Rocks in the U.S. Gulf Coast

Francisca E. Oboh-Ikuenobe and Carlos A. Jaramillo

ABSTRACT
Climate, vegetation, sediment supply, and burial conditions play important roles in the diversity and abundance of palynomorphs transported into and preserved in depositional sequences. Preserved palynological assemblages (spores, pollen, dinoflagellate cysts, and palynofacies) can be used to recognize past vegetation patterns, climates, and sea-level fluctuations, and such a model is used here to interpret the upper Eocene to lowermost Oligocene Pachuta and Shubuta Members of the Yazoo Formation, and the lower Oligocene Vicksburg Group in southeastern Mississippi and southwestern Alabama. Of the 260 spore and pollen taxa identified, eleven groups representing about 50% of the taxa have been recognized on the basis of abundance, occurrence in most samples and ecological significance. The most prominent groups are *Sequoiapollenites-Cupressacites*, *Momipites*, *Carya*, *Quercoidites*, *Cupuliferoipollenites*, *Cyrillaceaepollenites*, and *Siltaria*. Statistical analysis on presence-absence data produced five significant cluster groups that were dominated by species of the 11 groups, as well as by *Ulmipollenites thompsonianus* and *Salixipollenites parvus*. These cluster groups show some ecological significance because they appear to represent upland, swamp, and marsh taxa that were either preserved in the nearshore sediments or dispersed into the shallow-marine environment. The prominence of *Quercoidites*, *Sequoiapollenites*, *Cupressacites*, and *Cyrillaceaepollenites* in the rich pollen assemblage, which contains some tropical to subtropical elements, such as *Psilatricolporites operculatus* and *Nyssapollenites pulvinus*, suggests a warm-temperate paleoclimate during the early Oligocene in the study area. Although *Momipites* was generally more abundant in the Eocene and *Quercoidites* increased during the Oligocene, there was no significant paleofloristic change across the Eocene/Oligocene boundary. The major floral turnover along the Gulf Coast occurred well before the deposition of the Vicksburg units during the early Oligocene, probably in the middle Eocene. This observation is confirmed by dinoflagellate cyst (dinocyst) data that do not show any significant change in standing diversity, or in turnover rates across the Eocene/Oligocene boundary. A combination of dinocyst paleoecology, palynofacies, and lithofacies formed the basis of sequence stratigraphic interpretations. In southern Mississippi, we identified a latest Eocene maximum flooding surface in the middle of the Shubuta Clay, but this surface was identified at the top of the Shubuta Clay in southern Alabama, where the Eocene/Oligocene boundary was placed within a condensed interval. This condensation, which was identified at the Shubuta Clay–Vicksburg Group contact, is equivalent to the accumulation of the upper Shubuta Clay and Red Bluff Clay in southern Mississippi.

INTRODUCTION

The late Eocene to early Oligocene interval has received much attention because of a rapid cooling during the late Eocene, followed by a thermal minimum at the beginning of the Oligocene, as documented by microfaunas from deep-sea cores (Shackleton and Kennett, 1975; Wei, 1991). Fossil records of plant megafossils, pollen, and spores from the Gulf Coast and western North America have also

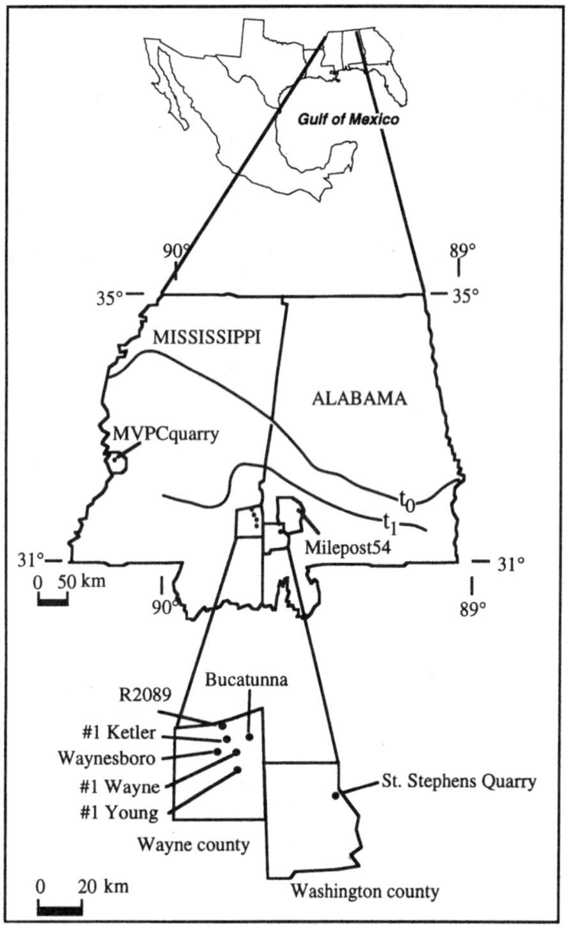

Figure 15.1. Map of Mississippi and Alabama showing the studied localities in the southern parts of the states. t_0 indicates the paleoshoreline during the accumulation of the Shubuta Clay and equivalent units, while t_1 represents the paleoshoreline during the accumulation of the Red Bluff Clay and equivalent units (paleoshorelines after Tew and Mancini, 1995).

indicated changes in vegetation that could be associated with paleoclimatic cooling toward the end of the Eocene or the beginning of the Oligocene (Wolfe, 1971, 1992, 1994; Elsik, 1974; Frederiksen, 1988). Outcrops of some of the most complete upper Eocene to lower Oligocene sedimentary rock sequences in North America are located in Mississippi and Alabama (figure 15.1) in the U.S. Gulf Coast. This study uses palynological information (pollen, spores, dinoflagellate cysts, and dispersed organic matter) from nine sections in southern Mississippi and southern Alabama to document paleoecological and paleofloristic patterns, and to make paleoclimatic inferences throughout the Eocene-Oligocene transition.

The diversity and abundance of palynomorphs and dispersed organic matter transported into and preserved in depositional sequences are dependent on climate, vegetation, sediment supply, and burial conditions. Paleovegetation patterns, paleoclimates, and sea-level fluctuations can be recognized from preserved palynological assemblages. These events have been reconstructed for the upper Eocene Pachuta and Shubuta Members of the Yazoo Formation, and the lower Oligocene Vicksburg Group in southeastern Mississippi and southwestern Alabama. We have also integrated dinocyst biostratigraphy, dinocyst paleoecology, and palynofacies, in order to obtain chronostratigraphic and paleobathymetric data to interpret the sequence stratigraphy and paleobathymetric fluctuations of upper Eocene to lower Oligocene strata in the area.

The few well-exposed Eocene to Oligocene sedimentary rocks in southern Mississippi and Alabama have been extensively studied. These units are more or less continuous and have provided field data for several paleogeographic and biostratigraphic studies (Hazel et al., 1980; Siesser, 1983; Tew and Mancini, 1995). The St. Stephens Quarry in southwestern Alabama has been the most studied, providing controversial information on the position of the Eocene/Oligocene boundary and its sequence stratigraphy (Mancini, 1979; Bybell and Poore, 1982; Siesser, 1983; Keller, 1985; Pasley and Hazel, 1990, 1995; Miller et al., 1993; Jaramillo and Oboh-Ikuenobe, 1999).

GEOLOGIC SETTING

The Cenozoic history of the Gulf of Mexico is characterized by a rapid sediment input and thick prograding depositional sequences that intertongue with interdeltaic shelf-edge sediments (Galloway, 1989). These units were deposited in the shallow-marine to marginal-marine environments of a passive margin (Hazel et al., 1980; Pasley and Hazel, 1995; Tew and Mancini, 1995). Upper Eocene sediments are characterized by fine-grained sediment deposition, while the lower Oligocene sediments constitute one of the great progradational wedges in the region (Galloway, 1989). Latest Eocene time in southern Mississippi and Alabama is represented by two lithostratigraphic units, the Pachuta Marl and Shubuta Clay of the Jackson Group (figure 15.2). The lower Oligocene strata are composed of eight lithostratigraphic units, all of which have formation status in Mississippi. In

Alabama, four of them have only member status. In ascending order, the units are the Bumpnose Limestone, the Red Bluff Clay, the Forest Hill Sand, the Mint Spring Marl (or member of the Marianna Limestone), the Marianna Limestone, the Glendon Limestone (or member of the Byram Formation), the Byram Formation (or informal Byram "marl" member of the Byram Formation), and the Bucatunna Clay (or member of the Byram Formation). The Bumpnose Limestone, Red Bluff Clay and Forest Hill Sand are lateral equivalents of one another, although the Forest Hill Sand is thicker in Mississippi. The Bumpnose and Red Bluff intertongue in some areas in southwest Alabama (Tew and Mancini, 1995) (figure 15.2).

MATERIALS AND METHODS

We studied nine localities in Mississippi and Alabama. In southern Mississippi (figure 15.1) there were 3 coreholes: #1 Young, #1 Ketler, #1 Wayne; and four outcrop sections: 0.4 km northeast of Bucatunna Creek (Bucatunna); Type Red Bluff (R2089); road cut east of Waynesboro, U.S. Route 84 (Waynesboro); and Mississippi Valley Portland Cement Quarry, Warren County (MVPC Quarry). Two outcrop sections, the St. Stephens Quarry and Milepost 54 on U.S. Rt. 84, east of Whatley, Clarke County, were studied in southwestern Alabama (figure 15.1). Approximately 160 palynological samples were used for this study. The R2089 and St. Stephens Quarry sections were analyzed for pollen, spores, and dinoflagellate contents. However, the #1 Young, #1 Ketler, and #1 Wayne cores have dinocyst and palynofacies data, while the Bucatunna, Waynesboro, and Mississippi Valley Portland Cement Quarry sections have only pollen and spore data.

The outcrop and core samples were prepared at the U.S. Geological Survey (Reston, Virginia) and by Global Geolab (Alberta, Canada), respectively, using standard palynological techniques, including mild oxidation in nitric acid or Schultze solution, and heavy liquid separation in zinc chloride. The residues were sieved through 10 μm nylon meshes before permanent slides were made. Illustrations and count data of spores and pollen can be found in Oboh and Reeves Morris (1994, 1995) and Oboh et al. (1996), and those for the dinocysts are in Jaramillo and Oboh-Ikuenobe (1999). The slides from the outcrop samples will eventually be deposited at the U.S. Geological Survey (Reston, Virginia), and the Geological Museum of the University of California at Riverside.

Average linkage cluster analysis, using Euclidean distance on presence-absence palynomorph data, was used to analyze the pollen and spore data. We used presence-absence data instead of abundance data in order to reduce the effect of dominant taxa in the analysis. A nonmetric multidimensional scaling (MDS) analysis with Spearman rank-order coeffi-

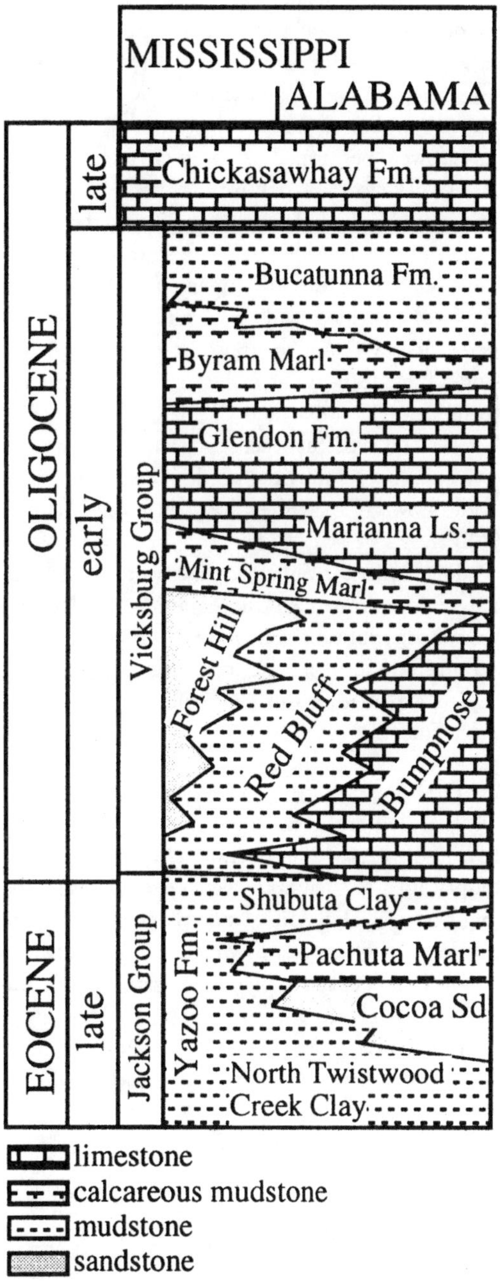

Figure 15.2. Simplified stratigraphic correlation chart of the Mississippi-Alabama region (modified from Galloway et al., 1991; Pasley and Hazel, 1995).

cient was used to analyze the dinocyst distribution. This method is best suited to finding a major terrestrial-marginal marine gradient along the first and second axes, and seems to be least vulnerable to distortion from high data diversity, nonnormality, and nonlinearity (Pielou, 1984). A set of 32 dinocyst taxa with high abundances and those with recognized paleoenvironmental significance were selected for this analysis (Jaramillo and Oboh-Ikuenobe, 1999). A number of statistical techniques were also used to analyze dinocyst diversity patterns across the Eocene-Oligocene transition. We use the word "diversity" to indicate number of species. Most of the palynological analyses were performed using Systat 5.0 and MVSP 3.0 <www.kovcomp.co.uk/mvsp/>.

Unoxidized dispersed organic matter of #1 Young samples provided the data set for the palynofacies analysis. At least 400 organic particles per sample were identified with a Zeiss transmitted-light microscope. Organic matter results were analyzed using a Euclidean-distance cluster analysis with average linkage (Jaramillo and Oboh-Ikuenobe, 1999). This method is useful because it averages all distances between pairs of objects in different clusters and decides how far apart they are (Sokal and Michener, 1958).

The chronostratigraphic framework used to date the sections was derived using graphic correlation on dinocyst distribution, and it was complemented by foraminifera and magnetic stratigraphy data. A detailed explanation can be found in Jaramillo and Oboh-Ikuenobe (1999).

POLLEN AND SPORE DISTRIBUTION

Two hundred and sixty spore and pollen taxa were identified. From the overall assemblage, eleven groups (50% of total taxa) were selected based upon a high relative abundance and recognized ecological significance. These groups were used to identify any relationship between pollen and spore distribution and the stratigraphy in the study area. These groups are spores, bisaccates, *Sequoiapollenites-Cupressacites, Momipites, Carya, Quercoidites, Fraxinoipollenites, Cupuliferoidaepollenites, Cupuliferoipollenites, Cyrillaceaepollenites,* and *Siltaria* (figure 15.3).

The Eocene/Oligocene boundary is associated with a slight increase in *Momipites* and *Quercoidites* whereas *Sequoiapollenites* decreases in abundance (figure 15.3). In the Vicksburg Group, increases up to the order of 7–30% were recorded for *Siltaria, Quercoidites,* and *Cyrillaceaepollenites* over their abundances in the uppermost Eocene Shubuta Clay. With the exception of a slight increase in the Red Bluff Clay, *Momipites* is generally more abundant in the Shubuta Clay (Jackson Group) than the younger Vicksburg formations. Elsik (1974) and Frederiksen (1980; 1988) have shown that most of the taxa occurring in our data have their first occurrence datums before the deposition of the Shubuta Clay and ranged into the Oligocene. Therefore, in our study area, the palynoflora at the Eocene/Oligocene boundary did not experience any abrupt floristic change. The palynofloral patterns preserved in the strata were controlled by the stratal architecture of the stratigraphic sequence, the vegetation on adjacent land areas, an abundant production of certain species, and their easy dispersal into offshore environments.

In southwestern Alabama and southeastern Mississippi, *Sequoiapollenites-Cupressacites* and *Quercoidites* are the most important components in the Vicksburg units. In southwest Alabama, *Sequoiapollenites-Cupressacites* are more prominent in the transgressive and highstand deposits of Mint Spring Marl, Marianna Limestone, and Glendon Limestone (figures 15.1 and 15.3). The floristic pattern in the Mint Spring Marl and Glendon Limestone in southwestern Mississippi, where the Marianna Limestone has been replaced laterally by the Mint Spring, is different from that in the eastern part of the study area: the *Sequoiapollenites-Cupressacites* group is a minor element whereas spores (figure 15.3) are more prominent in the assemblage. The Byram palynofloras are similar to those of the Bucatunna and Chickasawhay in southwestern Alabama. The coniferous pollen *Sequoiapollenites* is an ubiquitous element in the palynomorph assemblage. It is dominant (albeit in low numbers) in samples from the transgressive and highstand systems tracts of the Marianna and Glendon Limestones in southeastern Mississippi and southwestern Alabama where other taxa are relatively few, probably as a result of diagenetic alteration (Oboh and Reeves Morris, 1995). It is also dominant in the more palyniferous Mint Spring Marl in the area. The significance of *Sequoiapollenites* in these samples is purely ecological because it is not an important element in southwestern Mississippi, where spores, bisaccates, and other gymnosperms are more important. The genus probably experienced abundant production on adjacent land areas and was easily dispersed into the nearshore (Mint Spring) and offshore environments (Marianna and Glendon) where it was preserved (Oboh et al., 1996).

Five cluster groups containing 18 taxa (or

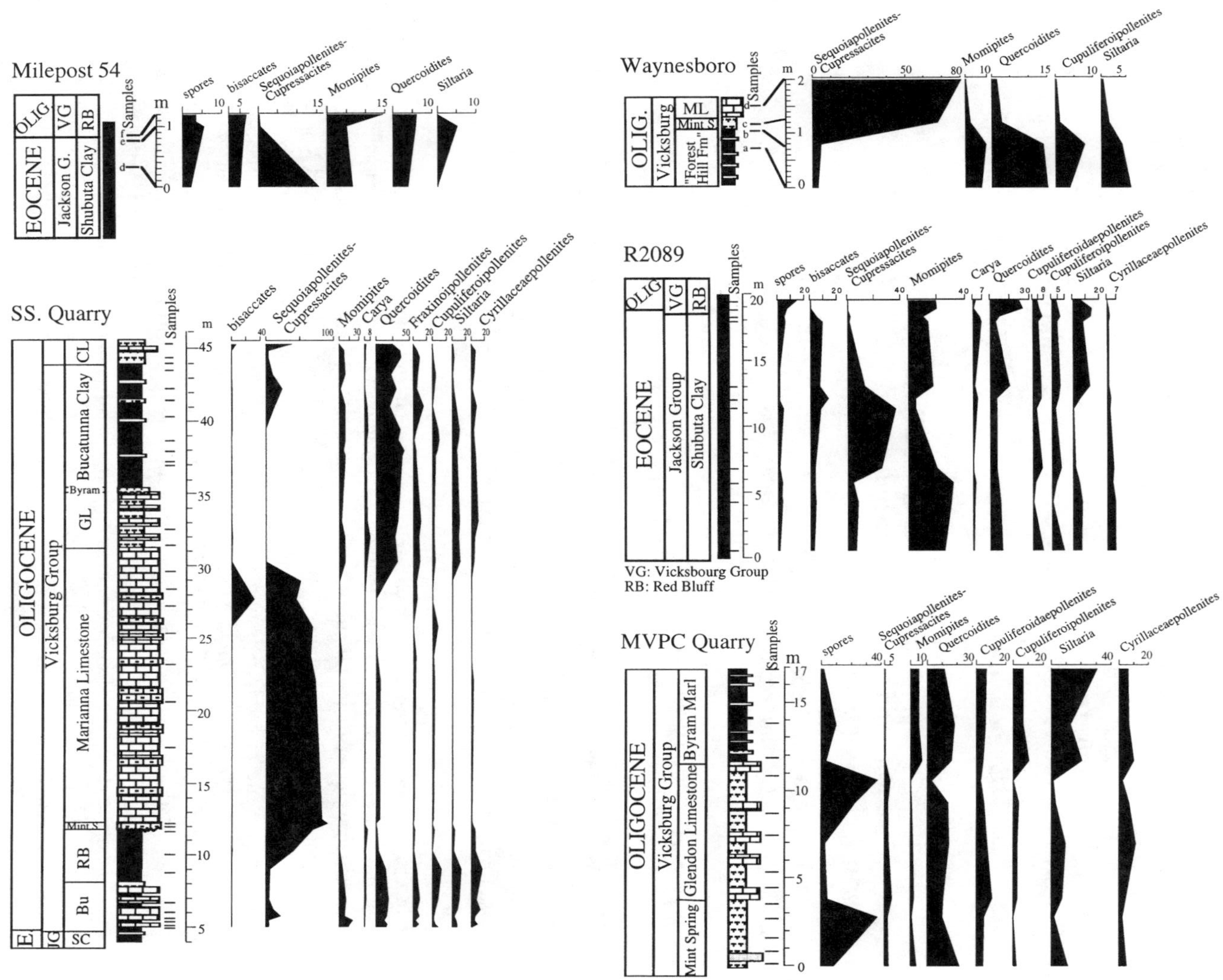

Figure 15.3. Relative frequencies (in percent) of eleven groups of spores and pollen for five of the sections analyzed in this study: Mississippi Valley Portland Cement Quarry (MVSP Quarry); St. Stephens Quarry (SS Quarry); Milepost 54 on U.S. Route 84, Clarke County, Alabama (Milepost 54); Type Red Bluff Section (R2089); and east of Waynesboro, Mississippi (Waynesboro). Groups with less than 5% were not plotted. Bu, Bumpnose; CL, Chickasawhay Limestone; JG, Jackson Group; GL, Glendon Limestone; ML, Marianna Limestone; RB, Red Bluff; VG, Vicksbourg Group. See lithology labels in figure 15.6.

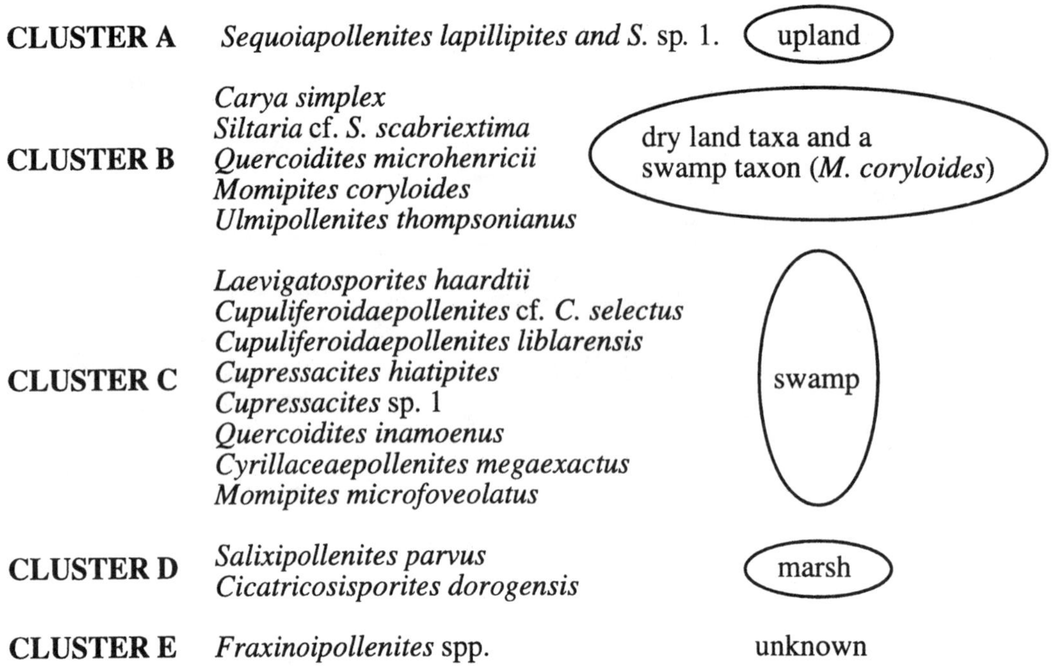

Figure 15.4. Taxonomic composition and ecological significance of five spore and pollen cluster groups containing 18 taxa found in the presence-absence cluster analysis.

approximately 7% of total assemblage), which are listed in Oboh and Reeves Morris (1994), were found in the presence-absence cluster analysis (figure 15.4). These 18 taxa contain mostly anemophilous taxa with a few zoophilous and hydrophilous elements. The five cluster groups are probably reflecting distinct plant assemblages. Group A (*Sequoiapollenites* species) probably represents an upland assemblage, group B contains both dry land taxa (*Carya*, *Quercoidites microhenrici*, *Ulmipollenites*) and a swamp taxon (*Momipites coryloides*) (Judith Gennett, written comm. 1994). Group C is dominated by such swamp taxa as *Momipites microfoveolatus*, *Cupressacites*, and *Cupuliferoidaepollenites*. Group D contains marsh elements such as *Salixipollenites* (Frederiksen, 1981). Group E contains only one taxon (*Fraxinoipollenites*) which is produced by several plant families (Frederiksen, 1980).

The rise in dominance of *Quercoidites* over *Momipites* was used by Elsik (1974) and Frederiksen (1988) to suggest a cooling and drying trend that took place either at the beginning of the Oligocene, or toward the end of the Eocene in the Gulf Coast. The prominence of pollen related to *Quercus*, *Sequoia*, *Carya*, *Cyrilla*, *Taxodium*, *Ulmus*, and *Fraxinus* in this palynomorph assemblage, however, is indicative of a warm-temperate climate in the eastern Gulf during Vicksburgian time. This observation is supported by (1) the rarity of cool temperate to cold temperate palynomorphs related to *Picea*, *Abies*, *Tsuga*, and *Betula*, (2) the low abundance of *Ephedra*, *Abies*, Gramineae, and other pollen which are typical of dry environments, and (3) the presence of some tropical to subtropical palynomorphs such as *Psilatricolporites operculatus*, and *Nyssapollenites pulvinus* (Germeraad et al., 1968; Groot, 1991). This early Oligocene warm-temperate paleoclimate was clearly cooler than the early and middle Eocene subtropical conditions in the Gulf Coast. However, the major floral turnover occurred well before the deposition of the lower Oligocene Vicksburg units.

DINOCYST DIVERSITY

We use the word "diversity" to denote number of species. Dinocyst diversity can be affected by three main factors: sample size (dinocysts per sample), sampling density for a given stratigraphic interval, and depositional environment. A number of techniques were used to take these factors into account: (1) The unbiased Simpson Index, or SI (Rosenzweig,

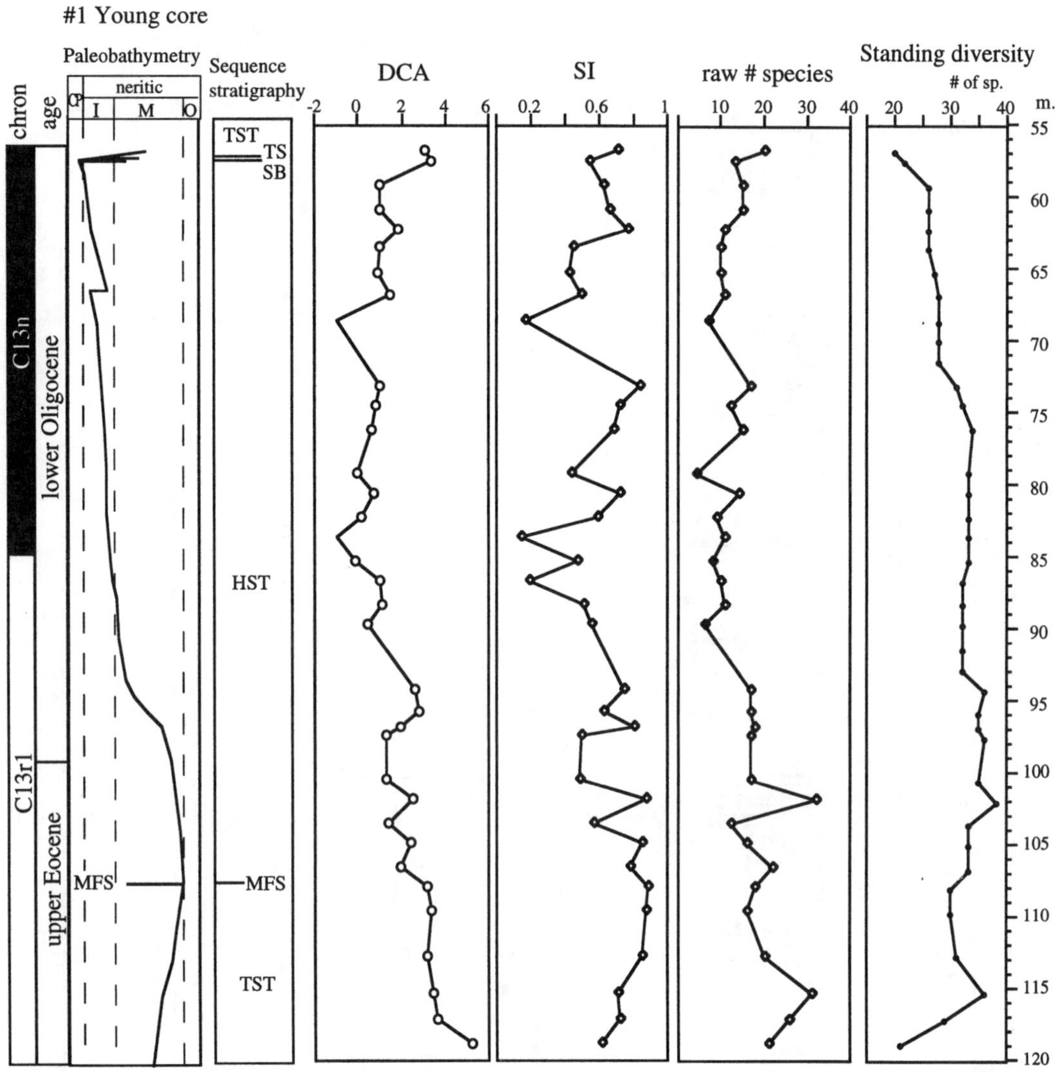

Figure 15.5. Detrended correspondence analysis (DCA), unbiased Simpson Index (SI) expressed as $(-\log_e(SI))$, raw number of species, and standing diversity (after range-through method was applied) for the #1 Young core. Paleobathymetric curve, sequence stratigraphy, and dating of the section after Jaramillo and Oboh-Ikuenobe (1999). CP, coastal plain; I, inner neritic; M, middle neritic; O, outer neritic; HST, highstand systems tract; TST, transgressive systems tract; LST, lowstand systems tract; TS, transgressive surface; MFS, maximum flooding surface; SB, sequence boundary; CS, condensed section.

1995), was calculated to estimate underlying diversity independently of sample size. This index adequately estimates diversity of small samples regardless of sample size, and when used as $(-\log_e(SI))$, it increases as the number of species increases (Rosenzweig, 1995); (2) The range-through method (Boltovskoy, 1988) was used to estimate standing diversity. This method assumes that a taxon is present in a sample if it is present below and above the sample examined, and it takes into account facies-related fossils and differences in capture probability for each taxon; (3) The overall dinocyst composition throughout the section was observed using detrended correspondence analysis, or DCA (Hill and Gauch, 1980). DCA summarizes the variations in the composition of the assemblages in a small number of dimensions (Wing, 1998). The ordination was performed on the raw abundance data excluding samples with fewer than 40 dinocysts per sample. Dinocyst data from the #1 Young core were selected for the diversity study

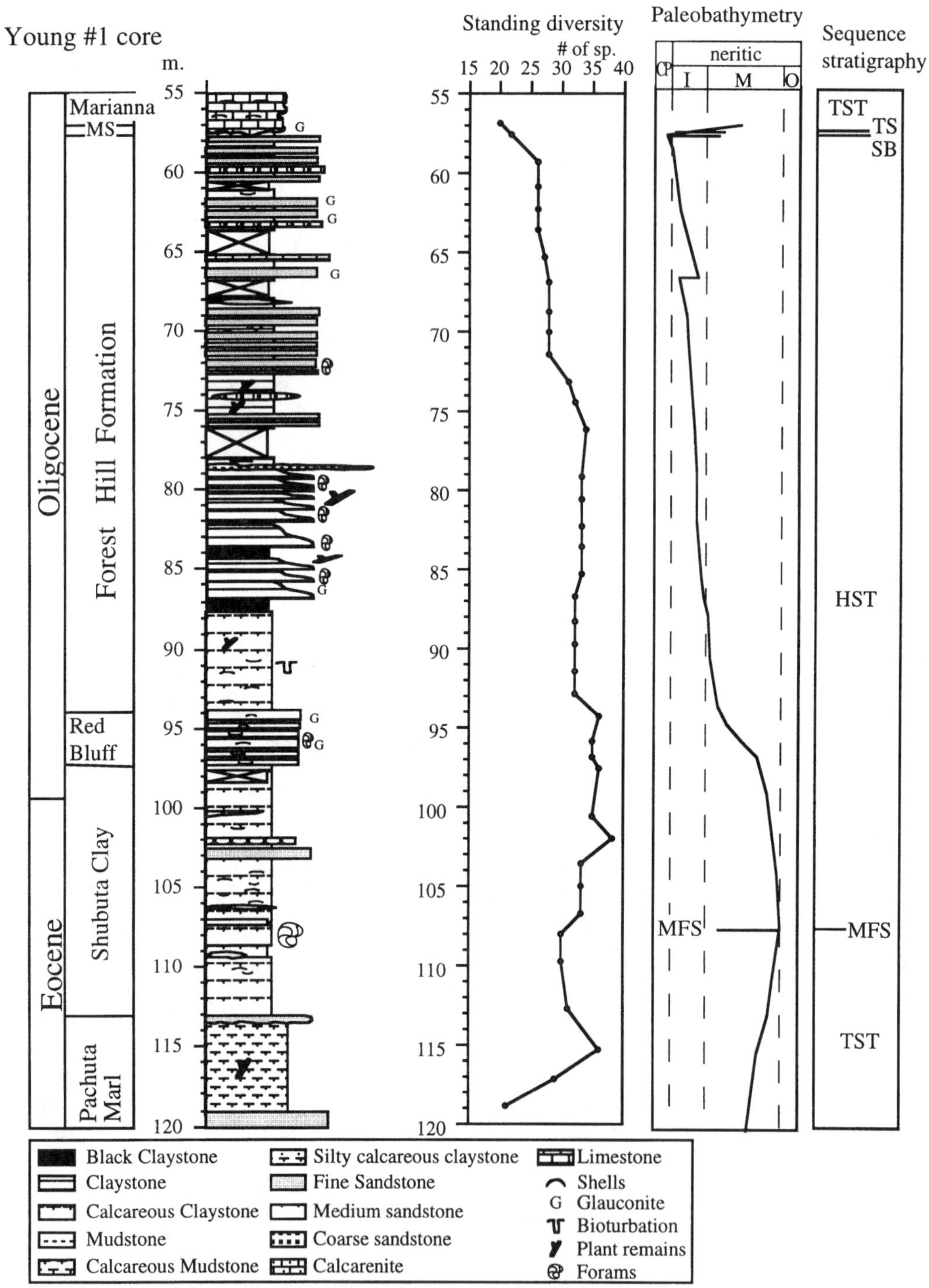

Figure 15.6. Comparison of dinocyst standing diversity patterns with lithofacies, paleobathymetry (derived from palynofacies and dinocyst paleobathymetric index), and sequence stratigraphy for the #1 Young core. Note how dinocyst diversity decreases as environments become shallower in the early Oligocene highstand system tract. Paleobathymetric curve, sequence stratigraphy, and dating of the section after Jaramillo and Oboh-Ikuenobe (1999). See label legends in figure 15.5.

because it is the section with the highest sampling density and best quality of recovery. Furthermore, it encompasses the Eocene-Oligocene transition.

The DCA, unbiased Simpson Index, and standing diversity show a constant dinocyst diversity throughout the Eocene-Oligocene transition (figure 15.5), although there is declining diversity later in the early Oligocene. This trend may be the product of the facies change toward more nearshore deposits (figures 15.5 and 15.6). This trend of decreasing diversity from middle offshore toward near-inshore deposits is observed in modern dinocyst distributions (e.g., Wall et al., 1977; Dale, 1996). The last and first occurrence patterns do not show any unusual event associated with the Eocene-Oligocene transition (figure 15.7). There is only one high first occurrence event that seems to be above background extinction and speciation rates (figure 15.7). However, this cluster of first occurrence datums at the maximum flooding surface is probably related to an artifact of sequence architecture (Holland, 1995), and does not indicate a true speciation event.

SEQUENCE STRATIGRAPHY

Paleobathymetric curves for each individual section, which constitute the basis for the sequence stratigraphic analysis, were derived from palynofacies analyses, dinocyst paleobathymetric indexes, and lithofacies. For a detailed explanation of techniques, results, and discussion of the curves, as well as sequence stratigraphy for each individual section, see Jaramillo and Oboh-Ikuenobe (1999). A correlation of all the studied sections (figure 15.8) shows that a maximum flooding surface in all the sections, except the St. Stephens Quarry, is isochronous and approximately 7.5 composite units (~0.1 Ma) older than the Eocene/Oligocene boundary. At the St. Stephens Quarry section, this surface is three composite units (~0.04 Ma) younger than the other sections. This maximum flooding surface occurs in the middle part of the Shubuta Clay in the four southern Mississippi sections and at the Shubuta–Bumpnose contact in the St. Stephens Quarry section. This contact represents a minor flooding surface in southern Mississippi. The Pachuta Marl and the lower Shubuta Clay (the entire Shubuta in Alabama) constitute a transgressive systems tract, with the upper Shubuta Clay and Bumpnose/Red Bluff/Forest Hill Formations representing the overlying highstand systems tract. The stratigraphic condensation at the St. Stephens Quarry section corresponds to the accumulation of the early Oligocene highstand in southern Mississippi. Figure 15.8 shows that the continuation of normal sedimentation in the St. Stephens Quarry area appears to correspond to the accumulation of the Forest Hill in southern Mississippi.

Up in the section, the Forest Hill–Mint Spring contact is interpreted as a sequence boundary that coincides with a transgressive surface. The Mint Spring Marl and Marianna Limestone represent a transgressive system tract. However, a possible lowstand deposit (forced regression) is interpreted in the #1 Wayne core section where sandy deposits occur in Mint Spring (figure 15.8).

Our interpretation of sequence stratigraphy is in general agreement with several previous studies (in particular, Tew and Mancini, 1995), although there are some distinct differences. First, the maximum flooding surface of the Shubuta–Forest Hill sequence occurs in the middle of the Shubuta Clay in southern Mississippi, but at the Shubuta-Bumpnose contact in the St. Stephens Quarry section. Second, the age of this maximum flooding surface is latest Eocene. Third, the lower 60 cm of the Bumpnose Limestone in the St. Stephens Quarry section constitute a condensed section that accumulated at the same time that the upper Shubuta and Red Bluff were being deposited in southern Mississippi. Fourth, the lowermost Oligocene sections constitute a highstand systems tract and not a lowstand systems tract as interpreted by Pasley and Hazel (1995).

REGIONAL AND GLOBAL PERSPECTIVES

The Eocene-Oligocene transition has received much attention because marine microfossils, most notably planktonic foraminifera and calcareous nannofossils, indicate thermal minimum at the boundary (Wei, 1991; Zachos et al., 1994). However, long-term cooling appears to have occurred long before the boundary itself, sometime during the middle to late Eocene. Several groups of organisms, such as mollusks in Oregon and Washington States (U.S.A.), document a decline in diversity and a shift from warm taxa to cold taxa during the late Eocene, just before the Eocene/Oligocene boundary (Hickman, this volume). In the U.S. Gulf Coast, our palynological data, which are limited to the Eocene-Oligocene transition, do not record distinct turnovers at the boundary itself, but note the dominance of the cooler-climate indicator *Quercoidites* over *Momipites* in the palynofloral assemblage. *Momipites* pollen is more common in warmer climates and it was more dominant during the Eocene (Elsik, 1974; Frederiksen, 1988, 1991; Oboh et al., 1996). Frederiksen (1991) was

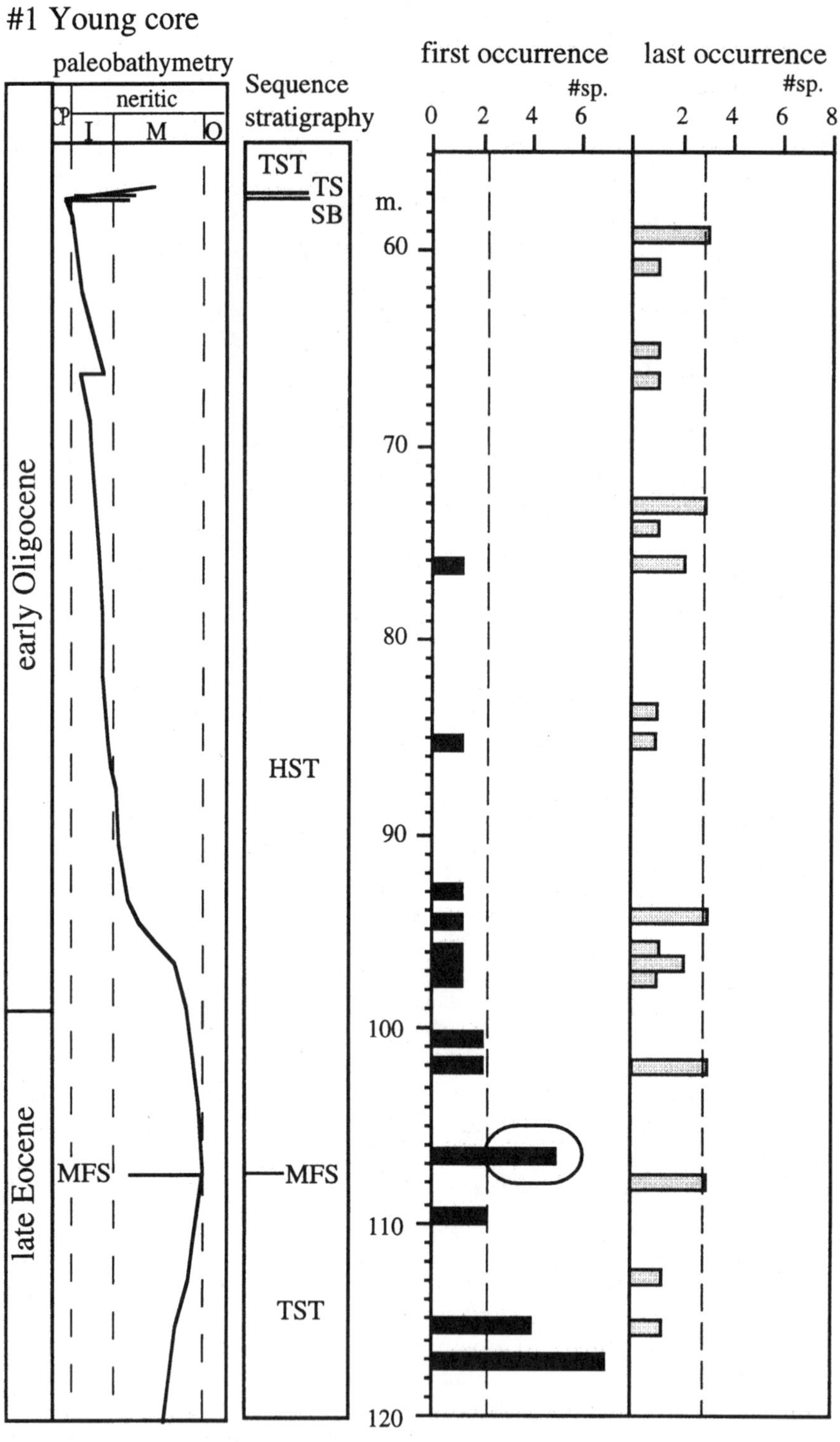

Figure 15.7. First and last occurrence datums per sample throughout the Eocene-Oligocene transition in the #1 Young core. Note how peak of first occurrence datum correlates with the maximum flooding surface, suggesting that the peak is an artifact of sequence architecture rather than an evolutionary event. Paleobathymetric curve, sequence stratigraphy, and dating of the section after Jaramillo and Oboh-Ikuenobe (1999). See label legends in figure 15.5.

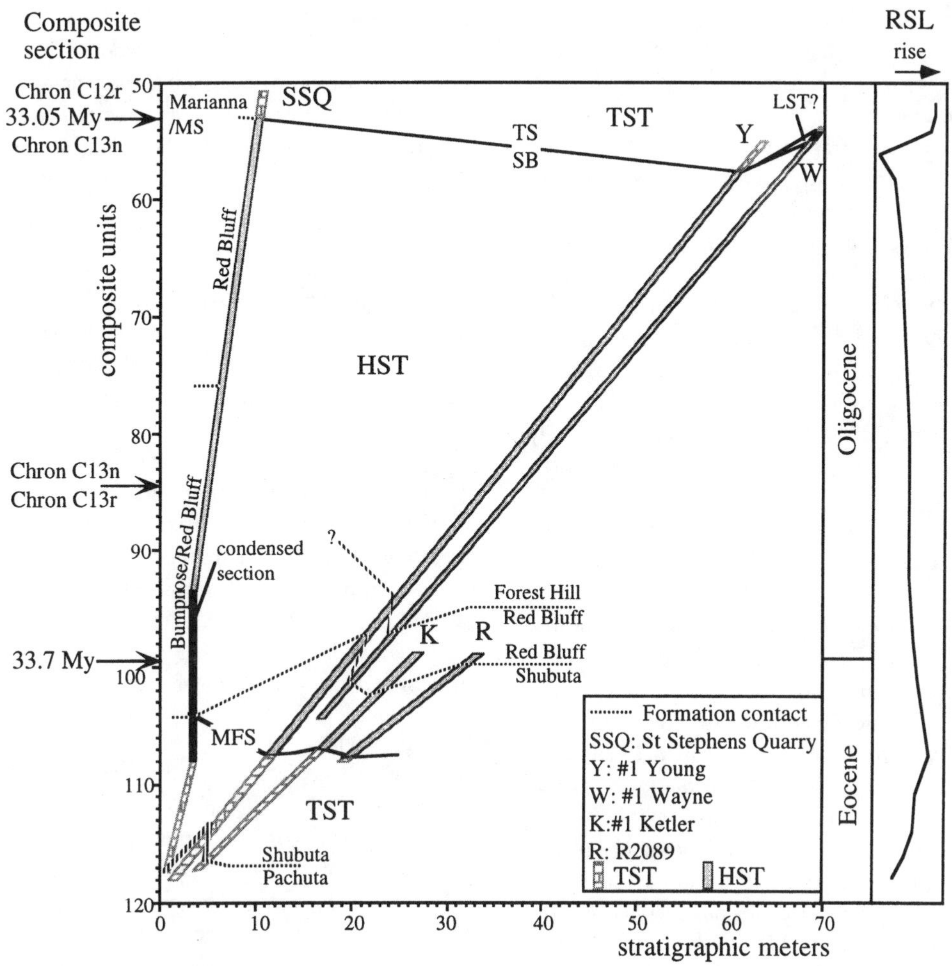

Figure 15.8. Time (composite section) versus stratigraphic thickness and sequence stratigraphy for the studied sections containing dinocyst data. Each composite unit is equivalent to 0.014 My. The Eocene/Oligocene boundary in the #1 Wayne corehole is at 55.7 m, 1.9 m above the Shubuta-Red Bluff contact, while the C13n-C12r magnetoboundary occurs at 7.1 m, 2 m above the Mint Spring-Marianna contact. In the #1 Ketler corehole, the Eocene/Oligocene boundary is located at 7.5 m, 3.7 m. above the Shubuta-Red Bluff contact. At the R2089 section, the Eocene/Oligocene boundary is located at 13.3 m, 0.1 m above the Shubuta-Red Bluff contact. At St. Stephens Quarry a 80 cm-condensed section containing the maximum flooding surface and the Eocene/Oligocene boundary begins 20 cm below the Shubuta-Bumpnose contact and ends 60 cm above the contact. See label description in legend of figure 15.5. MS, Mint Spring; RSL, relative sea level. Composite section after Jaramillo and Oboh-Ikuenobe (1999).

able to correlate similar pollen events in the Gulf Coast with those in southern California. However, Yancey et al. (this volume) document that *Quercoidites* abundance occurred earlier in Texas, during the late Eocene as opposed to the early Oligocene in the eastern Gulf Coast. Our early Oligocene palynofloral assemblage is indicative of a warm temperate climate in comparison with more tropical elements in early and middle Eocene assemblages. A similar observation was also made in Australia by Gammon et al. (this volume) in his study of Southern Hemisphere biosiliceous deposits. Our dinocyst diversity data (figure 15.5) do not record a rapid decline at the Eocene-Oligocene transition, and therefore support the spore and pollen data. The slight decrease in dinocyst diversity during the early Oligocene was likely due to a facies change from marine sediments to nearshore sediments.

Intensive debates have been going on for several years about the synchronicity of the sequence boundaries of the Haq et al. (1988) curve and its use for worldwide chronostratigraphic correlation (e.g., Miall, 1992). The Haq curve shows a sequence boundary at the Eocene/Oligocene boundary (TA 4.3/4.4 sequence boundary), which some people have argued should be used to define the Eocene/Oligocene boundary worldwide (e.g., Brinkhuis and Visscher, 1995). The 1992 Kyoto International Geological Congress defined the Eocene/Oligocene boundary by the last occurrence of the foraminifera *Hantkenina* spp. at the Massignano section in Italy (Berggren et al., 1995). This datum is located in the youngest part of the magnetostratigraphic C13r (C13r.14) Chron (Berggren et al., 1995). In the Priabonian type section, Brinkhuis (1994) indicated that the Eocene/Oligocene boundary (as defined by the last occurrence of *Hantkenina*) would be located 15 m below the lithostratigraphic position of a sequence boundary that is interpreted as the TA4.3/TA4.4 sequence boundary of the Haq et al. (1988) curve. Furthermore, this sequence boundary in the Priabonian type section is in the lower part of Chron C13n (Brinkhuis and Visscher, 1995:298, figure 4). The correlation of the Priabonian sequence boundary with the sequence boundary at the Forest Hill–Mint Spring contact is not possible because the sequence boundary in the U.S. Gulf Coast occurs at the C13n-C12r boundary (figure 15.8). Therefore, the Priabonian and Forest Hill-Mint Spring sequence boundaries are not coeval, and the TA 4.3/TA4.4 sequence boundary of the Haq et al. (1988) curve may not be worldwide in occurrence and should not be used as a time line.

CONCLUSIONS

The prominence of *Quercoidites, Sequoiapollenites, Cupressacites,* and *Cyrillaceaepollenites* in the rich pollen assemblage, which contains some tropical to subtropical elements, and lacks indicators of cool-temperate to cold-temperate or dry conditions, suggests a warm temperate paleoclimate during the beginning of the early Oligocene in the study area. Although *Momipites* was generally more abundant in the Eocene, and *Quercoidites* increased during the Oligocene, there was no significant paleofloristic change across the Eocene/Oligocene boundary. The major floral turnover along the Gulf Coast occurred well before the deposition of the Vicksburg units during the early Oligocene, probably at the end of the middle Eocene. Dinocyst data also do not show evidence of either a major species turnover or a change in standing diversity, suggesting that major changes in sea surface water in the area did not occur at the Eocene/Oligocene boundary. Sequence stratigraphic interpretations are in agreement with most previous studies.

ACKNOWLEDGMENTS

We thank Norman Frederiksen and Thomas Yancey for a detailed review of this paper. We also thank Norman Frederiksen and Lucy Edwards (U.S. Geological Survey) for loan of the outcrop samples, Larry Fearn and Mobil Exploration and Producing for providing the Mississippi corehole samples, and Mobil for permission to publish the results from the Mississippi coreholes. Lisa Reeves Morris and Oscar Yepes contributed data to this study. This research was funded by the Petroleum Research Fund, administered by the American Chemical Society, and by NSF Grant No. EAR-9405154 to FEO. Carlos Jaramillo thanks Colciencias for its support.

REFERENCES

Berggren, W. A., Kent, D. V., Swisher, C. C., III, and Aubry, M.-P., 1995, Revised Cenozoic geochronology and chronostratigraphy: SEPM Special Publication, v. 54, p. 129–212.

Boltovskoy, D., 1988, The range-through method and first-last appearance data in paleontological surveys: Journal of Paleontology, v. 62, p. 157-159.

Brinkhuis, H., 1994, Late Eocene to early Oligocene dinoflagellate cysts from the Priabonian type-area (Northeast Italy): biostratigraphy and paleoenvironmental interpretation: Palaeogeography, Palaeoclimatology, Palaeoecology, v. 107, p. 121-163.

Brinkhuis, H., and Visscher, H., 1995, The upper boundary of the Eocene Series: a reappraisal based on dinoflagellate cyst biostratigraphy and sequence stratigraphy: SEPM Special Publication, v. 54, p. 295-304.

Bybell, L. M., and Poore, R. Z., 1982, Late Eocene to early Oligocene calcareous nannofossils in Alabama and Mississippi: Transactions of the Gulf Coast Association of Geological Societies, v. 32, p. 295-302.

Dale, B., 1996, Dinoflagellate cyst ecology: modeling and geological applications, *in* Jansonius, J., and MacGregor, D. C., eds., Palynology: principles and applications: Dallas, American Association of Stratigraphic Palynologists Foundation, v. 3, p. 1249-1275.

Elsik, W. C., 1974, Characteristic Eocene palynomorphs in the Gulf Coast, U.S.A.: Palaeontographica Abteilung B, v. 149, p. 90-111.

Frederiksen, N. O., 1980, Sporomorphs from the Jackson Group (Upper Eocene) and adjacent strata of Mississippi and western Alabama: U.S. Geological Survey Professional Paper, v. 1084, p. 1-75.

Frederiksen, N. O., 1981, Middle Eocene to early Oligocene plant communities in the Gulf Coast, *in* Gray, J., Boucot, A. J., and Berry, W. B. N., eds., Communities of the past: Stroudsburg, Pennsylvania, Hutchinson Ross, p. 493-549.

Frederiksen, N. O., 1988, Sporomorph biostratigraphy, floral changes, and paleoclimatology, Eocene and earliest Oligocene of the eastern Gulf Coast: U.S. Geological Survey Professional Paper, v. 1448, p. 1-68.

Frederiksen, N. O., 1991, Pulses of middle Eocene to earliest Oligocene climatic deterioration in southern California and the Gulf Coast: Palaios, v. 6, p. 564-571.

Galloway, W. E., 1989, Genetic stratigraphic sequences in basin analysis II: application to Northwest Gulf of Mexico Cenozoic Basin: American Association of Petroleum Geologists Bulletin, v. 73, p. 143-154.

Galloway, W. E., Bebout, D. G., Fisher, W. L., Dunlap, J. B., Cabrera-Castro, R., Lugo-Riviera, J. E., and Scott, T. M., 1991, Cenozoic, *in* Salvador, A., ed., The Gulf of Mexico Basin: The Geology of North America: Boulder, Colorado, Geological Society of America, p. 245-324.

Germeraad, J. H., Hopping, C. A., and Muller, J., 1968, Palynology of Tertiary sediments from tropical areas: Review of Palaeobotany and Palynology, v. 6, p. 189-348.

Groot, J. J., 1991, Palynological evidence for Late Miocene, Pliocene and early Pleistocene climatic changes in the middle U.S. Atlantic Coastal Plain: Quaternary Science Reviews, v. 10, p. 147-162.

Haq, B. U., Hardenbol, J., and Vail, P. R., 1988, Mesozoic and Cenozoic chronostratigraphy and eustatic cycles, *in* Wilgus, C. K., Hastings, B. S., Posamentier, H., Van Wagoner, J., Ross, C. A., and Kendall, C. G. S. C., eds., Sea-level changes: an integrated approach: SEPM Special Publication, v. 42, p. 71-109.

Hazel, J. E., Mumma, M. D., and Huff, W. J., 1980, Ostracode biostratigraphy of the Lower Oligocene (Vicksburgian) of Mississippi and Alabama: Transactions of the Gulf Coast Association of Geological Societies, v. 30, p. 361-401.

Hill, M. O., and Gauch, H. G., 1980, Detrended correspondence analysis, an improved ordination technique: Vegetatio, v. 42, p. 47-58.

Holland, S. M., 1995, The stratigraphic distribution of fossils: Paleobiology, v. 21, p. 92-109.

Jaramillo, C. A., and Oboh-Ikuenobe, F. E., 1999, Sequence stratigraphic interpretations from palynofacies, dinocyst and lithological data of upper Eocene-lower Oligocene strata in southern Mississippi and Alabama, U.S. Gulf Coast: Palaeogeography, Palaeoclimatology, Palaeoecology, v. 145, p. 259-302.

Keller, G., 1985, Eocene and Oligocene stratigraphy and erosional unconformities in the Gulf of Mexico and Gulf Coast: Journal of Paleontology, v. 59, p. 882-903.

Mancini, E. A., 1979, Eocene/Oligocene boundary in southwest Alabama: Transactions of the Gulf Coast Association of Geological Societies, v. 29, p. 282-286.

Miall, A. D., 1992, The Exxon global cycle chart: an event for every occasion?: Geology, v. 20, p. 787-790.

Miller, K. G., Thompson, P. R., and Kent, D. V., 1993, Integrated late Eocene-Oligocene stratigraphy of the Alabama coastal plain: correlation of hiatuses and stratal surfaces to glacioeustatic lowerings: Paleoceanography, v. 8, p. 313-331.

Oboh, F. E., and Reeves Morris, L. M., 1994, Lower Oligocene palynosequences from the eastern Gulf Coast: Palynology, v. 18, p. 213-235.

Oboh, F. E., and Reeves Morris, L. M., 1995, Correlation between Sequoia type pollen and Lower Oligocene transgressive deposits in the eastern Gulf Coast: Palaios, v. 10, p. 371-382.

Oboh, F. E., Jaramillo, C. A., and Reeves Morris, L. M., 1996, Late Eocene-early Oligocene paleofloristic patterns in southern Mississippi and Alabama, U.S. Gulf Coast: Review of Palaeobotany and Palynology, v. 91, p. 23-34.

Pasley, M. A., and Hazel, J. E., 1990, Use of organic petrology in graphic correlation of biostratigraphic data in sequence stratigraphic interpretations: example from the Eocene/Oligocene boundary section, St. Stephens Quarry, Alabama: Gulf Coast Association of Geological Societies Transactions, v. 40, p. 661-683.

Pasley, M. A., and Hazel, J. E., 1995, Revised sequence stratigraphic interpretation of the Eocene/Oligocene boundary interval, Mississippi and Alabama, Gulf Coast Basin, U.S.A.: Journal of Sedimentary Research B, v. 65, p. 160-169.

Pielou, E. C., 1984, The interpretation of ecological data: New York, John Wiley and Sons, 263 p.

Rosenzweig, M. L., 1995, Species diversity in space and time: Cambridge, Cambridge University Press, 433 p.

Shackleton, N. J., and Kennett, J. P., 1975, Paleotemperature history of the Cenozoic and the initiation of Antarctic glaciation: oxygen and carbon isotope analyses in DSDP Sites 277, 279, and 281: Initial Reports of the Deep Sea Drilling Project, v. 29, p. 743-755.

Siesser, W. G., 1983, Paleogene calcareous nannoplankton biostratigraphy: Mississippi, Alabama and Tennessee: Mississippi Bureau of Geology Bulletin, v. 125, p. 1-61.

Sokal, R. R., and Michener, C. D., 1958, A statistical method for evaluating systematic relationships: University of Kansas Sciences Bulletin, v. 38, p. 1409-1438.

Tew, B. H., and Mancini, E. A., 1995, An integrated lithostratigraphic, biostratigraphic, and sequence stratigraphic approach to paleogeographic reconstruction: examples from the Upper Eocene and Lower Oligocene of Alabama and Mississippi: Palaios, v. 10, p. 133-153.

Wall, D., Dale, B., Lohmann, G. P., and Smith, W. K., 1977, The environmental and climatic distribution in modern marine sediments from regions in the North and South Atlantic Oceans and adjacent seas: Marine Micropaleontology, v. 2, p. 121-200.

Wei, W., 1991, Evidence for an earliest Oligocene abrupt cooling in the surface waters of the Southern Ocean: Geology, v. 19, p. 780-783.

Wing, S. L., 1998, Late Paleocene–early Eocene floral and climatic change in the Bighorn Basin, Wyoming, *in* Berggren, W., Aubry, M. P., and Lucas, S., eds., Late Paleocene–early Eocene biotic and climatic events: New York, Columbia University Press, p. 371-391.

Wolfe, J. A., 1971, Tertiary climatic fluctuations and methods of analysis of Tertiary floras: Palaeogeography, Palaeoclimatology, Palaeoecology, v. 9, p. 27-57.

Wolfe, J. A., 1992, Climatic, floristic, and vegetational changes near the Eocene/Oligocene boundary in North America, *in* Prothero, D. R., and Berggren, W. A., eds., Eocene-Oligocene climatic and biotic evolution: Princeton, New Jersey, Princeton University Press, p. 421-436.

Wolfe, J. A., 1994, Tertiary climatic changes at middle latitudes of western North America: Palaeogeography, Palaeoclimatology, Palaeoecology, v. 108, p. 195-205.

Zachos, J. C., Stott, L. D., and Lohmann, K. C., 1994, Evolution of early Cenozoic marine temperatures: Paleoceanography, v. 9, p. 353-387.

CHAPTER 16

Late Eocene-Early Oligocene Benthic Foraminifera in the Gulf Coastal Plain: Regional vs. Global Influences

Richard H. Fluegeman

ABSTRACT

The turnover in benthic foraminiferal faunas through the Eocene-Oligocene transition in the Gulf Coastal Plain has been considered abrupt by many workers. Although distinct Jackson and Vicksburg faunas can be identified, the transition between the two is gradual. This gradual transition makes recognition of important biohorizons in the benthic foraminiferal fauna possible. In the Gulf Coast section, numerous lowest and highest occurrence horizons have long been employed in subsurface correlation throughout the Gulf Coastal Plain. The appearances of distinct Vicksburg species such as *Lenticulina vicksburgensis* and *Cibicidoides cookei* are interpreted to occur at the Eocene/Oligocene boundary in the well-known sections of southwestern Alabama and eastern Mississippi. Results from the Mossy Grove core in west-central Mississippi, however, show that the appearance of the Vicksburg fauna occurs sometime after the Eocene-Oligocene biostratigraphic boundary.

Eocene-Oligocene benthic foraminiferal faunas from the Caribbean islands have little in common with those of the Gulf Coastal Plain because Caribbean assemblages are typically deep-water faunas. Their turnover pattern is similar to that found in deep-sea cores worldwide and is related to gradual global cooling through the Eocene.

Although faunal turnover patterns in neritic sections are generally influenced by regional processes such as sedimentation rates, local tectonics, and relative sea-level fluctuation, the Eocene-Oligocene benthic foraminiferal pattern in the Gulf Coastal Plain may be related to global phenomena. A similar turnover pattern in Eocene-Oligocene neritic benthic foraminiferal faunas occurs in southern Australia. This similarity may indicate that the gradual turnover of benthic foraminiferal faunas seen in the Eocene-Oligocene of the Gulf Coastal Plain is a response to global cooling of the late Eocene.

INTRODUCTION

The Gulf Coastal Plain of the United States has produced a wealth of data on upper Eocene and lower Oligocene benthic foraminifera over the years. The works of Cushman (1923, 1935), Howe and Wallace (1933), and Bandy (1949) are four of the most prominent. Cushman (1923) described foraminifera of the Oligocene Vicksburg Group from localities in western Alabama, eastern Mississippi, and northern Florida. Howe and Wallace (1933) described a foraminiferal fauna from the upper Eocene sediments exposed at Danville Landing, Louisiana. Cushman's (1935) wide-ranging study described foraminifera of the Eocene Jackson Group throughout the Gulf and southern Atlantic Coastal Plains, including samples from North Carolina, South Carolina, Georgia, Florida, Alabama, Mississippi, and Texas. These papers demonstrated distinct differences between the benthic foraminiferal faunas of the Jackson Group and the overlying Vicksburg Group. Bandy (1949) studied the distribution of Eocene and Oligocene foraminifera in a section along Little Stave Creek in Clarke County, Alabama. Although primarily taxonomic, Bandy's work showed changes in benthic foraminiferal faunas from the middle Eocene through the lower Oligocene.

The above works made great strides toward characterizing the upper Eocene and lower Oligocene foraminiferal faunas of the Gulf Coastal Plain. Although the distribution of the faunas was described geographically (Cushman 1923, 1935) and to a degree stratigraphically (Bandy, 1949), no attempt was made to correlate the upper Eocene and lower Oligocene rocks of the Gulf Coastal Plain with the European standard sections. Rather, they focused on the foraminifera of the major lithostratigraphic units: the Jackson Group and the Vicksburg Group. In the Gulf Coastal Plain, the Eocene/Oligocene boundary had long been placed at the Jackson-Vicksburg contact (Cooke, 1923). To be sure, the fossils of the Jackson were Eocene and those of the Vicksburg were Oligocene. No attempts had been made, however, to define this boundary in the Gulf Coastal Plain. References to changes in foraminiferal faunas at the Eocene/Oligocene boundary were therefore using this boundary in a lithologic rather than a chronostratigraphic context. The biostratigraphic work of Mancini (1979), Siesser (1983), and Mancini and Waters (1986) defined the Eocene/Oligocene boundary in the eastern Gulf Coastal Plain and correlated it to standard European sections.

Cheetham (1957) and Deboo (1965) recorded the ranges of foraminifera and other fossils at the Eocene/Oligocene boundary from sections in southwestern Alabama and southeastern Mississippi. Both papers attempted to improve recognition and correlation of the Eocene/Oligocene boundary within the eastern Gulf Coastal Plain. Using these and other published data on benthic foraminifera, Gaskell (1991) identified four major extinction cycles in benthic foraminifera within the Paleogene of the Gulf Coastal Plain. One of these extinction cycles coincides with the Eocene/Oligocene boundary, that is, the Jackson-Vicksburg lithologic contact. Following Mancini et al. (1987) Gaskell assumed that the Jackson-Vicksburg contact was conformable. A thorough discussion of the stratigraphic nature of the Jackson-Vicksburg contact in the eastern Gulf Coastal Plain can be found in Pasley and Hazel (1995). If this contact is unconformable, as indicated by Dockery (1990) and many others, the precise timing of the Eocene-Oligocene benthic foraminiferal extinction cycle of Gaskell (1991) is uncertain.

Despite the stratigraphic impression, there is a major faunal turnover in benthic foraminiferal faunas of the Gulf Coastal Plain sometime in the late Eocene or early Oligocene (Gaskell, 1991). It is the goal of this paper to review the faunal succession of benthic foraminifera in the upper Eocene and lower Oligocene rocks of the Gulf Coastal Plain, to identify as precisely as possible the time of the benthic foraminiferal extinction cycle, and to determine whether the faunal turnover is sudden or gradual with respect to benthic foraminifera.

STUDY AREA

The general focus of this investigation is the upper Eocene–lower Oligocene succession of the Gulf Coastal Plain of the United States, including benthic foraminiferal faunas from Texas to Florida. The detailed focus of this study will be on the well-exposed, extensively studied sections of Mississippi and southwestern Alabama. This region contains well-defined stratigraphic units in a framework of stable lithostratigraphic nomenclature.

It is in this region where the most reliable biostratigraphic control exists (Siesser, 1983; Mancini and Waters, 1986; Mancini and Tew, 1991), thus enabling correlation of key sections to the global geochronologic scale of Berggren et al. (1995). In this region are the important sections at Little Stave Creek, Clarke County, Alabama (Bandy, 1949); St. Stephens Quarry, Washington County, Alabama (Mancini and Copeland, 1986); the Chickasawhay River sections of Clarke County, Mississippi (Deboo, 1965; Dockery, 1980); the Mossy Grove core of Hinds County, Mississippi (Fluegeman, 1996) as well as many discontinuous but richly fossiliferous sections of the Jackson and Vicksburg Groups.

While there is general agreement on regional lithostratigraphy and biostratigraphy, there are wide differences concerning the sequence stratigraphic interpretation. The Eocene/Oligocene boundary is a particular point of difference among various workers. Dockery (1990) and Coleman and Galloway (1990) recognized the Eocene/Oligocene boundary at the Jackson-Vicksburg contact in the western Gulf Coastal Plain (Texas, Louisiana, and Mississippi) and interpret it as a sequence boundary. Baum and Vail (1988) and Mancini and Tew (1991) also interpreted the Eocene/Oligocene boundary at the Jackson-Vicksburg contact in Alabama and eastern Mississippi but considered it to be a condensed section. Fluegeman et al. (1994), based on a study of foraminifera from the Mossy Grove core of Mississippi, interpreted the Jackson-Vicksburg contact to be a sequence boundary, but suggested that it was not at the Eocene/Oligocene boundary. Current ideas about the sequence stratigaphy of the Eocene-Oligocene transition in the Gulf Coastal Plain are discussed by Mancini and Tew (1991).

BENTHIC FORAMINIFERAL FAUNAL SUCCESSION OF THE GULF COASTAL PLAIN

Overview

The Paleogene benthic foraminiferal faunas of the Gulf Coastal Plain can be informally separated into five general "faunas." These are recognizable by their species compositions and are separated stratgraphically by the extinction cycles of Gaskell (1991). These faunas bear the names of the lithostratigraphic units from which they were first described. These faunas are in ascending order: the Midway fauna (Paleocene and lower Eocene), the Claiborne fauna (middle Eocene), the Jackson fauna (upper Eocene and lowermost Oligocene), the Vicksburg fauna (lower Oligocene), and the Chikasawhay fauna (middle Oligocene).

The Eocene-Oligocene transition in the Gulf Coastal Plain is recorded by the transition from the Jackson fauna to the Vicksburg fauna. The Jackson fauna, as generally described by Cushman (1935), is recognized by species such as *Siphonina jacksonensis* Cushman and Applin, *S. danvillensis* Howe and Wallace, *Cibicidina walli* Bandy, *Uvigerina cocoaensis* Cushman, *Bulimina jacksonensis* Cushman, *Cibicidoides truncatus* (Bandy) (= *Cibicides truncatus*, 1949), and *Marginulina cocoaensis* Cushman. The Vicksburg fauna, as described by Cushman (1922), is distinguished from the underlying Jackson fauna by the appearance of *Lenticulina vicksburgensis* (Cushman) (= *Cristellaria vicksburgensis*, 1922), *Spiroplectammina latior* Bandy, *Cibicidoides cookei* (Cushman and Garret) (= *Cibicides cookei*, 1938), *Uvigerina vicksburgensis* Cushman and Ellisor, and *Eponides ellisorae* Garret.

The transition between the Jackson and the Vicksburg faunas is gradual. At Little Stave Creek, Bandy (1949) recognized that several species of benthic foraminifera continued from the Yazoo Clay (Jackson Group) into the overlying Red Bluff Clay (Vicksburg Group). The Red Bluff Clay at Little Stave Creek contains species characteristic of the lowermost Vicksburg fauna, including the distinctive species *Lenticulina vicksburgensis*.

Tipsword (1962) identified a general pattern of benthic foraminiferal succession in the Jackson and Vicksburg Groups of the Gulf Coastal Plain. Although he focused on identifying marker species for lithostratigraphic units rather than a continual pattern of faunal succession, his results indicate that a pattern is present in upper Eocene and lower Oligocene rocks of the Gulf Coastal Plain and that it may have biostratigraphic utility.

The succession of benthic foraminifera can also provide information about changing paleoecologic and paleoceanographic conditions. The extinction cycle in shallow-water benthic foraminiferal faunas at the Eocene/Oligocene boundary identified by Gaskell (1991) may be a response to changing global conditions. The Eocene-Oligocene transition is characterized by a distinct global cooling (Miller et al., 1991) which Thomas (1992) relates to gradual changes in deep-sea benthic foraminifera. In order to examine the effect of this global cooling on the neritic realm, it is necessary to study the succession of benthic foraminifera of the Jackson and Vicksburg Groups with detailed focus on several key sections.

The Claiborne-Jackson Group Faunal Transition

This transition is represented by the lithologic contact between the Claiborne and Jackson Groups. In southwestern Alabama, these groups are represented by the Gosport and Moodys Branch formations. The Gosport Formation is best known from exposures at Little Stave Creek, Clarke County, Alabama. Samples from the Gosport were included by Bandy (1949) in his study of foraminifera from this section. Benthic foraminiferal faunas from the Gosport at this locality are moderately diverse. They contain species typical of the Claiborne fauna, including *Siphonina claibornensis* Cushman, *Cibicidina subminuens* Bandy, and *Eponides mexicanus* (Cushman) (= *Pulvinulina mexicana*, 1925). Gardner (1957) additionally identified the stratigraphically important benthic foraminiferan *Nonionella cockfieldensis* Cushman and Ellisor from the middle and upper parts of the Gosport Formation at Little Stave Creek. Gosport samples examined for this study did not contain *Nonionella cockfieldensis*, but this species was identified from samples in the lower part of the Moodys Branch Formation. Other sections used in this study also contained *Nonionella cockfieldensis*: the lower Moodys Branch Formation at Claiborne Bluff on the Alabama River, Monroe County, Alabama; the upper Cockfield–lower Moodys Branch section at the Heard Cemetary cut on the Chickasawhay River, Clarke County, Mississippi (Dockery, 1980); the lower Moodys Branch Formation at La Fleur's Bluff (Riverside Park), Jackson, Mississippi; the Mossy Grove core, Hinds County, Mississippi; and the basal Jackson at Montgomery Landing, Sabine Parish, Louisiana.

The benthic foraminifera from the Moodys Branch Formation at Little Stave Creek include species typical of the Jackson fauna including *Siphonina danvillensis, S. jacksonensis,* and *Cibicidina*

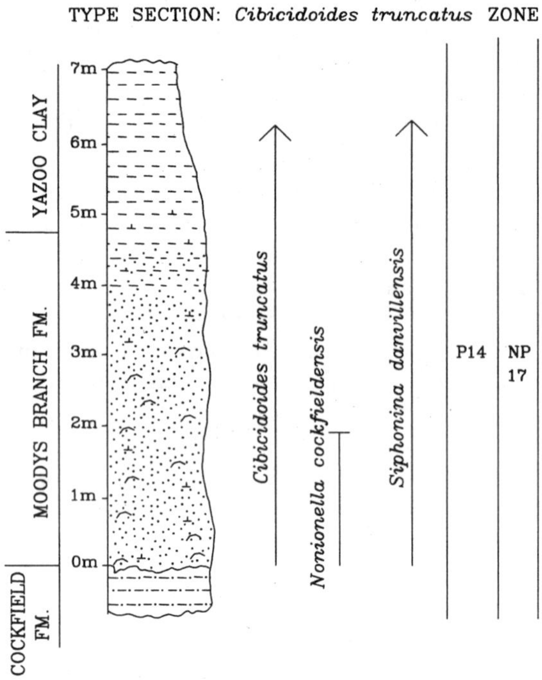

Figure 16.1. Measured section of the Moodys Branch Formation and lower Yazoo Clay at LaFleur's Bluff (Riverside Park), Jackson, Mississippi showing the distribution of biostratigraphically important benthic

walli. Additionally, sections of the Moodys Branch Formation at LaFleur's Bluff, in the Mossy Grove core in Mississippi, and at Montgomery Landing in Louisiana contain *Spiroplectammina mississippiensis* (Cushman) (= *Textularia mississippiensis,* 1922), *Uvigeriana jacksonensis* Cushman, and *Bolivina jacksonensis* Cushman and Applin. Within the Moodys Branch at all localities, there are benthic foraminiferal species common to the Claiborne fauna. Hence, in the Gulf Coastal Plain, the transition from the Claiborne fauna to the Jackson fauna is a gradual one.

Beyond the appearance of indicator species of the Jackson fauna, one important benthic foraminiferal biostratigraphic horizon can be identified within the Moodys Branch Formation. The highest occurrence of *Nonionella cockfieldensis* occurs within the lower part of the formation at the top of the *Scutella* bed as defined by Dockery (1980).

The section at La Fleur's Bluff, Jackson, Mississippi, is shown in figure 16.1.

The Yazoo Clay

The Yazoo Clay contains a diverse benthic foraminiferal fauna in the eastern Gulf Coastal Plain. Much of the knowledge of the Yazoo fauna comes from limited outcrops exposing only portions of the formation, which it is poorly exposed throughout the region. The Little Stave Creek section (Bandy, 1949) is an exception. This section provides important biostratigraphic control on the highest and lowest occurrences of many benthic species. Important subsurface sections such as the Mossy Grove core (Fluegeman, 1996) and the Mobil-Mississippi cores of Wayne County, Mississippi (Root et al., 1996), have provided supplemental information on the ranges of benthic foraminifera in the Yazoo Clay.

The Yazoo section at Little Stave Creek contains numerous highest and lowest occurrences of species. The difficulty lies in identifying those horizons that persist regionally and do not represent simply a change in local paleoecologic conditions.

In the Little Stave Creek section, the highest occurrence of *Cibicidoides truncatus* may be of regional significance. Bandy (1949) reports the last specimens of this species immediately above the top of the Moodys Branch Formation. In samples studied for this paper, the highest occurrence of *C. truncatus* is 1.5 m above the top of the Moodys Branch Formation in the North Twistwood Creek Clay member of the Yazoo. The lower part of the Yazoo Clay at Little Stave Creek and key benthic foraminiferal species are shown in figure 16.2.

The lowest occurrence of *Uvigerina cocoaensis* Cushman is another significant biohorizon within the Yazoo Clay. At Little Stave Creek, this species is first found at the base of the Pachuta Marl Member of the Yazoo and is a conspicuous part of the benthic foraminiferal fauna of the Yazoo in Mississippi. The lowest occurrence of this species is an important biostratigraphic marker in subsurface Jackson sections of Texas and Louisiana (Ellisor, 1933).

One especially important biohorizon is the lowest occurrence of *Marginulina cocoaensis* Cushman. This distinctive benthic foraminiferan is present in the uppermost part of the Shubuta Clay Member at Little Stave Creek. Its lowest occurrence coincides with the lowest occurrences of numerous species of benthic foraminifera at the Little Stave Creek section. Bandy (1949) did not extend the range of *Marginulina cocoaensis* to the Yazoo–Red Bluff contact. Samples from just below the Yazoo–Red Bluff contact examined in this study, however, do contain *Marginulina cocoaensis.* The precise stratigraphic range of this species is difficult to identify in the Little Stave Creek section because its lowest occurrence is just after an extended covered interval. Its highest

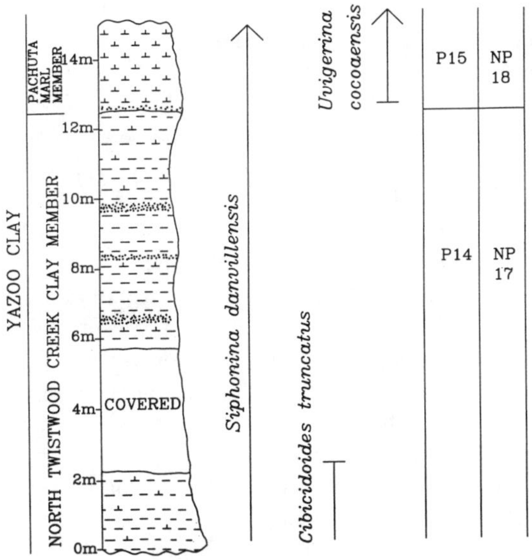

Figure 16.2. Measured section of the middle portion of the Yazoo Clay at Little Stave Creek, Clarke County, Alabama showing the location of important biohorizons.

occurrence at Little Stave Creek, as well as at the St. Stephens Quarry section and the Chickasawhay River section, is less problematic. Although there is some debate as to the nature of the Yazoo–Red Bluff contact in these sections, *Marginulina cocoaensis* does not occur in any samples from the Red Bluff Clay.

The Jackson-Vicksburg Group Transition

Benthic foraminifera from the Jackson-Vicksburg Group transition are best studied in southwestern Alabama and eastern Mississippi. The Yazoo–Red Bluff contact is best exposed in the Little Stave Creek and the St. Stephens Quarry sections in Alabama and in the Chickasawhay River section in Mississippi. In these sections, the Jackson-Vicksburg Group contact is the biostratigraphic Eocene/Oligocene boundary (Mancini and Waters, 1986; Siesser, 1983). In all three sections, the transition between the Jackson and the Vicksburg faunas is gradual. Many species present in the Yazoo Clay persist into the Red Bluff Clay. Despite the gradational transition, there are some noticeable changes in the benthic foraminiferal fauna across this contact. Some important species present in the uppermost Yazoo Clay but not in the Red Bluff are *Uvigerina cocoaensis, Marginulina cocoaensis, Siphonina jacksonensis,* and *Discorbis cocoaensis.* Benthic foraminifera with their lowest occurrences in the Red Bluff Clay include *Cibicidoides cookei, Eponides ellisorae, Lenticulina vicksburgensis,* and *Uvigerina vicksburgensis.* The Yazoo-Red Bluff interval at Little Stave Creek and key benthic foraminiferal species are shown in figure 16.3.

Both the studied samples and the published literature indicate that there is a gradual faunal turnover in benthic foraminifera at the Jackson-Vicksburg contact. If this turnover is to be related to any other faunal turnover events of the late Eocene or early Oligocene (Keller et al., 1992; Aubry, 1992; Thomas, 1992; McGowran et al., 1992), better biostratigraphic resolution is needed.

A Case Study: The Mossy Grove Core of Mississippi

The Mossy Grove core of Hinds County, Mississippi, drilled in 1991, penetrated a continuous section of the Yazoo Clay, 144 m (468 feet) thick, and a thick section of the underlying Moodys Branch Formation, 3.7 m (12 feet) thick. There are also some thin foraminifera-bearing beds in the Cockfield Formation below the Moodys Branch in the core. Details of the location and stratigraphy of the Mossy Grove core can be found in Dockery et al. (1991).

Fluegeman et al. (1994) and Fluegeman (1996) examined the foraminifera from this core. The Moodys Branch and Yazoo Clay section was sampled at 0.6-m (2-foot) intervals. Samples were also collected from the fossiliferous beds of the Cockfield

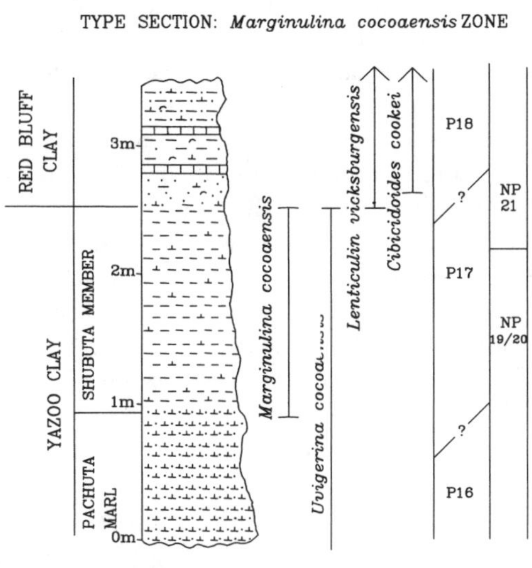

Figure 16.3. Measured section of the upper Yazoo and lower Red Bluff Clay at Little Stave Creek, Clarke County, Alabama showing the location of important biohorizons.

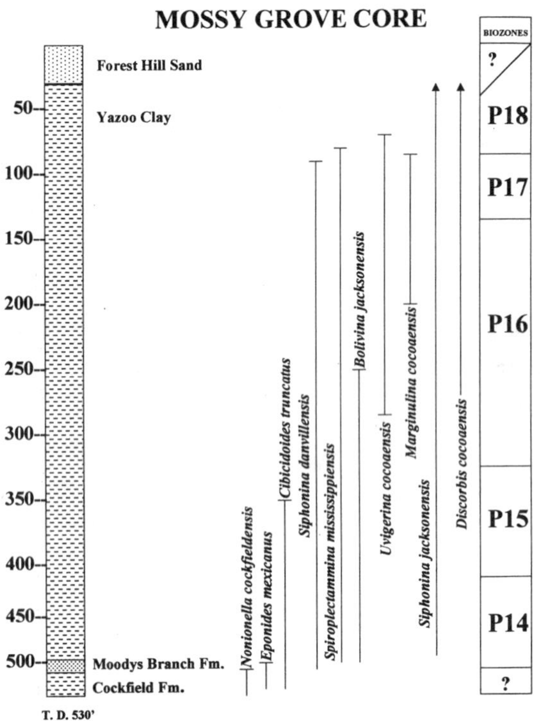

Figure 16.4. Distribution of key benthic foraminifera in the Mossy Grove core, Hinds County, Mississippi. Biostratigraphic zonation from Fluegeman (1996).

Formation at the base of the core. Planktonic foraminifera present enabled Fluegeman (1996) to identify the biozones of Berggren et al. (1995) in the core. He also identified benthic foraminiferal assemblages and was able to locate lowest and highest occurrences of key benthic foraminiferal species from the upper Eocene (figure 16.4). These biohorizons were related to planktonic foraminiferal biozones, in order to enhance stratigraphic control.

Of particular interest is Fluegeman's (1996) interpretation of the Eocene/Oligocene boundary. Using planktonic foraminifera, the boundary was identified at the 92-foot mark within the core, some 16.6 meters (54 feet) below the Jackson-Vicksburg contact and well within the Yazoo Clay. The Yazoo Clay above this boundary contained abundant benthic foraminifera, including *Uvigerina cocoaensis, Siphonina jacksonensis,* and *Discorbis cocoaensis* (figure 16.4). These species are characteristic of the Jackson fauna in the Gulf Coastal Plain. Even *Marginulina cocoaensis* occurs above the Eocene/Oligocene boundary in the core. No Vicksburg species have been identified in any of the samples.

The results from the Mossy Grove core indicate that the Jackson-Vicksburg benthic foraminiferal faunal turnover is not related to an event of the late Eocene or an Eocene/Oligocene boundary event. Rather, this turnover is likely a response to an early Oligocene event, possibly the Oi1 glacial episode of Miller et al. (1991).

EOCENE-OLIGOCENE BENTHIC FORAMINIFERA FROM OTHER REGIONS

The Caribbean Islands

Studies of Eocene and Oligocene foraminifera from the Caribbean Islands have focused on larger rather than smaller benthic foraminifera. Smaller benthic foraminifera of late Eocene and early Oligocene have been described from several islands, but few have been described from continuous sections.

Late Eocene foraminifera have been described from isolated localities in Cuba. Parker and Bermudez (1937) described species of *Bulimina* from Cuba. Bronnimann and Rigassi (1963) described in detail the stratigraphic successsion of Ciudad de La Habana, listing predominantly planktonic and larger benthic foraminifera. Of special interest to this study is the list of foraminifera reported from the San Luis Formation of eastern Cuba (former Oriente province) by Lewis and Straczek (1955). The foraminifera identified have little in common with the Eocene-Oligocene benthic foraminifera of the Gulf Coastal Plain, with two exceptions: *Spiroplectammina mississippiensis* and *Marginulina cocoaensis.* At present, this unit is poorly constrained biostratigraphically so little more can be said about the occurrence of these species of the Jackson fauna. Thin-sections of the San Luis Formation provided by G. Kysar of George Washington University reveal the foraminiferal fauna to be predominantly planktonics. It seems likely that any neritic benthic foraminifera in the San Luis were transported downslope, but the formation is worthy of further study.

Oligocene rocks were described from various localtities throughout Cuba by Iturralde-Vinent (1972). He documented occurrences of both planktonic and benthic foraminifera from his studied sections. Of the benthic foraminifera listed, none occurs in the Gulf Coastal Plain.

There are no detailed studies of late Eocene to early Oligocene benthic foraminifera from the island of Hispaniola. In southern Puerto Rico, Todd and Low (1976) examined well cuttings and described Oligocene benthic foraminifera. Although these are younger than the Vicksburg Group in the Gulf

Coastal Plain, their fauna did contain specimens of *Eponides ellisorae* and *Spiroplectammina mississippiensis.* They described the Eocene/Oligocene boundary in Puerto Rico as unconformable. Montgomery et al. (1991) also recognized a significant hiatus at the Eocene/Oligocene boundary in the Toa Baja well in northern Puerto Rico.

Late Eocene or early Oligocene benthic foraminifera have not been described in detail from Jamaica. Robinson and Wright (1993) and Robinson (this volume) have described larger foraminifera from the Eocene-Oligocene transition in Jamaica. It is not clear how the benthic foraminiferal fauna of these shallow water carbonates will compare with the predominantly clastic sediments of the Gulf Coastal Plain. Of interest is the deeper water Bonny Gate Formation. This unit is a planktonic foraminiferal chalk and appears to span the Eocene-Oligocene transition (Robinson and Wright, 1993). A late Eocene sample of the Bonny Gate examined for this study consisted of predominntly planktonic foraminifera with a small percentage of benthics. The benthic foraminifera are indicative of deep water conditions and are not likely to have much in common with the Eocene-Oligocene benthic foraminiferal faunas of the Gulf Coastal Plain. If suitable sections of the Bonny Gate can be located however, Jamaica may be an excellent place to study the Eocene-Oligocene transition.

The most comprehensive study of benthic foraminifera across the Eocene-Oligocene transition in the Caribbean region is that of Wood et al. (1985). They studied benthic foraminifera from the Oceanic Formation in Barbados. This fauna was interpreted as representing a water depth of between 2,500 and 3,000 m. They did not identify a strong faunal turnover at the Eocene-Oligocne boundary. Rather, new benthic foraminiferal species appear in the lower Oligocene part of the Oceanic Formation. There is no taxonomic similarity between the benthic foraminifera of the Oceanic Formation identified by Wood et al. (1985) and the Gulf Coast Section. Beckmann (1953) in his study of Oceanic Formation foraminifera, did identify *Cibicidoides cookei* from the base of the Oligocene in Barbados. This species is found in the Red Bluff Clay of Alabama and Mississippi and in the Vicksburg of the subsurface throughout the Gulf Coastal Plain. It is not present in the Oligocene part of the Yazoo Clay in the Mossy Grove core.

Benthic foraminifera from the Eocene-Oligocene transition have been described from Trinidad by Bolli et al. (1994). They report an unconformity between the Oligocene Cipero Formation and the upper Eocene San Fernando Formation. They identify the benthic foraminiferal fauna as deep water but feel the fauna indicates a strong shallowing through the San Fernando Formation (Bolli et al., 1994:232). The overall pattern of faunal turnover is a gradual one, similar to that seen in Barbados (Wood et al., 1985). There is little taxonomic similarity with the benthic foraminifera of the Gulf Coastal Plain. Two species of benthic foraminifera that occur in the Gulf Coastal Plain are reported from the Eocene-Oligocene of Trinidad. *Bulimina jacksonensis* occurs in the Eocene Navet and San Fernando Formations while *Cibicidoides cookei* occurs at the base of the Oligocene Cipero Formation. This is an interesting occurrence as *Bulimina jacksonensis* is one of the species that does not continue into the Vicksburg Group in the Gulf Coastal Plain. It does continue above the Eocene/Oligocene boundary in the Yazoo Clay of the Mossy Grove core.

DISCUSSION

The pattern of benthic foraminiferal faunal turnover through the Eocene-Oligocene transition in the Gulf Coastal Plain is a gradual one. It is attractive to relate this pattern to the global cooling identified in the late Eocene and early Oligocene (Miller et al, 1991; Miller, 1992). A gradual turnover in deep-sea benthic foraminiferal faunas related to this cooling has been identified by Thomas (1992). Bathyal and abyssal benthic foraminiferal turnover patterns, however, are not always in phase with changes in the neritic benthic foraminiferal faunas. The abrupt turnover in deep-sea benthic foraminifera at the Paleocene /Eocene boundary (Tjalsma and Lohmann, 1983; Thomas, 1990; and many others) is an oceanwide event in response to the Late Paleocene Thermal Maximum (LPTM). Yet neritic benthic foraminifera of the Gulf Coastal Plain are unaffected by the LPTM (Fluegeman, 2000).

The benthic foraminiferal turnover pattern of the Gulf Coastal Plain could be related to regional processes within the Gulf of Mexico Basin. The overprint of such regional processes as fluctuating sea level, sedimentation rates, river discharge, and local tectonics can mask the effect of global cooling on the benthic foraminiferal fauna. If the pattern of benthic foraminiferal turnover seen in the Eocene-Oligocene transition of the Gulf Coastal Plain is related to the global cooling, a similar pattern should be present in neritic benthic foraminiferal faunas of other regions. The Eocene-Oligocene benthic foraminiferal faunas of the Caribbean provide little help toward resolving this. The faunas have little taxonomic similarity to those of the Gulf Coastal Plain and the geologic setting of the Caribbean Basin ensures that most, if not all, Eocene-Oligocene benthic foraminifera will be bathyal or abyssal faunas.

McGowran and Beecroft (1986a,b) and McGowran et al. (1992) studied the faunal succession of Eocene-Oligocene neritic benthic foraminifera from various locations in southern Australia. They examined faunal turnover as well as benthic foraminiferal assemblages and their relationship to sequence stratigraphy. The interval they studied ranged in age from planktonic foraminiferal biozones P14 through P18. This is roughly correlative to the Moodys Branch through Red Bluff interval of the Gulf Coastal Plain. The turnover pattern they recognized was a gradual one with incoming species (lowest occurrences) far outnumbering outgoing species (highest occurrences). This is similar to the pattern seen in the Jackson fauna of the Gulf Coastal Plain. A sharp faunal turnover was identified near the P17/P18 planktonic foraminiferal biozone boundary (Eocene/Oligocene boundary). This boundary in southern Australia is an unconformity at the base of the Chinaman Gully Formation. A similar if less abrupt faunal turnover occurs at the Jackson-Vicksburg contact throughout the Gulf Coastal Plain, although in the Mossy Grove core, there is no faunal turnover in benthic foraminifera at the P17/P18 boundary. McGowran et al. (1992) relate the pattern of faunal turnover in benthic foraminifera in southern Australia to oceanographic changes brought about by late Eocene global cooling.

There is a similarity in the turnover pattern in benthic foraminiferal faunas in southern Australia to those in the Gulf Coastal Plain. This strongly suggests that late Eocene global cooling is a factor in the benthic foraminiferal succession in the Gulf Coastal Plain. More work is needed on the benthic foraminiferal of both regions to verify that the faunal turnover patterns are related to global cooling rather than local processes. Detailed work is currently underway on the Jackson benthic foraminifera of the Mossy Grove core and on the Red Bluff benthic foraminifera of the Mobil-Mississippi cores. These studies will further clarify the pattern of benthic foraminiferal succession across the Eocene-Oligocene transition in the eastern Gulf Coastal Plain.

REFERENCES

Aubry, M.-P., 1992, Late Paleogene calcareous nannoplankton evolution: a tale of climatic deterioration, *in* Prothero, D. R., and Berggren, W. A., eds., Eocene-Oligocene climatic and biotic evolution: Princeton, New Jersey, Princeton University Press, p. 272-309.

Bandy, O. L., 1949, Eocene and Oligocene foraminifera from Little Stave Creek, Clarke County, Alabama: Bulletins of American Paleontology, v. 32, no. 131, 210 p.

Baum, G. R., and Vail, P. R., 1988, Sequence stratigraphic concepts applied to Paleogeneoutcrops, Gulf and Atlantic basins: SEPM Special Publication, v. 42, p. 309-327.

Beckmann, J. P., 1953, Die Foraminiferen der Oceanic Formation (Eocaen-Oligocaen) von Barbados, Kleinen Antillen: Eclogae Geologicae Helvetiae, v. 46, p. 301-412.

Berggren, W. A., and Aubert, J., 1975. Paleocene benthonic foraminiferal biostratigraphy, paleobiogeography, and paleoecology of the Atlantic-Tethyan region: the Midway-type fauna: Paleogeography, Paleoclimatology, Paleoecolecology, v. 18, p. 73-192.

Berggren, W. A., Kent, D. V., Swisher, C. C., III, and Aubry, M.-P., 1995, A revised Cenozoic geochronology and chronostratigraphy: SEPM Special Publication, v. 54, p. 129-212.

Bolli, H. M., Beckmann, J.-P., and Saunders, J. B., 1994, Benthic foraminiferal biostratigraphy of the south Caribbean region: Cambridge, Cambridge University Press, 408 p.

Bronniman, P., and Rigassi, P., 1963, Contribution to the Geology of the area of the city of La Habana and its surroundings: Eclogae Geologicae Helvetiae, v. 56, p. 193-480.

Cheetham, A. H., 1957, Eocene/Oligocene boundary, eastern Gulf Coast region: Gulf Coast Association of Geological Societies Transactions, v. 7, p. 89-97.

Coleman, J., and Galloway, W. E., 1990, Sequence stratigraphic analysis of the lower Oligocene Vicksburg Formation of Texas, *in* Armentrout, J. M., Coleman, J. M., Galloway, W. E., and Vail, P. R., eds., Sequence stratigraphy as an exploration tool—concepts and practices in the Gulf Coast: 11th Annual Research Conference, Gulf Coast Section, SEPM Foundation Publication, p. 99-112.

Cooke, C. W., 1923, The correlation of the Vicksburg Group: U.S. Geological Survey Professional Paper, v. 133, p. 1-10.

Cushman, J. A., 1922, The Byram calcareous marl of Mississippi and its foraminifera: U.S. Geological Survey Professional Paper, v. 129-E, p. 79-122.

Cushman, J. A., 1923, The foraminifera of the Vicksburg Group: U.S. Geological Survey Professional Paper, v. 133, p. 11-71.

Cushman, J. A., 1935, Upper Eocene foraminifera of the southeastern United States: U.S. Geological Survey Professional Paper, v. 188, 88 p.

Cushman, J. A., and Garret, J. B., 1938, Three new rotaliform Foraminifera from the lower Oligocene and upper Eocene of Alabama: Contributions from the Cushman Laboratory of Foraminiferal Research, v. 14, pt. 3, p. 62-66.

DeBoo, P. B., 1965, Biostratigraphic correlation of the type Shubuta Member of the Yazoo Clay and the Red Bluff Clay with their equivalents in southwestern Alabama: Geological Survey of Alabama Bulletin, v. 80, 84 p.

Dockery, D. T., III, 1980, The invertebrate macropaleontology of the Clarke County, Mississippi, area: Mississippi Department of Natural Resources Bureau of Geology Bulletin, v. 122, 387 p.

Dockery, D. T., III, 1990, The Eocene/Oligocene boundary in the northern Gulf—a sequence boundary, *in* Armentrout, J. M., Coleman, J. M., Galloway, W. E., and Vail, P. R., eds., Sequence stratigraphy as an exploration tool—concepts and practices in the Gulf Coast: 11th Annual Research Conference, Gulf Coast Section, SEPM Foundation Publication, p. 141-150.

Dockery, D. T., III, Stover, C. W., Weathersby, P., Stover, C. W., Jr., and Ingram, S. L., 1991, A continuous core through the undifferentiated Yazoo Clay (late Eocene, Jackson Group) of central Mississippi: Mississippi Geology, v. 12, p. 21-27.

Ellisor, A. C., 1933, Jackson Group of formations in Texas with notes on Frio and Vicksburg: American Association of Petroleum Geologists Bulletin, v. 17, p.1293-1350.

Fluegeman, R. H., 1996, Preliminary paleontological report on the foraminifera of the Mossy Grove core, Hinds County, Mississippi: Mississippi Geology, v. 17, p. 9-15.

Fluegeman, R. H., 2000, Faunal succession of late Paleocene-middle Eocene neritic benthic foraminifera of the Gulf Coastal Plain of the United States: Geolgiska Foereningens i Stockholm Foerhandlinger (GFF), v. 122, p. 48-49.

Fluegeman, R. H., Berggren, W. A., and Briskin, M., 1990, Paleocene benthonic foraminiferal biostratigraphy of the eastern Gulf Coastal Plain: Micropaleontology, v. 36, p. 56-64.

Fluegeman, R. H., Denney, J. E., Finley, S. J., Hall, R. E., Rose, A. R., Voisard, V. L., Diaz, J. R., and Schwegman, M. E., 1994, Eocene-Oligocene benthic foraminiferal paleoecology and sea-level change records from the Mossy Grove core, Hinds County, Mississippi: Proceedings of the 7th International Symposium on the Observation of the Continental Crust Through Drilling, p. 189-192.

Gardner, J., 1957, Little Stave Creek—a paleoecologic study, *in* Ladd, H. S., Treatise on marine ecology and paleoecology: Geological Society of America Memoir 67, v. 2, p. 573-588.

Gaskell, B. A., 1991, Extinction patterns in Paleogene benthic foraminiferal faunas: relationship to climate and sea level: Palaios, v. 6, p. 2-16.

Howe, H. V., and Wallace, W. E., 1933, Foraminifera of the Jackson Eocene at Danville Landing on the Ouachita, Catahoula Parish, Louisisana: Louisiana Department of Conservation Bulletin, v. 2, p. 127-232.

Iturralde-Vinent, M. A., 1972, Principal characteristics of Oligocene and lower Miocene stratigraphy of Cuba: American Association of Petroleum Geologists Bulletin, v. 56, p. 2369-2379.

Keller, G., MacLeod, N., and Barrera, E., 1992, Eocene-Oligocene faunal turnover in planktic foraminifera, and Antarctic glaciation, *in* Prothero, D. R., and Berggren, W. A., eds., Eocene-Oligocene climatic and biotic evolution: Princeton, New Jersey, Princeton University Press, p. 218-244.

Lewis, G. E., and Straczek, J. A., 1955, Geology of south-central Oriente, Cuba: U.S. Geological Survey Bulletin 975-D, 336 p.

Mancini, E. A., 1979, Eocene/Oligocene boundary in southwestern Alabama: Gulf Coast Association of Geological Societies Transactions, v. 29, p. 282-289.

Mancini, E. A., and Copeland, C. W., 1986, St. Stephens Quarry (Lone Star Cement Company Quarry), St. Stephens, Washington County, Alabama, where a near complete Oligocene section, including the Eocene/Oligocene boundary, is exposed, *in* Neathery, T. L., ed., Centennial

Field Guide, v. 6: Southeastern Section of the Geological Society of America, p. 373-378.

Mancini, E. A., and Tew, B. H., 1991, Relationships of Paleogene stage and planktonic foraminiferal zone boundaries to lithostratigraphic and allostratigraphic contacts in the eastern Gulf Coastal Plain: Journal of Foraminiferal Research, v. 21, p. 48-66.

Mancini, E. A., and Waters, L. A., 1986, Planktonic foraminiferal biostratigraphy of upper Eocene and lower Oligocene strata in southern Mississippi and southwestern and south-central Alabama: Journal of Foraminiferal Research, v. 16, p. 24-33.

Mancini, E. A., Tew, B. H., and Waters, L. A., 1987, Eocene/Oligocene boundary in southeastern Mississippi and southwestern Alabama: A stratigraphically condensed section of a type 2 depositional sequence, *in* Ross, C. A., and Haman, D., eds., Timing and depositional history of eustatic sequences: constraints on seismic stratigraphy: Cushman Foundation for Foraminiferal Research Special Publication 24, p. 41-50.

McGowran, B., and Beecroft, A., 1986a, Foraminiferal biofacies in a silica-rich neritic sediment, late Eocene, South Australia: Palaeogeog-raphy, Palaeoclimatology, Palaeoecology, v. 52, p. 321-345.

McGowran, B., and Beecroft, A., 1986b, Neritic southern extratropical foraminifera and the Terminal Eocene Event: Palseogeography, Palseoclimatology, Palseoecology, v. 52, p. 321-345.

McGowran, B., Moss, G., and Beecroft, A., 1992, Late Eocene and early Oligocene in southern Australia: local neritic signals of global oceanic changes, *in* Prothero, D. R., and Berggren, W. A., eds., Eocene-Oligocene climatic and biotic evolution: Princeton, New Jersey, Princeton University Press, p. 178-201.

Miller, K. G., 1992, Middle Eocene to Oligocene stable isotopes, climate, and deep-water history: the terminal Eocene event? *in* Prothero, D. R., and Berggren, W. A., eds., Eocene-Oligocene climatic and biotic evolution: Princeton, New Jersey, Princeton University Press, p. 160-177.

Miller, K. G., Wright, J. D., and Fairbanks, R. G., 1991, Unlocking the ice house: Oligocene-Miocene oxygen isotopes, eustasy, and margin erosion, *in* Cloetingh, S., ed., Long-term sea level changes: Journal of Geophysical Research, B, p. 6829-6848.

Montgomery, H., Robinson, E., Saunders, J., and Van Den Bold, W., 1991, Paleontology of the Toa Baja well, Puerto Rico: Geophysical Research Letters, v. 18, p. 509-512.

Pasley, M. A., and Hazel, J. E., 1995, Revised sequence stratigraphic interpretation of the Eocene/Oligocene boundary interval, Mississippi and Alabama, Gulf Coast Basin, U.S.A.: Journal of Sedimentary Research B, v. 65, p. 160-169.

Parker, F. L., and Bermudez, P. J., 1937, Eocene species of the genera *Bulimina* and *Buliminella* from Cuba: Journal of Paleontology, v. 11, p. 513-516.

Robinson, E., and Wright, R. M., 1993, Jamaican Paleogene larger foraminifera, *in* Wright, R. M., and Robinson, E., eds., Biostratigraphy of Jamaica: Geological Society of America Memoir, v. 182, p. 283-345.

Root, S. A., Fearn, L. B., Rodgers, B. K., Armentrout, J. A., and Dockery, D. T., III, 1996, Biofacies expression of depositional sequences in a late Eocene–early Oligocene subtropical siliciclastic setting: a preliminary model, *in* Repetski, J. E., ed., Sixth North American Paleontological Convention Abstracts of Papers, p. 327.

Siesser, W. G., 1983, Paleogene calcareous nannoplankton biostratigraphy: Mississippi, Alabama, and Tennessee: Mississippi Department of Natural Resources Bureau of Geology Bulletin, v. 125, 61 p.

Thomas E., 1990, Late Cretaceous-early Eocene mass extinctions in the deep sea, *in* Sharpton, V. L., and Ward, P. D., eds., Global catastrophes in earth history: Geological Society of America Special Paper, v. 247, p. 481-495.

Thomas, E., 1992, Middle Eocene–late Oligocene bathyal benthic foraminifera (Weddell Sea): faunal changes and implications for ocean circulation, *in* Prothero, D. R., and Berggren, W. A., eds., Eocene-Oligocene climatic and biotic evolution: Princeton, New Jersey, Princeton University Press, p. 245-271.

Tipsword, H. L., 1962, Tertiary foraminifera in Gulf Coast petroleum exploration and development, *in* Rainwater, E. H., and Zingula, R. P., Geology of the Gulf Coast and central Texas: Gulf Coast Association of Geological Societies Guidebook, p. 16-57.

Tjalsma, R. C., and Lohmann, G. P., 1983. Paleocene-Eocene bathyal and abyssal benthic foraminifera from the Atlantic Ocean: Micropaleontology Special Publication, no. 4, 90 p.

Todd, R., and Low, D., 1976, Smaller foraminifera

from deep wells on Puerto Rico and St. Croix: U.S. Geological Survey Professional Paper, v. 863, 32 p.

Wood, K. C., Miller, K. G., and Lohmann, G. P., 1985, Middle Eocene to Oligocene benthic foraminifera from the Oceanic Formation, Barbados: Micropaleontology, v. 31, p. 181-197.

CHAPTER 17

Upper Paleogene Larger Foraminiferal Succession on a Tropical Carbonate Bank, Nicaragua Rise, Caribbean Region

Edward Robinson

ABSTRACT

The stratigraphic distribution of the larger foraminifera of the middle Eocene to lower Oligocene platform carbonates of the Nicaragua Rise suggest a stepwise extinction pattern. Major extinctions occur at the end of the middle Eocene and at or near the Eocene/Oligocene boundary. Less widespread extinctions occur within the upper Eocene. Whether the latter comprise more than one extinction event has not been ascertained, and the extreme rarity of planktonic species has, until now, inhibited a precise determination of the various extinction events (some of which may be environmentally controlled). Three strontium isotope determinations on assemblages near the Eocene/Oligocene boundary suggest that the larger foraminiferal genus *Fallotella*, a traditional Eocene marker, may have a range extending into the lowest part of the Oligocene.

INTRODUCTION

The Paleogene Period saw the development of a number of carbonate bank and platform systems in the Gulf of Mexico and Caribbean regions (figure 17.1). Some, such as the Bahamas-Florida platform and the Yucatan platform, had been in existence since the Mesozoic. Others, such as the Nicaragua Rise, developed over the eroded remains of a Mesozoic island arc and disrupted terrane from Central America (Eva and McFarlane, 1985). On the Caribbean plate itself, the two most extensive Paleogene carbonate buildups are those forming the Aves Ridge in the east and northeast, and the Nicaragua Rise in the northwest. The Aves Ridge is completely submerged except for the small shoal of Aves Island, and little is known about its litho- and biostratigraphy, except for exposure of Eocene rocks at its northern extremity, in the islands of St. Bartholomew and St Martin, from two wells drilled on Saba Bank, and from dredge samples (Droxler et al., 1998). The Nicaragua Rise is similarly a buried sequence but is rather better known from exposures in Jamaica (Robinson, 1993).

This paper examines the stratigraphic distribution of some larger foraminifera in the middle Eocene to Oligocene limestones of the Nicaragua Rise. The larger foraminifera are an artificial grouping of foraminiferidan protozoans that secrete calcareous or agglutinated tests, and range in size from less than a millimeter to as much as 60 mm across. The term "larger" is not so much a description of the size, as an expression of the fact that they usually have complex internal skeletal structures, requiring splitting, or thin-sectioning, for analysis and identification. Many unrelated families and genera are grouped under the umbrella term "larger foraminifera." Most extant and extinct groups are thought to have contained algal symbionts in life, and to have flourished in tropical to subtropical marine environments, at depths usually much less than 100 m (Ross, 1974; Hallock, 1984). These K-strategist foraminifera survive and diversify under tropical, oligotrophic conditions, but are adversely affected, even to extinction, when nutrient fluxes increase (Hallock, 1987; Hallock et al., 1991; Hottinger, 1997).

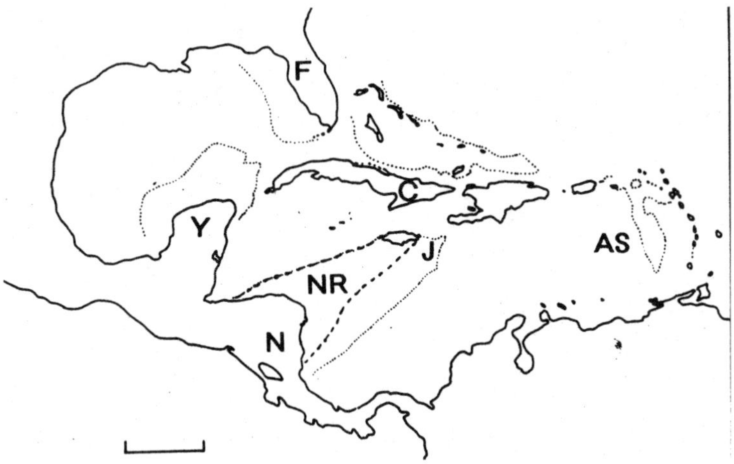

Figure 17.1. The Caribbean region, showing the location of the Nicaragua Rise (NR). AS, Aves Swell; C, Cuba; F, Florida; J, Jamaica; N, Nicaragua; Y, Yucatan. Bar scale is 500 km.

THE NICARAGUA RISE

Surface outcrops of the older Tertiary rocks of the Nicaragua Rise are limited to the island of Jamaica, where Pliocene-Quaternary uplift and erosion has exposed a succession through the Cenozoic carbonates down into the underlying Mesozoic island arc limestones and volcanics. Drilling by petroleum companies on the shallow parts of the Rise, particularly offshore Nicaragua and Honduras, has revealed an early Tertiary sequence essentially similar to that seen in Jamaica (Caceres Avila et al., 1984; Mills and Hugh, 1974; Robinson, 1993). In Jamaica, the upper middle Eocene to Oligocene section is represented by the lower part of the White Limestone Group (Hose and Versey, 1956; Eva and McFarlane, 1985; Robinson and Mitchell, 1999).

Two plate tectonic reconstructions (figure 17.2), modified from Pindell (1994), show the probable paleogeographic positions of the Nicaragua Rise in the middle Eocene and the Oligocene. At this time, the Rise formed one of the more southerly of an extensive complex of dominantly carbonate banks and platforms in the northern part of the Caribbean. Pindell did not provide paleolatitudes for his reconstructions. Although reconstructions (e.g., by Meschede and Frisch, 1998; Mann, 1999) have suggested that the Nicaragua Rise and the eastern Gulf Coast of North America occupied about the same latitudinal positions in the Eocene as they do today, Acton et al. (2000) indicate that the Paleogene position of the Rise may have been some five degrees south of its present position. At least in the earliest part of the middle Eocene, the Rise formed a partly subaerial extension of the Chortis Block, with Jamaica being connected to the North American landmass (Domning et al., 1997).

LARGER FORAMINIFERAL BIOSTRATIGRAPHY

The larger foraminifera considered here comprise species that lived within the back-reef and platform areas of carbonate buildups, in water normally much less than 40 m deep (Assemblage II species of Robinson, 1993). They are distinguished from the more cosmopolitan groups (Assemblage I of Robinson, 1993), including some 40 to 50 recorded species of nummulitids, lepidocyclinids, and orthophragmines (discocyclinids of some authors), that lived mainly at the shelf edge and on the upper slope of the tropical Western Hemisphere seafloor (Eva, 1976).

The distribution chart (figure 17.3) was compiled from observations of well records from the western end of the Nicaragua Rise, including two coreholes through the middle Eocene, and a corehole, numerous surface exposures, and other exploration wells on and offshore Jamaica. Control from calcareous nannofossils is sparse, as might be expected in very shallow carbonate environments. In addition, much of the Paleogene carbonate succession is well lithified, precluding determination of the nannoflora. Planktonic foraminifera and calcareous nannofossils provide indirect control through their co-occurrence with species of the lepidocyclinids in the outer parts of the platform environments and through interbedding, off the platform, of turbidites with hemipelagic chalks and micrites. The lepidocydinids are a probably polyphyletic larger foraminiferal group whose evolution through the Eocene provides key ties with

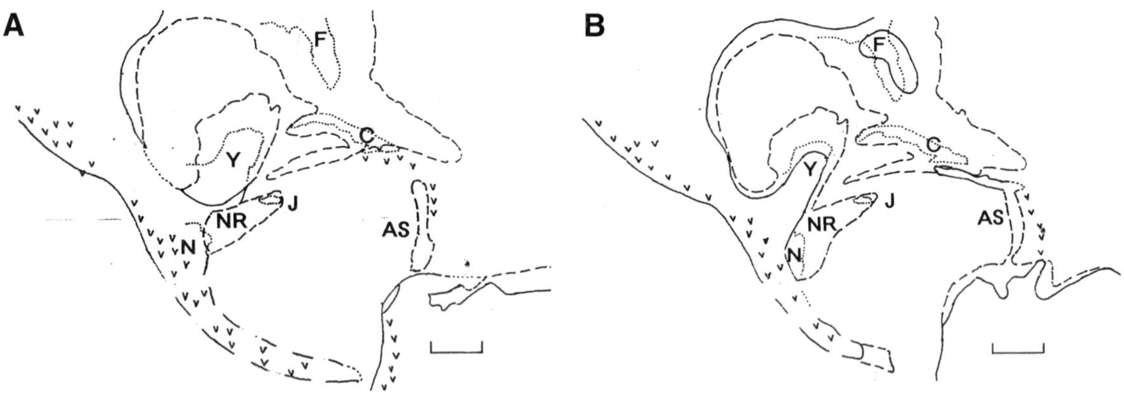

Figure 17.2. (A) Paleogeographic reconstruction for the middle Eocene. (B) Paleogeographic reconstruction for the early Oligocene, based on maps from Pindell (1994). Locality letters as for figure 17.1. Solid line outlines probable land areas, dashed line outlines shallow marine platforms and shelves; dotted lines show positions, in the reconstructions, of Florida, Cuba, Jamaica, Nicaragua and Yucatan. V symbols indicate regions of volcanic activity, some of which probably consisted of volcanic islands rather than continuous land areas. Bar scale is 500 km.

the platform faunas (Robinson, 1996; Robinson and Jiang, 1995; Robinson and Mitchell, 1999).

The reported overall diversity of the Assemblage II larger foraminiferal populations of the Eocene of the Rise is greater than that of the more northerly Bahamas and Florida platforms, with a number of distinctive species being endemic to the Rise or the immediately surrounding area (figure 17.3). Also, none of the 21 Assemblage II species listed on the distribution chart have been reported from South America, where carbonate platform construction was limited (Robinson, 1993). As figure 17.2 suggests, marine connections between the Caribbean Sea and the Florida–Gulf of Mexico region were probably more restricted than they are today. The Caribbean current and Loop current of the Gulf of Mexico did not exist in the middle Cenozoic, while there was relatively free east-west flow through the Central American seaway (Droxler et al., 1998). These oceanographic conditions may have inhibited lower to higher latitude faunal dispersal from the tropical Nicaragua Rise to the marginally tropical region of the Gulf Coast and Florida, and prevented dispersal toward South America. Eocene larger foraminiferal faunas recorded from the West Coast of the United States are closely related to Assemblage I species of the Caribbean region (Blondeau and Brabb, 1983), suggesting free connections at that time and supporting the concept of an open seaway through what is now the Panama Isthmus.

EOCENE/OLIGOCENE BOUNDARY

Some major Nicaragua Rise events shown on figure 17.3 include widespread extinction at the top of the Bartonian (sensu Berggren et al., 1995 and Serra-Kiel et al., 1998) and possible extinctions within the Priabonian (e.g. the imperforate conical genus *Cushmania*). Further extensive extinction of shallow water species occurred near the end of the Eocene.

In Jamaica, the Eocene imperforate conical genus *Fallotella* (*Dictyoconus* in part, of some authors) has been observed to extend consistently into stratigraphic horizons above the main end–Eocene extinction level of Assemblage II larger foraminifera. However, the genus does not appear to overlap upward with the characteristic Oligocene species *Praerhapydionina delicata* Henson (Robinson, 1995). Thus the *Fallotella* interval could be either uppermost Eocene or lowest Oligocene.

Lower Oligocene occurrences of *Fallotella cookei* and *F. floridana* have been reported from Cuba (Beckmann, 1958), Florida (Applin and Jordan, 1945) and Jamaica (Robinson and Wright, 1993). In Europe, the range of *Praerhapydionina delicata* is reported to extend throughout, and to be limited to the Oligocene, although Eocene occurrences have been recorded outside Europe (Cahuzac and Poignant, 1997). If this range also proves to hold true for the Caribbean, then it might be logical to retain the interval with *Fallotella*, but without *Praerhapydionina delicata* in the Eocene. As yet, Eocene/Oligocene

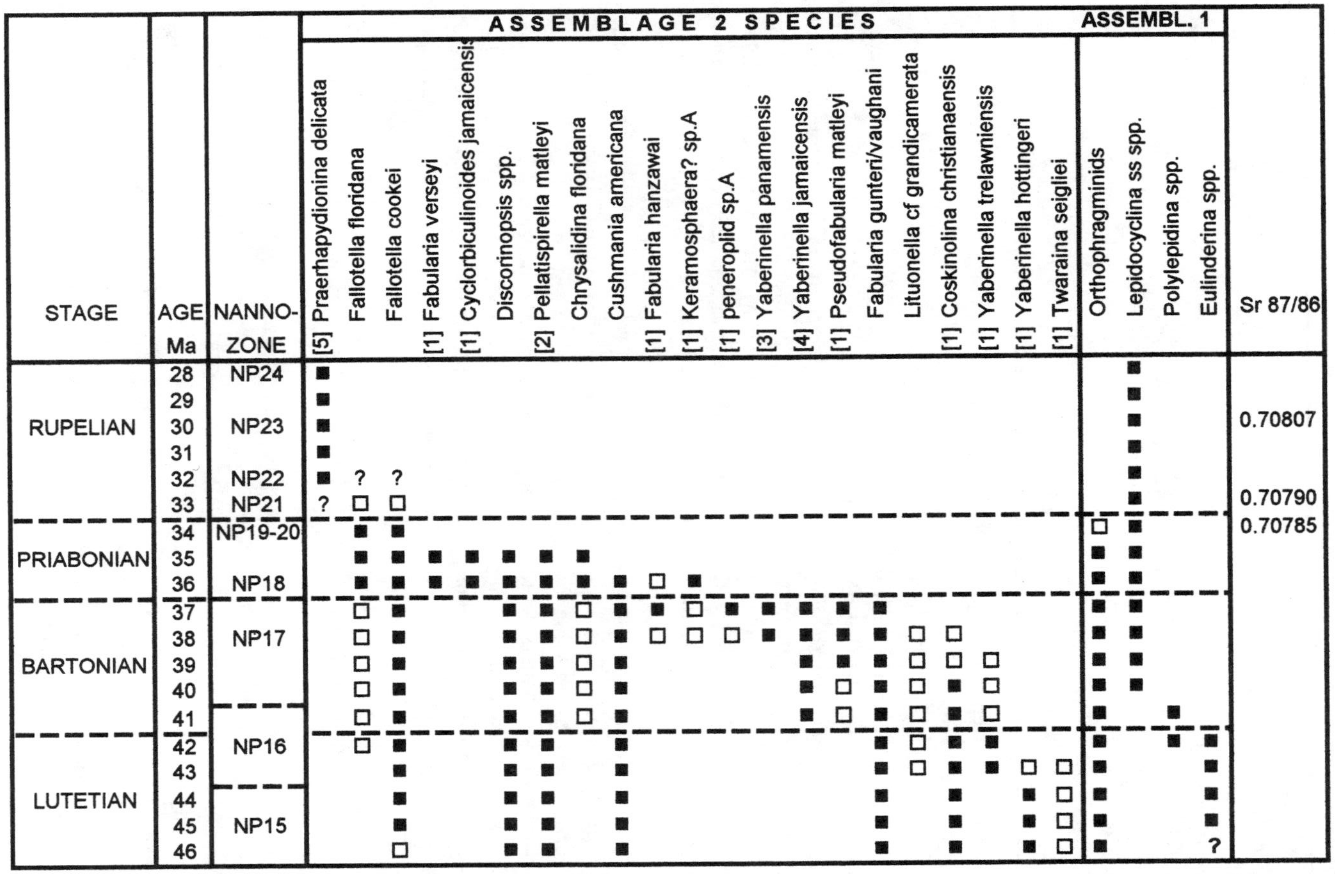

Figure 17.3. Stratigraphic distribution of larger foraminiferal species from Paleogene platform carbonates of the Nicaragua Rise (Assemblage II species of Robinson, 1993). Solid squares, known occurrences; open squares, occurrences with poor stratigraphic control; question marks, possible occurrences. [1] endemic to the Rise, [2] also reported from Cuba, southern Mexico and Hispaniola, [3] also reported from Panama, [4] also reported from Panama, Costa Rica and Oman, [5] widely reported from the Tethyan-Indo Pacific region but not from the Western Hemisphere, except Jamaica. The other species in the figure occur over most of the northern Caribbean and Gulf of Mexico. Selected Assemblage I genera (Robinson, 1993) are included for comparison. Time scale (Ma) after Berggren et al. (1995). Calcareous nannofossil zonation after Martini (1971). $^{87/86}$Sr stratigraphic positions are approximate.

Table 17.1. Larger Foraminifera in Samples Analyzed for their $^{87}Sr/^{86}Sr$ Isotopic Ratios

Sample	Assemblage	87/86 Sr ratio	Indicated age
*ER2585	*Peneroplis* sp. "*Archaias*" sp. *Neorotalia mexicana*	0.70807	Oligocene
ER2566	*Fallotella cookei* *Rotalia* sp.	0.70790	Lower Oligocene
ER2572	*Fabularia verseyi* *Fallotella cookei* *Chrysalidina floridana*	0.70785	Upper Eocene

*Collected above the lowest occurrence of *Praerhapydionina delicata*. See Robinson and Mitchell, 1999, figures 7 and 8 for details.

boundary sections have not been adequately studied in the Jamaican platform carbonates by detailed section measuring and sampling. Furthermore, the only records of *P. delicata* in the Western Hemisphere to date are from Jamaica (McFarlane, 1977; Robinson and Wright, 1993), so that comparative studies with other Caribbean sections of similar facies are of limited value.

Among Assemblage I genera (figure 17.3), the extinction level of the orthophragmines has been used, among the larger foraminifers, to mark the top of the Eocene in Europe and elsewhere (Pomerol and Premoli-Silva, 1986). However, in the type area of the Priabonian stage, the highest levels with orthophragmines (*Asterocyclina* beds), traditionally assigned to the Upper Eocene (Barbin and Bignot, 1986), contain palynomorphs assigned to the Gse dinoflagellate cyst zone. This zone falls within the basal Oligocene interval of the Massignano section, containing the global Eocene/Oligocene boundary point (GSSP; Brinkhuis, 1994; Brinkhuis and Visscher, 1995; Mietto, 2000). Thus, controversy persists over the placement of the Eocene/Oligocene boundary on the basis of larger foraminiferal assemblages.

The orthophragmines and even nummulites are absent, for environmental reasons, from the shallow carbonate platform sequences of central Jamaica. However, three preliminary strontium isotope ratio determinations from that part of the Jamaican platform section near the Eocene/Oligocene boundary were carried out on shell material taken from well-lithified limestone (P. Mueller, pers. commun. August 5, 1997) collected from field sections containing *Fallotella* (Robinson, 1995; Robinson and Mitchell, 1999). These are inserted on figure 17.3 and the larger foraminifers associated with the samples are given in table 17.1. The isotope ratio values are positioned approximately in figure 17.3, based on comparison with published biostratigraphic and interpreted magnetostratigraphic positions of Sr isotope values (Miller et al., 1991, 1993; McArthur, 1994; De Paolo and Ingram, 1985). Although the integrated chronostratigraphic scale of Hazel (1989) contains Sr values that correspond to a late Eocene age for that sample from the *Fallotella* interval of figure 17.3 ($^{87}Sr/^{86}Sr$ ratio of 0.70790), the biostratigraphic data of Miller et al. (1991) do in fact suggest an early Oligocene age for this horizon (e,g,, Miller et al., 1993). The lowest $^{87}Sr/^{86}Sr$ value (0.70785 of figure 17.3) is from a sample containing the *Fabularia verseyi* larger foraminiferal assemblage, whose biostratigraphic age has always been regarded as Eocene (Robinson and Mitchell, 1999; but see Chart 2 in De Graciansky et al., 1998, in which similar values range up into basal Oligocene). It is likely that all three Sr ratio values have been affected by diagenesis.

CONCLUSIONS

The extinctions of larger foraminifera inhabiting the carbonate platform of the Nicaragua Rise in the late Eocene appear to follow a stepwise pattern, with major events occurring at about the Bartonian/Priabonian boundary, and near the Eocene/Oligocene boundary. At least one lesser event occurred during the upper Eocene, but whether this, or the others, are the same events as those summarized by Prothero (1989) remain to be verified. The problems of correlating platform carbonate larger foraminiferal assemblages with the standard planktonic biostratigraphic zones require more work, emphasizing the integration of the sequence stratigraphy of platform assemblages with magnetic reversal and Sr isotope stratigraphy.

Documentation of Jamaican sections near the Eocene/Oligocene boundary is continuing.

ACKNOWLEDGMENTS

Paul Mueller kindly ran Sr isotope ratio determinations on the three samples listed in figure 17.3. Mark Jiang has provided published and unpublished calcareous nannofossil data used in constructing figure 17.3. A separate paper, documenting these data, is in preparation. The author thanks Jan Bartholdy and Jon Bryan for constructive reviews of the original manuscript.

REFERENCES

Acton, G. D., Galbrun, B., and King, J. W., 2000, Paleolatitude of the Caribbean Plate since the Late Cretaceous, *in* Leckie, R.M., Sigurdsson, H., Acton, G. D., and Draper, G., eds., Proceedings of the Ocean Drilling Program, Scientific Results, v. 165, p. 149-173.

Adams, C. G., and Racey, A., 1992, The occurrence and palaeobiogeographical significance of the foraminiferid *Yaberinella* from the Eocene of Oman: Palaeontology, v. 35, p. 237-245.

Adams, C. G., Genry, A. W., and Whybrow, P. J., 1983, Dating the terminal Tethyan event, in Meulen-kamp, J. E., ed., Reconstructioin of marine paleoenvironments: Utrecht Micropaleontology Bulletins, v. 30, p. 273-298.

Applin, E. R., and Jordan, L., 1945, Diagnostic Foraminifera from subsurface formations in Florida: Journal of Paleontology, v. 19, p. 129-148.

Barbin,V., and Bignot, G., 1986. New proposal for an Eocene/Oligocene boundary according to microfacies from the Priabonian type-section, *in* Pomerol, C., and Premoli-Silva, I., eds., Terminal Eocene events: Developments in Paleontology and Stratigraphy, v. 9, p. 49-52.

Beckmann, J. P., 1958, Correlation of pelagic and reefal faunas from the Eocene and Paleocene of Cuba: Eclogae Geologicae Helvetiae, v. 51 (2), p. 416-422.

Berggren, W. A., Kent, D. V., Swisher, C. C., III, and Aubry, M.-P., 1995, A revised Cenozoic geochronology and chronostratigraphy: SEPM Special Publication, v. 54, p. 129-212.

Blondeau, A., and Brabb, E. E., 1983, Large foraminifers of Eocene age from the Coast Ranges of California, *in* Brabb, E. E., ed., Studies in Tertiary stratigraphy of the California Coast Ranges, U.S. Geological Survey Professional Paper, v. 1213, p. 41-48.

Brinkhuis, H., 1994, Late Eocene to Early Oligocene dinoflagellate cysts from the Priabonian type-area (Northeast Italy): biostratigraphy and paleoenvironmental interpretation: Paleogeography, Paleoclimatology, Paleoecology, v. 107, p. 121-163.

Brinkhuis, H., and Visscher, H., 1995, The upper boundary of the Eocene Series: a reappraisal based on dinoflagellate cyst biostratigraphy and sequence stratigraphy: SEPM Special Publication, v. 54, p. 295-304.

Caceres Avila, F., Tappmeyer, D. M., Aves, H. S., Gillett, M., and Klenk, C. D., 1984, Recent studies of basins are encouraging for future exploration in Honduras: Oil and Gas Journal, v. 17, p. 139-149.

Cahuzac, B., and Poignant, A., 1997, Essai de biozonation de l'Oligo-Miocène dans le bassins européens à l'aide des grands foraminifères néritiques: Bulletin de la Société géologique de France, v. 168, no. 2, p. 155-169.

Cole, W. S., 1941, Stratigraphic and paleontologic studies of wells in Florida: Florida Geological Survey Bulletin, v. 19, p. 1-53.

De Graciansky, P.-C., Hardenbol, J., Jacquin, Th., and Vail, P. R., eds., 1998, Mesozoic and Cenozoic sequence stratigraphy of European basins: SEPM Special Publication 60.

De Paolo, D. J., and Ingram, B. L., 1985, High-resolution stratigraphy with strontium isotopes: Science, v. 227, p. 938-941.

Domning, D. P., Emry, R. J., Portell, R. W., Donovan, S. K., and Schindler, K. S., 1997, Oldest West Indian land mammal: rhinocerotoid ungulate from the Eocene of Jamaica: Journal of Vertebrate Paleontology, v. 17, p. 638-641.

Droxler, A. W., Burke, K., Cunningham, A. D., Hine, A. C., Rosencrantz, E., Duncan, D. S., Hallock, P., and Robinson, E., 1998, Caribbean constraints on circulation between Atlantic and Pacific Oceans over the past 40 million years, *in* Crowley, T. J., and Burke, K., eds., Tectonic boundary conditions for climate reconstructions: Oxford, Oxford University Press, p. 169-191.

Eva, A. N., 1976, The paleoecology and sedimentology of middle Eocene larger foraminifera in Jamaica: Maritime Sediments, Special Publication, no. 1, p. 467-475.

Eva, A. N. and McFarlane, N. A., 1985, Tertiary to early Quaternary carbonate facies relationships in Jamaica: Transactions of the 4th Latin American Geological Conference, Port-of-Spain, Trinidad and Tobago, 1979, v. 1, p. 210-219.

Hallock, P., 1984, Distribution of selected species of living algal symbiont-bearing foraminifera on two Pacific coral reefs: Journal of Foraminiferal Research, v. 14, p. 250-261.

Hallock, P., 1987, Fluctuations in the trophic resource continuum: a factor in global diversity cycles?: Paleoceanography, v. 2, p. 457-471.

Hallock, P., Premoli Silva, I., and Boersma, A., 1991, Similarities between planktonic and larger foraminiferal evolutionary trends through Paleogene paleoceanographic changes: Palaeogeography, Palaeoclimatology, Palaeoecology, v. 83, p. 49-64.

Hazel, J. E., 1989, Chronostratigraphy of upper Eocene microspherules: Palaios, v. 4, p. 318-329.

Hose, H. R., and Versey, H. R., 1956, Palaeontological and lithological divisions of the Lower Tertiary limestones of Jamaica: Colonial Geology and Mineral Resources, v. 6, p. 19-39.

Hottinger, L., 1997, Shallow benthic foraminiferal assemblages as signals for depth of their deposition and their limitations: Bulletin de la Société géologique de France, v. 168, p. 491-505.

Mann, P., 1999, Caribbean sedimentary basins: classification and tectonic setting from Jurassic to Recent, *in* Mann, P., ed., Caribbean basins: Sedimentary basins of the world, v. 4, p. 3-31.

Martini, E., 1971. Standard Tertiary and Quaternary calcareous nannoplankton zonation, *in* Farinacci, A., ed., Proceedings II Planktonic Conference, Roma, 1970, v. 2, p. 739-785.

McArthur, J. M., 1994, Recent trends in strontium isotope stratigraphy: Terra Nova, v. 6 (4), p. 331-358.

McFarlane, N. A., 1977, Some Eocene and Oligocene faunas from central Jamaica: Memorías del Segundo Congreso Latinoamericano de Geología, v. 3, p. 1393-1411.

Meschede, M., and Frisch, W., 1998, A plate-tectonic model for the Mesozoic and early Cenozoic history of the Caribbean plate: Tectonophysics, v. 296, p. 269-291.

Mietto, P., 2000, The Priabonian in the type locality (Vicentinian Prealps, NE Italy), *in* Bassi, D., Cosovic, V., Less, G., Mietto, P., Papazzoni, C. A., Trevisani, E., and Ungaro, S., eds., Biostratigraphy and paleoecology at the middle-upper Eocene boundary: the Venetian area, *in* Field Trip Guide to Shallow Water Benthic Communities at the Middle-Upper Eocene Boundary: Fifth Meeting of the IGCP 393, IUGS-UNESCO, Ferrara and Vicenza, 18-31 July 2000, p. 64-73.

Miller, K. G., Berggren, W.A., Zhang, J., and Palmer-Julson, A. A., 1991, Biostratigraphy and isotope stratigraphy of upper Eocene microtektites at site 612: how many impacts?: Palaios, v. 6, p. 17-38.

Miller, K. G., Thompson, P. R., and Kent, D. V., 1993, Integrated late Eocene-Oligocene stratigraphy of the Alabama coastal plain: correlation of hiatuses and stratal surfaces to glacioeustatic lowerings: Paleoceanography, v. 8 (2), p. 313-331.

Mills, R. A., and Hugh, K. E., 1974, Reconnaissance geologic map of Mosquitia region, Honduras and Nicaragua Caribbean Coast: American Association of Petroleum Geologists Bulletin, v. 58, p. 189-207.

Pindell, J. L., 1994, Evolution of the Gulf of Mexico and the Caribbean, *in* Donovan, S. K., and Jackson, T. A., eds., Caribbean geology: an introduction: UWIPA, Kingston, Jamaica, p. 13-39.

Pomerol, C., and Premoli-Silva, I., eds., 1986, Terminal Eocene events: Developments in Paleontology and Stratigraphy, v. 9, p. 1-420.

Prothero, D. R., 1989, Stepwise extinctions and climatic decline during the later Eocene and Oligocene, *in* Donovan, S. K., ed., Mass extinction: processes and evidence: New York, Columbia University Press, p.217-234.

Robinson, E., 1974, Some larger foraminifera from the Eocene limestones at Red Gal Ring, Jamaica: Verhandlungen Naturforschende Geselschaft in Basel, v. 84, p. 281-292.

Robinson, E., 1993. Some imperforate larger foraminifera from the Paleogene of Jamaica and the Nicaragua Rise: Journal of Foraminiferal Research, v. 23, p. 47-65.

Robinson, E., 1995, Larger foraminiferal assemblages from Oligocene platform carbonates, Jamaica: Tethyan or Caribbean?: Marine Micropaleontology, v. 26, p. 313-318.

Robinson, E., 1996, Using larger foraminifers in high-resolution biostratigraphy: an example from the Eocene of the Gulf of Mexico and northern Caribbean: Palaios, v. 11, p. 220-229.

Robinson, E., and Jiang, M.-J., 1995, Evolution of the foraminiferal genus *Lepidocyclina* in the middle Eocene and its implication for Gulf Coast stratigraphy: Gulf Coast Association of Geological Societies Transactions, v. 45, p. 509-517.

Robinson, E., and Mitchell, S. F., 1999, Middle Eocene stratigraphy and palaeogeography in Jamaica: a window on the Nicaragua Rise: Contributions to Geology, University of the West Indies, Mona, no. 4, 47 p.

Robinson, E., and Wright, R. M., 1993, Jamaican Palaeogene larger Foraminifera, *in* Wright, R. M., and Robinson, E., eds., Biostratigraphy of Jamaica: Geological Society of America Memoir 182, p. 283-345.

Ross, C. A., 1974, Evolutionary and ecological significance of large, calcareous Foraminiferida (Protozoa), Great Barrier Reef: Second International Coral Reef Symposium, v. 1, p. 327-333.

Serra-Kiel, J., and 18 others, 1998, Larger foraminiferal biostratigraphy of the Tethyan Paleocene and Eocene: Bulletin de la Société géologique de France, v. 169, p. 281-299.

APPENDIX: NOTES ON FORAMINIFERA OF FIGURE 17.3

Note that species are listed in the order in which they appear on the distribution chart. A reference is given in each case to a more complete discussion of the species.

Assemblage II Species

Praerhapydionina delicata Henson
First recorded from Jamaica by McFarlane (1974), this species is locally common in the Oligocene limestone of central Jamaica (McFarlane, 1974; Robinson and Wright, 1993; Robinson, 1995). Western Hemisphere records are limited to Jamaica, but the species is widely distributed in the middle Tertiary of the Old World (Adams et al., 1983; Cahusac and Poignant, 1997).

Fallotella floridana (Cole), *Fallotella cookei* (Moberg)
Although these species have been retained separately on the distribution chart, they frequently occur together, and *F. floridana* may merely be a variant of *F. cookei*, with fewer marginal partitions (Robinson and Wright, 1993).

Fabularia verseyi Cole
This distinctive species has been recorded only from the Nicaragua Rise, where it is locally abundant in the upper Eocene (Robinson and Wright, 1993). It differs from typical *Fabularia* in the frequent possession of up to five chambers in the final whorl.

Cyclorbiculinoides jamaicensis Robinson
This is a distinctive upper Eocene complex peneroplid that occurs commonly in the *F. verseyi* assemblage, and is limited to the Nicaragua Rise (Robinson and Wright, 1993).

Discorinopsis spp.
This genus is widespread in the Caribbean, Central America and Florida. Most occurrences should probably be placed in *D. gunteri* Cole (1941).

Pellatispirella matleyi (Vaughan)
This species group of large elphidiids first appears near the base of the middle Eocene and is common to abundant on the Nicaragua Rise, sometimes occurring in rock building quantities (Robinson and Wright, 1993).

Chrysalidina floridana (Cole)
Most references in the literature are as *Pseudochrysalidina floridana* (e.g., Robinson and Wright, 1993), but there appears to be no basis for a generic distinction from *Chrysalidina* and *Pseudochrysalidina* is here regarded as a junior synonym of *Chrysalidina*. In Jamaica, the species occurs in the upper part of the Eocene, but its stratigraphic distribution is not known in detail.

Cushmania americana (Cushman)
A widespread species group in the northern Caribbean and Florida, *C. americana* has been regarded as a marker for the top of the Florida middle Eocene (Cole, several well reports), but its upper range there, and in Jamaica, may be environmentally controlled. On the Nicaragua Rise, its range extends up into assemblages regarded as upper Eocene (Robinson and Wright, 1993).

Fabularia hanzawai Robinson
Distinguished from the *F. gunteri*/*F. vaughani* group by a thickened basal layer, this species may be ancestral to *F. verseyi* (Robinson, 1993).

Keramosphaera? sp. A
A small but distinctive complex milioliform species, seen only in thin sections, it appears to be restricted to the upper Eocene and possibly, the highest part of the middle Eocene (Robinson, 1974)

Peneroplid sp. A
An undescribed peneroplid with simple, annular chambers has been noted at the top of the middle Eocene section from a well at the western end of the Nicaragua Rise, and from Jamaica (Eva, 1976).

Yaberinella panamensis Robinson
This species occurs at the top of the range of *Yaberinella* on the Nicaragua Rise. It differs from *Y. jamaicensis* in its abbreviated spire (Robinson, 1993).

Yaberinella jamaicensis Vaughan
The species is widespread and locally abundant as a rock builder on the Nicaragua Rise. Its unusual distribution (figure 17.3) requires further study (Adams and Racey, 1992).

Pseudofabularia matleyi (Vaughan)
This monospecific genus appears to be of stratigraphic significance, apparently being limited to the top of the middle Eocene (Robinson and Wright, 1993).

Fabularia gunteri Applin and Jordan / *F. vaughani* Cole
This is the most widespread species group of *Fabularia* in the tropical American Eocene, but is probably limited to the early to middle Eocene, at least on the Nicaragua Rise (Robinson and Wright, 1993). Most Nicaragua Rise examples resemble *F. gunteri*.

Lituonella cf *L. grandicamerata* Cole
Several species of *Lituonella* may be present in the tropical American Eocene, but have not been investigated in detail (Robinson and Wright, 1993).

Coskinolina christianaensis Robinson
Records from the Nicaragua Rise (Robinson, 1993) indicate this species to be a middle middle to upper middle Eocene form.

Yaberinella trelawniensis Vaughan
The specific characteristics of this form are incompletely known, but it appears to be intermediate in its morphology and in stratigraphic range, between *Y. jamaicensis* and *Y. hottingeri* (Robinson, 1993).

Yaberinella hottingeri Robinson
This earliest recorded species of *Yaberinella* has not been reported away from the Rise (Robinson, 1993).

Twaraina seigliei Robinson
This monospecific, complex peneropliform genus has been noted only in wells at the western end of the Nicaragua Rise (Robinson, 1993).

Assemblage I Species
The biostratigraphic position of *Eulinderina* and *Polylepidina* spp. has been discussed by Robinson and Jiang (1995) and Robinson (1996). Other species of *Lepidocyclina* sensu lato and the orthophragmines (discocyclinids) of the Western Hemisphere still require detailed biostratigraphic and taxonomic revision. Other Assemblage I species, not on figure 17.3, of considerable potential use in correlating strata near the Eocene/Oligocene boundary include the *Heterostegina ocalana* group (Robinson and Wright, 1993) and the nummulitids.

CHAPTER 18

Molluscan Faunas Across the Eocene/ Oligocene Boundary in the North American Gulf Coastal Plain, with Comparisons to Those of the Eocene and Oligocene of France

David T. Dockery III and Pierre Lozouet

ABSTRACT

Distribution data is given for 947 molluscan taxa of late Eocene (Moodys Branch Formation) to early Oligocene (Byram Formation) age in the North American Gulf Coastal Plain. Only 26 of these taxa range across the Eocene/Oligocene boundary; the apparent extinction rate in the latest Eocene Yazoo Formation is 88.6%. The succession fauna of the lower Oligocene Vicksburg Group contains Tethyan elements having related species in the Stampian of the Aquitaine Basin in southwestern France; several related species are noted. Eocene/Oligocene boundary extinctions are attributed to falling sea level and climatic cooling. New appearances of related early Oligocene species on both sides of the Atlantic Ocean are attributed to the transport of plantotrophic larvae by ocean currents.

INTRODUCTION

A nearly continuous record of late Eocene and early Oligocene marine deposition is present in the North American Gulf Coastal Plain. This section contains diverse molluscan faunas above and below the boundary as well as excellent planktonic nannofossil and foraminiferal assemblages and radiometrically dated bentonites near the boundary interval. Dockery (*in* MacNeil and Dockery, 1984) proposed the disconformable contact of the late Eocene Shubuta Clay Member of the Yazoo Formation with the overlying early Oligocene Red Bluff Formation as exposed along the Chickasawhay River in Wayne County, Mississippi, as a stratotype for the Eocene/Oligocene boundary, a site where molluscan-rich early Oligocene sediments overlie pelagic clays of the latest Eocene. A stratotype for this boundary was eventually selected in the Marche-Umbria Basin of Italy in a pelagic section with a good microfossil record, but without a significant benthic molluscan record. The radiometric age of the boundary stratotype was interpolated by Montanari et al. (1988) as 33.7 ± 0.5 Ma. This age matches that of sanidine ages from bentonites near the top of the Yazoo Formation in coreholes at Society Ridge (33.70 ± 0.19 Ma, Obradovich and Dockery, 1996) and Mossy Grove (33.7 Ma, Fluegeman, 1996), both in northwestern Hinds County, Mississippi.

The molluscan record of western Europe is uneven across the Eocene/Oligocene boundary interval. The late middle Eocene Bartonian Stage of the Paris Basin contains some 1,300 to 1,400 species of mollusks with 1,000 species of gastropods alone (Lozouet, 1997). However, the molluscan faunas of the southwestern European late Eocene type Priabonian section in Italy are not well preserved; these faunas are best preserved in Germany where Cavelier (1979) documented some 800 species of gastropods. Above the boundary, the early Oligocene Stampian section of France's Aquitaine Basin contains (counting only the gastropods and bivalves) 577 molluscan species (unpublished data and Lozouet, 1997, 1999), while the Stampian of the Paris Basin contains 289 molluscan species (unpublished data and Gitton et al., 1986). In comparison, the late Eocene (Priabonian) section within the Mississippi Embayment of the North American Gulf Coast contains 426 molluscan taxa in the Jackson Group, and the early Oligocene section contains 547 species in the Vicksburg Group. The appendix lists a total of

947 molluscan taxa, including species, subspecies, varieties, and unnamed species, recorded in the Jackson and Vicksburg groups of the north-central Gulf Coastal Plain, only 26 of which range across the Eocene/Oligocene boundary.

It is interesting to note that while mollusks are not an important component of the Eocene/Oligocene boundary stratotype in Italy, they were the means by which Charles Lyell (1830) subdivided Europe's Tertiary section into Eocene, Miocene, and Pliocene epochs. Lyell chose the name Eocene to represent the "dawn of the recent," an epoch following the extinction of many Cretaceous molluscan taxa (as well as the dinosaurs and other creatures) and with taxa resembling those of recent faunas, but with few living molluscan species. The Miocene was a name meaning "less recent [living molluscan species]," and the Pliocene was a name meaning "more recent [living molluscan species]." Thus, Lyell constructed a Tertiary timetable based on percentages of extant molluscan species. At that time there were no planktonic microfossil P zones and NP zones, and benthic mollusks proved their worth in regional correlations of geologic time.

Lyell traveled to America in 1841–1842 and in 1845–1846, in part to see if his Tertiary epochs applied to the American Coastal Plain section as well as to that of Europe. On March 19, 1846, Lyell arrived at Jackson, Mississippi, by rail and was taken to what would become the type Jackson Group locality, an exposure of the Moodys Branch Formation at Town Creek (only a ten-minute walk from the train station). Lyell was immediately able to identify the exposure as Eocene in age based on the fossil mollusks he collected. He also recognized the Jackson exposure as younger than that of the type Claiborne Group (Bartonian) section he had visited earlier in Alabama and as older than the type Vicksburg Group section he had visited the day before. Having only Eocene, Miocene, and Pliocene divisions, Lyell recognized the Vicksburg mollusks to be intermediate between the Eocene and Miocene faunas and placed them as upper Eocene. Conrad followed this placement when he published the Vicksburg mollusks in 1848. In 1854, Beyrich named the Oligocene Epoch, meaning "little recent," and Conrad recognized the Vicksburg mollusks to be of Oligocene age in his faunal lists of 1865 and 1866.

SEQUENCE STRATIGRAPHY

Distribution data for 2,816 Paleocene and Eocene molluscan taxa (including species, subspecies, and unnamed species) for the North American Gulf and Atlantic coastal plains show that 2,261, or 80%, of these occur in four adjoining states: Texas, Louisiana, Mississippi, and Alabama (Dockery, 1986). These states straddle a plunging syncline and ancient seaway known as the Mississippi Embayment. Distribution data for 547 early Oligocene mollusks from the Vicksburg Group show all these taxa to occur within (but not necessarily restricted to) Mississippi between the Chickasawhay River on the east and the Mississippi River on the west. Thus, the important marine record for fossil mollusks for both the late Eocene and early Oligocene is centered in the northern Gulf within the Mississippi Embayment

Palmer (1979) recognized the importance of the molluscan record in understanding the history of sea-level or "rhythm of the Paleocene-Eocene seas" in the central Gulf Coastal Plain. She recognized a correlation between regional stratigraphic stages (groups), molluscan faunas, and global sea-level cycles. Vail et al. (1977) and Vail and Hardenbol (1979) also depended heavily on the Gulf Coastal Plain record in developing their eustatic sea-level charts for the Tertiary Period. In these charts, the late Eocene Priabonian Stage was placed in cycle TE3, and the early Oligocene Rupelian Stage was placed in cycle TO1. Baum and Vail (1988) later broke the correlation between stratigraphic stages and sea-level cycles when they recognized three cycles in the Priabonian Stage of the Gulf and Atlantic coastal plains. The first, cycle TE3.1 included the Gosport Sand, and thus extended across the Claiborne/Jackson Group/Stage boundary. The third, cycle TE3.3, included the Red Bluff and Forest Hill formations, and thus extended across the Jackson/Vicksburg Group boundary. The placement of early Vicksburg sediments in cycle TE3.3 was strongly influenced by the Eocene/Oligocene boundary section at St. Stephens Quarry in Washington County, Alabama, a downdip section with deep-water clay and limestone facies. Here Baum and Vail (1988) placed a condensed section at the Jackson-Vicksburg Group contact. Mancini and Tew (1991) restricted cycle TE3.1 to include only the Gosport Sand and placed only lower Jackson Group sediments of the Moodys Branch Formation and the North Twistwood Creek Clay Member of the Yazoo Formation in cycle TE3.2. However, they followed Baum and Vail (1988) in extending cycle TE3.3 into the Vicksburg Group, placing within it the Cocoa Sand, Pachuta Marl, and Shubuta Clay members of the Yazoo Formation along with the Red Bluff and Forest Hill formations.

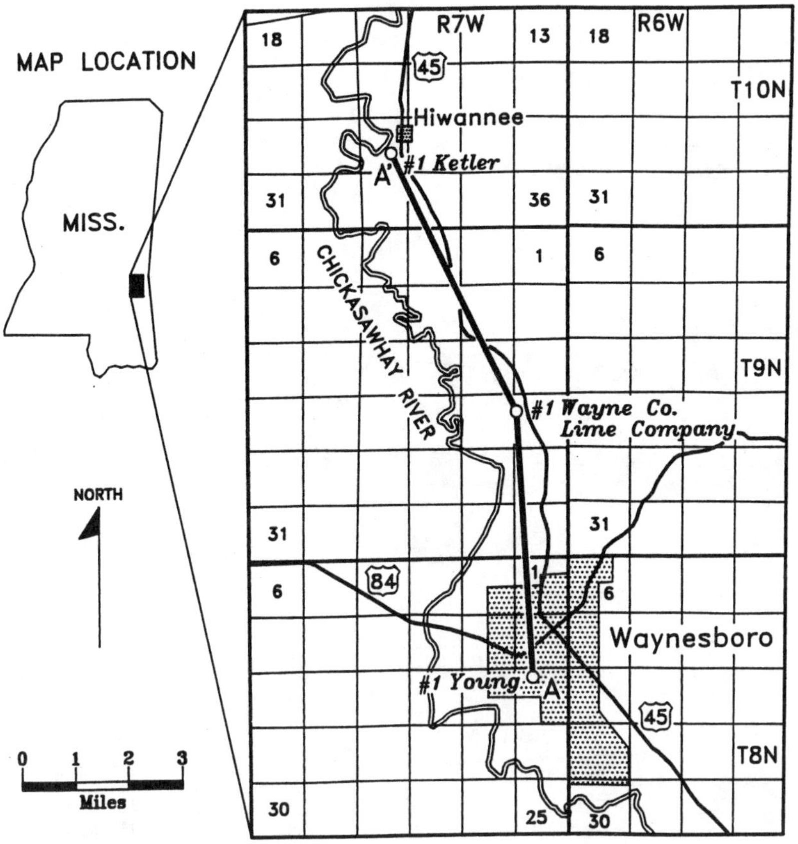

Figure 18.1. Locations of Mississippi Office of Geology-Mobil Exploration and Producing Technical Center coreholes along the Chickasawhay River in eastern Mississippi. From Dockery et al. (1994).

To address these problems and to study how systems tracts might be recognized from microfossil occurrences, Mobil Exploration and Producing Technical Center contracted with the Mississippi Office of Geology to drill three coreholes along the Chickasawhay River in eastern Mississippi in the late summer and fall of 1993. Figure 18.1 shows the location of these coreholes. Figure 18.2 is a correlation of units in the #1 Young corehole with those in a composite of the #1 Wayne County Lime Company and #1 Ketler coreholes. This stratigraphic cross section is hung on the Gosport–Cockfield contact with the Gordon Creek Shale Member of the Cook Mountain Formation.

The Cook Mountain Formation, as shown at the base of the #1 Young and #1 Ketler coreholes, contains a diverse record of middle Eocene marine mollusks from Texas to Alabama and up the Atlantic Coastal Plain to North Carolina and Virginia. This formation is included in the sequence stratigraphy cycle TE2.4 of Mancini and Tew (1991). The overlying Gosport Sand (cycle TE3.1 of Mancini and Tew, 1991) contains a diverse molluscan record in southwestern Alabama and grades northwestward into the deltaic/fluvial sands of the Cockfield Formation, a transition that occurs between the #1 Young and #1 Ketler coreholes. The upper Cockfield Formation contains an estuarine fauna in its Creola Member with a low-diversity assemblage of fossils, containing species also found in the overlying Moodys Branch Formation. The Creola Member has a spotty occurrence from eastern Mississippi to its type locality at Creola Bluff on the Red River in Louisiana.

The lower Jackson Group of cycle TE3.2 includes the Moodys Branch Formation, a transgressive systems tract, and highstand deposits of the overlying North Twistwood Creek Member of the Yazoo Formation. Mancini and Tew (1991) placed both of these units in the *Truncorotaloides rohri* planktonic foraminiferal Interval Zone of late Bartonian age. They recognized the upper Jackson Group cycle TE3.3 to contain the Cocoa Sand Member of the Yazoo Formation (which they placed as a shelf margin sand within the *Porticulasphaera semiinvoluta* Interval Zone of Priabonian age) and the overlying Pachuta Marl Member and Shubuta Clay Member (as the

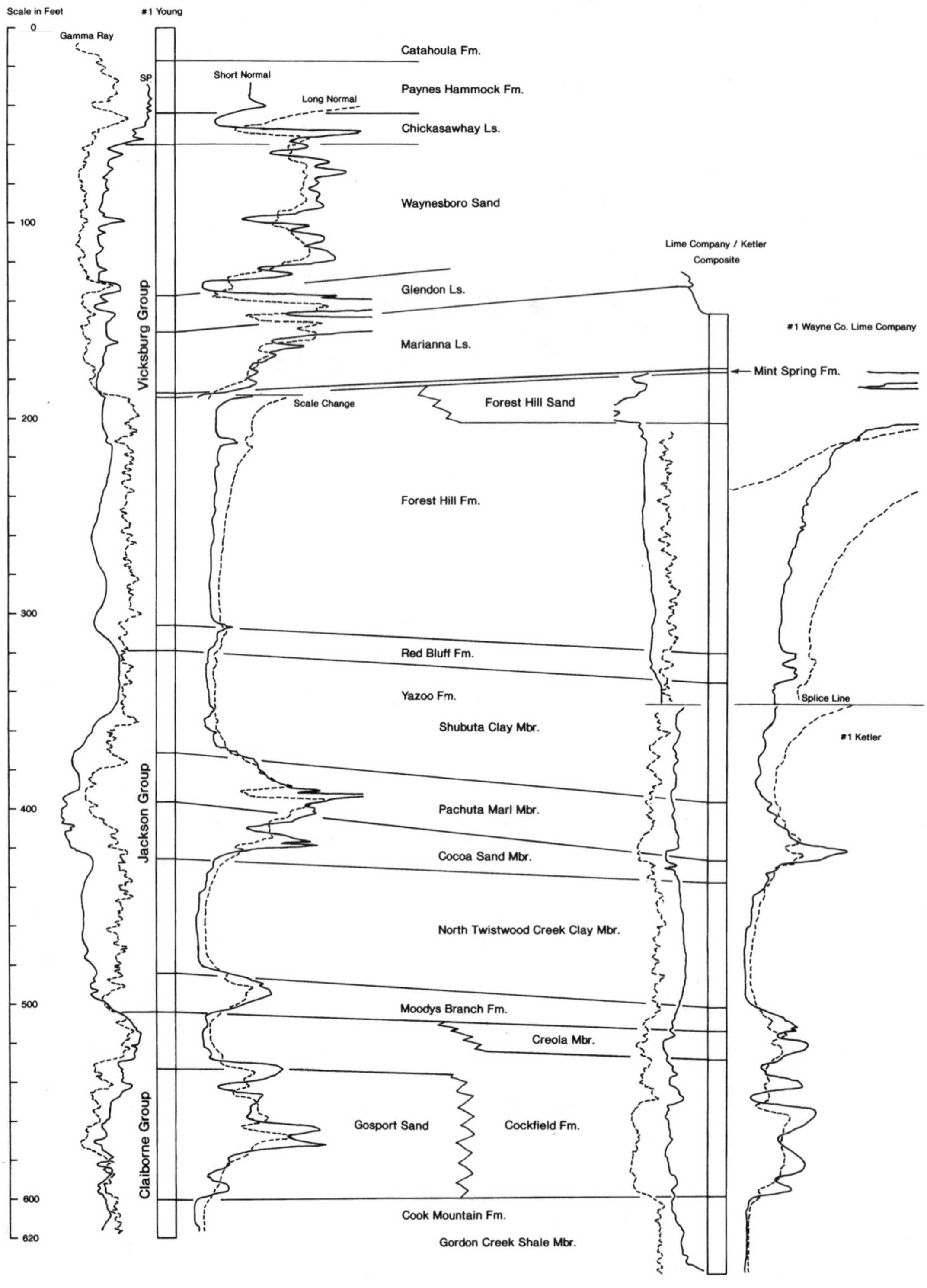

Figure 18.2. Stratigraphic cross section of the #1 Young corehole and a composite of the #1 Wayne County Lime Company and #1 Ketler coreholes hung on the Gosport Sand/Cockfield-Cook Mountain contact. From Dockery et al. (1994).

transgressive systems tract for the cycle and within the *Globorotalia cerroazulensis* Interval Zone of Priabonian age).

The electric signatures of the #1 Ketler and #1 Wayne County Lime Company geophysical logs (figure 18.2) show the Red Bluff Formation to be more resistive on the short and long normal curves than the Shubuta Clay below and the clays of the Forest Hill Formation above. This is due to the sandy, glauconitic, and fossiliferous nature of the Red Bluff clays and to the frequent ironstone concretions and burrow fillings associated with erosional surfaces (diastems) within the formation. The Red Bluff Formation appears to be the transgressive systems tract for an early Oligocene depositional cycle rather than the highstand systems tract for the late Eocene cycle TE3.3. Mancini and Tew (1991) placed the Red Bluff, Forest Hill, Mint Spring, Marianna, and Glendon formations of the Vicksburg Group within the *Pseudohastigerina micra* Interval Zone of Rupelian age.

The writers believe the disconformable contact between the Shubuta Clay and the Red Bluff Formation to be a Type 1 sequence boundary (see also Dockery, 1990; Coleman and Galloway, 1990); the discussion over the nature of this boundary has continued for some time. The deep-water blue-gray (also called turquoise) clays of the Shubuta maintain a similar lithology across Mississippi, while the overlying dark-gray, glauconitic, fossiliferous clays of the Red Bluff Formation grade northwestward along the outcrop belt into lignites and sands of the basal Forest Hill Formation. This indicates a prominent lowering of sea level at the Jackson-Vicksburg Group boundary and a subsequent sea-level rise in which deltaic facies were predominant in central and western Mississippi and the Red Bluff transgressive systems tract was restricted to southeastern Mississippi.

Coreholes through the Jackson-Vicksburg Group boundary north and south of Jackson, Mississippi, show a basal lignite in the Forest Hill Formation to rest on marine clay at the top of the Yazoo Formation. One corehole at the Forest Hill type locality in south Jackson showed carbonized roots extending from the lignite into the Yazoo clay. South of the Forest Hill type locality, localized channel-like sands in the basal Forest Hill Formation provide the sole aquifer for the water supply of the town of Terry in southeastern Hinds County. Water from Forest Hill aquifers generally has a reddish-brown color due to the presence of lignite.

The Mississippi Petrified Forest at the town of

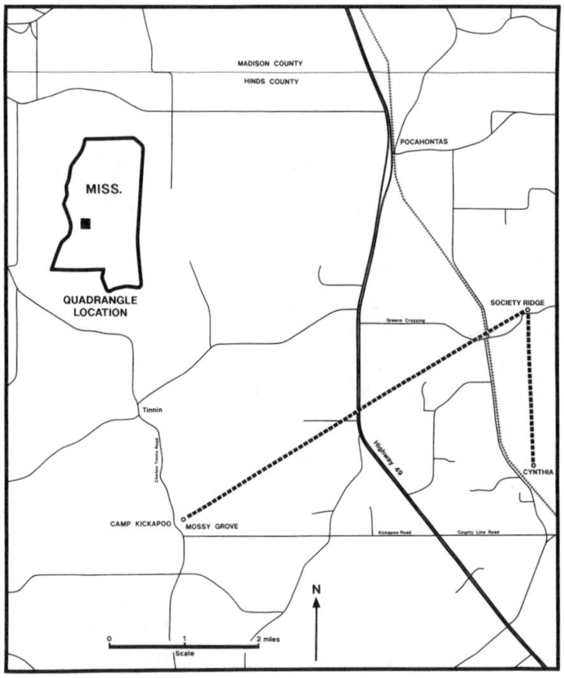

Figure 18.3. Location of core/test holes at and near the Miss Lite clay pit at Cynthia in northwestern Hinds County, central Mississippi. From Dockery et al. (1991).

Flora, northwest of Jackson in Madison County (a Registered National Natural Landmark promoted as the only petrified forest east of the Mississippi River), is an accumulation of large petrified logs in the basal channel sands of the Forest Hill Formation. Wood types at the Mississippi Petrified Forest and at Sand Hill to the south in northwestern Hinds County (near the Mossy Grove corehole of figure 18.3) include the conifers *Cupressinoxylon* and *Staphyleoxylon*, extinct genera comparable to those in the families Podocarpaceae and Cupressaceae, which now live in the Southern Hemisphere (Blackwell, 1983). Fossil woods in the Forest Hill Formation lack well-defined growth rings indicative of seasonal growth in temperate climates.

Pasley and Hazel (1990) noted an increase in the percentage of terrestrial organic matter above the Eocene/Oligocene boundary (at the Shubuta–Red Bluff contact) at St. Stephens Quarry in Washington County, Alabama (the section that led to the interpretation of a condensed section at the boundary). Later Pasley and Hazel (1995) determined that this section did not contain all the components of a depositional sequence and explained the boundary as a merged surface of maximum starvation and overlying sequence boundary. This seems reasonable, as the Shubuta Clay Member is only 5 m (16 feet) thick at St. Stephens Quarry and is 20 m (60 feet) thick along the Chickasawhay River in eastern Mississippi.

Echols et al. (this volume) present detailed analyses of the Mississippi Office of Geology/Mobil Exploration coreholes in Wayne County, Mississippi. They recognize a relative fall in sea level across the Yazoo-Red Bluff contact and a change from pelagic sedimentation in the Yazoo (Shubuta Clay Member) to the current-winnowed glauconite-shell-organic-rich sediments of the Red Bluff Formation. Their explanation for this apparent drop in sea level is the filling of accommodation space by sedimentation. However, in the #1 Young and #1 Ketler cores, only 8 m (25 feet) of sediment (or less) separate the maximum flooding surface of the Shubuta Clay from its upper contact with the Red Bluff Formation. This thickness of sediment is inconsequential and certainly insufficient to account for such a dramatic and consistent change in facies across the Yazoo–Red Bluff contact. Also, there is no sediment fill separating the lower Red Bluff contact from the maximum flooding surface at the top of the Shubuta Clay in the St. Stephens Quarry section.

Echols et al. (this volume) also believe the Yazoo–Red Bluff contact to be diachronous between the #1 Young, #1 Wayne County Lime Company, and #1 Ketler coreholes and correlate the upper Shubuta Clay of the #1 Young with the lower Red Bluff Formation of the #1 Wayne County Lime Company and #1 Ketler. According to their model, terrigenous clays bypassed the shallower Red Bluff shelf to be deposited downdip (over a distance of only 8 km, or five miles) as the upper Shubuta Clay. An argument against this is that the Red Bluff Formation is largely a gray organic-rich clay with glauconitic sands and shells concentrated above diastems within the lower part of the formation. The color contrast across the Yazoo–Red Bluff contact is readily apparent in all outcrops. It is unlikely that gray-colored prodelta clays associated with the Red Bluff Formation would lose their organic content and turn into the "turquoise-blue" Shubuta Clay just 8 km (five miles) downslope.

A better solution is to recognize the Red Bluff Formation as the transgressive systems tract and condensed section of the first Oligocene stratigraphic sequence, and the overlying deltaic section of the Forest Hill Formation as the highstand component of that sequence. The writers agree with Mancini and Tew (1991) in the placement of the next sequence boundary at the base of the Mint Spring Formation, a transgressive systems tract, with the Marianna and Glendon limestones as condensed section and highstand systems tract carbonate facies, but would recognize this sequence at TO1.2 rather than TO1.1. Also, as is the case with the Creola Member of the Cockfield Formation and the Moodys Branch Formation, the sands in the upper Forest Hill Formation in the #1 Wayne County Lime Company corehole in eastern Mississippi, which were correlated as NP21 where exposed on Highway 84 in Wayne County near the Alabama state line (Siesser, 1983), contain marine fossils like those in the Mint Spring Formation of central and western Mississippi (see appendix), which were correlated as NP22 (Siesser, 1983). These sands represent the initial flooding of estuarine environments preceding the open marine incursion associated with the diachronous (NP21–NP22) Mint Spring transgressive systems tract.

Mancini and Tew (1991) placed a sequence boundary between the Glendon Limestone and the overlying Byram Formation and recognized the Byram as a transgressive systems tract with the overlying Bucatunna Formation as the highstand component. While the Bucatunna contains mineral-rich estuarine clays, the Byram is also a shallower-water facies than the Glendon Limestone below it. If the Mint Spring Formation is the transgressive systems tract leading to the widespread carbonate deposition (stretching from the Mississippi River to Florida) found in the Marianna and Glendon limestones, then the Byram Formation seems to be its retrograde (regressive) equivalent. Also, the diverse molluscan faunas of the Mint Spring Formation can be found as molds in the Marianna Limestone, and the diverse faunas of the Byram Formation can be found as molds in the Glendon Formation. One indication that the Byram Formation is a nearshore, shallow-shelf environment is the findings of two large land mammal remains in the formation at Edwards, Mississippi. These remains include the skull of the amynodont *Metamynodon planifrons,* as reported by Manning et al. (1985), and the lower jaws of the rhinoceros *Subhyracodon occidentalis,* as reported by Manning (1997), and tie the early Oligocene marine section with the Orellan North American Land Mammal Age.

The Byram and Bucatunna formations do not appear in the cross section of figure 18.2 because they have been cut out by channel sands of the Waynesboro Sand, a lowstand fluvial deposit at the base of Super-cycle Te at the early/late Oligocene boundary. Dissolution associated with the Waynesboro Sand produced the "horse-bone weathering" of the underlying Glendon Limestone, including large voids filled with fresh-water sparry calcite. Overlying the Waynesboro Sand are the Chickasawhay Limestone and Paynes Hammock Formation of late Oligocene age and the Catahoula Formation of Oligocene to Miocene age.

Figure 18.3 shows the location of three test/coreholes northwest of Jackson, Mississippi, where the Yazoo Formation is an undifferentiated clay of some 133-150 m (400 to 450) feet in thickness. In comparison, the Yazoo Formation in the #1 Young (figure 18.2) along the Chickasawhay River is only 53 m (165 feet) thick. Figure 18.4 is a structural cross section, showing the Miss Lite clay pit where late Eocene molluscan faunas occur. The bentonite at the top of the Yazoo Formation in the #1 Society Ridge core/test hole has a radiometric age of 33.70 ± 0.19 Ma (Obradovich and Dockery, 1996), which agrees with the age of the bentonite (also 33.70 Ma) in the upper Yazoo Formation in the #1 Mossy Grove corehole (Obradovich, cited in Fluegeman, 1996:12). This date places an upper limit on the age of the late Eocene molluscan faunas of the Miss Lite clay pit and also on the last occurrence of the archaeocete whales. Three archaeocete whale species are known from the Miss Lite clay pit, including *Basilosaurus cetoides*, *Zygorhiza kochi*, and *Pontogeneus priscus* (as recently identified by Uhen, in press). Fluegeman et al. (1994) recognized the sparsely fossiliferous section of Yazoo clay above the bentonite in the Mossy Grove corehole as Oligocene in age. Based on another bentonite in the upper Yazoo Formation at Sartaria, Mississippi, which yielded sanidine ages of 34.3 Ma (Obradovich et al., 1993), the late Eocene molluscan faunas of the Miss Lite clay pit and equivalent faunas at Danville Landing on the Ouachita River in Catahoula Parish, Louisiana, are between 34.3 and 33.70 Ma. These faunas account for a large majority of 159 taxa listed as occurring in the Yazoo Formation in the appendix.

PUNCTUATED MOLLUSCAN SUCCESSION IN THE NORTHERN GULF

Figure 18.5 is a correlation chart of Paleogene units in the northern Gulf from Texas to Alabama. Bold lines bracket formations comprising molluscan faunal zones. These zones appear in figure 18.6 under the Horizon Column, where they take the name of the formation with the most significant molluscan record (i.e., that with the most species). The number of molluscan taxa for each horizon, including species, subspecies, and unnamed species occurring in the states of Texas, Louisiana, Mississippi, and Alabama, is shown in the bar graph to the right. The fraction of species restricted to each horizon is indicated by the shaded region of each bar and by the "Restricted Species Percentage" in the right column.

Major sources for the Paleogene molluscan record include Palmer and Brann (1965, 1966), Dockery (1982), and MacNeil and Dockery (1984). The work of Palmer and Brann is particularly important as it summarized a large volume of literature on over 2,800 Paleocene and Eocene molluscan species, updating generic placement and placing in synonymy a number of redundant species names. This is also a reliable catalogue as the senior author was a credible field paleontologist and systematist familiar with both the faunas and stratigraphy. Additional reports on Paleocene and Eocene mollusks used in the database for figures 18.6 and 18.7 include Toulmin (1977), Dockery (1977, 1980), Dockery and Siesser (1984), and several papers in *Tulane Studies in Geology* and *Mississippi Geology*, bringing the species count to 2,979.

The greatest molluscan diversity within the lower Oligocene Vicksburg Group occurs within the transgressive systems tract of the Mint Spring Formation with 325 species. This number is very close to that for a similar transgressive systems tract in the Eocene Moodys Branch Formation of the underlying Jackson Group, which has 346 species. The latest Eocene molluscan record found in the upper Yazoo Formation at the Miss Lite clay pit in central Mississi-ppi and in the Danville Landing Member at its type locality at Danville Landing on the Ouachita River in Catahoula Parish, Louisiana, contains a deep-water fauna of only moderate diversity. Still this fauna contains such important Eocene taxa as *Venericardia (Venericor)*, *Calyptraphorus*, *Pseudoliva*, *Athleta*, and *Belosaepia*.

Figure 18.7 shows the percentage of species with first appearance data (FAD) and last appearance data (LAD) for each faunal horizon and, in a third column, the net difference between the two. For the three horizons that make up the Paleocene Midway Group, the basal Clayton Formation has a dominance in FAD species. This changes slightly in favor of LAD species in the Porters Creek Formation, and then to a large dominance in LAD species in the Naheola Formation. Above a Type 1 sequence

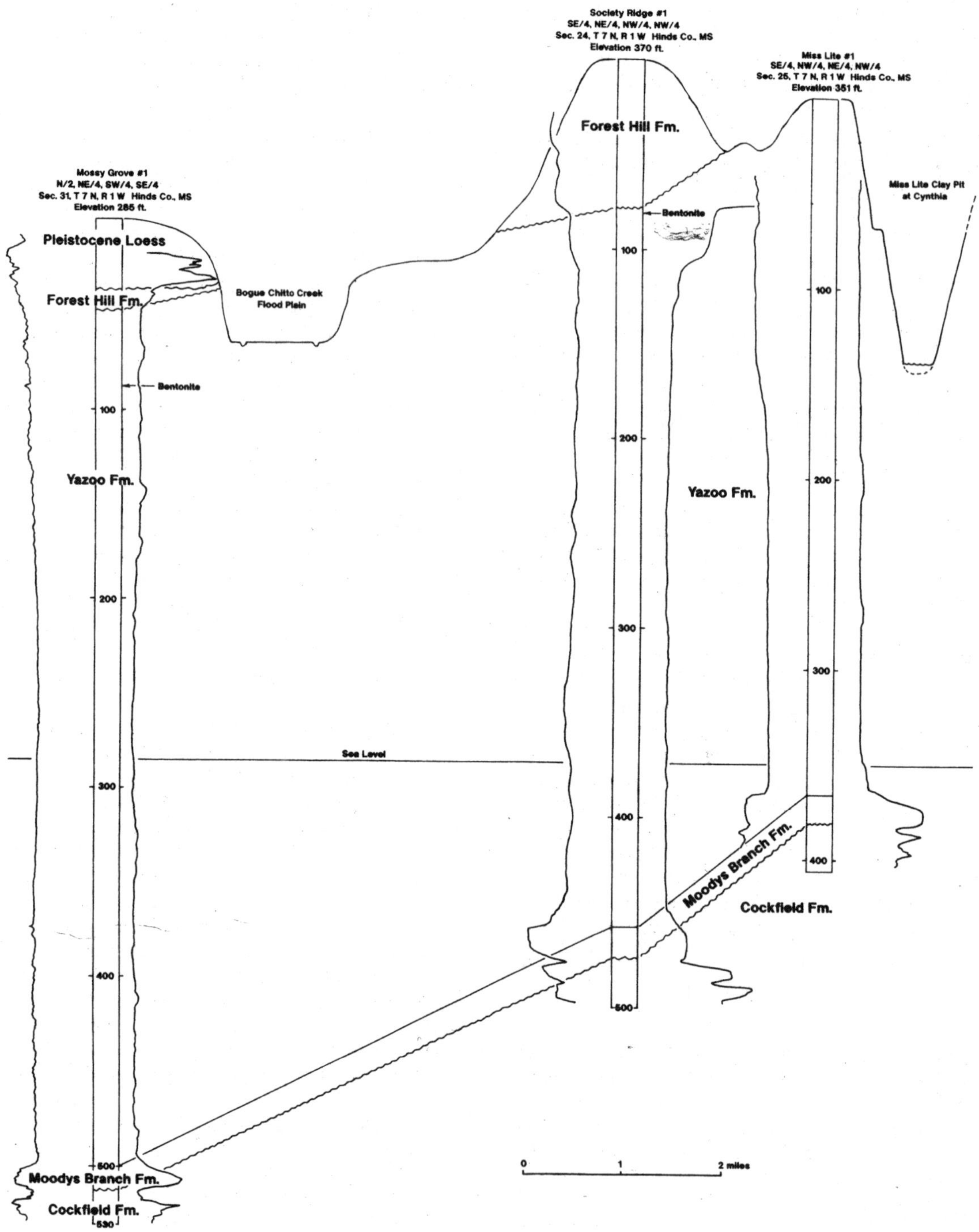

Figure 18.4. Structural cross section of core/test holes on the northwestern flank of the Jackson Dome in Hinds County. These holes penetrate thick sections of undifferentiated Yazoo Formation. Bentonites near the top of the Yazoo Formation in the Mossy Grove and Society Ridge coreholes have radiometric dates of 33.7 Ma. From Dockery et al. (1991).

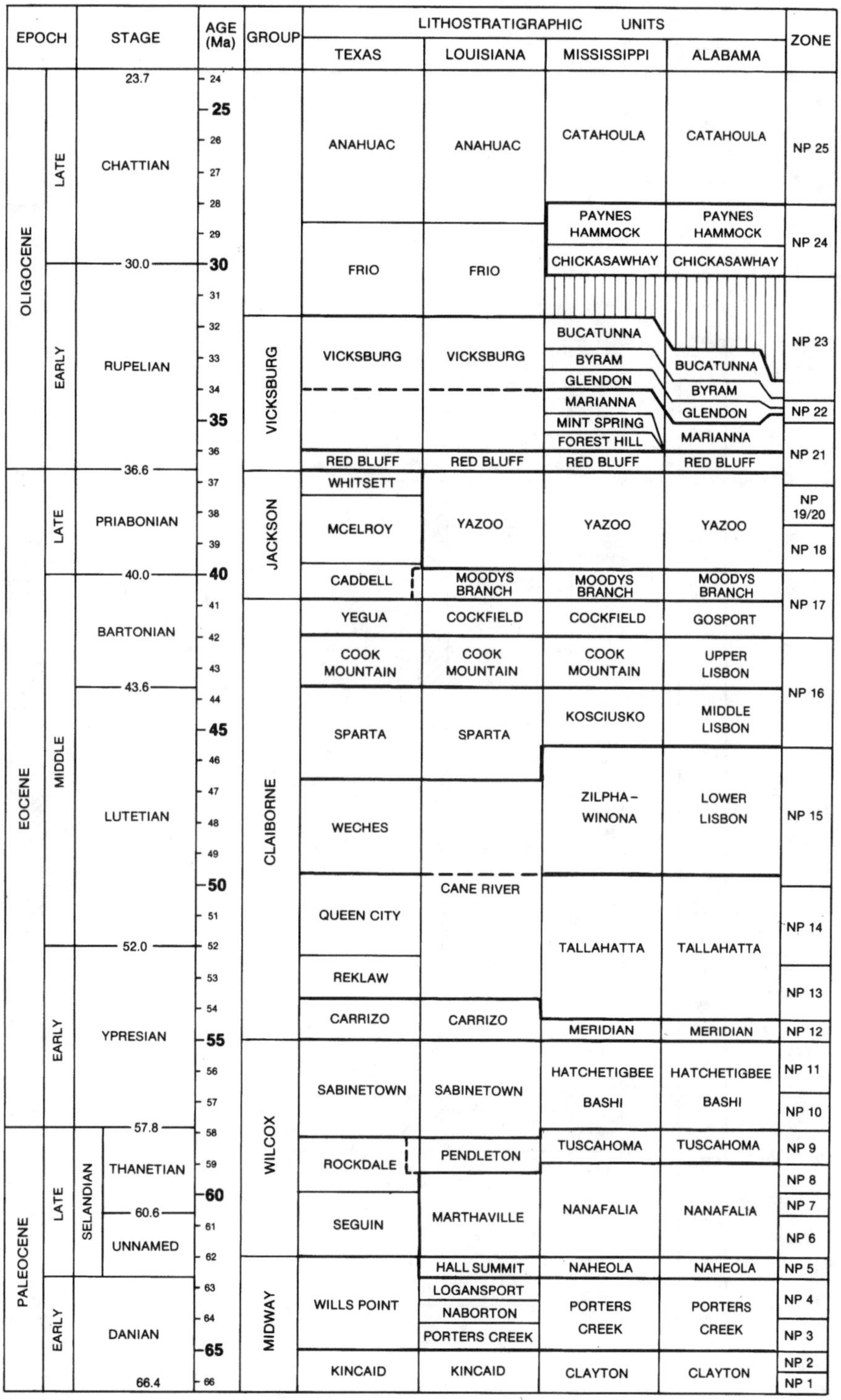

Figure 18.5. Correlation chart of Paleogene formations in Texas, Louisiana, Mississippi, and Alabama. The Texas and Louisiana sections are modified from a chart by Paleo-Data, Inc. (no date), and the Mississippi and Alabama sections are from Siesser et al. (1985). Heavy lines separate stratigraphic horizons given in figures 6 and 7. The time scale and synchronism of calcareous nannoplankton zonation follow Berggren et al. (1985). From Dockery (1986).

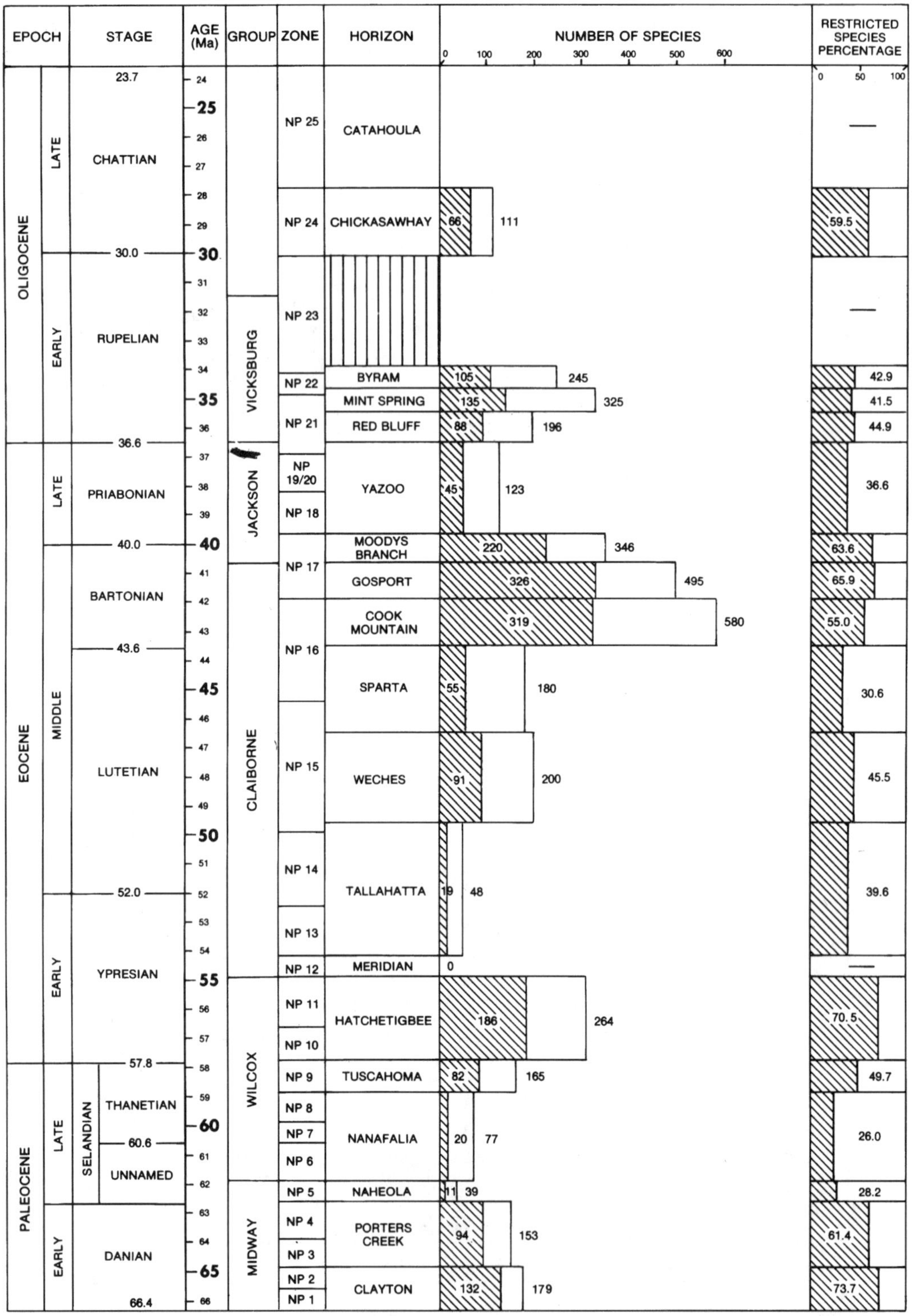

Figure 18.6. Composite section for the Paleogene of the northern Gulf with bar graph showing the number of molluscan species (including species, subspecies, and unnamed forms) recorded for each horizon. Shaded areas of the graph indicate the number of species restricted to each horizon. The percentage of the fauna restricted to each horizon is given in the right-hand column. From Dockery (1986).

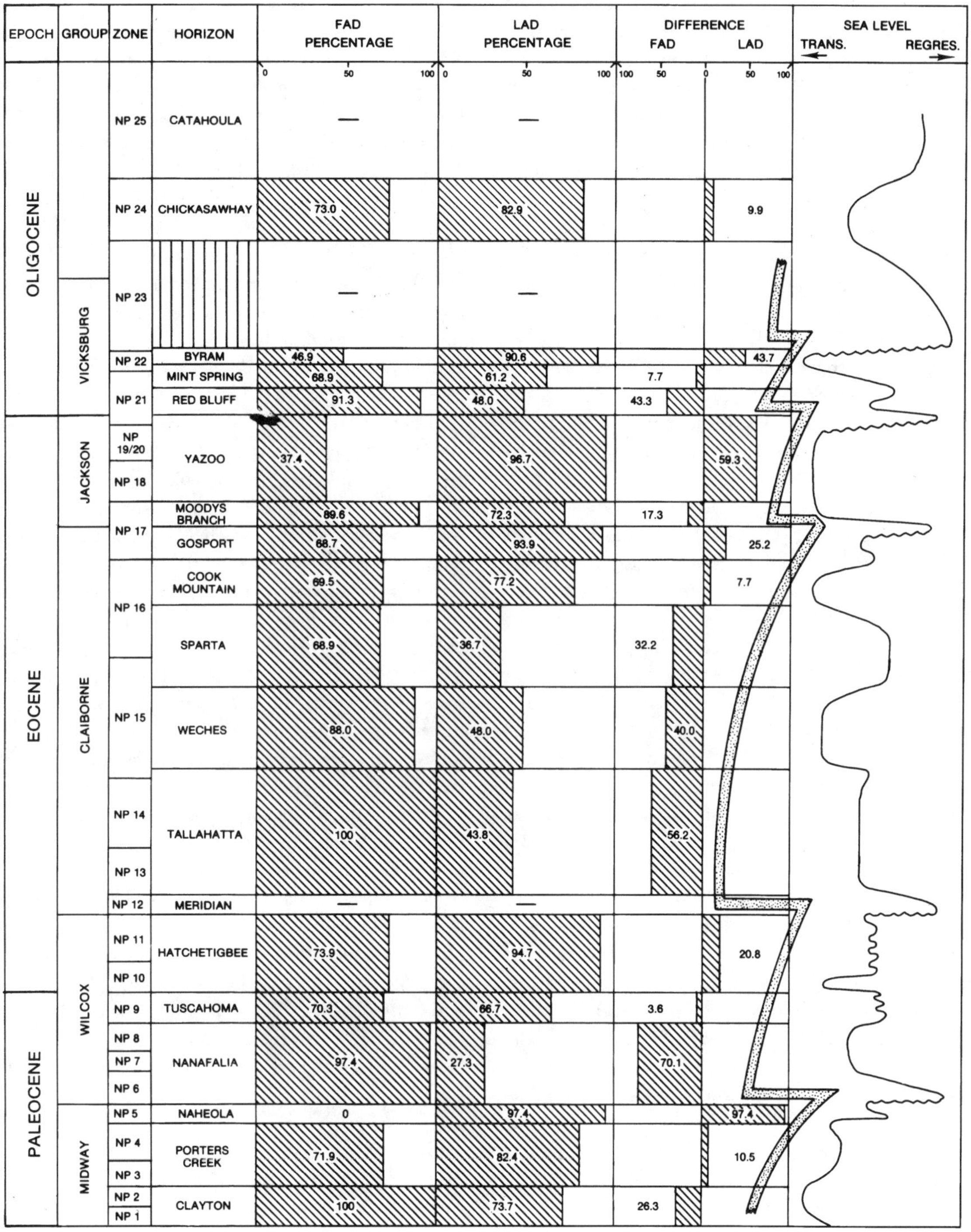

Figure 18.7. Composite Paleogene section with percentages of molluscan species at each horizon having first appearance data (FAD), last appearance data (LAD), and the difference between FAD and LAD percentages given in adjacent columns. A stippled curve follows trends indicated in the difference column. The right-hand column contains a sea-level curve for the northern Gulf showing transgressive and regressive cycles. From Dockery (1986).

boundary at the base of the Wilcox Group is a large dominance of FAD species in the Nanafalia Formation, and the pattern continues as before to the contact with the overlying Claiborne Group. From this figure it is evident that major faunal turnovers occur at group/stage boundaries, which coincide with Type 1 sequence boundaries. The Eocene/Oligocene boundary at the Jackson–Vicksburg Group contact would be an exception to this if the boundary were considered a condensed section. However, with more evidence to prove it a sequence boundary (in the writers' opinion), there is a perfect correlation between major sea level falls and major molluscan turnover events. The strength of this correlation depends on where one places the sequence boundaries in the stratigraphic record. A sea-level curve in the right-hand column of figure 18.7 shows the writers' interpretation.

The nature of molluscan turnovers across the Eocene/Oligocene boundary and the environmental influences that caused them have been a matter of discussion. Hansen (1987a, 1987b) favored a stepped extinction model associated with bolide impacts, while Dockery (1987) favored a single mass extinction associated with falling sea level. This discussion continued (Dockery, 1988a; Hansen, 1988) following the publication of Hut et al. (1987). Though subsequent work has not led to a consensus on the cause of molluscan extinctions, there is agreement concerning the timing of extinctions across the Eocene/Oligocene boundary. After collecting bulk samples from the upper Yazoo Formation at the Miss Lite clay pit and from the Red Bluff Formation in eastern Mississippi, Haasl and Hansen (1996) found 87 molluscan taxa in the Yazoo clay, of which 70 were identified to species (Dockery and Siesser, 1984, listed only 56 species from this site based on surface collections). Of those 70 species, 52 (74%) occurred in the Moodys Branch Formation below, while only 8 species occurred above the Eocene/Oligocene boundary in the Red Bluff Formation, giving an apparent extinction rate of 88.6%. This data supports Dockery's (1987) abrupt pattern of extinction at the Eocene/Oligocene boundary.

Extinctions comprise only half the equation of faunal turnovers at group/stage boundaries in the Gulf Coast. The sudden appearance of replacement faunas in transgressive systems tracts above these boundaries must also be explained. In the case of early Oligocene faunas, several Tethyan elements suggest that these had an eastern source and may have been transported across the Atlantic as larvae by ocean currents. One obstacle to the transport of Tethyan molluscan larvae into the northern Gulf was the Suwannee Current, which flowed through the Gulf Trough (of the Georgia coastal plain) into the Atlantic Ocean. The strength of this current was controlled by eustasy and was strongest during the highest sea levels (Popenoe et al., 1987). The Gulf Trough reached its maximum development in the late Eocene; however, its current volume and velocity were diminished at the end of the Eocene (Huddlestun, 1993:98).

The Gulf Trough and Suwannee Current formed a barrier between the provincial molluscan faunas of the northern Gulf and the more Tethyan faunas (i.e., containing the gastropods *Gisortia* and *Velates* in the late Eocene) of the Florida Platform Province (Dockery, 1988b). A fall in sea level at the end of the Eocene diminished the Suwannee Current and allowed an influx of Tethyan planktotrophic larvae, thus explaining the appearance of similar species in the Vicksburg Group of the Gulf Coast and the Stampian of the Aquitaine Basin on the Atlantic Coast of southwestern France. Eustasy-related faunal turnovers were also cited by Lozouet (1997) for changes in the Oligocene of the Aquitaine Basin, where there are some 1,200 late Oligocene gastropods, of which only 70, or 5.8%, occur in the underlying Stampian deposits. Figure 18.8 shows important Paleogene molluscan localities in southwestern France including: (1) the famous early Eocene (Cuisian) locality at the Tuilerie Quarry at Gan near Pau, (2) early Oligocene (Stampian) localities in the Lagouarde and Espibos parishes of Dax (sited under the name Dax), (3) late Oligocene (Chattian) localities in the Abesse and Estoti parishes of Dax and the Mineur Parish of Pontonx, and (4) an early Miocene locality in the Moulin de Carreau Parish of Corbieu. The calcareous nannoplankton biostratigraphy of these localities was studied by Siesser and Dockery (1985) and the paleoecology by Dolin et al. (1985).

MOLLUSCAN FAUNAL TURNOVERS AT THE EOCENE/OLIGOCENE BOUNDARY

The average rate of extinction for molluscan species between formations in the Gulf Coast Paleogene is 70%; this rate increases to 95% at group/stage boundaries (Dockery, 1988a). These numbers were not available to Lyell in 1846 when he recognized Eocene marine sediments at Jackson, Mississippi. What was important to Lyell in the recognition of geologic age were the shells of *Cardita planicosta* [now *Venericardia (Venericor) planicosta*]. These were the

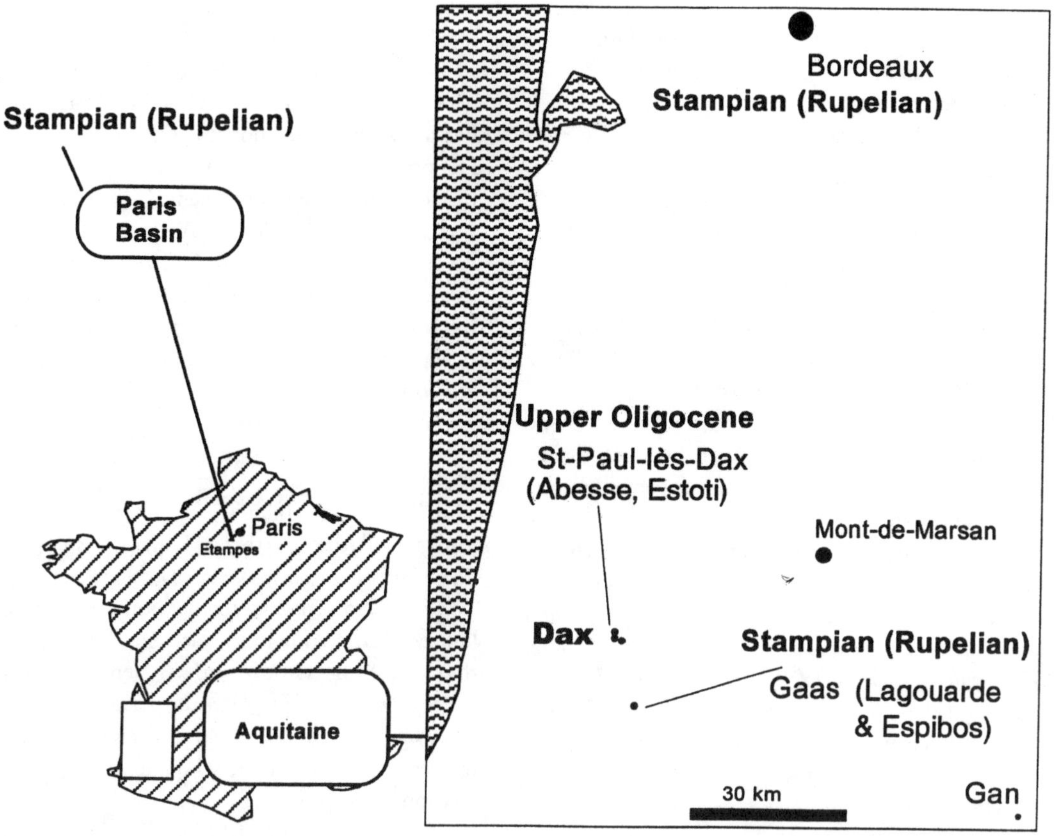

Figure 18.8. Paleogene molluscan localities in France, including the early Oligocene (Stampian) at Etampes in the Paris Basin (*left*) and, in the Aquitaine Basin (*right*), the Early Eocene (Cuisian) locality at the Tuilerie Quarry at Gan, the early Oligocene (Stampian) localities at Lagouarde and Espibos south of Dax, and the late Oligocene (Chattian) localities at Abesse and Estoti just northwest of Dax.

shells of planicostate venericard clams common to Eocene beds in England and France and later to be found worldwide (Gardner and Bowles, 1939). Lyell had also seen these clams in the Eocene of the American Atlantic Coastal Plain and at Claiborne, Alabama. Also present at Jackson was the volutid gastropod *Athleta*, which is common where planicostate venericards are found.

Extinctions at the Eocene/Oligocene boundary in the Gulf Coastal Plain are more than a bump in the background extinction rate from 70% to above 95%; it is the extinction of certain molluscan taxa that characterize both Paleocene and Eocene faunas worldwide. In comparison, extinctions at the Paleocene/Eocene boundary in the Gulf Coast at the contact between the Tuscahoma and Bashi formations are largely a non-event (Dockery, 1998). Weighting the impact of extinction events based only on species counts misses the mark in that most published species are rare; half the molluscan species of the Gulf Coast Paleogene are known from a single formational horizon (Dockery, 1987). In contrast, planicostate venericards are either common or dominant in Gulf Coast faunas of late Paleocene to late Eocene age and are absent in all species-rich environments of Oligocene age.

Palmer and Brann (1965) listed 27 species of planicostate venericards of the subgenus *Venericor* for the Paleogene of the Atlantic and Gulf coastal plains, with combined ranges extending from the basal Paleocene Clayton Formation to the latest Eocene Yazoo Formation and with localities from Maryland to Texas. Counting the last planicostate venericard species in the late Eocene, *Venericardia (Venericor) klimacodes*, as "1" in an extinction count is tantamount to counting the last dinosaur as "1" at the end of the Cretaceous without giving the event any other meaning.

The molluscan extinctions in the Gulf Coast must be measured in two ways, by the number of species affected and by the significance of the species affected. Taxa that are recurrent common elements of Paleogene molluscan faunas should carry more weight in extinction event appraisals. The "heavy weight" list of Paleogene taxa that make their last appearance just below the Eocene/Oligocene boundary in the Gulf Coast includes planicostate venericards, the gastropods *Calyptraphorus, Athleta, Lapparia, Pseudoliva, Cornulina,* and *Bullia,* and the cephalopod *Belosaepia. Athleta* continues into the Stampian of the Aquitaine Basin of France and the Miocene of Europe and is still living along the coast of South Africa, while *Pseudoliva* survives along the West Coast of Africa, but still the disappearance of these genera in the North American Gulf Coast is noteworthy. Of the many species listed in their catalogue of Paleocene and Eocene mollusks, Palmer and Brann (1965:8) recognized conspicuous evolutionary lines among *Venericardia (Venericor), Ostrea, Athleta, Calyptraphorus, Lapparia, Pseudoliva, Caricella,* and *Bullia.* Of these lines, only *Ostrea* and *Caricella* survive the Eocene/Oligocene boundary in the Gulf Coast region.

The composition of certain higher molluscan taxa in the Gulf Coastal Plain includes notable changes across the Eocene/Oligocene boundary, and, for some of these taxa, the change is synchronous in the European section. The first appearance of closely related species in the lower Oligocene of North America and Europe points to the distribution of planktotrophic larvae by ocean currents (Dockery, 1996). The observation of planktotrophic larvae within the following molluscan groups is based in part on Lozouet's examination of some 400 Paris Basin Eocene species and 1,452 Aquitaine Basin species, ranging in age from lower Oligocene to upper Miocene. The following discussion notes changes across the Eocene/Oligocene boundary among families and superfamilies of the Bivalvia and Gastropoda and among solitary and hermatypic corals. Many of these observations are also cited in Dockery (1982, 1996), MacNeil and Dockery (1984), and Lozouet (1997).

BIVALVIA

Crassatellidae

The *Crassatellas* of the Vicksburg Group replace the Eocene *Venericors* as the large, thick-shelled, shallow-burrowing bivalves of the early Oligocene in the Gulf Coast. The common Vicksburg *Crassatellas, C. lirasculpta* Dockery and *C. mississippiensis megacostata* Dockery of the Mint Spring Formation and *C. mississippiensis* Conrad of the Byram Formation appear to be related to *C. ponderosa* (Gmelin, 1791) of the Paris Basin Eocene and not to *C. alta* or *C. texanus* Heilprin of the Gulf Coast Eocene. *C. lirasculpta* is very close to the late Oligocene *C. emiliae* Cossmann and Peyrot of the French Aquitaine Basin, and *C. mississippiensis* is similar to the early Oligocene (Stampian) *C. pseudotumida* Benoist of the Aquitaine Basin. It is also interesting to note that the genus *Bathytormus,* which is common in the Claiborne and Jackson groups, does not occur in the Vicksburg Group.

Cardiacea

The Cardiacea, which are rather boring in the late Eocene of the northern Gulf Coast with only *Nemocarium nicolleti* (Conrad) and the rare species *Laevicardium gardnerae* Cooke (known only from the holotype) occurring in the Jackson Group, blossom in the Vicksburg Group with 6 genera (*Agnocardia, Trachycardium, Trigoniocardia, Nemocardium, Laevicardium,* and *Dinocardium*) and eight species. It is interesting to note that, while absent or rare in the Jackson Group of the northern Gulf, the genera *Agnocardia, Trachycardium, Trigoniocardia,* and *Laevicardium* are present in the Claiborne Group and in the middle and late Eocene Tethyan faunas of carbonate units in South Carolina (Santee Limestone) and Florida (Avon Park Limestone and Inglis Formation). Perhaps they re-inhabited the Gulf after a low sea-level stand at the Eocene/Oligocene boundary.

Veneridae

In both the Eocene and Oligocene of the Gulf Coast, the most common venerid bivalves are *Callista* and *Pitar.* However, it's interesting to note the new appearance of the venerid *Ventricolaria ucuttana* (Dall) in the Red Bluff Formation at about the same time that *V. praecursor* (Mayer) appears in the Stampian of France. Also the Gulf Coast early Oligocene venerids resemble modern faunas in having a diversity of the subfamily Chioninae including *Chione (s.s.) craspedonia* Dall, *C. (s.s.) perbrevisformis* Dockery, *C. (s.s.) bainbridgensis* Dall, *C. (Lirophora) victoria* Dall, and *Chamelea mississippiensis* (Conrad).

GASTROPODA

Stromboidea

The superfamily Stromboidea are largely tropical to

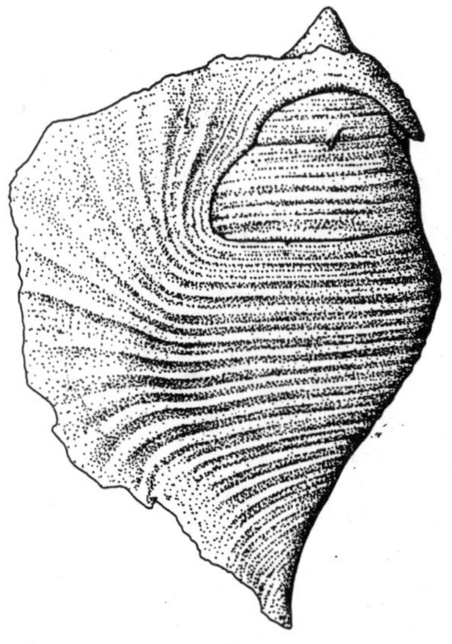

Figure 18.9. *Playtoptera extenta* (Conrad, 1854) from the late Eocene Moodys Branch Formation in Jackson, Mississippi. One of several elegant strombid taxa, which, collectively, characterize Eocene marine deposits worldwide.

subtropical gastropods typified by shells with flaring or modified outer lips. In the upper Eocene Moodys Branch Formation, this superfamily contains the genera *Seraphs*, *Dientomochilus (Dasyostoma)*, *Calyptraphorus*, and *Platyoptera* (figure 18.9). Of these, only *Seraphs* survives into the Oligocene (but not in the Gulf Coastal Plain). The extinction of *Calyptraphorus* is particularly noteworthy as it ranges from Cretaceous to late Eocene, is common in Paleocene and Eocene marine environments in the Gulf Coast, and is one of the most common large gastropods in the upper Eocene Moodys Branch Formation. *Calyptraphorus* was placed as a subgenus of *Tibia* (*Tibia s.s.* is currently living in the Indo-Pacific realm) by Wenz (1938) and is one of several *Tibia*-like taxa to become extinct by the end of the Eocene. The only member of the Stromboidea present in the early Oligocene of both the northern Gulf and the Aquitaine and Paris basins of France is *Aporrhais (Goniocheila)*, a member of the stromboid family Aporrhaidae. An extant species of this family, *Aporrhais occidentalis* Beck, inhabits arctic to temperate waters from the Arctic Canada to North Carolina (Abbott, 1974).

The Eocene Stromboidea of western Europe, according to Le Renard and Pacaud's (1995) revision of Cossmann and Pissarro (1911), includes *Rimella*, *Cyclomolops*, *Strombolaria*, *Ectinochilus*, *Dientomochilus s.s.*, *Dientomochilus* (*Digitolabrum*), *Tibia (Eotibia)*, *Tibia (Amplogladius)*, *Tibia (Hippochrenes)*, *Tibia (Wateletia)*, *Tibia* (*Semiterebellum*), *Tibia (Chedevillia)*, *Seraphs s.s.*, *Seraphs (Miniseraphs)*, *Seraphs (Diameza)*, and *Paraseraphs*. Of these, only *Seraphs s.s.* survives in the Tethyan realm of the southern European Oligocene with *Aporrhais* occurring in the Paris Basin and northern European Oligocene. The tropical early Oligocene faunas of the Aquitaine Basin also contain large specimens of the strombid genera *Strombus* and *Oostrombus*. So great is the loss of stromboid taxa at the Eocene/Oligocene boundary that the only other comparable extinction with this group is at the Cretaceous/Tertiary boundary. In fact, *Calyptraphorus* survived the K-T extinction event only to perish at the Eocene/Oligocene boundary. Such a loss of tropical taxa suggests that climatic cooling as well as eustasy played a role in their demise.

Cypraeoidea

The Cypraeoidea of the Moodys Branch Formation includes *Sphaerocypraea*, *Sulcocypraea*, *Cypraedia*, *Eucypraedia*, *Cypraeacties*, *Cypraeorbis*, *Jenneria*, and *Transovula*. Also, in the upper Eocene Crystal River Formation (an equivalent to the Yazoo Formation) of peninsular Florida is the giant cypraeid *Gisortia*. In contrast to this diversity, the overlying Oligocene beds contain only *Sulcocypraea*, *Cypraeorbis*, and *Simnia*. Extinct at the Eocene/Oligocene boundary are *Gisortia* and *Transovula*.

Ranellidae (=Cymatiidae)

In the northern Gulf, the early Oligocene ranellid gastropods are more abundant and diverse than those of the late Eocene. In the Moodys Branch Formation there is only *Sassia (Personella)*. In the early Oligocene Vicksburg Group are *Cymatium (Septa)*, *Cymatium (Ranularia)*, *Sassia s.s.*, *Sassia (Personella)* [Beu, 1988:85, placed *Byramia* MacNeil, 1984, as a synonym of *Personella*], and *Distorsio s.s.* It is interesting to note that similar species of *Distorsio s.s.* occur on both sides of the Atlantic in the early Oligocene, *Distorsio (s.s.) crassidens* (Conrad) occurring in the Mint Spring and Byram formations of the Gulf Coast and *Distorsio (s.s.) tortuosa* (Borson) in the Stampian of the Aquitaine Basin of France. The same is true for *Sassia s.s.* where *Sassia (s.s.) conradiana* (Aldrich) occurs in the Red Bluff Formation of Mississippi, *Sassia (S.) conradiana menthafons* MacNeil occurs in the Mint Spring Formation of

Mississippi, and *Sassia (s.s.) subspinosum* (Grateloup) occurs in the Stampian of the Aquitaine Basin of France. *Cymatium amnicretum* MacNeil from the Mint Spring Formation of Mississippi is very close to *C. subcorrugatum* (D'Orbigny) from the Stampian of Gaas (Aquitaine Basin).

Cassidae

The present systematics of the cassid gastropods is not in a satisfactory state to make meaningful stratigraphic correlations of species. However, some observations can be made. *Galeodaria* Conrad, 1865, is presented by the type species *G. petersoni* (Conrad) in the Moodys Branch Formation of the Gulf Coast and continues into the Oligocene with *G. shubutensis* (Aldrich) in the Red Bluff Formation, *G. shubutensis gardnerae* MacNeil in the Mint Spring Formation, and *G. tricarinata* (Conrad) in the Byram Formation.

Gardner (1939) named *Gomphopages* as a cassid subgenus of *Galeodea*, including within it the type species *Galeodea (Gomphopages) turneri* Gardner (see Garvie, 1996, pl. 14, figures 6-7, for best illustration) from the lower Eocene Recklaw Formation of Texas and questionably *Galeodea (Gomphopages) millsapsi* Sullivan and Gardner from the Moodys Branch Formation of Mississippi. This group becomes extinct at the Eocene/Oligocene boundary but unfortunately contains (in the writers' opinion) *Cassidaria planotecta* Meyer and Aldrich from the middle Eocene Cook Mountain Formation of Mississippi, which Gardner (1939) named as the type species for *Galeodea (Mambrinia)*. If the type species is ignored, *Mambrinia* includes: *Galeodea koureos* Gardner (which should have been the type species) from the upper Paleocene Tuscahoma Formation and Early Eocene Bashi Formation of Alabama and the upper Paleocene Pendleton Formation of Texas; a group of seven species from the Paleocene-Eocene of the Paris Basin, including *G. (M.) nodosa carinata* (Bruguière), *G. (M.) endis* Deshayes, *G. (M.) gallica* Wrigley, *G. (M.) petiosa* Deshayes, *G. (M.) nodosa retusa* Deshayes, *G. (M.) singularis* Deshayes, and *G. (M.) textiliosa* Deshayes as listed by Le Renard and Pacaud (1995); and one species from the earliest Oligocene of the Gulf Coast (Red Bluff Formation), *G. (M.) brevidentata* (Aldrich). *Galeodea (Mambrinia) brevidentata* of Mississippi's early Oligocene has been recognized in the Stampian of Gaas (Aquitaine Basin) as *Mambrinia* cf. *brevidentata*.

Other late Eocene cassids from the Gulf Coast include *Phalium taitii johnsoni* Palmer and *Galeodea planotecta jacksonia* Palmer. In the earliest Oligocene, several cassid lineages begin in the Red Bluff Formation, including: (1) *Galeodaria shubutensis* (Aldrich)—*G. shubutensis gardnerae* MacNeil (Mint Spring Formation)—*G. tricarinata* (Conrad) (Byram Formation), (2) *Sconsia prelintea* MacNeil—*S. lintea* (Mint Spring and Byram formations), and (3) *Phalium (Menthafontia)* sp.—*P. (M.) menthafons* MacNeil (Mint Spring Formation)—*P. (M.) mississippiensis* (Conrad) (Byram Formation). Also in the Mint Spring and Byram formations is *Cassis flintensis* Mansfield, which is very close to *Cassis elegans* Grateloup from the Stampian of the Aquitaine Basin of France (see MacNeil and Dockery, 1984, pl. 49 for comparisons).

Harpidae

The species *Oniscidia harpula* (Conrad) from the Mint Spring and Byram formations of Mississippi is very close to *Oniscidia neglecta* (Michelotti) of the Stampian of the Aquitaine Basin (both the generic names *Oniscidia* and *Morum* have been applied to these species).

Turritellidae

One molluscan taxon that is surprisingly rare in early Oligocene Gulf Coast faunas is the gastropod *Turritella*. *Turritella* is common in many marine environments from Cretaceous to Recent in age and is one of the most common taxa in upper Eocene sediments of the Moodys Branch Formation and in the uppermost Eocene sediments of the Yazoo Formation. Yet it is uncommon in the diverse molluscan fauna of the lowermost Oligocene Red Bluff Formation, is uncommon in the upper Forest Hill Formation, is common at only one locality in the Mint Spring Formation, and is the dominant element at one locality in the Byram Formation (Allmon and Dockery, 1992), while being uncommon at most other Byram localities. *Turritella* populations rebound in the late Oligocene to form the *Turritella* limestones of the Chickasawhay Limestone (May et al., 1974:95-96).

Globulariidae

The family Globulariidae is represented by species in the middle and late Eocene of both the Gulf Coast and Europe. Of interest is the genus *Ampullinopsis*, which is restricted to the Oligocene and appears at the same time on both sides of the Atlantic, being represented in the Red Bluff, Mint Spring, and Byram formations of the Vicksburg Group by the species *Ampullinopsis mississippiensis* (Conrad).

Xenophoridae
The Subgenus *Stellaria* is represented in the Lower Oligocene of the Gulf Coast by *Xenophora* (*Stellaria*) *conica* Dall and in the Oligocene, Miocene, and Pliocene of Europe by *X.* (*S.*) *testigera* (Bronn). Ponder and Cooper (1983:14) suggested that *X.* (*S.*) *conica* was ancestral to *X.* (*S.*) *testigera* and believed the latter species to be conspecific with *X.* (*S.*) *digitata* Martens, a Recent species living off the west coast of Africa.

Muricidae
Vokes (1970) believed *Pterynotus* (*Pterynotus*) *burnsii* (Aldrich) of the Red Bluff Formation in the Gulf Coast to be a direct descendant of the middle and late Eocene Tethyan species *Pterynotus* (*Pterynotus*) *tripteriodes* (Lamarck). However, *P. burnsii* is perhaps more closely related to *Pterynotus consobrinus* (d'Orbigny) of the Stampian of the Aquitaine Basin. Both Koenen (1889) and Wrigley (1930) recognized the Red Bluff species *Pterynotus* (*Pterochelus*) *angelus* (Aldrich) to be closely related to the Tethyan species *Pterynotus* (*Pterochelus*) *bispinosus* (J. de C. Sowerby), which occurs in the late Eocene of England, Germany, and Belgium. Again, an even closer relationship exists between *P. angelus* and the Stampian species *Pterynotus magnei* (Vergneau-Saubade) of the Aquitaine Basin. The generic name *Ponderia* is also used for this species group.

Melongenidae
The first true *Melongena* appear in the early Oligocene on both sides of the Atlantic with *M. crassicornuta* Conrad of the Byram Foramtion in Mississippi being very close to *M. laxecarinata* Michelotti of the Stampian of the Aquitaine Basin.

Pteropods (Thecosomata)
The planktonic gastropods known as pteropods show a higher survival rate across the Eocene/Oligocene boundary than do benthic mollusks. Of the 6 species occurring in the Jackson Group, two are also found in the Vicksburg Group. One of these, the curved *Creseis*-like taxon *Bovicornu eocenense* Meyer, forms a tight range zone bounding the boundary and occurring within NP21-age sediments of the upper Yazoo Clay and the Red Bluff Formation.

ANTHOZOA

Corals
Corals are commonly associated with Paleogene molluscan faunas in the Gulf Coast with the solitary coral *Flabellum* being very common in marine sediments of middle and late Eocene age. This is an important association as certain gastropods such as the Cypraeoidea prey almost exclusively upon corals. Others, such as the Conidae, prey on the Cypraeoidea, which prey on corals (Barclay, 2000). A striking feature of early Oligocene faunas in the northern Gulf Coastal Plain is the total absence of *Flabellum* even though the genus is living today and is common in the latest Eocene faunas of the Yazoo Formation. *Flabellum* is the only large common fossil below the Eocene/Oligocene boundary where the Shubuta Clay Member of the Yazoo Formation is in contact with the overlying Red Bluff Formation in eastern Mississippi. Here the overlying Red Bluff Formation contains an abundance of the solitary coral *Balanophyllia elongata* Vaughan and rare branches of the hermatypic coral *Archohelia vicksburgensis* (Conrad). *A. vicksburgensis* becomes the dominant coral in the Byram Formation.

Hermatypic corals were also important in the Oligocene of the European tropical to subtropical Tethyan realm with small coral reefs appearing in the Aquitaine Basin. The Paris Basin and the rest of northern Europe contained warm temperate molluscan faunas and almost no corals in the Oligocene, an evidence of climatic cooling. This is in contrast to the very similar molluscan and coral faunas that existed across Europe in the more tropical climates of Eocene times.

CONCLUSIONS

The ranges of several common and long-ranging molluscan taxa found in the Paleocene and Eocene marine sediments of the North American Gulf Coastal Plain are truncated at the Eocene/Oligocene boundary. Replacement taxa of similar standing in the lower Oligocene Vicksburg Group are similar to species in the Stampian of the Aquitaine Basin of France. The disconformity that separates Gulf Coast Eocene and Oligocene faunas is noted by all but interpreted as both a condensed interval and as the opposite, a Type 1 unconformity. If the latter is true as the writers believe, then a fall and subsequent rise in sea level separates the extinction of Eocene species and the appearance of Oligocene species. Such an event would explain the extinction of species, not only by both the loss of habitat and a change in the nature of the remaining available habitat, but also by altering ocean currents, which would explain the arrival of new taxa having no ancestral species in the Gulf Coast Eocene.

The extinction and appearance of molluscan species on both sides of the Atlantic affect the same groups, the majority of which probably have planktotrophic larvae. The extinction of tropical strombid taxa at the Eocene/Oligocene boundary is particularly notable, along with the subsequent first appearance of true *Strombus* in the Oligocene of the Tethyan realm of southwestern France and the absence of all Strombidae in the Oligocene of the Paris Basin and the rest of northern Europe. Eustacy-related events coupled with climatic cooling best explain the timing and mechanisms of extinctions and appearances of molluscan faunas across the Eocene/Oligocene boundary in both North America and Europe.

REFERENCES

Abbott, R. T., 1974, American Seashells, 2nd ed.: New York, Van Nostrand Reinhold, 663 p.

Allen, J. E., 1968, New species of Sepiida (Mollusca, Cephalopoda) from the Eocene of the Gulf Coast: Tulane Studies in Geology, v. 6, no. 1, p. 33-37.

Allen, J. E., 1970, New species of Eocene Mollusca from the Gulf Coast: Tulane Studies in Geology and Paleontology, v. 8, no. 1, p. 69-78.

Allmon, W. D., 1996, Systematics and evolution of Cenozoic American Turritellidae (Mollusca: Gastropoda) I: Paleocene and Eocene coastal plain species related to "*Turritella mortoni* Conrad" and "*Turritella humerosa* Conrad": Palæontographica Americana, no. 59, 134 p.

Allmon, W. D., and Dockery, D. T., III, 1992, A turritelline gastropod-dominated bed in the Byram Formation (Oligocene) of Mississippi: Mississippi Geology, v. 13, no. 2, p. 29-35.

Barclay, D. E., 2000, Cone wars: American Conchologist, v. 27, no. 4, p. 28-30, continued in v. 28, no. 1, p. 25-30.

Baum, G. R., and Vail, P. R., 1988, Sequence stratigraphic concepts applied to Paleogene outcrops, Gulf and Atlantic basins: SEPM Special Publication, v. 42, p. 309-327.

Berggren, W. A., Kent, D. W., Flynn, J. J., and Van Couvering, J. S., 1985, Cenozoic geochronology: Geological Society of America Bulletin, v. 96, p. 1407-1418.

Beu, A. G., 1987, Taxonomy of gastropods of the families Ranellidae (=Cymatiidae) and Bursidae. Part 2. Descriptions of 14 new modern Indo-West Pacific species and subspecies, with revisions of related taxa: New Zealand Journal of Zoology, v. 13, p. 273-355.

Beu, A. G., 1988, Taxonomy of gastropods of the families Ranellidae (=Cymatiidae) and Bursidae. Part 5. Early history of the families, with four new genera and recognition of the Family Personidae: Saito Ho-on Kai Special Publication (Professor Tamio Kotaka Commemorative Volume), p. 69-96.

Blackwell, W. H., 1983, Fossil wood from "Sand Hill," western central Mississippi: Bulletin of the Torrey Botanical Club, v. 110, no. 1, p. 63-69.

Cavelier, C., 1979, La limite Eocène-Oligocène en Europe occidentale: Science Géologiques, Mémoire 54, 280 p.

Cernohorsky, W. O., 1985, The taxonomy of some Indo-Pacific Mollusca. Part 12. With remarks on two American gastropod species: Records of the Auckland Institute and Museum, v. 22, p. 47-67.

Coleman, J., and Galloway, W. E., 1990, Sequence stratigraphic analyses of the lower Oligocene Vicksburg Formation of Texas, p. 99-112, *in* Sequence stratigraphy as an exploration tool, concepts and practices in the Gulf Coast: Eleventh Annual Research Conference, Gulf Coast Section, SEPM Foundation, December 2-5, 1990, Houston, Texas, 409 p.

Conrad, T. A., 1848a, Observations on the Eocene formation, and descriptions of one hundred and five new fossils of that period, from the vicinity of Vicksburg, Mississippi, with an Appendix: Academy of Natural Sciences of Philadelphia, Proceedings for 1847, v. 3, p. 280-299.

Conrad, T. A., 1848b, Observations on the Eocene formation, and descriptions of one hundred and five new fossils of that period, from the vicinity of Vicksburg, Mississippi, Mississippi, with an Appendix: Academy of Natural Sciences of Philadelphia, Journal, Second series, v. 1, pt. 2, p. 111-134.

Conrad, T. A., 1865, Catalogue of the Eocene and Oligocene Testacea of the United States: American Journal of Conchology, v. 1, no. 1, p. 1-35.

Conrad, T. A., 1866, Check list of invertebrate fossils of North America Eocene and Oligocene: Smithsonian Miscellaneous Collection, v. 7, p. 1-41.

Cossmann, M., and Pissarro, G., 1911, Iconographie complète des coquilles fossiles de l'Éocène des environs de Paris, Tome II, parsó In-4°: Paris (Hermann), p. 26-45.

Dockery, D. T., III, 1977, Mollusca of the Moodys Branch Formation, Mississippi: Mississippi Geological Survey, Bulletin 120, 212 p.

Dockery, D. T., III, 1980, The invertebrate macropaleontology of the Clarke County, Mississippi, area: Mississippi Bureau of Geology, Bulletin 122, 387 p.

Dockery, D. T., III, 1982, Lower Oligocene Bivalvia of the Vicksburg Group in Mississippi: Mississippi Bureau of Geology, Bulletin 123, 261 p.

Dockery, D. T., III, 1984, Crisis events for Paleogene molluscan faunas in the southeastern United States: Mississippi Geology, v. 5, no. 2, p. 1-7.

Dockery, D. T., III, 1986, Punctuated succession of Paleogene mollusks in the northern Gulf Coastal Plain: Palaios, v. 1, no. 6, p. 582-589.

Dockery, D. T., III, 1987, Eocene-Oligocene molluscan extinctions: Comment & Reply: Comment: Palaios, v. 2, no. 6, p. 620-621.

Dockery, D. T., III, 1988a, Molluscan extinction rates in question: Nature, v. 331, p. 123.

Dockery, D. T., III, 1988b, The influence of the ancient Gulf Stream on Paleogene molluscan provinces in the northern Gulf and Atlantic coastal plains (abstract): Abstracts with Programs of the 37th Annual Meeting of the Southeastern Section of the Geological Society of America, v. 29, no. 4, p. 261.

Dockery, D. T., III, 1990, The Eocene/Oligocene boundary in the northern Gulf—a sequence boundary, *in* Sequence Stratigraphy as an Exploration Tool, Concepts and Practices in the Gulf Coast: Eleventh Annual Research Conference, Gulf Coast Section, SEPM Foundation, December 2-5, 1990, Houston, Texas, p. 141-150,

Dockery, D. T., III, 1996, Eocene, Oligocene, and Pleistocene (pre-loess alluvium) stratigraphy and fossil localities of west-central Mississippi: Geological Society of America, Southeastern Section Field Trip, March 13, 1996, 33 p.

Dockery, D. T., III, 1998, *Carditamera williamsi*, a new species from the late Eocene of Mississippi: Mississippi Geology, v. 18, no. 4, p. 57-60

Dockery, D. T., III, 1998, Molluscan faunas across the Paleocene/Eocene series boundary in the North American Gulf Coastal Plain, *in* Aubry, M.-P., Lucas, S., and Berggren, W. A., eds., Late Paleocene-Early Eocene climatic and biotic events in the marine and terrestrial records: New York, Columbia University Press, p. 296-322,

Dockery, D. T., III, and Siesser, W. G., 1984, Age of the upper Yazoo Formation in central Mississippi: Mississippi Geology, v. 5, no. 1, p. 1-10.

Dockery, D. T., III, Stover, C. W., Weathersby, P. W., Stover, C. W., Jr., and Ingram, S. L., 1991, A continuous core through the undifferentiated Yazoo clay (late Eocene, Jackson Group) of central Mississippi: Mississippi Geology, v. 12, no. 3-4, p. 21-27.

Dockery, D. T., III, Thompson, D. E., and Ingram, S. L., 1994, The Mobil-Mississippi Office of Geology core-hole project: Mississippi Geology, v. 15, no. 1, p. 8-15.

Dolin, L., 1991a, *Cypraeacites blowi* n. sp.; first occurrence of the genus in the New World. Studies on Paleogene Cypraeoidea (Mollusca: Gastropoda) from the Gulf Coast Basin—II: Mississippi Geology, v. 11, no. 4, p. 6-8.

Dolin, L., 1991b, *Cypraedia (Eucypraedia) multicarinata* (Dall, 1890); a late Eocene Ovulidae from Florida, Mississippi, Colombia, and Peru. Studies on Paleogene Cypraeoidea (Mollusca: Gastropoda) from the Gulf Coast Basin—III: Mississippi Geology, v. 12, nos. 3-4, p. 28-31.

Dolin, L., and Dolin, C., 1981, *Sphaerocypraea jacksonensis* (Johnson) from the Moodys Branch Formation (Eocene), Mississippi: Mississippi Geology, v. 2, no. 2, p. 17-19.

Dolin, C., Dolin, L., and Lozouet, P., 1985, Paleoecology of some classic Tertiary localities in the Aquitaine and Paris Basins of France: Mississippi Geology, v. 5, no. 4, p. 4-13.

Emerson, W. K., 1993, A new species of columbellid gastropod from the Old World tropics: The Nautilus, v. 106, no. 4, p. 147-151.

Fluegeman, R. H., 1996, Preliminary paleontological report of the Foraminifera of the Mossy Grove core, Hinds County, Mississippi: Mississippi Geology, v. 17, no. 1, p. 9-15.

Fluegeman, R. H., Denney, J. E., Finley, S. J., Hall, R. E., Rose, A. R., Voisard, V. L., Diaz, J. R., and Schwegman, M. E., 1994, Eocene-Oligocene benthic foraminiferal paleoecology and sea-level change records from the Mossy Grove core, Hinds County, Mississippi: Proceedings Seventh International Symposium on the Observation of the Continental Crust Through Drilling, Santa Fe, New Mexico, p. 189-192.

Gardner, J., 1939, Notes on fossils from the Eocene of the Gulf Province. I. The annelid genus *Tubulostium*. II. The gastropod families Cassididae, Ficidae, and Buccinidae: U.S. Geological Survey, Professional Paper, v. 193-B, p. 17-44.

Gardner, J., and Bowles, E., 1939, The *Venericardia*

planicosta group in the Gulf Province: U.S. Geological Survey, Professional Paper, v. 189-F, p. 143-215.

Garvie, C. L., 1996, The molluscan macrofauna of the Reklaw Formation, Marquez Member (Eocene: Lower Claibornian), in Texas: Bulletins of American Paleontology, v. 111, no. 352, 117 p., 23 pl.

Gitton, J. L., Lozouet, P., and Maestrati, P., 1986, Biostratigraphie et paléoécologie des gisements types du Stampien de la région d'Etampes (Essonne): Éditions du Bureau de Recherches Géologiques et Minières, Extrait de Géologie de la France No. 1, 101 p.

Haasl, D. M., and Hansen, T. A., 1996, Timing of latest Eocene molluscan extinction patterns in Mississippi: Palaios, v. 11, no. 5, p. 487-494.

Hansen, T. A., 1987a, Extinction of late Eocene to Oligocene molluscs: Relationship to shelf area, temperature changes, and impact events: Palaios, v. 2, no. 1, p. 69-75.

Hansen, T. A., 1987b, Eocene-Oligocene Molluscan Extinctions: Comment & Reply: Reply: Palaios, v. 2, no. 6, p. 621-622.

Hansen, T. A., 1988, Molluscan extinction rates in question: Hansen replies: Nature, v. 331, January 14, 1988, p. 123.

Harry, H. W., and Dockery, D. T., III, 1983, Notes on the lower Oligocene oysters of Mississippi: Mississippi Geology, v. 4, no. 2, p. 7-14.

Hodgkinson, K. A., Garvie, C. L., and Bé, A. W. H., 1992, Eocene euthecosomatous Pteropoda (Gastropoda) of the Gulf and Eastern Coasts of North America: Bulletins of American Paleontology, v. 103, no. 341, 62 p.

Huddlestun, P. F., 1993, A revision of the lithostratigraphic units of the coastal plain of Georgia: Georgia Geological Survey Bulletin, v. 105, 152 p.

Hut, P., Alvarez, W., Elder, W. P., Hansen, T., Kauffman, E. Keller, G. G., Shoemaker, E. M., and Weissman, P. R., 1987, Comet showers as a cause of mass extinctions: Nature, v. 329, p. 118-126.

Koenen, A. von, 1889, Das norddeutche Unter-Oligocän und seine Mollusken-Fauna. Lief. 1: Strombidae—Muricidae—Buccinidae: Abhandlungen zur Geologischen Specialkarte von Preussen und den Thüringischen Staaten, v.10, 280 p.

Le Renard, J., and Pacaud, J.-M., 1995, Révision des Mollusques Paléogènes du Bassin de Paris II—Liste des références primaires des espèces: Cossmanniana, v. 3, no. 3, p. 65-132.

Lozouet, P., 1997, Le domaine Atlantique Européen au Cénozoïque moyen: diversité et évolution des gastéropodes: Thèse de Doctorat du Muséum National d'Histoire Naturelle, 309 p. (microfilm: ANRT, Grenoble, no. 75872).

Lozouet, P., 1999, Nouvelles espèces de gastéropodes (Mollusca: Gastropoda) de l'Oligocène et du Miocène d'Aquitaine (Sud-Ouest de la France). Partie 2: Cossmanniana, v. 6, nos. 1-2, p. 1-68, 20 pls.

Lyell, C., 1830, Principles of geology: London, John Murray, v. 1.

MacNeil, F. S., and Dockery, D. T., III, 1984, Lower Oligocene Gastropoda, Scaphopoda, and Cephalopoda of the Vicksburg Group in Mississippi: Mississippi Bureau of Geology, Bulletin 124, 415 p.

Mancini, E. A., and Tew, B. H., 1991, Relationships of Paleogene stage and planktonic foraminiferal zone boundaries to lithostratigraphic and allostratigraphic contacts in the eastern Gulf Coastal Plain: Journal of Foraminiferal Research, v. 21, no. 1, p. 48-66.

Manning, E. M., 1997, An early Oligocene rhinoceros jaw from the marine Byram Formation of Mississippi: Mississippi Geology, v. 18, no. 2, p. 13-31.

Manning, E., Dockery, D. T., III, and Schiebout, J. A., 1985, Preliminary report of a *Metamynodon* skull from the Byram Formation (Lower Oligocene) in Mississippi: Mississippi Geology, v. 6, no. 2, p. 1-16.

May, J. H., Baughman, W. T., McCarty, J. E., Glenn, R. C., and Hall, W. B., 1974, Wayne County Geology and Mineral Resources: Mississippi Geological Survey, Bulletin 117, 293 p.

Montanari, A., Deino, A. L., Drake, R. E., Turin, B. D., DePaolo, D. J., Odin, G. S., Curtis, G. H., Alvarez, W., and Bice, D. M., 1988, Radioisotopic dating of the Eocene/Oligocene boundary in the pelagic sequence of the northeastern Appenines, *in* Premoli-Silva, I., Coccioni, R., and Montanari, A., eds., The Eocene/Oligocene boundary in the Marche-Umbria Basin (Italy): Ancona (Italy), p. 195-208.

Obradovich, J. D., and Dockery, D. T., III, 1996, Revisions to the geochronology of the uppermost Yazoo Formation, central Mississippi, and a new estimate for the age of the Eocene/Oligocene boundary (abstract): Journal of the Mississippi Academy of Sciences, v. 41, no. 1, p. 54.

Obradovich, J. D., Dockery, D. T., III, and Swisher, C. C., II, 1993, ^{40}Ar-^{39}Ar ages of bentonite beds in the upper part of the Yazoo Formation (upper Eocene), west-central Mississippi: Mississippi Geology, v. 14, no. 1, p. 1-9.

Paleo-Data, Inc. (New Orleans, Louisiana), no date, Gulf Coast Cenozoic stratigraphic chart, 1 sheet.

Palmer, K. V. W., 1979, Rhythm of the Paleocene-Eocene seas of the central Gulf Coast, as defined by fossils and sediments: Tulane Studies in Geology and Paleontology, v. 15, no. 3, p. 69-74.

Palmer, K. V. W., and Brann, D. C., 1965, Catalogue of the Paleocene and Eocene Mollusca of the southern and eastern United States. Part I. Pelecypoda, Amphineura, Pteropoda, Scaphopoda, and Cephalopoda: Bulletins of American Paleontology, v. 48, no. 218, p. 1-466.

Palmer, K. V. W., and Brann, D. C., 1966, Catalogue of the Paleocene and Eocene Mollusca of the southern and eastern United States. Part II. Gastropoda: Bulletins of American Paleontology, v. 48, no. 218, p. 471-1057.

Pasley, M. A., and Hazel, J. E., 1990, Use of organic petrology and graphic correlation of biostratigraphic data in sequence stratigraphic interpretations: example from the Eocene/Oligocene boundary section, St. Stephens Quarry, Alabama: Gulf Coast Association of Geological Societies, Transactions, v. 40, p. 661-683.

Pasley, M. A., and Hazel, J. E., 1995, Revised sequence stratigraphic interpretation of the Eocene/Oligocene boundary interval, Mississippi and Alabama, Gulf Coast Basin, U.S.A.: Journal of Sedimentary Research B, v. 65, no. 1, p. 160-169.

Ponder, W. F., and Cooper, J., 1983, A revision of the Recent Xenophoridae of the World and of the Australian fossil species (Mollusca: Gastropoda): Australian Museum, Sydney, Memoir 17, 126 p.

Popenoe, P., Henry, V. J., and Idris, F. M., 1987, Gulf Trough—the Atlantic connection: Geology, v. 15, p. 327-332.

Richards, H. G. and Palmer, K. V. W., 1953, Eocene molluks from Citrus and Levy counties, Florida: Florida Geological Survey, Bulletin 35, 95 p.

Robinson, J. E., 1983, New and interesting Eocene molluscan species in the Moodys Branch Formation: Mississippi Geology, v. 3, no. 3, p. 9-12.

Robinson, J. E., and Dockery, D. T., III, 1981, New occurrences of molluscan species in the Moodys Branch Formation at Jackson, Mississippi: Mississippi Geology, v. 1, no. 3, p. 13-15.

Schiebout, J. A. and Bold, W. V. D., eds., 1986, Montgomery Landing Site, marine Eocene (Jackson) of central Louisiana: Proceedings of a Symposium, 1986 Annual Meeting, Baton Rouge, Louisiana, October 22, 1986, Gulf Coast Association of Geological Societies, 238 p.

Siesser, W. G., 1983, Paleogene calcareous nannoplankton biostratigraphy: Mississippi, Alabama and Tennessee: Mississippi Bureau of Geology, Bulletin 125, 61 p.

Siesser, W. G., and Dockery, D. T., III, 1985, Calcareous nannoplankton biostratigraphy of selected Tertiary localities in the Paris, Adour, and Bearn basins of France and their correlation with the North American Gulf Coast Tertiary sequence: Mississippi Geology, v. 5, p. 9-19.

Siesser, W. G., Fitzgerald, B. G., and Kronman, D. J., 1985, Correlation of Gulf Coast provincial Paleogene stages with European standard stages: Geological Society of America Bulletin, v. 96, p. 827-831.

Toulmin, L. D., 1977, Stratigraphic distribution of Paleocene and Eocene fossils in the eastern Gulf Coast region: Geological Survey of Alabama Monograph, v. 13, 602 p.

Tucker, J. K., 1994, The genus *Conorbis* with the description of a new species from the Eocene of Mississippi: Mississippi Geology, v. 15, no. 3, p. 43-55.

Uhen, M. D., *Pontogeneus priscus*, the largest dorudontine archaeocete whale: Mississippi Geology, in press.

Vail, P. R., and Hardenbol, J., 1979, Sea-level changes during the Tertiary: Oceanus, v. 22, p. 71-79.

Vail, P. R., Mitchum, R. M., Jr., Todd, R. G., Widmier, J. M., Thompson, S., III, Sangree, J. B., Bubb, J. N., and Hatlelid, W. G., 1977, Seismic stratigraphy and global changes of sea level, *in* Seismic stratigraphy—applications to hydrocarbon exploration: American Association of Petroleum Geologists, Memoir 26, p. 49-212.

Vokes, E. H., 1968, Cenozoic Muricidae of the western Atlantic region: Part IV—*Hexaplex* and *Murexiella*: Tulane Studies in Geology, v. 6, no. 3, p. 85-126.

Vokes, E. H., 1970, Cenozoic Muricidae of the western Atlantic region. Part V—*Pterynotus* and *Poirieria*: Tulane Studies in Geology and Paleontology, v. 8, no. 1, p. 1-50.

Wenz, W., 1938, Teil 1, Allgemeiner Teil and Prosobranchia Lfg. 1: Handbuch der Paläozoologie. Band 6. Gastropoda, Berlin, p. 1-240.

Wrigley, A., 1930, Notes on English Eocene Mollusca, with descriptions of new species: Malacological Society of London, Proceedings 1930, v. 19, pt. 3, p. 91-115.

APPENDIX:

First appearance data (FAD) for Eocene-Oligocene mollusks of the Jackson and Vicksburg Groups in Mississippi, Louisiana, Arkansas, Texas, and Alabama. Stratigraphic units include the Claiborne Group (CL), Moodys Branch Formation (MB), Yazoo Formation (YA), Red Bluff Formation (RB), Forest Hill Formation (FH), Mint Springs Formation (MT), Marianna Limestone (MR), Glendon Limestone (GN), and Byram Formation (BY). Bold indicates type horizon; occurrences labeled as "cf.," "aff.," and "?" are included. Numbers indicate occurrences in: (1) Mississippi, (2) Louisiana, (3) Arkansas, (4) Alabama, (5) Texas, (6) Mississippi and Louisiana, (7) Mississippi and Arkansas, (8) Mississippi, Louisiana, and Arkansas, (9) Mississippi, Louisiana, Arkansas, and Texas, (10) Mississippi, Louisiana, Texas, and Alabama, (11) Louisiana, Alabama, and Texas, (12) Florida and Alabama, (13) Mississippi and Texas, (14) Mississippi, Louisiana, and Texas, (15) Louisiana and Alabama, (16) Mississippi, Louisiana, Arkansas, and Alabama, (17) Mississippi and Alabama, (18) Mississippi, Louisiana, and Alabama, (19) Mississippi, Arkansas, and Texas, (20) Louisiana and Arkansas, (21) Mississippi and Florida, (22) Mississippi, Louisiana, and Florida, (23) Mississippi, Louisiana, Arkansas, Texas, and Alabama, (24) Mississippi, Alabama, and Texas, (25) South Carolina, (26) Mississippi, Louisiana, Alabama, and Florida, (27) Mississippi, Louisiana, Arkansas, Alabama, and Florida, (28) Alabama, Georgia, and South Carolina, (29) Alabama and Georgia, (30) Florida, (31) Mississippi, Alabama, Texas, and South Carolina, (32) Arkansas and Alabama, (33) Mississippi, Alabama, and Florida, (34) Louisiana, Alabama, Florida, and Georgia, (35) Mississippi and South Carolina, (36) Georgia, (37) Mississippi, Florida, and Georgia, and (38) Mississippi, Alabama, and Georgia. Eocene occurrence records and nomenclature are from Richards and Palmer (1953), Palmer and Brann (1965, 1966), Toulmin (1977), Dockery (1977, 1980, 1998), Dockery and Siesser (1984), E. Vokes (1968, 1970), Allen (1968, 1970), Robinson and Dockery (1981), Robinson (1983), Schiebout and Bold (1986), Dolin and Dolin (1981), Dolin (1991a, 1991b), Hodgkinson et al. (1992), Tucker (1994), Haasl and Hansen (1996), and Allmon (1996). Oligocene records and nomeclature are from Dockery (1982), Harry and Dockery (1983), MacNeil and Dockery (1984), Cernohorsky (1985), Beu (1987, 1988), and Emerson (1993).

	EOCENE			OLIGOCENE					
Species	**CL**	**MB**	**YA**	**RB**	**FH**	**MT**	**MR**	**GN**	**BY**
Gastropoda with FAD in CL									
1. *Acteon idoneus* Conrad, 1833	**CL10**	MB7							
2. *Architectonica (Stellaxis) alveata* (Conrad, 1833)	**CL11**	MB6	YA2						
3. *Architectonica (Architectonica) bellistriata* Conrad *in* Wailes, 1854	CL5	**MB6**	YA6						
4. *Athleta symmetricus* (Conrad *in* Wailes, 1854)	CL5	**MB8**							
5. *Bullia altilis* (Conrad, 1832)	**CL4**	MB7							
6. *Bullia altilis subglobosa* (Conrad, 1832)		**CL4**	MB3						
7. *Calyptraea (Trochita) aperta* (Solander *in* Brander, 1766)	CL4	MB14	YA15	RB1		MT1			
8. *Cerithiella* sp. B	CL17	MB1							

9. *Cirsotrema (Coroniscala) nassulum* (Conrad, 1833)	CL17	MB1					
10. *Creseis simplex* (Meyer, 1886)	CL4	MB1					
11. *Eopleurotoma carya carola* Harris, 1937	CL25	MB2					
12. *Glyptotoma crassiplicata* (Gabb, 1860)	CL14	MB1					
13. *Hipponix pygmaeus* Lea, 1833	CL24	MB6				MT1	
14. *Mesalia vetusta* (Conrad, 1833)	CL4	MB8					
15. *Mitrella (Columbellopsis) parva* (Lea, 1841)	CL4	MB1					
16. *Niso umbilicata* (Lea, 1833)	CL17	MB1	YA1				
17. *Polinices eminulus* (Conrad, 1833)	CL4	MB3					
18. *Pseudoliva vetusta* (Conrad, 1833)	CL24	MB8					
19. *Pyramidella bastropensis* Harris, 1895	CL5		YA1				
20. *Pyramimitra terebraeformis* (Conrad, 1848)	CL11		YA1				
21. *Sablea minuta* Allen, 1970	CL2	MB6		RB1		MT1	
22. *Seila constricta* (Lea, 1841)	CL4	MB1					
23. *Sinum (Sinum) beatricae* Palmer, 1937	CL17			RB1		MT1	BY1
24. *Streptochetus limulus* (Conrad, 1833)	CL17	MB1					
25. *Sveltella parva* (Lea, 1833)	CL18	MB1					
26. *Tornatellaea lata* (Conrad *in* Morton, 1834)	CL2	MB17	YA1				
27. *Trypanotoma terebriformis curta* Harris, 1937	CL1		YA2				
28. *Umbraculum planulatum* (Conrad *in* Wailes, 1845)	CL1	MB6					
Bivalvia with FAD in CL							
29. *Alveinus minutus* Conrad, 1865	CL4	MB1					
30. *Barbatia (Acar) aspera* (Conrad *in* Wailes, 1854)	CL24	MB1					
31. *Barbatia (Cucullaearca) ludoviciana* (Harris, 1899)	CL14	MB6	YA1				
32. *Chlamys cawcawensis* (Harris, 1919)	CL28	MB4					
33. *Chlamys (Aequipecten) deshayesi* (Lea, 1833)	CL4	MB29					
34. *Crassostrea alabamiensis* (Lea, 1833)	CL4	MB1					
35. *Crenella isocardioides* (Lea, 1833)	CL4	MB1					
36. *Glycymeris (Glycymeris) idonea* (Conrad, 1833)	CL4	MB6					
37. *Hilgardia multilineata* (Conrad *in* Wailes, 1854)	CL1	MB16	YA4				
38. *Pitar (Katherinella) trigoniata* (Lea, 1833)	CL28	MB7					
39. *Spisula jacksonensis* Cooke, 1926	CL4	MB1					
40. *Spisula (Symmorphomactra) praetenuis* (Conrad, 1833)	CL4	MB8					
41. *Tellina (Eurytellina) vaughani* Cooke, 1926	CL4	MB1					
Scaphopoda with FAD in CL							
42. *Dentalium (Antalis) minustriatum* Gabb, 1860	CL31	MB2					
Cephalopoda with FAD in CL							
43. *Belemnosella floweri* (Palmer, 1937)	CL4	MB2					
Gastropoda with FAD in MB							
44. *Abderospira oviformis* (Meyer, 1886)		MB1	YA1				
45. *Acrilla unilineata* (Heilprin, 1880)		MB1	YA2				
46. *Acirsa? solumcostata* Dockery, 1977		MB1					
47. *Acteon annectens* Meyer, 1885		MB6	YA2				
48. *Agaronia media* (Meyer, 1885)		MB9					
49. *Agaronia* sp. A		MB1					
50. *Apiotoma palmerae* Dockery, 1977		MB1					
51. *Architectonica (Solariaxis) acuta* Conrad *in* Wailes, 1854		MB6	YA2				
52. *Architectonica (Architectonica) billmoorei* Dockery, 1977		MB1					
53. *Architectonica (Granosolarium) meekana subsplendida* Palmer, 1947		MB6	YA6				
54. *Architectonica (Architectonica) meliconae* Robinson and Dockery, 1981		MB1			MT1		

Species						
55. *Architectonica (Granosolarium) ornata jacksonia* Palmer, 1947	MB6					
56. *Architectonica (Architectonica) trilirata palmeri* MacNeil, 1984	MB6			MT1		
57. *Athleta haleanus jacksonius* Palmer, 1947	MB2					
58. Cf. *Athleta triplicatus* (Meyer, 1887)	MB1					
59. *"Bellatara" floridana* Palmer, 1953	MB12					
60. *Bittium koeneni* Meyer, 1886	MB1	YA1				
61. *Bonellitia jacksonica* (Cooke, 1926)	MB6	YA1				
62. *Bovicornu gracile* Meyer, 1887	MB1					
63. *Buccitriton jacksonense* (Cooke, 1926)	MB1					
64. *Bullata semen jacksonensis* (Meyer, 1885)	MB13	YA1				
65. *Caecum (Micranellum) alterum* Meyer, 1887	MB1					
66. *Calyptraea alta* (Conrad *in* Wailes, 1854)	MB6	YA2				
67. *Calyptraea glandaria* Dockery, 1977	MB1					
68. *Calyptraphorus stamineus* (Conrad, 1856)	MB16	YA6				
69. *Capulus americanus* Conrad *in* Wailes, 1854	MB14	YA2	RB1			
70. *Capulus cassis* Allen, 1970	MB2					
71. *Capulus (Hyalorisia) nettlesi* Robinson, 1983	MB1					
72. *Caricella (Atraktus) fenestra* Dockery, 1977	MB1					
73. *Caricella giganta* Dockery, 1977	MB1					
74. *Caricella howei* Palmer, 1947	MB6	YA6				
75. *Caricella polita* Conrad *in* Wailes, 1854	MB1					
76. *Caricella subangulata* Conrad *in* Wailes, 1854	MB8					
77. *Caricella turneri* Palmer, 1947	MB6					
78. *Cerithiella jacksonensis* (Meyer, 1886)	MB6					
79. *Circulus ottonius* Palmer, 1947	MB1					
80. *Cirsotrema (Coroniscala) danvillense* Palmer, 1947	MB2	YA18				
81. *Cirostrema (Coroniscala) nassulum creolum* Palmer, 1947	MB6	YA17				
82. *Cirostrema (Coroniscala) ranellinum* (Dall, 1900)	MB2	YA17				
83. *Cirostrema (Coroniscala)* sp. B	MB1					
84. *Cirostrema*? sp. D	MB1					
85. *Clavilithes humerosus* Conrad *in* Wailes, 1854	MB16					
86. *Cochlespira bella polita* (Harris, 1937)	MB6	YA1				
87. *Cochlespira columbaria* Aldrich, 1886	MB6					
88. *Colubraria vokesae* Allen, 1970	MB2					
89. *Conomitra hammakeri* (Harris, 1984)	MB20	YA6				
90. *Conomitra jacksonensis* Cooke, 1926	MB8	YA2				
91. *Conorbis alatoideus* Aldrich, 1885	MB6					
92. *Conorbis dolini* Tucker, 1994	MB1					
93. *Conus tortilis* Conrad *in* Wailes, 1854	MB16	YA18				
94. *Coptostoma ulmulum secutorum* Palmer, 1947	MB2					
95. *Cordiera ludoviciana* (Vaughan, 1896)	MB6	YA1				
96. *Cornulina dalli* (Harris, 1894)	MB19					
97. *Cornulina dalli cetaria* Palmer, 1947	MB20					
98. *Coronia childreni* (Lea, 1833) var. B	MB20					
99. *Coronia conjuncta* (Casey, 1904)	MB6	YA6				
100. *Coronia montgomeryensis* (Harris, 1937)	MB1	YA6				
101. *Coronia nodulina* (Casey, 1904)	MB1					
102. *Coronia nodulina bunkerensis* (Harris, 1937)	MB2	YA2				
103. *Coronia weisbordi* (Harris, 1937)	MB					
104. *Creseis corpulenta* (Meyer, 1887)	MB1					BY1
105. *Creseis hastata* (Meyer, 1886)	MB1		RB1		MT1	BY1
106. *Crommium jacksonense* (Harris, 1897)	MB1					
107. *Cylichna (Cylichnella) bitruncata* (Meyer, 1886)	MB1					
108. *Cymatosyrinx dorseyi* (Cooke, 1926)	MB1					
109. *Cymatosyrinx palmerae* Harris, 1947	MB1	YA6				
110. *Cypraeacites blowi* Dolin, 1991	MB2					
111. *Cypraedia fenestralis* Conrad *in* Wailes, 1854	MB6					
111. *Cypraedia (Eucypraedia) multicarinata* (Dall, 1890)	MB21					
114. *Cypraeorbis towncreekensis* (Dockery, 1977)	MB1					
112. *Cypraeorbis ventripotens* (Cossmann, 1903)	MB6	YA2				
115. *Dientomochilus (Dasyostoma) rugostomum* (Johnson, 1899)	MB6					

116. *Diodora tenebrosa veatchi* Palmer, 1947	MB6				
117. *Dolicholatirus leaensis* Harris, 1897	MB9	YA2			
118. *Ectinochilus (Vaderos) stenzeli* Palmer, 1947	MB2	YA2			
119. *Eoclathurella jacksonica* Casey, 1904	MB6	YA2			
120. *Eoclathurella obesula* Casey, 1904	MB6				
121. *Eoclathurella ornata* Dockery, 1977	MB1				
122. *Eoclathurella*? sp.	MB1				
123. *Eodaphne powelli* Allen, 1970	MB2				
124. *Eopleurotoma albirupsis* Harris, 1947	MB3				
125. *Eopleurotoma carya* Harris, 1937 "var."	MB2				
126. *Eopleurotoma julia* (Cooke, 1926)	MB1				
127. *Eopleurotoma*? sp.	MB1				
128. *Eucheilodon crenocarinata* Heilprin, 1880	MB6	YA6			
129. *Eulima jacksonensis* de Gregorio, 1890	MB1	YA1			
130. *Euspira jacksonensis* Palmer, 1947	MB14	YA2			
131. *Ficus filia* (Meyer, 1885)	MB6				
132. *Ficus merita* Palmer, 1947	MB6				
133. *Fusinus insectoides* (Harris, 1897)	MB6				
134. *Galeodaria petersoni* (Conrad *in* Wailes, 1845)	MB19				
135. *Galeodea (Gomphopages) millsapsi* Sullivan and Gardner, 1939	MB1				
136. *Galeodea planotecta jacksonia* Palmer, 1947	MB1				
137. *Globularia morgani* (Johnson, 1899)	MB1				
138. *Glyptotoma crassiplicata montgomeryensis* Harris, 1937	MB2				
139. *Harpa (Eocithara) jacksonensis* Harris, 1897	MB6				
140. *Haustator perdita* Conrad, 1865	MB6				
141. *Haustator perdita jacksonensis* Cooke, 1926	MB6	YA1			
142. *Haustator rivurbana* Cooke, 1926	MB1				
143. *Hemisinus jacksonius* Palmer, 1947	MB20				
144. *Hexaplex (Hexaplex) katherinae* Vokes, 1968	MB6				
145. *Hexaplex (Hexaplex) marksi* (Harris, 1894)	MB8				
146. *Hexaplex (Hexaplex) supernus* Palmer, 1947	MB6	YA6			
147. *Infracoronia ludoviciana normani* (Harris, 1937)	MB2	YA1			
148. *Jenneria ludoviciana* Johnson, 1899	MB6				
149. *Lapparia dumosa* (Conrad *in* Wailes, 1854)	MB14				
150. *Lapparia dumosa exiqua* Palmer, 1937	MB6				
151. *Lapparia dumosa* (Conrad *in* Wailes, 1854) var.	MB1				
152. *Lapparia fasciola* Dockery, 1977	MB1				
153. *Latirus humilior* (Meyer, 1885)	MB1				
154. *Latirus humilior cognatus* Palmer, 1947	MB2				
155. *Latirus humilior jacksonensis* (Aldrich, 1885)	MB1				
156. *Latirus humilior urbanus* Palmer, 1947	MB9				
157. *Latirus liratus* Dockery, 1977	MB1				
158. *Latirus suturalis* Johnson, 1899	MB1				
159. *Levifusus branneri* Harris, 1894	MB8	YA2			
160. *Levifusus fulguriparens* Maury, 1909	MB2	YA2			
161. *Levifusus moodianus* Cooke, 1926	MB1				
162. *Levifusus mortoniopsis* (Gabb, 1860) subsp.	MB2				
163. *Levifusus* sp. D	MB2				
164. *Lithophysema grande* (Aldrich, 1886)	MB10	YA2			
165. *Litiopa spirata* (Meyer, 1887)	MB1				
166. *Mathilda regularis* (Meyer, 1886)	MB1	YA1	RB1	MT1	BY1
167. *Mazzalina inaurata humerosa* Harris, 1894	MB7				
168. *Mazzalina inaurata oweni* Dall, 1890	MB7	YA1			
169. *Metula gentilicia* Palmer, 1947	MB1				
170. *Metula johnsoni* (Vaughan, 1896)	MB2				
171. *Metula subgracilis* Johnson, 1899	MB6				
172. *Microdrillia cossmanni* (Meyer, 1887)	MB1				
173. *Microdrillia ouachitae* Harris, 1937	MB1	YA6			
174. *Microdrillia* sp. C	MB1				
175. *Microsurcula nucleola* Casey, 1904 var.	MB1				
176. *Mitra (Fusimitra) millingtoni* Conrad *in* Wailes, 1854	MB23	YA2			
177. *Mnestia meyeri* (Cossmann, 1893)	MB1	YA1			
178. *"Murex" angulatus* Meyer, 1886	MB1				

Species	MB	YA	RB	FH	MT	BY
179. *Natica permunda* Conrad *in* Wailes, 1854	MB14	YA6				
180. *Odostomia jacksonensis* Dockery	MB1	YA1				
181. *Oliva (Strephonella) affluens* (Casey, 1903)	MB1			FH1	MT1	BY1
182. *Ottoina kinkelini* (Meyer, 1887)	MB1					
183. *Palmerella alveata* (Conrad *in* Wailes, 1854)	MB6					
184. *Palmerella arenicola* (Conrad, 1865)	MB16	YA2				
185. *Palmerella clevelandia* Harris, 1894	MB8	YA2				
186. *Palmerella creola* Palmer, 1947	MB2					
187. *Papillina dumosa* (Conrad *in* Wailes, 1854)	MB8					
188. *Petaloconchus transcostatus* Dockery, 1977	MB1					
189. *Phalium brevicostatum creolum* Palmer, 1947	MB2					
190. *Phalium taitii johnsoni* Palmer, 1947	MB6					
191. *Philine (Megistostoma) dockeryi* Allen, 1970	MB1					
192. *Platyoptera extenta* (Conrad *in* Wailes, 1854)	MB18	YA2				
193. *Pleurofusia collaris* (Casey, 1903)	MB17					
194. *Pleurofusia danvicola parahilgardina* Harris, 1947	MB2					
195. *Pleurofusia fluctuosa* (Harris, 1937)	MB6					
196. *Pleurofusia hilgardi* (Casey, 1903)	MB1					
197. *Pleurofusia subservata* Harris, 1947	MB2					
198. *Pleuroliria jacksonella* Casey, 1904	MB6					
199. *"Pleurotoma" exsculpata* Meyer, 1885	MB1					
200. *Pliciscala cribrum* (Cooke, 1926)	MB1					
201. *Pliciscala pearlensis* (Meyer, 1887)	MB1					
202. *Polinices weisbordi* Palmer, 1937	MB16	YA1				
203. *Pseudoliva vetusta perspectiva* Conrad *in* Gabb, 1860	MB8	YA6				
204. *Pseudotoma aldrichi* (Meyer, 1887)	MB1					
205. *Pseudotoma axeli* Harris, 1947	MB1					
206. *Pseudotoma heilprini* (Aldrich, 1885)	MB17					
207. *Pseudotoma heilprini* (Aldrich, 1885) "var."	MB2	YA6				
208. *Pterynotus (Pterynotus) weisbordi* Palmer, 1947	MB2					
209. *Puncturella jacksonensis* Meyer, 1887	MB1					
210. *Pyramidella (Montigma) crassispirata* (Meyer, 1887)	MB1					
211. *Pyramidella (Syrnola) meyeri* Cossmann, 1893	MB1	YA1				
212. *Pyramimitra quadralirata* Dockery, 1977	MB1					
213. *Retusa (Cylichnina) jacksonensis* (Meyer, 1886)	MB9	YA1				
214. *Retusa (Cylichnina) jacksonensis exta* Harris, 1894	MB3					
215. *Rissoina mississippiensis* Meyer, 1886	MB1					
216. *Sassia (Personella) jacksonensis* (Meyer, 1885)	MB22					
217. *Scaphander jacksonensis* Palmer, 1947	MB1	YA1				
218. *Scobinella famelica* Casey, 1903 subsp.	MB2					
219. *Scobinella? jacksonensis* (Meyer, 1886)	MB1					
220. *Scobinella louisianae* Harris, 1937	MB6					
221. *Scobinella newtonensis* Aldrich, 1911 subsp.	MB6					
222. *Scobinella transitionalis* Harris, 1937	MB2					
223. *Seraphs* sp.	MB1					
224. *Serpulorbis chavani* Palmer, 1947	MB6					
225. *Simnia* sp.	MB1					
226. *Sinistrella americana* (Aldrich, 1885)	MB1	YA6				
227. *Sinistrella meyeri* Allen, 1970	MB2					
228. *Sinum jacksonense* Dockery, 1977	MB6					
229. *Siphonalia jacksonia* Harris, 1897	MB1					
230. *Siphonalia sullivani* Palmer, 1947	MB2	YA2				
231. *Solariella cancellata jacksonia* Palmer, 1947	MB1					
232. *Solariorbis quadrangularis* (Meyer, 1887)	MB1	YA1				
233. *Solariorbis subangulatus* (Meyer, 1886)	MB1					
234. *Sphaerocypraea jacksonensis* (Johnson, 1899)	MB6					
235. *Sulcocypraea conradi* (Schilder, 1927)	MB1					
236. *Sulcocypraea healeyi* (Aldrich, 1923)	MB6		RB1	FH1	MT1	
237. *Sullivania perexilis* (Aldrich, 1886)	MB1					
238. *Teinostoma moodiense* Palmer, 1947	MB1					
239. *Teinostoma (Idioraphe) verrilli* Meyer, 1885	MB1	YA1	RB1			
240. *Tenuiscala aspersa* (Meyer, 1887)	MB1					
241. *"Terebra" abditiva* Palmer, 1947	MB2					
242. *Terebra (Mirula) jacksonensis* Cooke, 1926	MB8					
243. *Tornus infraplicatus* Johnson, 1899	MB1		RB1			

244. *Transovula (Oxycypraea) producta* Dockery, 1977	**MB1**						
245. *Trigonostoma selectum* Palmer, 1947	**MB2**						
246. *Tritiaria albirupina* (Harris, 1894)	**MB20**	YA1					
247. *Tritiaria hilli* (Harris, 1894)	**MB8**	YA2					
248. *Tritiaria jacksonensis* (Johnson, 1899)	**MB1**						
249. *Tritiaria magnocostata* (Johnson, 1899)	**MB6**						
250. *Tritonoatractus montgomeriensis* (Vaughan, 1896)	**MB8**	YA6					
251. *Tritonoatractus pearlensis* (Aldrich, 1885)	**MB18**	YA2					
252. Cf. *Tudicla* sp. A	MB1						
253. *Turbonilla (Strioturbonilla) major* Meyer, 1887	**MB1**	YA1					
254. *Typhis (Rugotyphis) dentatus* Johnson, 1899	**MB1**						
255. *Unitas pearlensis* (Meyer and Aldrich, 1887)	**MB1**						
256. *Uromitra grantensis* (Johnson, 1899)	**MB6**						
257. *Vasum humerosum* Vaughan, 1896	**MB6**						
258. *Viviparus erronis* Allen, 1970	**MB2**						
259. *Xenophora humilis* (Conrad, 1848)	MB18					**BY1**	
260. *Xenophora reclusa* (Conrad *in* Wailes, 1854)	**MB6**				MT1		
Bivalvia with FAD in MB							
261. *Abra (Abra) nitens jacksonica* Harris, 1946	**MB1**						
262. *Anodontia (Eophysema) ozarkana* (Harris, 1897)? "var."	MB1	YA2					
263. *Arcoperna filosa* Conrad, 1865	**MB1**						
264. *Atrina jacksoniana* Dall, 1898	**MB6**	YA4					
265. *Atrina* sp. A	MB2						
266. *Barbatia corvamnis* Harris, 1946	**MB3**						
267. *Barbatia (Cucullaearca) cuculloides* (Conrad, 1833)	**MB18**						
268. *Barbatia (Barbatia) seraperta* Harris, 1946	**MB1**						
269. *Bathytormus clarkensis postclarkensis* (Harris, 1919)	**MB7**	YA2					
270. *Bathytormus flexurus* (Conrad *in* Wailes, 1854)	**MB14**						
271. *Bathytormus flexurus productus* (Conrad, 1863)	**MB1**						
272. *Biocorbula pearlensis* (Meyer, 1886)	**MB1**						
273. *Caestocorbula wailesiana* (Harris *in* Dall, 1898)	**MB18**	YA17					
274. *Callista (Callista) annexa* (Conrad, 1865)	**MB26**	YA4					
275. *Callista (Costacallista) pearlensis* (Harris, 1897)	**MB1**						
276. *Calorhadia (?Litorhadia) albirupina* (Harris, 1946)	**MB3**						
277. *Carditamera williamsi* Dockery, 1998	**MB1**						
278. *Chama (Psilopus) mississippiensis* Conrad, 1848	MB6		RB1	FH1	MT1		**BY1**
279. *Chlamys (Aequipecten) clinchfieldensis* Harris, 1951	**MB29**	YA12					
280. *Chlamys (Aequipecten) cocoana* Dall, 1898	MB29	YA29	**RB17**				
281. *Chlamys nupera* (Conrad *in* Wailes, 1854)	**MB18**						
282. *Corbula (Caryocorbula) densata* Conrad *in* Wailes, 1854	MB27	YA4					
283. *Corbula (Caryocorbula) willistoni* Meyer, 1885	**MB7**	YA18					
284. *Corbula (Caryocorbula) willistoni arkansia* Harris, 1946	**MB7**						
285. *Crassetella* sp. I	MB1						
286. *Crassinella pygmaea* (Conrad, 1865)	**MB1**						
287. *Crenella? tenuis* (Meyer, 1887)	**MB1**						
288. *Diplodonta (Diplodonta) ungulina yazoocola* Harris, 1946	**MB1**						
289. *Eburneopecten (Eburneopecten) frontalis* (Dall, 1898)	**MB1**						
290. *Eburneopecten (Eburneopecten) scintillatus* Conrad, 1865	**MB17**	YA4					
291. *Eburneopecten (Eburneopecten) subminutus* (Aldrich, 1903)	MB1		**RB1**	FH1	MT1		BY1
292. *Erycina zitteli* Meyer, 1887	**MB1**						
293. *Felaniella palmerae* Dockery, 1980	**MB1**						
294. *Gari (Gobraeus) jacksonensis* Harris, 1946	**MB22**						
295. *Gastrochaena (Gastrochaena) mississippiensis* Harris, 1946	**MB1**						
296. *Glycymeris (Glycymeris) filosa* (Conrad *in* Wailes, 1854)	**MB18**						
297. *Gryphaeostrea plicatella* (Morton, 1833)	MB17	**YA17**					

298. *Hormomya hamatoides* (Call, 1891)	**MB**3						
299. *Kelliella boettgeri* Meyer, 1886	**MB**6	YA1					
300. *Laevicardium gardnerae* Cooke, 1926	**MB**1						
301. *Limopsis (Pectunculina) radiata* Meyer, 1885	**MB**6						
302. *Lirodiscus (Lirodiscus) jacksonensis* (Meyer, 1885)	**MB**27	YA30					
303. *Lirodiscus (Lirodiscus) pretriangulata* (Dockery, 1977)	**MB**6						
304. *Lirodiscus (Lirodiscus)* sp.	MB1						
305. *Lithophage (Lithophage)* sp.			MB1				
306. *Lucina (Callucina?) curta* (Conrad, 1865)	**MB**9	YA1					
307. *Lucina (Callucina?) subcurta* Harris, 1946	**MB**6						
308. *Mactra inornata* Meyer, 1885	**MB**1						
309. *Margaritaria inexpectens* Vokes, 1964	**MB**2						
310. *Miltha (Plastomiltha?) gaufia* Harris, 1946	**MB**18						
311. *Nemocardium (Nemocardium) nicoletti* (Conrad, 1841)	**MB**18	YA6					
312. *Nucula (Nucula) magnifica yazooensis* Harris, 1946	**MB**6						
313. *Nucula (Nucula) spheniopsis* Conrad, 1865	**MB**18	YA4					
314. *Nuculana linifera* Conrad, 1865	**MB**1						
315. *Ostrea falco* Dall, 1896	**MB**18	YA17					
316. *Pachecoa corvamnis* (Harris, 1946)	**MB**3						
317. *Panopea (Panopea) oblongata* Conrad, 1848	MB1	YA4		MT1	MR1	GN1	**BY**1
318. *Periploma claibornense parvum* Harris, 1885	**MB**1						
319. *Periploma collardi turgidum* Harris, 1946	**MB**3						
320. *Periploma equalum* Dockery, 1977	**MB**1						
321. *Pitar (Pitar) securiformis* (Conrad, 1865)	**MB**16	YA18					
322. *Pleuromeris inflatior jacksonensis* Meyer, 1885	**MB**1						
323. *Pleuromeris quadrata* Dockery, 1980	**MB**8						
324. *Plicatula creola* Allen, 1970	**MB**2						
325. *Poromya mississippiensis* Meyer and Aldrich, 1887	**MB**1						
326. *Pteria limula vanwinkleae* Harris, 1946	**MB**18	YA4					
327. *Pycnodonta (Pycnodonta) trigonalis* (Conrad *in* Wailes, 1854)	**MB**18	YA12					
328. *Pycnodonta vicksburgensis* (Conrad, 1848) var.	MB4	YA18					
329. *Solemya (Archarax) dolinorum* Robinson, 1983	**MB**1						
330. *Spisula mississippiensis* (Conrad, 1848) subsp.	MB1						
331. *Spondylus (Spondylus) dumosus* (Morton, 1834)	MB4		**RB**17				
332. *Spondylus hollisteri* Harris, 1951	MB4	YA12					
333. *Tellina (Arcopagia) eburneopsis* Conrad, 1865	**MB**8						
334. *Tellina (Eurytellina) linifera* Conrad, 1865	**MB**18						
335. *Tellina pearlensis* Meyer, 1887	**MB**1						
336. *Tellina (Arcopaginula) raveneli weisbordi* Harris, 1946	**MB**2						
337. *Tellina (Eurytellina) spillmani* Dall, 1900	**MB**1						
338. *Tellina (Eurytellina) spillmani corvia* Harris, 1946	**MB**3						
339. *Tellina (Arcopagia) trumani garlandica* Harris, 1946	**MB**1						
340. *Tellina (Eurytellina) vaughani* Cooke, 1926 var.	MB1						
341. *Tellina vicksburgensis moodiana* Cooke, 1926	**MB**1						
342. *Teredo mississippiensis* Conrad *in* Wailes, 1854	**MB**6						
343. *Timothynus bulla* (Conrad, 1865)	**MB**8						
344. *Timothynus deflatus* Harris, 1946	**MB**1						
345. *Venericardia (Venericor) apodensata* Gardner and Bowles, 1939	**MB**23						
346. *Venericardia (Rotundicardia) diversidentata* Meyer, 1885	**MB**18	YA1					
347. *Venericardia (Rotundicardia) diversidentata* Meyer, 1885 var.	MB6						
348. *Venericardia praecisa* Dall, 1903	**MB**3						
349. *"Venericardia"* sp. C	MB4						
350. *"Venus" jacksonensis* Meyer, 1887	**MB**1						
351. *Verticordia (Verticordia) cossmanni* Dall, 1903	**MB**1						
352. *Yoldia (Calorhadia) mater* (Meyer, 1885)	**MB**19	YA4					
353. *Yoldia (Calorhadia) reginajacksonis* (Harris, 1897)	**MB**8	YA1					
354. *Yoldia (Calorhadia) reginajacksonis* (Harris, 1897) "var."	MB1						

Taxon	MB	YA	RB	MT	BY
355. *Yoldia (Orthoyoldia) rubamnis* (Harris, 1946)	**MB**7				
356. *Yoldia (Orthoyoldia)* sp. C	MB2				
Scaphopoda with FAD in MB					
357. *Cadulus (Polyschides) jacksonensis* Meyer, 1885	**MB**1				
358. *Cadulus juvenis* Meyer, 1886	**MB**1				
359. *Cadulus (Polyschides) margarita* Palmer, 1947	**MB**14				
360. *"Dentalium bitubatum"* Meyer, 1886	**MB**1				
361. *Dentalium (Antalis) danvillense* Palmer, 1947	MB6	YA6			
362. *Dentalium (Antalis) mississippiense jacksonense* Palmer, 1947	**MB**18	YA2			
363. *Dentalium (Antalis) vincense* Palmer, 1947	**MB**3				
364. *Dentalium* sp. E.	MB32				
365. *Fustiaria (Laevidentalium) danai* (Meyer, 1885)	**MB**6				
366. *Fustiaria (Laevidentalium) subcompressa* (Meyer, 1885)	**MB**1				
Cephalopoda with FAD in MB					
367. *Aturia alabamensis* (Morton, 1834)	**MB**33	YA1	RB1		
368. *Belemnosella palmerae* Allen, 1968	**MB**2				
369. *Belosaepia* sp. A	MB1				
Gastropoda with FAD in YA					
370. *Abderospira leblanci* Palmer, 1947		**YA**2			
371. *Asthenotoma danvitexa* Harris, 1947		**YA**6			
372. *Asthenotoma* sp. A		YA2			
373. *Asthenotoma* sp. B		YA2			
374. *Athleta* sp. F		YA1			
375. *Bovicornu eocenense* Meyer, 1886		YA1	**RB**1		
376. *Cantharus* sp.		YA1			
377. *Cerithiella ouachitensis* Palmer, 1947		**YA**2			
378. *Coronia genitiva* (Casey, 1904) "var."		YA2			
379. *Coronia? ludonorma* Harris, 1947		**YA**2			
380. *Coronia nodulina* (Casey, 1904) "var."		YA2			
381. *Coronia nodulina ouachitensis* (Harris, 1937)		**YA**6			
382. *Coronia? plentopsis* (Harris, 1947)		**YA**2			
383. *Eodrillia?* sp.		YA1			
384. *Eopleurotoma ouachitensis* Harris, 1937		**YA**15			
385. *Infracoronia ludoviciana normani* Harris, 1937 var.?		YA1			
386. *Latirus* sp.		YA1			
387. *Levifusus spiniger* (Conrad, 1848)		YA1	RB1	MT1	BY1
388. *Limacina* sp.		YA1			
389. *Niso* sp. B		YA2			
390. *Ottoina?* sp.		YA1			
391. *Palmerella arenicola danvillensis* Stenzel and Turner, 1940		**YA**6			
392. Cf. *Penion penrosei* (Palmer, 1947)		**YA**2			
393. *Pleurofusia danvicola* Harris, 1937		**YA**6			
394. *Pleurofusia danvicola* Harris, 1937 var.		YA2			
395. *Pleurofusia evanescens* (Casey, 1903)		**YA**2			
396. *Pseudotoma heilprini gibsoni* Harris, 1937		**YA**2			
397. Cf. *Rhizorus* sp.		YA2			
398. *Ringicula* sp.		YA1			
399. *Scaphander* sp.		YA1			
400. *Sinum (Sigaretotrema) danvillense,* Palmer, 1947		**YA**2	RB1		
401. *Siphonalia sullvani ouachitae* Harris, 1947		**YA**6			
402. *Tritonoatractus pearlensis danvillensis* Palmer, 1947		**YA**6			
403. *Turricula plutonica weisbordi* Harris, 1937		**YA**6			
404. *Turritella bunkerhillensis* Palmer, 1947		**YA**2			
405. *Turritella* sp. R.		YA1			
Bivalvia with FAD in YA					
406. *Barbatia* sp. B		YA1			
407. *Callista* sp. A		YA4			
408. *Chamys danvillensis* Weisbord *in* Tucker-Rowland, 1936		**YA**18			
409. *Chlamys (Aequipecten) spillmani* (Gabb, 1860)		**YA**33			

410. *Chlamys* sp. D	YA1				
411. *Limopsis* sp.	YA1				
412. *Lirodiscus* sp.	YA1				
413. *Modiolus cretaceus* (Conrad, 1935)	**YA4**				
414. *Modiolus*? sp.	YA1				
415. *Nuculana* sp.	YA1				
416. *Periploma* sp.	YA1				
417. *Plicatula? louisiana* Harris, 1946	**YA2**				
418. *Plicatula* sp. C	YA34				
419. *Pycnodonta vicksburgensis mortoni* (Gabb, 1862)	**YA4**				
420. *Saxolucina (Plastomiltha)* sp.?	YA1				
421. *Spheniopsis mississippiensis* (Meyer, 1887)	YA1	**RB1**		MT1	BY1
422. *Trinacria*? sp.	YA1				
423. *Venericardia (Venericor) klimacodes* Gardner and Bowles, 1939	**YA35**				
424. *Yoldia* sp. A	YA1				
425. *Yoldia* sp. B	YA1				
Scaphopoda with FAD in YA					
426. *Cadulus* sp.	YA1				
Cephalopoda with FAD in YA					
427. *Belosaepia* sp.	YA1				
Gastropoda with FAD in RB					
428. *Aclis* sp.? B		RB1			
429. *Acteocina crassiplica involuta* MacNeil, 1984		RB1		MT1	**BY1**
430. *Acteon (Acteon) meyeri* MacNeil, 1984		RB1	FH1	**MT1**	BY1
431. *Acteon (Acteon) subaldrichi* MacNeil, 1984		**RB1**			
432. *Agatrix mississippiensis* (Conrad, 1848)		RB1	FH1	MT1	**BY1**
433. *Alabina menthafontis* MacNeil, 1984		RB1		**MT1**	
434. *Ampullinopsis mississippiensis* (Conrad, 1848)		RB1		**MT1**	BY1
435. *Architectonica (Architectonica) fuscicava* MacNeil, 1984 var. ?		RB1			
436. *Architectonica (Granosolarium) hargeri* (Meyer, 1886)		**RB1**			
437. *Architectonica (Architectonica) textilina caseyi* MacNeil, 1984		**RB1**			
438. *Architectonica (Architectonica)* sp.		RB1			
439. *Bathytoma rhomboidea* MacNeil, 1984		**RB1**			
440. *Bathytoma rhomboidea lyrata* MacNeil, 1984		**RB1**			
441. *"Bittium" acuta* (Meyer, 1886)		**RB1**			
442. *Bittum (Argyropeza?) ottoi* MacNeil, 1984		RB1	FH1	**MT1**	BY1
443. *Calyptraea (Trochita)* sp.		RB1		MT1	
444. *Capulus planus* Dockery, 1984		**RB1**			
445. *Caricella (Atraktus) reticulata* (Aldrich, 1885)		**RB1**		MT1	
446. *Cerithiella langdoni* (Aldrich, 1885)		**RB1**	FH1	MT1	
447. *Cerithiella langdoni* (Aldrich, 1885) var.		RB1			
448. *Cerithiella nassuloides* MacNeil, 1984		RB1		**MT1**	
449. *Chicoreus (Phyllonotus) mississippiensis* (Conrad, 1848)		RB1		MT1	**BY1**
450. *Chicoreus (Phyllonotus) stetopus* (de Gregorio, 1890)		RB1		**MT1**	
451. *Clathurella meyeri* MacNeil, 1984		**RB1**			
452. *Clavidrupa anita* (Aldrich, 1885)		**RB1**			
453. *Clavilithes longiformis* Dockery, 1984		**RB1**			
454. *Clavilithes vicksburgensis* (Conrad, 1849)		RB1		**MT1**	
455. *Cochlespira cookei rubracollis* MacNeil, 1984		**RB1**			
456. *Confusiscala (Funiscala) durhami* MacNeil, 1984		**RB1**			
457. *Conomitra crenulata* Dockery, 1984		**RB1**			
458. *Conomitra vicksburgensis laevigata* Dockery, 1984		RB1	FH1	**MT1**	
459. *Conus alveatus* Conrad, 1865		RB1	FH1	MT1	**BY1**
460. *Conus protractus* Meyer, 1885		RB1		MT1	**BY1**
461. *Coronia (Coroniopsis) ancilla* (Casey, 1903)		**RB1**			
462. *Crenilabium altispira* MacNeil, 1984		RB1	FH1		
463. "*Cyclostremiscus*" sp.		RB1			BY1
464. *Cylichna* sp.		RB1			
465. *Cypraeorbis* sp.?		RB1			
466. *Dermomurex (Takia) cookei* Vokes, 1975		**RB1**		MT1	
467. *Diodora mississippiensis* (Conrad, 1848)		RB1	FH1	MT1	**BY1**

Species	RB	FH	MT	BY
468. *Dolicholatirus cervicrassus* Dockery, 1984	**RB1**			
469. *Epitonium (Sthenorytis) whitfieldi* (Aldrich, 1885)	**RB1**			
470. *Eulimella clearyensis* MacNeil, 1984	RB1		**MT1**	BY1
471. *Eulimella* sp.	RB1			
472. *Eumetula vicksburgella* MacNeil, 1984	**RB1**		MT1	
473. *Euspira vicksburgensis* (Conrad, 1848)	RB1		MT1	**BY1**
474. *Euspira* sp.?	RB1			
475. *Ficus mississippiensis* Conrad, 1848	RB1		MT1	BY1
476. *Galeodaria shubutensis* (Aldrich, 1885)	**RB1**			
477. *Gemmula amica* (Casey, 1903)	**RB1**			
478. *Gemmula amica* (Casey, 1903) var. A	RB1			
479. *Latirus aldrichi* Dockery, 1984	**RB1**		MT1	
480. *Latirus indistinctus* Aldrich, 1884	**RB1**			
481. *Latirus mississippiensis* (Conrad, 1848)	RB1			**BY1**
482. *Latirus protractus* (Conrad, 1848)	RB1		MT1	**BY1**
483. *Limacina cf. L. inflata* (d'Orbigny, 1847)	RB1			
484. *Limacina* sp.	RB1			
485. *Lyria (Enaeta) isabellae modesta* Dockery, 1984	RB1	**FH1**		BY1
486. *Lyria (Lyria) nestor* Casey, 1903	**RB1**			
487. *Mambrinia brevidentata* (Aldrich, 1885)	**RB**17			
488. *Mathilda inaequistriata* (Meyer, 1886)	**RB1**			
489. *Melanatria serratiodes* (Aldrich, 1894)	**RB1**			
490. *Melanella amnicreta* MacNeil, 1984	RB1		**MT1**	
491. *Melanella postnotata* MacNeil, 1984	RB1		**MT1**	
492. *Metula (Metula) dockeryi* Emerson, 1993	**RB1**			
493. *Metula (Metula) fastidiosa* Casey, 1903	**RB1**	FH1		
494. *Metula (Caseyella) hiwanneensis* Dockery, 1984	**RB1**			
495. *Metula (Caseyella) neptuneiformis* MacNeil, 1984	**RB1**			
496. *Microdrillia brevis allo* MacNeil, 1984	**RB1**			
497. *Microdrillia infans* (Meyer *in* Smith, 1886)	**RB1**			
498. *Microdrillia vicksburgella* Casey, 1903	RB1	FH1	MT1	**BY1**
499. *Mitra (Fusimitra) conquisita* Conrad, 1848	RB1	FH1	MT1	**BY1**
500. *Mitra (Fusimitra) mississippiensis* Conrad, 1848	RB1			**BY1**
501. *Mitrella (Columbellopsis) fuscicava* MacNeil, 1984	RB1	FH1	**MT1**	
502. *Murexiella (Murexiella) vaughani* MacNeil, 1984	**RB1**			
503. *Natica (Naticarius) acuticallosa* MacNeil, 1984	RB1		MT1	**BY1**
504. *Natica (Naticarius) aff. N. (N.) alazana* Cooke, 1928	RB1		MT1	
505. *Natica (Natica) caseyi* MacNeil, 1984	RB1		MT1	**BY1**
506. *Niso fuscicava* MacNeil, 1984	RB1		**MT1**	
507. *Odostomia (Odostomia) boettgeri* Meyer, 1887	RB1		**MT1**	
508. *Odostomia (Odostomia) byramensis* MacNeil, 1984 var.?	RB1			
509. *Odostomia (Odostomia)* sp.	RB1			
510. *Olssonella elongata* Dockery, 1984 var.?	RB1			
511. *Phalium (Menthafontia)* sp.	RB1			
512. *Phandella transemma* MacNeil, 1984	RB1		**MT1**	
513. *Phandella* sp.	RB1			
514. *Pleurofusia clarkeana* (Aldrich, 1894)	**RB1**			
515. *Pleurofusia (Xestosurcula) fessa* MacNeil, 1984	**RB1**		MT1	
516. *Pleurofusia hiwanneensis* MacNeil, 1984	**RB1**			
517. *Pleurofusia oblivia* (Casey, 1903)	**RB1**			
518. *Pleuroliria subsimilis* Casey, 1904	**RB1**			
519. *Pleuroliria tenuis* MacNeil, 1984	**RB1**			
520. *Pliciscala (Nodiscala?) caseyi* MacNeil, 1984	RB1		**MT1**	BY1
521. *Pterynotus (Pterochelus) angelus* (Aldrich, 1886)	**RB1**			
522. *Pterynotus (Pterynotus) burnsii* (Aldrich, 1894)	**RB1**			
523. *Ringicula (Ringiculella) mississippiensis petila* MacNeil, 1984	RB1	FH1	MT1	**BY1**
524. *Sassia (Sassia) conradiana* (Aldrich, 1885)	**RB1**			
525. *Scalina rubricollis* MacNeil, 1984	**RB1**			
526. *Scaphander (Scaphander) primus* Aldrich, 1885	**RB1**			
527. *Scobinella caelata* Conrad, 1848	RB1			**BY1**
528. *Scobinella pluriplicata* Casey, 1903	**RB1**			
529. *Sconsia prelintea* MacNeil, 1984	**RB1**			
530. *Serpulorbis* sp.	RB1		MT1	BY1
531. *Siphonochelus (Laevityphis) curvirostratus* (Conrad, 1848)	RB1		MT1	**BY1**
532. *Spiradaphne lowei refugium* MacNeil, 1984	**RB1**			

533. *Strombiformis caseyi* MacNeil, 1984	RB1		**MT1**			
534. *Syntomodrillia collarubra* MacNeil, 1984	**RB1**					
535. *Terebra (Strioterebrum) alaba* MacNeil, 1984	RB1		**MT1**			
536. *Terebra (Terebrellina) divisura clearyensis* MacNeil, 1984	RB1		**MT1**			
537. *Terebra (Terebrellina) hiwanneensis* MacNeil, 1984	**RB1**					
538. *Tornatellaea* sp.?	RB1					
539. *Triphora (Triphora) bilineata* (Meyer, 1886)	**RB1**	FH1				BY1
540. *Triphora (Triphora) meridionalis* (Meyer, 1886)	**RB1**		MT1			
541. *Tritiaria falsus* (Casey, 1903)	**RB1**					
542. *Tritiaria macilenta* (Casey, 1903)	**RB1**					
543. *Tritiaria meyeri* MacNeil, 1984 n. subsp.?	RB1					
544. *Tritiaria scapulistriata* MacNeil, 1984	**RB1**					
545. *Tropisurcula caseyi* (Aldrich, 1903)	**RB1**	FH1	MT1			BY1
546. *Turbonilla mississippiensis* Meyer, 1886	**RB1**					
547. *Turricula (Orthosurcula) longiforma* (Aldrich, 1885)	**RB1**					
548. *Turritella caseyi* MacNeil, 1984	RB1		MT1			**BY1**
549. *Turritella premimetes* MacNeil, 1984	RB1		**MT1**			
550. *Turritella rubricollis* MacNeil, 1984	**RB1**					
551. *Typhis (Typhina) mississippiensis* Gertmann, 1969	**RB1**					
552. *Umbraculum* sp.	RB1					
553. *Urosalpinx? aspinosus* (Meyer, 1886)	**RB1**					
554. *Varicobela smithii* (Aldrich, 1885)	**RB1**		MT1			
555. *Vexillum (Costellaria) lintoidea* (Aldrich, 1894)	**RB1**					
556. *Vitrinella (Vitrinella) laevis* (Meyer, 1886)	**RB1**					BY1
557. *Vitrinella (Vitrinellops)* sp.	RB1					
558. *Xenophora (Stellaria) conica* Dall, 1892	**RB1**		MT1			

Bivalvia with FAD in RB

559. *Abra pectorosa* (Conrad, 1848)	RB1					**BY1**
560. *Agnocardia glebosum* (Conrad, 1848)	RB1	FH1	MT1			BY1
561. *Anodontia (Anodontia) mississippiensis* (Conrad, 1848)	RB1		MT1			BY1
562. *Astarte triangulata* Meyer *in* Smith, 1886	**RB1**					BY1
563. *Barbatia paradiagona* Dockery, 1982	RB1		**MT1**			
564. *Brevinucula pseudopunctata* Dockery, 1982	**RB1**					
565. *Callista (Callista) sobrina* (Conrad, 1848)	RB1	FH1	MT1			**BY6**
566. *Chama (Chama) pappiladerma* Dockery, 1982	RB1					
567. *Chione (Lirophora) victoria* Dall, 1903	RB1		**MT1**			
568. *Chlamys (Anatipopecten) anatipes* (Morton, 1833)	RB12	FH1	MT1	MR36	GN4	BY4
569. *Corbula (Caryocorbula) engonata* Conrad, 1848	RB1		MT1			BY1
570. *Corbula (Vokesula) rufaripa* Dockery, 1982	**RB1**		MT1			
571. *Dimya rufaripa* Vokes, 1979	**RB1**					
572. *Eburneopecten (Eburneopecten) subminutus* (Aldrich, 1903)	**RB1**	FH1	MT1			BY1
573. *Glycymeris intercostata* (Gabb, 1860)	RB1		MT1			
574. *Halirus (Halirus) quadrangularis* (Aldrich, 1903)	**RB1**					
575. *Jouannetia (Pholadopsis) triguetra* (Conrad, 1848)	RB1		MT1			BY1
576. *Kelliella rufaripa* Dockery, 1982	**RB1**		MT1			
577. *Myrtea (Myrtea) scopularis* (Casey, 1903)	**RB1**					
578. *Nemocardium (Nemocardium) eocenense* (Meyer, 1887)	**RB1**					
579. *Nucula tallahalaensis* Dockery, 1982	RB1					**BY1**
580. *Nucula vicksburgensis* Conrad, 1848	RB1	FH1	MT1			BY1
581. *Nuculana triangulata* (Meyer *in* Smith, 1886)	**RB1**					
582. *Pecten (Pecten) perplanus* Morton, 1833	**RB**33	FH1		MR4		
583. *Pitar aldrichi* Dockery, 1982	**RB1**	FH1	MT1			
584. *Plectodon intastriata* (Conrad, 1848)	RB1		**MT1**			
585. *Plicatula variplicata* Dockery, 1982	RB1	FH1				**BY1**
586. *Pteria argentea* (Conrad, 1848)	RB1		MT1			BY1
587. *Pycnodonta paroxis* (Lesueur MS *in* Dockery, 1982)	RB1	FH1	**MT1**			BY1
588. *Pycnodonta vicksburgensis* (Conrad, 1848)	RB1	FH1	MT1	MR1	GN1	BT1
589. *Scapharca (Scapharca) invidiosa* (Casey, 1903)	**RB1**					
590. *Scapharca (Scapharca)* sp.	RB1					
591. *Septifer (Septifer) probolus* Dockery, 1982	**RB1**					
592. *Spondylus (Spondylus) dumosus* (Morton, 1834)	**RB**17					
593. *Spondylus (Spondylus) granulocostatus* Dockery, 1982	**RB**4					
594. *Tellina lintea* (Conrad, 1848)	RB1	FH1	MT1			BY1
595. *Timothynus turgida* (Conrad, 1848)	RB1	FH1	MT1.			BY1

596. *Tiza* sp.	RB1					
597. *Trinacria menthifontis* Dockery, 1982	RB1		**MT1**			
598. *Venericardia (Rotundicardia) carsonensis* Dall, 1903	**RB1**	FH1	MT1			BY1
599. *Ventricolaria ucuttana* (Dall, 1903)	**RB1**					BY1
600. *Verticordia (Verticordia) dalliana* Aldrich, 1903	**RB1**	FH1	MT1			BY1
601. *Yoldia clydoniona* Dockery, 1982	**RB1**	FH1				
Scaphopoda with FAD in RB						
602. *Cadulus (Polyschides) corpulentus* Meyer *in* Smith, 1886	**RB1**					
603. *Cadulus (Polyschides) quadriturritus* Meyer *in* Smith, 1886	**RB1**	FH1	MT1			
604. *Cadulus (Polyschides) vicksburgensis* Meyer, 1885	RB1		MT1			**BY1**
605. *Dentalium polygonuum* Casey, 1903	**RB1**					
606. *Dentalium zephyrinum* Casey, 1903	**RB1**					
607. *Fustiaria (Rhabdus)* sp.	RB1		MT1			
Gastropoda with FAD in FH						
608. *Acteocina crassiplica* (Conrad, 1848)		FH1	MT1		GN1	**BY1**
609. *Acteocina crassiplica altispira* MacNeil, 1984		FH1	**MT1**			
610. *Acteon (Acteon) menthafons* MacNeil, 1984		FH1	**MT1**			BY1
611. *Acteon (Acteon) pretextilis* MacNeil, 1984		FH1	**MT1**			
612. *Atys (Atys) pinguis* MacNeil, 1984		FH1	**MT1**			
613. *Bathytoma congesta fontis* MacNeil, 1984		FH1	**MT1**			
614. *Cerithioderma*? sp.		FH1				
615. *Conomitra crenulata modesta* Dockery, 1984		FH1	**MT1**			
616. *Cylichna nida* MacNeil, 1984		FH1	**MT1**			BY1
617. *Melongena (Myristica) crassicornuta* Conrad, 1848		FH2				**BY1**
618. *Sassia (Sassia) conradiana menthafons* MacNeil, 1984		FH1	**MT1**			
619. *Scaphander (Coeloscapha) hilgardi* MacNeil, 1984		FH1	**MT1**			BY1
620. *Semivertagus silvacollinis* Dockery, 1984		**FH1**				
621. *Sigatica conradii* (Dall, 1892)		FH1	MT1			**BY1**
622. *Vexillum (Costellaria) cervilirata* Dockery, 1984		**FH1**				
623. *Vexillum (Costellaria) laevicostata* Dockery, 1984		FH1	**MT1**			
624. *Vexillum (Costellaria) multicostata* Dockery, 1984		**FH1**	MT1			
625. *Volvulella subspinosa* MacNeil, 1984		FH1	**MT1**			BY1
Bivalvia with FAD in FH						
626. *Arca (Arca) subprotracta* Heilprin, 1882		FH1	**MT1**			
627. *Astarte menthifontis* Dockery, 1982		FH1	**MT1**			BY1
628. *Chama* sp.		FH1				
629. *Chione (Chione) craspedonia* Dall, 1903		FH1	**MT1**			
630. *Chlamys (Aequipecten) redwoodensis* Dockery, 1982		FH1				**BY1**
631. *Crassinella variablis* Dockery, 1982		FH1	MT1			**BY1**
632. *Crenella fenestra* Dockery, 1982		FH1	**MT1**			BY1
633. *Diplodonta (Diplodonta) eburnea* (Conrad, 1848)		FH1	MT1			BY1
634. *Donax funerata* Conrad, 1848		FH1	**MT1**			
635. *Ervilia exterolaevis* Dockery, 1982		**FH1**	MT1			
636. *Limaria* sp.		FH1				
637. *Lucina (Parvilucina) posteocurta* Dockery, 1982			FH1	**MT1**		
BY1						
638. *Myrtea (Myrtea) vicksburgensis* (Casey, 1903)		FH1	**MT1**			BY1
639. *Nemocardium (Nemocardium) diversum* (Conrad, 1848)		FH1	MT1			**BY1**
640. *Pitar (Lamelliconcha) megacostata* Dockery, 1982		FH1	**MT1**			
641. *Scapharca (Scapharca) delicatula* (Casey, 1903)		FH1	**MT1**			
642. *Semelina pilsbryi* (Casey, 1903)		FH1	**MT1**			BY1
643. *Spisula inaequilateralis* (Meyer *in* Smith, 1886)			FH1	**MT1**		
644. *Spondylus (Spondylus) filiaris* Dall, 1916		FH1	MT1			BY1
645. *Tellina subprotexta* Dockery, 1982		FH1	**MT1**			
646. *Trachycardium eversum* (Conrad, 1848)		FH1	MT1			BY1
647. *Trigoniocardia (Americardia) silvacollina* Dockery, 1982		**FH1**				
Scaphapoda with FAD in FH						
648. *Dentalium mississippiense* Conrad, 1848		FH1				**BY1**

Gastropoda with FAD in MT			
649. *Aclis matsoni* MacNeil, 1984 var.	MT1		
650. *Acrilla (Acrilloscala) palmerae* MacNeil, 1984	MT1		
651. *Acteon (Acteon) aldrichi* MacNeil, 1984	MT1		
652. *Acteon (Acteon) preluculi* MacNeil, 1984	MT1		
653. *Acteon (Kleinacteon) puteatus* MacNeil, 1984	MT1		
654. *Acteon (Acteon)* sp. A	MT1		
655. *Admetula inflata* Dockery, 1984	MT1		
656. *Amyssodropa clearyensis* MacNeil, 1984	MT1		
657. *Aporrhais (Goniocheila) menthafontis* MacNeil, 1984	MT1		
658. *Architectonica (Architectonica) fuscicava* MacNeil, 1984	MT1		BY1
659. *Architectonica (Pseudotorinia) julia* MacNeil, 1984	MT1		
660. *Architectonica (Architectonica) menthafontis* MacNeil, 1984	MT1		
661. *Atys (Roxaniella) caseyi* MacNeil, 1984	MT1		
662. *Atys (Roxaniella) caseyi* MacNeil, 1984 var.	MT1		
663. *Bittium (Argyropeza?) caseyi* MacNeil, 1984	MT1		
664. *Caecum solitarium* Meyer, 1886	MT1	GN1	
665. *Calyptraea (Trochita) conradi* MacNeil, 1984	MT1		BY1
666. *Caricella (Atraktus) demissa* Conrad, 1848 var. B	MT1		
667. *Caricella (Atraktus)* sp.	MT1		
668. *Cassis flintensis* Mansfield, 1940	MT37		BY1
669. *Cerithiella leafensis* MacNeil, 1984	MT1		
670. *Clathurella meyeri sylvarenensis* MacNeil, 1984	MT1		
671. *Clavilithes* sp. A	MT1		
672. *Cochlespira cookei* MacNeil, 1984	MT1		
673. *Coronia (Coroniopsis) tenella antetenella* MacNeil, 1984	MT1		
674. *Crenilabium paucicrenulatus* MacNeil, 1984	MT1		
675. *Crucibulum (Dispotaea) hyalonama* MacNeil, 1984	MT1		
676. *Cyclostremiscus menthafons* MacNeil, 1984	MT1		
677. *Cylichna acutiscapulae* MacNeil, 1984	MT1		
678. *Cylichna acutiscapulae corrugata* MacNeil, 1984	MT1		
679. *Cymatium (Septa) amnicretum* MacNeil, 1984	MT1		
680. *Cymia (Tritonopsis) subalveata* (Conrad, 1848)	MT1		BY1
681. *Cypraeorbis sphaeroides* (Conrad, 1848)	MT1		BY1
682. *Diodora postantica* MacNeil, 1984	MT1		
683. *Discopsis pilsbryi* MacNeil, 1984	MT1		
684. *Distorsio (Distorsio) crassidens* (Conrad, 1848)	MT1		BY1
685. *Dolicholatirus perexilis* (Conrad, 1848)	MT1		BY1
686. *Epitonium (Gryoscala)* n. sp.	MT1		
687. *Epitonium* sp.?	MT1		
688. *Euspira vicksburgensis cookei* MacNeil, 1984	MT1		
689. *Fusiturricula ichusa* MacNeil, 1984	MT1		
690. *Galeodaria shubutensis gardnerae* MacNeil, 1984	MT1		
691. *Gemmula mediosa* MacNeil, 1984	MT1		
692. *Kurtziella protatrostyla* MacNeil, 1984	MT1		
693. *Levifusus nodulatum* (Conrad, 1849)	MT1		
694. *Litiopa meyeri* MacNeil, 1984	MT1		
695. *Lyria (Lyria) mississippiensis* (Conrad, 1848)	MT1		BY1
696. *Mathilda aff. M. plexita* Dall, 1896	MT1		
697. *Melanella* sp.	MT1		
698. *Microdrillia brevis gemma* MacNeil, 1984	MT1		
699. *Microsurcula mentha* MacNeil, 1984	MT1		BY1
700. *Microsurcula mentha* MacNeil, 1984 var. A	MT1		
701. *Mitodrillia harmonica* (Casey, 1903)	MT1		
702. *Mitodrillia pharus* MacNeil, 1984	MT1		
703. *Mitodrillia pharus crassispiropsis* MacNeil, 1984	MT1		
704. *Odostomia (Odostomia) aff. O. (O.) angularis* Dall and Bartsch, 1904	MT1		
705. *Odostomia (Miralda) menthafons* MacNeil, 1984	MT1		
706. *Oliva (Strephonella) mississippiensis* Conrad, 1848	MT1		BY1
707. *Olivella (Callianax) vicksburgensis* Dockery, 1984	MT1		
708. *Olssonella elongata* Dockery, 1984	MT1		
709. *Oniscidia harpula* (Conrad, 1848)	MT1		BY1
710. *Onoba* sp.?	MT1		
711. *Persicula vicksburgensis* Dockery, 1984	MT1		

712. *Phalium (Menthafontia) menthafons* MacNeil, 1984	MT1		
713. *Phos (Strongylocera) caseyi* MacNeil, 1984	MT1		
714. *Phos (Antillophos) hopkinsi* MacNeil, 1984	MT1		
715. *Phos (Strongylocera) vicksburgensis* (Aldrich, 1885)	MT1		
716. *Pleurofusia clarkeana fascia* MacNeil, 1984	MT1		
717. *Pleurofusia clarkeana juba* MacNeil, 1984	MT1		
718. *Pleurofusia elegantula* MacNeil, 1984	MT1		BY1
719. *Pleurofusia menthafons* MacNeil, 1984	MT1		
720. *Pleurofusia menthafons* MacNeil, 1984 var. A	MT1		
721. *Pleurofusia (Xestosurcula) plutonica* (Casey, 1903)	MT1		
722. *Pleurofusia servata* (Conrad, 1848)	MT1		BY1
723. *Pleurofusia servata* (Conrad, 1848) var. B	MT1		
724. *Pleurofusia wythei* MacNeil, 1984	MT1		
725. *Pleuroliria cochlearis* (Conrad, 1848)	MT1		BY1
726. *Pleuroliria cochlearis vetula* MacNeil, 1984	MT1		
727. *Pleuroliria longirostropsis bicollaris* MacNeil, 1984	MT1		
728. *Pliciscala (Punctiscala) cookei* MacNeil, 1984	MT1		
729. *Pliciscala (Nodiscala?)* n. sp.	MT1		
730. *Pyramidella (Voluspa) chavanoidea* MacNeil, 1984	MT1		
731. *Pyramidella (Voluspa) leafensis* MacNeil, 1984	MT1		
732. *Pyramidella (Voluspa?) microcosta* MacNeil, 1984	MT1		
733. *Rictaxis andersoni* (Conrad, 1848)	MT1	GN1	BY1
734. *Ringicula (Ringiculella) crassata* MacNeil, 1984	MT1	GN1	
735. *Ringicula (Ringiculella) irrasa* MacNeil, 1984	MT1		
736. *Ringicula (Ringiculella) mississippiensis* Conrad, 1848 subsp.?	MT1		
737. *Sassia (Personella) abbreviata* (Conrad, 1848) var.	MT1		BY1
738. *Sassia (Personella) caseyi* MacNeil, 1984	MT1		
739. *Sassia (Cymatiella) fuscicava* MacNeil, 1984	MT1		
740. *Scalina menthafontis* MacNeil, 1984	MT1		
741. *Scalina trigintanaria hopkinsi* MacNeil, 1984	MT1		
742. *Scalina trigintanaria* (Conrad, 1848) var.?	MT1		BY1
743. *Scobinella pluriplicata subpluriplicata* MacNeil, 1984	MT1		BY1
744. *Sconsia lintea* (Conrad, 1848)	MT1		BY1
745. *Seila* ap.	MT1		
746. *Semivertagus methafontis* MacNeil, 1984	MT1		
747. *Sinum (Sigaretotrema) mississippiensis* (Conrad, 1848)	MT1		BY1
748. *Solariella clearyensis* MacNeil, 1984	MT1		
749. *Solariella fragum* MacNeil, 1984	MT1		
750. *Solariella laevifunda* MacNeil, 1984	MT1		
751. *Solariella methafontis* MacNeil, 1984	MT1		BY1
752. *Solariobis* sp.	MT1		
753. *Spiradaphne lowei refugium* MacNeil, 1984	MT1		
754. *Sulcocypraea lintea methafons* MacNeil, 1984	MT1		
755. *Syntomodrillia funis* MacNeil, 1984	MT1		
756. *Syntomodrillia funis* MacNeil, 1984 var. A	MT1		
757. *Teinostoma (Teinostoma) caseyi* MacNeil, 1984	MT1		
758. *Telescopium leafensis* MacNeil, 1984	MT1		
759. *Terebra (Laeviterebrum) spinula* MacNeil, 1984	MT1		
760. *Terebra (Strioterebrum) vincta* MacNeil, 1984	MT1		BY1
761. *Tornatellaea brevispira* MacNeil, 1984	MT1		
762. *Triphora (Euthymella) fuscicava* MacNeil, 1984	MT1		
763. *Triphora (Triphora) menthafons* MacNeil, 1984	MT1		
764. *Tritiaria menthafons* MacNeil, 1984	MT1		
765. *Tritiaria meyeri* MacNeil, 1984	MT1		
766. *Tritiaria mississippiensis cookei* MacNeil, 1984	MT1		
767. *Tritiaria refugensis* MacNeil, 1984	MT1		
768. *Tritiaria vaughani* MacNeil, 1984	MT1		BY1
769. *Turbonilla caseyi* MacNeil, 1984	MT1		
770. *Turbonilla leafensis* MacNeil, 1984	MT1		BY1
771. *Turritella boycensis* MacNeil, 1984	MT1		
772. *Turritella caelatura* Conrad, 1848	MT1		
773. *Turritella caelatura* Conrad, 1848 var.	MT1		
774. *Turritella carota* MacNeil, 1984	MT1		BY1
775. *Vetidrillia palmerae* MacNeil, 1984	MT1		
776. *Vexillum (Costellaria) multicostata* Dockery, 1984 var.?	MT1		

777. *Vitrinella (Vitrinella) meyeri* MacNeil, 1984	**MT1**			
778. *Vitrinella (Vitrinella) vicksburgensis* (Meyer, 1886)	**MT1**			
Bivalvia with FAD in MT				
779. *Anomia microstriata* Dockery, 1982	**MT1**			
780. *Arcoperna linteata* Dockery, 1982	**MT1**			
781. *Astarte planilamella* Dockery, 1982	**MT1**			
782. *Atrina argentea* (Conrad, 1848)	MT1			BY1
783. *Barbatia mississippiensis* (Conrad, 1848)	MT1			BY1
784. *Brachidontes mississippiensis* (Conrad, 1848)	MT1			BY1
785. *Callista (Callista) goniopisthus* Dockery, 1982	**MT1**			
786. *Carditella aldrichi* (Casey, 1903)	**MT1**			BY1
787. *Chlamys (Lyropecten) menthifontis* Glawe, 1970	**MT1**	MR1		
788. *Coralliophaga (Coralliophaga) corrugata* Dockery, 1982	**MT1**			
789. *Crassatella (Crassatella) lirasculpta* Dockery, 1982	**MT1**			
790. *Crassatella (Crassatella) mississippiensis megacosta* Dockery, 1982	**MT1**			
791. *Corbula (Varicorbula) laqueata* Casey, 1903	MT1			**BY1**
792. *Crassostrea* sp.	MT1			
793. *Diplodonta (Diplodonta) elatia* Dockery, 1982	MT1			**BY1**
794. *Divaricella (Divalinga) subrigaultiana* (Meyer *in* Smith, 1886)	**MT1**			
795. *Donax* sp.?	MT1			
796. *Felaniella (Felaniella) compacta* Dockery, 1982	**MT1**			
797. *Gari (Gari) papyria* (Conrad, 1848)	MT1			BY1
798. *Glycymeris arctata* (Conrad, 1848)	MT1			**BY6**
799. *Glycymeris suwannensis* Mansfield, 1937	**MT**21			
800. *Laevicardium leptorimum* Dockery, 1982	**MT1**			
801. *Limaria (Limaria) staminea* (Conrad, 1848)	**MT1**			
802. *Limopsis (Limopsis)* sp.	MT1			
803. *Lucina (Callucina) choctavensis* Meyer *in* Smith, 1886	MT1			BY1
804. *Lucina (Lucina) fimbripallium* Dockery, 1982	**MT1**			BY1
805. *Lucina (Cavilinga) imbricolamella* Dockery, 1982	**MT1**			
806. *Lucina (Cavilinga) triloba* Dockery, 1982	**MT1**			
807. *Lucina (Lucinisca) varisculpta* Dockery, 1982	**MT1**			BY1
808. *Lucina* sp.	MT1			
809. *Pecten (Pecten) poulsoni* Morton, 1834	MT1	**MR**17	GL4	
810. *Periploma macneili* Dockery, 1982	**MT1**			BY1
811. *Pitar (Lamelliconcha) astartiformis* (Conrad, 1848)	MT1			BY1
812. *Pitar (Lamelliconcha) calcanea* (Dall, 1903)	**MT**37			
813. *Pitar (Lamelliconcha) protena* Dockery, 1982	**MT1**			BY1
814. *Pitar (Lamelliconcha) silicifluvia* (Dall, 1916)	MT37			
815. *Scapharca (Scapharca) chordicosta* Dockery, 1982	MT1			**BY1**
816. *Semele mississippiensis* (Conrad, 1848)	MT1			BY1
817. *Semele staminea* (Conrad, 1848)	MT1			BY1
818. *Solecurtus vicksburgensis* Aldrich, 1885	MT1			BY1
819. *Spisula (Mactomeris) mississippiensis* (Conrad, 1848)	MT1		GL1	BY1
820. *Sportella oblonga* (Conrad, 1848)	MT1			BY1
821. *Tellidorella interlacinia* Dockery, 1982	**MT1**			
822. *Tellina vicksburgensis* Conrad, 1848	MT1			**BY1**
823. *Teredo* sp.	MT1	MR1		
824. *Thracia (Thracia) vicksburgiana* (Dall, 1903)	**MT1**			
825. *Tiza alta* (Conrad, 1848)	**MT1**			BY1
826. *Trachycardium planicostata* Dockery, 1982	**MT1**			
827. *Yoldia serica* (Conrad, 1848)	MT1			BY1
Scaphopoda with FAD in MT				
828. *Dentalium opaculum* Casey, 1903	**MT1**			
829. *Dentalium varicostata* Dockery, 1984	**MT1**			
830. *Fustiaria (Episiphon) menthifonta* Dockery, 1984	**MT1**			
831. *Fustiaria (Fustiaria)* sp.	MT1			
Bivalvia with FAD in MR				
832. *Chlamys (Lyropecten) duncanensis* Mansfield, 1934		MR36	**GN33**	BY4
833. *Spondylus (Spondylus)* sp.		MR1		

Bivalvia with FAD in GN		
834. *Pecten (Pecten) byramensis* Gardner, 1945	GN26	**BY17**
835. *Pholadomya* sp.	GN1	
Cephalopoda with FAD in GN		
836. *Aturia berryi* Stenzel, 1940	**GN1**	
Gastropoda with FAD in BY		
837. *Aclis matsoni* MacNeil, 1984		**BY1**
838. *Aclis* sp. A		BY1
839. *Acteon (Acteon)* sp. B		BY1
840. *Admetula regularia* Dockery, 1984		**BY1**
841. *Alaba blakneyensis* MacNeil, 1984		**BY1**
842. *Alaba macneili* Dockery, 1984		**BY1**
843. *Anticlimax byramensis* MacNeil, 1984		**BY1**
844. *Aporrhais (Goniocheila) lirata* (Conrad, 1848)		**BY1**
845. *Architectonica (Architectonica) trilirata* (Conrad, 1848)		**BY1**
846. *Architectonica (Architectonica) vicksburgensis* (Dall, 1892)		**BY1**
847. *Arene (Arene) nodosa* Dockery, 1984		**BY1**
848. *Bathytoma congesta* (Conrad, 1848)		**BY1**
849. *Calliostoma* sp.?		BY1
850. *Capulus (Brocchia) langdoni* MacNeil, 1984		**BY1**
851. *Caricella (Atraktus) demissa* Conrad, 1848		**BY1**
852. *Caricella (Atraktus) demissa* Conrad, 1848 var. A		BY1
853. *Cerithiella* sp.		BY1
854. *Chicoreus (Phyllonotus) dormani* (Vokes, 1967)		**BY1**
855. *Clathurella blakneyensis* MacNeil, 1984		**BY17**
856. *Clathurella* sp.		BY1
857. *Clavilithes lesueuri* Dockery, 1984		**BY1**
858. *Clavilithes* sp. B		BY1
859. *Cochlespira cristata* (Conrad, 1848)		**BY1**
860. *Conomitra staminea* (Conrad, 1848)		**BY1**
861. *Conomitra vicksburgensis* (Conrad, 1848)		**BY1**
862. *Conorbis porcellanus* (Conrad, 1848)		**BY1**
863. *Conus alveatus spiralis* Dockery, 1984		**BY1**
864. *Coronia (Coroniopsis) tenella* (Conrad, 1848)		**BY1**
865. *Crassispira (Crassispirella) abundans* (Conrad, 1848)		**BY1**
866. *Crassispira (Crassispirella) lyopleura* MacNeil, 1984		**BY1**
867. *Cyclostremiscus quadracordata* Dockery, 1984		**BY1**
868. *Cymatium (Ranularia) vicksburgense* Dockery, 1984		**BY1**
869. *Cymia (Tritonopsis) subalveata* (Conrad, 1848) subsp.?		BY1
870. *Dolicholatirus exilis confertus* Dockery, 1984		**BY1**
871. *Galeodaria tricarinata* (Conrad, 1848)		**BY1**
872. *Gemmula rotaedens* Conrad, 1848		**BY1**
873. *Harpa vicksburgiana* Dockery, 1984		**BY1**
874. *Marginella* sp.?		BY1
875. *Metula (Caseyella) blakneyensis* MacNeil, 1984		**BY1**
876. *Metula (Metula) fragilis* Casey, 1903		**BY1**
877. *Microdrillia brevis* (Meyer, 1886)		**BY1**
878. *Microsurcula intacta* (Casey, 1903)		**BY1**
879. *Microsurcula intacta jayensis* MacNeil, 1984		**BY4**
880. *Microsurcula intacta* (Casey, 1903) var. A		BY1
881. *Mitrella (Columbellopsis) aff. M. (C.) oryzoides* Gardner, 1947		BY1
882. *Mitrolumna (Cymakra) tallahalaensis* (Dockery, 1984)		**BY1**
883. *Odostomia (Odostomia) byramensis* MacNeil, 1984		**BY1**
884. *Odostomia (Odostomia) vicksburgella* Dockery, 1984		**BY1**
885. *Olivella (Callianax) vicksburgensis* Dockery, 1984 var.		BY1
886. *Phalium (Menthafontia) mississippiensis* (Conrad, 1848)		**BY1**
887. *Phandella monroensis* MacNeil, 1984		**BY4**
888. *Phandella nepionica* Casey, 1903		**BY1**
889. *Pleurofusia decliva* (Conrad, 1848)		**BY1**
890. *Pleurofusia decliva asper* MacNeil, 1984		**BY1**
891. *Pleurofusia longirostropsis* de Gregorio, 1890		**BY1**
892. *Pleurofusia servata* (Conrad, 1848) var. A		BY1
893. *Pleurofusia trichorda* MacNeil, 1984		**BY1**

894. *Pliciscala (Nodiscala?) byramensis* MacNeil, 1984	BY1
895. *Poirieria (Panamurex) macneili* Vokes, 1970	BY1
896. *Pseudofulgur lirata* Dockery, 1984	BY1
897. *Pseudofulgur vicksburgensis* (Conrad, 1848)	BY1
898. *Pyramidella (Voluspa)* n. sp.	BY1
899. *Pyrunculus laevipyrum* MacNeil, 1984	BY1
900. *Ringicula (Ringiculella) mississippiensis* Conrad, 1848	BY1
901. *Ringicula (Ringiculella) mississippiensis nuda* MacNeil, 1984	BY1
902. *Sassia (Personella) abbreviata* (Conrad, 1848)	BY1
903. *Sassia (Personella) mississippiensis* (Conrad, 1848)	BY1
904. *Scalina trigintanaria* (Conrad, 1848)	BY1
905. *Scobinella famelica* Casey, 1903	BY1
906. *Scobinella macer* Casey, 1903	BY1
907. *Semicassis caelatura* (Conrad, 1848)	BY1
908. *Simnia (Calpurna) cookei* MacNeil, 1984	BY1
909. *Spiradaphne lowei* MacNeil, 1984	BY1
910. *Solariella tallahalaensis* Dockery, 1984	BY1
911. *Sulcocypraea lintea* (Conrad, 1848)	BY1
912. *Syntomodrillia tantula* (Conrad, 1848)	BY1
913. *Syntomodrillia tantula* (Conrad, 1848) var. A	BY17
914. *Teinostoma (Idiorapha) minuta* MacNeil, 1984	BY1
915. *Terebra (Terebrellina) divisura* Conrad, 1848	BY1
916. *Terebra (Strioterebrum) tantula* Conrad, 1848	BY1
917. *Terebra (Strioterebrum)* sp.	BY1
918. *Tritiaria mississippiensis* (Conrad, 1848)	BY1
919. *Turbinella wilsoni* Conrad, 1848	BY1
920. *Turricula (Orthosurcula) byramensis* MacNeil, 1984	BY1
921. *Turritella mississippiensis* Conrad, 1848	BY38
922. *Turritella mundula* MacNeil, 1984	BY1
923. *Turritella aff. T. planigyrata* Guppy, 1867	BY1
924. *Vexillum (Costellaria) cellulifera* (Conrad, 1848)	BY1
925. *Varicobela aldrichi* Dockery, 1984	BY1
926. *Volema hopkinsi* MacNeil, 1984	BY1

Bivalvia with FAD in BY

927. *Barbatia (Cucullaearca) lima* (Conrad, 1848)	BY1
928. *Bornia tallahalaensis* Dockery, 1982	BY1
929. *Chamelea mississippiensis* (Conrad, 1848)	BY1
930. *Chione (Chione) bainbridgensis* Dall, 1916	BY37
931. *Chione (Chione) perbrevisformis* Dockery, 1982	BY1
932. *Crassatella (Crassatella) mississippiensis* Conrad, 1848	BY1
933. *Cuspidaria (Tropidomya?)* sp.	BY1
934. *Dinocardium vicksburgense* (Conrad, 1848)	BY1
935. *Ervilia lamelloexteria* Dockery, 1982	BY1
936. *Glycymeris mississippiensis* (Conrad, 1848)	BY1
937. *Gryphaeostrea* sp.	BY1
938. *Nuculana akidonta* Dockery, 1982	BY1
939. *Ostrea* sp.	BY1
940. *Pitar (Lamelliconcha) imitabilis* (Conrad, 1848)	BY1
941. *Pitar (Lamelliconcha) perbrevis* (Conrad, 1848)	BY1
942. *Pitar (Hyphantosoma) semipunctata* (Conrad, 1848)	BY1
943. *Scapharca (Scapharca) lesueuri* Dall, 1898	BY17
944. *Spisula funerata* (Conrad, 1848)	BY1
945. *Spisula* sp.	BY1
946. *Tellina serica* Conrad, 1848	BY1

Scaphopoda with FAD in BY

947. *Dentalium strenuum* Casey, 1903	BY1

Biotic Patterns in Eocene-Oligocene Mollusks of the Atlantic Coastal Plain, U.S.A.

David C. Campbell and Matthew R. Campbell

ABSTRACT

The mid-Eocene to early Oligocene was a time of major faunal change, but the exact patterns and their connections to global climate change and impact events are debated. Although the Paleogene molluscan faunas of the Carolinas remain poorly documented, some preliminary patterns may be observed. The middle Eocene (Claibornian, Bartonian) faunas of the Santee Limestone and lower Castle Hayne Limestone have the highest faunal richness. All studied faunas have very high genus-level affinities with the U.S. Gulf Coast and moderate to high species-level affinities. At the genus level, the affinity with tropical, Tethyan faunas is also consistently high, but the species-level affinities are lower. Uppermost middle Eocene to possibly late Eocene (upper Bartonian to Priabonian, Jacksonian equivalent) beds in the Castle Hayne Limestone and "Cross Formation" of authors have lower faunal richness but also poorer preservation and poorer documentation. The Oligocene fauna shows somewhat higher faunal richness than the later Eocene, but with almost entirely new species relative to the Eocene faunas.

All taxa endemic to the Carolinas are also short-lived, not surviving from one unit to the next. Survival of non-endemics is moderate to high at the genus level. Species survival is much lower. Substantial species-level extinctions across both the approximate Claiborne/Jackson boundary (mid-Bartonian) and the Eocene/Oligocene boundary suggest a gradual or stepwise decline rather than an abrupt Eocene-Oligocene extinction, although the boundary itself is not represented in the region. This overall pattern resembles that in the Gulf Coast. The low local survival may reflect the proximity of the Chesapeake Bay impact, but problems of taphonomy and documentation cannot yet be eliminated.

INTRODUCTION

Some of the earliest described Paleogene deposits from the United States. were in the Carolinas (Finch, 1824). Early interest in these deposits waned with the discovery of well-preserved faunas in the Gulf Coast. Most publications on the Paleogene of the Carolinas are stratigraphic rather than paleobiological in focus. As a result, the faunas remain poorly known, with many taxa awaiting discovery or description. Rossbach and Carter (1991) provide the only thorough Paleogene molluscan monograph for the Carolinas, on the Oligocene fauna from the lower River Bend Formation, although Kellum (1926) also described several taxa from the Castle Hayne and Belgrade formations.

The prevalence of limestone in the region has made study more difficult. Although well-lithified limestones preserve detailed molds, they require collection of large blocks to provide adequate material. Additionally, several beds are marly and do not preserve molds well. Thus, taphonomic effects may obscure the original biological patterns. Erosion provides another problem. Eocene exposures include erosional remnants scattered across the Coastal Plain, and the Oligocene exposures are extremely local. Despite these difficulties, some biotic patterns are evident. The carbonate environment, with the influence of the warm Suwannee current (Popenoe et al., 1987), distinguished this region from the nearby Gulf Coast, creating potential barriers to migration. The

Paleogene exposures in Florida are also mostly carbonates, but represent a shallow offshore bank, more tropical than the Carolinas and separated by the Gulf Trough. In Florida, the Claibornian exposure area is very limited, with no pre-Claibornian exposure. The Floridian faunas show strong tropical, Tethyan influence (Palmer, 1967; Givens, 1989; Nicol, 1991). These molluscan faunas also remain poorly documented. With the distinctive environmental features and geographic barriers, the faunas of the Carolinas have certain similarities to those of Florida and other Tethyan regions and to the Gulf Coast, as well as endemic components. This distinctive mix of taxa makes the Carolinas an important comparison for faunal patterns based on other regional faunas. In addition, the updip Orangeburg District Bed, an informal unit, contains a diverse silicified fauna with high similarity to the Cook Mountain and upper Lisbon Formation and the Gosport Sand. Some taxa from this unit have been described in Harris (1919) and Palmer (1937); a more extensive taxonomic list is in Dockery and Nystrom (1992). It is relatively similar to the Gulf Coast both environmentally and faunally and is not further considered here.

Eocene nearshore deposits near Raleigh, North Carolina, and Columbia, South Carolina, indicate that the ocean extended at least 50 to 100 km inland beyond the studied localities at times of maximum highstand. The presence of glauconite and phosphatized hardgrounds and the abundance of epibionts (such as bryozoans, siliquariid and vermetid snails, oysters, and serpulid annelids) and endolithic organisms (such as clionid sponges, boring annelids, lithophaginids, gastrochaenids, and acrothoracican barnacles) indicate slow or intermittent sedimentation during the deposition of the Santee and Castle Hayne limestones. Epifauna are also common in the "Cross Formation" of authors, but endolithic organisms seem less abundant. The general absence of abraded specimens in all of these deposits indicates a low-energy, subtidal environment. Otte (1986) suggested a maximum water depth for the Castle Hayne Limestone of over 100 m, though most beds were deposited in shallower water. The Castle Hayne Limestone exposures near Wilmington, North Carolina, show temporal variation between his deep, quiet-water and shallow-water, high-energy facies.

Paleogeography, fauna, and benthic algae flora indicate a predominantly middle- to outer-shelf environment. No Eocene terrestrial vertebrates are known from the Carolinas, whereas terrestrial vertebrate remains occasionally occur in Paleocene and Neogene marine deposits of the Carolinas and Paleogene marine deposits of the Gulf Coast. Similarly, the lower abundance of Eocene terrigenous sporomorphs in South Carolina relative to the Gulf Coast suggests a greater distance from land (Frederiksen, 1980). Although the Ocala Limestone was deposited much farther offshore, the presence of fossil seagrasses (Ivany et al., 1990) indicates shallow water. The only benthic photosynthetic organisms yet found in the Eocene limestones of the Carolinas are calcareous red algae, which tolerate lower light levels than most other plants and algae. Also, pleurotomariids, brachiopods, spatangoid echinoids, crinoids, and other elements of the modern outermost shelf to upper slope faunas are particularly common and diverse at the localities that would have been farther offshore, such as Georgetown, South Carolina, and Wilmington, North Carolina. The predominance of these taxa suggests that most of the deposition occurred in deeper water, even if sea level varied significantly during this time. Popenoe et al. (1987) considered the Eocene limestones of South Carolina to represent middle to outer shelf deposits, with coeval North Carolina deposits generally shallower. This contradicts the faunal trends observed in the present study. Further study of more definitive depth indicators, such as benthic foraminifera, would help resolve this issue.

Some previous studies on the Eocene of the eastern United States have used mollusks as temperature indicators, though the limited data for the Carolinas has prevented much assessment for this region (Palmer, 1974; Hansen, 1987a, 1988b). As data on modern temperature tolerances are limited, they were only able to classify taxa into vague "warm" or "cool" categories. The criteria for assigning genera to these categories were not explicit, and Hansen considers a few of Palmer's warm genera to be cool-tolerant. Better data on modern temperature preferences are required before these studies can be refined.

Dockery (1984, 1986, 1987, 1988, 1998) and Hansen (1987a, 1987b, 1988a, 1988b, 1992) have discussed molluscan extinction patterns in the Paleogene of the Gulf Coast. As the molluscan faunas were major factors in defining the stages, not surprisingly the stage boundaries are times of high levels of extinction, including the Claibornian/Jacksonian boundary and the Jacksonian/Vicksburgian boundary. The Claibornian/Jacksonian boundary falls within the Bartonian (Mancini and Tew, 1994), whereas the Jacksonian/Vicksburgian boundary is just after the Eocene/Oligocene boundary (Hansen, 1992). No richly fossiliferous beds occur between the end of the Eocene and the end of the Jacksonian, so the exact time of the extinction relative to the Eocene/Oligocene boundary remains unclear (Hansen, 1992). Within the stages, the diversity generally gradually increases. In several publications, Dockery and Hansen debated both the method of enumerating taxa (whether to include or exclude poorly known taxa) and the correlations to particular environmental factors. Dockery (1984) argued that diversity tended to correlate with shelf area and suggested that the terrigenous detrital influx during lowstands was a major causative factor in the extinctions. Hansen (1987a) argued that temperature was a stronger correlative factor. The presence of similar patterns in the Carolinas and the Gulf would tend to support a more global causative factor (such as temperature) rather than a Gulf-specific factor. However, increased erosion of the southern Appalachians could simultaneously affect sedimentation in both areas. Although the Carolinas retained predominantly carbonate deposition throughout the interval, the proportion of terrigenous input does vary. Thus, comparison of patterns between the regions is suggestive but not definitive for assessing the two models. Also, both authors noted that regression is correlated with other environmental factors such as cooling, and Dockery (1998) also noted the possibility of ranges truncated by major unconformities rather than actual time of extinction. Further work on sedimentology and sediment tolerances of the fauna will better test the models. The temperature curve used by Hansen (1992) is a bit difficult to interpret. It apparently represents a consensus of the Gulf terrestrial and North Atlantic marine curves in Wolfe and Poore (1982). The switching of two figures in Wolfe and Poore (1982) adds to the confusion. The correlation of temperature and extinctions may be tested by further regional climatic study, especially methods such as stable isotopic measurement that can be performed directly on samples from the mollusk-bearing units.

The Eocene-Oligocene transition is a time of particular interest for study of faunal change. The initiation of large-scale Antarctic glaciation led to global climatic change, and large bolide impacts also occurred during this time. However, the exact nature of the biotic response has been uncertain. Particular issues are the timing of faunal turnover and the gradual, stepwise, or abrupt nature of the changes. The distinctive faunal and environmental features of the Carolinas make them an important source of comparisons to the better-known Gulf Coast and European faunas. Also, the proximity to the Chesapeake Bay crater (Poag et al., 1994) makes the Carolinian faunas particularly important for assessing its biological impact.

The Eocene stratigraphy of the region remains debated. The present study recognized three informal Eocene biostratigraphic units and one Oligocene informal unit. Unit I includes the Moultrie Member of the Santee Limestone and the lower Castle Hayne Limestone (New Hanover and Comfort Members), including exposures near Wilmington, North Carolina. Unit II is the Cross Member of the Santee Limestone of Ward et al. (1978). The stratotype at the Martin Marietta Berkeley Quarry is faunally and lithologically distinct from the Eocene beds in the Giant Portland and other quarries that have been referred to the Cross Member or the Cross Formation. The confusion reflects both previous uncertainty as to the correlation of the units and somewhat different use of the term "Cross Formation" in Baum et al. (1979). These younger "Cross Formation" beds share several taxa with the Spring Garden Member of the Castle Hayne Limestone and an unnamed soft white limestone that underlies the Spring Garden at New Bern. Unit III includes these younger "Cross Formation" beds, the Spring Garden Member, and the soft white limestone. Unit I is tentatively correlated with the upper Claiborne Cook Mountain and Lisbon formations of the Gulf Coast. Unit II is tentatively correlated to the Gosport Sand (uppermost Claibornian), and Unit III is probably Jacksonian, based on macrofossil biostratigraphy. However, Ward (pers. commun.) considers the Cross Member to be equivalent to the uppermost Lisbon Formation and Unit III to be Gosport equivalent. The chronostratigraphic relationships of Unit IV, the Oligocene lower River Bend Formation, remain uncertain. It probably was deposited near the Vicksburgian/Chickasawhayan boundary (Rossbach and Carter, 1991). Further subdivision of the Eocene should be possible in future studies with more stratigraphic data.

The present study compiles published and unpublished data on the molluscan faunas of the upper middle Eocene through middle Oligocene to provide a preliminary assessment of the biogeographic and temporal patterns. In particular, the relative affinities with Tethyan versus Gulf Coast faunas and the changes through time were poorly known based on previously published data.

Table 19.1. Affinities of Studied Taxa

	Unit I	Unit II	Unit III	Unit IV
Number of genera	85	9	37	45
Number of species	102	5	28	39
Endemic genera	1.2%	0.0%	0.0%	2.2%
Tethyan genera	89.4%	100.0%	94.6%	84.4%
Gulf Coast genera	91.8%	100.0%	94.6%	91.1%
Claibornian genera	—	—	91.9%	73.3%
Jacksonian genera	94.1%	89.9%	—	77.8%
Oligocene genera	84.7%	89.9%	91.9%	—
Endemic species	20.6%	0.0%	17.9%	38.5%
Tethyan species	12.7%	0.0%	17.9%	15.4%
Gulf Coast species	69.6%	100.0%	82.1%	56.4%
Claibornian species	—	—	53.6%	5.1%
Jacksonian species	36.3%	40.0%	—	7.7%
Oligocene species	1.0%	0.0%	3.6%	—

Note: Units I and II were defined as Claibornian, Unit III was defined as Jacksonian, and Unit IV was defined as Oligocene, so the correlations are 100.0% by definition.

METHODS

Compilation of published records and examination of collections yielded a list of molluscan taxa for the Santee Limestone, the Castle Hayne Limestone, the lower River Bend Formation, and other deposits of debated stratigraphic nomenclature. These faunas were then compared at the genus and species levels to those of other regions. Paleogene occurrences in Florida, the Paris Basin and southern Europe, and other tropical regions were considered Tethyan. Additionally, Palmer (1967), Givens (1989) and Nicol (1991) identified certain taxa as Tethyan. These assignments were also used in assessing the affinities of the faunas. To gauge extinction and origination patterns, the time of origin and last occurrence of taxa were noted if within the late middle Eocene to early Oligocene.

Occurrences in the present study were included as biostratigraphic extensions. This caused the affinities of each unit with its own time interval to be 100% despite the presence of taxa not otherwise known from coeval deposits. Range-through was also extrapolated, so that a taxon documented only from the lower Eocene and the Oligocene would also be coded as Claibornian and Jacksonian. Geographic occurrence was coded irrespective of time; thus, taxa that briefly invaded one region are consistently recorded as occurring in that region. Examples include the Gulf Coast occurrences of *Glyptoactis alticostata* and *Crassatella alta*, which are confined to the Gosport Sand despite their long ranges in the Carolinas. A genus represented from the same unit by both specifically uncertain material and recognizable material was reported under the recognized species. Thus, *Crassostrea* sp. and *Crassostrea gigantissima* would be merged into one record, but *Crassostrea* sp. nov. would remain separate.

DATA

Although few taxa were previously reported, 85 molluscan genera and 102 species were documented from Unit I, 9 genera and 5 species from Unit II, 37 genera and 28 species from Unit III, and 45 genera and 39 species from Unit IV (appendix). Major sources of data on taxa include Campbell (1995a, 1995b) for the Santee Limestone, Palmer and Brann (1965-1966) and Cossmann and Pissarro (1904-1913) on Eocene ranges, Rossbach and Carter (1991) and Carter and Rossbach (1992) for the River Bend Formation, and Dockery (1982) and MacNeil and Dockery (1984) for Oligocene ranges. The appendix lists the taxa from the Carolinas, and table 19.1 gives the biogeographic and biostratigraphic affinities.

RESULTS

The middle Eocene (Claibornian, Bartonian) faunas of Unit I have the highest faunal richness. All studied faunas have very high genus-level affinities with those of the U.S. Gulf Coast and moderate to high species-level affinities. At the genus level, the affinity with the more tropical Tethyan faunas of Florida, the Paris Basin, and other regions is also consistently high, but the species-level affinities with the Tethyan faunas are lower. The Cross Member fauna is very poorly known, with strong similarity to the Gulf Coast for identified taxa. Units II and III have lower faunal richness than Unit I but also poorer preservation and poorer documentation, with generally softer matrix that does not preserve molds as well. Unit IV has somewhat higher faunal richness than Units II and III, with almost entirely new species relative to the Eocene faunas. No taxa endemic to the Carolinas are known to survive from one unit to the next. Survival of non-endemics is moderate to high at the genus level, but at least 10% of the genera that survive globally become extinct in the Carolinas at each boundary. At the species level, survival is much lower, with only one cosmopolitan species surviving from the Eocene into the Oligocene.

For most units, the number of reported genera exceeds the number of reported identified species. This reflects the combined effects of poor preservation, limited collection, and lack of taxonomic review of most groups. It suggests that much further work remains to be done on these faunas.

DISCUSSION

Substantial species-level extinction across both the Claibornian/Jacksonian boundary (mid-Bartonian) and the Eocene/Oligocene boundary suggest a gradual or stepwise extinction rather than abrupt Eocene-Oligocene decline. Both intervals are times of high faunal turnover in the Gulf Coast (Dockery, 1998; Hansen, 1992). The lack of continuous sections across these boundaries in the Carolinas, and the poor documentation for Unit II, prevent precise determination of the timing of change. The low local survival in the Carolinas around the Eocene/Oligocene boundary interval may reflect the proximity of the Chesapeake Bay impact, but problems of taphonomy and lack of documentation cannot presently be ruled out. The level of similarity between the faunas of the Units I–III and Claibornian and Jacksonian faunas is much higher than for any of these units with Oligocene faunas. Likewise, the affinity of Unit IV with Claibornian and Jacksonian faunas is low. This suggests that the faunas changed more extensively around the Eocene/Oligocene boundary than in the late middle Eocene. However, the magnitude of the late middle Eocene extinction seems higher, as reflected in the larger drop in taxonomic richness. The higher similarity of Unit III with the Claibornian than Unit I with the Jacksonian shows that Unit III includes mostly survivors from the Claibornian, but many Unit I taxa had disappeared by the Jacksonian. This pattern was also noted by Dockery (1998) for the Gulf Coast, with peak diversity in the Claibornian, followed by a major extinction, but greater change in the character of the fauna at the Eocene/Oligocene boundary, with fewer surviving species and disappearance of several typical Eocene lineages. The higher species richness in Unit IV relative to Unit III may reflect either climatic amelioration in the early Oligocene or diversification of eurythermal lineages, if it is not merely a monographic artifact.

Although the documented levels of species richness closely match the levels of study for each unit, the decrease in diversity from the first unit to the later ones seems real based on examination of large samples from different units. The number of species in any one sample of Unit I generally exceeds that for a comparable sample of another unit. Only rarely does one species dominate an entire sample from Unit I, whereas often one species dominates the younger samples from Units II and III.

Apart from the poorly documented Unit II, the Carolinian faunas have moderate species-level endemism (18% to 39%). This suggests that the Carolinas represented a distinct biogeographic unit from the Gulf Coast throughout the studied time interval. The biogeographic relationship with Florida is less clear. Many distinctive Tethyan taxa in Florida have not yet been found in the Carolinas, but several others do occur in both regions. Faunal documentation for both regions is poor enough to make conclusions tentative. In particular, the similarity to Floridian faunas is likely to increase with better documentation of those faunas. In general, the Carolinian faunas appear intermediate between the Floridian and Gulf Coast faunas. No endemic species have yet been recognized for the Cross Member. This partly reflects the poor documentation of the unit. However, it may reflect a true pattern. Both *Crassatella alta* and *Glyptoactis alticostata* occur in the Carolinas before and after the Cross Member, but in the Gulf they occur only in the Gosport Sand. If the Cross Member and the Gosport Sand represent a time of exceptionally warm conditions in the Gulf, when Carolinian taxa invaded that region (Dockery and Nystrom, 1992), then the Carolinian endemism would be much less than usual.

No strong trends were noted in the proportion of Tethyan elements in the faunas. Extinctions caused by cooling might be expected to substantially decrease the affinity between more northern and southern faunas, as is observed in the Pliocene of this region (Campbell, 1998). The continued high Tethyan component, despite global cooling, may indicate a greater influence of regional factors such as sea-level change (Dockery, 1984 and following) or shifts in circulation patterns. However, simultaneous climatic change in the Tethyan and Carolinian regions could also produce continued high similarity despite cooling, by leaving eurythermal taxa in both areas. The limited exposure and few studies of coeval faunas from north of the Carolinas in the western Atlantic make comparison with cooler provinces difficult.

Differences between the Carolinian and Gulf faunas reflect a variety of environmental factors. Temperatures may have been slightly warmer in the Carolinas (Dockery and Nystrom, 1992) and temperature variations would have been milder in the offshore environment of the Carolinas than in the nearshore regions of the Gulf Coast. Also, the offshore setting would have been less turbid. The substrate difference of carbonate sand, mud, and hardgrounds versus clastic sediments would affect some species, as would water depth. The Floridian faunas had similar substrates but warmer, shallower water.

Throughout the studied time interval, many species attained greater sizes in the Carolinas than in the Gulf Coast. The biological significance of this difference is not yet certain. Many mollusks have indeterminate growth. Abnormally small size suggests stressful conditions limiting growth. Abnormally large size may reflect better conditions for growth, such as high levels of food or lower mortality rates and thus more time to grow. Conversely, size can reflect the relative investment of energy in growth versus in reproduction, so that small size may reflect good conditions for reproduction and large size may reflect unfavorable conditions for reproduction. The difficulty of this comparison lies in determining whether the Gulf Coast or the Carolinian individuals were normal in size.

CONCLUSIONS

The two major molluscan extinctions during the Eocene-Oligocene transition in the Carolinas appear to fall in the late middle Eocene and near the Eocene/Oligocene boundary. These approximately coincide with extinctions at the Claibornian/Jacksonian and Jacksonian/Vicksburgian boundaries in the United States Gulf Coast. Although the faunas are not yet well correlated with the impact events, the lack of endemic holdovers from the late Eocene into the Oligocene may reflect the Chesapeake Bay impact. The faunas have strongest affinities with those of the Gulf Coast, but also show affinities with Tethyan faunas and have some endemic components.

The documented species richness of Unit I, the middle Eocene Santee Limestone and lower Castle Hayne Limestone, is far higher than for any of the later units, but the preservation, exposure area, and documentation also are better.

Thus, mollusks in this region appear to have gone through stepwise extinctions, similar to the pattern reported for the Gulf Coast mollusks and several other faunas. However, preservational factors and monographic bias may play important roles in producing the apparent pattern.

ACKNOWLEDGMENTS

We thank Donald Prothero for the invitation to participate in the Penrose Conference on the marine Eocene-Oligocene transition. Wally Fallaw and Victor Zullo first directed our attention to the Paleogene of the Carolinas. Joseph Carter, Tim Bralower, and Dan Textoris reviewed the senior author's master's thesis, from which much of the Eocene information is derived. Lauck Ward and Roger Portell reviewed this paper and provided many helpful comments.

REFERENCES

Baum, G. R., Harris, W. B., and Zullo, V. A., 1979, Historical review of Eocene to early Miocene stratigraphy, North Carolina, *in* Baum, G. R., Harris, W. B., and Zullo, V. A., eds., Structural and stratigraphic framework for the Coastal Plain of North Carolina: Raleigh, Carolina Geological Society and Atlantic Coastal Plain Geological Association, Field Trip Guidebook, p. 1-15.

Campbell, D. C., 1995a, New molluscan faunas from the Eocene of South Carolina: Tulane Studies in Geology and Paleontology, v. 27, p. 119-151.

Campbell, D. C., 1995b, Gastropods from the Eocene limestones of the Carolinas [unpubl. M.S. thesis]: Chapel Hill, University of North Carolina, 190 p.

Campbell, M. R., 1998, Plio-Pleistocene Bivalvia of the western Atlantic Ocean: temporal and taxonomic resolution and the anatomy of an extinction [unpubl. M.S. thesis]: Chapel Hill, University of North Carolina, 68 p.

Carter, J. G., and Rossbach, T. R., 1992, Molluscan biostratigraphy of the River Bend Formation, New Bern and Belgrade, North Carolina, *in* Dennison, J. M., and Stewart, K. G. eds., Geologic Guides to North Carolina and adjacent areas: Chapel Hill, North Carolina, Department of Geology, University of North Carolina at Chapel Hill, Geologic guidebook no. 1, p. 139-143.

Cossmann, A. E. M., and Pissarro, G., 1904-1913, Iconographie complète des coquilles fossiles de l'Éocène des environs de Paris: Paris, Société Geologique de la France. 2 vols.

Dockery, D. T., III, 1982, Lower Oligocene Bivalvia of the Vicksburg Group in Mississippi: Mississippi Department of Natural Resources Bureau of Geology Bulletin 123, 261 p.

Dockery, D. T., III, 1984, Crisis events for Paleogene molluscan faunas in the southeastern United States: Mississippi Geology, v. 5, p. 1-7.

Dockery, D. T., III, 1986, Punctuated succession of Paleogene mollusks in the northern Gulf Coastal Plain: Palaios, v. 1, p. 582-589.

Dockery, D. T., III, 1987, Eocene-Oligocene molluscan extinctions: Comment: Palaios, v. 2, p. 620-621.

Dockery, D. T., III, 1988, Molluscan extinction rates in question: Comment: Nature, v. 331, p. 123.

Dockery, D. T., III, 1998, Molluscan faunas across the Paleocene/Eocene Series boundary in the North American Gulf Coastal Plain, *in* Aubry, M.-P., Lucas, S., and Berggren, W. A., eds., Late Paleocene–early Eocene climatic and biotic events in the marine and terrestrial records: New York, Columbia University Press, p. 296-322.

Dockery, D. T., III, and Nystrom, P. G., 1992, The Orangeburg District molluscan fauna of the McBean Formation: a new, diverse, silicified fauna within the *Glyptoactis (Claibornicardia) alticostata* zone above the occurrence of *Cubitostrea sellaeformis*, an interval zone of Gosport age, *in* Zullo, V. A., Harris, W. B., and Price, V., eds., Proceedings of the Second Bald Head Island Conference on Coastal Plains Geology: Savannah River Region: Transition Between the Gulf and Atlantic Coastal Plains, November 1990: Wilmington, N.C., UNC-Wilmington, p. 90-96.

Finch, J., 1824, Geological essay on the Tertiary formations in America: American Journal of Science and Arts, v. 7, p. 31-43.

Frederiksen, N. O., 1980, Paleogene sporomorphs from South Carolina and quantitative correlations with the Gulf Coast: Palynology, v. 4, p. 125-179.

Givens, C. R., 1989, First record of the Tethyan genus *Volutilithes* (Gastropoda: Volutidae) in the Paleogene of the Gulf Coastal Plain, with a discussion of Tethyan molluscan assemblages in the Gulf Coastal Plain and Florida: Journal of Paleontology, v. 63, p. 852-856.

Hansen, T. A., 1987a, Extinction of Late Eocene to Oligocene molluscs: relationship to shelf area, temperature change, and impact events: Palaios, v. 2, p. 69-75.

Hansen, T. A., 1987b, Eocene-Oligocene molluscan extinctions: Reply: Palaios, v. 2, p. 621-622.

Hansen, T. A., 1988a, Molluscan extinction rates in question: Reply: Nature, v. 331, p. 123.

Hansen, T. A., 1988b, Early Tertiary radiation of marine molluscs and the long-term effect of the Cretaceous-Tertiary extinction: Paleobiology, v. 14, p. 37-51.

Hansen, T. A., 1992, The patterns and causes of molluscan extinction across the Eocene/Oligocene boundary, *in* Prothero, D. R., and Berggren, W. A., eds., Eocene-Oligocene climatic and bio-tic evolution: Princeton, New Jersey, Princeton University Press, p. 341-348.

Harris, G. D., 1919, Pelecypoda of the St. Maurice and Claiborne Stages: Bulletins of American Paleontology, v. 6, no. 31, p. 1-261, 59 pls.

Ivany, L. C., Portell, R. W., and Jones, D. S., 1990, Animal-plant relationships and paleobiogeography of an Eocene seagrass community from Florida: Palaios, v. 5, p. 244-258.

Kellum, L. B., 1926, Palentology and stratigraphy of the Castle Hayne and Trent Marls in North Carolina: U.S. Geologica Survey Professional Paper 143, 56 p.

MacNeil, F. S., and Dockery, D. T., III, 1984, Lower Oligocene Gastropoda, Scaphopoda, and Cephalopoda of the Vicksburg Group in Mississippi: Mississippi Geological Survey, Bulletin 124, 415 p.

Mancini, E. A., and Tew, B. H. 1994. Claiborne-Jackson Group contact (Eocene) in Alabama and Mississippi: Transactions, Gulf Coast Association of Geological Societies, v. 44, p. 431-439.

Nicol, D., 1991, Tethyan molluscs of the middle and late Eocene of peninsular Florida: Papers in Florida Paleontology, no. 4, p. 1-6.

Otte, L. J., 1986, Regional perspective on the Castle Hayne Limestone, *in* Textoris, D. A., ed., SEPM,

Third midyear, Field Guidebooks, Southeastern United States: Tulsa, SEPM, p. 270-276.

Palmer, K. V. W., 1937, The Claibornean Scaphopoda, Gastropoda, and dibranchiate Cephalopoda of the southern United States: Bulletins of American Paleontology, v. 7, no. 32, p. 1-732.

Palmer, K. V. W., 1967, A comparison of certain Eocene molluscs of the Americas with those of the Western Tethys, *in* Adams, C. G., and Ager, D. V., eds., Systematics Association Publication 7: Aspects of Tethyan Biogeography: London, Systematics Association, p. 183-193.

Palmer, K. V. W., 1974, Composition with relationships of Paleocene and Eocene molluscan fauna of the East Americas: Verhandlungen der Naturforshenden Gesellschaft in Basel, v. 84, p. 468-482.

Palmer, K. V. W., and Brann, D. C., 1965-1966, Catalogue of the Paleocene and Eocene Mollusca of the Southern and Eastern United States: Bulletins of American Paleontology, v. 48(218), p. 1-1057.

Poag, C. W., Powars, D. S., Poppe, L. J., and Mixon, R. B., 1994, Meteoroid mayhem in Ole Virginny: source of the North American tektite strewn field: Geology, v. 22, p. 691-694.

Popenoe, P., Henry, V. J., and Idris, F. M., 1987, Gulf Trough—the Atlantic connection: Geology, v. 15, p. 327-332.

Rossbach, T. J., and Carter, J. G., 1991, Molluscan biostratigraphy of the lower River Bend Formation at the Martin Marietta Quarry, New Bern, North Carolina: Journal of Paleontology, v. 65, p. 80-118.

Ward, L. W., Lawrence, D. R., and Blackwelder, B. W., 1978, Stratigraphic revision of the middle Eocene, Oligocene, and lower Miocene—Atlantic Coastal Plain of North Carolina: U.S. Geological Survey Bulletin 1457-F, p. F1-F23.

Wolfe, J. A., and Poore, R. Z., 1982, Tertiary marine and nonmarine climatic trends, *in* Berger, W. H., and Crowell, J. C., eds., Climate in Earth History: Washington, D.C., National Academy Press, p. 154-158.

APPENDIX: TAXA

Key: EG, endemic genus; TG, Tethyan genus; GG, Gulf Coast genus; CG, Claibornian genus; JG, Jacksonian genus; OG, Oligocene genus; ES, endemic species; etc.

Unit I, Moultrie Member of Santee Limestone and Lower Castle Hayne Limestone
Entemnotrochus nixus (Tuomey, 1853) TG, CG, JG, OG, ES, CS
Entemnotrochus sp. nov. TG, CG, JG, OG, ES, CS
Emarginula (*Arginula*) sp. nov. TG, GG, CG, JG, OG, ES, CS
Puncturella (*Altrix*) *altior* (Meyer and Aldrich, 1886) GG, CG, JG, OG, GS, CS
Diodora penderensis (Kellum, 1926) TG, GG, CG, JG, OG, GS, CS
Diodora tenebrosa (Conrad, 1833) TG, GG, CG, JG, OG, GS, CS
Tiburnus eboreus (Conrad, 1833) GG, CG, JG, OG, GS, CS
Solariella louisiana Dall, 1893 TG, GG, CG, JG, OG, GS, CS
Solariella stalagmium "*modesta*" (Meyer and Aldrich, 1886) TG, GG, CG, JG, OG, GS, CS
Solariella tricostata (Conrad, 1835) TG, GG, CG, JG, OG, GS, CS
Haustator arenicola (Conrad, 1865) TG, GG, CG, JG, OG, GS, CS, JS
Haustator ghigna (de Gregorio, 1890) TG, GG, CG, JG, OG, GS, CS, JS
Mesalia claibornensis Harris, 1895 TG, GG, CG, JG, OG, GS, CS
Tenagodus vitis (Conrad, 1833) TG, GG, CG, JG, OG, GS, CS
Serpulorbis squamulosus (Conrad, 1834) TG, GG, CG, JG, OG, GS, CS, JS
Xenophora sp. TG, GG, CG, JG, OG
Cheilea boutillieri (Cossmann, 1888) TG, CG, JG, OG, TS, CS
?*Cheilea bernayi* (Cossmann, 1885) TG, CG, JG, OG, TS, CS
Capulus americanus (Conrad, 1854) TG, GG, CG, JG, OG, GS, CS, JS
Calyptraea (*Trochita*) "*aperta*" (Solander, 1766) TG, GG, CG, JG, OG, TS, GS, CS, JS, OS
Cypraeorbis sp. TG, GG, CG, JG, OG
Cypraedia spp. TG, GG, CG, JG
"*Natica*" "*Naticarius*" *semilunata* Lea, 1833 GG, CG, JG, OG, GS, CS
cf. *Ampullinopsis citrinensis* Palmer, 1953 TG, GG, CG, JG, OG, TS, CS, JS
Phalium (?) *taiti* (Conrad, 1834) TG, GG, CG, JG, OG, GS, CS, JS
Phalium (*Galeodosconsia*?) *brevicostatum* (Conrad, 1834) TG, GG, CG, JG, OG, GS, JS
Ficus affinis Van Winkle, 1919 TG, GG, CG, JG, OG, GS, CS, JS
Ficopsis penita (Conrad, 1833) TG, GG, CG, GS, CS
Cirsotrema (*Coroniscala*) *aequabile* (Kellum, 1926) TG, GG, CG, JG, OG, ES, CS
Cirsotrema (*Coroniscala*) sp. nov. TG, GG, CG, JG, OG, ES, CS
Pterynotus (*Pterynotus*) sp. aff. *P. sabinola* (Palmer, 1960) TG, GG, CG, JG, OG, ES, CS
Penion bellus (Conrad, 1833) TG, GG, CG, JG, OG, GS, CS
Papillina sp. cf. *P. dumosa trapaquara* Harris, 1895 TG, GG, CG, JG, GS, CS, JS
"*Clavilithes abruptus*" (Tuomey, 1853) of Kellum, 1926 TG, GG, CG, JG, OG, ES, CS
Clavilithes penrosei (Heilprin, 1891) TG, GG, CG, JG, OG, GS, CS
Clavilithes regexus Harris, 1895 TG, GG, CG, JG, OG, GS, CS
Mitrella (*Clinurella*) *bucciniformis* (Heilprin, 1879) GG, CG, JG, OG, GS, CS
Persicula larvata (Conrad, 1833) TG, GG, CG, JG, OG, GS, CS
Prunum columba (Lea, 1833) TG, GG, CG, JG, OG, GS, CS
Mitra (*Fusimitra*) *millingtoni* Conrad, 1854 TG, GG, CG, JG, OG, TS, GS, CS, JS
Mitra (*Fusimitra*) *perexilis* Conrad, 1833 TG, GG, CG, JG, OG, GS, CS
Mitra (*Fusimitra*) *neta* Gardner, 1945 TG, GG, CG, JG, OG, GS, CS
Agaronia alabamensis (Conrad, 1833) TG, GG, CG, JG, OG, TS, GS, CS, JS
Lapparia conradi Palmer, 1953 TG, GG, CG, JG, TS, CS, JS
Volutocorbis sp. nov. TG, GG, CG, JG, OG, ES, CS
Lyrischapa sp. nov. TG, GG, CG, ES, CS

Caricella (*Caricella*) *bolaris* (Conrad, 1833) TG, GG, CG, JG, OG, GS, CS
Caricella (*Caricella*) *pyruloides* s.l. (Conrad, 1832) TG, GG, CG, JG, OG, GS, CS, JS
Caricella (*Atraktus*) sp. TG, GG, CG, JG, OG, ES, CS
Voluticella levensis Palmer, 1953 TG, CG, JG, TS, CS, JS
Santeevoluta wilmingtonensis (Brown and Pilsbry, 1912) EG, CG, ES, CS
Conomitra fusoides (Lea, 1833) TG, GG, CG, JG, OG, GS, CS, JS
Admetula jacksonica (Cooke, 1926) TG, GG, CG, JG, OG, GS, CS, JS
Conus spp. TG, GG, CG, JG, OG
Eopleurotoma sayi (Lea, 1833) TG, GG, CG, JG, GS, CS
Surculoma subequalis var.? (Conrad, 1835) TG, GG, CG, JG, OG, GS, CS
Nucula sp. TG, GG, CG, JG, OG
Nuculana trumani (Harris, 1919) TG, GG, CG, JG, OG, GS, CS
Yoldia (*Calorhadia*) *reginajacksonis* (Harris, 1897) TG, GG, CG, JG, OG, GS, CS, JS
Barbatia sp. nov. TG, GG, CG, JG, OG, ES, CS
Barbatia rhomboidella (Lea, 1833) TG, GG, CG, JG, OG, GS, CS
Pachecoa sp. GG, CG, JG
Arcoperna sp. TG, GG, CG, JG, OG
Modiolus cawcawensis Harris, 1919 TG, GG, CG, JG, OG, ES, CS
Lithophaga sp. TG, GG, CG, JG, OG
Glycymeris idonea (Conrad, 1833) TG, GG, CG, JG, OG, GS, CS, JS
Atrina sp. TG, GG, CG, JG, OG
Pteria sp. TG, GG, CG, JG, OG
Cubitostrea sellaeformis (Conrad, 1832) TG, GG, CG, JG, OG, GS, CS
Pycnodonte ludoviciana (Harris, 1919) TG, GG, CG, JG, OG, GS, CS
Nayadina (*Exputens*) spp. TG, CG, JG, TS, CS, JS
Plicatula filamentosa Conrad, 1833 TG, GG, CG, JG, OG, TS, GS, CS, JS
Ctenoides sp. nov. TG, GG, CG, JG, OG, GS, CS
Eburneopecten calvatus (Morton, 1833) GG, CG, JG, OG, ES, CS
"*Chlamys*" *cawcawensis* (Harris, 1919) TG, GG, CG, JG, OG, ES, CS
"*Chlamys*" *membranosa* (Morton, 1833) TG, GG, CG, JG, OG, ES, CS
"*Chlamys*" *wahtubbeana* Dall, 1898 TG, GG, CG, JG, OG, GS, CS
Spondylus lamellacea Kellum, 1926 TG, GG, CG, JG, OG, GS, CS
Venericardia (*Venericor*) *claiboplata* Gardner and Bowles, 1939 TG, GG, CG, JG, OG, GS, CS
Venericardia (*Rotundicardia*) *diversidentata diversidentata* Meyer, 1885 TG, GG, CG, JG, OG, GS, CS, JS
Venericardia (*Rotundicardia*) *eutawcolens* Harris *in* Van Winkle and Harris, 1919 TG, GG, CG, JG, OG, ES, CS
Venericardia (*Rotundicardia*) *rotunda* Lea, 1833 TG, GG, CG, JG, OG, GS, CS
Glyptoactis (*Claibornicardia*) *alticostata* (Conrad, 1833) TG, GG, CG, JG, OG, GS, CS, JS
Glyptoactis (*Claibornicardia*) *complexicostata* (Meyer and Aldrich, 1886) TG, GG, CG, JG, OG, GS, CS
Pleuromeris parva (Lea, 1833) GG, CG, JG, OG, GS, CS, JS
Lirodiscus smithvillensis (Harris, 1895) TG, GG, CG, JG, GS, CS
Lirodiscus (*Crustuloides*) sp. nov. TG, GG, CG, JG, ES, CS
Crassatella alta Conrad, 1832 TG, GG, CG, JG, OG, GS, CS, JS
Crassatella eutawcolens (Harris *in* Van Winkle and Harris, 1919) TG, GG, CG, JG, OG, TS, CS, JS
Crassatella texalta Harris, 1895 TG, GG, CG, JG, OG, GS, CS
Crassatella willcoxi (Brown and Pilsbry, 1912) TG, GG, CG, JG, OG, ES, CS
Gastrochaena (*Gastrochaena*) sp. TG, GG, CG, JG, OG
Eosolen sp. TG, GG, CG
Semele australina Harris, 1919 TG, GG, CG, JG, OG, GS, CS
Semele linosa (Conrad, 1833) TG, GG, CG, JG, OG, GS, CS
Gari eborea (Conrad, 1833) TG, GG, CG, JG, OG, GS, CS
Agnocardia claibornensis (Aldrich, 1911) GG, CG, JG, OG, GS, CS

Nemocardium sp. TG, GG, CG, JG, OG
Chama monroensis Aldrich, 1903 TG, GG, CG, JG, OG, GS, CS
Diplodonta ungulina (Conrad, 1833) TG, GG, CG, JG, OG, GS, CS
Meiocardia carolinae Harris *in* Van Winkle and Harris, 1919 TG, CG, JG, OG, ES, CS
Callista aequora (Conrad, 1833) TG, GG, CG, JG, OG, GS, CS
Callista perovata (Conrad, 1833) TG, GG, CG, JG, OG, GS, CS
Callista perovata lisbonensis (Harris, 1919) TG, GG, CG, JG, OG, GS, CS
Pitar poulsoni (Conrad, 1833) TG, GG, CG, JG, OG, GS, CS
Spisula praetenuis (Conrad, 1833) TG, GG, CG, JG, OG, TS, GS, CS, JS
Spisula praetenuis australina Harris, 1919 TG, GG, CG, JG, OG, CS
Spisula decisa palmaris Harris, 1919 TG, GG, CG, JG, OG, GS, CS
Corbula (*Caryocorbula*) *willistoni* Meyer, 1885 TG, GG, CG, JG, OG, GS, CS, JS
Pholadomya harrisi Gardner, 1927 TG, GG, CG, JG, OG, GS, CS
Pecchiolia dalliana Harris, 1919 TG, CG, JG, OG, ES, CS
Eutrephoceras carolinensis Kellum, 1926 TG, GG, CG, TS, CS
Aturia sp. TG, GG, CG, JG, OG

Unit II, Cross Member of Santee Limestone
Haustator arenicola (Conrad, 1865) TG, GG, CG, JG, OG, GS, CS, JS
Xenophora sp. TG, GG, CG, JG, OG
Crepidula sp. TG, GG, CG, JG, OG
Sinum bilix (Conrad, 1833) TG, GG, CG, JG, OG, GS, CS
Ficopsis penita (Conrad, 1833) TG, GG, CG, GS, CS
Glycymeris staminea (Conrad, 1833) TG, GG, CG, JG, OG, GS, CS
Pycnodonte sp. TG, GG, CG, JG, OG
Cubitostrea sp. TG, GG, CG, JG, OG
Crassatella alta Conrad, 1832 TG, GG, CG, JG, OG, GS, CS, JS

Unit III, Upper "Cross Formation" and Upper Castle Hayne Limestone
Entemnotrochus sp. TG, CG, JG, OG
Haustator arenicola (Conrad, 1865) TG, GG, CG, JG, OG, GS, CS, JS
Haustator ghigna (de Gregorio, 1890) TG, GG, CG, JG, OG, GS, CS, JS
cf. *Serpulorbis chavani* Palmer, 1947 TG, GG, CG, JG, OG, GS, CS, JS
Calyptraea (*Trochita*) "*aperta*" (Solander, 1766) TG, GG, CG, JG, OG, TS, GS, CS, JS, OS
Cypraeorbis sp. TG, GG, CG, JG, OG
Euspira sp. GG, CG, JG, OG
Crommium jacksonense (Harris, 1897) TG, GG, CG, JG, OG, TS, GS, CS, JS
cf. *Sassia* sp. nov. TG, GG, JG, OG, ES, JS
Ficus filia (Meyer, 1885) TG, GG, CG, JG, OG, GS, CS, JS
Ficus merita Palmer, 1947 TG, GG, CG, JG, OG, GS, JS
Scalina sp. TG, GG, CG, JG, OG
Agaronia sp. TG, GG, CG, JG, OG
Caricella (*Caricella*) *pyruloides* s.l. (Conrad, 1832) TG, GG, CG, JG, OG, GS, CS, JS
Caricella (*Caricella*) *subangulata* Conrad *in* Wailes, 1854 TG, GG, CG, JG, OG, GS, JS
Admetula jacksonica (Cooke, 1926) TG, GG, CG, JG, OG, GS, JS
Conus spp. TG, GG, CG, JG, OG
Architectonica sp. TG, GG, CG, JG, OG
Tornatellaea lata (Conrad *in* Morton, 1834) TG, GG, CG, JG, OG, GS, CS, JS
Scaphander jacksonensis Palmer, 1947 TG, GG, CG, JG, OG, GS, JS
Nucula sp. cf. *magnifica* Conrad, 1833 TG, GG, CG, JG, OG, GS, CS, JS
Glycymeris sp. TG, GG, CG, JG, OG
Limopsis aviculoides (Conrad, 1833) TG, GG, CG, JG, OG, GS, CS, JS

Lithophaga sp. TG, GG, CG, JG, OG
Pycnodonte trigonalis (Conrad *in* Wailes, 1854) TG, GG, CG, JG, OG, TS, GS, JS
Batequeus ducenticostatus Campbell, 1995 TG, CG, JG, ES, JS
"*Aequipecten*" sp. nov. TG, GG, CG, JG, OG, ES, JS
cf. "*Myrtea*" *curta* (Conrad, 1865) TG, GG, JG, OG, GG, JG
Cavilinga pomilia (Conrad, 1833) TG, GG, CG, JG, OG, GS, CS, JS
Glyptoactis (*Claibornicardia*) *alticostata* (Conrad, 1833) TG, GG, CG, JG, OG, GS, CS, JS
Bathytormus sp. TG, GG, CG, JG, OG
Crassatella alta Conrad, 1832 TG, GG, CG, JG, OG, ES, GS, CS, JS
Crassatella eutawcolens (Harris *in* Van Winkle and Harris, 1919) TG, GG, CG, JG, OG, TS, CS, JS
Nemocardium nicolletti (Conrad, 1841) TG, GG, CG, JG, OG, GS, JS
Tellina (*Eurytellina*) *linifera* (Conrad, 1865) TG, GG, CG, JG, OG, GS, JS
Panopea sp. TG, GG, CG, JG, OG
Callista neusensis (Harris, 1919) TG, GG, CG, JG, OG, ES, JS
Corbula (*Caryocorbula*) *densata* Conrad *in* Wailes, 1854 TG, GG, CG, JG, OG, TS, GS, CS, JS
Caestocorbula wailesiana (Harris *in* Dall, 1898) TG, GG, CG, JG, GS, JS
Antalis danvillensis Palmer *in* Harris and Palmer, 1947 TG, GG, CG, JG, OG, GS, JS
Aturia sp. TG, GG, CG, JG, OG

Unit IV, Lower River Bend Formation
Diodora mississippiensis (Conrad, 1848) TG, GG, CG, JG, OG, TS, GS, OS
"*Turritella*" *caelatura alani* Rossbach and Carter, 1991 TG, GG, CG, JG, OG, ES, OS
"*Turritella*" *neusensis* Rossbach and Carter, 1991 TG, GG, CG, JG, OG, ES, OS
Scalina aff. *Scalina menthfontis* MacNeil in MacNeil and Dockery, 1984 TG, GG, CG, JG, OG, ES, OS
Calyptraea (*Trochita*) cf. *C.* (*T.*) *aperta* (Solander in Brander, 1766) C. (T.) aperta TG, GG, CG, JG, OG, TS, GS, CS, JS, OS
Xenophora sp. TG, GG, CG, JG, OG
Euspira vicksburgensis (Conrad, 1848) GG, CG, JG, OG, GS, OS
Euspira vicksburgensis var.? Rossbach and Carter, 1991 GG, CG, JG, OG, ES, OS
Sinum sp. TG, GG, CG, JG, OG
Apiocypraea sp. OG
Galeodaria britti Rossbach and Carter, 1991 GG, JG, OG, ES, OS
Phalium newbernensis Rossbach and Carter, 1991 TG, GG, CG, JG, OG, ES, OS
Semicassis caelatura (Conrad, 1848) TG, GG, OG, GS, OS
Ficus aff. *F. harrisi* (Martin, 1904) TG, GG, CG, JG, OG, OS
Ficus mississippiensis Conrad, 1848 TG, GG, CG, JG, OG, TS, GS, OS
Cymatium planinodum Rossbach and Carter, 1991 TG, GG, CG, JG, OG, ES, OS
Sassia sp. TG, GG, CG, JG, OG
Distorsio (*Distorsio*) *crassidens* (Conrad, 1848) TG, GG, CG, JG, OG, TS, GS, OS
Oocorys vadosus Rossbach and Carter, 1991 OG, ES, OS
Dermomurex (*Takia*) *portelli* Vokes, 1992 TG, GG, OG, TS, OS
Ecphora wheeleri Rossbach and Carter, 1991 TG, GG, OG, ES, OS
Streptochetus indistinctus (Aldrich, 1894) TG, GG, CG, JG, OG, GS, OS
Levifusus spiniger (Conrad, 1848) TG, GG, CG, JG, OG, GS, JS, OS
Lyria (*Lyria*) *concinna* Rossbach and Carter, 1991 TG, GG, CG, JG, OG, ES, OS
Scaphella saintjeani Rossbach and Carter, 1991 TG, GG, OG, ES, OS
Oniscidia harpula (Conrad, 1848) TG, GG, CG, JG, OG, GS, OS
Olivella sp. TG, GG, CG, JG, OG
Agatrix mississippiensis (Conrad, 1848) TG, GG, JG, OG, GS, OS
Terebra (*Terebrellina*) *divisura divisura* Conrad, 1848 TG, GG, CG, JG, OG, GS, OS
Conus alveatus Conrad, 1865 TG, GG, CG, JG, OG, GS, OS
Pleurofusia menthafons MacNeil in MacNeil and Dockery, 1984 TG, GG, CG, JG, OG, GS, OS

Turricula (*Orthosurcula*) *aequa* Rossbach and Carter, 1991 TG, GG, CG, JG, OG, ES, OS
Clavidrupa anita (Aldrich, 1885) GG, OG, GS, OS
Architectonica fungina (Conrad, 1833) TG, GG, CG, JG, OG, GS, CS, OS
Tornatellaea sp. TG, GG, CG, JG, OG
Yoldia serica (Conrad, 1848) TG, GG, CG, JG, OG, GS, OS
Glycymeris aff. *Glycymeris cookei* Dall, 1916 TG, GG, CG, JG, OG, TS, GS, OS
Rebeccapecten trentensis Harris, 1919 EG, OG, ES, OS
Pycnodonte paroxis (Dockery, 1982 after Lesueur, 1829) TG, GG, CG, JG, OG, GS, OS
Stewartia micraulax Rossbach and Carter, 1991 OG, ES, OS
Astarte sp. TG, GG, CG, JG, OG
Panopea sp. TG, GG, CG, JG, OG
Nemocardium diversum (Conrad, 1848) TG, GG, CG, JG, OG, GS, OS
Pitar (*Lamelliconcha*) *imitabilis* (Conrad, 1848) TG, GG, CG, JG, OG, GS, OS
Callista sp. TG, GG, CG, JG, OG
Ventricolaria judithae Jones, 1984 TG, GG, OG, ES, OS
Bicorbula sp. TG, GG, CG, JG, O
Dentalium misissippiense Conrad, 1848 TG, GG, CG, JG, OG, GS, JS, OS
Dentalium opaculum? Casey, 1904 TG, GG, CG, JG, OG, GS, OS

CHAPTER 20

Diversity Patterns in Eocene and Oligocene Echinoids of the Southeastern United States

Burchard D. Carter

ABSTRACT

The echinoids of the southeastern United States show an increase in species (alpha) diversity from the Bartonian to the Priabonian, despite a decrease from two to one in biogeographic regions sampled (loss of gamma diversity). This is attributed to an increase in beta diversity resulting from an increased number of carbonate facies available for sampling. This correlation between alpha and beta diversity is even more pronounced when examining successive Priabonian biozones. Subsequent stages and substages show steadily declining diversity, and even when changes in beta diversity are considered, apparently imply a real alpha diversity loss from the Eocene to the Oligocene. Other regions of the world also show a general diversity decline from the Eocene to the Oligocene, though the exact pattern at the stage and substage level differs from place to place. Most of these differences also can be hypothesized to have resulted from unique beta diversity patterns. The genera present in Eocene and Oligocene deposits of the southeastern United States suggest that the major change in temperature from tropical to temperate did not begin before the Chattian, and was not complete until the Miocene. If declining temperature had any effect on the fauna, it was simply to decrease diversity, not allow replacement of a warm-water fauna by a cooler-water one. In this region, survival rates of echinoid species from stage to stage do not support hypotheses of unusual extinction rates.

INTRODUCTION

Echinoids are among the most conspicuous components of the Paleogene fossil fauna of the southeastern United States, particularly in the carbonate rocks. This paper describes the temporal trends in species diversity of this important taxon, relating it to changes in geographical and ecological diversity through the later Paleogene and into the Neogene.

Past work on diversity trends in echinoids of the region has focused on two potential explanations for observed trends. McKinney and Oyen (1989) and McKinney et al. (1992) correlated general trends for the entire Cenozoic to paleotemperature curves. The latter paper suggested that declining temperature was the best correlate for diversity in the southeastern United States, but that preservational and ecological factors had substantial control on echinoid diversity in other regions. Carter and Hamza (1994) and Carter (1997) elaborated further upon the integral role of environment and of preservation of appropriate environments upon echinoid diversity.

For this paper, data on the stratigraphic and geographic distribution of echinoids are taken from Cooke (1941, 1942, 1959), Toulmin (1977), Kier (1968, 1980, 1997), McKinney and Zachos (1987), Carter (1987, 1989, 1990), Carter and Hammack (1989), Carter and McKinney (1992), collections of the Florida Museum of Natural History and the Alabama Geological Survey, and from personal collections. Interpretations of substrate preferences for late Eocene species in the southeastern Coastal Plain are taken from Carter et al. (1989). Those for other times

Table 20.1. Stage and Substage Diversity of Southeastern Echinoids from the Middle Eocene to the Late Oligocene

		Lithofacies Distribution	Total
	CHATTIAN	4/4/0	8
	RUPELIAN 2	5/4/0	9
	RUPELIAN 1	6/6/3	15
	PRIABONIAN 3	17/17/9	43
26/17/9 ---- 52	PRIABONIAN 2	13/7/6	26
	PRIABONIAN 1	16/6/1	23
	BARTONIAN	25/8/1	34

Note: The number of ambiguously interpreted species appears to become larger as the proportion of mud-tolerant species in the fauna increases, for the most part reflecting the preferred preservation of regular echinoids in the quiet water environments implied by the mud. The last column of numbers is simply the sum of the three in the preceding column, giving the total diversity for that interval. The Priabonian and Rupelian have been divided into subintervals for this table, but the numbers for the entire stages are also provided at the bottom of the table.

and regions have been made using the same techniques outlined therein.

DATA

Danian echinoids are fairly diverse in the limestones of the Clayton Formation of the Gulf Coast. Thanetian, Ypresian, and Lutetian faunas are of very low diversity and generally of poor preservation in the siliciclastics of that region. Thus, the following discussion will focus on the Bartonian to Chattian faunas. Table 20.1 illustrates the diversity of echinoids in several stages and substages.

The Priabonian Stage has the highest diversity of all, with the maximum value in the uppermost part of the interval. The Bartonian value derives mostly from the fauna of the Castle Hayne Limestone in North Carolina. I have treated the upper two units of Kier (1980) as belonging in this stage, as is conventional. However, some workers (e.g., Zullo and Harris, 1986) consider part of this interval, including the richest echinoid-bearing horizons, to be Priabonian. Should this interpretation prove correct, the Bartonian number would decline from 34 to 19, and the Priabonian number would rise accordingly. Clearly, overall diversity rose between the Bartonian and Priabonian, after which it declined sharply in the lower Rupelian and continued to do so more gradually thereafter. The remainder of this section focuses on the geographic and environmental distributions of the species within each time interval, to allow us to track changes in beta and gamma diversity that might contribute to this pattern of alpha diversity change.

Bartonian

Figure 20.1 shows the distributions of Bartonian taxa and their substrate preferences in the southeastern United States. The heavy dashed line across southern Georgia and northern Florida, on this and all other maps, is the location of the Suwannee Strait, a deep-water feature that exercised considerable control on facies distributions in the region (Carter and McKinney, 1992). The lighter dotted line approximates the boundary between predominantly carbonate rocks to the east and predominantly siliciclastic ones to the west.

The very low diversity faunas of the Gulf Coastal Plains reflect sampling of only tidal flat and very shallow-water facies (Florida: Zachos, 1978; Ivany et al., 1990), poor sampling of fossils available only from well cuttings and cores (Georgia: Kier, 1968), or dominance of siliciclastic sediments (Alabama to Texas.) For reasons that have yet to be explained, echinoids often appear to prefer carbonate over siliciclastic environments (Telford and Mooi, 1986; Carter, 1995; Carter and McGee, 1995). Total diversity in this region is 12 species, practically all of which preferred sandy bottoms.

In contrast, the carbonates of the Castle Hayne Formation of the Carolinas have a richer echinoid fauna of 23 species, with apparently no species preferring mud bottoms. The higher diversity of the Carolina fauna probably reflects both better exposure and sampling of carbonate sedimentary rocks compared to the Gulf Coast. The overall high diversity for the Bartonian results from the provinciality of the faunas. Only three species are shared between the

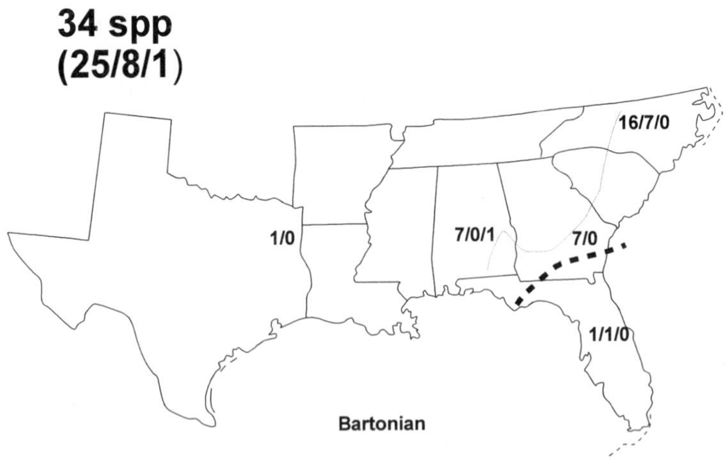

Figure 20.1. Distributions and diversity of Bartonian echinoids. Numbers separated by slashes are numbers of sand dwelling species, species of uncertain preference, and mud dwelling species from left to right. Those at the top of the diagram are for the entire region, those within the map are for local faunas. Heavy dashed line shows position of the Suwannee Strait; lighter dotted line separates predominantly siliciclastic rocks to the west from carbonates to the east. (These conventions are used on all other figures unless otherwise noted.)

Carolinas and the Gulf Coast, and only an additional two are shared between the western Gulf Coast and southeastern Georgia. This high gamma diversity is in contrast to all other intervals discussed in this report.

Priabonian

The Priabonian has the highest overall diversity of the Paleogene stages, as well as the highest proportion of mud-tolerant (and ambiguously interpreted) species at 50%. Though there is some taxonomic distinction between the faunas north and south of the Suwannee Strait, Carter and McKinney (1992) determined that this results from facies and stratigraphic mismatching of the regions rather than real biogeographic distinctiveness. Thus we can assume that gamma diversity has no effect on our taxic counts. Stratigraphic resolution for this interval allows subdivision into three substages, and an interesting ecological correlate to diversity emerges upon examination at this scale.

Figure 20.2A shows diversity and distribution data for the early Priabonian (*Oligopygus phelani* zone). Overall diversity is 23 species, with only 1 species having test characters to allow it to inhabit a muddy substrate. There is little geographic faunal distinction, beyond a few species occurring either north or south of the Suwannee Strait (but not both). The diversity is comparable to that of the Carolina carbonates in the Bartonian, which were also dominated by sand-dwelling species.

Figure 20.2B is a map showing the distributions and diversity of middle Priabonian (*Oligopygus haldemani* zone) echinoids. Note that the carbonate/siliciclastic boundary has shifted well westward, so that all the echinoids are found in carbonate rocks. Diversity is modestly higher in this zone than the previous one, with the gains reflecting the few mud-tolerant species found near the northern edge of the Suwannee Strait.

Figure 20.2C shows the distributions and diversity of late Priabonian (*Oligopygus wetherbyi* zone) species. Again, most species are found in the carbonates in the eastern part of the Gulf Coast, though a few are known from the siliciclastics of Texas and western Louisiana. Diversity in this interval reaches its maximum value of 43 species. The number of mud-tolerant and ambiguous species is accordingly high, both in the vicinity of the Suwannee Strait and farther from it.

Carter (1990) attributed the upsection increase in carbonate mud to the marine transgression suggested by several sedimentological studies (Randazzo and Saroop, 1976; Zachos, 1978; Fenk, 1978; Sharpe, 1980) and global sea-level curves. Figure 20.3 summarizes the facies evolution of the carbonates through the Priabonian. In the *Oligopygus phelani* zone, the low diversity faunas inhabited widespread sandy bottoms. By the time of the *O. haldemani* zone, some mixed sand and mud bottoms occurred in extreme northern Florida and some predominantly

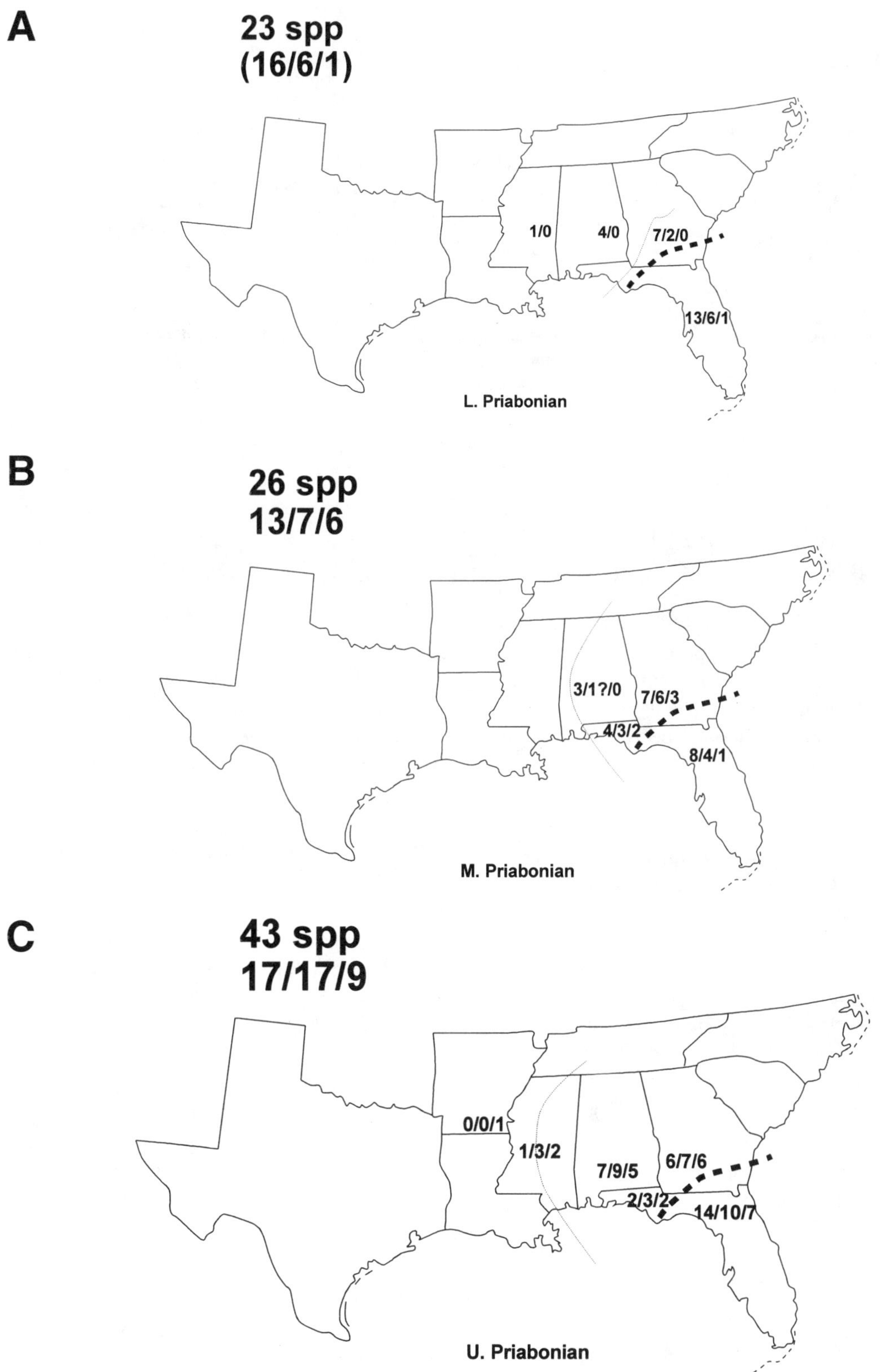

Figure 20.2. Distributions and diversity of early Priabonian echinoids: (A) early Priabonian echinoid diversity; (B) middle Priabonian echinoid diversity; (C) late Priabonian echinoid diversity.

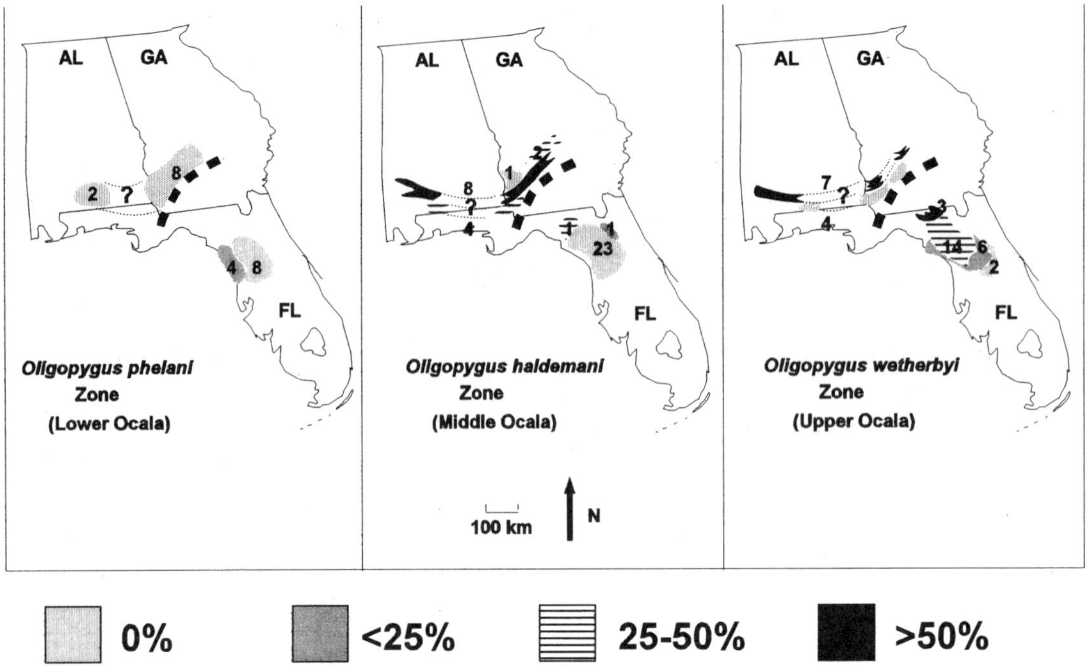

Figure 20.3. Lithofacies patterns of late Eocene carbonates in the eastern Gulf Coastal Plain. The percentages and symbols represent the proportion of mud dwelling species in the fauna. Note that the increasing diversity of echinoids through this interval tracks the development of more complex lithofacies patterns, controlled by the location of the Suwannee Strait. Numbers on the map indicate the numbers of individual localities per substrate type upon which the lithofacies model is based (modified from Carter and McKinney, 1992).

muddy ones in southern Georgia and Alabama, with a modest diversity increase accompanying them. By the time of the *O. wetherbyi* zone, very muddy bottoms persisted in southern Georgia and Alabama, and mixed and predominantly muddy ones had become more widespread in northern and central Florida. This was the time of maximal echinoid diversity in the region. It seems likely that prolonged inundation by continually rising sea level may have allowed a progressively more complex suite of facies to develop throughout the Priabonian, providing progressively more resource diversity for echinoids to partition (Carter, 1995).

Rupelian

Figure 20.4A illustrates the distribution and diversity of lower Rupelian echinoids. As in the later Priabonian, all species are from the carbonate facies. Diversity in these units is lower than in the later Priabonian, and the proportion of mud-dwellers is generally lower as well. More habitat complexity existed than the ratio of mud-dwellers to mud/sand-dwellers suggests; the Bridgeboro Limestone of southwest Georgia, southeast Alabama, and west Florida is primarily a coralgal conglomeratic facies, and contains a fairly diverse fauna of echinoids (Manker and Carter, 1987).

Later Rupelian faunas throughout the southeast are completely dominated by sand-dwelling echinoids, with diversity dropping to the lowest level yet seen (figure 20.4B).

Chattian

Figure 20.4C shows the distributions and diversity values for Chattian species. As in the later Rupelian, overall diversity is very low, reflecting the complete dominance of species that preferred sand bottoms.

GLOBAL PATTERNS

Southeastern United States

Table 20.2 summarizes the diversity trends in the southeastern United States and compares them with several other regions. Diversity rises between the Bartonian and the Priabonian, reaching its highest value, then drops precipitously in the Rupelian and continues to decline throughout that interval and

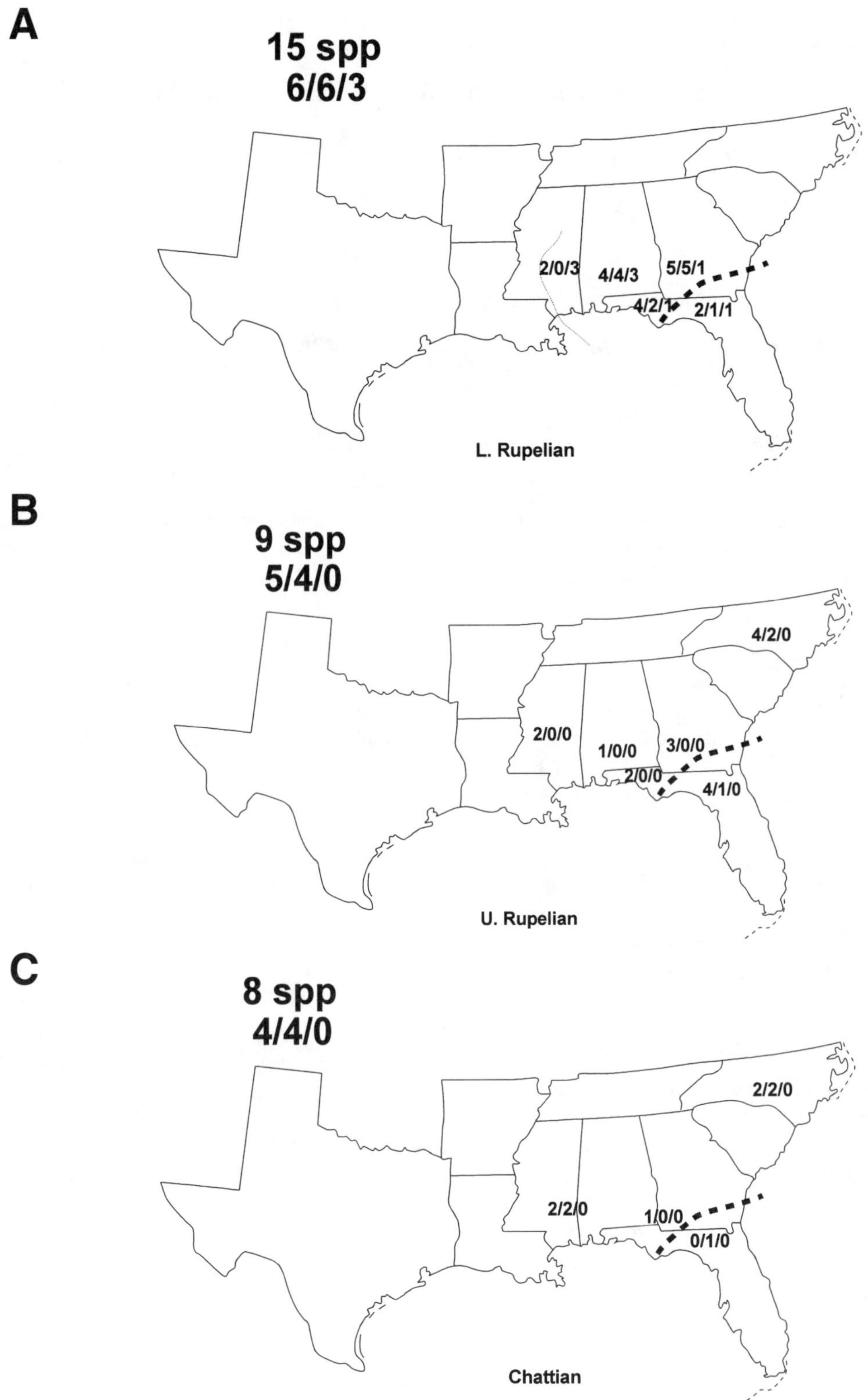

Figure 20.4. Distributions and diversity of Oligocene echinoids: (A) early Rupelian echinoid diversity; (B) late Rupelian echinoid diversity; (C) Chattian echinoid diversity.

Table 20.2. Comparison of Changes in Diversity from the Middle Eocene to the Early Oligocene in the Southeastern United States, Jamaica, Southwestern France, Northeastern Italy, Egypt, and Pakistan

	SE USA	JAMAICA	SW FR	VENETO	EGYPT	PAK
RUPELIAN 2	9/4?	12/3?	17/2?	23/9	0	- -?- -?- - 14/4?
RUPELIAN 1	15/9?		? ↕ ? 17/2?	25/11	0	
PRIABONIAN	52/26	21/8	17/2?	70/25	11/4?	- -?- -?- - 37/17
BARTONIAN	34/9?	26/10	62/16	87?/35	38/27	

Note: The numbers represent the number of sand-dwelling species and the number of *potentially* mud-dwelling ones, combining the uncertain ones and the more obviously mud-loving ones. Lack of personal examination of many of these species—particularly from Jamaica, Italy, and Pakistan—makes this a conservative course.

into the Chattian. Rupelian diversity is low because of the prevalence of sandy substrata; but more interesting, it is lower than in previous intervals that were also dominated by such sediments. Recall that in the Bartonian of the Carolinas and the earliest Priabonian of the Gulf Coast, diversities were around 25 species. This contrasts sharply with the 9 species of the sandy sediments of the upper Rupelian or the 8 species of the Chattian (table 20.1). Thus there appears to be a decline in alpha diversity from the Eocene to the Oligocene beyond what can be accounted for by any change in beta diversity.

Jamaica

Data for the faunas of Jamaica were taken from Donovan (1993, 1994, 1996) and Dixon and Donovan (1994, 1998). These authors often warned that poor preservation in the upper Eocene and, to a lesser extent, the Oligocene precluded certainty about diversity, but the general pattern is interesting nevertheless. Diversity in Jamaica appears to have been highest in the Bartonian, with a gradual decline into the Priabonian, and finally a precipitous drop in the Oligocene. The Rupelian drop is almost certainly artifactual in part (Dixon and Donovan, 1994). The apparent trend differs from the southeastern United States in lacking the Priabonian increase in diversity. It is open to question what better preservation and retrieval of fossils from the upper Eocene in Jamaica would do to this pattern. As in the United States, diversity declines were accompanied by decreases in the proportion of mud-tolerant and potentially mud-tolerant species.

Southwestern France

The Aquitaine Basin and the Bordelaise of southwest France have rich Paleogene echinoid faunas. Data were taken from a number of monographic works, notably Castex (1930), Castex and Lambert (1920), Cotteau (1885-1889, 1889-1894), Lambert (1912), and Roman and Cahuzac (1992). In these regions, Bartonian diversity has by far the highest value of any interval. There is a sharp drop into the Priabonian, which is dominated by siliciclastic sedimentary rocks. In either younger Priabonian or early Rupelian rocks (the local Tongrian Stage) diversity remains at its low Priabonian level, and this persists into the later Rupelian. Chattian diversity, though not tabulated here, is even lower than in the Rupelian. As in the United States and Jamaica, the dramatic drop in diversity coincided with a similarly dramatic drop in the mud dwelling species.

Northeastern Italy

In the Veneto region of northern Italy, echinoid diversity is consistently higher through the later Eocene and early Oligocene than in any other region considered so far. The data for this region were taken from a large number of monographs cited in Carter (1997). The rocks of the Veneto appear unusual in that a much larger proportion of mud-dwelling taxa are preserved here in the Oligocene. Comparison of the Oligocene numbers for the Veneto with those of other regions, for instance the United States or Jamaica, makes clear that both total diversity and number of possible mud-dwellers in the Oligocene of the Veneto are comparable to those for the Eocene of the other regions.

Egypt

Egypt has a highly diverse middle Eocene echinoid fauna. Data for these values come from numerous monographs and museum collections, cited in Carter and Hamza (1994). Separation of Lutetian and Bartonian taxa is very difficult, and the majority of species range into both intervals; the value given is a conservative estimate of species restricted to the latter. Note that diversity is comparable to that of the Bartonian in the United States, Jamaica, or Pakistan, and that the proportion of mud-dwelling species is exceptionally high in comparison with any other value in table 20.2. The complete lack of Oligocene echinoids in Egypt results from the lack of any marine rocks of Oligocene age in that country. Carter and Hamza (1994) argued that echinoid diversity at the stage level in Egypt tracked the number of distinct carbonate facies throughout the Eocene.

Pakistan

Stratigraphic distributions of echinoids in the Paleogene of Pakistan are very uncertain. Data are from Duncan and Sladen (1882-1886), who assigned the taxa to either the Kirthar Group or the Nari Group. Kumar (1985) assigned the former to the middle Eocene ("Lutetian") to Priabonian, and the latter to the Oligocene (both lower and upper). Later work separated the Kirthar as understood by Duncan and Sladen into an upper unit (Kirthar) and a lower one (the Laki, which is Ypresian according to Kumar). Collections of fossils from this region in the Museum of Paleontology at Berkeley, along with careful cross-referencing of collecting localities, allow a good estimate of which taxa belong in which of these intervals. Note that high middle Eocene diversity in this region is accompanied by a high proportion of mud-tolerant species, and that the decline in diversity into the Oligocene is accompanied by a decline in the proportion of such species.

General Patterns

All the regions outlined above show a marked diversity decline from the Eocene into the Oligocene, though the pathway by which this proceeded is not the same in all cases. In some areas (France, Egypt, and Pakistan) the Bartonian appears to have had the highest diversity with rapid decline thereafter. In others (Italy and Jamaica) the middle to late Eocene decline was not so marked, whereas in the United States, there seems to have been an actual diversity increase into the late Eocene. In all cases, declining overall diversity tracked a decline in the number of taxa that could have survived in fine-grained sediments. The diversity decline into the Oligocene, though still obvious, was much less severe in Italy, the one place where abundant fine-grained sediments persisted into the Oligocene.

I suggest that the specific pattern of diversity decline is a consequence not of real loss of taxa, but rather of loss of beta diversity available for sampling. In the most extreme case (Egypt), the echinoids *appear* to have become extinct at the end of the Eocene. Clearly, this is an illusion, and results from the unavailability of appropriate rocks for collecting the fossils. It is tempting to wonder what might be the effect on apparent diversity in southwestern France, for example, if Priabonian rocks containing abundant carbonate muds were exposed in that region.

Nevertheless, there are two reasons for supposing that the overall decline in species richness from the Eocene into the Oligocene is real. First, in the Veneto region, where proportions of mud-dwellers remained high throughout the Oligocene, diversity still declined to roughly one-third its Eocene level. Second, in the United States, diversity in the calcarenite lithofacies of the later Rupelian was less than half what it had been in similar facies in the Bartonian (of North Carolina) and the early Priabonian (of the eastern Gulf Coast).

THE ROLE OF TEMPERATURE

The diversity decline from the Eocene to the Oligocene in the southeastern United States and elsewhere may have resulted from changing global temperature, as McKinney and Oyen (1989) and McKinney et al. (1992) have suggested, but the pattern in the fossil record is not as straightforward in this regard as it could be. Table 20.3 shows the extant irregular echinoid genera with congeneric Paleogene relatives present in this region and compares them to modern geographic distributions of those genera (data from Ghiold, 1988/1989). Examination of reg-

Table 20.3. Temperature Preferences of Modern Genera with Relatives in Paleogene Rocks of the Southeastern United States

	Upper Eocene	Lower Oligocene	Upper Oligocene	Miocene*
Genera that presently live at high latitudes	*Echinocyamus*	NONE	*Echinocyamus*	*Echinocyamus* *Leodia* *Spatangus* *Echinocardium*
Genera that presently live primarily in the tropics, but with few temperate representatives	*Eurhodia* *Echinolampas* *Clypeaster* *Schizaster (Schizaster)* *Brissopsis* *Brissus*	*Clypeaster* *Cyclaster* *Lovenia* *Brissus*	*Echinolampas*	*Clypeaster* *Encope* *Lovenia* *Brissus*
Genera that presently live only in the tropics	*Fibularia* *Rhyncholampas* *Schizaster (Paraster)* *Agassizia* *Plagiobrissus* *Eupatagus*	*Rhyncholampas* *Schizaster (Paraster)* *Agassizia*	*Agassizia*	*Agassizia*

***Excludes species from southern Florida, which still has a tropical fauna.**

Note: The shift from predominantly tropical to predominantly temperate appears to have happened in the Miocene, though poor data for the Chattian do not allow rejection of the hypothesis that it was under way in that interval.

ular echinoid taxa gives the same sense, but the taxonomic assignments are rather more uncertain for fossil regular echinoids, and so they are excluded. It is clear from Table 20.3 that upper Eocene to lower Oligocene faunas are dominated by genera whose extant species live entirely or predominantly in tropical waters. The upper Oligocene is ambiguous because in its small fauna only three irregular echinoid genera have extant relatives (and all the regular echinoid genera known from the interval are extinct). Though probably not a representative sample, these taxa seem to suggest a fairly even distribution of tropical, temperate, and high-latitude species, perhaps marking the onset of the transition from tropical to temperate conditions. Because of the predominance of extratropical genera in its fauna, the Miocene appears to have been the time when the southeastern United States first acquired an unambiguously temperate echinoid fauna. If changing temperature were responsible for echinoid diversity loss across the Eocene/Oligocene boundary, it did not make itself obvious in the taxonomic/ecological makeup of the faunas that survived it.

EXTINCTIONS

Figure 20.5 reiterates the echinoid diversities of the four stages around the Eocene/Oligocene boundary in the southeastern United States, and also shows the number of species known to survive across the three boundaries. Some identifications are uncertain, particularly of species near the Bartonian/Priabonian boundary. Survival at this boundary was at least 21% and perhaps as high as 41% of the species comprising Bartonian diversity. From the Priabonian to the Rupelian, species survival was around 14–16% of Priabonian diversity (or 33–38% of the smaller Rupelian diversity). Fourteen percent of Rupelian species survived into the Chattian (38% of the smaller Chattian fauna). Though similar tabulations have not been made for other regions, anecdotally I suspect these are fairly standard survival rates for most stage boundaries. Two critical observations emerge from this figure. First, the highest survival rate is from the Bartonian to the Priabonian, suggesting that any extinction event at the end of the middle Eocene had no inordinate effect on southeastern echinoids. Second, percentage species survival at the end of the

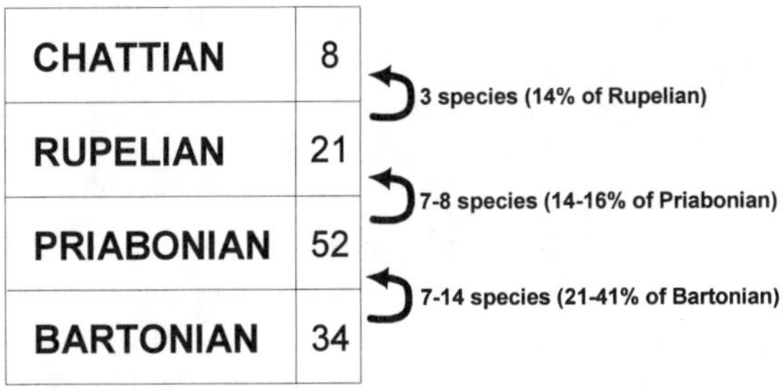

Figure 20.5. Survival rates of echinoid species between successive stages of the Paleogene, southeastern United States.

Priabonian is almost exactly the same as at the Rupelian/Chattian boundary, suggesting that any terminal Eocene extinction event also had no unusual effect upon echinoids.

CONCLUSIONS

1. Southeastern echinoids show an increase in species diversity from the Bartonian to the Priabonian, despite a decrease from two to one in biogeographic regions sampled (loss of gamma diversity). This alpha diversity increase is attributed to an increase in beta diversity resulting from an increased number of carbonate facies available for sampling. This correlation between alpha diversity and beta diversity is even more pronounced when examining successive Priabonian stages.

2. Subsequent stages and substages show steadily declining diversity, even when changes in beta diversity are considered, apparently implying a real alpha diversity loss from the Eocene to the Oligocene.

3. Other regions of the world also show a general diversity decline from the Eocene to the Oligocene, though the exact pattern at the stage and substage level differs from place to place. Most of these differences can also be hypothesized to have resulted from unique beta diversity patterns.

4. The genera present in Eocene and Oligocene deposits of the southeastern United States suggest that the major change in temperature from tropical to temperate did not begin before the Chattian, and was not complete until the Miocene. If declining temperature had any effect on the fauna, it was simply to decrease diversity near the Eocene/Oligocene boundary, not to allow replacement of a warm-water fauna by a cooler-water one.

5. In the southeastern United States, survival rates of echinoid species from stage to stage do not support hypotheses of unusual extinction rates.

ACKNOWLEDGMENTS

I profusely thank Don Prothero for again providing a context in which to consider the diversity of my favorite beasts during my favorite historical time interval, and an excuse for getting out of the Southeast during the droughty August of 1999. Lois Elms organized and ran a great week in Olympia. As for this manuscript, Warren Allmon suggested the addition of some material that greatly improved it. Natalie Jordan's senior thesis on the echinoids of Pakistan provided a basic data set that allowed me easily to extract good estimates of stratigraphic distribution of Eocene echinoids in that country. Yonnie Williams, as usual, saved you a lot of trouble by translating my first draft into real English. Steve Donovan updated my discussion of Jamaican echinoids, suggested several sensible modifications to the manuscript, and delivered much needed coups de grâce to several superfluous commas that had stubbornly refused to die in previous edits.

REFERENCES

Carter, B. D., 1987, Paleogene echinoid distributions in the Atlantic and Gulf Coastal Plains: Palaios, v. 2, p. 390-404.

Carter, B. D., 1989, Echinoid biofacies and lithofacies distributions in the upper Eocene of the Dougherty Plain, southwestern Georgia: Southeastern Geology, v. 30, p. 175-191.

Carter, B. D., 1990, Late Eocene echinoid biofacies of Florida: Palaios, v. 5, p. 176-183.

Carter, B. D., 1995, Temporal variations in diversity of the Echinoidea through the Paleogene, eastern Gulf Coastal Plain: the effects of substrate: Geological Society of America Abstracts with Programs, v. 27, no. 2, p. 41-42.

Carter, B. D., 1997, Inferring substrate preferences from test morphology in echinoids, and interpreting spatial and temporal patterns of diversity, *in* Waters, J. A., and Maples, C. G., eds., Geobiology of echinoderms: Paleontological Society Papers 3, p. 121-145.

Carter, B. D., and Hammack, R. E., 1989, Stratigraphic distribution of late Eocene (Priabonian, Jacksonian) echinoids in Georgia: suggested correlations with Florida and the Carolinas: Palaios, v. 4, p. 86-91.

Carter, B. D., and Hamza, F., 1994, Substrate preferences and biofacies distributions of Egyptian Eocene echinoids: Palaios, v. 9, p. 237-253.

Carter, B. D., and McGee, R. Q., 1995, Comparison of echinoid and mollusc taphonomy in the Late Eocene of the Gulf coast: greater diversity does not equate with a better record: Geological Society of America Abstracts with Programs, v. 27, no. 2, p. 42.

Carter, B. D., and McKinney, M. L., 1992, Eocene echinoids, the Suwannee Strait, and biogeographic taphonomy: Paleobiology, v. 18, p. 299-325.

Carter, B. D., Beisel, T. H., Branch, W. B., and Mashburn, C. M., 1989, Substrate preferences of late Eocene (Priabonian, Jacksonian) echinoids of the eastern Gulf Coast: Journal of Paleontology, v. 63, p. 495-503.

Castex, L., 1930, Révision des Échinides du Nummulitique du Department des Landes: Actes de la Société Linnaean de Bordeaux, v. 82, p. 1-69.

Castex, L., and Lambert, J., 1920, Révision des Échinides des falaises de Biarritz: Actes de la Société Linnaean de Bordeaux, v.71, p. 117-200.

Cooke, C. W., 1941, Cenozoic regular echinoids of eastern United States: Journal of Paleontology, v. 15, p. 1-20.

Cooke, C. W., 1942, Cenozoic irregular echinoids of eastern United States: Journal of Paleontology, v. 16, p. 1-62.

Cooke, C. W., 1959, Cenozoic echinoids of eastern United States: U.S. Geological Survey Professional Paper 321, 106 p.

Cotteau, G. H., 1885-1889, Terraines tertiares Éocène v. 1, *in* d'Orbigny, A., Paléontologie Française: Paris, G. Masson, 672 p.

Cotteau, G. H., 1889-1894, Terraines tertiares Éocène v. 2, *in* d'Orbigny, A., Paléontologie Française: Paris, G. Masson, 789 p.

Dixon, H. L., and Donovan, S. K., 1994, Local extinction patterns and the decline of the Jamaican Paleogene echinoid fauna: Palaios, v. 9, p. 506-511.

Dixon, H. L., and Donovan, S. K., 1998, Oligocene echinoids of Jamaica: Tertiary Research, v. 18, p. 95-124.

Donovan, S. K., 1993, Jamaican Cenozoic echinoids, *in* Wright, R. M. and Robinson, E., eds., Biostratigraphy of Jamaica: Geological Society of America Memoir 182, p. 371-412.

Donovan, S. K., 1994, Middle to upper Eocene echinoids of Jamaica, *in* David, B., Guille, A., Feral, J.-P., and Roux, M., eds., Echinoderms through time: Rotterdam, A.A. Balkema, p. 629-635.

Donovan, S. K., 1996, A regular echinoid from the Walderston Formation (lower Oligocene) of Jamaica: Caribbean Journal of Science, v. 32, p. 78-82.

Duncan, P. M., and Sladen, W. P., 1882–1886, Tertiary and Upper Cretaceous fauna of western India, v. 1, pt. 3, Fossil Echinoidea of western Sind and the coast of Baluchistan and of the Persian Gulf, from the Tertiary formations: Palaeontologica Indica, series 14, v. 1, p. 1-392.

Fenk, E. M., 1978, Sedimentology and stratigraphy of middle and upper Eocene carbonate rocks, Lake, Hernando, and Levy Counties, Florida: M.S. thesis, University of Florida, Gainesville, 133 p.

Ghiold, J., 1988/1989 Species distributions of irregular echinoids: Biological Oceanography, v. 6, p. 79-162.

Ivany, L. C., Portell, R. W., and Jones, D. S., 1990, Animal-plant relationships and paleobiogeography of an Eocene seagrass community from Florida: Palaios, v. 5, p. 244-258.

Kier, P. M., 1968, Echinoids from the middle Eocene Lake City Formation of Georgia: Smithsonian Miscellaneous Collections, v. 153, no. 2, 45 p.

Kier, P. M., 1980, The echinoids of the middle Eocene Warley Hill Formation, Santee Limestone, and Castle Hayne Limestone of North and South Carolina: Smithsonian Contributions to Paleobiology 39, 102 p.

Kier, P. M., 1997, Oligocene echinoids of North Carolina: Smithsonian Contributions to Paleobiology 83, 37 p.

Kumar, R., 1985, Fundamentals of historical geology and the stratigraphy of India: New York, John Wiley, 254 p.

Lambert, J., 1912, Revision des Échinides fossiles du Bordelais: Actes de la Société Linnaean de Bordeaux, v. 66, p. 45-120.

Manker, J. P., and Carter, B. D., 1987, Sedimentology and paleoecology of a lower Oligocene rhodolith facies in south Georgia and north Florida: Palaios, v. 2, p. 181-188.

McKinney, M. L., McNamara, K. J., Carter, B. D., and Donovan, S. J., 1992, Evolution of Paleogene echinoids: a global and regional view, *in* Prothero, D. R., and Berggren, W. A., eds., Eocene-Oligocene climatic and biotic evolution: Princeton, New Jersey, Princeton University Press, p. 349-367.

McKinney M. L., and Oyen, C. W., 1989, Causation and nonrandomness in biological and geological time series: temperature as a proximal control of extinction and diversity: Palaios, v. 4, p. 3-15.

McKinney, M. L., and Zachos, L. G., 1987, Echinoids in biostratigraphy and paleoenvironmental reconstruction: a cluster analysis from the Eocene Gulf Coast (Ocala Limestone): Palaios, v. 2, p. 420-423.

Randazzo, A. F., and Saroop, H. C., 1976, Sedimentology and paleoecology of middle and upper Eocene carbonate shoreline sequences, Crystal River, Florida, U.S.A.: Sedimentary Geology, v. 15, p. 259-291.

Roman, J., and Cahuzac, B., 1992, Une riche faune d'échinoides de l'Éocène d'Angoumé (sud-Aquitaine, France): Geobios, v. 14, p. 113-121.

Sharpe, C. L., 1980, Sedimentological interpretation of Tertiary carbonate rocks from west central Florida: [unpubl. M.S. thesis], University of Florida, Gainesville, 170 p.

Telford, M., and Mooi, R., 1986, Resource partitioning by sand dollars in carbonate and siliceous sediments: evidence from podial and particle dimensions: Biological Bulletin, v. 171, p. 197-207.

Toulmin, L. D., 1977, Stratigraphic distribution of Paleocene and Eocene fossils in the eastern Gulf Coastal Plain: Alabama Geological Survey Monograph 13, 602 p.

Zachos, L. G., 1978, Stratigraphy and petrology of two shallow wells, Citrus and Levy Counties, Florida: [unpubl. M.S. thesis], University of Florida, Gainesville, 105 p.

Zullo, V. A., and Harris, W. B., 1986, Introduction: sequence stratigraphy, lithostratigraphy, and biostratigraphy of the North Carolina Eocene carbonates, *in* Textoris, D.A., ed., Southeastern United States: SEPM Guidebooks, Third Annual Midyear Meeting, Raleigh, North Carolina, p. 257-263.

CHAPTER 21

The Eocene/Oligocene Transition in Marine Vertebrates of the Gulf Coastal Plain

Earl M. Manning

ABSTRACT

A series of problems relating to differences in area of exposure, number of published faunas, reliability of species identifications, differential habitat preservation, and relative amount of diagenesis make it difficult at this point to make precise comparisons of diversity counts in the marine vertebrate faunas across the Eocene/Oligocene boundary of the Gulf Coastal Plain (GCP). Generally, however, it appears that aside from a few first and last appearances, the changes are mostly very gradual—often at a subspecific level. This contrasts strikingly with vertebrate faunal changes at the early/middle Eocene and early/late Oligocene boundaries, which appear stepwise. The major faunal changes, however, are probably caused only by the poor records of the intermediate faunas.

A few species, notably several Tethyan forms, such as the giant sea snakes (possibly victims of cooling Oligocene seawater), the early whale *Basilosaurus* (possibly surviving into the earliest Oligocene, in the uppermost Jackson Group), and the sawfish *Propristis*, die out at the end of the Eocene in the GCP. Several taxa show only subspecific change (*Carcharhinus gibbesi*, *Carcharodon auriculatus*, *Sphyraena*, *Lophius*?, etc.) or no change (*Galeocerdo alabamensis*, *Hemipristis wyattdurhami*, *Plinthicus* n. sp., etc.) across the Eocene/Oligocene boundary. A few taxa (notably, the daggernose shark *Isogomphodon acuarius*) first appear in the early Oligocene of the GCP.

Some taxa showed regional differences along the Eocene-Oligocene Gulf coastline. The coastline itself was radically changed at the Eocene/Oligocene boundary, as the Mississippi Embayment was largely lost.

INTRODUCTION

History

Even though fossil mollusks from what are now identified as (early Oligocene) Vicksburg Group beds were reported from the Redwood area, northeast of Vicksburg, Warren County, Mississippi, as early 1784 (Hutchins, 1784:55), and from what are now upper middle to uppermost Eocene Jackson Group beds from a well north of Jackson, in southern Madison County, Mississippi, as early as 1819 (Cornelius, 1819, p. 327), Jackson and Vicksburg Group fossils of the Gulf Coastal Plain (GCP) didn't become widely known until the early 1830s.

What later became the Marianna Limestone of the Vicksburg Group of the GCP was described by Conrad (1832:14) as a "very white friable Secondary limestone, full of *Nummulites*, and containing *Gryphoea*, &c." (later called the "nummulite beds," incorrectly, for the larger foraminiferan later named *Lepidocyclina*). This bed was incorrectly correlated with the Cretaceous (then, as "Secondary"), and incorrectly thought to lie below (now the upper middle Eocene Gosport Formation, upper Claiborne Group) "Claiborne beds" at Claiborne Landing, Alabama (Conrad, 1832:14). Conrad based his correlation on material sent to Philadelphia by local collectors, a year before he had seen the beds for himself. The latter bed was correctly assigned to the "Middle Tertiary" "London Clay" of England, or "Calcaire grossier" of France, by Conrad (1832:13), even before the term "Eocene" (then also including the Paleocene and Oligocene) was named by Lyell in 1833. The first described Vicksburg Group invertebrates (including "*Nummulites*" *mantelli*, later *Lepidocyclina*) were reported from the friable limestone near Claiborne, Alabama, by Morton (1833). Fossils from what later became the Yazoo Formation of the Jackson Group of the GCP were first described by Harlan (1834), from northeastern Louisiana, in reporting *Basilosaurus* (now a specialized early whale, but then considered an Eocene plesiosaur). As it was soon after also reported from Alabama, then Mississippi, these beds came to be known as the "Zeuglodon beds," for an 1839 improper emendation of Harlan's original name. The problem of Conrad's (1832) incorrect age and stratigraphic position assignments was greatly compounded by Morton's influential (1834) review of American Cretaceous fossils. Morton considered both Jackson Group fossils, such as the newly named "*Nautilus*" (now *Aturia*) *alabamensis* (1834:33); and Vicksburg Group fossils, including "*Nummulites*" *mantelli* (1834:45); to be "newer Cretaceous" (later "First Group," "Upper Division," or "Upper Cretaceous"; in comparison to the now late Paleocene "Medial Cretaceous" fossils, and the now late Cretaceous "Lower Cretaceous" fossils—Morton, 1842). Both Jackson and Vicksburg beds were incorrectly thought to lie beneath the Eocene Claiborne beds.

The proper Eocene age (sensu lato, including the Oligocene) of the Zeuglodon and nummulite beds wasn't corrected until Conrad (1846:209-210); and the true stratigraphic order of nummulite beds (Vicksburg Group) over Zeuglodon beds (Jackson Group) over Claiborne beds (Claiborne Group) wasn't corrected until the publications by Lyell (1846:314) (Manning and Bograd, 2000:60).

Although Harlan had reported and figured a vertebrate from what is now the Jackson Group of the GCP as early as 1834, vertebrates from what is now the Vicksburg Group (shark teeth from the Vicksburg area) from the GCP were not reported until Wailes (1845). Knowledge of the marine vertebrate faunas on either side of the Eocene/Oligocene boundary in the Gulf Coastal Plain has come sporadically ever since.

Eocene-Oligocene Deposition in the Gulf Coastal Plain

Generally, the record of marine vertebrate faunas across the Eocene/Oligocene boundary is much more complete in the GCP than in many other areas of the world, notably, in comparison to faunas of the North American Atlantic and Pacific Coastal Plains. The reason for this is tectonic, and not entirely clear. The Pacific Coastal Plain was the active plate margin, with only a very narrow coastal plain, close to a steep shelf-edge. Many of the contemporaneous known deposits are of far greater water depths than those of the GCP. On the Atlantic Coastal Plain, Eocene/Oligocene boundary deposits are often unusually thin in comparison to either the earlier Eocene or later Miocene beds (Poag, 1992:figure 6-5). Why so little sediment was being shed by the Appalachians in the late Eocene and Oligocene, in contrast to periods just before and after that time, has been little discussed. The timing of the increased rate of deposition on the passive margin might suggest that activation of the Appalachians was tied to West Coast tectonic activity—the Laramide compression before, and the Basin and Range extension after. The transmission of stress from one end of the North American continent to the other seems unlikely, but some explanation is required for the dramatic changes in the Atlantic Coastal Plain deposition rate.

One of the few moderately thick Jackson Group–equivalent deposits of the Atlantic Coastal Plain is the Chickahominy Formation, which filled in the top of the Chesapeake Bay crater (see Poag, this volume). It is, unfortunately, entirely a subsurface unit. Some beds of the equivalent age have been preserved in structural lows northeast and southwest of the Cape Fear Arch, near the North Carolina-South Carolina eastern border. The upper Eocene-lower Oligocene beds of the Atlantic Coastal Plain are generally thin and often covered by younger deposits.

The reasons why the contemporaneous beds in the GCP are so much thicker than those of the Atlantic and Pacific coasts are fairly clear. From the middle Cretaceous until the early late Eocene (the timing is discussed further later), a huge aulacogen created a topographic low over a graben in the area of the lower Mississippi Valley. This permitted the Gulf of Mexico to extend far north of the main Gulf coastline (forming the Mississippi Embayment), and caused increased sedimentation in the area of the structural low. Related to this feature, a large river delta (possibly an early Mississippi River) was restricted to the area, and increased deltaic sedimentation, causing downwarping of the Gulf continental margin, which also greatly increased the deposition rate. As a result, the stratigraphic record of the Eocene-Oligocene transition is unusually complete in the Gulf Coastal Plain.

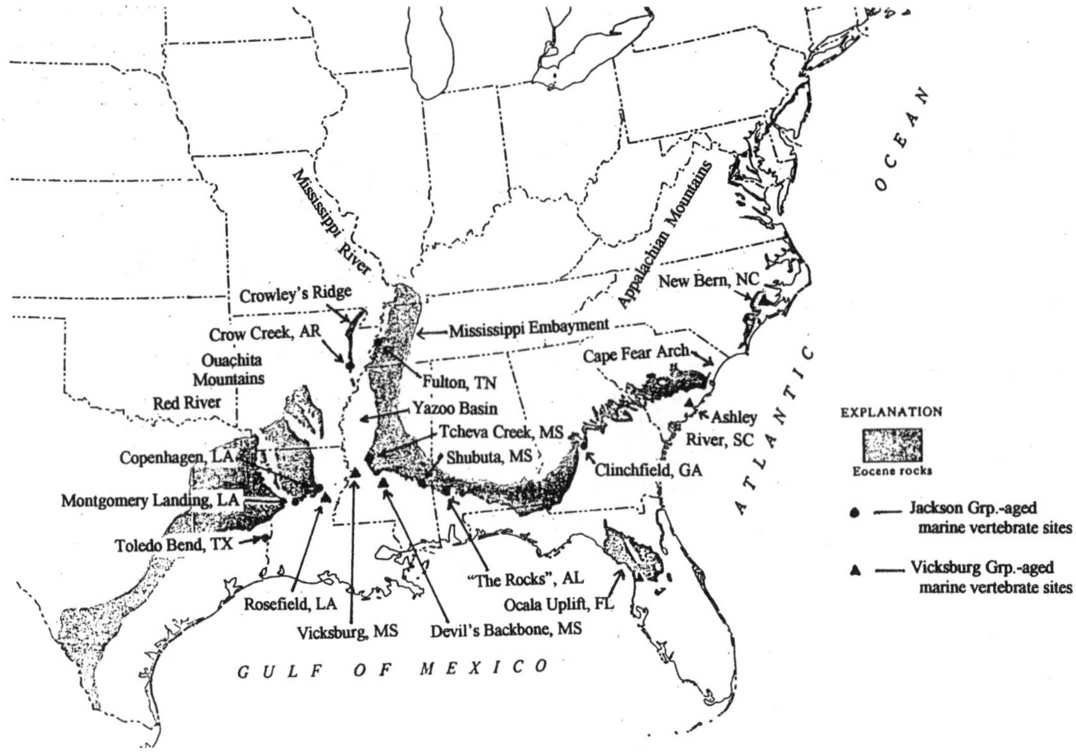

Figure 21.1. Some Jackson (late middle to latest Eocene) and Vicksburg (early Oligocene) Group aged sites with marine vertebrate faunas of the Gulf Coastal Plain and southern Atlantic Coastal Plain. The Vicksburg Group outcrop is a narrow band seaward of the Eocene exposure. Sites shown are mainly the major ones, discussed in the text.

Abbreviations

MGS	paleontology collection, Mississippi Office of Geology (formerly the Mississippi Geological Survey); Jackson
MMNS VP	vertebrate paleontology collection, Mississippi Museum of Natural Science; Jackson
TU–VP	vertebrate paleontology collection, Tulane Museum of Natural History; Belle Chasse, Louisiana

PROBLEMS WITH MAKING DIVERSITY COUNTS

There are several difficulties with making precise diversity counts of marine vertebrates across the Eocene/Oligocene boundary in the GCP at the present time.

Known Jackson and Vickburg Faunas

Because the Jackson Group is far more broadly exposed in the GCP, and because it has been more extensively collected for fossils than has the Vicksburg Group, far more Jackson (late middle to latest Eocene) faunas are known than Vicksburg (early Oligocene) ones. As has been shown statistically by Peters and Foote (2002), a false "extinction event" can be created by comparing the fauna of a widely exposed bed to that of a narrowly exposed one. That has almost certainly been done in places in the Eocene-Oligocene interval. The Jackson also appears to include more vertebrate concentrations than does the Vicksburg. These biases make the Jackson (late Eocene) diversity seem to be higher than that of the Vicksburg (early Oligocene).

The following brief review mentions only a few of the broader studies of the late Eocene and early Oligocene marine vertebrate faunas of the Gulf Coastal Plain. Many reports discuss individual elements of the faunas and are too numerous to cover here. For more specific studies of Louisiana and Mississippi, see Domning (1969); of Mississippi, see Daly (1992); and of Alabama, see Thurmond and Jones (1981). Figure 21.1 shows the locations of most of the sites discussed in this chapter. The sites are mentioned in geographic order, from southwest to northeast.

Jackson Group and Jackson Equivalents

Texas. A diverse Jackson Group vertebrate fauna has been reported (Manning, 1990:535-538) from a very peculiar situation in easternmost Texas. The late Eocene Toledo Bend marine vertebrate fauna is entirely reworked—recovered from an early Miocene river channel deposit. Because of the reworking, it is possible that some early Oligocene taxa may be misinterpreted as late Eocene.

Louisiana. Good Jackson vertebrate faunas have been reported from Montgomery Landing (Manning and Standhardt, 1986) and Copenhagen (Stringer, 1979). The former includes both Moodys Branch and lower Yazoo Formation material, while the latter is in the upper Yazoo Formation.

Arkansas. Despite the scarcity of Jackson Group surface exposure in the state, an excellent Moodys Branch-equivalent vertebrate fauna is known from Crow Creek (Westgate, 1984, 2001), near the southern end of the Crowley's Ridge erosional remnant. In addition to this material, new material (with some unreported taxa) from Crow Creek has been collected by William Parks (MGS collection). A second fauna, the Lover's Leap fauna from Copperas Creek, has recently been reported by Westgate (2001).

Tennessee. The Fulton fauna (briefly noted in Corgan, 1976:76-77, and Corgan and Breitberg, 1997:119, and currently under study by Parks and the author) is a Moodys Branch-equivalent fauna in western Tennessee. It is of special interest, as the northernmost Eocene marine fauna of the Mississippi Embayment.

Mississippi. Lists of Jackson Group vertebrates from Mississippi have a long history (Wailes, 1854:276-278; Harper, 1857:142-144, 152, 160-161, 157-158; Hilgard, 1866:126-135). Jackson Group teleost otoliths were also reported early from the state (Koken, 1888), and have been more recently studied by Frizzell and Dante (1965). More recent work, on several sites, has been done by Leriche (1942), and on the Techeva Creek fauna by Breard (1991).

Alabama. The Alabama Jackson vertebrates have likewise long been known, and most recently reviewed by Leriche (1942), White (1956:123-147; primarily studying the early-collected fauna of "The Rocks"), and Thurmond and Jones (1981).

Florida. A small late Eocene marine vertebrate fauna is known from the Ocala Uplift, of northern peninsular Florida (Tessman, 1969). Most of the fauna is from the Crystal River Limestone, or reworked from it.

Georgia. At the eastern end of the GCP, an excellent Jackson Group vertebrate fauna has been reported from near Clinchfield (Case, 1981). A second large fauna has recently been reported from the Hardie Kaolin Mine near Gordon (Westgate, 2001).

South Carolina. At the southern end of the Atlantic Coastal Plain, in a small tectonic basin below the Cape Fear Arch, is another important fauna composed entirely of reworked material. The Ashley River fauna, near Charleston, was one of the first Eocene-Oligocene shark and ray faunas studied in America. It was reported in monograph form by Gibbes (1848, 1849, 1850). To an even greater extent than the Toledo Bend fauna, the reworked Ashley River fauna is difficult to divide into individual Eocene and Oligocene faunas.

Northern Atlantic Coastal Plain. Because Eocene-Oligocene transition beds are generally thin on the Atlantic Coastal Plain, there are few vertebrate faunas of comparable age to those of the GCP.

Vicksburg Group and Vicksburg Equivalents

Mississippi. Although excellent figures of a variety of Vicksburg Group marine vertebrates (including shark teeth, bony fish otoliths, and sea cow rib fragments) from the Vicksburg area were prepared for an unpublished manuscript as early as 1829 (Lesueur, 1829), the illustrations weren't printed until 153 years later (Dockery, 1982a:8, pl. 2 and 3; 1982b, appendix 2, pl. 2 and 3). The fauna was listed early (Hilgard, 1866:142), but not studied in detail until later. The otolith fauna was reported early (Koken, 1888), and further studied by Frizzell and Dante (1965). More recent work on Wayne County material has been done by Arata and Jackson (1965:175).

Alabama. Otoliths from the St. Stephens area were studied by Koken (1888), and of several sites in the state by Frizzell and Dante (1965). A small fauna from St. Stephens quarry is noted in Whetstone and Martin (1978), and later by Thurmond and Jones (1981:6, 95).

South Carolina. The reworked Ashley River fauna (Gibbes, 1848, 1849, 1850) probably includes more Oligocene elements than Eocene ones, though age determinations are difficult.

North Carolina. An important fauna has been reported (Case, 1980) from the southern Atlantic Coastal Plain, from the small basin north of the Cape

Fear Arch, in New Bern. Although the site was originally reported as early Miocene, and from the "Trent" Formation, it has been argued that the usage of the term "Trent" has been so ambiguous that it should be abandoned (Ward et al., 1978). The vertebrate fauna appears to be from what they term the River Bend Formation. The mollusks of the Vicksburg-equivalent lower River Bend Formation have been reported (Rossbach and Carter, 1991). The Case (1980) New Bern vertebrate fauna is reinterpreted here as Vicksburg-equivalent. The early Miocene (Aquitanian) age assignment may have been made because Case (1982:265) mistakenly felt that there was no Oligocene marine sedimentary record in the United States. The New Bern fauna is critical to the study of the Vicksburg Group vertebrates.

General

As presently known, far more Jackson Group marine vertebrate faunas are known from the Gulf Coastal Plain than of the Vicksburg Group. Furthermore, the Jackson Group faunas (as reported) are more diverse, and better studied, than are the Vicksburg Group faunas. Adding to the problem, the best Vicksburg-equivalent fauna of the Atlantic Coastal Plain, the New Bern fauna, is recorded in the literature as early Miocene. The result is that the late Eocene Jackson fauna gives the false impression of being far more diverse than the smaller, less well-studied, and sometimes misidentified early Oligocene Vicksburg fauna.

UNPUBLISHED FAUNAS

Aside from the previously mentioned sketchy faunal lists, otolith studies, and sea cow work (e.g. Whetstone and Martin, 1978), reviews of the Vicksburg Group marine vertebrate faunas of Louisiana, Mississippi, and Alabama have never been published. Good screen-washed faunas of isolated marine vertebrate material (notably of the Rosefield, Louisiana, fauna and the Byram Formation fauna of the Vicksburg, Mississippi, area in the LSU Museum of Natural Sciences; and the Mint Springs Formation fauna of the Devil's Backbone site, Mississippi, in the MMNS VP collection) have been collected, but only partly studied (author's research). Only a few of the important points from that work can be included here, much remains unpublished. The effect of comparing the well-published Jackson Group marine vertebrate fauna to the largely unpublished Vicksburg Group one is, again, to make the late Eocene fauna seem more diverse.

UNRELIABLE SHARK SPECIES COUNTS

A particular difficulty with measuring early Tertiary shark diversity is determining the true number of species present in a given fauna, when based on isolated teeth. Species counts are often artificially raised by counting upper and lower teeth, distinctive teeth (symphyseals, intermediates, posteriors, etc.), or adult and juvenile teeth (shark teeth are replaced frequently, and succeeding sets vary in morphology and are progressively larger) as separate species. Two faunas with identical species counts can be reported with greatly different numbers, if one is studied by a taxonomic splitter and the other by a lumper.

Manning and Standhardt (1986:145) suggested that, unlike the modern tiger shark *Galeocerdo cuvieri*, the early Tertiary *G. alabamensis* (figure 21.3, no. 3) appears to have narrower primary cusps on the lower teeth than on the uppers. The upper teeth appear to have been described as a separate species, *G. clarkensis* White, 1956. Likewise, contrary to Purdy et al. (2001: 146-147), I consider the Miocene *G. "contortus"* Gibbes, 1849, to be the lower teeth of *G. aduncus* Agassiz, 1843.

Applegate (1965) pointed out that teeth from various positions in a single shark's jaws can often look nothing like one another. This has led to teeth from unusual positions being assigned to separate species. An extreme case of this was noted by Ward (1988). He showed that the neotype of *Carcharias hopei* (here as *Odontaspis hopei*), a phosphatic nodule containing many associated teeth of the same individual, spanned the morphology of at least three other commonly used shark species of another genus—*Hypotodus.*

Besides often being far smaller than adult shark teeth, young juvenile shark teeth sometimes lack the distinctive features (e.g., lingual costae, accessory cusps) of the adults and can have much narrower outlines.

A persistent problem in comparing early Tertiary shark faunas based on isolated teeth (as nearly all faunas are in the Gulf Coastal Plain) is the use of outdated sources for identification. Long-ago-corrected errors reappear again and again in faunal lists and are difficult to root out. An example is the use of *Lamna elegans* for the lower teeth of *Odontaspis macrotus.* This happens when a report with excellent illustrations but outdated identifications (such as Fowler, 1911) is used as a source.

In the author's opinion, an extreme example of oversplitting of shark species is the recent study of the important late Paleocene and early Eocene faunas of the Red Hot Truck Stop, of Meridian, Mississippi (Case, 1994), the most thoroughly studied faunas of those ages of the GCP. In the author's opinion, the sand tiger shark species *Odontaspis macrotus* has there been described as fifteen separate species (four new), in seven genera (*Odontaspis*, ?*Jaekelotodus*, *Pseudodontaspis*, *Striatolamia*, *Carcharias*, *Anomotodon*, and *Cretolamna*), based on variants in tooth position, ontogeny, and of the population.

As a result of these problems, if all faunas are not reviewed by the same people, faunal species counts can not be reliably compared.

DIVERSITY PROBLEMS RELATED TO SEQUENCE STRATIGRAPHY

Only recently have vertebrate paleontologists taken into account how the cyclic nature of coastal marine sedimentary deposits affect the nature of the vertebrate faunas contained in them. Geologists have long been aware that concentrations of fossil shark teeth occur in certain beds (Harper, 1857:81-82, 282), but it wasn't clear that the same sedimentary environment was being repeated, or even that the teeth were being concentrated secondarily by reworking from underlying beds (the beds were seen, at times, as mass shark die-offs). With the recent application of sequence stratigraphy to coastal vertebrate paleontology, a clearer pattern has emerged.

Where in a Sequence Vertebrate Faunas Are Found

Only two beds of each marine sequence, reflecting opposite shelf habitats, usually contain good vertebrate faunas. The bulk of the Coastal Plain marine record includes very few vertebrate remains. The upper, regressive, part of each sequence—the highstand systems tract (HST)—is usually poor in vertebrates. Even in the middle of the transgression, the middle lowstand systems tract (LST), which often contains good invertebrate faunas, there is often only rare, scattered vertebrate material (for the early Tertiary, these remains commonly include shark, ray, and bony fish teeth; shark and ray calcified vertebral centra; bony fish vertebrae and miscellaneous bones; sea snake vertebrae; sea turtle shell fragments; whale and sea cow teeth, vertebrae, and rib fragments). The gaps between faunas result in a relatively episodic Coastal Plain vertebrate record, often of alternating paleoecologies.

The base of the sequence. It has long been noted in the sequence stratigraphic literature (Baum and Vail, 1988:313) that the base of a sequence can often be identified in updip deposits by a thin (often less than a foot thick) bed of phosphatic pebbles with an abundance of "fish" teeth (mainly shark teeth). This transgressive lag bed, usually marking the start of a transgression, is often critical to sequence studies, as it is sometimes the *only* bed with a sharp unconformity (at its base), making it ideal as a basal sequence boundary marker. This marine lag is formed when sea level is lowest, under very high energy, in shallow-shelf conditions—the rising sea permits fast coastal currents to rip up soft, previously deposited beds, and causes the more durable particles in those beds to be concentrated, as the finer particles are washed out to sea (much as in the process of river lag formation).

These lags are often the most vertebrate-rich bed of each sequence—the diversity is artificially raised as thinly scattered remains from several beds are concentrated in a single thin unit. On the other hand, a large variety of forms with delicate hard parts are also lost in these very abrasive conditions, reducing the raised diversity considerably. Additionally, taxa that can not be identified from fragmentary remains become unidentifiable. Paleoecologically, these lag faunas (they aren't "local faunas," despite their discrete stratigraphic horizons, as they include a longer-than-usual span of time) are very difficult to interpret, as they frequently include reworked deep-water marine, estuarine, fluviatile, and often terrestrial elements along with their primary shallow-marine fauna. Traditionally, vertebrate paleontologists interested in anatomy or systematics have avoided these rich beds (even as shark tooth collectors flock to them), as the material is usually fragmentary and badly worn from mechanical wear (rolling along the seafloor in a pebble bed), and rarely includes the well-preserved articulated skeletons necessary for anatomical and systematic research. This bias has sometimes made it difficult to interpret what the shallow-marine fauna of a region really was. What the lag faunas lack in quality, they make up for in quantity, and richly reward serious screen-washing.

Eocene-Oligocene examples of the GCP. In the central Gulf Coastal Plain, the Jackson Group includes two basal sequence beds—the base of the upper Bartonian Moody Branch Formation and base of the lower Priabonian Cocoa Sand Member of the Yazoo Formation (Tew and Mancini, 1995:figures 2, 4, and 5). The basal Moodys Branch includes a rich vertebrate fauna at several localities of the GCP. So far as is known, the basal Cocoa Sand (a shelf margin, rather than transgressive systems tract— Tew and Mancini, 1995) does not. The Vicksburg Group also includes two basal sequence beds—the middle Rupelian basal

Mint Springs Formation and basal Byram Formation (Tew and Mancini, 1995), both of which are transgressive systems tract units, and both of which contain good vertebrate faunas (that of the Mint Springs is richer). In contrast, Dockery and Lozouet (this volume) place the Vicksburg sequence boundaries at the base of the Red Bluff and above the Byram Formation (in addition to that at the base of the Mint Springs), an interpretation not followed here.

The middle of the sequence. The vertebrate fauna from the deeper-marine part of the sequence (probably not much deeper than outer shelf depths, about 200 m, for surface deposits), surrounding the Maximum Flooding Surface (MFS), or turnaround point, where the transgression turns to regression, is less well known in the early Tertiary of the GCP. The comparable deep-water chalk beds of the Mississippi and Alabama Upper Cretaceous have a far better known vertebrate fauna. They're famous for their beautifully preserved marine vertebrate skeletons (bony fish, sea turtles, mosasaurs, etc.). These beds were deposited under very low-energy conditions (either near deltas, in the fine pro-deltaic clays carried farthest from land; or away from deltas, in planktonic foraminiferal ooze). In contrast to that from the base of a sequence, vertebrates found in midsequence are generally little worn, often articulated, and rarely reworked. In really deep-water conditions, such as the chalk beds, bottom conditions are often anoxic, so the fauna is mostly pelagic—taxa drifted down to a scavenger-free bottom (GCP early Tertiary surface beds rarely got that deep). Though there aren't many early Tertiary midsequence vertebrates, those specimens found in this part of the sequence are often unusually well preserved, making for a small, but well-identified fauna. It is sometimes thought that vertebrate remains concentrate on the seafloor under conditions of sediment starvation (the MFS is considered a "condensed section" because very little sediment is deposited in the deepest water, when farthest from shore), simply because they accumulate on the seafloor and do not get buried quickly. To test this idea, several years ago the author collected and screen-washed a large sample from the latest Eocene MFS "pebble bed" at the top of the Shubuta Member, upper Yazoo Formation, at St. Stephens Quarry, southwestern Alabama. The "pebbles" turned out to be small pyrite nodules (indicative of the low-oxygen conditions of the deep seafloor), associated with a sparse invertebrate hash, containing only a few shark and bony fish teeth. The MFS bed itself seems not to be a good source of vertebrate remains.

Eocene-Oligocene examples of the GCP. In the GCP, the Jackson Group includes midsequence beds in the upper Moodys Branch Formation and late Priabonian upper Shubuta Member of the Yazoo Formation (Tew and Mancini, 1995:figures 2, 4, and 5). The upper Moodys Branch does not appear to have been deposited in very deep-water, and includes few articulated vertebrates, but partial skeletons of the derived early whale *Basilosaurus cetoides* (see Fordyce, this volume) have been recovered from the upper Yazoo in several states, including the site of the generic type specimen, from northeast Louisiana (Harlan, 1834).

The Yazoo Formation of southeastern Mississippi and southwestern Alabama includes a distinctive deep-water marine fauna—with the billfish *Cylindracanthus rectus*, the mako shark *Isurus praecursor*, and (to a lesser extent) the great white shark *Carcharodon auriculatus*. A down-dropped area seems to have existed in the area in the late Eocene. A comparable deep-water fauna is not known from the Vicksburg Group.

A remarkably complete squirrelfish skeleton, *Holocentrites ovalis*, has been reported from the late Eocene Crystal River Limestone (probably a shallow-marine midsequence) of the Florida panhandle (Conrad, 1941). The species has also been reported from the Vicksburg Group of Mississippi (Stewart, 1992), and from three unreported partial skeletons from the Glendon Limestone of Mississippi (MMNS VP 773, and uncatalogued specimens in the Mississippi Office of Geology [Smith Coll.] and LSU Museum of Natural Sciences collection).

Like the Jackson Group, the Vicksburg Group includes two midsequence beds—in the middle Rupelian middle Marianna Limestone and middle Byram Formation (Tew and Mancini, 1995: figures 2, 4, and 5). Relatively few deep-water marine elements have been recovered from the Vicksburg Group, suggesting that the deepest Vicksburg bed is considerably shallower than the deepest Jackson bed—perhaps the difference between middle- and outer-shelf depths. This probably reduced the total diversity of the Vickburg Group, again making its fauna appear less diverse than that of the Jackson Group.

Comparing Faunas from Different Parts of the Sequence

To early coastal biostratigraphers, the marine record of the Coastal Plain seemed particularly easy to read—just a stacked series of faunas. Sequence stratigraphy shows that it is not that simple. It is unfair to compare a shallow-marine fauna to the deep-marine fauna that overlies it in a transgression, even if they are closest chronologically. This is particularly true of the two main Jackson Group vertebrate faunas—the shallow-water fauna of the basal Jackson Group basal Moodys Branch Formation has more in common with the shallow-water lower Oligocene basal Mint Springs Formation of the lower Vicksburg Group, than with the deep-water upper Eocene upper Yazoo Formation of the upper Jackson Group. Beyond the paleoecological factors, the high diversity of a lag concentration compares more closely, in preservational terms, with that of another lag fauna than to the low-diversity, lower-energy, deep-water fauna made up largely of scattered pelagic taxa drifted down over an anoxic, benthic-poor bottom. To compare the GCP vertebrate faunas on either side of the Eocene/Oligocene boundary gives a false disparity.

Terrestrial Faunas

Although the emphasis of this paper is on marine vertebrate faunas, mention should be made of the contrast in terrestrial vertebrate remains found associated with the two faunal types noted above, from the base and middle of sequences (land and freshwater vertebrate concentrations are also found in fluviatile channel deposits in the upper parts of sequences, the highstand systems tract, in the GCP). As can be seen from the review of land mammals in marine beds in Mississippi in Manning (1997:28), their preservation depends greatly on the part of the sequence they are from. Like the marine vertebrates, land mammal remains from the basal sequence lag beds are generally isolated teeth, jaw fragments, and bone fragments; while the deeper-water midsequence remains tend to be far more complete. Basal sequence examples from the early Tertiary of the GCP include an early Eocene primate jaw fragment from Mississippi (Beard and Tabrum, 1991) and two mammal jaw fragments from the late middle Eocene basal Jackson Group of Arkansas (Westgate and Emry, 1985). Midsequence examples include a nearly complete middle Eocene titanothere skull and jaw from Mississippi (Gazin and Sullivan, 1942), a nearly complete early Oligocene amynodont skull (Manning et al., 1986), and a nearly complete early Oligocene rhinoceros jaw (Manning, 1997) from the same upper Vicksburg Group site in Mississippi. The reworked (often several times) basal sequence material is usually scrappy and poorly preserved, while the deeper-water midsequence remains appear to have simply dropped off bloated floating carcasses while far off shore, resulting in unusually good preservation (Manning, 1997:24-25).

Terrestrial mammals from either type of fauna are especially significant in biostratigraphy, as they permit correlation of marine stages to Land Mammal "Ages," for example, the late Bartonian to the early Uintan (Westgate and Emry, 1985), or the middle Rupelian to the early Orellan (Manning, 1997).

Diagenesis

Secondary loss of diversity caused by in situ alteration of sediment is often ignored when comparing coastal faunas. Because its effects depend on the type of bed in which the fossils are buried, faunas whose diversity was originally comparable can end up very different.

The basal sequence transgressive lag beds have often been affected by both phosphate and carbonate diagenesis. As the vertebrate remains are concentrated, the less stable original phosphatic mineral, carbonate hydroxyl apatite (dahlite—found in all vertebrate bones and teeth when alive) is flushed out of the remains and is deposited as the more stable mineral fluorapatite in concretions (often formed around crustaceans), rip-up clasts, phosphatized wood, mollusk steinkerns, and the like (author's research). Less substantial vertebrate hard parts are lost completely, removing taxa from the fauna. Later carbonate diagenesis occurs as, first, aragonitic and, later, calcitic remains are dissolved, as groundwater preferentially flows through these coarse clastic lag beds (they are often primary aquifers). This results in the loss of all bony fish otoliths and nearly all original mollusk remains (leaving just the phosphatic steinkerns to indicate their presence).

The midsequence marl, chalk, and limestone beds also suffer carbonate diagenesis, but of a less severe form. Generally, only the less stable aragonitic forms are lost, the more stable calcitic forms remain (for invertebrate faunas, this can mean that a high diversity mollusk fauna can, after partial dissolution, look like a low-diversity oyster bed). In particularly deep-water, low-oxygen, midsequence beds (more typical of the Cretaceous chalks than the Eocene/Oligocene marls), pyrite sometimes encrusts the vertebrate remains, making them difficult to identify. Black shales are uncommon in the GCP, but when marine vertebrates are found in them near the surface, sulphur from the pyrite often mixes with the carbonate from dissolved aragonitic mollusks to produce selenite, which infiltrates the vertebrate remains and can make them unrecognizable (as in many Pierre Shale vertebrates from the American West).

Because of both the greater mechanical wear and the greater diagenetic alteration, the faunas of the lag

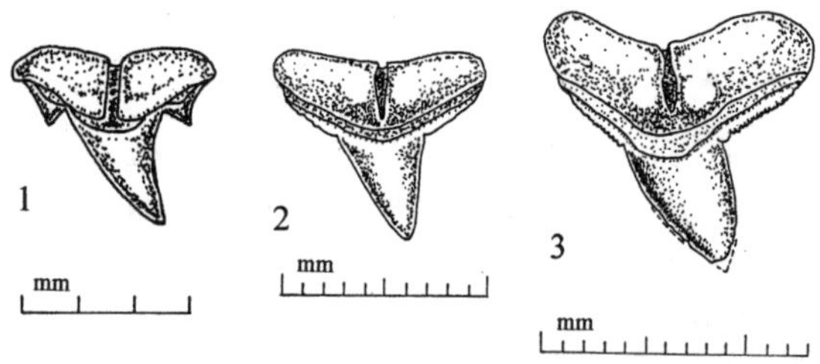

Figure 21.2. Time ranges given are for the Gulf Coastal Plain only. Scale bar units are in millimeters, except for figures 21.9 and 21.10-1 (in centimeters). Carcharhinid (reef shark) right upper lateral teeth, in lingual view. (1) *Abdounia minutissima* (range: late Paleocene to early Eocene). MGS 1905. Early Eocene, Bashi Fm.; B & B Truck Sales, Meridian, Lauderdale Co., Miss. (2) *Carcharhinus gibbesi gilmorei* (range: middle Eocene to late Oligocene). TU-VP 1010. Late middle Eocene, Moodys Branch Fm.; Town Creek, Jackson, Hinds Co., Miss. (3) *Carcharhinus gibbesi gibbesi* (range: early to late Oligocene). TU-VP 1017. Late early Oligocene, Byram Fm.; Vicksburg, Warren Co., Miss.

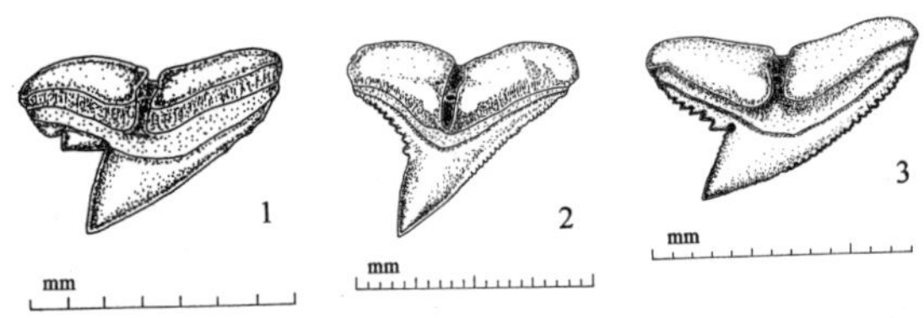

Figure 21.3. *Galeocerdo* (tiger shark) left upper teeth, in lingual view. (1) *G. minor* (range: late Paleocene to early Eocene). MGS 1906. Early Eocene, Bashi Fm.; B & B Truck Sales, Meridian, Lauderdale Co., Miss. (2) *G. latidens* (range: middle Eocene). TU-VP 1007. Late middle Eocene, Gosport Fm.; Little Stave Creek, Clarke Co., Ala. (3) *G. alabamensis* (range: late middle Eocene to early Oligocene). TU-VP 1011. Late middle Eocene, Moodys Branch Fm.; Claiborne Bluff, Monroe Co., Ala.

beds suffer a proportionately greater diversity loss than those of the midsequence beds. Because the late Eocene Jackson Group includes both types of vertebrate faunas, while the early Oligocene Vicksburg Group vertebrate fauna is primarily known from lag beds, the Vicksburg fauna again appears far less diverse than that of the Jackson.

GENERAL TRENDS IN DIVERSITY CHANGE

Although, for the reasons just given, it is not yet possible to cite specific numbers in the diversity change of the marine vertebrate fauna of the Gulf Coastal Plain (GCP) across the Eocene/Oligocene boundary, it is possible to provide some general impressions of the faunal change of that time and place.

First, despite the impression given by published works, the early Oligocene fauna probably wasn't much less diverse than that of the late Eocene. A false impression is conveyed by the relative number of published versus unpublished faunas, area of outcrop, variety of habitats preserved, degree of diagenesis, and so on, that the late Eocene fauna is far more diverse.

Second, the vertebrate diversity change across the Eocene/Oligocene boundary (in contrast to the mollusk record of the same beds) appears to be very gradual, mostly on a subspecific level.

Third, in contrast to the Eocene/Oligocene boundary faunal change, those of the early/middle Eocene and early/late Oligocene boundaries appear to be major (stepwise) faunal changes. These changes, however, are probably "more apparent than real," as both seem to be the result of long unfossiliferous periods between the two levels. This is especially likely for the radically changed late Oligocene fauna, as a result of nondeposition during the important middle Oligocene sea-level lowstand.

SOME SPECIFIC CHANGES

Only a few of the specific changes in the marine vertebrate fauna of the Eocene and Oligocene of the GCP can be cited here. The changes are discussed in chronological order.

The Early/Middle Eocene Boundary

A major (stepwise) faunal change seems to have taken place between the Wilcox and Claiborne groups of the GCP. This is likely because of the poor vertebrate faunas of the upper Wilcox and lower Claiborne. The best Wilcox vertebrate fauna is that of the basal Bashi Formation of the lower upper Wilcox Group (Case, 1994a, b). The best Claiborne vertebrate faunas are in the upper Claiborne—the Cook Mountain and basal Gosport formations. This is a gap between foraminiferal Zones NP10 and NP16, about 12 million years (from about 42 to 54 Ma). A comparable gap in vertebrate faunas is present in the mid-Atlantic Coastal Plain of Maryland and Virginia, between the early Eocene Nanjemoy and late middle Eocene Piney

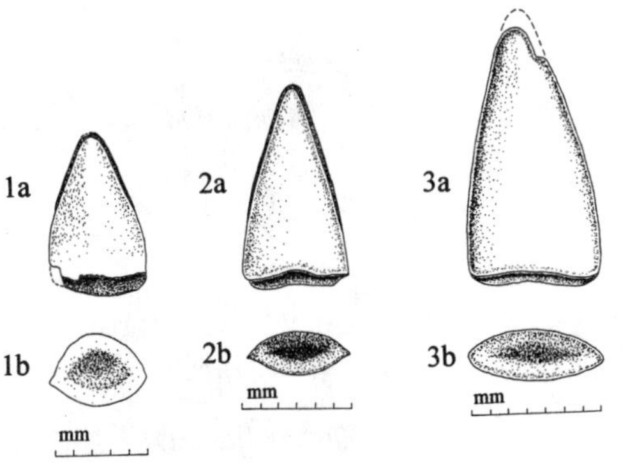

Figure 21.4. *Scomberomorus* (mackerel) and *Sphyraena* (barracuda) lateral teeth, in (a) lateral and (b) basal views. (1) *Scomberomorus proosti* (range: late Paleocene to early Eocene). MGS 1907. E. Eocene, Bashi Fm.; B & B Truck Sales, Meridian, Lauderdale Co., Miss. (2) *Sphyraena* sp. (range: late middle to late Eocene). MGS 1908. Late middle Eocene, Moodys Branch Fm.-equivalent; Crow Creek, St. Francis Co., Ark. (3) *Sphyraena* sp. (range: early Oligocene). MMNS VP 772. E. Oligocene, Mint Springs Fm.; Devil's Backbone, near Cleary, Rankin Co., Miss.

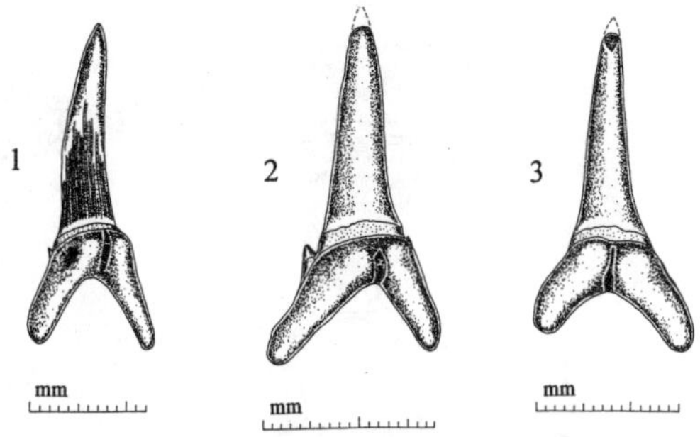

Figure 21.5. *Odontaspis* (sand tiger shark) lower anterior teeth, in lingual view. (1) *O. macrotus* (range: late Paleoc. to middle Eocene). TU-VP 1006. Late middle Eocene, Gosport Fm.; Little Stave Creek, Clarke Co., Ala. (2) *O. hopei* (range: late middle to late Eocene). MGS 1909. Late middle Eocene, Moodys Branch Fm.-equivalent; Crow Creek, St. Francis Co., Ark. (3) *O. cuspidatus*? (range: early Oligocene to Mioc.). MMNS VP 770. E. Oligocene, Mint Springs Fm.; Devil's Backbone, near Cleary, Rankin Co., Miss.

Point formations (Ward and Wiest, 1990).

Among the notable changes at this boundary are those in carcharhinid and tiger sharks, and in predaceous bony fish. The small, primitive carcharhinid *Abdounia minutissima* (Winkler, 1873) (here considered senior synonym of the type species, *Eugaleus beaugei* Arambourg, 1935) (figure 21.2, no. 1) is replaced by a small species of the modern reef shark *Carcharhinus*, *C. gibbesi gilmorei* (Leriche, 1942) (figure 21.2, no. 2). The latter had, for many years, been incorrectly placed in the lemon shark *Negaprion* until corrected by Cappetta (1987:121).

The tiger shark *Galeocerdo* is not usually recorded as appearing in the GCP until the middle Eocene, but a small, early Eocene species usually placed in *Galeorhinus* (the soupfin sharks), *G. minor* (Agassiz, 1843) (figure 21.3, no. 1), could as easily be identified as a small, primitive member of *Galeocerdo*. An argument could even be made that the *Galeocerdo* lineage goes back to the Late Cretaceous *Paracorax jaekeli* (Woodward, 1895, originally placed in *Galeocerdo*), despite its current placement in the Anacoracidae by Cappetta (1987:108). The typical middle Eocene tiger shark of the GCP, *Galeocerdo latidens* (Agassiz, 1843) (figure 21.3, no. 2), is a large species with distinctive teeth whose primary cusp is not as sharply angled as usual for the genus.

There is a curious turnover in predaceous bony fish of the GCP at the early/middle Eocene boundary. Teeth of the mackerel *Scomberomorus proosti* (Storms, 1897) (*Cybium* in earlier literature) (figure 21.4, no. 1) are common in the early Eocene and typically have a thick base (figure 21.4, no. 1b). For reasons unknown, they are almost entirely replaced in the middle Eocene (and thereafter in the Tertiary of the GCP) by teeth of a small form of the barracuda *Sphyraena* (figure 21.4, no. 2), which have a similar lateral outline but are slimmer at the base (figure 21.4, no. 2b). Most bony fish teeth are simple, elongate cones, so mackerel and barracuda teeth tend to stand out.

The Claiborne/Jackson Group Boundary

The main change in the marine vertebrate fauna of the GCP middle Eocene (a smaller change than that of the early/middle Eocene) is at the Claiborne/Jackson Group boundary, near the end of the Bartonian. Here there is a notable exchange of Tethyan elements (i.e., the giant sea snake *Pterosphenus*, the whale *Basilosaurus*, and the primitive sawfish *Propristis*—all of which had also existed in earlier forms in the upper Claiborne).

Distinctive changes occur at this time in the lineages of the sand tiger shark *Odontaspis* and the

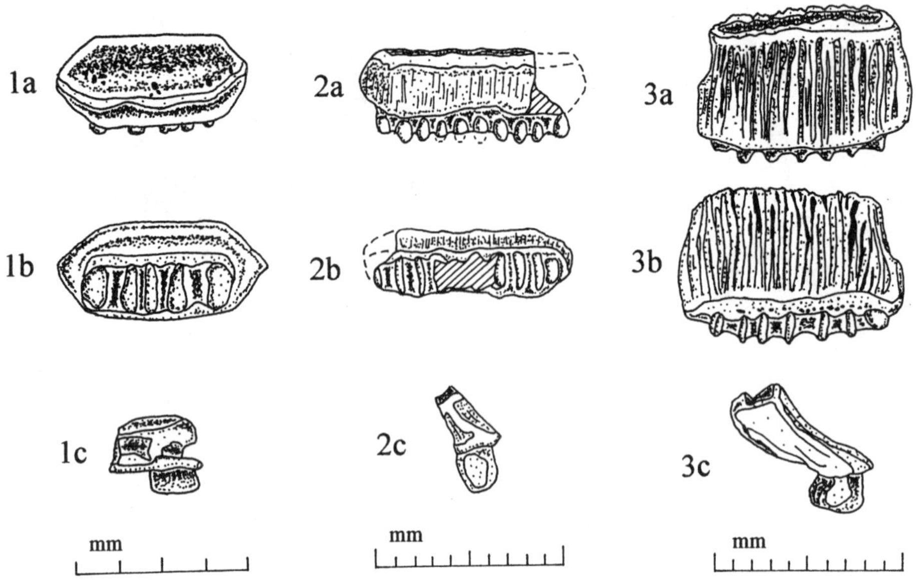

Figure 21.6. *Burnhamia/Plinthicus* (extinct ray lineage) medial teeth, in (a) occlusal, (b) basal, and (c) lateral views. (1) *Burnhamia daviesi* (range: early to middle Eocene). MGS 1910. E. Eocene, Bashi Fm.; B & B Truck Sales, Meridian, Lauderdale Co., Miss. (2) *Plinthicus* n. sp. (range: late Eocene to early Oligocene). TU-VP 1016. Late early Oligocene, Byram Fm.; Byram, Hinds Co., Miss. (3) *Plinthicus stenodon* (range: middle Mioc.). TU-VP 1019. M. Mioc., Pungo River Fm.; Lee Creek Mine, Buford Co., N.C.

extinct ray *Plinthicus. Odontaspis* Agassiz, 1838 (here treated as including the primitive *Striatolamia* Gluckman, 1964; and *Carcharias* Rafinesque, 1810 [*Odontaspis* is to be given precedence over *Carcharias* when the two are considered synonymous—ICZN Opinion 1459], which is often treated as a separate GCP lineage) is generally the most abundant shark in early Tertiary marine vertebrate sites of the GCP. In most early Tertiary lag deposits of the GCP, it is the genus that makes up the bulk of the shark tooth beds. Late Paleocene, early Eocene, and middle Eocene Claiborne Group sites generally include *O. macrotus* (Agassiz, 1843) (figure 21.5, no. 1), while late middle to late Eocene Jackson Group sites generally include *O. hopei* (Agassiz, 1843) (figure 21.5, no. 2). Adult lower teeth of *O. macrotus* have distinctive costae on the basal half of the tooth crown on the lingual side (these ridges are usually called "striations," even though they aren't grooves in a smooth surface), while the costae in *O. hopei* are reduced or lost. Like a few other shark species, O. *hopei* appears later in the GCP than in the higher-latitude European sites.

The lineage of the unusual ray *Plinthicus* Cope, 1869 is represented in the early Eocene and middle Eocene Claiborne Group of the GCP by *Burnhamia daviesi* (Woodward, 1889) (figure 21.6, no. 1). Unlike the teeth of most advanced rays (*Rhinoptera, Myliobatis, Aetobatus*, etc.), members of this lineage have a deep crease in the occlusal surface of their elongate "pavement" teeth, rather than the typical flat crushing surface. This seems to suggest that they ate far softer food than the others. In *Burnhamia*, the tooth crown is low and the occlusal surface is positioned directly over (or under, in the upper teeth) the roots. In *Plinthicus*, the tooth crown is high (figure 21.6, nos. 2c and 3c), and the occlusal surface is positioned posterior to the roots. *Plinthicus* is represented in the Jackson and Vicksburg groups by an undescribed small species (figure 21.6, no. 2), whose crown hasn't been "stretched" to the extent of the typical Miocene *P. stenodon* (figure 21.6, no. 3).

The Bartonian/Priabonian Boundary

There is no significant faunal change in the marine vertebrate fauna of the GCP at the Bartonian/Priabonian (middle/late Eocene) boundary, within the lower Yazoo Formation of the Jackson Group. The same is true of the middle Yazoo Formation, at approximately the time of the Chesapeake Bay bolide impact (see Poag, this volume). It does not appear that there was a major faunal change in the GCP as a result of the impact.

The Eocene/Oligocene Boundary

Late Eocene extinctions. A few marine vertebrate taxa become extinct at the end of the Eocene (the Priabonian/Rupelian boundary) in the GCP. It is notable that several of these (the giant sea snakes, *Basilosaurus, Propristis*) are forms that took part in the Tethyan exchange at the start of the Jackson Group. It is possible that the giant sea snakes (the boid-derived Palaeophidae), as reptiles, were affected by the cooler winter seawater temperatures at this time (Ivany et al., 2000; this volume), but it is impossible to be certain. The group was highly successful in the Eocene (Holman et al., 1991), so their absence in the Oligocene is striking. It is worth noting that the extinction of the derived early whale *Basilosaurus* (as well as the less derived Jackson whales *Zygorhiza kochii* and "*Pontogeneus priscus*") in the GCP, and the absence of whales in the early Oligocene of the GCP, is purely local—whales as a group cross the boundary easily (see Fordyce, this volume). It is even possible that *Basilosaurus cetoides* (Owen, 1839) survived just past the Eocene/Oligocene boundary in the GCP. Jaramillo and Oboh-Ikuenobe (1999) indicate that the Eocene/Oligocene boundary is, at certain sites, as much as 2 m below the top of the Jackson Group (upper Shubuta Clay, upper Yazoo Formation). As *Basilosaurus* remains are reported throughout the Jackson Group, it is possible that some may be from the earliest Oligocene part. As few marine vertebrates are recorded from this uppermost level, Jackson Group faunas are treated as Eocene here. Eocene-Oligocene comparisons here are mostly between Jackson and Vicksburg Group faunas.

Although these few extinctions help define the Jackson Group vertebrate fauna, they do not define a major faunal change in the GCP. This is in striking contrast with the mollusk extinction rate at the Eocene/Oligocene boundary in the same beds, as discussed by Dockery and Lozouet (this volume). They find an extinction rate of above 95%, a nearly complete extinction (only 26 species and subspecies cross the boundary, of 426 Jackson mollusk taxa). Why the mollusks should be affected so much more than the vertebrates of the same time and place is unclear, though none of the vertebrates are as strictly benthic as the mollusks.

Changes through the Eocene/Oligocene boundary. Most marine vertebrates of the GCP show little or no change through the Eocene/Oligocene boundary. Taxa showing no change include the tiger shark *Galeocerdo alabamensis* Leriche, 1942 (figure 21.3, no. 3), the snaggletooth shark *Hemipristis wyattdurhami* White, 1956 (figure 21.10, no. 7), and the previously noted undescribed small species of the extinct ray *Plinthicus* (figure 21.6, no. 2). *Galeocerdo alabamensis* does not appear to be derived from the middle Eocene *G. latidens*, as it has the more typical primary cusp angle.

Several marine vertebrates of the GCP show only very minor change across the Eocene/Oligocene boundary. These include changes in the sand tiger shark *Odontaspis*, the reef shark *Carcharinus*, the great white shark *Carcharodon*, the barracuda *Sphyreana*, the drum fish *Sciena*, and the goosefish(?) *Lophius*(?). The difference between the Jackson Group (figure 21.5, no. 2) and Vicksburg Group (figure 21.5, no. 2) *Odontaspis* is slight. The Vicksburg Group material is mostly very worn lag specimens, making identification even more difficult. The Vicksburg form appears to be *O. cuspidatus* (Agassiz, 1844), but the identification is uncertain.

The Jackson and Vicksburg Group *Carcharhinus* are intergradational, although the Vicksburg form tends to be slightly larger, have a coarser serration on the heels of the crown, and the base of the primary cusp is slightly more "pinched" than in the Jackson form. "*Sphyrna*" *gilmorei* Leriche, 1942 (originally thought to be a hammerhead shark) from the Yazoo Clay of Choctaw County, Alabama, fits the Jackson form well. White (1956:143-144) interpreted *Carcharias* (*Aprionodon*) *gibbesi* Woodward, 1889, originally described by Gibbes (1849:192) from South Carolina, as including *gilmorei* as a subspecies (as a lemon shark, *Negaprion gibbesi gilmorei*). Although he considered *N. g. gilmorei* as "slightly larger" than *N. g. gibbesi* (White, 1956:144), his description of *N. g. gibbesi* upper teeth as more clearly serrate than *N. g. gilmorei*, and his illustrations showing a "pinching" of the base of the primary cusp (White, 1956: figures 79-80) show that the typical *N. gibbesi* is actually the larger, Oligocene form. White considered the South Carolina type as being Eocene, but it could as easily be Oligocene, and it is interpreted that way here. The *Carcharhinus g. gilmorei* illustrated here (figure 21.2, no. 2) is typical of the Jackson Group material, but the *C. g. gibbesi* shown (figure 21.2, no. 3) is a somewhat extreme form, to illustrate the full range of variation. The two subspecies are fully intergradational, justifying White's (1956) including them both in the same species. Although *C. g. gilmorei* is not uncommon throughout the Oligocene, *C. g. gibbesi* (which also ranges throughout the Oligocene) does not occur in the Eocene beds with *C. g. gilmorei*. Because indicator

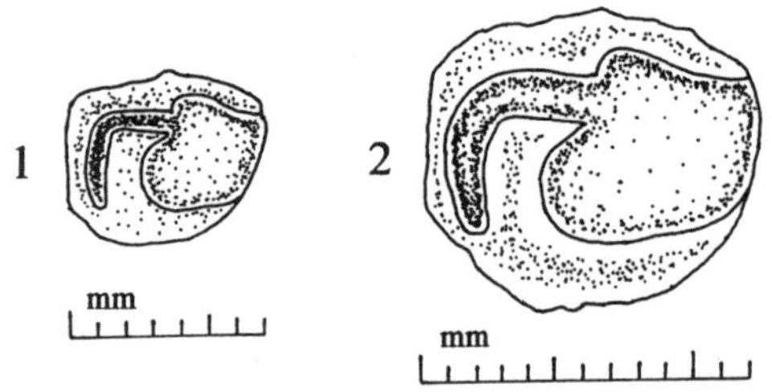

Figure 21.7. Sciaenid (drum fish) otoliths (left sagittae), in inner view. (1) *Corvina intermedia* (range: late middle to late Eocene). MGS 1911. Late middle Eocene, Moodys Branch Fm.-equivalent; Crow Creek, St. Francis Co., Ark. (2) *Sciaena gemma* (range: early Oligocene). MGS 1912. E. early Oligocene, Red Bluff Fm.; west of Hiwannee, Wayne Co., Miss.

vertebrate taxa are so rare at the Eocene-Oligocene transition in the GCP, the occurrence of *C. g. gibbesi* is a helpful indicator of Oligocene beds.

In a review of the largest shark teeth from Mississippi, Dockery and Manning (1986:14) noted that the teeth of the great white shark *Carcharodon auriculatus* (de Blainville, 1818) from the Vicksburg Group are slightly larger, and have slightly more reduced accessory cusps than do those from the Jackson Group. It could be argued that the Oligocene forms should be treated as a separate subspecies, perhaps as *Carcharodon auriculatus angustidens* (Agassiz, 1835). It is my opinion that Purdy et al. (2001:123-128) have shown definitively that these forms belong to the genus *Carcharodon* Smith (*in* Miller and Henle, 1838), and not in the paraphyletic "*Carcharocles*" Jordan and Hannibal, 1923, or "*Procarcharocles*" Casier, 1960.

The large size of certain bony fish teeth is also a helpful indicator of the Oligocene side of the Eocene/Oligocene boundary, although the changes are again gradational. Teeth of the barracuda *Sphyraena* (a genus rarely seen in the cooler-water sites of the Atlantic Coastal Plain) tend to be larger in the early Oligocene Vicksburg Group (figure 21.4, no. 3) than those of the late Eocene Jackson Group (figure 21.4, no. 2) (the mean of the population is larger, though the smaller Oligocene specimens overlap). This increasing size trend appears to continue through the late Oligocene and Miocene. The otoliths of the early Oligocene lower Vicksburg Group drum fish [notably *Sciaena gemma* (Koken, 1888)—figure 21.7, no. 2] are strikingly larger than those of closely-related late Eocene Jackson Group ones [such as *Corvina intermedia* (Koken, 1888)—figure 21.7, no. 1], so much so that the Oligocene ones are often obtained in surface collecting (a rarity with otoliths, which are usually collected in screen-washing only). Less well known than the barracuda teeth or drum fish otoliths in the early Tertiary of the GCP, are needle-like bony fish teeth, possibly the teeth of the goose fish *Lophius* (whose jaws often wash up on Gulf Coast beaches today). Those of the Jackson Group *Lophius*(?) (figure 21.8, no. 1) are considerably smaller than those of the Vicksburg Group (figure 21.8, no. 2).

Early Oligocene appearances. A very few vertebrate taxa seem to make their first appearance in the GCP in the early Oligocene Vicksburg Group, and do not occur in closely related form in the late Eocene Jackson Group. One notable Vicksburg first appearance datum is the teeth of the daggernose shark *Isogomphodon*, a form distinctive for its smooth heels and elongate narrow primary cusps (with insufficient attention to detail, it is often mistaken for *Odontaspis*). It seems likely that this is the species identified as *Aprionodon acuarius* (Probst, 1879) by Case (1980:90), and placed in *Isogomphodon* by Cappetta (1987:121). Case likely placed it in the early Miocene European species as he identified the

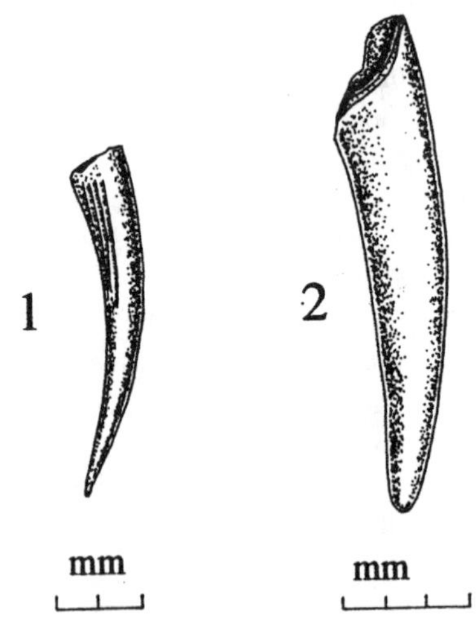

Figure 21.8. *Lophius*? (goosefish?) teeth, in lateral view. (1) *L*? sp. (range: late middle to late Eocene). MGS 1913. Late middle Eocene, Moodys Branch Fm.-equivalent; Fulton, Lauderdale Co., Tenn. (2) *Lophius*? sp. (range: early Oligocene). MMNS VP 771. Early Oligocene, Mint Springs Fm.; Devil's Backbone, near Cleary, Hinds Co., Miss.

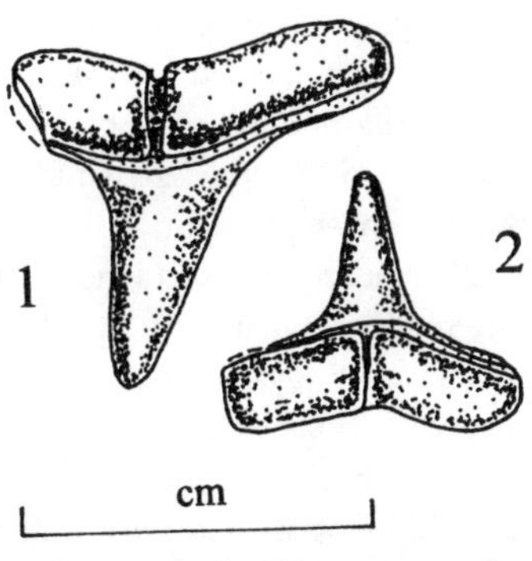

Figure 21.9. *Isogomphodon* (daggernose shark) teeth, in lingual view. (1 and 2) *I. acuarius* (range: early Oligocene to early Mioc.). MGS 1914. (1) Upper left, and (2) lower right teeth. Early Oligocene, Mint Springs Fm.; Devil's Backbone, near Cleary, Rankin Co., Miss.

New Bern site as of that age. Here, it is identified as Vicksburg-equivalent early Oligocene. *I. acuarius* (figure 21.9, nos. 1 and 2) is probably the shark most useful for determining an early Oligocene age for GCP faunas.

The Early/Late Oligocene Boundary

Like the early/middle Eocene vertebrate faunal change in the GCP (and unlike the Eocene/Oligocene boundary change), that of the early/late Oligocene is a major, stepwise, faunal change. In a sense, it is the time of the appearance of a new "Miocene-type" marine fauna, like those of the phosphate beds (excluding the uppermost, Pliocene, part) of Florida. This major change, as noted earlier, probably results in large part from the missing marine beds of the "middle" Oligocene (late Rupelian in a two-part Oligocene) sea-level lowstand. It is this lowstand that created the major unconformity between the Vicksburg and Chickasawhay groups of the GCP.

Two examples of the changeover to a "Miocene-type" marine vertebrate fauna in the GCP are in the snaggletooth shark *Hemipristis* and in the extinct ray *Plinthicus*. *Hemipristis wyattdurhami* (figure 21.10, no. 1) is found in both the Jackson and Vicksburg groups, but is replaced by the more derived *H. serra* Agassiz, 1843 (figure 21.10, no. 2) in the late Oligocene (typically, the medial serration of the upper teeth is better developed in *H. serra*, and adult size is increased). The typical *Plinthicus stenodon* Cope, 1869 (figure 21.6, no. 3) of the late Oligocene to middle Miocene has a crown far more "stretched out" than that of the *P.* n. sp. (figure 21.6, no. 2) of the Jackson and Vicksburg groups. In *P. stenodon*, the crowns of succeeding teeth are "shingled" on each other. This makes for an exceedingly weak tooth, in comparison to those of more typical derived rays, such as *Myliobatis* or *Aetobatus*. Rather than using their teeth for crushing hard-shelled prey (like mollusks or crustaceans), *Plinthicus* seems to have used its grooved occlusal tooth surfaces to gently squeeze very soft prey. Although Cappetta (1987:177) has placed *Plinthicus* and *Burnhamia* in the Mobulidae (devil and manta rays), based primarily on a similarity to certain female *Mobula* teeth, the basic construction of the teeth suggests a derivation from the cownose ray *Rhinoptera* (a conclusion also reached by Purdy et al., 2001:96). *Plinthicus stenodon* does not appear to extend into the late Miocene, making it useful in distinguishing middle from late Miocene faunas.

Although sea cows have not been emphasized in this study, it is worth noting that, at the early/late Oligocene boundary, the two main groups—the dugongs and manatees—appear to diverge and speciate. The same could be said of the two main whale groups—the toothed and baleen whales. As already noted, the stepwise change in the marine vertebrate fauna of this period is likely more apparent than real—a result of missing intermediary faunas.

GEOGRAPHIC DIFFERENCES AT THE EOCENE-OLIGOCENE BOUNDARY

Faunal Differences Along the Coastline

The Upper Mississippi Embayment. Although the marine vertebrate faunas of the Jackson Group of the southern Mississippi Embayment, of southeast Arkansas and northwest Mississippi, differ little from

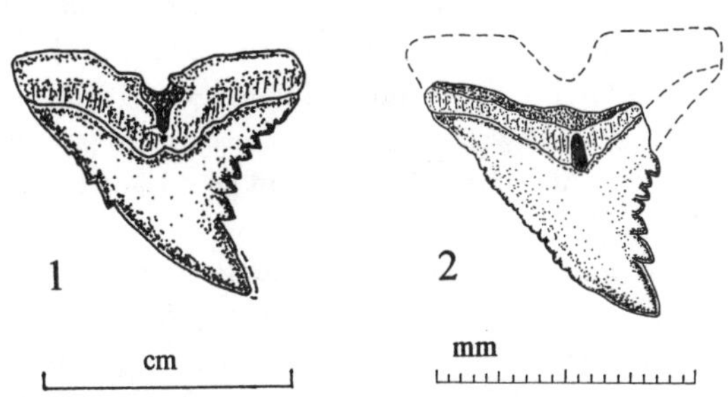

Figure 21.10. *Hemipristis* (snaggletooth shark) right upper lateral teeth, in lingual view. (1) *H. wyattdurhami* (range: late middle Eocene to early Oligocene). MGS 1915. Late middle Eocene, Moodys Branch Fm.-equivalent; Crow Creek, St. Francis Co., Ark. (2) *H. serra* (range: late Oligocene to early Pleist.). TU-VP 1018, partial tooth missing most of root. Late Oligocene, upper River Bend Fm.; Silverdale, Onslow Co., N.C.

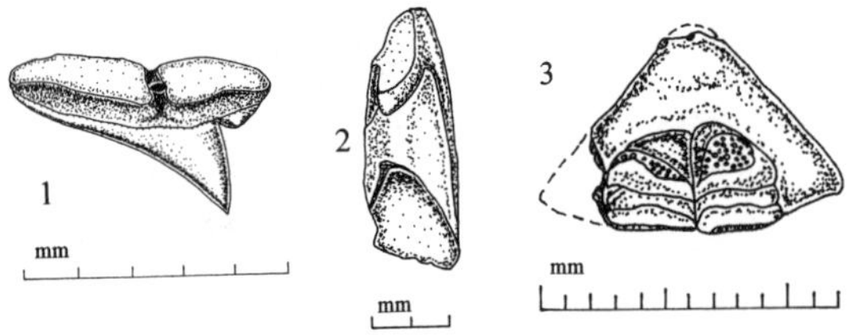

Figure 21.11. Upper Mississippi Embayment (1 and 2), and eastern carbonate zone (3) elements. (1) *Rhizoprionodon* (sharpnose shark) sp. lateral tooth, in lingual view. MGS 1916. Late middle Eocene, Moodys Branch Fm.-equivalent; Fulton, Lauderdale Co., Tenn. (2) Unident. ballistid (trigger fish) anterior tooth, in medio-lingual view. MGS 1917. Late middle Eocene, Moodys Branch Fm.-equivalent; Crow Creek, St. Francis Co., Ark. (3) *Chilomycterus* (burrfish) sp. toothplate, in occlusal view. TU-VP 1008. Late middle Eocene, Gosport Fm.; Little Stave Creek, Clarke Co., Ala.

those of other Jackson faunas (there are no known Vicksburg Group faunas of the Embayment), there are a few important differences in the northern Embayment. These are known through the research of William Parks, on the Crow Creek fauna of Arkansas and Fulton fauna of Tennessee (figure 21.1). The Parks collection (MGS) shows that the two faunas are contemporaneous (both Moodys Branch Formation, lower Jackson Group, equivalents) and generally similar, yet with distinctive elements (author's research). The Fulton fauna, the northernmost Eocene marine fauna of the Embayment, includes abundant teeth of the small sharpnose shark *Rhizoprionodon* (figure 21.11, no. 1), while the nearby Crow Creek fauna includes none. Likewise, the Crow Creek fauna includes anterior teeth of the triggerfish (figure 21.11, no. 2), not seen at Fulton. These differences suggest that the northern Embayment had a more nearshore habitat and fauna than the southern end.

The eastern clearwater fauna. Because of the location of the Mississippi delta, the sediment of the western GCP Jackson and Vicksburg groups tends to be dominated by clays, while the more eastern part, in southwest Alabama, tends to be more carbonate-rich (Tew and Mancini, 1995). A few of the eastern marine vertebrates also suggest a more clearwater (less turbid) habitat. One of these elements is the toothplate of the burrfish (a relative of the more derived porcupinefish) *Chilomycterus* (figure 21.11, no. 3). The more common elements, such as the teeth of the sand tiger shark *Odontaspis* and the eagle ray *Myliobatis*, remain the same.

The high-latitude fauna. One would expect the fauna of the northern Atlantic Coastal Plain to differ greatly from that of the GCP, as higher-latitude faunas lived in cooler water. Although the marine vertebrate faunas of the northern Atlantic CP aren't well known, the early Tertiary faunas, like the better-known late Cretaceous faunas of the same region, show only subtle differences from the warmer-water faunas of the GCP. It is noticeable in the chimaeroids (which are far more common in the northern faunas, notably in the middle Eocene of New Jersey [Fowler, 1911] and England [Kemp et al., 1990]) and in three families of sharks (the Squalidae [dogfish], Hexanchidae [cow sharks], and Heterodontidae [bullhead, or horn sharks]), all of which are noticeably rarer in the GCP. Some of these taxa are picked up in the cold-water faunas of unusually deep-water faunas of the GCP. A few warm-water taxa, notably the barracuda *Sphyraena*, become far rarer in the high-latitude fauna.

General. Taken as a whole, the geographic differences along the Gulf and Atlantic Coastal Plain are generally relatively minor and are not such that they would make difficult the identification of age-equivalents. Still, they need to be considered when comparing faunas across the Eocene/Oligocene boundary, so that all differences aren't considered age-related.

CHANGE IN THE MISSISSIPPI EMBAYMENT ACROSS THE EOCENE/OLIGOCENE BOUNDARY

Recent research on the marine faunas of the northern Mississippi Embayment has indicated that the Embayment had been greatly reduced (to a very small bay) by the early Oligocene. The collection of Jackson Group fossils and sediment samples from the gravel bars of the Mississippi River in northwest Mississippi and southeast Arkansas by John Connaway and Lonnie Looper has revealed details of the final stages of the Mississippi Embayment. A small Jackson fauna (largely mollusks, *Odontaspis* teeth and *Basilosaurus* rib fragments) has been found on gravel bars throughout the area, contrary to the commonly held notion

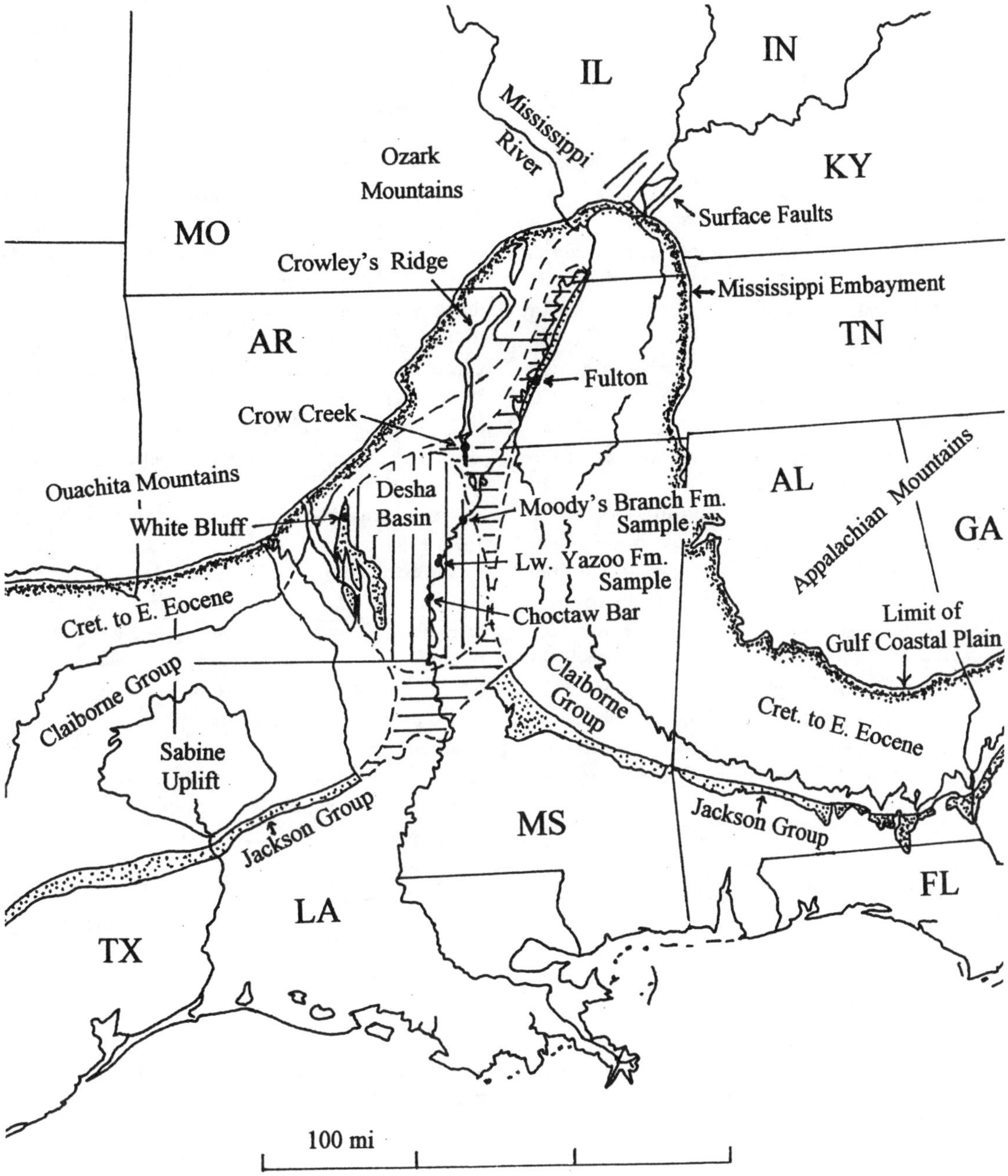

Figure 21.12. Mississippi Embayment portion of the central Gulf Coastal Plain. Jackson Group (late middle to latest Eocene) fauna and sample sites of the embayment are located (see text). Area of parallel lines is subsurface Jackson Group. The Desha Basin (vertical lines) is the area of thickest deposition in the Jackson Group of the Mississippi Embayment.

that the Jackson in that area is deeply buried below Pleisto-cene and Holocene sediments. Some of these collecting sites, significantly, are located atop the Desha Basin (figure 21.12), the thickest part of the Mississippi Embayment deposits, and the last part of the bay to fill in. Abundant specimens of the Jackson Group bivalve *Venericardia apodensata* have been recovered from Choctaw Bar (Looper collection). A small block of glauconitic sandstone with a Moodys Branch Formation–equivalent microfauna has been collected (Connaway collection) in the northeast part of the Desha Basin (figure 21.12). Most critical of all, a lithified marl sample with a lower Yazoo Formation pteropod fauna (Looper collection) has been collected near the center of the Desha Basin (figure 21.12). The latter suggests that the youngest marine beds in the Mississippi Embayment may be no more recent than early late Eocene (early Priabonian). Despite the sea-level highstand at the midsequence at the end of the Eocene, there is no evidence that the Mississippi Embayment, as a major feature of the Gulf coastline, existed at that time. It is unlikely that, in the early Oligocene Vicksburg Group time, the Embayment consisted of more than a very small bay in the GCP.

The loss of this large, central bay, and change in the Gulf coastline in the early Oligocene could have had an effect on the fauna at the Eocene/Oligocene boundary.

CONCLUSIONS

1. Diversity counts of marine vertebrate species across the Eocene/Oligocene boundary in the Gulf Coastal Plain are highly affected by several factors: the number of known faunas, how well studied the faunas are, by the reliability of the shark and ray identifications, the relative differences in diagenetic alteration of the sediment, differences in deposition rates, and a series of factors resulting from the sequence statigraphic position of the faunas. These effects make the Jackson Group (late middle to latest Eocene) vertebrate fauna appear to be far more diverse than that of the early Oligocene Vicksburg Group.

2. In contrast to the mollusk fauna, the actual change in the marine vertebrate fauna across the Eocene/Oligocene boundary in the GCP was probably mostly very gradual (often on a subspecific level); in contrast to the major, stepwise faunal changes of the early/middle Eocene and early/late Oligocene boundaries, which are probably due to missing faunas of intermediate age.

3. The Mississippi Embayment was largely lost across the Eocene/Oligocene boundary.

ACKNOWLEDGMENTS

Several large collections of Gulf Coastal Plain vertebrates were studied for this report. David Dockery (Mississippi Office of Geology, Jackson) permitted the use of material he had collected, as well as the recently donated Park Collection material from Fulton, Tennessee, and Crow Creek, Arkansas. Eleanor Daly (Mississippi Museum of Natural Science, Jackson) tolerantly permitted the use of the important Frazier Devil's Backbone collection at the difficult time of a move to a new building. Henry Bart (Tulane Museum of Natural History, Belle Chase, Louisiana) permitted the use of the extensive Vokes Coastal Plain vertebrate fossil collection. The GCP vertebrate fossil collection of the Louisiana State University Museum of Natural Sciences in Baton Rouge, much of it collected by Henry Howe, was also studied extensively while the author worked there, 1982-1991.

Robert Wiest (College Park, Maryland) kindly donated the *Plinthicus stenodon* specimen figured here. John Connaway (Mississippi Department of Archives and History, Clarksdale) and Lonnie Looper (Greenville, Mississippi) permitted the study of rare Jackson Group fossils from northwestern Mississippi in their collections. William Parks (Tennessee Division of Geology, retired; Cordova, Tennessee) permitted the author to see unpublished manuscripts he had written on his Fulton and Crow Creek collections (now in Jackson) Wylie Poag (U.S. Geological Survey, Woods Hole, Massachusetts) provided a helpful discussion of the Jackson Group-equivalent beds of the Atlantic Coastal Plain. James Westgate (Lamar University, Beaumont, Texas) provided a particularly detailed and helpful review of the manuscript. David Dockery and Eleanor Daly also kindly reviewed it.

Thanks are particularly extended to Donald Prothero (Occidental College, Los Angeles) for the invitation to participate in the conference and volume.

REFERENCES

Applegate, S. P., 1965, Tooth terminology and variation in sharks, with special reference to the sand shark, *Carcharias taurus* Rafinesque: Los Angeles County Museum, Contributions in Science, no. 86, 18 p.

Arata, A. A., and Jackson, C., 1965, Cenozoic vertebrates from the Gulf Coastal Plain—1: Tulane Studies in Geology, v. 3, no. 3, p. 175-177.

Baum, G. R., and Vail, P. R., 1988, Sequence stratigraphic concepts applied to Paleogene outcrops,

Gulf and Atlantic basins, *in* Wilgus, C., Hastings, B. S., Kendall, C. G. St. C., Posamentier, H. W., Ross, C. A., and Van Wagoner, J. C., eds., Sea-level changes: an integrated approach: SEPM, Special Publication, v. 42, p. 309-327.

Beard, K. C., and Tabrum, A. R., 1991, The first early Eocene mammal from eastern North America: an omomyid primate from the Bashi Formation, Lauderdale County, Mississippi: Mississippi Geology, v. 11, no. 2, 6 p.

Breard, S. Q., Jr., 1991, Paleoecology of a late Eocene (Bartonian) vertebrate fauna, Moodys Branch Formation, Techeva Creek, Mississippi: Gulf Coast Association of Geologic Societies, Transactions, v. 41, p. 43-55.

Cappetta, H., 1987, Chondrichthyes II, Mesozoic and Cenozoic Elasmobranchii, *in* Schultz, H.-P., ed., Handbook of paleoichthyology, v. 3B: New York, Gustav Fischer Verlag, 193 p.

Case, G. R., 1980, A selachian fauna from the Trent Formation, lower Miocene (Aquitanian) of eastern North Carolina: Palaeontographica, Abteilung A, v. 171, p. 75-103.

Case, G. R., 1981, Late Eocene selachians from south-central Georgia: Paleontographica, Abteilung A, v. 176, p. 52-79.

Case, G. R., 1982, A pictorial guide to fossils: New York, van Nostrand Reinhold Co., 514 p.

Case, G. R., 1994a, b, Fossil fish remains from the late Paleocene Tuscahoma and early Eocene Bashi formations of Meridian, Lauderdale County, Mississippi: Paleontographica, Abteilung A, v. 230, no. 4-6, a—Part I. Selachians. P. 97-138, 15 pls.; b—Part II. Teleosteans. P. 139-153, 2 pls.

Conrad, G. M., 1941, A fossil squirrel-fish from the upper Eocene of Florida: Florida Geological Survey, Geological Bulletin, v. 22, 25 p.

Conrad, T. A., 1832, Fossil shells of the Tertiary formations of North America: printed for the author by Judah Dobson, Philadelphia, v. 1, no. 1, 20 p. Reprinted: Ithaca, NY, Paleontological Research Institution, 1893 and 1963, p. 9-34, 6 pls.

Conrad, T. A., 1846, Observations on the Eocene Formation of the United States, with descriptions of species of shells, &c. occurring in it: American Journal of Science, series 2, v. 1, nos. 2 and 3, in two parts: p. 209-221 (no. 2), p. 395-405 (no. 3).

Corgan, J. X., 1976, Vertebrate fossils of Tennessee: Tennessee Division of Geology, Bull. 77, 100 p.

Corgan, J. X., and Breitburg, E., 1997, Tennessee's prehistoric vertebrates: Tennessee Division of Geology, Bulletin 84, 170 p. (dated 1996, but distributed in 1997).

Cornelius, E., 1819, On the geology, mineralogy, scenery, and curiosities of parts of Virginia, Tennessee, and the Alabama and Mississippi territories, &c. with miscellaneous remarks, in a letter to the editor: American Journal of Science, series 1, v. 1, p. 214-226, 317-331 (in two parts).

Daly, E., 1992, A list, bibliography and index of the fossil vertebrates of Mississippi: Mississippi Office of Geology, Bulletin 128, 47 p.

Dockery, D. T., III, 1982a, Lesueur's Walnut Hills fossil shells: Mississippi Geology, v. 2, no. 3, p. 7-13, 12 pls.

Dockery, D. T., III, 1982b, Lower Oligocene Bivalvia of the Vicksburg Group in Mississippi: Mississippi Bureau of Geology, Bulletin 123, 261 p..

Dockery, D. T., III, and Manning, E. M., 1986, Teeth of the giant shark *Carcharodon auriculatus* from the Eocene and Oligocene of Mississippi: Mississippi Geology, v. 7, no. 1, p. 7-19.

Domning, D. P., 1969, A list, bibliography, and index of the fossil vertebrates of Louisiana and Mississippi: Gulf Coast Association of Geological Societies, Transactions, v. 19, p. 385-422.

Fowler, H. W., 1911, A description of the fossil fish remains of the Cretaceous, Eocene and Miocene formations of New Jersey: Geological Survey of New Jersey, Bulletin, v. 4, 192 p.

Frizzell, D. L., and Dante, J. H., 1965, Otoliths of some early Cenozoic fishes of the Gulf Coast: Journal of Paleontology, v. 39, no. 4, p. 687-718.

Gazin, C. L., and Sullivan, J. M., 1942, A new titanothere from the Eocene of Mississippi, with notes on the correlation between the marine Eocene of the Gulf Coastal Plain and continental Eocene of the Rocky Mountain region: Smithsonian Miscellaneous Collections, v. 101, no. 13, 13 p.

Gibbes, R. W., 1848, Monograph of the fossil Squalidae of the United States: Academy of Natural Sciences of Philadelphia, Journal, v. 1, ser. 2, pt. 2, p. 139-147.

Gibbes, R. W., 1849, Monograph of the fossil Squalidae of the United States: Academy of Natural Sciences of Philadelphia, Journal, v. 1, ser. 2, pt. 3, p. 191-206.

Gibbes, R. W., 1850, New species of *Myliobatis* from the Eocene of South Carolina, with other genera not heretofore observed in the United States:

Academy of Natural Sciences of Philadelphia, Journal, v. 1, ser. 2, pt. 4, p. 299-300.

Harlan, R., 1834, Notice of fossil bones found in the Tertiary Formation of the state of Louisiana: Academy of Natural Sciences of Philadelphia, Transactions, new ser., v. 4, no. 12, p. 397-403.

Harper, L., 1857, Preliminary report on the geology and agriculture of the state of Mississippi: Jackson (printed in New York for), E. Barksdale, 350 p.

Hilgard, E. W., 1866, Report on the geology and agriculture of the state of Mississippi: Jackson, E. Barksdale, 391 p. (dated 1860, but see Hilgard, 1901, p. 305).

Hilgard, E. W., 1901, A historical outline of the geological and agricultural survey of the state of Mississippi: American Geologist, v. 27, p. 284-311.

Holman, J. A., Dockery, D. T., III, and Case, G. R., 1991, Paleogene snakes of Mississippi: Mississippi Geology, v. 11, no. 1, 12 p.

Hutchins, T., 1784, An historical narrative and topographical description of Louisiana and West-Florida: Philadelphia, Robert Aitken, 5-95 p. 1968 facimile reproduction, Gainesville, University of Florida Press, 5-95 p.

Ivany, L. C., Patterson, W. P., and Lohmann, K. C., 2000, Cooler winters as a possible cause of mass extinctions at the Eocene/Oligocene boundary: Nature, v. 407, no. 6806, p. 887-890.

Jaramillo, C. A., and Oboh-Ikuenobe, F. E., 1999, Sequence stratigraphic interpretations from palynofacies, dinocyst and lithological data of upper Eocene–lower Oligocene strata in southern Mississippi and Alabama, U.S. Gulf Coast: Palaeogeography, Palaeoclimatology, Palaeoecology, v. 145, nos. 1-3, p. 259-302.

Kemp, D. J., Kemp, E., and Ward, D. J., 1990, An illustrated guide to the British middle Eocene vertebrates: London, privately published by D. J. Ward, 59 p.

Koken, E., 1888, Neue Untersuchungen an tertiaren Fisch-otolithen: Deutsche geologische Gesellshaft, Zeitschrift, v. 40, p. 274-305.

Leriche, M., 1942, Contribution a l'étude des faunes ichthyologiques marines des terrains Tertiaires de la plaine cotière Atlantique et du centre des Etats-Unis: Société Géologique de France, Mémoire, v. 20, no. 45, p. 5-110.

Lesueur, C.-A., 1829 manuscript, The geology and paleontology of Walnut Hills, Mississippi: Lesueur Collection, no. 45120-45178: Le Havre, France, Museum of Natural History.

Lyell, C., 1833, Principles of geology, v. 3: London, J. Murray.

Lyell, C., 1846, [Notes on the geology of Alabama and Georgia]: American Journal of Science, ser. 2, v. 1, no. 2, p. 313-315.

Manning, E. M., 1990, The late early Miocene Sabine River: Gulf Coast Association of Geological Societies, Transactions, v. 40, p. 531-549.

Manning, E. M., 1997, An early Oligocene rhinoceros jaw from the marine Byram Formation of Mississippi: Mississippi Geology, v. 18, no. 2, 13-31 p.

Manning, E. M., and Bograd, M. B. E., 2000, Annotated biliography of the geology of Mississippi to 1850: Mississippi Geology, v. 20, no. 4, p. 49-72.

Manning, E. M., and Standhardt, B. R., 1986, Late Eocene sharks and rays of Montgomery Landing, Louisiana, *in* Schiebout, J. A., and van den Bold, W., eds., Montgomery Landing site, marine Eocene (Jackson) of central Louisiana: Gulf Coast Association of Geological Societies, Symposium Proceedings, p. 133-161.

Manning, E. M., Dockery, D. T., III, and Schiebout, J. A., 1985, Preliminary report of a *Metamynodon* skull from the Byram Formation (lower Oligocene) in Mississippi: Mississippi Geology, v. 6, no. 2, 16 p.

Morton, S. G., 1833, Supplement to the "Synopsis of the organic remains of the Ferruginous Sand Formation of the United States": American Journal of Science, series 1, v. 23, no. 2, p. 288-294.

Morton, S. G., 1834, Synopsis of the organic remains of the Cretaceous Group of the United States: Philadelphia, Key and Biddle, 88 p.

Morton, S. G., 1842, Description of some new species of organic remains of the Cretaceous Group of the United States: with a tabular view of the fossils hitherto discovered in this formation: Academy of Natural Sciences of Philadelphia, Journal, ser. 1, v. 8, p. 207-227.

Peters, S. E., and Foote, M., 2002, Determinants of extinction in the fossil record: Nature, v. 416, no. 6874, p. 420-424.

Poag, C. W., 1992, U. S. Middle Atlantic continental rise: provenance, dispersal, and deposition of Jurassic to Quaternary sediments, *in* Poag, C. W., and de Graciansky, P. C., eds., Geologic evolution of Atlantic continental rises: New York, Van Nostrand Reinhold, p. 100-156.

Purdy, R. W., Schneider, V. P., Applegate, S. P., McLel-lan, J. H., Meyer, R. L., and Slaughter, B. H., 2001, The Neogene sharks, rays, and bony fishes from the Lee Creek mine, Aurora, North Carolina, *in* Ray, C. E., and Bohaska, D. J., eds., Geology and paleontology of the Lee Creek mine, North Carolina: Smithsonian Contributions to Paleobiology, no. 80, p. 71-202.

Rossbach, T. J., and Carter, J. G., 1991, Molluscan biostratigraphy of the lower River Bend Formation at the Martin Marietta Quarry, New Bern, North Carolina: Journal of Paleontology, v. 65, no. 1, p. 80-118.

Stewart, J. D., 1992, First Mississippi records of *Holocentrites* (Beryciformes, Holocentridae), and confirmation of its myripristine affinities: Journal of Vertebrate Paleontology, v. 12, supplement to no. 3 (Abstracts), p. 53A.

Stringer, G. L., 1979, A study of the upper Eocene otoliths of the Yazoo Clay in Caldwell Parish, Louisiana: Tulane Studies in Geology and Paleontology, v. 15, no. 3, p. 95-103.

Tessman, N. T., 1969, The fossil sharks of Florida [Unpubl. M.S. thesis]: Gainesville, University of Florida, 132 p.

Tew, B. H., and Mancini, E. A., 1995, An integrated stratigraphic method for paleogeographic reconstruction: examples from the Jackson and Vicksburg groups of the eastern Gulf Coastal Plain: Palaios, v. 10, no. 2, p. 133-153.

Thurmond, J. T., and Jones, D. E., 1981, Fossil vertebrates of Alabama: Tuscaloosa, University of Alabama Press, 244 p.

Wailes, B. L. C., 1845, On the geology of Mississippi: Association of American Geologists and Naturalists, Proceedings, Abstract, v. 6, p. 80-81.

Wailes, B. L. C., 1854, Report on the agriculture and geology of Mississippi: Philadelphia, Lippincott, Grambo and Co., for E. Barksdale, Jackson, 371 p., 17 pls.

Ward, D. J., 1988, *Hypotodus verticalis* (Agassiz 1843), *Hypotodus robustus* Leriche (1921) and *Hypotodus heinzelini* (Casier 1967), Chondrichthyes, Lamniformes, junior synonyms of *Carcharias hopei* (Agassiz 1843): Tertiary Research (Leiden), v. 10, no. 1.

Ward, D. J., and Wiest, R. L., 1990, A checklist of Paleocene and Eocene sharks and rays (Chondrichthyes) from the Pamunkey Group, Maryland and Virginia, USA: Tertiary Research (Leiden), v. 12, no. 2, p. 81-88.

Ward, L. W., Lawrence, D. R., and Blackwelder, B. W., 1978, Stratigraphic revision of the middle Eocene, Oligocene, and lower Miocene—Atlantic Coastal Plain of North Carolina: U.S. Geological Survey Bulletin 1475-F, 23 p.

Westgate, J. W., 1984, Lower vertebrates from the late Eocene Crow Creek Local Fauna, St. Francis County, Arkansas: Journal of Vertebrate Paleontology, v. 4, no. 4, p. 536-546.

Westgate, J. W., 2001, Paleontology and biostratigraphy of marginal marine Gulf Coast Eocene vertebrate localities, *in* Gunnell, G., ed., Eocene faunas: unusual occurrences and rarely sampled habitats: New York, Plenum Press, p. 268-295.

Westgate, J. W., and Emry, R. J., 1985, Land mammals of the Crow Creek Local Fauna, late Eocene, Jackson Group, St. Francis County, Arkansas: Journal of Paleontology, v. 59, no. 1, p. 242-248.

Whetstone, K., and Martin, L. D., 1978, An Oligocene (Orellan) sirenian from the Bucatunna Formation of Alabama: Tulane Studies in Geology and Paleontology, v. 14, no. 4, p. 161-163.

White, E. I., 1956, The Eocene fishes of Alabama: Bulletins of American Paleontology, v. 36, no. 156, p. 123-152.

CHAPTER 22

A Brief Account of the Evolution of the Caribbean Seaway: Jurassic to Present

Manuel A. Iturralde-Vinent

ABSTRACT

This paper presents a summary of the history of the Caribbean Seaway, from its Lower to Middle Jurassic Pangaean epicontinental sea precursor (rift valleys), its debut as an oceanic seaway in the Oxfordian, and its subsequent evolution until the Present. Since the Oxfordian, marine connectivity between the Tethys–Atlantic and the Pacific Oceans along the Caribbean Seaway have been shaped by the occurrence of different topographic barriers, sometimes as isolated islands and submarine highs, sometimes as continuous land barriers. This scenario can be used to evaluate possible biotic interchange, either by land between the American continents or by water between the Pacific and Tethys–Atlantic Oceans. During the Eocene-Oligocene transition (35–33 Ma), there was a land barrier in the Caribbean (Gaarlandia), that precluded free ocean-water circulation between the Pacific and the Atlantic–Caribbean, probably contributing to the climatic change that took place at this time interval. This barrier was progressively drowned during the Oligocene, restoring the circum-tropical current that warmed up temperature within the Caribbean Sea and the Tropical Pacific.

INTRODUCTION

It is well known that, due to changing Earth paleogeography, global ocean-water circulation undergoes change, which is controlled by the configuration of the continents, islands, oceans, seas, and shallow submarine areas, as well as atmospheric and water composition and dynamic factors (Barron and Peterson, 1989). As a consequence, present-day patterns of marine circulation cannot be extrapolated to evaluate paleobiogeographic scenarios (Berggren and Hollister, 1974; Frakes et al., 1992; Parrish, 1992). In this paper, I briefly explore the origin and evolution of the Caribbean area, using the database compiled by Iturralde-Vinent and MacPhee (1999), as well as the paleogeographic reconstructions published recently (Iturralde-Vinent and MacPhee, 1999; Kerr et al., 1999; Iturralde-Vinent, 1998, Lawver et al., 1999). This is a necessary step toward the understanding of the history of marine biotic exchange between the Pacific and Tethyan/Atlantic oceans.

Figure 22.1 illustrates the history of topographic barriers between North America and South America, and shows how they affected the marine flow across the area. This history can be subdivided into the following major steps: (1) the western Pangaean epicontinental seaway; (2) the early opening of the Caribbean Seaway; (3) the Mesozoic evolution of the Caribbean Seaway; (4) the early Cenozoic evolution; and (5) the latest Eocene to Recent evolution of the seaway.

THE WESTERN PANGAEAN EPICONTINENTAL SEAWAY

Similarities among several groups of marine animals (bivalves, ammonites, marine reptiles) that occur both in the western Tethys and the East Pacific suggest that since the Pliensbachian-Hettangian (nearly 190 ± 5 Ma) an aquatic connection existed between these regions (Smith, 1983; Damborenea and Maceñido, 1979; Riccardi, 1991; Gasparini, 1992, 1996). Although the connection, theoretically, may have taken place eastward across the eastern Tethys into the

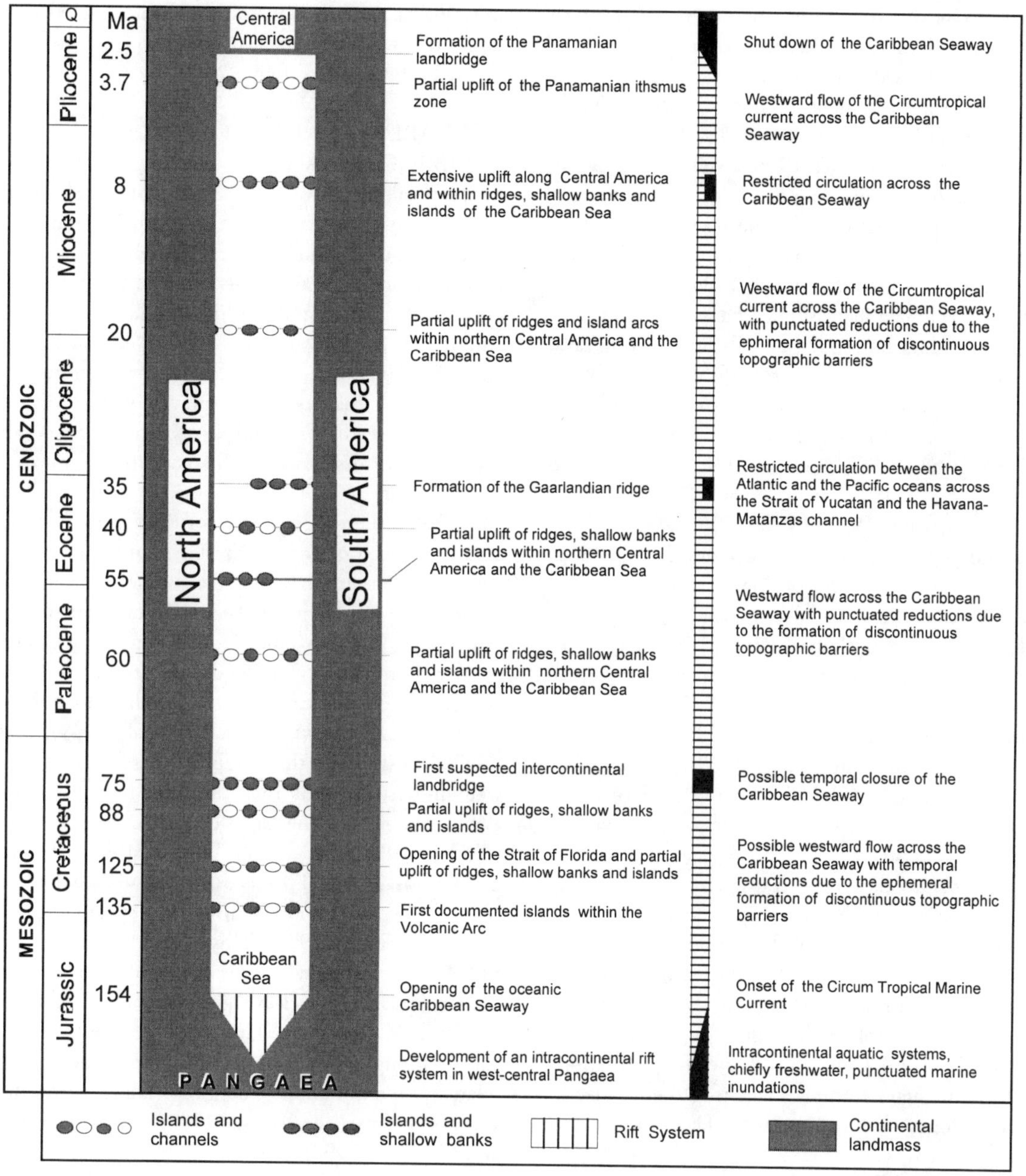

Figure 22.1. Major events in the evolution of the Caribbean Seaway from Jurassic to Recent. Data from Iturralde and MacPhee (1999:figure 12 and tables 1–4). On the left part of the graphic, the dotted vertical strips are the permanent continental masses of North and South America (NA and SA), the central patternless zone represent the Caribbean marine area, and the bubble rows indicate topographic barriers within the Caribbean. The right part of the graphic provides information about the topographic barriers and the marine seaway. The thin vertical strip within the textual area is intended to show in black major shutdowns of the flow along the Caribbean Seaway, and with horizontal pattern the time lapses when water exchange was possible between the Pacific and Atlantic/Tethyan oceans. Chronology not to scale.

Pacific Ocean, similarities between the western Tethyan and the East Pacific taxa suggest that the connection passed through the central part of the Pangaean supercontinent (Gasparini, 1978; Smith, 1983; Riccardi, 1991; Gasparini, 1978, 1992, 1996; Gasparini and Fernández 1996; Damborenea, 2000, Aberham, 2001). This connection first yielded a limited exchange of shallow marine taxa (Smith, 1983; Damborenea, 2000, Aberham, 2001), but with time the waterway widened (Pindell, 1994; Lawver et al., 1999) allowing an increasing exchange of marine animals, including both open and shallow marine taxa, limited during the Bathonian-Oxfordian, and extensive since the Oxfordian (Westermann, 1981; Westermann, 1992; Hilldebrandt, 1981; Hilldebrandt et al., 1992: Imlay, 1984; Sandoval and Westermann, 1986; Bartok et al., 1985; Riccardi, 1991; Gasparini, 1978, 1992, 1996; Gasparini and Fernández, 1996; Iturralde-Vinent and Norell, 1996; Damborenea, 2000). According to recent paleogeographic reconstructions, this passageway was established before the breakup of Pangaea, probably across a dominantly shallow siliciclastic interior sea located within western Pangaea (figure 22.1; Salvador, 1987; Pindell, 1994; Pindell and Tabbutt, 1995).

The fact that large ocean-dwelling saurians are recorded from the Middle Jurassic of the southeastern Pacific (Gasparini, 1992; Gasparini and Fernández, 1996) suggest that the western Pangaean epicontinental seaway may have been transected by some deep-water channels, probably along grabens and semigrabens produced by crustal extension (rift valleys). This hypothesis is supported by the existence of "continental margin and marine facies" in the pre-Oxfordian (probably Bathonian-Bajocian) siliciclastic rocks of western Cuba (San Cayetano Formation), with sedimentary features suggesting the occurrence of paleoslopes from fluvio-marine facies into marine environments (Haczewski, 1976; Pszczolkowski, 1978). It is also supported by the occurrence of Jurassic rift grabens and semigrabens in the eastern and southern continental margin of North America (Salvador, 1987; Pindell, 1994; Marton and Buffler, 1994, Bullard et al., 1965; Klitgord et al., 1988; Gradstein et al., 1990; Jones et al., 1995, Pindell and Talbut, 1995; Milani and Tomas Filho, 2000).

The Pangaean epicontinental sea should not be thought of as the proto-Caribbean basin itself, but as a precursor situated within western Pangaea (Iturralde-Vinent and MacPhee, 1999). For this reason, I propose to avoid the confusing terms "Hispanic Corridor" or "Caribbean Corridor" to designate this seaway and refer to it instead by a more descriptive term, the western Pangaean epicontinental seaway and rift valley system (figure 22.1).

THE EARLY OPENING OF THE CARIBBEAN SEAWAY

As the break-up of western Pangaea completely separated Laurasia and Gondwana, a new paleogeographic feature formed: the Caribbean Seaway. This seaway is here understood as a truly oceanic basin located between the North and South American continental landmasses. It represents a linkage between the Pacific and the Atlantic–Tethyan Oceans. In order to explore the origin of this basin, two sources of information are needed: earliest Mesozoic occurrence of ocean crustal indicators, and the first occurrence of true marine environments.

The earliest occurrence of Mesozoic oceanic crustal indicators within the Mesoamerican realm comprise the pillow basalts and associated marine sediments reported from the Andean terranes of northern South America (Bajocian–Bathonian Siquisique basalts: Bartok et al. 1985) and from the Guaniguanico terrane of western Cuba (Oxfordian El Sábalo basalts: Pszczolkowski, 1994; Kerr et al., 1999). The original position of the Siquisique basalts is not yet well understood, but they probably belong to the Pacific margin of Pangaea, as do the Andean terranes (Pindell, 1994; Pindell and Tabbutt, 1995, Aleman and Ramos, 2000). If this is true, they predate the formation of the early Caribbean oceanic crust (Iturralde-Vinent, 1998). This is not the case for the El Sábalo basalts, which probably formed on the Mayan (Yucatan) block margin facing the Caribbean (Iturralde-Vinent, 1994; Hutson et al., 1999), so they can be thought of as representing a magmatic event related to the initiation of ocean spreading within the Caribbean (Iturralde-Vinent, 1998; Iturralde-Vinent and MacPhee, 1999; Kerr et al., 1999). This inference is corroborated by the occurrence of an important assemblage of Oxfordian marine animals (bivalves, ammonites, ganoid fish and marine reptiles) in the Guaniguanico terrane of western Cuba (Iturralde-Vinent and Norell, 1996) that are clearly related to both the western Tethys and the eastern Pacific faunas (Ricardii, 1991; Gasparini, 1992; Gasparini and Fernández, 1996; Fernández and Iturralde-Vinent, 2000).

The opening of the Caribbean Seaway represents an important paleoceanographic golden spike, at which point the Mesozoic deep circum-equatorial ocean-water circulation that linked the Tethys, the central Atlantic, and the Pacific Oceans may have begun.

THE MESOZOIC EVOLUTION OF THE CARIBBEAN SEAWAY

It is generally understood that active seafloor spreading between North and South America lasted from Late Jurassic through late Albian, widening the gap between the continents (Pindell, 1994). Subsidence of the crust produced abyssal water-depth sections within the area, as early as the Tithonian (Pszczolkowski, 1978; 1982; Iturralde-Vinent, 1998; Iturralde-Vinent and MacPhee, 1999). These processes generally improved the chances for interoceanic marine communication across the Caribbean Seaway. Such a paleoceanographic scenario allowed extensive biotic exchange between the Tethys, the Caribbean, and the Pacific, as recorded by the distribution of several mollusks (Sohl and Kollmann, 1985; Ricardii, 1991; Rojas et al., 1996; Boitrón-Sánchez and López-Tinajero, 1996).

Some numerical models of Cretaceous ocean circulation across the Tethys suggest that the main circulation was directed eastward (Barron and Peterson, 1989). This hypothesis is contradicted by the pattern of ammonite (cephalopod), acteonelid (gastropod) and rudist (bivalve) dispersion from east to west (Sohl and Kollmann, 1985; Ricardii, 1991; Rojas et al., 1996; Boitrón-Sánchez and López-Tinajero, 1996).

Marine connectivity along the Caribbean Seaway was eventually moderated by local and regional uplift within the Caribbean (figure 22.1). Iturralde-Vinent and MacPhee (1999) have compiled abundant information concerning the occurrence of locals and regional unconformities and hiatuses, as well as the presence of shallow-water banks, indicating both subaerial exposure and/or submarine topographic highs during the time interval under consideration. These positive topographies must have ultimately interrupted deep-ocean circulation, and forced circulation along deep trenches and channels. However, current knowledge precludes accurate reconstruction of the specific paleoceanographic scenarios.

The Bahamas are an important permanent barrier that occurs within the Caribbean Seaway. By the Late Jurassic-Early Cretaceous, this shallow-water structure included the present-day Florida Peninsula and the Bahamas, but since the Aptian this megaplatform has been subdivided into smaller platforms separated by a network of abyssal channels (Buffler and Hurts, 1985; Iturralde-Vinent, 1994; 1998). This general barrier allowed restricted deep-marine circulation through the channels, but it concentrated the main circulation across the Strait of Florida and the central Atlantic Ocean. Some other barriers probably occurred as islands chains (erosional surfaces and hiatuses within the volcanic sections) and shallows (interbedded shallow marine limestones) along the active volcanic arcs of the Caribbean realm (Iturralde-Vinent and MacPhee, 1999), but during the Cretaceous these topographic features were mostly located on the western periphery of the Caribbean (Pindell, 1994; Iturralde-Vinent, 1998). Within these volcanic terranes, major uplift events are recorded during the Neocomian, Aptian-Albian, and Coniancian-Santonian; but these barriers were surely non-permanent, because subsidence and deposition of marine rocks are also recorded in them after each uplift (Iturralde-Vinent and MacPhee, 1999).

The documented interchange of large terrestrial tetrapods between North and South America, late in the Cretaceous, strongly suggests that a land bridge linked these continents (Gingerich, 1985; Gayet et al., 1992). Iturralde-Vinent and MacPhee (1999) suggested that this land bridge probably coincided in time with the regional uplift that took place by late Campanian–early Maastrichtian along the Antillean Cretaceous volcanic arc (figure 22.1). This event would have closed the Caribbean Seaway for a period of 1–3 million years, producing a major change in ocean-water circulation and global climate, like the one recorded after the formation of the Panamanian land bridge in the Pliocene. Prior to the late Campanian. topographic barriers within the Cretaceous volcanic arcs were ephemeral and probably not continuous landmasses and shallows separated by local deep-water channels (Iturralde-Vinent and MacPhee, 1999), so their effect on the marine flow across the Caribbean Seaway would have been less drastic (figure 22.1).

THE EARLY CENOZOIC EVOLUTION OF THE CARIBBEAN SEAWAY

The end of the Cretaceous in the Caribbean realm was accompanied by strong modifications in the tectonic regime of the area. Rock deformation and collisional tectonics associated with important vertical movements played an outstanding role since the latest Campanian, and the formation and destruction of topographic barriers was a common phenomenon during the Cenozoic (Khudoley and Meyerhoff, 1971; Pszczolkowski and Flores, 1986; Droxler et al., 1998; Iturralde-Vinent and MacPhee, 1999). This dynamic of the crustal movements within the

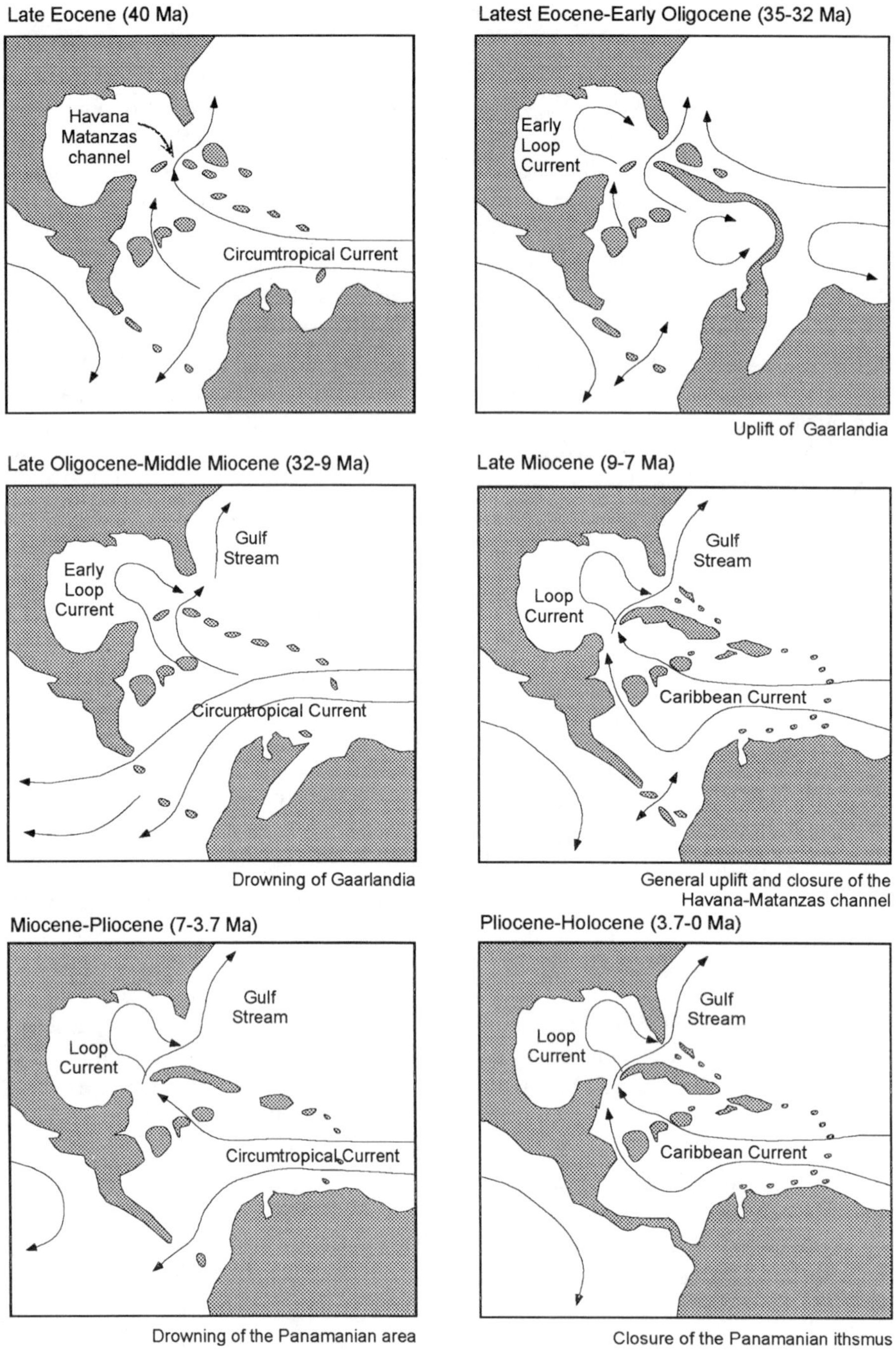

Figure 22.2. Main scenarios of late Eocene to Recent marine surface circulation across the Caribbean Seaway. After data from Prothero (1994), Coates and Obando (1996), Droxell et al. (1998), Iturralde-Vinent and MacPhee (1999: figure 10). Since the Late Eocene the course of westward-flowing currents in the mid-Atlantic/Tethys across the Caribbean Seaway has been greatly affected by the appearance and disappearance of various topographic barriers. Thus, the circumtropical current, which originally passed into the western Pacific (late Eocene) was temporarily disrupted during the Eocene-Oligocene transition by the emergence of Gaarlandia, during the late Miocene by the partial emergence of Central America, and was permanently disrupted by the uplift of Central America in the late Pliocene.

Caribbean also produced a more complicated landscape, as several volcanic chains and ridges were present, with a clear tendency toward the reduction of the size of the marine basins (Iturralde-Vinent and MacPhee, 1999:figures 6-8). Nevertheless, those barriers did not prevent marine currents from flowing across the Caribbean Seaway (figure 22.1), as Paleogene and Neogene biotas from the Pacific and Atlantic–Tethys intermingled in the Caribbean and/or migrated between these oceans (Squires, this volume; Dockery and Lozouet, this volume; Duque-Caro, 1990; Collins, 1996; Iturralde-Vinent et al., 1997). This biotic interchange was particularly marked during the periods of highstand, recorded by the frequent occurrence of deep-water environments within Caribbean early Cenozoic stratigraphic sections (Iturralde-Vinent and MacPhee, 1999).

Consequently, during the Paleocene and Eocene, the so-called circum-global equatorial current (Droxler et al., 1998) flowed between the Tethys– Atlantic and the Caribbean–Pacific across the Caribbean Seaway. Figure 22.2 (60–35 Ma) illustrates, in a very general manner, the Caribbean paleogeography during this time interval.

THE LATEST EOCENE TO RECENT EVOLUTION OF THE CARIBBEAN SEAWAY

The transition between the end of the Eocene and the beginning of the Oligocene (Zones P16 to P18 of Berggren et al., 1995) coincides with the Pyrenean phase of tectogenesis (Leonov and Khain, 1987), the effects of which are well represented in the Caribbean Region (MacPhee and Iturralde-Vinent, 1995; Iturralde-Vinent and MacPhee, 1999). During this phase, regional tectonic uplift coincided with a major eustatic sea level drop at about 35 Ma (Miller et al., 1996). As a result, subaerial exposure within the Caribbean Region was probably more extensive then than at any other time in the Cenozoic, including the late Quaternary. Figure 22.2 (35-33 Ma) reflects this fact. However, it is important to compare this map with other "Oligocene" reconstructions which represent this period as one of overall minimum land exposure (e.g., González de Juana et al., 1980; Galloway et al., 1995; Macellari, 1995). The explanation for the difference in treatment lies in the fact that most Oligocene reconstructions depict the paleogeography of the mid- to later Oligocene, by which time the Pyrenean orogenic phase had terminated (Iturralde-Vinent and MacPhee, 1999).

Evidence for Pyrenean uplift can be seen in stratigraphic sections as well as submarine dredge samples, drill cores, and seismic lines recovered from many parts of the Caribbean and surrounding continental borderlands (table 22.1). Many stratigraphic sections consistently lack marine sediments of latest Eocene-early Oligocene age, with this time interval represented instead by hiatuses, red beds, and other kinds of terrestrial deposits (Iturralde-Vinent and MacPhee, 1999). In many sedimentary basins located near former topographic highs, latest Eocene-early Oligocene sediments carry abundant land-derived debris, chiefly very coarse conglomerates (matrix- or clast-supported) and sandstones. This type of sedimentary unit is so common in the Caribbean realm that Iturralde-Vinent and MacPhee (1999) dubbed it the "Eocene-Oligocene transition conglomerate event". In places far removed from terrestrial sediment sources, such as the Colombian, Venezuelan, and Yucatan basins, deep-marine deposition was condensed (Edgar et al., 1973; Sigurdsson et al., 1997).

The Aves Ridge is considered to have constituted the southeastern continuation of the Greater Antilles Ridge in latest Eocene-early Oligocene time. Iturralde-Vinent and MacPhee (1999) argue that exposure of the ridge crest created, for a short time around 33-35 Ma, a series of large, closely-spaced islands or possibly a continuous peninsula (as part of Gaarlandia) stretching from northern South America to the Puerto Rico–Virgin Islands to Central Cuba. It is noteworthy that during the Eocene-Oligocene transition, western Cuba was uplifted but separated by the Havana-Matanzas deep-water channel from central Cuba (actually from Garrlandia), and by the Strait of Yucatan from the Yucatan Peninsula (figure 22.2; Iturralde-Vinent et al., 1996). Some positive topographic features characteristic of the present Caribbean seafloor did not exist in Eocene-Oligocene times (and are therefore not depicted in figure 22.2). These are the Cayman Trench, Mona Channel, Anegada Trough, and so on (see discussion by MacPhee and Iturralde-Vinent, 1994, 1995).

During the latest Eocene (37 Ma), after the collision of the Antillean arc with the Bahamas, general uplift started all along the axis of the present day Greater Antilles (Iturralde-Vinent and MacPhee, 1999). Short after this event, during the Eocene-Oligocene transition (about 35-33 Ma), a pronounced topographic barrier (Gaarlandia) reduced to a minimum water exchange between the Atlantic and the Caribbean–Pacific, and a special pattern of ocean-water circulation similar to present day was established for a short time (figure 22.2), probably con-

Table 22.1. Stratigraphic Indicators for Land, Shallow-Marine, and Deep-Marine Environments During the Latest Eocene-Early Oligocene (35-33 Ma, Zones P16-18) Interval Within the Caribbean Realm and Its Surroundings

Location	*Land indicators* (hiatuses, red beds, alluvia, paleosls, terrestrial plant and animal fossils, nearshore conglomerates)	*Shallow marine and costal indicators* (lagoonal sediments, carbonate shelf, etc.)	*Deep marine indicators* (Globigerine ooze, etc.)
Florida Peninsula	X	X	-
Bahamas Platform	X	X	X
Southern Mexico	X	X	X
Yucatan Peninsula	X	X	X
Northern Central America	X	-	-
Southern Central America	-	X	X
Northeastern South America	X	X	X
Northeastern South America shelf and islands	X	X	-
Nicaragua Rise	X	X	-
Cayman Ridge	X	X	-
Beata Ridge	-	X	-
Aves Ridge	X	X	-
Cuba	X	X	X
Jamaica	X	X	X
Hipaniola	X	X	X
Puerto Rico-Virgin Islands	X	X	-
Lesser Antilles	X	X	?
Yucatan Basin and Cayman Rise	-	?	X
Colombia and Venezuela Basins	-	-	X
Grenada Basin	-	X	X

Note: Some present-day geographic areas are not listed because they did not exist at this time interval (Cayman Trench, Windward Passage, Mona Passage, Anegada Trough, etc.). Compiled from Iturralde-Vinent and MacPhee, 1999, table 2 and appendix 1.

tributing to the climatic change recorded at the Eocene-Oligocene transition (Prothero, 1994).

Since the early Oligocene, Gaarlandia progressively drowned. Inundation of terrestrial environments began as early as Zone P19 (Berggren et al., 1995) and continued into Zone P22, as it is indicated by the widespread occurrence of transgressive Oligocene marine sediments in the Caribbean Region (Iturralde-Vinent and MacPhee, 1999: table 3, appendix 1). Marine communication between the Atlantic and Pacific was re-established, and the circum-tropical current warmed up the water temperature within the Caribbean Seaway and the equatorial Pacific (Droxler et al., 1998). This scenario lasted until the middle Miocene (figure 22.2: 33-9 Ma), and by the late Miocene, water circulation across the Caribbean Seaway was restricted again by regional uplift (figure 22.2: 9-7 Ma; Coates and Obando, 1996; Iturralde-Vinent and MacPhee, 1999). This is suggested by a foraminiferal faunal turnover in the Panamanian area, by the abundance of high latitude Californian elements in the Panamanian fauna, and bu the occurrence of South American land mammals in the late Miocene-Pliocene of North America (Gingerich, 1985; Duque-Caro, 1990; Collins, 1996). The Panamanian barrier was drowned at the end of the Miocene (figure 22.2: 7-3.7 Ma), and soon arose again, first as a shallow submarine ridge, later as a land bridge between North and South American continents (figure 22.2 3.7-0 Ma; Duque-Caro, 1990; Collins, 1996; Coates and Obando, 1996).

CONCLUSIONS

The history of the Caribbean Seaway is very complicated, but can be generally explored with the available data. Figure 22.1 is a summary of the history of marine connections and topographic barriers within the Caribbean area. This scenario can be used to evaluate possible biotic interchange, either by land between the American continents or by water between the Pacific and Tehys/Atlantic oceans. Figure 22.2 illustrates the Paleocene to Recent evolution of the seaway, which is the main focus of the this volume. In general, the evolution of the Caribbean Seaway must have had a strong bearing on the global patterns of ocean circulation and climate change in the past, but this is an issue that remains to be studied in more details.

ACKNOWLEDGMENTS

This is a contribution to UNESCO/IUGS IGCP Project 433 Caribbean Plate Tectonics. The author wish to thanks his colleague Ross MacPhee (American Museum of Natural History), and an anonymous reviewer, for important suggestions to the original typescript and drawings. Attending the GSA Penrose Conference on The Marine Eocene-Oligocene Transition (Olympia, 1999) was possible thanks to a grant from the Cuba Group of SSRC. Field work in the Greater Antilles for this paper was partially supported by a NGS Grant to the author. Laboratory work at the Institute for Geophysics of the University of Texas at Austin (UTIG) was possible thanks to grants from the Institute for Latin American Studies (UTEXAS) and UTIG.

REFERENCES

Aberhan, M., 2001, Bivalve paleogeography and the Hispa-nic corridor: time of opening and effectiveness of a proto-Atlantic seaway: Paleogeography, Paleoclimatology, Paleoecology, v. 165 p. 375-394.

Aleman, A., and Ramos, V. A., 2000, Northern Andes, *in* Cordani, U. C., Milani, E. J., Thomaz Filho, A., and Campos, D. A., eds., Tectonic evolution of South America: 31 International Geological Congress, p. 453-480.

Barron, E. J., and Peterson, W. H., 1989. Model simulation of the Cretaceous Ocean circulation: Science, v. 244, p. 684-686.

Bartok, P. E., Renz, O., and Westermann, G. E. G., 1985, The Siquisique ophiolites, Northern Lara State, Venezuela: A discussion on their Middle Jurassic ammonites and tectonic implications: Geological Society of America Bulletin, v. 96, p. 1050-1055.

Berggren, W. A., and Hollister, C. D., 1974, Paleogeography, paleobiogeography and the history of circulation in the Atlantic Ocean, *in* Hay, W. W., ed., Studies in paleoceanography: SEPM Special Publication, v. 20 p. 126-186.

Berggren, W. A., Kent, D. V., Swisher, C. C., III, and Aubry, M.-P., 1995, A revised Cenozoic geochronology and chronostratigraphy: SEPM Special Publication, v. 54, p. 129-212.

Boitrón-Sánchez, B., and López-Tinajero, Y., 1996. Mollusk gastropods in a Lower Cretaceous rudist-bearing formation of Jalisco, west central Mexico: Revistas Mexicana de Ciencias Geológicas, v. 12, no. 2, p. 157-168.

Buffler, R. T., and Hurst, J. M., 1995, Disintegration of the Jurassic-Lower Cretaceous Megabank, Cuba-Florida-Bahamas (abstract): St. Petersburg,

Florida, 1st SEPM Congress on Sedimentary Geology, Congress Program and Abstracts, p. 35.

Coates, A. G., and Obando, J. A., 1996. The geologic evolution of the Central American isthmus. *in* Jackson, J. B. C., Budd, A. F., and Coates, A. G., eds., Evolution and environment in Tropical America: Chicago, University of Chicago Press, p. 21-56.

Collins, L. S., 1996. Environmental changes in the Caribbean shallow waters relative to closing tropical American Seaway, *in* Jackson, J. B. C., Budd, A. F., and Coates, A. G., eds., Evolution and environment in Tropical America: Chicago, Univer-sity of Chicago Press, p. 130-167.

Damborenea, S. E., 2000, Hispanic corridor: its evolution and the biogeography of bivalve molluscs, *in* Hall, R. L., and Smith, P. L., eds., Advances in Jurassic Research 2000: Switzerland, Transtech Publications, v. 6 p. 369-380.

Damborenea, S. E., and Maceñido, M. O., 1979. On the paleogeographical distribution of pectinid genus *Weyla* (Bivalvia, Lower Jurassic): Palaeogeography, Palaeoclimatology, Palaeoecology, v. 27, p. 85-102.

Droxler, A. W., Burke, K., Cunningham, A. D., Hine, A. C., Rosencrantz, E., Duncan, D. S., Hallock, P., and Robinson, E., 1998, Caribbean constraints on circulation between Atlantic and Pacific oceans over the past 40 million years, *in* Crowley, T., and Burke, K., eds., Tectonic boundary conditions for climate reconstruction: Oxford, Oxford University Press, p. 160-191.

Duque-Caro, H., 1990. Major Neogene events in Panamanian South America, *in* Tsuchi, R., ed., Pacific Neogene events, their timing, nature, and interrelationships: Tokyo, Proceedings of the International Geological Congress, p. 101-114,.

Fernández, M., and Iturralde-Vinent, M., 2000, An Oxfordian ichthyosaur (Reptilia) from Viñales, western Cuba: paleobiogeographic significance: Journal of Vertebrate Paleontology, v. 20, no. 1, p. 191-193.

Frakes, L. A., Francis, J. E., and Syktus, J. I., 1992, Climate modes of the Phanerozoic: the history of Earth's climate over the past 600 million years: Cambridge, Cambridge University Press, 274 p,

Galloway, W. E., Bebout, D. G., Fisher, W. L., Dunlap, J. B., Jr., Cabrera-Castro, R., Lugo Rivera, J. E., and Scott, T. M., 1991. Cenozoic, *in* Salvador, A., ed., The Geology of North America, The Gulf of Mexico basin, v. J, pp. 245-324.

González de Juana, C., Iturralde de Arozena, J. M., and Picard Cadillat, X., 1980. Geología de Venezuela y de sus cuencas petrolíferas: Caracas, Ediciones Foninves, 2 vols., 1031 p.

Gasparini, Z., 1978, Consideraciones sobre los Metrorhynchidae (Crocodylia, Mesosuchia): su origen, taxonomía y distribución geográfica: Obra del Centenario del Museo de La Plata, Paleontología, v. 5, p. 1-9.

Gasparini, Z., 1992, Marine reptiles of the circum-Pacific region, *in* Westermann, G. E. G., ed., The Jurassic of the circum-Paciific, world, and regional geology: London, Cambridge University Press, v. 3, p. 361-364.

Gasparini, Z., 1996, Biogeographic evolution of the South American crocodilians, *in* Arratia, G., ed., Contributions of South America to Vertebrate paleontology: Münchner Geowissenschaften Abhandlung A, v. 30, p. 159-184.

Gasparini, Z., and Fernández, M., 1996, Biogeographic affinities of the Jurassic marine reptile fauna of South America, *in* Riccardi, A. C., ed., Advances in Jurassic Research: Switzerland, Transtec Publications, GeoResearch Forum, v. 1-2 p. 443-449.

Gayet, M., Rage, J. C., Sampere, T., and Gagnier, P. Y., 1992. Modalité des exchanges de vertebrés continentaux entre l'Amerique du Nord et l'Amerique du Sud au Cretacé superior et au Paleocene: Bulletin de Société Géologique de France, v. 163, p. 781-791.

Gingerich, P. D., 1985. South American mammals in the Paleocene of North America, *in* Stehli, F. G., and Webb, S. D., eds., The great American biotic interchange: New York, Plenum Press, pp. 123-137.

González de Juana, C., Iturralde de Arozena, J. M., and Picard Cadillat, X., 1980, Geología de Venezuela y de sus cuencas petrolífera: Caracas, Ediciones, 2 vol.

Gradstein, F. M., Jansa, L. F., Srivastava, S. P., Williamson, M. A., Bonham Carter, G., and Stam, B., 1990, Aspects of North Atlantic paleo-oceanography, *in* Keen, M. J., and Williams, G. L., eds., Geology of the continental margin of Canada: Geological Survey of Canada, Geology of Canada, v. 2, p. 351-389.

Haczewski, G., 1976, Sedimentological reconnaissance of the San Cayetano Formation: an accumulative continental margin in the Jurassic: Acta Geologica Polonica, v. 26, no. 2, p. 331-353

Hutson, F., Mann, P., and Renne, P., 1999. ^{40}Ar / ^{39}Ar

dating of single muscovite grains in Jurassic siliciclastic rocks (San Cayetano Formation): constraints on the paleoposition of western Cuba: Geology, v. 26, p. 83-86.

Iturralde-Vinent, M., 1994. Cuban geology: a new plate tectonic synthesis: Journal of Petroleum Geology, v. 17, p. 39-70.

Iturralde-Vinent, M., 1998, Sinopsis de la constitución geológica de Cuba, *in* Melgarejo, J. C., and Proenza, J. A., eds., Geología y Metalogénia de Cuba: una introducción: Acta Geológica Hispánica, v. 33, no. 1-4, p. 9-56.

Iturralde-Vinent, M., and Norell, M., 1996, Synoptic catalog of Late Jurassic marine reptiles from Cuba: American Museum Novitates, no. 3164, p. 1-17.

Iturralde-Vinent, M., Hubbell, G., and Rojas, R., 1996, Catalog of Cuban fossil Elasmobranchii (Paleocene-Pliocene) and paleogeographic implications of their lower to middle Miocene occurrence: Journal of the Geological Society of Jamaica, v. 31, p. 7-22.

Iturralde-Vinent, M. A., and MacPhee, R. D. E., 1999, Paleogeography of the Caribbean region: implications for the Cenozoic biogeography: Bulletin of the American Museum of Natural History, v. 238, p. 1-95.

Jones, E. J. W., Cande, S. C., and Spathopoulos, F., 1995, Evolution of a major oceanographic pathway: the equatorial Atlantic, *in* Scrutton, R. A., Stoker, M. S., Shimmield, G. B., and Tudhope, A. W., eds., The tectonics, sedimentation and paleooceanography of the North Atlantic region: Geological Society of London Special Publication, v. 9 p. 199-213.

Kerr, A. C., Iturralde-Vinent, M., Saunders, A. D., Babbs, T. L., and Tarney, J., 1999, A geochemical reconnaissance of Cuban Cretaceous volcanic rocks: implications for plate tectonic models of the Caribbean: Geological Society of America Bulletin, v. 111, p. 1581-1599.

Khudoley, K., and Meyerhoff, A., 1971, Paleogeography and geological history of Greater Antilles: Geological Society of America Memoir, v. 129, p. 1-199

Klitgord, K. D., Hutchinson, D. R., and Schouten, H., 1988, U.S. Atlantic continental margin; structural and tectonic framework, *in* Sheridan, R. E., and Grow, J. A., eds., The Geology of North America, v. 1-2, The Atlantic continental margin, Geological Society of America, p. 19-55.

Lawver, L. A., Coffin, M. F., Dalziel, I. W. D., Gahagan, L. M., and Schmitz, R. M., 1999, Plates 1999, Atlas of paleogeographic reconstructions: Austin, Texas, University of Texas Institute for Geophysics, PLATES Progress Report No. 35-0799, 84 p.

Leonov, Yu. G., and Khain, V. E., 1987, Problems of a systematic approach to the study and correlation of tectonic phases, *in* Leonov, Yu. G., and Khain, V.E., eds., Global correlation of tectonic movements: New York, John Wiley, pp. 1-12.

Macellari, C. E., 1995.Cenozoic sedimentation and tectonics of the southwestern Caribbean pull-apart basin, Venezuela and Colombia, *in* Tankard, A. J. R., Suarez Soruco, R., and Welsink, H. J., eds., Petroleum basins of South America: American Association of Petroleum Geologists Memoir, v. 62, p. 757-780.

MacPhee, R. D. E., and Iturralde-Vinent, M. A., 1994, First Tertiary land mammal from Greater Antilles: an early Miocene sloth (Xenarthra, Megalonychidae) from Cuba: American Museum Novitates, no. 3094, p. 1-13.

MacPhee, R. D. E., and Iturralde-Vinent, M. A., 1995, Origin of the Greater Antillean land mammal fauna, 1: New Tertiary fossils from Cuba and Puerto Rico: American Museum Novitates, no. 3141, p. 1-31.

Marton, G., and Buffler, R. T., 1994, Jurassic reconstruction of the Gulf of Mexico basin: International Geology Review, v. 36, p. 545-586.

Miller, K. G., Mountain, G. S., et al., 1996, Drilling and dating New Jersey Oligocene-Miocene sequences: ice volume, global sea level, Exxon record: Science, v. 271, p. 1092-1095.

Parrish, J. T., 1992, Climatology and oceanology, *in* Westermann, G. E. G., ed., The Jurassic of the circum-Pacific, world, and regional geology: London, Cambridge University Press, p. 365-379.

Pindell, J., 1994, Evolution of the Gulf of Mexico and the Caribbean, *in* Donovan, S. K., and Jackson, T. A., eds., Caribbean geology, an introduction: Jamaica, The University of West Indians Publishers Association, p.13-40.

Prothero, D. R., 1994, The Eocene-Oligocene transition: paradise lost: New York, Columbia University Press, 281 p.

Pszczolkowski, A., 1978, Geosynclinal sequences of the Cordillera de Guaniguanico in western Cuba; their lithostratigraphy, facies development and paleogeography: Acta Geologica Polonica, v. 28, no. 1, p. 1-96.

Pszczolkowski, A., 1982, Cretaceous sediments and paleogeography in the western part of the Cuban miogeosyncline: Acta Geologica Polonica, v. 32, no. 1-2, p. 135-161.

Pszczolkowski, A., 1994, Lithostratigraphy of Mesozoic and Paleogene rocks of Sierra del Rosario, western Cuba: Studia Geologica Polonica, v. 105, p. 39-66.

Pszczolkowski, A., and Flores, R., 1986, Fases tectónicas del Cretácico y del Paleógeno en Cuba occidental y central: Bulletin of the Polish Academy of Sciences, Earth Sciences, v. 34, no. 1, p. 95-111.

Ricardii, A. C., 1991, Jurassic and Cretaceous marine connections between the southeast Pacific and Tethys: Palaeogeography, Palaeoclimatology, Palaeoecology, v. 87, p. 155-189.

Rojas, R., Iturralde-Vinent, M., and Skelton, P., 1995, Stratigraphy, composition and age of Cuban rudist-bearing deposits: Revista Mexicana de Ciencias Geológicas, v. 12, no. 12, p. 272-291.

Salvador, A., 1987, Late Triassic-Jurassic paleogeography and origin of the Gulf of Mexico basin: Bulletin of the American Association of Petroleum Geologists, v. 71, no. 4, p. 419-451.

Smith, P. L., 1983, The Pliensbachian ammonite *Dayiceras dayiceroides* and Early Jurassic paleogeography: Canadian Journal of Earth Sciences, v. 20, p. 86-91.

Sohl, N. F., and Kollmann, H. A., 1985, Cretaceous actaeonellid gastropods from the western hemisphere. U.S. Geological Survey Professional Papers, v. 1304, p. 1-104.

CHAPTER 23

The Eocene-Oligocene Transition in the Equatorial Atlantic (ODP Site 925): Paleoproductivity Increase and Positive $\delta^{13}C$ Excursion

Liselotte Diester-Haass and Jim Zachos

ABSTRACT

To decipher the Eocene-Oligocene evolution of paleoproductivity in the western equatorial Atlantic, we have measured the isotopic composition and biogenic sediment properties of a sequence of pelagic sediments from ODP Site 925, Ceara Rise. This includes the stable carbon isotopic ratios of bulk carbonate, the percent coarse fraction, and benthic foraminiferal abundances that were used to compute benthic foraminiferal accumulation rates (BFAR). The carbon isotope record and biostratigraphy provide stratigraphic control while the BFAR, carbonate dissolution, radiolarian accumulation rates constrain changes in paleoproductivity. On the basis of changes in the various indicies, four distinct units are distinguished:

1. Unit 1, the middle Eocene, characterized by low productivity as reflected in low BFAR and low carbonate dissolution. Relatively high radiolarian accumulation rates reflect on higher concentrations of dissolved Si in regional bottom waters.

2. Unit 2, the late Eocene, characterized by BFAR values comparable to those in unit 1 but with lower radiolarian accumulation rates. Carbonate dissolution is decoupled from productivity and appears to be sensitive to terrigenous input and decay of terrestrial organic matter.

3. Unit 3, the earliest Oligocene, characterized by an abrupt excursion in $\delta^{13}C$ and paleoproductivity to values two to three times higher than Eocene. Paleoproductivity remains high for about 400,000 years, then declines to levels higher than in the Eocene, varying rhythmically throughout. The BFAR, carbonate dissolution, and radiolarian accumulation rates covary.

4. Unit 4, upper early Oligocene, is characterized by low paleoproductivity and the complete absence of radiolaria. The two- to threefold increase in paleoproductivity values in the earliest Oligocene is attributed to enhanced upwelling during the period of rapid expansion of Antarctic ice sheets. The simultaneous 0.9‰ increase in $\delta^{13}C$ is due to a substantial increase in organic carbon to inorganic carbon burial rates, which may have contributed to the earliest Oligocene cooling via drawdown of $pCO_{2\ atm}$.

INTRODUCTION

The Eocene-Oligocene transition marks a profound change in global climate and oceanography. A sharp shift in marine oxygen isotope ratios toward heavier values reflects a global cooling by 3-4° and the growth of an Antarctic ice sheet with a mass at least half that of the present day ice sheet (Zachos et al., 1994; Miller et al., 1987; Lear et al., 2000). Marine faunal and floral assemblages indicate a global decrease in diversity in association with the climate shift (Prothero, 1994; Prothero and Berggren, 1992), and in the Southern Ocean, marine biological productivity increases by a factor of >2–3 times (Diester-Haass and Zahn, 1996, 2001; Diester-Haass, 1996; Salamy and Zachos, 1999). The issue of how this major climatic event is temporally linked to the major plate reorganizations in the Southern Hemisphere, such as the opening of the Drake Passage (at 37 Ma, Diester-Haass and Zahn, 1996) and Tasman seaway at about 32.5 Ma (Lawver and Gahagan, 1997, see here for

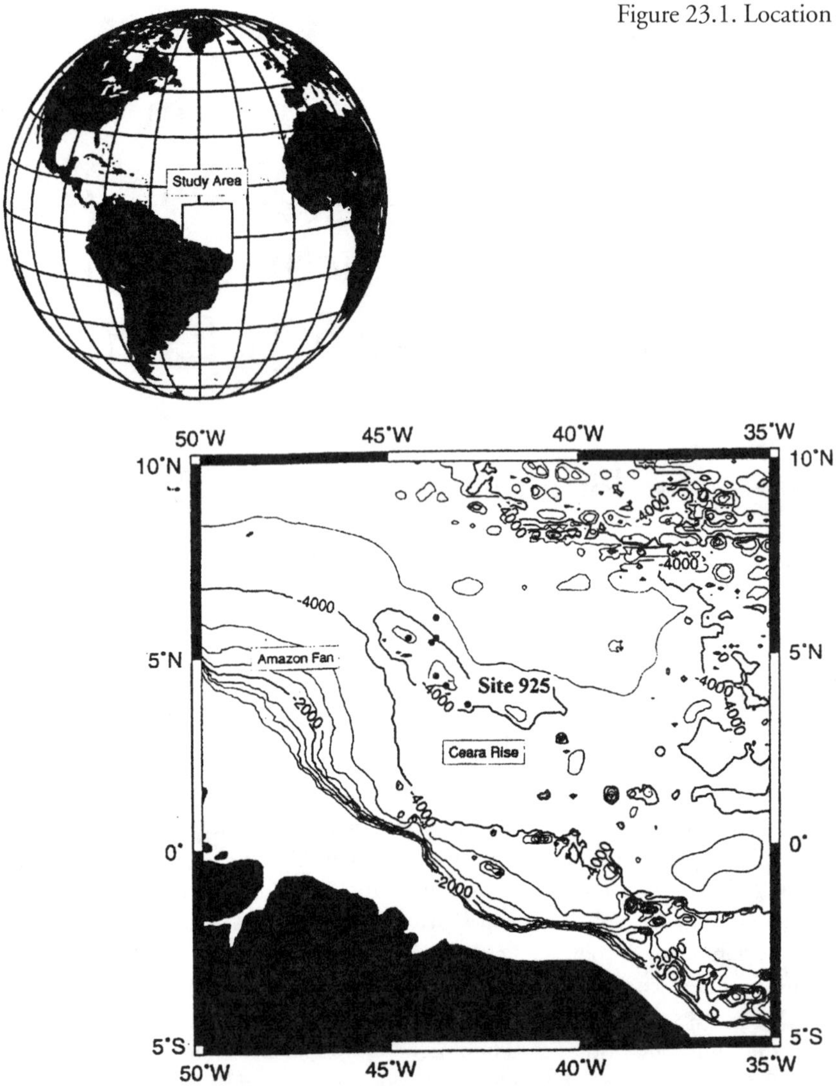

Figure 23.1. Location of Site 925 on Ceara Rise.

discussion of earlier results; Diester-Haass and Zahn, 2001) or the closure of the Indian-Tethyan seaway (Lawver and Gahagan, 1997) remains highly controversial. Other forcing mechanisms may have also played a role, including changes in marine surface and deep circulation, vertical mixing, albedo changes related to ice expansion, reductions in atmospheric methane (Sloan et al., 1992) and CO_2 content (Arthur et al., 1991; Freeman and Hayes, 1992; Sloan and Rea, 1995). In particular, pCO_2 levels were on the decline due to the effects of increased silica weathering (Ruddiman et al., 1997), and reduced seafloor spreading (Berner, 1990). Recent investigations, however, question the relation between temperature and ice volume on the one hand and atmospheric pCO_2 on the other hand (Pagani et al., 1999; Flower, 1999; Pearsson and Palmer, 1999; 2000). On short-time scales greater marine biological production and sequestering of organic carbon in marine sediments also contributed to lowering pCO_2 (Diester-Haass and Zahn, 1996). The focus of this paper is primarily on the marine biological productivity response, and possible effects on atmospheric CO_2 concentrations. More specifically, we are interested in how biogenic sediment accumulation rates in the western equatorial Atlantic (ODP Site 925) changed during the Eocene-Oligocene transition. Our goal is to determine whether the increased fertility observed in the Southern Ocean at the Eocene-Oligocene transition was a local phenomenon associated with regional effects of ice sheet buildup and circulation change, or whether productivity increased at low latitudes as

Table 23.1. Biostratigraphy of Site 925 from Pearson and Chaisson (1997), Based on Planktonic Foraminifera

mbsf	Age Ma	LSR (cm/ky)
652.20	30.3	3.85
663.75	30.6	3.40
728.30	32.5	3.87
776.65	33.75	1.03
779.22	34.0	1.28
840.77	38.8	2.87
880.97	40.2	

well. If so, the related effects on regional C_{org}/C_{carb} burial rates might have contributed to the postulated pCO_2 lowering at the Eocene-Oligocene transition.

MATERIALS AND METHODS

ODP Site 925, Hole A, is located on Ceara Rise in the equatorial Atlantic (4°12.249'N, 43°29.334'W), at 3042 m water depth (figure 23.1) (Curry et al., 1995). It is the shallowest of 5 sites drilled on Ceara Rise during ODP Leg 154. The Eocene-Oligocene transition occurs in sediments consisting of nannofossil chalks (above 700 mbsf) and limestones (below 700 mbsf). Calcite recrystallization is pervasive ranging from spar-filled foraminifers to calcite cement. (Curry et al., 1995). Samples were taken at 50–150-cm intervals, which given linear sedimentation rates provide a mean resolution of 40,000 years per sample. The paleodepth at 35 Ma is estimated to be roughly 2,500 m (recommended by K.Hinz, pers. commun., 1998). The sediments were freeze-dried, weighed, soaked in 10% H_2O_2 for several hours and washed over 40 and 63 µm sieves, dried and sieved into the following subfractions: 63–125,125–250, 250–500, 500–1000, >1000 µm. About 25 biogenic, terrigenous, and authigenic components were distinguished in each fraction, and percent composition of total sand (>63 µm) or coarse fraction (CF) and 40–63 µm fraction calculated. Using linear sedimentation rates and bulk density data (DBD)(Curry et al., 1995) accumulation rates of components (AR comp.) were calculated according to

AR comp. =
LSR* DBD * %comp./ 100*1000 (mg/cm^2*ky),

where LSR is linear sedimentation rate constrained by biostratigraphic datums as identified by Pearson and Chaisson (1997) (table 23.1) and calibrated to the time scale of Berggren et al. (1995).

The Site 925 stable isotope stratigraphy is based on values of whole-rock samples. Roughly 300–500 µg of sample was powdered and roasted under vacuum in preparation for analysis. The powdered sample was then reacted in an Autocarb common acid bath at 90°C. Gas analysis was carried out on an OPTIMA gas source mass spectrometer. External analytical precision based on replicate analyses of two standards NBS-19 and Carrera Marble, the lab working standard, was better than ±0.1‰ for both O and C isotopes. Unless otherwise stated, all isotope values discussed in this paper are reported in the per mil notation ‰ relative to the PDB standard.

RESULTS

Linear sedimentation rates increase from 1.4 up to 3.8 cm/1,000 years at the Eocene/ Oligocene boundary (figure 23.2A). Dry bulk density (DBD) is rather constant, but drops at 690 mbsf, in the same interval where per cent coarse fraction drops (figure 23.2A). This coincides with the chalk to limestone transition at 690 mbsf. Below this level foraminifer tests are usually filled with secondary calcite. Per cent CF varies between 2 and 25% (figure 23.2B). The CF consists of >90% planktonic foraminifers (in 800–820 mbsf, where radiolaria occur [figure 23.3B], planktonic foraminifer percentage decreases to 65%) and 0.4-4% benthic foraminifers. Ostracods, echinoderms, fish debris, and terrigenous particles are minor constituents (<1% of the CF). In several intervals pyrite represents up to 2% of the CF.

The accumulation rates of planktonic foraminifers increase from <500 to 1700 mg/cm^2*ky in the earliest Oligocene, and decrease at 690 mbsf, where % CF also decreases (figure 23.4A). Benthic foraminifers show a similar trend. Absolute values increase from <10 in the middle / late Eocene to up to 25 mg/ cm^2*ky in the earliest Oligocene (figure 23.4A). The mean signal amplitude increases from the Eocene to the Oligocene. Radiolaria are present in the middle late Eocene and lowermost Oligocene with accumulation rates as high as 40 mg/ cm^2*ky (figure 23.4B). They are absent between 790 and 770 mbsf and above 700 mbsf. Accumulation rates of echinoderms and ostracods, which are absent in some intervals, are relatively low (figure 23.5A, B). Nonetheless, echinoderms show a marked increase in the earliest Oligocene. Accumulation rates of fish debris (figure 23.5C) are low in the Eocene and increase by a factor of 4 in the earliest Oligocene before declining above 720 mbsf.

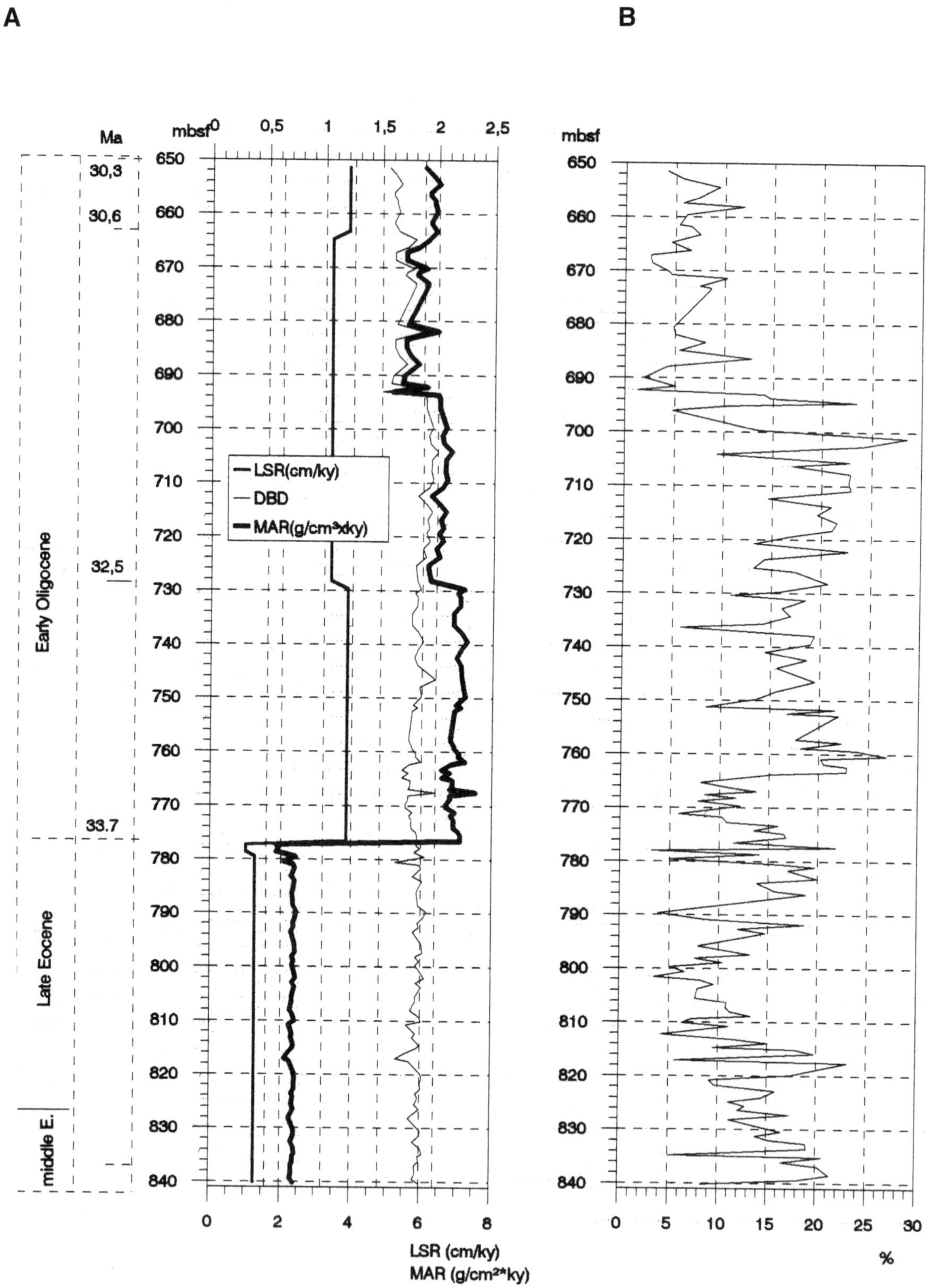

Figure 23.2. (A) Linear sedimentation rates, dry bulk density, and mass accumulation rates. See text for stratigraphy. (B) Percent sand fraction of total sediment.

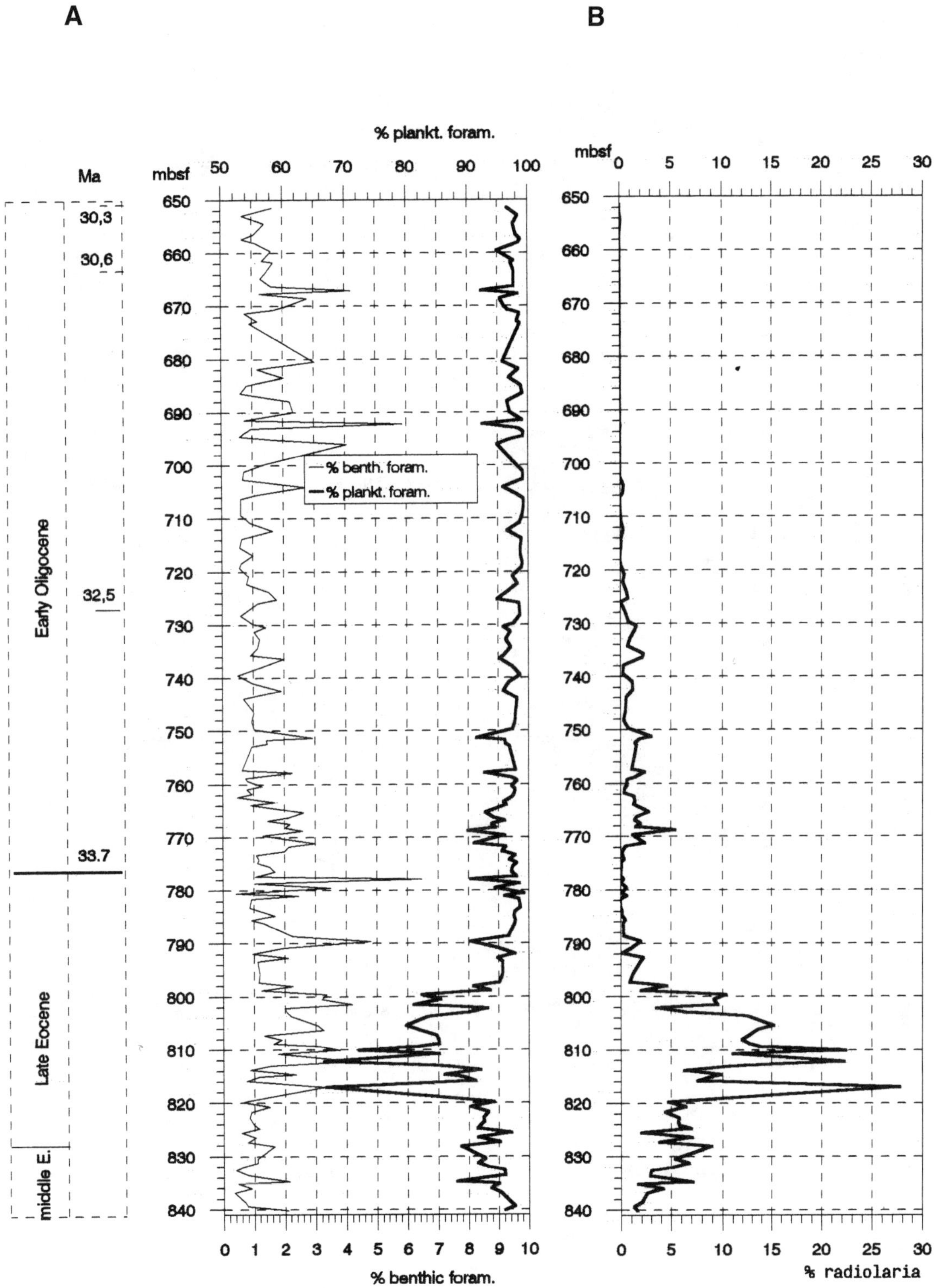

Figure 23.3. (A) Percent benthic and percent planktonic foraminifers of sand fraction. (B) Percent radiolaria of sand fraction.

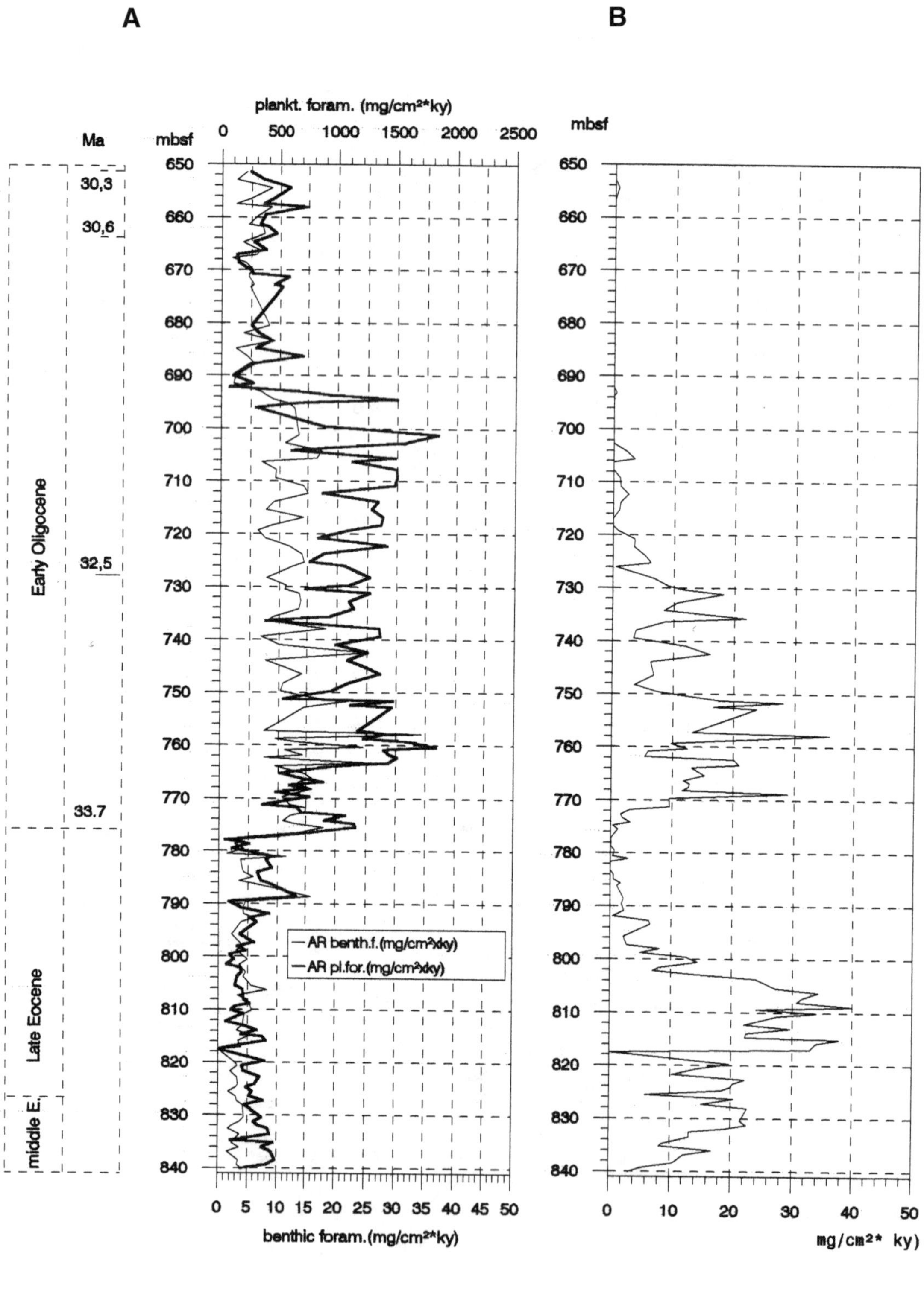

Figure 23.4. (A) Accumulation rates of benthic (>63 μm sized) and planktonic foraminifers in mg/cm² *ky. (B) Accumulation rates of radiolaria.

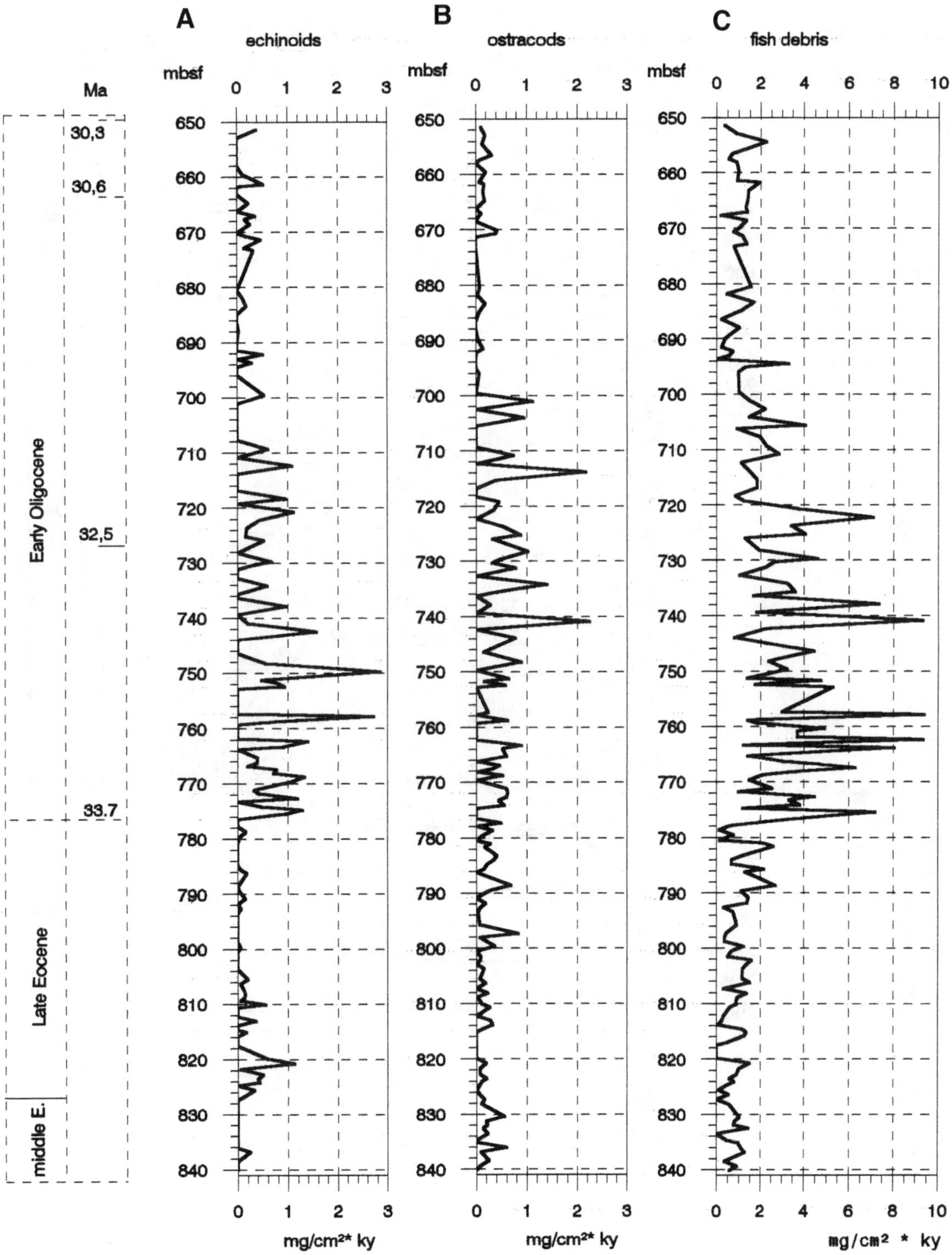

Figure 23.5. (A) Accumulation rates of echinoids. (B) Accumulation rates of ostracods. (C) Accumulation rates of fish debris.

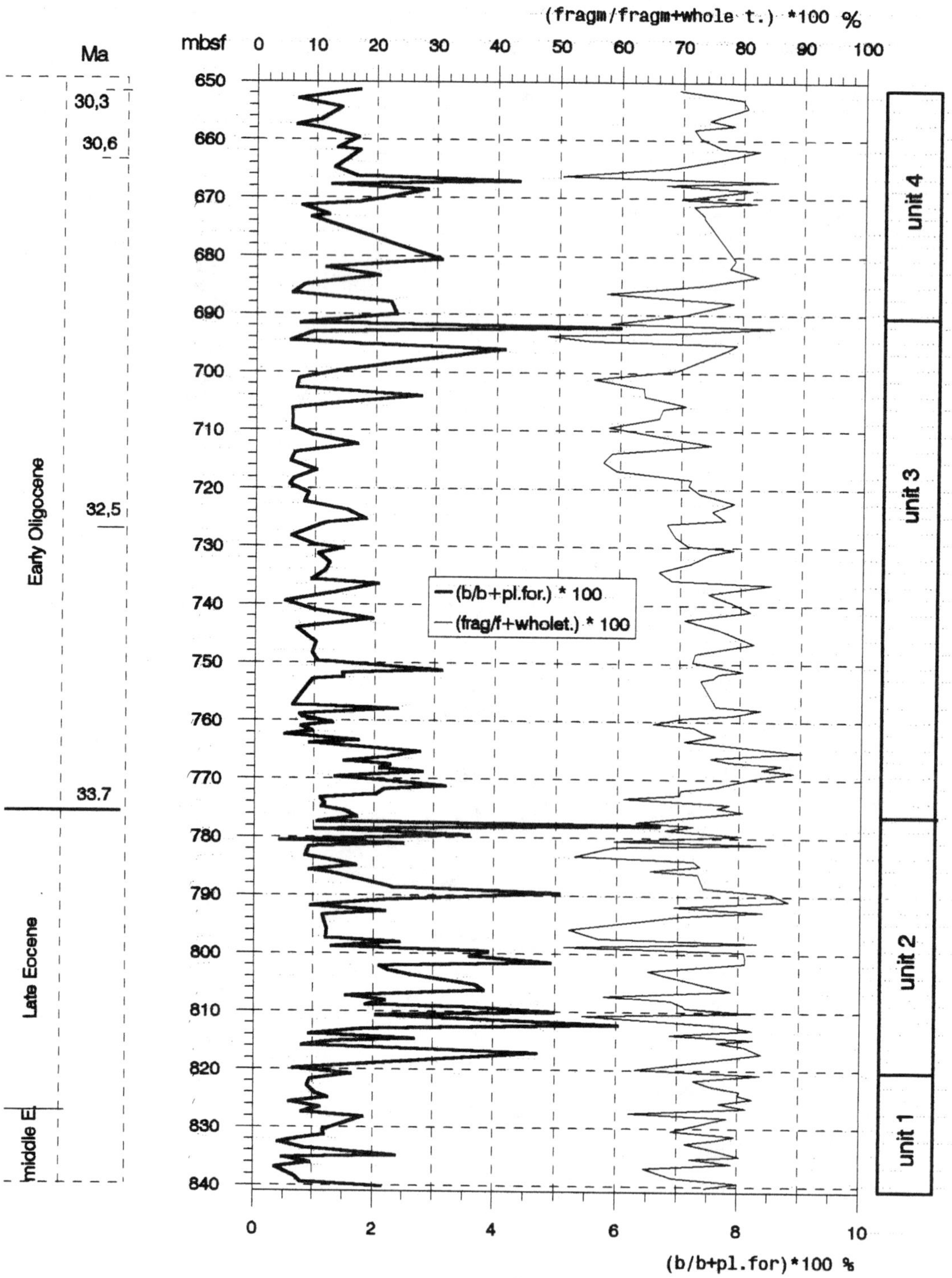

Figure 23.6. Carbonate dissolution established by means of benthic /planktonic foraminiferal ratios and fragmentation of planktonic foraminifers (for calculation see text) unit 1-4: see discussion

Carbonate dissolution indices were established by computing the benthic/planktonic foraminiferal ratio (% benthics/% benthics + % planktonics)*100) and percent fragmentation of planktonic foraminifers (% fragments/% fragments + % whole tests)*100) (figure 23.6). In general, the carbonate dissolution indices show high variability, between 0.5% and 6% benthics of the benthic/planktonic ratio and 50–90% fragmentation (figure 23.6). Benthic/planktonic foraminiferal ratios increase from 0.5–2 up to 6% at 820 mbsf, in the early upper Eocene, and drop again at about 760 mbsf to values <2%. As expected, there is an inverse correlation between % CF and plankton foraminiferal accumulation rates and carbonate dissolution. The increased dissolution tends to lower the % CF, and hence, the plankton foraminiferal accumulation rates. Similarly, in the Oligocene, ostracods tend to be rare or absent in intervals marked by strong carbonate dissolution.

Paleoproductivity rates can be qualitatively estimated by means of benthic foraminiferal accumulation rates (BFAR, calculated as numbers per cm^2 and ky) (Herguera and Berger, 1991; Herguera, 2000; Nees, 1997; Yasuda, 1997). We calculated BFAR as numbers of benthic foraminifers >125 µm/ cm^2*ky. (Note that Herguera and Berger [1991], and Yasuda [1997] used the >150 µm size fraction) In general, the BFAR record (figure 23.7) shows a trend similar to the accumulation of >63 µm sized foraminifers (figure 23.4A) with a minimum in values at the middle/late Eocene transition, a gradual increase in the late Eocene, and a rapid excursion in the lowermost Oligocene, where values increase by a factor of about 5 followed by declines at 758 and 700 mbsf. Bulk carbonate $\delta^{13}C$ values vary between 0.6 and 2‰ (figure 23.7). Abrupt shifts to higher values occur at the middle/late Eocene boundary, at the Eocene/Oligocene transition and in the upper early Oligocene. The early Oligocene $\delta^{13}C$ peak corresponds with a paleoproductivity increase, whereas elsewhere in the Oligocene the opposite is the norm.

The coarse fraction component of 925 sediments contains two minor nonbiogenic components: mica (figure 23.8A) and pyrite (figure 23.8B). Mica is present primarily in the uppermost middle Eocene and upper Eocene. Pyrite is rare in the middle Eocene, and increases in the upper Eocene where carbonate dissolution is high. In the Oligocene each pyrite peak corresponds to a maximum in carbonate dissolution. Magnetic susceptibility data (Curry et al., 1995) (figure 23.9A) show a trend that is comparable to that of benthic/planktonic foraminiferal ratios (figure 23.6): low values in the upper middle Eocene, an increase in the late Eocene–earliest Oligocene, a decrease above 760 mbsf. Per cent $CaCO_3$ (Curry et al., 1995) shows an opposite trend with lowest values of about 45–60% $CaCO_3$ where magnetic susceptibility is highest and highest percent $CaCO_3$ values of 60–85% in the uppermost part of the investigated sequence (figure 23.9B).

DISCUSSION

Proxies of Productivity

Quantifying past variations in marine productivity continues to be one of the more difficult challenges in paleoceanography. Many methods have been developed to reconstruct production rates, or some aspect of production (Berger and Herguera, 1992; Berger et al., 1994; and numerous references herein). Export production can be semi-quantitatively evaluated by means of benthic foraminiferal accumulation rates. The BFAR is controlled by several factors related to production: the main factor is supply of particulate organic matter from the photic zone (Herguera and Berger, 1991; Herguera, 2000). This is a function of several factors, including water depth, which remains relatively constant over this period, primary production, export production, as well as the rate of oxidation of organic matter in the water column, which is temperature dependent (Thomas et al., 2000). Other factors that control accumulation rates of benthic foraminifers are in situ dissolution, which is also minor above the CCD, and evolutionary adaptation to changing climate (Corliss et al., 1984; Thomas, 1992; Thomas et al., 1995). Analyses of Miocene-Pliocene benthic foraminiferal assemblages on Ceara Rise (Yasuda, 1997) as well as modern specimens (Thomas et al., 1995) have shown that phytodetritus species abundances vary in direct response to the availability of fresh organic matter. In most of the ocean, the flux of organic matter to the seafloor is highest during short but concentrated phytoplankton blooms. As the organic fluff arrives at the seafloor, it is rapidly populated by benthic foraminifers. The same processes likely operated in the Eocene-Oligocene period as well (Thomas and Gooday, 1996). Furthermore, Yasuda (1997) has shown that the accumulation rate of phytodetritus species has a high correlation coefficient with the number of total individuals, although phytodetritus species account for only one-third of the total assemblage. These and other data strongly imply that accumulation rates of benthic foraminifers are controlled mainly by food supply, and thus export production. We assume that

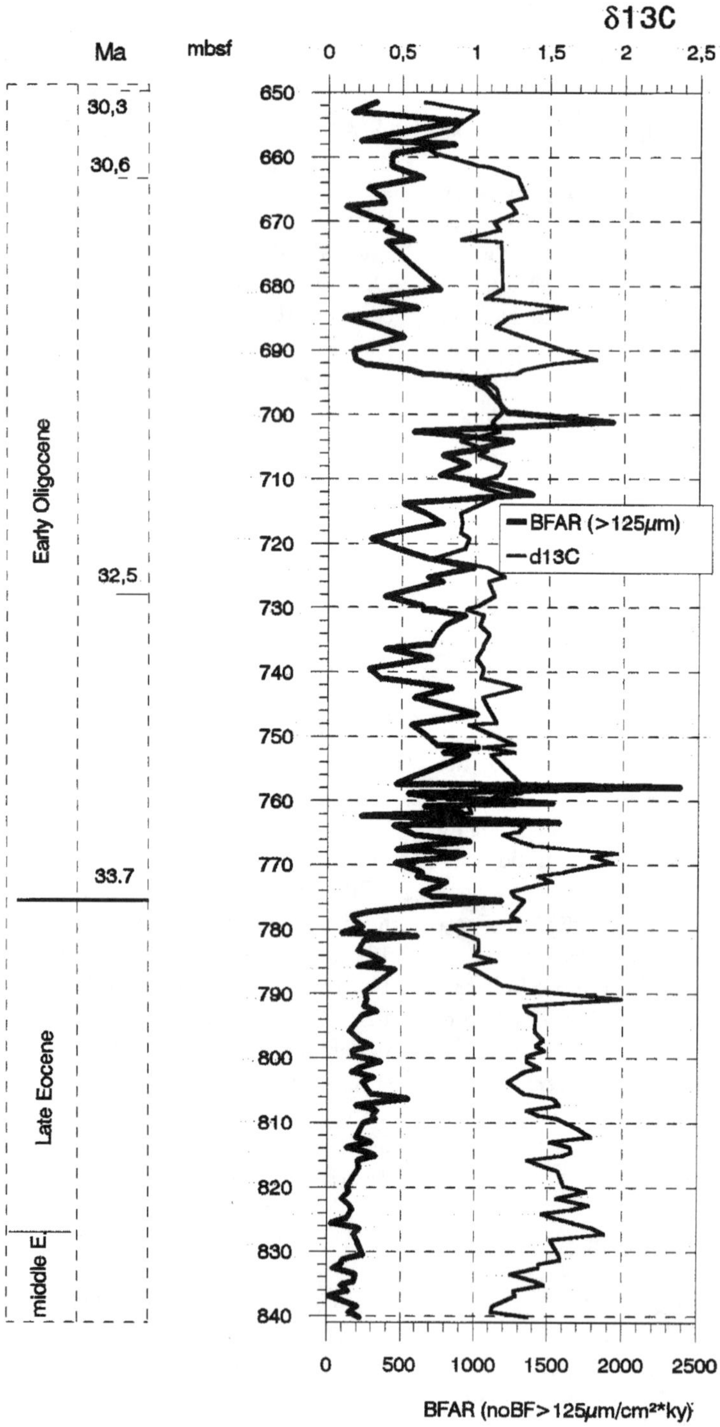

Figure 23.7. Benthic foraminiferal accumulation rates (BFAR): calculated as numbers of benthic foraminifers >125 µm per cm^2 and ky and $\delta^{13}C$.

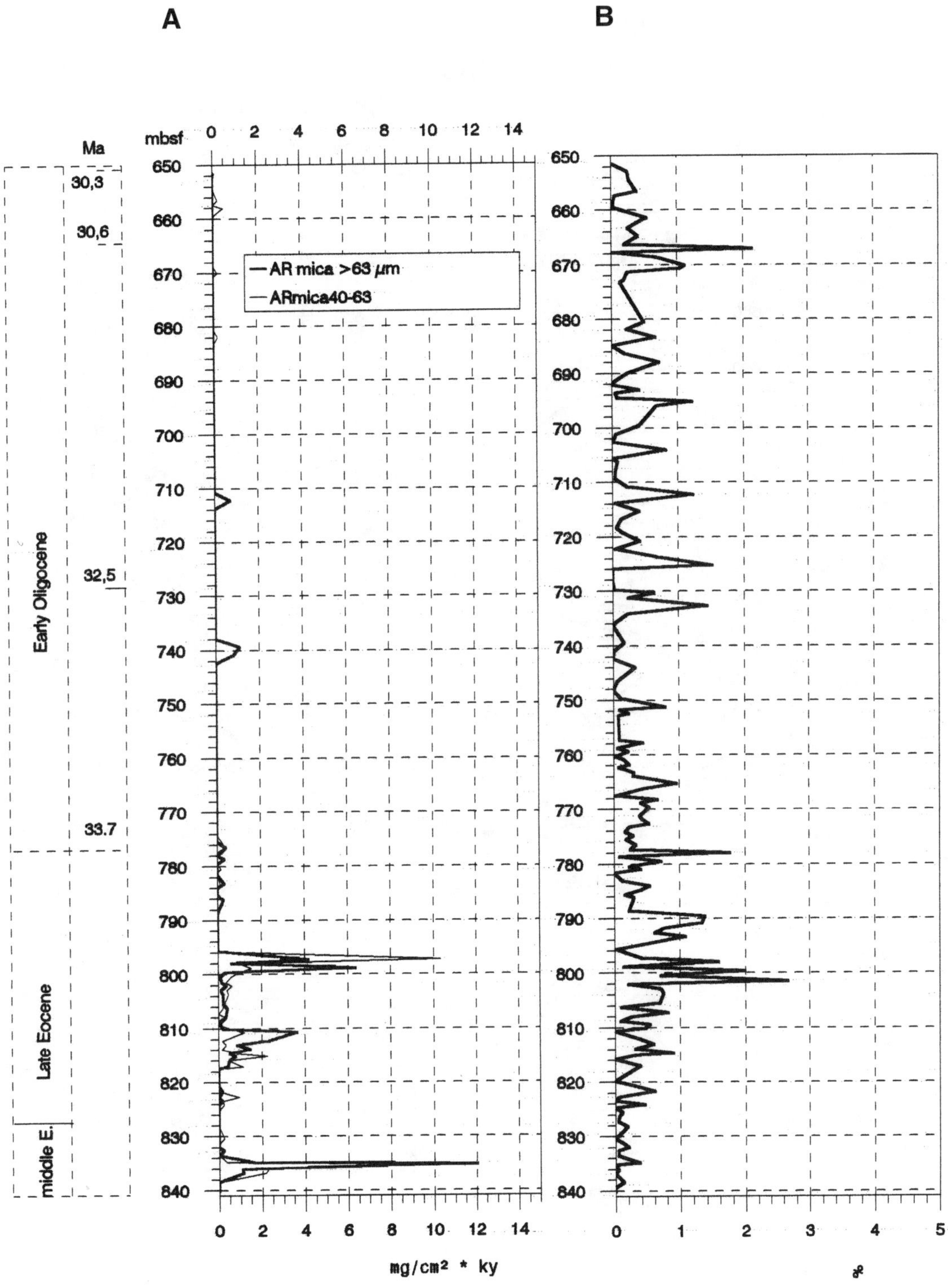

Figure 23.8. (A) Accumulation rates of mica in >63 and 40-63 μm fraction. (B) Percent pyrite in sand fraction.

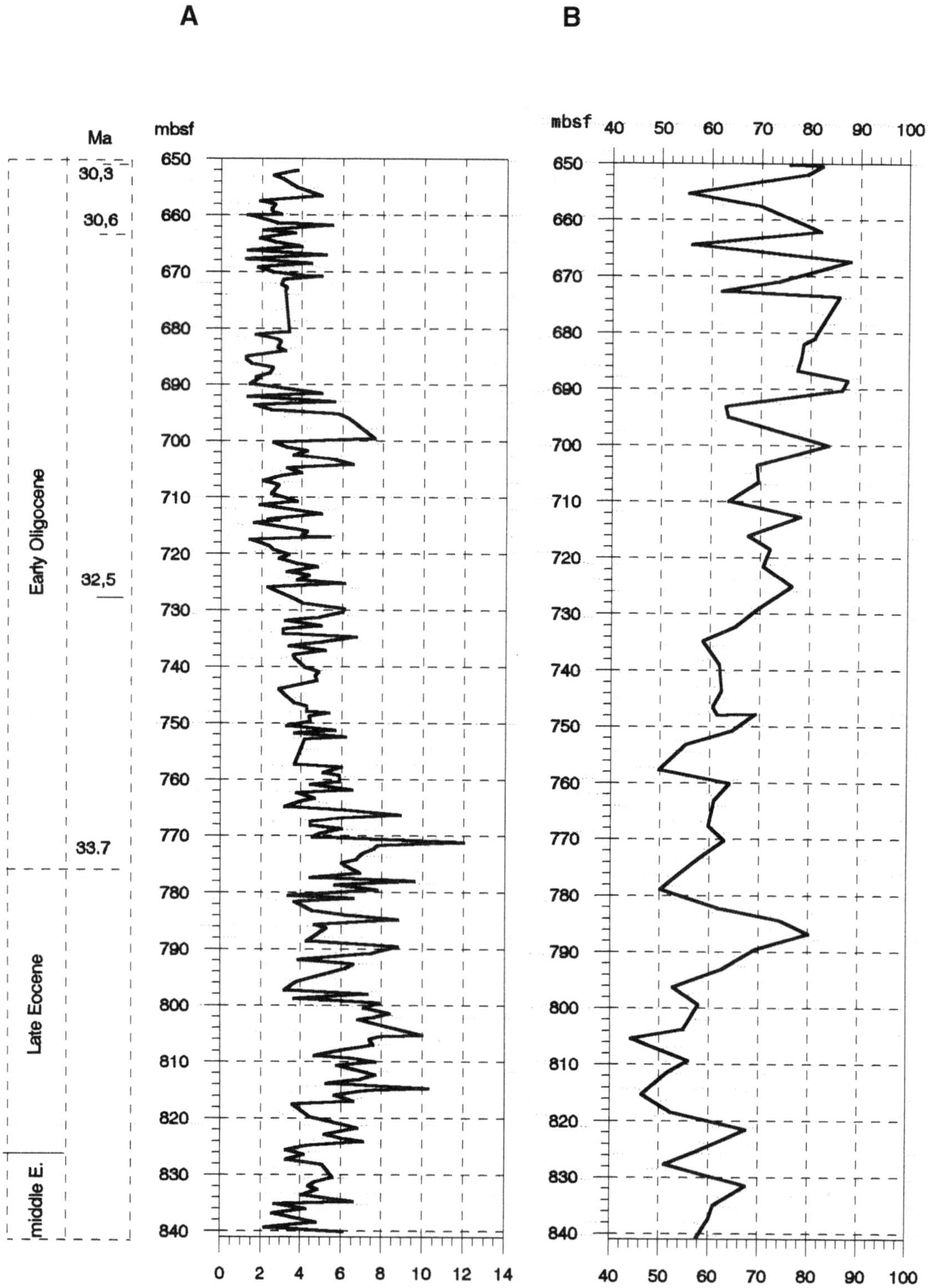

Figure 23.9. (A) Magnetic susceptibility. (B) Percent $CaCO_3$ of total sediment from Curry et al. (1995).

the food flux–foram accumulation relationship holds for the Eocene-Oligocene period. Thus the increased abundance of benthic foraminifers (BFAR) in the earliest Oligocene probably reflects on more intense seasonal production of organic matter in the western equatorial Atlantic. One might assume that the increased export production in the earliest Oligocene should be reflected in higher organic carbon contents of the sediments. This is not the case: in these highly oxidized Paleogene sediments, the organic carbon is gradually consumed after burial. As with most pelagic sediments, organic matter content is low throughout (Curry et al., 1995)

The planktonic foraminifer dissolution index may also vary as a function of paleoproductivity. In sediments deposited at relatively shallow depths (i.e., above the lysocline) changes in the degree of planktonic foraminiferal dissolution are driven mainly by changes in the $C_{org}/CaCO_3$ rain ratio. Because this rain ratio generally tends to increase with higher rates of primary production (Archer and Maier-Reimer, 1994), dissolution and the fragmentation of planktonic foraminifers as well as the ratio of benthics to planktonics also tend to increase (Berger, 1976; Diester-Haass, 1995; 1996). In general, our Site 925 BFAR and PF dissolution data exhibit a positive relationship (figures 23.6 and 23.7) suggesting that both reflect on changes in organic carbon fluxes. Besides BFAR and plankton foraminiferal dissolution index, opal accumulation may also serve as a qualitative proxy of productivity (Calvert, 1974; Heath, 1974; Berger, 1976).

The accumulation rates of radiolaria (>63 μm; no diatoms present) at Site 925 are supposed to covary with bulk opal content, a phenomenon that has been observed in modern sediments (Leinen et al., 1986). At Site 925, the radiolarian accumulation record shows several peaks (figure 23.4B), one of which coincides with the maximum in BFAR in the early Oligocene. Seawater is generally undersaturated with respect to silica, and as a result, opal skeletons are readily dissolved. There are several factors, however, that can either slow the rate of silica dissolution during sinking, or increase burial rate, thereby enhancing the probability of preservation in sediments. Such factors include temperature (lower temperatures enhance opal preservation) and trace element chemistry (Nelson et al., 1995). Still, the most important factor appears to be the rate of production. A high rate of silica supply in a short time span enhances the probability of preservation. For example, during a diatom bloom, the formation of aggregates or digestion of opal organisms by copepods or other metazoans produces large particles (fecal pellets) that rapidly transport silica to the seafloor (Nelson et al., 1995). In short, opal-rich sediments tend to occur where diatom production rates and fluxes are high.

On geologic time scales, additional factors capable of influencing the supply of opal to sediments must also be considered. For one, Fe supply is important in terms of maintaining high diatom production rates where nutrients are abundant (for discussion, see Takeda, 1998). Most important, the flux of SiO_2 from continental weathering to the ocean largely controls the rate at which opal is buried (Froelich et al., 1992). For example, Schneider et al. (1997) show that changing climatic conditions during the late Quaternary altered the weathering and silica supply from equatorial Africa to the eastern Atlantic in the Congo Basin, essentially decoupling the opal burial signal from the upwelling signal. Similarly, increased volcanism and hydrothermal activity (Owen and Rea, 1985) as well as widespread areas with lateritic weathering during the Paleogene may have increased the flux of dissolved silica to the ocean thereby triggering widespread opal deposition (McGowran, 1989). Furthermore, the cycling of silica between and within ocean basins can lead to a decoupling of the preservation signal of diatoms from productivity (Lange and Berger, 1993).

Still, despite the myriad of factors that contribute to opal production and preservation, large-scale changes in ocean and atmosphere dynamics have the greatest potential to alter patterns of primary production and, hence, opal accumulation. Over the last 100 Ma, the most consequential changes in circulation occurred during the Paleogene. Baldauf and Barron (1990) and Barron and Baldauf (1989) have proposed that these circulation changes were responsible for the observed changes in global opal accumulation patterns. In the early Paleogene, polar regions were warmer and latitudinal temperature gradients were minimal, resulting in sluggish surface circulation. Under these conditions, advection of nutrient-enriched subsurface waters was geographically widespread,which resulted in diffuse production. In the early Oligocene, as the poles cooled, the reverse pattern developed. Upwelling and areas of high productivity become regionally focused, a pattern that is consistent with findings of modeling simulations (Heinze and Crowley, 1997). Finally, ignorance of opal diagenesis makes interpretation of opal accumulation rates problematic (Archer et al., 1993).

Summarizing these thoughts, one can conclude

that the opal (diatom and radiolarian) content of sediments, or any proxy alone for that matter, cannot be used to characterize paleoproductivity. Clearly, the application of several proxies simultaneously is required to minimize the uncertainties in reconstructing paleoproductivity (Berger et al., 1994).

EOCENE-OLIGOCENE PALEOPRODUCTIVITY VARIATIONS.

The following characterization of Eocene-Oligocene paleoproductivity in the western equatorial Atlantic is based on a suite of proxy records from Site 925. This includes the % CF, BFAR, fragmentation of planktonic foraminifers, and radiolaria accumulation records. We also consider the bulk $\delta^{13}C$, more in terms of the global C-isotope budget and carbon burial. On the basis of sediment composition, accumulation rates, and inferred variations in productivity, four distinct units are recognized in the Eocene-Oligocene interval at Site 925 (figure 23.6):

1. Unit 1. The middle Eocene (840–820 mbsf): BFAR varies between <100 and 300 (figure 23.7), benthic/planktonic foraminiferal ratios are low, pointing to weak dissolution (figure 23.6). Variations in inferred productivity and dissolution (fragmentation of planktonic foraminifers) are synchronous. Radiolaria are present (figure 23.4B).

2. Unit 2. Late Eocene (820–776 mbsf): strong increase in benthic/planktonic foraminiferal ratios points to an increase in carbonate dissolution; dissolution proxies vary in parallel (figure 23.6), accumulation rates of planktonic foraminifers are lower than in unit 1 (figure 23.4A), radiolarian accumulation rates are high in 815-800 mbsf (figure 23.4B). There is no relation between the very strong carbonate dissolution and BFAR. In the uppermost part of unit 2: radiolaria decrease to nearly zero, BFAR increases.

3. Unit 3. Earliest Oligocene (776–690 mbsf): dissolution is weaker than in unit 2, highest BFAR and high radiolarian accumulation rates (but less than in unit 2). Parallel changes in BFAR, radiolarian accumulation rates, and carbonate dissolution and drop in $\delta^{13}C$ values with increasing productivity.

4. Unit 4. Late early Oligocene (690–652 mbsf): dissolution similar or higher than in unit 3, no opal, BFAR slightly lower than in unit 3.

These four units reflect on a fundamental difference between Eocene and Oligocene sediments. In the Oligocene units, 3 and 4, the paleoproductivity proxies such as dissolution of carbonates, BFAR, opal accumulation rates, and $\delta^{13}C$ all covary, whereas in the Eocene there is a decoupling of these proxies. This suggests that one or more of the four productivity proxies was strongly influenced by factors other than primary production.

In the middle Eocene unit 1, although fragmentation is high, carbonate dissolution appears to be minimal. Benthic/planktonic foraminiferal ratios vary between 0.5 and 2.5%, values typical for depths of 2,500 m (Berger and Diester-Haass, 1988), where dissolution is minimal. The high fragmentation likely reflects postburial diagenetic breakdown of planktonic foraminiferal shells. Because BFAR vary in parallel with carbonate dissolution, we suspect that the subtle variations in calcite dissolution are the result of productivity-driven changes of in situ C_{org} oxidation and CO_2 release (Archer and Maier-Reimer, 1994). Radiolarian accumulation rates, on the other hand, do not show a relation to either dissolution or paleoproductivity variations. It therefore appears that the higher radiolarian abundances are the result of either a higher mean ocean silica content of water (Owen and Rea, 1985; McGowran, 1989; Berger, 1991) that led to a widespread silica deposition (Barron and Baldauf, 1989), or to a better preservation because of more heavily silicified tests (Schmitz, 1987; Barron and Baldauf, 1989). This interpretation is somewhat in conflict with previous ideas on the nature of middle Eocene biosiliceous sediments, or cherts, in the equatorial Atlantic, North Atlantic, and the Pacific that have been attributed to increased surface-water productivity (Baldauf and Barron, 1990). Our data on BFAR and dissolution, at least for this equatorial site indicate a period of low primary production and a decoupling of the opal signal from the export production signal.

In unit 2 (late Eocene), most of the proxies exhibit patterns that are opposite to those observed in unit 1. The most prominent change is in dissolution (~820 mbsf) which increases significantly. Radiolarian accumulation rates are high and tend to covary with BFAR, suggesting common forcing. Carbonate dissolution oscillations increase in amplitude in this unit but appear to be decoupled from cycles in the two productivity proxies. An increase in the accumulation of terrigenous clastic material, mainly mica, in the >40 μm fraction (figure 23.8A) points to a possible late Eocene change in run-off and weathering conditions in South America. Today, and in the Quaternary, much of the terrigeneous material at Ceara Rise derives from the Amazon River basin (Curry et al., 1995). This was probably true in the Eocene as well, although the rate of supply was much lower. Shipboard magnetic susceptibility records reach a maximum in late Eocene unit 2 (e.g., minima in $CaCO_3$) (figures 23.9), an observation that is also consistent with a change in terrigenous fluxes at this time. The radiolarian maximum from 830 mbsf to 800 mbsf also corresponds to this magnetic susceptibility maximum. We suspect that the periodic increased delivery of terrigenous material resulted from periodic shifts from cool, dry conditions to warmer, wetter conditions in South America. This would also have increased the dissolved silica supply, thereby enhancing radiolarian production and/or preservation (comparable to the late Quaternary processes of Zaire [Schneider et al., 1997]). The lack of a direct relationship between higher carbonate dissolution and variations in either radiolarian accumulation rates or export productivity as inferred from BFAR indicates the change might be caused by an increase in the flux and oxidation of terrestrial plant organic matter, as observed in the Quaternary (Bickert et al., 1997). Each mica peak corresponds to a dissolution spike (figures 23.8A and 23.6), although not all dissolution spikes correspond to a peak in mica. In situ decay of terrestrial plant material, like marine organic matter, may have lowered pore water carbonate ion concentration thereby increasing dissolution (Bickert et al., 1997).

In unit 3 (early Oligocene), the most prominent changes are in the bulk $\delta^{13}C$ and BFAR records. The 0.9‰ increase in $\delta^{13}C$ values accompanied by two- to threefold increases in BFAR, and fish debris and echinoderms suggest a significant change in regional fertility. Echinoderm abundances as well as benthic foraminifera vary in response to food supply from the photic zone (Gooday and Turley, 1990). Following the excursion, productivity remains relatively high varying rhythmically. In this unit, carbonate dissolution and productivity strongly covary with increased dissolution coinciding with high productivity as inferred from BFAR. As a result, planktonic foraminiferal accumulation rates vary inversely to productivity. The small-scale rhythmic changes in radiolarian accumulation rates are in parallel with BFAR and dissolution variations. In short, all proxies clearly point to a sharp increase in paleoproductivity in the earliest Oligocene. This is consistent with previous findings of an abrupt increase in low-latitude marine productivity during the earliest Oligocene (Fenner, 1977; Aubry, 1992; Mikkelsen and Barron, 1997).

The base of unit 4 is marked by a decline in paleoproductivity (figures 23.6 and 23.7). In addition, radiolaria are absent while dissolution is comparable to that in unit 3. Variations in BFAR tend to be rhythmic and are accompanied by dissolution changes. We attribute the variations in radiolarian accumulation rates, and their disappearance at 700 mbsf, to a local decline in productivity as well as a reduction in deep-water silica concentrations. The latter may reflect on late Oligocene circulation changes (i.e., increased basin/basin fractionation due to intermittent formation of proto-NADW; Baldauf and Barron, 1990; Miller et al., 1987) and/or in the supply of silica to the ocean.

To summarize, Site 925 benthic foraminiferal accumulation rates, which are controlled mainly by primary productivity in the photic zone, show evidence of a significant but brief (~400,000 year) increase in paleoproductivity in the earliest Oligocene (figure 23.10). Carbonate dissolution, which at this shallow site is driven primarily by in situ organic carbon from primary production and terrestrial plant material, increases in the late Eocene. Carbon isotope values, which reflect on regional and local changes in surface-water/thermocline carbon cycling as well as global changes in burial of organic and inorganic carbon, show a sharp positive excursion in the earliest Oligocene, parallel to the increase in paleoproductivity. Radiolarian accumulation rates, which are controlled by three factors—silica concentrations of seawater, continental weathering and local riverine silica supply, and local variations in fertility through thermocline mixing—are highest in the middle Eocene and early late Eocene suggesting presence of more silica-rich waters.

The increase in paleoproductivity from the late Eocene to the Oligocene and the sharp earliest Oligocene excursion are clearly global phenomena, the latter having been observed in the Southern Ocean (Diester-Haass, 1996; Diester-Haass and Zahn, 1996; Salamy and Zachos, 1998; Diester-Haass and Zachos, submitted) as well as in middle and subtropical latitudes (Diester-Haass and Zahn, 2001). This shift has been attributed to climatic/circulation changes triggered by the gradual closure of Indian-Tethyan seaway (Lawver and Gahagan, 1997) and opening of Drake and Tasman seaways (Lawver and Gahagan, 1997). The onset of a circum-Antarctic surface current at about 37 Ma (Diester-Haass and Zahn, 1996) and growth of Antarctic glaciers at the Eocene-Oligocene transition (Zachos et al., 1994) is thought to have stimulated the developments of frontal systems comparable to those of today (Diester-Haass and Zahn, 1996). Increased meridional temperature gradients and enhanced wind velocities

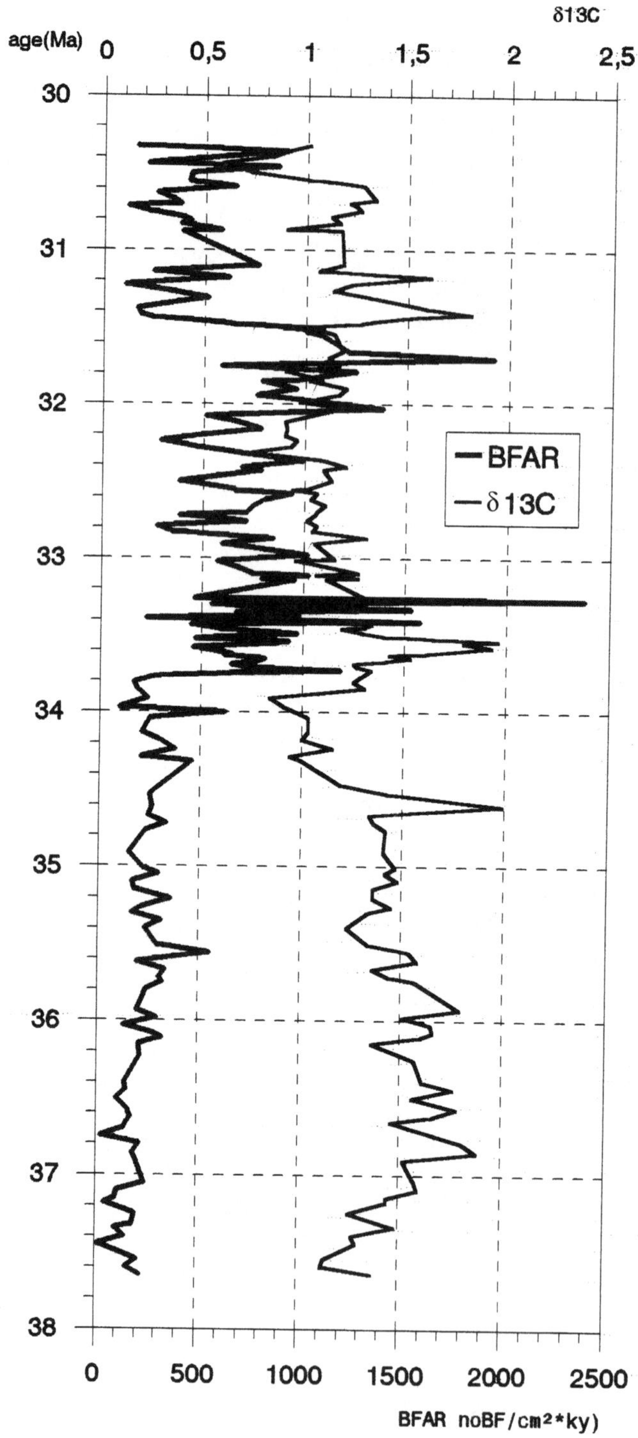

Figure 23.10. $\delta^{13}C$ and benthic foraminiferal accumulation rates (BFAR) versus age (Ma).

enhanced thermocline mixing, thus increasing productivity in the photic zone during the late Eocene until the Oligocene.

The cause of the sudden excursion to higher paleoproductivity (two to three times) in the earliest Oligocene and the subsequent decline is unknown. One possibility involves the sudden deep sea cooling and expansion of ice sheets on Antarctica. Zachos et al. (1996) proposed that high-latitude cooling and the appearance of a large ice sheet significantly enhanced overturn of the ocean, particularly in regions of wind-driven upwelling. The higher paleoproductivity persisted for only 400,000 years (figure 23.10) as the nutrient levels dropped to some steady state level. Despite the drop, paleoproductivity stabilized at a higher level than in the Eocene. The 0.9‰ increase in $\delta^{13}C$ at Site 925 during the earliest Oligocene paleoproductivity shift, which is observed also at southern high latitudes and at ODP Sites 592 and 763, is attributed to a substantial increase in organic carbon to inorganic carbon burial rates that resulted from enhanced marine biological productivity (Zachos et al., 1996; Diester-Haass and Zahn, 1996; Diester-Haass and Zahn, 2001; Salamy and Zachos, 1998). Sequestration of organic carbon in marine sediments leads to a lowering of oceanic and atmospheric CO_2 levels and may be responsible for the extreme but transient cooling of the earliest Oligocene.

CONCLUSION

Several proxies including BFAR (benthic foraminiferal accumulation rates), carbonate dissolution, and radiolarian accumulation rates were used to reconstruct paleoproductivity variations at ODP Site 925 during the Eocene-Oligocene transition. These records document a prominent increase in productivity in the equatorial Atlantic in the earliest Oligocene. This productivity increase, on the basis of bio- and carbon isotope stratigraphy, is correlated to similar increases in the Southern Ocean and in midlatitudes. This widespread increase in productivity may have lead to enhanced carbon export and burial, which further contributed to the global cooling in the earliest Oligocene via extraction of CO_2 from the global carbon ocean–atmosphere carbon reservoir.

ACKNOWLEDGMENTS

We thank the Ocean Drilling Program for providing samples and the Deutsche Forschungsgemeinschaft and National Science Foundation for financial assistance. We thank several collegues for stimulating discussions: Wolfgang Berger, James Kennett, Stephan Pekar, EllenThomas and Rainer Zahn and all the participants of the Penrose Conference on the Marine Eocene-Oligocene Transition, especially Don Prothero for having invited us to this exciting meeting. We thank Wolfgang Berger and Rainer Zahn for reviews of the manuscript and Kara Kuvakas, Melissa Harvey, and Geoff Koehler for technical support.

REFERENCES

Archer, D. and Maier-Reimer, E., 1994, Effect of deep-sea sedimentary calcite preservation on atmospheric CO_2 concentration: Nature, v. 367, p. 260-263.

Archer, D., Lyle, M., Rodgers, K., and Froelich, P., 1993, What controls opal preservation in tropical deep-sea sediments?: Paleoceanography, v. 8, no. 1, p. 7-21.

Arthur, M. A., Hinga, K. R., Pilson, M. E. Q., Whitaker, E., and Allard, D., 1991, Estimates of pCO_2 for the last 120 Ma based on the $\delta^{13}C$ of marine phytoplanktonic organic matter (abstract): EOS, v. 72, p. 166.

Aubry, M.-P., 1992, Late Paleogene calcareous nannoplankton evolution: a tale of climatic deterioration, *in* Prothero, D. R., and Berggren, W. A., eds., Eocene-Oligocene climatic and biotic evolution: Princeton, New Jersey, Princeton University Press, p.272-309.

Baldauf, J. G., and Barron, J. A., 1990, Evolution of biosiliceous sedimentation patterns—Eocene through Quaternary: paleoceanographic response to polar cooling, *in* Bleil, U., and Thiede, J., eds., Geological History of th Polar Oceans, Arctic versus Antarctic: Norwell, Massachusetts, Kluwer Academic Publishers, p. 575-607.

Barron, J. A., and Baldauf, J. G., 1989, Tertiary cooling steps and paleoproductivity as reflected by diatoms and biosiliceous sediments, *in* Berger, W. H., Smetacek, V., and Wefer, G., eds., Productivity of the oceans: present and past: New York, John Wiley, p. 341-354.

Berger, W. H.,1976, Biogenous deep-sea sediment preservation and interpretation: Chemical Oceanography, v. 5, p. 265-388.

Berger, W. H., 1991, Produktivität des Ozeans aus geologischer Sicht: Denkmodelle und Beispiele: Zeitschrift Deutsche Geologische Gesellschaft, v. 142, p. 149-178.

Berger, W. H., and Diester-Haass, L., 1988, Paleoproductivity: the benthic/planktonic ratio in foraminifera as a productivity index: Marine Geology, v. 81, p.15-25.

Berger, W. H., and Herguera, J. C., 1992, Reading the sedimentary record of the oceans productivity, *in* Falkowski, P. D., and Woodhead, A. D., eds., Primary productivity and biogeochemical cycles in the sea: New York, Plenum Press, p. 455-486.

Berger, W. H., Herguera, J. C., Lange, C. B., and Schneider, R., 1994, Paleoproductivity: flux proxies versus nutrient proxies and other problems concerning the Quaternary productivity record, *in* Zahn, R., et al., eds., Carbon cycling in the glacial ocean: constraints on the ocean's role in global change: Springer Verlag, NATO ASI Series, v. 117, p. 385-412.

Berggren, W. A., Kent D. V., Swisher, C. C., II, and Aubry, M.-P., 1995, A revised Cenozoic geochronology and chronostratigraphy: SEPM Special Publication, v. 54, 212 p.

Berner, R. A., 1990, Atmospheric carbon dioxide levels over Phanerozoic time: Science, v. 249, p. 1382-1386.

Bickert, T., Curry, W. B., and Wefer, G., 1997, Late Pliocene to mid-Pleistocene (2.6-1.0 Ma) carbonate dissolution in the western equatorial Atlantic: results of Leg 154, Ceara Rise, *in* Shackleton, N. J., et al., eds., Proceedings of the ODP, Scientific Results, v. 154, p. 229-238.

Calvert, S. E., 1974, Deposition and diagenesis of silica in marine sediments: Special Publication of the International Association of Sedimentologists, v. 1, p. 273-299.

Corliss, B. H., Aubry, M.-P., Berggren, W. A., Fenner, J., Keigwin, L. D., and Keller, G., 1984, The Eocene-Oligocene boundary event in the deep-sea: Science, v. 226, p. 806-810.

Crowley, T. J. and North, G. R., 1988, Abrupt climate change and extinction events in earth history: Science, v. 240, p. 996-1002.

Curry, W. B., Shackleton, N. J., Richter, C., et al., 1995, Proceedings Ocean Drilling Program, Initial Reports, Leg 154: College Station, Texas (Ocean Drilling Program), 1111 p.

Diester-Haass, L., 1995, Middle Eocene to early Oligocene paleoceanography of the Antarctic Ocean (Maud Rise, ODP Leg 113, Site 689): change from a low to a high productivity ocean: Paleogeography, Paleoclimatology, Paleoecology, v. 113, p. 311-334.

Diester-Haass, L., 1996, Late Eocene-Oligocene paleoceanography in the southern Indian Ocean (ODP Site 744): Paleogeography, Paleoclimatology, Paleoecology., v. 130, p. 99-119.

Diester-Haass, L. and Zahn, R., 1996, The Eocene-Oligocene transition in the Southern Ocean: history of water masses, circulation, and biological productivity inferred from high resolution records of stable isotopes and benthic foraminiferal abundances (ODP Site 689): Geology, v. 24, no. 2, p.16-20.

Diester-Haass, L., and Zahn, R., 2001, Paleoproductivity increase at the Eocene-Oligocene climatic transition: ODP/DSDP Sites 763 and 592: Paleogeography, Paleoclimatology, Paleoecology, v. 172, p. 153-170.

Diester-Haass, L., and Zachos, J., submitted, The Eocene-Oligocene climatic transition: Paleoproductivity and its impact on the marine carbon cycle: International Journal of Earth Sciences, Spring-Verlag.

Fenner, J., 1977, Cenozoic diatom biostratigraphy of the equatorial and southern Atlantic Ocean, *in* Supko, P. R., Perch-Nielsen, K. et al., Initial Reports of the Deep Sea Drilling Project, v. 39, p. 491-624.

Flower, B. P., 1999, Warming without high CO_2?: Nature, v. 399, p. 313-314.

Freeman, K. H., and Hayes, J. M., 1992, Fractionation of carbon isotopes by phytoplankton and estimates of ancient CO_2 levels: Global Biogeochemical Cycles, v. 6, no. 2, p. 185-198.

Froelich, P. N., Blanc, V., Mortlock, R. A., Chillrud, S. N., Dunstan, W., Udomkit, A., and Peng, T.-H., 1992, River fluxes of dissolved silica to the ocean were higher during glacials: Ge/Si in diatoms, rivers and oceans: Paleoceanography, v. 7, no. 6, p.739-767.

Gooday, A. J., and Turley, C. M.,1990, Responses by benthic organisms to inputs of organic material to the ocean floor: a review: Philosophical Transactions of the Royal Society of London, ser. A, v. 331, p. 119-138.

Gooday, A. J., and Rathburn, A. E., 1999, Temporal variability in living deep-sea benthic foraminifera: a review: Earth-Science Reviews, p. 187-212.

Heath, G. R., 1974, Dissolved silica and deep-sea sediments, *in* Hay, W. W., ed., Studies in paleoceanography: SEPM Special Publication, v. 20, p. 77-93.

Heinze, C., and Crowley, T., 1997, Sedimentary response to ocean gateway circulation changes: Paleoceanography, v.12, no. 6, p. 742-754.

Herguera, J. C., 2000. Last glacial productivity patterns in the eastern equatorial Pacific: benthic foraminifera records: Marine Micropaleontology, v. 40, p. 259-275.

Herguera, J. C., and Berger, W. A., 1991, Paleoproduc-tivity from benthic foraminifera abundance: glacial to postglacial change in the west-equatorial Pacific: Geology, v. 19, p.1173-1176.

Lawver, L. A., and Gahagan, L. M., 1997, Opening of Drake Passage and its impact on Cenozoic ocean circulation, *in* Crowley, T., and Burke, K. eds., Tectonic boundary conditions for ocean circulation: Oxford University Press, p. 213-224.

Lange, C. B., and Berger, W. H., 1993, Diatom productivity and preservation in the western equatorial Pacific: the Quaternary record, *in* Berger, W. H., Kroenke, L. W., Mayer, L. A., et al., Proceedings of the Ocean Drilling Program, Scientific Results, v. 130, p.509-523.

Lear, C. H., Elderfield, H., and Wilson, P. A., 2000, Palaeotemperature and ocean chemistry records for the early Paleogene from Mg/Ca and Sr/Ca in benthic foraminiferal calcite: Science, v. 14, no. 287, p. 269-272.

Leinen, M., Cwienk, D., Heath, R. G., Biscaye, P., Kolla, V., Thiede, J., and Dauphin, J. P., 1986, Distribution of biogenic silica and quartz in recent deep-sea sediments: Geology, v. 14, p. 199-203.

McGowran, B., 1989, Silica burp in the Eocene Ocean: Geology, v. 17, p. 857-860.

Mikkelsen, N., and Barron, J. A., 1997, Early Oligocene diatoms on the Ceara Rise and the Cenozoic evolution of biogenic silica accumulation in the low-latitude Atlantic, *in* Shackleton, N. J., Curry, W. B., Richter, C., and Bralower, T. J. eds., Proceedings of the Ocean Drilling Program, Scientific Results, v. 154, p. 483-490.

Miller, K. G., Fairbanks, R. G., and Mountain, G. S., 1987, Tertiary oxygen isotope synthesis, sea-level history and continental margin erosion: Paleoceanography, v. 2, p.1-19.

Nelson, D. M., Treguer, P., Brzezinski, M. A., Leynaert, A., and Queguiner, B., 1995, Production nad dissolution of biogenic silica in the ocean: revised global estimates, comparison with regional data and relatioship to biogenic sedimentation: Global Biogeochemial Cycles, v. 9, no. 3, p. 359-372.

Nees, S., 1997, Late Quaternary paleooceanography of the Tasman Sea: the benthic foraminiferal view: Palaeogeography, Palaeoclimatology, Palaeoecology, v. 131, p. 365-389.

Owen, R. M., and Rea, D. K., 1985, Sea-floor hydrothermal activity links climate to tectonics: the Eocene carbon dioxide greenhouse: Science, v. 227, p. 166-169.

Pagani, M., Arthur, M. A., and Freeman, K. H., 1999, Miocene evolution of atmospheric carbon dioxide: Paleoceanography, v.14, p. 273-292.

Pearson, P. N., and Palmer, M. R., 1999. Middle Eocene seawater pH and atmospheric carbon dioxide concentrations: Science, v. 284, p. 1824-1826.

Pearson, P. N., and Palmer, M. R., 2000. Atmospheric carbon dioxide concentrations over the past 60 million years: Nature, v. 406, p. 695-699.

Pearson, N. P., and Chaisson, W. P., 1997, Late Paleocene to middle Miocene planktonic foraminifer biostratigraphy of the Ceara Rise, *in* Shackleton, N. J., Curry, W. B., Richter, C., and Bralower, T. J., eds., Proceedings of the Ocean Drilling Program, Scientific Results, v. 154, p. 33-68.

Prothero, D. R. 1994, The Eocene-Oligocene transition: paradise lost: New York, Columbia University Press, 291 p.

Prothero, D. R., and Berggren, W. A., eds., 1992, Eocene-Oligocene climatic and biotic evolution: Princeton, New Jersey, Princeton University Press., 568 p.

Ruddiman, W. F., Raymo, M. E., Prell, W. L., and Kutzbach, J. E., 1997, The uplift-climate connection: a synthesis, *in* Ruddiman, W. F., ed., Tectonic uplift and climate change: New York, Plenum Press, p. 471-515.

Salamy, K. A., and Zachos, J. C., 1999, The effect of late Eocene–early Oligocene climate change on Southern Ocean fertility: inferences from sediment accumulation and stable isotope data: Palaeogeography, Palaeoclimatology, Palaeoecology, v. 145, p.79-93.

Schmitz, B., 1987, Barium, equatorial high productivity, and the northward wandering of the Indian continent: Paleoceanography, v. 1, no. 2, p. 63-77.

Schneider, R. R., Price, B., Müller, P. J., Kroon, D., and Alexander, I., 1997, Monsoon related variations in Zaire (Congo) sediment load and influence of fluvial silicate supply on marine productivity in the east equatorial Atlantic during the last 200.000 years: Paleoceanography, v. 12, no.3, p. 463-481.

Sloan, L. C., and Rea, D. K., 1995, Atmospheric carbon dioxide and early Eocene climate: a general circulation modeling sensitivity study: Palaeogeography, Palaeoclimatology, Palaeoecol-ogy, v.119, p. 275-292.

Sloan, L. C., Walker, J. C. G., Moore, T. C., Rea, D. K., and Zachos, J. C., 1992, Possible methane-induced polar warming in the early Eocene: Nature, v. 357, p.320-322.

Takeda, S., 1998, Influence of iron availability on nutrient consumption ratio of diatoms in oceanic waters: Nature, v. 393, p. 774-777.

Thomas, E., 1992, Middle Eocene-late Oligocene bathyal benthic foraminifera (Weddell Sea): Faunal changes and implications for ocean circulation, *in* Prothero, D. R., and Berggren, W. A., eds., Eocene-Oligocene climatic and biotic evolution: Princeton, New Jersey, Princeton University Press, p. 245-272.

Thomas, E., Booth, L., Maslin, M., and Shackleton, N. J., 1995, Northeastern Atlantic benthic foraminifera during the last 45.000 years: Changes in productivity seen from the bottom up: Paleoceanography, v. 10, p. 545-562.

Thomas, E., and Gooday, A. J., 1996., Cenozoic deep-sea benthic foraminifera: tracers for changes in oceanic productivity?: Geology, v. 24, p. 355-358.

Thomas, E., Zachos, J. C., and Bralower, T. J., 2000, Deep-sea environments on a warm earth: latest Paleocene-early Eocene, *in* Wing, S., ed., Warm climates in earth history: New York, Cambridge University Press, p. 132-160.

Van Andel, Tj. H., Heath, G. R., and Moore, T. C., 1975, Cenozoic tectonics, sedimentation and paleoceanography of the central equatorial Pacific: Geological Society of America Memoir v.143, p. 47-61.

Yasuda, H., 1997, Late Miocene–Holocene paleoceanography of the western equatorial Atlantic: evidence from deep-sea benthic foraminifera, *in* Shackleton, N. J., Curry, W. B., Richter, C., and Bralower, T. J., eds., Proceedings of the Ocean Drilling Program, Scientific Results, v. 154, p. 395-432.

Zachos, J. C., Stott, L. D., and Lohmann, K. C, 1994, Evolution of early Cenozoic marine temperatures: Paleoceanography, v. 9, p.353-387.

Zachos, J. C., Quinn, T. M., and Salamy, K. A., 1996, High-resolution (10^4 years) deep-sea foraminiferal stable isotope records of the Eocene-Oligocene climate transition: Paleoceanography, v. 11, p. 251-266.

Part III

THE NORTH SEA AND TETHYS

CHAPTER 24

The Eocene/Oligocene Boundary in the North Sea Area: A Sequence Stratigraphic Approach

Noël Vandenberghe, Henk Brinkhuis, and Etienne Steurbaut

ABSTRACT

The study of the Doel 2b borehole in northern Belgium enables the identification of four depositional sequences in the Eocene-Oligocene transition interval. These sequences, which have occurred consistently throughout the North Sea Basin, are calibrated to the standard sequence chronostratigraphic scales and the oxygen isotope events on the basis of calcareous nannofossil and dinoflagellate cyst correlations. The Bassevelde 1 sequence is bound at its base by the Pr1 boundary, which coincides with the Epi-1 cooling event. The lower boundary of the Bassevelde 2 sequence, probably of less importance, corresponds to the Pr2 sequence boundary. The Bassevelde 3 sequence falls within nannofossil Zone NP21 and is correlative of the marine Tongrian stratotype. The lower boundary of this sequence is characterized by dinocyst events occurring at the Eocene-Oligocene GSSP (Global Stratotype Section and Point), whose age is estimated to be 33.7 Ma. It is suggested that this boundary equates the Pr3 sequence boundary and corresponds to the isotope event Oi1 (Miller) or O1a (Zachos). The boundary between the Bassevelde 3 and the overlying Ruisbroek-Boom-1 sequence represents a major cooling. It is correlated with the Ru1 sequence boundary and the isotope event Oi1a (Miller) or O1b (Zachos). This major cooling event is situated well within the Oligocene if calibrated against the Hantkeninidae extinction event, although it coincides or, at least, is very close to the boundary between the Priabonian and Rupelian unit stratotypes.

Since the Bassevelde 3 sequence is a lateral equivalent of the marine Tongrian stratotype and also the lowermost Oligocene sequence, it is worth considering reintroducing the threefold subdivision of the Oligocene.

INTRODUCTION

The Eocene-Oligocene transition represents a critical phase in earth history reflecting the change from a greenhouse into an icehouse world caused by the interaction of large-scale tectonic, oceanic, and climatic processes. The evolution of the Antarctic continental ice sheets invoked major eustatic sea-level changes. In recent years, a reasonably detailed and well-calibrated record of the climatic and sea-level history of the Eocene-Oligocene transition has become available from the Atlantic margins and various European basins (e.g. Haq et al., 1987; Brinkhuis and Visscher, 1995; Pekar and Miller, 1996; Zachos et al., 1996; de Graciansky et al., 1998).

A key area for the study of this stratigraphic interval is northern Belgium, part of the Cenozoic North Sea Basin and home of the lower Oligocene Rupelian Stage. Vandenberghe et al. (1998) have provided a detailed overview of the sequence stratigraphic architecture of the marginal-marine strata and recognized third-order sequences in the Priabonian-Rupelian interval comparable to sequences recognized elsewhere. The interpretation of the terminal Eocene and earliest Oligocene sequences remained tentative, however, due to the absence of suitably documented sections. A continuously cored interval comprising the Eocene-Oligocene transitional strata has now become available in the recently drilled Doel 2b borehole, north of Antwerp (figure 24.1).

The aim of this paper therefore is to provide an

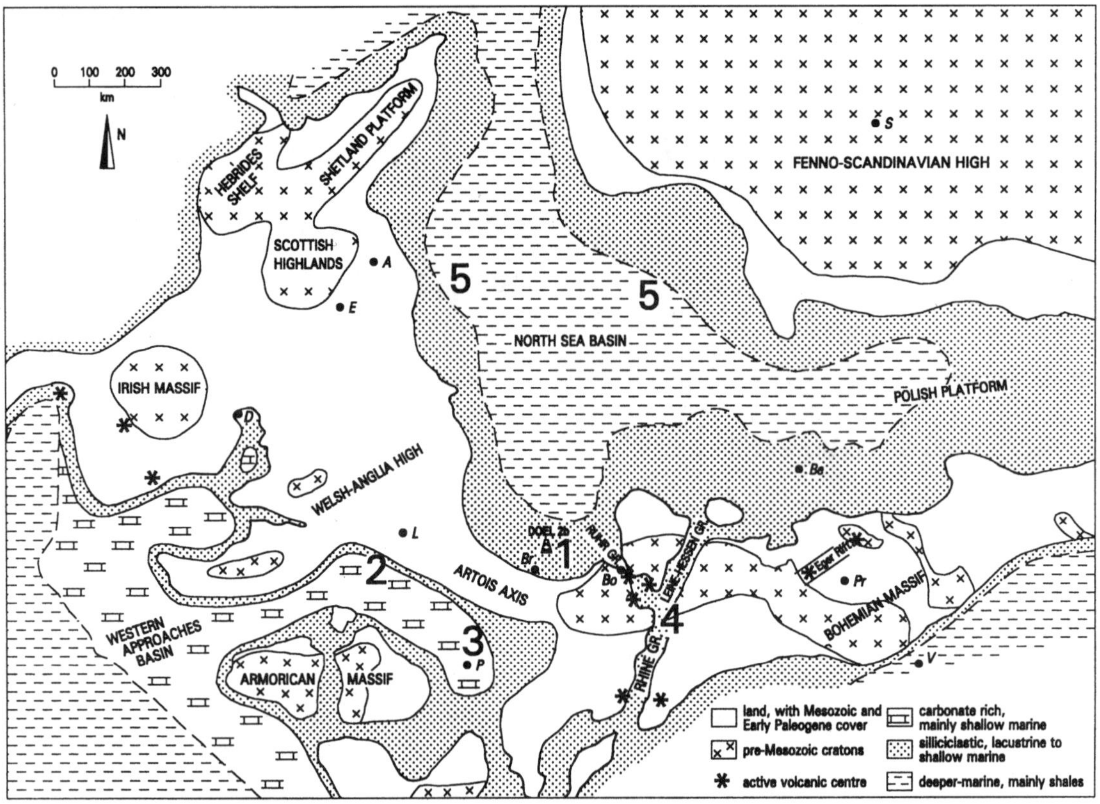

Figure 24.1. The Oligocene paleogeography (modified after Ziegler, 1982) with indication of the areas studied in the paper: Doel 2b borehole, central and eastern Belgium (1), Isle of Wight (2), Paris Basin (3), Rhine Graben area (4), North Sea subsurface (5). The capital letters are the initials of the major cities in the area: Aberdeen, Berlin, Bonn, Brussels, Dublin, Edinburgh, London, Paris, Prague, Vienna, and Stockholm.

accurate age-assessment of the Doel 2b Eocene-Oligocene succession, provide an integrated sequence stratigraphic interpretation, provide comparisons to the sequence stratigraphy established in other parts of the European North Sea Basin, and briefly discuss the relationship with central and northeastern classic Italian sections, including the Massignano Eocene-Oligocene GSSP.

THE SEQUENCE STRATIGRAPHIC INTERPRETATION OF THE DOEL 2B BOREHOLE IN THE UPPER EOCENE AND LOWER OLIGOCENE INTERVAL

The data for the stratigraphic interpretation of the Doel 2b well are shown in figure 24.2. A comparison of the core lithology with the resistivity geophysical well log shows that the resistivity log is acceptable as a proxy for the grain-size evolution, except in the interval between 129 and 145 m where the resistivity log apparently is strongly influenced by the occurrence of thin alternating clay and sand layers and maybe also by high mica content in the sediment. Therefore in this interval in figure 24.2, the core description—not the well log— is shown. Otherwise, the resistivity log is represented. Palynology, including dinoflagellate cyst analysis, is represented in figure 24.3 and calcareous nannoplankton analysisin figure 24.4.

The Bartonian-Priabonian Boundary

The top of the Onderdijke Clay at 173.25-meter depth is the deepest lithostratigraphic reference level (figure 24.2). The Onderdijke Clay is a continuous and widely recognized layer in the region, the top of which is generally assumed to mark the Bartonian/Priabonian boundary (Vandenberghe et al., 1998). In the absence of a GSSP for the base of the Priabonian, the NP17/NP18 zonal boundary, defined by the first appearance of *Chiasmolithus oamaruensis*, is generally accepted to represent the Bartonian/Priabonian boundary (Berggren et al., 1995). This nannofossil event cannot be exactly pinpointed in Belgium, as the

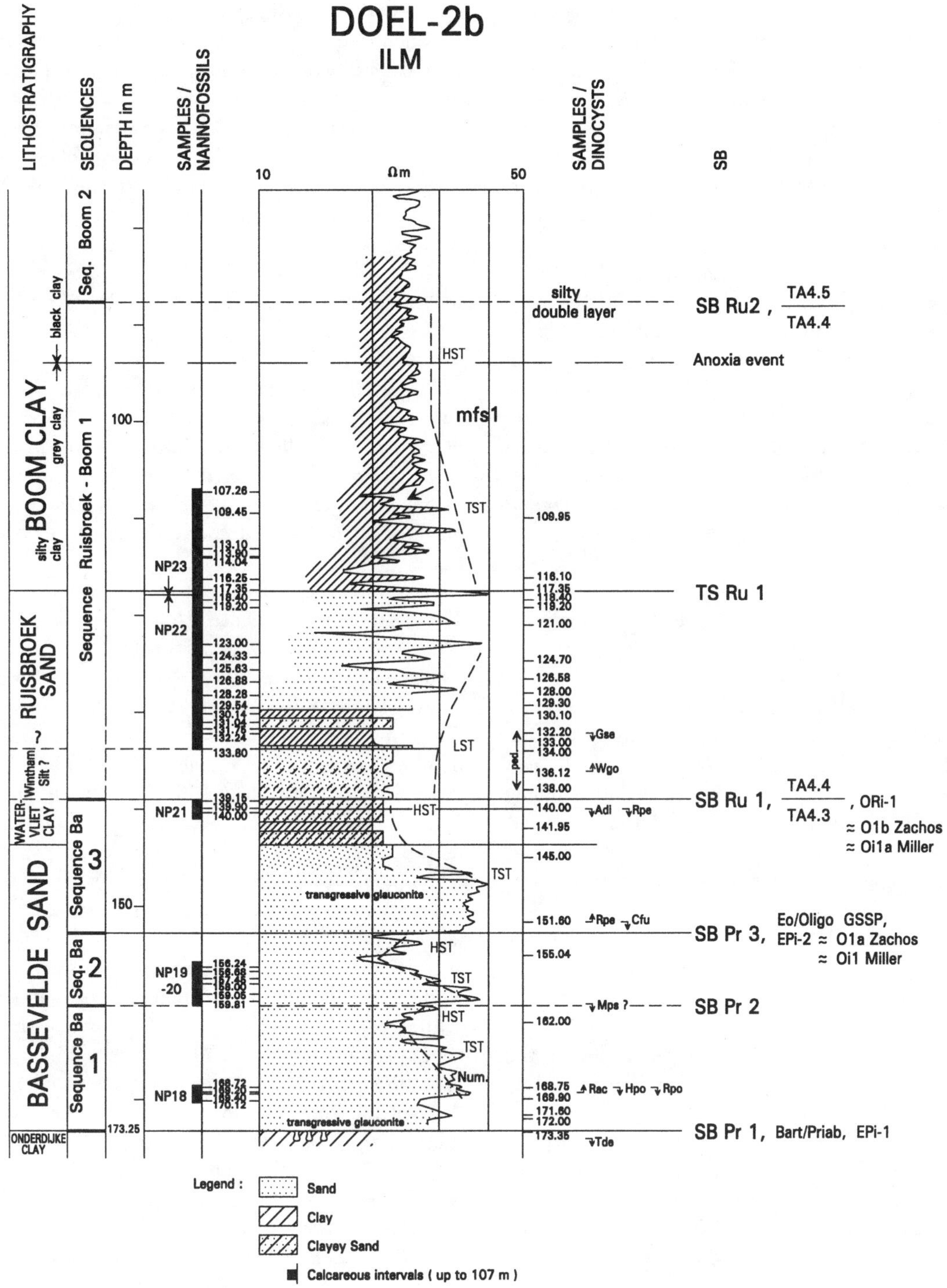

Figure 24.2. Synoptic stratigraphic information obtained from lithologies, calcareous nannoplankton and dinoflagellate analyses in the Doel 2b borehole, together with the sequence stratigraphic interpretation and the proposed correlation with the isotope stratigraphy and the sequence chronostratigraphy in de Graciansky et al. (1998). Dinocyst abbreviations: Adi, *Areosphaeridium diktyoplokum*; Cfu, *Cordosphaeridium funiculatum*; Gse, *Glaphyrocysta semitecta*; Hpo, *Heteraulacacysta porosa*; Mps, *Melitasphaeridum pseudorecurvatum*; Rac, *Reticulosphaera actinocoronata*; Rpe, *Rhombodinium perforatum*; Rpo, *Rhombodinium porosum*; Tde, *Thalassiphora delicata*; Wgo, *Wetzeliella gochtii*.

Onderdijke Clay and the basal part of the overlying Bassevelde Sand are noncalcareous (Steurbaut, 1986). In the Onderdijke Clay of the Doel 2b borehole, the last occurrence of the dinoflagellate cyst (hereafter, dinocyst) *Thalassiphora delicata* marks the end of the Bartonian, following Williams et al. (1998). Samples in the basal part of the overlying Bassevelde Sand still contain some typical Bartonian dinocysts, such as *Rhombodinium porosum* and *Heteraulacacysta porosa.* This indicates either that deposition of the base of the overlying Bassevelde Sand (below NP18) still occurred during the end of the Bartonian, or that a local reworking of Bartonian material took place at the beginning of the Priabonian. This last option is probable, based on the unconformity between the Onderdijke Clay and the Bassevelde Sand (Vandenberghe et al., 1998) and it is also supported by the presence of the Priabonian dinocyst *Rhombodinium perforatum* (Williams et al., 1998) at the very base of the Bassevelde Sand. It should also be noted, however, that in de Graciansky et al. (1998) the Bartonian/Priabonian boundary is shown to be 0.1 million years younger than the Pr1 sequence boundary.

The Bassevelde 1 Sequence (173.25–160 m)

In the Doel 2b borehole the upper Eocene Bassevelde Sand on top of the Onderdijke Clay can be split into three cycles based on the grain-size evolution trends (figure 24.2). The base of each cycle is characterized by coarse sediment (173.25 m, 160 m and 152.40 m). The basal transgressive parts of the lower and the upper cycles contain in both cases abundant glauconite but not the base of the middle cycle. Glauconitic greensands in the area generally occur at the transgressive base of a sequence throughout the Cenozoic deposits (Vandenberghe et al., 1998). The oldest cycle is henceforth called Bassevelde 1, the middle cycle Bassevelde 2, and the youngest cycle Bassevelde 3 (figure 24.2).

In the Bassevelde 1 cycle, a few meters above the lower transgressive glauconite section, a layer of *Nummulites* occurs, the youngest level known from the Cenozoic of the region. The same *Nummulites* level is also found in the stratotype area of the Bassevelde Sand (Steurbaut, 1986) and in the nearby Kallo borehole (Gulinck, 1969) but unfortunately the *Nummulites* species has no particular stratigraphic value (Drooger, 1969). This carbonate-containing level, however, contains nannoplankton in the Doel 2b well (170.32–168.52) indicative of Zone NP18. The appearance of *Reticulatosphaera actinocoronata* at 168.75 m therefore falls within NP18 and may indicate a distinct cooling of the seawater although the nannofossils *Neococcolithes minutus, Pontosphaera prava* and *Pemma* spp. still indicate relatively warm surface waters.

The sedimentary cycle shows a fining upward trend, expressed by the resistivity values, followed by a short coarsening upward trend. These grain-size trends are interpreted, respectively, as the transgressive systems tract and the highstand systems tract of a depositional sequence, with the maximum flooding surface at the reversal of the trend at about 162 m. It can be concluded that the sequence boundary at the base of the Bassevelde Sand corresponds in time to the regionally important sequence boundary Pr1 (de Graciansky et al., 1998: chart 1; see also Stover and Hardenbol, 1994: figure 3) coinciding with a marked oxygen isotope shift known as the Epi-1 cooling event (Abreu and Anderson, 1998). The NP19/20 nannoplankton interval already occurs in the base of the Bassevelde 2 sequence.

The Bassevelde 2 Sequence (160–152.40 m)

The Bassevelde 2 cycle starts at its base with a sharp increase in resistivity, reflecting a jump to coarser grain-size. Similarly to the Bassevelde 1 sequence, a fining upward trend followed by a coarsening upward trend is recorded in the middle cycle. Also based on the grain-size trends, this middle cycle is interpreted as a sequence with transgressive and highstand systems tracts separated by a maximum flooding surface at the reversal of the grain-size trends. The absence of a glauconitic base probably points to the lesser importance and shorter duration of the sea-level fluctuation at this boundary, in view of the necessary time required for glauconite formation. The base and the central part of this sequence are correlatable to nannoplankton Zones NP19/20 (figure 24.2). The nannofossil associations can be differentiated from these of the underlying Bassevelde sequence 1 by a substantial decrease in abundance and in species diversity. They are furthermore characterized by the disappearance of *Sphenolithus celsus,* the sudden increase of *Discoaster saipanensis,* and the entry of *Isthmolithus recurvus.* In the Bassevelde sequence 2, fewer acritarchs and prasinophytes but more bisaccate pollen are present than in the sequence below (figure 24.3). These marked differences support the differentiation of the Bassevelde 1 and 2 sequences. In the nearby Doel 14E-194 borehole, which is located at the same site with only about 1 m difference between comparable levels in the Doel 2b well, Stover and Hardenbol (1994) reported the presence of

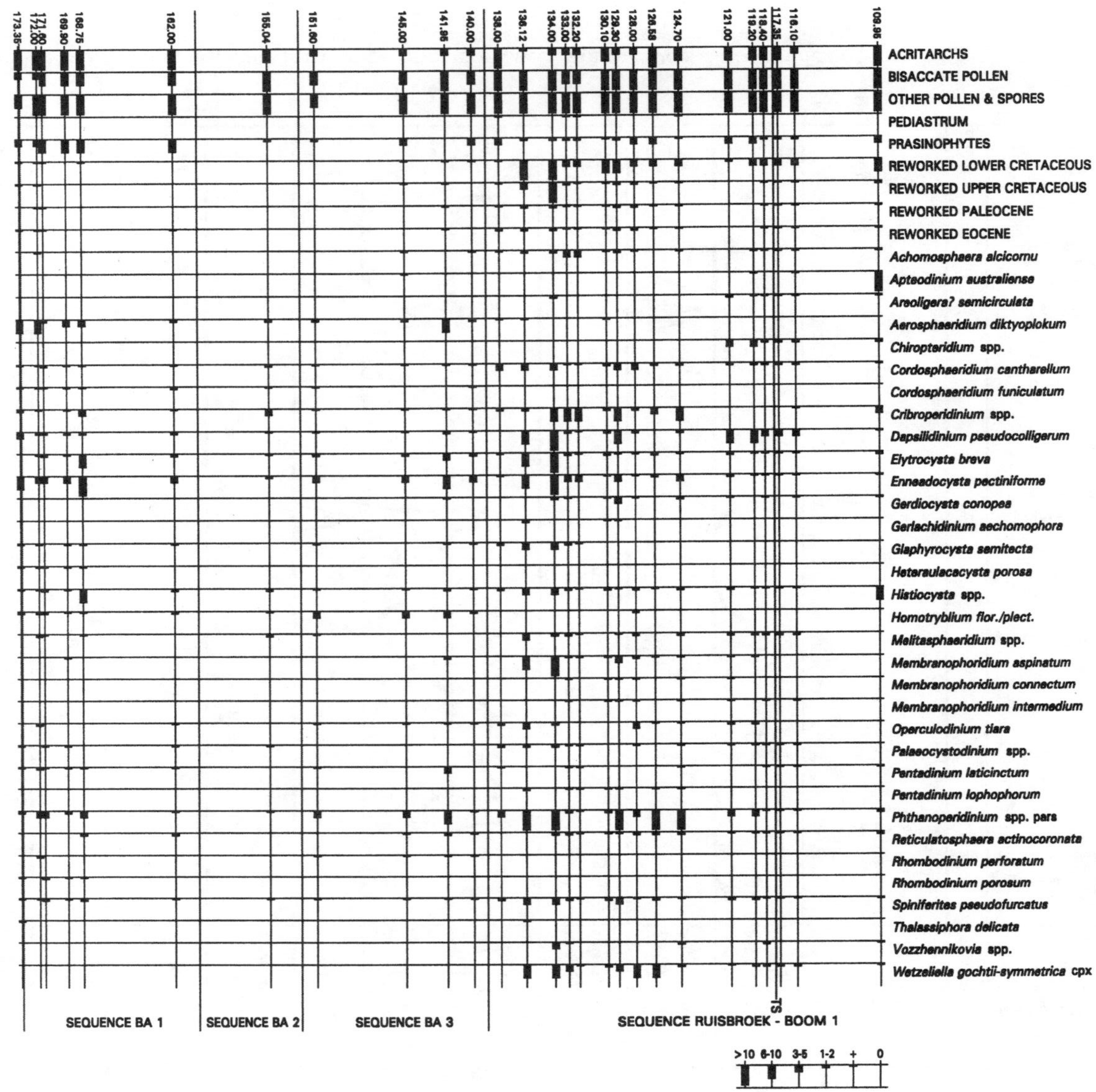

Figure 24.3. Summarizing diagram of the palynology of the Doel 2b borehole, including distribution of selected dinoflagellate cysts.

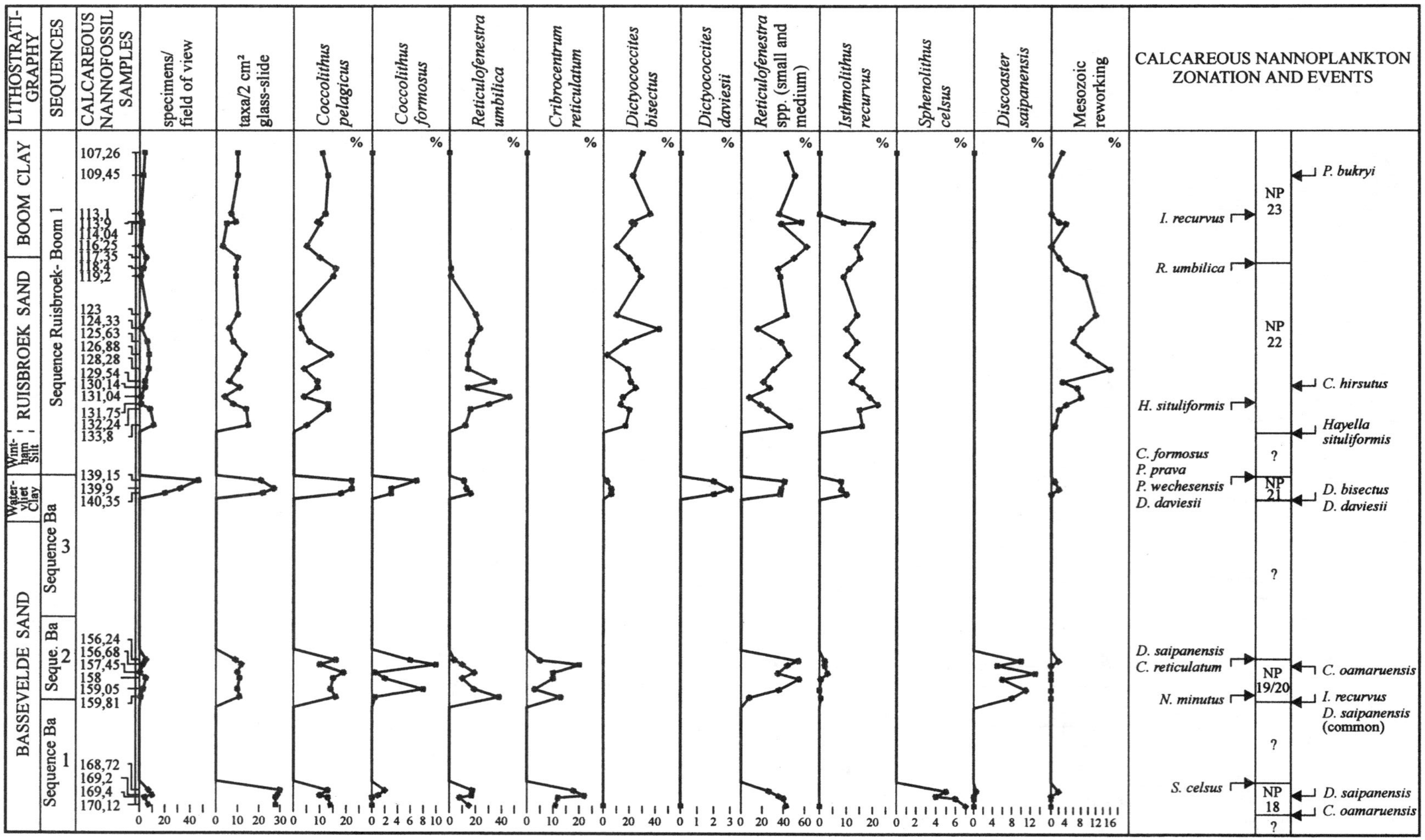

Figure 24.4. Calcareous nannoplankton zonation, events, and quantitative distribution of marker species in the Doel 2b borehole.

Melitasphaeridium pseudorecurvatum at 159.90 m but not in the overlying samples (139.90 m and higher). From the present dinocyst analysis it can be concluded that the top of the Mps dinocyst Zone (*M. pseudorecurvatum*) of Brinkhuis and Biffi (1993) occurs at the base of the Bassevelde 2 sequence. Besides, the top of the Mps dinocyst Zone in the Massignano section in central Italy is also characterized by a strong decrease in prasinophytes (Brinkhuis and Biffi, 1993: figure 6) as is the case at the sequence boundary between Bassevelde 1 and 2 (figure 24.3). It can be concluded that the sequence boundary at the base of Bassevelde 2 corresponds very closely in time to the Pr2 sequence boundary at the NP18/19 zonal boundary (de Graciansky et al., 1998), that the sequence Bassevelde 2 is represented only by 7.60 m of sediment, and that it is correlated to sequence TA4.2 (Haq et al., 1987).

The transition between the sequences Bassevelde 2 and 3 is marked by the first consistent occurrence (=FCO) of *Rhombodinium perforatum*, the presence of which upwards is restricted to the sequence Bassevelde 3. The top of the consistent occurrence (=LCO) of *Cordosphaeridium funiculatum* is located at 151.6 m, precisely in the base of the Bassevelde 3 sequence. These dinocyst events are associated with the foraminiferal family Hantkeninidae extinction event characterizing the Eocene/Oligocene boundary in Italy (Brinkhuis and Biffi, 1993). Therefore the boundary age between the Bassevelde 2 and 3 sequences is considered to be 33.7 Ma.

The Bassevelde 3 Sequence (152.40 – 139.00 m)

The Bassevelde 3 sequence starts at the base with a few meters of glauconite-rich sands, continues upwards in a fining grain-size trend which is reversed in a coarsening upward trend at about 142 m. It represents the upper part of the Bassevelde Sand and it is particularly rich in coarse mica flakes. In the Doel 2b borehole these sands do not contain calcareous nannoplankton, but Zone NP21 is recognized in the clay-rich upper part (141–139 m) of the sequence. However, as the dinocyst events situated at the base of the sequence Bassevelde 3, namely the LCO of *Cordosphaeridium. funiculatum* and the FCO of *Rhombodinium perforatum*, are time equivalent of the Eocene-Oligocene GSSP Hantkeninidae event they occur in Zone NP21 (Brinkhuis and Biffi, 1993). Hence, it can reasonably be assumed that the whole Bassevelde 3 is situated within Zone NP21. The nannofossil associations of the Bassevelde 3 sequence differ from these of the underlying Bassevelde 2 sequence by a substantial increase in abundance and species diversity, indicating high-productivity surface waters. The Bassevelde 3 associations are furthermore marked by the disappearance of warm-water taxa such as discoasters, by the appearance of *Dictyococcites bisectus* and *D. daviesii*, and by a pronounced increase in *Isthmolithus recurvus*, which are all cool-water taxa (Wei and Wise, 1990).

The sandy sedimentation of the transgressive systems tract rapidly turns into an alternation of mainly clayey sediments at about 144 m (figure 24.2). In this area the Bassevelde Sands are commonly overlain by the Watervliet Clay. This clay unit, however, is mainly known from well log responses and its signature, including that of its top, is not well defined. Therefore, it may have different meanings in the geological descriptions of different boreholes by different authors and hence correlation by lithostratigraphy alone can be deceptive at the level of stratigraphic resolution required for sequence stratigraphy as applied in this paper. This explains small differences in the positions of the lithostratigraphic units interpreted in the Doel 14E-194 borehole by Stover and Hardenbol (1994: figure 22) compared to our analysis of the nearby Doel 2b well and the difficulty of comparing in detail with the dinocyst analysis by De Coninck (1995a, 1999) in the area.

The sediments above the 139 m level are generally sandier than the clays below but remain clayey. A first lower coarse horizon comes in at 133.5–133.8 m but it is overlain again by mainly clayey to silty sediments and followed by a second upper coarse horizon at 129.65–128 m. The overlying sediments still contain some clay layers, but gradually the sand becomes coarser and dominates the lithology. It is probable that in practice, during routine log interpretation and borehole sample examination, the clays in the top of the Bassevelde sequence 3 (139–144 m) and the clayey sands above, until the lower or upper coarse horizon, would be described as Watervliet Clay. However, the clayey sandy sediments between the clay and the coarse levels could equally well be interpreted as the Wintham Silt (figure 24.2), a lithostratigraphic member introduced by Steurbaut (1992). The sands above the coarse level (133.80 m) routinely would be described as Ruisbroek Sand. The last interpretation, including the Wintham Silt, is applied in figure 24.2.

In the Doel 2b borehole a major change in the organic-walled microfossil content occurs at 139 m where the clays again turn to more sandier sediment (figure 24.2). The dinocyst abundance is sharply decreased and the lowest of all samples studied. More

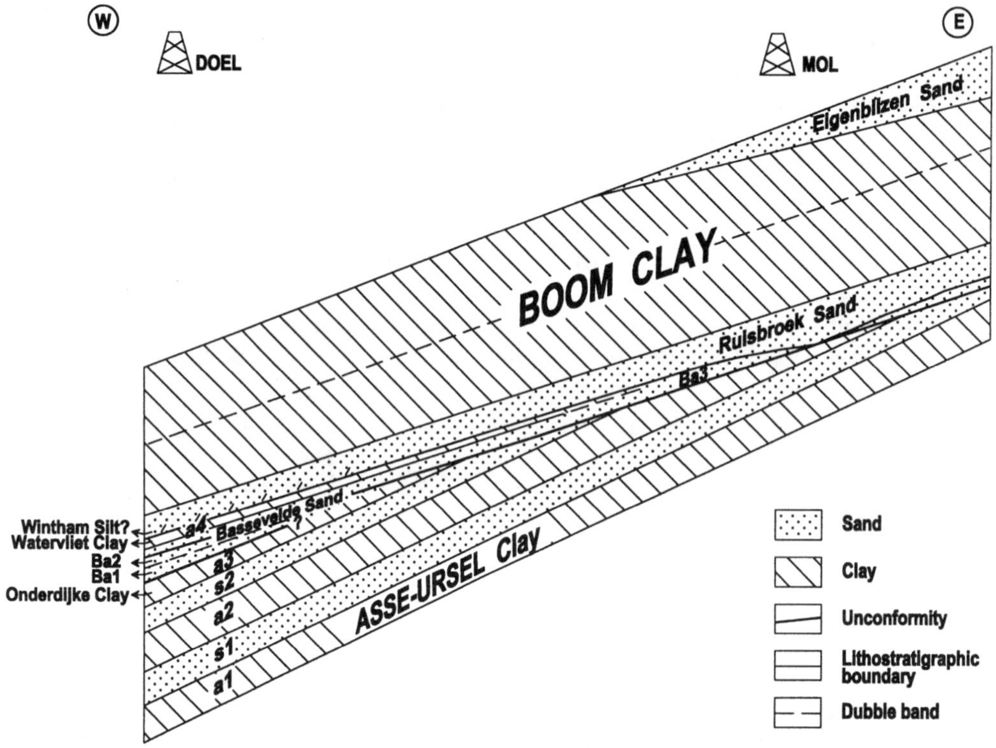

Figure 24.5. Schematic cross-section through the upper Eocene to lower Oligocene of northern Belgium. The vertical scale can be estimated from figure 24.2 and the horizontal distance between Doel and Mol is approximately 60 km. Note the erosive character of the Ruisbroek Sand and the major unconformity underlying the Bassevelde 3 sequence. The precise geometric relationships between the Bassevelde 1, 2 and the Onderdijke Clay is not well known. In earlier stratigraphic literature and practice the Onderdijke Clay and Watervliet Clay are called respectively a3 and a4 (a for argile, French for clay) and the Bassevelde Sand and Ruisbroek Sand respectively s3 and s4 (s for *sable*, French for "sand"). These units are subdivisions of the Kallo Complex, an obsolete stratigraphic term introduced by Gulinck (1969) in his description of the nearby Kallo well for the succession of clay and sand layers between the Asse Clay (upper Lutetian) and the Boom Clay (Rupelian), two easily identifiable thicker clay units in northern Belgium.

bisaccate pollen as well as other pollen and spores occur above this level. In addition, from this depth on, reworked Eocene dinocysts occur and a couple of meters higher large numbers of reworked Lower and Upper Cretaceous and Paleocene dinocysts appear. The consistent presence of *Areosphaeridium diktyoplokum* also ends at 139 m. Its disappearance has been inferred to be related to a distinct cooling of the seawater well known from Chron C13N (Brinkhuis and Visscher, 1995). At the same level the Priabonian dinocyst *Rhombodinium perforatum* disappears. Also at 139 m disappears the *Homotryblium floripes/plectilum* plexus, which is present from the bottom of the Doel 2b section, although in important numbers only from the base of the Bassevelde sequence 3. Its temporal disappearance may be related to the cooling of the seawater, since it represents a tropical taxon (Brinkhuis, 1994). The interval above 139 m is characterized by the sharp increase in *Cordosphaeridium cantharellum*, *Cribroperidinium* spp., *Achomosphaera alcicornu*, *Dapsilidinium pseudocolligerum*, *Glaphyrocysta semitecta*, *Operculodinium tiara*, *Palaeocystodinium* spp. and *Spiniferites pseudofurcatus,* and by the first appearance of *Wetzeliella gochtii*. The latter may be considered to be a typical Oligocene species. The warm-water *Coccolithus formosus*, frequently represented at 139 m, is absent in the next calcareous level at 133.8 m, pointing to a cooling of the surface waters.

Because of the marked change in the dinocyst

assemblage, which probably indicates a minor hiatus, and the change in facies from marine to lagoonal sediment (see below), the transition from clay to sandy silt at 139 m is interpreted as the top of the Bassevelde 3 sequence. Its age is most convincingly indicated by the disappearance of *Areosphaeridium diktyoplokum*, defining the boundary between the Adi and Rac dinoflagellate zones of Brinkhuis and Biffi (1993) in central Italy (within NP21, in the basal part of P18).

The Ruisbroek-Boom Sequence

The base of the sequence. The first clayey to silty sand unit above the Bassevelde 3 sequence, between 139 m and 133.5 m, is characterized by very abundant bisaccate pollen, by many reworked dinocysts from the Lower Cretaceous through the Eocene, and by the presence to frequent presence of *Corodosphaeridium cantharellum, Elytrocysta breva, Enneadocysta pectiniforme, Gerdiocysta conopea, Glaphyrocysta semitecta, Histiocysta* spp., *Melitasphaeridium* spp., *Membranophoridium aspinatum, Pentadinium lacticinctum,* and *Wetzeliella gochtii.* The presence of *Gerdiocysta conopea, Wetzeliella gochtii, Phthanoperidinium* spp., many bisaccate and other pollen, spores, and the freshwater alga *Pediastrum* point to a lagoonal rather than shallow marine environment for this interval. The same holds for the clayey sediments between the two coarse levels above (133.5 m and 129.65 m; see figure 24.2): pollen and spores content remains high, *Pediastrum* remains present, reworked dinocysts are less numerous and are now dominated by a Lower Cretaceous provenance, and the continued presence of *Glaphrocysta semitecta, Membranophoridium aspinatum, M. connectum* and *M. intermedium.* The two coarse glauconite containing sediment horizons indicate flooding of the lagoonal sediments by marine incursions. The simultaneous appearance of *Chiropteridium* spp. and disappearance of *Membranophoridium aspinatum* toward the top of the Ruisbroek Sand also points to the installation of an open-marine environment.

At the base of the Ruisbroek Sands in the nearby Kallo well (106 m), a colder-water influx is indicated by the presence of the dinocyst *Svalbardella* sp. (De Coninck, 1999), supporting the interpretation for a marked cooling at the end of sequence Bassevelde 3 already discussed above. Also the appearance of frequent *Glaphyrocysta semitecta* and *Achomosphaera alcicornu* in the area at the base of the Ruisbroek-Boom sequence can be explained by the cooling pulse, which forces these species to migrate southward (Brinkhuis and Biffi, 1993).

The Ruisbroek Sand. The Ruisbroek Sand, as well as the underlying lagoonal deposits, is clearly deposited in a broad erosive space appearing from a basal unconformity in the mapping (see figure 24.5). These sands continue to contain many reworked dinocysts, in particular from a Lower Cretaceous source pointing to the uplift of the Artois axis in the south (figure 24.1), and also reworked Mesozoic calcareous nannofossils (figure 24.4). Their sand-size evolution is systematically coarsening upward. For all these reasons the Ruisbroek Sand is considered as a lowstand systems tract (Vandenberghe et al., 1998), overlain by the transgressive Boom Clay. These sands also have a distinct brownish grey hue compared to the greenish-gray color of the Bassevelde Sands. The lagoonal sediments at the base of the sequence belong also to the lowstand systems tract. The presence of marine shallow-water sediments above lagoonal sediments can be considered as the evolution to more open but still very shallow marine conditions when the sea level starts slowly rising. This effect of more marine conditions toward the top of a lowstand systems tract is certainly helped by the start of important subsidence in the area creating space for the deposition of the relatively deeper water (50–100 m) Boom Clay. It is worth noting that in a similar way more shoreward the equivalent lowstand deposits of the same sequence are exclusively non-marine but lagoonal and estuarine (formerly called continental Tongrian) with at the top incoming coastal deposits just before the onset of the main transgressive flooding surface (Vandenberghe et al., 1998; Verbeeck et al., 1998).

The calcareous nannoflora has drastically changed between the top of the Watervliet Clay at 139 m and the base of the Ruisbroek Sand at 133.8 m, the intermediate interval being noncalcareous. Between these levels there is a major reduction in abundance and in species diversity related to cooling and to the presence of restricted environments. Warm-water taxa, among which are *Coccolithus formosus, Pontosphaera prava* and *P. wechesensis* permanently disappear, whereas cool-water forms first appear (*Cyclococcolithus hirsutus*) or bloom (*Isthmolithus recurvus, Dictyococcites bisectus*). As is generally the case in the area (Steurbaut, 1986, 1992), including the Doel 2b borehole, calcareous nannofossils in the Ruisbroek Sand at the marine part of the lowstand systems tract range from the base of Zone NP22, including *Hayella situliformis* and the last *Lanternithus minutus*, to the latest NP22 or earliest NP23 at the very top (NP22 between 133.85 and 118.40 m and NP23 from 117.35 m onward, see figures 24.2 and 24.4). Most important, it should be noted that the lagoonal base of the lowstand deposits between 139 and 134 m is devoid of calcareous nannoplankton, so that no exact age of the interval can be given. The top of the underlying Bassevelde 3 sequence does contain nannoplankton indicative of NP21 and is associated with the LO of *Aerosphaeri-dium diktyoplokum.* Usually the latter event occurs within Zone NP21. Therefore, either the 139–134 m interval still belongs to NP21, or the erosion shown by the start of the reworking of the dinocysts and ac-companying the low sea level at the sequence boundary at 139 m, has eroded the top NP21 sediments.

The dinoflagellate marker *Wetzeliella gochtii* has its first occurrence at 136.12 m, a few meters above the base of the Ruisbroek-Boom sequence. The last occurrence of *Glaphrocysta semitecta* is situated just above the lower coarse glauconitic marine influx at the base of the marine Ruisbroek Sand: its last abundant occurrence is at 133 m and its last occurrence is at 132.2 m. It defines the Rac/Cin dinoflagellate zonal boundary of Brinkhuis and Biffi (1993). Note also that the Ruisbroek Sands contain some pure clay layers toward the top (figure 24.2) unlike in their type area to the south. This is probably due to the deeper distal setting of the Doel area, with respect to the type area.

The transition to the Boom Clay Formation. Above the Ruisbroek Sands occurs the Boom Clay (figure 24.5). It is well exposed along the Rupel River and was originally meant to be the unit stratotype of the Rupelian (Dumont, 1849). The base of the Boom Clay is marked in the exposure area in the "Land van Waas" by a phosphatic level consisting of phosphatized fossils, bones, trace-fossils, and internal molds of mollusks, originally cementing the top of the Ruisbroek Sand. This level has subsequently been slightly reworked by erosion due to the important transgression cutting new beach profiles upon its way southwards. It is the transgressive surface delimiting the underlying lowstand systems tract sediments from the overlying transgressive type Rupelian Boom Clay. In the Doel 2b borehole the sedimentary transition between Ruisbroek Sand and Boom Clay is less abrupt than in the outcrop area to the south, probably due to the already deeper depositional environment in the Doel area. The boundary between both units in the Doel 2b borehole is selected based on the position of the reversal in the resistivity and gamma ray trends, a well-proven procedure in the log correlation of the Rupelian in north Belgium, the Netherlands, and the lower Rhine area in Germany (Vandenberghe et al., 2001).

Reworking of older dinocysts, in particular those derived from Lower Cretaceous rocks, continues at the base of the Boom Clay pointing to the continued uplift, exposure and erosion of the Artois axis. The abundance of *Apteodinium australiense* a few meters above the base of the Boom Clay indicates warmer sea surface temperatures than before.

The Rupelian Boom Clay

The Boom Clay is a banded silty clay displaying cycles of different periodicities (figure 24.6). The banding is caused by obliquity and dominant eccentricity orbital cycles (Van Echelpoel and Weedon, 1990). Third-order sequences are reflected by the vertical trends in the grain-size and geochemical properties (Vandenberghe et al., 1997, 1998; Laenen, 1998) and dinoflagellate distribution (Stover and Hardenbol, 1994) (figure 24.6). The thickness evolution of the individual cycles, reflecting sediment supply, shows a 400,000-year cyclicity; other longer duration cycles are detected by geochemistry, stable isotopes, and clay minerals (Laenen, 1998; Vandenberghe et al., 1997).

A third-order sequence boundary within the Boom Clay is expressed by a very silty zone located at about 88 m in the Doel 2b borehole. Its age can be estimated (figure 24.6) from paleomagnetic information, from dinocyst and planktonic foraminiferal zonations, and from counting Milankovitch cycles from boundary levels of biozones and magnetozones with known ages from Berggren et al. (1995). It occurs certainly above the C12r/C12n magnetochron boundary (31 Ma) and probably already within the C11r magnetochron (30.1–30.4 Ma). Using the equal time duration of the clay-silt cycles and the age of the foraminiferal boundary levels, an age of 30.3 Ma can be calculated while the dinoflagellate events around this sequence boundary identified by Stover and Hardenbol (1994) point to an age between 30.7 and 31.5 Ma (LO of *Fibrocysta axialis* at 30.7 Ma situated just above the sequence boundary and the FO of *Apteodinium spiridoides* at 31.5 Ma just below the sequence boundary). This age estimate falls in between the Ru2 and Ru3 sequence boundary ages of, respectively, 32 Ma and 29.4 Ma in de Graciansky et al. (1998). Since this sequence boundary within the Boom Clay is the first major boundary above the Ru1 it is suggested that it represents the Ru2 sequence boundary of de Graciansky et al. (1998). Slightly older than this sequence boundary, at 30.8 Ma start

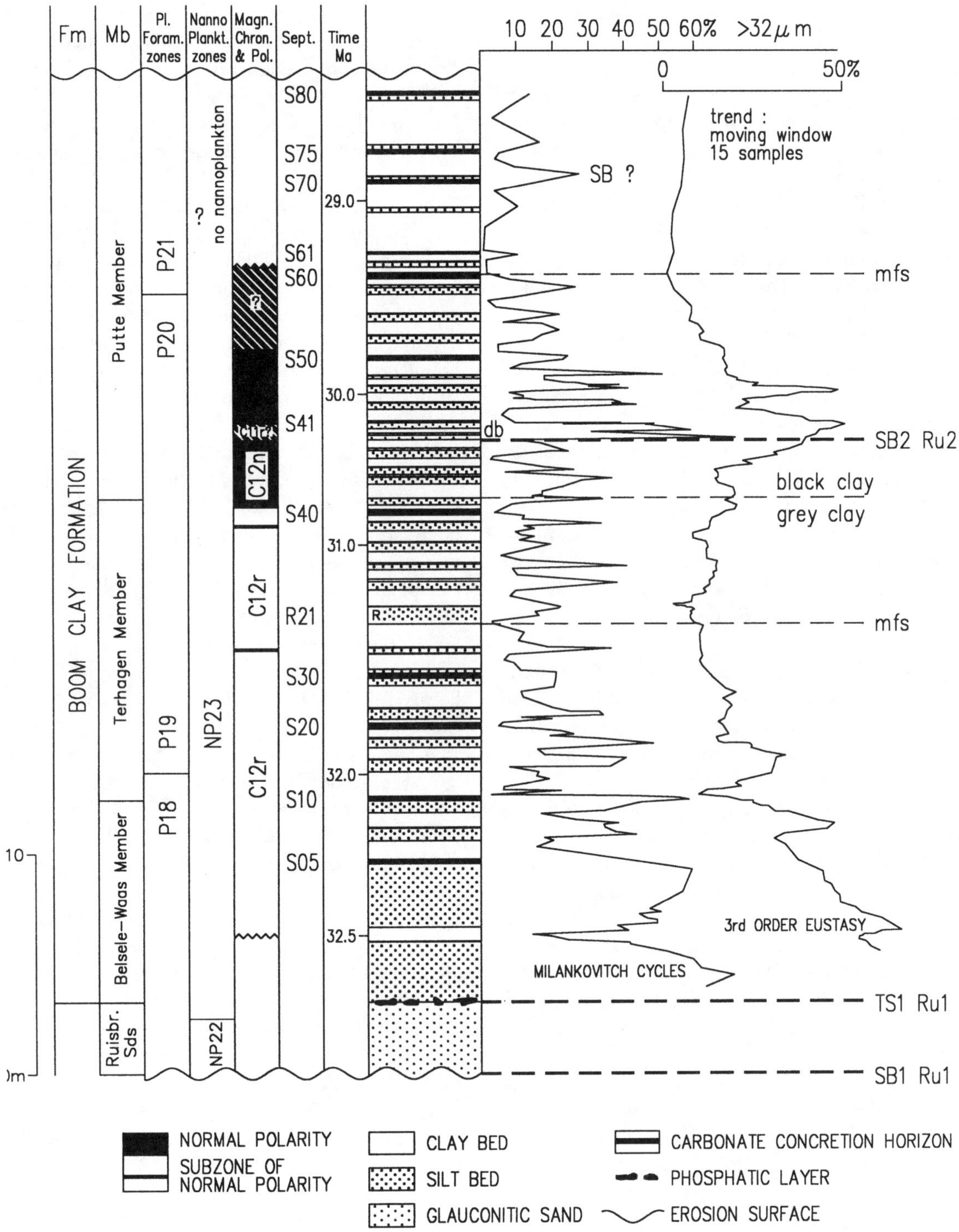

Figure 24.6. Synthesis of the Rupelian stratigraphy in the unit stratotype area (modified from Vandenberghe et al., 1997).

regionally widespread subanoxic conditions expressed by higher organic matter preservation in black clays.

DISCUSSION

Four depositional sequences have been identified in the Doel 2b borehole between 173 m and 88 m and the age of their boundaries determined by calcareous nannoplankton and dinoflagellates. The lowermost sequence boundary occurs at or just below the Bartonian-Priabonian boundary. It is correlative to the Pr1 sequence boundary of de Graciansky et al. (1998), the base of TA4.1 of Haq et al. (1987) and to the Epi-1 event of Abreu and Anderson (1998).

The sequence above (Bassevelde 1) is expressed by grain-size trends typical for a transgressive followed by a highstand systems tract. At the transgressive base occurs a glauconite layer. The sequence boundary at the top may be correlated to Pr2 of de Graciansky et al. (1998) and TA4.2 of Haq et al. (1987). Since no glauconite concentration is found at the base, this sequence boundary is considered to be of minor importance, in accordance with the observation that according to Abreu and Anderson (1998) no important isotope fluctuation occurs at this level.

The overlying Bassevelde 2 sequence is also expressed by a lower fining upward and an upper coarsening upward grain-size trend. It is bounded above by a sequence boundary characterized by dinocyst events that occur at the Eocene/Oligocene boundary as defined at Massignano in Italy.

The overlying Bassevelde 3 sequence, including also the Watervliet Clay, coincides approximately with biochron NP21 and is bounded at its top (139 m) by an erosive surface marking a major sequence boundary. At this time the Artois axis to the south is uplifted and eroded and the marine basin to the north subsides. The importance of this sequence boundary is further indicated by the presence in the overlying Ruisbroek-Boom-1 sequence (figure 24.2) of a well-developed lowstand systems tract, already in Zone NP22 with a few meters at its very base that might still be NP21, and more particularly by a major change in dinocysts and nannoplankton at the sequence boundary, pointing to a strong cooling pulse. This sequence boundary is therefore correlative to Ru1 of de Graciansky et al. (1998), and to the TA4.3/4 boundary of Haq et al. (1987), since it is associated with the major isotope event Ori-1 (Abreu and Anderson, 1998). Besides, the Ru1 sequence boundary, as identified at the base of the Ruisbroek Sands (sensu lato) by Vandenberghe and Van Echelpoel (1987), Stover and Hardenbol (1994) and Vandenberghe et al. (1998) occurs just below the Rupelian unit stratotype in the Boom Clay along the Rupel River and is meant to be the Ru1 at the Eocene/Oligocene boundary in de Graciansky et al. (1998) and the TA4.3/4.4 of Haq et al. (1987). However, the cooling and hence Ru1 in the Doel 2b borehole, occurring at the top of the Adi dinocyst Zone and very near the NP21 to NP22 transition across an erosive surface, is situated well above the Eocene/Oligocene boundary as associated with the extinction of the Hantkeninidae planktonic foraminifera. The problem was already recognized and discussed by Stover and Hardenbol (1994: figure 4). They demonstrated that the Ru1 sequence boundary at the base of the Ruisbroek Sands (sensu lato) did occur at or just above the top of the Priabonian type section as defined respectively by Barbin (1988) and Hardenbol (1968). Obviously the Eocene-Oligocene GSSP defined at Massignano in the Marche-Umbria Basin in Italy is older than the boundary between the Priabonian and the Rupelian unit stratotypes as shown by Brinkhuis and Visscher (1995) and Berggren et al. (1995). Therefore the Ru1 sensu de Graciansky et al. (1998) has to be positioned within the lower Oligocene in the upper part of the Zone NP21 and above the Adi/Rac transition between 32.8 Ma (base NP22) and 33.5 Ma (top Adi).

The sequence boundary between the Bassevelde 2 and 3 sequences, which is associated with the Eocene/Oligocene boundary at 33.7 Ma probably corresponds to the Pr3 sensu de Graciansky et al. (1998), and the TA4.2/3 boundary of Haq et al. (1987), and to the important isotope cooling event Epi-2 of Abreu and Anderson (1998). The two isotope cooling events at the beginning of the Oligocene according to Miller et al. (1991), Pekar and Miller (1996) and Zachos et al. (1996), Oi1, or O1a and Oi1a, or O1b, respectively occur at 33.5 Ma and near the NP21/NP22 boundary at 32.8 Ma. The last one indeed corresponds with the Ru1 S.B. as defined in this analysis of the Doel 2b well. It is suggested that the former one corresponds approximately to the boundary between Bassevelde 2 and 3 sequences. The strong cooling at the Ru1 sequence boundary undoubtedly caused an important sea-level lowering. The preservation of lagoonal and shallow-marine lowstand deposits in such a situation requires strong tectonic subsidence of the area preceding and allowing the thick marine Boom Clay deposition. Sequence analysis of the Paleogene nearshore sediments in this slowly subsiding area (Vandenberghe et al., 1998) has revealed that most lithological units consist of relating thin units separated by unconformities, lacking an LST. LST sediments could only be identified in areas with incised valleys, which had created the additional space for LST sediments to be preserved.

The sequence boundary within the Boom Clay at about 88-m depth in the Doel 2b well is a well-known level in the Boom Clay and it corresponds to the mid-Rupelian sequence boundary TA4.4/4.5 (see also Stover and Hardenbol, 1994). As it is the immediate younger third-order sequence boundary above Ru1 in a sedimentary succession lacking significant erosive surfaces, it is logical to correlate this boundary with Ru2. However, it seems to have an age between the Ru2 and Ru3 ages reported by de Graciansky et al. (1998: chart 1) and therefore it might be different from Ru2 as well.

Counting down the orbital cycles from the P18/19 planktonic foraminiferal zonal boundary leads to an estimated minimum age of about 32.5 Ma and maximum age of 32.8 Ma for the transgressive surface TSRu1 (uncertain due to the silty base of the Belsele Waas Member where the clear clay-silt couples are absent). Being situated at the base of NP23, this transgressive surface has an age of about 32.30 Ma (latitude dependent boundary, see Berggren et al., 1995). This is compatible with the age of the Ru1 sequence boundary age between 32.8 Ma and 33.5 Ma as discussed before.

The Stratigraphic Sequences in the Eocene-Oligocene Transitional Strata of the North Sea Basin

The Tongrian deposits in Belgium. The Tongrian is a regional stage traditionally regarded as the lower division of the former threefold Oligocene subdivision (e.g., Vinken, 1988). Sometimes, however, the Tongrian Stage is incorporated in the top Eocene (e.g., Haq and Van Eysinga, 1987), but sometimes only the marine lower Tongrian, dated as NP21, is included in the Eocene, and the overlying continental Tongrian, belonging to Zone NP22, in the Oligocene (e.g., Cavelier, 1979; Pomerol, 1982). The base of the marine Tongrian (Grimmertingen Sand) is a very marked regional unconformity. Based on the presence of NP21 in the marine Tongrian, it is correlated with the Bassevelde 3 sequence in the Doel 2b borehole. The continental Tongrian deposits are correlated with the Ruisbroek Sand in the Doel 2b well, based on the common presence of *Gerdiocysta conopea* and the presence of nannoplankton Zone NP22 in some marine incursion layers. Therefore the continental Tongrian represents the landward lowstand systems tract of the Ruisbroek-Boom-1 sequence above Ru1 (see Vandenberghe et al., 1998). Just above the Ru1 S.B. (in the Hoogbutsel complex) occurs a tetrapod fauna of Zone MP21, locating Stehlin's Grande Coupure (GC) between the Bassevelde 3 and the Ruisbroek-Boom-1 sequences.

The Isle of Wight section. The stratigraphy of the Solent Group, which spans the Eocene-Oligocene transitional interval in the Isle of Wight, has been described by Hooker (1992), Collinson (1992), and King and Hooker (1995). The Colwell Bay Member is a marine unit within the dominantly nonmarine Headon Hill Formation. It contains nannoplankton indicating the NP19/20 interval and therefore a correlation with the Bassevelde 2 sequence is proposed. The Colwell Bay Member is overlain by mainly freshwater deposits, including the upper part of the Headon Hill Formation, the Bembridge Limestone Formation, and the lower part of the Bouldnor Formation (Bembridge Marls Member), which are deposited in meso- to polyhaline environments (King and Hooker, 1995). The Bembridge Marls belong to the MP20 mammal zone and therefore could correspond to the higher sea level during the Bassevelde 3 or marine Tongrian sequence in Belgium. The lower part of the Hamstead Member above the "*Nematura* Band" contains the oldest MP21 fauna, as in the base of the continental Tongrian in Belgium. Also, the green clays and the mollusks of the Middle and Upper Hamstead beds bear a strong resemblance to the facies of the continental Tongrian in Belgium. The boundary between the Bembridge Marls Member and the Hamstead Member, which is marked by a paleosol, should therefore coincide with the Ru1 boundary in Belgium, a conclusion in accordance with the dinocyst study of De Coninck (1995b) and the very marked floral change at this boundary as described by Machin (1971). The onset of higher salinities is documented in the overlying Cranmore Member and seems to represent the Rupelian transgression.

The Ludian and Sannoisian facies of the Paris Basin. The Eocene Ludian facies of the central Paris Basin consist of gypsum and marls, the Oligocene Sannoisian facies of green clays and limestones. The former is correlated with the marine Tongrian and the latter with the continental Tongrian in Belgium (Pomerol, 1982).

Collinson (1992) presented a detailed correlation scheme of the Eocene-Oligocene strata between the Hampshire Basin and the Paris Basin. The Upper Gypsum mass, indicating a warm and dry climate,

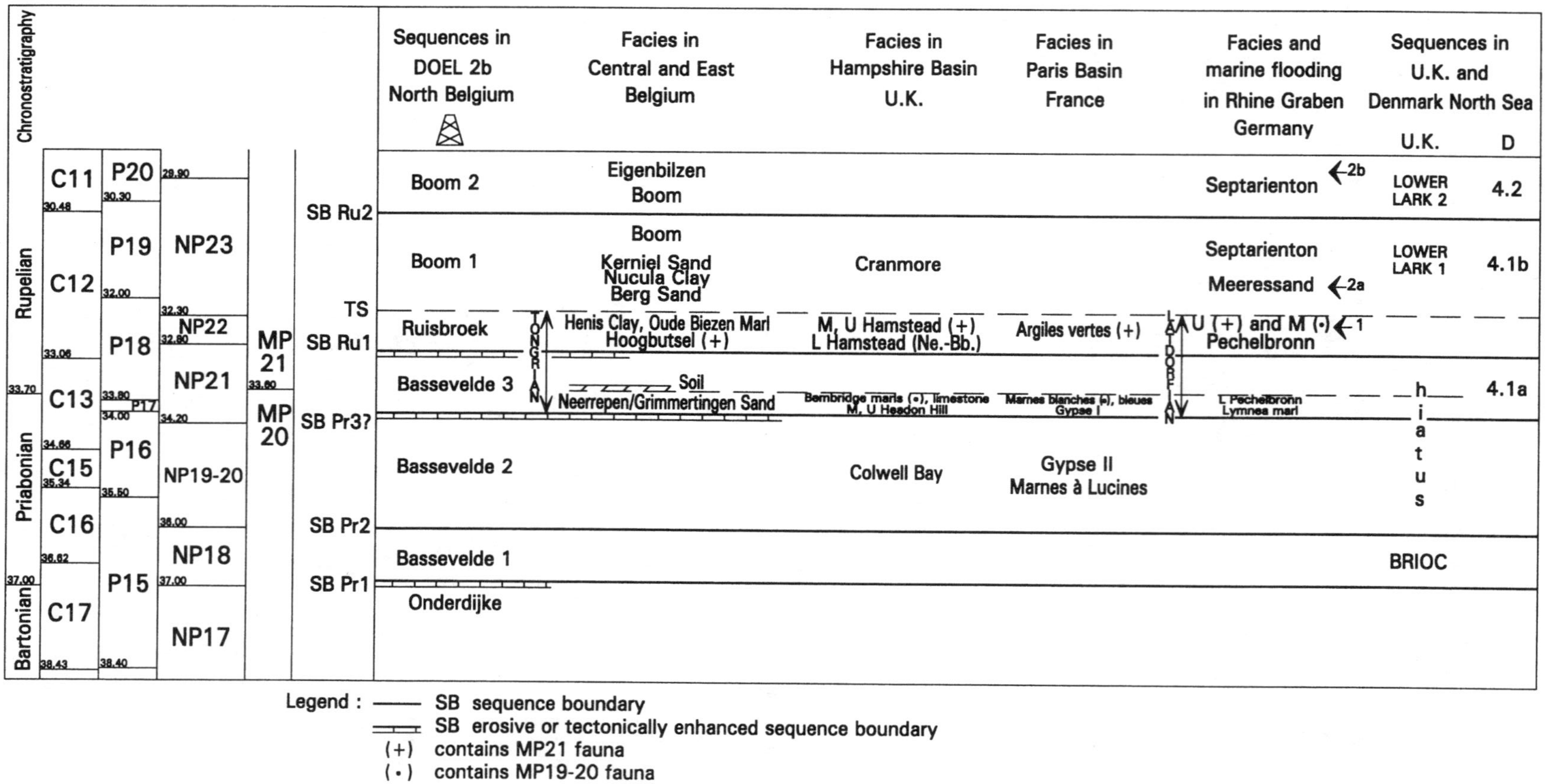

Figure 24.7. Comparison of the Doel 2b sequences with the different facies, events and sequences in the Eocene/Oligocene boundary interval of central and eastern Belgium, the Hampshire Basin, the Paris Basin, the Rhine area, and the North Sea subsurface, based on literature. The correlations are discussed and references given in the text. The biostratigraphy is from Berggren et al. (1995).

and the overlying *Marnes bleues d'Argenteuil*, which contain a tropical flora, correspond respectively to the Bembridge Limestone Formation and the lower part of the Bembridge Marls. The overlying *Marnes Blanches de Pantin* show a floral change with the loss of several tropical taxa and the appearance of more temperate taxa. They still contain MP19/20 faunas. The Grande Coupure occurs in the base of the overlying *Argile verte de Romainville*. This green clay resembles the clays in the Hamstead Member. It belongs to nannoplankton Zone NP22 and to mammal Zone MP21, while the flora has returned to tropical. This situation is very similar to the sequences described in Belgium where the Neerrepen Sand in the top of the Bassevelde 3 or marine Tongrian sequence also shows an increase in temperate conifer pollen, while the Henis green clay facies of the continental Tongrian, just above the Ru1 sequence boundary, shows a return of warm and tropical pollen.

The Latdorfian strata in Germany. In the Rhine Graben, Martini (1990) recognized in the Middle Pechelbronn strata, containing nannoplankton Zone NP22, a marine incursion (arrow 1 in figure 24.7) that he correlates with the major flooding of the TA4.4 sequence of Haq et al. (1987) and another marine incursion (arrow 2 in figure 24.7) in overlying sediments of NP23 age, that he correlates with the flooding of the TA4.5 sequence of Haq et al. (1987). A younger marine incursion (arrow 3 in figure 24.7) occurs in the base of NP24. We suggest that the NP22 and NP23 incursions represent floodings during the lowstand and during the transgressive systems tract of the Ruisbroek-Boom-1 sequence above Ru1 (TA4.4) and the younger one at the base of NP24 represents the Boom-2 sequence above Ru2. The Middle Pechelbronn strata occupy the upper part of the Latdorfian, which is a regional stage name for the lower Oligocene in Germany. The latter is often correlated with the Tongrian and the Sannoisian, other regional stage names for the lower part of the Oligocene in the former threefold subdivision of the Oligocene in which the Rupelian represented the middle Oligocene (e.g. Martini et al., 1986). Unlike in the other areas discussed, in the Rhine area the GC occurs between the NP22 and NP23 boundaries (Martini and Rothe, 1998), a situation discussed in detail by Hooker (1992).

The North Sea subsurface. Michelsen (1994) and Michelsen et al. (1998) identify three sequences in the lower Oligocene, with the upper one continuing into the upper Oligocene. The lower boundary of their lowermost sequence (4.1) is located within nannoplankton Zone NP21, within agglutinating foraminiferal Zone NSA6b and just below the top of the *Areosphaeridium diktyoplokum* highest occurrence. These sequences contain more silt and mica than the clayey sequences below, which is probably controlled by the start of Late Cenozoic uplift (Michelsen, 1994). Michelsen (1996) and Danielsen et al. (1997) could subdivide the lower sequence into two individual sequences (4.1a and b) and suggested a correlation of the lowermost one with the Viborg Formation in onshore Denmark. The upper sequence (4.1b) is situated partly in nannoplankton Zones NP22 and NP23 (Danielsen et al., 1997: figure 4). The lower boundary of sequence 4.2 lies within NP23 and the lower boundary of sequence 4.3 within NP24. It is therefore suggested that sequence 4.1a corresponds to the marine Tongrian or Bassevelde 3 sequence in Belgium, the 4.1b sequence to the Ruisbroek-Boom-1 sequence, and the middle Lower Oligocene sequence (4.2) to the Boom-2 sequence above the Ru2 sequence boundary. It should be noted that Neal (1996: figure 5) also correlates the Viborg Formation in Denmark with the marine Tongrian in Belgium.

In the Moray Firth, Central Graben, and South Viking Graben, Neal (1996) and Neal et al. (1998) have identified three sequences in the Priabonian and the Rupelian. The lower sequence (Brioc) contains nannoplankton Zone NP18 and consequently corresponds to the Bassevelde 1 sequence in Belgium. In this area, an important hiatus is followed by sequence Lower Lark 1, which starts after the disappearance of *A. diktyoplokum* and therefore can be correlated with the Ruisbroek-Boom-1 sequence in Belgium. The Lower Lark 2 sequence is situated within NP23 and is correlated by Neal (1996) to the Boom-2 sequence above Ru2 and can also be correlated with the middle lower Oligocene (4.2) sequence of Michelsen et al. (1998).

The Eocene-Oligocene in central Italy. In central Italy, from dinoflagellate studies, Brinkhuis and Biffi (1993) associate the major low eustatic sea-level position corresponding to the first major Antarctic ice sheet growth with the Adi/Rac dinocyst boundary. This level corresponds to Ru1 in the Doel 2b borehole. The slightly older sea-level low identified in Italy occurs just before or at the Hantkeninidae-defined Eocene/Oligocene boundary and can therefore be associated in the Doel 2b borehole with the sequence boundary at the base of Bassevelde 3 or the marine Tongrian. With some minor modification, the sequence stratigraphic interpretation of the Priabona and Bressana sections in Brinkhuis and Visscher

(1995) and of the Massignano area in van Mourik and Brinkhuis (1999) also fits the stratigraphic position of these sea-level lows.

CONCLUSIONS

A detailed analysis of the Eocene-Oligocene transitional interval in the Doel 2b borehole at the southern rim of the North Sea Basin allows the identification of several depositional sequences. A comparison with the facies evolution and the biostratigraphy in the same stratigraphic interval in neighboring classical outcrop sections and in the subsurface of the North Sea shows the sequences to be of general occurrence in the North Sea Basin (figure 24.7). The ages of the sequence boundaries based on dinocyst and nannoplankton biostratigraphic datings allow correlation with the sequences identified in the Eocene-Oligocene GSSP area in central Italy and with the sequences described in Haq et al. (1987), and in de Graciansky et al. (1998) and the isotope events described by Abreu and Anderson (1998), Pekar and Miller (1996), and Zachos et al. (1996).

The Bassevelde 1 sequence is bound at its base by the Pr1 boundary, coinciding with the Epi-1 cooling event. The nannofossils in this sequence still indicate warm surface waters, although dinocysts suggest already a cooling. The Bassevelde 2 sequence, probably of less importance, corresponds with the Pr2 sequence boundary. The Bassevelde 3 sequence falls within the NP21 Zone and is correlative of the marine Tongrian. The lower boundary of the sequence is characterized by dinocyst events occurring at the Eocene-Oligocene GSSP. The nannofossils in sequence Bassevelde 3 indicate a distinct cooling of the surface waters. It is suggested that its base correlates with the Pr3 sequence boundary and with the isotope event Oi1 (Miller) or O1a (Zachos).

The boundary between the Bassevelde 3 and the overlying Ruisbroek-Boom-1 sequence represents a major cooling, as indicated both by the dinocysts and the calcareous nannoplankton. It occurs in the top of the NP21 or the base of the NP22 Zone. It is correlated with the Ru1 sequence boundary and the isotope event Oi1a (Miller) or O1b (Zachos). This major cooling event is situated well within the Oligocene if calibrated against the Hantkeninidae extinction event but it does correspond very closely to the boundary between the Priabonian and Rupelian unit stratotypes. Since the Bassevelde 3 sequence (the marine Tongrian), is the lowermost Oligocene sequence, it is worth considering to reintroduce the threefold subdivision of the Oligocene. After this major cooling event, the dinocysts show a reappearance of warmer conditions. The next sequence boundary within the Rupelian Boom Clay represents the TA4.4/5 boundary of Haq et al. (1987) and should logically be Ru2 but it has a slightly younger age than Ru2 in de Graciansky et al. (1998).

ACKNOWLEDGMENTS

We thank NIRAS-ONDRAF and SCK-CEN for permission and facilities to study the borehole. We thank R. Coccioni for examining the foraminifera of the Doel 2b borehole, which unfortunately did not contribute to the problems studied in this paper. We sincerely thank O. Michelsen and C. King for their constructive reviewing of the paper, leading to several improvements. This study is part of a project on the Eocene-Oligocene transition in Belgium funded by the Fonds voor Wetenschappelijk Onderzoek (Fund for Scientific Research)—Vlaanderen (research project No. G.0093.01).

REFERENCES

Abreu, V. S., and Anderson, J. B., 1998, Glacial eustasy during the Cenozoic: Sequence stratigra-phic implications: American Association of Petro-leum Geologists Bulletin, v. 82, p. 1385-1400.

Barbin, V., 1988, The Eocene-Oligocene transition in shallow-water environment: the Priabonian stage type area (Vicentin, northern Italy), *in* Premoli-Silva, I., Coccioni, R., and Montanari, A., eds., The Eocene/Oligocene boundary in the Marche-Umbria Basin (Italy): International Union of Geological Sciences, International Subcommission on Paleogene Stratigraphy, Ancona, Italy, Special Publication II-7, p. 163-171.

Berggren, W. A., Kent, D. V., Aubry, M.-P., and Hardenbol, J., 1995, Geochronology, time scales and global stratigraphic correlation: SEPM Special Publication 54, 386 p.

Brinkhuis, H., 1994, Late Eocene to early Oligocene dinoflagellate cysts from the Priabonian type area (northwest Italy): biostratigraphy and palaeoenvironmental interpretation: Palaeogeography, Palaeoclimatology, Palaeoecology, v. 107, p. 121-163.

Brinkhuis, H., and Biffi, U., 1993, Dinoflagellate cyst stratigraphy of the Eocene/Oligocene transition in central Italy: Marine Micropaleontology, v. 22, p. 131-183.

Brinkhuis, H., and Visscher, H., 1995, The upper boundary of the Eocene series: a reappraisal based on dinoflagellate cyst biostratigraphy and sequence stratigraphy, *in* Berggren, W. A., Kent,

D. V., Aubry, M.-P., and Hardenbol, J., eds., Geochronology, time scales and global stratigraphic correlation: SEPM Special Publication 54, p. 295-304.

Cavelier, C., 1979, La limite Eocène-Oligocène en Europe occidentale: Sciences Géologiques Institut de Géologie Strasbourg Mémoire 54, p. 1-280.

Collinson, M. E., 1992, Vegetational and floristic changes around the Eocene/Oligocene boundary in western and central Europe, *in* Prothero, D. R., and Berggren, W. A., eds., Eocene-Oligocene climatic and biotic evolution: Princeton, New Jersey, Princeton University Press, p. 437-450.

Danielsen, M., Michelsen, O., and Clausen, O. R., 1997, Oligocene sequence stratigraphy and basin development in the Danish North Sea sector based on log interpretations: Marine and Petroleum Geology, v. 14, no. 7/8, p. 931-950.

De Coninck, J., 1995a, Microfossiles à paroi organique du Bartonien, Priabonien et Rupélien inférieur dans le sondage de Kallo; espèces significatives dans les sondages de Woensdrecht, Kallo et Mol: Mededelingen Rijks Geologische Dienst, v. 53, p. 65-105.

De Coninck, J., 1995b, Corrélations entre les dépôts du Lutétien au Rupélien du Bassin belge, et des Bassins de Hamsphire et de Paris: Mededelingen Rijks Geologische Dienst, v. 53, p. 107-118.

De Coninck, J., 1999, Organic-walled phytoplankton biostratigraphy of the Eocene-Oligocene transition in the Kallo borehole and the Rupelian stratotype area (northwestern Belgium): Bulletin Belgische Vereniging voor Geologie, t. 105 (1996)., no. 3-4, p. 171-209

de Graciansky, P.-Ch., Hardenbol, J., Jacquin, Th., and Vail, P. R., 1998, Mesozoic and Cenozoic sequence stratigraphy of European basins: SEPM Special Publication 60, 786 p.

Drooger, C. W., 1969, Le sondage de Kallo (au Nord-Ouest d'Anvers). Microfauna close to the Eocene/Oligocene boundary in the boring at Kallo: Mémoires Explication Cartes Géologiques et Minières de la Belgique, v. 11, p. 9-38.

Dumont, A., 1849, Rapport sur la carte géologique du Royaume: Académie Royale de Belgique Bulletin, v. XVI, IIe partie, p. 351-373.

Gulinck, M., 1969, Le sondage de Kallo (au nord-ouest d'Anvers). I. Coupe résumée des terrains traversés au sondage de Kallo et profil géologique NS passant par Woensdrecht-Kallo-Halle: Mémoires Explication Cartes Géologiques et Minières de la Belgique, v. 11, p. 3-7.

Haq, B. U., and Van Eysinga, F. W. B., 1987, Geological time table: Elsevier, 4th edition.

Haq, B. U., Hardenbol, J., and Vail, P. R., 1987, Chronology of fluctuating sea levels since the Triassic: Science, v. 235, p. 1156-1166.

Hardenbol, J., 1968, The Priabonian type section (a preliminary note): BRGM Mémoire, v. 58, p. 629-635.

Hooker, J. J., 1992, British mammalian paleocommunities across the Eocene-Oligocene transition and their environmental implications, *in* Prothero, D. R., and Berggren, W. A., eds., Eocene-Oligocene climatic and biotic evolution: Princeton, New Jersey, Princeton University Press, p. 494-515.

King, C., and Hooker, J. J., 1995, The Paleogene of the Isle of Wight (Hampshire Basin, England), *in* RCNPS/RCNNS Meeting, Southampton, England, Field Excursion 6-8 September 1995, 16 p.

Laenen, B., 1998, The geochemical signature of relative sea-level cycles recognised in the Boom Clay: Aardkundige Mededelingen, v. 9, p. 61-82.

Machin, J., 1971, Plant microfossils from the Tertiary deposits of the Isle of Wight: New Phytologist, v. 70, p. 851-872.

Martini, E., 1990, The Rhinegraben system, a connection between northern and southern seas in the European Tertiary: Veröff. Übersee-Mus., v. A10, p. 83-98.

Martini, E., and Rothe, P., 1998, Sieblos an der Wasserkuppe: Forschungsbohrungen in einem alttertiären See, *in* Martini, E., and Rothe, P., eds., Die alttertiäre Fossilslagerstätte Sieblos an der Wasserkuppe/Rhön: Geologische Abhandlung Hessen, v. 104, p. 7-27.

Martini, E., Fahlbusch, V., and Hagn, H., 1986, The Eocene/Oligocene boundary and the Latdorfian (lower Oligocene): Newsletters in Stratigraphy, v. 17, no. 1, p. 37-43.

Michelsen, O., 1994, Stratigraphic correlation of the Danish onshore and offshore Tertiary successions based on sequence stratigraphy: Bulletin Geological Society of Denmark, v. 41, p. 145-161.

Michelsen, O., 1996, Late Cenozoic basin development of the eastern North Sea Basin: Bulletin Geological Society Denmark, v. 43, p. 9-21.

Michelsen, O., Thomsen, E., Danielsen, M., Heilmann-Clausen, C., Jordt, H., and Laursen, G.V., 1998, Cenozoic sequence stratigraphy in the Eastern North Sea, *in* de Graciansky, P.-Ch., Hardenbol, J., Jacquin, Th., and Vail, P .R., eds.,

Mesozoic and Cenozoic Sequence Stratigraphy of European Basins: SEPM Special Publication, v. 60, p. 91-118.

Miller, K. G., Wright, J. D., and Fairbanks, R. G., 1991, Unlocking the ice house: Oligocene-Miocene oxygen isotopes, eustasy, and margin erosion: Journal of Geophysical Research, v. 96, no. B4, p. 6829-6848.

Neal, J. E., 1996, A summary of Paleogene sequence stratigraphy in northwest Europe and the North Sea, *in* Knox, R. W. O'B., Corfield, R. M., and Dunay, R. E., eds., Correlation of the early Paleogene in Northwest Europe: Geological Society Special Publication 101, p. 15-42.

Neal, J. E., Stein, J. A., and Gamber, J. H., 1998, Nested stratigraphic cycles and depositional systems of the Paleogene Central North Sea, *in* de Graciansky, P.-Ch., Hardenbol, J., Jacquin, Th., and Vail, P. R., eds., Mesozoic and Cenozoic Sequence Stratigraphy of European Basins: Society for Sedimentary Geology Special Publication, v. 60, p. 261-288.

Pekar, S., and Miller, K. G., 1996, New Jersey Oligocene "Icehouse" sequences (ODP Leg 150X) correlated with global $\delta^{18}O$ and Exxon eustatic records: Geology, v. 24, p. 567-570.

Pomerol, C., 1982, The Cenozoic Era: Tertiary and Quaternary: Horwood Series in Geology, 272 p.

Steurbaut, E., 1986, Late middle Eocene to middle Oligocene calcareous nannoplankton from the Kallo well, some boreholes and exposures in Belgium and a description of the Ruisbroek Sand Member: Contributions to Tertiary and Quaternary Geology , v. 23, no. 2, p. 49-83.

Steurbaut, E., 1992, Integrated stratigraphic analysis of Lower Rupelian deposits (Oligocene) in the Belgian Basin: Annales de la Société Géologique de Belgique, t. 115, fasc. 1, p. 287-306.

Stover, L. E., and Hardenbol, J., 1994, Dinoflagellates and depositional sequences in the Lower Oligocene (Rupelian) Boom Clay Formation, Belgium: Bulletin Belgische Vereniging voor Geologie, 1993, v. 102, no. 1-2, p. 5-77.

Vandenberghe, N., and Van Echelpoel, E., 1987, Field guide to the Rupelian stratotype: Bulletin Belgische Vereniging voor Geologie, v. 96, no. 4, p. 325-337.

Vandenberghe, N., Laenen, B., Van Echelpoel, E., and Lagrou, D., 1997, Cyclostratigraphy and climatic eustasy: example of the Rupelian stratotype: Acad. Sci. Paris, Sciences de la Terre et des Planètes, C.R., v. 325, p. 305-315.

Vandenberghe, N., Laga, P., Steurbaut, E., Hardenbol, J., and Vail, P. R., 1998, Tertiary sequence stratigraphy at the southern border of the North Sea Basin in Belgium, *in* de Graciansky, P.-Ch., Hardenbol, J., Jacquin, Th., and Vail, P .R., eds., Mesozoic and Cenozoic Sequence Stratigraphy of European Basins: SEPM Special Publication, v. 60, p. 119-154.

Vandenberghe, N., Hager, H., van den Bosch, M., Verstraelen, A., Leroi, S., Steurbaut, E., Prüfert, J., and Laga, P., 2001, Stratigraphic correlation by calibrated well logs in the Rupel Group between North Belgium, the Lower-Rhine area in Germany and Southern Limburg and the Achterhoek in The Netherlands: Proceedings 7th Joint Biannual Regional Committees on Northern Paleogene and Neogene Stratigraphy, Aardkundige Mededelingen 11, Leuven University Press.

Van Echelpoel, E., and Weedon, G. P., 1990, Milankovitch cyclicity and the Boom Clay Formation, an Oligocene siliciclastic shelf sequence in Belgium: Geological Magazine, v. 127, no. 6, p. 599-604.

van Mourik, C. A., and Brinkhuis, H., 1999, Late Eocene to early Oligocene organic walled dinoflagellate cysts of the Massicore, Central Italy: Journal of Conference Abstracts, EUG 10, Strasbourg, France, Cambridge Publications 4-1, p. 734-735.

Verbeeck, K., Vandenberghe, N., and Ottenburgs, R., 1998, On the presence of sepiolite in the Borgloon Formation (Tongeren Group) in East Brabant, Belgium: Aardkundige Mededelingen, v. 9, p. 121-124.

Vinken, R., ed., 1988, The Northwest European Tertiary Basin: Geologisches Jahrbuch, Reihe A, H. 100, 508 p.

Wei, W., and Wise, S. W., 1990, Biogeographic gradients of middle Eocene-Oligocene calcareous nannoplankton in the South Atlantic Ocean: Palaeogeography, Palaeoclimatology, Palaeoecology, v. 79, p. 29-61.

Williams, G. L., Brinkhuis, H., Bujak, J. P., Damassa, S. P., Hochuli, P. A., de Verteuil, L., and Zevenboom, D., 1998, Dinoflagellates, *in* de Graciansky, P.-Ch., Hardenbol, J., Jacquin, Th., and Vail, P. R., eds., Mesozoic and Cenozoic sequence stratigraphy of European basins: SEPM Special Publication, v. 60, p. 764-765

Zachos, J. C., Quinn, T. M., and Salamy, S., 1996, High resolution (10^4 yr) deep-sea foraminiferal stable isotope records of the earliest Oligocene climate transition: Paleoceanography, v. 9, p. 353-387.

Ziegler, P. A., 1982, Geological atlas of Western and Central Europe: Elsevier, Amsterdam, 130 p.

CHAPTER 25

Deep-Water Benthic Foraminiferal Events from the Massignano Eocene/Oligocene Boundary Stratotype, Central Italy

Rodolfo Coccioni and Simone Galeotti

ABSTRACT

We carried out a quantitative and semi-quantitative analysis of late Eocene–earliest Oligocene deep-water benthic foraminifera from the lower bathyal Massignano Global Stratotype Section and Point. Species distribution and faunal parameters (planktonic/total foram ratio, faunal density, diversity, the proportion of agglutinating and infaunal forms, and the dissolved oxygen index) enabled us to do the following:

1. Recognize and calibrate several, discrete bioevents against a well-established bio-, chrono-, and magnetostratigraphy.

2. Characterize the timing, and the paleoecological and paleoceanographic conditions in the western Tethys that accompanied the end of the well-documented, global Eocene-Oligocene deep-water foraminiferal faunal turnover. The timing of this faunal turnover, which is known to be time-transgressive, is here suggested to follow a latitudinal gradient with areas close to the Antarctic continent showing earlier taxonomic turnover.

3. Recognize that *Nuttallides truempyi* disappeared well before cold bottom waters reached the western Tethys, indicating that this event was not related to a general cooling and the ensuing enhanced bottom-water corrosivity and oxygenation. A mechanism invoking changes in paleoproductivity regimes possibly related to increased seasonality, is a more likely explanation of the observed pattern.

INTRODUCTION

The Eocene-Oligocene transition contains the intertwined history of global cooling and changes in oceanic circulation (see Prothero, 1994, for a review) and represents one of the three major intervals of turnover in Cenozoic deep-sea benthic foraminifera (see Miller et al., 1992). In contrast to other periods of dramatic and sudden extinction in benthic foraminiferal assemblages (i.e., Paleocene/Eocene boundary), a series of first and last appearances occurred over a period of several millions of years from the late middle Eocene into the earliest Oligocene (see Tjalsma and Lohmann, 1983; Berggren and Miller, 1989; Miller et al., 1992; Thomas, 1992a, b). This faunal turnover reflects major oceanographic changes related to a long-term global cooling and resulted in a species-richness gradient between high and low latitudes as a consequence of decreased diversity among high-latitude faunas during the late Eocene (Thomas and Gooday, 1996). An explanation for such a decrease in diversity at high latitudes has taken into account the physiochemical properties of deep-water masses including high $CaCO_3$ corrosivity and/or oxygenation (Thomas, 1992a; Kaiho, 1994a).

Most of the studies on benthic foraminiferal paleoecology across the Eocene-Oligocene transition are based on lower bathyal to abyssal assemblages (Miller et al., 1992). However, no quantitative, high-resolution study on benthic foraminiferal assemblages across the Eocene-Oligocene transition has been carried out in the western Tethys. The Massignano Global Stratotype Section and Point (GSSP) offers the opportunity to calibrate the lower bathyal benthic foraminiferal events to the calcareous plankton bio-, magneto-, and chronostratigraphy through the late Eocene (ca. 36 Ma)–earliest Oligocene (ca. 33 Ma).

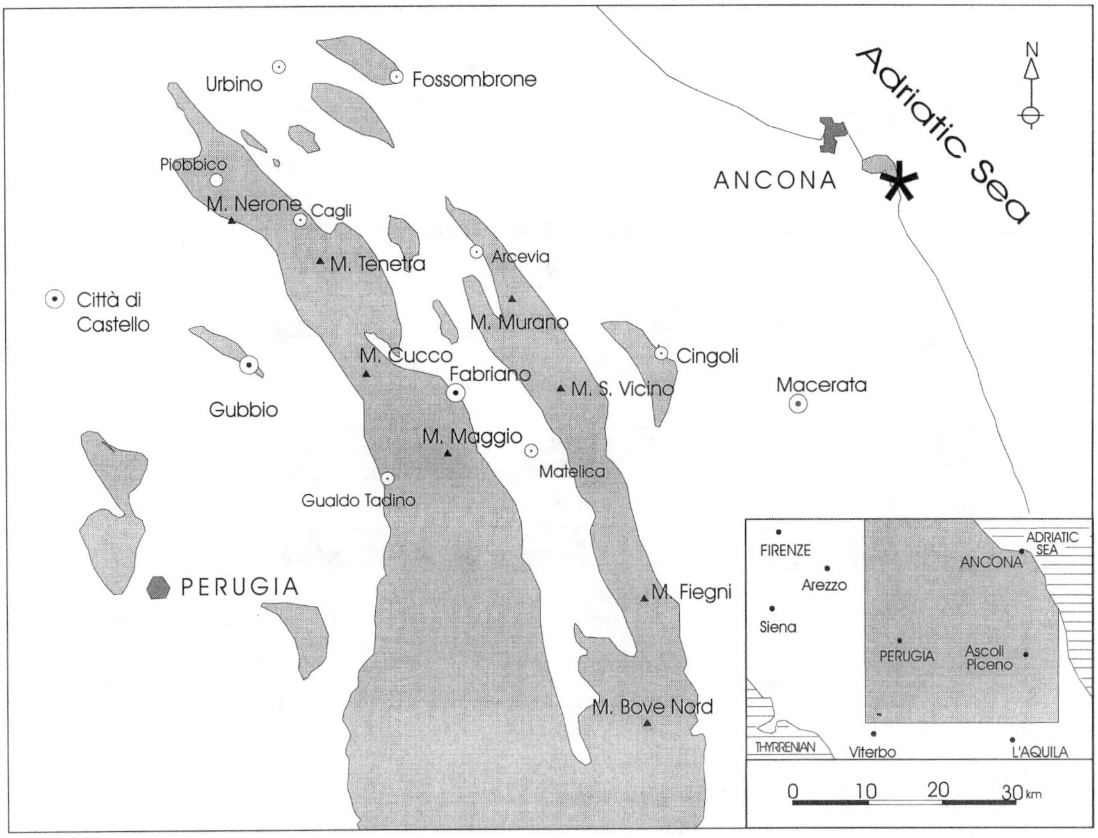

Figure 25.1. Tectonic sketch-map of the Umbria-Marche Apennines where the shaded areas represent the Meso-Cenozoic orogenic belt. The location of the Massignano GSSP is marked by an asterisk.

THE MASSIGNANO EOCENE/OLIGOCENE BOUNDARY GSSP

The Massignano section, located in the Umbria-Marche Basin (figure 25.1), is the currently accepted type section for the Eocene/Oligocene boundary (Premoli Silva and Jenkins, 1993). The exposed sequence, which covers the latest Eocene to earliest Oligocene, is 23 m thick and consists of reddish and greenish-gray hemipelagic marls and calcareous marls of the Scaglia Variegata and Scaglia Cinerea formations with some intercalations of volcanic biotite-rich layers (figure 25.2). Detailed micropaleontological and biostratigraphic (Coccioni et al., 1986, 1988; Coccioni, 1988; Nocchi et al., 1988; Parisi and Coccioni 1988; Bellagamba and Coccioni, 1990; Brinkhuis and Biffi, 1993), magnetostratigraphic (Bice and Montanari, 1988; Lowrie and Lanci, 1994), sedimentological, geochemical and geochronological (Capo and DePaolo 1988; Montanari et al., 1988, 1993; Odin et al., 1988) studies have been performed on this section. Vonhof (1998) suggested that the $\delta^{18}O$ record at Massignano is diagenetically altered and it is here disregarded. The Massignano GSSP spans calcareous nannofossil Zones NP18 to NP21 of Martini (1971) and CP15a to CP16 of Okada and Bukry (1980), planktonic foraminiferal Zones P15 to P18 of Blow (1969, 1979) and Berggren and Miller (1988), and Zones *Porticulosphaera semiinvoluta* to *Turborotalia cerroazulensis–Pseudohastigerina* spp. of Berggren et al. (1995), and Chrons C16n to C13n. Following the time scale of Berggren *et al.* (1995), the Massignano GSSP spans 3 million years (36.2 to 33.2 Ma) with a mean sedimentation rate of 7.6 m/myr.

MATERIAL AND METHODS

Data are presented from 58 samples. Sample preparation included gentle crushing, immersion in a hydrogen peroxide solution (10%), and sieving through a 63 μm fine mesh. A quantitative analysis based on approximately 300 specimens per sample was carried

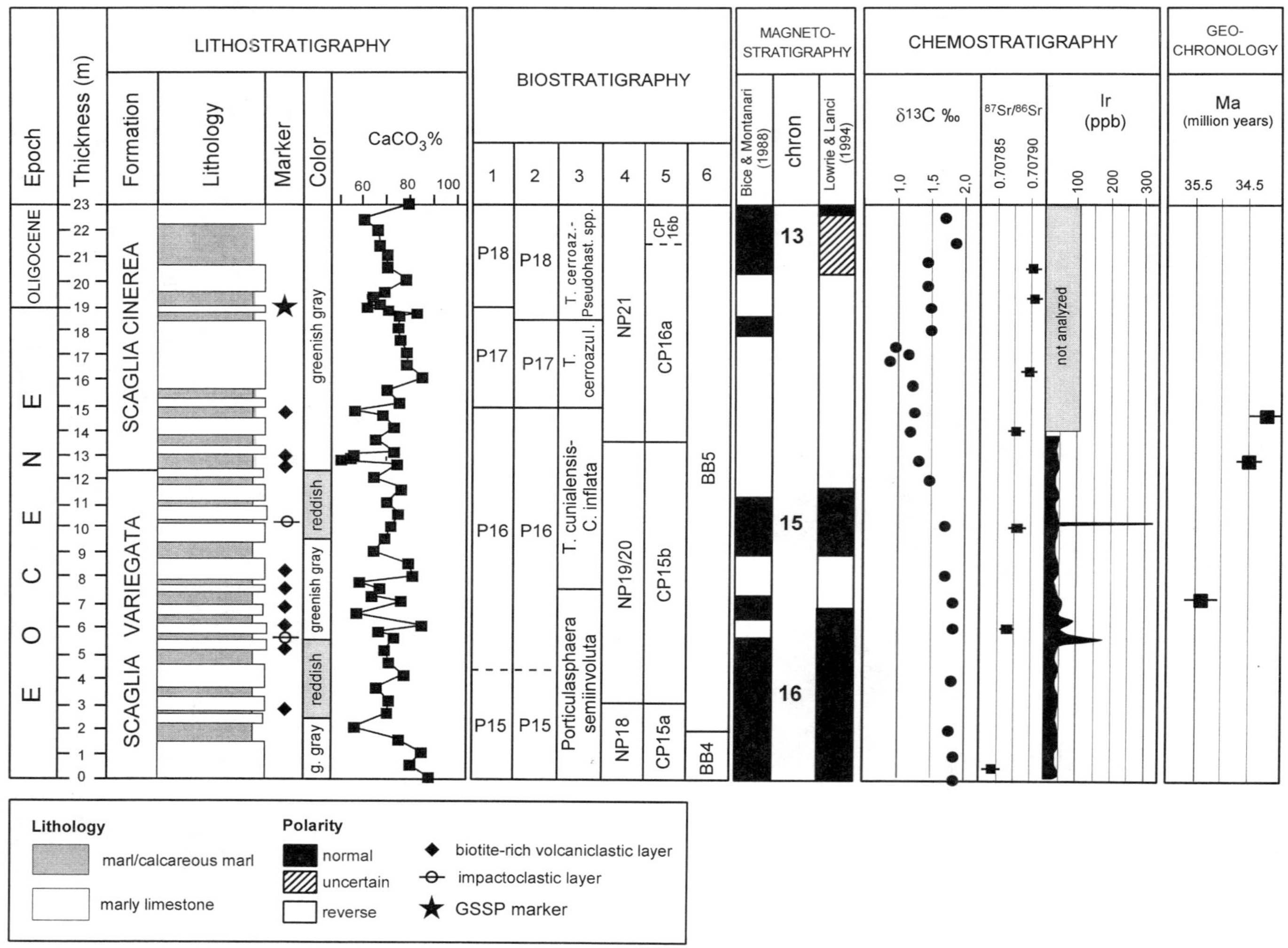

Figure 25.2. Integrated stratigra-phy of the Massignano GSSP. All the available data are reported excluding the oxygen isotope record which is diagenetically altered and is here disregarded. Litho- and biostratigraphy (columns 1, 4, and 5) after Coccioni et al. (1988). Different biozonal schemes according to their authors are reported. Planktonic foraminifera: (1) Blow, 1969, 1979; (2) Berggren and Miller, 1988; (3) Berggren et al., 1995. Calcareous nannoplankton: (4) Martini, 1971; (5) Okada and Bukry, 1980. Benthic foraminifera: (6) Berggren and Miller, 1989. Carbon isotope record after Odin et al. (1988). Strontium isotope record after Capo and DePaolo (1988). Iridium concentration record after Montanari et al. (1993). Geochronology after Montanari et al. (1988) and Odin et al. (1988).

out on the fraction greater than >125 μm. A semi-quantitative survey conducted on the fraction >63 μm allowed to check the presence of rare and/or small-sized species. The generic classification followed herein is mainly based on that outlined by Loeblich and Tappan (1988).

To check how benthic foraminiferal assemblages responded to paleoenvironmental changes, the following faunal parameters have been calculated: % planktonic forams (planktonic/total foraminifera); faunal density (number of foraminifera >63 μm per g of dry sediment); diversity (Fisher-a index); Ag/Ag + CH x 100 ratio where Ag and CH are the numbers of agglutinating and calcareous-hyaline forms, respectively; I/I + E x 100 ratio where I and E are the numbers of infaunal and epifaunal forms, respectively (morphotypes based on Corliss [1985], Bernhard [1986], and Corliss and Chen [1988]); Dissolved Oxygen Index (DOI) (O/O + D x 100 ratio where O and D are the numbers of Oxic and Dysoxic forms, respectively). Assessment of identified species as oxic or dysoxic taxa have been carried out largely following Kaiho (1991, 1994a). The time scale adopted in this study is that of Berggren et al. (1995).

RESULTS

Benthic foraminiferal assemblages from the Massignano GSSP are diverse and well preserved. One hundred and 67 species belonging to 70 genera were identified. Cosmopolitan, long-ranging benthic foraminifera dominate, including genera *Cibicidoides*, *Hanzawaia*, *Osangularia*, *Oridorsalis*, *Bulimina*, *Gyroidinoides* and, subordinately, by *Karreriella*, *Vulvulina*, *Dorothia*, *Planulina*, *Dentalina*, and *Uvigerina* (figure 25.3). The latter genus is mainly represented by costate taxa (*biserialis*, *havanensis*, *jacksonensis*, *multistriata*, *rippensis*, and *spinicostata*) and, subordinately, by hispid forms (*ecuadorensis*, *farinosa*, *mantaensis*). In the lower part of the section, *Nuttallides truempyi* represents an important component (up to 20%) of the assemblages. Small-sized *Globocassidulina subglobosa* are common in the size fraction <125 μm.

Some specific Last Occurrences (LO) and First Occurrences (FO) are observed at different stratigraphic levels (figure 25.4): (1) FO of *Cibicidoides subhaidingerii* at 1.5 m, (2) LO of *Cibicidoides truncanus* at 2 m, (3) Last Common Occurrence (LCO) of *Nuttallides truempyi* at 4 m, (4) LO of *Nuttallides truempyi* at 9 m, and (5) FO of *Cibicidoides cookei* at 23 m. Other significant faunal events are: (1) the general increase in abundance of *Oridorsalis* and *Gyroidinoides* starting close to the LCO of *Nuttallides truempyi*, (2) the increase in abundance of *Globocassidulina subglobosa* in the size fraction <125 μm from 12 m upward, (3) the decrease in abundance of buliminids, represented by *Bulimina jarvisi*, followed by the temporary loss of this group at 18.5 m, and 4) a remarkable bloom of small-sized (<125 μm) bolivinids (i.e., *Brizalina* ex gr. *antegressa*) between 15 and 19 m.

All the faunal parameters taken into consideration show large fluctuations throughout the Massignano GSSP (figure 25.5). The P/P+Bx100 ratio is typically in the range of open-marine settings fluctuating from 91% to 99%. The faunal density ranges from 47 to 78. The Fisher-a index varies from 3.7 to 20. Calcareous-hyaline forms largely dominate the assemblages with a relative abundance always higher than 75%. Increasing values of the Ag/Ag+CHx100 ratio are observed throughout the Massignano GSSP. The proportion of infaunal forms ranges from 10% to 40%. The DOI fluctuates between 60 and 98.

Three discrete Benthic Foraminiferal Intervals (BFIs) have been distinguished based on patterns and faunal parameters (figure 25.5).

BFI-I (0-9 m)

This interval encompasses the upper part of Zone *Porticulosphaera semiinvoluta* and the lower part of Zone *Turborotalia cunialensis-Cribrohantkenina inflata* of Berggren et al. (1995). P/P+Bx100 ratio remains fairly stable throughout. Diversity, Ag/Ag+CHx100, and I/I+Ex100 ratios increase upsection, whereas DOI values gradually decrease. The faunal density increases up to 6 m, then rapidly decreases in the upper part of BFI-I.

BFI-II (9-19 m)

This interval corresponds to most of the Zone *Turborotalia. cunialensis–Cribrohantkenina inflata*, the entire Zone *Turborotalia cerroazulensis*, and the lowermost part of Zone *Turborotalia cerroazulensis–Pseudohastigerina* spp. of Berggren et al. (1995). Faunal parameters exhibit relatively large sample-to-sample fluctuations without any clear trend in the average values. However, DOI falls to the lowest observed value at 13 m, then recovers to previous values by 16 m and remains fairly constant in the upper third of BFI-II.

BFI-III (19-23 m)

This interval falls within the lower part of Zone *Turborotalia cerroazulensis–Pseudohastigerina* spp. of

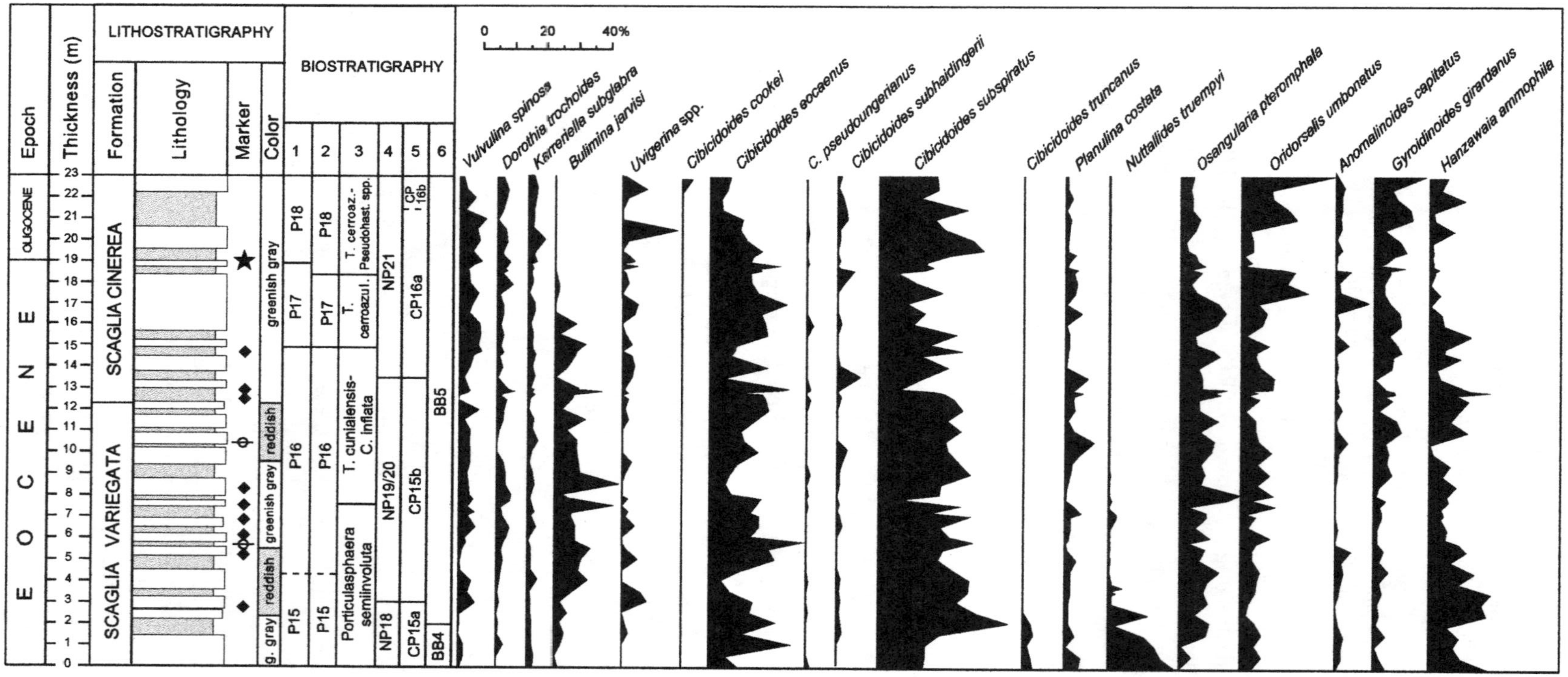

Figure 25.3. Relative abundance of selected benthic foraminiferal taxa from the size fraction >125 μm in the Massignano GSSP arranged in taxonomic order. Litho- and biostratigraphy as in figure 25.2.

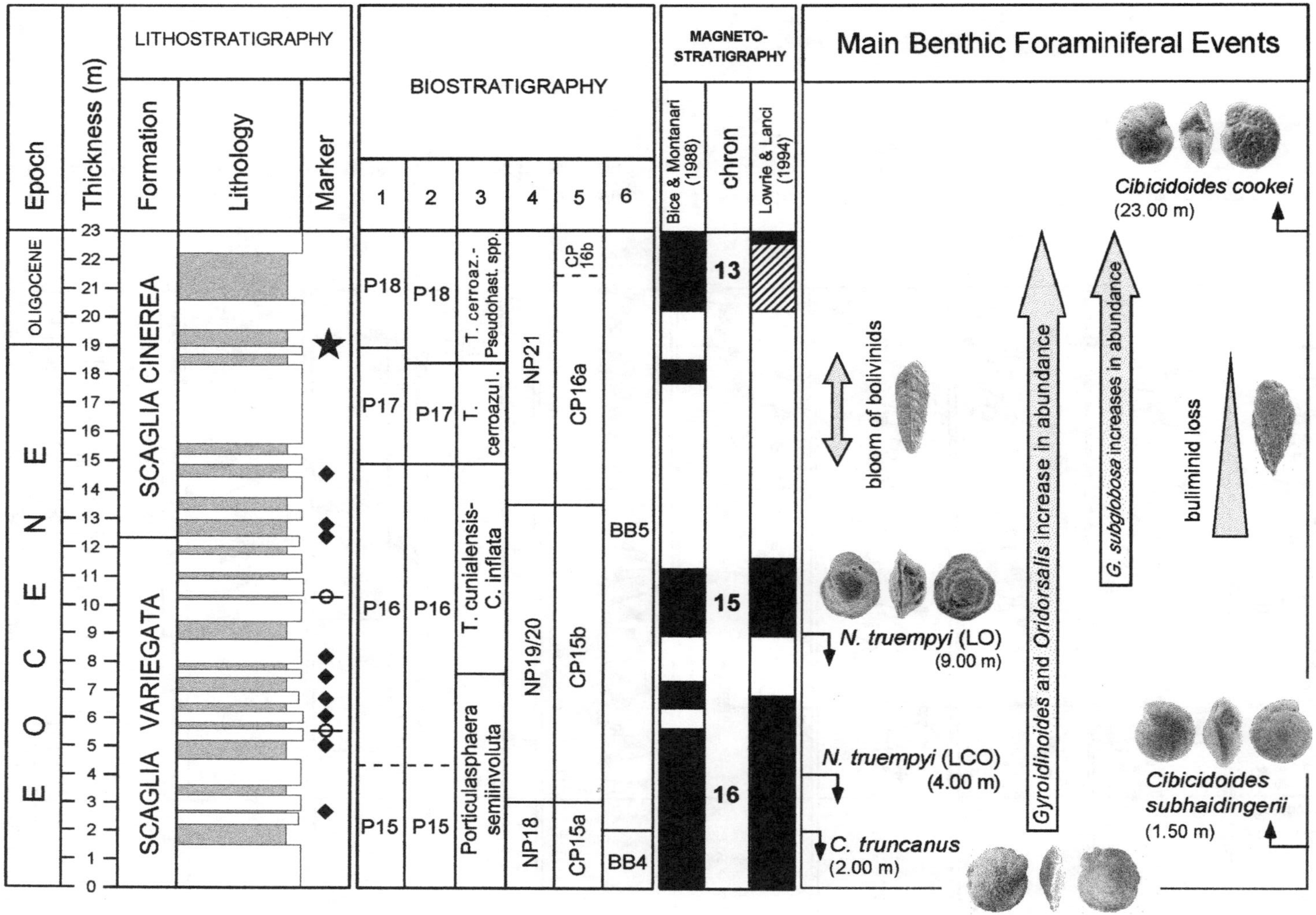

Figure 25.4. Main events recognized in deep-water benthic foraminiferal assemblages of the Massignano GSSP. Litho- and biostratigraphy as in figure 25.2.

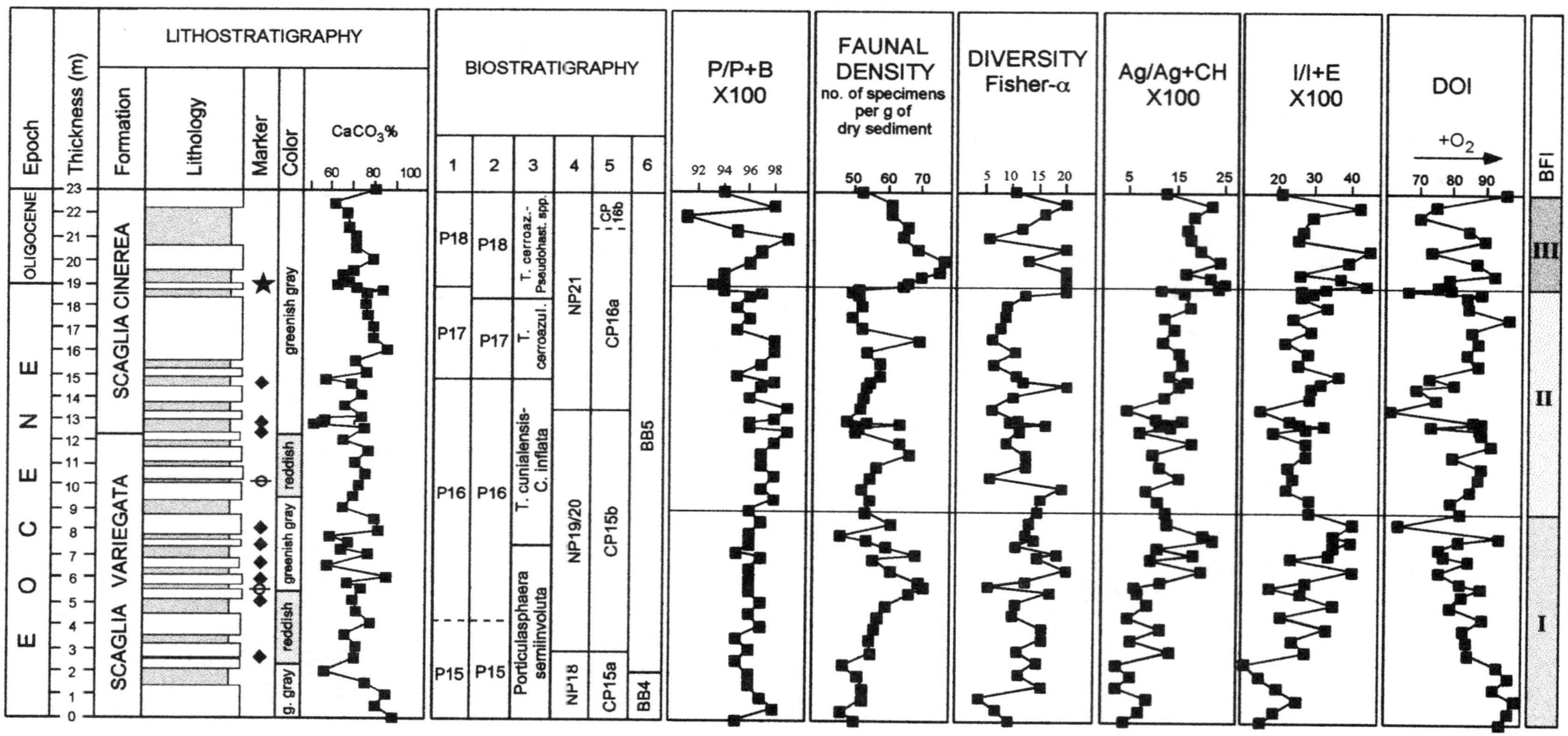

Figure 25.5. Benthic foraminiferal faunal parameter record in the Massignano GSSP. The Benthic Foraminiferal Intervals (BFI) defined according to faunal parameters record are shown as I to III from bottom to top. Litho- and biostratigraphy as in figure 25.2.

Berggren et al. (1995) and is characterized by very large, sample-to-sample fluctuations of several faunal parameters.

PALEOBATHYMETRY

Paleobathymetric assessment is based on the distribution of index forms as reported by van Morkhoven et al. (1986). The lack of faunal components typical of abyssal depth provides a lower depth limit of 2,000 m. The absence of taxa generally restricted to <600 m (e.g., rectuvigerinids) and <1,000 m (e.g., some uvigerinids and cibicids) in indicative of an upper depth limit of 1,000 m. Accordingly, the high abundance of *Anomalinoides capitatus, Bulimina jarvisi, Cibicidoides eocaenus, Cibicidoides subspiratus, Hanzawaia ammophila, Nuttalides truempyi,* and *Planulina costata* indicates a lower bathyal depositional setting at a paleodepth of 1,000–2,000 m. The absence of *Cibicidoides praemundulus,* which was primarily a lower bathyal and abyssal form showing greatest abundances at abyssal depth (van Morkhoven et al., 1986) and dominating lower bathyal to abyssal upper Eocene-Oligocene biofacies (Miller et al., 1992), restricts the paleodepth estimation to 1000-1500 m (upper part of the lower bathyal setting). Although some of the bathymetric indices show large fluctuactions in relative abundance or even local disappearance, no remarkable paleobathymetric change occurs throughout the Massignano GSSP by the general composition of the assemblages. Accordingly, the P/P+Bx100 ratio shows constant values typical for a bathyal depositional environment.

BIOSTRATIGRAPHY

The value of Cenozoic benthic foraminifera as biomarkers for long-distance correlations has been evidenced by Berggren and Miller (1989). These authors developed two discrete biozonations based on bathyal and abyssal taxa, respectively, and calibrated them to the P-Zones of Berggren and Miller (1988).

Following the calibration of Berggren and Miller (1989), the lower bathyal Massignano GSSP spans a time interval containing the BB4/BB5 zonal boundary, which is defined by the LO of *C. truncanus.* This event is reported by Berggren and Miller (1989) to occur at the P17/P18 zonal boundary of Berggren and Miller (1988) corresponding to the *Turborotalia cerroazulensis*/ *T. cerroazulensis– Pseudohastigerina* spp. zonal boundary of Berggren et al. (1995). In the Massignano GSSP, the LO of *Cibicidoides truncanus* is observed 2 m above the base of the section (figure 25.4), therefore falling in the upper part of Zone P15 of Blow (1969, 1979) and of Zone *Porticulosphaera semiinvoluta* of Berggren and Miller (1988) that is within the *Porticulosphaera semiinvoluta* Zone of Berggren et al. (1995) at about 36.1 Ma (figure 25.6). In terms of calcareous nannofossil biozonation, this bioevent occurs in the uppermost part of Zones NP18 and CP15a of Martini (1971) and Okada and Bukry (1980), respectively. Our data suggest that, in the western Tethys, the LO of *C. truncanus* might have occurred much earlier than previously observed at other sites (figure 25.6).

According to Berggren and Miller (1989), the LO of *Nuttalides truempyi,* defining the AB7/AB8 boundary of their abyssal zonation, takes place at the same time of the LO of *Cibicidoides truncanus.* In the lower bathyal Massignano GSSP, the LO of *Nuttalides truempyi* is observed 9 m above the base of the section post-dating its LCO (4 m) of about 0.7 Ma and falling in the middle part of Zone P16 of Blow (1969, 1979) and Berggren and Miller (1988) that is in the lower part of the *Turborotalia cunialensis* Zone of Berggren et al. (1995) at about 35.1 Ma (figures 25.4 and 25.6). In terms of calcareous nannofossil biozonation, this event occurs in the middle part of Zone NP19/20 and CP15b of Martini (1971) and Okada and Bukry (1980), respectively (figure 25.4).

A comparison with benthic foraminiferal record from other sites (figure 25.6), suggested that other bioevents of possible biostratigraphic use for the Eocene-Oligocene transition would be the FO of *Cibicidoides subhaindigerii* and of *Cibicidoides cookei* which, in the Massignano GSSP, occur at 1.5 m and 23 m, respectively (figures 25.4 and 25.6). Accessory benthic foraminiferal events reported by Berggren and Miller (1989) to occur in the late Eocene are the migrations into the bathyal realm from the neritic realm of planulinids and siphoninids. The former has been reported to occur in the middle part of Zone P15 and the latter in the lower-middle part of Zone P16 of Berggren and Miller (1988). In the lower bathyal setting of the Massignano GSSP, planulinids (i.e., *Planulina costata* and *Planulina renzi*) are already consistently present from the base of the section, that is from the upper part of Zones P15 of Blow (1969, 1979) and of Berggren and Miller (1988). The migration of siphoninids from the neritic environment to bathyal settings, which, according to Berggren and Miller (1989), took place in the lower middle part of Zone P16 of Berggren and Miller (1988), is not observed in the Massignano section, suggesting that this group experienced a different timing of migration in the western Tethys.

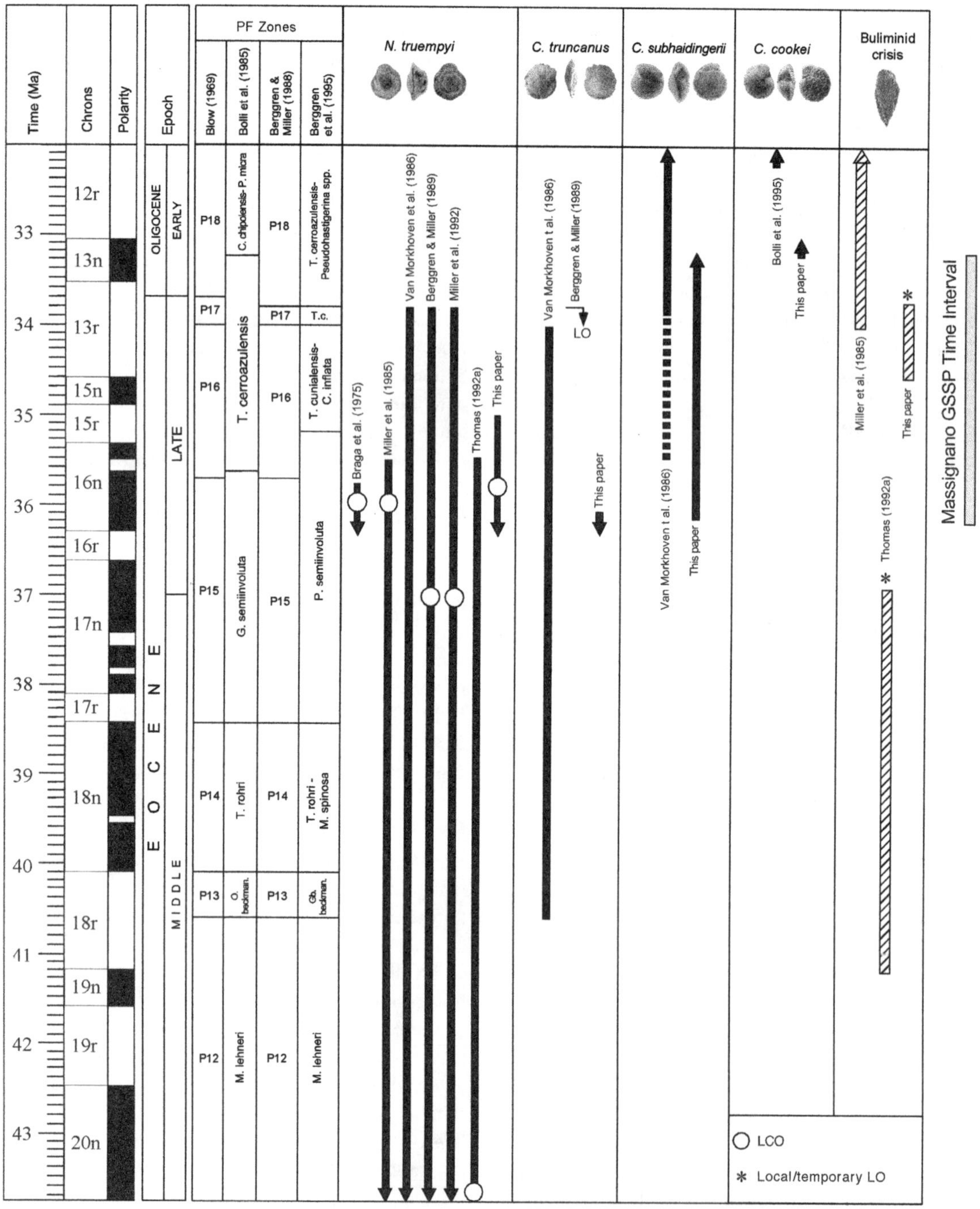

Figure 25.6. Stratigraphic ranges of biostratigraphically significant deep-water benthic foraminifera from the middle middle Eocene-earliest Oligocene interval plotted against the different planktonic foraminiferal (PF) Zones. Time scale after Beggren et al. (1995).

FAUNAL EVENTS AND PALEOECOLOGY

The disappearance of *N. truempyi* is the most important benthic foraminiferal event recorded in the surveyed interval. The LO of *N. truempyi* marks the end of a stepwise taxonomic reorganization affecting deep-sea benthic foraminifera in the late Eocene, whereby the LCO of this taxon marks the beginning of this gradual turnover (Miller et al., 1992). In bathyal to abyssal settings, a biofacies dominated by *Nutallides truempyi* was replaced by a biofacies dominated by *Globocassidulina subglobosa*, *Oridorsalis* spp., *Gyroidinoides* spp., and *Cibicidoides praemundulus* (see Miller et al., 1992 with references therein). In more general terms, the LCO of *Nutallides truempyi* marks the onset of the change to a transitional assemblage, in turn giving rise to "Modern Faunas" by middle–late Miocene time (see Thomas, 1992b). Based on data from several sites of the North and South Atlantic, Gulf of Mexico, and Caribbean Sea, Miller et al. (1992) stated that the LCO of *Nutallides truempyi* could have occurred either in the late middle Eocene or early late Eocene. However, at Site 689 (Weddelll Sea, Antarctica) the LCO of *Nutallides truempyi* is recorded in older stratigraphic levels occurring in the middle middle Eocene, close to the CP13/CP14a nannofossil zonal boundary, that is at the base of Zone *Morozovella lehneri* of Berggren et al. (1995) (Thomas, 1992a) (figure 25.6). On the other hand, as already pointed out by Miller et al. (1992), the timing of this event is poorly correlated to the geomagnetic polarity time scale and uncertain, although it is well understood that it is time-transgressive. In the Massignano GSSP, the LCO of *Nutallides truempyi* occurs well within the late Eocene, in the upper part of Chron C16n2n, that is in the uppermost part of Zone *Porticulosphaera semi-involuta* of Berggren et al. (1995). Following the time scale of Berggren et al. (1995), the numerical age of the *Nutallides truempyi* LCO at Massignano would be about 35.8 Ma. Taking into consideration available data (Braga et al., 1975; Miller et al., 1985, 1992; van Morkhoven et al., 1986; Berggren and Miller, 1989; Thomas, 1992a), the timing of *Nutallides truempyi* LCO occurred much earlier in areas close to the Antarctic continent, and later in lower-latitude regions (i.e., Bay of Biscay, Southern Alps, and Northern Apennines).

At Massignano, the LCO and the subsequent disappearance (8.5 m) of *Nutallides truempyi* fall within BFI-I in a period of markedly decreasing oxygenation on the seafloor as testified by the DOI record (figure 25.5). At the same time, an increase in faunal density and relative abundance of infaunal forms is observed. High proportion of infaunal forms have been found to be associated with higher fluxes of organic matter to the seafloor in Recent assemblages (Corliss and Chen, 1988). Moreover, the abundance of modern deep-sea foraminifera, as indicated by both biomass of living populations and the accumulation rate of dead tests is closely related to food flux to the seafloor, with higher numbers under more eutrophic conditions (Altenbach, 1988, Altenbach and Sarnthein, 1989; Herguera and Berger, 1991; Herguera, 1992; Berger and Herguera, 1992; Sjoersdma and van der Zwaan, 1992). Accordingly, benthic foraminiferal record suggests that the LCO of *Nutallides truempyi* occurred in a time of decreased ventilation on the seafloor and increasing organic fluxes in the low-latitude, lower bathyal setting of the Massignano GSSP. However, no changes are observed in the $\delta^{13}C$ record are observed in the same interval, suggesting that low oxygenation rather than increasing organic fluxes was responsible for changes observed in the faunal parameter record.

Following the disappearance of *Nutallides truempyi*, the rise of *Gyroidinoides* and *Oridorsalis* and, subsequently, that of *Globocassidulina subglobosa* observed in our material well fits within the framework of the deep-sea benthic foraminiferal taxonomic reorganization occurring by middle–late Eocene time worldwide (see Miller et al., 1992). Associated with such a faunal reorganization, a decline in the relative abundance of buliminids is reported to occur from Sites 549 (Goban Spur Region, Miller et al., 1985) and 689 (Weddell Sea, Thomas, 1992a,b). According to the latter author, the buliminid crisis could be, at least partly, related to the increasing levels of oxygenation of the deep waters in the Southern Ocean. This increased oxygenation could have resulted in more thorough oxidation of organic material, leaving less food available for the benthic foraminiferal faunas (Thomas, 1992a). At Massignano, similar to the events observed by Miller et al. (1985) and Thomas (1992a, b), a crisis of buliminids follows the loss of *Nutallides truempyi* and accompanies the increase in abundance of *Oridorsalis* spp., *Gyroidinoides* spp., and *Globocassidulina subglobosa,* which are the dominant taxa of the Oligo-Miocene transitional faunas (figures 25.4 and 25.6). Moreover, as observed at Site 689, the buliminid crisis occurred at Massignano in a period of increasing oxygenation as testified by increasing DOI values in the upper part of BFI-II (figures 25.4 and 25.5). It has often been suggested that the distribution of buliminids and bolivinids is controlled by the interplay of organic flux, oxygen and competition (see Lutze and Coulborn, 1984; Gooday, 1994; Thomas and Gooday, 1996; among others). These forms are known to proliferate under conditions of raised organic flux and/or a decreased oxygen content (see a review in van der Zwaan et al., 1999). The loss of buliminids well fits with the decrease of primary productivity indicated by the $\delta^{13}C$ record which shows a gradual ~1‰ decrease from 10 m to 18 m, that is in correspondence to the BFI-II. Interestingly, such a gradual decrease starts in correspondence to the relatively large iridium anomaly observed by Montanari et al. (1993) at 10.25 m. However, the low resolution of $\delta^{13}C$ record does not allow any further discussion, and further analyses are required to better constrain the possible correspondence of the observed trends. Surprisingly, a remarkable bloom of bolivinids is associated with the buliminid crisis at Massignano. It is possible that these forms did not strictly show the same ecological preferences, at least in terms of oxygenation, and/or that different forms belonging to the bolivinids group might not show the same tolerance to dysaerobic environment.

Another group often associated with the bolivinids and buliminids in terms of preference for eutrophic environments and tolerance to dysaerobic conditions is the uvigerinids (see van der Zwaan et al., 1999). The record of *Uvigerina* species from the Massignano GSSP does not show any remarkable change in abundance and only a minor increase in abundance is observed just above the Eocene/Oligocene boundary. A general indication with respect to the trophic conditions on the seafloor can be traced by considering the distribution of different morphotypes belonging to this group. In particular, costate forms would be associated with higher organic fluxes (Borsetti et al., 1986). At Massignano, a general dominance of costate forms would suggest that relatively high organic flux regimes prevailed on the seafloor. However, in agreement with the faunal parameters record, consistently low abundances of this group indicate that truly eutrophic and/or dysaerobic conditions never developed on the seafloor at this site.

Following the buliminid crisis and the bloom of bolivinids, large sample-to-sample fluctuations of faunal parameters are observed in BFI-III. This possibly testifies to the development of a more dynamic environment characterized by rapidly changing conditions. Accordingly, increased diversity values in this interval indicate enhanced heterogeneity in the deep sea (see Grassle and Morse-Porteous, 1987; Grassle and Maciolek, 1992).

PALEOCEANOGRAPHY

Although not associated with a major decrease of diversity, changes observed at Massignano are comparable to the well-documented worldwide taxonomic reorganization of deep-water benthic foraminifera. Similar to other locations (see Miller et al., 1985, 1992; Berggren and Miller, 1989; Thomas, 1992a; Thomas and Gooday, 1996), the extinction of *Nutallides truempyi* is followed by an increase in the relative abundance of *Gyroidinoides, Oridorsalis,* and *Globocassidulina subglobosa* that is, in turn, associated with a crisis of buliminids at Massignano. Changes in the late Eocene-earliest Oligocene deep-sea benthic foraminiferal faunas have been related to deep-water cooling and ensuing increased oxygenation (Kaiho, 1994b) or $CaCO_3$ corrosivity (Thomas, 1992a) related to Antarctic Bottom Water (AABW) flow. Contrary to that observed at high latitudes, changes recorded at Massignano are accompanied by decreasing ventilation on the seafloor at least in the BFI-I (i.e., up to the base of Zone *Turborotalia cunialensis-Cribrohantkenina inflata* of Berggren et al., 1995, that is in Chron C15r, at an estimated age of around 34.9 Ma). It is possible that until this time, the relatively "young" deep-water masses (i.e., in equilibrium with the atmosphere only a short time before reaching the site and, thus, enriched in oxygen, poor in CO_2 and nutrients) forming at high latitudes in the late Eocene (Miller et al., 1987) did not reach the western Tethys lower bathyal settings. Accordingly, a high-resolution integrated stratigraphic study carried out across an Ir-rich horizon falling within BFI-I (from 4 m to 8 m above the base of the section) reveals that decreasing DOI values in benthic foraminifera correspond to cooler surface-water conditions as testified by planktonic foraminiferal and dinoflagellate cyst distribution suggesting that a thermohaline circulation regime existed in the western Tethys in the late Eocene (Galeotti, 1998; Coccioni et al., 2000). The change observed in the DOI record at top of BFI-I (figure 25.5) might suggest that, later on, western Tethys bathyal settings started to be invaded by better-oxygenated deep waters, possibly from the Atlantic. Interestingly, the onset of increased intensity of abyssal circulation associated with the initial entry of bottom waters (though originally thought to originate in the Northern Hemisphere) has been reported to occur at the same time by Miller et al. (1985) in the northeastern Atlantic Site 549, at a paleolatitude comparable to that of the Umbria-Marche Basin. At Site 549, the dominance of buliminids up to the P16/P17 zonal boundary of Blow (1969, 1979) is a further indication that a vigorous circulation of younger and colder bottom waters was fully established only at a time falling in our BFI-II where the shift of the DOI record is observed.

The disappearance of *Nutallides truempyi* as observed in the western Tethys (see also Braga et al., 1975) slightly precedes such a change in the deep-water circulation in this region, and therefore cannot be related to either increased ventilation or increased corrosivity of bottom waters. However, the global significance of *Nutallides truempyi* disappearance invokes a global cause to explain this event. The influence on benthic foraminifera of physiochemical properties of bottom waters (temperature, oxygen content, salinity, total dissolved carbonate) as compared to changes in food supply from surface primary production is still a matter of intense discussion (see review in van der Zwaan et al., 1999). According to some authors, oxygenation is the major factor controlling the benthic foraminiferal distribution (see Kaiho, 1994a). However, data on living faunas suggest that food supply is of prime importance, oxygen concentration being a major factor only under high organic input conditions leading to oxygen depletion (Sen Gupta and Machain-Castillo, 1993; Gooday, 1994; Schmiedl, 1995). Accordingly, the loss of *Nutallides truempyi* and associated taxonomic reorganization occurring worldwide in the middle late Eocene is better explained in terms of a primary productivity regime change. According to Thomas and Gooday (1996), the global change in benthic foraminiferal faunas was related to the onset of a seasonally fluctuating, more unpredictable food supply. These authors suggested that such a mechanism would also be responsible for the species-richness latitudinal gradient developing in the middle late Eocene as a result of more developed seasonality at high latitudes. Besides this, the influence of seasonality likely started at low latitudes much later than in high-latitude areas, explaining the difference observed in the timing of the *Nutallides truempyi* LCO at different sites which, according to the time scale of Berggren *et al.* (1995), would span a time interval of about 7 million years.

CONCLUSIONS

High-resolution analysis of late Eocene to earliest Oligocene deep-water benthic foraminiferal assemblages from the lower bathyal Massignano GSSP enabled us to recognize and calibrate several discrete bioevents against well-established bio-, chrono-, and magnetostratigraphy.

The analysis of faunal parameters allowed us to characterize the paleoecological and paleoceanographic conditions that accompanied the closing of the Eocene-Oligocene deep-water benthic foraminiferal faunal turnover in the Western Tethys.

The timing of this faunal turnover shows large differences (up to 7 myr) from site to site and is here suggested to follow a latitudinal gradient with areas close to the Antarctic continent showing earlier taxonomic turnovers. western Tethys bathyal settings were not invaded by relatively young bottom waters forming at high latitudes prior to the loss of *Nutallides truempyi.* It is then unlikely that this event was either related to increased oxygenation or to corrosivity of bottom waters. A mechanism related to changes in paleoproductivity regimes due to increased seasonality is more appropriate to explain the observed pattern.

ACKNOWLEDGMENTS

We thank M. E. Katz and M. A. Kaminski for their useful comments which greatly improved this paper. Funding was provided by MURST Ricerca Scientifica (ex 60%) and CNR (grant 97.00242CT05) to R. C.

REFERENCES

Altenbach, A. V., 1988, Deep sea benthic Foraminifera and flux rate of organic carbon. Revue de Paléobiologie, Special Volume 2, p. 719-720.

Altenbach, A. V., and Sarnthein, M., 1989, Productivity record in benthic foraminifera, *in* Berger, W. H., Smetacek V. S., and Wefer, G., eds., Productivity of the oceans: past and present: Chichester, Wiley, p. 255-269.

Bellagamba, M., and Coccioni, R., 1990, Deep-water agglutinated foraminifera from the Massignano section (Ancona, Italy), a proposed stratotype for the Eocene/Oligocene boundary, *in* Hemleben, Ch., Kaminski, M. A., Khunt, W., and Scott, D., eds., Paleoecology, biostratigraphy, paleoceanography and taxonomy of agglutinated Foraminifera: NATO Advance Studies Institute Series, v. 327, p. 883-921.

Berger, W. H., and Herguera, J. C., 1992, Reading the sedimentary record of the ocean's productivity, *in* Falkowski, P. G., and Woodhead, A. D., eds., Primary productivity and biogeochemical cycles in the aea: New York, Plenum Press, p. 455-486.

Berggren, W. A., and Miller, K. G., 1988, Paleogene tropical planktonic foraminiferal biostratigraphy and magnetobiochronology: Micropaleontology, v. 34, p. 362-380.

Berggren, W. A., and Miller, K. G., 1989, Cenozoic bathyal and abyssal calcareous benthic foraminiferal zonation: Micropaleontology, v. 35, p. 308-320.

Berggren, W. A., Kent, D. V., Swisher, C. C., and Aubry, M.-P., 1995, A revised Cenozoic geochronology and chronostratigraphy, *in* Berggren, W. A., Kent, D. V., Aubry, M.-P,. and Hardenbol, J., eds., Geochronology, Time Scales and Global Stratigraphic Correlation: SEPM Special Publication, v. 54, p. 129-212.

Bernhard, J. M., 1986, Characteristic assemblages and morphologies of benthic foraminifera from anoxic, organic-rich deposits: Jurassic through Holocene: Journal of Foraminiferal Research, v. 16, p. 207-215.

Bice, D., and Montanari, A., 1988, Magnetic stratigraphy of the Massignano section across the Eocene/Oligocene boundary, *in* Premoli Silva, I., Coccioni, R., and Montanari, A., eds., The Eocene/Oligocene boundary in the Marche-Umbria basin (Italy): International Subcommission on Paleogene Stratigraphy, Industrie Grafiche Filli Aniballi, Ancona, Special Publication, p. 111-117.

Blow W. H., 1969, Late middle Eocene to Recent planktonic foraminiferal biostratigraphy, *in* Brönniman, P., and Renz, H. H., eds., Proceedings of the First International Confe-rence on Planktonic Microfossils, Geneva, 1967: Leiden, Brill, v. 1, p. 199-422.

Blow W. H., 1979, The Cenozoic Globigerinida: a study of the morphology, taxonomy, evolutionary relationships and the stratigraphical distribution of some Globigerinida (mainly Globigerinacea): Leiden, Brill, 3 vol., 1413 p.

Bolli, H. M., Beckmann, J.-P., and Saunders, J. B., 1985, Benthic foraminiferal biostratigraphy of the south Caribbean region: Cambridge University Press, 408 p.

Bolli, H. M., Saunders, J. B., and Perch-Nielsen, K., eds., 1985, Plankton stratigraphy: Cambridge University Press, 1032 p.

Borsetti, A. M., Iaccarino, S., Jorissen, F. J., Poignant, A., Sztrakos, K., van der Zwaan, G. J., and Verhallen, P. J. J. M., 1986, The Neogene development of *Uvigerina* in the Mediterranean, *in* van der Zwaan, G. J., Jorissen, F. J., Verhallen, P. J. J. M., and von Daniels, C. H., eds., Atlantic-European Oligocene to Recent *Uvigerina*: Utrecht Micropaleontology Bulletin, v. 35, p. 183-235.

Braga, G. P., De Biase, R., Grünig, A., and Proto Decima, F., 1975, Foraminiferi bentonici del Paleocene ed Eocene della Sezione di Possagno, *in* Bolli, H. M., ed., Monographia micropaleontologica sul Paleocene e l'Eocene di Possagno, Provincia di Treviso, Italia: Schweizerischze Paläontologische Abhandlungen, v. 97, p. 85-198.

Brinkhuis, H., and Biffi, U., 1993, Dinoflagellate cyst stratigraphy of the Eocene/Oligocene transition in Central Italy: Marine Micropaleontology, v. 22, p. 131-183.

Capo, R. C., and De Paolo, D. J., 1988, Sr isotopic analysis of marine carbonates from the Massignano section across the Eocene/Oligocene boundary, *in* Premoli Silva, I., Coccioni, R., and Montanari, A., eds., The Eocene/Oligocene boundary in the Marche-Umbria basin (Italy): International Subcommission on Paleogene Stratigraphy, Ancona, Special Publication, p. 189-192.

Coccioni, R., 1988, The genera *Hantkenina* and *Cribrohantkenina* (Foraminifera) in the Massignano section (Ancona, Italy), *in* Premoli Silva, I., Coccioni, R., and Montanari, A., eds., The Eocene/Oligocene boundary in the Marche-Umbria basin (Italy): International Subcommission on Paleogene Stratigraphy, Ancona, Special Publication, p. 81-96.

Coccioni, R., Monaco, P., Monechi, S., Nocchi, M., and Parisi, G., 1986, The Eocene/Oligocene boundary at Massignano (Ancona, Italy): biostratigraphy based on calcareous nannofossils and planktonic foraminifera: Bulletin of Liaison and Information, IGCP Project 196, v. 6, p. 37-44.

Coccioni, R., Monaco, P., Monechi, S., Nocchi, M., and Parisi, G., 1988, Biostratigraphy of the Eocene/Oligocene boundary at Massignano (Ancona, Italy), *in* Premoli Silva, I., Coccioni, R., and Montanari, A., eds., The Eocene/Oligocene boundary in the Marche-Umbria basin (Italy): International Subcommission on Paleogene Stratigraphy, Industrie Grafiche F.lli Aniballi, Ancona, Special Publication, p. 59-80.

Coccioni R., Basso, D., Brinkhuis, H., Galeotti, S., Gardin, S., Monechi, S., and Spezzaferri, S., 2000, Marine biotic signals across a late Eocene impact layer at Massignano, Italy: evidence for long-term environmental perturbations?: Terra Nova, v. 12, p. 258-263.

Corliss, B. H., 1985, Microhabitats of benthic foraminifera within deep-sea sediments: Nature, v. 314, p. 435-438.

Corliss, B. H., and Chen, C., 1988, Morphotype patterns of Norwegian Sea deep-sea benthic foraminifera and ecological implications: Geology, v. 16, p. 716-719.

Galeotti, S., 1998, Crisi biologiche e foraminiferi bentonici attraverso i limiti Cretacico/Terziario, Paleocene/Eocene e Eocene/Oligocene [Unpubl. Ph.D Thesis]: University of Parma, 205 p.

Gooday, A. J., 1994, Deep-sea benthic foraminiferal species which exploit phytodetritus: characteristic features and controls on distribution: Marine Micropaleontology, v. 22, p. 187-205.

Grassle, J. F., and Maciolek, N. J., 1992, Species diversity in deep-sea communities: Trends in Ecology and Evolution, v. 4, p. 12-15.

Grassle, J. F., and Morse-Porteous, L. S., 1987, Macrofaunal colonization of disturbed deep-sea environments and the structure of deep-sea benthic communities: Deep-Sea Research, v. 34, p. 1911-1950.

Herguera, J. C., 1992, Deep-sea benthic foraminifera and biogenic opal: Glacial to postglacial productivity changes in the western equatorial Pacific: Marine Micropaleontology, v. 19, p. 79-98.

Herguera, J. C., and Berger, W. H., 1991, Paleoproductivity from benthic foraminifera abundance: glacial to interglacial change in the west-equatorial: Geology, v. 19, p. 1173-1176.

Kaiho, K., 1991, Global changes of Paleogene aerobic/anaerobic benthic foraminifera and deep-sea circulation: Palaeogeography, Palaeoclimatology, Palaeoecology, v. 83, p. 65-85.

Kaiho, K., 1994a, Benthic foraminiferal dissolved-oxygen index and dissolved oxygen levels in the modern ocean: Geology, v. 22, p. 719-722.

Kaiho, K., 1994b, Planktonic and benthic foraminiferal extinction events during the last 100 m.y.: Palaeogeography, Palaeoclimatology, Palaeoecology, v. 111, p. 45-71.

Loeblich, A. F., and Tappan, H., 1988, Foraminiferal genera and their classification: New York, Van Nostrand Reinhold, v. 1 and v. 2, 970 p., 212 p.

Lowrie, W., and Lanci, L., 1994, Magnetostratigraphy of Eocene/Oligocene boundary sections in Italy: no evidence for short subchrons within chrons 12R and 13R: Earth and Planetary Sciences Letters, v. 126, p. 247-258.

Lutze, G. H., and Coulborn, W. T., 1984, Recent benthic foraminifera from the continental margin of northwest Africa: community structure and distribution: Marine Micropaleontology, v. 8, p. 361-401.

Martini, E., 1971, Standard Tertiary and Quaternary calcareous nannoplankton zonation, *in* Farinacci, A., ed., Proceedings Second Planktonic Conference on Microfossils, Roma 1970: Tecnoscienza, v. 2, p. 739-785.

Miller, K. G., Curry, W. B., and Ostermann, D. R., 1985, Late Paleogene (Eocene to Oligocene) benthic foraminiferal oceanography of the Goban Spur region, Deep Sea Drilling Project Leg 80, *in* de Graciansky, P. C., Poag, C. W., et al., Initial Reports DSDP, v. 80, p. 505-538.

Miller, K. G., Fairbanks, R. G., and Mountain, G. S., 1987, Tertiary oxygen isotope synthesis, sea level history, and continental margin erosion: Paleoceanography, v. 2, p. 1-9.

Miller, K. G., Katz, M. E., and Berggren, W. A., 1992, Cenozoic deep-sea benthic Foraminifera: a tale of three turnovers: Studies in Benthic Foraminifera, BENTHOS 90, Sendai: Tokai University Press, p. 67-75.

Montanari, A., Deino, A. L., Drake, R. E., Turrin, B. D., DePaolo, D. J., Odin, G. S., Curtis, G. H., Alvarez, W., and Bice, D.M., 1988, Radioisotopic dating of the Eocene/Oligocene boundary in the pelagic sequence of the Northern Apennines, *in* Premoli Silva I., Coccioni R., and Montanari A., eds, The Eocene/Oligocene boundary in the Marche-Umbria basin (Italy): International Subcommi-sion on Paleogene Stratigraphy, Industrie Grafiche F.lli Aniballi, Ancona, Special Publication, p. 195-208.

Montanari, A., Asaro, F., Michel, H. V., and Kennett, J. P., 1993, Iridium anomalies of late Eocene age at Massignano (Italy), and ODP Site 689B (Maud Rise, Antarctic): Palaios, v. 8, p. 420-437.

Nocchi, M., Monechi, S., Coccioni, R., Madile, M., Monaco, P., Orlando, M., Parisi, G., and Premoli Silva, I., 1988, The extinction of Hantkeninidae as a marker for recognizing the Eocene/Oligocene boundary: a proposal, *in* Premoli Silva, I., Coccioni, R., and Montanari, A., eds., The Eocene/Oligocene boundary in the Marche-Umbria basin (Italy): International Subcommission on Paleogene Stratigraphy, Industrie Grafiche F.lli Aniballi, Ancona, Special Publication, p. 249-252.

Odin, G. S., Clauser, S., and Renard, M., 1988,

Sedimentological and geochemical data on the Eocene/Oligocene boundary at Massignano (Apennines, Italy), *in* Premoli Silva, I., Coccioni, R., and Montanari, A., eds., The Eocene/Oligocene boundary in the Marche-Umbria basin (Italy): International Subcommission on Paleogene Stratigraphy, Ancona, Special Publication, p. 175-186.

Odin, G. S., Guise, P., Rex, D. C., and Kreuzer, H., 1988, K-Ar and $^{40}Ar/^{39}Ar$ geochronology of Late Eocene biotites from northeastern Apennines, *in* Premoli Silva, I., Coccioni, R., and Montanari, A., eds., The Eocene/Oligocene boundary in the Marche-Umbria basin (Italy): International Subcommission on Paleogene Stratigraphy, Industrie Grafiche F.lli Aniballi, Ancona, Special Publication, p. 175-186.

Okada, H., and Bukry, D., 1980, Supplementary modification and introduction of code numbers to the "Low latitude coccolith biostratigraphy zonation" (Bukry, 1973, 1975): Marine Micropaleontology, v. 5, p. 321-325.

Parisi, G., and Coccioni, R., 1988, Deep-water benthic Foraminifera at the Eocene/Oligocene boundary in the Massignano section (Ancona, Italy), *in* Premoli Silva, I., Coccioni, R., and Montanari, A., eds., The Eocene/Oligocene boundary in the Marche-Umbria basin (Italy): International Subcommission on Paleogene Stratigraphy, Industrie Grafiche F.lli Aniballi, Ancona, Spec. Publ., p. 97-109.

Premoli Silva, I., and Jenkins, D. J., 1993, Decision on the Eocene/Oligocene boundary stratotype: Episodes, v. 16, p. 379-381.

Prothero, D. R., 1994, The Eocene-Oligocene transition: paradise lost: New York, Columbia University Press, 291 p.

Schmiedl, G., 1995, Rekonstruktion der spaetquartaeren Tiefenwasserzirkulation und Produktivitaet im oestlichen Suedatlantik anhand von benthischen Foraminiferenvergesellschaftungen: Berichte zur Polarforschung, v. 160, 207 p.

Sen Gupta, B. K., and Machain-Castillo, M. L., 1993, Benthic foraminifera in oxygen-poor habitats: Marine Micropaleontology, v. 20, p. 183-201.

Sjoersdma, P. G., and van der Zwaan, G. J., 1992, Simulating the effect of changing organic flux and oxygen content on the distribution of benthic foraminifera: Marine Micropaleontology, v. 19, p. 163-180.

Thomas, E., 1992a. Middle Eocene-late Oligocene bathyal benthic foraminifera (Weddell Sea): faunal changes and implications for ocean circulation, *in* Prothero, D. R., and Berggren, W. A., eds., Eocene-Oligocene climatic and biotic evolution: Princeton, New Jersey, Princeton University Press, p. 245-271.

Thomas, E., 1992b, Cenozoic deep-sea circulation: evidence from deep-sea benthic foraminifera: American Geophysical Union, Antarctic Research Series, v. 56, p. 141-165.

Thomas, E., and Gooday, A. J., 1996, Cenozoic depsea benthic foarminifers: Tracers for changes in oceanic productivity?: Geology, v. 24, p. 355-358.

Tjalsma, R. C., and Lohmann, G. P., 1983, Paleocene-Eocene bathyal and abyssal benthic foraminifera from the Atlantic Ocean: Micropaleontology Special Publication, v. 4, 90 p.

van der Zwaan, G. J., Duijnstee, I. A. P., den Dulk, M., Ernst, S. R., Jannink, N. T., and Kouwenhoven, T. J., 1999, Benthic foraminifers: proxies or problems? A review of paleoecological concepts: Earth-Science Reviews, v. 46, p. 213-236.

Van Morkhoven, F. P. C. M., Berggren, W. A., and Edwards, A. S., 1986, Cenozoic cosmopolitan deep-water benthic Foraminifera: Bulletin Centre Recherces Exploration-Production, Elf-Aquitaine, Memoire 11, 421 p.

Vonhof, H. B., 1998, The strontium isotope stratigraphic record of selected geologic events [Unpubl. Ph.D. Thesis]: University of Amsterdam, 138 p.

Widmark, J. G. V., and Speijer, R. P., 1997, Benthic foraminiferal ecomarker species of the terminal Cretaceous (late Maastrichtina) deep-sea Tethys: Marine Micropaleontology, v. 31, p. 135-155.

CHAPTER 26

Stratigraphy and Benthic Foraminiferal Events Across the Middle-Late Eocene Transition in the Western Negev, Israel

Roberto Barbieri, Chaim Benjamini, Simonetta Monechi, and Viviana Reale

ABSTRACT

Upper middle through upper Eocene rocks (Zones P14–P15; NP17–NP19–20) from the western Negev, Israel (southern margin of the central Tethys), contain benthic foraminiferal assemblages that largely consist of cosmopolitan, lower to upper bathyal taxa, with many elements shared with other low-latitude areas. There is a gradual change both in the taxonomic structure, mainly established by first rather than last appearances, and in the species abundances that allows identification of three distinct assemblages. Faunal turnover especially affects trochospiral taxa, so that the wide-scale events marking the middle/late Eocene boundary include the last appearance or sharp decrease of *Nuttallides truempyi*, *Cibicidoides hettneri*, and *C. truncanus*. Although the diversity indices are generally high and the taxonomic composition suggests oligotrophic and oxic environments, just above the middle/late Eocene boundary there is an increase in relative abundance of species and morphogroups that appears to reflect a temporary oxygen shortage. These conditions persisted for more than the time span represented by Zone NP18, and seem to have influenced the distribution of planktonic taxa (foraminifera and calcareous nannofossils), with some effect on their local biostratigraphic ranges. An approximately coeval faunal perturbation recorded from the northern Tethyan margin produced bolivinid/uvigerinid blooms. Thus, although relative sea-level shallowing (from lower to middle bathyal) is indicated at the base of the late Eocene interval, this deep-sea foraminiferal change could be a faunal response to changes in the origin and type of the Tethyan deep waters.

INTRODUCTION

Outcrops of middle and upper Eocene rocks are widespread in Israel and primarily consist of chalks, chalk-like limestones and marls deposited in pelagic settings, with some nummulitic interruptions (Benjamini, 1980; Schaub et al., 1995). The abundant, diverse biota in these deposits has allowed establishment of an accurate biostratigraphy, including correlation between nummulite, planktonic foraminiferal, and calcareous nannofossil zonal schemes (Schaub et al., 1995). In contrast to recent focus on earlier Paleogene deposits, especially across the Paleocene/Eocene boundary, few studies have been devoted to smaller benthic foraminifera of the later Eocene of Israel, although benthic foraminifera are effective proxies for many paleoceanographic/ paleobathymetric parameters. The stratigraphically continuous nature and good exposures of sections, including the middle-late Eocene transition, offers a means for evaluation of the southern Tethyan pelagic microfaunal and -floral structure at a time of complex global paleoclimatic and paleoceanographic change. These changes have been related to gradual cooling of the oceans (Berggren and Prothero, 1992; Miller, 1992) following the detachment of Australia from Antarctica (Kennett, 1977). Nevertheless, many open questions remain on the nature of this cooling, on its effects on oceanic circulation patterns and thermal gradients (Thomas, 1992a), and on the development of the first Antarctic ice cover. Growing evidence for the latter suggests its initiation by the late middle Eocene (Barron et al., 1991; Browning et al., 1997).

In general, extinctions at the middle/late Eocene boundary were severe. Significant extinctions among

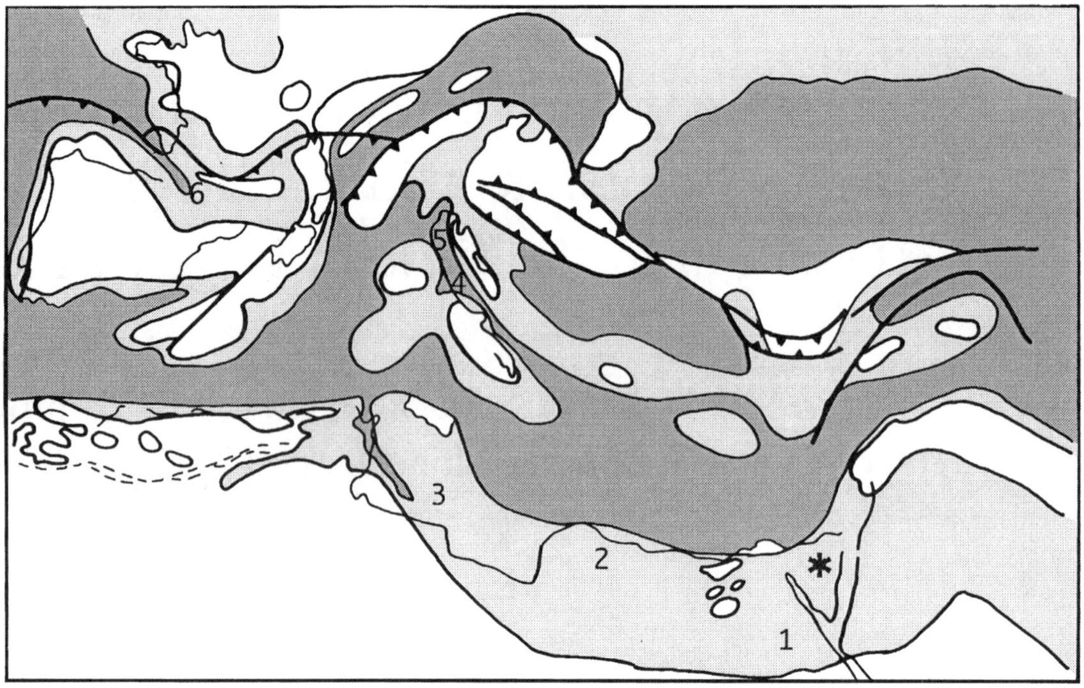

Figure 26.1. Paleogeographic map of the Mediterranean Tethys (modified after Dercourt et al., 1993) during middle Eocene (Lutetian) time. Patterns indicate the following environments: gray, slope and deep basin; light gray, shallow carbonate platform and terrigenous shelf; white, lands above sea level. Asterisk locates the study area. Numbers refer to localities/sections cited in text: 1, Gebel Aweina section, Egypt (Speijer and Schmitz, 1998); 2, northern Cyrenaica, Libya (Barr and Berggren, 1980); 3, northwestern Libya offshore (Barbieri, 1990); 4, Umbro-Marchean region, Italy (Nocchi et al., 1988; Parisi and Coccioni, 1988; Parisi et al., 1988); 5, Possagno section, Italy (Braga et al., 1975; Grünig and Herb, 1980); 6: Biarritz, southwestern France (Mathelin and Sztràkos, 1993).

planktonic foraminifera reflect disappearance of tropical epipelagic taxa (Keller, 1982; Boersma et al., 1987; Hallock et al., 1991; Keller et al., 1992), but large-scale extinction events also affected shallow-water larger foraminifera (Hallock et al., 1991) and deep-sea benthic foraminifera (Müller-Merz and Oberhänsli, 1991; Thomas, 1992b; Nomura, 1995). Neritic smaller benthic foraminifera were comparatively less affected, presumably because of their greater environmental tolerance.

We present here the stratigraphy and benthic foraminiferal introduction/extinction patterns of the Nahal Nizzana section, northwestern Negev, Israel. This is the type section for the Qezi'ot Formation (Benjamini, 1980) and includes a more or less complete chronostratigraphic profile across the middle/late Eocene transition. The lithology and quality of preservation of the microfossil record of this section has permitted a biostratigraphically controlled, foraminiferal-based paleoenvironmental/paleodepth reconstruction of the region across this interval.

MATERIALS AND METHODS

Lithology

The Nahal Nizzana section consists of some 23 m of well-exposed, undisturbed whitish chalk and light grey to whitish, chalky marls of the Qezi'ot Formation. It is located at the base of the NW-SE trending ridge south of the settlement of Qezi'ot, close to the Israel-Egypt border (figures 26.1 and 26.2). Benjamini (1980) established the type section of the Qezi'ot Formation at this location, which is the best exposure but may not comprise the entire thickness of the unit. The composite section (figure 26.3) measured was from two adjacent sections, less than 50 meters away one from the other, each approximately of the same thickness. Weathering has prevented a proper distinction in the field of the subtle lithologic differences so that most of the exposure appears of massive and unbedded nature. In the middle part of the lower section, however, a chalk bedding plane is recognizable and has allowed a measurement of dip of about 10°NW, which was assumed for the entire

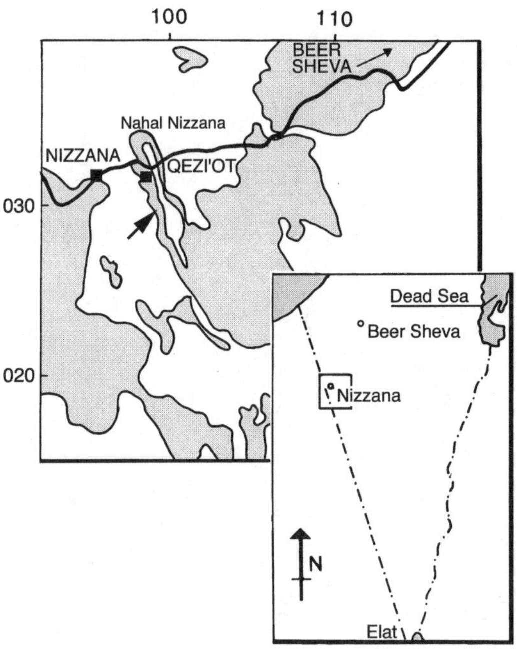

Figure 26.2. Location map of the Nahal Nizzana section (arrow). The gray pattern indicates the Eocene outcrops (after Benjamini, 1980).

exposure. Field continuity between the lower and upper portions of the section cannot be precisely established; however, the above considerations of local outcrop geology suggest that any possible gap or overlap should be very small and is in any case beneath biostratigraphic resolution. In its type area, the Qezi'ot Formation unconformably overlies the lower Eocene Nizzana Formation and is erosively truncated by Neogene conglomerates. Percent $CaCO_3$ values (figure 26.3), rather than visual inspection in the field, show fluctuations in calcium carbonate content, ranging from 60% to 80%, which may reflect either carbonate dilution by terrigenous sediment or changes in primary productivity. The carbonate component consists especially of calcareous nannofossils; foraminifera represent up to 1% in weight, as calculated from the washed residues (>63 µm) in the understanding, however, that many tests have some infilling material (nannofossils and calcite crystals). Only one marked, positive % carbonate excursion (a planktonic foraminiferal wackestone) was observed in the lower part of the upper section. There is no indication of significant carbonate dissolution and with few exceptions calcareous microfossil preservation ranges from good (planktonic foraminifera and nannoplankton) to excellent (benthic foraminifera).

Sample Preparation and Data Collection

Forty-five samples were collected along the type section of the Qezi'ot Formation at approximately 50-cm intervals irrespective of the subtle lithological changes. Shallow holes about 30–40-cm deep were dug to collect unweathered and noncontaminated rock. All samples were used for planktonic foraminifera and calcareous nannofossil determinations, and 40 were used for quantitative estimates on benthic foraminifera.

To extract the foraminifera, sample aliquots of about 100 g dry weight (oven dried at 60°C) were mechanically crushed and disaggregated according to the solvent (kerosene) method described by Thomas and Murney (1985) and washed through a 63 µm sieve. For further cleaning of the washed residues, short exposure (from one to several minutes) to ultrasonic vibration provided good results for most of the samples. Half of each sample residue was then sent to Beer Sheva for planktonic foraminiferal analysis. The >63 µm fraction was used for the quantitative benthic foraminiferal analysis. Splits of 250 to nearly 400 specimens were used from each sample, then identified, counted, and stored on cardboard slides in the micropaleontological collection of the Capellini Paleontological Museum, University of Bologna, Italy. The benthic curve (figure 26.3) represents the percentages of benthic foraminifera of the total (benthic + planktonic) foraminifera and are expressed as $100*B/(B + P)$.

Nannofossil smear slides were prepared using standard techniques and investigated through a light microscope at magnification of 1,560X. Qualitative and semiquantitative methods were used in order to investigate first and last occurrences. Semiquantitative data were collected by counting 300 specimens, and in order to document the presence of rare species, at least four traverses of each slide were examined. Relative species abundances of selected species were calculated as percentages on the 300 species counted. Calcareous nannofossils are common to abundant throughout the section, with the exception of the upper part of the lower portion, where the assemblages are less common and poorly preserved. Nannofossil reworking is present in a few samples and consists of rare specimens of Cretaceous species.

BIOSTRATIGRAPHY

Planktonic foraminifera and calcareous nannoplankton were used to establish the stratigraphic age of the section (figure 26.3). Biozone definitions in the present study follow Berggren et al. (1995), for plankton-

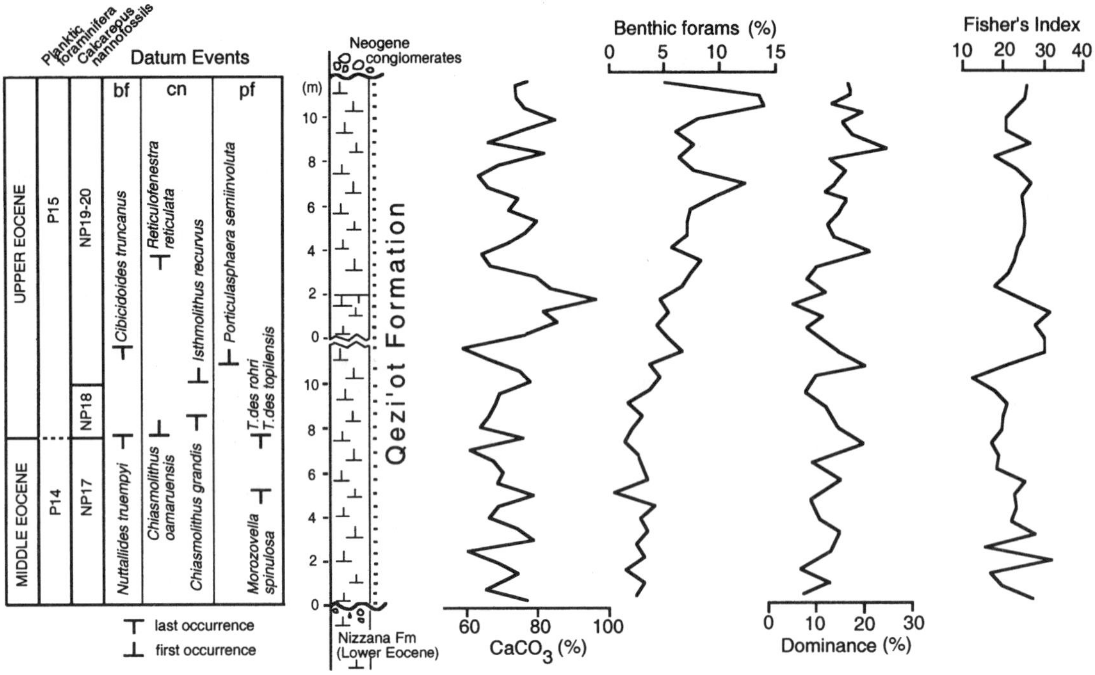

Figure 26.3. Planktonic foraminiferal and nannofossil biostratigraphy, datum events (bf, benthic foraminifera; cn, calcareous nannofossils; pf, planktonic foraminifera), percent benthic foraminifera, diversity indices (dominance and Fisher's index), and percent $CaCO_3$, Nahal Nizzana section.

ic foraminifera (P-zones), and Martini (1971) for calcareous nannofossils (NP-zones).

In terms of the planktonic foraminiferal zonation, the Nahal Nizzana section ranges from the upper part of Zone P14 (*Truncorotaloides rohri–Morozovella spinulosa* Partial Range Zone) to the upper part of Zone P15 (*Porticulasphaera semiinvoluta* Interval Zone). Between these two zones lies a poorly defined interval, about 2 m thick, largely due to poor preservation (compaction has led to flattening of the larger planktonic foraminiferal tests) and consequent absence of recognizable bioevents. The base of this interval coincides with the middle/late Eocene boundary, established on the basis of the nannofossil NP17/NP18 zonal boundary. Similar to sections of the northern Tethyan margin (Nocchi et al., 1988; Parisi et al., 1988), the highest occurrence (HO) of *Morozovella spinulosa* (5.1 m from the section base) significantly predates those of *Truncorotaloides rohri* and *T. topilensis* (7.7 m from the section base). Although small acarininids referable to *Acarinina rugosoaculeata* persist well into Zone NP18, a marked change in the planktonic foraminiferal composition occurs at 10.9 m above the base, established by the quantitative increase of *Globigerinatheka* (especially *G. index* group) and the extinction of the last muricate taxa. The gradual, stepwise fall-off of the muricate forms, with larger, more ornate forms going out first, and smaller, more generalized forms exiting last, suggests that the small thickness of Zone NP18 here, with an estimated duration of 1 million years (Berggren et al., 1995), is not the result of a gap in sedimentation, but may rather be the result of condensation. Some incompleteness near the base of the interval, however, cannot be unequivocally excluded.

The uppermost 6 m of the Nahal Nizzana section is placed close to the top of Zone P15. In terms of dominant taxa, this upper Zone P15 assemblage is characterized by abundant, diverse globigerinathekids (*Globigerinatheka index* group, *G. subconglobata* group, and *P. semiinvoluta*) and turborotaliids (*Turborotalia cerroazulensis cerroazulensis*). Relatively young features include the presence of advanced forms of the *Subbotina praeturritilina*/*Subbotina gortanii* lineage, *Globorotaloides suteri*, and five-chambered *Globigerina ouachitaensis*. Hantkeninids are generally rare, suggesting some environmental exclusion perhaps related to their being positively sensitive to oxygen minimum conditions (Boersma and Premoli

Silva, 1986; Boersma et al., 1987), and only broken chambers with spines were recovered.

The calcareous nannofossil biostratigraphy of the section ranges from Zone NP17 to Zone NP19-20. The upper boundary of NP17 (*Discoaster saipanensis* Zone) is defined by the lowest occurrence (LO) of *Chiasmolithus oamaruensis*, at 7.7 m from the base of the lower section. A further useful event to approximate the middle/late Eocene boundary is the LO of *C. grandis*, which defines the upper boundary of Okada and Bukry's (1980) Subzone CP14b. Contrary to the correlation scheme of Berggren et al. (1995), the above two ranges overlap; this also appears to occur in other northern Tethyan and Pacific sections (e.g., Proto Decima et al., 1975; Nocchi et al., 1986; Bralower and Mutterlose, 1995). However, for *C. grandis*, showing a continuous distribution up to 4.1 m and scattered occurrences above this level, we cannot exclude the effect of some reworking. In fact, it is generally difficult to recognize zonal boundaries based on *Chiasmolithus* in low–latitude areas, where this genus is rare or absent (Perch-Nielsen, 1985). The LO of *Isthmolithus recurvus* at 10.1 m defines the boundary between Zones NP18 and NP19–20. In the studied material this cold-water taxon has a very short range, being limited to the uppermost part of the lower section.

The upper portion of the Nahal Nizzana section is characterized by the HO of *Reticulofenestra reticulata*, at 3.8 m above the base of the second portion of the section. The few upsection records of this taxon are probably due to reworking, although Bralower and Mutterlose (1995) found at the tropical ODP Site 865 an extended distribution of *R. reticulata*, with last occurrence well above that of *Calcidiscus protoannulus*, in the Oligocene interval. According to Berggren et al. (1995), however, the disappearance of *R. reticulata* is a diachronous event between low-middle and high-latitude regions. In the Massignano section, Italy (stratotype for the Eocene/Oligocene boundary), for example, the HO of *R. reticulata* has been recorded just above the LO of *I. recurvus* but below the HO of *C. protoannulus* (Coccioni et al., 1988). The combined NP19–20 Zone proposed by Martini (1976) is here used because the LO of *Sphenolithus pseudoradians*, which defines the base of the NP20 Zone, is a biostratigraphically unreliable event and only a patchy occurrence was found in the study material.

In summary, correlation between the planktonic foraminifera and calcareous nannofossil zones indicate some differences from Berggren et al. (1995). The NP17/NP18 zonal boundary correlates to the tentative boundary placed between P14 and P15 Zones, while in the Berggren and co-workers zonal scheme it correlates with the middle part of the P15 Zone. At the Possagno section, Italy, however, Proto Decima et al. (1975) found the same correlation between the two zonal schemes. In addition, it should be pointed out that the *Chiasmolithus oamaruensis* first appearance datum is difficult to detect because of the presence (lower part of its range) of transitional forms with *C. expansus*. We believe that the establishment of meso/eutrophic conditions at the base of the late Eocene interval, as recognized through the benthic foraminiferal analysis (see below), may have somewhat altered the distribution of the calcareous plankton taxa. Interestingly, *C. oamaruensis* has been regarded as an opportunistic taxon adapted to eutrophic conditions (Aubry, 1992).

FAUNAL RESULTS

Structure of the Benthic Foraminiferal Assemblages

All the split aliquots (from 245 to 389 specimens/sample) of benthic foraminifers were identified at the species level, so that the taxonomic census accounts for a large number of rare taxa (species present with 1 specimen/sample); these comprise between one-fourth to two-fifths of the species present in each sample. In all, the benthic foraminifera from Nahal Nizzana comprise nearly 250 species belonging to 90 genera. Most of the species identified have calcareous hyaline tests, as agglutinated and porcelaneous taxa show a sample abundance from 1% to 14% and from 0% to 6%, respectively. The preservation of benthic foraminifera is generally very good, including those upsection intervals in which flattened larger planktonic tests abound.

Different indices, such as those for diversity (simple diversity and Fisher-α) and dominance (expressed as the more abundant species/sample, in %), reflect persistent taxonomic diversity throughout the section. Diversity expressed by Fisher-α values generally exceeds 20 (figure 26.3) and corresponds to simple diversities (number of species/sample) that average 58 and 65 in the lower and upper part of the section, respectively. The Fisher-α curve, which allows a better comparison with the diversity calculated for other sections/stratigraphic intervals, shows a negative correlation with the dominance curve, which, in turn, mostly ranges between 10% and 20%. The main contributions to the diversity of the assemblages derive from uniserial taxa (more than 30 species), coiled vaginulinids (25 species), parrelloidids

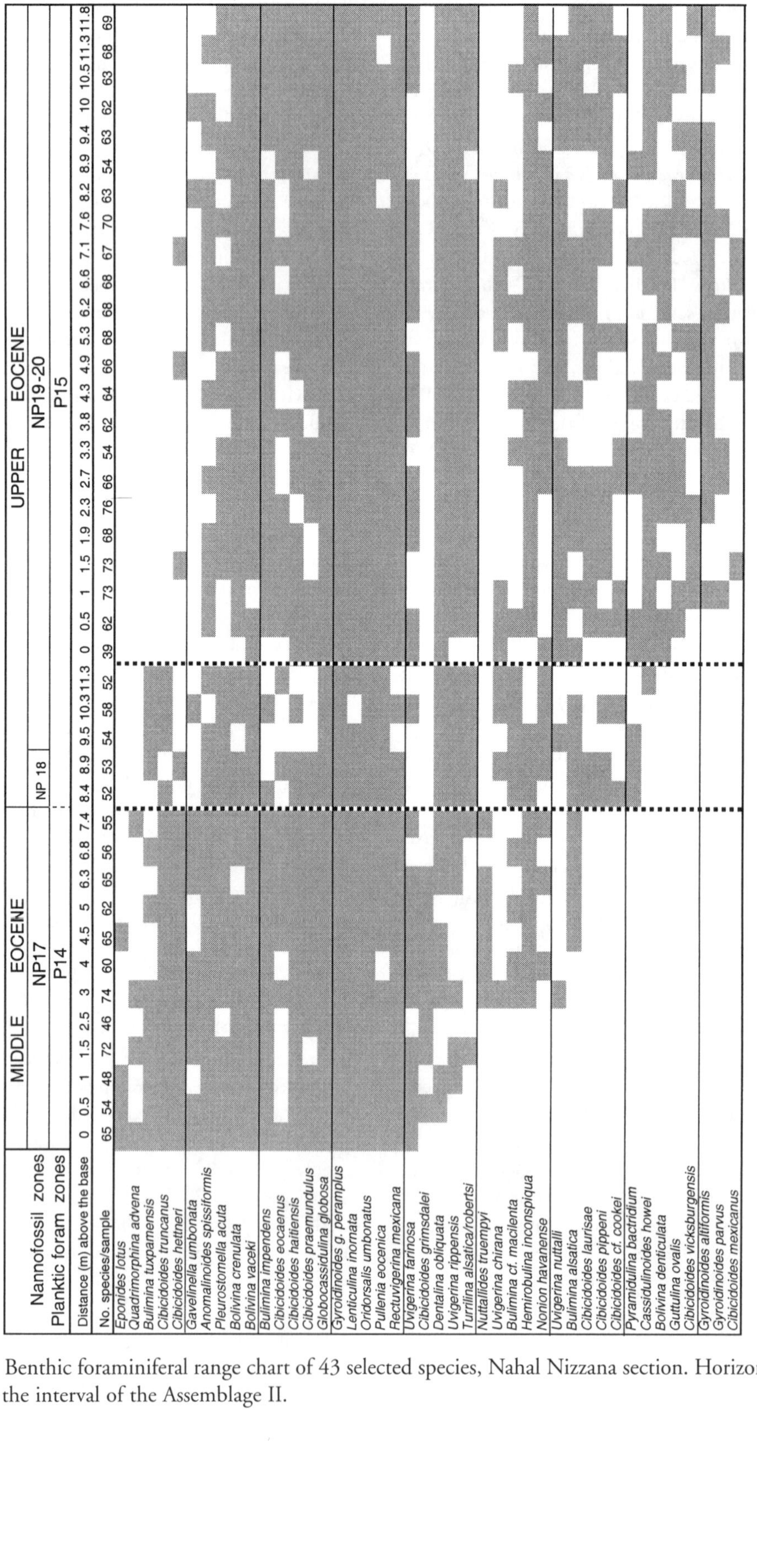

Figure 26.4. Benthic foraminiferal range chart of 43 selected species, Nahal Nizzana section. Horizontal broken lines bound the interval of the Assemblage II.

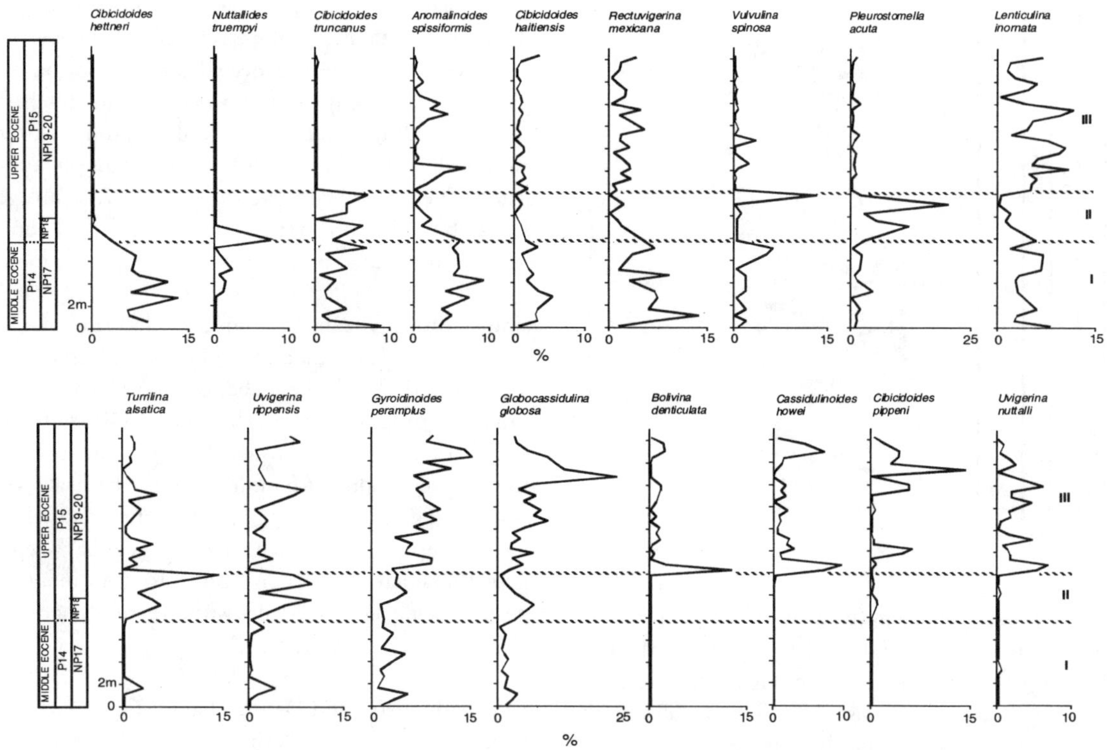

Figure 26.5. Relative species abundance of benthic foraminifera, Nahal Nizzana section. Horizontal pattern and Roman numerals mark the position of the benthic assemblages. Note the different percentage scales.

(14 species), pleurostomellids (13 species), and bolivinids (9 species).

First rather than last occurrence datums characterize the faunal changes at Nahal Nizzana (figure 26.4). First occurrences among taxa having a persistent record (i.e., species recorded from one-third to more than two-thirds of the study samples) and significant abundance (at least 1–2% per sample) are gradual and include some 17 taxa, such as *Cassidulinoides howei, Nonion affinis, Turrilina alsatica*, buliminids, hemirobulinids, gavelinellids, and parrelloidids. Last occurrences mostly comprise trochospiral species, including *Cibicidoides grimsdalei, C. hettneri, C. truncanus, Eponides lotus, Gavelinella umbonata, Nuttallides truempyi*, and *Quadrimorphina advena.* These changes, associated with remarkable variations in frequency distribution of the common taxa (figure 26.5), allow identification of three assemblages, distinguished in figures 26.5 and 26.6 by Roman numerals. Assemblage I occurs in the lower part of the section, up to the NP17/NP18 zonal boundary. Assemblage II approaches the lower part of the NP19–20 combined Zone, including the interval with the main planktonic foraminiferal change, and Assemblage III encompasses the remaining part of the section.

The upper boundary of Assemblage I is marked by the disappearance of *Nuttallides truempyi*, and of *Cibicidoides hettneri* as a persistent and abundant faunal element. Taxa showing a concomitant reduction of frequency include *Anomalinoides spissiformis, Cibicidoides haitiensis*, and *Rectuvigerina mexicana.* A few other species whose range is limited to this lower assemblage include *Cibicidoides grimsdalei, Eponides lotus*, and *Quadrimorphina advena*, whereas *Gavelinella umbonata* is restricted to this assemblage as a persistent faunal element. Several first occurrences, such as those of *Cibicidoides laurisae, C. pippeni, C.* cf. *cookei*, and *Pyramidulina bactridium*, characterize the base of Assemblage II, whose upper boundary is established by the disappearance of *Cibicidoides truncanus* and *Bulimina tuxpamensis.* This assemblage is also typified by a slightly lower simple diversity (average 52 species/sample) and a high frequency, with respect to the underlying and overlying assemblages, of *Nodogenerina atlantisae, Pleurostomella acuta, T. alsatica*, and *Uvigerina rippensis*, and a low frequency of *A. spissiformis, Bolivina vaceki*, and *Lenticulina*

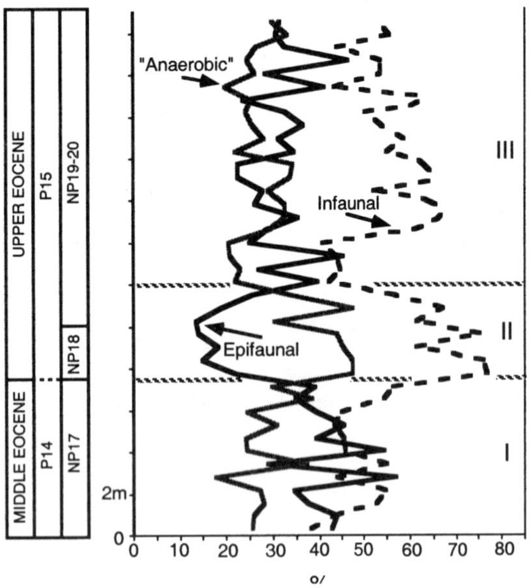

Figure 26.6. Proportion (percent) of the foraminiferal specimens assigned to infaunal, epifaunal, and "anaerobic" taxa as estimated through their test morphology. Roman numerals and horizontal lines mark the position of the benthic assemblages.

inornata. Assemblage III is distinguished through the appearance and/or sharp increase of *Bolivina denticulata*, *Cassidulinoides howei*, *Globocassidulina globosa*, *Gyroidinoides peramplus*, *Uvigerina nuttalli*, and an increase of the benthic versus planktonic foraminiferal proportion. The upper 5 m of the section records a further increase of the benthic percentage, increase in relative abundance of *G. globosa* and *Cibicidoides pippeni*, and decrease of *Karreriella subglabra* (figure 26.5).

The relationship between test morphology and occupation of specific microhabitats, as recognized for extant benthic foraminiferal species, has been extended to the fossil record, although recent indications (e.g., Moodley et al., 1997) that a direct relation between shape of test and epi- or infaunal mode of life is often loose. However, since the proportion of infaunal versus epifaunal taxa (table 26.1, as defined since Corliss, 1985; see also van der Zwaan et al., 1999 for a recent review) in a given assemblage should represent a compromise between faunal density (lower in infaunal microhabitat) and resource availability (higher at the interface), this proportion may be a useful parameter for recognizing broad faunal perturbations whose precise nature remains unknown. Faunal proportions plotted in figure 26.6 show that at Nahal Nizzana, specimens with an inferred infaunal mode of life outnumber the epifaunal ones. This figure is expected for deep-sea paleocommunities largely made up of taxa with inferred deposit-feeding strategies (trophic groups after Koutsoukos and Hart, 1990), such as those investigated. The interval corresponding to the Assemblage II exhibits changes in the epifaunal/infaunal ratio leading to infaunal proportions that exceed by more than 50% the epifaunal ones. Fisher-α values and simple diversity exhibit a modest decrease in the interval, with fewer than 20 and 40–55 species/sample, respectively. Such a change in the epifaunal/infaunal proportions at the NP17/NP18 zonal boundary corresponds to a paleodepth fluctuation, from lower middle to middle upper bathyal (see below), and this may suggest some cause-and-effect relationship. However, the reestablishment of the epifaunal/infaunal proportions prior to this infaunal peak (Assemblage III) occurs without any recognized change in paleobathymetry.

Paleobathymetry and Depth-Diagnostic Species

It may be that true isobathyal benthic foraminiferal species hardly exist; however, use of benthic foraminifera as proxies for paleobathymetry follows the evidence that certain assemblages have been related to paleodepth ranges in geographically wide areas. In general, however, environmental factors that determine benthic foraminiferal distribution and abundance, especially organic flux and oxygenation (see Gooday, 1994; van der Zwaan et al., 1999, for reviews), are only partially controlled by depth and their fluctuations may cause effective limitations in paleobathymetric reconstructions. Foraminiferal assemblages at Nahal Nizzana allow inference of significant paleodepth changes (shallowing) from the lower to the upper part of the succession. Evidence is derived from the proportion of the benthic versus planktonic specimens (figure 26.3) and from a number of paleodepth markers (figure 26.7), many of them characterized by a cosmopolitan distribution.

The percentage of benthic foraminifera in the total foraminiferal fauna can be reliably applied throughout the section because of a substantial absence of dissolution. However, since benthic/planktonic proportions are dependent not only on depth of deposition but also on paleoproductivity (Berger and Diester-Haass, 1988; Herguera and Berger, 1991) and on distance from the coast, we used this faunal index as a proxy with care, only in conjunction with changes in the distribution patterns

Table 26.1. Infaunal and Epifaunal Benthic Foraminiferal Taxa Grouped According to the Morphologic Concepts by Corliss (1985), Jones and Charnock (1985), Koutsoukos and Hart (1990), and Gooday (1994)

Infaunal taxa	Epifaunal taxa
Bolivina crenulata	*Anomalinoides spissiformis*
Bolivina denticulata	*Anomalinoides* spp.
Bolivina vaceki	*Cibicidoides eocaenus*
Bolivina spp.	*Cibicidoides grimsdalei*
Bulimina alazanensis	*Cibicidoides haitiensis*
Bulimina alsatica	*Cibicidoides hettneri*
Bulimina impendens	*Cibicidoides laurisae*
Bulimina tuxpamensis	*Cibicidoides mexicanus*
Bulimina spp.	*Cibicidoides pippeni*
Cassidulina spp.	*Cibicidoides praemundulus*
Dentalina spp.	*Cibicidoides truncanus*
Globobulimina spp.	*Cibicidoides* spp.
Globocassidulina globosa	*Eponides lotus*
Laevidentalina spp.	*Falsoplanulina ammophila*
Melonis spp.	*Gavelinella* spp.
Nodogenerina spp.	*Gyroidinoides peramplus*
Nonion havanense	*Gyroidinoides* spp.
Orthomorphina havanensis	*Nuttallides truempyi*
Pleurostomella spp.	*Oridorsalis umbonatus*
Pullenia spp.	*Osangularia mexicana*
Pyramidulina spp.	*Planulina* spp.
Rectuvigerina mexicana	Coiled vaginulinids
Siphonodosaria spp.	
Turrilina alsatica	
Uvigerina nuttalli	
Uvigerina rippensis	
Uvigerina spp.	
Elongate, tapered agglutinated	
Uncoiled, cylindrical vaginulinids	

of the benthic foraminifera. Benthic ratios in the lower part of the section (NP17 Zone and the lower NP18 Zone) are uniform, with values between 1% and 3%. A regular increase to values of about 9% begins within the upper NP18 Zone up to the uppermost part of the section, in which excursions in benthic ratios range from 6% to nearly 14%. These data are consistent with generally bathyal (or deeper) paleodepths (Gibson, 1989; van der Zwaan et al., 1990).

Frequently occurring, cosmopolitan benthic taxa useful in the paleobathymetric definition are presented in figure 26.7. Their paleodepth ranges have been earlier determined by Tjalsma and Lohmann (1983); Miller et al. (1985); Miller and Katz (1986); van Morkhoven et al. (1986); Berggren and Miller (1989); and Müller-Merz and Oberhänsli (1991). The bathymetric zones we have used are after van Morkhoven et al. (1986) and Berggren and Miller (1989). Assemblages from the lower part of the section, including most of the middle Eocene interval, up to the top of the Zone P14, are largely composed of bathyal–upper abyssal species, such as *Anomalinoides spissiformis, Cibicidoides haitiensis, C. hettneri, Nuttallides truempyi,* and *Rectuvigerina mexicana.* In combination with *Cibicidoides grimsdalei* (up to 5% relative abundance), whose upper depth limit is within the lower bathyal zone (van Morkhoven et al., 1986), these faunal components suggest some 1,000 m paleodepth. *C. grimsdalei,* a truly cosmopolitan and readily recognizable taxon, is also a persistent faunal element in middle–late Eocene sections of the northern Tethyan region (Parisi and Coccioni, 1988; Nocchi et al., 1988) that were paleobathymetrically assigned to lower middle bathyal settings. The exit of *N. truempyi* at Nahal Nizzana seems unrelated to paleobathymetry; in fact, this taxon was recovered from late Paleocene–early Eocene, upper bathyal deposits of Libya (Barr and Berggren, 1980) and Egypt (Speijer, 1994; Speijer and Schmitz, 1998). In general, however, this global extinction/sharp reduction event has been related to paleodepth changes and/or increase in $\delta^{18}O$ (Miller et al., 1992; E. Thomas, pers. commun. 1999). Moreover, data from the Indian Ocean have allowed a correlation between the extinction of *N. truempyi* (at the end of the Eocene) and a positive shift of $\delta^{13}C$ (Nomura, 1995), thus suggesting a relationship with (lowered) food resources.

Assemblages of the middle–upper part of the Nahal Nizzana succession, from Zone NP18 upsection, suggest shallower paleodepths. The following distributional data are relevant to paleobathymetric interpretation: (1) the disappearance of the deep-dwelling *Cibicidoides grimsdalei*; (2) the concurrence of *Bulimina impendens, C. laurisae, C. mexicanus*; (3) the appearance of *Uvigerina nuttalli.* The assumed upper depth limit of *Bulimina. impendens* was about 500 m (Berggren and Aubert, 1983) and although this taxon can hardly be distinguished from its quasi-homeomorphic ***B.*** *trinitatensis* (only this latter taxon, to our knowledge, is mentioned from sections of the Mediterranean Tethys), any problem of identification is paleobathymetrically negligible in consideration of their similar upper depth limits (see in Barr and Berggren, 1980, for *B. trinitatensis*). The association of *B. impendens* with *C. mexicanus,* which has been viewed as the deeper (middle to lower bathyal) member of a bathymetric cline (van Morkhoven et al., 1986), and *C. laurisae,* a predominantly middle bathyal-abyssal taxon (van Morkhoven et al., 1986),

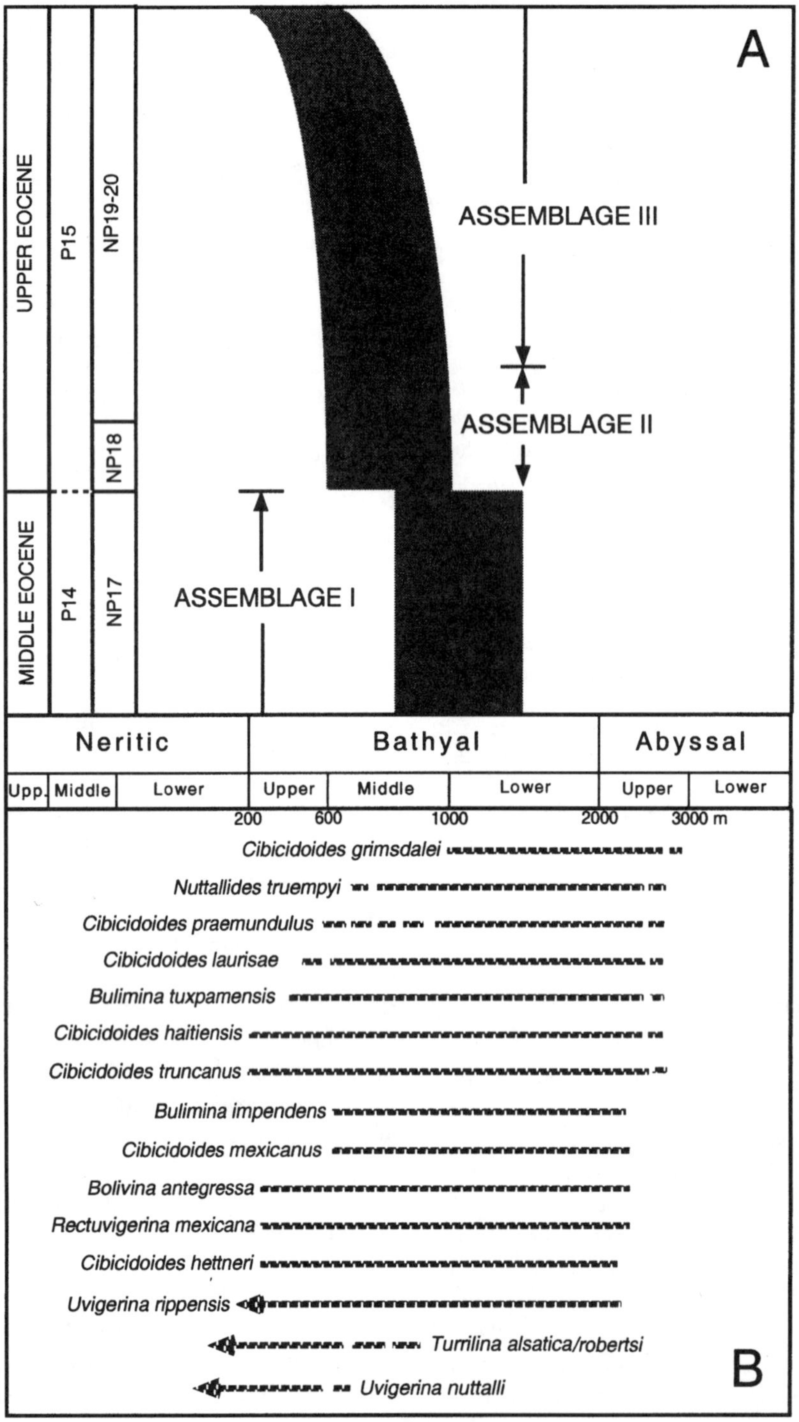

Figure 26.7. (A) Paleodepth evolution of the Nahal Nizzana section (*shaded boxes*); bathymetric zones are after van Morkhoven et al. (1986) and Berggren and Miller (1989). (B) Paleobathymetric distribution of commonly occurring species in the Nahal Nizzana section, for references see the text. *B. antegressa* is here considered a synonym of *B. vaceki*; *C. hettneri* corresponds to *Cibicidoides* sp. 14 of van Morkhoven et al. (1986).

which is recorded from the shallower paleodepths in middle to lower bathyal, NW Atlantic sites (Miller and Katz, 1986), suggests assignation of this part of the section to a paleobathymetry approximately between 500–600 m and 1,000 m.

The development of the costate *U. nuttalli* throughout the upper portion of the Nahal Nizzana section inserts a shallower (upper bathyal to outer neritic, Boersma, 1984) component. There seems to be some positive relationship between *U. nuttalli* morphotypes and the type (carbonate) of sediment (Barr and Berggren, 1980; Borsma, 1984). However, because no significant change occurs along the lithologic log of the section between the lower and upper portions, the entry of *U. nuttalli* with significant numbers (up to nearly 7%) is presumably related to changes in paleodepth. In Libyan middle Eocene this species is recorded from outer neritic deposits (Barr and Berggren, 1980; Barbieri, 1990), although with accompanying species substantially different from those recovered at Negev. A trend toward further shallowing, well into the upper bathyal, can be envisaged in the uppermost part of the section. Together with benthic relative abundance in excess of 10%, the high frequency (up to 14%) of *Cibicidoides pippeni* (figure 26.5), which is considered a neritic–upper bathyal ecotype of the *C. mexicanus* plexus (van Morkhoven et al., 1986), supports this statement.

These paleodepths somewhat conflict with paleogeographic reconstructions that assign the investigated area to a shallow terrigenous/carbonate shelf (figure 26.1), with position of shelf/slope transition during the middle Eocene similar to where it is today. Interestingly, however, the regional (central Israel) work by Buchbinder et al. (1988), in which recognition of mass-transport features in early middle Eocene pelagic chalk was used as a mean for locating the position of the continental slope, shows an inferred position of the shelf/slope transition 10–35 km inland from the present-day Mediterranean coast. These mass-transport features took place on the flanks of mostly submerged folds of the Syrian Arc fold belt. These folds developed in the transition zone between Tethyan spreading to the west and the compressive regime where the Arabian Plate approached the Taurus part of the Alpine orogen. Nummulitic shoals and local emergence characterized Syrian Arc structural highs, and are found in relative proximity to lower neritic to bathyal chalks, and mass-transport phenomena testify to the steep local gradients on the fold flanks. The actual shoreline of the southern continent was very distant (figure 26.1), therefore, continent-derived clastics or organic detritus were cut off from the northwestern Negev both by distance and by the fold system.

Comparison with Global Foraminiferal Changes Across the Middle/Late Eocene Transition

Global distribution patterns for Eocene deep-sea foraminifera indicate that gradual changes affected what were remarkably cosmopolitan assemblages. Foraminiferal extinction patterns from the middle-late Eocene up to the Eocene/Oligocene boundary show selectivity, principally involving low-latitude planktonic (Keller et al., 1992) and larger benthic foraminiferal taxa (Banerjee and Boyajian, 1996, 1997). The extinction curve proposed by Kaiho (1994) for the late Eocene interval shows a long-term pattern and a medium extinction percentage (relatively to foraminiferal groups) for the deep-water benthic taxa. In general, changes include both the structure of assemblages, with the succession/overturn of different faunal groups, and the degree of diversity.

There is a large set of data indicating faunal reorganization around the middle/late Eocene boundary, although explanations for its causes are made difficult by the superposition of local effects (i.e. shallowing) over more general ones. Faunal evidences from the northern Atlantic Ocean region and the Caribbean (Tjalsma and Lohmann, 1983; Wood et al., 1985) off New Jersey (Miller and Katz, 1986), and from the Goban Spur (Miller et al., 1985), indicate a major change close or just above the middle/late Eocene boundary, approximately concomitant with the quantitative decline of *N. truempyi*. Similar data were also recorded from low-latitude, lower bathyal DSDP sites of the southern Atlantic Ocean (Müller-Merz and Oberhänsli, 1991), in which the most abrupt faunal change approximately coincides with the middle/late Eocene boundary, and from upper to middle bathyal, low-latitude ODP sites of the Indian Ocean (Nomura, 1995).

In general, at bathyal depths, *N. truempyi* and associated faunas characterize the middle Eocene assemblages; these are then replaced by a *Globocassidulina globosa/subglobosa–Gyroidinoides* spp.-dominated, late Eocene assemblage (Tjalsma and Lohmann, 1983; Wood et al., 1985; Miller and Katz, 1986). In the Negev, *N. truempyi* is associated with the bathyal marker *Cibicidoides hettneri* (= *Cibicidoides* sp. 14 of van Morkhoven et al., 1986). Their disappearance (or sharp decrease) close to the top of P14 Zone and NP18 Zone, respectively, is only partially replaced by the entry of *Cibicidoides laurisae*, another plano-convex taxon, which is a lower to middle bathyal marker.

In the Negev material, cosmopolitan taxa overwhelmingly dominate the faunal stocks and determine most of the first/last occurrences and relative changes in abundance. A large number of these species belong to low trochospiral (epifaunal) and cylindrical (infaunal) species. In general, the benthic assemblages from paleobathymetrically comparable North Atlantic sequences (DSDP/ODP sites and land sections) bear close resemblance to the those described in this study; for example, the middle bathyal assemblages reported by Miller and Katz (1986), from DSDP Site 612, and by Miller et al. (1985), from DSDP Site 548, in which a number of taxa from about 50% to 75%, respectively, are shared (once removed problems by synonyms) with the Negev material. A similar pattern, but limited to the calcareous trochospiral taxa, is also from the lower bathyal Oceanic Formation (data from the range charts by Bolli et al., 1994).

A characteristic taxon in the lower half of the Negev succession is *C. truncanus* (= *C. parki Auct.*), a truly cosmopolitan species and useful stratigraphic marker for the late middle to late Eocene time span (van Morkhoven et al., 1986). At the northern margin of the Mediterranean Tethys, this species is associated with a *N. truempyi*-dominated assemblage in the early late Eocene of the Marna di Possagno Formation, Italy (Braga et al., 1975; Grünig and Herb, 1980). In areas where lower–middle bathyal conditions persist up to the Eocene/Oligocene boundary, such as in the Massignano section (Italy), *C. truncanus* disappears within the upper part of Zone P16 (Parisi and Coccioni, 1988), suggesting some bathymetric relationship with its Mediterranean range. To a wider extent, the assemblages of Nahal Nizzana show marked similarities with those from the lower to middle bathyal successions of the Umbro-Marchean region, Italy (Nocchi et al., 1988), where more than 50% of the recorded in situ taxa are shared with those from Negev. The diversity indices calculated in that region, however, would indicate little to moderate diversity (average Fisher-α = 5–10; average simple diversity = 15–30, Parisi and Coccioni, 1988; Parisi et al., 1988) compared with those from the Negev. We attribute the reason for this divergence to differences in lithology and inconsistencies in the diversity estimate (e.g., small number of picked specimens/sample). Late Eocene neritic sediments overlying a middle-earliest late Eocene bathyal succession at Possagno (Grünig and Herb, 1980) display a marked regional signature and a consequent loss of the cosmopolitan character exhibited by the underlying deeper deposits. This is consistent with the poor similarity between assemblages of the northern Tethyan margin with the coeval, neritic faunas from the New Jersey Coastal Plain (Browning et al., 1997).

An absence that is difficult to explain is that of *Buliminella grata*, a cosmopolitan, long ranging taxon that is common in the Atlantic (Tjalsma and Lohmann, 1983) and rare in Mediterranean regions (Braga et al., 1975; Grünig and Herb, 1980). In the southern Atlantic *B. grata* is more common at abyssal sites (Müller-Merz and Oberhänsli, 1991); perhaps its absence from the Nahal Nizzana section may thus depends on factors related to paleodepth.

Although bolivinid stocks can be compared on the species level from one location to another with difficulty, because the oxygen concentration is able to influence their abundance, shape, and ornamentation (Sen Gupta and Machain-Castillo, 1993), they represent a primary faunal component at Nahal Nizzana, as the most abundant species, *Bolivina vaceki*, comprises up to more than 20% (average 8%) of the entire assemblage. Similarly, this strongly compressed, striate taxon dominates the bolivinid stock of the middle–late Eocene of the North and South Atlantic (given as *Bolivina antegressa*), where it is limited to the shallower (bathyal) sites (Miller and Katz, 1986; Müller-Merz and Oberhänsli, 1991). In addition, the Negev bolivinid stocks, disregarding species with patchy and inconsistent distribution, also consist of *Bolivina crenulata* (up to more than 7%, average 2%), occuring throughout the section, and *Bolivina denticulata,* which is present (up to more than 12%) only in its upper part. Whereas *B. crenulata* characterizes the bolivinid stock of the bathyal, late Eocene deposits of Biarritz, Gulf of Biscay (Mathelin and Sztràkos, 1993), bolivinids from the South Atlantic with a similar ornamentation are represented by *Bolivina huneri,* which, however, is more common at abyssal sites (Müller-Merz and Oberhänsli, 1991). *B. denticulata* is reported from the northern Tethyan domain in the upper bathyal interval of the Marna di Possagno Formation (Braga et al., 1975; Grünig and Herb, 1980) and in the Häringer Schichten, northern Alps (Lindenberg, 1965). It was also recorded by Bandy (1949) from the lower middle Eocene Tallahatta Formation (stratigraphy after Gaskell, 1991), Gulf Coast, United States. It would appear that paleobathymetric constraints limited the distribution of *B. denticulata* to the shallower conditions of Zone NP19-20 in the Nahal Nizzana section.

DISCUSSION

The complete census of deep-sea benthic foraminifera of middle through late Eocene age from the Nahal Nizzana section records a number of faunal events that may be recognized from deep water sections world-wide. Because of more favorable preservational conditions when compared with isochronous, isobathyal sequences of the northern Tethyan margin, the deposits from Negev have yielded assemblages that in terms of abundance and preservation allow a better comparison with those of the Caribbean region, where an exhaustive faunal record of the middle–late Eocene transition is known. A number of somewhat disparate faunal events may be reconciled when considering that they occur during the faunal perturbation coinciding with the interval of distribution of the Assemblage II, at the base of the late Eocene time span.

Together with marked changes in the abundance for a number of taxa, Assemblage II shows the main change in the epifaunal/infaunal ratio (figure 26.6). Infaunal dominance (to about 75%) is primarily due to the increase of cylindrical morphotypes belonging (mostly) to *T. alsatica* (up to 14% of the total benthic fauna), *P. acuta* (up to 20%), *Uvigerina rippensis* (up to 10%), and uniserial taxa (up to 23%). Increased abundance of inferred infaunal representatives (and consequent decrease of epifaunal ones) is commonly attributed to high flux of organic carbon and/or low oxygen concentration (see the review by Gooday, 1994). Although the admittedly less direct relationship between test morphology and environmental factors would seem to reduce its utility for paleoecological reconstructions, a dramatic increase of the infaunal versus epifaunal taxa has been recorded in critical periods. This is the case at the Paleocene/Eocene boundary in southern high latitudes, for which the suggested cause is a change in the thermal structure of deep-water circulation with a consequent decrease of oxygenation (Thomas, 1990, 1992b). Similarly, in the central Tethyan realm across the Paleocene/Eocene interval, the deep-water benthic foraminiferal extinction event has been related to marked oxygen deficiency (Speijer, 1994; Speijer et al., 1997). The drop of the diversity indices generally registered at the end of Paleocene (e.g., Thomas, 1992b) is absent in the early late Eocene of the Negev, where diversity shows only a modest decrease (compare with Fisher-α values in the bathyal, Paleocene/Eocene boundary Wadi Nukhl section [Sinai], Speijer, 1994). However, a relative increase of abundance of the "anaerobic indices" (sensu Kaiho, 1991) was recorded in the Assemblage II. These forms, which comprise up to about 45% of the entire benthic fauna (figure 26.6), are inferred according to their slender, cylindrical/flattened shapes and reduced or absent ornamentation. In the remaining Assemblages I and III, the average values for the entire stock of these forms range from 20% to 30%. To these "anaerobic" forms were included the uniserial taxa with smooth and/or weakly ornamented/hispid tests, *Pleurostomella* spp., *Bolivina* spp., and smooth to weakly ornamented buliminids. The Assemblage II is also typified by increased abundance of species that may be related positively with raised nutrient availability/reduced oxygenation, such as *U. rippensis*, if ecologic behavior similar to that of *Uvigerina peregrina*, its extant homeomorph, is assumed, and the meso-eutrophic turrilinids (Koutsoukos and Hart, 1990; Speijer and Schmitz, 1998, for the turrilinid *T. brevispira*). With this respect, a notable exception is represented by *Lenticulina inornata* (by far the most abundant lenticulinid) which records a decrease of relative abundance approximately corresponding to the Assemblage II (figure 26.6). As well as in recent settings, *Lenticulina* spp. have been associated with high food and low oxygen levels in several southern Tethyan Paleogene sections (Kouwenhoven et al., 1997; Speijer and Schmitz, 1998). The other two benthic foraminiferal assemblages at the Nahal Nizzana section show relatively stable taxonomic structures, reflected in slow, gradual turnovers and with epifaunal/infaunal and anaerobic species abundances consistent with the inferred paleobathymetry (compare epifaunal/infaunal proportions at the lower bathyal, South Atlantic sites by Oberhänsli et al., 1991): prevailing oligotrophic and oxic environments are there indicated.

To summarize, the observed endobenthic + "anaerobic" faunal peak suggests that the deep-water paleoenvironments at the base of the late Eocene, in comparison with the lower and upper intervals, experienced some oxygen depletion. For such a faunal setting, causes connected with the bathymetric shallowing event located at the boundary between Assemblages I and II are probably of limited importance. Continent-derived (runoff) nutrients, for example, which might increase surface primary production and phytodetritus delivery to the deep waters, can be hardly envisaged because of the distance from the shoreline (figure 26.1) and the SW-NE–oriented structures of the Syrian Arc fold system (Buchbinder et al., 1988). A combined effect of low oxygen and high nutrients can instead be produced by a surface fertility increase as a result of some coastal upwelling, which was active in the area at least since late Cretaceous and during Paleocene–early Eocene time (Almogi-Labin et al., 1993; Charisi and Schmitz, 1998), caused by atmospheric contrast between Tethys and the southern continent. Since the Santonian, in fact, the offshore, parallel-to-coast, south Tethyan upwelling belt was turning on-and-off, with a periodical expansion or contraction of the oxygen-minimum zone. These ecological conditions may have affected surface as well as bathyal waters and, rather than paleobathymetry by itself, could have caused the early disappearance (ecologic exclusion by stressful, oxygen-depleted conditions) of the aerobic *N. truempyi.* In the northern Tethyan margin, for example, the exit of this species takes place at younger levels, close to the P15/P16 zonal boundary and at the base of the NP19–20 Zone, in areas with lower–middle bathyal paleodepths, such as in the Scaglia Variegata Formation of the Umbro-Marchean area (Nocchi et al., 1988).

Tethyan paleogeographic maps, such as those by Dercourt et al. (1993) (figure 26.1), show large and deep Eocene passageways between western and central Tethys and the world ocean that allow a potentially large water exchange. However, the persistent, open discussion on the origin of Tethyan water masses and on their contribution to the world ocean during the Eocene is mainly caused by inadequate paleontological and geochemical information, especially compared with those from the Atlantic and Indian Oceans. A development of Antarctic cooling should have led to an upper middle Eocene increase of latitudinal thermal gradients and to production of corrosive, well-oxygenated deep-water bodies as observed in the Atlantic and Indian Oceans (Oberhänsli et al., 1991; Nomura, 1995). In the middle Eocene of the South Atlantic, a good relationship is established between oxygen availability, changing ratios of epifaunal/infaunal assemblages, and stable isotope (carbon) gradients, but not during the late Eocene (Oberhänsli et al., 1991). Oberhänsli and co-workers argue that late Eocene changes in the infaunal proportions are not necessarily related solely to the oxygenation of the bottom waters, but could also be related to their corrosiveness, as indicated by increase in relative abundance of infaunal indices with corrosive and well-oxygenated bottom waters. It is so far impossible to assess whether the faunal perturbation evidenced by Assemblage II originated solely from regional events (e.g., upwelling) or as a faunal response of the global paleoceanographic reorganization related to the changes in origin (high- or lower-latitude sources) of deep waters throughout the middle Eocene. The absence of very low percentages of planktonic foraminifera (a proxy for corrosive water bodies in deep settings), such as those registered in south Atlantic, late Eocene bathyal sites (Oberhänsli et al., 1991), however, should contrast a direct influence by corrosive water. In addition, *Oridorsalis umbonatus*, an epifaunal species susceptible to dissolution (Corliss, 1983), attains similar (although fluctuating) abundances throughout the section. Perhaps the bolivinid and uvigerinid blooms recorded at the middle/late Eocene boundary in the Umbro-Marchean area (Nocchi et al., 1988) were caused by similar poorly oxygenated paleoenvironmental conditions. These conditions might reflect the development of (western Tethys-wide?) sluggish circulation that did not evolve into true anoxic conditions, such as those experienced in central Tethys at the Paleocene/Eocene boundary (Speijer et al., 1998), and that did not determine changes in the gradual faunal turnover of deep-water benthic foraminifera, but a change in epifaunal/infaunal species proportions just after the middle/late Eocene boundary.

CONCLUSIONS

1. Planktonic foraminifera and calcareous nannoplankton have allowed the Nahal Nizzana section to be subdivided into two and three zones, respectively, included between the upper middle Eocene and the middle late Eocene. The stratigraphic record is largely continuous although condensation and/or a possible gap, between Zones NP17 and NP18 (middle/late Eocene boundary), cannot be unequivocally excluded.

2. The diverse benthic foraminiferal assemblages are overwhelmingly dominated by cosmopolitan taxa and bear a close resemblance to lower–middle bathyal, coeval faunas recorded from the northern Tethyan margin and north Atlantic. Assemblages show a relatively stable taxonomic structure, and there is a gradual turnover, straddling the middle/late Eocene boundary, in which appearances outnumber the disappearances. Paleobathymetric analysis leads to suggested paleodepths ranging from lower bathyal, in the lower part of the section (Zones P14, NP17), to middle and then upper bathyal for the remaining part of the section.

3. The exit of *N. truempyi* at the base of late Eocene seems unrelated to paleobathymetry. Rather, it may have resulted from the establishment of some oxygen depletion. Such conditions are suggested by

changes in relative proportions of infaunal versus epifaunal taxa and low-oxygen-tolerant species. It seems improbable that this infaunal peak (and concomitant epifaunal drop) was caused by enhanced organic carbon flux. If a relationship is established with broadly coeval (middle/late Eocene boundary) acme of certain infaunal taxa registered from the northern margin of the western Tethys, then we may speculate on a change in deep-water circulation leading to a relatively wide and long-lasting (more than 1 million years) period of relatively low oxygen concentrations.

ACKNOWLEDGMENTS

We thank Tanja Kouwenhoven and Robert Speijer, for their review and useful comments, Remo Gamberini and Piero Trentini, for technical assistance (sample preparations and calcium carbonate analyses). This work was funded by the NATO Collaborative Research Grant CRG 972157 and MURST Cofin 1997-1998.

REFERENCES

Almogi-Labin, A., Bein, A., and Sass, E., 1993, Late Cretaceous upwelling system along the southern Tethys margin (Israel): Interrelationship between productivity, bottom water environments, and organic matter preservation: Paleoceanography, v. 8, p. 671-690.

Aubry, M.-P., 1992, Late Paleogene calcareous nannoplankton evolution: a tale of climatic deterioration, *in* Prothero, D. R., and Berggren, W. A., eds., Eocene-Oligocene climatic and biotic evolution: Princeton, New Jersey, Princeton University Press, p. 272-309.

Bandy, O. L., 1949, Eocene and Oligocene foraminifera from Little Stave Creek, Clarke County, Alabama: Bulletin of American Paleontology, v. 32, p. 31-210.

Banerjee, A., and Boyajian, G. E., 1996, Changing biological selectivity of extinction in the Foraminifera over the past 150 m.y.: Geology, v. 24, p. 607-610.

Banerjee, A., and Boyajian, G. E., 1997, Selectivity of foraminiferal extinction in the late Eocene: Paleobiology, v. 23, p. 347-357.

Barbieri, R., 1990, L'Eocene medio e superiore del bacino di Tripolitania (Libia nord occidentale): biostratigrafia e paleoecologia: Bollettino della Societa' Paleontologica Italiana, v. 29, p. 253-271.

Barr, F. T., and Berggren, W. A., 1980, Lower Tertiary biostratigraphy and tectonics of northeastern Libya, *in* Salem, M. J., and Busrewil, M. T., eds., The geology of Libya: London, Academic Press, v. 1, p. 163-192.

Barron, J. A., Larsen, B., and Baldauf, J. G., 1991, Evidence for late Eocene to early Oligocene Antarctic glaciation and observations of late Neogene glacial history of Antarctica: results from Leg 119, *in* Barron, J. A., Larsen, B., et al., eds., Proceedings of the Ocean Drilling Program, Scientific Results, v. 119, p. 869-891.

Benjamini, C., 1980, Stratigraphy and foraminifera of the Qezi'ot and Har'Aqrav Formations (latest middle to late Eocene) of the Western Negev, Israel: Israel Journal of Earth Science, v. 29, p. 227-244.

Berger, W. H., and Diester-Haass, L., 1988, Paleoproductivity: the benthic/planktonic ratio in foraminifera as a productivity index: Marine Geology, v. 81, p. 15-25.

Berggren, W. A., and Aubert, J., 1983, Paleogene benthic foraminiferal biostratigraphy and paleobathymetry of the Central Coast Ranges of California, *in* Brabb, E. E., ed., Studies in Tertiary Stratigraphy of the California Coast Ranges: U.S. Geological Survey Professional Paper 1213, p. 4-21.

Berggren, W. A., and Miller, K. G.,1989, Cenozoic bathyal and abyssal calcareous benthic foraminiferal zonation: Micropaleontology, v. 35, p. 308-320.

Berggren, W. A., and Prothero, D. R., 1992, Eocene-Oligocene climatic and biotic evolution: an overview, *in* Prothero, D. R., and Berggren, W. A., eds., Eocene-Oligocene climatic and biotic evolution: Princeton, New Jersey, Princeton University Press, p. 1-28.

Berggren, W. A., Kent, D. V., Swisher, C. C., III, and Aubry, M.-P., 1995, A revised Cenozoic geochronology and chronostratigraphy: SEPM Special Publication 54, p. 129-212.

Boersma, A., 1984, Handbook of common Tertiary *Uvigerina*: Stony Point, New York, Microclimates Press, p. 207.

Boersma, A., and Premoli Silva, I., 1986, Terminal Eocene events: planktonic foraminifera and isotopic evidence, *in* Pomerol, Ch., and Premoli Silva, I., eds., Terminal Eocene Events, Developments in Palaeontology and Stratigraphy 9: Amsterdam, Elsevier, pp. 213-223.

Boersma, A., Premoli Silva, I., and Shackleton, N. J., 1987, Atlantic Eocene planktonic foraminiferal paleohydrographic indicators and stable isotope paleoceanography: Paleoceanography, v. 2, p. 287-331.

Bolli, H. M., Beckmann, J.-P., and Saunders, J. B., 1994, Benthic foraminiferal Biostratigraphy of the south Caribbean Region: Cambridge, Cambridge University Press, 408 p.

Braga, G., De Biase, R., Grünig, A., and Proto Decima, F., 1975, Foraminiferi bentonici del Paleocene ed Eocene della sezione di Possagno: Schweizerische Paläontologische Abhandlungen, v. 97, p. 85-111.

Bralower, T. J., and Mutterlose, J., 1995, Calcareous nannofossil biostratigraphy of Site 865, Allison Guyot, Central Pacific Ocean: a tropical Paleogene reference section, *in* Winterer, E. L., Sager, W. W., et al., eds.: Proceedings of the Ocean Drilling Program, Scientific Results, v. 143, p. 31-74.

Browning, J. V., Miller, K. G., and Bybell, L. M., 1997, Upper Eocene sequence stratigraphy and the Absecon Inlet Formation, New Jersey Coastal Plain, *in* Miller, K.G., and Snyder, S.W., eds.: Proceedings of the Ocean Drilling Program, Scientific Results, v. 150X, p. 243-266.

Browning, J. V., Miller, K. G., Van Fossen, M., Liu, C., Pak, D. K., Aubry, M.-P., and Bybell, L. M., 1997. Lower to middle Eocene sequences of the New Jersey coastal plain and their significance for global climate change, *in* Miller, K. G., and Snyder, S. W., eds.: Proceedings of the Ocean Drilling Program, Scientific Results, v. 150X, p. 229-242.

Buchbinder, B., Benjamini, C., Mimran, Y., and Gvirtzman, G., 1988, Mass transport in Eocene pelagic chalk on the northwestern edge of the Arabian platform, Shefela area, Israel: Sedimentology, 35, 257-274.

Charisi, S. D., and Schmitz, B., 1998, Paleocene to early Eocene paleoceanography of the Middle East: the $\delta^{13}C$ and $\delta^{18}O$ isotopes from foraminiferal calcite: Paleoceanography, v. 13, p. 106-118.

Coccioni, R., Monaco, P., Monechi, S., Nocchi, M., and Parisi, G., 1988, Biostratigraphy of the Eocene-Oligocene boundary at Massignano (Ancona, Italy), *in* Premoli Silva, I., Coccioni, R., and Montanari, A., eds., The Eocene-Oligocene boundary in the Marche-Umbria Basin (Italy). Interdisciplinary studies presented at "The Eocene-Oligocene boundary ad hoc meeting," Ancona, October 1987: Ancona, F.lli Anniballi, p. 59-80.

Corliss, B. H., 1983, Distribution of Holocene deep-sea benthonic foraminifera in the southwest Indian Ocean: Deep-Sea Research, v. 30(2A), p. 95-117.

Corliss, B. H., 1985, Microhabitats of benthic foraminifera within deep-sea sediments: Nature, v. 314, p. 435-438.

Dercourt, J., Ricou, L. E., and Vrielynck, B., eds., 1993, Atlas Tethys paleoenvironmental maps: Paris, Gauthier-Villars, p. 307.

Gaskell, B. A., 1991, Extinction patterns in Paleogene benthic foraminiferal faunas: relationship to climate and sea level: Palaios, v. 6, p. 2-16.

Gibson, T. G., 1989. Planktonic benthonic foraminiferal ratios: modern patterns and Tertiary applicability: Marine Micropaleontology, v. 15, p. 29-52.

Gooday, A. J., 1994, The biology of deep-sea foraminifera: a review of some advances and their application in paleoceanography: Palaios, v. 9, p. 14-31.

Grünig, A., and Herb, R., 1980, Paleoecology of late Eocene benthonic foraminifera from Possagno (Treviso—northern Italy): Cushman Foundation Special Publication 18, p. 68-85.

Hallock, P., Premoli Silva, I., and Boersma, A., 1991, Similarities between planktonic and larger foraminiferal evolutionary trends through Paleogene paleoceanographic changes: Palaeogeography, Palaeoclimatology, Palaeoecology, v. 83, p. 49-64.

Herguera, J. C., and Berger, W. H., 1991, Paleoproductivity from benthic foraminifera abundance: Glacial to postglacial change in the west-equatorial Pacific: Geology, v. 19, p. 1173-1176.

Jones, R. W., and Charnock, M. A., 1985, "Morphogroups" of agglutinating foraminifera. Their life positions and feeding habits and potential applicability in (paleo)ecological studies: Revue de Paléobiologie, v. 4, p. 311-320.

Kaiho, K., 1991, Global changes of Paleogene aerobic/anaerobic benthic foraminifera in deep-sea circulation: Palaeogeography, Palaeoclimatology, Palaeoecology, v. 83, p. 65-85.

Kaiho, K., 1994, Planktonic and benthic foraminiferal extinction events during the last 100 m.y.: Palaeogeography, Palaeoclimatology, Palaeoecology, v. 111, p. 45-71.

Keller, G., 1982, Paleoclimatic analyses of middle Eocene through Oligocene planktonic foraminiferal faunas: Palaeogeography, Palaeoclimatology, Palaeoecology, v; 43, p. 73-94.

Keller, G., MacLeod, N., and Barrera, E., 1992, Eocene-Oligocene faunal turnover in planktonic foraminifera and Antarctic glaciation, *in* Prothero, D. R., and Berggren, W. A., eds., Eocene-Oligocene climatic and biotic evolution: Princeton, New Jersey, Princeton University Press, p. 218-244.

Kennett, J. P., 1977, Cenozoic evolution of Antarctic glaciation the circum-Antarctic Ocean and their impact of global paleoceanography: Journal of Geophysical Research, v. 82, p. 3843-3860.

Koutsoukos, E. A. M., and Hart, M. B., 1990, Cretaceous foraminiferal morphogroup distribution patterns, palaeocommunities and trophic structures: a case study from the Sergipe Basin, Brazil: Transactions of the Royal Society of Edinburgh: Earth Sciences, v. 81, p. 221-246.

Kouwenhoven, T. J., Speijer, R. P., van Oosterhout, C. W. M., and van der Zwaan, G. J., 1997, Benthic foraminiferal assemblages between two major extinction events: the Paleocene El Kef section, Tunisia: Marine Micropaleontology, v. 29, p. 105-127.

Lindenberg, H. G., 1965, Die Bolivinen (Foram.) der Häringer Schichten. Mikropaläontologische Untersuchungen im Alttertiar des Unterinntal—Gebietes: Bollettino della Società Paleontologica Italiana, v. 4, p. 64-160.

Martini, E., 1971, Standard Tertiary and Quaternary calcareous nannoplankton zonation, *in* Farinacci, A., ed., Proceedings Second Planktonic Conference, Roma 1970: Roma, Tecnoscienza, p. 739-785.

Martini, E., 1976, Cretaceous to Recent calcareous nannoplankton from Central Pacific Ocean (DSDP Leg 33), *in* Schlanger, S.O., Jackson, E.D., et al., eds.: Initial Reports of the Deep Sea Drilling Project, v. 33, p. 383-423.

Mathelin, J.C., and Sztràkos, K., 1993, L'Eocène de Barritz (Pyrénées Atlantiques, SW France). Stratigraphie et paléoenvironment. Monographie des foraminifères: Cahiers de Micropaléontologie, v. 8, p. 5-182.

Miller, K. G., 1992, Middle Eocene to Oligocene stable isotopes, climate, and deep-water history: the terminal Eocene event? *in* Prothero, D. R., and Berggren, W. A., eds., Eocene-Oligocene climatic and biotic evolution: Princeton, New Jersey, Princeton University Press, p. 160-177.

Miller, K. G. and Katz, M. E., 1986, Eocene benthic foraminiferal biofacies of the New Jersey transect, *in* Poag, C. W., Watts, A. B., et al., eds.: Initial Reports of the Deep Sea Drilling Project, v. 95, p. 267-298.

Miller, K. G., Curry, W. B., and Ostermann, D. R., 1985, Late Paleogene (Eocene to Oligocene) benthic foraminiferal oceanography of the Goban Spur region, Deep Sea Drilling Project Leg 80, *in* Graciansky, P. C. de, Poag, C. W., et al., eds.: Initial Reports of the Deep Sea Drilling Project, v. 80, p. 505-538.

Miller, K. G., Katz, M. E., and Berggren, W. A., 1992, Cenozoic deep-sea benthic foraminifera: a tale of three turnovers, *in* Takayanagi, Y., and Saito, T., eds., Studies in Benthic Foraminifera (Benthos '90, Sendai 1990): Tokay University Press, p. 67-75.

Moodley, L., van der Zwaan, G. J., Herman, P. M. J., Kempers, L., and van Breugel, P., 1997, Differential response of benthic meiofauna to anoxia with special reference to Foraminifera (Protista: Sarcodina): Marine Ecology Progress Series, v. 158: 151-163.

Müller-Merz, E., and Oberhänsli, H., 1991, Eocene bathyal and abyssal foraminifera from a South Atlantic transect at 20–30°S, *in* Premoli Silva, I., and Boersma, A., eds., The oceans of the Paleogene: Palaeogeography, Palaeoclimatology, Palaeoecolology, v. 83, p. 117-171.

Nocchi M., Parisi, G. Monaco P., Monechi S., Madile M., Napoleone G., Ripepe M., Orlando M. Premoli Silva I., and Bice, D.M., 1986, The Eocene-Oligocene boundary in the Umbrian pelagic sequence (Italy), *in* Pomerol, C., and Premoli Silva, I., eds., Terminal Eocene Events, Developments in Palaeontology and Stratigraphy 9: Amsterdam, Elsevier, p. 1-24.

Nocchi, M., Parisi, G., Monaco, P., Monechi, S., Madile, M., 1988, Eocene and Early Oligocene micropaleontology and paleoenvironments in SE Umbria, Italy: Palaeogeography, Palaeoclimatology, Palaeoecology, v. 67, p. 181-244.

Nomura, R., 1995, Paleogene to Neogene deep-sea paleoceanography in the eastern Indian Ocean: benthic foraminifera from ODP Sites 747, 757 and 758: Micropaleontology, v. 41, p. 251-290.

Oberhänsli, H., Müller-Merz, E., Oberhänsli, R., 1991, Eocene paleoceanographic evolution at 20–30°S in the Atlantic Ocean, *in* Premoli Silva, I., and Boersma, A., eds., The oceans of the Paleogene: Palaeogeography, Palaeoclimatology, Palaeoecology, v. 83, p. 173-215.

Okada, H., and Bukry, D., 1980, Supplementary modification and introduction of code numbers to the low-latitude coccolith biostratigraphic zonation: Marine Micropaleontology, v. 5, p. 321-325.

Parisi, G., and Coccioni, R., 1988, Deep-water benthic foraminifera at the Eocene-Oligocene boundary in the Massignano section (Ancona, Italy), *in* Premoli Silva, I., Coccioni, R., and Montanari, A., eds., The Eocene-Oligocene boundary in the Marche-Umbria Basin (Italy). Interdisciplinary studies presented at "The Eocene-Oligocene boundary ad hoc meeting," Ancona, October 1987: Ancona, F.lli Anniballi, p. 97-109.

Parisi, G., Guerrera, F., Madile, M., Magnoni, G., Monaco, P., Monechi, S., and Nocchi, M., 1988, Middle Eocene to early Oligocene calcareous nannofossil and foraminiferal biostratigraphy in the Monte Cagnero section, Piobbico (Italy), *in* Premoli Silva, I., Coccioni, R., and Montanari, A., eds., The Eocene-Oligocene boundary in the Marche-Umbria Basin (Italy). Interdisciplinary studies presented at "The Eocene-Oligocene boundary ad hoc meeting," Ancona, October 1987: Ancona, F.lli Anniballi, p. 119-135.

Perch-Nielsen, K., 1985, Cenozoic calcareous nannofossils, *in* Bolli, H .M., Saunders, J. B., and Perch-Nielsen, K., eds., Plankton stratigraphy: Cambridge, Cambridge University Press, p. 427-554.

Proto Decima, F., Roth, P. H., and Todesco, M. L., 1975, Nannoplancton calcareo del Paleocene e dell'Eocene della sezione di Possagno: Schweizerische Paläontologische Abhandlungen, v. 97, p. 35-55.

Schaub, H., Benjamini, C., and Moshkovitz, S., 1995, The biostratigraphy of the Eocene of Israel: Schweizerische Paläontologische Abhandlungen, v. 117, p. 58.

Sen Gupta, B., and Machain-Castillo, M. L., 1993, Benthic foraminifera in oxygen-poor habitats: Marine Micropaleontology, v. 20, p. 183-201.

Speijer, R. P., 1994, Extinction and recovery patterns in benthic foraminiferal paleocommunities across the Cretaceous/Paleogene and Paleocene/Eocene boundaries: Geologica Ultraiectina, v. 124, p. 191.

Speijer, R. P., and Schmitz, B., 1998, A benthic foraminiferal record of Paleocene sea level and trophic/redox conditions at Gebel Aweina, Egypt: Palaeogeography, Palaeoclimatology, Palaeoecology, v. 137, p. 79-101.

Speijer, R. P., Schmitz, B., and van der Zwaan, G. J., 1997, Benthic foraminiferal extinction and repopulation in response to latest Paleocene Tethyan anoxia: Geology, v. 25, p. 683-686.

Thomas, E., 1990, Late Cretaceous-early Eocene mass extinctions in the deep sea, *in* Sharpton, V. L., and Ward, P. D., eds., Global catastrophes in earth history: an interdisciplinary conference on impacts, volcanism, and mass mortality: Geological Society of America Special Paper 247, p. 481-495.

Thomas, E., 1992a, Middle Eocene–late Oligocene bathyal benthic foraminifera (Weddell Sea): faunal changes and implications for ocean circulation, *in* Prothero, D. R., and Berggren, W. A., eds., Eocene-Oligocene climatic and biotic evolution: Princeton, New Jersey, Princeton University Press, p. 245-266.

Thomas, E., 1992b, Cenozoic deep-sea circulation: evidence from deep-sea benthic foraminifera, *in* Kennett, J. P., and Warnke, D. A., eds., The Antarctic paleoenvironment: a perspective on global change: Antarctic Research Series 56, p. 141-165.

Thomas, F. C., and Murney, M. G., 1985, Techniques for extraction of foraminifers and ostracodes from sediment Samples: Canadian Technical Report of Hydrography and Ocean Sciences 54, p. 24.

Tjalsma, R. C., and Lohmann, G. P., 1983, Paleocene-Eocene bathyal and abyssal benthic Foraminifera from the Atlantic Ocean: Micropaleontology Special Publication 4, p. 90.

van der Zwaan, G. J., Jorissen, F. J., and de Stigter, H. C., 1990, The depth dependency of planktonic/benthic foraminiferal ratios: constraints and applications: Marine Geology, v. 95, p. 1-16.

van der Zwaan, G. J., Duijnstee, I. A. P., den Dulk, M., Ernst, S. R., Jannink, N. T., and Kouwenhoven, T. J., 1999, Benthic foraminifers: proxies or problems? A review of paleoecological concepts: Earth-Science Reviews, v. 46, p. 213-236.

van Morkhoven, F. P. C. M., Berggren, W. A., and Edwards, A. S., 1986, Cosmopolitan deep-Water benthic Foraminifera: Bulletin des Centres de Recherches Exploration-Production Elf-Aquitaine Mem. 11, p. 421.

Wood, K. C., Miller, K. G., and Lohmann, G. P., 1985, Middle Eocene to Oligocene foraminifera from the Oceanic Formation, Barbados: Micropaleontology, v. 31, p. 181-197.

CHAPTER 27

The Development of Facies Patterns of Middle Eocene to Lower Oligocene Circum-Alpine Shallow-Water Carbonate Environments

James H. Nebelsick, Michael Rasser, and Davide Bassi

ABSTRACT

Middle Eocene to lower Oligocene shallow-water carbonate facies from the circum-alpine area are compared with respect to dominating biogenic components and their distributions in space and time. The study areas (northeastern Italy, northern Slovenia, Austria, and southern Bavaria) were situated at the junction of the Tethys seaway, the developing Paratethys and the boreal North Sea area, and were influenced by plate tectonic movements and formation of the Alpine chain. Fourteen different major facies types (MFTs) are distinguished; these are dominated by coralline algae, large and small foraminifera, corals, and bryozoans. The distribution of MFTs is controlled not only by global events (such as the extinction of larger foraminifera) but also by local variations of environmental parameters. The most dramatic changes occur within the larger foraminifera, with three extinction patterns that have a direct influence on facies patterns. Large alveolinids disappear in the uppermost middle Eocene, large *Nummulites* and *Orbitolites* become extinct at the middle Eocene/late Eocene boundary, Orthophragminids disappear at the late Eocene/Oligocene boundary. These developments in carbonate environments during the change from a "greenhouse" to "icehouse" world mark the transition to the carbonate facies types as we know them today.

INTRODUCTION

Shallow-water carbonates contain a wealth of information concerning environmental and paleoclimatic gradients in marine benthic environments. Carbonates are also sensitive to environmental changes resulting in corresponding variations in constituent components as well as facies distributions. The different character of carbonates can be used, for example, to recognize distinct biogeographic or climatic provinces based on component distributions (e.g., the distribution and diversity of symbiont-bearing organisms such as larger foraminifera and corals; Nelson, 1988; Betzler et al., 1997; James, 1997).

The climatic transition between the earliest Eocene and late Oligocene is known to have had a dramatic effect on a wide variety of different organisms (Berggren and Prothero, 1992). This transition between the "greenhouse" and "icehouse" worlds has been ultimately linked to opening of the circum-Antarctic seaway and an increase in Antarctic ice volume. The early Eocene is recognized as the warmest part of the Cenozoic, with a reduced latitudinal sea-surface temperature (SST) gradient relative to today and warmer deep water with about half the maximum vertical temperature gradient of modern oceans (e.g., Shackleton and Boersma, 1981; Miller et al., 1987, 1991; Zachos et al., 1994). In the early to middle Eocene, global climates start to cool and by the Oligocene glacial conditions had become established in the high latitudes of the Southern Hemisphere. This resulted in a dramatic decrease of deep water as well as high-latitude SST from the early Eocene to the Oligocene. Although a total decrease of 12°C has been postulated for high-latitude temperatures during this time, there was apparently little change in low-latitude SST (Zachos et al., 1994).

In this paper we concentrate on the middle Eocene to lower Oligocene, circum-alpine carbonates. These include a compilation of modern studies on

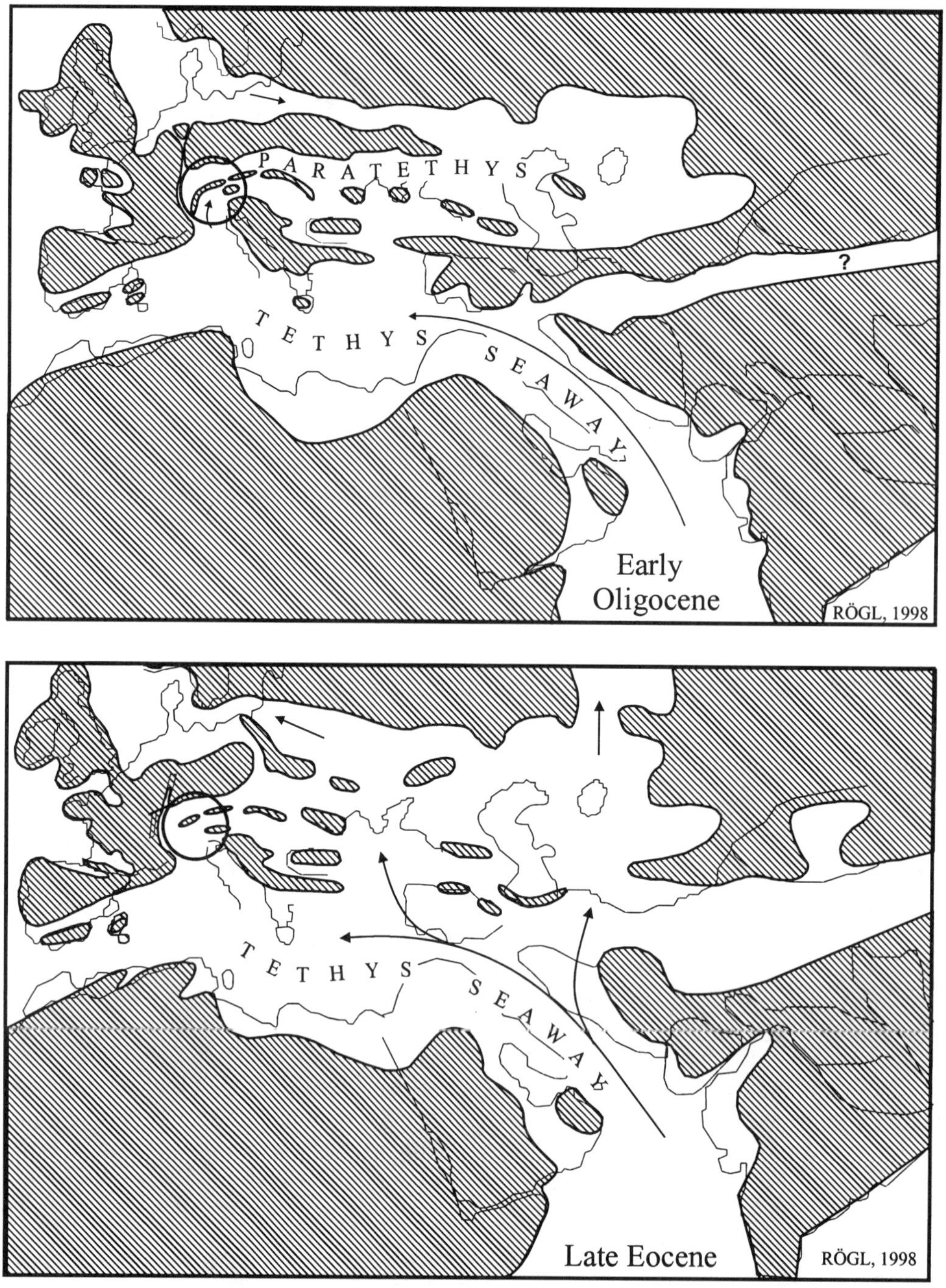

Figure 27.1. Paleogeographic development of the western Tethys during the late Eocene and early Oligocene (after Rögl, 1998). Circle indicates general study area.

historically well-known localities that have been very important for the characterizing of Eocene and Oligocene carbonate successions in general. The time frame was chosen to include major changes in component and facies composition that started to take place in the middle Eocene. These carbonates are dominated by larger foraminifera, coralline algae, and corals. The most dramatic faunal change in Eocene and Oligocene shallow-water carbonates occurs in the diversity and distribution of larger foraminifera. More larger foraminiferal genera existed during the early-middle Eocene than at any other time during the Cenozoic. The coexistence of the larger foraminiferal groups of orbitolitids, alveolinids, nummulitids, and orthophragminids is a unique event in earth history restricted to the period of late Paleocene to the end of the middle Eocene (Hottinger, 1983a). The large variety of different taxa present is reflected not only in numerous monographs (e.g., Schweighauser, 1953; Hottinger, 1960; Drobne, 1977; Schaub, 1981; Less, 1987), but also in studies concerning their biodiversity and biogeography. There is a pronounced provincialism of larger foraminifera within the Indo-West Pacific, Tethyan, and American biogeographic realms (Adams et al., 1983, 1986, 1990). Hottinger (1983a) pointed out the correlation between larger foraminiferal assemblages and facies types from the Late Cretaceous to the Recent and recognized the repetition of larger foraminiferal morphologies in similar facies developments.

Paleocene to lower Oligocene reefs are among the rarest in all of the Phanerezoic (Kiessling et al., 1999). The abundance and geographic distribution of reefs, however, increased along with coral diversity during the middle to late Eocene, with many modern reef-building coral genera appearing during this time (Flügel and Flügel-Kahler, 1992). Coral communities show extinction in species, but not in genera at the Eocene/Oligocene boundary (Veron, 1995). The early Oligocene is marked by a decrease of coral diversity. The late Oligocene, however shows the maximum reef development of the Paleogene at least in the Caribbean (Frost and Langenheim, 1974; Frost et al., 1983).

Coralline algae are the least-known major components in Paleogene limestones with respect to diversity, abundance, and distribution. This is, in part, certainly due to the lack of a uniform taxonomic basis needed for the study of diversity gradients and biogeography. There is, however, a noticeable increase of coralline algae diversity during the Paleogene, with the floras showing distinctly modern aspects (Flügel and Flügel-Kahler, 1992).

The geographic position of the study areas (figure 27.1) allows for the investigation of facies changes both across chronostratigraphic boundaries, and along latitudinal gradients from "tropical" to "boreal" settings. The position of the study areas marks the junction of three different biogeographic provinces: the Tethys Ocean to the south, the North Sea area to the north, and, in between, the developing Paratethys. The Tethys Seaway, situated between the African-Arabian plate to the south and the European plate to the north, was part of a circumglobal seaway in the Paleogene, become narrower during the Cenozoic and finally being cut off in the early Miocene (Rögl, 1998). The Paratethys was separated from the Tethys through the developing Alpine chain and became a distinct paleogeographic unit in the early Oligocene (Báldi, 1984). A number of localities in carbonates of northern Italy are important with respect to the middle Eocene/late Eocene as well as Eocene/Oligocene boundaries (e.g., Castellarin and Cita, 1969; Cita, 1969; Barbin, 1986, 1988, 1992; Pomerol and Premoli Silva, 1986; Premoli Silva and Jenkins, 1993; Brinkhuis, 1994; Brinkhuis and Visscher, 1995; Aubry et al., 1999).

Only a few studies have specifically commented on possible paleoclimatic influences of floral/faunal composition in the study areas, these being restricted to northern Italy. Bosellini and Russo (1988) correlated an increase in diversity of Rupelian coral communities from the Eocene/Oligocene boundary onward to climatic amelioration during the Oligocene. Barbin (1992), using palynoflora, showed no evidence for climatic deterioration across the Eocene/Oligocene boundary, suggesting that terrestrial conditions remained warm-humid tropical to subtropical through this interval. Brinkhuis (1994), using dinoflagellate cysts, suggested sustained relatively warm tropical to subtropical environments throughout the Eocene-Oligocene transition although the influence of colder waters (early Oligocene cooling event) could be tentatively detected. Local environmental influences in the restricted neritic setting, however, may be more important than SSTs in influencing the observed floral compositions (Brinkhuis 1994).

GEOLOGIC SETTING

The general geologic setting of the studied localities is very complex due to plate and structural reorganization of the Alpine orogeny. The Alpine orogeny was caused by an interplay of movements and collisions of the African and European plates and intervening microplates (e.g., Ziegler, 1987; Wagreich, 1995). This led to widely different depositional settings and subsequent fates of Paleogene sediments. The carbonate facies analyzed here originate from eight areas that show different depositional histories (figure 27.2): the eastern and western Lessini Shelf, as well as eastern Colli Berici from northeastern Italy; the Gornji Grad area in northern Slovenia; the Central Alpine Gosau Basin of the Krappfeld; the Inner Alpine Basins as

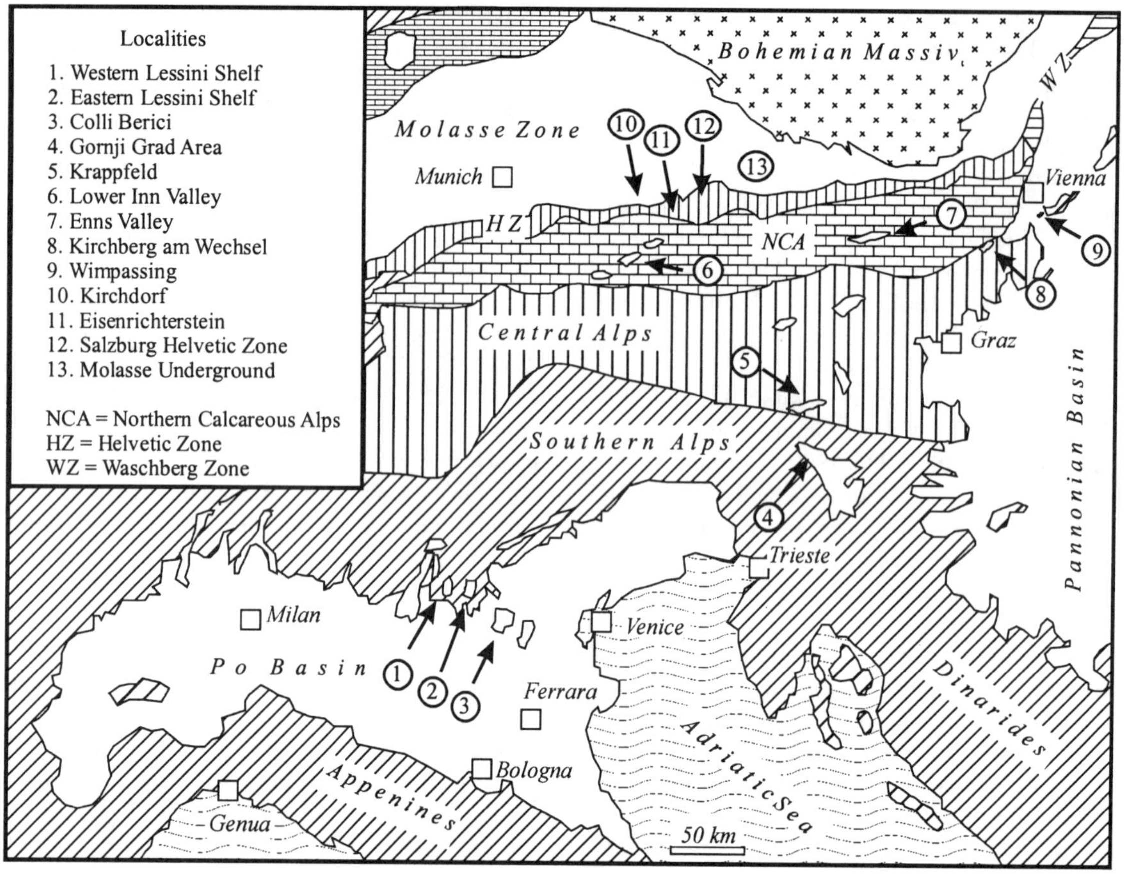

Figure 27.2. Distribution of the study areas.

well as the Helvetic Unit and autochthonous Molasse underground from Austria and Bavaria.

Northeastern Italy

The Lessini Shelf is a paleogeographic unit of the Southern Alps and represents the Paleogene, shallow-water carbonate successions of the Venetian area (Monti Lessini, north-eastern Italy). The Lessini Shelf is superimposed onto the Liassic Trento Platform (Castellarin, 1972; Bosellini, 1989), a major structural and paleogeographic unit of the Adriatic plate. During the complex collision between Europe and the Adriatic plate that generated the Alpine chain, the tectonic pattern in this area was largely controlled by inherited Permian-Jurassic tectonic structures (Doglioni and Bosellini, 1987; Cassinis and Castellarin, 1988). Paleogene, shallow-water carbonate sedimentation of the Lessini Shelf was initiated on topographic highs that resulted from block-faulting of the former Trento Platform (Bosellini, 1989). The Lessini Shelf can be divided into an "eastern" and "western" regions that show different facies developments, especially in the early Oligocene.

The Colli Berici, located at the south Vicenza plain, are geologically interpreted as an isolated, flattened, slightly bent anticline along a NE-SW axis. The Colli Berici are thought to be the southwestern continuation of the Monti Lessini, as they show stratigraphic analogies with the Paleogene formations of the Lessini Shelf (Mietto, 1988). The stratigraphical succession is mainly represented by carbonate successions and igneous rocks, ranging from the Late Cretaceous to early Miocene. The volcanic successions in the Monti Lessini and Colli Berici are mainly represented by middle Eocene basaltic volcanics, tuffs, and minor lava flows, which can alternate with nummulitic calcarenites (Ungaro and Bosellini, 1965; Piccoli, 1966; De Vecchi and Sedea, 1995).

Slovenia

The Gornji Grad area (northern Slovenia, Southern Alps) represents a transition zone between the northern most extension of the Tethys to the south and the developing Paratethys. A number of Paleogene sections are present in a geologically complex setting in the surroundings of Gornji Grad. The formation of this basin is related to tectonic movements along the Periadriatic Lineament that separates the Central Alps from the Southern Alps (Fodor et al., 1998; Placer, 1999).

Eastern Alps and Alpine Foreland

During most of the Alpine orogeny, a northern shelf represented by the southern rim of the European plate (Helvetic Unit, autochthonous Eocene of Molasse), and a southern shelf along the northern margin of the Alpine archipelago (northern Alpine Gosau Basins, Inner Alpine/parautochthonous Molasse basins) can be differentiated. These two shelves were separated by the Penninic Ocean. The Alpine nappe system formed an archipelago, on which the Gosau basins of the Central Alps were situated. The Gosau basins represent small-scale basins formed during erosional phases of Alpine orogeny. In the current study we group the Northern Alpine Gosau Basins and the Inner Alpine/Parautochthonous Molasse basins into the "Inner Alpine basins".

The late Eocene was characterized by orogenic activities resulting from the final closure of the Penninic Ocean. This was followed by overthrusting of the Rhenodanubian Flysch Zone and the Helvetic Unit by the end of the Eocene (Wagreich, 1995). The northward movement of the basinal axis in the Alpine Foreland caused a northward shifting of the northern shelf area from the Helvetic Zone to the autochthonous Molasse underground (Hagn, 1981; Tollmann, 1985). The closure of the Penninic Ocean marks the beginning of the Molasse sedimentation in the Paratethys now separated from the Tethys by the rising Alpine chain. The Alpine foredeep of the Molasse zone separated the northern and the southern shelves and accommodated the erosional products from the uprising Alps. Piggy-back basins (Inner Alpine Molasse basins) were formed along the northern Alpine thrust front (Steininger et al., 1986).

STRATIGRAPHIC OVERVIEW

Monti Lessini and Colli Berici

The Upper Cretaceous–early middle Eocene successions of the Lessini Shelf and of the Colli Berici are represented by deep-water deposits (Scaglia rossa, Scaglia cinerea; Massari et al., 1976; Lehner et al., 1987); the middle Eocene–Oligocene successions by shallow-water deposits (Mietto, 1988; Antonelli et al., 1988). Four complete depositional sequences were revealed by sequence stratigraphy analysis of the Paleogene successions from the western (Luciani, 1989) and eastern (Trevisani, 1997) margins of the Lessini Shelf: (E1) early-middle Eocene, (E2) middle-late Eocene, (O1) late Eocene-early Oligocene, and (O2) late Oligocene. These depositional sequences, which consist mainly of carbonate platform-to-basin depositional systems, are bounded by unconformity surfaces.

The middle Eocene shallow-water carbonate successions consist of large nummulitic calcarenites and coralline-foraminiferal packstones that generally pass upsection into smaller miliolid packstones/grainstones (Ungaro, 1969; Garavello and Ungaro, 1982; Papazzoni and Sirotti, 1995). The uppermost middle Eocene (Bartonian), a regressive phase, caused an emergence with brackish conditions in some areas of the Monti Lessini (Frascari, 1963) and the western Colli Berici (Ungaro and Bosellini, 1965; Ungaro, 1969). The Priabonian carbonate deposits are mainly represented by small *Nummulites* calcarenites, coralline packstones, orthophragminid packstones/marls, and bryozoan marls (Ungaro, 1969; Antonelli et al., 1990; Bassi, 1998). Larger foraminifera gradually disappear at the uppermost middle upper Eocene, while during the Priabonian, frequent occurrences of small *Nummulites* and "*Operculina*" persist. Simultaneously, coralline algae increase in abundance and become facies dominating.

At the Eocene/Oligocene boundary, a regressive phase is documented in the western Lessini by a disconformity surface of the subaerially exposed Calcare di Nago platform, which in turn is overlain by the basin sediments of the upper Eocene-lower Oligocene "Marne di Bolognano" (Castellarin and Cita, 1970; Luciani, 1989). At the eastern Lessini Shelf, the Priabonian/Rupelian boundary is marked by regressive deposits that consist of bryozoan marls, sandstones, and silty marls (Cita, 1969; Setiawan, 1983; Barbin, 1988; Trevisani, 1997). In the Colli Berici, the Eocene/Oligocene boundary is marked by bryozoan marls and marly calcarenites. Although the number of sedimentary hiatuses increase around the Eocene/Oligocene boundary both on the shelf and in ocean basins (Pomerol and Premoli Silva, 1986), continuous marine sequences across the boundary have been recorded in Venetian neritic settings (Ungaro, 1978; Garavello and Ungaro, 1982).

Coral, coralline, miliolid, and small nummulite facies characterize the Oligocene deposits of the eastern Lessini Shelf and the eastern Colli Berici. The lower-middle Oligocene deposits are represented by the Calcareniti di Castelgomberto which overlie the Marne di Priabona both in the Lessini Shelf and in eastern Colli Berici. They are characterized by shallow-water, massive or irregularly bedded limestones, and irregularly bedded calcarenites. The fossils are represented by nummulitids, soritids, small benthic miliolid foraminifera, massive and branched corals, mollusks, and echinoids (Coletti et al., 1973; Geister and Ungaro, 1977; Ungaro, 1978; Frost, 1981; Bosellini and Trevisani, 1992; Bosellini and Stemann, 1996); coralline algae are also present. Both the Lessini Shelf and the Colli Berici areas emerged towards the uppermost Rupelian.

Slovenia

The transgressive sedimentary succession of Paleogene sediments begins with Eocene sediments. These are followed by the marine Gornji Grad Beds, which include the limestones containing the carbonate facies discussed here. These Gornji Grad beds are followed by foraminifera-rich marls and finally tuffites (Schmiedl et al., 2002). Drobne et al. (1985) recognized an early Oligocene age for the Gornji Grad Beds using larger foraminifera. The limestones are dominated by coralline algae (Bassi and Nebelsick, 2000; Bassi et al., 2000; Nebelsick et al., 2000), small *Nummulites*, small benthic foraminifera, corals and bivalves.

Eastern Alps and Alpine Foreland (South to North)

Central Alpine Gosau basin of the Krappfeld. The Krappfeld Gosau basin represents the southernmost Paleogene Gosau Basin. Due to its southern position, it forms a transitional element between the northern calcareous Alps and the southern Alps (Tollmann, 1977; Wilkens, 1989, 1991; Rasser, 1994). The succession starts with the Paleocene continental Holzer Formation and grades into upper Paleocene to lower Eocene sandstones and nummulite marls of the Sittenberg Formation. The overlying Dobranberg Formation marks a transition to very pure larger foraminiferal limestones. Basal parts of this formation are characterized by intercalations of *Alveolina* and nummulitic limestones with local intercalations of orthophragminid foraminifera (Wilkens, 1989, 1991; Hillebrandt, 1993). The nummulitid foraminifera are early Eocene in age (Hillebrandt, 1993). Coralline algae and encrusting foraminifera dominate the middle part of the succession where they form huge accumulations of rhodoliths and acervulinid macroids (Rasser, 1994) that are thought to be of Ypresian to Lutetian in age (Wilkens, 1989; see also Moussavian, 1984). The succession from larger foraminifera to coralline algal limestones is interpreted as a deepening upward, shallow marine environment (Rasser, 1994). The pure nummulite limestones are interpreted by Wilkens (1989, 1991) to have been formed by in-situ winnowing sensu Aigner (1985).

Northern Calcareous Alps. Paleogene neritic carbonates within the Northern Calcareous Alps are characteristically small-scaled occurrences, sometimes autochthonous, but often known only from poorly dated, reworked clasts in Miocene conglomerates. These occurrences are usually separated into Gosau basins and Inner Alpine/parautochthonous Molasse basins (here grouped together into the Inner Alpine basins) along the Alpine thrust front. The Paleogene carbonates of the lower Inn Valley are represented by lower Oligocene coralline algae-dominated facies, which overlie the siliciclastic Häring Beds and are part of the "Zementmergel" Beds (Hagn, 1981; Stingl, 1990; Krois, 1992; Ortner, 1996). Coralline algae, branched corals, amphistigenids, and bryozoans can be abundant. Paleogene components of fluvial conglomerates near Radstadt/Pongau (Enns Valley) contain upper Paleocene coral limestones and lower to middle Eocene detritical carbonates rich in nummulitids (Trauth, 1918; Moussavian, 1984). Parautochthonous boulders of Paleogene carbonates overlay the crystalline basement at Kirchberg am Wechsel in lower Austria. They are dominated by large nummulitids and also contain orthophragminids and coralline algae; larger foraminifera indicate a late Eocene age (Kramer, 1962; Ebner et al., 1991). Finally, upper Eocene coralline algal limestones are known from Wimpassing in Burgenland (Trauth, 1918; Papp, 1958).

Helvetic Zone. The Helvetic Zone is a tectonic unit of the Alpine orogenic system. It forms a narrow zone extending from Switzerland to eastern Austria between the Rhenodanubian Flysch Zone to the south and the Molasse zone to the north (figure 27.2). The best outcrops of Paleogene carbonates of the Helvetic Zone are near Salzburg (Gohrbandt, 1963; Vogeltanz, 1970; Rasser and Piller, 1999a, b) where coralline algal limestones of the uppermost Paleocene overlie Paleocene continental series. The overlying Ypresian to lowermost Bartonian sandstones and limestones are characterized by large nummulitids and orthophragminids. This succession is followed by the middle to late Eocene so-called "Stockletten," which contains intercalations of coralline algal olistoliths in the late Eocene. The occurrence of coralline algae within the deep-marine "Stockletten" is seen to reflect the presence of olistoliths originating from a shallow-marine environment, perhaps from the autochthonous late Eocene of the Molasse underground (Darga, 1992).

Molasse zone and autochthonous underground. The first marine sediments of the Molasse Zone are known from the late Eocene, which are called "pre-Molasse," "Molasse underground" or "autochthonous Eocene of the Alpine Foreland" which underlie the Molasse sediments sensu strictu (Steininger et al., 1986). These sediments, designated as the Perwang-Group by Wagner (1996), can be separated into a northern and a southern facies realm (Wagner, 1980; Rasser, 2000). The northern facies, which were deposited near the coast, contains siliciclastic series (*Cerithium* beds, Amphing Sandstone, Voidsdorf Formation) and diverse coralline algal facies ("*Lithothamnion* Limestone"). The latter are dominated by coralline branches, rhodoliths, and coralline bindstones. In addition, peyssonneliaceans, nummulitids, and branching corals can be abundant. The southern facies are characterized by a succession from nummulite sandstones to orthophragminid marls, and finally to bryozoan marls (Rasser et al., 1999; Rasser, 2000). Upper Eocene sediments, which were dated by means of larger foraminifera and calcareous nannoplankton are overlain by Oligocene deep-water facies. Facies successions are expected to express the ongoing subsidence of the foreland due to Alpine orogeny. The Oligo-Miocene Molasse sediments of the Molasse zone are dominated by clastics with restricted carbonate developments in the upper Oligocene (Roetzel et al., 1991) and lower Miocene (Nebelsick, 1989).

COMPARISON OF CARBONATE FACIES

The following Major Facies Types (MFTs) are considered to represent the most important types within the middle Eocene to Oligocene carbonates under consideration (table 27.1). This subdivision is seen to represent an objective differentiation from the authors' own work as well as from other published data. The MFTs are defined using the most abundant taxa and exclude subordinate or transitional facies types as well as special floral/faunal developments within the specific study areas. The distribution of MFTs in different areas is summarized in table 27.2, their distribution through time in table 27.3.

A generalized ramp morphology is chosen as the general carbonate depositional environment in the schematic distribution of MFTs shown in figure 27.3. Due to the flexural downbending of the crust, caused by tectonic loading, carbonate platforms of foreland basins are usually homoclinal ramps (Dorobek, 1995). Burchette and Wright (1992) differentiated between inner ramp (upper shore face to fair-weather wave base [FWWB]), middle ramp (FWWB to storm-wave base), and outer ramp (below normal SWB). This subdivision certainly applies to the late Eocene of the Alpine Foreland. A carbonate ramp geometry with a gentle slope from the shoreline to the adjacent basin can be argued for the western (Luciani et al., 1988; Bassi, 1998) and the eastern (Trevisani, 1997) Lessini Shelf as well as the Colli Berici (Geister and Ungaro, 1977; Ungaro, 1978; Papazzoni, 1994) (but see Frost, 1971; Bosellini, 1989, for an interpretation as a fringing, protruding carbonate platform scattered with reefs, lagoons, islands and volcanoes and interpreted reefs on its margins).

MAJOR FACIES TYPES

MFT-1: Maerl Facies

Maerl is the term used for sediments composed of coralline algal branches, rhodoliths, and their detritus. Today, they are known from temperate and subarctic environments (Adey, 1986; Freiwald et al., 1991; Freiwald, 1994). The dominant components in the study area are fragmented and unfragmented coralline algal branches, which can contribute up to 90% of the allochems; rhodoliths can be locally abundant. Nummulitids, smaller miliolid and textulariid foraminifera, and peyssonneliacean algae are subordinate; branching corals and arenitic siliciclasts are also present. This facies occurs as a coralline algal rudstone with grainstone or packstone matrix; sorting is usually poor, sometimes nearly bimodal. This facies shows massive bedding, and its fabric lacks any preferred orientation or gradation.

Maerl has a restricted occurrence in the late Eocene of the Molasse underground (Rasser, 2000) and in the early Oligocene of Northern Italy (see table 28.1). The only detailed interpretation of this facies from the study area is from the Eocene of the Molasse underground where the coralline algae are represented by *Phymatolithon* sp. and an undetermined melobesioid coralline algal species (Rasser, 2000). This facies was compared with the present-day maerl from the Mediterranean Sea. It is interpreted to have formed in higher-energy banks or megaripples, which

Table 27.1. Middle Eocene-Lower Oligocene Circum-Alpine Major Facies Types

No.	Major Facies Types	Dominant Components	Subordinate Components	Texture	Fabric
1	Maerl Facies	Coralline algal branches and detritus	Nummulitids, smaller miliolids textularid foraminifera	Rudstones with grainstone and packstone matrix	Massive and unbedded, no orientations or gradations
2	Rhodolith Facies	Non-geniculate coralline algae, peyssonneliacean algae, encrusting acervulinid foraminifera	Serpulids, *Haddonia*, bryozoans	Rudstones with grainstone and packstone matrix	Various rhodolith shapes: spherical, ellipsoidal, discoidal and boxwork; encrusting and protuberant algal growth forms
3	Crustose Coralline Algal Facies	Encrusting coralline algae, rhodoliths, encrusting corals, nummulitids locally abundant	Smaller miliolid and textulariid foraminifera, bryozoans, peyssonneliacean algae, echinoderms, serpulids	Meter thick bindstones composed of 0.5 to 1 mm thick algal crusts	Algal crusts usually horizontally orientated
4	Coralline Algal Debris Facies	Coralline algal debris, larger foraminifera, bryozoans, corals	Peyssonneliacean algae, molluscs, smaller benthic foraminifera (rotaliids, textulariids), peloids, siliciclastics	Rudstones with grainstone or packstone matrix	No bedding or grading, various types of sorting
5	Peyssonneliacean Facies	*Polystrata alba* forming rounded aggregates	Coralline algae	Rudstones with packstone matrix	No grading, sorting or orientation
6	Large *Nummulites* Facies	Various species of large *Nummulites*	Other larger and small small benthic foraminifera, molluscs, bryozoans, decapods, echinoids	Rudstones with packstone matrix	Both orientated and chaotic fabrics present
7	Small *Nummulites* Facies	Various species of small *Nummulites*,	Coralline algal debris other larger and smaller benthic foraminifera, bivalves, echinoids, brachiopods, corals	Packstones and rudstones with packstone/ grainstone matrix or siliciclastic matrix	Graded bedding can occur
8	Orthophrag -minid Facies	Large, thin, disc-shaped and saddle shaped orthophragminids	Larger and smaller benthic foraminifera (rotaliids and textulariids), bivalves, coralline algal crusts, planktic foraminifera can be present	Rudstones with wacke- to packstone matrix	Horizontal orientation usually present
9	*Orbitolites* Facies	*Orbitolites* tests and small miliolid foramini -fera	Peneroplid foraminifera, bivalves and gastropods	Rudstones with pack- to grainstone matrix	Graded bedding present
10	Smaller Miliolid Facies	Diverse small benthic miliolid foraminifera, peneroplid, and alveolinid foraminifera	Textulariid foraminifera, *Sphaerogypsina*, bivalves, echinoderms, geniculate and non-geniculate coralline algae	Packstones and grainstones	No grading, sorting or orientation
11	Alveolinid Facies	Alveolinids, small miliolids, asterigerinids, nummulitids	Small benthic foraminifera, echinoids, coralline algae	Packstones and grainstones	Both oriented and chaotic fabrics
12	Acervulinid Facies	Acervulinid macroids formed by *Acervulina ogormani* and *A. linearis*, tubular aggregates of *A. multiformis*	Macroids also contain coralline algae, serpulids and homotrematid foraminifera, larger and smaller miliolid foraminifera	Rudstones	Macroids up to 10 cm in diameter with laminar-encrusting growth-forms, no grading, sorting, or orientation
13	Coral Facies	Branching and encrusting, corals, coralline ting,	Small nummulitids, bryozoans, brachiopods, miliolids	Rudstones with pack- to grainstone matrix	Both branching and encrusting coral growth forms
14	Bryozoan Facies	Diverse bryozoans	Larger foraminifera, smaller benthic rotaliid foraminifera, coralline algae	Rudstones with wackestone matrix, marly packstones	Diverse bryozoan growth forms present, components mostly horizontally orientated

Table 27.2. Major Facies Types: Middle Eocene-Lower Oligocene Circum-Alpine Distributions

No.	Major Facies Types	Middle Eocene					Late Eocene						Early Oligocene			
		ELS	WLS	ECB	KF	HZ	ELS	WLS	ECB	IAB	HZ	MOL	ELS	ECB	GGB	IAB
1	Maerl											x	x	x		
2	Rhodolith	x	x	x	x		x	x	x		x	x	x	x		
3	Crustose Coralline								x			x		x		
4	Coralline Algal Debris	x	x	x	x		x	x	x		x	x	x	x	x	x
5	Peyssonneliacean											x				
6	Large *Nummulites*	x	x	x	x	x										
7	Small *Nummulites*	x	x	x	x		x	x	x	x		x	x	x	x	
8	Orthophragminid	x	x	x			x	x	x	x		x				
9	*Orbitolites*	x														
10	Smaller Miliolid	x	x	x			x				x		x	x	x	
11	Alveolinid	x	x													
12	Acervulinid	x			x		x				x	x				
13	Coral							x		x	x	x	x	x	x	x
14	Bryozoan						x		x			x				x

ELS = eastern Lessini Shelf, WLS = western Lessini Shelf, ECB = eastern Colli Berici, KF = Krappfeld, HZ = Helvetic Zone, IAB = Inner Alpine Basins, MOL = Mollase underground, GGB = Gornji Grad Beds.

were cut by sand channels. This maerl is composed of interlocking branches and represents an unstable substrate that is not conducive for the settlement of sessile communities and frameworks. Different rates of hydrodynamic energy are reflected by the amount of fragmentation of coralline branches and the composition of matrix.

MFT-2: Rhodolith Facies

Rhodoliths consist of unattached nodules with a size larger than 2 cm (Adey, 1986, but see Bosence, 1983, for another size classification) that consist predominantly (>50%) of non-geniculate coralline red algae (Bosence, 1983). Rhodoliths are characteristic features of many Cenozoic shallow-water carbonates. The rhodoliths that dominate this facies consist predominately of non-geniculate coralline algae although other encrusting organisms, such as peyssonneliacean algae (*Polystrata alba* [Pfender] Denizot) and encrusting foraminifera (*Acervulina ogormani* [Douville] and *Acervulina linearis* [Hanzawa]) can be important contributors. Serpulids, bryozoans and other sessile foraminifera such as *Haddonia* sp. can also be important. Size and shape of the rhodoliths in the study area are highly variable. They can reach up to 10 cm and show spherical, ellipsoidal, discoidal, or boxwork shapes; the algal thalli can be encrusting or show protuberances.

This facies shows a very wide distribution throughout the study area (see table 27.2). There are no distinct changes of rhodolith size and shape within the studied successions; there are, however, noticeable differences concerning their inner arrangement and composition. The composition of middle Eocene rhodoliths of the Krappfeld is interpreted to reflect water depth (Rasser, 1994). The vertical succession from pure coralline algal rhodoliths, which upsection show an increasing abundance of peyssonneliacean algae and acervulinid foraminifera, and finally grade into an acervulinid macroid facies, are interpreted as a deepening upward trend. upper Eocene rhodoliths from the western Lessini Shelf show a significant increase in size and in constructional voids with increasing water depth (Bassi, 1998).

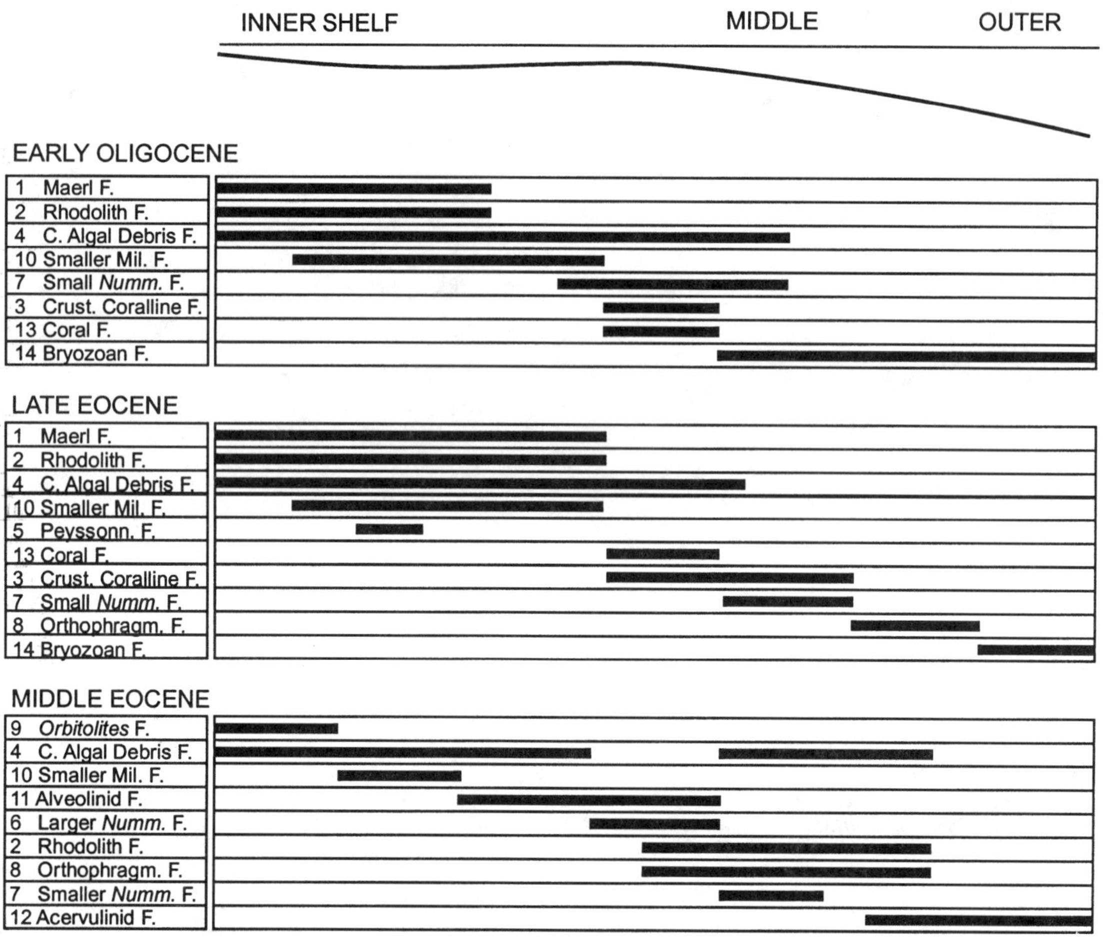

Figure 27.3. Schematic distribution of Major Facies Types (MFT) during the middle Eocene, late Eocene and lower Oligocene in the circum-alpine region.

The upper Eocene rhodolith facies of the Molasse underground (Rasser, 2000) show a high diversity of coralline algal species, lack acervulinids, but can contain abundant peyssonneliaceans (Rasser, 2000). These features point to lower hydrodynamic energy and relatively rapid smothering by sediment. upper Eocene rhodolith facies of the Helvetic Zone are intercalated with deep-water Stockletten facies and contain few peyssonneliaceans and *Acervulina linearis.* These probably represent olistoliths within a deeper-marine setting, which originated from a now unpreserved shallow-water environment (Darga, 1992). The rhodolith facies is also known from the middle Eocene of northern Italy (Bassi et al., 1998), upper Eocene eastern Lessini Shelf (Garavello and Ungaro, 1982) and eastern Colli Berici (Bassi, 1995) as well as the lower Oligocene of the eastern Lessini Shelf (Bosellini and Trevisani, 1992) and Colli Berici (Ungaro, 1978).

MFT-3: Crustose Coralline Algal Facies

Crustose corallines can show millimeter-scale lamination due to binding. Bindstone-forming coralline algae are abundant in present-day tropical (e. g., Steneck et al., 1997) and nontropical (e.g., Bosence, 1985) environments. They can form a structural relief above the surrounding substrate (see Rasser, 2000, for review). The bindstones are dominated by encrusting coralline algae; rhodoliths, encrusting corals and large foraminifera (mainly nummulitids) can be locally abundant

This facies has a restricted distribution (see table 27.2). In the autochthonous upper Eocene of the Molasse underground, crusts are formed by *Phymato-*

Table 27.3. Middle Eocene-Lower Oligocene Major Facies Types: Circum-Alpine Stratigraphic Distribution

No.	Major Facies Types	Middle Eocene	Late Eocene	Early Oligocene
1	Maerl		x	x
2	Rhodolith	x	x	x
3	Crustose Coralline		x	x
4	Coralline Algal Debris	x	x	x
5	Peyssonneliacean		x	
6	Large *Nummulites*	x		
7	Small *Nummulites*	x	x	x
8	Orthophragminid	x	x	
9	*Orbitolites*	x		
10	Smaller Miliolid	x	x	x
11	Alveolinid	x		
12	Acervulinid	x	x	
13	Coral		x	x
14	Bryozoan		x	x

lithon sp. and *Neogoniolithon* sp. Horizontally orientated, 0.5- to 1-mm-thick individual crusts can form bindstones of up to 4 m thickness (Rasser, 2000). In the Colli Berici, upper Eocene coralline bindstones are formed by *Lithothamnion* spp. and *Mesophyllum* sp.; *Nummulites, Operculina, Pellatispi-ra, Spiroclypeus* and orthophragminids also occur (Bassi, 1995). Both occurrences are additionally characterized by small (up to 3 cm), ellipsoidal, laminar rhodoliths with a high percentage of constructional voids. In the Colli Berici, this facies passes laterally into nummulitid-, orthophragminid- or peyssonneliacean-dominated facies. They occurred in an open-platform setting and were subjected to sporadic occurrences of higher hydrodynamic energy; micrite was deposited during periods of lower hydrodynamic energy. Subordinate rhodoliths in this facies show a very high percentage of voids (Bassi, 1995). Some of these bindstones are comparable with the present-day "Coralligene de plateau" (e.g., Bosence, 1985) and with Miocene representatives from Malta (Bosence and Pedley, 1982).

MFT-4: Coralline Algal Debris Facies

This facies is dominated by the debris of various species of coralline algae as well as larger foraminifera, bryozoans, and corals. A number of subordinate components are also present (table 27.1). This facies is very common and occurs in almost all of the studied areas. It is, however, not recorded from the late Eocene of the Inner Alpine basins (e. g., Papp, 1958; Moussavian, 1984) and in the middle Eocene of the Helvetic Zone (Darga, 1992). This facies can occur in a wide variety of environmental settings and all of the above-mentioned coralline algal facies can contribute to this facies. It usually represents an unstable substrate detrimental for the development of sessile communities and frameworks.

MFT-5: Peyssonneliacean Facies

This facies is dominated by peyssonneliacean red algae, which are distinguished from the coralline algae by aragonitic cell walls (Bassi, 1997). The dominant components are *Polystrata alba* (Pfender) Denizot, which forms small, monospecific subspherical aggregates (usually 2–10 mm in diameter), and crusts, which account for up to 77% of the total components. Coralline algae can also be abundant. This facies is known only from the late Eocene of the Molasse underground and is interpreted to have been deposited in a very low-energy environment with a muddy substrate, probably in the deeper photic zone (Rasser, 2000).

MFT-6: Large *Nummulites* Facies

This facies is dominated by one or several species of large-sized *Nummulites* tests with subordinate smaller nummulite species, orthophragminids, coralline algae, and other organisms (Matteucci and Pignatti, 1990). It shows a wide geographic distribution, but is restricted to the middle Eocene. This facies is characterized by *Nummulites gizehensis* (Forskål), *N. millecaput* Boubée, *N. brogniarti* D'Archiac and Haime, *N. perforatus* (De Montfort), *N. lyelli* D'Archiac and Haime, and *N. biedai* Schaub in the middle Eocene of the eastern (Schaub, 1981; Garavello and Ungaro, 1982) and the western Lessini Shelf (Papazzoni and Sirotti, 1995) and eastern Colli Berici (Schweighauser, 1953; Ungaro, 1969; Papazzoni and Sirotti, 1995). According to von Hillebrandt (1993), the middle Eocene large *Nummulites* facies of the Krappfeld is characterized by *Nummulites praeaturicus* Schaub, *N. hilarionis* Schaub, *N. beneharnensis* (De La Harpe), *N. boussaci* (Rozlozsnik), *N. praediscorbinus* Schaub, and *N. millecaput* (Boubée).

The interpretation of this facies in both the Southern Alps (Schaub, 1981) and the Krappfeld area (Wilkens, 1989) have followed the general interpretations as parautochthonous accumulations which formed shoals or bars (Arni, 1965; Matteucci and Pignatti, 1990) due to in situ winnowing (sensu Aigner, 1985). More recently, Racey (1995) used the ratio of A to B forms to distinguish autochthonous (dominated by A forms) from allochthonous (dominated by B forms) accumulations.

MFT-7: Small *Nummulites* Facies

The main components of this facies are small nummulites such as *Nummulites discorbinus* (Schlotheim), *N. bullatus* Azzaroli, *N. beaumonti* D'Archiac and Haime, *N. ptukhiani* Kacharava, *N. striatus* (Bruguière), *N. variolarius/incrassatus*, *N. chavannesi* De La Harpe, *N. fabianii* (Prever), *N. stellatus* Roveda, *N. pulchellus* Hantken, *N. retiatus* Roveda, *N. vascus* Joly and Leymerie and *N. fichteli* Michelotti. Subordinate components are small orthophragminids, *Heterostegina*, *Assilina*, *Operculina*, *Pellatispira*, *Biplanispira*, *Spiroclypeus*, small orthophragminids, small benthic foraminifera (e.g., asterigerinids, textulariids, miliolids, gypsinids, *Cibicides*, *Eponides*, *Chapmanina*, *Halkyardia*, *Europertia*, *Fabiania*, etc.) and other components including abundant coralline debris produced by encrusting and branching growth forms.

This facies is widely distributed throughout the study area and well documented from the middle Eocene of the western (Castellarin and Cita, 1969) and the eastern (Schaub, 1981; Garavello and Ungaro, 1982) Lessini Shelf, eastern Colli Berici (Schweighauser, 1953), and the Krappfeld (Wilkens, 1989; 1992); the late Eocene of the western (Castellarin and Cita, 1969; Luciani, 1989), eastern (Cita, 1975; Garavello and Ungaro, 1982) Lessini Shelf, eastern Colli Berici (Ungaro, 1969), the Inner Alpine basins (Papp, 1957), and the Molasse underground (Rasser et al., 1999); early Oligocene from the eastern Lessini Shelf (Bosellini and Trevisani, 1992), eastern Colli Berici (Ungaro, 1978), and Gornji Grad beds in Slovenia (Nebelsick et al., 2000).

The small nummulitid facies has been interpreted as a shallow-water inner-platform deposit (Arni, 1965) and may represent parautochthonous, reworked and resedimented deposits (see Aigner, 1985; Matteucci and Pignatti, 1990). Bioerosion by endolithic algae suggest a long surface residence time of the tests (Accarie, 1987). These resedimented deposits are usually characterized by smaller *Nummulites* species that reach a minimum size during the late Eocene and Oligocene. Siliciclastic nummulite sandstones of the Molasse underground represent the base of a succession from nummulite facies to orthophragminid facies and to bryozoan facies. They represent the beginning of a deepening upward sequence (Rasser et al., 1999; Rasser, 2000).

MFT-8: Orthophragminid Facies

Large thin disc-shaped and saddle-shaped orthophragminids (sensu Less, 1987) dominate this facies that occurs in middle and upper Eocene deposits including the middle Eocene of the eastern (Setiawan, 1983) and western Lessini Shelf (Castellarin and Cita, 1969), and eastern Colli Berici (Schweighauser, 1953), as well as the late Eocene of the eastern (Setiawan, 1983), and the western Lessini Shelf (Bassi, 1998), eastern Colli Berici (Schweighauser, 1953), Inner Alpine basins (Darga, 1992) and the autochthonous Molasse underground (Rasser, 2000). While orthophragminids of the Molasse underground can be accompanied by small nummulitids or bryozoans, those of the Helvetic Zone are accompanied by coral debris and encrusted by coralline algae.

Orthophragminid assemblages are considered to represent open-shelf deposits (deeper part of the photic zone) (Fermont, 1982; Buxton and Pedley, 1989; Eichenseer and Luterbacher, 1992). *Discocyclina*-dominated assemblages have been regarded to be adapted to shallower waters and lower turbulence than the *Asterocyclina*-dominated ones (Sirotti, 1978: Less, 1987). There is, however, insufficient information concerning the palaeoecology of the orthophragminids that became important constituents of the Priabonian successions after the decrease in the *Nummulites* diversity. The bathymetric interpretation of these assemblages is based on the knowledge of modern nonrelated forms showing similar morphologies (Ferràndez-Cañadell and Serra-Kiel, 1992). Priabonian orthophragminid facies have also been described from the uppermost midramp setting of the western Lessini Shelf (Bassi, 1998).

MFT-9: *Orbitolites* Facies

The *Orbitolites* Facies, dominated by larger foraminifera *Orbitolites*, is restricted to uppermost middle Eocene of the eastern Lessini Shelf (Garavello and Ungaro, 1982). This facies is associated with accumulations of large bivalves and gastropods as well as conglomerates. The environment of this facies has been interpreted as a shallow-water ("beach deposits" of Eichenseer and Luterbacher, 1992) or possibly hypersaline setting subject to frequent salinity and temperature variations (Hottinger, 1983a).

MFT-10: Smaller Miliolid Facies

The packstones and grainstones of the smaller miliolid facies are dominated by a wide range of foraminifera including the name-giving miliolids, peneroplids, and alveolinids (*Borelis* and *Glomalveolina*) as well as *Silvestriella*, *Cibicides*, *Austrotrillina*, asterigerinids, and rare textulariids. In Eocene examples, the Alveolinidae *Alveolina* and *Glomalveolina* also occur together with *Austrotrillina*. This facies is present in the uppermost middle Eocene from the western (Castellarin and Cita, 1969) and the eastern Lessini Shelf (Cita, 1975; Garavello and Ungaro, 1982), eastern Colli Berici (Ungaro, 1969); the late Eocene of the Colli Berici (Barbin and Keller-Gruenig, 1991) and Priabonian type locality (eastern Lessini; Setiawan, 1983), the late Eocene of the Helvetic Zone (Darga, 1992) and the early Oligocene from the eastern Lessini Shelf (Bosellini and Trevisani, 1992), eastern Colli Berici (Ungaro, 1978), and Gornji Grad area (Bassi and Nebelsick, 2000). This facies represents a shallow-water setting with high hydrodynamic and high light intensity (Buxton and Pedley, 1989).

MFT-11: Alveolinid Facies

This facies is dominated by large-sized alveolinid tests that form massive to bedded packstones to rudstones in the middle Eocene at the western and the eastern Lessini Shelf (Hottinger, 1960). Small miliolids, asterigerinids, and *Nummulites* are also common. Small benthic foraminifera, coralline algae and echinoderms are subordinate. Alveolinid-dominated facies were interpreted by Arni (1965) to occur in different ecological settings ranging from restricted hypersaline shelf through shoals to the open marine inner shelf.

MFT-12: Acervulinid Facies

Acervulinid foraminifera can show growth-form features comparable to that of coralline algae (Perrin, 1992). Acervulinid macroids are unattached nodules predominantly composed of acervulinid foraminifera, comparable to rhodoliths (Hottinger, 1983b). The acervulinid macroids are mostly formed by *Acervulina ogormani* and *A. linearis*. *A. linearis* and *A. multiformis* Trauth can also form small spherical crusts with a central part filled with sparite. The macroids can reach sizes up to 10 cm in diameter and show a laminar-encrusting growth form. Smaller aggregates of *A. multiformis* can show half-tube growth forms, which are characteristically U-shaped in thin section, with a size of a few millimeters, frequently forming grainstones.

This facies occurs in the middle Eocene of the Krappfeld (Rasser, 1994); middle and late Eocene of the eastern Lessini Shelf (Ungaro, 1996), and upper Eocene Molasse underground (Rasser, 2000); and in olistoliths of the Helvetic Zone (Darga, 1993). The acervulinid macroids of the Krappfeld are interpreted to have formed in deeper-water environments than the underlying rhodolith facies and therefore express a deepening upward trend. The tubular and half-tubular growth forms of acervulinids are interpreted to reflect the bio-immuration of a nonpreserved fleshy algae or seagrass (Ungaro, 1996).

MFT-13: Coral Facies

Branching and encrusting corals dominate this facies together with coralline algae. It shows a wide distribution in the late Eocene and early Oligocene. Published occurrences include the late Eocene of the western Lessini Shelf (Bosellini, 1998) and the Inner Alpine Basins (Eisenrichterstein, Bavaria; Darga, 1992), the Helvetic Zone (Kirchdorf; Darga, 1992), and the Molasse underground (Rasser, 2000); the early Oligocene of the lower Inn Valley, the eastern Lessini Shelf (Bosellini and Russo, 1988; Bosellini and Trevisani, 1992; Bosellini and Stemann, 1996), eastern Colli Berici (Geister and Ungaro, 1977; Ungaro, 1978; Frost, 1981) and the Gornji Grad area (Bassi and Nebelsick, 2000).

Corals of the Eisenrichterstein and Kirchberg

localities of the late Eocene of the Helvetic Zone are characterized by fragmented corals, which do not form a framework (Darga, 1992), and which can be accompanied by encrusting coralline algae. Corals of the uppermost Eocene Molasse underground (Rasser, 2000) are represented by encrusting corals, which can rarely dominate coralline algal bindstones, as well as by branching corals forming biostromes, which are accompanied by encrusting and branching coralline algae. lower Oligocene corals are dominated by branching corals, which are comparable to those of the upper Eocene Molasse underground. Upper Eocene coral facies of the western Lessini Shelf show a change with depth in taxonomic composition, in abundance and proportion within the framework, and in growth forms (Bosellini, 1998).

Coral limestones of the upper Eocene Helvetic Zone (Kirchdorf) are interpreted as olistoliths originating from a shallow-water area, which is not preserved today; also those of the inneralpine basin (Eisenrichterstein) are frequently fragmented. Darga (1992) discussed reef models from the Eisenrichterstein; however, he also stated that no true reefal limestones occur. The "reef limestones" of Darga are characterized by abundant coral fragments, which in fact do not form any sort of framework (Darga, 1992). Upper Eocene olistoliths of the deep-water facies of the Helvetic Zone are partly dominated by coral fragments; the shallow-water area, from which these olistoliths came, is not preserved. Darga reconstructed a fringing reef, although there are no indications for any reef fabric (Darga, 1992). upper Eocene coral facies of the western Lessini Shelf have been interpreted as shelf-edge patch reefs developed on the shelf margin (Bosellini, 1998).

The lower Oligocene corals of the eastern Lessini Shelf proliferated in shallow-water muddy bottoms during low-energy intervals, while bioclastic sands were swept across the shelf during high-energy intervals (Bosellini and Trevisani, 1992). Bosellini and Russo (1988) show an increase in diversity of Rupelian coral communities from the Eocene/Oligocene boundary onward. The early Oligocene of the eastern Lessini Shelf is characterized by the poritid coral *Actinacis rollei* which grows on muddy substrates; it shows mainly branching forms even though massive globose and laminar colonies are present (Bosellini and Trevisani, 1992). The lower Oligocene coral facies from the eastern Colli Berici show lens- to dome shape structures composed mainly of massive hermatypic scleractinians, and extensive coral carpets formed predominantly by branching forms (Geister and Ungaro, 1977; Ungaro, 1978). Frost (1981) recognized three different coral assemblages in his analysis of the early Oligocene of the Colli Berici. The barrier reef model proposed by Frost (1981) for the eastern Colli Berici, however, is not substantiated by the detailed facies analysis conducted by Geister and Ungaro (1977) and Ungaro (1978) in this area. Coral facies are also recognized in the early Oligocene of Slovenia (Nebelsick et al., 2000) and the lower Inn Valley. Here again, isolated corals in muddy environments or present in coral thickets are recognized; buildups in the sense of a reef structure are not present.

MFT-14: Bryozoan Facies

The bryozoan facies occurs in the late Eocene of the Molasse underground (Rasser, 2000), lower Oligocene of the early Inn Valley, and late Eocene of the eastern Lessini Shelf (Setiawan, 1983) and eastern Colli Berici (Ungaro, 1969, 1978). The bryozoan facies of the lower Inn Valley follow algal- and foraminifera-dominated facies and are interpreted as indicating increasing water depth. Two alternative interpretations were made for bryozoan facies of the upper Eocene Molasse underground: the succession from orthophragminid facies to bryozoan facies can be interpreted by either a deepening upward or an increasing eutrophication (Rasser, 2000). A sequence stratigraphic approach for the upper Eocene bryozoan facies of the eastern Lessini Shelf (Priabonian type locality) suggests that it represents regressive deposits, which characterize the lowstand systems tracts of the overlying lower Oligocene depositional sequence (Trevisani, 1992). Brinkhuis (1994), however, ascribed the same deposits ("bryozoan limestones and the bryozoan marls") to the early Oligocene and interpreted them as inner-neritic deposits representing a highstand systems tract.

DISCUSSION

The most dramatic changes within the studied facies patterns occur within the larger foraminifera with three extinction patterns that have a direct influence on facies patterns (table 27.3): (1) large alveolinids (*Alveolina*) (MFT-11) disappear in the early Bartonian (uppermost middle Eocene) along with the acervulinid *Acervulina ogormani* (MFT-12); (2) large *Nummulites* (MFT-6) and *Orbitolites* (MFT-9) become extinct at the middle Eocene/late Eocene boundary; (3) orthophragminids disappear (MFT-8) at the Eocene/Oligocene boundary. The ecological space left by these facies are occupied by others (see figure 27.3). For example the open platform environment occupied by the large *Nummulites* facies (MFT-6) in the middle Eocene are "replaced" by the maerl facies (MFT-1) in the late Eocene. Other major facies types such as the rhodolith (MFT-2), small *Nummulites* (MFT-7) and miliolid facies (MFT-10) do not show any specific restrictions in distribution and are also known before (compare Taberner and Bosence 1985) as well as after the time frame discussed here. The specific occurrence of other facies are obviously influenced by local current patterns. For example, the presence of the bryozoans facies (MFT-14) is correlated to higher nutrient input due to coastal runoff and/or upwelling patterns.

All the facies types in the study areas that are still present today already existed in the middle Eocene. Orthophragminids are the last facies in the study area to disappear (at the Eocene/Oligocene boundary). Conversely, all the Oligocene facies present in the study areas are still extant today. This emphasizes, on the one hand, the importance of the "greenhouse-ice-house" transition period in the development of "modern" facies types as we know them today. On the other hand, it becomes clear, at least from the standpoint of major facies types, that this transition occurred not suddenly, but gradually over at least three important biotic turnovers characterized by faunal extinctions and facies replacements.

These facies represent, in part, global development of faunal and floral successions that can also be followed outside the study area. The facies thus correspond in part to those discussed by Buxton and Pedley (1989) in their generalized cross sections of carbonate ramp depositional setting in the Paleogene. Important changes in biotic components and major facies types do not seem to be immediately related to dramatic variations in the temperatures of either terrestrial environments or SSTs in the immediate study areas (compare Barbin, 1992, and Brinkhuis, 1994). This may be due to the fact that (1) changes in temperature gradients in the Northern Hemisphere are not so strong as in the Southern Hemisphere, or (2) an open Tethys seaway with low-latitude "tropical" affinities persisted throughout the Eocene and Oligocene, which may have precluded any dramatic climatic changes in the study area. In fact, the most important biogeographic change on a global scale immediately affecting the study areas first occurs in the early Miocene with the closing of the Tethys Seaway (see Rögl, 1998).

As stated by Adams et al. (1990), however, it is not safe to infer climatic changes from faunal distribution patterns seen only in one region, because distribution patterns can be altered locally by tectonic and plate movements. Veron (1995) also pointed out the difficulties in using coral distributions to track SSTs, as they are often limited by short-term environmental fluctuations. Extending the comparative approach using major facies types expanded upon in this paper to other areas within the Tethys and other biogeographic provinces (for example, to carbonates in equatorial settings) should further help to reveal major environmental influences on Eocene and Oligocene biotic and facies distributions.

ACKNOWLEDGMENTS

J. N. was supported by the DFG (German Science Foundation) SFB275 (Sonderforschungsbereich): Climate-coupled Processes in Mesozoic and Cainozoic Geological Systems. D. B. was supported by a M.U.R.S.T. ex-40% (C. Loriga) at the University of Ferrara, Italy. M. R. was supported by the Jubiläumsfond der Österreichischen Nationalbank. We thank Martyn Pedley (University of Hull) and an anonymous reviewer for their comments on the manuscript.

REFERENCES

Accarie, H., 1987, Dynamique sédimentarie et structurale au passage plate-forme bassin. Les faciés crétacés et tertiaires: Massif de la Maiella (Abruzzes, Italie): École des Mines, Mémoires de Sciences de la Terre, p. 1-162.

Adams, C. G., Gentry, A., and Whybrow, P.J., 1983, The terminal Tethyan event: Utrecht Micropaleontological Bulletin, v. 30, p. 273-298.

Adams, C. G., Butterlin, J., and Samantra, B. K., 1986, Larger foraminifera and events at the Eocene/Oligocene boundary in the Indo-West Pacific region, *in* Pomerol, C., and Premoli-Silva, I., eds., Terminal Eocene Event: Amsterdam, Elsevier, p. 237-252.

Adams, C. G., Lee, D. E., and Rosen, B. R., 1990. Conflicting isotopic and biotic evidence for tropical sea-surface temperatures during the Tertiary: Palaeogeography, Palaeoclimatology, Palaeoecology, v. 77, p. 289-313.

Adey, W. H., 1986, Coralline algae as indicators of sea-level, *in* van de Plasche, O., ed., Sea-level research: a manual for the collection and evaluation of data: Norwich, Free University Geo Book, p. 229-280.

Aigner, T., 1985, Biofabrics as dynamic indicators in nummulite accumulations: Journal of Sedimentary Petrology, v. 55, p. 131-134.

Antonelli, R., Barbieri, G., Dal Piaz, G.V., Dal Pra, A., De Zanche, V., Grandessa, P., Mietto, P., Sedea, R., and Zanferani, A., 1988, Note Illustrative della Carta Geologica del Veneto 1: 250.000. Regione Veneto, Dipartimento di Geologia, Paleontologia e Geofisica, Università di Padova, p. 1-31.

Arni, P., 1965, L'évolution des Nummulitinae en tant que facteur de modification des dépots littoraux: Mémoires du Bureau de Recherches Géologique et Minieres, v. 32, p. 7-20.

Aubry, M.-P., Berggren, W. A., Van Couvering, J. A., and Steininger, F., 1999, Problems in chronostratigraphy: stages, series, unit and boundary stratotypes, global stratotype section and point and tarnished golden spikes: Earth-Science Reviews, v. 46, p. 99-148.

Báldi, T., 1984, The terminal Eocene and early Oligocene events in Hungary and the separation of an anoxic, cold Paratethys: Eclogae Geologicae Helvetiae, v. 77, p. 1-27.

Barbin, V., 1988, The Eocene-Oligocene transition in shallow-water environment: the Priabonian stage type area (Vicentin, northern Italy), *in* Premoli Silva I., ed., E/O meeting, Ancona, October 1987: Special Publication of the International Subcommission on Paleogene Stratigraphy, v. 2, p. 163-171.

Barbin, V., 1992, Paleoenvironment of the Bredola area (late Priabonian stage stratotype, northern Italy) during the Eocene-Oligocene transition: Neues Jahrbuch für Geologie Paläontologie Mitteilungen, v. 1992, p. 221-230.

Barbin V., and Keller-Gruenig A., 1991, Benthic foraminiferal assemblages from the Brendola section (Priabonian stage stratotype area, northern Italy): Distribution, palaeoenvironment and palaeoecology: Marine Micropaleontology, v. 17, p. 237-254.

Bassi, D., 1995, Crustose coralline algal pavements from late Eocene Colli Berici of northern Italy: Rivista Italiana di Paleontologia e Stratigrafia, v. 101, p. 81-92.

Bassi, D., 1997, Vegetative anatomy and paleoecology of *Polystrata alba* (Pfender) Denizot, 1968 (Cryptonemiales, Peyssonneliaceae) from the upper Eocene of northern Italy: Revue de Paléobiologie, v. 16, p. 309-320.

Bassi, D., 1998, Coralline algal facies and their palaeoenvironments in the late Eocene of northern Italy (Calcare di Nago): Facies, v. 39, p. 179-202.

Bassi, D., and Nebelsick, J. H., 2000, Calcareous algae from the lower Oligocene Gornji Grad Beds of Northern Slovenia: Rivista Italiana di Paleontologia e Stratigrafia, v. 106, p. 99-102.

Bassi, D., Loriga, C., and Ungaro, S., 1998, Large foraminiferal facies in the middle and upper Eocene of northeastern Italy (Colli Berici, Vicenza): Tertiary to Recent larger foraminifera: their depositional environments and importance as petroleum reservoirs: Conference-Workshop, ARCO-BG-Chevron, 15–19 Feb., Kingston Surrey, U.K., Abstracts, p. 4.

Bassi, D., Woelkerling, W. J., and Nebelsick, J. H., 2000, Taxonomic and biostratographical reassessment of *Subteranniphylum* Elliott (Corallinales, Rhodophyta): Palaeontology, v. 43, p. 405-425.

Berggren, W. A., and Prothero, D. R., 1992, Eocene-Oligocene climatic and biotic evolution: an overview, *in* Prothero, D. R., and Berggren, W. A., eds., Eocene-Oligocene climatic and biotic evolution: Princeton, New Jersey, Princeton University Press. p. 1-28.

Betzler, C., Brachert, T., and Nebelsick, J. H., 1997, The warm temperate carbonate province: facies, zonations and delimitations: Courier Forschungsinstitut Senckenberg, v. 201, p. 83 - 100.

Bosellini, A., 1989, Dynamics of Tethyan carbonate platforms, *in* Crevello, P. D., Wilson, J. L., Starg, J. F., and Read, J. F., eds., Controls on carbonate platform and basin development: SEPM, Special Publications, v. 44, p. 3-13.

Bosellini, F. R., 1998, Diversity, composition and structure of late Eocene shelf-edge coral associations (Nago Limestones, northern Italy): Facies, v. 39, p. 35-40.

Bosellini, F .R., and Russo, A., 1988, The Oligocene *Actinacis* coral community of the southern Alps (Italy): temperature vs. terrigenous control: Proceedings of the 6th International Coral Reef Symposium, Townsville, Australia, p. 385-391.

Bosellini, F. R., and Stemann, T. A., 1996, Autoecological significance of growth form in the scleractinian *Actinacis rollei* Reuss (Oligocene, Lessini Mountains, northern Italy), *in* Cherchi, A., ed., Autoecology of selected fossil organisms: achievements and problems: Bollettino della Societá Italiana di Paleontologia, Special Volume 3, p. 31-43.

Bosellini, F. R., and Trevisani, E., 1992, Coral facies and cyclicity in the Castelgomberto Limestone (early Oligocene, eastern Lessini Mountains, northern Italy): Rivista Italiana di Paleontologia e Stratigrafia, v. 98, p. 339-352.

Bosence, D. W. J., 1983, Description and classification of rhodoliths, *in* Peryt, T. M., ed., Coated grains: Berlin-Heidelberg, Springer Verlag, p. 217-224.

Bosence, D. W. J., 1985, The "Coralligène" of the Mediterranean—a Recent analog for Tertiary coralline algal limestones, *in* Toomey, D. F., and Nitecki, M. H., eds., Paleoalgology: contemporary research and applications: Berlin, Springer Verlag, p. 216-225

Bosence D. W. J., and Pedley, H. M., 1982, Sedimentology and palaeoecology of a Miocene coralline algal biostrome from the Maltese Islands: Palaeogeography, Palaeoclimatology, Palaeoecology, v. 38, p. 9-43.

Brinkhuis, H., 1994, Late Eocene to early Oligocene dinoflagellate cysts from the Priabonian type-are (northeast Italy): biostratigraphy and paleoenvironmental interpretation: Palaeogeography, Palaeoclimatology, Palaeoecology, v. 107, p. 121-163.

Brinkhuis, H., and Visscher, H., 1995, The upper boundary of the Eocene Series: a reappraisal based on dinoflagellate cyst biostratigraphy and sequence stratigraphy: SEPM Special Publications, v. 54, p. 295-304.

Burchette, T. P., and Wright, V. P., 1992, Carbonate ramp depositional systems: Sedimentary Geology, v. 79, p. 3-57.

Buxton, M. W. N., and Pedley, H. M., 1989, A standardized model for Tethyan Tertiary carbonate ramps: Journal of the Geological Society, v. 146, p. 746-748.

Cassinis, G., and Castellarin, A., 1988, Il significato delle linee della Gallinera e della Giudicarie sud nella geologia dell'Adamello e zone circostanti: Atti Ticinensi di Scienze della Terra, v. 31, p. 446-462.

Castellarin, A., 1972, Evoluzione paleotettonica sinsedimentaria del limite tra "Piattaforma veneta" e "Bacino lombardo" a nord di Riva del Garda: Giornale di Geologia, v. 38, p. 11-212.

Castellarin, A., and Cita, M. B., 1969, La coupe priabonienne de Nago (Prov. Trento) et la limite Éocène-Oligocene: Mémoires du Bureau de Recherches Géologique et Minieres, v. 69, p. 93-117.

Cita, M. B., 1969, Le Paléocène et l'Eocène de l'Italie du nord: Mémoires du Bureau de Recherches Géologique et Minieres, v. 69, p. 417-429.

Cita, M. B., 1975, Stratigrafia della Sezione di Possagno, *in* Bolli, H. M., ed., Monografia micropaleontologica sul Paleogene e l'Eocene di Possagno, Provincia di Treviso, Italia: Schweizerische Paläontologische Abhandlungen, v. 97, p. 9-32.

Coletti, F., Piccoli, G., Sambugar, B., and Vendemiati Dei Medici, M. C., 1973, I molluschi fossili di Castelgomberto e il loro significato nella paleoecologia dell'Oligocene veneto: Memorie dell' Istituto di Geologia e Mineralogia dell'Università di Padova, v. 28, p. 1-32.

Darga, R., 1992, Geologie, Paläontologie und Palökologie der südostbayerischen unter-priabonen (Ober-Eozän) Riffkalkvorkommen des Eisenrichtersteins bei Hallthurm (Nördliche Kalkalpen) und des Kirchbergs bei Neuberern (Helvetikum): Münchner Geowissenschaftliche Abhandlungen (A), v. 23, p. 1-166.

De Vecchi, G., and Sedea, R., 1995, The Paleogene basalts of the Veneto region (NE Italy): Memorie di Scienze Geologiche dell'Università di Padova, v. 47, p. 253-274.

Doglioni, C., and Bosellini, A., 1987, Eoalpine and mesoalpine tectonics in the Southern Alps: Geologische Rundschau, v. 77, p. 734-754.

Dorobek, S. L., 1995, Synorogenic carbonate platforms and reefs in foreland basins: controls on stratigraphic evolution and platform/reef morphology, *in* Dorobek, S. L., and Ross, G. M., eds., Stratigraphic evolution of foreland basins: SEPM Special Publications, v. 52, p. 127-147.

Drobne, K., 1977, Alvéolines paléogènes de la Slovénie et de l'Istrie: Schweizerische Paläontologische Abhandlungen, v. 99, p. 1-132.

Drobne, K., Pavlovec, R., Drobne, F., Cimerman, F., and Sikic, L., 1985, Some larger foraminifera from the upper Eocene and basal Oligocene beds in north Slovenia: Geologica Glasnik, Herald Geological Bulletin, v. 28, p. 77-117.

Ebner, F., Sachsenhofer, F., and Schwendt, A., 1991, Das Tertiär von Kirchberg am Wechsel: Mitteilungen des naturwissenschaftliches Vereins der Steiermark, v. 121, p. 119-127.

Eichenseer H., and Luterbacher H. P., 1992, The marine Paleogene of the Tremp region (NE Spain): depositional sequences, facies history, biostratigraphy and controlling factors: Facies, v. 27, p. 119-152.

Ferràndez-Cañadell, C., and Serra-Kiel, J., 1992, Morphostructure and paleoecology of *Discocyclina* Gümbel, 1870: Journal of Foraminiferal Research, v. 22, p. 147-165.

Flügel, E., and Flügel-Kahler, E., 1992, Phanerozoic reef evolution: basic questions and data base: Facies, v. 26, p. 167-277.

Fodor, L., Jelen, B., Márton, E., Skaberne, D., Car, J., and Vrabec, M., 1998, Miocene-Pliocene tectonic evolution of the Slovenian Periadriatic fault: implications for Alpine-Carpathian extrusion models: Tectonics, v. 17, p. 690-709.

Frascari, F., 1963, La trasgressione priaboniana nel Vicentino orientale: Giornale di Geologia, v. 30, p. 219-231.

Freiwald, A., 1994, Sedimentological and biological aspects in the formation of branched rhodoliths in northern Norway: Beiträge zur Paläontologie von Österreichs, v. 20, p. 7-19.

Freiwald, A., Henrich, R., Schäfer, P., and Willkomm, H., 1991, The significance of high-boreal to subarctic maerl deposits in northern Norway to reconstruct Holocene climatic changes and sea level oscillations: Facies, v. 25, p. 315-340.

Frost, S. H., 1981, Oligocene reef coral biofacies of the Vicentin, northern Italy, *in* Toomey, D.F., ed., European fossil reef models: SEPM, Special Publications, v. 30, p. 483-539.

Frost, S. H., and Langenheim, R. L., 1974, Cenozoic reef biofacies: De Kalb, Northern Illinois University Press, 388 p.

Frost, S. H., Harbour, J. L, Realini, M. J., and Harris, P. M., 1983, Oligocene reef tract development southeastern Puerto Rico. Part I. Text: Sedimenta, v. 9, p. 1-104.

Garavello, A. M., and Ungaro, S., 1982, Studio biostratigrafico e paleoecologico della serie eocenica di Pradipaldo nella zona pedemontana meridionale dell'Altipiano di Asiago (Vicenza): Geologica Romana, v. 21, p. 655-675.

Geister, J., and Ungaro, S., 1977, The Oligocene coral formations of the Colli Berici: Eclogae geologicae Helvetiae, v. 70, p. 811-823.

Gohrbandt, K., 1963, Paleozän und Eozän des Helvetikums nördlich von Salzburg: Verhandlung Geologische Bundesanstalt, Sonderheft F, *in* Grill, R., Kollmann, K., Küppers, H., and Oberhauser, R., eds., Exkursionsführer für das Achte Europäische Mikropaläontologische Kolloquium in Österreich: Vienna, Geologische Bundesanstalt, p. 47-57.

Hagn, H., ed., 1981, Die Bayerischen Alpen und ihr Vorland in mikropaläontologischer Sicht: Geologica Bavarica, v. 82, p. 1-408.

Hillebrandt, A. von, 1993, Nummuliten und Assilinen aus dem Eozän des Krappfeldes in Kärnten (Österreich): Zitteliana, Hagn/Herm-Festschrift, v. 20, p. 277-293.

Hottinger, L., 1960, Recherches sur les Alvéolines du Paléocène et de l'Éocène: Schweizerische Paläontologische Abhandlungen, v. 75-76, p. 1-243.

Hottinger, L., 1983a, Processes determining the distribution of larger foraminifera in space and time: Utrecht Micropaleontological Bulletin, v. 30, p. 239-253.

Hottinger, L., 1983b, Neritic macroid genesis, an ecological approach, *in* Peryt, T.M., ed., Coated grains: Berlin-Heidelberg, Springer Verlag, p. 38-55.

James, N. P., 1997, The cool-water carbonate depositional realm, *in* James, N. P., and Clarke, J. A. D., eds., Cool water carbonates: SEPM Special Publications, v. 56, p. 1-22.

Kiessling, W., Flügel, E., and Golonka, J., 1999, Paleoreef maps: evaluation of a comprehensive database on Phanerozoic reefs: American Association of Petroleum Geologists Bulletin, v. 83, p. 1552-1587.

Kramer, H., 1962, Das Eozänvorkommen am Goldberg bei Kirchberg Wien [unpubl. Ph.D. diss.]: Vienna, University of Vienna, 110 p.

Krois, P., 1992, Aspekte zur Sedimentgeologie und Fazies des inneralpinen Tertiärs (Oligozän) im Raum Häring- Angerberg, Tirol [Unpubl. Ph.D. diss.]: Innsbruck, University Innsbruck, 158 p.

Lehner, B. L., Knappertsbush, M. W., and Heer, P. H., 1987, Biostratigraphy, lithostratigraphy and sedimentology of the Maiolica Lombarda and the Scaglia Lombarda on the west side of Lake Garda (northern Italy): Memorie di Scienze Geologiche, v. 34, p. 1-35.

Less, G., 1987, Paleontology and stratigraphy of the European orthophragminae: Geologica Hungarica (Palaeontologica), v. 51, p. 1-373.

Luciani, V., 1989, Stratigrafia sequenziale del Terziario nella catena del Monte Baldo (Province di Verona e Trento): Memorie di Scienze Geologiche, v. 41, p. 263-351.

Luciani, V., Barbujani, C., and Bosellini, A., 1988, Facies e cicli del Calcare di Nago (Eocene superiore. Trentino meridionale): Annali dell' Università di Ferrara, Nuova Serie 1, v. 4, p. 47-62.

Massari, F., Medizza, F., and Sedea, R., 1976, L'evoluzione geologica dell'Area euganea tra il Giurese superiore e l'Oligocene inferiore: Memorie dell'Istituto di Geologia e Mineralogia, Università di Padova, v. 30, p. 174-197.

Matteucci, R., and Pignatti, J. S., 1990, The taphonomy of *Nummulites*, *in* Robba, E., ed., Proceedings, 4th Symposium: Ecology and paleoecology of the benthic communities, Sorrento 1988: Turin, Museo Regionale di Scienze Naturali di Torino, p. 183-198.

Mietto, P., 1988, Aspetti geologici dei Monti Berici, *in* Signum, P., ed., I Colli Berici. Natura e civiltà: Padua, Edizioni, p. 13-23.

Miller, K. G., Fairbanks, R. G., and Mountain, G. S., 1987. Tertiary oxygen isotope synthesis, sea-level history and continental margin erosion: Paleoceanography, v. 2, p. 1-19.

Miller, K. G., Wright, J. D., and Fairbanks, R. G., 1991, Unlocking the ice house: Oligocene-Miocene oxygen isotopes, eustasy, and margin erosion: Journal of Geophysical Research, v. 96, p. 6829-6848.

Moussavian, E., 1984, Die Gosau- und Alttertiär-Gerölle der Angerberg-Schichten (Höheres Oligozän, Unterinntal, Nördliche Kalkalpen): Facies, v. 10, p. 1-86.

Nebelsick, J. H., 1989, Temperate water carbonate facies of the early Miocene Paratethys (Zogelsdorf Formation, lower Austria): Facies, v. 21, p. 11-40.

Nebelsick, J. H., Drobne, K., and Bassi, D., 2000, Microfacies analysis and palaeoenvironmental interpretation of lower Oligocene, shallow-water carbonates (Gornji Grad Beds, Slovenia): Facies, v. 43, p. 157-176.

Nelson, C. S., 1988. An introductory perspective on non-tropical shelf carbonates: Sedimentary Geology, v. 60, p. 3-12.

Ortner, H., 1996, Deformation und Diagenese im Unterinntaler Tertiär (zwischen Rattenberg und Durchholzen) und seinem Rahmen. [Unpubl. Ph.D. diss.]: Innsbruck, University Innsbruck, 234 p.

Papazzoni, C. A., 1994, Macroforaminifera and paleoenvironments near the middle-upper Eocene boundary in the Mossano section (Berici mts., Vicenza, northern Italy), *in* Matteucci, R., Carboni, M. G., and Pignatti, J. S., eds., Studies on ecology and paleoecology of benthic communities: Bollettino della Società Paleontologica Italiana, Special Publication v. 2, p. 203-212.

Papazzoni, C. A., and Sirotti, A., 1995, Nummulite biostratigraphy at the middle/upper Eocene boundary in the northern Mediterranean area: Rivista Italiana di Paleontologia e Stratigrafia, v. 101, p. 63-80.

Papp, A., 1958, Vorkommen und Verbreitung des Obereozäns in Österreich: Mitteilungen der Geologischen Gesellschaft in Wien, v. 50, p. 251-270.

Piccoli, G., 1966, Studio geologico sul vulcanismo paleogenico veneto: Memorie dell'Istituto di Geologia e Mineralogia dell'Università di Padova, v. 26, p. 1-100.

Placer, L., 1999, Contribution to the macrotectonic subdivision of the border region between Southern Alps and External Dinarides: Geologija, v. 41, p. 223 - 255.

Pomerol, C., and Premoli-Silva, I., eds., 1986, Terminal Eocene events: Amsterdam, Elsevier, v. 9, p. 1-414.

Premoli Silva, I., and Jenkins, D.G., 1993, Decision on the Eocene/Oligocene boundary stratotype: Episodes, v. 16, p. 379-382.

Racey, A., 1995, Palaeoenvironmental significance of larger foraminiferal biofabrics: implications for petroleum exploration, in Al-Husseini M. I., ed., Middle East Petroleum Geosciences, GEO'94, v. 2, p. 793-820.

Rasser, M. W., 1994, Facies and palaeoecology of rhodoliths and acervulinid macroids in the Eocene of the Krappfeld (Austria): Beiträge zur Paläontologie, v. 19, p. 191-217.

Rasser, M. W., 2000, Coralline red algal limestones of the late Eocene Alpine Foreland Basin in upper Austria: component analysis, facies and palecology: Facies, v. 42, p. 59-92.

Rasser, M. W., and Piller, W. E., 1999a, Lithostratigraphische Neugliederung im Paläogen des österreichischen Helvetikums: Abhandlungen der Geologischen Bundesanstalt Wien, v. 56, p. 699-712.

Rasser, M. W., and Piller, W. E., 1999b, Die klassischen Profile des Paläogen des Haunsberg-Helvetikums (Salzburg): Die Typusprofile von Olching-Formation, Kroisbachgraben-Subformation, und Frauengrube-Subformation: Abhandlungen der Geologischen Bundesanstalt Wien, v. 56, p. 713-722.

Rasser, M. W., Less, G., and Báldi-Beke, M., 1999. Biostratigraphy and facies of the late Eocene of the Western Austrian Molasse Zone with special reference to the larger foraminifera: Abhandlungen der Geologischen Bundesanstalt Wien, v. 56, p. 679-698.

Roetzel, R., Steininger, F. F., and Piller, W., 1991, Steyregg, Sandgrube Treul, *in* Roetzel, R., and Steininger, F. F., eds., Exkursion im Tertiär Österreichs: Vienna, Österreichische Paläontologische Gesellschaft, p. 47-49.

Rögl, F., 1998, Palaeogeographic considerations for Mediterranean and Paratethys Seaways (Oligocene to Miocene): Annalen des Natur-historisches Museum Wien, v. 99A, p. 297-310.

Schaub, H., 1981, Nummulites et Assilines de la Téthys Paléogène: Taxonomie, phylogénèse et biostratigraphie: Schweizerische Paläontologische Abhandlungen, v. 104-106, p. 1-236.

Schmiedl, G., Scherbacher, M., Bruch, A., Jelen, B., Nebelsick, J. H., Hemleben, H., Mosbrugger, V., and Rifelj, H., 2002, Paleoenvironmental evolution of the Paratethys in the Slovenian Basin during the late Paleogene: International Journal of Earth Sciences, v. 91, p. 123-132.

Schweighauser, J., 1953, Mikropaläontologische und stratigraphische Untersuchungen im Paleocaen und Eocaen des Vicentin (Nord-Italien): Schweizerische paläontologische Abhandlungen, v. 70, p. 1-97.

Setiawan, J. R., 1983, Foraminifera and microfacies of type Priabonian: Utrecht Micropaleontological Bulletin, v. 29, p. 1-161.

Shackleton, N. J., and Boersma, A., 1981, The climate of the Eocene ocean: Journal of the Geological Society of London, v. 138, p. 153-157.

Steininger, F. F., Wessely, G., Rögl, F., and Wagner, L., 1986, Tertiary sedimentary history and tectonic evolution of the Eastern Alpine Foredeep: Giornale di Geologia, ser. 3, v. 48, p. 285-297.

Steneck, R. S., Macintyre, I. G., and Reid, R. P., 1997, A unique algal ridge system in the Exuma Cays, Bahamas Coral Reefs, v. 16, p. 29-37.

Stingl, V., 1990, Die Häringer Schichten vom Nordrand des Unterinntaler Tertiär Beckens (Angerberg/Tirol): Fazies, Sedimentpetrographie und beckengenetische Aspekte: Geologisch-Paläontologische Mitteilungen Innsbruck, v. 17, p. 31-38.

Taberner, C., and Bosence, D. W. J., 1985, Ecological succession from corals to coralline algae in Eocene Patch Reefs, Northern Spain, *in* Toomey, D. F., and Nitecki, M. H., eds., Paleoalgology: contemporary research and applications: Berlin, Springer Verlag, p. 226-236.

Tollmann, A., 1977, Geologie von Österreich. Bd. I: Vienna, Deuticke, 766p.

Tollmann, A., 1985, Geologie von Österreich. Bd. 2: Vienna, Deuticke, 523p.

Trauth, F., 1918, Das Eozänvorkommen bei Radstadt im Pongau und seine Beziehung zu den gleichaltrigen Ablagerungen bei Kirchberg am Wechsel und Wimpassing am Leithagebirge: Denkschriften der Kaiserlich-Königliche Akademie der Wissenschaften Wien, Mathematisch-naturwissenschaftliche Klasse, v. 95, p. 171-278.

Trevisani, E., 1997, Stratigrafia sequenziale e paleogeografia del margine orientale del Lessini Shelf durante l`Eocene superiore (Prealpi Venete, province di Vivenza e Treviso) Studu Trentini di Scienze Naturali: Acta Geologica, v. 71, p. 145-168.

Ungaro, S., 1969, Étude micropaléontologique et stratigraphique de l'Éocene supérieur (Priabonian) de Mossano (Colli Berici): Mémoires du Bureau de Recherches Géologique et Minieres, v. 69, p. 267-280.

Ungaro, S., 1978, L'Oligocene dei Colli Berici: Rivista Italiana di Paleontologia, v. 84, p. 199-278.

Ungaro, S., 1996, Adaptive morphological strategy of *Gypsina* (encrusting Foraminifer), *in* Cherchi A., ed., Autecology of selected fossil organisms: achievements and problems: Bollettino della Società Paleontologica Italiana, special v. 3, p. 233-241.

Ungaro, S., and Bosellini, A., 1965, Studio micropaleontologico e stratigrafico sul limite Eocene-Oligocene nei Colli Berici occidentali: Annali dell'Università di Ferrara, Nuova Serie, sezione 9, v. 3, p. 157-183.

Veron, J. E. N., 1995, Corals in space and time: The biogeography and evolution of the Scleractinia: Sydney, University of New South Wales Press, 321p.

Vogeltanz, R., 1970, Sedimentologie und Paläogeographie eines eozänen Sublitorals im Helvetikum von Salzburg (Österreich): Verhandlung der Geologische Bundesanstalt, v. 1970, p. 373-451.

Wagner, L., 1980, Geologische Charakteristik der wichtigsten Erdöl- und Erdgasträger der oberösterreichischen Molasse Teil I: Die Sandsteine des Obereozäns: Erdöl Erdgas Zeitschrift, v. 96, p. 338-346.

Wagner, L., 1996, Stratigraphy and hydrocarbons in the upper Austrian Molasse Foredeep (active margin), *in* Wessely, G., and Liebl, W., eds., Oil and gas in Alpidic thrustbelts and basins of central and eastern Europe: Special Publications, v. 5, p. 217-235.

Wagreich, M., 1995, Subduction tectonic erosion and late Cretaceous subsidence along the northern Austroalpine margin (Eastern Alps, Austria): Tectonophysics, v. 242, p. 63-78.

Wilkens, E., 1989, Entstehung von Großforaminiferen-Akkumulationen, Biofabric-Entwicklung und Bioklastaggregaten im Alttertiär des Sonnberges: Arbeitstagung Geologischen Bundesanstalt, p. 100-106.

Wilkens, E., 1991, Das zentralalpine Paläogen der Ostalpen: Stratigraphie, Sedimentologie und Mikrofazies Großforaminiferen reicher Sedimente [Unpubl. Ph.D. diss.]: Hamburg, University of Hamburg, 227 p.

Zachos, J. C., Scott, L. D., and Lohmann, K. C., 1994, Evolution of early Cenozoic marine temperatures: Paleoceanography, v. 9, p. 353-387.

Ziegler, P. A., 1987, Late Cretaceous and Cenozoic intra-plate compressional deformations in the Alpine foreland—a geodynamic model: Tectonophysics, v. 137, p. 389-420.

Part IV

CAUSES AND CONSEQUENCES

Late Eocene Impacts: Geologic Record, Correlation, and Paleoenvironmental Consequences

C. Wylie Poag, Edward Mankinen, and Richard D. Norris

ABSTRACT

We present new magnetostratigraphic and stable isotopic ($\delta^{18}O$, $\delta^{13}C_{carb}$) data to help improve correlations among three late Eocene impact craters and their inferred breccia and ejecta deposits. Our analyses also shed light on potential global environmental consequences attributable to the impacts. The new data come from a continuously cored interval of the subsurface Chickahominy Formation, which lies conformably above the Chesapeake Bay impact crater in southeastern Virginia. The new magnetostratigraphic data indicate that the Chesapeake Bay impact took place in Chron C16n.2n, the same magnetochron that encompasses the late Eocene ejecta layer at Massignano, Italy. This correlation places both the Chesapeake Bay impact and the Massignano ejecta at ~35.6 Ma, and resolves a previous miscorrelation between these two sites based on planktonic foraminifera and calcareous nannofossils. The new magnetostratigraphic correlations also suggest that the published magnetostratigraphic framework for ejecta-bearing late Eocene strata at ODP Site 689B (Maud Rise) is incorrect, due to an incomplete section.

New $\delta^{18}O$ data (single species of benthic foraminifera) from the same Chickahominy section at Chesapeake Bay indicate that successional intervals of warm oceanic bottom-water may be characteristic of the late Eocene. We infer that the warm intervals correlate with successive episodes of greenhouse warming, triggered in part by a comet shower, which produced the Chesapeake Bay, Toms Canyon, Popigai, and presumably additional (as yet undiscovered) late Eocene impact craters.

We also demonstrate that a marked negative excursion of $\delta^{13}C_{carb}$ persists through the upper half of the Chickahominy Formation. This excursion, also recorded at Massignano, at Bath Cliff, Barbados, and at other widespread localities in the world ocean, may be additional evidence of global-scale, long-term environmental disturbances related to the bolide impacts. As such, this $\delta^{13}C$ signal may be useful for global subdivision of the late Eocene stratigraphic record.

INTRODUCTION

The Problem

How many bolides struck Earth in the late Eocene, where are their respective ejecta deposits, what are their geochronological relationships, and what paleoenvironmental effects are attributable to them? Definitive answers to these questions have been elusive, but three impact sites have been identified to date, and can be correlated with proximal and distal breccias and far-field ejecta deposits (figure 28.1). Geochemical evidence, such as spikes of ^{3}He and Ir enrichment, and the presence of multiple ejecta layers (possibly as many as six), have been used to infer that several additional bolides struck the Earth in the late Eocene. Two of the three documented late Eocene bolides struck the North American continental shelf and formed craters now covered by hundreds to thousands of meters of water and marine sediments (Chesapeake Bay and Toms Canyon craters; Poag et al., 1994; Poag, 1997a; Poag and Poppe, 1998; figure 28.1). The third late Eocene bolide excavated a subaerial crater, which is exposed on the Anabar shield near the town of Popigai, northern Siberia (Masaitis

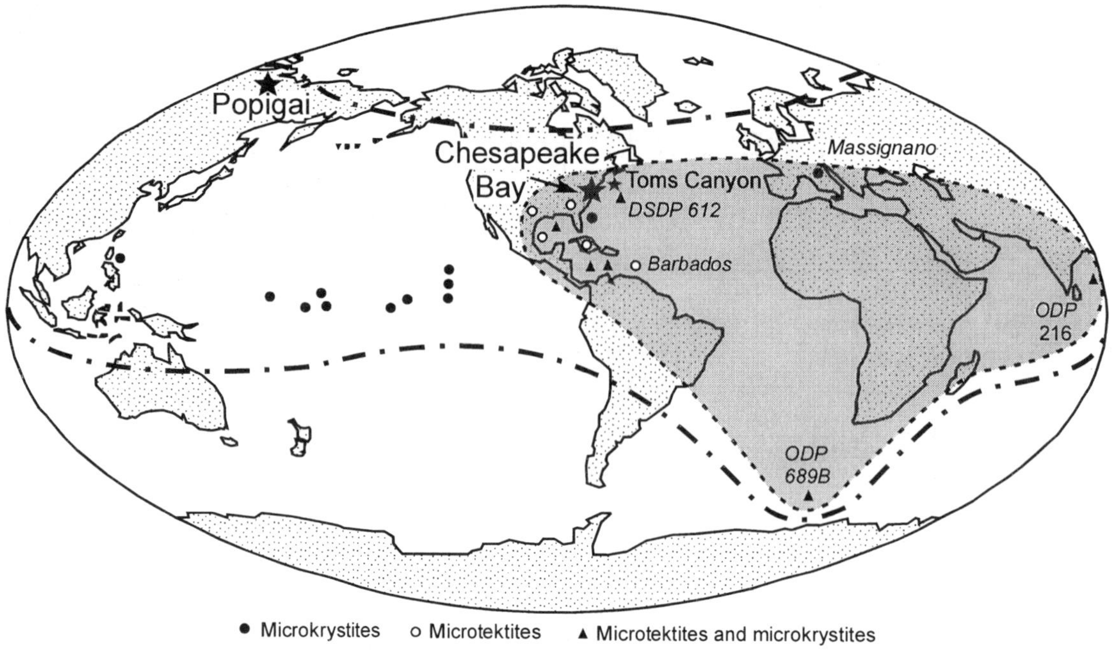

Figure 28.1. Location of documented late Eocene impact craters and related impactites. Note that North American tektite strewn field (*shaded*), correlated with Chesapeake Bay impact, is currently not known in Pacific and Arctic regions; presence of microtektite layer at Massignano is inferred, but not yet documented. Microkrystite strewn field (*dash-dot line*), correlated with Popigai impact, appears to be globally distributed.

et al., 1975, 1999; Masaitis 1994; figure 28.1). The structure and morphology of all three craters are well established by drilling and by gravity and/or seismic reflection surveys.

Ejecta and/or crater-filling breccia from these three impacts have been identified at numerous localities (D'Hondt et al., 1987; Glass, 1989; Glass and Koeberl, 1999a, b; Langenhorst, 1996; Masaitis, 1994; Montanari et al., 1993; Pierrard et al., 1998; Poag et al., 1994; Masaitis et al., 1999; Vonhof and Smit, 1999; Whitehead et al., 2000), but other upper Eocene ejecta deposits, especially the well-studied layer at Massignano, Italy, currently lack firmly established source craters. Biochronology, magnetochronology, and radioisotopic chronology provide approximate correlations between ejecta deposits and their respective craters, but various extrapolations, interpolations, and inferences are still required to make the correlations plausible.

Estimated Crater Ages

The breccia deposits of the Chesapeake Bay and Toms Canyon craters, as well as the youngest deposits overlying them, belong to a thin biostratigraphic interval in which the upper part of planktonic foraminiferal Zone P15 overlaps the lower part of calcareous nannoplankton Zone NP19–20 (Poag and Aubry, 1995; Poag, 1997b). Their biochronological ages are, therefore, nearly identical at 35.2–36.0 Ma, according to the time scale of Berggren et al. (1995). No radioisotopic date has yet been derived directly from these two North American craters.

In contrast, the age of the Popigai crater has been determined only from radioisotopes. An $^{40}Ar/^{39}Ar$ analysis of associated tagamites (melt-rocks generated by the impact; Masaitis et al., 1975, 1999) yielded an estimated age of 35.7 ± 0.2 Ma (Bottomley et al., 1997). Eocene sedimentary rocks are absent at the Popigai site, which precludes obtaining a biochronological age there.

Ages of Distal Impactites

Ejecta deposits biochronologically equivalent to the Chesapeake Bay and Toms Canyon craters are widespread in the western North Atlantic, Caribbean, and Gulf of Mexico, including the coastal plains of Georgia and Texas. Ejecta from 10 sites in this 9 million km^2 region formerly delineated the North American tektite strewn field (NATF) (figure 28.1; Glass, 1989; Koeberl, 1989). Vonhof and Smit (1996,

1999) and Glass and Koeberl (1999a, b), however, have recently expanded the distribution of the NATF to include Ocean Drilling Program (ODP) Site 689B on Maud Rise, near Antarctica (figure 28.1; Leg 113; Barker, Kennett, et al., 1990) and ODP Site 216 on Ninetyeast Ridge in the eastern Indian Ocean (figure 28.1; Leg 22; von der Borch, Sclater, et al., 1974). The radioisotopic age of the NATF is estimated from $^{40}Ar/^{39}Ar$ plateau analyses of tektite fragments sampled at two locations. A northern location, cored at DSDP Site 612, located 20 km south of the Toms Canyon structure and 330 km northeast of the Chesapeake Bay crater yielded an age of 35.2–35.5 ± 0.3 Ma (Obradovich et al., 1989; Poag and Aubry, 1995; figure 28.1). A southern location, sampled at Bath Cliff, Barbados, 3,500 km southeast of Chesapeake Bay, yielded an essentially identical age of 35.4 ± 0.6 Ma (figure 28.1; Glass et al., 1986). On the basis of similar ages and geochemistry, the NATF is now believed to have been produced mainly by the Chesapeake Bay impact (Poag et al., 1994; Koeberl et al., 1996; Glass et al., 1998; Whitehead et al., 2000), though some of the Site 612 ejecta may have come from the Toms Canyon crater (Poag and Aubry, 1995; Poag and Poppe, 1998).

At deep-sea sites containing ejecta of the NATF, the deposit typically contains microtektites, tektite fragments, shocked quartz and feldspar, coesite, and stishovite (Glass and Wu, 1993; Glass et al., 1998; figure 28.1). This microtektite layer is often underlain by a thin layer of microkrystites (spherules containing clinopyroxene crystals and/or nickel-rich spinels; Glass et al., 1985; Glass and Burns, 1987; Pierrard et al., 1999; Whitehead et al., 2000; figure 28.1) associated with a moderate Ir anomaly (Sanfilippo et al., 1985; Montanari et al., 1993). These two layers are generally separated by 2–25 cm of sediment (equivalent to 10,000–20,000 years; Glass et al., 1998; Glass and Koeberl, 1999a, b), and both layers reside in the P15/NP19-20 overlap interval (Poag, 1997b). Some sites in the NATF, however, contain just one of these layers, and only the microkrystite layer has been compellingly documented at Pacific sites (figure 28.1).

A particularly important occurrence of late Eocene ejecta has been reported from the world stratotype of the Eocene/Oligocene boundary at Massignano, Italy (Odin and Montanari, 1989; Berggren et al., 1995; figure 28.1). Geographically, the Massignano section appears to be within the NATF (near its northern limit). The composition of the Massignano ejecta, however, resembles a mixture of the microtektite and microkrystite layers. Nickel-rich spinels, shocked quartz, and an Ir anomaly have been documented (Montanari et al., 1993; Clymer et al., 1996; Langenhorst, 1996; Pierrard et al., 1998), and the ^{3}He content is unusually high (Farley et al., 1998). The presence of the nickel-rich spinels at Massignano has been taken as evidence of its correlation with the microkrystite layer (Vonhof and Smit, 1999). Though the Massignano ejecta appears to occur in a single layer, and microtektites have not yet been identified, we speculate that this occurrence may represent a condensation of the microtektite and microkrystite layers whose separate identities have been obscured by bioturbation (Vonhof and Smit, 1999) and low deposition rates in this relatively thin (12 m) upper Eocene section.

Magnetostratigraphic studies of the Massignano section reveal that the ejecta deposit resides within the upper part of Chron C16n.2n (Bice and Montanari, 1988; Lanci et al., 1996), and is, therefore, >35.6 million years old. A radioisotopic age of 35.7 ± 0.4 Ma for the Massignano ejecta has been derived indirectly by complicated interpolations between (1) a weighted mean age estimate from an ash bed at Massignano (35.4 ± 0.3 Ma), and (2) a similarly derived estimate (36.4 ± 0.4 Ma) from an ash bed located ~80 km away in the Contessa Highway section, which was extrapolated by means of magnetochronology to the Massignano section (summarized by Montanari et al., 1993).

Correlation Problems

The biostratigraphic position of the Massignano ejecta was reported by Coccioni et al. (1988) and Montanari et al. (1993) to be within planktonic foraminiferal Biozone P16, whose biochron spans the interval from 34.0 Ma to 35.2 Ma (Berggren et al., 1995). If true, the Massignano ejecta would be biochronologically younger, but radioisotopically older, than the Chesapeake Bay and Toms Canyon impacts (and their related impactites). However, it has been compellingly demonstrated that the stratigraphic ranges of key species of planktonic foraminifera and calcareous nannofossils are diachronous between the Massignano section, Site 612, and Chesapeake Bay (Exmore breccia and Chickahominy Formation; Miller et al., 1991; Montanari et al., 1993; Poag and Aubry, 1995).

Biochronological correlations with other pertinent upper Eocene sections are hampered by poor specimen preservation in deep-sea sediments (including Bath Cliff, Barbados) and by significant faunal and floral disparities as a result of biogeographic differences between tropical micro- and nannofossil assemblages (Caribbean and Gulf of Mexico), mixed tropical and temperate assemblages (Site 612, Chesapeake Bay), Tethyan assemblages (Massignano), and Antarctic assemblages (Site 689B). Inconsistent upper Eocene biostratigraphic interpretations at two additional core holes near Site 612 (ODP Sites 903, 904; figure 28.1; McHugh et al. 1993; Snyder et al., 1993; Aubry, 1993) have further confused correlations with the offshore New Jersey sites. For example, at Site 904A, these three groups of authors place the Zone P15–P14 boundary (planktonic foraminifera), the Zone NP 19/20 boundary (calcareous nannofossils), and the Zone NP16 boundary (calcareous nannofossils) at significantly different core depths.

Although the magnetostratigraphic records at Massignano and Site 689B have helped to resolve some of the biostratigraphic and radioisotopic correlation problems, critical inferences, extrapolations, and interpolations are still required, which weakens some of the correlations. We present new magnetostratigraphic and stable isotopic data ($\delta^{18}O$, $\delta^{13}C_{carb}$) from the Chickahominy Formation (figure 28.2), which (1) further constrain the age of the Chesapeake Bay crater; (2) clarify correlations between Chesapeake Bay, Massignano, Barbados, and Antarctic ODP Site 689B; (3) aid in future refinements of late Eocene correlations; (4) provide stratigraphic constraints that will assist in the search for additional occurrences of late Eocene impact ejecta; and (5) improve our understanding of potential paleoenvironmental perturbations arising from the impacts.

ANALYTICAL METHODS

We took 32 samples for paleomagnetic study from a 66-m section of the Chickahominy Formation cored continuously at a site near Kiptopeke, Virginia (figure 28.2; Poag et al., 1994; Poag and Aubry, 1995; Poag, 1997b). The Chickahominy Formation is a moderately consolidated, silty clay unit, 60-150 m thick, which accumulated in relatively deep marine water (~200 m paleodepth). The Chickahominy is restricted to the subsurface of southeastern Virginia (essentially present only over the impact crater), and is the only documented deposit in the Atlantic Coastal Plain that represents the final 1.5–2.3 million years of Eocene marine deposition (subsequent to the impact). We sampled the Kiptopeke core at ~2-m intervals, which represent ~25-38 kyr. The core was oriented only with respect to top and bottom, so the geomagnetic polarity of each sample is based on magnetic inclination. The natural remanent magnetization (NRM) of each sample was measured with a superconducting magnetometer housed in a magnetically shielded room (USGS, Menlo Park, Calif.). The NRM was very weak, with a geometric mean intensity of 0.055 mA^2/kg for 31 samples. Progressive alternating-field demagnetization experiments on a few selected samples indicated that this method would be ineffective for recovering the primary remanence within this core. Remaining samples, therefore, were subjected to progressive thermal demagnetization. Thermal demagnetization was performed in a partial vacuum (50 µm) to remove any gases emitted during the experiments and to inhibit chemical alteration. Ovens used for the heating experiments were enclosed in a Rubens coil array, which maintained an ambient field of approximately 40 nT. All samples began to chemically alter by about 400°C.

Because of the low signal-to-noise ratio in these samples, it often was difficult to ascertain whether magnetization had reached a stable end point, or to allow us to rely entirely upon principal component analysis (Kirschvink, 1980). Directional trends during the experiments, however, generally were unambiguous and allowed us to determine the geomagnetic polarity with some degree of confidence. Samples in which the magnetic direction was inclined below the horizontal are considered to be of normal polarity; those with inclinations above the horizontal are interpreted to be of reversed polarity.

We performed stable-isotope analyses for oxygen and carbon on 40 samples taken from the same cored interval sampled for the paleomagnetic study (figure 28.2). We used monospecific samples (~3–20 individuals of the benthic foraminifer *Cibicidoides speciosus* [Cushman and Cederstrom], 1945) from the >63 µm grain-size fraction. Mass spectrometry was performed using a Finnigan MAT 252 instrument with an on-line automated carbonate reaction Kiel device (Woods Hole Oceanographic Institution). Analytical precision based on repeated analysis of standards (NBS-19, Carrara Marble, and B-1 marine carbonate) was better than ±0.03‰ for $\delta^{13}C$ and 0.08‰ for $\delta^{18}O$ relative to the Peedee belemnite (PDB) standard.

We used the magnetobiochronological framework (synthesis of radioisotopes, geomagnetic polarity, planktonic foraminifera, calcareous nannofossils) of Berggren et al. (1995) to guide interpretation and correlation of our biostratigraphic, magnetostratigraphic, and chemostratigraphic results.

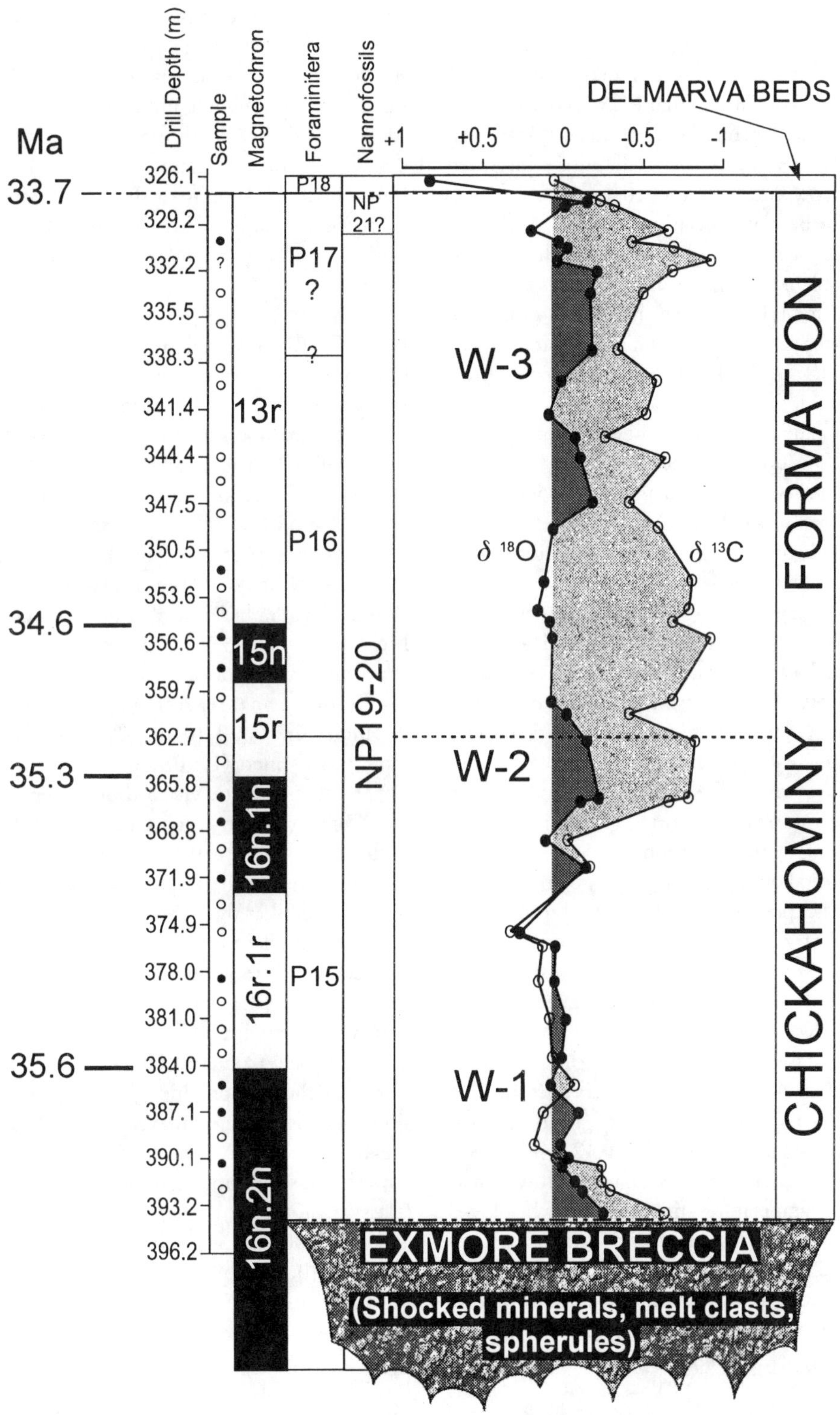

Figure 28.2. Geochronology and stratigraphy of Chickahominy Formation in Kiptopeke corehole, Delmarva Peninsula, Virginia. W-1, W-2, and W-3 indicate intervals of relatively warm climate and bottom water, as inferred from analysis of $\delta^{18}O$ in benthic foraminifer *Cibicidoides*. Solid circles in sample column represent normal magnetochrons; open circles represent reversed magnetochrons.

RESULTS

Magnetostratigraphic Record

The geomagnetic polarity record derived from the Chickahominy Formation (figure 28.2) is based on our requirement that one or more adjacent samples must indicate identical polarity. Thus, we interpret our data to represent three reversed and three normal intervals. The best-fit correlation with the Chickahominy biostratigraphic record assumes that the boundary between planktonic foraminiferal Zones P15 and P16 (~362.7 m; 1,190 ft) is in the middle of Chron C15r (Berggren et al., 1995). This places the superjacent 2 normal samples in Chron C15n, above which the stratigraphically highest 12 samples represent Chron C13r. The next 4 samples below Chron C15r represent Chron C16n.1n, the next lower 6 samples represent Chron C16r.1r, and the lowest 5 samples represent Chron C16n.2n.

The Chickahominy geomagnetic polarity record also displays a few single-sample reversals. Although some of these may represent actual geomagnetic field behavior (see, for example, the "tiny wiggles" or "cryptochrons" of Cande and Kent, 1992), others may be spurious results. Misinterpretation of a sample's magnetic polarity can result from misorientation of a core segment on the drilling rig, from mechanical disturbance within the core, or from secondary components of magnetization (overprints) not removed during demagnetization.

The Chickahominy geomagnetic polarity record indicates that deposition of the Exmore breccia and, thus, excavation of the Chesapeake Bay crater, took place within Chron C16n.2n, which constrains its age to somewhat older than 35.6 Ma (figure 28.2).

Stable Isotopic Record

The benthic foraminiferal $\delta^{18}O$ record from the Chickahominy Formation shows three relatively distinct negative excursions of around -0.12 to -0.25‰, interpreted to represent pulses of relatively warmer bottom-water temperatures (figure 28. 2). The oldest warm pulse, W-1, correlates with the upper part of Chron C16n.2n and perhaps the lower part of Chron C16r.1r; W-2 correlates with Chron C16n.1n and the lower half of Chron C15r; the youngest, W-3, correlates with the upper two-thirds of Chron C13r. A distinct positive excursion to +0.81‰ occurs in the lowest sample of the Delmarva beds, a lower Oligocene marine sand deposit, which rests unconformably above the Chickahominy.

The $\delta^{13}C_{carb}$ record from the Chickahominy Formation displays a short negative excursion (-0.3 to -0.7‰) near the base, correlative with Chron C16n.2n. Above that, a positive interval (+0.3 to -0.2‰) extends to the base of Chron C16n.1n, followed by a prominent negative excursion (-0.25 to -0.9‰), which extends upward to the top of the Eocene section, correlative with the interval from upper Chron C16n.1n to the upper part of Chron C13r (figure 28.2). The $\delta^{13}C$ signal switches to positive values again (+0.06‰) in the lower Oligocene section.

NEW CORRELATIONS

Our interpretations indicate that the Chesapeake Bay impact, deposition of the Exmore breccia, deposition of the North American tektite-strewn field, and deposition of the impactite at Massignano (5.61 m above the base of the section) all took place within the upper part of Chron C16n.2n (figure 28.3A). This clarifies the biostratigraphic problem at Massignano, caused by misplacement of the impactite there within Biozone P16, rather than P15 (Coccioni et al., 1988; Montanari et al., 1993).

Our new paleomagnetic data also have interesting implications regarding the ejecta cored at Antarctic Site 689B (figure 28.3B). According to Stott and Kennett (1990), Speiss (1990), and Miller (1992), the Antarctic ejecta (both layers of the doublet) correlates with either Chron C16n.1n or C15r, both of which are ~ 0.1–0.9 million years younger than the late Eocene impacts. Either interpretation would imply that the Site 689B ejecta doublet represents two unidentified, closely spaced, late Eocene impacts, which took place between deposition of the Massignano ejecta layer and a younger iridium-enriched layer (within Chron C13r; Montanari et al., 1993) in the Contessa Highway section in Italy. However, the stratigraphic relationships and compositions of the two ejecta layers at Site 689B strongly suggest correlation with the Chesapeake Bay and Popigai impacts (upper and lower layers, respectively; Glass and Koeberl, 1999a; Vonhof and Smit, 1999), and cast doubt on the published magnetostratigraphy of Site 689B. On this basis, we infer that one or more unconformities have interrupted the stratigraphic succession at Site 689B (figure 28.3B).

PALEOENVIRONMENTAL IMPLICATIONS

$\delta^{18}O$ Data

Along with other evidence of warm late Eocene paleoclimates (Haq and Lohmann, 1976; Wolfe, 1978, 1992; Wolfe and Poore, 1982; Aubry, 1992; Keller et al., 1992; Prothero, 1994; Bestland et al., 1996), Poag

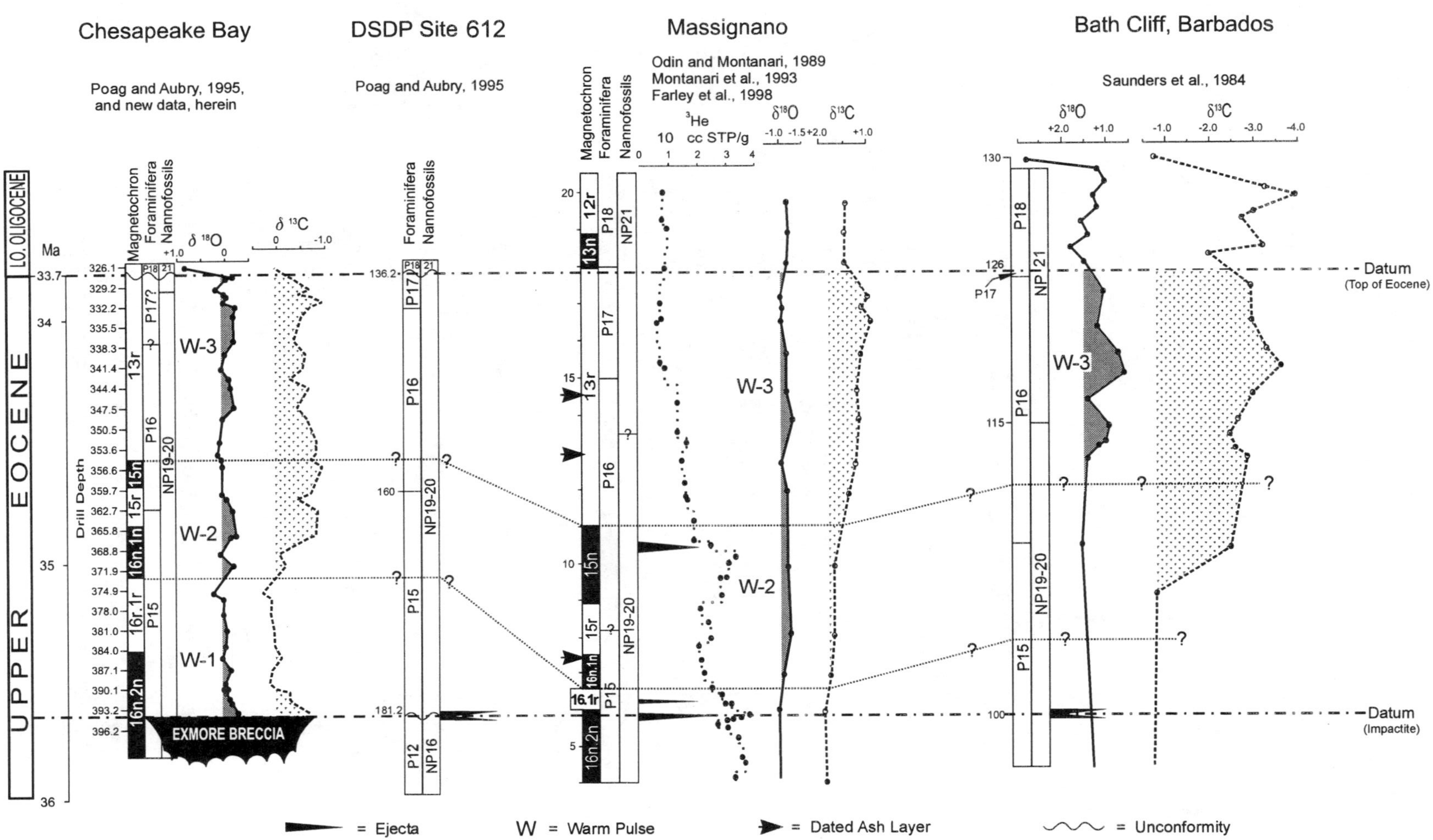

Figure 28.3. Correlation of late Eocene impactites (breccia and ejecta) based on integrated magnetochronology, biostratigraphy, and stable isotopic stratigraphy. All stratigraphic sections measured in meters. (A) Correlation of Chesapeake Bay stratigraphy with DSDP Site 612, Massignano, Italy, and Bath Cliff, Barbados.

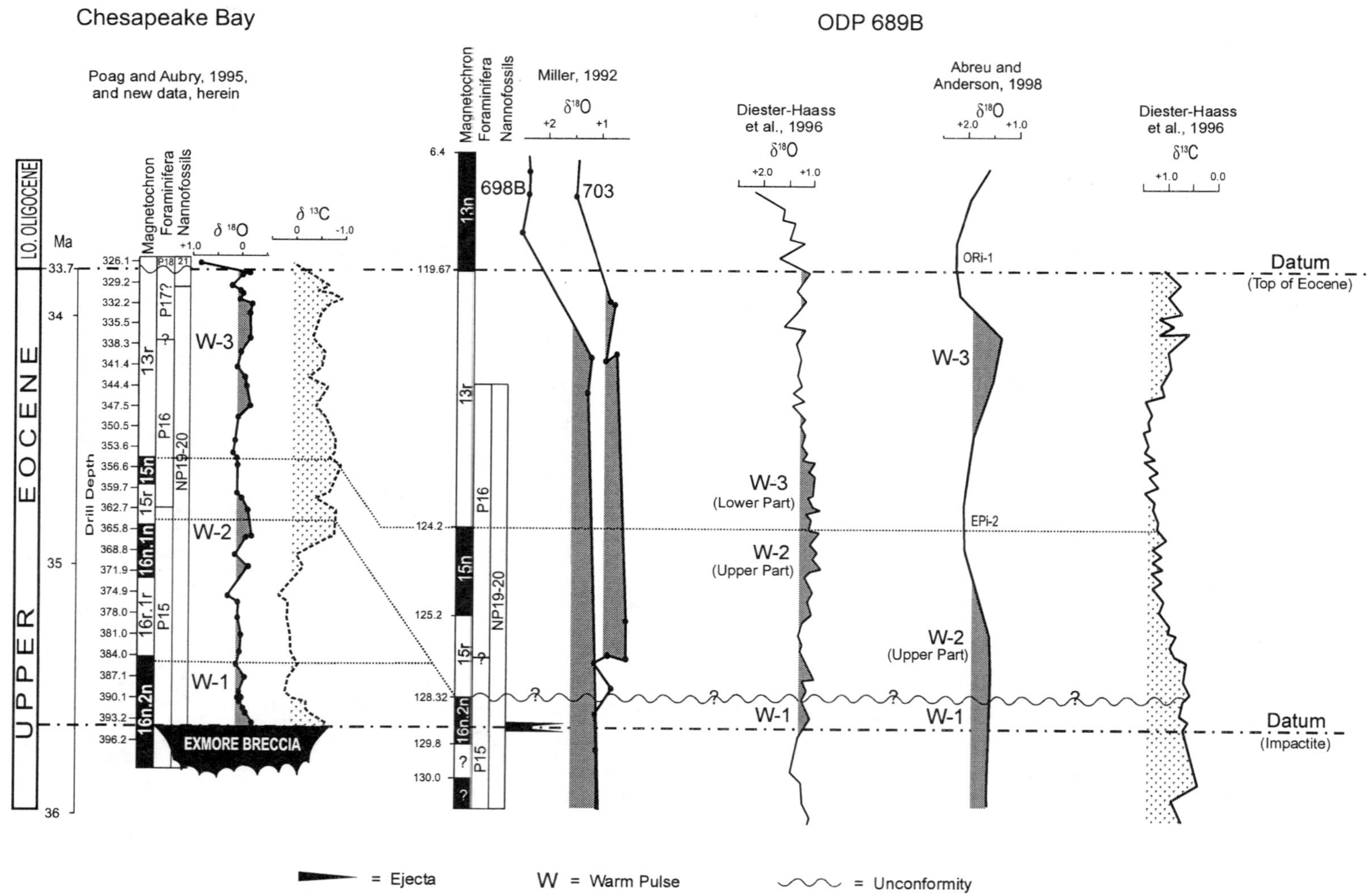

Figure 28.3. (B) Correlation of Chesapeake Bay with three different studies of ODP Site 689B.

(1997b) used Miller's (1992) isotopic data (benthic foraminifera) from Southern Ocean ODP Sites 689B and 703 to support the speculation that the three documented bolide impacts produced overriding greenhouse effects and triggered a long-term (~2 million year) pulse of late Eocene climatic warmth. Miller (1992) used a few widely spaced samples to infer an interval of relatively negative $\delta^{18}O$ (warm bottom water) between the ejecta doublet in core 689B and the Eocene/Oligocene boundary (figure 28.3B).

The $\delta^{18}O$ record in the Chickahominy Formation, however, comes from a thicker, more rapidly deposited upper Eocene sedimentary section than any other impactite site studied (66 m thick vs 45 m at Site 612, 25 m at Bath Cliff, 14 m at 689B, and 12 m at Massignano). This expanded record shows a threefold subdivision of the inferred late Eocene warm pulse (figures 28.2 and 28.3A). The oldest warm subdivision, W-1, is expressed by negative $\delta^{18}O$ in Chron C16n.2n and the lower part of Chron C16r.1r. Subpulse W-1 correlates with the Chesapeake Bay, Toms Canyon, and Popigai impacts, but has not been identified at Massignano or Bath Cliff (insufficient data; figure 28.3A).

The next youngest Chickahominy warm subpulse, W-2, coincides with Chron C16n.1n and the lower two-thirds of Chron C15r (figures 28.2 and 28.3). Subpulse W-2 correlates with the lower part of the lower warm interval at Massignano, but has not been identified at Bath Cliff (insufficient data; figure 28.3A).

The youngest Chickahominy warm subpulse, W-3, occupies the middle to upper part of Chron C13r, and correlates with the upper warm interval at Massignano and the single warm interval identified at Bath Cliff (figure 28.3A).

Relatively positive $\delta^{18}O$ values in the Chickahominy record during Chron C15n and early Chron C13r indicate cooler water, and correlate with a similar interval at Bath Cliff. This interval, however, is difficult to recognize at Massignano (figure 28.3A).

Upon further consideration of the published record, we conclude that the presence of at least two to three late Eocene warm pulses also can be inferred from Antarctic cores (figure 28.3B). First of all, the isotopic record at Site 689B is more complicated than indicated by Miller's (1992) coarse sampling. For a subsequent isotopic study (benthic foraminifera) of the same core, Diester-Haass and Zahn (1996) sampled at 20-cm intervals, and produced a much different $\delta^{18}O$ profile from that of Miller (1992). Warm subpulses W-1 and W-2 appear to be represented (figure 28.3B), and perhaps the lower part of W-3. There is no clear separation of W-2 and the lower part of W-3, however, and the upper part of W-3 is only faintly represented.

Abreu and Anderson (1998) composited and smoothed the late Eocene $\delta^{18}O$ record in 19 Southern Ocean sites (including Site 689B) and presented a third paleoenvironmental interpretation. They recognized two relatively negative intervals (ice retreat; warmer bottom water) separated by a distinct positive excursion (ice advance; colder bottom water) during late Chron C15n and early Chron C13, which they designated as event EPi-2 (figure 28.3B). The $\delta^{18}O$ correlations between the Kiptopeke core, Massignano, and 689B suggest that the magnetostratigraphy at 689B is correct in the interval interpreted to represent Chrons C15r to C13r. If one assumes that the ejecta doublet represents the upper part of Chron C16n.2n, however, then Chrons C16n.1n and C16r.1r appear not to be represented at Site 689B (figure 28.3B). We infer, then, that the stratigraphically lower warm subpulse of Abreu and Anderson (1998) is, in reality, two warm subpulses separated by an unconformity. Thus, three warm pulses appear to be represented in the Antarctic region, although not all can be distinguished at every site studied.

A broad, late Eocene warm pulse also can be inferred from the $\delta^{18}O$ records at numerous other localities in the Atlantic, Pacific, and Indian Oceans (e.g., Miller et al., 1985, 1987; Keigwin and Corliss, 1986; Shackleton, 1986; Steckler et al., 1999; figures 28.4 and 28.5), which attests to its near global occurrence. The significance of this warm pulse (a succession of at least three subpulses) has been overshadowed in all previous studies, however, by the spectacular cooling event that followed in the early Oligocene.

$\delta^{13}C_{carb}$ Data

The late Eocene $\delta^{13}C$ record has not received as much attention as the $\delta^{18}O$ record because of the latter's importance in defining the early Oligocene ice-advance signal and initial development of the oceanic cryosphere. The Chickahominy $\delta^{13}C$ curve contains easily distinguishable subdivisions, however, and holds promise for identifying fundamental late Eocene environmental changes and for improving correlations with other upper Eocene sections and related impactites (figures 28.2 and 28.3A–B).

The most obvious characteristic of the late

Eocene $\delta^{13}C$ record is the persistence of distinctly negative values in the upper two-thirds of the section at Chesapeake Bay, Massignano and Bath Cliff (encompassing Chron C16n.1n through most of Chron C13r; figure 28.3A). A coincident negative $\delta^{13}C$ excursion characterizes the late Eocene record at many additional deep-sea sites (often the same sites showing the negative $\delta^{18}O$ excursion; figures 28.4 and 28.5).

A second, older, less prominent, negative $\delta^{13}C$ excursion is present at Chesapeake Bay (upper part of Chron C16n.2n) and can be seen in some (e.g., Shackleton, 1986) but not all the other sites discussed herein (figures 28.3A–B and 28.4).

Several notable negative excursions of $\delta^{13}C$ have been documented in other parts of the geologic record, also coincident, or nearly so, with negative $\delta^{18}O$ excursions. Perhaps the most dramatic is the Late Paleocene Thermal Maximum (Norris and Röhl, 1999). The origins of such rapid $\delta^{13}C$ excursions have been attributed mainly to either a catastrophic release of biogenic methane from global dissociation of submarine gas hydrates (Bains et al., 1999; Norris and Röhl, 1999) or a massive injection of atmospheric

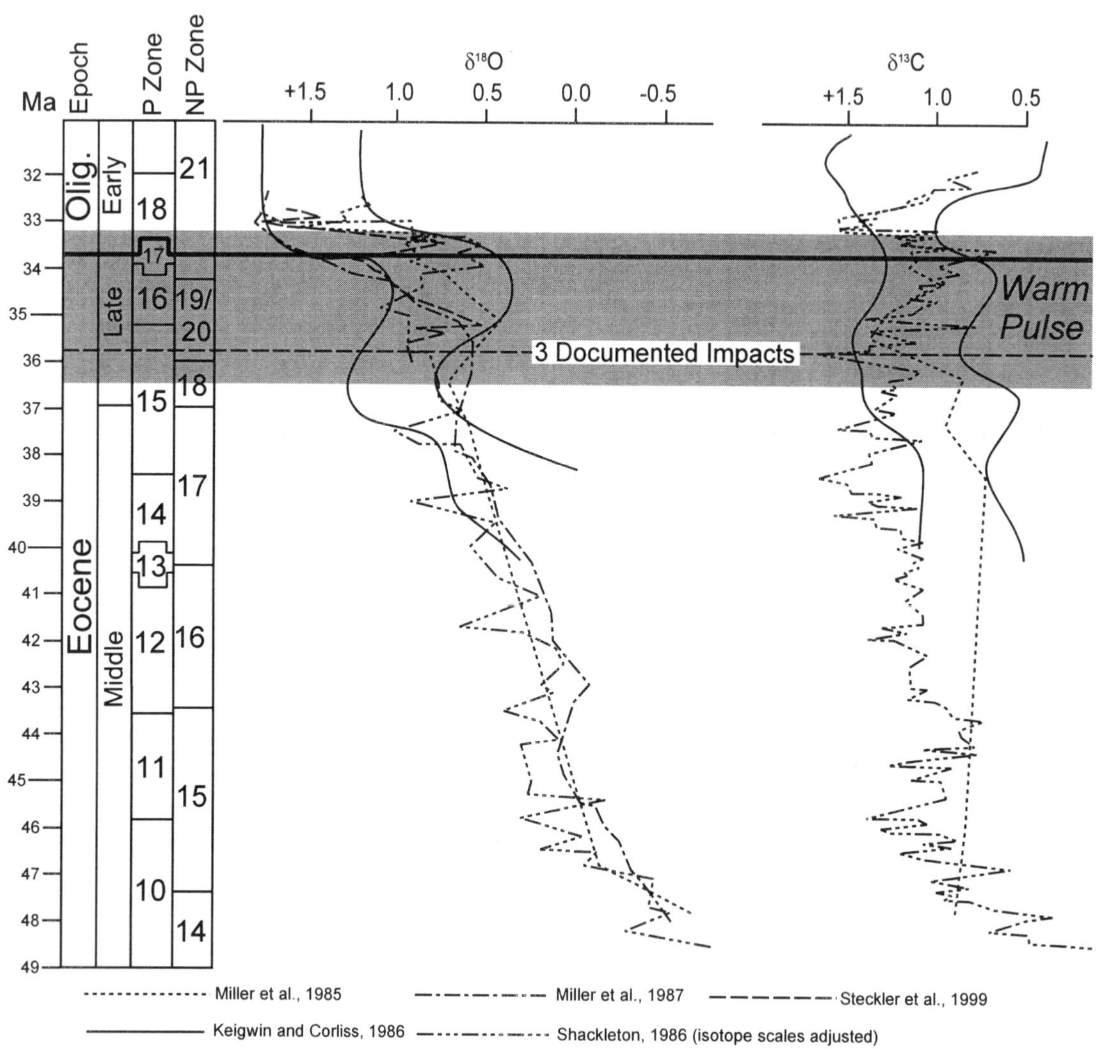

Figure 28.4. Stratigraphic correlation of Eocene-Oligocene isotopic records ($\delta^{18}O$ and $\delta^{13}C_{carb}$) from 26 sites in the Pacific, Atlantic, Indian, and Southern Oceans. The curves from Keigwin and Corliss (1986) represent a broad envelope bounding their data points.

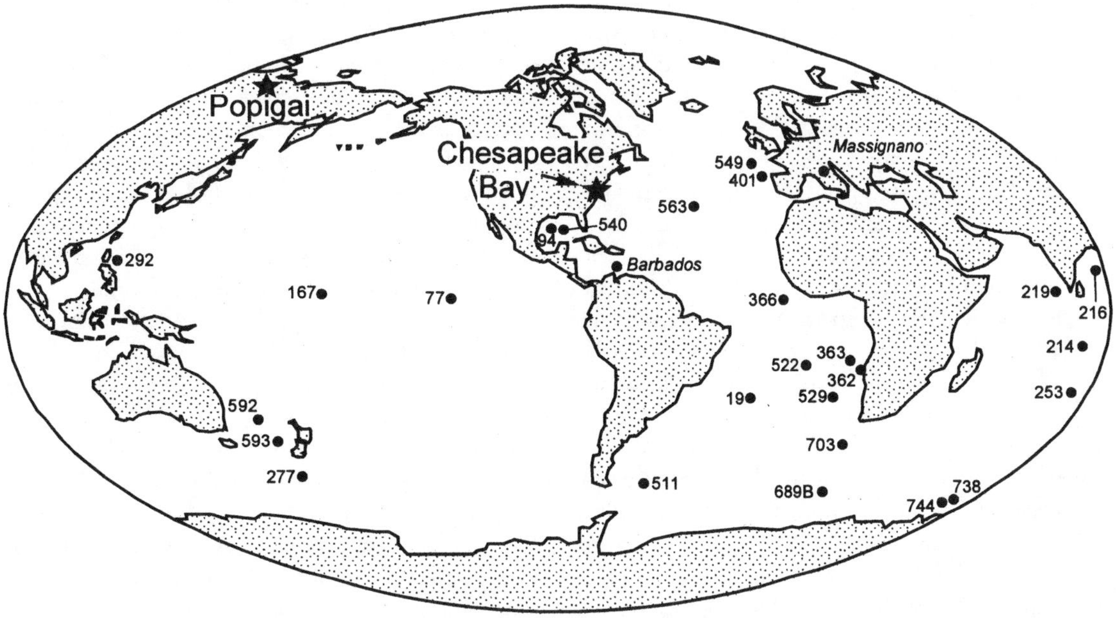

Figure 28.5. Stable isotopic data from these 28 core sites indicate significant negative excursions of $\delta^{18}O$ and $\delta^{13}C_{carb}$ in the late Eocene, from which we infer at least three subpulses of climatic warmth and two perturbations of the global carbon budget.

CO_2 by sustained, intense, volcanic degassing (Zachos et al., 1993; Thomas and Shackleton, 1996). The late Eocene negative excursion of $\delta^{13}C$, however, was a more sustained event, and has been interpreted in a paleoceanographic context to indicate an incursion of colder, more vigorous bottom waters (Miller et al., 1985), and increased paleoproductivity (Diester-Haass and Zahn, 1996).

Theoretically, atmospheric disturbances caused by dust and particulates from a single projectile the size of the Chesapeake Bay impactor (3–5 km) would last, at most, for only a few decades (Toon et al., 1994). However, individual late Eocene impacts could be expected to also change atmospheric gas composition (e.g., the global carbon budget) by injecting massive volumes of CO_2 into the atmosphere from vaporized seawater (e.g., Chesapeake Bay and Toms Canyon) and carbonate-rich target rocks (e.g., Popigai: Masaitis et al., 1975; and Toms Canyon: Poag and Poppe, 1998). A succession of impacts (accompanied by numerous atmospheric bursts) during a 2-million-year-long comet shower, on the other hand, could plausibly sustain atmospheric perturbations for most of the late Eocene.

BIOTIC PERTURBATIONS

At this writing, no mass extinction is known to have coincided with the three known late Eocene impacts (Prothero and Berggren, 1992). On the other hand, this lack of immediate biotic response does not imply that no significant biological changes arose from the impacts. We believe that our new magnetostratigraphic and chemostratigraphic data support the concept that late Eocene, impact-generated, greenhouse warming interrupted a progressive early Eocene to early Oligocene cooling trend (Poag, 1997b). A number of marine and terrestrial paleoclimatic analyses provide independent evidence of late Eocene climatic warming. Examples include (1) a 6-8° temperature rise over North America, as indicated by leaf-margin analysis of land plants (Wolfe, 1978, 1992, 1994); (2) warming of oceanic surface waters, which permitted low-latitude nannofloras to migrate into higher latitudes (Haq and Lohman, 1976); and (3) maintenance of humid, subtropical climate in midlatitude western North America (Bestland et al., 1996).

The late Eocene warmth was manifested in a succession of warm subpulses, during which a complex succession of minor extinction events, species originations, biogeographic shifts, and faunal turnovers took place. The warmth abated rapidly in the early Oligocene, accompanied by a dramatic biotic crash (Corliss et al., 1984; Miller et al., 1985; Sanfilippo et al., 1985; Keller, 1986; MacLeod et al., 1990; Wei and Wise, 1990; Aubry, 1992; Keller et al., 1992; Thomas, 1992; Poag and Aubry, 1995; Prothero and Emry, 1996; Prothero and Heaton, 1996; Meng and McKenna, 1998). Thus, the inferred succession of late Eocene impacts may have created, or contributed to, the particularly steep temperature gradient between the late Eocene greenhouse and the early Oligocene icehouse, thereby creating threshold conditions necessary for the consequent early Oligocene mass extinction. Ironically, this impact-derived disruption of the global cooling trend may have postponed a pending late Eocene mass extinction until the early Oligocene.

SUMMARY AND CONCLUSIONS

New magnetostratigraphic and stable isotopic data ($\delta^{18}O$, $\delta^{13}C_{carb}$) from the late Eocene Chickahominy Formation (Kiptopeke core), which overlies the Chesapeake Bay impact crater, help to improve correlations between three known late Eocene impact craters and their respective impactites. The Chesapeake Bay bolide struck during the late part of Chron C16n.2n, a little more than 35.6 Ma. That same magnetostratigraphic interval contains a widespread doublet of distal ejecta characterized by (1) an upper layer containing microtektites, tektite fragments, shocked quartz, coesite, and stishovite (North American tektite strewn field); and (2) a lower layer containing microkrystites (containing clinopyroxene crystals and/or nickel-rich spinels; commonly called the cpx layer), accompanied by an enrichment in Ir. Both ejecta layers are correlative with an interval notable for its unusual enrichment in extraterrestrial ^{3}He. Distinct stratigraphic subdivisions in the stable isotopic record in the Kiptopeke core further enhance correlations between Chesapeake Bay, Massignano, Barbados, and Antarctic cores, and promise better ties to other upper Eocene sections. The Chickahominy record of $\delta^{18}O$, for example, permits a threefold subdivision of late Eocene bottom-water temperature. From this record, we infer at least three consecutive subpulses of warm climate, derived from long-term greenhouse effects produced by successive bolide impacts during a comet shower. The more compressed (thinner) stratigraphic sections at Bath Cliff, Massignano, and Maud Rise record less complete records of late Eocene warmth.

A notable negative excursion of $\delta^{13}C$ in the upper half to two-thirds of late Eocene time is strongly correlative between Chesapeake Bay, Massignano, and Bath Cliff. This excursion implies a substantial change in the global carbon balance, which also is attributable to successional bolide impacts. A less prominent negative excursion of $\delta^{13}C$, is noted at the base of the Chickahominy section but has not been widely recognized elsewhere, probably because of sparse sampling in that stratigraphic interval.

ACKNOWLEDGMENTS

We are grateful to Bill Glass and Chris Koeberl for discussions of various aspects of impactite distribution and correlation. T. Scott Bruce and the Virginia Department of Environmental Quality provided samples from the Chickahominy Formation. John Bratton, Larry Poppe, Bill Glass, and Philippe Claeys provided helpful reviews of earlier versions of the manuscript.

REFERENCES

Abreu, V. S., and Anderson, J. B., 1998, Glacial eustasy during the Cenozoic: sequence stratigraphic implications: American Association of Petroleum Geologists Bulletin, v. 82, p. 1385-1400.

Aubry, M.-P., 1992, Late Paleogene calcareous nannoplankton evolution: A tale of climatic deterioration, *in* Prothero, D. R., and Berggren, W. A., eds., Eocene-Oligocene climatic and biotic evolution: Princeton, New Jersey, Princeton University Press, p. 272-309.

Aubry, M.-P., 1993, Data report: Eocene to upper Miocene calcareous nannofossil stratigraphy, *in* Mountain, G. S., Miller, K. G., Blum, P., Poag, C. W., and Twichell, D. C., eds., Proceedings of the Ocean Drilling Program, Scientific Results, Volume 150: College Station, Texas, Ocean Drilling Program, p. 435-437.

Bains, S., Corfield, R. M., and Norris, R. D., 1999, Mechanisms of climate warming at the end of the Paleocene: Science, v. 285, p. 724-727.

Barker, P. F., Kennett, J. P., et al., 1990, Proceedings of the Ocean Drilling Program, Scientific Results, Volume 113: College Station, Texas, Ocean Drilling Program, p. 1-785.

Berggren, W. A., Kent, D. V., Swisher, C. C., III, and Aubry, M.-P., 1995, A revised Cenozoic geochronology and chronostratigraphy: SEPM Special Publication, v. 54, p. 129-212.

Bestland, E. A., Retallack, G. J., Rice, A. E., and

Mindszenty, A., 1996, Late Eocene detrital laterites in central Oregon: Mass balance, geochemistry, depositional setting, and landscape evaluation: Geological Society of America Bulletin, v. 108, p. 285-302.

Bice, D. M., and Montanari, A., 1988, Magnetic stratigraphy of the Massignano section across the Eocene/Oligocene boundary, *in* Premoli Silva, I., Coccioni, R., and Montanari, A., eds., The Eocene-Oligocene boundary in the Marche-Umbria Basin: International Commission on Paleogene Stratigraphy Special Publication p. 111-117.

Bottomley, R., Grieve, R. A. F., York, D., and Masaitis, V., 1997, The age of the Popigai impact event and its relation to events at the Eocene/Oligocene boundary: Nature, v. 388, p. 365-368.

Cande, S. C., and Kent, D. V., 1992. A new geomagnetic polarity time-scale for the Late Cretaceous and Cenozoic: Journal of Geophysical Research, v. 97, p. 13,917-13,951.

Clymer, A. K., Bice, D. M., and Montanari, A., 1996, Shocked quartz from the late Eocene: impact evidence from Massignano, Italy: Geology, v. 24, p. 483-486.

Coccioni, R., Monaco, P., Monechi, S., and Parisi, G., 1988, Biostratigraphy of the Eocene/Oligocene boundary at Massignano (Ancona, Italy), *in* Premoli Silva, I., Coccioni, R., and Montanari, A., eds., The Eocene-Oligocene boundary in the Marche-Umbria Basin (Italy): International Subcommission on Paleogene Stratigraphy, Special Publication II, p. 59-80.

Corliss, B. H., Aubry, M.-P., Berggren, W. A., Fenner, J. M., Keigwin, L. D., and Keller, G., 1984, The Eocene/Oligocene boundary event in the deep sea: Science, v. 226, p. 806-810.

Cushman, J. A., and Cederstrom, D. J., 1945, An upper Eocene foraminiferal fauna from deep wells in York County, Virginia: Virginia Geological Survey, Bulletin 67, 58 p.

Diester-Haass, L., and Zahn, R., 1996, Eocene-Oligocene transition in the Southern Ocean: history of water mass circulation and biological productivity: Geology, v. 24, p. 163-166.

D'Hondt, S. L., Keller, G., and Stallard, R. F., 1987, Major element compositional variation within and between different late Eocene microtektite strewn fields: Meteoritics, v. 22, p. 61-79.

Farley, K. A., Montanari, A., Shoemaker, E. M., and Shoemaker, C. S., 1998, Geochemical evidence for a comet shower in the late Eocene: Science, v. 280, p. 1250-1253.

Glass, B. P., 1989, North American tektite debris and impact ejecta from DSDP Site 612: Meteoritics, v. 24, p. 209-218.

Glass, B. P., and Burns, C. A., 1987, Late Eocene crystal-bearing spherules: Two layers or one?: Meteoritics, v. 22, p. 265-279

Glass, B. P., and Koeberl, C., 1999a, Ocean Drilling Project Hole 689B spherules and upper Eocene microtektite and clinopyroxene-bearing spherule strewn fields: Meteoritics and Planetary Science, v. 34, p. 197-208.

Glass, B. P., and Koeberl, C., 1999b, North American microtektites in the Indian Ocean?: Meteoritics and Planetary Science, v. 34, Supplement, p. A43-A44.

Glass, B. P., and Wu, J., 1993, Coesite and shocked quartz discovered in the Australasian and North American microtektite layers: Geology, v. 21, p. 435-438.

Glass, B. P., Hall, C. M., and York, D., 1986, $^{40}Ar/^{39}Ar$ laser probe dating of North American tektite fragments from Barbados and the age of the Eocene/Oligocene boundary: Chemical Geology, v. 59, p. 181-186.

Glass, B. P., Burns, C. A., Crosbie, J. R., and DuBois, D. L., 1985, Late Eocene North American microtektites and clinopyroxene-bearing spherules: Journal of Geophysical Research, v. 90, p. D175-D196.

Glass, B. P., Koeberl, C., Blum, J. D., and McHugh, C. M. G., 1998, Upper Eocene tektite ejecta layer on the continental slope off New Jersey: Meteoritics and Planetary Science, v. 33, p. 229-241.

Haq, B. U., and Lohmann, G. P., 1976, Cenozoic calcareous nannoplankton biogeography of the Atlantic Ocean: Marine Micropaleontology, v. 1, p. 119-194.

Keigwin, L. D., and Corliss, B. H., 1986, Stable isotopes in late middle Eocene to Oligocene foraminifera: Geological Society of America Bulletin, v. 97, p. 335-345.

Keller, G., 1986, Stepwise mass extinctions and impact events: Late Eocene to early Oligocene: Marine Micropaleontology, v. 10, p. 267-293.

Keller, G., MacLeod, N., and Barrera, E., 1992, Eocene-Oligocene faunal turnover in planktic foraminifera, and antarctic glaciation, *in* Prothero, D. R., and Berggren, W. A., eds., Eocene-Oligocene climatic and biotic evolution: Princeton, New Jersey, Princeton University Press, p. 218-243.

Kirschvink, J. L., 1980, The least-squares line and plane and the analysis of paleomagnetic data: Geophysical Journal of the Royal Astronomical Society, v. 62, p. 699-718.

Koeberl, C., 1989, New estimates of area and mass for the North American tektite strewn field: Proceedings of the 19th Lunar and Planetary Science Conference: Houston, Texas, Lunar and Planetary Institute, p. 745-751.

Koeberl, C., Poag, C. W., Reimold, W. U., and Brandt, D., 1996, Impact origin of the Chesapeake Bay structure and the source of the North American tektites: Science, v. 271, p. 1263-1266.

Lanci, L., Lowrie, W., and Montanari, A., 1996, Magnetostratigraphy of the Eocene/Oligocene boundary in a short drill-core: Earth and Planetary Science Letters, v. 143, p. 37-48.

Langenhorst, F., 1996, Characteristics of shocked quartz in the late Eocene impact ejecta from Massignano (Ancona, Italy): clues to shock conditions and source crater: Geology, v. 24, p. 487-490.

Masaitis, V. L., 1994, Impactites from Popigai crater, *in* Dressler, B. O., Grieve, R. A. F., and Sharpton, V. L., eds., Large Meteorite Impacts and Planetary Evolution: Geological Society of America Special Paper 293, p. 153-162.

Masaitis, V., Mikhailov, M. V., and Selivanovskaya, T. V., 1975, Popigai meteorite crater [in Russian]: Nauka Press, Moscow, 124 p.

Masaitis, V. L., Naumov, M. V., and Mashchak, M. S., 1999, Anatomy of the Popigai impact crater, Russia, *in* Dressler, B. O., and Sharpton, V. L., eds., Large Meteorite Impacts and Planetary Evolution II, Geological Society of America Special Paper 339, p. 1-17.

McHugh, C. M. G., Snyder, S. W., Deconinck, J.-F., Saito, Y., Katz, M. E., and Aubry, M.-P., 1993, Upper Eocene tektites of the New Jersey continental margin, Site 904, *in* Mountain, G.S., Miller, K. G., Blum, P., Poag, C. W., and Twichell, D. C., eds.: Proceedings of the Ocean Drilling Program, Scientific Results, Volume 150: College Station, Texas, Ocean Drilling Program, p. 241-265.

McLeod, N., Keller, G., and Kitchell, J. A., 1990, Progenesis in late Eocene populations of *Subbotina linaperta* (Foraminifera) from the western Atlantic: Marine Micropaleontology, v. 16, p. 219-240.

Meng, J., and McKenna, M. C., 1998, Faunal turnovers of Palaeogene mammals from the Mongolian Plateau: Nature, v. 394, p. 364-367.

Miller, K. G., 1992, Middle Eocene to Oligocene stable isotopes, climate, and deep-water history: The terminal Eocene event? *in* Prothero, D. R., and Berggren, W. A., eds., Eocene-Oligocene climatic and biotic evolution: Princeton, New Jersey, Princeton University Press, p. 160-177.

Miller, K. G., Curry, W. B., and Ostermann, D. R., 1985, Late Paleogene (Eocene to Oligocene) benthic foraminiferal oceanography of the Goban Spur region, Deep-Sea Drilling Project Leg 80, *in* Graciansky, P. C. de, Poag, C. W., et al., Initial Reports of the Deep Sea Drilling Project, Volume 80: Washington, D.C., U.S. Govern-ment Printing Office, p. 505-531.

Miller, K. G., Fairbanks, R. G., and Mountain, G. S., 1987, Tertiary oxygen isotope synthesis, sea level history, and continental margin erosion: Paleoceanography, v. 2, p. 1-19.

Miller, K. G., Berggren, W. A., Zhang, J., and Palmer-Julson, A. A., 1991, Biostratigraphy and isotope stratigraphy of late Eocene microtektites at Site 612: How many impacts?: Palaios, v. 6, p. 17-38.

Montanari, A., Asaro, F., Michel, H. V., and Kennett, J. P., 1993, Iridium anomalies of late Eocene age at Massignano (Italy) and ODP Site 689B (Maud Rise, Antarctic): Palaios, v. 8, 420-437.

Norris, R. D., and Röhl, U., 1999, Carbon cycling and chronology of climate warming during the Paleocene/Eocene transition: Nature, v. 401, p. 775-778.

Obradovich, J., Snee, L. W., and Izett, G. A., 1989, Is there more than one glassy impact layer in the late Eocene?: Geological Society of America, Abstracts with Programs, v. 21, p. 134.

Odin, G. S., and Montanari, A., 1989, Age radiométrique et stratotype de la limite Éocène Oligocène: C.R. Acadamie Sciences Paris, v. 309, p. 1039-1945.

Pierrard, O., Robin, E., Rocchia, R., and Montanari, A., 1998, Extraterrestrial Ni-rich spinel in late Eocene sediments from Massignano: Geology, v. 26, p. 307-310.

Pierrard, O., Robin, E., Rocchia, R., Lefevre, I., Smit, J., and Vonhof, H., 1999, Late Eocene Ni-rich spinel from LL44-GPC3 (Central Pacific), ODP Site 689B (Maud Rise, Antarctic), DSDP Sites 94 (Gulf of Mexico) and 612 (U.S. East Coast): Lunar and Planetary Science XXX, Abstract # 1674, Lunar and Planetary Institute, Houston.

Poag, C. W., 1997a, The Chesapeake Bay bolide impact: a convulsive event in Atlantic Coastal Plain evolution: Sedimentary Geology Special Issue, v. 108, p. 45-90.

Poag, C. W., 1997b, Roadblocks on the kill curve: testing the Raup hypothesis: Palaios, v. 12, p. 582-590.

Poag, C. W., and Aubry, M.-P., 1995, Upper Eocene impactites of the U.S. East Coast: Depositional origins, biostratigraphic framework, and correlation: Palaios, v. 10, p.

Poag, C. W., and Poppe, L. J., 1998, The Toms Canyon structure, New Jersey outer continental shelf: A possible late Eocene impact crater: Marine Geology, v. 145, p. 23-60.

Poag, C. W., Powars, D. S., Poppe, L. J., and Mixon, R. B., 1994, Meteoroid mayhem in Ole Virginny: Source of the North American tektite strewn field: Geology, v. 22, p. 691-694.

Prothero, D. R., 1994, The Eocene-Oligocene transition: paradise lost: New York, Columbia University Press, 291 p.

Prothero, D. R., and Berggren, W. A., eds., 1992, Eocene-Oligocene climatic and biotic evolution: Princeton, New Jersey, Princeton University Press, 568 p.

Prothero, D. R., and Emry, R. J., 1996, The Terrestrial Eocene-Oligocene transition in North America: Cambridge, Cambridge University Press, 688 p.

Prothero, D. R., and Heaton, T. H., 1996, Faunal stability during the early Oligocene climatic crash: Palaeogeography, Palaeoclimatology, Palaeoecology, v. 127, p. 257-283.

Sanfilippo, A., Riedel, W. R., Glass, B. P., and Kyte, F. T., 1985, Late Eocene microtektites and radiolarian extinctions on Barbados: Nature, v. 314, p. 613-615.

Saunders, J. B., Bernoulli, D., Müller-Merz, E., Oberhänsli, H., Perch-Nielsen, K., Riedel, W. R., Sanfilippo, A., and Torrini, R., Jr., 1984, Stratigraphy of the late middle Eocene to early Oligocene in the Bath Cliff section, Barbados, West Indies: Micropaleontology, v. 30, p. 390-425.

Shackleton, N. J., 1986, Paleogene stable isotope events: Palaeogeography, Palaeoclimatology, and Palaeoecology, v. 57, p. 91-102.

Snyder, S. W., Miller, K. G., and Saperson, E., 1993, Paleogene and Neogene planktonic foraminiferal biostratigraphy of the New Jersey continental slope: Sites 902, 903, and 904 (Leg 150), *in* Mountain, G. S., Miller, K. G., Blum, P., Poag, C. W., and Twichell, D. C., eds., Proceedings of the Ocean Drilling Program, Scientific Results, Volume 150: College Station, Texas, Ocean Drilling Program, p. 3-15.

Speiss, V., 1990, Cenozoic magnetostratigraphy of Leg 113 drill sites, Maud Rise, Weddell Sea, Antarctica, *in* Barker, P. F., Kennett, J. P., et al., Proceedings of the Ocean Drilling Program, Scientific Results, Volume 113: College Station, Texas, Ocean Drilling Program, p. 261-318.

Steckler, M. S., Mountain, G. S., Miller, K. G., and Christie-Blick, N., 1999, Reconstruction of Tertiary progradation and clinoform development on the New Jersey passive margin by 2-D backstripping: Marine Geology, v. 154, p. 399-420.

Stott, L. D., and Kennett, J. P., 1990, Antarctic Paleogene planktonic foraminifer biostratigraphy: ODP Leg 113, Sites 689 and 690, *in* Barker, P. F., Kennett, J. P., et al., Proceedings of the Ocean Drilling Program, Scientific Results, Volume 113: College Station, Texas, Ocean Drilling Program, p. 549-569.

Thomas, E., 1992, Middle Eocene–late Oligocene bathyal benthic foraminifera (Weddell Sea): faunal changes and implications for ocean circulation, *in* Prothero, D. R., and Berggren, W. A., eds., Eocene-Oligocene climatic and biotic evolution: Princeton, New Jersey, Princeton University Press, p. 245-266.

Thomas, E., and Shackleton, N. J., 1996, The Paleocene-Eocene benthic foraminiferal extinction and stable isotope anomalies, *in* Knox, R. W., Corfield, R. M., and Dunay, R. E., eds., Correlation of the early Paleogene in northwest Europe: Geological Society Special Publications 101, p. 401-441.

Toon, O. B., Zahnle, K., Turco, R. P., and Covey, C., 1994, Environmental perturbations caused by asteroid impacts: *in* Gehrels, T., ed., Hazards Due to Comets and Asteroids: Tucson, University of Arizona Press, p. 791-826.

von der Borch, C. C., Sclater, J. G., et. al., 1974, Initial Reports of the Deep Sea Drilling Project, Volume 22: Washington, D.C., U.S. Government Printing Office, 890 p.

Vonhof, H. B., and Smit, J., 1996, Late Eocene impact ejecta at DSDP Site 689B (Maud Rise, Antarctica): stretching up the strewn fields of the late Eocene impacts, *in* Drobne, K., Gorican, S., and Kotnik, B., eds.: The role of impact processes in the geological and biological evolution of Planet Earth: Ljubljana, Slovenia, p. 97.

Vonhof, H. B., and Smit, J., 1999, Late Eocene microkrystites and microtektites at Maud Rise (Ocean Drilling Project Hole 689B; Southern Ocean) suggest a global extension of the approximately 35.5 Ma Pacific impact ejecta strewn field: Meteoritics and Planetary Science, v. 34, p. 747-756.

Wei, W., and Wise, S. W., Jr., 1990, Biogeographic gradients of middle Eocene-Oligocene calcareous nannoplankton in the South Atlantic Ocean: Palaeogeography, Palaeoclimatology, Palaeoecology, v. 79, p. 26-61.

Whitehead, J., Spray, J. G., Grieve, R. A. F., Papanastassiou, D. A., Ngo, H. H., and Wasserburg, G. J., 2000, Rb-Sr and Sm-Nd of upper Eocene microtektites: a potential Popigai source: Lunar and Planetary Science XXXI, CD-ROM, Abstract # 1373.

Wolfe, J. A., 1978, A paleobotanical interpretation of Tertiary climates in the Northern Hemisphere: American Scientist, v. 66, p. 694-703.

Wolfe, J. A., 1992, Climatic, floristic, and vegetational changes near the Eocene/Oligocene boundary in North America, *in* Prothero, D. R., and Berggren, W. A., eds., Eocene-Oligocene climatic and biotic evolution: Princeton, New Jersey, Princeton University Press, p. 421-436.

Wolfe, J. A., 1994, Tertiary climatic changes at middle latitudes of western North America: Palaeogeography, Palaeoclimatology, Palaeoecology, v. 108, p. 195-205.

Wolfe, J. A., and Poore, R. Z., 1982, Tertiary marine and nonmarine climatic trends, *in* Berger, W., and Crowell, J. C., eds., Climate in earth history: Washington, D.C., National Academy of Science, p. 154-158.

Zachos, J. C., Lohmann, K. C, Walker, J. C. G., and Wise, S. W., 1993, Abrupt climate change and transient climates during the Paleogene: a marine perspective: Journal of Geology, v. 101, p. 191-213.

CHAPTER 29

Boundaries, Turnover, and the Causes of Evolutionary Change: A Perspective from the Cenozoic

Warren D. Allmon

ABSTRACT

Biostratigraphic boundaries have been fundamental to stratigraphic geology since its founding, but their biological interpretation has lagged. After taking into account the effects of preservation, such boundaries, particularly those marked by coincident appearance and disappearance of taxa, suggest that evolutionary origination and extinction are not just coincidentally associated, but are dynamically connected as evolutionary processes. The incompleteness of the geological record will always limit our resolution on such questions, but we may nevertheless be able to learn about evolution at these boundaries if we examine faunal turnover using explicit evolutionary models as hypotheses for testing. Specifically, we may be able to understand more about the relationships among environmental change, speciation, and extinction. Environmental change may dynamically link the processes of speciation and extinction through the effects of disturbance, which at intermediate levels can stimulate speciation but at higher levels can increase extinction.

Boundaries between epochs of the Cenozoic (such as the Paleocene/Eocene, Eocene/Oligocene, and Pliocene/Pleistocene) are frequently marked by high rates of apparently roughly coincident appearance and disappearance of many marine taxa, especially mollusks, and may therefore offer good cases for such analysis. These were clearly times of accelerated environmental change in at least parts of the marine realm, and also perhaps accelerated origination and extinction in several higher taxa of marine organisms.

Such insights may be of particular relevance to understanding the causes of diversity patterns, reaction of the biota to massive environmental change, recovery from mass extinction, and putative examples of "coordinated stasis."

INTRODUCTION

Biologically defined boundaries—times of appearance and/or disappearance of taxa—have been fundamental to stratigraphic paleontology and geology for almost 200 years (Berry, 1968; Hancock, 1977; Schoch, 1989). Defining such boundaries has traditionally been one of the principal tasks of paleontology—it is crucial to the establishment of all types of biostratigraphic units or zones (Hedberg, 1976)—and recognition of boundaries remains an extremely important element of integrated approaches to chronostratigraphy today (Remane et al., 1996).

Only relatively recently, however, have paleontologists devoted comparable attention to exploring what may be happening biologically and/or geologically at these boundaries. That is, why exactly do taxa appear or disappear? It has long been acknowledged that biostratigraphic boundaries may be created by both the evolutionary processes of origination and extinction as well as nonevolutionary phenomena such as changes in facies and degree of preservation (e.g., Neaverson, 1955; Ager, 1973). Thus any observed pattern of first and last appearances in the stratigraphic record may potentially be amenable to a variety of causal interpretations. If we can by whatever means control for the nature of the stratigraphic record, and interpret at least part of the observed patterns of appearance and disappearance as representative of the actual processes of evolutionary origination (speciation) and extinction, then biostratigraphic

Table 29.1. Relationship Between Observed Patterns of First Appearance Datum (FAD) and Last Appearance Datum (LAD), with Possible Explanations

Explanation/Observation	FAD	LAD	FAD + LAD
Hiatus/incomplete preservation	Record incomplete	Record incomplete	Record incomplete
Environmental change	Immigration	Local extinction	"Turnover"
Evolutionary change	Radiation	Global extinction	"Turnover"

boundaries represent potentially rich sources of insight on how these processes work.

Considerable recent work has focused on boundaries marked by abundant last appearances and their interpretation as episodes of mass extinction (e.g., Ryder et al., 1996; Sepkoski, 1996). Attention has also focused on intervals of abundant first appearances and their interpretation as evolutionary radiations (e.g., Erwin, 1992). Comparatively less biological attention has been given to boundaries marked by numerous first and last appearances—that is, times that may have witnessed abundant origination roughly coincidental with abundant extinction. Such boundaries may be common in the geological record. The papers in this volume indicate that the Eocene/Oligocene boundary may have been such a time.

These episodes are often referred to as "turnovers," but the exact evolutionary mechanisms underlying them have remained obscure. Stanley (1986, 1990) has considered potential causal connections between speciation and extinction. Vrba's (1993, 1995) "turnover pulse hypothesis" specifically addresses the connection between environmental change and faunal turnover. The pattern of "coordinated stasis" (Brett et al., 1996), in which groups of coexisting species display coincident stability over extended intervals separated by episodes of relatively rapid change, implies a coincidence of origination and extinction at boundaries between blocks of faunal equilibrium (Ivany, 1996a).

Exactly or even approximately how common such coincident appearance and disappearance of taxa at biostratigraphic boundaries is remains an open question. The growing literature on coordinated stasis suggests that although coincidence does occur at some boundaries, boundary events are frequently marked by either appearances or disappearances, but not both (e.g., Patzkowsky and Holland, 1997, 1999). More studies will be required to adequately document relative frequency of these alternatives (Ivany, 1996a, 1999).

This paper explores the implications of such patterns of coincident appearance and disappearance for our understanding of the evolutionary process. If we can have confidence that a series of first and last appearances does represent actual evolutionary originations and extinctions, what does this mean? What does it suggest for our understanding of how evolution works that, in at least some cases, origination and extinction are closely linked in time and space?

NATURE OF THE DATA

Biostratigraphic data consist essentially of "first appearance datums" (FADs) and "last appearance datums" (LADs) (Van Couvering and Berggren, 1977; Schoch, 1989:201ff). Biozonal boundaries can be defined by either FADs or LADs, but in practice many formal zonal designations are based on definition of the lower boundary by one or more FADs (Schoch, 1989; Blatt et al., 1991). Clustered or coincident FADs and LADs may be used to define biostratigraphic units such as "concurrent range zones" and "Oppel zones" (Hedberg, 1976). Boundaries marked by large numbers of FADs and LADs may be called "turnover events" or "bioevents" (e.g., Walliser, 1986, 1996). When the faunas of entire formations or zones are considered, the number of FADs and LADs at such boundaries can be large (e.g., Dockery, 1986).

The occurrence of FADs and LADs is determined by a number of factors, chief among them the nature and completeness of the fossil record, but also including how data from the record are accumulated. Consideration of all of these factors is necessary in order to make valid conclusions about the causes for observed patterns (table 29.1).

Causes of Patterns of Appearance

1. Completeness of the record. The factors that influence our knowledge of and confidence in the fossil record have been well reviewed (e.g., McKinney, 1991; Foote, 1996, 1997; Donovan and Paul, 1998; Foote and Sepkoski, 1999; Holland, 2000). FADs are almost certainly later than the actual first appearance of a taxon, and LADs are almost certainly earlier than the actual last occurrence; the question is, by how much? The occurrence of LADs and FADs adjacent to unconformities may be particularly problematic. If FADs or LADs are coincident with a local or regional unconformity, then the actual timing and location of the true first and last appearances may be highly uncertain (Haq and Worsley, 1982; Schoch, 1989; McKinney and Allmon, 1995; Holland, 1996, 2000). Such "pileups" of FADs and LADs may indeed be taken under some circumstances to be indicative of otherwise unrecognizable unconformities (Schoch, 1989, p. 204). It should be noted, however, that the environmental changes that often accompany unconformities may themselves have been related to the inferred evolutionary changes (Patzkowsky and Holland, 1993; McKinney and Allmon, 1995; Holland, 1996).

2. Facies changes. When a biostratigraphic boundary is marked by a significant change in facies, then the likelihood increases that the pattern of first and last appearances is due as much to environmental as to evolutionary change (Jablonski, 1980; Holland, 1996). Comparing turnover between two adjacent formations, for example, that represent very different facies may simply be recording a change of environment, such as a marine transgression, rather than any particular pattern of origination and extinction. Multiple outcrop studies that cut across contemporary paleoenvironmental gradients can lessen the effects of such biases (McKinney and Allmon, 1995; Holland, 1996).

3. Monographic effects. Our knowledge of the fossil record is uneven not only because of physical incompleteness of preservation, but also because of differing levels of scientific attention devoted to different horizons and faunas; some boundaries and some faunas have been much better studied than others. Faunal monographs are not usually done in stratigraphic order, and this may leave crucial gaps in knowledge of a series of formations. Even when all relevant faunas have been examined, this may not have been done recently, and modern workers may be at least partly dependent on older taxonomic monographs, some of which may be better than others. The massive work of Woodring (1957–1982) on the Neogene of Panama, for example, appears to be very incomplete (Jackson et al., 1993, 1999), whereas the enormous catalog of Palmer and Brann (1965–1966) on the Paleocene and Eocene of the Gulf and Atlantic Coastal Plains appears to be much more adequate in its coverage (Dockery and Hansen, 1987, 1988), perhaps because it represents essentially a meta-analysis of many monographs, combined with critical judgment and additions by one of the leading monographers involved. Other factors being equal, it is also likely that tropical faunas, being larger, will almost always be less well known than higher latitude faunas (e.g., Jackson et al., 1999).

There are also biases that may affect our understanding before data are even published. Within faunas, more attention may be paid to larger or more attractive taxa; this "ugly fossil syndrome" (Tang, 2000) may lead to certain taxa being consistently understudied. Even if taxa or faunas are studied in detail, the results may never be published (Teichert et al., 1987).

Monographic effects can be minimized in a number of ways, including restudying and, ideally, resampling classic localities and faunas, and giving more attention to analysis of sampling (in)adequacy and to obtaining more adequate samples (e.g., CoBabe and Allmon, 1994; Jackson et al., 1999). Perhaps the most important point is to approach published data critically but not overly so; it is not all guaranteed to be comprehensive, but neither is it all guaranteed to be unusable.

4. Museum collections. Museum collections are important sources of data on stratigraphic distributions of fossil taxa (White and Allmon, 2000), but they should be used carefully. Unstudied museum collections may be the only available sources of information to address some problems created by monographic effects (above). Museum collections representing large, bulk, or stratigraphic collections may be particularly valuable in this regard. In poorly collected areas, or in sparsely fossiliferous units, on the other hand, museum collections may represent only the larger or more conspicuous elements of a fauna (e.g., Jackson et al., 1999), and be underrepresentative of rare or small taxa. Large faunas, in particular, must be sampled repeatedly in order to adequately sample rare taxa (Valentine, 1989; CoBabe and Allmon, 1994; Jackson et al., 1999), and this may not have taken place in the reconnaissance/exploration style in which many museum collections were made.

5. Phylogenetic analysis. Careful phylogenetic

analysis of a clade can reveal the putative existence of lineages that are not preserved in the fossil record. These "ghost lineages"—taxa that must have been present during a particular stratigraphic interval based on a cladistic analysis but that are not known from fossils—can have a significant effect on hypotheses of temporal pattern of first and last appearance (e.g., Norell, 1992; MacFadden, 1992; Schneider, 1998). The vast majority of clades in the fossil record have not been analyzed phylogenetically, a situation unlikely to change in the near future, and it remains an open question to what degree this shortcoming renders invalid conclusions about evolutionary originations.

Interpretations of Patterns of Appearance

Assuming some control for the geological quality of the record, first appearances (FADs) in the record may represent genuine evolutionary appearances or immigration events. Schoch (1989:201) suggests that the "tacit assumption, barring any evidence to the contrary, is generally that a FAD represents the immigration of a species shortly after its origination (evolution) in some unspecified area." Depending on the completeness of the record, a FAD may come shortly or long after and geographically close to or far removed from the speciation event that produced the taxon. Last appearances (LADs) are generally interpreted as indicative of local extinction and may or may not approximate the global extinction of the taxon (e.g., Signor and Lipps, 1982).

The interpretation of FADs and LADs is really a matter of estimating the relative order of events with some confidence interval that depends in large part on temporal or stratigraphic resolution. FADs represent latest possible times for the origin of a taxon; LADs represent earliest possible times for the extinction of a taxon. We judge the "actual" relative sequence of FADs and LADs by the confidence we have that they match the real first and last appearance, respectively. If the resolution is lower than the temporal gap between two events, we will not be able to resolve them and they will appear simultaneous. Some records will be much better for examining these patterns than others (e.g., Marshall, 1997).

EVOLUTIONARY CONNECTIONS BETWEEN FIRST AND LAST APPEARANCES

If it can be determined with some confidence that a pattern of first and last appearances is due mostly to patterns of evolutionary origination and extinction, what does this mean? To explore the relationship between speciation and extinction, it is helpful to be as explicit as possible about how at least some instances of speciation actually come about. Every occurrence of allopatric speciation can be seen as occurring in three stages (Allmon, 1992): (1) an isolated population must form—that is, become separated from the parental population; (2) it must persist long enough to differentiate genetically into a new species; and (3) it must actually undergo that differentiation (figure 29.1). (Allopatric speciation is clearly not the only mode of speciation, but it is arguably the dominant mode in most animal groups that form the bulk of the fossil record [e.g., Futuyma and Mayer, 1980; Berlocher, 1998].)

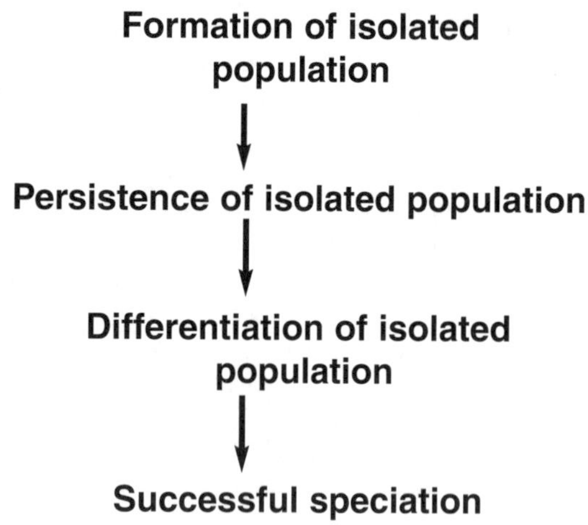

Figure 29.1. A three-stage framework for examining the process of allopatric speciation (from Allmon, 1992).

This three-stage framework suggests one or more explicit connections between the environment and the speciation process, and perhaps also between speciation and extinction. Specifically, it suggests that environmental disturbance may, at lower intensities, lead to increased speciation by increasing the rate of isolate formation, whereas at higher intensities it may lead to decreased speciation and increased extinction by decreasing the rate of isolate persistence; thus, for a given clade and a given type of disturbance, speciation should be most frequent at an intermediate level of that disturbance (McKinney and Allmon, 1995; Allmon et al., 1998) (figure 29.2).

In order to examine what might occur at a stratigraphic boundary marked by environmental change, origination, and extinction, we may add a second clade to the analysis (figure 29.3). If (as seems likely)

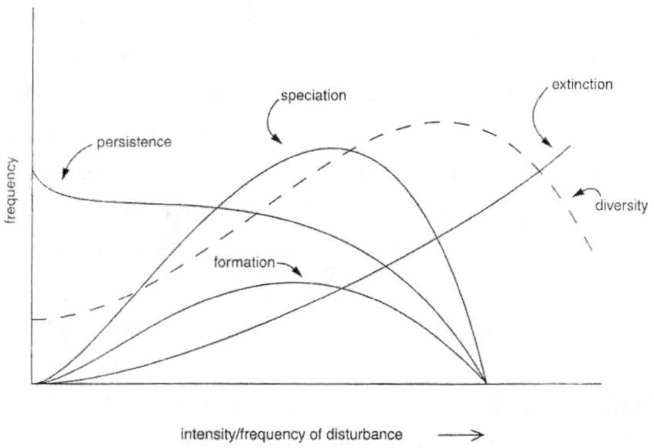

Figure 29.2. Qualitative relationships among two of the three processes (isolate formation and isolate persistence) leading to speciation in a single clade and intensity/frequency of environmental disturbance, together with successful speciation, extinction of established species, and total net speciesrichness or diversity (from Allmon, 1992).

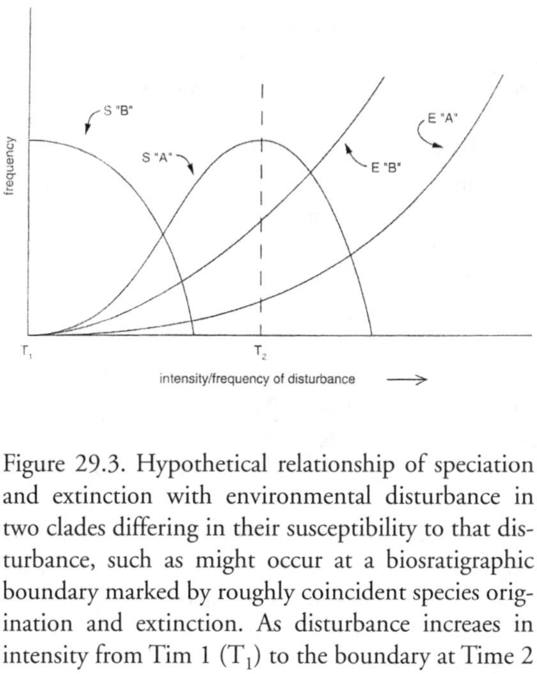

Figure 29.3. Hypothetical relationship of speciation and extinction with environmental disturbance in two clades differing in their susceptibility to that disturbance, such as might occur at a biosratigraphic boundary marked by roughly coincident species origination and extinction. As disturbance increaes in intensity from Tim 1 (T_1) to the boundary at Time 2 (T_2), speciation rate (S) quickly decreases in the more susceptible clade "B," while rising in clade "A." Extinction rate (E) rises more quickly in clade "B" than in clade "A." At the T_2 boundary, clade "B" will have suffered greater extinction, while clade "A" experienced greater speciation, and an episode of "turnover" results.

the two clades differ in their sensitivity and reaction to environmental change, they may pass differentially through this continuum of disturbance. One clade begins to experience the effects of disturbance via increased rates of isolate formation. As the disturbance increases, rate of origination declines as isolates have a difficult time persisting long enough to differentiate genetically. But the isolates that have differentiated and are now successful species survive and are now the newly originated diversity. As disturbance increases, at least to a point, they do not suffer enhanced extinction rates. The second clade, however, is more susceptible to the environmental changes, either because its isolates do not persist long enough to form successful new species or because it has fewer isolates and/or a smaller overall geographic range (McKinney and Allmon, 1995). It experiences only increased extinction. In this model, as the two clades pass through an interval of environmental change, origination in one group will precede extinction in another. At the end, we will call it "turnover."

Other models of faunal change with very different predictions are, of course, possible. The most obvious is what may be called the "extinction-replacement model," in which environmental disturbance causes extinction in one clade, and the lost diversity is subsequently replaced by origination (or immigration) in another clade (e.g., Harries et al., 1996). The fact that such models have alternative patterns of temporal ordering of events suggests that it may be possible to test among them, given records of appropriate stratigraphic resolution.

EXAMPLES FROM THE CENOZOIC

Paleogene of the Coastal Plain

Paleogene mollusk faunas of the U.S. Gulf and Atlantic Coastal Plains have long been known to exhibit a "stepped" or "punctuated" pattern of origination and extinction; blocks of relative faunal stability are interspersed with intervals in which as much as 90% of the fauna turns over, that is, disappears and is replaced (Dockery, 1986). Although all of these turnover events coincide with major regional unconformities, it is clear that a large proportion (if not the majority) of the taxa involved do show actual origination and extinction approximately at these times (McKinney and Allmon, 1995). The exact causes of these turnover events have remained controversial, but they are assumed to be correlated with significant environmental change. Dockery (1986), for example, has argued that sea level changes are responsible, while Hansen (1987, 1992) has argued for temperature changes (cooling) or extraterrestrial impact as the primary cause (see also Dockery and Hansen, 1987, 1988).

In a global context, changes at the Paleocene/Eocene boundary (between the Wilcox and Claiborne Groups in the Coastal Plain; Dockery, 1998) might be associated with the events of global warming and decrease in oceanic primary productivity (or at least an expansion of oligotrophic habitat) at this time (Rea et al., 1990; Berggren et al., 1998; Boersma et al., 1998). Changes at the Eocene/Oligocene boundary (between the Jackson and Vicksburg Groups; Dockery and Lozouet, this volume) may have been associated with abrupt global cooling and increase in productivity.

A common feature in the putative environmental changes at these Paleogene boundaries is disturbance. It therefore seems reasonable to examine evolutionary mechanisms within this context. Disturbances—whether caused by change in water temperature, nutrient levels, or sea level—could have led to both origination and extinction at around the same time via the model described above. In order to test this idea further, more data are needed on (1) the details of environmental change at stage/group boundaries in the Coastal Plain, and (2) environmental tolerances of the taxa that show FADs and LADs at each boundary. Testable hypotheses can then be set up and examined critically.

For example, it is highly suggestive that at the Paleocene/Eocene boundary in the Gulf Coastal Plain, essentially all of the large turritellid gastropod species become extinct (Allmon, 1996), whereas at the Eocene/Oligocene boundary they show an expansion of diversity (MacNeil and Dockery, 1984). Turritellids are typically suspension feeders and might be expected to show enhanced extinction when primary productivity declined and enhanced speciation when it increased. This pattern contrasts, however, with that shown by other clades in the region; venericard bivalves, for example, show both extinction and origination at around the Paleocene/Eocene boundary (Gardner and Bowles, 1939). What is different about the biology of these two clades that might account for what appears to be a differential reaction to environmental change?

Although the apparent coincidence of environmental and faunal change at these boundaries points to a connection between the two, a broader analysis of the patterns of environmental and faunal changes throughout the Paleogene of the Coastal Plain suggests that the relationship may be complex. Ivany and Morris (Morris and Ivany, 1994; Ivany, 1996b; Morris, unpublished data) compared patterns of molluscan faunal similarity among formations in the Paleocene and Eocene of the Gulf Coastal Plain to the most detailed temperature and sea-level records available. They found no overall correlation between change in sea level and change in the mollusk fauna, and an ambiguous relationship between change in temperature and faunal change (there was a significant correlation with magnitude but not with sign of temperature change). Major episodes of faunal turnover (such as the Paleocene/Eocene and Eocene/Oligocene boundaries) do coincide with major environmental changes, but similar environmental changes occur at other times when there is little or no faunal change. Such a nonlinear relationship between environmental and faunal change does not mean that they are not related; it suggests that this relationship is complex and likely to involve threshold effects and other elements of nonlinear dynamics (Ivany, 1996a,b; see also Morris et al., 1995).

Pliocene of Florida and the Caribbean

One of the larger known regional turnover events occurred among marine faunas in the western Atlantic in the late Pliocene (Allmon, 2001). At around 2.5 Ma, 25–50% of the bivalves and 50–70% of the gastropods of southern and central Florida turned over with little change in total diversity (Allmon et al., 1996a). In Costa Rica and Panama around 2.0 Ma, around 50% of the mollusk genera and subgenera turned over (Jackson et al., 1993), also with little change in diversity. The precise timing of origination vs. extinction has not been determined for the vast majority of these mollusk taxa.

Among hermatypic corals in the Caribbean (Budd and Johnson, 1999), origination peaked between 3 and 4 Ma, mostly among living species; extinction peaked between 1 and 2 Ma, mostly involving species that had originated prior to 5 Ma. The extinction was further selective in that species with larger colonies and massive or platy shapes were more likely to survive than smaller, free-living, or branching colonies. "Species with small massive colonies tend to live in small, short-lived populations with highly fluctuating recruitment rates" (Johnson et al., 1995)—species with small massive colonies originated at higher rates prior to turnover.

The causes of these faunal changes have also been controversial (see Allmon, 2001; Allmon et al., 1996a,b). It is reasonable, however, that environmental disturbance, probably associated with a decline in nutrients, and so in marine productivity, associated with the oceanographic consequences of the closure of the Central American Seaway (CAS) by formation of the Central American Isthmus (CAI), was largely responsible (Allmon, 2001; Allmon et al., 1996b). A causal scenario much like the following is consistent with the available evidence:

Beginning around 5.0 Ma and culminating between 3.0 and 2.0 Ma, oceanic circulation changed across much of the western Atlantic, primarily in response to the formation of the CAI and closure of the CAS; these changes in circulation led to a decline in marine nutrients and primary productivity across much of the region. This led initially to an increase in formation of isolated populations due either to the disruption and decrease in food supplies or to the instability of food and nutrient conditions. The latter situation may have been particularly important for zooxanthellate corals, which are especially sensitive to changes in nutrient conditions. This increase in isolate formation led to an increase in rate of speciation. Disturbance, in the form of continued changes in nutrients and productivity conditions, continued to increase, and this led eventually to a decrease in the rate of persistence of isolated populations, a decrease in species origination, and an increase in rate of species extinction. For some taxa, extinction continued to increase, eventually exceeding origination.

DISCUSSION

Patterns of first and last appearances in the stratigraphic record should not be taken at face value, but neither should they be ignored. Examining boundaries characterized by coincident or approximately coincident FADs and LADs within an explicit framework for the mechanisms of extinction and speciation may allow us to push the available fossil record farther as a source of insight on the relationship between faunal and environmental change. To the degree that "coordinated stasis" is a general pattern (Brett et al., 1996; but see Miller, 1997; Patzkowsky and Holland, 1997), then such boundaries may well prove to be loci of a large proportion of evolutionary change, and deserving of detailed mechanistic analysis.

Viewed in this context, the Cenozoic examples cited here point to more refined research questions than have usually been asked of these records before, such as (1) Why do populations of certain taxa (those showing high extinction) show less resistance to environmental change than other taxa (those showing high origination)? (2) Are faunas responding to the pace of change in the environment, or do threshold effects play a role (Ivany, 1996a,b)? (3) Are environmental changes themselves characterized by threshold effects (Ivany, 1996a,b)? The nonlinear relationship between environmental and faunal change observed in the Coastal Plain Paleogene (Morris and Ivany, 1994; Ivany, 1996b) suggests that it is the faunas themselves displaying threshold effects.

But faunas do not speciate; individual species do. A nonlinear response of a fauna to a particular environmental change must involve responses of the constituent species of that fauna. In neither the Paleogene nor the Neogene of the western Atlantic do we currently have adequately detailed understanding of the environmental tolerances of most of the mollusk taxa involved in these turnover events. Such understanding will be required to advance this analysis. Knowledge of corals in the Plio-Pleistocene is more developed in this regard, and recent analyses are consistent with the model proposed here (e.g., Johnson et al., 1995; Budd and Johnson, 1999).

Perhaps most important, turnover events at boundaries marked by significant environmental change should not be seen as either "black boxes," into and out of which lineages simply pass. Just correlating environmental change with faunal change does not really tell us much about the evolutionary process. The papers in this volume reveal how much better our knowledge of patterns of environmental and biotic change in the Cenozoic is becoming. We should now match this with similar improvements in the specificity of the processes behind these patterns.

CONCLUSIONS

These examples suggest that we should look at these boundaries as events of environmental change that different taxa respond to differently—specifically in their propensity to form, differentiate, and preserve isolated populations and emerging new species. This point of view can set us off toward investigating specific hypotheses of environmental change and biotic response in individual species or clades.

As we examine boundaries marked by coincident extinction and origination, we often say "turnover" and leave it at that. We are often correctly cautious about the completeness of the stratigraphic and paleontological record at such boundaries and reasonably conclude that we can say nothing of biological significance. We should at least keep in our minds as we examine such boundaries, however, the possibility that there are true signals in the record about the connection between environmental and evolutionary change. Applying explicit models of the process of speciation may not be fully possible in all such cases, but it may alert us to signals and connections that we would otherwise miss.

ACKNOWLEDGMENTS

I am grateful to Linda Ivany, Liz Nesbitt, and Don Prothero for inviting me to participate in the Penrose Conference, and to Linda Ivany, Mike McKinney, and particularly Mark Patzkowsky for comments on an earlier draft of the manuscript.

REFERENCES

Ager, D. V., 1973, The nature of the stratigraphical record: London, Macmillan, 114 p.

Allmon, W. D., 1992, A causal analysis of the stages of allopatric speciation: Oxford Surveys in Evolu-tionary Biology, v. 8, p. 219-257.

Allmon, W. D., 1996, Evolution and systematics of Cenozoic American Turritellidae (Gastropoda). I. Paleocene and Eocene species related to "*Turritella mortoni* Conrad" and "*Turritella humerosa* Conrad" from the U.S. Gulf and Atlantic Coastal Plains: Palaeontographica Americana, no. 59, p. 1-134.

Allmon, W. D., 2001, Nutrients, temperature, disturbance, and evolution: a model for the late Cenozoic marine record of the Western Atlantic: Palaeogeography, Palaeoclimatology, Palaeoecology, v. 166, p. 9-26.

Allmon, W. D., Morris, P. J., and McKinney, M. L., 1998, An intermediate disturbance hypothesis of maximal speciation, *in* McKinney, M. L., and Drake, J. A., eds., Biodiversity dynamics: turnover of populations, taxa, and communities: New York, Columbia University Press, p. 349-376.

Allmon, W. D., Rosenberg, G., Portell, R. W., and Schindler, K., 1996a, Diversity of Pliocene-Recent mollusks in the western Atlantic: extinction, origination, and environmental change, *in* Jackson, J. B. C., Budd, A. F., and Coates, A. G., eds., Evolution and environment in tropical America: Chicago, University of Chicago Press, p. 271-302.

Allmon, W. D., Emslie, S. D., Jones, D. S., and Morgan, G. S., 1996b, Late Neogene oceanographic change along Florida's west coast: evidence and mechanisms: Journal of Geology, v 104, p. 143-162.

Berggren, W. A., Lucas, S., and Aubry, M.-P., 1998, Late Paleocene-early Eocene climatic and biotic evolution: an overview, *in* Aubry, M.-P., Lucas, S. G., and Berggren, W. A., eds., Late Paleocene-early Eocene climatic and biotic events in the marine and terrestrial records: New York, Columbia University Press, p. 1-17.

Berlocher, S. H., 1998, Can sympatric speciation via host or habitat shift be proven from phylogenetic and biogeographic evidence? *in* Howard, D. J., and Berlocher, S. H., eds., Endless forms: species and speciation: New York, Oxford University Press, p. 99-113.

Berry, W. B. N., 1968, Growth of a prehistoric time scale: San Francisco, Freeman, 158 p.

Blatt, H., Berry, W. B. N., and Brande, S., 1991, Principles of stratigraphic analysis: Boston, Blackwell Scientific Publications, 512 p.

Boersma, A., Premoli-Silva, I., and Hallock, P., 1998, Trophic models for the well-mixed and poorly mixed warm oceans across the Paleocene/Eocene epoch boundary, *in* Aubry, M.-P., Lucas, S. G., and Berggren, W. A., eds., Late Paleocene-early Eocene climatic and biotic events in the marine and terrestrial records: New York, Columbia University Press, p. 204-213.

Brett, C. E., Ivany, L. C., and Schopf, K .M., 1996, Coordinated stasis: an overview: Palaeogeography, Palaeoclimatology, Palaeoecology, v. 127, p. 1-20.

Budd, A. F., and Johnson, K. G., 1999, Origination preceding extinction during late Cenozoic turnover of Caribbean reefs: Paleobiology, v. 25, p. 188-200.

CoBabe, E. A., and Allmon, W. D., 1994, Effects of sampling on paleoecologic and taphonomic analyses in high-diversity fossil accumulations: an example from the Eocene Gosport Sand, Alabama: Lethaia, v. 27, p. 167-178.

Dockery, D. T., III, 1986, Punctuated succession of Paleogene mollusks in the northern Gulf Coastal Plain: Palaios, v. 1, p. 582-589.

Dockery, D. T., III, 1998, Molluscan faunas across the Paleocene/Eocene series boundary in the North American Gulf Coastal Plain, *in* Aubry, M.-P., Lucas, S. G., and Berggren, W. A., eds., Late Paleocene-early Eocene climatic and biotic events in the marine and terrestrial records: New York, Columbia University Press, p. 296-322.

Dockery, D. T., III, and Hansen, T. A., 1987, Eocene-Oligocene molluscan extinctions: Comment and reply: Palaios, v. 2, p. 620-622.

Dockery, D. T., III, and Hansen, T. A., 1988, Mollus-can extinction rates in question (Comment and reply): Nature, v. 331, p. 123.

Donovan, S. K., and Paul, C. R. C., eds., 1998, The adequacy of the fossil record: Chichester, John Wiley and Sons, 312 p.

Erwin, D. H., 1992, A preliminary classification of evolutionary radiations: Historical Biology, v. 6, p. 133-148.

Foote, M. J., 1996, On the probability of ancestors in the fossil record: Paleobiology, v. 22, p. 141-151.

Foote, M. J., 1997, Estimating taxonomic durations and preservation probability: Paleobiology, v. 23, p. 278-300.

Foote, M. J., and Sepkoski, J. J., Jr., 1999, Absolute measures of the completeness of the fossil record: Nature, v. 398, p. 415-417.

Futuyma, D. J., and Mayer, G. C., 1980, Non-allopatric speciation in animals: Systematic Zoology, v. 29, p. 254-271.

Gardner, J. A., and Bowles, E., 1939, The *Venericardia planicosta* group in the Gulf Province: U.S. Geological Survey Professional Paper 189-F, p. 141-215.

Hancock, J. M., 1977, The historic development of biostratigraphic correlation, *in* Kauffman, E. G., and Hazel, J. E., eds., Concepts and methods of biostratigraphy: Stroudsburg, Pennsylvania, Dowden, Hutchinson, and Ross, p. 3-22.

Hansen, T. A., 1987, Extinction of late Eocene to Oligocene molluscs: relationship to shelf area, temperature changes, and impact events: Palaios, v. 2, p. 69-75.

Hansen, T. A., 1992, The patterns and causes of molluscan extinction across the Eocene/Oligocene boundary, *in* Prothero, D. R., and Berggren, W. A., eds., Eocene-Oligocene climatic and biotic evolution: Princeton, New Jersey, Princeton University Press, p. 341-348.

Haq, B. U., and Worsley, T. R., 1982, Biochronology—biological events in time resolution, their potential and limitations, *in* Odin, G. S., ed., Numerical dating in stratigraphy: Chichester, John Wiley and Sons, p. 19-35.

Harries, P. J., Kauffman, E. G., and Hansen, T. A., 1996, Models for biotic survival following mass extinction, *in* Hart, M. B., ed., Biotic recovery from mass extinction events: London, The Geological Society, p. 41-60.

Hedberg, H. D., ed., 1976, International stratigraphic guide—a guide to stratigraphic classification, terminology and procedure: New York, John Wiley and Sons, 200 p.

Holland, S. M., 1996, Recognizing artifactually generated coordinated stasis: implications of numerical models and strategies for field tests: Palaeogeography, Palaeoclimatology, Palaeoecology, v. 127, p. 147-156.

Holland, S. M., 2000, The quality of the fossil record: a sequence stratigraphic perspective, *in* Erwin, D. H., and Wing, S. L., eds., Deep time. Paleobiology's perspective: Lawrence, Kansas, The Paleontological Society, p. 148-168.

Ivany, L. C., 1996a, Coordinated stasis or coordinated turnover? Exploring instrinsic vs. extrinsic controls on pattern: Palaeogeography, Palaeoclimatology, Palaeoecology, v. 127, p. 239-256.

Ivany, L. C., 1996b, Coordinated faunal stasis: a new twist on the biological record of environmental change [abs.]: EOS (Transactions, American Geophysical Union), v. 77, p. 403.

Ivany, L. C., 1999, So . . . Now what? Thoughts and ruminations about coordinated stasis: Palaios, v. 14, p. 297-300.

Jablonski, D., 1980, Apparent versus real biotic effects of transgressions and regressions: Paleobiology, v. 6, p. 397-407.

Jackson, J. B. C., Jung, P., Coates, A. G., and Collins, L. S., 1993, Diversity and extinction of tropical American mollusks and emergence of the Isthmus of Panama: Science, v. 260, p.1624-1626.

Jackson, J. B. C., Jung, P., and Fortunato, H., 1996, Paciphilia revisited: transisthmian evolution of the *Strombina* group (Gastropoda: Columbellidae), *in* Jackson, J. B. C., Budd, A. F., and Coates, A. G., eds., Evolution and environment in tropical America: Chicago, University of Chicago Press, p. 234-260.

Jackson, J. B. C., Todd, J. A., Fortunato, H., and Jung, P., 1999, Diversity and assemblages of Neogene Caribbean Mollusca of lower Central America, *in* Collins, L. S., and Coates, A. G., eds., A paleobiotic survey of Caribbean faunas from the Neogene of the Isthmus of Panama: Bulletins of American Paleontology, no. 357, p.193-230.

Johnson, K. G., Budd, A. F., and Stemann, T. A., 1995, Extinction selectivity and ecology of Neogene Caribbean reef corals: Paleobiology, v. 21, p. 52-73.

MacFadden, B. J., 1992, Interpreting extinctions from the fossil record: methods, assumptions, and case examples using horses (Family Equidae), *in* Novacek, M. J., and Wheeler, Q. D., eds., Extinction and phylogeny: New York, Columbia University Press, p. 17-45.

MacNeil, F. S., and Dockery, D. T., III, 1984, Lower Oligocene Gastropoda, Scaphopoda, and Cephalopoda of the Vicksburg Group in Mississippi: Mississippi Geological Survey Bulletin 124, 415 p.

Marshall, C. R., 1997, Confidence intervals on stratigraphic ranges with nonrandom distributions of fossil horizons: Paleobiology, v. 23, p. 165-173.

McKinney, M. L., 1991, Completeness of the fossil record: an overview, *in* Donovan, S.K., ed., The processes of fossilization: New York, Columbia University Press, p. 66-83.

McKinney, M. L., and Allmon, W. D., 1995, Metapopulations and disturbance: from patch dynamics to biodiversity dynamics, *in* Erwin, D. H., and Anstey, R. L., eds., New approaches to speciation in the fossil record: New York, Columbia University Press, p. 123-183.

Miller, A. I., 1997, Coordinated stasis or coincident relative stability? Paleobiology, v. 23, p. 155-164.

Morris, P. J., and Ivany, L. C., 1994, Biotic response to environmental change in light of ecological locking [abs.]: EOS (Transactions, American Geophysical Union), v. 75, p. 80.

Morris, P. J., Ivany, L. C., Schopf, K. M., and Brett, C. E., 1995, The challenge of paleoecological stasis: reassessing sources of evolutionary stability: Proceedings of the National Academy of Sciences of the USA, v. 92, p.11269-11273.

Neaverson, E., 1955, Stratigraphical palaeontology. A study of ancient life-provinces: Oxford, Clarendon Press, 806 p.

Norell, M. A., 1992, Taxic origin and temporal diversity: the effect of phylogeny, *in* Novacek, M. J., and Wheeler, Q. D., eds., Extinction and plylogeny: New York, Columbia University Press, p. 89-118.

Palmer, K. V. W., and Brann, D. C., 1965–66, Catalogue of the Paleocene and Eocene Mollusca of the southern and eastern United States: Bulletins of American Paleontology, v. 48, 1057 p.

Patzkowsky, M. E., and Holland, S. M., 1993, Biotic response to a Middle Ordovician paleoceanographic event in eastern North America: Geology, v. 21, p. 619-622.

Patzkowsky, M. E., and Holland, S. M., 1997, Patterns of turnover in Middle and Upper Ordovician brachiopods of the eastern United States: a test of coordinated stasis: Paleobiology, v. 23, p. 420-443.

Patzkowsky, M. E., and Holland, S. M., 1999, Biofacies replacement in a sequence stratigraphic framework: Middle and Upper Ordovician of the Nashville Dome, Tennessee, USA: Palaios, v. 14, p. 301-323.

Rea, D. K., Zachos, J. C., Owen, R. M., and Gingerich, P. D., 1990, Global change at the Paleocene/Eocene boundary: climatic and evolutionary consequences of tectonic events: Palaeogeography, Palaeoclimatology, Palaeoecology, v. 79, p. 117-128.

Remane, J., Bassett, M. G., Cowie, J. W., Gohrbandt, K. H., Lane, H. R., Michelsen, O., and Naiwen, W., 1996, Revised guidelines for the establishment of global chronostratigraphic standards by the International Commission on Stratigraphy (ICS): Episodes, v. 19, p. 77-81.

Ryder, G., Fastovsky, D., and Gartner, S., eds., 1996, The Cretaceous-Tertiary event and other catastrophes in Earth history: Geological Society of America Special Paper 307, 569 p.

Schneider, J. A., 1998, Phylogeny of stem-group Eucardiidae (Bivalvia: Cardiidae) and the significance of the transitional fossil *Perucardia*: Malacologia, v. 40, p. 37-62.

Schoch, R. M., 1989, Stratigraphy: principles and methods: New York, Van Nostrand Reinhold, 375 p.

Sepkoski, J. J., Jr., 1996, Patterns of Phanerozoic extinction: a perspective from global data bases, *in* Walliser, O. H., ed., Global bio-events and event-stratigraphy: Berlin, Springer-Verlag, p. 35-51.

Signor, P. W., III, and Lipps, J. H., 1982, Sampling bias, gradual extinction patterns, and catastrophes in the fossil record: Geological Society of America Special Paper 190, p. 291-296.

Stanley, S. M., 1986, Population size, extinction, and speciation: the fission effect in Neogene Bivalvia: Paleobiology, v. 12, p. 89-110.

Stanley, S. M., 1990, The general correlation between rate of speciation and rate of extinction: fortuitous causal linkages, *in* Ross, R. M., and Allmon, W. D., eds., Causes of evolution: apaleontological perspective: Chicago, University of Chicago Press, p. 103-127.

Tang, C., 2000, Ugly fossil syndrome: Palaios, v. 15, p 175-176.

Teichert, C., Sweet, W. C., and Boucot, A. J., 1987, The unpublished fossil record: Senckenbergiana Lethaea, v. 68, p.1-19.

Valentine, J. W., 1989, How good was the fossil record? Clues from the California Pleistocene: Paleobiology, v. 15, p. 83-94.

Van Couvering, J. A., and Berggren, W. A., 1977, Biostratigraphical basis of the Neogene time scale, *in* Kauffman, E. G., and Hazel, J. E., eds., Concepts and methods of biostratigraphy: Stroudsburg, Pennsylvania, Dowden, Hutchinson, and Ross, p. 283-306.

Vrba, E. S., 1993, Turnover-pulses, the Red Queen, and related topics: American Journal of Science, v. 293a, p. 418-452.

Vrba, E. S., 1995, On the connections between paleoclimate and evolution, *in* Vrba, E. S., Denton, G. H., Partridge, T. C., and Burckle, L. H., eds., Paleoclimate and evolution, with emphasis on human origins: New Haven, Yale University Press, p. 24-48.

Walliser, O. H., ed., 1986, Global bio-events. A critical approach: Berlin, Springer-Verlag, 442 p.

Walliser, O. H., ed., 1996, Global bio-events and event-stratigraphy: Berlin, Springer-Verlag, 333 p.

White, R. D., and Allmon, W. D., eds., 2000, Standards and guidelines for care and computerization of invertebrate paleontology collections: Paleontological Society Special Publications, v. 10, 260 p.

Woodring, W. P., 1957–1982, Geology and paleontology of Canal Zone and adjoining parts of Panama: U.S. Geological Survey Professional Paper 306, 759 p.

CHAPTER 30

The Marine Eocene-Oligocene Transition: A Synthesis

Linda C. Ivany, Elizabeth A. Nesbitt, and Donald R. Prothero

ABSTRACT

Progressive global cooling and increasing seasonality brought on by a combination of tectonic and oceanographic factors led to a series of worldwide pulsed turnovers through the middle Eocene and into the Oligocene. Real patterns of evolutionary turnover are often obscured in shelf settings by regional tectonism and facies changes, but, as seen in this volume, progress is being made at seeing the global pattern through the regional overprint. Most taxonomic groups show significant perturbations at or near the middle/late Eocene boundary and near or shortly after the Eocene/Oligocene boundary. Other events were noted in some groups in the late middle Eocene, within the late Eocene, and later into the Oligocene. Difficulties in establishing good age control and correlating widely separated sections limit our ability to determine the degree to which these events were isochronous, but the strong similarity of records from disparate regions, environments, and taxonomic groups suggests that these turnovers were global and hence triggered by changes occurring on a global basis, such as climate cooling. Extinction seems to have played a more important role in low-latitude turnovers, while origination was a significant component at high latitudes. Warm-water taxa were, without exception, hit harder than cool-water taxa, particularly at the middle/late Eocene boundary. While cosmopolitan taxa are more likely to survive turnovers, at high latitudes the reverse is true, and at the Eocene/Oligocene boundary itself taxa seem to have been affected indiscriminately.

INTRODUCTION

It has been well over a decade since the 1989 Penrose Conference on the Eocene-Oligocene transition was convened (Prothero and Berggren, 1992). Since that time, much new information has come to light, and work that was initiated as a result of that first conference has come to fruition. This volume, based on the 1999 Penrose Conference (Prothero et al., 2000), offers the most up-to-date views on the nature of the Eocene-Oligocene transition from climatic, oceanographic, and paleontologic perspectives. The majority of the papers presented here are chosen to represent the varied and ongoing work in Eocene-Oligocene marine *shelf* sections, from which much of our macrofossil record comes, and to draw comparisons between these data and those from open-ocean and terrestrial sources. Additional papers focus on deep-sea and coastal-lowland-forest environments. As editors, we intend that this selection provide a comprehensive look at the patterns and processes that characterize the time period from the middle Eocene through the Oligocene, spanning geographic, taxonomic, and methodological boundaries to allow for a global, multidisciplinary perspective.

The transition from global "greenhouse" conditions of the early and middle Eocene to global "icehouse" conditions of the early Oligocene marks a turning point in Cenozoic Earth history. The early Eocene world is characterized by palm trees growing in Alaska (Scotese, 2002) and the northern Rocky Mountains (Greenwood and Wing, 1995), crocodiles living on Ellesmere Island well above the Arctic Circle

(Markwick, 1994), forests blanketing much of Antarctica (Case, 1988), and a host of unusual archaic taxa scattered across the marine shelves and continents (papers in Prothero and Berggren, 1992; Prothero, 1994). By the start of the Oligocene, changes in ocean circulation brought about by the opening of the passage between Australia and Antarctica, and the Drake Passage between South America and Antarctica, had resulted in progressive global cooling (Zachos et al., 1994), the extinction of hundreds of warm-water taxa (Dockery, 1986; Haasl and Hansen, 1996), and the development of a continent-scale ice sheet on Antarctica (Zachos et al., 1992). Events associated with the Eocene-Oligocene transition in particular ushered in more modern climatic and biologic systems. The general oceanic and atmospheric circulation patterns of today were first established at this time. Groups that survived and radiated following the close of the Eocene are those that, for the most part, are still extant. During this interval, large shifts toward both warmer and cooler conditions occurred on both gradual and episodic time scales (Kennett and Stott 1991; Zachos et al., 1992; Zachos et al., 1993; Zachos et al., 1994; Lear et al., 2000; Ivany et al., 2000). Not surprisingly, the faunal and floral response to such changes was variable in both space and time, and some groups were more sensitive to change than others. Our goal here is to relate the latest information concerning climatic and oceanographic change associated with this transition, and to examine geographic and taxonomic patterns in biotic turnover that provide clues about where, when, and how fast these environmental changes are happening.

The papers in this volume are organized by geographic region. However, a number of recurring themes through the papers highlight the timing and magnitude of both climatic and biotic change. We felt it would be useful to synthesize that information in one place, and this summary chapter is structured in such a manner. We discuss, in sequence:

—The tectonic and paleogeographic setting of the Paleogene

—Specific issues related to the stratigraphy of shelf deposits, the topic of most of these studies

—Advances in recognizing and correlating boundary sections

—Trends in the expression of climate change

—Patterns of faunal and floral turnover, and

—A synthesis of pattern by latitude and environment

By necessity, all the papers offered in this volume are concerned with one or more of these issues, and they offer a range of potential solutions to the problems of comparing sections both in space and in time through the Eocene-Oligocene transition.

This chapter is not meant to be a comprehensive literature review, but a summary of the topics covered in this volume. Clearly, there are many seminal papers published earlier on many of these subjects, and the reader is referred to papers in this volume for the relevant literature reviews.

TECTONIC AND PALEOGEOGRAPHIC SETTING FOR THE EOCENE AND OLIGOCENE

Paleogeographic and paleoclimatic reconstructions (Scotese, 2002) portray the Eocene world with a wide Tethys Sea, which extended from the western Indian Ocean, through the Mediterranean, to between North and South America. During this interval, India collided with the south-central Asian coastline, coinciding with realignment of the Pacific midocean ridge system. South America, Australia and Antarctica were still connected as remnants of Gondwana, and North America and Europe were intermittently connected by the Bering land bridge. Thus both the southern and northern hemispheric oceanic circulations were significantly different from today's pattern. Eustatic sea levels were high (Vail et al., 1991), and the only circumglobal circulation occurred around the equator. Across-latitude circulation was severely restricted. A drop in global sea level around the Eocene/Oligocene boundary resulted in the Tethys becoming more restricted. But more significantly, Antarctica separated from both Australia and South America, generating the first circum-Antarctic current and the deep-water circulation originating from the Antarctic (Berggren and Prothero, 1992; Diester-Haass and Zahn, 1996; Exon et al., 2002).

In this volume, papers that focus on changes in seaway paleogeography and oceanic circulation patterns bear particular relevance to the Gulf of Mexico, Caribbean Sea, and Atlantic Ocean. Iturralde-Vinent (this volume) uses stratigraphic techniques to understand the changes in the Caribbean Seaway during the Eocene-Oligocene transition. He documents shifts from free exchange between Tethyan and Pacific faunas earlier in the Paleogene to more restricted exchange near the boundary, related to tectonism and uplift in the Caribbean. The biogeography of benthic foraminiferal assemblages allows Robinson (this volume) to add more data supporting the progressive restriction of circulation between the Gulf of Mexico and the Caribbean Sea during the Eocene. In addition, Florida was more insulated from the rest of the Gulf Coast by the northeast-flowing Suwannee Current, which prevented many Tethyan mollusks from crossing the gap between Florida and the central Gulf. Dockery and Lozouet (this volume) record the arrival of characteristic Tethyan forms in western Alabama and Mississippi near the Eocene/Oligocene boundary, which indicates breaching of that divide as sea level fell.

Another line of evidence for a rapid change in oceanic circulation is illustrated by Scholl et al. (this volume) from the depositional patterns of the deep-sea Meiji sediment body in the northernmost Pacific Ocean. The pelagic sediments of the Meiji body accumulated along the northern flank of the Emperor Seamounts chain beginning in Oligocene time. From its bathymetric setting, depositional form, and textures, Scholl et al. (this volume) conclude that the deposition of the Meiji body was initiated by changing oceanographic circulation in the early Oligocene that brought cold, dense, and salty abyssal waters into the North Pacific.

THE NATURE OF SHELF DEPOSITS

In comparison with records from the deep sea, shelf sections can be difficult to work with due to their time-transgressive nature and regional variation of sedimentation rates, and the lithologic heterogeneity and intense bioturbation of many deposits. Eustatic sea-level fluctuations and active tectonics result in substantial and frequently undetermined disconformities (Ager, 1993; Posamentier and Allen, 1993). Faunal correlation along shelves, and with the deep-sea record, is also complicated by provinciality and the very poor record of planktonic taxa in nearshore environments. The next step, interpreting apparent faunal turnover across temporal units and with continental deposits, is often fraught with difficulties and inconsistencies. Sequence boundaries offer consistent, traceable surfaces with which to correlate sections, but they may not be the same age everywhere, particularly in updip versus downdip sections. In addition, the amount of time missing at sequence boundaries varies from place to place depending on where on the shelf the section is located (Christie-Blick, 1991).

Condensed sections, which are isochronous everywhere on the contiguous shelf, offer one solution to the problem of correlating disparate shelf sequences (Loutit et al., 1988). In addition, biostratigraphically useful pelagic fossils are more frequently preserved in deeper-water condensed strata, adding to their value in correlation (e.g., Echols et al., this volume; Robinson, this volume). Terrestrial palynomorphs can also be preserved in marine shelf sediments (e.g., Oboh-Ikuenobe and Jaramillo, this volume; Yancey et al., this volume), offering a way to correlate terrestrial with marine records. Despite their limitations, therefore, shelf sediments can provide a valuable (and often the only) link between terrestrial and deep-sea sections.

Interpretation of apparent faunal turnover events can be complicated by correlative changes in lithology and hence paleoenvironment, making it unclear whether the turnover is real or simply facies-controlled (Holland, 1995). Collecting assemblages from a range of localities up and down a contiguous shelf can allow comparison of faunas drawn from the same lithology across a probable transition, thereby minimizing this potential bias, but this is an option only when both updip and downdip sections are available. The issue of facies change becomes particularly important when boundaries are recognized based on faunal turnovers but those faunas are facies-specific. The dramatic middle-late Eocene turnover in mollusks of the Pacific Northwest provides a good illustration: the transition from the warm shallow-marine faunas of the Cowlitz Formation to the cooler, deeper-water faunas of the Lincoln Creek and Keasey formations was also accompanied by a large facies shift toward fine-grained tuffaceous sediments (Hickman, this volume; Nesbitt, this volume). It is unclear in this case how much of the turnover was brought on by global cooling and how much simply reflected a facies shift due to regional tectonics. Papers by Fluegeman (this volume) and Echols et al. (this volume) offer an additional example from the U.S. Gulf Coast, where the Eocene/Oligocene boundary had been traditionally placed at the lithologic and faunal break between the Jackson and Vicksburg Groups. While the faunas of the Jackson and Vicksburg Groups are distinct, the contact between them has recently been shown to be time transgressive (Jaramillo and Oboh-Ikuenobe, 1999). With close examination of the faunas across the boundary in several geographic areas, these authors found that the boundary as defined by the disappearance of *Hantkenina* spp. occurs below the Jackson-Vicksburg contact in some areas and above it in others.

Rare and unusual facies also pose correlation dilemmas. Gammon et al. (this volume) describe geographically extensive biosiliceous beds within late Eocene shallow-shelf limestones in Western Australia. These beds consist of bioclastic concentrations of siliceous sponge spicules that have Mesozoic-Tethyan taxonomic affinity. Today such lithistid-hexactinellid assemblages occur only in deep water. However, in the Eocene shelf of Australia they flourished at neritic depths because of elevated amounts of silica-rich, land-derived nutrients and calm, warm waters. Likewise, taxa relegated to the deep sea today are found in Eocene shallow-marine deposits of the Antarctic Peninsula (Aronson et al., 1997).

ADVANCES IN IDENTIFYING AND CORRELATING BOUNDARY SECTIONS

A significant limitation with respect to studies of the Eocene-Oligocene transition is our ability to (1) agree on a defining characteristic for particular chronostratigraphic boundaries; (2) recognize them in our respective sections; and (3) correlate our sequences with others in different areas. These concerns, of course, are not unique to the Eocene-Oligocene interval, but characterize all studies based on stratigraphic sections. Berggren et al. (1992) presented a substantial review of the issues surrounding Paleogene geochronology and made strides toward an age reassessment of the Eocene/Oligocene boundary. Those estimates were fine-tuned in Berggren et al. (1995), with an age assessment of the boundary at 33.7 Ma, and their revised and updated Cenozoic geochronology is the time scale we use in this volume.

The currently accepted marker for the Eocene/Oligocene boundary in a stratigraphic section is the last appearance of the planktonic foraminifer *Hantkenina* spp., as seen in the type section of Massignano, Italy. However, Brinkhuis (1992) showed that the relative rarity of *Hantkenina* caused this criterion to place portions of the type upper Eocene Priabonian Stage into the Oligocene, as per the Signor-Lipps effect (Signor and Lipps, 1982). Consequently, Berggren et al. (1995) suggested using the Last Appearance Datum (LAD) for the *Turborotalia cerroazulensis/cunialensis* group instead, or in addition, because of the higher probability of their being better preserved in a section. While the two events may eventually be shown to have been synchronous, at present the LAD of the *T. cerroazulensis/cunialensis* group, defining the P17/P18 planktonic foraminiferal zonal boundary, is about 0.1 million years younger than the LAD of *Hantkenina* spp. that defines the actual Eocene/Oligocene boundary.

The long-recognized concern associated with biostratigraphically recognized boundaries, that the first (FAD) and last appearances of taxa may not be geographically synchronous, is expanded upon in a number of the papers in this volume. Coccioni and Galeotti (this volume), for example, make clear that the disappearance of the benthic foraminifer *Nuttallides truempyi* was diachronous and occurred earlier at high latitudes than at low. This also was true of calcareous nannofossils, which appear to be temperature-limited and hence not globally synchronous in their FADs and LADs (Wei and Wise, 1990; Vandenberghe et al., this volume). Since the Eocene-Oligocene transition marks a profound cooling trend, temperature limitation and the resulting diachronous nature of faunal events become particularly significant problems.

Magnetic stratigraphy provides an entirely independent check on other types of correlation. Using magnetostratigraphy pinned to radiometric dates, and to known microfossil zones, Prothero (this volume) presents a higher resolution chronostratigraphy of shelf sediments along the western North American coast than previously available. Without this resolution, the studies of faunal and floral turnover, extinction and originations would be difficult to place in a global context.

Chemostratigraphy is another approach that is becoming more and more useful for the Cenozoic. While geochemical tools have become widely accepted as useful for correlation, workers seem to have resisted employing them to define particular chronostratigraphic intervals. This is changing with the recent acceptance of the carbon isotope anomaly formerly known as the LPTM (late Paleocene thermal maximum) as the recognized marker for the Paleocene-Eocene boundary (Aubry et al., 1998; Röhl et al., 2000). For the Eocene-Oligocene transition, the large step toward more positive oxygen isotopic compositions of marine carbonate has been recognized across a globally wide range of sections, latitudes, and environments. This is the so-called Oi1 event (Miller, 1992), which marks the rapid expansion of ice on Antarctica in what is now considered the earliest Oligocene (Zachos et al., 1992). Recognition of this isotopic shift in boundary sections, and the extent to which it reflects ice growth and not temperature change, can help constrain the age of particular deposits and correlate disparate sections. Geochemical excursions may well prove to be the ideal tool for recognizing global isochronous time planes on which to hang our sections and avoid the problem of diachroneity of biostratigraphic events. Strontium isotope stratigraphy, too, is especially useful in marine sections from the late Eocene into the Oligocene because the steep slope of the marine reservoir curve can allow for very precise age determinations, assuming no problems with diagenesis or freshwater influence (DePaolo and Ingram, 1985; McArthur et al., 2001).

Radiometric dates bracketing fossiliferous horizons are the decisive factor in refining chronologies. New $^{40}Ar/^{39}Ar$ dates accompanied by paleomagnetic and floral correlations have resulted in the first reliable chronostratigraphy of both interior and lowland paleofloras that span the Eocene-Oligocene transition in the Pacific Northwest (Myers, this volume). Climate data obtained from whole-flora leaf morphologies produced a sequence of temperature and precipitation data that can be directly compared with the shelf and deep-marine record. Latest middle Eocene (39–38 Ma) lowland-forest data indicate a cooling of 3–4°C was followed by an unusual period of warming around 35 Ma. The terminal Eocene cooling produced ~6°C decline in mean average temperatures combined with a pronounced seasonality of rainfall (Myers, this volume). This climate change was not as severe as previously inferred (Wolfe, 1994), but it does correspond closely with the shallow-marine invertebrate record from the northeastern Pacific.

CLIMATE CHANGE ASSOCIATED WITH THE EOCENE-OLIGOCENE TRANSITION

We will not attempt to review here the voluminous and growing literatures on the deep-marine record of changes across the Eocene-Oligocene transition. We instead refer the reader to useful reviews by Miller (1992) and Zachos et al. (1993, 1994) and to papers in the previous Eocene-Oligocene Penrose Conference volume (Prothero and Berggren, 1992). Oceanic paleotemperatures are generally derived from the stable isotopic compositions of biogenic carbonate, which presumably reflect the temperature of the water in which the organism grew, and/or from biologic inference based on the affinities of organisms present at any given time in a sequence. When such information is derived from a marine source, a recorded change in the temperature of water may reflect a true global pattern of global cooling and/or the growth of polar ice caps, or it may be associated with regional changes in circulation. These factors are clearly not unrelated on a large scale, nevertheless, local and regional changes in upwelling or circulation can produce dramatic changes in the temperature of the water in a given area that may not be indicative of what is happening elsewhere. Buening et al. (1998) present an example of this phenomenon in their reconstruction of paleotemperatures from southeastern New Zealand based on the oxygen isotope values of fossil brachiopod shells. The cooling they see across the Eocene/Oligocene boundary is more than they expected, given the presence of warmer-water taxa on the north and west sides of the island. They reconcile the difference by proposing that the cooling reflected the development of a cool current that flowed up from Antarctica and past the east side of the island, while the north and west remained warm. The current itself was likely due to the presence of ice on Antarctica, and so was related to global climate change, but the difference between the east and west sides of the island was more a function of regional variation in ocean circulation.

High-resolution $\delta^{18}O$ data obtained from foraminifera along the Gulf Coast shelf produced unusual and intriguing results. Hurley and Fluegeman (this volume) compared their isotope curves with the planktic:benthic foraminifera ratio curve over the 1.7 million years of the P14 planktonic foraminiferal zone. They found an inverse similarity between the two curves and concluded that the isotope pattern did not result from changes in seawater temperatures per se, but from fluctuating ice volumes in the late middle Eocene reflecting 17 inferred glacio-eustatic cycles.

The range of seasonal variation is particularly important in controlling the biogeographic distributions of organisms and yet this variable is generally are not accessible from oceanographic data sources. Advances in our ability to microsample accretionary biogenic materials, such as bivalve shells and fish otoliths, have allowed for reconstruction of seasonal variation from high-resolution stable isotopic analysis. Ivany et al. (2000; this volume) recognize an increase in the seasonal range of temperature variation across the Eocene/Oligocene boundary in the U.S. Gulf Coastal Plain as inferred from microsampling and oxygen isotopic analysis of otolith carbonate, and they infer a link with the associated faunal turnover. Data from shelf sections such as these will become increasingly important for constraining this variable, as long-lived marine taxa are generally not recovered from deep-sea sections.

Ultimately, the Eocene-Oligocene transition can most likely be traced back to tectonics—plate motion and the effects it had on oceanic circulation, and mountain building and the effects it had on atmospheric composition. The conventional story, and it is undoubtedly more complicated than presented here, is that the progressive isolation of Antarctica caused by the rifting away of Australia, Tasmania, and South America led to the development of circum-Antarctic circulation in the Southern Ocean. Recent analyses suggest that this isolation began with the opening of the Drake Passage, possibly at the middle/late Eocene boundary at 37 Ma (Diester-Haass and Zahn, 1996), The final event was the separation of East Antarctica from Tasmania at 33.5 Ma (Exon et al., 2002), which allowed for ocean waters to continuously circulate around Antarctica, rather than be deflected up to warmer climes. This caused the water and the continent it circled to cool down, eventually leading to the growth of continental ice sheets. It also generated the psychrosphere, the cold bottom-water currents that sink from the Antarctic surface waters and flow at abyssal depths all the way to the North Atlantic and North Pacific.

The second part of the story comes with the proposed links among tectonics, continental weathering, and climate change. The hypothesis proposed by Raymo et al. (1988) is that the uplift of the Tibetan Plateau led to increased rates of chemical weathering of silicate minerals, thus removing large quantities of CO_2 from the atmosphere. Reduced atmospheric CO_2 produced a significant and geologically rapid cooling of the planet. The drawback of this explanation is that the process, while probably important in the Neogene, is not likely to have played a significant role in Eocene-Oligocene cooling because the Himalayan orogeny had only just gotten underway. Zachos et al. (1999) propose an alternative mechanism to explain the trend toward more radiogenic strontium in the marine reservoir, a strong piece of evidence used by Raymo and colleagues to support their hypothesis. Zachos et al. (1992) suggest instead that Eocene-Oligocene cooling and ice growth on Antarctica increased the rate of chemical weathering itself, thereby causing the observed trend in strontium isotopes and, in theory, also causing further climate cooling by the drawdown of CO_2. The issue remains largely unresolved, and is perhaps one of the most interesting in the earth sciences today (see also papers in Chemical Geology volume 161). Clearly, though, tectonics, ocean circulation, and atmospheric composition all play a role in the long-term Cenozoic cooling trend that began in the middle Eocene.

Did the late Eocene impact events have any demonstrable effects on global climate? Poag et al. (this volume) make a case for periods of greenhouse warming associated with impact events. They base their arguments on the oxygen isotopic compositions of benthic foraminifera taken from a deep shelf section in Virginia immediately overlying the Chesapeake Bay impact structure. In addition, they document a negative $\delta^{13}C$ signal in many sections worldwide that they speculate may be related to impact event perturbations of the global carbon cycle. This is in contrast to data obtained from the 35.5 Ma impact horizon in an ODP core from Maude Rise (high latitude, Southern Ocean) and an equivalent horizon from the midlatitude, type section at Massignano, Italy (Vonhof et al., 2000). Oxygen isotope values from the impact horizon indicate accelerated global cooling, within an already deteriorating climate trend. This is accompanied by a marked peak in the relative abundance of the dinoflagellate *Thalassiphora pelagica* at both Maud Rise and Massignano, indicating cooling and increased productivity.

It is interesting to note that these two impact hypotheses predict opposite effects: Poag et al. (this volume) suggest that an impact caused a brief warming episode (or at least that it temporarily suspended global cooling), while Vonhof et al. (2000) argued that it caused global cooling. Disentangling potential causes for an observed climatic trend can clearly be difficult, and simple correlation of events in time is not enough to establish causality. In fact, it remains unclear how impacts of the size of Popigai and Chesapeake Bay could cause such long-term and global effects (either warming or cooling). While there were undoubtedly severe short-term regional effects from these impacts, in all likelihood they produced relatively insignificant long-term global effects in comparison to the overall climatic trends of the Eocene-Oligocene transition.

Recently, Fawcett and Boslough (2002) have suggested that the cooling of the late Eocene could have been due to a ring of impact debris around the Earth, which casts a shadow and shades the tropics for about 100,000 years. There are severe problems with this hypothesis. First, the post-impact late Eocene was marked by *warming*, not cooling (which Poag et al., this volume, attribute to impact debris). Also, the effects take longer than 100,000 years to develop after the impact, and persist for longer than 100,000 years.

PATTERNS, TIMING, AND INFERRED CAUSES OF FAUNAL TURNOVER

Climatic change during the Eocene-Oligocene transition had profound effects on global biotic communities, as many of the papers in this volume demonstrate. The nature and magnitude of these biotic changes can often be even more informative than physical or geochemical indicators in understanding the nature and timing of the climatic changes.

The Biotic Record

Because many higher taxa are characterized by particular and distinct ecologies (e.g., mode of life, feeding strategy, developmental pathway, and so on), establishing the taxonomic expression of a turnover interval can be helpful in determining the proximal and ultimate cause(s) of the event. Here we discuss the major groups represented within this volume and the patterns they typify across the Eocene-Oligocene transition.

In the record of Eocene-Oligocene benthic foraminiferal assemblages, pulses of extinction can be noted at the end of the middle Eocene and/or at or close to the Eocene/Oligocene boundary in the Tethys and Paratethys in Italy (Coccioni and Galeotti, this volume), Israel (Barbieri et al., this volume), the circum-Alpine region (Nebelsick et al., this volume), southern Australia (McGowran et al., 1992), in the Caribbean (Robinson, this volume), and the U.S. Gulf Coast (Gaskell, 1991; Fluegeman, this volume). In both the Caribbean and the Gulf Coast, there are indications that the turnover generally associated with the boundary may in fact have been taking place in the earliest Oligocene (Fluegeman, this volume; Robinson, this volume), and therefore perhaps have been associated with Oi1 cooling.

Mollusks are abundant, ecologically diverse, and well skeletonized and therefore well preserved. Thus they are the primary marine-shelf faunal component for which there are good data across the Eocene-Oligocene transition. The U.S. Gulf Coastal Plain records arguably one of the best sections of fossiliferous Paleogene sediments in the world, and the mollusks there have been intensively studied for more than a century and a half (synthesized in the works of Palmer and Brann, 1965, 1966; Dockery, 1982; and MacNeil and Dockery, 1984). One place in the world that rivals the Gulf in the nature of its molluscan fossils is the Paris Basin and associated regions in France. Dockery and Lozouet (this volume) present a comparison of these two spectacularly diverse sequences. In the Gulf Coast, there is substantial molluscan taxonomic turnover at group-level boundaries throughout the Paleogene (Dockery, 1986). In both regions, the greatest turnover is near the Eocene/Oligocene boundary, with more than 95% of Gulf Coast mollusk species going extinct, among them several long-ranging and abundant Paleocene and Eocene forms. In the U.S. Atlantic Coastal Plain, limited outcrops and poor preservation have made molluscan faunas difficult to record and analyze. Using what is available, Campbell and Campbell (this volume) found close taxonomic affinities of these faunas to the Gulf Coast assemblages, and noted the similar stepwise extinction patterns across the middle/late Eocene and late Eocene/Oligocene boundaries. In addition they observe that no taxa endemic to the region survive from one stratigraphic unit to the next.

A trio of papers on the western North American molluscan faunas shows a similar pattern of pulsed faunal turnover at the stage level, with significant events at the middle/late Eocene boundary, near the Eocene/Oligocene boundary, and at the early/late Oligocene boundary (Hickman, this volume; Nesbitt, this volume; Squires, this volume). As in the Gulf Coast, tropical taxa were particularly hard hit during turnovers. Farther north, in the Kamchatka Peninsula, the same middle/late Eocene and terminal Eocene turnovers occurred, and once again warm-water taxa sufferred preferentially. Interestingly, while Alaskan faunas responded similarly to middle-late Eocene events, they were seemingly unaffected by Eocene/Oligocene boundary extinctions (Oleinik and Marincovich, this volume). In all regions, there was a notable radiation of new taxa that either migrated from higher latitudes or are endemic. The Eocene/Oligocene boundary marks the end of the characteristic shallow-marine Tethyan fauna in North America and the northern Pacific Ocean.

Ostracodes, too, can be useful indicators of oceanographic change. Schellenberg (1999) documents significant drops in species richness associated with the Eocene/Oligocene boundary at several sites in the Southern Ocean. Associated with the diversity drop is a transition from dominantly deposit-feeding to dominantly filter-feeding forms. This indicates a trend toward low-oxygen condition, for filter feeders are better able to move water over their bodie and hence excange oxygen more efficientlys.

The pattern of echinoderm diversity through the Eocene and into the Oligocene does not seem to consistently follow that of the mollusks and benthic foraminifera. Nearly all regions considered show an overall decline in echinoid diversity from the middle Eocene to the Oligocene, but the specifics of the patterns differ among the regions (Carter, this volume). Unlike that of other invertebrate groups, the diversity of echinoids in the southeastern United States actually increased from the middle to the late Eocene, then declined into the early Oligocene (Carter, this volume). Carter suggests that the record of echinoid diversity across the shelf was primarily controlled by facies effects, specifically the relative proportions of carbonate versus siliciclastic sediment and the balance between muddy and sandy substrates. Conversely, in the Pacific Northwest, Burns and Mooi (this volume) point out that sand dollars in the Oligocene were significantly different from those in the Eocene, implying an Eocene/Oligocene boundary turnover. Based on paleobiogeography, they suggest that the Oligocene fauna may have had its origins at higher latitudes, consistent with the molluscan data. Other echinoderms have a less complete record due to poorer preservation potential, and so patterns can be only tenuous. The diversity and abundance of crinoids, asteroids, and ophiuroids in marine shelf sediments actually seem to have increased at high southern latitudes through the Eocene. Aronson et al. (1997) report unusual echinoderm-rich faunas near the top of the Eocene section on Seymour Island on the Antarctic Peninsula, potentially in response to the decrease in predation pressure associated with climatic cooling. In the Pacific Northwest, however, crinoids disappeared from shallow water at the end of the Eocene (Burns and Mooi, this volume).

The record of terrestrial vertebrates through the Eocene-Oligocene transition was covered in some detail by papers in Prothero and Berggren (1992) and Prothero and Emry (1996). Here, we add contributions regarding the record of marine vertebrates. Manning (this volume) records an unexpected pattern of FADs and LADs for marine vertebrates along the U.S. Gulf Coast, although he cautions that both the temporal and the stratigraphic record is poor. He noted an end-Eocene extinction of several Tethyan forms, such as the archeocete whale *Basilosaurus*, giant sea snakes, and the sawfish, *Propristis*. However, all other marine vertebrate groups show a gradual change, often at subspecific level, across the boundary or no change at all. On a global scale, Fordyce (this volume) reviews whale diversity, extinctions, and originations across the boundary. At the family level, there was an initial rapid appearance of grades and clades in the Ypresian-Lutetian inteval (early to middle Eocene), and a second radiation near the Eocene/Oligocene boundary. Eustatic sea-level fall substantially reduces the available record of marine mammals for the early Oligocene. No archaeocetes ranged into the Oligocene, and new data show a complex pattern for the Oligocene radiation of cetaceans around the Pacific Ocean, including the oldest odontocetes from the Priabonian in Washington. Fordyce (this volume) concludes that this diversity pattern resulted from the restructuring of food resources during a period of oceanic cooling.

PALEOBIOLOGY OF TURNOVER

Turnover events are characterized by some combination of disappearance of old forms and/or appearance of new forms (Allmon, this volume), and hence can affect taxonomic diversity in several ways. It is conceivable that a boundary could be marked by 100% turnover of the fauna but have no associated change in diversity (number of taxa) across that boundary. Stepwise turnover of shelf communities through the Eocene and Oligocene seems to characterize nearly all the records reported here, with particular pulses at the middle/late Eocene boundary and near the Eocene/Oligocene boundary. In order to understand these events better, it is important to consider diversity in combination with either origination or extinction (or both). As an example, Campbell and Campbell (this volume) point out that the record of molluscan diversity in the Atlantic Coastal Plain tends to be characterized more by pulses of extinction than by pulses of origination. The Gulf Coast seems to be distinguished by both (Dockery, 1986). But in Alaska, origination actually exceeds extinction (Oleinik and Marincovich, this volume). All these turnovers, however, may have been roughly of the same magnitude.

A common thread that seems to be present in the studies of Atlantic, Gulf, and European mollusks is that *diversity loss* is most significant at the middle/late Eocene boundary, yet the *turnover* in the fauna is greater near the Eocene/Oligocene boundary (Dockery and Lozouet, this volume; Campbell and Campbell, this volume). This implies, in the framework of faunal turnovers presented by Allmon (this volume), that while extinction was greatest at the middle/late Eocene boundary, it was not accompanied by much in the way of origination. The late Eocene fauna is simply a whittled-down version of the middle Eocene biota. At the end of the Eocene, turnover in the fauna was more complete such that not only did many forms suffer extinction, but many new taxa appeared for the first time in the early Oligocene. The result is that the Oligocene fauna is recognizably and distinctly different from the Eocene one, despite the fact that extinction alone might have been greater at the middle/late Eocene boundary.

On the Pacific Coast, the end of the middle Eocene is similar in that it marks a very large drop in molluscan diversity, but here turnover as well is virtually complete—taxonomic overlap between the units above and below is near zero (Hickman, this volume; Nesbitt, this volume; Squires, this volume). Near the Eocene/Oligocene boundary, turnover is comparatively severe and the percent extinction larger (Hickman, this volume)

The greater turnover experienced in the Pacific Northwest at the middle/late Eocene boundary in comparison with the Gulf Coast may be a function of its higher latitude, for warm, equable, environments there in the middle Eocene may have been more severely affected by cooling than those at lower latitudes. Farther north in the Pacific, on the Kamchatka Peninsula, middle-late Eocene turnover was also significant, but first appearances were more important than extinctions (Oleinik and Marincovich, this volume)—the exact opposite of the pattern from the Gulf Coast. Likewise, the Eocene/Oligocene boundary marks the most significant diversity drop there. In contrast, as mentioned earlier, the middle/late Eocene and Eocene/Oligocene boundaries in Alaska (Oleinik and Marincovich, this volume) are not so much marked by extinction, but by origination!

An ever-present question associated with any given turnover is how much of the taxonomic change is evolutionary (true origination and extinction) and how much is simply migration out of or into a given area. Dockery and Lozouet (this volume) consider this issue for the U.S. Gulf Coast and find that in many cases the extinctions at the Eocene/Oligocene boundary were truly extinctions (at least in the Western Hemisphere), rather than emigrations. In addition, they maintain that turnover was rapid enough near the Eocene/Oligocene boundary and without a significant loss in diversity, that the first appearances must have included a substantial amount of immigration from elsewhere, rather than evolution in situ. Interestingly as well, the new taxa that appeared in both the Gulf and western Europe are very similar, if not identical (Dockery and Lozouet, this volume), suggesting an open biogeographic connection at that time. The presence of these characteristic Tethyan forms in the basal Oligocene of Mississippi suggests to them that the source of larvae was to the east, in the vicinity of the Paris and Aquitaine Basins in France. Biogeography is a factor that is relevant in other ways, too. Campbell and Campbell (this volume), in their comparison of turnover in mollusk genera restricted to the Carolinas versus those more globally distributed, not surprisingly find that endemic genera were more likely than widespread ones to suffer extinction during turnovers. At the species level, however, even widespread taxa suffered greatly near the Eocene/Oligocene boundary (only a single species survived the transition), suggesting a severe and global event. Interestingly, at high latitudes in the North Pacific, the more cosmopolitan taxa seem to have been more strongly affected by extinction than those with restricted distributions (Oleinik and Marinco-vich, this volume). Cosmopolitan taxa, in fact, were replaced by more endemic genera throughout the Eocene-Oligocene transition. In this case, the widespread taxa tended to have ranges that extended to warmer, lower latitudes. Their selective loss suggests that thermal tolerance may play a more important role than degree of endemism in determining survival probability at high latitudes.

CAUSES FOR TURNOVER?

In narrowing down the cause for a particular turnover, one of the possible strategies would be to ask whether the turnover itself is local or global. Do faunal events recognized in shelf sections match those noted on other continents or in open ocean (benthic or planktonic) assemblages? Fluegeman (this volume) points out that the pattern of turnover in benthic foraminifera, for example, is often not in phase with that in the deep sea because neritic taxa tend to be more sensitive to local and regional forcing factors, while the driving forces behind change in deep-sea benthics are more likely to be global. In the U.S. Gulf Coast, however, the similarity of the benthic foraminiferal record (e.g., Gaskell, 1991; Fluegeman, this volume) to that in southern Australia discussed in McGowran et al. (1992) suggests that this may not be the case for Eocene-Oligocene turnovers (Fluegeman, this volume). The similarity of molluscan extinction patterns across the middle/late Eocene and the Eocene/Oligocene boundaries in the Gulf and Atlantic Coastal Plains, Europe (Campbell and Campbell, this volume; Dockery and Lozouet, this volume), and the Pacific (Nesbitt, this volume; Hickman, this volume; Squires, this volume; Oleinik and Marincovich, this volume) also indicates a more global event. Nevertheless, several authors point out the importance of facies changes in influencing the patterns recovered in their sections. Care should be taken to constrain the turnovers as carefully as possible in time and space so that local versus global control can be evaluated.

Selectivity of extinction in terms of the type of fauna affected can be especially useful for determining the cause(s) of a particular episode or interval of turnover. A recurrent theme within the molluscan records of the Eocene-Oligocene transition seems to be that warm-water taxa suffered preferentially in comparison with cool-water taxa, particularly at the middle/late Eocene boundary turnover. Hansen

(1987, 1992) argues for a correlation between Gulf Coastal Plain turnovers and temperature based on the selective loss of warm-water taxa. Dockery and Lozouet (this volume) acknowledge temperature as a factor in the extinctions as well, but imply it to be less important than sea level. For the Pacific Coast, Nesbitt (this volume), Hickman (this volume), and Squires (this volume) point out the strong bias against warm-water and tropical taxa at the middle/late Eocene boundary in particular, and the same is true in the North Pacific (Oleinik and Marincovich, this volume).

Dockery (1986) and Dockery and Lozouet (this volume) also note that the molluscan faunal turnovers in the U.S. Gulf Coast correlate well with Type 1 sequence boundaries, and on that basis suggest a correlation between major sea-level falls and molluscan turnovers.Gaskell (1991) documents the same relationship for Gulf Coast benthic foraminifera. Their inferred mechanism by which sea level would affect diversity is through loss of habitable area and change in circulation patterns that force biogeographic change. While it is clear that the relationship with sea level exists in the Gulf, the widespread selective loss of warm-water taxa through the middle-late Eocene and into the Oligocene in a number of other areas strongly argues for temperature being a dominant causal factor in the extinctions globally.

In bathyal environments, the productivity regime and degree of oxygenation of the bottom seem to be particularly important factors in faunal turnover. Barbieri et al. (this volume) note a significant transition at the middle/late Eocene boundary in benthic foraminiferan assemblages that they link to a shift from oxygenated, oligotrophic conditions to oxygen-limited and nutrient-rich conditions. The same trend toward highly productive surface waters is seen near the Eocene/Oligocene boundary in bathyal (Coccioni and Galeotti, this volume) and abyssal sediments based on benthic foraminiferal abundances and carbon isotopic work (Diester-Hass, 1995; Thomas and Gooday, 1996; Salamy and Zachos, 1999; Diester-Hass and Zahn, 2001; Diester-Haass and Zachos, this volume). Increases in paleoproductivity of ocean waters was likely linked to cooling and concomitant changes in ocean circulation, leading to an increase in upwelling intensity.

As to whether the documented late Eocene bolide impacts (e.g., Poag et al., this volume) caused any response in the faunas at the time, there does not seem to be any obvious correlation between impacts and extinction peaks in the U.S. Gulf Coast (Hansen, 1987). In the Carolinas, where the effect plausibly would have been more severe, the problem of poor preservation makes the issue difficult to resolve (Campbell and Campbell, this volume). Robinson (this volume) notes an extinction of lesser magnitude in the benthic foraminiferan record in the Caribbean during the late Eocene. However, poor age constraints do not allow precise correlation with impact or other events. Given that the papers in this volume and others suggest major pulses of extinction at the end of the middle Eocene (well before the impacts) and the early Oligocene (well after), it is unlikely that the Chesapeak Bay or Popigai impacts had a significant effect on the global biota.

NEW AVENUES FOR RESEARCH

Since the proceedings volume of the first Penrose Conference on the Eocene-Oligocene transition was published (Prothero and Berggren, 1992), new avenues for research have opened up through the development of more refined techniques for investigating processes across an epoch boundary. Higher-resolution chronostratigraphy utilizing radiometric dating techniques, chemo-, magneto-, and sequence stratigraphy, microsampling of shell material for stable isotope investigations, and leaf morphology analysis have allowed us to focus on biotic and climatic changes occurring on progressively fine time scales. However, much work remains to be done in correlating boundary sequences across paleobiogeographic gaps. We all aspire to better age constraints so that we can place a virtual golden spike to mark the Eocene-Oligocene transition in each of our respective sections. Such precision will finally allow us to compare data across provincial boundaries, across oceans, and between taxa.

We see a number of new fruitful directions for research:

—Correlate marine and continental data to better resolve how climatic change affects coastal plain and inland/upland communities and determine whether turnovers are synchronous

—Refine the time scale of climatic change in different regions to determine the degree to which they correlate and how the magnitude of change varies geographically

—Explore latitudinal diversity trends, and how they respond to global climate change

—Investigate ecological and taxonomic selectivity across transitions in order to learn more about cause

—Deconvolve the relative contributions of in situ evolutionary change, true extinction, and migration

—Collect more faunal and environmental data in geographical regions that are less well represented
—Solve the ice volume versus temperature conundrum of oxygen isotopes, perhaps with progress being made using trace elements (e.g., Lear et al., 2000)
—Use biogeography of assemblages to better understand ocean circulation and thermal gradients

As Allmon (this volume) states, "Biologically defined boundaries—times of appearances and/or disappearances of taxa—have been fundamental to stratigraphic paleontology and geology for almost 200 years." This observation sets the stage for topics explored in this volume. Stratigraphic boundaries are nearly always associated with or defined by local and regional extinctions and originations, which themselves are often the result of environmental change. Recognizing the geologic cause(s) of such change, refining its temporal and geographic expression, and understanding the biological response to it, are our goals. As investigations move toward better temporal resolution of events using nonbiotic signatures, we are allowed freedom to disentangle the biotic events from their climatic and/or tectonic causes. Precise stratigraphic correlations continue to be the blueprint from which we work, and as paleontologists we can now utilize new geochemical tools to investigate faunal and floral responses to climate change. We hope this will help us to better understand the complexities associated with the study of climate change and the associated biological response to the past, as well as in our own immediate future.

REFERENCES

Ager, D. V., 1993, The nature of the stratigraphical record: Chichester, Wiley, 151p.

Aronson, R. B., Blake, D. B., and Oji, T., 1997, Retro-grade community structure in the late Eocene of Antarctica: Geology, v. 25, p. 903-906.

Aubry, M.-P., Lucas, S., and. Berggren, W. A., eds., 1998, Late Paleocene-early Eocene biotic and climatic events in the marine and terrestrial records: New York, Columbia University Press.

Baum, J. S., Baum, G. R., Thompson, P. R., and Humphrey, J. D., 1994, Stable isotopic evidence for relative and eustatic sea-level changes in Eocene to Oligocene carbonates, Baldwin County, Alabama: Geological Society of America Bulletin, v. 106, p. 824-839.

Berggren, W. A., Kent, D. V., Obradovich, J. D., and Swisher, C.C., II. 1992, Toward a revised Paleogene geochronology, *in* Prothero, D. R., and Berggren, W. A., eds., Eocene-Oligocene climatic and biotic evolution: Princeton, New Jersey, Princeton University Press, p. 29-45.

Berggren, W. A., Kent, D. V., Swisher, C. C., II., and Aubry, M.-P., 1995, A revised Cenozoic geochronology and chronostratigraphy: SEPM Special Publication, v. 54, p. 129-212.

Berggren, W. A., and Prothero, D. R., 1992, Eocene-Oligocene climatic and biotic evolution: an overview, *in* Prothero, D. R., and Berggren, W. A., eds., Eocene-Oligocene climatic and biotic evolution: Princeton, New Jersey, Princeton University Press, pp. 1-28.

Case, J., 1988, Paleogene floras from Seymour Island, Antarctic Peninsula, *in* Feldmann, R. M., and Woodburne, M. O., eds., Geology and paleontology of Seymour Island, Antarctic Peninsula: Geological Society of America Memoir 169, pp. 523-530.

Christie-Blick, N., 1991, Onlap, offlap, and the origin of unconformity-bounded depositional sequences: Marine Geology, v. 97, pp. 35-56.

DePaolo, D. J., and Ingram, B. L., 1985, High-resolution stratigraphy with strontium isotopes: Science, v. 227, p. 938-941.

Diester-Haass, L., 1995, Middle Eocene to early Oligocene paleoceanography of the Antarctic Ocean (Maud Rise, ODP Leg 113, Site 689): change from a low to a high productivity ocean: Palaeogeography, Palaeoclimatology, Palaeoecology, v. 113, p. 311-334.

Diester-Haass, L., and Zahn, R., 1996, The Eocene-Oligocene transition in the Southern Ocean: history of water masses, circulation, and biological productivity inferred from high resolution records of stable isotopes and benthic foraminiferal abundances (ODP Site 689): Geology, v. 24, p.16-20.

Diester-Haass, L., and Zahn, R., 2001, Paleopro-ductivity increase at the Eocene-Oligocene climatic transition: ODP/DSDP sites 763 and 592: Palaeogeography, Palaeoclimatology, Palaeoecology, v. 172, p. 153-170.

Dockery, D. T., III, 1982, Lower Oligocene Bivalvia of the Vicksburg Group: Mississippi Department of Natural Resources Bureau of Geology Bulletin, v. 123, p. 1-261.

Dockery, D. T., III, 1986, Punctuated succession of Paleogene mollusks in the Northern Gulf Coastal Plain: Palaios, v. 1, p. 582-589.

Exon, N., and 20 others, 2002, Drilling reveals climatic consequences of Tasmanian gateway opening: EOS, v. 83, no. 23, p. 253-259.

Fawcett, P. J., and Boslough, M. B. E., 2002, Climatic effects of an impact-induced equatorial debris ring: Journal of Geophysical Research, v. 107, no. D15, p. 10129-10146.

Gaskell, B. A., 1991, Extinction patterns in Paleogene benthic foraminiferal faunas: Relationship to climate and sea level: Palaios, v. 6, p. 2-16.

Greenwood, D. R., and Wing, S. L., 1995, Eocene continental climates and latitudinal temperature gradients: Geology, v. 23, p. 1044-1048.

Haasl, D. M., and Hansen, T. A., 1996, Timing of latest Eocene molluscan extinction patterns in Mississippi: Palaios, v. 11, p. 487-494.

Hansen, T. A., 1987, Extinction of late Eocene to Oligocene molluscs: relationship to shelf area, temperature changes, and impact events: Palaios, v. 2, p. 69-75.

Hansen, T., 1992, The patterns and causes of molluscan extinction across the Eocene/Oligocene boundary, *in* Prothero, D. R., and Berggren, W. A., eds., Eocene-Oligocene climatic and biotic evolution: Princeton, New Jersey, Princeton University Press, pp. 341-348.

Holland, S. M., 1995, The stratigraphic distribution of fossils: Paleobiology, v. 21, p. 92-109.

Ivany, L. C., Patterson, W. P., and Lohmann, K. C, 2000, Cooler winters as a possible cause of mass extinctions at the Eocene/Oligocene boundary: Nature, v. 407, p. 887-890.

Jaramillo, C., and Oboh-Ikuenobe, F. E., 1999, Sequence stratigraphic interpretations from palynofacies, dinocyst and lithological data of upper Eocene-lower Oligocene strata in southern Mississippi and Alabama, US Gulf Coast: Palaeogeography, Palaeoclimatology, Palaeoecology, v. 145, pp. 259-302

Kennett, J. P., and Stott, L. D., 1991, Abrupt deep-sea warming, palaeoceanographic changes and benthic extinctions at the end of the Palaeocene: Nature, v. 353, p. 225-229.

Lear, C. H., Elderfield, H., and Wilson, P. A., 2000, Cenozoic deep-sea temperatures and global ice volumes from Mg/Ca in benthic foraminiferal calcite: Science, v. 287, p. 269-272.

Loutit, T. S., Hardenbol, J., and Vail, P. R., 1988, Condensed sections: the key to age determination and correlation of continental margin sequences: SEPM Special Publication, v. 42, pp. 183-213.

MacNeil, F. S., and Dockery, D. T., III, 1984, Lower Oligocene Gastropoda, Scaphopoda, and Cephalopoda of the Vicksburg Group in Mississippi: Mississippi Department of Natural Resources Bureau of Geology Bulletin, v. 124, p. 1-415.

Markwick, P. J., 1994, "Equability," continentality, and Tertiary "climate": the crocodilian perspective: Geology, v. 22, p. 613-616.

McArthur, J. M., Howarth, R. J., and Bailey, T. R., 2001, Strotiium isotope stratigraphy: LOWESS version 3: best fit to the marine Sr-isotope cuve for 0-509 Ma and accompanying look-up table for deriving numerical age: Journal of Geology, v. 109, p. 155-170.

McGowran, B., Moss, G., and Beecroft, A., 1992, Late Eocene and Early Oligocene in Southern Australia: Local neritic signals of global oceanic changes, *in* Prothero, D. R., and Berggren, W. A., eds., Eocene-Oligocene climatic and biotic evolution: Princeton, New Jersey, Princeton University Press, pp. 178-201.

Miller, K. G., 1992, Middle Eocene to Oligocene stable isotopes, climate, and deep-water history: the Terminal Eocene Event? *in* Prothero, D. R., and Berggren, W. A., eds., Eocene-Oligocene climatic and biotic evolution: Princeton, New Jersey, Princeton University Press, pp. 160-177.

Palmer, K. V. W., and Brann, D. C., 1965-1966, Catalogue of the Paleocene and Eocene Mollusca of the Southern and Eastern United States: Part I. Pelecypoda, Amphineura, Pteropoda, Scaphopoda, and Cepholopoda, and Part II. Gastropoda: Bulletins of American Paleontology, v. 48, p. 1=1057.

Posamentier, H. W., and Allen, G. P., 1993, Variability of the sequence stratigraphic model: effects of local basin factors: Sedimentary Geology, v. 86, p. 91-109.

Prothero, D. R., 1994, The Eocene-Oligocene transition: paradise lost: New York, Columbia University Press, 291 p.

Prothero, D. R., and Berggren, W. A., eds., 1992, Eocene-Oligocene climatic and biotic evolution: Princeton, New Jersey, Princeton University Press, 568 p.

Prothero, D. R., and Emry, R. J., eds., 1996, The terrestrial Eocene-Oligocene transition in North America: New York, Cambridge University Press.

Prothero, D., Ivany, L., and Nesbitt, E., 2000, Penrose Conference Report: The marine Eocene-Oligocene transition: GSA Today, v. 10, no. 7, p. 10-11.

Raymo, M. E., Ruddiman, W. F., and Froelich, P. N., 1988, Influence of late Cenozoic mountain building on ocean geochemical cycles: Geology, v. 16, p. 649-653.

Röhl, U., Bralower, T. J., Norris, R. D. and Wefer, G., 2000, New chronology for the late Paleocene thermal maximum and its environmental implications: Geology, v. 28; p. 927-930.

Salamy, K. A., and Zachos, J. C., 1999, Latest Eocene-early Oligocene climate change and Southern Ocean fertility: inferences from sediment accumulation and stable isotope data: Palaeogeography, Palaeoclimatology, Palaeoecology, v. 145, p. 61-77.

Schellenberg, S. A., 1999, Deep-ocean ostracode response to the Eocene-Oligocene transition: high-resolution records from the Southern Ocean (ODP Sites 744 and 689): Geological Society of America Abstracts with Programs, v. 31, no. 7 , p. 310.

Scotese, C. R., 2002, Paleomap Project, http://www.cscotese.com/climate.htm, version 4/20/02

Signor, P., and Lipps, J. H., 1982. Sampling bias, gradual extinction patterns, and catastrophes in the fossil record: Geological Society of America Special Paper, v. 190, p. 291-296.

Thomas, E., and Gooday, A. J., 1996, Cenozoic deep-sea benthic foraminifera: tracers for changes in oceanic productivity?: Geology, v. 24, p. 355-358.

Vail, P. R., Audemard, F., Bowring, S. A., Eisner, P. N., and Perez-Cruz, C., 1991. The stratigraphic signatures of tectonics eustacy and sedimentology—an overview, *in* Einsele, G., Ricken, W., and Seilacher, A., eds., Cycles and events in stratigrpahy: Berlin, Springer-Verlag, p. 617-659.

Vonhof, H. B., Smit, J., Brinkhuis, H., Montanari, A., and Nederbracht, A. J., 2000, Global cooling accelerated by early-late Eocene impacts? Geology, v. 28, p. 687-690.

Wei, W., and Wise, S. W., Jr., 1990. Biogeographic gradients of middle Eocene-Oligocene calcareous nannoplankton in the South Atlantic Ocean: Palaeogeography, Palaeoclimatology, Palaeoecology, v. 79, p. 29-61.

Wolfe, J. A., 1994, Tertiary climatic changes at middle latitudes of western North America: Palaeogeography, Palaeoclimatology, Palaeoecology, v. 105, p. 195-205.

Zachos, J. C., Breza, J. R., and Wise, S. W., 1992, Early Oligocene ice-sheet expansion on Antarctica: Stable isotope and sedimentological evidence from Kerguelen Plateau, southern Indian Ocean: Geology, v. 20, p. 569-573.

Zachos, J. C., Lohmann, K. C, Walker, J. C. G., and Wise, S. W., 1993, Abrupt climate change and transient climates during the Paleogene: A marine perspective: Journal of Geology, v. 101, p. 191-213.

Zachos, J. C., Opdyke, B. N., Quinn, T. M., Jones, C. E., and Halliday, A. N., 1999, Early Cenozoic glaciation, antarctic weathering, and seawater $^{87}Sr/^{86}Sr$: is there a link?: Chemical Geology, v. 161, p. 165-180.

Zachos, J. C., Stott, L. D., and Lohmann, K. C, 1994, Evolution of early Cenozoic marine temperatures: Paleoceanography, v. 9, p. 353-387.

Contributors

Warren D. Allmon
Paleontological Research Institution
1259 Trumansburg Road
Ithaca, NY 14850

John M. Armentrout
20060 SE Highway 224
Clackamas, OR 97015

Roberto Barbieri
Dipartimento di Scienze della Terra e Geologico-Ambientali
Università di Bologna
Via Zamboni 67
40126 Bologna, Italy

Davide Bassi
Dipartimento di Scienze, Geologiche e Paleontologiche
Universita degli Studi di Ferrara
Corso Ercole I d'Este
I-44100 Ferrara, Italy

Chaim Benjamini
Department of Geological and Environmental Sciences
Ben Gurion University of the Negev
P.O. Box 653
Beer Sheva 84105, Israel

Yvonne Bone
Department of Geology and Geophysics
University of Adelaide
Adelaide, SA. 5005, Australia

Henk Brinkhuis
Laboratory of Paleobotany and Palynology
Utrecht University
Budapestlaan 4
NL-3584 CD Utrecht, The Netherlands

Casey Burns
The Burke Museum, Box 353010
University of Washington, Seattle, WA 98195

David C. Campbell
Department of Biological Sciences, Box 870345
University of Alabama
Tuscaloosa, AL 35487

Matthew R. Campbell
Department of Geological Sciences
Indiana University
1001 East Tenth Street
Bloomington IN 47405

Burchard D. Carter
Department of Geology and Physics
Georgia Southwestern State University
Americus, Georgia 31709

Jonathan D. A. Clarke
Department of Geology
Australian National University
Canberra, ACT. 0200, Australia

Rodolfo Coccioni
Istituto di Geologia and Centro di Palinologia
Libera Universita degli Studi di Urbino
Campus Scientifico
Localita Crocicchia
61029 Urbino, Italy

James C. Cooke
6139 Annapolis
Dallas, TX 75214

Liselotte Diester-Haass
Zentrum fur Umweltforschung
Universitat des Saarlandes
66041 Saarbrucken, Germany

David T. Dockery III
Mississippi Office of Geology
Jackson, MS 39289

Ronald J. Echols
1616 West Seventh Street
Irving, TX 75060

William C. Elsik
The MycoStrat Connection
P.O. Box 549
Snook, TX 77878

Larry B. Fearn
7024 Spanish Oaks Drive
N. Richmond Hills, TX 75234

Richard H. Fluegeman
Department of Geology
Ball State University
Muncie, IN 47306

R. Ewan Fordyce
Department of Geology
University of Otago
Dunedin, New Zealand

Simone Galeotti
Istituto di Geologia and Centro di Palinologia
Libera Universita degli Studi di Urbino
Campus Scientifico, Localita Crocicchia
61029 Urbino, Italy

Paul R. Gammon
Department of Geology and Geophysics
Adelaide University
Adelaide, 5005, Australia

Carole S. Hickman
Department of Integrative Biology and Museum of Paleontology
University of California
Berkeley, CA 94720

John V. Hurley
6103 St. Albion Way, I-311
Mountlake Terrace, WA 98043

Manuel A. Iturralde-Vinent
Museo Nacional de Historia Natural
Obispo No. 61, Plaza de Armas
Havana 10100 Cuba

Linda C. Ivany
Department of Earth Sciences
Syracuse University
Syracuse, NY 13244

Noel P. James
Department of Geological Sciences
Queen's University
Kingston, ON K7L 3N6, Canada

Carlos A. Jaramillo
Instituto Colombiano del Petroleo, A.A. 4185
Bucaramanga, Colombia

Kyger C Lohmann
Department of Geological Sciences
University of Michigan
Ann Arbor, MI 48109

Pierre Lozouet
Musée National d'Histoire Naturelle
8 rue du Buffon
75005 Paris, France

Edward Mankinen
U.S. Geological Survey, 345 Middlefield Road
Menlo Park, CA 94025

Earl M. Manning
Department of Geology
Tulane University
New Orleans, LA 70118

Louie Marincovich Jr.
Department of Invertebrate Zoology and Geology,
California Academy of Sciences
San Francisco, CA 94118

Simonetta Monechi
Dipartimento di Scienze della Terra
Università di Firenze, Via La Pira 4
50121 Firenze, Italy

Richard Mooi
Department of Invertebrate Zoology and Geology
California Academy of Sciences
San Francisco, CA 94118

Jeffrey A. Myers
Department of Earth and Physical Sciences
Western Oregon University
Monmouth, OR 97361

James H. Nebelsick
Institut und Museum für Geologie und Paläontologie
Sigwartstrasse 10
D-72076, Tübingen, Germany

Elizabeth A. Nesbitt
The Burke Museum
Department of Earth and Planetary Sciences
Box 353010
University of Washington
Seattle, WA 98195-3010

Marlene A. Noble
U.S. Geological Survey
345 Middlefield Road
Menlo Park, CA 94025

Richard D. Norris
Woods Hole Oceanographic Institution
Woods Hole, MA 02543

Francisca E. Oboh-Ikuenobe
Department of Geology and Geophysics
University of Missouri-Rolla
Rolla MO 65409

Anton E. Oleinik
Florida Atlantic University
Department of Geography and Geology
Boca Raton, FL 33431

William P. Patterson
Department of Earth Sciences
Syracuse University
Syracuse, NY 13244

C. Wylie Poag
U.S. Geological Survey
384 Woods Hole Road
Woods Hole, MA 02543

Donald R. Prothero
Department of Geology
Occidental College
Los Angeles, CA 90041

Michael Rasser
Institut für Paläontologie
Geozentrum
Althanstrasse 14, A-1090
Vienna, Austria

David K. Rea
Department of Geological Sciences
University of Michigan
Ann Arbor, MI 48109

Viviana Reale
Dipartimento di Scienze della Terra
Università di Firenze
Via La Pira 4
50121 Firenze, Italy

Edward Robinson
Department of Geography & Geology
University of the West Indies
Mona, Kingston 7, Jamaica

B. Kirby Rodgers
Global Biostrat
3312 Dorado Beach Drive
Farmers Branch, TX 75234

Stephan A. Root
Global Biostrat
3312 Dorado Beach Drive
Farmers Branch, TX 75234

Recep H. Sancay
Turkiye Petrolleri
Anasirma Merkezi
Mustafa Kemal Mah. 2, Cad. No. 86
06520 Sogutozu Turkey

David W. Scholl
Department of Geophysics
Stanford University
Stanford, CA 94305

Richard L. Squires
Department of Geological Sciences
California State University
Northridge, California 91330

Etienne Steurbaut
K.U. Leuven
Historische Geologie
Redingenstraat 16, B-3000
Leuven, Belgium

Andrew J. Stevenson
U.S. Geological Survey
345 Middlefield Road
Menlo Park, CA 94025

Peter R. Thompson
Global Biostrat
3312 Dorado Beach Drive
Farmers Branch, TX 75234

Noël Vandenberghe
K.U. Leuven
Historische Geologie
Redingenstraat 16
B-3000 Leuven, Belgium

Thomas E. Yancey
Department of Geology and Geophysics
Texas A&M University
College Station, TX 77843

Jim Zachos
Earth Sciences Department
University of California
Santa Cruz, CA 95064

Index